# GAMMA-RAY BURST AND AFTERGLOW ASTRONOMY 2001

# Related Titles from AIP Conference Proceedings

**637** Classical Nova Explosions: International Conference on Classical Nova Explosions
Edited by Margarita Hernanz and Jordi José, November 2002, 0-7354-0092-X

**599** X-Ray Astronomy: Stellar Endpoints, AGN, and the Diffuse X-Ray Background
Edited by Nicholas E. White, December 2001, 0-7354-0043-1

**587** Gamma 2001: Gamma-Ray Astrophysics 2001
Edited by Steven Ritz, Neil Gehrels, and Chris R. Shrader, October 2001, 0-7354-0027-X
CD-ROM: 0-7354-0030-X

**586** Relativistic Astrophysics: 20th Texas Symposium
Edited by J. Craig Wheeler and Hugo Martel, October 2001, 0-7354-0026-1

**565** Young Supernova Remnants: Eleventh Astrophysics Conference
Edited by Stephen S. Holt and Una Hwang, May 2001, 0-7354-0001-6

**558** High Energy Gamma-Ray Astronomy: International Symposium
Edited by Felix A. Aharonian and Heinz J. Völk, April 2001, 1-56396-990-4

**556** Explosive Phenomena in Astrophysical Compact Objects: First KIAS Astrophysics Workshop
Edited by Heon-Young Chang, Chang-Hwan Lee, Mannque Rho, and Insu Yi, March 2001, 1-56396-987-4

**526** Gamma-Ray Bursts: 5th Huntsville Symposium
Edited by R. Marc Kippen, Robert S. Mallozzi, and Gerald J. Fishman, June 2000,
CD-ROM included, 1-56396-947-5

**522** Cosmic Explosions: Tenth Astrophysics Conference
Edited by Stephen S. Holt and William W. Zhang, June 2000, 1-56396-943-2

**516** 26th International Cosmic Ray Conference: ICRC XXVI, Invited, Rapporteur, and Highlight Papers
Edited by Brenda L. Dingus, David B. Kieda, and Michael H. Salamon
May 2000, 1-56396-939-4

**515** GeV-TeV Gamma Ray Astrophysics Workshop: Towards a Major Atmospheric Cherenkov Detector VI
Edited by Brenda L. Dingus, Michael H. Salamon, and David B. Kieda, May 2000, 1-56396-938-6

**510** The Fifth Compton Symposium
Edited by Mark L. McConnell and James M. Ryan, March 2000, 1-56396-932-7

**433** Workshop on Observing Giant Cosmic Ray Air Showers from $>10^{20}$ eV Particles from Space
Edited by John F. Krizmanic, Jonathan F. Ormes, and Robert E. Streitmatter,
June 1998, 1-56396-788-X

**428** Gamma-Ray Bursts: 4th Huntsville Symposium
Edited by Charles A. Meegan, Robert D. Preece, and Thomas M. Koshut, May 1998,
1-56396-766-9

To learn more about these titles, or the AIP Conference Proceedings Series, please visit the webpage http://proceedings.aip.org/proceedings

# GAMMA-RAY BURST AND AFTERGLOW ASTRONOMY 2001

A Workshop Celebrating the First Year of the HETE Mission

Woods Hole, Massachusetts   5–9 November 2001

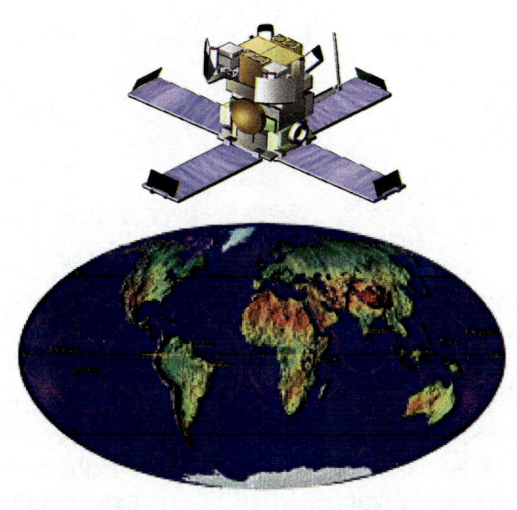

EDITORS
G. R. Ricker
R. K. Vanderspek

Massachusetts Institute of Technology
Cambridge, Massachusetts

Melville, New York, 2003
AIP CONFERENCE PROCEEDINGS ■ VOLUME 662

**Editors:**

George R. Ricker
Roland K. Vanderspek

Center for Space Research
Massachusetts Institute of Technology
77 Massachusetts Avenue
Cambridge, MA 02139-4307
USA

E-mail: grr@space.mit.edu
roland@space.mit.edu

The articles on pp. 143 and 465-468 were authored by U.S. Government employees and are not covered by the below mentioned copyright.

Authorization to photocopy items for internal or personal use, beyond the free copying permitted under the 1978 U.S. Copyright Law (see statement below), is granted by the American Institute of Physics for users registered with the Copyright Clearance Center (CCC) Transactional Reporting Service, provided that the base fee of $20.00 per copy is paid directly to CCC, 222 Rosewood Drive, Danvers, MA 01923. For those organizations that have been granted a photocopy license by CCC, a separate system of payment has been arranged. The fee code for users of the Transactional Reporting Service is: 0-7354-0122-5/03/$20.00.

© 2003 American Institute of Physics

Individual readers of this volume and nonprofit libraries, acting for them, are permitted to make fair use of the material in it, such as copying an article for use in teaching or research. Permission is granted to quote from this volume in scientific work with the customary acknowledgment of the source. To reprint a figure, table, or other excerpt requires the consent of one of the original authors and notification to AIP. Republication or systematic or multiple reproduction of any material in this volume is permitted only under license from AIP. Address inquiries to Office of Rights and Permissions, Suite 1NO1, 2 Huntington Quadrangle, Melville, NY 11747-4502; phone: 516-576-2268; fax: 516-576-2450; e-mail: rights@aip.org.

L.C. Catalog Card No. 2003102745
ISBN 0-7354-0122-5
ISSN 0094-243X

Printed in the United States of America

# CONTENTS

**Preface** .................................................................................................. xiii
**Banquet Photographs** ........................................................................... xv

## I. THE HETE SATELLITE MISSION

**The High Energy Transient Explorer (HETE): Mission and Science Overview** ........................ 3
  G. R. Ricker, J.-L. Atteia, G. B. Crew, J. P. Doty, E. E. Fenimore, M. Galassi, C. Graziani, K. Hurley,
  J. G. Jernigan, N. Kawai, D. Q. Lamb, M. Matsuoka, G. Pizzichini, Y. Shirasaki, T. Tamagawa,
  R. Vanderspek, G. Vedrenne, J. Villasenor, S. E. Woosley, and A. Yoshida and the HETE Science Team

**In-Flight Performance and First Results of FREGATE** ................................................ 17
  J.-L. Atteia, M. Boer, F. Cotin, J. Couteret, J.-P. Dezalay, M. Ehanno, J. Evrard, D. Lagrange, M. Niel,
  J.-F. Olive, G. Rouaix, P. Souleille, G. Vedrenne, K. Hurley, G. Ricker, R. Vanderspek, G. Crew, J. Doty,
  and N. Butler

**In-Orbit Performance of the WXM (Wide-Field X-Ray Monitor)** ....................................... 25
  N. Kawai, A. Yoshida, M. Matsuoka, Y. Shirasaki, T. Tamagawa, K. Torii, T. Sakamoto, D. Takahashi,
  E. Fenimore, M. Galassi, T. Tavenner, D. Q. Lamb, C. Graziani, T. Donaghy, R. Vanderspek,
  M. Yamauchi, K. Takagishi, I. Hatsukade, and the HETE-2 Science Team

**An Overview of the HETE Soft X-Ray Camera** .......................................................... 33
  J. Villasenor, R. Dill, J. P. Doty, G. Monnelly, R. Vanderspek, S. Kissel, G. Prigozhin, G. B. Crew,
  and G. R. Ricker

**HETE: Novel Technology for a Small Space Mission** ................................................... 38
  J. Doty, R. Dill, G. B. Crew, T. Brady, J. Francis, G. Huffman, J. Roberts, R. Vanderspek,
  and the HETE Science Team

**HETE-II and the Interplanetary Network** ............................................................... 42
  K. Hurley, J.-L. Atteia, G. Crew, G. Ricker, J. Doty, G. Monnelly, R. Vanderspek, J. Villasenor,
  and T. Cline

**Using Wavelets to Detect HETE Untriggered Bursts** ................................................... 45
  N. Butler and J. Doty

**HETE Soft X-Ray Camera Imaging: Calibration, Performance, and Sensitivity** ......................... 49
  G. Monnelly, J. N. Villasenor, J. G. Jernigan, G. Prigozhin, R. Vanderspek, G. B. Crew, J. Doty,
  A. Levine, and G. Ricker

**GRB 010921: Discovery of the First HETE Afterglow** .................................................. 56
  P. A. Price for the REACT GRB Collaboration

**The E-Peak Distribution of the GRBs Detected by HETE FREGATE Instrument** ......................... 59
  C. Barraud, J.-L. Atteia, J.-F. Olive, J.-P. Dezalay, D. Q. Lamb, N. Kawai, A. Yoshida, Y. Shirasaki,
  T. Sakamoto, T. Tamagawa. K. Torii, M. Matsuoka, E. E. Fenimore, M. Galassi, T. Tavenner,
  T. Q. Donaghy, and C. Graziani

**A Study of the Background in the HETE Equatorial Orbit** ............................................. 63
  N. Butler, A. Dullighan, J.-L. Atteia, N. Kawai, A. Yoshida, and T. Tamagawa

**Real-Time Message from the HETE Spacecraft via the HETE Burst Alert Network** ..................... 66
  G. B. Crew, J. Villasenor, G. Monnelly, J. Doty, R. Vanderspek, and G. Ricker

**HETE Mission Operations** ............................................................................. 70
  G. B. Crew, R. Vanderspek, J. Doty, J. Villasenor, G. Monnelly, N. Butler, G. Prigozhin, and G. Ricker

**Sensitivity of the FREGATE Experiment** ............................................................... 73
  J.-P. Dezalay, J.-F. Olive, J.-L. Atteia, J.-P. Lestrade, C. Barraud, N. Butler, G. Crew, J. Doty, K. Hurley,
  G. Ricker, and R. Vanderspek

**Localization of GRBs by Bayesian Analysis of Data from the HETE WXM** .............................. 76
  C. Graziani, D. Q. Lamb, and the HETE Science Team

**A "Spiffy" Trigger for Gamma-Ray Bursts** ............................................................. 79
  C. Graziani

**FREGATE Observation of a Strong Burst from SGR1900+14** ............................................ 82
  J.-F. Olive, K. Hurley, J.-P. Dezalay, J.-L. Atteia, C. Barraud, N. Butler, G. B. Crew, J. Doty, G. Ricker,
  R. Vanderspek, D. Q. Lamb, N. Kawai, A. Yoshida, Y. Shirasaki, T. Sakamoto, T. Tamagawa, K. Torii,
  M. Matsuoka, E. E. Fenimore, M. Galassi, T. Tavenner, T. Q. Donaghy, and C. Graziani

**In-Flight Verification of the FREGATE Spectral Response** ............................................. 88
  J.-F. Olive, J.-P. Dezalay, J.-L. Atteia, C. Barraud, N. Butler, G. B. Crew, J. Doty, and R. Vanderspek

**In-Flight Characterization of the HETE Soft X-Ray CCD Cameras** ......... 91
  G. Prigozhin, J. Villasenor, R. Vanderspek, G. Jernigan, J. Doty, G. Crew, and G. Ricker

**X-Ray Bursts Observed by the HETE-2 Satellite** ......... 94
  T. Sakamoto, D. Takahashi, N. Kawai, A. Yoshida, Y. Shirasaki, T. Tamagawa, K. Torii, M. Matsuoka, E. Fenimore, M. Galassi, D. Q. Lamb, C. Graziani, and the HETE-2 Team

**The Effectiveness of the HETE-2 Triggering Algorithm** ......... 97
  T. Tavenner, E. Fenimore, M. Galassi, R. Vanderspek, B. Preger, C. Graziani, D. Lamb, N. Kawai, A. Yoshida, Y. Shirasaki, and T. Tamagawa

**What HETE Sends to the GCN** ......... 101
  R. Vanderspek, G. B. Crew, S. Barthelmy, and the HETE Science Team

**First Year Operations of the HETE Burst Alert Network** ......... 107
  J. Villasenor, G. Crew, G. Monnelly, J. Doty, R. Foster, G. Ricker, R. Vanderspek, N. Kawai, A. Yoshida, M. Boer, K. Hurley, R. Manchanda, G. Pizzichini, J. Braga, and G. Azzibrouck

**Soft Gamma-Ray Repeaters as Observed by the HETE-2 WXM** ......... 111
  K. Torii, A. Yoshida, N. Kawai, Y. Shirasaki, T. Tamagawa, T. Sakamoto, M. Matsuoka, E. Fenimore, M. Galassi, J.-L. Atteia, K. Hurley, N. Butler, D. Q. Lamb, C. Graziani, T. Donaghy, R. Vanderspek, G. Ricker, and the HETE-2 Science Team

**Astrometric Calibration and Estimate of the Systematic Error in WXM Localizations Obtained by the Chicago Bayesian Method** ......... 114
  C. Graziani, Y. Shirasaki, T. Donaghy, E. Fenimore, M. Galassi, N. Kawai, D. Q. Lamb, T. Sakamoto, D. Takahashi, T. Tamagawa, T. Tavenner, K. Torii, A. Yoshida, and R. Vanderspek

**Astrometric Calibration and Estimate of the Systematic Error in HETE WXM Localizations Obtained by the RIKEN Cross Correlation Method** ......... 117
  Y. Shirasaki, N. Kawai, A. Yoshida, M. Matsuoka, T. Tamagawa, K. Torii, T. Sakamoto, E. Fenimore, M. Galassi, R. Vanderspek, and the HETE-2 Science Team

## II. GLOBAL PROPERTIES OF GAMMA-RAY BURSTS

**Spectral Properties of GRBs Observed with BeppoSAX** ......... 123
  E. Costa and F. Frontera

**The Use of $E_{pk}$ Decay Rates to Find Standard Candles in GRBs** ......... 130
  D. Kocevski and E. Liang

**Kinematics of Gamma-Ray Bursts and Their Relationship to Afterglows** ......... 134
  J. D. Salmonson

**An Observational Evidence for the Difference between the Short and Long Gamma-Ray Bursts** ......... 137
  L. G. Balázs, P. Mészáros, Z. Bagoly, I. Horváth, and A. Mészáros

**The Statistical Determination of the Burst Energy Distribution** ......... 140
  D. L. Band

**Seven-Year Konus GRB Catalog** ......... 143
  T. L. Cline, S. D. Barthelmy, P. S. Butterworth, T. B. Sheets, R. I. Aptekar, D. D. Frederiks, S. V. Golenetskii, V. N. Il'inskii, E. P. Mazets, and V. D. Pal'shin

**Comments on Anisotropic Distributions of Faint BATSE GRBs** ......... 144
  J. Hakkila, T. W. Giblin, W. S. Paciesas, R. J. Roiger, D. J. Haglin, and C. A. Meegan

**The Internal Luminosity Functions of BATSE 5B Gamma-Ray Bursts** ......... 147
  J. Hakkila, T. W. Giblin, T. M. Freismuth, K. C. Young, A. J. Sprague, A. D. Stallworth, D. J. Haglin, R. J. Roiger, and W. S. Paciesas

**The AMANDA Search for High Energy Neutrinos from Gamma-Ray Bursts** ......... 150
  R. Hardtke for the AMANDA Collaboration

**A Gamma-Ray Burst Bibliography, 1973–2001** ......... 153
  K. Hurley

**Explaining GRB Lag with Spectral Evolution** ......... 156
  D. Kocevski and E. Liang

**Observational Appearance of Relativistic Outflows in the Presence of Neutron Component** ......... 159
  A. A. Belyanin, E. V. Derishev, V. V. Kocharovsky, and Vl. V. Kocharovsky

**The Results of Statistical Tests of the Angular Distribution of Gamma-Ray Bursts** ......... 163
  R. Vavrek, L. G. Balázs, A. Mészáros, I. Horváth, and Z. Bagoly

**The Dynamics of Magnetized Outflows in GRBs** .................................................. 166
    N. Vlahakis and A. Königl
**Hard X-Ray Afterglows of Short GRBs** ............................................................. 169
    D. Lazzati, E. Ramirez-Ruiz, and G. Ghisellini
**Constraints on the Bulk Lorentz Factor of GRB 990123** .......................................... 172
    A. M. Soderberg and E. Ramirez-Ruiz
**BATSE 5B Sky Exposure and Trigger Efficiency** ................................................... 176
    J. Hakkila, G. N. Pendleton, C. A. Meegan, M. S. Briggs, R. M. Kippen, and R. D. Preece
**A Comparison of Unsupervised Classifiers on BATSE Catalog Data** ............................... 179
    J. Hakkila, R. J. Roiger, D. J. Haglin, T. W. Giblin, and W. S. Paciesas

## III. GAMMA-RAY BURST PROGENITORS AND PROMPT EMISSION

### Sources and Central Engines

**The Central Engines of Gamma-Ray Bursts** ........................................................ 185
    S. E. Woosley, W. Zhang, and A. Heger
**Merging Neutron Star–Black Hole Binaries** ....................................................... 193
    M. Ruffert and H.-T. Janka
**Comparing GRB Progenitors** ...................................................................... 199
    C. L. Fryer
**Collapsar Disks and Winds** ...................................................................... 202
    A. I. MacFadyen
**Erupting Fireballs, Nozzles, and Precursors** .................................................... 206
    E. Ramirez-Ruiz, A. I. MacFadyen, and D. Lazzati
**Gamma-Ray Bursts: The Tip of the Iceberg?** ..................................................... 210
    M. H. P. M. van Putten
**On the Progenitors of Collapsars** ............................................................... 214
    A. Heger and S. E. Woosley
**Gamma-Ray Bursts from Accreting Black Holes** ................................................... 217
    W. H. Lee and E. Ramirez-Ruiz
**Neutron Star Binaries as Central Engines of GRBs** .............................................. 220
    S. Rosswog
**A Model for Short Gamma-Ray Bursts: Heated Neutron Stars in Close Binary Systems** ............. 223
    J. D. Salmonson and J. R. Wilson
**Relativistic Jets in Collapsars** ................................................................ 226
    W. Zhang, S. E. Woosley, and A. I. MacFadyen

### Prompt Emission

**X-Ray Flashes and X-Ray Counterparts of Gamma-Ray Bursts** ...................................... 229
    J. Heise
**Analytic Modeling of GRB Spectral and Temporal Evolution** ...................................... 237
    E. Liang and D. Kocevski
**Observations of the Highest Energy Gamma Rays from Gamma-Ray Bursts** ........................... 240
    B. L. Dingus
**Spectral Characteristics of X-Ray Flashes Compared to Gamma-Ray Bursts** ........................ 244
    R. M. Kippen, P. M. Woods, J. Heise, J. J. M. in 't Zand, M. S. Briggs, and R. D. Preece
**Spectral Properties of Short Gamma-Ray Bursts** .................................................. 248
    W. S. Paciesas, M. S. Briggs, R. D. Preece, and R. S. Mallozzi
**Towards an Understanding of Prompt GRB Emission** ............................................... 252
    N. M. Lloyd-Ronning
**Luminosity and Variability of Collimated Gamma-Ray Bursts** ...................................... 260
    S. Kobayashi, F. Ryde, and A. MacFadyen
**Determining Bolometric Corrections for BATSE Burst Observations** ............................... 264
    L. Borgonovo, F. Ryde, M. de Val Borro, and R. Svensson

**BATSE–EGRET Combined Spectral Fits** ................................................................. 267
    M. M. González, Y. Kaneko, R. Preece, and B. L. Dingus

**Spectral Analysis of Bright Gamma-Ray Bursts** ........................................................ 270
    G. Ghirlanda, A. Celotti, and G. Ghisellini

**Extended Power-Law Decays in BATSE Gamma-Ray Bursts: Signatures of External Shocks?** ......... 273
    T. W. Giblin, V. Connaughton, J. van Paradijs, R. D. Preece, M. S. Briggs, C. Kouveliotou,
    R. A. M. J. Wijers, and G. J. Fishman

**RXTE All-Sky Monitor Observations of GRB Light Curves** .............................................. 276
    A. M. Levine, H. Bradt, R. Remillard, and D. A. Smith

**Temporal Properties of Short and Long Gamma-Ray Bursts** ............................................ 280
    S. McBreen, F. Quilligan, B. McBreen, L. Hanlon, and D. Watson

**Low-Energy Study of Gamma-Ray Bursts Using Two BATSE Spectroscopy Detectors** ............... 283
    M. J. Pangia and R. D. Preece

**Analytical Descriptions of Gamma-Ray Burst Pulses** ................................................... 286
    F. Ryde, D. Kocevski, and E. Liang

**The Expected Thermal Precursors of Gamma-Ray Bursts in the Internal Shock Model** ............... 289
    F. Daigne and R. Mochkovitch

**The Self-Consistent keV to TeV Spectra of Gamma-Ray Bursts Produced by the Synchrotron-Self-Compton Emission in Relativistic Shocks** ...................................................... 292
    E. V. Derishev, V. V. Kocharovsky, Vl. V. Kocharovsky, and P. Mészáros

**Inverse Compton Model of Gamma-Ray Burst Spectra** ................................................ 295
    E. Liang, M. Boettcher, and D. Kocevski

**The Physics of Pulses in GRBs** ........................................................................... 299
    F. Daigne and R. Mochkovitch

## IV. AFTERGLOWS OF GAMMA-RAY BURSTS

### Afterglow Theory

**Properties of Gamma-Ray Burst Jets Obtained from Afterglow Modelling** .......................... 305
    A. Panaitescu and P. Kumar

**GRB Remnants** ............................................................................................. 313
    T. Piran and S. Ayal

**Optical and X-Ray Afterglows in the Cannonball Model** ............................................. 319
    A. De Rújula

**Afterglows with High Inferred Values of $\varepsilon_e$ and $\varepsilon_B$: A Clue to the Gamma-Ray Burst Environment?** ........................................................................ 327
    A. Königl and J. Granot

**Gamma-Ray Burst Afterglows: Two Post-Standard Effects** ........................................... 331
    B. Zhang and P. Mészáros

**Afterglow Lightcurves, Viewing Angle, and the Jet Structure of Gamma-Ray Bursts** ............... 335
    E. Rossi, D. Lazzati, and M. J. Rees

### Radio, Infrared, and Optical Observations

**On the Color Indices and Absolute Brightnesses of the Optical Afterglows of GRB** ............... 338
    V. Šimon, R. Hudec, N. Masetti, and G. Pizzichini

**SCUBA Sub-Millimeter Observations of Gamma-Ray Bursters** ...................................... 342
    I. A. Smith, R. P. J. Tilanus, R. A. M. J. Wijers, N. Tanvir, P. Vreeswijk, E. Rol, and C. Kouveliotou

**A Serendipitous Search for GRB Afterglows by Subaru/Suprime-Cam: A Test of GRB Beaming** ... 346
    T. Totani, S. Miyazaki, Y. Mizumoto, R. Ogasawara, T. Takada, N. Yasuda, M. Doi, W. Kawasaki,
    N. Kawai, A. Yoshida, and Y. Urata

**GRB Afterglows and Other Transients in the SDSS**.................................................349
    B. C. Lee, D. E. Vanden Berk, D. Lamb, B. Wilhite, D. E. Reichart, J. F. Beacom, D. L. Tucker,
    B. Yanny, K. Abazajian, J. Adelman, J. Annis, B. Chen, M. Harvanek, A. Henden, K. Hurley, Z. Ivezic,
    R. Kehoe, S. Kleinman, R. Kron, J. Krzesinski, D. Long, T. McKay, R. McMillan, E. H. Neilsen,
    P. R. Newman, A. Nitta, P. Palunas, D. P. Schneider, S. Snedden, J. Wren, D. York, J. W. Briggs,
    J. Brinkmann, I. Csabai, G. S. Hennessy, S. Kent, R. Lupton, H. J. Newberg, and C. Stoughton

**Gamma-Ray Burst Optical Afterglow Observations at Nyrola Observatory**.........................352
    A. Oksanen, H. Hyvönen, and M. Moilanen

**The AAVSO International GRB Network**.........................................................355
    A. Price

**Colour–Colour Diagram as a Tool for Prompt Search of GRB Afterglows; the Discovery of the GRB 001011 Optical/Near-Infrared Counterpart**.........................................357
    J. Gorosabel, J. U. Fynbo, P. Møller, J. Hjorth, H. Pedersen, L. Christensen, B. L. Jensen, M. I. Andersen,
    C. Wolf, J. Afonso, M. A. Treyer, G. Mallén-Ornelas, A. J. Castro-Tirado, A. Fruchter, J. Greiner,
    S. Klose, C. Kouveliotou, N. Masetti, E. Palazzi, F. Frontera, E. Pian, N. Tanvir, P. M. Vreeswijk, E. Rol,
    I. Salamanca, L. Kaper, E. van den Heuvel, and R. A. M. J. Wijers

**Searches for Orphan Optical Afterglows**.......................................................360
    R. Hudec, V. Ríkal, J. Polcar, and F. Hroch

**Optical Follow-Up Observations of GRBs at the Klet' Observatory**................................363
    J. Polcar, R. Hudec, M. Topinka, J. Tichá, M. Tichy, N. Masetti, and G. Pizzichini

**Super-LOTIS and LOTIS for HETE-2 GRB Triggers**...............................................366
    H. S. Park, G. G. Williams, E. Ables, S. D. Barthelmy, T. Cline, N. Gehrels, D. Hartmann, K. Hurley,
    K. Lindsay, R. Nemiroff, W. Pereira, and D. Perez-Ramirez

**GRB 980329: Determining Density without a Redshift**...........................................369
    S. A. Yost

## *X-Ray Afterglows*

**X-Ray Spectroscopy of Gamma-Ray Bursts**......................................................372
    L. Piro

**BATSE Observations of GRB 991216**............................................................379
    V. Connaughton, T. W. Giblin, R. M. Kippen, R. D. Preece, M. S. Briggs, and C. A. Meegan

**A Radiative Recombination Edge in X-Ray Afterglow of GRB 970828, and Non-Equilibrium Ionization States**...........................................................................383
    D. Yonetoku, T. Murakami, A. Yoshida, K. Masai, N. Kawai, and M. Namiki

**The Prompt and Afterglow Emission of GRB 001109 Measured by BeppoSAX**.......................387
    L. Amati, F. Frontera, J. M. Castro Cerón, E. Costa, M. Feroci, G. Gandolfi, P. Giommi, C. Guidorzi,
    N. Masetti, E. Montanari, L. Piro, P. Soffitta, and J. J. M. in 't Zand

## V. ASSOCIATIONS WITH GAMMA-RAY BURSTS: SUPERNOVAE, SOURCE ENVIRONMENTS, AND HOST GALAXIES

## *Supernova Connection*

**GRB 000911: Evidence for an Associated Supernova?**............................................393
    S. Covino, D. Lazzati, G. Ghisellini, D. Fugazza, S. Campana, P. Saracco, P. A. Price, E. Berger,
    S. Kulkarni, E. Ramirez-Ruiz, A. Cimatti, M. Della Valle, S. di Serego Alighieri, A. Celotti, F. Haardt,
    G. L. Israel, and L. Stella

**Signs and Consequences of Supernova - Gamma-Ray Burst Association**...........................396
    R. A. M. J. Wijers

## *Source Environment*

**Evidence for Circumburst Extinction of Gamma-Ray Bursts with Dark Optical Afterglows and Evidence for a Molecular Cloud Origin of Gamma-Ray Bursts**...............................403
    D. E. Reichart and P. A. Price

X-Ray Spectroscopy of Gamma-Ray Bursts: The Path to the Progenitor........................ 411
    D. Lazzati, R. Perna, and G. Ghisellini
The Role of Dust in GRB Afterglows................................................... 415
    D. Q. Lamb
Observational Properties of Afterglows from Mergers of Compact Objects.................... 417
    R. Perna and K. Belczynski

## Host Galaxies

Radio, Sub-mm, and X-Ray Studies of Gamma-Ray Burst Host Galaxies......................... 420
    E. Berger
The Search for the Afterglow of the Dark GRB 001109 ..................................... 424
    J. M. Castro Cerón, J. Gorosabel, A. J. Castro-Tirado, V. V. Sokolov, V. L. Afansiev, T. A. Fatkhullin,
    S. N. Dodonov, V. N. Komarova, A. M. Cherepashchuk, K. A. Postnov, J. Greiner, S. Klose, J. Hjorth,
    H. Pedersen, E. Rol, J. Fliri, M. Feldt, G. Feulner, M. I. Andersen, B. L. Jensen, F. J. Vrba, A. A. Henden,
    and G. Israelian
Unveiling the Progenitors of Gamma-Ray Bursts Through Observations of Their
Host Galaxies........................................................................ 428
    R.-R. Chary

## VI. GAMMA-RAY BURSTS AND COSMOLOGY

Gamma-Ray Bursts as a Probe of Cosmology.............................................. 433
    D. Q. Lamb
Measuring $\Omega_M$, $\Omega_\Lambda$, and the SFR with Class III GRBs................. 438
    A. Balastegui, P. Ruiz-Lapuente, and R. Canal
Gamma-Ray Bursts and Cosmic Radiation Backgrounds...................................... 442
    D. H. Hartmann, T. M. Kneiske, K. Mannheim, and K. Watanabe
Estimation of the Redshifts for Long Gamma-Ray Bursts.................................. 446
    Z. Bagoly, I. Csabai, A. Mészáros, P. Mészáros, I. Horváth, L. G. Balázs, and R. Vavrek
Determining the GRB (Redshift, Luminosity)-Distribution Using Burst Variability........ 450
    T. Donaghy, D. Q. Lamb, D. E. Reichart, and C. Graziani
What Are Luminosity Indicators Telling Us?............................................ 454
    N. M. Lloyd-Ronning, C. L. Fryer, and E. Ramirez-Ruiz
Consequences of a Dependence of GRB Properties on Local Metallicity................... 457
    E. Ramirez-Ruiz, A. W. Blain, and D. Lazzati
Determining the Gamma-Ray Burst Rate as a Function of Redshift........................ 460
    N. Weinberg, C. Graziani, D. Q. Lamb, and D. E. Reichart

## VII. INSTRUMENTATION AND TECHNIQUES

## Current and Future Satellite Missions

The Swift GRB MIDEX Mission.......................................................... 465
    N. Gehrels on behalf of the Swift Team
The GLAST Burst Monitor.............................................................. 469
    C. Meegan, G. Lichti, M. Briggs, R. Diehl, G. Fishman, R. Kippen, C. Kouveliotou, A. von Kienlin,
    W. Paciesas, R. Preece, and V. Schönfelder
The Current Performance of the Third Interplanetary Network.......................... 473
    K. Hurley, T. Cline, I. Mitrofanov, E. Mazets, S. Golenetskii, F. Frontera, E. Montanari, C. Guidorzi,
    and M. Feroci
Proposed Next Generation GRB Mission: EXIST........................................ 477
    J. Grindlay, N. Gehrels, F. Harrison, R. Blandford, G. Fishman, C. Kouveliotou, D. H. Hartmann,
    S. Woosley, W. Craig, and J. Hong

**ECLAIRs: A Microsatellite to Observe the Prompt Optical and X-Ray Emission of Gamma-Ray Bursts** ................................................................. 481
    D. Barret

**The *Swift* X-Ray Telescope** ................................................................. 488
    D. N. Burrows, J. E. Hill, J. A. Nousek, A. Wells, A. Short, M. Turner, O. Citterio, G. Tagliaferri, and G. Chincarini

**The Trigger Algorithm for the Burst Alert Telescope on Swift** ........................... 491
    E. E. Fenimore, D. Palmer, M. Galassi, T. Tavenner, S. Barthelmy, N. Gehrels, A. Parsons, and J. Tueller

**Lobster Eye: New Approach to Monitor GRBs in X-Rays** ................................. 494
    R. Hudec, A. Inneman, L. Pina, and V. Hudcová

**Afterglow Studies with the Swift UV/Optical Telescope** .................................. 497
    S. D. Hunsberger, P. W. A. Roming, J. A. Nousek, and K. O. Mason

**Searches for Hard X-Ray Gamma-Ray Burst Afterglows with the BAT on Swift** ............ 500
    H. A. Krimm, L. M. Barbier, S. D. Barthelmy, A. Eftekharzadeh, E. E. Fenimore, N. Gehrels, D. D. Hullinger, C. Markwardt, D. M. Palmer, A. M. Parsons, H. Ozawa, J. Tueller, and G. Weidenspointner

**The Development of GRAPE, a Gamma-Ray Polarimeter Experiment** ...................... 503
    M. L. McConnell, J. R. Ledoux, J. R. Macri, and J. M. Ryan

**Afterglow and Transient Astronomy with the Swift GRB Explorer** ........................ 506
    J. A. Nousek, M. M. Chester, F. E. Marshall, and the Swift Team

**High Resolution Spectroscopy of the X-Ray Emission of GRBs by IMBOSS on the ISS** ..... 509
    L. Colasanti, L. Piro, L. Pacciani, E. Costa, G. Gandolfi, P. Soffitta, F. Gatti, D. Pergolesi, M. Razeti, R. Vaccarone, G. Testera, M. Pallavicini, A. Ferrari, E. Trussoni, M. Orio, D. Mc Cammon, T. Sanders, M. Galeazzi, A. Szymkowiak, S. Porter, and R. Kelley

**Performance of the Swift X-Ray Telescope (XRT) Mirror/Detector Combination** ........... 511
    A. D. Short, R. M. Ambrosi, I. B. Hutchinson, R. Willingale, A. F. Abbey, A. A. Wells, J. E. Hill, D. N. Burrows, G. Tagliaferri, and O. Citterio

## *Ground-Based Instruments*

**The ROTSE-IIIa Telescope System** ........................................................ 514
    D. Smith, C. Akerlof, M. C. B. Ashley, D. Casperson, G. Gisler, R. Kehoe, S. Marshall, K. McGowan, T. McKay, M. A. Phillips, E. Rykoff, W. T. Vestrand, P. Wozniak, and J. Wren

**Monitoring of the Prompt GRB Afterglow with the REM Telescope** ........................ 517
    S. Covino, F. Zerbi, G. Chincarini, G. Ghisellini, M. Rodonó, L. A. Antonelli, P. Conconi, G. Cutispoto, E. Molinari, L. Nicastro, and E. Palazzi

**BART — Recent Status** ................................................................. 520
    M. Jelínek, R. Hudec, P. Kubánek, M. Nekola, J. Soldán, I. Stoklasová, M. Topinka, R. Smída, L. Svéda, F. Hroch, J. Polcar, and A. J. Castro-Tirado

**Analyses of GRBs on Astronomical Emulsions** ............................................ 523
    R. Hudec, V. Hudcová, F. Krolupper, and P. Kroll

**Simultaneous and Quasisimultaneous Optical Data for GRBs** .............................. 526
    R. Hudec, I. Stoklasová, M. Jelínek, R. Smída, L. Svéda, and P. Kroll

**VHE Observations of GRB with Milagro** ................................................. 529
    J. E. McEnery for the Milagro Collaboration

**Rapid Notification of TeV Transients with the Milagro Telescope** ......................... 532
    M. F. Morales for the Milagro Collaboration

**Discovery Mode Search Techniques for Gamma-Ray Telescopes** .......................... 535
    M. F. Morales for the Milagro Collaboration

**Optical Transient Monitor (OTM) for BOOTES Project** ................................... 538
    P. Páta, M. Bernas, A. J. Castro-Tirado, and R. Hudec

**Rapid Identification of Optical Afterglows: Bright Prospects** .............................. 541
    P. A. Price, B. P. Schmidt, and T. S. Axelrod

**Secondary Science with ROTSE Data** ................................................... 544
    J. Štrobl, R. Hudec, M. Jelínek, V. Šimon, F. Hroch, C. Akerlof, and the ROTSE Team

**Searching for Optical Transients in Real-Time: The RAPTOR Experiment** ................. 547
    W. T. Vestrand, K. Borozdin, S. P. Brumby, D. Casperson, E. Fenimore, M. Galassi, G. Gisler, K. McGowan, S. Perkins, W. Priedhorsky, D. Starr, R. White, P. Wozniak, and J. Wren

**A System for Photon-Counting Spectrophotometry of Prompt Optical Emission from Gamma-Ray Bursts** .................................................................. 550
    W. T. Vestrand, K. Albright, D. Casperson, E. Fenimore, C. Ho, W. Priedhorsky, R. White, and J. Wren

**Recent Developments in the BOOTES Experiment** ......................................................... 553
    A. de Ugarte Postigo, T. J. Mateo Sanguino, J. M. Castro Cerón, P. Páta, M. Bernas, M. Jelinek, R. Hudec, S. McBreen, J. Á. Berná, C. E. García Dabó, J. Gorosabel, J. M. Más-Hesse, T. Soria, B. A. de la Morena, J. Torres Riera, and A. J. Castro-Tirado

## *Techniques*

**An Update on the GRB ToolSHED Project Status** .............................................................. 556
    J. Hakkila, D. J. Haglin, R. J. Roiger, T. W. Giblin, W. S. Paciesas, and C. A. Meegan

## VIII. SOFT GAMMA REPEATERS

**The Effects of Burst Activity on Soft Gamma Repeater Pulse Properties and Persistent Emission** ............................................................................................................... 561
    P. M. Woods

**An Extended Burst Tail from SGR 1900+14 with a Thermal X-Ray Spectrum** ............. 570
    G. T. Lenters, P. M. Woods, J. E. Goupell, C. Kouveliotou, E. Göğüs, K. Hurley, D. Frederiks, and S. Golenetskii

**The Environments of SGRs: A Brief & Biased Review** ..................................................... 574
    S. S. Eikenberry

**Mid-Infrared Observations of the SGR 1900+14 Error Box** ............................................. 579
    S. Klose, B. Stecklum, D. H. Hartmann, F. J. Vrba, A. A. Henden, and A. Bacmann

**Search for Optical Activity of SGR 1806-20 at the SAO 6-m Telescope** ........................ 583
    G. Beskin, V. Debur, A. Panferov, I. Panferova, V. Plokhotnichenko, A. Pozanenko, V. Loznikov, M. Boer, J.-L. Atteia, A. Klotz, and G. Ricker

**Symposium Participants** ........................................................................................................ 585
**Author Index** ........................................................................................................................... 589

# PREFACE

Gamma-ray Burst Astronomy has developed rapidly as an exciting new discipline of high energy astrophysics, following on from the discovery of counterparts at redshifts reaching z=4.5. Pioneering studies using the long-lived BATSE instruments provided the first evidence that gamma-ray bursts (GRBs) are cosmological in nature. Accurate GRB locations distributed to the astronomical community within a day by the BeppoSAX satellite led to the breakthrough discovery of radio, optical, and X-ray afterglows. GRBs are now widely recognized not only as very energetic, ultra relativistic phenomena that are likely linked to the deaths of massive stars, but also as cosmological beacons that mark the end of the "dark ages" in the early universe. Successful studies by the Chandra and XMM-Newton Observatories of BeppoSAX and IPN localizations are providing the first detailed spectra of the late X-ray afterglows of GRBs.

The HETE mission, launched on 9 October 2000, has dramatically quickened the pace of GRB discoveries. By disseminating accurate burst coordinates to the astronomical community within a minute of the burst onset, HETE has made possible the panchromatic observation of the early, bright phases of the relativistic fireball expansion in several GRB afterglows. For the very first time, high dispersion spectroscopy and accurate spectropolarimetry of afterglows with 10 meter class optical telescopes are being obtained in HETE-enabled early measurements, when the afterglows are 100-1000 times brighter. Further progress will be expected from the SWIFT, AGILE, GLAST, and ECLAIRS missions that will be launched later in this decade.

To celebrate the first year of the HETE mission, and to present new results from both HETE and other observatories, a workshop was held in a unique setting in the village of Woods Hole, Massachusetts, on Cape Cod, during the week of 5-9 November 2001. More than 150 participants attended. The scientific program covered the rapidly expanding range of GRB research, with roughly half of the sessions being devoted to prompt burst emission, and half to burst afterglows, both from observational and theoretical perspectives. Special sessions were devoted to future missions and instrumentation.

The meeting was organized by the HETE Science Team, with contributions from members in the US, France, Japan, Brazil, India, and Italy. The program was developed and organized by the Scientific Organizing Committee (SOC). The members of the SOC were:

Jean-Luc Atteia (Obs. Midi-Pyrenees)
Thomas Cline (GSFC/NASA)
Enrico Costa (IAS-CNR)
Brenda Dingus (U. Wisconsin)
Ed Fenimore (LANL)
Jerry Fishman (MSFC/NASA)
Dale Frail (NRAO)
Andy Fruchter (STSCI)
Neil Gehrels (GSFC/NASA)
Jochen Greiner (MPE)
John Heise (SRON)
Kevin Hurley (UC Berkeley)
Nobu Kawai (TITech)

Marc Kippen (LANL)
Chryssa Kouveliotou (MSFC/NASA)
Shri Kulkarni (Cal Tech)
Don Lamb (U Chicago)
Avi Loeb (Harvard)
Masaru Matsuoka (NASDA)
Chip Meegan (MSFC/NASA)
Peter Meszaros (Penn State)
Luigi Piro (IAS-CNR)
George Ricker, Chairman (MIT)
Marco Tavani (IFC-CNR)
Ralph Wijers (U. Amsterdam)
Stan Woosley (UC Santa Cruz)

A special session devoted to soft gamma repeaters (SGRs) was held on the final day of the conference. Organized by Chryssa Kouveliotou (NASA/MSFC) and Kevin Hurley (UC/Berkeley), the SGR Workshop reviewed recent observations and emerging theories in this important field that has traditionally been related to gamma-ray bursts, but has now emerged as an independent discipline within galactic high energy astronomy.

The Local Organizing Committee was chaired by Roland Vanderspek, and was comprised by members of the HETE Science Team at MIT. A multitude of contributions from Jean Farewell of the MIT Center for Space Research, and by Trish Dobson, Ruth Paglierani, and John Vallerga of the Conference Connection ensured a smoothly-run meeting. The staff and management of the Woods Hole Marine Biology Laboratory were gracious in extending their hospitality; the site proved to be one that encouraged informal discussions, with excellent regional cuisine and exceptionally pleasant November weather. In the aftermath, assistance in the refereeing of papers was provided by Nat Butler, John Doty, Allyn Dullighan, Al Levine, and Joel Villasenor.

In this volume, we have assembled the written papers from both the poster and oral presentations from the workshop. The papers have been arranged by topical sections, although many papers contain research results relevant to several topics. In particular, the twenty-six papers presented in the section on the HETE Satellite Mission comprise both a tutorial on the spacecraft and its instruments, as well as a summary of early scientific results. In the period since the workshop, the full potential of the HETE mission has been amply demonstrated in the wealth of new sources it has discovered. At the time of this writing, HETE had accurately localized 33 GRBs. Of these localizations, ten had led to the detection of X-ray, optical, or radio afterglows; seven of the afterglows had established redshifts. Of the GRBs localized by HETE, approximately forty percent have been "X-ray rich" events. Optical, IR, and radio follow-up observations of HETE GRBs commencing within less than a minute of the prompt gamma-ray emission appear to have revealed the nature of heretofore mysterious "dark bursts", and have greatly enhanced our understanding of extragalactic X-ray flashes. Routinely, the soft X-ray cameras (SXC) on HETE had produced prompt localizations accurate to better than one arc minute, enabling the use of sensitive narrow field instruments on the very largest ground-based telescopes, as well as follow-up observations by the Chandra and HST Observatories.

We gratefully acknowledge supplemental support from NASA Headquarters (HQ) and from the Swift project at GSFC. Don Kniffen and Lou Kaluzienski of NASA HQ have been especially generous in their devotion to and assistance of the HETE Mission. Continuing support for the HETE program in the USA is provided under NASA contract NASW-4690.

*Cambridge, MA, USA*  
*February 2003*

*George R. Ricker*  
*Roland K. Vanderspek*

# I. THE HETE SATELLITE MISSION

# The High Energy Transient Explorer (HETE): Mission and Science Overview

G. R. Ricker*, J-L. Atteia[†], G. B. Crew*, J. P. Doty*, E. E. Fenimore**, M. Galassi**, C. Graziani[‡], K. Hurley[§], J. G. Jernigan[§], N. Kawai[¶], D. Q. Lamb[‡], M. Matsuoka[‖], G. Pizzichini[††], Y. Shirasaki[‡‡], T. Tamagawa[§§], R. Vanderspek*, G. Vedrenne[¶¶], J. Villasenor*, S. E. Woosley*** and A. Yoshida[†††]

*MIT Center for Space Research, Cambridge, MA 02139, USA*
[†]*Laboratoire d'Astrophysique, Observatoire Midi-Pyrenees, France*
**Los Alamos National Laboratory, Los Alamos, NM, USA*
[‡]*Department of Astronomy & Astrophysics, University of Chicago, Chicago, IL 60637, USA*
[§]*UC Berkeley Space Sciences Laboratory, Berkeley, CA 94720 USA*
[¶]*Tokyo Institute of Technology, Tokyo, Japan*
[‖]*NASDA, Tokyo, Japan*
[††]*Consiglio Nazionale Delle Ricerche, Italy*
[‡‡]*National Astronomical Observatory, Tokyo, Japan*
[§§]*The Institute of Physical and Chemical Research (RIKEN), Tokyo, Japan*
[¶¶]*Centre D'Etude Spatiale des Rayonnements, France*
***Department of Astronomy & Astrophysics, University of California, Santa Cruz, CA 95064, USA*
[†††]*Aoyama University, Tokyo, Japan*

**Abstract.** The High Energy Transient Explorer (*HETE*) mission is devoted to the study of gamma-ray bursts (GRBs) using soft X-ray, medium X-ray, and gamma-ray instruments mounted on a compact spacecraft. The *HETE* satellite was launched into equatorial orbit on 9 October 2000. A science team from France, Japan, Brazil, India, Italy, and the US is responsible for the *HETE* mission, which was completed for $\sim 1/3$ the cost of a NASA Small Explorer (SMEX). The *HETE* mission is unique in that it is entirely "self-contained," insofar as it relies upon dedicated tracking, data acquisition, mission operations, and data analysis facilities run by members of its international Science Team.

A powerful feature of *HETE* is its potential for localizing GRBs within seconds of the trigger with good precision ($\sim 10'$) using medium energy X-rays and, for a subset of bright GRBs, improving the localization to $\sim 30''$ accuracy using low energy X-rays. Real-time GRB localizations are transmitted to ground observers within seconds via a dedicated network of 14 automated "Burst Alert Stations," thereby allowing prompt optical, IR, and radio follow-up, leading to the identification of counterparts for a large fraction of *HETE*-localized GRBs. *HETE* is the only satellite that can provide near-real time localizations of GRBs, and that can localize GRBs that do not have X-ray, optical, and radio afterglows, during the next two years. These capabilities are the key to allowing *HETE* to probe further the unique physics that produces the brightest known photon sources in the universe.

To date (December 2002), *HETE* has produced 31 GRB localizations. Localization accuracies are routinely in the $4'$-$20'$ range; for the five GRBs with SXC localization, accuracies are $\sim 1$-$2'$. In addition, *HETE* has detected $\sim 25$ bursts from soft gamma repeaters (SGRs), and >600 X-ray bursts (XRBs).

## INTRODUCTION

As has long been recognized, accurate locations and rapid follow-up observations in many wavelengths are central to understanding the nature of gamma-ray bursts. For this reason, a strategy evolved in the late 1970's and early 1980's (e.g., Woosley et al. 1982) to detect GRBs, not only in gamma-rays, but also in emitted X-rays and optical light. While detection of short transients at the lower energies posed observational challenges, and the strength of the optical signal was unknown, the possibility of arc minute localizations deduced from the X-rays

**FIGURE 1.** *HETE* satellite undergoing balance testing prior to launch. The 4 solar panels are shown fully deployed; they span $\sim 2.5$ meters tip-to-tip. The solar cells are facing down in the picture. The science instruments are situated at the top of the spacecraft, $\sim 1$ meter from the plane of the solar panels.

that were known to be present was very appealing. This strategy was implemented in the *HETE-1* (High Energy Transient Explorer) satellite, which was unfortunately lost due to a rocket failure on 1996 November 4, and in the highly successful *BeppoSAX* Mission [56].

The *HETE-2* satellite (henceforth simply "*HETE*", see Figure 1), which was successfully launched into equatorial orbit on 9 October 2000, is the first space mission entirely devoted to the study of gamma-ray bursts (GRBs). *HETE* utilizes a matched suite of low energy X-ray, medium energy X-ray, and gamma-ray detectors mounted on a compact spacecraft. A unique feature of *HETE* is its potential for localizing GRBs with $\sim 1$-$10'$ accuracy in real time aboard the spacecraft. GRB locations are transmitted, within seconds to minutes, directly to a dedicated network of telemetry receivers at 14 automated "Burst Alert Stations" (BAS) sited along the satellite ground track (Villasenor et al 2002). The BAS network then re-distributes the GRB locations world-wide to all interested observers via Internet and the GRB Coordinates Network (GCN) in $\approx 1$ s [15]. Thus, prompt optical, IR, and radio follow-up identifications can be anticipated for a large fraction of *HETE* GRBs.

The *HETE* mission had produced 18 gamma-ray burst (GRB) localizations as of May 2002, with accuracies of 6 - 20 arcminutes. Eleven localizations were made between August 2001, and May 2002, corresponding to a rate of $\approx 15$ yr$^{-1}$. (This rate of localizations is $\approx 2$ times the mean rate achieved by the BeppoSAX satellite in its six years of operations). Four of the 10 most recent localizations had yielded optical afterglows (with 2 redshifts measured); one more yielded probable X-ray and radio afterglows. Three Chandra Target of Opportunity (ToO) observations had been carried out based on *HETE* localizations. *HETE* had detected a dozen short ($<2$s) duration GRBs and localized one of them. *HETE* also discovered 17 unusually "X-ray rich" GRBs, for which more than 30% of the energy is emitted in the 2 - 10 keV band. Of these 17 "X-ray rich" GRBs, 8 had been localized. (In total, *HETE* had detected $\approx 100$ GRBs of all types.)

However, despite the successes of the *HETE* mission in localizing GRBs, the scientific yield from the first year of the *HETE* mission was less than had been anticipated. Even now – with instrumental and operational challenges overcome – the GRB localization rate, while significantly higher than the BeppoSAX mean rate, is $\approx 1/3$ the rate that had been expected for *HETE* prior to launch. Although 6 SGR and $>30$ XRB Wide-Field X-Ray Monitor (WXM) localizations have gone out in near-real time, no accurate near-real time GRB localizations had been circulated as of May 2002.

Several factors contributed to the first year problems of the *HETE* mission. Difficulties of the kind that most missions experience during their Performance Verification phase had to be overcome, including issues involving optical aspect, spacecraft system reboots, an unreliable Cayenne Primary Ground Station, and implementation and testing of the sophisticated WXM flight software that localizes bursts. All of the above problems are now fixed, but because of manpower shortages in the first year arising from initial underfunding of the MO&DA effort, fixing them took much longer than had been expected (ie, $\approx 9$ months rather than the $\approx 3$ months anticipated).

The rate of *HETE* GRB localizations is lower than was expected prior to launch due to two factors: 1) a much smaller than expected "live time" for the instruments; and 2) a lower-than-expected rate of detection of GRBs by the WXM. It has been necessary to operate FREGATE and WXM only from terminator to terminator and often only from orbit "dusk" to orbit "dawn," in order to safeguard the health of these instruments because of solar activity (solar maximum occurred in January 2001 and an unexpected secondary maximum has occurred recently). This has reduced the WXM localization rate by a factor of about 1.5 compared to that expected prior to launch.

Three additional factors affected the rate: 1) the BATSE GRB rate, from which we scaled the predicted *HETE* rate, was reduced by a factor of 0.82 in the 4B catalog (Paceisas et al 1999); 2) it is becoming apparent that short GRBs, which have very hard gamma-ray spectra, may also be X-ray poor; and 3) the average ratio of $L_x/L_\gamma$ for GRBs is lower than was expected. While these three factors – taken together – reduce the expected rate of *HETE* GRB localizations by a factor of about 1.9 compared to the rate expected before launch, their quantification also represents new knowledge about GRBs that has important implications for future GRB missions. The result of the lower than expected "live time" and the re-

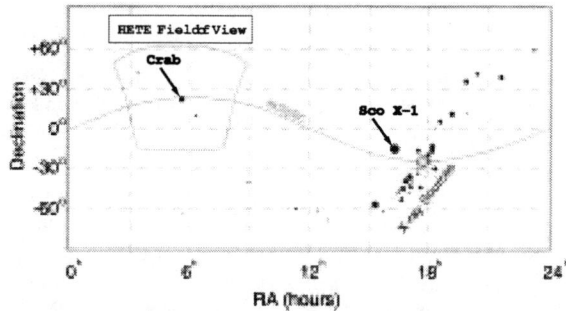

**FIGURE 2.** Map in celestial coordinates of the viewing region accessible to the *HETE* WXM and SXC instruments. Since *HETE* generally points anti-sun, the center of the instruments' field-of-view (FOV) traces along the ecliptic as the year progresses, always 12 hours distant in RA from the Sun. The date chosen for this figure was December 15, when the Crab Nebula is nearly centered in the *HETE* WXM FOV, shown as a red trapezoid. Approximately six months later, *HETE* would be viewing the region near Sco X-1 and the Galactic Center.

duction in the localizable population is a reduction in the overall *HETE* GRB localization rate of a factor of about 1.5 x 1.9 = 2.8 compared to the rate that was expected prior to launch.

What about the future? With a 2-year extended mission, the 4-year *HETE* mission can be expected to produce totals of ≈60 localizations, ≈25 optical afterglows, and $\gtrsim 25$ redshifts. The actual numbers are likely to be larger, as solar activity declines and the "live time" of the FREGATE and WXM instruments can be safely increased from 40 minutes to as much as 55 minutes per orbit. In addition, the demonstrated sensitivity of the SXC holds out the possibility for a number of near real-time $\lesssim 30''$ localizations and the detection of X-ray emission lines in the spectra of a few very bright GRBs. All in all, with the mysteries that still surround GRBs and the (currently) unique capabilities of the *HETE* FREGATE, WXM and SXC instruments, the scientific promise of the extended *HETE* mission is great. The extended *HETE* mission will yield new knowledge about GRBs and will continue to serve as a means of developing and testing flight software that is similar to the flight software for the BAT on the upcoming *Swift* mission; both will enhance the scientific productivity of the *Swift* mission.

## STATUS OF THE MISSION

In this section, we review the status of the *HETE* instruments, spacecraft and *HETE* ground station network. After a lengthy startup phase, all the components of

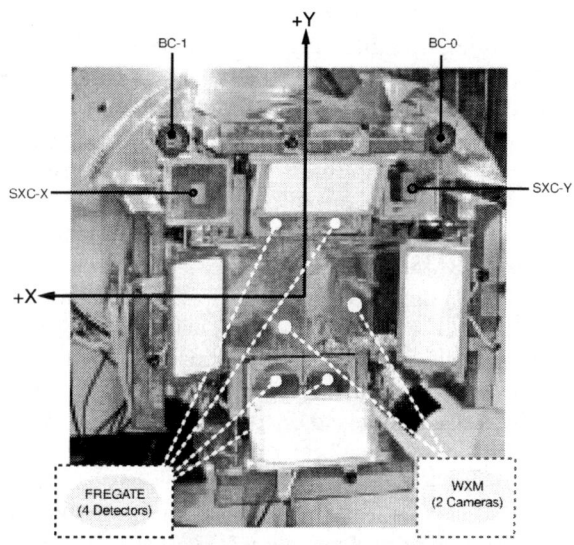

**FIGURE 3.** Instrument face of the *HETE* satellite. In this view, the two pairs of circular detectors comprising FREGATE are labeled "FREGATE (4 Detectors);" the two 1-D masks and detectors comprising the WXM are located in the center of the picture and are labeled "WXM (2 cameras);" and the two square units comprising the SXC are located at the upper left and upper right corners of the picture and are labeled "SXC-X" and "SXC-Y."

the fully-integrated *HETE* system are currently operating well. The anti-sun orientation of the *HETE* satellite dictates the region of the sky viewed by the science instruments as the year progresses (see Figure 2). The *HETE* spacecraft, and the science instruments mounted thereon, are shown in Figure 3.

## Spacecraft

All spacecraft subsystems are operating nominally, including the command and data communications system, the attitude control system, and the power system.

**Communications system.** The spacecraft RF system showed sensitivity to noise sources on board that were not detected in ground testing, resulting in lost or garbled commands. Sometimes, the extra noise bits caused enough extra interrupts to upset the spacecraft controller and cause a full reboot of the flight software. These problems were the biggest source of operational inefficiency in the early part of the mission. In May 2001 we modified flight and ground RF software to reduce the sensitivity of the RF system to these noise sources. We no longer see command errors due to noise: the changes have reduced the error rate by five orders of magnitude.

The current RF system is very robust and capable of downloading 150-200 Mbytes of data per day.

**Attitude Control System.** The attitude control system displayed its authority when *HETE* attained its on-station orientation, in its nominal anti-solar orientation, within three hours of launch. Since then, it has performed well, both in normal operations and after rare spacecraft anomalies. We have recently improved orbit night stability to $<2'$ RMS.

**Power system.** Before launch, the spacecraft was advertised to be "walk-away safe", in that the power system was able to fully charge the spacecraft batteries regardless of the orientation of the spacecraft. The power system has shown this to be true in a handful of anomalous situations and has been rock-solid for the duration of the mission. Degradation of the batteries has been negligible. Solar panel performance has been well above expectations: in normal operation *HETE* has a power margin of about 50% above the nominal orbit cycle requirements.

**TABLE 1.** Omnidirectional γ-ray Spectrometer

| | |
|---|---|
| Built by | CESR (France) |
| Instrument type | Cleaved NaI(Tℓ) |
| Energy Range | 6 keV to $>400$ keV |
| Timing Resolution | 10 $\mu$s |
| Spectral Resolution | $\sim$13% @ 81 keV, |
| | $\sim$10% @ 356 keV |
| Effective Area | 160 cm$^2$ |
| Background (25-100 keV) | 100 c/s/detector |
| Sensitivity, 6 σ, 50 - 300 keV | $\sim 1 \times 10^{-7}$ erg/cm$^2$/s |
| Passively collimated FOV | $\sim$4 steradians |

## Gamma-Ray Detectors

The prime objectives of the French Gamma Telescope (FREGATE) are the detection and spectroscopy of GRBs, SGR bursts, and XRBs, and monitoring of other variable sources. Table 1 lists its properties. FREGATE has functioned flawlessly since launch [13]; its sensitivity is more than a factor of two better than was expected prior to launch. Its GRB detection rate is $\approx 44$ bursts per year coincident with other spacecraft, plus over 50 additional GRBs per year detected only by *HETE*. Of the $\approx 44$ coincident bursts per year, about 11 originate from outside the passively collimated field of view of the detector, a fact which enhances FREGATE's value to the Interplanetary Network [17]. (The passive collimator reduces the low energy background by excluding the diffuse cosmic component and the contributions from soft sources in the Galactic center region, but is semi-transparent to GRBs).

**TABLE 2.** Wide Field X-ray monitor

| | |
|---|---|
| Built by | RIKEN (Japan) and LANL |
| Instrument type | Coded Mask with PSPC |
| Energy Range | 2 to 25 keV |
| Timing Resolution | 1 ms |
| Spectral Resolution | $\sim$ 25% @ 20 keV |
| Detector QE | 90% @5 keV |
| Effective Area | $\sim$175 cm$^2$ (each of two units) |
| Sensitivity (10 σ) | $\sim 8 \times 10^{-9}$ erg/cm$^2$/s |
| | (2-10 keV) |
| Field of View | 1.6 steradians (FWZM) |
| Localization resolution | $19'$ (5σ burst) |
| | $2.7'$ (22σ burst) |

The FREGATE spectral response was extensively calibrated prior to launch, and the calibrations have been confirmed in flight using observations of the pulsed spectrum of the Crab. The in-flight energy resolution is excellent and the lower energy threshold is 6 keV, due to the use of cleaved crystals. For each burst whose arrival direction is known, a response matrix is constructed and models of the incident photon spectrum are being fit to the data using XSPEC. The combination of broad energy range and good energy resolution is a powerful one for the discovery and study of spectral features.

## WXM

The primary objectives of the Wide-Field X-Ray Monitor (WXM) are the detection, localization, and spectra of GRBs, SGR bursts, and XRBs. Table 2 lists its properties. The WXM has performed well since launch. The sensitivity of the WXM is equal to that expected prior to launch.

The localization accuracy of the WXM is better than expected prior to launch (systematic error $<2.7'$ rather than $>5'$), as calibrated using observations of the Crab, Sco X-1, SGRs 1806-20 and 1900+14, and numerous known X-ray burst sources [18] [21] [14]. The WXM has successfully localized nearly all of the GRBs that it has detected as of May, 2002. The flight algorithm had correctly localized 7 GRBs in real time. Unfortunately, in most of these cases the burst occurred near full moon; consequently, there was no real-time aspect and therefore no GCN Notice was sent out to the community. One SGR burst and two X-ray bursts that the WXM flight software correctly localized in near-real time were sent out to the astronomical community. An additional 6 SGR bursts and 25 X-ray bursts that the WXM flight software correctly localized in near-real time were not sent out to the astronomical community, in order not to overwhelm the community with non-GRB burst alerts, or, because of a lack of real-time aspect, were distributed with imprecise positions.

The WXM spectral response was extensively calibrated prior to launch, and the calibrations have been confirmed and refined in flight using observations of the Crab nebula. The in-flight spectral resolution is equal to that expected. For each burst that is detected, a response matrix is being constructed and models of the incident photon spectrum are being fit to the data.

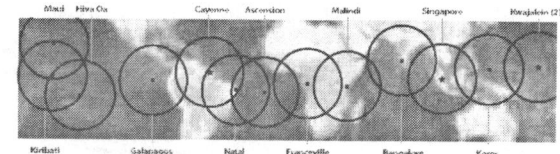

**FIGURE 4.** The 14 *HETE* Burst Alert Stations (BAS) are distributed in longitude near the equator in a manner that *HETE* is within range of a BAS over 99% of its orbit.

## SXC

The *HETE* soft X-ray cameras (SXC; [16] [20]) were designed to provide fine ($<1'$) localization of bursts, based on their 1-10 keV emission. The SXC consists of two orthogonally-oriented, one-dimensional coded aperture units, each comprised by a fine ($\sim 45$ $\mu$m) featured slit mask, separated by 10 cm from an X-ray CCD detector plane, with 15 $\mu$m pixels. The pre-launch and current properties of the SXC are given in Table 3.

To prevent contamination of the CCD detectors by non X-ray photons, the SXC was provided with two redundant optical blocking filter (OBF) systems: 1) very thin (0.05 $\mu$m) aluminized plastic films, covering the entire aperture, and 2) Be foil (25 $\mu$m) filters, covering 1/2 of the detector area. The thin film OBF system operated for 3 months (Y-camera) and 5 months (X-camera), respectively. After those times, the films were apparently lost due to erosion by atomic oxygen.[1]

The foil OBFs have remained intact throughout the mission. However, the foil OBFs protect only 1/2 of the detectors, and are subject to edge leakage during bright moon conditions. As a result, the SXC sensitive area is now 1/2 the value at launch, and the SXC can only be used during the "lunar dark" of the month (*i.e.*, 24 days, or $\sim$80% of each lunar month).

After the loss of the outer OBFs, an extensive re-write of the SXC flight software was required to correct for residual light leakage. The revised flight software was uploaded and tested in July 2001, which successfully revived the SXC. Prior to launch, we had estimated that the SXC would localize about 1/3 as many GRBs as would the WXM (albeit with 100x smaller error circles), based on a solid angle-area-time product comparison. In fact, with the SXC acting as a "vernier" for the WXM, the usable SNR for SXC localizations was improved by $\sim$1.5 x over pre-launch estimates (see Table 3). With the reduction in sensitive area because of the Outer OBF loss, and accounting for the reduced operating time per month, we estimate that the SXC will localize 1/5 as many GRBs as will the WXM. For a WXM localization rate of 15 events yr$^{-1}$, we now expect an SXC localization rate of $\sim$3 events yr$^{-1}$.

During July-August 2001, the SXC localized 9 galactic transients, produced by 4 XRBs and 1 SGR [20], to an accuracy of $< 1'$. During the first 9 months after the SXC had been restored to operation, all 10 of the WXM-localized GRBs occurred either at times of bright moon, were measured to be faint in the WXM, or were outside the useful FOV of the SXC. Thus, no GRB localizations by the SXC had occurred as of May 2002. (During August–December 2002, five GRB localizations by the SXC occurred).

### Primary Ground Stations

The *HETE* primary ground stations (PGS) are located at Singapore, at the Kwajalein Atoll in the Republic of the Marshall Islands, and at Cayenne, French Guinea. All three stations are currently operating at full capacity, downlinking up to 9 Mbytes of spacecraft data per contact for up to 15 contacts per day. Because commanding of the *HETE* flight instruments is carried out by automated ground software, the reliability of the PGS network is paramount to the routine operation of the spacecraft. Although the operation of the PGS network is now relatively routine, there was a significant learning curve with all three stations, with some requiring more time and effort to achieve stable operation.

## BURST ALERTS

**Burst Alert Network.** The *HETE* Burst Alert Network (BAN) currently consists of fourteen burst alert stations (BAS) sited at locations along the equator (see Figure 4). There was a significant debugging period re-

---

[1] Atomic oxygen densities at 600 km apparently increase by $10^3$-$10^5$ times during intense solar flares [51]. *HETE* experienced a series of such flares in late 2000 – early 2001, shortly before the loss of the SXC outer OBFs. The existence of this strong dependence of atomic oxygen on solar activity at the *HETE* orbital altitude was not known at the time the SXC underwent its design reviews in 1998.

**TABLE 3.** Soft X-ray Camera

|  | At launch (both OBFs intact) | At present (Inner OBF only) |
|---|---|---|
| Built by | MIT CSR | |
| Instrument type | 4 CCID-20 | 2 CCID-20 |
| Camera dimensions | 10cm × 10cm × 17.5cm | |
| Energy Range | 500 eV to 14 keV | 1.3 to 14 keV |
| Timing Resolution | 1.2s | |
| Spectral Resolution | 46 eV @ 525 eV | ∼300 eV |
| | 129 eV @ 5.9 keV | @ all energies |
| Dectector QE | 93% @ 5 keV, | 93% @ 5 keV, |
| Dectector QE > 20% | 0.5–14 keV | 1.3–14 keV |
| Effective Area | 74.4 cm$^2$ | 37.2 cm$^2$ |
| Source Sens (Crab) | 2.3 $t^{-1/2}$ (5.5$\sigma$) | 1.9 $t^{-1/2}$ (3.5$\sigma$) |
| Burst Sens (cts/cm$^2$/s) | 1.0 (5.5$\sigma$) | 0.8 (3.5$\sigma$) |
| FOV (FWZM) | 1.88 sr | 1.29 sr* |
| Focal Plane scale | 33″ per CCD pixel | |
| Localization | 40″(systematics limit) | |

* There is an additional 0.54 sr FOV for one-sided localizations.

quired to get most of the stations to operate automatically: all stations in the BAN, with the exception of Kiribati, are now operating routinely, except for instances of power or Internet interruptions. (Difficulties at Kiribati have been ameliorated by the addition of the new Maui station, which covers much of the Kiribati footprint). The average real-time coverage of *HETE*'s orbit has improved from ∼ 40% at the beginning of 2001 to the current value of ∼ 90%.

Notification of *HETE* bursts proceeds in two steps: first, the results of the on-board analysis of burst data are distributed via the Burst Alert Network in near real time; then, once the full burst data set is available on the ground, the results of ground analyses may be distributed.

**Real-Time Alerts.** Information about GRBs detected by *HETE* is transmitted to the BAN in real time, immediately relayed to the *HETE* Mission Control Center at MIT, and then promptly reformatted and sent to the GRB Coordinates Distribution Network (GCN); the GCN then distributes the information in the form of email, pager and/or Internet socket messages to observers around the world (see Figure 5). The first message received by observers includes the time, energy range, duration, and significance of the detected burst; subsequent messages can include burst localizations as calculated by the flight software, if the burst was in the WXM and SXC fields-of-view.

The response time of the *HETE*-to-GCN system depends strongly on the presence of a functioning BAS with Internet access within receiving range of *HETE* at the time of message transmission. When this is the case,

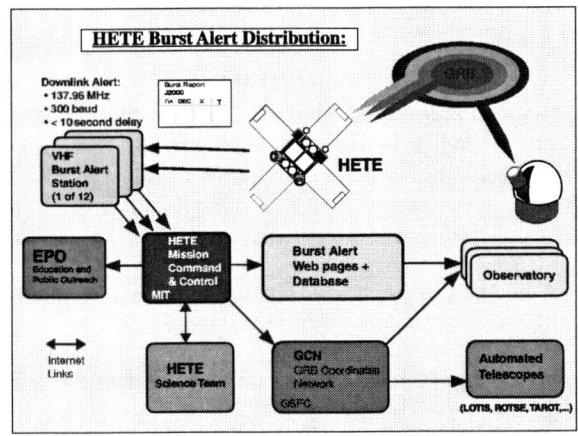

**FIGURE 5.** Flow chart for dissemination of a *HETE* burst alert. As soon as a burst localization is calculated by *HETE*, it is transmitted to the Burst Alert Stations (BAS) at VHF frequencies. The localizations are then routed from the BAS to a collection point at MIT via the Internet. Next, the localizations are posted to the GCN, to the MIT *HETE* web pages, and to the *HETE* Science Team. Several of the links shown are served in parallel by socket connections, email, and beepers.

the time between burst detection and receipt by the observer is typically under 20s; in other cases, the delay can be as long as 15 minutes. Details are given by Crew et al. [25].

**Results of Ground Analyses.** Once the detailed burst data are collected by a PGS, typically 20–60 minutes after the burst, more detailed analyses of the burst data can be performed. In typical cases, a more precise

**TABLE 4.** Eighteen GRBs localized by the *HETE* Satellite between launch and May 2002.

| GRB | *HETE* ID | GCN Circular | FREGATE SNR* | WXM SNR* | Error Radius | Alert Delay | Localization Delay (hrs) | Comments |
|---|---|---|---|---|---|---|---|---|
| 010110 | untrig | - | - | 23.3 | [∼10′] | N/A | N/A | Fregate was off |
| 010213 | untrig | 934 | 9.8 | 69.4 | 3.5′ | 36h | 36 | X-ray rich |
| 010225 | H1491 | - | 15.9 | 19.8 | [∼10′] | N/A | N/A | X-ray rich; no aspect |
| 010326B | H1496 | 1018 | 19.7 | 11.2 | 18′ | 4.75h | 4.75 | Soft spectrum |
| 010612 | H1546 | 1065 | 23.2 | 9.1 | 36′ | 15s | 69 | *HETE* +IPN |
| 010613 | H1547 | 1067 | 70.2 | 12.1 | 36′ | 102s | 57 | Double-peaked |
| 010629B | H1573 | 1075 | 62.3 | 21.8 | 15′ | 78s | 6.5 | Double-peaked |
| **010921** | **H1761** | **1096** | **83.3** | **7.3** | **10′** | **17s** | **5.2** | **Optical ID** |
| 010928 | H1770 | 1103 | 48.9 | 6.3 | 6′ | 203s | 6.2 | 1-sided error box |
| 011019 | untrig | 1109 | 10.1 | 14.1 | 35′ | 12.1h | 12.1 | X-ray rich |
| 011130 | H1864 | 1165 | 4.1 | 9.7 | 7.6′ | 4.4h | 4.4 | X-ray rich (Chandra TOO) |
| 011212 | untrig | 1194 | 7.4 | 8.9 | 11′ | 6.3h | 6.3 | X-ray rich |
| **020124** | **H1896** | **1220** | **41.1** | **11.4** | **12′** | **11s** | **1.4** | **Optical ID; Long** |
| **020127** | **H1902** | **1229** | **37.9** | **17.1** | **8′** | **96s** | **1.8** | **Radio ID (Chandra TOO)** |
| **020305** | **H1939** | **1262** | **7.2** | **3** | **25′** | **42s** | **10** | **Optical ID; double-peaked** |
| 020317 | H1959 | 1280 | 11.8 | 8.8 | 18′ | 34s | 0.9 | X-ray rich |
| **020331** | **H1963** | **1315** | **27.2** | **9.7** | **8′** | **55s** | **0.7** | **Optical ID; Long** |
| *020531* | *H2042* | *1399* | *22.4* | *7.0* | *35′* | *178s* | *1.5* | *Short GRB (Chandra TOO)* |

* SNR = signal-to-noise ratio in the energy ranges of 2-25 keV and 8-85 keV for WXM and FREGATE, respectively.

burst localization with a smaller error radius than any calculated on board is available within 10 minutes of receipt of data. Such positions are distributed as a GCN Notice as soon as the on-duty burst analysis team agrees on the results of the analyses. Bursts H1896, H1902, H1959, and H1963 are examples of this process (refer to Table 4).

In some cases, the ground burst analysis is more complex: the spacecraft aspect is difficult to calculate due to contamination by moonlight, the burst shows strong spectral evolution, or the results of independent analyses of the same data show large discrepancies. In these situations, the refined burst localization is delayed by several hours, while the burst-analysis team analyzes the full data set. Bursts H1761, H1770, and H1864 are examples of these cases.

Finally, routine inspections of survey data have revealed several untriggered bursts. These analyses are typically only completed several hours after the burst occurred, so significant delays are incurred in the analyses of the data. Bursts GRB011019 and GRB011212 are examples of this type of burst.

## LESSONS LEARNED

In its first year, *HETE* did not reach full operational status as quickly and did not achieve as high a rate of GRB localizations as both the *HETE* team and the community had expected prior to launch.

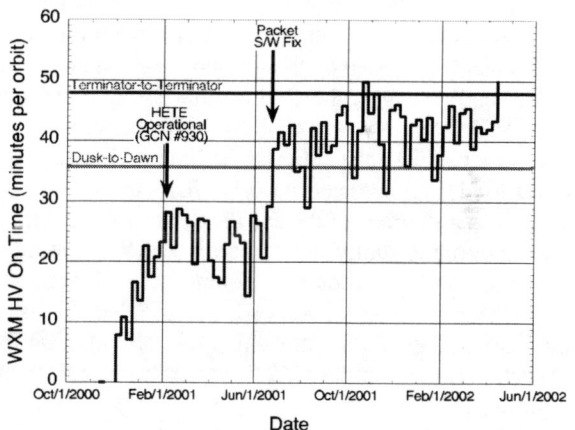

**FIGURE 6.** The mean WXM HV on time as a function of time through the *HETE* mission, which is a good indicator of the overall operational efficiency of *HETE* during its mission. Indicated are the time at which *HETE* was declared capable of detecting and localizing GRBs and the date on which spacecraft communications became more robust. The mean operation efficiency of *HETE* effectively doubled after July 1, 2001, when compared with the months immediately after declaration of operations.

The delay in reaching full operational status was, in large part, a consequence of inadequate funding of the *HETE* Mission Operations and Data Analysis (MO&DA) effort, below even the minimal budget proposed at the outset of the *HETE-2* mission. This funding level left the

*HETE* team under-manned and with insufficient reserves to solve, in a timely fashion, the usual problems that surface during the Performance Verification phase of any mission.

We describe here some of the major difficulties we encountered; some lessons we learned that may be valuable to other missions - especially those studying transients (e.g. *Swift*); and the reasons why we think that in a full 4-year mission, *HETE* will still be able to fulfill its original mission objectives.

**Shortfall in MO&DA Funding**. As a consequence of curtailed MO&DA funding, the *HETE* operations and instrument teams were seriously under-manned. Team members who had the ability and the knowledge to fix the initial glitches in spacecraft and instrument software had to operate the spacecraft and the ground stations instead.

A pertinent example was frequent rebooting of the spacecraft during the first months of the mission. This was due to rare errors in the command packets received by the satellite which could cause the spacecraft processor to reboot. A second example was the difficulty achieving stable spacecraft aspect. All of the links in the software chain between the star tracking cameras, the sun sensors, and the control software for the momentum wheel had been tested on the ground prior to launch, but end-to-end testing of the chain could not be done until the *HETE* satellite was in orbit. A third example was the inaccuracy of the ephemerii (the TLEs) provided by NORAD for *HETE*'s equatorial orbit. This meant that the directions and times of the *HETE* passes over the three primary ground stations were poorly known, leading to failed up- and down-loads of commands and data. The *HETE* team was forced to solve this problem by using the experimental GPS unit (which was provided by our French partners) on the *HETE* satellite to calculate the required TLEs.

Each of these problems could have been solved in a timely way with adequate staff. NASA provided supplementary MO&DA funding in May 2001, which allowed the *HETE* project to hire several temporary additional personnel, who were able to address these problems. However, the underfunding of the initial period of the *HETE* primary mission had increased the Performance verification phase of the *HETE* mission from the $\approx 3$ months that had been planned to $\approx 9$ months.

**Cayenne Primary Ground Station**. The Cayenne Primary Ground Station (PGS) plays a key role in the ability of the *HETE* mission to uplink commands to the spacecraft and instruments, and to download the large data masses produced by burst triggers to the ground in a timely fashion. This is particularly the case because the Cayenne PGS is widely separated in longitude from the other two primary ground stations, which are located in Singapore and at Kwajalein. Unfortunately, the Cayenne primary ground station, which was provided by our French partners, was unreliable for the first year of the *HETE* mission. As a result of extensive modifications made by our partners to the Cayenne PGS, it is now working well. This experience lends added emphasis to the well-known fact that missions involving contributions by a variety of international partners are more complicated, and frequently more difficult to manage and operate.

**Complexity of Autonomous Operations**. Any mission that relies on on-board systems for autonomous operation and source localization are likely to require more than 30 days of check out and on-orbit calibration. System interactions are extremely complex, and one cannot simulate all of them on the ground. Interference from scattered optical light, changing backgrounds, power management in unscheduled modes, and unplanned science events that interrupt pre-planned sequences by the satellite contribute to the complexity of the task. Because of these factors, the challenge of producing localizations in near real-time is much greater than was appreciated prior to launch; every link in a long chain of criteria must be satisfied: the burst must be bright enough, it must occur within roughly 30 deg of the center of the FOV of the WXM, the flight software must correctly select burst and background time intervals despite the tremendous diversity of burst time histories, accurate and stable optical aspect must be available, the secondary ground station must be active and functioning properly, the GCN network must be active and functioning properly, etc.

**Lower than Expected GRB Localization Rate**. The current rate of GRB localizations is lower than was expected prior to launch due primarily to two factors: (1) a much smaller than expected "live time" for the instruments; and (2) a lower than expected rate of detection of GRBs by the WXM. It has been necessary to operate the French Gamma-Ray Telescope (FREGATE) and WXM only from terminator to terminator (48 minutes per orbit) and often only from "dusk" to "dawn" (35 minutes per orbit), in order to safeguard the health of these instruments because of solar activity (solar maximum occurred in January 2001 and an unexpected secondary maximum occurred in late 2001). These factors reduced the WXM localization rate by a factor of about 1.5 compared to that expected prior to launch (see Figure 6).

Three additional factors affected the rate: 1) the BATSE GRB rate, from which we scaled the predicted *HETE* rate, was reduced by a factor of 0.82 in the 4B catalog (Paceisas et al 1999); 2) it is becoming apparent that short GRBs, which have very hard gamma-ray spectra, may also be X-ray poor; and 3) the average ratio of

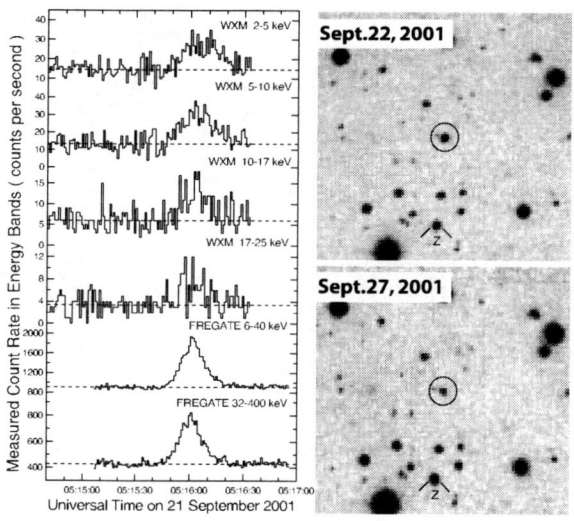

**FIGURE 7.** At left, the light curve of GRB010921 as seen by FREGATE and the WXM. At right, images of the fading optical counterpart of GRB010921 [19]

$L_x/L_\gamma$ for GRBs is lower than was expected. While these three factors – taken together – reduce the expected rate of *HETE* GRB localizations by a factor of about 1.9 compared to the rate expected before launch, their quantification also represents new knowledge about GRBs that has important implications for future GRB missions. The result of the lower than expected "live time" and the reduction in the localizable population is a reduction in the overall *HETE* GRB localization rate of a factor of about 1.5 x 1.9 = 2.8 compared to the rate that was expected prior to launch.

## HETE SCIENTIFIC RESULTS

After a longer than anticipated startup phase, the *HETE* spacecraft and instruments are now working well. Recently, the mission has produced an extensive body of scientific results for transient phenomena. In this section, we describe some of the scientific highlights of the first year's results.

### Gamma-Ray Bursts

FREGATE is detecting GRBs at a rate of $\approx 100$ bursts $yr^{-1}$, comprised of $\approx 44$ confirmed GRBs $yr^{-1}$ and over 50 unconfirmed GRBs $yr^{-1}$. About 75% of the confirmed GRBs originate within the FOV of the FREGATE collimators. The WXM had localized 18 GRBs as of May 2002, with accuracies of 6 - 20 arcminutes. Table 4 lists the properties of these 18 events.

**GRB Afterglows and Redshifts.** In Figure 7, we show the lightcurves in 6 X-ray and γ-ray energy bands, and the fading optical afterglow, of GRB010921, the first *HETE* burst for which an optical afterglow was detected. This burst was relatively bright and has a redshift $z = 0.450$, making it a favorable candidate for detecting a SN component in the lightcurve and spectrum of the optical afterglow. However, no such component was detected. During the 9 months prior to May 2002, the *HETE* mission has produced 11 localizations, corresponding to a rate of $\approx 15$ $yr^{-1}$. Four of these 11 localizations yielded optical afterglows (with 2 redshifts measured); one more yielded a probable X-ray and radio afterglow. Figure 8 shows the *HETE* WXM error circle, IPN annulus, and location of the optical transient for the 4 bursts with optical afterglows.

The dissemination of a number of *HETE* GRB localizations within 0.7-2 hours of the burst (see Table 4) allowed optical follow-up observations to probe the brightness distribution of optical afterglows at early times. Figure 9 shows that, if some GRB optical afterglows are very bright, these near-real-time localizations will allow robotic telescopes to probe the transition from the burst itself to the afterglow.

**X-Ray Rich GRBs.** One of the very first GRBs localized by *HETE*, the burst 010213, was an "X-ray rich" GRB (see Figure 10). As of May 2002, FREGATE had detected 17 "X-ray rich GRBs" (i.e., a burst for which $L_x/L_\gamma \equiv F_{2-10keV}/F_{50-300keV}$ is $> 0.3$) (see Figure 11); six of these occurred within the WXM FOV and were localized by the WXM prior to May 2002 (see Table 4). Thus, $\approx 40\%$ of the GRBs detected by the *HETE* mission are "X-ray rich" bursts. This is similar to the recent report that 30-50% of the GRBs seen by the WFC's on BeppoSAX are "X-ray rich" [50].

Are X-ray rich GRBs nearby or exceptionally far away? Are any associated with supernovae? The *HETE* mission has disseminated the localizations of three X-ray rich GRBs within 0.9-6.3 hours of the burst (see Table 4). Yet no radio or optical afterglow has been detected from these or any other X-ray rich GRBs, raising the question of whether such bursts have unusually faint or no optical and radio afterglows. Finding the counterpart of an X-ray rich GRB would therefore be a breakthrough. During the *HETE* extended mission from 2002-2004, the WXM can be expected to localize an additional 12 or so "X-ray rich" GRBs, which will provide opportunities for follow-up observations to detect or set severe upper limits on optical or radio afterglows from these bursts.

**Short GRBs.** Short GRBs (duration $< 2$ s, [40]) remain an enigma. At present, nothing is known about the distance to nor the nature of the short GRBs, despite extensive efforts. *BATSE* localized several short GRBs in near-real time, but the areas of the error circles were typ-

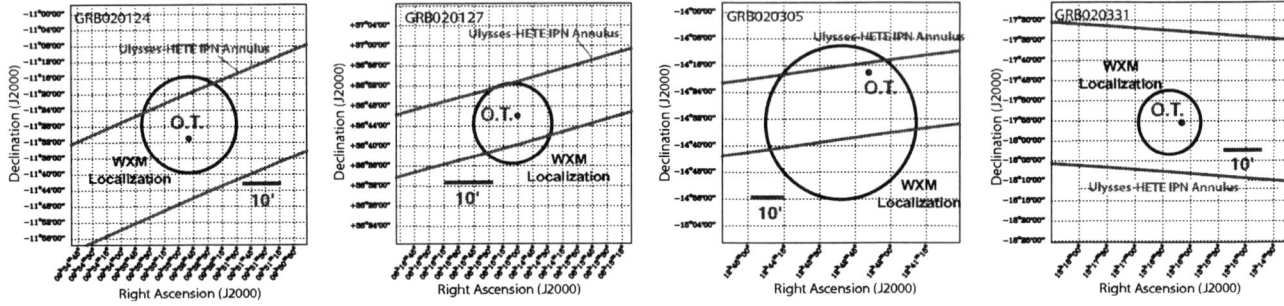

**FIGURE 8.** *HETE* error circles, IPN annuli, and locations of optical afterglows for four GRBs.

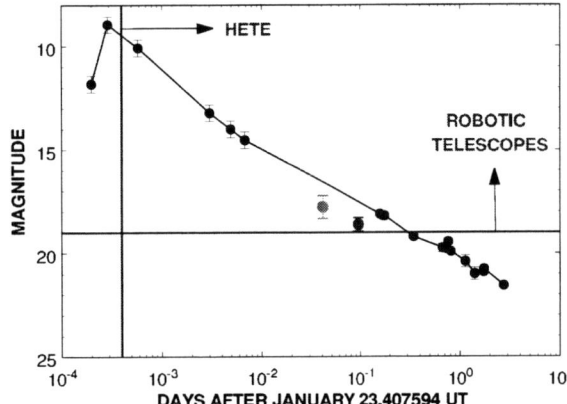

**FIGURE 9.** The magnitude of the optical emission associated with GRB990123 as a function of time [35]. The two points slightly below the GRB990123 decay line are measurements for two recent *HETE* bursts (green: GRB020331 [9]; red: GRB020124 [10]). The vertical line at 35s indicates a expected delay to issue the first real-time *HETE* localization notice. The horizontal line shows the sensitivities of currently operating robotic telescopes such as Super-LOTIS, ROTSE III, and TAROT-2.

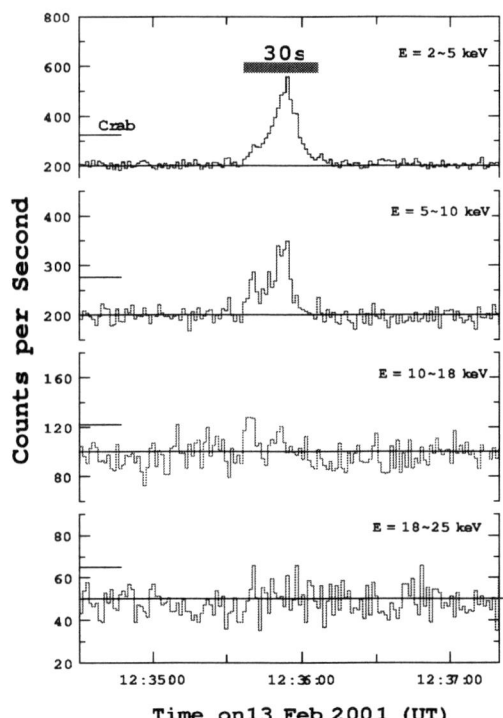

**FIGURE 10.** WXM Light curves for the "X-ray rich" burst GRB010213 discovered by *HETE*. A strong peaking of the light curve below 20 keV is particularly prominent for this burst.

ically $\sim$ 30 square degrees, making radio, optical, and X-ray follow-up observations difficult. The Third Interplanetary Network (IPN) derived localizations for four short GRBs (000607, 001025B, 001204, and 010119) with delays of 15-65 hours. However, in three of these cases, the opportunity for follow-up observations was compromised either by the burst being close to the sun (000607, 65 degrees) or close to the Galactic plane (001025B, b $\sim$ 4 degrees; 010119, b $\sim$ 5 degrees). Only one burst (001204) was optimally placed on the sky for followup observations. However, in this case the delay (65 hours) in deriving a localization for the burst hampered follow-up efforts.

FREGATE has detected a dozen such bursts (corresponding to $\approx$ 20% of the GRBs that it has detected). Figure 12 shows the time history of GRB010628, one of the short bursts detected by FREGATE. The localiza-

tion by *HETE* of the short-hard burst GRB020531 (duration $\sim$ 300ms in the FREGATE 30-400 keV band [53]) promises to enlighten us a great deal as to the nature of short-hard bursts [54]. The prompt reporting and localization of GRB020531 by *HETE* and the IPN enabled a far-flung multiwavelength campaign to be mounted, including multi-epoch ToO observations by the Chandra X-ray Observatory [55] and the Hubble Space Telescope.

**Emission Lines in GRB X-Ray Afterglows.** The report [45] of hydrogen-like emission lines from a series of low-Z elements (Mg XII, Si XIV, S XVI, Ar XVIII,

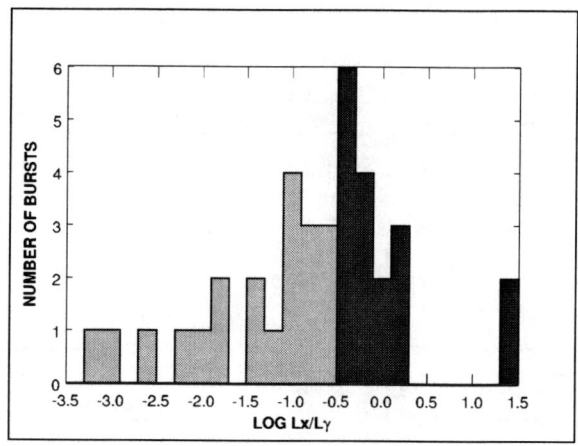

**FIGURE 11.** The X-ray (2-10 keV) to gamma-ray (30-400 keV) fluence ratios of HETE bursts. X-ray rich bursts are shaded dark. Some outliers have been omitted for clarity. Note the tail of low $L_x/L_\gamma$ bursts.

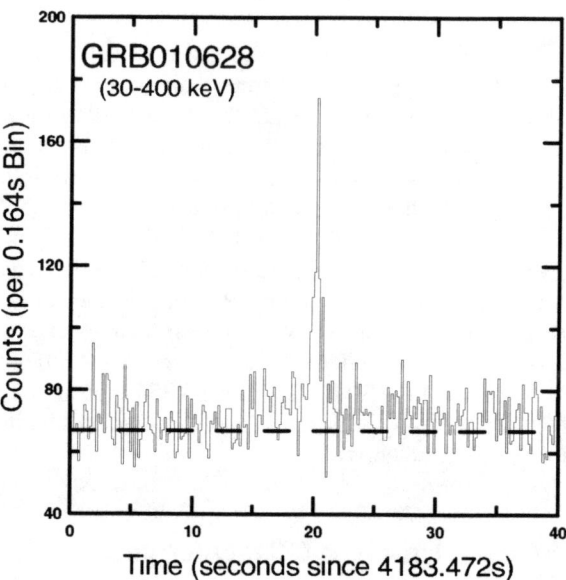

**FIGURE 12.** A short GRB detected by FREGATE: for this burst, $T_{50}=0.5$s, $T_{90}=1.0$.

and Ca XX, and possibly Ni XXVIII) in the X-ray afterglow of GRB011211 demonstrates that that for GRBs at redshifts z<2.5, both the location and the redshift of the GRB can be accurately determined from X-ray observations in the 0.5-10 keV band alone. X-ray observations of such emission lines can strengthen the association between GRBs and the deaths of massive stars, provide unique information about the nucleosynthetic yield of Type Ib/Ic supernovae, and reveal the dynamics of the GRB explosion [49] [45] [42].

GRB011211 lies at a redshift $z = 2.14$, so 4 of the

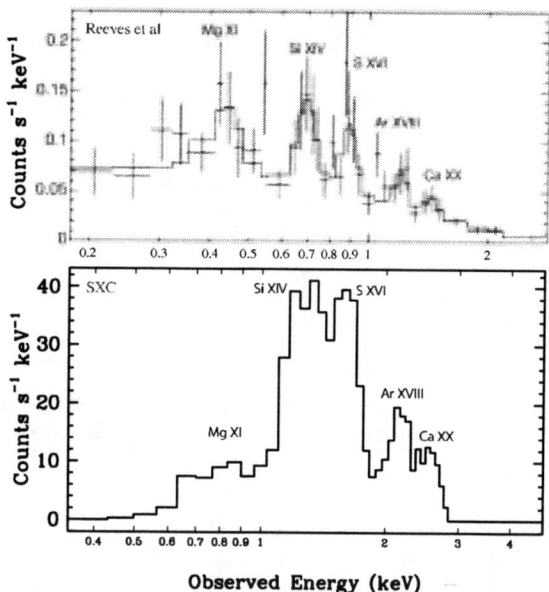

**FIGURE 13.** In the upper panel, the XMM spectrum of GRB011211 [45]; in the lower panel, the simulated spectrum of a bright (250 Crab-s) GRB as seen in the SXC.

5 emission lines lie below the SXC window cutoff (at 1.3 keV), and thus would not be detectable by the SXC. However, for GRBs with $z < 0.8$ (median value for the 25 GRBs with known redshift), the five lines would be accessible to SXC observations. To evaluate the possibility that the SXC might be able to detect such emission lines, we have modeled the prompt emission from a plasma with the properties reported for GRB011211 by Reeves et al, but at a redshift of $z = 0.8$. In Figure 13, we show the spectrum that we would expect to measure. Our simulation assumes that the burst has a flux $\sim 5\times$ Crab and a duration of 50s. A burst with this brightness could be expected to fall within the FOV of the SXC a couple of times per year, at favorable periods in the lunar month. As can be seen in Figure 13, the emission line structure is quite evident in the simulation. Thus there is a possibility that during the two-year *HETE* extended mission the SXC will be able to detect X-ray emission lines in the spectra of a few very bright GRBs.

## Soft Gamma Repeaters

*HETE* is a sensitive SGR monitor. Its anti-solar pointing direction places the three known Galactic SGRs, and the fifth possible SGR, within the fields of view of the WXM and FREGATE for several months each year. This was the case in June/July 2001, when SGR1900 and SGR1806 became active: FREGATE detected a total of 19 bursts from these SGRs; at the same time, the WXM

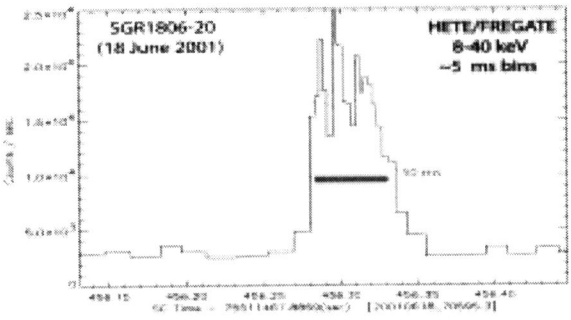

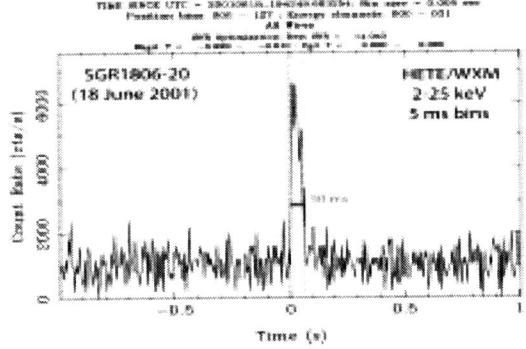

**FIGURE 14.** Light curve for a soft gamma repeater burst from SGR1806-20 detected by *HETE* on 18 June 2001. The upper panel shows the FREGATE light curve in the 8-40 keV band, while the lower panel shows the WXM light curve in the 2-25 keV band. The burst duration of $\sim 50$ ms is indicated by a horizontal bar in each panel; due to plot scale choices, more background data is plotted before and after the burst for the WXM.

detected 6 of these bursts and 20 other unconfirmed SGR bursts, most of which are likely to be from these sources [7] [8]. In Figure 14, we show a sample short-duration ($\sim$ 50ms) burst from SGR1806-20, observed by both the WXM and FREGATE. In addition, FREGATE has detected SGR bursts up to 130 degrees off-axis, which means that it can monitor much of the sky for activity from any of the known SGRs, as well as from any new SGRs, for much of the year.

*HETE*'s excellent energy resolution and low-energy sensitivity have already revealed new, unexpected spectral behavior in an intense burst from SGR1900+14 on 2001 July 2 [23]. Despite the fact that SGR1900+14 has been observed to burst hundreds of times, the unusual spectrum we observed has never been detected before, due to the fact that previous experiments lacked *HETE*'s combination of broad energy coverage and excellent resolution.

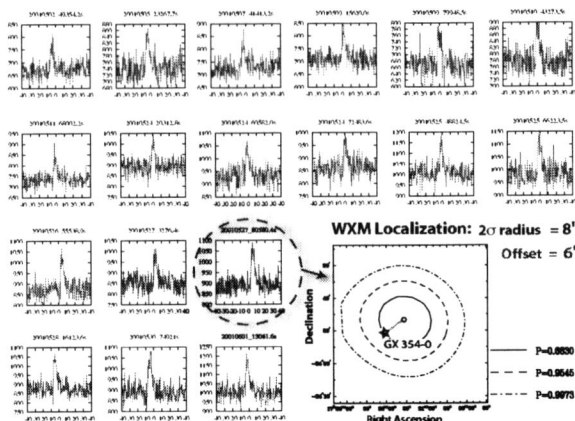

**FIGURE 15.** The time histories of 18 of the 30 XRBs from GX354-0 observed by FREGATE along with the WXM localization obtained for one of these bursts.

## X-ray Bursts

The galactic plane is in the FOV of FREGATE and the WXM for approximately three months per year (mid-May through mid-August). *HETE* is able to detect many X-ray bursts (XRBs) during these months, and to monitor particular XRB sources for an extended period of time.

The WXM detected about 130 XRBs from 15 XRB sources in 2001 [22]. Of these XRBs, 53 were triggered events and 77 were untriggered events. Two XRBs could not be associated with known XRB sources; they may be XRBs from previously unknown XRB sources or extra-galactic X-ray rich GRBs. Figure 15 shows the time histories of 18 of the 30 XRBs from GX354-0 observed by FREGATE, along with the WXM localization obtained for one of these bursts.

## Unsolved GRB Questions

Despite the rapid progress made during the *BATSE* and *Beppo SAX* eras, it would be premature to pronounce the GRB mystery solved. There is no consensus even on the nature of the central engine powering the subclass of long bursts. Is it a newly formed black hole undergoing accretion or the collapse of a rapidly spinning magnetar? If accretion, is the jet energized by black hole rotation, MHD processes in the disk, or neutrinos? Do the GRB and the supernova it produces happen simultaneously, or are they separated by days or even years [39]?

**Diversity of GRBs.** There is also growing evidence that what we call GRBs may be a catch-all term for a diverse family of high-energy transients. Short bursts (duration $< 2$s [40]) remain mysterious; the first prompt localization by *HETE* of GRB020531 may shed light on such short GRBs [54]. To these we must now add "X-ray rich GRBs" [41] and "X-ray flashes" [50]. Is each of these a different kind of event, or is there some underlying unified model capable of producing all of them (*e.g.*, depending on the observer's viewing angle)?

Theory also predicts other classes of GRBs having longer durations than ordinary GRBs. Some may come from a first generation of stars which produced an abundance of black holes with 100 - 200 $M_\circ$ [43] [44], others from variations on the black hole theme (black holes made by fall back in supernovae, black hole helium core mergers, etc). It would be surprising if nature has found only one way of making brilliant flashes of X rays and gamma rays, even at cosmic distances.

**GRB–SN Connection**. One possible breakthrough would be proof that long GRBs are accompanied by Type Ib/Ic supernovae. An optical spectrum taken at the right epoch can confirm that the excess emission in the afterglow is supernova and not a light echo or some other modulation of the afterglow emission itself. Properties of the supernova will be constraining on the progenitor of the GRB, the total energy of the GRB, and the explosion mechanism.

**GRBs as a Probe of Cosmology**. *HETE* can detect GRBs out to $z \approx 10-20$ [38] [26], and may therefore provide the first identification of a very high ($z > 5$) redshift GRB. In addition, the dissemination by *HETE* of GRB localizations in near-real time will make it possible to bring to bear the Chandra X-ray Observatory, as well as large-aperture ground-based optical telescopes (*e.g.*, Keck, Magellan, Subaru, and the VLT) soon after the GRB. This is necessary in order to achieve high signal-to-noise, high resolution spectra of GRB X-ray and optical afterglows – a requirement for achieving the promise of GRBs for cosmology.

With a mission duration extending from 2002-2004, *HETE* can be expected to produce totals of $\approx 60$ localizations, $\approx 25$ optical afterglows, and $\gtrsim 25$ redshifts. The actual numbers are likely to be larger, as solar activity declines and the "live time" of the FREGATE and WXM instruments can be safely increased from 40 minutes to as much as 55 minutes per orbit. All in all, with the mysteries that still surround GRBs and the (currently) unique capabilities of the *HETE* FREGATE, WXM and SXC instruments, the scientific promise of the ongoing *HETE* mission is great.

## CONCLUSIONS

The *HETE* mission has achieved the original goals set for the program more than a decade ago, despite its highly constrained funding and manpower limits. The harvest of *HETE* is proving to be a rich one: as of December 2002, *HETE* had localized 31 GRBs with its WXM instrument. Of these, 5 had refined ($\sim 1'$) localizations from the SXC instrument, 11 were "X-ray rich", one was of short duration, and 8 had optical identifications. For the optical identifications, two had been established within minutes of the GRB. Until the advent of the Swift mission in 2004, *HETE* should continue to serve the GRB observer community as the premiere source of accurate, prompt source localizations.

## ACKNOWLEDGMENTS

The combined efforts over almost two decades by many talented and dedicated individuals made the *HETE* mission possible. In particular, the efforts of Bob Dill, Tye Brady, Dave Breslau, Gus Comeyne, Francis Cotin, Jim Francis, Greg Huffman, Frank LaRosa, Fred Miller, Jerry Roberts, and Fuyuki Tokanai were particularly notable. Participation by our scientific colleagues in Brazil (led by Joao Braga) and India (let by Ravi Manchanda) has been essential to HETE's success.

The *HETE* mission is supported in the USA by NASA Contract NASW-4690; in Japan, in part by the Ministry of Education, Culture, Sports, Science, and Technology Grant-in-Aid 13440063; in France, by CNES Contract 793-01-8479. K. Hurley is grateful for *Ulysses* support under Contract JPL 958059, and for *HETE* support under Contract MIT-SC-R-293291. G. Pizzichini acknowledges support by the Italian Space Agency (ASI).

## REFERENCES

1. Cline, T., *et al.* 2000, Ap. J. 531, 407
2. Fuchs, Y., *et al.*, 1999, Astron. Astrophys. 350, 891
3. Gaensler, B., *et al.*, 2001, Ap. J. 559, 963
4. Kouveliotou, C., *et al.*, 1998, Nature, 393, 235
5. Kouveliotou, C., *et al.*, 1999, Ap. J. 510, L115
6. Hurley, K., *et al.* 1999a, ApJS, 120, 399
7. Ricker, G., *et al.*, GCN 1068, 2001
8. Ricker, G., *et al.*, GCN 1073, 2001
9. Kato *et al.*, GCN 1363
10. Torii *et al.*, GCN 1378
11. Thompson, C., and Duncan, R., 1995, MNRAS 275, 255
12. Vrba, F., *et al.*, 2000, Ap. J. 533, L17
13. Atteia, J-L. *et al.*, 2002, in Gamma-Ray Burst and Afterglow Astronomy 2001: A Workshop Celebrating the First Year of the HETE Mission, eds. G. R. Ricker and R. K. Vanderspek (New York: AIP), in press
14. Kawai, N., et al. *ibid.*
15. Vanderspek, R.K., et al. *ibid.*
16. Villasenor, J.N. *et al.*, *ibid.*
17. Hurley, K. *et al.*, *ibid.*
18. Graziani, C. *et al.*, *ibid.*
19. Price, P. A. *et al.*, *ibid.*
20. Monnelly, G.P. *et al.*, *ibid.*
21. Shirasaki, Y. *et al.*, *ibid.*

22. Sakamoto, T. *et al.*, *ibid.*
23. Olive, J.-F. *et al.*, *ibid.*
24. Crew, G.B., et al. *HETE Mission Operations*, *ibid.*
25. Crew, G.B., et al. *Real-Time Messages from the HETE Spacecraft via the HETE Burst Alert*, *ibid.*
26. Bromm, V. & Loeb, A. 2002, ApJ, in press, astro-ph/0201400
27. Ruffert, M., & Janka, H. 1999, Astr. Astroph. 344, 573
28. MacFadyen, A., and Woosley, S., Ap. J. 524, 262, 1999
29. Usov, V., Nature 357, 472, 1992
30. Wheeler, C. *et al.*, Ap. J. 537, 810, 2000
31. Bloom, J. *et al.*, Nature 401, 453, 1999
32. Bloom, J., *et al.*, 2002, Ap. J., submitted
33. Reichart, D., Ap. J. 521, L111, 1999
34. Garnavich, P. M. *et al.*, 2002 Ap. J., submitted
35. Galama, T. *et al.*, Nature 398, 394.
36. Galama, T. *et al.*, Ap. J. 536, 185, 2000
37. Frail, D., *et al.*, Ap. J. 562, L55, 2001
38. Lamb, D., and Reichart, D., Ap. J. 536, 1, 2000
39. Vietri, M., and Stella, L., Ap. J. 527, L43, 1999
40. Kouveliotou, C., *et al.* Ap. J. Lett. 413, L101, 1993
41. Heise, J., *et al.*, 2001, ESO Astrophysics Symposia, Springer (Berlin).
42. Lazzati, D., Ramirez-Ruiz, E. & Rees, M.J., 2002, ApJ (submitted)
43. Abel, T., *et al.* Science 295, 93, 2002
44. Fryer, C., *et al.* Ap. J. 550, 372, 2001
45. Reeves, J. *et al.*, Nature 416, 512, 2002
46. Cen, R., Haiman, Z. & Lamb, D. 2002, ApJ (submitted)
47. Savagli, S., Fall, S. & Fiore, F. 2002, ApJ (submitted)
48. Reichart, D. E. and Yost, S. A. 2002, ApJ, in press
49. Piro, L. *et al.* 2000, Science, 290, 955
50. Heise, J. 2002, talk at the 200th AAS meeting
51. Dooling, D. and Finckenor, M., NASA/TP-1999-209260.
52. Bloom, J.S., et al. 2001, GCN Circ. 1135.
53. Ricker G.R., et al. 2002, GCN Circ. 1399.
54. Lamb D.Q., et al. 2002, ApJ, submitted.
55. Butler, N.R., et al. 2002, GCN Circ. 1415.
56. Costa, E., Frontera, F., Heise, J., et al. 1997, Nature, **387**, 783.
57. Djorgovski, S.G., et al. 2001, GCN Circ. 1108.
58. Heise, J., in't Zand, J., Kippen, R., and Woods, P., *X-Ray Flashes and X-Ray Rich Gamma-Ray Bursts*, in Gamma-Ray Bursts in the Afterglow Era, (Rome, Italy, 17-20 October 2000), ESO Astrophysics Symposia, Springer (Berlin), p. 16, 2001.
59. Hurley, K. 1992, *Gamma-Ray Burst Observations: Past and Future*, in Gamma-Ray Bursts, Eds. W. Paciesas and G. Fishman, AIP Conf. Proc. 265 (AIP Press- New York), 3.
60. Hurley, K., et al. 2001, GCN Circ. 1097.
61. Kouveliotou, C., et al. 1993, Ap. J. Lett., **413**, L101.
62. Kulkarni, S., et al. 2002, in preparation.
63. Paciesas, W.S. et al. 1999, ApJS, **122**, 465P.
64. Park, H. S., et al. 2001a, GCN Circ. 1114.
65. Park, H. S., et al. 2002, submitted to Ap. J. Letters (astro-pf/0112397).
66. Price, P.A., et al. 2001a, GCN Circ. 1107.
67. Price, P.A., et al. 2002, submitted to Ap. J. Letters (astro-pf/0201399).
68. Ricker, G.R., et al. 2001a, GCN Circ. 1096.
69. Shirasaki, Y., et al. 2000, Proc. SPIE, **4012**, pp 166-177.
70. Woosley, S. E. et al. 1984, The High Energy Transient Explorer (HETE), in *High Energy Transients in Astrophysics* (Santa Cruz, CA 1983), ed. S. E. Woosley, A.I.P. Conf. Proc. 115 (AIP Press-New York), p. 709.

# In-Flight Performance and First Results of FREGATE

J-L. Atteia*, M. Boer*, F. Cotin*, J. Couteret*, J-P. Dezalay*, M. Ehanno*, J. Evrard*, D. Lagrange*, M. Niel*, J-F. Olive*, G. Rouaix*, P. Souleille*, G. Vedrenne*, K. Hurley[†], G. Ricker**, R. Vanderspek**, G. Crew**, J. Doty** and N. Butler**

*C.E.S.R., 9 Avenue du Colonel Roche, 31028 Toulouse Cedex 4, FRANCE
[†]UC Berkeley Space Sciences Laboratory, Berkeley, CA 94720-7450
**M.I.T. Center for Space Research, 70 Vassar St., Cambridge, MA 02139

**Abstract.** The gamma-ray detector of HETE-2, called FREGATE, has been designed to detect gamma-ray bursts in the energy range 6-400 keV. Its main task is to alert the other instruments of the occurrence of a gamma-ray burst (GRB) and to provide the spectral coverage of the GRB prompt emission in hard X-rays and soft gamma-rays. FREGATE was switched on on October 16, 2000, one week after the successful launch of HETE-2, and has been continuously working since then. We describe here the main characteristics of the instrument, its in-flight performance and we briefly discuss the first GRB observations.

## INTRODUCTION

The HETE-2 spacecraft [13, 5] has been designed to distribute gamma-ray burst localizations within several seconds of the burst detection. The localization process and the distribution of rapid alerts is a complex chain of events which starts when a GRB is detected and identified as such. HETE-2 (hereafter HETE for simplicity) carries three experiments: the Soft X-ray Camera (SXC), The Wide-Field X-ray Monitor (WXM) and the FREnch GAmma TElescope (FREGATE). The latter is a traditonal gamma-ray detector operating in the energy range 6-400 keV, which was built by the Centre d'Etude Spatiale des Rayonnements (CESR) in Toulouse, France. FREGATE has three main goals:

- Detecting count rate increases and qualifying them as gamma-ray burst candidates.
- Performing GRB spectroscopy over a broad energy range (complementing and overlapping the WXM energy range).
- Monitoring the activity of galactic transient sources like the Soft Gamma Repeaters.

This paper describes FREGATE and explains how it is operated and calibrated in-flight. It also contains a brief discussion of the types of events detected during the first year of operation and a highlight of the preliminary scientific results of the mission.

**FIGURE 1.** FREGATE in the laboratory before its integration on the spacecraft. The four detectors on the left are 20 cm high.

## DESCRIPTION OF THE INSTRUMENT

The FREGATE hardware consists of four identical detectors and one electronics box (see Figure 1). The instrument weights 14 kg and consumes 9 watts of electrical power. The four detectors are co-aligned on the spacecraft in order to share the same field of view. Contrary to the other HETE instruments (the WXM and the SXC), FREGATE has no localization capability, except the abil-

**TABLE 1.** FREGATE performances

| | |
|---|---|
| Energy range | 6 - 400 keV |
| Effective area (4 detectors, on axis) | 160 cm$^2$ |
| Field of view (FWZM) | 70° |
| Sensitivity (50 - 300 keV) | $10^{-7}$ erg cm$^{-2}$ s$^{-1}$ |
| Dead time | 10 $\mu$sec |
| Time resolution | 6.4 $\mu$sec |
| Maximum acceptable photon flux | $10^3$ ph cm$^{-2}$ sec$^{-1}$ |
| Spectral resolution at 662 keV | $\sim$ 8 % |
| Spectral resolution at 122 keV | $\sim$ 12 % |
| Spectral resolution at 6 keV | $\sim$ 42 % |

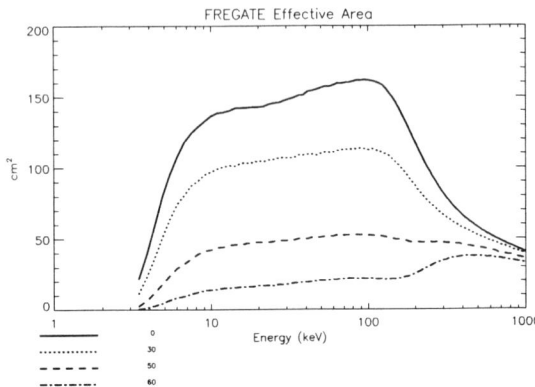

**FIGURE 3.** FREGATE effective area for on-axis and off-axis sources.

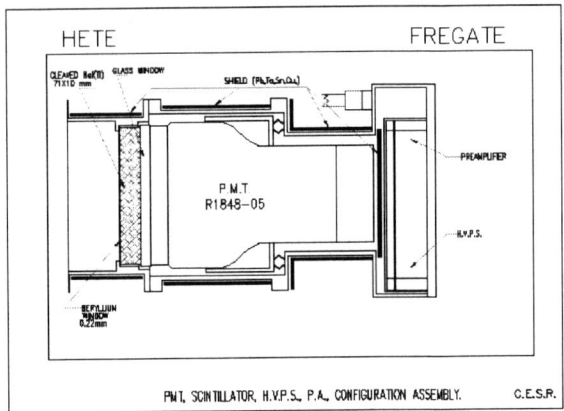

**FIGURE 2.** Schematic drawing of one FREGATE detector. The crystal has a diameter of 71mm and a thickness of 10mm. The length of the detector is 20cm.

## Detectors

Each FREGATE detector consists of a cleaved NaI crystal (a cylinder 10mm thick and 71mm in diameter) read by a photomultiplier (Hamamatsu 1848, figure 2). In order to extend the spectral coverage of FREGATE at low energies, we chose cleaved crystals which have no dead layer. For the same reason the crystals are encapsulated in a beryllium housing which reduces the absorption of low energy gamma-rays. The thickness of the housing on the front side of the detectors is 0.8 mm and the overall transmission of the entrance window is larger than 65% at 6 keV and reaches 85% at 10 keV (see figure 3). At high energies the electronics limits the energy range to photons with energies below 400 keV. The geometric area of the sum of the four detectors for a source on-axis is nearly 160 cm$^2$. In addition there are two on-board radioactive sources of Barium 133 which illuminate the detectors from the outside with photons

ity to recognize whether a transient occurred within or outside its field of view. The main characteristics of FREGATE are described in table 1.

at 81 and 356 keV, allowing to monitor the gain of the detectors from the ground (see subsection "In-flight calibration" for more details).

## Shield and Collimator

As shown in figure 2, the body of the detectors is surrounded by a graded shield made of lead, tantalum, tin, copper, and aluminium (0.8 mm of lead, 0.3 mm of tantalum, 0.7 mm of tin, 0.3 mm of copper, and 0.8 mm of aluminum). The aim of the shield is to prevent the photons from outside the field of view from reaching the crystal. The transparency of the shield is 5.5% at 150 keV, 55% at 300 keV, and 75% at 500 keV.

A peculiarity of the FREGATE detectors is that the shield goes beyond the front face of the crystal, reducing the field of view (FOV) of the instrument and acting as a collimator. The angular response of FREGATE with its collimator is shown in figure 4. While this collimator decreases the number of GRBs that FREGATE can detect, it plays an essential role for a mission like HETE where a synergy must be found between three sets of instruments with different properties and constraints. In the context of FREGATE the collimator provides the following advantages (in order of increasing importance):

- It increases the fraction of FREGATE GRBs which are within the FOV of the WXM.
- It decreases by a factor of two the count rate from the diffuse X-ray background.
- It restricts to 4 months per year the transit time of galactic sources within the field of view of FREGATE (instead of nearly 6 months).

While GRBs which are outside the field of view of the WXM cannot be localized, it was thought that the FOV of FREGATE should nevertheless be wider than

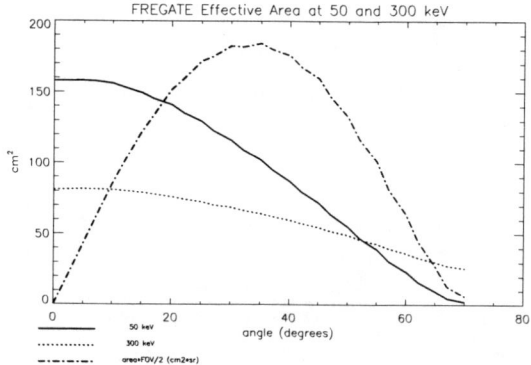
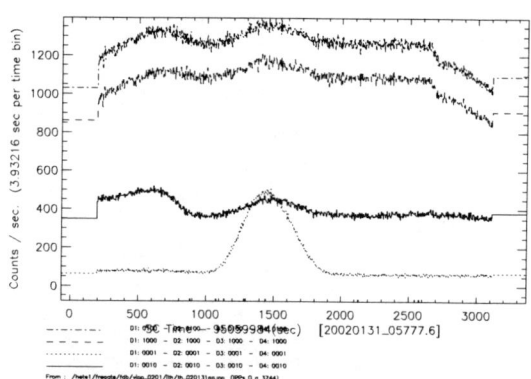

**FIGURE 4.** FREGATE response to off-axis sources. Solid line: angular response at 50 keV where the collimator is opaque. dotted line: angular response at 300 keV where the collimator is mostly transparent. Dash-dotted line: expected number of GRBs as a function of their angle of arrival.

**FIGURE 5.** Continuous data generated by FREGATE during nighttime. From top to bottom, the light curves in the 4 energy bands B, A, C, and D (see text) for the sum of the 4 detectors. The data have been regrouped in 4 second bins for clarity, the actual resolution of the data is 0.16 seconds. The large peak in energy band D is due to protons trapped in the SAA.

the FOV of the WXM. The main reason for this choice was that a wider field of view of FREGATE ensured that all the GRBs detected by the WXM would illuminate at least 60 cm$^2$ of FREGATE detectors (as was the case for GRB010921, [14]). Moreover FREGATE-only GRBs are useful for broadband spectroscopic studies because they increase the statistics for rare events (like X-Ray Flashes or short GRBs) and they can be localized by the IPN.

### Readout electronics

Each detector has its own analog and digital electronics. The analog electronics contain a discriminator circuit with four adjustable channels and a 14-bit PHA whose output is regrouped into 512 evenly-spaced energy channels (each approximately 0.8 keV wide). The (dead) time needed to encode the energy of each photon is 14 $\mu$s for the PHA and 9 $\mu$s for the discriminator. The digital electronics process the individual pulses, to produce the following output :

- Time histories in 4 energy channels (every 20ms).
- 128-channel energy spectra spanning the range 0-400 keV (every 80 ms).
- A circular buffer containing the most recent 65536 photons tagged in time (resolution 6.4 $\mu$s) and in energy (resolution 0.8 keV)..

### On-board software

The main tasks of the on-board software include the configuration of the instrument, the acquisition of data from the electronics, their packaging for the telemetry, and the search for excesses in the count rate.

*Configuration of the instrument.* The output of FREGATE depends on a number of adjustable parameters such as the settings of the high voltages, the limits of the energy channels, the trigger criteria, or the on-board data compression. These parameters are described in a configuration file which can be uploaded when the spacecraft is in contact with one of the three primary ground stations (PGS) [3]. When a new configuration file is uploaded, the on-board software modifies the configuration of FREGATE accordingly.

*Data packaging.* Every 20 ms the FREGATE Digital Signal Processor (DSP) reads the data from the electronics and prepares the data for the telemetry. The data products generated by FREGATE are described in the next Section.

*Search for excesses.* The on-board software also scans the data in real time to search for sudden increases in the count rate recorded by the instrument. This work is described in details in Section "FREGATE triggers" below.

## DATA TYPES

FREGATE generates 4 types of data: Housekeeping (HK), light curves and spectra generated by the FREGATE DSP, light curves generated by one of the on-board tranputers (the so-called X-$\gamma$ transputer), and burst data. Housekeeping data are produced continuously; light curves and spectra are produced when the high voltages are on; and burst data are produced only after a trig-

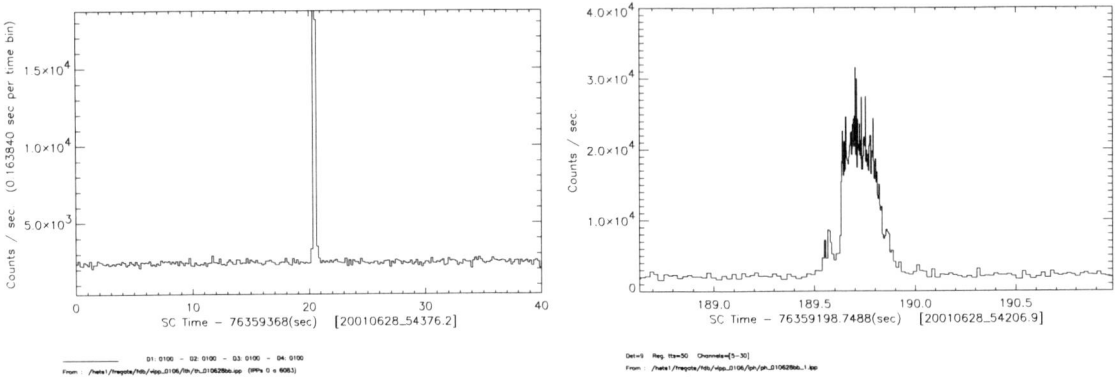

**FIGURE 6.** A burst from SGR 1900+14 as seen in the continuous data (left) and in the burst data (right).

ger.

The HK data and the light curves produced by the on-board transputer will not be discussed here, however, the continuous data and the burst data generated by the FREGATE DSP are explained in detail below.

*Light curves and spectra generated by the DSP*

During nighttime, when the high voltages are on, FREGATE produces continuous light curves and spectra.

The light curves represent the count rates measured by the 4 detectors in 4 broad energy channels with time resolutions of 0.16 and 0.32 s. The limits of the energy channels are usually set to

- 6 - 40 keV for channel A.
- 6 - 80 keV for channel B.
- 32 - 400 keV for channel C.
- > 400 keV for channel D.

An example of these data is shown in figure 5.

Simultaneously FREGATE generates four 128-channel energy spectra covering the energy range 0-400 keV every 5 or 10 seconds (but the electronics threshold and the absorption of the beryllium window reduce the effective energy range to 6-400 keV).

*Burst data*

When a trigger occurs, burst data are generated in addition to the continuous data. The burst data consist of 256k photons (64k per detector) tagged in time (with a resolution of 6.4 $\mu$s) and in energy (256 energy channels spanning the range 0-400 keV). These burst data allow detailed studies of the spectro-temporal evolution of bright GRBs. An example of the gain provided by the burst data is shown in figure 6.

# FREGATE OPERATION

FREGATE operations are driven by alternating nighttime and daytime periods. Because HETE instruments always point in the antisolar direction they have the earth in their field of view during about 45 min per orbit (the duration of one orbit is 90 min); this is the daytime period.

During daytime the High Voltages of the detectors are switched off, partly to reduce the amount of data produced by the spacecraft and partly to avoid the triggers due to solar X-ray flares reflected on the atmosphere of the Earth (see figure 11).

During nighttime the high voltages are switched on. The detectors continuously record the gamma-ray flux in four energy bands and these data are processed by the DSP to search for excesses due to GRBs (see next section).

*In-flight calibration*

Since FREGATE records only the time and the energy of the photons, we need to calibrate only the FREGATE timing and energy scales.

The FREGATE time is directly derived from the HETE time and the calibration of FREGATE time is, in reality, the issue of the spacecraft time calibration which is not discussed here.

As mentioned above, the energy calibration of the detectors is made possible by two on-board radioactive sources of Barium 133 which illuminate the detectors from the outside ($^{133}$Ba emits gamma-ray lines at 81 and 356 keV, with a half-life of 10.5 years) . A typical background spectrum, accumulated for 1200 seconds, is

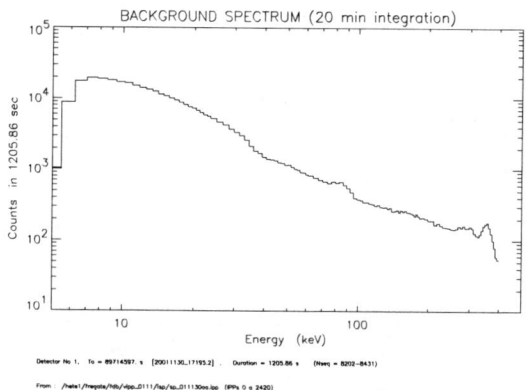

**FIGURE 7.** Background spectrum showing the two calibration lines at 81 keV and 356 keV.

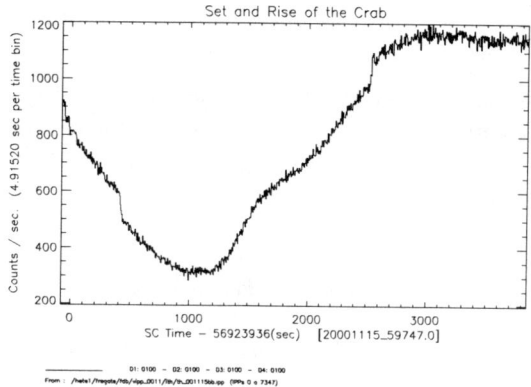

**FIGURE 9.** Crab occultation: the two steps in the light curve are due to the transit of the Earth in front of the Crab nebula. The light curve, which has a temporal resolution of 5s, shows the total count rate measured by the four FREGATE detectors in the energy range 6-80 keV. The Crab is approximately 30° off-axis.

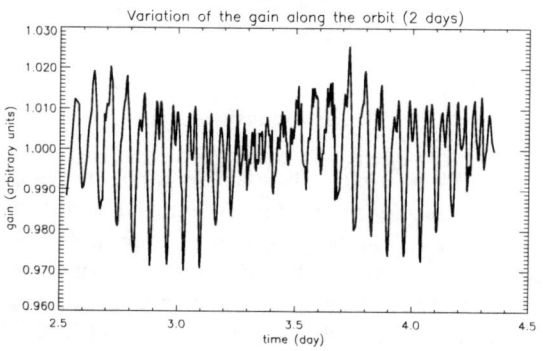

**FIGURE 8.** Gain fluctuations during 26 consecutive orbit. The gain fluctuations are due to the changing orientation of the magnetic field along the orbit.

shown in figure 7. These radioactive sources allow the monitoring of the gain of the four detectors of FREGATE with a time resolution of a few minutes (the gain is manifested in the response of the detectors to photons with a given energy). The gain fluctuates on two timescales: along one orbit (90 min) and on a longer period of several weeks.

On the short term, the gain changes with the orientation of the magnetic field along the orbit. This effect is due to the incomplete collection of the photoelectrons on the first anode of the PMT (which has no magnetic shield). A typical example of such variations is shown in figure 8.

The gain also exhibits a tendency to decrease on the long term. this trend is monitored on the ground and compensated by regular increases of the high voltages (by a few percent) every 2 months.

### Energy response

The in-flight energy response of a gamma-ray detector is the combination of its physical response, its intrinsic non-linearities, and the gain variations. The physical response of FREGATE detectors has been evaluated with detailed Monte Carlo simulations. The output of the simulation program has been checked against ground calibrations made with 9 radioactive sources having energies in the range 8 keV ($^{65}$Zn) to 1332 keV ($^{60}$Co). The same sources provide a measure of the intrinsic non-linearities of the detectors. Finally the gain fluctuations are measured on-board as explained above.

The quality of the spectral response of FREGATE has been evaluated via the deconvolution of the hard X-ray emission from the Crab nebula. The spectrum of the Crab nebula has been constructed from the amplitude of the Crab occultation steps observed at various energies, see figure 9. The deconvolved spectrum is fully compatible with the well known spectrum of the Crab nebula for energies between 10 keV and 200 keV and for angles between 0° and 50°. This procedure and the results obtained are described in details in [10].

### Ground operations

The operational tasks on the ground are very simple. In normal operation they include the following actions:

- Upload the HV ON/OFF sequences.
- Check the health of the detectors.
- Adjust the gains and update the configuration.

- Search for events which didn't fire the on-board trigger.

When an astronomical event is detected it is also necessary to update the FREGATE catalog(s) and to construct the response matrices if it is within the field of view.

## FREGATE TRIGGERS

### GRB detection

The data recorded by FREGATE are searched for GRBs and other astronomical transients both on-board and on the ground. Two real-time programs run on-board: the DSP trigger and the transputer trigger. The former is described below. The transputer trigger has been designed to add more flexibility and to search for excesses in the combined data from FREGATE and the WXM, it is described in [16].

The on-board GRB detection is completed by two programs which automatically process the data when they arrive at the ground. These programs are more efficient than the on-board processing for the detection of long or soft events, they are described in [2] and in [6].

*DSP trigger.* The DSP triggers when the count rate measured over a time interval $\Delta t$ exceeds the average count rate, measured over the last $T$ seconds, by more than $k$ standard deviations. Four timescales ($\Delta t$) are used for the trigger detection : 20 ms, 160 ms, 1.3 s, and 5.2 s. The duration of the background integration is an adjustable parameter, usually set to 30 s. The trigger thresholds $k$ are adjustable and currently set to values between 4.5 and 6. The trigger detection algorithm works in parallel in the energy channels B (6-80 keV) and C (30-400 keV). In order to decrease the rate of false triggers, due to electronic noise or particles, we discard the triggers which are detected by only one of the four detectors. When the DSP triggers it sends an alert message to the HETE trigger monitor [3, 17], which alerts the other instruments of HETE and the VHF transmitter. Simultaneously FREGATE starts to record burst data (see "DATA TYPES"). Additionally, FREGATE will go into burst mode when it receives a message from the on-board trigger monitor (e.g. following a WXM trigger). Because the high-energy sky is essentially variable, many types of events can trigger FREGATE. We summarize below the origin of the triggers detected during the first year of the mission.

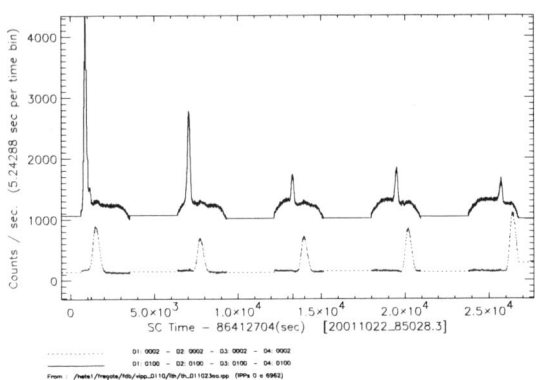

**FIGURE 10.** Electrons and protons trapped in the radiation belts. The lower curve shows the signal due to the protons in the South Atlantic Anomaly. The upper curve shows large peaks due to trapped electrons, they are detected a few minutes before the SAA.

### Non-astrophysical events

*Pre-SAA electrons.* In addition to the high fluxes of protons detected in the South Atlantic Anomaly (SAA), energetic electrons are sometimes trapped in electron radiation belts crossed by HETE a few minutes before the SAA. These populations of energetic electrons are highly variable and they reach a maximum in the days following large coronal mass ejections (CMEs) from the Sun. When HETE goes through these radiation belts (at longitudes between 80 and 100 degrees) FREGATE measures high count rates due the interaction of the electrons with the spacecraft (producing X-rays) and with the detectors (see figure 10) . These high count rates will sometimes trigger FREGATE.

*Solar flares reflected on the Earth's atmosphere.* FREGATE high voltages are usually switched off during daytime. However, there are occasions when we want FREGATE on with the Earth in the field of view. In this case the low energy threshold of FREGATE make it very sensitive to solar flares reflected by the atmosphere of the Earth. An example of such a flare (class M2.2) is shown in figure 11.

*Noise triggers.* The noise triggers are due to the statistical fluctuations of the background count-rate measured by the detectors. FREGATE triggers only when it detects two simultaneous excesses in the sum of detectors 1+2 AND in the sum of detectors 3+4, this strategy reduces the number of noise triggers to less than 1 per month.

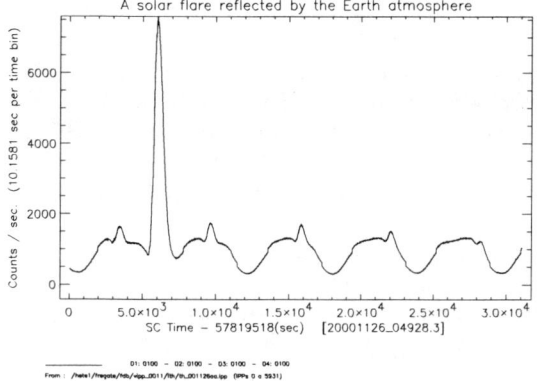

**FIGURE 11.** A solar flare reflected by the atmosphere of the Earth on November 26 2000.

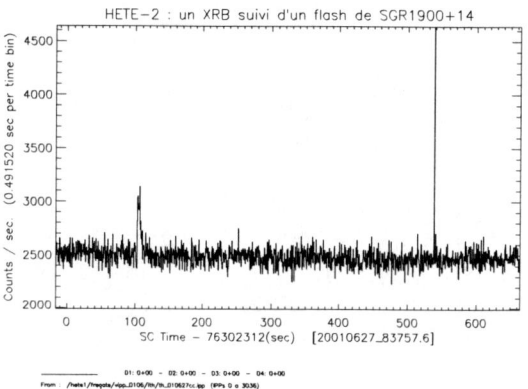

**FIGURE 13.** During summer time, when it has the galactic bulge within its field of view, FREGATE detects many galactic transients. This light curve (6-80 keV) shows an X-Ray Burst followed by a burst from SGR1900+14 about seven minutes later.

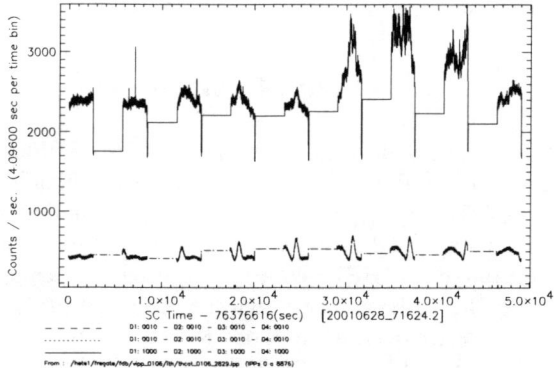

**FIGURE 12.** Count rates measured by FREGATE in two energy channels (top: 6-40 keV; bottom 30-400 keV) during 9 consecutive orbits with SCO X-1 in the the field of view (45 degrees off-axis). The flaring activity of SCO X-1 is clearly visible during orbits 6 to 9 in the low energy channel. Note also that SCO X-1 has no effect on the high energy channel (above 30 keV), which can be used to detect GRBs. The three short spikes in orbit 2 and 3 are X-ray bursts.

### GRBs and other astrophysical transients

**SCO X-1.** SCO X-1 is the brightest hard X-ray source in the sky. It exhibits rapid flaring (see figure 12) and generates many triggers when the trigger is enabled in energy channel B (6-80 keV). From the beginning of April to the end of July, when SCO X-1 is within the field of view of FREGATE, the trigger is disabled in channel B, completely suppressing SCO triggers.

**X-ray bursts.** X-ray bursts (XRBs) are due to thermonuclear explosions at the surface of accreting neutron stars in binary systems. Several dozen X-ray bursters are present in the galactic bulge. When FREGATE has the galactic bulge within its field of view it detects a few XRBs per day. These events do not trigger FREGATE, whose low energy trigger is disabled during summer time (when SCO X-1 is in the field of view, see above), but they are identified a posteriori by the ground processing. A sure way to identify XRBs is to associate their arrival direction with a known X-ray source. The WXM is well suited to do this job, but its field of view is only half the field of view of FREGATE, implying that the identification of the XRBs detected by FREGATE at large off-axis angles must rely solely on their spectro-temporal properties. An example of an XRB detected by FREGATE can be found in figure 13. A list of X-Ray Bursts detected and localized by the WXM during the summer 2001 is given in [15].

*Soft Gamma-ray Repeaters.* During the summer 2001, both SGR1900+14 and SGR1806-20 were active. From the beginning of June to the end of August FREGATE detected about 30 short bursts which can be attributed to these Soft Gamma Repeaters (figures 6 and 13). Six of these bursts were localized by the WXM [9], two were emitted by SGR1806-20 and four by SGR1900+14. On July 2 2001, FREGATE detected a high fluence burst from SGR1900+14 (GCN1078), a detailed analysis of this event is given in [11].

*Gamma-Ray Bursts.* Gamma-ray bursts constitute the main scientific target of HETE. Between October 2000 and September 2001, FREGATE has detected 32 confirmed GRBs and a few unconfirmed events (see [4] for a list). We estimated the sensitivity of FREGATE to be $10^{-7}$ erg cm$^{-2}$ in the energy range [50-300] keV [4]. Figure 15 shows the number of confirmed GRBs detected

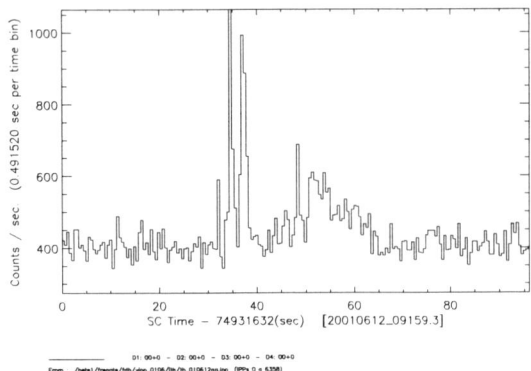

**FIGURE 14.** GRB010612 observed by FREGATE (30 to 400 keV), the light curve shows the total of the four detectors, the GRB is 13 degrees off-axis.

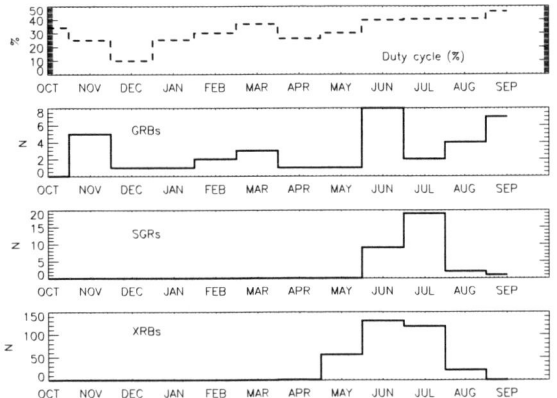

**FIGURE 15.** Monthly detection number of various types of high energy transients from October 2000 to September 2001. From May to August, the Galactic bulge was in the field of view of FREGATE. Note the ordinate of the XRB plot.

by FREGATE since it was turned on in October 2000. The higher number of GRBs per month after May 2001 is due to an increased observational efficiency, which should result in the detection of a larger number of GRBs in 2002. Some interesting results have already been obtained with the GRBs detected during the first year of FREGATE operation, the reader will find some of them in these proceedings and a short list is given in the Conclusion below.

## CONCLUSION

The first year of FREGATE operation shows that it fulfills the goals for which it has been designed : alerting HETE when a GRB occurs, performing the broadband spectroscopy of GRBs and other astrophysical transients (SGRs and XRBs), and providing a census of galactic and extragalactic high energy transients. Regarding this last issue, we note that, as a consequence of the anti-solar pointing strategy of HETE, FREGATE observes the same portion of the sky during more than $3 \times 10^6$ seconds per year. In addition FREGATE has been successfully integrated into the Interplanetary Network of gamma-ray burst detectors [8].

This volume contains some significant scientific results obtained by HETE and FREGATE. The followup of GRB010921 detected by FREGATE and localized with the WXM has led to the identification of the first HETE afterglow at a redshift z=0.45. [14, 12] The spectro-temporal evolution of a bright burst from SGR1900+14 in analysed in detail in [11]. Barraud et al. (2002) discuss the existence of very soft GRBs (probably similar to the X-Ray Flashes [7]) which have less than 10% of their fluence above 30 keV.

## ACKNOWLEDGMENTS

J-L Atteia acknowledges the support of the FREGATE team in CESR (F. Cotin, J. Couteret, M. Ehanno, J. Evrard, D. Lagrange, G. Rouaix and P. Souleille) who built a successful instrument and never lost enthusiasm after the loss of HETE-1. A special tribute is due to M. Niel, who designed and tested the prototype of the detectors. FREGATE is supported in France by the CNES under contract CNES 793-01-8479. None of the results presented here would have been obtained without the dedication of the HETE OPS team at MIT who manages FREGATE operations.

## REFERENCES

1. Barraud C., et al., these proceedings
2. Butler N., et al., these proceedings
3. Crew G., et al., 2002, these proceedings
4. Dezalay J-P., et al., 2002, these proceedings
5. Doty J., et al., 2002, these proceedings
6. Graziani C., et al., 2002, these proceedings
7. Heise J., et al., 2002, these proceedings
8. Hurley K., et al., 2002, these proceedings
9. Kawai N., et al., 2002, these proceedings
10. Olive J-F, et al., et al., 2002a, these proceedings
11. Olive J-F., et al., et al., 2002b, these proceedings
12. Price. P., et al., 2002, submitted to ApJ
13. Ricker G., et al., 2002a, these proceedings
14. Ricker G., et al., 2002b, submitted to ApJ
15. Sakamoto T., et al., 2002, these proceedings
16. Tavenner T., et al., 2002, these proceedings
17. Vanderspek R., et al., 2002, these proceedings

# In-Orbit Performance of WXM (Wide-Field X-Ray Monitor)

N. Kawai[*], A. Yoshida[†], M. Matsuoka[**], Y. Shirasaki[‡], T. Tamagawa[§], K. Torii[§], T. Sakamoto[*], D. Takahashi[†], E. Fenimore[¶], M. Galassi[¶], T. Tavenner[¶], D.Q. Lamb[‖], C. Graziani[‖], T. Donaghy[‖], R. Vanderspek[††], M. Yamauchi[‡‡], K. Takagishi[‡‡], I. Hatsukade[‡‡] and HETE-2 Science Team

[*]*Department of Physics, Tokyo Institute of Technology, Meguro-ku, Tokyo 152-8551, Japan*
[†]*Aoyama Gakuin University, Shibuya-ku, Tokyo 150-8366, Japan*
[**]*NASDA, Tsukuba, Ibaraki 305-8505, Japan*
[‡]*JST/NASDA, Tsukuba, Ibaraki 305-8505, Japan*
[§]*RIKEN, Wako, Saitama 351-0198, Japan*
[¶]*Los Alamos National Laboratory, Los Alamos, NM 87545, USA*
[‖]*Department of Astronomy & Astrophysics, University of Chicago, Chicago, IL 60637, USA*
[††]*MIT Center for Space Research, MIT, Cambridge, MA 02139, USA*
[‡‡]*Department of Engineering, Miyazaki University, Miyazaki, Japan*

**Abstract.** The Wide-field X-ray Monitor (WXM) is one of the three main scientific instruments on HETE-2, and is designed to measure the light curves, spectra, and locations of gamma-ray bursts (GRBs) and other transients in the energy range of 2–25 keV. It consists of Xe-filled 1-D position-sensitive proportional counters equipped with two 1-D coded apertures in orthogonal directions with a field of view of $40° \times 40°$. The sophisticated onboard processing allows the localization of GRBs in real time with $\sim 10'$ accuracy based on the alerts from FREGATE, the gamma-ray detector. The WXM also triggers on its own count time history with a flexible algorithm and can localize X-ray events on various time scales. We present the design and basic characteristics of the detectors, the handling of the data, the in-flight performance, and some of the early observations.

## INTRODUCTION

The primary goals of the HETE-2 mission [1] are the multiwavelength observation of gamma-ray bursts (GRBs) and the prompt distribution of precise GRB co-ordinates to the astronomical community for immediate follow-up observations. To achieve these goals, HETE-2 is equipped with three scientific instruments: Wide-field X-ray Monitor (WXM), Soft X-ray Camera (SXC; [2]) and French Gamma Telescope (FREGATE; [3]). They share a common field of view of $\sim 1.5$ steradians, and, together, are sensitive to photons in the energy range of 0.5 keV to over 400 keV. The two X-ray instruments have position-sensitive detectors equipped with coded apertures, allowing HETE to determine the location of a GRB. While the SXC has CCDs (MIT-LL CCID-20) as the detecting elements, and has high spatial resolution ($10''$), WXM has a larger effective area and thus is more sensitive to weaker bursts than SXC. If a GRB triggers either FREGATE or WXM itself, the position histogram representing the X-ray shadow of the coded aperture cast by the GRB is quickly analyzed on board, and the derived source location ($\sim 10'$ accuracy) is immediately reported to the ground.

In this paper we present a brief description of the WXM instrument, how it is operated, its performance in orbit, and a brief summary of its early results. Details of the WXM are given in [4].

## DESCRIPTION OF THE INSTRUMENT

The WXM consists of two identical units of position-sensitive X-ray detectors. Each unit, in turn, consists of two one-dimensional position sensitive proportional counters (PSPCs) sensitive to X-rays in the 2–25 keV energy range (Fig. 1).

Two units are placed in orthogonal directions to each other for measuring the X and Y directions independently (Fig. 1). The detectors are placed 187 mm below

**TABLE 1.** Wide Field X-Ray Monitor

| | |
|---|---|
| Built by | RIKEN and LANL |
| Instrument type | Coded Mask with PSPC |
| Energy Range | 2 to 25 keV |
| Timing Resolution | 1 ms |
| Spectral Resolution | ~25% @ 20 keV |
| Detector QE | 90% @5 keV |
| Effective Area | ~175 cm²* |
| Sensitivity (10 σ) | ~$8 \times 10^{-9}$ erg/cm²/s (2-10 keV) |
| Field of View | 1.6 steradians (FWZM) |
| Localization resolution | 19' (5σ burst) |
| | 2.7' (22σ burst) |

* each of two units

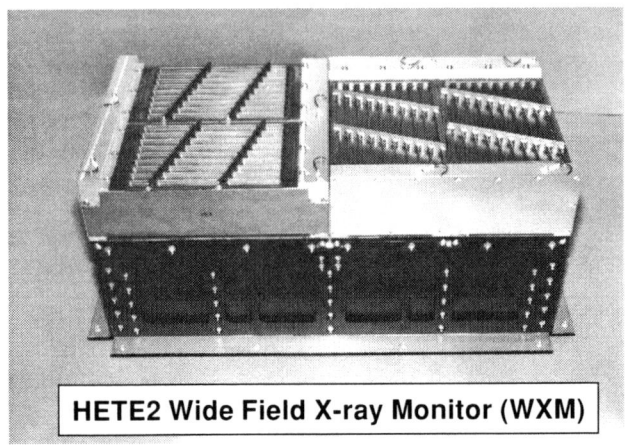

**FIGURE 1.** The WXM detector system consisting of four identical position-sensitive proportional counters and the electronics.

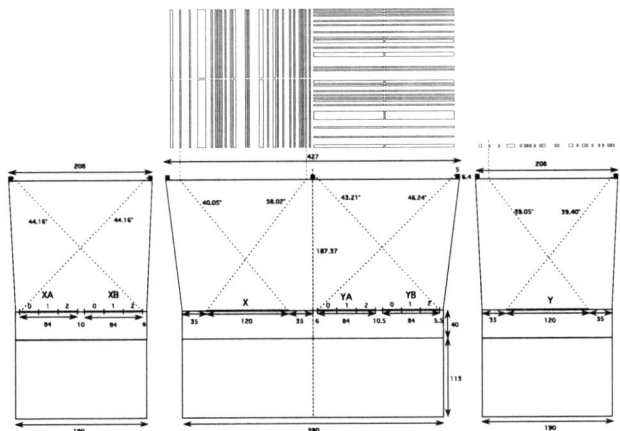

**FIGURE 2.** A schematic drawing of WXM. The coded mask is placed at 187.37 mm above the proportional counters and supported by an aluminum structure.

the coded aperture (Fig 2). The X unit has a field of view defined by $-38° < \theta_x < +40°$ and $-44° < \theta_y < +44°$, and the Y unit has a field of view $-46° < \theta_x < +43°$ and $-39° < \theta_y < +39°$.

The proportional counters and electronics were provided by RIKEN and Miyazaki University, while the coded aperture and on-board localization software is provided by Los Alamos National Laboratory.

### Coded Mask

The coded mask is made of a plate of aluminum (0.5mm thick) and gold (0.025mm thick) with a series of slits of randomly varying width (random mask). The location of the GRB is determined by measuring a set of two shift distances of the mask pattern in the X and Y directions. The mask pattern of the WXM was selected from 100,000 random patterns with the same open fraction and element size in order to optimize the localization accuracy for GRBs.

### Detectors

Each PSPC has three carbon fiber anode wires with 10μm diameter in its upper cells and four tungsten wires in its lower cells, and is filled with 1.4 atm Xenon gas with 3% $CO_2$ as quenching gas (Fig. 3). The three upper cells are used for X-ray detection. They have a depth of 25.5 mm and a width of ~27 mm, and are separated by the cathode wires placed at intervals of 3.4 mm. The four lower cells are used for rejecting charged particle events by anti-coincidence, and have a depth of 11.5 mm. The detection area of $120 \times 83.5$ mm² at the top side is sealed by 100 μm thick Be windows. The WXM has two high voltage (HV) power supplies, each of them connected to one detector in the X system, and one in the Y system. The voltage can be set to a value within a range of 0–1992 V in steps of 7.8 V. In the standard setting, a positive high voltage of 1650 V is applied to the anodes, which was chosen to provide acceptable position resolution (~1 mm at 8 keV) and energy resolution (~20 % at 5.9 keV) at the same time.

### Electronics and Processing

The position of an incoming X-ray is determined by the charge-division method. That is, $Q_L/(Q_L + Q_R)$ has a linear relation with the incident position, and $Q_L + Q_R$ measures photon energy. Here, $Q_L$ and $Q_R$ represent the pulse height values measured at the left and right ends of

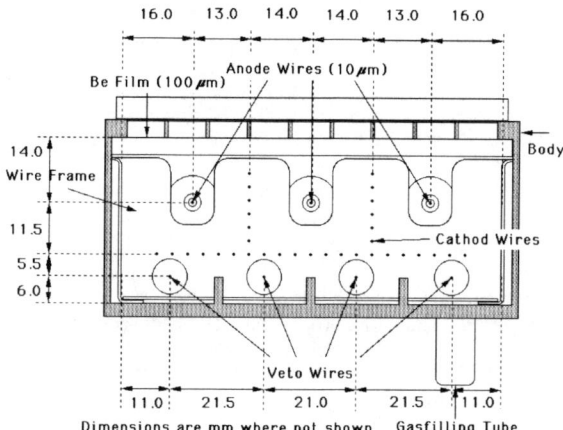

**FIGURE 3.** Cross section of a PSPC of WXM. Three position sensitive anode wires made of carbon fiber are placed in the three upper cells for X-ray detection. The lower cells form a veto layer for rejection of charged particle events. The entrance window has a area of $120 \times 83.5$ mm$^2$ and is made of 100 $\mu$m-thick Be foil.

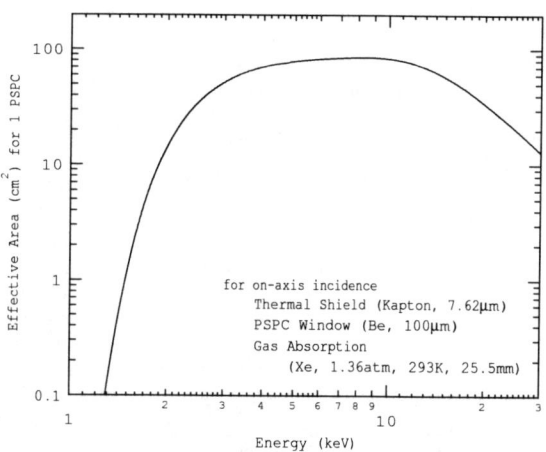

**FIGURE 4.** Effective area of one WXM PSPC for on-axis X-ray incidence as a function of the X-ray energy. The maximum is 85.4 cm$^2$ at 8.3 keV.

anodes respectively.

The signals from both ends of anodes are processed separately by charge-sensitive preamplifiers (Amptek, A225), go through the main amplifiers, and are provided to the discriminators and to 12-bit ADC with peak-hold logic. The gain of the main amplifier is adjustable in 32 levels with a dynamic range of $\sim 10$, and the discriminator has four levels. For a detected event, the 2-bit detector identifier, 4-bit anode-hit flag, and two 12-bit pulse height values from the left and right channels are packed in two 16-bits words and transmitted to the DSP in the spacecraft processor on the serial synchronous interface (SSI). WXM also transmits the HK (house keeping) words to the DSP on demand on the serial communication interface (SCI). The HK words contain the temperatures, measured high voltage values, electronics status bits, and the last commands accepted.

### Onboard Software

The data from WXM are handled in one of the four nodes in the spacecraft computer system.

The WXM and FREGATE use a common processor board (XG node) and each has a separate DSP (X-DSP and G-DSP) for handling the event data from the instruments and forming data products as well as commanding the instruments.

The WXM instrument provides the housekeeping data such as power status, HV setting, temperatures of an electronics board and the support structure wall as well as the photon data. Once the X-DSP receives those data, it counts the total number of the received events every 4 sec, assigns the photon data time stamps with 1 $\mu$s resolution, counts and calculates energies and incident positions. For the safety of the proportional counters, the X-DSP monitors the total received count rate. If the rate exceeds 20000 counts per 4 sec, a HV shutdown command is sent to the WXM. In the case that the count rate is too high and the observed signal becomes like a DC current, photon counting is not possible. In order to protect against such a situation, there is a mechanism to shut down the HV when the count rate is less than a given commandable value.

In processing photon data, the coincidence with the veto signal or another anode wire is checked and, if found, the data is rejected as a background event. If the energy is in the range of 2 to 25 keV, the data is accepted to form data products. The following data products are regularly generated at commandable time intervals:

1. the WXM housekeeping data and total received count rate every 4 s,
2. time history of four energy bands every 80 ms,
3. position histogram for two energy bands every 320 ms,
4. pulse height histogram every 4.9 s.

The count rates and position histograms are transferred to the XG-transputer and are used to search for a transient event and to localize the event, while the housekeeping data and pulse height histograms are directly transferred to the telemetry.

The XG-Transputer processes both of the WXM and FREGATE time histories to search for transient phenomena and, if one is found, provides a trigger to the other

**TABLE 2.** Summary of the WXM data products

| Type | Time resolution | Energy channels | Description |
|---|---|---|---|
| HK | 4.096 s | — | Temperature, Status, HV monitor etc. |
| SHK | 4.096 s | — | Total counts etc. |
| TH | 1.024 s | 4* | Time histories, three anodes summed for each detector |
| PH | 4.915 s | 32 | Pulse height spectra, each anode |
| POS | 6.55 s | 2† | Position histograms, 1 mm bins, anodes summed for each detector |
| RAW | 1 $\mu$s | 4096** | Raw data from the instrument |
| TAG | 256 $\mu$s | 32 | Time-tagged photons with 1-mm bin position, only for bursts |
| others | — | — | Various messages on burst trigger and localization, foreground and background position histograms, etc. |

\* 2–5/5–10/10–17/17–25 keV
† 2–10/10–25 keV
\*\* for each of left and right channels

transputer processes and localizes the event. The triggering time scales used as of April 2002 are from 80 ms to 14 s and the threshold levels are set to 4.7–6.3 $\sigma$, depending on the time scale.

In order to reduce the data mass downlinked to the ground, the low resolution time history and position histogram data are generated and transferred to the telemetry.

*Data Products*

In Table 2 is shown a list of currently implemented data products. Any values mentioned in the table are software variable, so we can change them in accordance with scientific profit.

The status of the detectors and temperatures are reported in **HK**, and the number of total events (both X-rays and charged particles) transmitted from the detectors to the DSP is reported in **SHK** both at the highest priorities.

For each GRB trigger, **TAG** data are downlinked to the ground with a medium priority. These are time-tagged photon data with 1-ms time resolution, 5-bit energy, and 7-bit position (1-mm bin) information. The duration of the data (including the background) is determined onboard according to the light curve of the event. The foreground and background position histograms used in the flight localization are also downlinked together with messages containing the information such as trigger time, trigger criteria met, foreground/background intervals, trigger significance, and image quality measure.

There are also three persistent streams of data. **TH** stands for "time history", and consists of the light curve histogram for each detector (XA, XB, YA, and YB) in four energy bands with a typical time resolution of 1.0 seconds. It also reports the number of events outside the normal energy bounds and the suspected particle events which hit multiple wires. **PHA** contains the energy spectrum histograms for each anode wire with 32 energy channels and time resolution of 4.9 seconds. **POS** contains the position histograms (1-mm bin) for each detector in two energy bands with a time resolution of 6.6 seconds. These histograms are downlinked to the ground with lower priority, and can be lost when there is a large mass of high priority data such as TAG.

In addition, **RAW**, a raw 32-bit event data from the detector (12-bit left and right pulse height with anode IDs) with 1-ms time resolution attached, can be downlinked on demand. This is mostly used for instrument calibration, and only three minutes of RAW data are taken regularly every day.

The onboard process also sends out messages which contain useful information when a trigger occurs.

These data products are downlinked to the ground via S-band at the contacts at primary ground stations[5]. A limited number of important health and status data, such as high voltage status and counting rate of each detector, are broadcast via VHF continuously[6]. At a burst trigger, a small number of essential pieces of information on the bursts, such as the trigger significance and localization results, are transmitted via VHF ([7]).

## OPERATIONS OF WXM

In November 2000, two weeks after the launch, the WXM detectors were briefly turned on to check out their performance. The position encoding, the position resolution, the energy scale (gain), and the energy resolution were checked using the onboard $^{55}$Fe calibration sources. No significant changes since the ground testing were observed.

The orbital period of HETE-2 is ~96 minutes, and the WXM is mostly operated based on the phase in the orbit. For one third of the orbit, the sun is occulted by

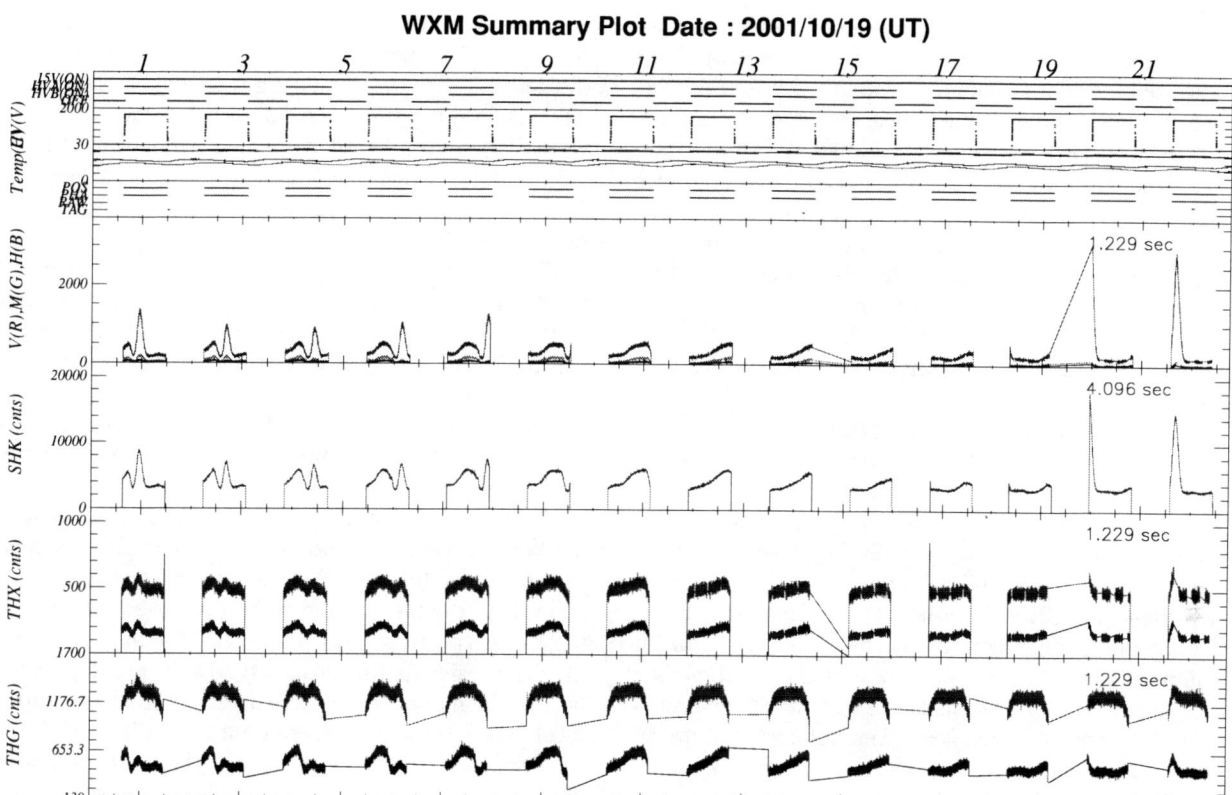

**FIGURE 5.** The summary plot of WXM on October 19, 2001, which shows a typical operation of WXM. The upper four panels show the status of the detectors. The fifth panel shows events that are likely due to charged particles, such as coincidence events; the sixth panel (labeled SHK) shows the total count rate; the seventh panel (THX) shows the time history of WXM in two bands; and the bottom panel (THG) shows the FREGATE time histories in Channel B (6–80 keV) and Channel C (32–400 keV). The instruments are turned on during the night-side half of the orbits.

the earth, and the spacecraft is in the "night". This is the prime time for observing GRB with WXM for three reasons: the field of view of WXM is completely clear of the earth, as it is centered on the anti-sun direction; the spacecraft attitude is locked with the optical star cameras, which perform best during the night time; and the X-ray instrument is free from hazardous impulsive solar flares. The field of view is clear of the earth for additional eight minutes on both sides of an orbit night. However, it is difficult to obtain onboard attitude in these regions, and there are risks for the instruments. In late 2000, the solar X-ray activity was extremely high, and we decided to take a conservative policy and turn on the high voltages of the WXM detectors mostly only during the orbit night.

In order to protect against hazardous conditions, such as high particle flux in the South Atlantic Anomaly (SAA), the count rate of the WXM detectors are monitored in real time. If the total incoming counts from the detectors (including both X-ray and non-X-ray events) exceeds the predefined threshold, the high voltage (HV) to the detectors is shut down automatically. This automatic protection mechanism has been working quite reliably, and we do not schedule HV shutdown commands for SAA.

The health and status of the WXM instrument is regularly monitored on ground by the WXM operation team. The histograms of the quiescent data products (count rate time history, position histograms, and pulse hight distribution) are updated at every downlink, and inspected by the duty operators. The energy/position linearity and resolution of the proportional counters are inspected at regular intervals with the calibration source using the RAW data, which are taken for three minutes every day.

A sample of a WXM summary plot for a typical day is shown in Figure 5. The status of WXM and the count rates from WXM and FREGATE are shown together. In this particular plot, the satellite passed through only the north rim of the SAA region in night, and therefore, there was no high voltage shutdown associated with SAA.

In December 2000 and January 2001, the Crab nebula was in the field of view, and we collected the data which was used to calibrate the alignment of the WXM coordinate system and that of the optical cameras. On January 11, 2001, a GRB was detected by the WXM and was localized. Unfortunately the optical camera system was still being tested, and an accurate spacecraft attitude was not available for this event. In February 2002, the attitude control using the optical system started to operate normally, and the first GRB localization by WXM was delivered to the community late in this month. With the improvement in the operations of the spacecraft as well as the ground systems, the observing efficiency of WXM gradually increased to 30–40% in April-May 2001.

In the period from late April to early September, the WXM field of view moved over to the galactic bulge region, where bright binary X-ray sources are present. In particular, Sco X-1, the brightest X-ray sources in the sky with strong variability, caused a significant loss of sensitivity for GRBs from late April to mid July 2001. On the other hand, these X-ray sources were useful for calibration. The Sco X-1 data were used to calibrate the alignment of the WXM coordinate system with respect to that of the optical star cameras. Similarly, most of the X-ray bursts and the bursts from soft gamma-ray repeaters (SGRs) were localized on the ground and identified with theier known source locations, and then used to cross-check the alignment calibration at various positions in the field of view, which was not possible with Sco X-1.

Some of the X-ray bursts triggered the burst detection and localization in the spacecraft onboard process, and were very useful as true end-to-end tests of the flight software. The details of the calibration of localization is given in [8] and [9]. In total, 135 X-ray bursts were localized with the WXM, and 133 of them were identified with catalogued X-ray burst sources. The complete catalog of the localized X-ray bursts are shown in [10]. Two bursts were localized to SGR 1806−06, and four bursts were localized to SGR 1900+14 [12].

In early August, the WXM field of view left the galactic bulge region, and the trigger criteria were optimized for GRBs. The observing efficiency in September 2001 was 43%, i.e., 10 observing hours per day on average.

## LOCALIZATION WITH WXM

The localization capability of WXM is the key to the HETE-2 mission.

The positional response of the WXM detectors were extensively calibrated on the ground. The non-linearity (so-called "S-curve") was characterized for each of the twelve anode wires to an accuracy of 0.1 mm. In addition, the position encoding is monitored in flight with the $^{55}$Fe calibration sources, which illuminate each anode wire at two positions ($\pm 4$ cm from the center) near the detector ends. We found slow, subtle, but systematic changes in the position encoding of all the wires, and we update the on-board correction parameters every 6 months. It is also calibrated using the position histograms of the Crab nebula and Sco X-1, which is the brightest X-ray source in the sky.

Three localization algorithms have been independently developed: onboard localization algorithm at Los Alamos [11], and ground analysis algorithms at RIKEN [8] and Chicago [9]. The consistency of localization with these methods were checked against each other using both simulations and real data from astrophysical sources such as X-ray bursts, gamma-ray bursts, and bright X-ray sources, and the results are found to be consistent.

Using the Crab nebula and Sco X-1, the alignment of the WXM coordinates with respect to the optical camera system was calibrated. With this calibration, the r.m.s. residual of the localization of Sco X-1 was $2.4'$ around the known position, and it was $2.7'$ for the Crab nebula. The residuals for these two sources are dominated by systematic uncertainties. The errors in X-ray bursts localization (ground analysis) are typically $6-10'$, and are dominated by statistical noise. The uncertainties in the onboard localizations were also evaluated with X-ray bursts. 41 out of 52 flight localizations of XRBs were correct with an r.m.s. uncertainty of $11.3'$. The localizations for these X-ray bursts are dominated by statistical erros, and the results are therefore consistent with those for Sco X-1 and the Crab nebula. See [8] for details.

## EARLY RESULTS

Since the launch until October 2001, a total of 152 bursts have been localized with WXM (Table 3).

### Gamma-Ray Bursts

In the first year, WXM localized 10 GRBs including three X-ray rich GRBs (or X-ray flashes). The list of

**TABLE 3.** 151 bursts localized by HETE-2 WXM

| Class | Remarks | Number |
|---|---|---|
| GRB | (including 3 X-ray rich GRBs) | 10 |
| SGR | SGR 1806−20 | 2 |
| | SGR 1900+14 | 4 |
| XRB | Total | 135 |
| | (triggered) | (53) |
| | (untriggered) | (82) |

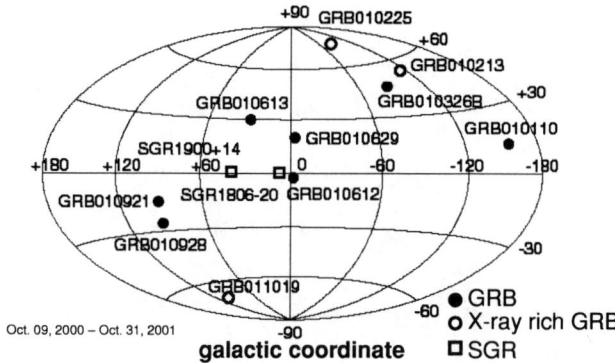

**FIGURE 6.** The spatial distribution of the gamma-ray bursts localized by WXM is shown in galactic coordinates. It roughly follows the ecliptic plane.

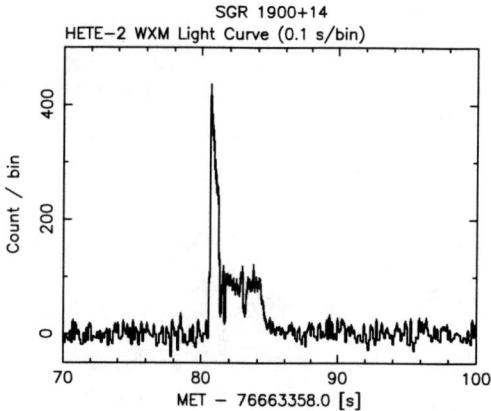

**FIGURE 8.** Light curve of a burst from SGR 1900+14 on July 2, 2001 (BID 1576) in 2–25 keV obtained with WXM. The dips in the light curve are artifacts due to the dead time.

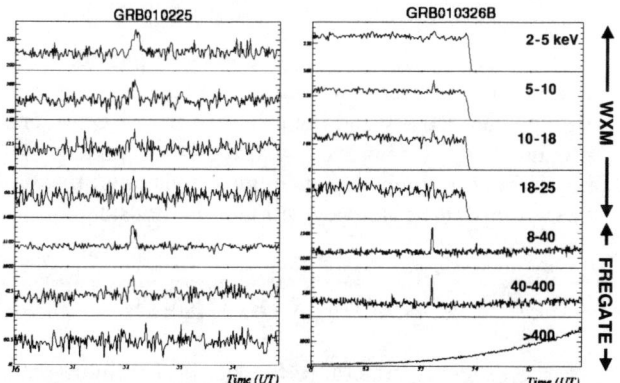

**FIGURE 7.** The light curves of two gamma-ray bursts (GRB 010225 and GRB 010326) obtained with HETE-2 WXM and FREGATE at various energy bands

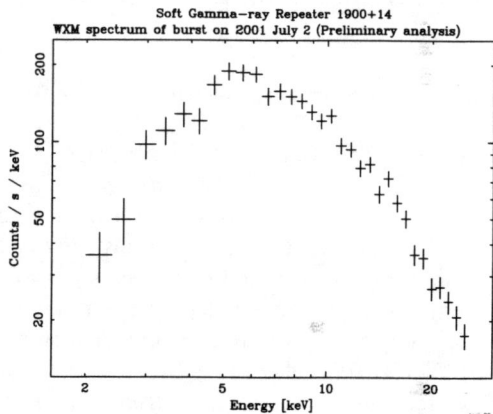

**FIGURE 9.** Energy spectrum of the burst (BID 1576) obtained with WXM.

the localized GRBs are shown in Table 4, and their locations in the sky are shown in Figure 6. The light curves of two GRBs are shown for the four energy bands in WXM and three energy bands in FREGATE in Figure 7. GRB010225 is a long and X-ray rich GRBs, while GRB010326B is a short GRB (duration ∼2 sec) with relatively hard spectrum.

*Soft Gamma-Ray Repeaters*

During the ∼60 days in June–August 2001, two bursts were localized to SGR1806−20 and four bursts were localized to SGR1900+14, including an intense burst from 1900+14 on 2 July 2001 (Fig.8,9). The complete list of the 27 candidate events from the SGRs detected by HETE-2, and the preliminary spectral analyses on the intense burst on 2 July 2001 are presented in [12].

*X-ray Bursts*

From May to September 2001, when the HETE-2 field of view passed through the galactic bulge region, 135 X-ray bursts (XRBs) were detected. 133 were localized to

**TABLE 4.** List of GRBs/XRFs localized by HETE-2 WXM in the first year of the mission

| GRB ID | Counterparts | Remarks |
|---|---|---|
| 010110 | – | Fregate was off |
| 010213 | $m_R > 17$ | X-ray rich |
| 010225 | – | X-ray rich |
| 010326B | $m_R > 22$ | |
| 010612 | – | Sco X-1 in FOV |
| 010613 | – | Sco X-1 in FOV |
| 010629 | $m_R > 20.5$ | Sco X-1 in FOV |
| 010921 | Yes | 1-dim + IPN |
| 010928 | – | 1-dim |
| 011019 | – | X-ray rich |

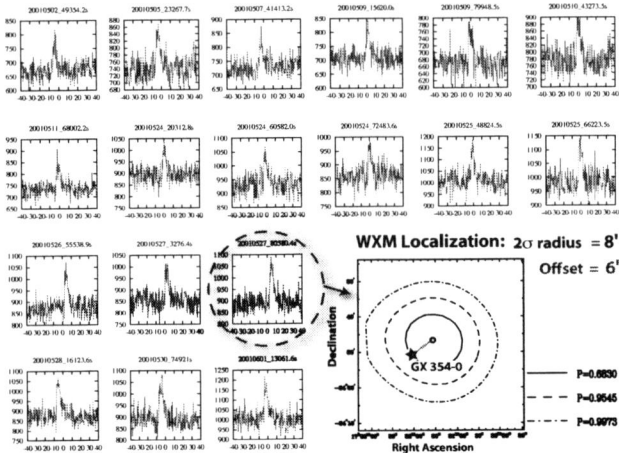

**FIGURE 10.** The time histories of 18 of the 30 XRBs from GX354-0 observed by FREGATE along with the WXM localization obtained for one of these bursts.

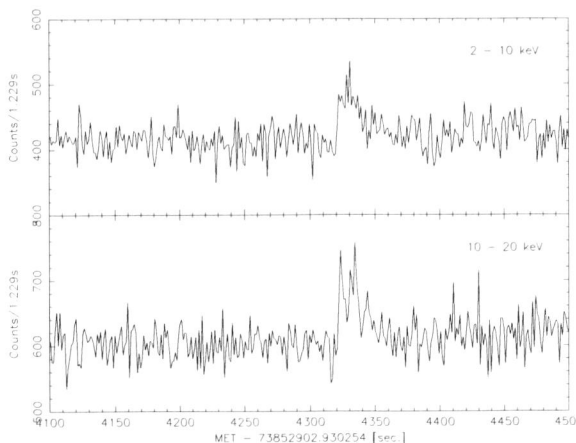

**FIGURE 11.** Light curve of an X-ray burst from X1812−121 in two energy bands observed with WXM on May 30, 2001.

known X-ray burst sources, which include 38 XRBs from GS 1826−238, 30 XRBs from X1728−34, and 17 XRBs from Aql X-1. The typical error of the ground localization was $6'$ for these source. Twenty-seven XRBs were correctly localized by the WXM flight code in near-real time, but only two of these localizations were sent out to the astronomical community; the remainder were not sent out in order not to overwhelm the community with non-GRB burst alerts, or were sent out with incorrect positions. Two XRBs could not be associated with known XRB sources; they may be XRBs from previously unknown XRB sources or extragalactic X-ray rich GRBs. Figure 10 shows the time histories of 18 of the 30 XRBs from GX354-0 observed by FREGATE, along with the WXM localization obtained for one of these bursts.

The complete list of the detected time and sources of the X-ray bursts are presented in [10].

## CONCLUSION

The performance of WXM in orbit shows that its design goal has been successfully achieved. We have had no degradation in detector performance since the launch (even since the ground testing), and its positional and spectral performance have been continuously monitored using the $^{55}$Fe calibration sources built in WXM.

Its localization capability has been demonstrated with localization of the 10 GRBs, 7 SGR bursts, and 135 X-ray bursts, and the typical uncertainties are $10'$ r.m.s. Its alignment with the spacecraft aspect system has been calibrated using the strong persistent sources such as Crab and Sco X-1.

Its observing efficiency in the first year has been limited due to unforeseeable constraints such as unusually strong solar activity and the problems in spacecraft and ground systems. However, with the better understanding of the instrument and operational constraints that has been gained, the efficiency will improve considerably.

## ACKNOWLEDGMENTS

We are specially grateful to the late Prof. Minoru Oda, who supported the HETE mission since its beginning. His encouragement at the loss of HETE-1 and his strong support for HETE-2 were invaluable to us.

The authors thanks the operation team at MIT for continuous support since the beginning of the original HETE-1 mission. We also thank to the support teams at the primary ground stations and the secondary ground stations, in particular Mr. Mak Choong Weng and other members of National University of Singapore, where largest mass of the data is downlinked.

## REFERENCES

1. Ricker, G., et al., 2002, these proceedings.
2. Villasenor, J., et al., 2002, these proceedings.
3. Atteia, J.-L., et al., 2002, these proceedings.
4. Shirasaki, Y., , et al., 2002, Proc. SPIE Vol. 4012, p. 166-177.
5. Crew, G., et al., 2002a, these proceedings.
6. Crew, G., et al., 2002b, these proceedings.
7. Vanderspek, R., et al., 2002, these proceedings.
8. Shirasaki, Y., et al., 2002, these proceedings.
9. Graziani, C., et al., 2002, these proceedings.
10. Sakamoto, T., et al., 2002, these proceedings.
11. Fenimore, E.E., et al., 2002, these proceedings.
12. Torii, K., et al., 2002, these proceedings.

# An Overview of the HETE Soft X-ray Camera

J.N. Villasenor*, R. Dill[†], J.P. Doty*, G. Monnelly*, R. Vanderspek*, S. Kissel*, G. Prigozhin*, G. B. Crew* and G.R. Ricker*

*MIT Center for Space Research, Cambridge, MA*
[†]*ITT, Fort Wayne, IN*

**Abstract.** A new type of imaging detector, the Soft X-ray Camera (SXC), is now flying on the HETE-2 satellite as part of the instrument suite to detect and localize GRBs. The low point spread function of CCDs combined with a finely ruled and highly aligned coded mask results in a compact instrument (~10 cm on a side) which can localize transients with high precision (~30 arcseconds) over a large field of view (~1 sr). We present an overview of the design, fabrication, and testing of the SXC. The in-flight performance and capabilities are then presented. Finally, the adverse effects of the space environment (in particular the micrometeorite flux and increased atomic oxygen concentrations during solar maximum) on the SXC, and the steps taken to mitigate these effects are discussed. Both GRBs and XRBs are being routinely localized with high accuracy by the SXC.

## INTRODUCTION

Following the loss of HETE-1 in November of 1996 due to a rocket failure, planning for a replacement mission spacecraft immediately commenced. It was recognized at that time that GRBs are indeed extra-galactic in origin and thus likely to be optically faint. Plausible estimates indicated that a fraction of GRBs should have a detectable soft X-ray flux; thus, replacing the UV CCDs on HETE-1 with X-ray CCDs would increase the chances of obtaining fine source localizations.

The Soft X-Ray Camera (SXC) was conceived to exploit the inherent advantages of X-ray CCDs. These devices have very small point spread functions, resulting in accurately-resolved single-photon positions. Combining X-ray CCDs with a 1-D coded mask possessing micron-sized features results in an instrument capable of fine angular resolution (~30 arcsecond). X-ray CCDs also have excellent spectral resolution and complement the WXM instrument by extending the spectral coverage down to 1 keV.

In view of the tight delivery schedule for HETE-2, it was necessary that the SXCs be direct replacements for the previous UV cameras. The available spacecraft mass and volume resources could accomodate two 1-D coded mask units, each paired with an optical CCD camera to determine accurate aspect. The SXCs were to be placed at two corners of the spacecraft, and utilized the same electronics and computer resources previously allocated for the HETE-1 UV system (Fig. 1).

The MIT CCD laboratory used its in-house facilities

**FIGURE 1.** SXC on HETE and the Pegasus rocket. The two SXC units occupy the top corners; the solar panels deploy (down) clear of the instruments.

to develop and test the SXCs and boresight cameras, completing such tasks as CCD calibration, mechanical alignment, thermal testing using an environmental vacuum chamber, and acoustic loading tests on an accelerated schedule. Development of the SXC began in 1997, and delivery and integration of the fully developed system to the spacecraft occurred in mid-1999.

## DESIGN CONSIDERATIONS

The general SXC parameters are shown in Table 1. The major challenges in fabricating this instrument are to

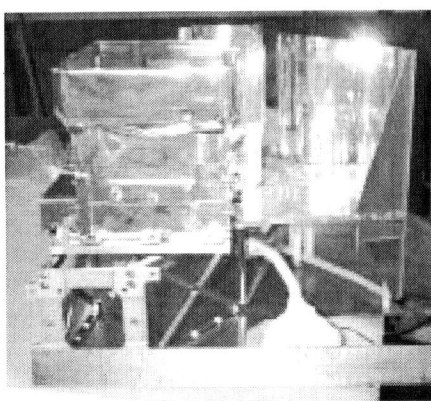

**FIGURE 2.** Side view of a SXC module. The cooling radiator is the curved attachment to the right. The boresight camera lens housing is also covered with thermal tape to allow heat dissipation.

**TABLE 1.** General SXC parameters

| | |
|---|---|
| Camera Size | $100 \times 100 \times 100$ mm |
| Instrument type | Coded Mask with CCID-20 |
| Field of View | 0.9 sr (FWHM) |
| Angular Resolution | 33 arcseconds per CCD pixel |
| Detector-Mask Distance | 95 mm |
| Mask Open fraction | 0.2 |
| Mask element size | 45 $\mu$m |
| Timing Resolution | 1.2 s |

create an accurately-aligned system of mask and focal plane elements and to control any shifts or distortions of these elements that might occur in orbit. The image cross correlation can be blurred by any of the following factors:

- Distortion in the mask or the focal plane.
- Rotation of mask slits relative to the CCD columns.
- Tilt of the focal plane with respect to the mask plane.

The first and second items smear the shadowing edge from one pixel to adjacent pixels on the focal plane. The third item is the equivalent of a rotation for off-axis images, and can be corrected for by additional ground processing. Effects due to the first two items cannot be removed and must be minimized by accurate fabrication and alignment.

Maintaining an isothermal SXC also minimizes thermal distortion; thus, the walls are covered with thermal tape both inside and out to distribute the heat evenly (Fig. 2). Thermal shifts during an orbit can cause motion of the reference cameras with respect to the SXC instrument. Hence, an optical camera is mounted adjacent to each SXC on a common baseplate.

# CODED MASK

Each SXC coded mask consists of a single electroformed sheet of 99% gold stretched over a stainless steel frame. This is the first time such a process has been used to transfer and replicate micron sized features over a large area (10 cm $\times$ 10 cm) with a minimum of distortion (non-parallel slits). The slit edges were measured to have deviations of less than 5 $\mu$m rms from an ideal ruled reference plane. The sheets were fabricated by Dynamics Research Corporation in Wilmington, MA.

The coefficient of thermal expansion of the stainless steel frame is lower than that of gold, hence the sheet is under tension and free from deformation as it cools to operational temperatures. The mesh thickness is 25 $\mu$m, which ensures better than 1:100 blocking factor over the 0.5-14 keV range.

The pattern consisted of 2100 coded elements (slits), with an element size of 45 $\mu$m. The pattern was chosen to minimize sidelobe and coding pattern noise for a given signal and background noise. For mechanical reasons, the pattern contained no single stops that would be flanked by open elements, which might lead to breakage due to launch stresses.

# FOCAL PLANE

The X-ray CCDs are front-side CCID-20 $2048 \times 4096$ arrays ($15 \times 15$ $\mu$m pixels) fabricated at MIT Lincoln Labs. Each device operates as a collection of 2048 long and narrow detectors which integrate events over 1.2 s. Two of these CCDs are mounted side by side on a common aluminum nitride (AlN) baseplate, separated by 1 mm. The parallel alignment of the two CCDs was within 5 $\mu$m over 6.1 cm. The relative displacement of the CCDs were measured and are shown in Fig. 3. Some warping of the CCD is evident, due to internal stresses developed during the silicon etching process; in-flight calibration is required to correct for the effect of this stress upon SXC localization accuracy.

The side walls have a 0.25 mm tin layer sandwiched beneath the thermal tape to limit off-axis X-ray penetration. To minimize power, the CCDs are passively cooled using a radiator plate attached to the baseplate. The design temperature was -50 C, achieved in flight.

# OPTICAL BLOCKING FILTER

To prevent contamination of the CCD detectors by non X-ray photons, the SXC was provided with two redundant optical blocking filter (OBF) systems: 1) very thin (0.05 $\mu$m) aluminized plastic films, covering the entire

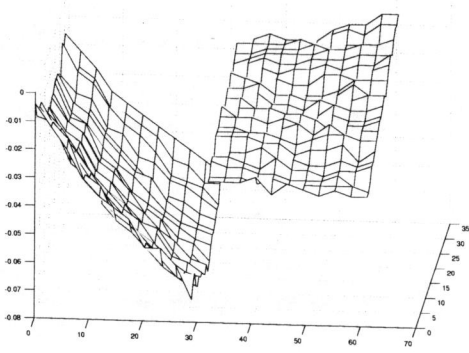

**FIGURE 3.** Measured surface map of the CCDs on one of the focal planes showing the relative orientation. All axes are in mm. The effect of the "tilt" of one CCD with respect to another can be corrected through in-flight calibration.

**TABLE 2.** CCID-20 parameters

| | |
|---|---|
| Pixel array | 2048×4096 |
| Pixel size | 15 $\mu$m × 15 $\mu$m |
| CCD Dimensions | 6.1 cm × 3.1 cm |
| Effective Area | 18.90 cm$^2$ |
| Depletion Depth | 30 $\mu$m |
| Readout Mode | 100 row summation |
| Energy Range | 1-14 keV |

aperture, and 2) Be foil (25 $\mu$m) filters, covering one half of the detector area.

The outer OBF was made of 5000 Å polyimide coated with 1500 Å of Aluminum. The visible light transmission was $10^{-8}$, sufficient to block out moonlight. The optical blocking filter design was based on the Chandra HRC filters, which had an equivalent surface area. A 25 $\mu$m beryllium film was placed on top of one CCD on each focal plane as insurance against micrometeorite impact and outer OBF failure. Side vents on the SXC walls were included to prevent the filters from bursting during launch; these had light baffles to minimize light leakage.

## ALIGNMENT AND MEASUREMENT

The CCD and mask alignment/uniformity were measured with a narrow depth-of-focus microscope and an XY translation stage (Fig. 5). A video camera system enhanced surface features and allowed precise measurement accurate to ±2 $\mu$m over a ~12 cm travel on all three axes. Mask distortion (non-parallelism of the slits) was measured by following mask edges using this setup. Likewise, the focal plane was mapped by noting the CCD gate features.

The tilt misalignment is minimized by careful design. The mask and frame were mounted onto the wall frame,

**FIGURE 4.** SXC focal plane showing the two CCID-20's mounted side by side on a common AlN baseplate.

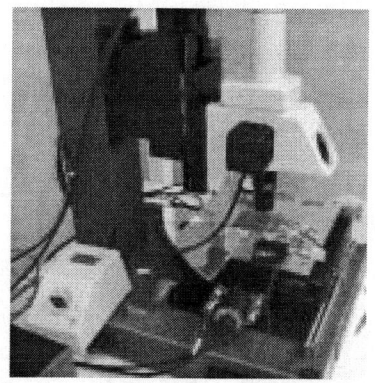

**FIGURE 5.** Alignment apparatus, showing the x-y stage, the microscope, and one focal plane being mapped.

which was machined to a uniform height to within tolerances of a few microns. Similarly, the baseplate and all relevant attachments and surfaces to the baseplate were lapped to preserve the flatness of the focal plane.

The rotational alignment of the mask/focal plane was measured by pivoting the mask (already mounted to the walls) with respect to the focal plane/baseplate combination, using one micrometer to rotate, and another to gauge movement.

## IN-FLIGHT PERFORMANCE

From the time of launch in October 2000 until approximately December 2000, both SXCs exhibited excellent in-flight performance characteristics, with stable baselines and energy resolution comparable to pre-launch values. In-flight measurements using the Crab Nebula exhibited narrow, tightly-collimated correlation peaks which were consistent with the pre-launch ground calibrations. Approximately three months (X camera) and five months (Y camera) post launch, the CCDs began to exhibit a saturation response that was indicative of se-

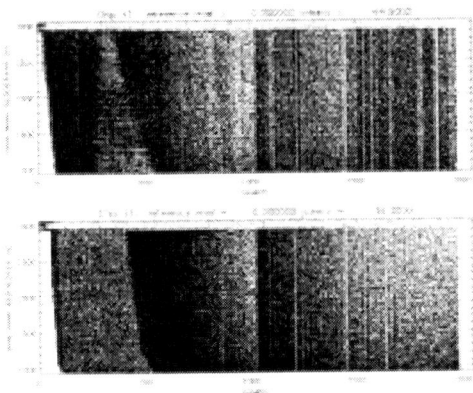

**FIGURE 6.** Operations with the moon close to the field of view. Each point is a photon event recorded over the width of the CCD, with time moving upwards. The period up to 2900s show the CCDs flooded with scattered light.

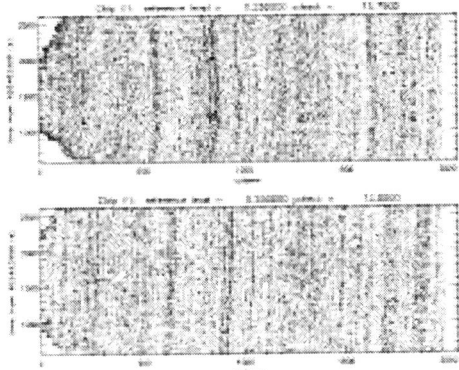

**FIGURE 7.** Tracks in the X and Y SXCs with the Crab in the field of view. The drift in the spacecraft pitch and yaw are evident.

vere light leakage. Comparision of the various CCD pairs indicated that the sensors most adversely affected were those which were protected by only the thin-film outer OBF. An extensive series of diagnostic tests showed that the outer OBFs appeared to have become gradually more transmissive, indicating an erosive loss mechanism (see discussion below).

The foil OBFs have remained intact throughout the mission. However, the foil OBFs protect only one half of the detectors and are subject to edge leakage during bright moon conditions. As a result, the SXC sensitive area is now one half of the value at launch, and the SXC can only be used during the "lunar dark" of the month (i.e., 24 days, or ~80% of each lunar month).

Using the Crab and Sco X-1 as astrometric reference sources, the SXC has been calibrated to provide localizations of better than 43 arcseconds at 90% confidence (Fig. 7). An extensive discussion of the localization accuracy can be found in [1].

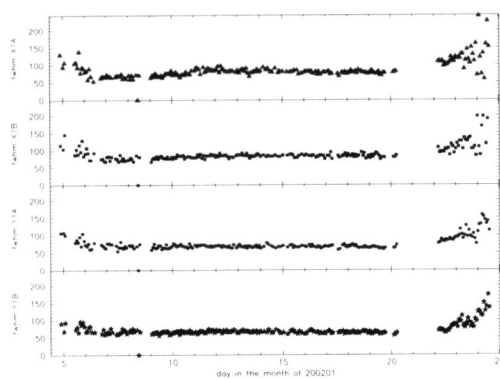

**FIGURE 8.** The gain of the detectors, showing the operational period of the instrument. The increased levels at the beginning and end are due to the higher background light levels.

The small Fe55 flight calibration source attached to the mask frame calibrates the energy gain and the spectral resolution of the CCDs. A side effect of the code fix allowing operations with background light variations (Fig. 8) was to diminish the spectral resolving capability of the SXCs, e.g., FWHM at 5.9 keV is now ~400 eV. The spectral performance of the SXC is described in more detail in [2].

The code running the SXC undergoes periodic revisions to improve performance. In addition, the improvement of spacecraft aspect will likely permit in-flight (prompt) localizations, centered on a region identified by WXM on-board localization.

## EFFECTS OF OBF LOSS

Atomic oxygen (AO) densities at 600 km apparently increase by $10^3 - 10^5$ times during intense solar flares [3]. HETE experienced a series of such flares in late 2000 – early 2001, shortly before the loss of the SXC outer OBFs. The existence of this strong dependence of atomic oxygen on solar activity at the HETE orbital altitude was not known at the time the SXC underwent its design reviews in 1998.

AO attacks carbon based films, including polyimide, at a rate proportional to the time integrated flux impinging on the surface. Polyimide has a recession rate of about 2.8 $\mu$m for a fluence of $10^{20}$ atoms/cm$^2$ [3]. For a polyimide thickness of 5000 Å, and an averaged AO density of $10^6$/cm$^3$ the OBF would be eroded in 40 days. This is in reasonable agreement with the time of loss of the two outer OBFs: the SXC-Y outer OBF was lost in December, 2000, and the SXC-X outer OBF was lost in February, 2001.

The flux of orbital debris (mainly $Al_2O_3$ particles from spent rocket stages) and micrometeorites may also have degraded the CCDs which had no Be filter. The outer OBF acts as a Whipple shield, vaporizing and fragmenting particles which could otherwise cause deep cratering. It is estimated that there may be as many as 10 impacts/year/CCD which could cause damaging craters without the outer OBF. That number is reduced to 1-2 events if the outer OBF is present [4]. Debris/micrometeorite cratering on the CCDs is not well understood. However, over a period of months the clocking voltage on the bare CCDs became insufficient to flush out the photon-generated charge, which could possibly indicate gate shorts caused by debris pitting.

## CONCLUSIONS

The design details for the SXC, a novel imaging instrument flown for the first time on HETE, have been presented. For the first time, the challenges of fabricating, calibrating, and flying an integrated mask-detector assembly requiring micron tolerances have been successfully met. Despite initial difficulteis with two of the four optical blocking filters, excellent positional locations ($< 40$ arcseconds RMS) for GRBs are being routinely achieved. [ Note added in proof: as of December 2002, five GRBs and more than 60 XRBs have been localized to sub-arcminute precision by the HETE SXC].

## ACKNOWLEDGMENTS

We thank Tyc Brady, David Breslau and Fred Miller for helping fabricate and test the SXCs.

## REFERENCES

1. Monnelly, G., *et al.*, these proceedings.
2. Prigozhin, G., *et al.*, these proceedings.
3. Dooling, D. and Finckenor, M. ,"Material Selection Guidelines to Limit Atomic Oxygen Effects on Spacecraft Surface", NASA/TP-1999-209260.
4. Pak, S. 1999, "Micrometeoroid, Orbital Debris, and Shielding Analysis for the SXC of The HETE-2 Spacecraft", MIT Aero Astro Masters thesis.

# HETE: Novel Technology for a Small Space Mission

J. Doty*, R. Dill*, G. B. Crew*, T. Brady[†], J. Francis*, G. Huffman**, J. Roberts[‡], R. Vanderspek* and the HETE Science Team[§]

*MIT Center for Space Research, Cambridge, MA, 02139 USA
[†]H.L. Draper Laboratories, Cambridge, MA, 02139 USA
**GMH Enterprises, Petaluma, CA, 94952 USA
[‡]Roberts Engineering, Arlington, MA, 02476 USA
[§]An international collaboration of institutions including MIT, LANL, U. Chicago, U.C. Berkeley, U.C. Santa Cruz (USA), CESR, CNES, Sup'Aero (France), RIKEN, NASDA (Japan), TESRE (Italy), INPE (Brazil), TIFR (India)

**Abstract.** The High Energy Transient Explorer is an unusually small and inexpensive satellite. We will discuss the technology that made this possible. HETE incorporates a robust and efficient power system, a 3-axis stabilized gyroless control system, and an unusually efficient communication system. We will discuss the advantages this technology offers for the HETE mission, as well as the constraints it imposes and the difficulties it has caused.

## INTRODUCTION

HETE-2 was launched on October 9, 2000 by a Pegasus rocket near Kwajalein atoll. It is in a nearly circular (apogee 620 km, perigee 580 km altitude) equatorial (inclination 1.95 degrees) orbit. It is a very low cost mission compared to other astrophysics missions: we estimate that the total cost, including very substantial contributions from our foreign collaborators, has been about $26 million. An overview of the mission is given in [1]; the instruments are described in [2, 3, 4]; the ground assets are described in [5, 6].

HETE-2 is intended to obtain gamma ray burst locations suitable for optical followup. For this reason, we decided to center its field of view on the antisolar direction. This allows us to use fixed solar panels, and places the primary science instruments on the cold side of the spacecraft. We thus need no active cooling for the CCD instruments: the SXC modules are maintained below -50 C by radiators and heat shields. Limiting the range of solar attitudes for normal operation also simplified the attitude control system design.

## HARDWARE

The key to selecting technology for a low cost mission is to start from that particular mission's needs. Irrelevant requirements must be avoided. It is essentially superstitious to insist on "space qualified" technology without

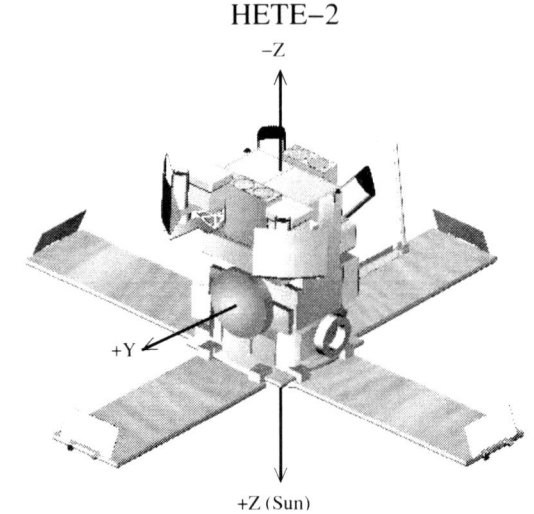

**FIGURE 1.** HETE-2: the instrument axis is -Z; the angular momentum axis is +Y.

critically examining whether it is the best match to your requirements. Low Earth orbit is not an especially harsh environment.

Most of the electronic parts we used to construct HETE-2 were ordinary mass-produced commercial parts ordered "off the shelf" from industrial suppliers. The most obvious advantage of this is price, but that is actually of very little importance: you can't save money by using inferior technology. The real advantage is that the parts selected met HETE's needs much better than "space qualified" parts would have. Prototype parts delivered overnight were the same as flight parts, avoiding nasty surprises. Commercial parts have higher quality than "space qualified" parts, avoiding rework of flight assemblies. The diversity of commercial technology helps when matching the parts to the requirements, simplifying circuits, reducing power consumption, reducing component stress, and increasing margins.

HETE-2 has six batteries, each composed of 23 "Energizer" brand rapid-charge NiCd cells (1.2 AH, 2/3 C size). To help insure long life, we measured the capacity of each cell and grouped cells of similar capacity together. We avoid deep discharges: in normal operation we only need about 20% of available capacity to power HETE through orbit night. With care to avoid over and under charge, the batteries have shown no measurable change in characteristics over more than 8000 cycles. These cells are extremely rugged: we severely abused some prototype batteries in early testing without significant loss of capacity. With six independent batteries, HETE can continue to function without difficulty if two are lost. This is one area where redundancy is inexpensive: battery mass and volume are proportional to total capacity so the number of batteries is a small consideration. Chargers are neither terribly complex nor bulky.

One reason we chose these cells is that available "space qualified" batteries were too large, making redundancy impossible. Hermetically sealed batteries also are brittle: if abused they lose their seals and electrolyte. HETE's vented cells may lose some electrolyte if abused, but the loss is controlled and should cease when the abuse ceases.

HETE's semiconductor parts are almost all plastic encapsulated, instead of hermetically sealed. While hermetic parts show better reliability in some kinds of tests at high humidity, there is no reason to believe they are any more reliable in actual service in space. Matching the capabilities of the parts to the requirements often leads one to select parts that are unavailable in hermetic packages. Choosing parts that are right for the application leads to simpler, more robust circuitry. We attempted to select parts made with processes that are reasonably radiation tolerant, and HETE-2 employs circuit-level defenses (circuit breakers, current limiters, error correcting codes) against radiation-induced upsets. This is generally simpler than working around the peculiarities of "radiation hardened" parts.

HETE has now operated in orbit for over 18 months, and we have seen no evidence of any electronic part failure.

The are places, however, where we judged that specialized space technology was desirable. For deploying solar panels, HETE-2 uses Starsys hot-wax actuators. These are simple, testable, and reliable. The solar cells themselves are "space qualified" Si cells, as we believe these will show better tolerance to the enormous radiation dose they are experiencing.

## SOFTWARE

HETE-2 uses a duplex S-band link for command uplink and bulk data downlink. This link employs a stream transport protocol, with handshaking, error checking, and retransmission of bad data ("backward error correction"). The protocol is a variant of SDLC, which is older and simpler than the more familiar Internet protocol, TCP. The downlink also uses 2:1 convolutional encoding for "forward error correction". Backward error correction greatly reduces error rates: even when the link quality is poor, data makes it through reliably, although when retransmission is frequent the data rate suffers. Forward error correction reduces the frequency of retransmission.

HETE-2 commands and data use a common packet format for uplink, downlink, within the spacecraft, and on the ground. This simplifies handling and routing of commands and data.

The HETE processor box contains twelve processors. Eight Motorola 56001 digital signal processors handle interfaces to the instruments and the spacecraft. These processors are very fast, but have limited memory capacity. Four Inmos T805 Transputer processors handle bulk data storage and processing code too elaborate for the smaller processors.

For the 56001 processors, we wrote a very simple operating system. It serves a single background task and interrupt handlers. For HETE-1 the Transputers used a commercial real time operating system. We found this to be overly complex: a large amount of application software was present only to correct for OS misfeatures. For HETE-2, we wrote our own simplified multitask operating system, simplifying the application code and improving performance and stability.

HETE software development uses the most widely available tools (CVS and the GNU development suite) as much as possible. Some proprietary tools from Inmos and Motorola were required. Most of the software is written in C. It is extremely advantageous to use standard, widely supported, familiar tools.

## ATTITUDE CONTROL

The attitude control system (ACS) stabilizes HETE-2 in all three rotational degrees of freedom. It is physically much simpler than most other "3 axis" stabilization systems. There are no gyros, and only a single momentum wheel. In normal operation, star trackers are the primary attitude sensors for about 45% of each orbit. When star data is available, the ACS is designed to anchor the spacecraft in inertial space using its attitude at the time of the first tracker solution of the series ("night" mode). Scattered sunlight blinds the star trackers for the other 55% of the orbit: in this case the ACS is designed to point the instrument (-Z) axis in the antisolar direction ("day" mode), using a pair of four pixel pinhole cameras to determine the location of the Sun.

The momentum wheel controls rotation about the Y axis. Magnetic torque coils control the angular momentum, using magnetometers to determine the required moment for the desired torque. In day mode, the coils control solar elevation relative to +Z and total angular momentum. In night mode, they control rotation about X and Z.

Despite its simplicity, ACS performance is good. Most of the time, night mode holds the attitude within 2 arc minutes of the target.

Day mode does not control rotation about Z, and night mode only holds its rate to zero. HETE-2 therefore tends to rotate very slowly about Z. If left uncontrolled for several weeks, this rotation generally ends with +Y near the south ecliptic pole, apparently because of some slight magnetic imbalance. We can actively control this by biasing the torque coils upon ground command: rotations of up to 40 degrees per day are possible. While this is not very fast, we have found it useful when calibrating using known X-ray sources. We also have used this technique to orient HETE-2 in its least vulnerable attitude during the Leonids.

We can also command HETE-2 to rapidly rotate about the Y axis, using the Sun as a reference to stabilize in day mode with the instrument axis in a selected position away from the antisolar point. If we do this before HETE-2 enters the Earth's shadow, night mode will then hold the attitude at this position. After sunrise, we command HETE-2 to rotate back to a nominal attitude to recharge batteries and keep the CCD instruments cool. We have successfully tested this as a way to keep moonlight out of the star trackers, and we intend to use this capability to keep the field of view away from Sco X-1 and the bright X-ray sources in the galactic bulge during spring and summer of 2002.

Day and night modes only work properly when the attitude and momentum are close to nominal values, and night mode requires star tracker data. For times when these preconditions have not been met, the ACS has a selection of simpler control laws. These modes use magnetometers to sense rotation rate, and coarse sun sensors (photodiodes on the spacecraft principal axes) to sense solar attitude. "Detumble" mode removes energy from spacecraft rotation until the rates about the X and Z axes are tolerably low. "Spinup" mode adds angular momentum around the Y axis until it is sufficient to support the higher modes. "Reorient" mode precesses the angular momentum vector until it is approximately perpendicular to the Sun. "Wheel Deploy" mode spins up the momentum wheel to its nominal rate. "Acquisition" mode uses wheel torques to point the Z axis at the sun. "Panel Deploy" mode releases the solar panels (HETE-2 has only entered this mode once since launch). The ACS automatically selects the most advanced mode whose preconditions have been met: there is no "safe hold" mode.

## ROBUSTNESS

HETE is "walk away safe": it autonomously handles situations that might endanger it. No ground intervention is required to safeguard HETE in an emergency.

The biggest hazard is power loss: orienting the solar panels toward the Sun is a complex task that can go wrong in a large number of ways. HETE defends against this with "battery wait" mode: if its main power bus voltage drops below a preset (adjustable) threshold, it shuts down nearly all spacecraft systems: only coarse timekeeping and battery chargers are powered in this mode. The minimum possible power from the solar panels is 6W, orbit average (from sunlight reflected from the Earth). Only 2W is consumed in battery wait mode, leaving at least 4W for battery charging.

HETE leaves battery wait mode when a majority of its six battery chargers indicates full charge. In practice, this means all batteries are essentially fully charged, as they tend naturally toward balance. If for some reason this condition cannot be reached, battery wait mode will end when a 38 hour timeout has expired. When battery wait mode ends, the power controller powers up the "spacecraft controller" processors.

A 16 second watchdog timer guards against crashes of the spacecraft controller processors, rebooting them if they appear to have failed. The power system timer will cycle power to the spacecraft if it is not reset from the ground for 38 hours (this has never happened outside of preflight testing).

## DIFFICULTIES

The reason that space flight is so expensive is that any mission must overcome many difficulties. Solutions to the primary difficulties create secondary difficulties. It is difficult to get this process to converge. To be successful, a low cost mission must choose approaches that create as little trouble as possible: convergence must be rapid. It is not possible to do this perfectly, so HETE-2 had a number of secondary difficulties to overcome.

The most serious technical difficulty HETE-2 had to overcome was noisy digital interfaces on board. High density digital circuitry saves mass, power, and money, but its high speed can be trouble when it is used at interfaces, even if the interface speed is low. We identified and fixed a number of noise problems before launch, but one did not manifest itself until after launch. There appears to be an intermittent problem with noise on the data clock between the S band transceiver and the DSP that serves it. This causes packet corruption on the uplink, and since uplink packet error checks are performed by the transceiver, these errors can go undetected. The result was that about 1% of all uplink messages were garbled. HETE-2 science operations are command intensive: it is typical to have 100 significant messages in the programmed uplink for a pass, so a 1% error rate is very serious.

In addition, the noise is sometimes so bad that the extra processor interrupts can significantly slow down the DSP. This caused frequent watchdog timeouts, reducing operational efficiency further.

We solved these problems with new software, first uplinked on 15 May, 2001. We added an additional checksum field to each uplink packet, effectively eliminating the undetected errors. Interrupting the downlink data flow when DSP processing falls behind schedule has almost completely eliminated watchdog timeouts. Neither of these causes any data loss, just a slight reduction in throughput.

On the management side, a small team can be more of a challenge than a large one. To keep costs down, nearly everyone will be juggling multiple activities. It is generally impossible to identify the critical path, because problems in one activity pull effort into that activity, delaying other activities. Management must be flexible, helping the team to self organize around trouble.

Design criticism is important, but very difficult to get in a low cost project. Design reviews are dysfunctional because reviewers are generally unfamiliar with low cost technology. This wasted a great deal of time on HETE-2.

Unfamiliar technology and practices look riskier than they really are. An irrational approach to risk exacerbates this problem: risk is not the same thing as failure probability. Risk (in an unmanned mission) should properly be assessed as financial expose times failure probability. A low cost mission should never be considered high risk.

## THE PAYOFF

It took us a while, but HETE-2 is working well. Low-cost spacecraft are the way to get space exploration back on the high technology path, where increased capability and reduced cost go hand in hand. Low cost missions are the place to develop technology: CCD measurement chain technology for Chandra and ASTRO-E were originally developed for HETE.

For scientific missions, scientists must take the lead, because knowledge of the application is even more important than knowledge of the technology. Low cost missions, led by scientists, are key to the future.

Is HETE novel technology? By most standards it is not. It is simply effective technology selected to meet the real needs of the mission. In space, that counts as novel.

## ACKNOWLEDGMENTS

Even a small space mission is an enormous collective endeavor. Many contributed to the success of HETE, and have my sincere thanks. Major contributers to the design and construction of HETE-2 include Joel Villasenor, Glen Monnelly, Nat Butler, Fred Miller, Janice Crisafulli, Jean Winter, Mike Vezie, Tim Jones, Scott McDermott, Steve Kissel, Dave Breslau, Pete Tappan, Francois Martel, Frank LaRosa, Chuck McCall, Richard Dynes, Richard Warner, and George Ricker

## REFERENCES

1. Ricker, G. R., et al., these proceedings.
2. Atteia, J-L., et al. 2002, these proceedings.
3. Kawai, N., et al. 2002, these proceedings.
4. Villasenor, J. N., et al. 2002, these proceedings.
5. Crew, G. B., et al. 2002, these proceedings.
6. Villasenor, J. N., et al. 2002, these proceedings.

# HETE-II and the Interplanetary Network

K. Hurley*, J.-L. Atteia[†], G. Crew**, G. Ricker**, J. Doty**, G. Monnelly**, R. Vanderspek**, J. Villasenor** and T. Cline[‡]

*UC Berkeley Space Sciences Laboratory, Berkeley, CA 94720-7450*
[†]*CESR, BP 4346, 31028 Toulouse Cedex 4, France*
**M.I.T. Center for Space Research, 70 Vassar St., Cambridge, MA 02139*
[‡]*NASA GSFC, Code 661, Greenbelt, MD 20771*

**Abstract.**
The FREGATE experiment aboard HETE-II has been successfully integrated into the Third Interplanetary Network (IPN) of gamma-ray burst detectors. We show how HETE's timing has been verified in flight, and discuss what HETE can do for the IPN and vice-versa.

## INTRODUCTION

The FREGATE experiment aboard HETE-II is an excellent complement to the detectors in the IPN. It has good sensitivity to bursts, thanks to its large surface area and its steady background in equatorial orbit. Its time resolution in both triggered and untriggered modes is sufficiently high for cross-correlating precisely with other spacecraft. And finally, its energy range is well matched to those of other instruments in the network. However, before any experiment can be added to the network, its timing must be verified in flight. In this paper, we will first explain the procedure we have used to demonstrate that the HETE timing is accurate. Then we will show how the IPN can be used to improve the location accuracy of HETE bursts, and how the FREGATE data are useful to the IPN.

## TIMING

In principle, we expect the timing of the HETE spacecraft to be good, both because it is in low Earth orbit, and because it is derived from an onboard GPS receiver. While there are several ways to confirm the timing of any spacecraft in flight, the best is to triangulate burst sources with known positions, either soft gamma repeaters, or gamma-ray bursts for which optical or radio counterparts are found. In its first year of operation, FREGATE detected numerous bursts from SGR1806-20 and 1900+14; seven of them were observed in conjunction with other spacecraft in the IPN. The positions of both SGRs are known to arcsecond or sub-arcsecond accuracy, and therefore serve as timing calibration sources.

In figures 1 and 2 we show HETE/Ulysses annuli for bursts from each SGR. From these triangulations we conclude that HETE timing errors are indeed negligible, and moreover, that HETE is a sensitive SGR detector when it is pointed towards the Galactic plane. Since this occurs for several months around December and June, it is clear that HETE will serve as a monitor of the known SGRs, and may also detect new ones when they become active.

## WHAT HETE BRINGS TO THE IPN

At present, the IPN consists of Ulysses, in heliocentric orbit, Mars Odyssey, now in orbit around Mars, and several missions close to Earth [1]. Prior to the launch of HETE, the two principal GRB detectors close to Earth were the BeppoSAX Gamma- Ray Burst Monitor (GRBM) and the Konus experiment aboard the Wind spacecraft. BeppoSAX has a high duty cycle, but as it is in low-Earth orbit, it misses roughly half of all GRBs due to Earth occultation. The GRBM has a coarse location capability (10-20$^o$). The mission will end some time in 2002. Wind is nominally at the $L_1$ Lagrange point, so no bursts are missed due to occultation, but the duty cycle is not as high as that of BeppoSAX, and this causes some events to be missed. Because of the placement of the two Konus detectors aboard the spacecraft, the experiment can be used to determine the ecliptic latitude of a burst to an accuracy of about 10$^o$ in many, but not all, cases. The Wind mission is now expected to be supported at least through 2002, at which time the launch of

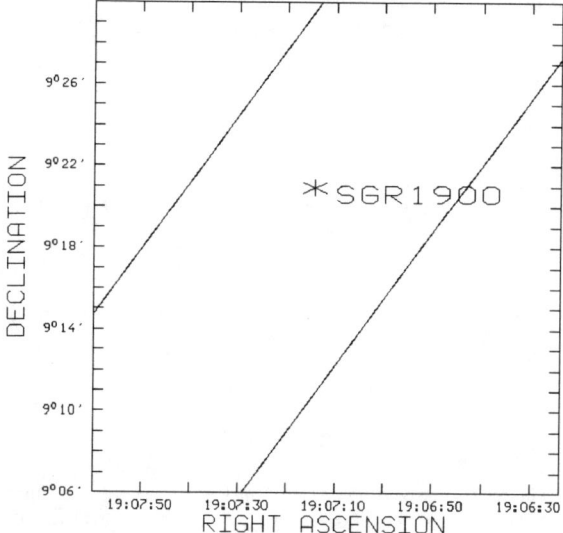

**FIGURE 1.** The FREGATE/Ulysses location annulus for a burst from SGR1900+14 on June 29 2001. The annulus width is 11'. The displacement between the known position of the SGR and the center line of the annulus is 31", which corresponds to a timing error of 93 ms. As this is much less than the 1 sigma statistical uncertainty in the cross-correlation, it indicates that the HETE timing errors are negligible.

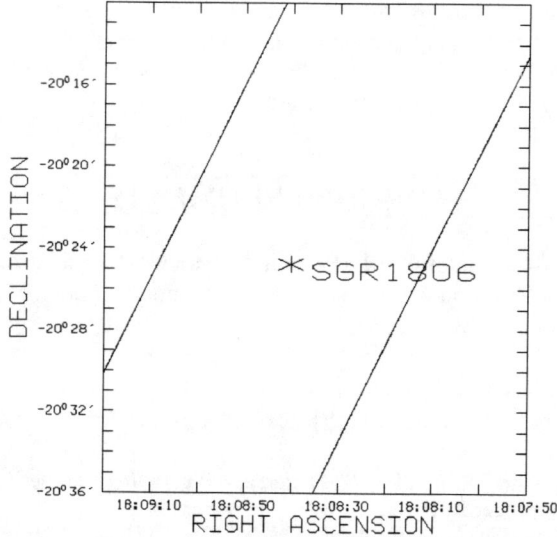

**FIGURE 2.** The FREGATE/Ulysses location annulus for a burst from SGR1806-20 on June 23 2001. The annulus width is 12'. The displacement between the known position of the SGR and the center line of the annulus is 1", which corresponds to a negligible timing error of 3 ms.

INTEGRAL should provide a replacement or near-Earth vertex for the IPN.

HETE complements these missions in three ways, as far as the IPN is concerned. First, it fill in the gaps by providing data on bursts missed by the other spacecraft. To date, about one burst per month has been detected by FREGATE and verified by Ulysses or Mars Odyssey, but has not been detected by Konus or by the GRBM. Second, it provides burst data in near-real time. The GRBM data are delayed by up to several hours, and the Konus data by half a day or more. Thus even when these experiments detect a burst, valuable time can be gained by receiving the HETE data and alerting the Ulysses and Mars Odyssey teams at JPL to expedite processing of the data in the small time window where the burst should occur. Finally, FREGATE data can be used to resolve the ambiguity in a 3 or 4 spacecraft localization. In the case where Ulysses, Mars Odyssey, and FREGATE observe a burst, FREGATE's coarse localization capability can serve to indicate which of two alternate triangulation positions is the correct one. In the case where Ulysses, Mars Odyssey, FREGATE, and Konus observe a burst, the separation between FREGATE and Konus will often be just enough to generate a third, independent annulus of location which again resolves the ambiguity. It should also be noted that the time resolution of FREGATE in the untriggered mode is 0.164 s, which is a factor of about 5 to 20 better than the untriggered modes of the GRBM and Konus, which results in more accurate cross-correlations.

## WHAT THE IPN DOES FOR HETE

The IPN can assist the HETE mission in four ways. First, it can confirm candidate GRBs and SGR bursts. Typically, these are events which are observed only by FREGATE. 24 such bursts have been confirmed to date, or a rate of about two per month.

Second, it can refine WXM error circles. Figure 3 shows one example, GRB010613. This is an extreme case, in that both the WXM error circle and the IPN annulus are rather large. However, there are numerous examples of smaller BeppoSAX WFC error circles which were refined by the IPN [2], and we expect most HETE-IPN bursts to be similar to them.

Third, it can transform "one dimensional" WXM error boxes which are too large to search for counterparts into small error boxes which are suitable for optical and radio follow-up observations. A "one-dimensional" WXM error box is one which results when a burst is detected and localized in only one of the two crossed WXM detectors, leading to a long, narrow error box. Figure 4 shows an example, GRB010921. Figure 5 shows an enlargement of the region where the IPN annulus intersects the WXM

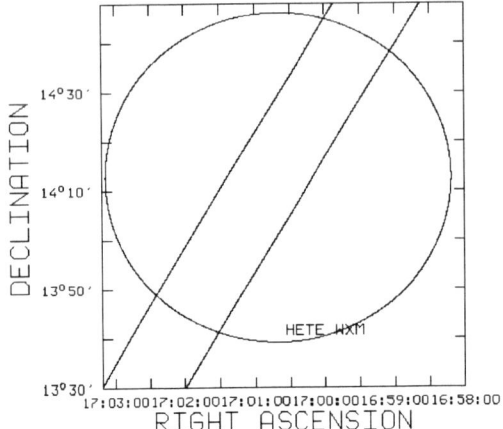

**FIGURE 3.** The 72' diameter WXM error circle for GRB010613, and the 15' wide FREGATE/Ulysses location annulus for it. The intersection of the annulus and the error circle forms an error box whose area is approximately 1000 square arcminutes, or about a factor of four smaller than the error circle alone. A similar annulus could also have been generated using the Konus-Wind data, had FREGATE not detected this burst for any reason.

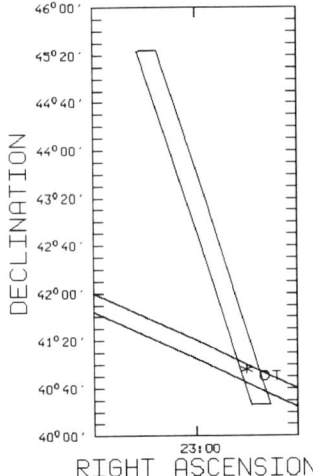

**FIGURE 4.** The 5.2° by 17' one-dimensional WXM error box for GRB010921, and the 14' wide FREGATE/Ulysses location annulus for it. The position of the optical transient is marked with an asterisk. A similar annulus could also have been generated using the Konus-Wind data, had FREGATE not detected this burst for any reason.

error box. This burst and its counterpart are described in detail elsewhere [3, 4, 5].

Finally, the IPN can, in principle, transform one dimensional SXC error boxes into small error boxes in the same way that it transforms WXM error boxes. However, this has not yet been done for any burst, since none has

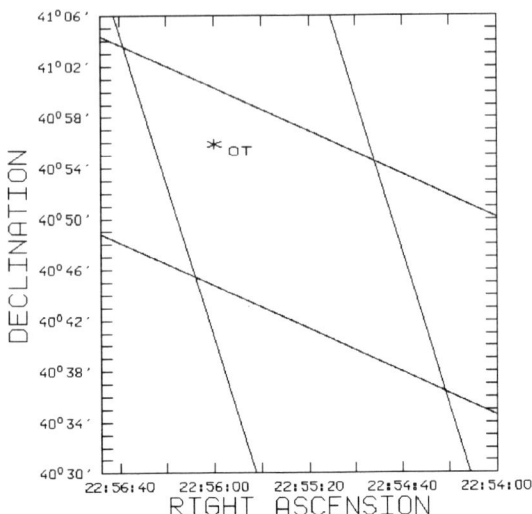

**FIGURE 5.** Expanded view of the 310 square arcminute IPN/WXM error box for GRB010921, and the position of the optical transient.

been localized by the SXC in this way.

## CONCLUSIONS

We have successfully integrated the HETE spacecraft into the third interplanetary network of gamma-ray burst detectors, and have demonstrated the value of HETE to the IPN and vice-versa. We are now looking forward to several more years of operations.

## ACKNOWLEDGMENTS

KH is grateful for IPN support under JPL Contract 958056, and for HETE-FREGATE support under MIT Contract SC-R-293291.

## REFERENCES

1. Cline, T. et al., *AIP Conference Proceedings*, **these proceedings** (2002).
2. Hurley, K. et al., *Ap. J.*, **534**, 258–264 (2001).
3. Ricker, G. et al., *Ap. J.*, **submitted** (2001).
4. Park, H.-S. et al., *Ap. J.*, **submitted** (2001).
5. Kulkarni, S. et al., *Ap. J.*, **submitted** (2001).

# Using Wavelets to Detect HETE Untriggered Bursts

N. Butler* and J. Doty*

*MIT CSR, Cambridge, MA, 02139*

**Abstract.** We have developed a simple scheme using a 1-dimensional discrete wavelet transform, to search for bursts in the ground analysis of *HETE* data. The method, which complements the on-board *HETE* triggering systems, has proven useful for verifying that these systems are functioning and for detecting untriggered bursts. We outline the method and discuss the results obtained.

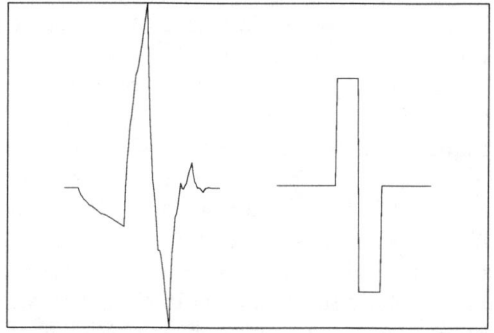

**FIGURE 1.** Two example wavelets. The wavelet on the right, the most compact in a series discovered by I. Daubechies, is used in the wavelet used in this study.

## INTRODUCTION

Wavelets are functions in a Hilbert space, much like the sines and cosines of Fourier analysis. But unlike sines and cosines, wavelets are localized in time as well as in frequency. Two examples are shown in Figure 1. In general, one trades smoothness for compactness in selecting a mother wavelet. Wavelets are normalized to zero so that they have zero response to a flat signal. In triggering, it is often also sensible to require zero response to a linearly rising or falling signal. The first wavelet in Figure 1, which is the wavelet used in this study, has this property. This, in addition to its relative compactness over smoothness, suggests that this wavelet is appropriate for triggering on bursts.

The discrete wavelet transform (DWT) is of particular utility for triggering on GRB's, where an a priori burst profile is not known. Although the initial choice of a wavelet basis is motivated by an anticipated burst profile, we can build arbitrarily complex profiles out of linear wavelet superpositions. (The wavelet basis is established by translating and stretching in time the mother wavelet.)

Traditional triggering methods, like those used on board *HETE*, can be viewed as wavelet decompositions. Consider the right most wavelet in Figure 1 (Haar wavelet), flipped top to bottom. Integrating this shape across a stream of data, we are essentially applying a one-side ("BATSE-like") trigger. The negative portion samples the background, whereas the positive portion samples the burst. More general triggering algorithms allow for wavelets with negative regions (i.e. background samples) both before and after the positive region. One runs several trials with different stretchings of the positive and negative regions, discovering the filter which optimally brings out a count excess. The wavelet coefficients we calculate (i.e. the S/N determinations from the separate trials) are not statistically independent and cannot in general be combined to find a (possibly) higher S/N.

The DWT is a fast linear operation, which returns a number of wavelet coefficient (i.e. correlations of the data with the basis wavelets) equal to the number of elements in the initial data array. A threshold applied to exclude coefficients below a certain significance then captures any jagged features in the data to a small number of wavelet coefficients. These are then combined, based on proximity in phase space, into triggers.

Additional background on wavelets, including transform algorithms, can be found in Press [1].

## THE TRANSFORM

For Poisson data, the signal variance is correlated with the background count rate. A square root transformation helps to remove this correlation and allows for a simple statistical understanding of the wavelet coefficients. We

find
$$y = \sqrt{x} + \sqrt{1+x} \quad (1)$$

to be a particularly well behaved form of the square root transformation for a small number of counts. This transformation takes a Poisson random variable to an approximately normal random variable with unit variance (independent of the mean). As the wavelet basis is orthonormal, we then get wavelet coefficients in $\sigma$ units. That is, for Poisson noise at a constant mean, the coefficients will on average be zero. Deviations from zero will follow a normal distribution with $\sigma = 1$. This interpretation breaks down at low count rates.

Consider a signal $\psi_s = \Sigma_i a_i \phi_i$, where $a_i$ are the wavelet coefficients and $\phi_i$ are the wavelets, buried within a noisy data set represented by $y = \Sigma_j a_j \phi_j$. Integrating the wavelets in the signal against the entire data set, we can define a signal to noise ratio:

$$S/N = \frac{\int y * \psi_s dt}{\sqrt{\int \psi_s^2 dt}} = \sqrt{\Sigma_i a_i^2} \quad (2)$$

Thus, to get the total S/N of an event decomposed into multiple wavelets, we just add the wavelet coefficients in quadrature. If we consider the background limited case with a counts excess of $N$ counts in $n_s$ bins over a flat background $N_b$ in each of $n_b$ bins, it is straight-forward to show that a $\psi_s$ in the shape of a one-side ("BATSE-like") trigger yields:

$$S/N \approx \frac{N}{\sqrt{n_s N_b + \frac{n_s^2}{n_b} * N_b}} \quad (3)$$

This is the usual expression for model variance (where we are attempting to disprove the null hypothesis). The S/N defined in equation (2) is, however, better behaved than model variance at very low and very high count rates. For wavelet shapes more general than that used to calculate (3) the S/N can be considered analogous to, but not strictly equal to, model variance. It should also be noted that the sum in (2) will not on average be equal to zero for Poisson noise. Therefore it is important to threshold out insignificant $a_i$ terms. We typically retain $a_i > 2.5\sigma$.

Figure 2 shows an example using the Daubechies wavelet of Figure 1. The line in the left-most plot is meant to represent the Poisson mean of counts in a detector. The ideal calculated S/N corresponding to the counts excess is 6. Notice that our calculation of the S/N degrades as we raise the significance threshold on allowed wavelet terms. The over-plotted curve represents the thresholded inverse wavelet transformation. With the addition of noise, thresholding would be important, and we would have calculated S/N~5 with 2 or 3 wavelet terms contributing. Also, notice that the excess is concentrated into a small number of terms. The situation will be better, producing higher fidelity S/N's, for excesses which more resemble the mother wavelet.

Figure 3 shows an X-ray burst in the Fregate instrument [2] on-board *HETE*, detected in June of 2001. To the left we show the four most significant wavelets over $3\sigma$, the sum of which is over-plotted on the actual data in the right-most plot. If we neglect the two medium scale terms with $\sigma \approx 3.5$, two terms dominate. These have $\sigma = 11.6$ and $\sigma = 10.6$, and reconstruct the burst S/N to better than 90% accuracy.

## ARCHIVAL SEARCHES

To further quantify the sensitivity of the method, we searched through *HETE* archival data from June 1 trough August 1 of 2001 for X-ray bursts. In this period, the *HETE* field of view passed through the Galactic plane. We find >140 bursts over S/N=10. From the cumulative distribution in Figure 4, it is apparent that the method is affected by excess noise at S/N≈5-6. This is largely a critique of the manner in which we assemble wavelet terms into events, differentiating signal from noise, after applying the DWT and thresholding. We have found the following simple scheme to be sufficient: neglecting wavelet terms with scales larger than our characteristic minimum duration background variations (~60s), we use the larger scale wavelet terms as windows which gather up temporally overlapping terms at smaller scales. We break up events where there is a sizeable gap in the run of scales.

Due to telemetry saturation, triggers on board *HETE* that would have caught the events described in the previous paragraph were disabled. Several other archival searches have been conducted to find untriggered bursts. In addition to the fourteen SGR bursts on which *HETE* did trigger in June and July of 2001 [3], we uncovered an additional eleven which were missed on-board due to the disabling of certain trigger criteria. A recent search for GRB's in the Fregate 32-400 keV band over the mission lifetime has found no unaccounted for events over S/N=6.

## GROUND TRIGGERING AT MIT

Data from *HETE* are searched for untriggered bursts upon reaching MIT. We compare counts rates in multiple instruments, detectors, and energy bands to find coincident excesses. We use the trigger algorithm described in Graziani [4] as well as the one described here. Notifications are sent out via email (typically ~1 hour after a burst). We are working to supplement this software with

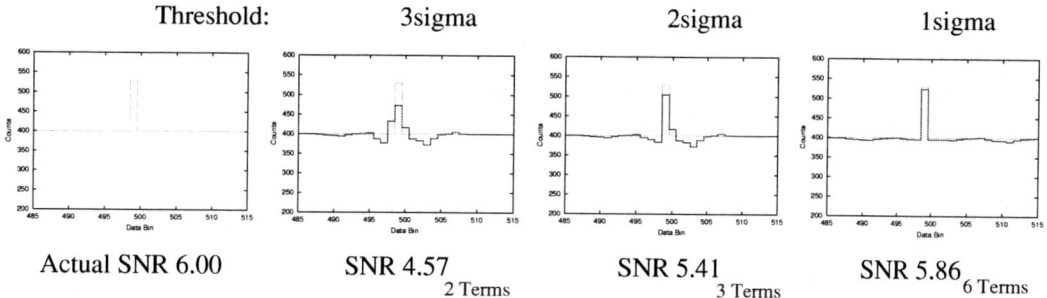

**FIGURE 2.** A contrived example: wavelet reconstruction of a single bin, S/N=6 count excess.

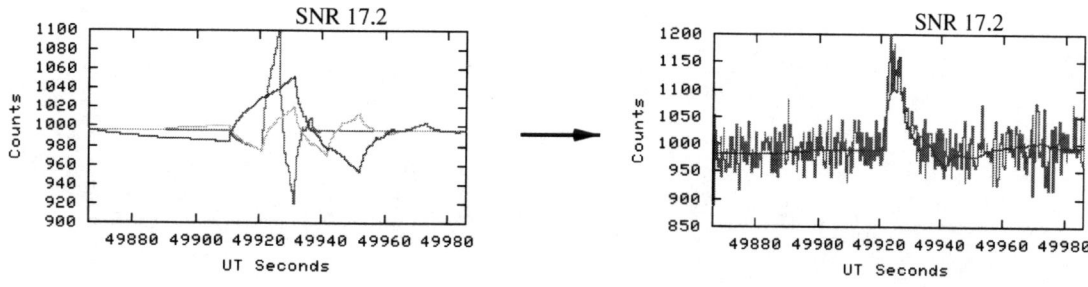

**FIGURE 3.** Wavelet transform over-plotted on *HETE* Fregate Data in the 8-85 keV band. Two wavelet components are sufficient to reconstruct the burst profile to 10% accuracy.

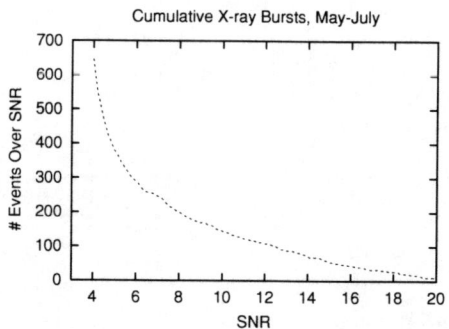

**FIGURE 4.** The cumulative distribution of detected X-ray bursts in the Fregate 8-85 keV band turns up at S/N$\approx$5-6.

automated localization routines in order to send notifications automatically through the GCN.

Ground triggering's first major success came with a faint, soft untriggered burst on the 19th of October, 2001 (GCN Circular #1109). GRB01212 was also discovered in this manner.

## SIGNAL TO NOISE RATIOS FOR HETE BURSTS

As a further diagnostic on the method, we display the calculated S/N's for all of *HETE*'s bursts (as of November 1, 2001), which were coincident with IPN bursts and/or were localized by the *HETE* WXM instrument. Of these 28 bursts, the 9 that were localized by the WXM are represented with filled-in boxes.

We note from the plots that the IPN appears biased against observing *HETE*'s soft bursts. Also, most of the bursts localized by the WXM exhibited larger S/N's at low energy.

## ACKNOWLEDGMENTS

Special thanks to the *HETE* operations team. Thanks to George Ricker for several discussions influencing this work.

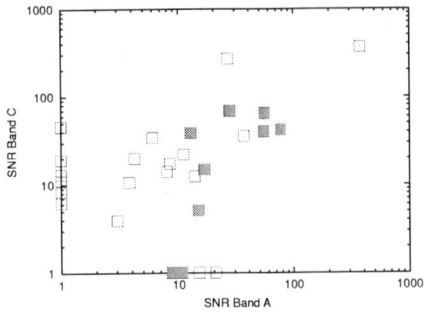

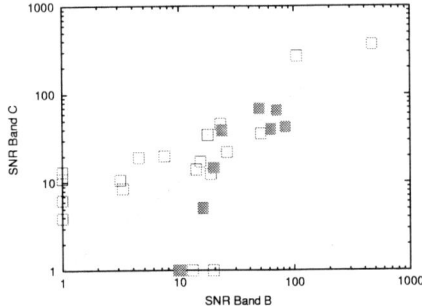

**FIGURE 5.** S/N's for *HETE* bursts. Filled-in boxes represent WXM localizations. The flight triggers operate on Fregate Bands B (8-85 keV) and C (30-400 keV). Band A is 5-40 keV.

# REFERENCES

1. Press, et. al., W. H., *Numerical Recipes in C*, Cambridge University Press, 1992, pp. 591–606.
2. Atteia, J.-L. et. al., "In-flight Performance and First Results from the Fregate Instrument on HETE", in *these proceedings.*, 2002.
3. Ricker, G. R. et. al., "High Energy Transient Explorer (HETE): Mission and Science Overview", in *these proceedings.*, 2002.
4. Graziani, C. et. al., "A Spiffy Trigger for GRBs", in *these proceedings.*, 2002.

# HETE Soft X-ray Camera Imaging: Calibration, Performance, and Sensitivity

G. Monnelly*, J. N. Villasenor*, J. G. Jernigan[†], G. Prighozin*, R. Vanderspek*, G. B. Crew*, J. Doty*, A. Levine* and G. Ricker*

*MIT Center for Space Research, Cambridge, MA, 02139 USA
[†]Space Sciences Laboratory, University of California, Berkeley

**Abstract.**
The HETE Soft X-ray Camera (SXC) uses X-ray CCDs and a micro-fabricated coded aperture mask to image X-ray sources with sub-arcminute accuracy over a steradian field of view. Calibration of imaging with observations of the Crab and Sco X-1 is described. The accuracy of SXC localizations is determined trough observations of known steady and transient galactic X-ray sources and sensitivity to GRBs is estimated through detections of steady and transient Galactic X-ray sources. [NB: As of October 2002, the SXC had localized 4 GRBs to arcminute accuracy].

## INTRODUCTION

The HETE Soft X-ray Camera (SXC) consists of two one-dimensional coded aperture cameras with orthogonal orientation (referred to as SXC-X and SXC-Y). Each has a coded aperture mask that is 10 centimeters square with a random pattern of slits with a 45 micron feature size. Each camera has two large format X-ray CCDs with 15 micron square pixels. However, one of the two CCDs in each SXC is not operational due to the loss of the outer optical blocking filter shortly after launch. For an overview of the SXC, see [2] and [3]. For details about the the in-flight performance of the CCDs, see [1].

The characteristic angular resolution of each SXC is $96''$, set by the mask element size (45 microns) and the mask to detector distance (96 mm). Calibration observations demonstrate a $43''$ radius, 90% confidence error circle (see section on Calibration, below). The mask pattern is over-sampled three times by the 15 micron CCD pixels. Each camera has a 1.6 steradian field of view (full width, zero response) that is slightly rectangular: 65° in the resolution direction, 79° in the non-resolution direction. The fields of view of SXC-X and SXC-Y have a 1.3 steradian area of overlap, shown in Figure 1.

The SXCs are designed to provide accurate localizations ($< 1'$) over a wide field a few. To achieve this, the CCDs and mask must maintain precise alignment throughout each orbit. This alignment can be tested by observing the pattern of illumination on the CCDs from a bright x-ray source such as Sco X-1. Any misalignment between the mask and CCD produces a blurring of the illumination pattern. Data collected in orbit verify that alignment has been maintained within an acceptable tolerance.

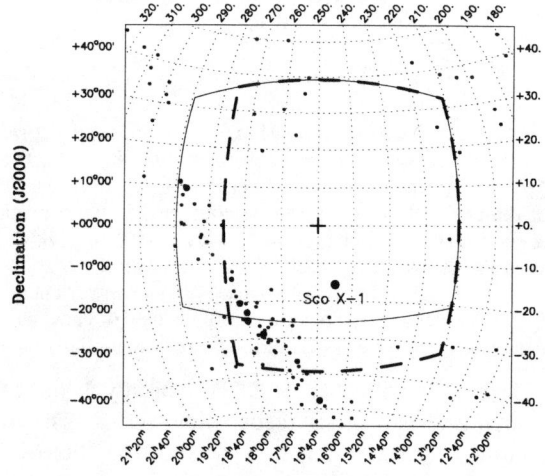

**FIGURE 1.** SXC fields of view: SXC-X *thick dashed*, SXC-Y *thin solid*. This view is pointing near Sco X-1, with the Galactic Center to the lower left. Each SXC has position resolution in only one direction; here, SXC-X determines right ascension, and SXC-Y determines declination.

# IMAGING

Localization of a point source results from a one dimensional cross-correlation of the CCD data with all possible angles in the sky. There are 2782 resolution elements, each 96″ in extent, across the 65° field of view. A celestial X-ray source will appear as a peak in the cross-correlation map, demonstrated in Figure 2. A detection in a single camera yields a long thin rectangle that is less than 1′ wide, and 79° long (the width of the field of view in the non-resolution direction); a detection in both SXC-X and SXC-Y can be combined to yield a celestial position with sub-arc minute precision. Further details of the SXC localization accuracy will be discussed below.

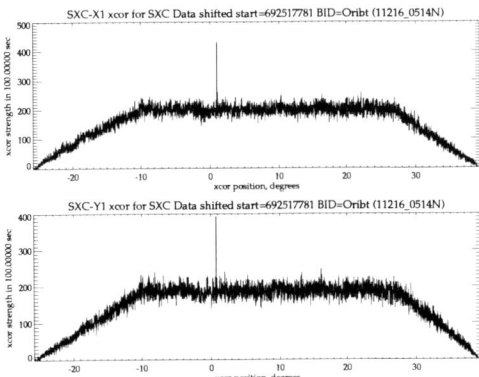

**FIGURE 2.** SXC one-dimensional image showing a correlation peak corresponding to the Crab in both the X (*top*) and Y (*bottom*) cameras for a 100 second observation. The two one-dimensional positions can be combined to yield a celestial coordinate pair. The plateau on which the Crab peak sits is due to the diffuse X-ray background and other X-ray sources, including the Crab. The noise plateau slants downward near the edges as it extends to the outer limits of the field of view that are shadowed by the SXC walls, and thus are only partially coded.

The pointing direction of HETE typically drifts at the arc minute level on 10 second time scales. The drift is sufficiently small that it does not effect the WXM, but can affect the SXC because the cross-correlation image feature size of 96″ is comparable to the aspect drift. This causes the peak from a point source, such as the Crab, to move across a cross-correlation image, as shown in Figure 3. Without further processing, the drift timescale limits the duration over which SXC data can be integrated to detect sources on long time scales. However, because the spacecraft aspect is measured by the optical star cameras, we can correct the SXC data for the aspect drift by shifting the cross-correlation image to compensate for the drift. The resulting aspect-corrected data, as shown in Figure 4, is coherent with time, and is thus more suitable for detection of faint signals on longer time scales.

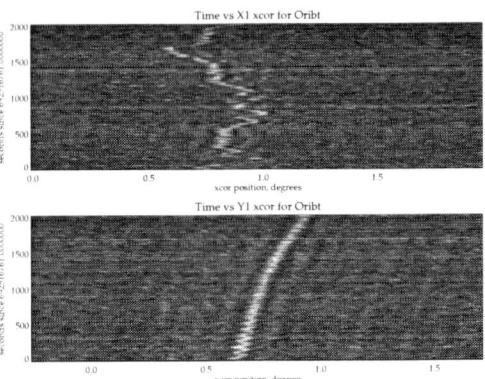

**FIGURE 3.** SXC image showing a localization peak corresponding to the Crab. In these plots, the y-axis spans 2000 seconds of data in 2 second steps, the x-axis labels the cross-correlation position in a restricted 2° portion of the field of view, and the intensity of the image represents the strength of the cross-correlation at each angle and time. Because of spacecraft pointing instability, the position of the Crab is seen to drift by a degree throughout the orbit.

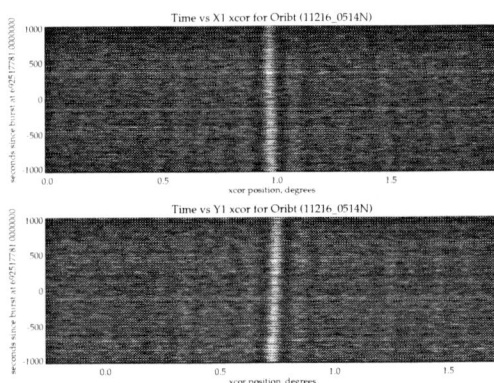

**FIGURE 4.** SXC image showing a localization peak corresponding to the Crab. These plots are the same as Figure 3, except that optical aspect data has been used to shift the cross-correlation to correct for spacecraft aspect drift at each point in time.

# CALIBRATION

To improve the accuracy of SXC localizations, and to estimate their uncertainties, we have performed a series of calibration observations using known cosmic X-ray sources. Calibration observations began with observations of the Crab in December 2000 – January 2001. The calibration dataset was expanded with observations of Sco X-1 in May – July 2001, and the Crab again in December 2001 – January 2002. During calibration observations, the spacecraft roll was adjusted so that the calibration source covered all areas of the SXC field of view, as shown in Figure 5.

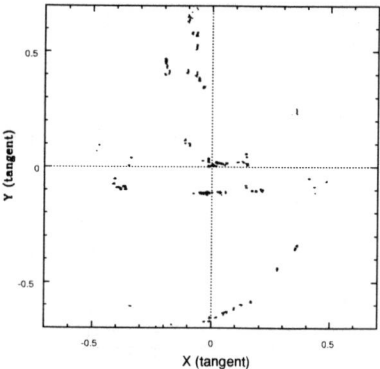

**FIGURE 5.** Coverage of the SXC field of view by calibration observations of the Crab and Sco X-1. The plot shows a tangent plane projection of the SXC field of view, covering an angular range of ±30°.

Calibration data is used to determine the parameters in a simplified geometrical model of the SXC which relates position in the cross-correlation map to location in celestial coordinates. The model parameters are determined with a joint fit of all calibration data with fit parameters representing the relative position of the CCD and mask, and the relative rotation between the SXC and spacecraft. The fitted solution is combined with spacecraft aspect information, obtained from optical star cameras, to calculate positions in celestial coordinates.

Empirical measurements show a 20″ radius RMS localization accuracy in the X and Y directions. This corresponds to a 90% confidence, two-dimensional error circle of 43″ radius. Figure 6 shows the localization accuracy from multiple independent observations of Sco X-1. The largest contributor to this error is likely to be mechanical distortions arising from periodic thermal fluctuations during each 90 minute orbit. With additional modelling, we should be able to further reduce this systematic thermal effect.

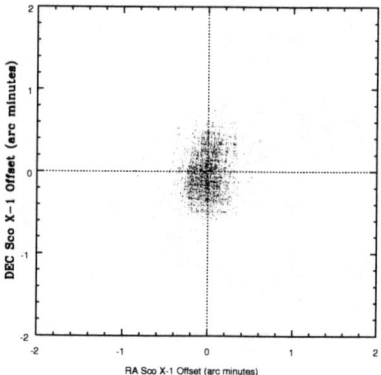

**FIGURE 6.** SXC astrometry: multiple independent localizations of Sco X-1 at different times and in different parts of the field of view showing 20″ radius RMS accuracy.

## SXC AS A *VERNIER* FOR THE WXM

Because the SXC has such a wide field of view with a fine angular resolution, resulting in a cross-correlation map with thousands of elements, only a very bright source is likely to produce a correlation peak that is brighter than all the background peaks. However, any source that the SXC could detect is very likely to be also detected and localized by the WXM. As a standard analysis procedure, we use the source position determined by the WXM to refine the region of the SXC cross-correlation map that is searched (Figures 8). Thus, the SXC serves as a *vernier* to refine the WXM localization. Typical WXM localizations are less than 15′ in radius, which is a tremendous reduction of the 65° SXC field of view (2782 resolution elements reduced to less than 20). This reduction greatly increases the chances that the true source has a higher cross-correlation strength than any background peak. However, when the search region is tightly restricted, it is difficult to tell *a priori* if the highest peak indeed corresponds to the source detected by the WXM and is not just a noise peak.

We address the problem of possible false SXC detection in a restricted search area by increasing the search region to ± 1° centered on the WXM position. If the source is detected by the SXC then it will with very high probability fall within the WXM error region. However, if it isn't detected, then the SXC cross-correlation peak can fall anywhere in the ± 1° search region. For example, if the WXM error region is 1/10 of the SXC search region in one dimension, and the SXC peak falls within the WXM error region in both X and Y cameras, then there is only a 1% chance that it is a false detection.

As of August 2002, the SXC had not yet detected X-ray emission from a Gamma Ray Burst. Of all the GRBs detected by the WXM, none were detected by the SXC because of its smaller field of view (Figure 1), its reduced duty cycle (due to full moon constraints) or its lower sensitivity. However, a number of Galactic X-ray sources, both steady and transient, were detected while HETE was pointed at the Galactic Center during the summers of 2001 and 2002. In June of 2002 a series of tests were initiated where HETE's on-board trigger system was configured to trigger on XRBs with HETE pointed at the Galactic Center (normally when the Galactic Center is in view, low energy trigger thresholds are set to exclude the ∼ 5 − 10 XRBs detected per day.) These tests verified the WXM's ability to generate accurate flight localizations that are sent out to the GCN with less then one minute of delay, and demonstrated the SXC's ability to provide refinements, based on ground analysis, of the WXM positions. During three days of testing, five XRBs were localized (Table 1). Four of these (H2050, H2052, H2057, and H2063) were two-axis detections, while one (H2055) was a one axis detection. H2050, H2052, and

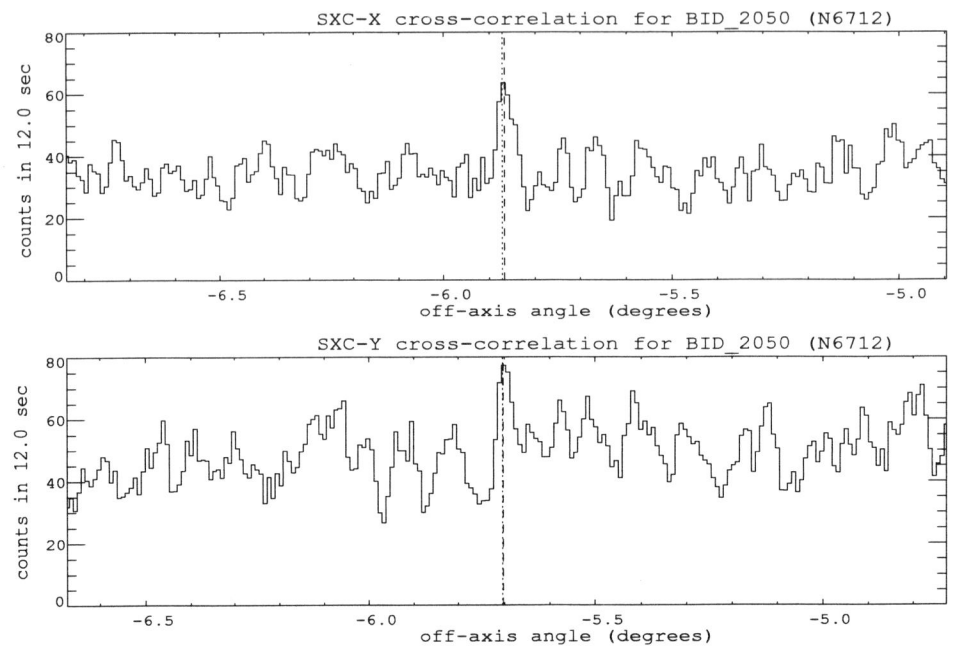

**FIGURE 7.** N6712 outburst on June 8, 2002, localized in SXC-X at a SNR of 3.9 and an offset of 16″ from the known source position and in SXC-Y at a SNR of 2.6 with an offset of 5.4″. The *short-dashed* line represents the known source position, while the *long-dashed* line represents the SXC localization.

**TABLE 1.** Five XRBs localized by the SXC during three days of observations of the Galactic Center in June of 2002.

| Burst ID | Date (2002) | Time (UT) | Source | Axis | localization offset | SNR | Delay (hrs) | Remarks |
|---|---|---|---|---|---|---|---|---|
| 2050 | 8 June | 1059 | N6712 | x | 16″ | 3.9 | 7.1 | Good aspect; |
|  |  |  |  | y | 5.4″ | 2.6 |  | Sent to GCN |
| 2052 | 8 June | 1602 | N6624 | x | 27″ | 4.8 | 3.7 | One aspect point; |
|  |  |  |  | y | 42″ | 4.1 |  | Sent to GCN |
| 2055 | 9 June | 1304 | N6624 | x | – | – | 1.6 | Good aspect |
|  |  |  |  | y | 14″ | 2.9 |  |  |
| 2057 | 9 June | 2045 | N6624 | x | 31″ | 2.9 | 3.8 | Good aspect; |
|  |  |  |  | y | 14″ | 3.9 |  | Sent to GCN |
| 2063 | 10 June | 1935 | N6712 | x | 19″ | 8.1 | 0.75 | High aspect drift; |
|  |  |  |  | y | 60″ | 7.2 |  | Sent to GCN, but GCN was down |

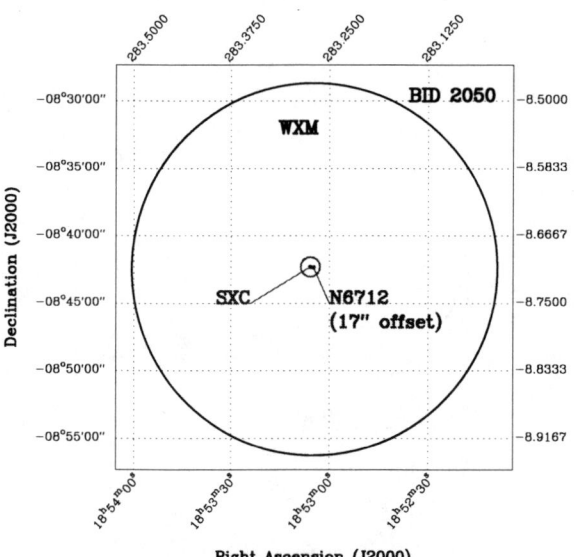

**FIGURE 8.** *Upper panels:* N6712 outburst on June 8, 2002, localized by SXC and WXM. The SXC-X and SXC-Y positions are combined to give celestial coordinates. Note that the SXC error circle (43″ radius) is considerably smaller than the WXM error circle (14′ radius).

H2063 were released as GCN Alerts with delays of 3-7 hours (H2050: Figure 8, H2052 and 2063: Figure 9). With improved automation, the typical delay will soon drop to less than two hours.

**TABLE 2.** Consistency of detecting the Crab during 2000 seconds (one orbit) of data taken with the Crab on-axis. This number is representative of the SXC's ability to detect a 1 Crab GRB with a duration equal to the integration time. The integrations listed in the "After Sept. 2002" column are adjusted to reflect flight-code improvements that yielded a cumulative 100% increase in event detection rate. For example, with the post-Sept 2002 sensitivity, a 2.5 Crab-sec source will be detected with > 90% reliability in both axes.

| Integration (sec) | | % correct | |
| --- | --- | --- | --- |
| Before Apr 2002 | After Sept 2002 | SXC-X | SXC-Y |
| 1 | 0.5 | 71 | 74 |
| 2 | 1.0 | 82 | 82 |
| 5 | 2.5 | 91 | 93 |
| 10 | 5.0 | 97 | 99 |

## SENSITIVITY

Typically the SXC count rate is dominated by the diffuse X-ray background (DXRB), which yields 15 cts/s in each of the two SXC cameras. During January and December, the Crab is present in the SXC field of view, increasing the count rate by 4 cts/s per camera when it is on axis. With the increase of 50% in X-ray detection efficiency achieved in April 2002, the Crab count rate is projected to be 6 cts/s per camera. Finally, with a software modification in September 2002, the Crab count rate was further increased to 8 cts/s per camera (NB: The DXRB rate was also increased, reaching 30 cts/s per camera after September 2002).

An analysis of 2000 seconds (one orbit) of SXC data from December 12, 2001 when the Crab was 1.19° from the SXC boresight was used to characterize the sensitivity to a 1 Crab source. Using the technique described above, we searched the SXC cross-correlation map within ±1° of the known Crab position to check how consistently the Crab was detected. A detection was defined as the brightest cross-correlation peak being located within 2′ of the known Crab position. Integration times of 1, 2, 5, and 10 seconds are used to simulate a Crab-like transient with these durations (Table 2). Because the SXC has a very consistent background count rate, the sensitivity to constant sources for any integration time should be very similar to the sensitivity to transient sources, such as GRBs, with duration equal to that integration time. The Crab data used in this estimate were taken before September 2002, when flight-code improvements yielded a 100% increase in event detection rate. Because the increase is approximately energy independent, the sensitivity can be re-calculated by simply noting that an integration of $x$ seconds before April 2002 is equivalent to a $x/2$ second integration after September 2002. The current sensitivity to a 1 Crab source, adjusted accordingly, is shown in Table 2.

The SXC sensitivity can also be characterized in terms of integration time and flux threshold for a given signal-to-noise ratio (SNR). We are able to set the SXC SNR threshold at 3.5σ, which by itself is a low significance, but there is additional information that makes this level reasonable, predicated on a concurrent, high-significance WXM detection. First, the time of the burst will be known from the WXM, and likely FREGATE as well. Second, the WXM-localization allows the SXC to act as a "vernier"; thus, it is only necessary to search a small part, and not the entirety, of the SXC field of view. Because we are using this additional information to restrict the SXC analysis, it is possible to set a low threshold of 3.5σ, which for a burst of duration $t$ seconds gives a sensitivity (compared to the Crab flux) of $1.9t^{-1/2}$ Crab. Additional details comparing the pre-launch predicted SXC performance to the post-September 2002 SXC perfor-

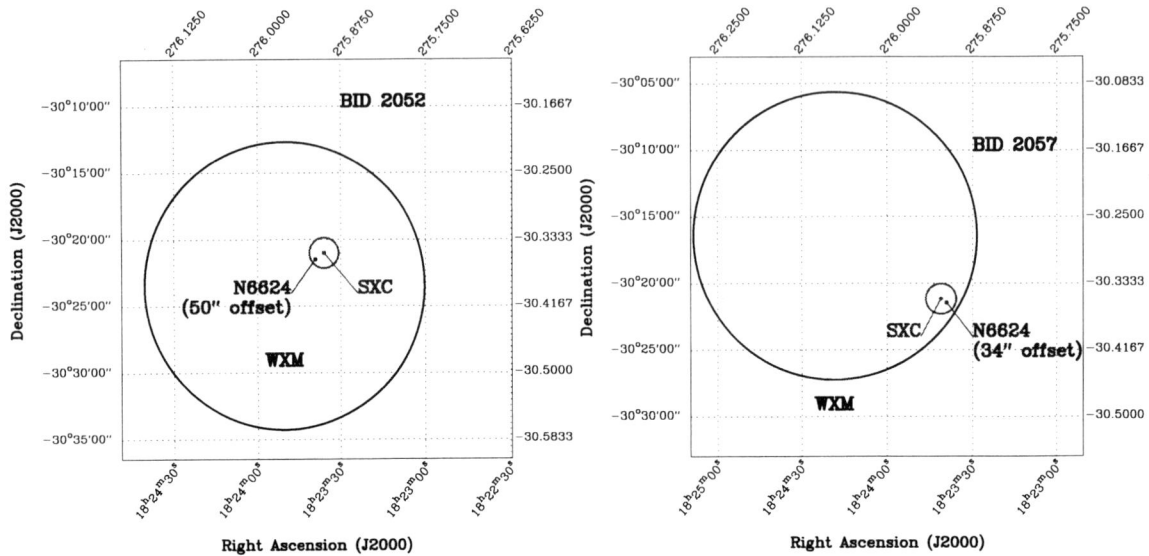

**FIGURE 9.** SXC and WXM error regions for two XRBs, as released in GCN Alerts.

**TABLE 3.** Soft X-ray Camera

|  | At launch (both OBFs intact) | At present (Inner OBF only) |
| --- | --- | --- |
| Built by | MIT CSR | MIT CSR |
| Instrument type | 4 CCID-20 | 2 CCID-20 |
| Camera dimensions | 10cm × 10cm × 17.5cm | 10cm × 10cm × 17.5cm |
| Energy Range | 500 eV to 14 keV | 1.3 to 14 keV |
| Timing Resolution | 1.2s | 1.2s |
| Spectral Resolution | 46 eV @ 525 eV | ∼300 eV |
|  | 129 eV @ 5.9 keV | @ all energies |
| Dectector QE | 93% @ 5 keV, | 93% @ 5 keV, |
| Dectector QE $> 20\%$ | 0.5–14 keV | 1.3–14 keV |
| Effective Area | 74.4 cm$^2$ | 37.2 cm$^2$ |
| Source Sens (Crab) | 2.3 $t^{-1/2}$ (5.5$\sigma$) | 1.9 $t^{-1/2}$ (3.5$\sigma$) |
| Burst Sens (cts/cm$^2$/s) | 1.0 (5.5$\sigma$) | 0.8 (3.5$\sigma$) |
| FOV (FWZM) | 1.88 sr | 1.29 sr* |
| Focal Plane scale | 33″ per CCD pixel | 33″ per CCD pixel |
| Localization | 40″(systematics limit) | 40″(systematics limit) |

* There is an additional 0.54 sr FOV for one-sided localizations.

mance are given in Table 3.

## SUMMARY, CONCLUSIONS, AND ONGOING WORK

The SXC is currently performing consistently, localizing sources with a $20''$ RMS accuracy, corresponding to a two dimensional 90 percent confidence error radius of $43''$. Because of the loss of the outer optical blocking filter, the SXC now operates with only half of the CCD collecting area, and cannot operate within $\sim 3$ days of the full moon. Nonetheless, more efficient operating modes have restored the SXC's effective operating sensitivity to the anticipated pre-launch values (see Table 3). As evidence for the restored effectiveness of the SXC, it recently detected and localized 4 GRBs (GRB020813, GRB020819, GRB020903, GRB021004).

## ACKNOWLEDGMENTS

We would like to thank David Breslau, Bob Dill, and Steve Kissel for help in design, construction, and testing of the SXC.

## REFERENCES

1. Prigozhin, G., *et al.*, these proceedings (2002).
2. R. Vanderspek, et al. GRB Observations with the HETE Soft X-ray Cameras, *A&AS*, 138:565–566, September 1999.
3. Villasenor, J. N., et al., these proceedings (2002).

# GRB 010921: Discovery of the First HETE Afterglow

Paul A. Price
for the REACT GRB Collaboration

*Palomar Observatory, Caltech 105-24, Pasadena, CA 91125.*

**Abstract.** We present the discovery of the optical and radio afterglow of GRB 010921, the first gamma-ray burst afterglow to be found from a localisation by the High Energy Transient Explorer satellite. Discovery of the afterglow enabled us to determine the redshift, $z = 0.45092$ from the host galaxy. We expect that the low redshift of this GRB will enable the strongest limits on the detection of an underlying supernova to be made with HST.

## DISCOVERY OF THE AFTERGLOW

GRB 010921 triggered the FREGATE instrument on board HETE-2 on 2001 September 21.21934 UT as HETE trigger number 1761. The GRB was detected by only one of HETE's two Wide-field X-ray Monitors, and so localisation was only possible in a single dimension. Localisation in the second dimension was done by the Inter-Planetary Network, and the resultant 250-square-arcmin error box was reported to the community fourteen hours after the GRB through the GRB Coordinate Network (GCN) Circulars [5].

We observed the error box with the Large Format Camera (LFC) on the Hale 200-inch telescope at Palomar Observatory, commencing 22 hours after the GRB [3] in Sloan $r'$. Even with LFC's 26-arcmin diameter field-of-view, three pointings were required to cover the entire error box. A quick comparison of the images with the DPOSS did not identify the afterglow within the large error box. Hoping to identify the afterglow by its variability, we re-observed the error box with LFC five days later.

Reduction of data from LFC is complicated by the large field-of-view of the instrument which leads to optical distortions in the focal plane, some of which can be removed in software, but coma at the edges of the field is unavoidable. We employed on-chip binning to reduce the file size of the images, but which also lowers the dynamic range of the images, making image comparison and photometry difficult. Comparison of the first- and second-epoch images was done through PSF-matched image subtraction [1]. Despite a subtraction that was imperfect due to the aforementioned problems, we identified two sources within the error box which were clearly variable.

In order to distinguish between the afterglow and a variable star, we re-observed the field 21 and 26 days after the GRB respectively. Photometry from these epochs revealed that one of the previously-identified variable sources had increased in brightness, while the other had faded further and appeared to be settling to a constant flux level, reminiscent of the behaviour of other optical afterglows.

We observed the field with the Very Large Array (VLA[1]) 26 days after the GRB, followed by several observations over the course of the following weeks. We identified in the first epoch a radio source at coordinates $22^h55^m59^s931 \pm 0.018$, $40°55'52''23 \pm 0.20$ (J2000), coincident with the optical afterglow candidate, with a spectral slope of $\beta = 0.35 \pm 0.19$ between 4.86 and 22.5 GHz. This value is consistent with that expected from the synchrotron emission of an afterglow ($\beta = 1/3$; Sari, Piran & Narayan [9]). Further observations indicate variability in the source, which is readily interpreted as due to interstellar scintillation, a phenomenon commonly seen in radio afterglows.

The $(B-V)$ and $(R-I)$ colours at 2001 September 22 identify the source as having a power-law spectrum, according to the colour-colour selection plots of Rhoads [8]. Finally, from images taken 1 day and 26 days after the burst (dominated by the optical transient and the galaxy respectively), we measure an offset of the afterglow from the host galaxy of $0.351 \pm 0.049$ arcsec, or $2.18 \pm 0.30$ kpc at the distance of the galaxy[2]. We therefore conclude that the transient is not an AGN, but rather

---

[1] The National Radio Astronomy Observatory (NRAO) is a facility of the National Science Foundation operated under cooperative agreement by Associated Universities, Inc. NRAO operates the VLA.
[2] We assume a standard flat cosmology with $H_0 = 65$ km s$^{-1}$ Mpc$^{-1}$, $\Omega_M = 0.3$ and $\Omega_\Lambda = 0.7$.

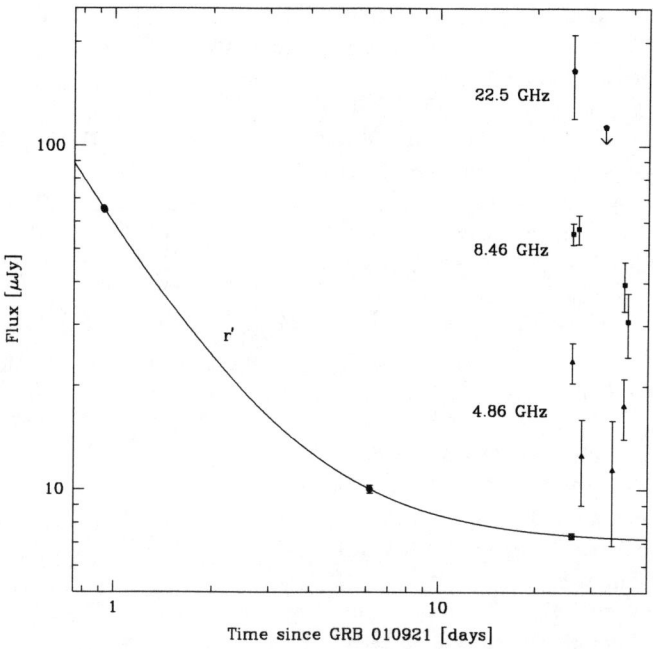

**FIGURE 1.** The optical and radio light curves of the afterglow of GRB 010921. The optical measurements displayed here have been corrected for Galactic extinction using $E_{(B-V)} = 0.148$. The radio measurements were multiplied by the following scale factors for presentation: 22.5 GHz 1/2; 8.46 GHz 1/4; 4.86 GHz 1/8.

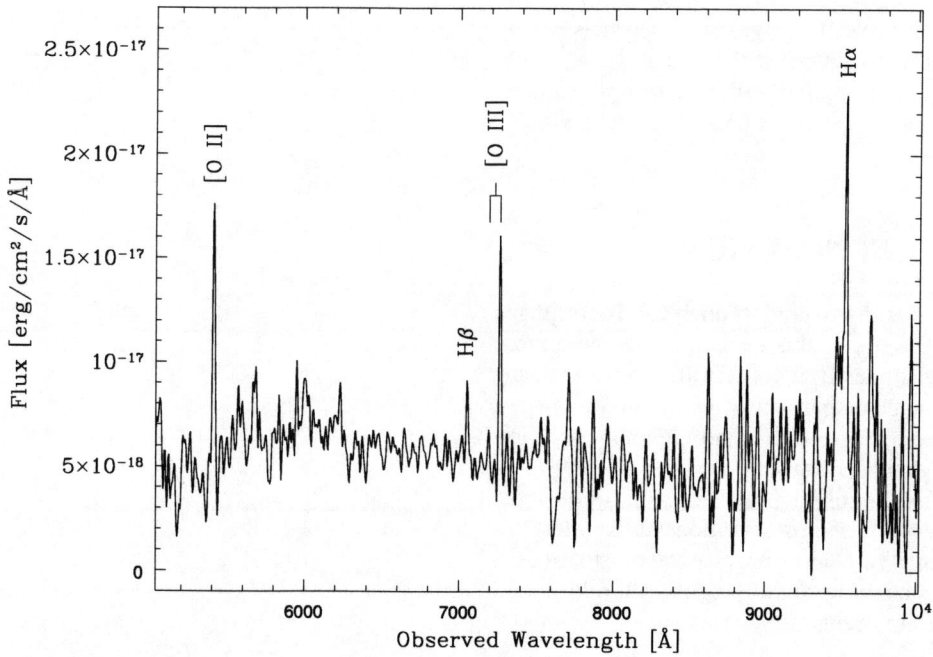

**FIGURE 2.** The spectrum of the host galaxy of GRB 010921. The spectrum has been smoothed with a Gaussian of FWHM = 12 Å, approximately the instrumental resolution. Emission lines corresponding to [O II], H$\beta$, [O III] and H$\alpha$ (labelled) are clearly detected, corresponding to a mean redshift of $z = 0.45092$.

the afterglow of GRB 010921, detected and localised by HETE-2.

Fits of a power-law temporal decay and power-law spectrum to the optical data corrected for Galactic extinction yield good fits ($\chi^2$/DOF = 10.9/11), with $\alpha = -1.579 \pm 0.058$ and $\beta = -2.00 \pm 0.13$. Demanding that the afterglow spectral slope match that predicted from the temporal slope (e.g. Price et al. [6]) requires extinction in the source frame with $A_V > 0.4$ mag, which likely indicates that this GRB is in a dusty environment.

## THE HOST GALAXY

We observed the host galaxy on 2001 Oct 17.25 with the Double Spectrograph on the Palomar 200-inch telescope. Several emission lines are detected, corresponding to [O II], H$\beta$, [O III] and H$\alpha$, all at a mean redshift of $z = 0.45092 \pm 0.00040$. The widths of the lines range from 10 – 13 Å, consistent with the instrumental resolution. From the observed Balmer decrement and an LMC extinction curve, we calculate $A_V$ for the galaxy to be 1.3 mag, perhaps indicative that the extinction observed in the afterglow is due to interstellar (not circumburst) dust.

From the FREGATE 8–400 keV fluence of $1.5 \times 10^{-5}$ erg/cm$^2$ [7], we derive the isotropic-equivalent prompt energy release as $9.0 \times 10^{51}$ erg. In order for this energy to be consistent with the Frail et al. [4] relation, we would require a jet opening angle of 19°, or a jet break time at approximately 36 days after the burst, consistent with the observed light-curve.

## CONCLUSION

The afterglow was discovered through the use of image differencing. The use of this technique may be a more robust method of identifying GRB afterglows than the traditional manual comparison with sky survey images, since it enables the detection of an afterglow superimposed on a bright host galaxy.

Since the afterglow of GRB 010921 is one of the most nearby afterglows, further observations of the afterglow with HST should be able to provide the most stringent limits on any supernova underlying the afterglow. And thus the HETE era begins....

## ACKNOWLEDGEMENTS

PAP gratefully acknowledges travel assistance from the Alex Rodgers Travelling Scholarship, and the advice and support from the members of the Caltech-NRAO-CARA GRB collaboration. We are grateful to the staff of the Palomar and US Naval Observatories for their expert help. INDNJC.

## REFERENCES

1. Alard, C., 2000, A&AS, 144, 363.
2. Berger, E. et al., 2001, ApJ, 556, 556.
3. Fox, D.W. et al., 2001, GCN Circular 1099.
4. Frail, D.A., et al., 2001, 2001, ApJ, 562, L55.
5. Hurley, K. et al., 2001, GCN Circular 1097.
6. Price, P.A. et al., 2001, ApJ, 549, L7.
7. Ricker, G.R., et al., 2001, in preparation.
8. Rhoads, J.E., 2001, ApJ, 557, 943.
9. Sari, R., Piran, T. & Narayan, R., 1998, ApJ, 497, L17.

# The E-Peak Distribution of the GRBs Detected by HETE FREGATE Instrument

C. Barraud[*], J. L. Atteia[*], J. F. Olive[*], J. P. Dezalay[*], D. Q. Lamb[†], N. Kawai[‡], A. Yoshida[§], Y. Shirasaki[¶], T. Sakamoto[||], T. Tamagawa[††], K. Torii[††], M. Matsuoka[‡‡], E. E. Fenimore[||], M. Galassi[||], T. Tavenner[||], T. Q. Donaghy[†] and C. Graziani[†]

[*]*Centre d'Etude Spatiale des Rayonnements-CNRS/UPS- 9 Av du Colonel Roche, BP 4346-31 028 Toulouse Cedex 4,FRANCE*
[†]*Department of Astronomy and Astrophysics, University of Chicago, 5640 South Ellis Avenue, Chicago, IL 60637.*
[**]*Department of Physics, Tokyo Institute of Technology, 2-12-1 Ookayama, Meguro-ku, Tokyo 152-8551, Japan.*
[‡]*RIKEN (Institute of Physical and Chemical Research),2-1 Hirosawa, Wako, Saitama 351-0198, Japan.*
[§]*Department of Physics, Aoyama Gakuin University, Chitosedai 6-16-1 Setagaya-ku, Tokyo 157-8572, Japan*
[¶]*National Astronomical Observatory, Osawa 2-21-1, Mitaka, Tokyo 181-8588 Japan.*
[||]*Los Alamos National Laboratory, P.O. Box 1663, Los Alamos, NM, 87545.*
[††]*RIKEN (Institute of Physical and Chemical Research), 2-1 Hirosawa, Wako, Saitama 351-0198, Japan.*
[‡‡]*Tsukuba Space Center, National Space Development Agency of Japan, Tsukuba, Ibaraki, 305-8505, Japan.*

**Abstract.** The FREGATE gamma ray detector of HETE-2 is sensitive to photons between 6 and 400 keV. This sensitivity range, extended towards low energies, allows us to explore the emission of GRBs in hard X-rays. We fit the spectra of 23 GRBs with Band's spectral function in order to derive the distribution of their peak energies (E-peak). This distribution is then compared with the E-peak distributions measured by BATSE and GINGA.

## INTRODUCTION

We present here a preliminary analysis of the spectral distribution of FREGATE's GRBs. FREGATE is the gamma-ray detector of HETE-2 (see [1] for a description of FREGATE). During its first year of operation (HETE was launched on October 9th 2000), FREGATE has detected 37 GRB candidates: for 13 of them, a position has been determined, 13 have no localization but are in the field of view and the remaining 11 are out of the field of view.

In this preliminary study, we focus on the distribution of E-peak which is the energy at which the $\nu F_\nu$ spectrum reaches a maximum. To obtain the E-peak of a GRB, we fit its energy spectrum with a simple model (the so called Band's function) using the Xspec software.

In addition, this analysis provides the fluences of the GRBs detected within the field of view of FREGATE.

## THE E-PEAK DISTRIBUTION

### Spectral fitting

The Xspec software requires 3 input components: the *observed spectrum* is obtained by subtracting the background from the signal. It is coded on 128 channels in energy for the four detectors of FREGATE. The *instrumental response* is computed from a Monte Carlo simulation of the instrument, it has been extensively tested with ground calibrations and inflight calibrations with the Crab nebula [2]. The *model spectrum* is the model GRBM from Xspec, based on Band's function [3]:

$N(E) = A.E^\alpha.e^{-E/E_o}$ for $E < (\alpha - \beta)E_o$
$N(E) = B.E^\beta$ otherwise.

We can then calculate the E-peak energy ($E_p$), the energy at which the $\nu F_\nu$ spectrum has a maximum. $E_p$ is given by:

$$E_p = E_o(2 + \alpha) \quad (1)$$

$E_p$ is defined when: $\alpha > -2$ and $\beta < -2$.

As an example the figure 1 shows the spectrum of GRB001225 which is well fitted by the Band's model.

In comparison, the figure 2 shows a spectrum which can be fitted by a simple power law.

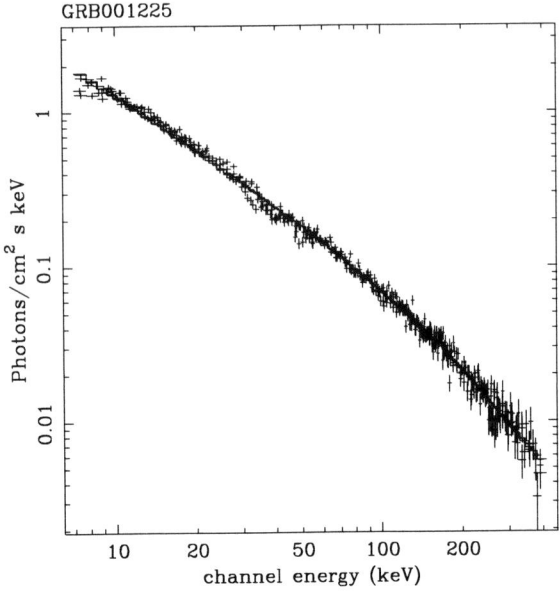

**FIGURE 1.** Spectrum of GRB001225 fitted with Band's function. $\alpha = -1.15, \beta = -1.90, E_o = 277 keV, E_p = 235 keV$

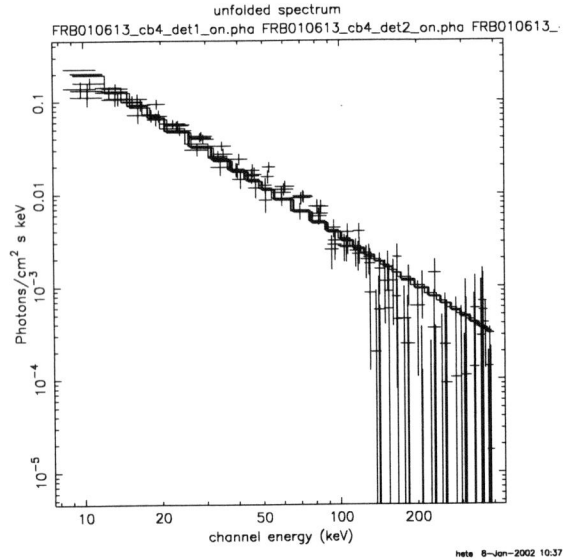

**FIGURE 2.** Spectrum of GRB010613 fitted with a power law, $\alpha = -1.78$

Table 1 lists the spectral parameters of 23 GRB candidates which occured within the FOV of FREGATE. Most of these events were confirmed by other GRB detectors on ULYSSES, KONUS, SAX-GRBM, RXT-ASM. In this preliminary analysis, we accept without question the spectral parameters provided by Xspec.

## The E-peak distribution

Figure 3 shows the E-peak distributions measured with the GRBs detected by FREGATE compared to BATSE and GINGA. The value of $E_P$ for FREGATE GRBs is obtained with Xspec, the distribution of BATSE comes from [4] and the GINGA's distribution was calculated using [5].

Some GRBs with few counts at high energies can be fitted with $\beta = -10$. We attribute this absence of constraints on $\beta$ to the lack of sensitivity of FREGATE at high energies. When this is the case, if $\alpha \geq -2$, we consider that the data do not allow us to derive $E_p$ and we provide no value (see GRB010613). if $\alpha \leq -2$, the $\nu F_\nu$ spectrum is steadily decreasing in the energy range of FREGATE and we arbitrarily set $E_p$ to 10 keV.

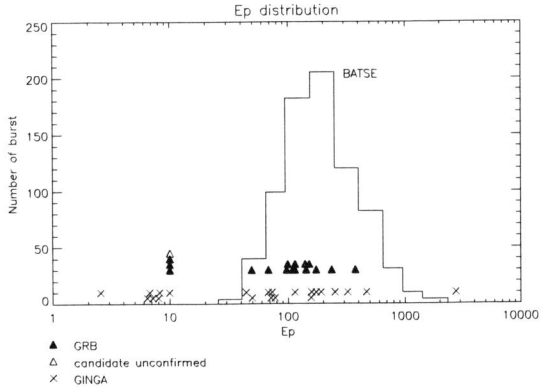

**FIGURE 3.** The E-peak distribution of the GRBs detected by FREGATE (triangles), GINGA (crosses) and BATSE.

The spectral fitting of FREGATE GRBs also allows us to compare the fluences in various energy ranges.

## The fluence

Figure 4 shows the fluence in the range 30-400 keV versus the fluence in the range 8-30 keV in units of $10^{-6} erg.cm^{-2}$. The line 50% represents the place where there is as much fluence in the range 8-30 keV as in the range 30-400 keV. These two fluences are well correlated for confirmed GRBs (with 80% of the fluence in the range 30-400 keV on average).

A few events are very soft, having most of their fluence below 30 keV (they are located below the line 50%). The nature of these events is still unclear: genuine GRBs, very soft GRBs... see also [6]. According to a suggestion by J.Heise (GCN 1138), we call them X-Ray Flashes (XRFs).

For comparison, a few typical X-ray bursts have been added to this figure.

**TABLE 1.** GRB candidates detected by FREGATE

| Bursts | angle | fluence* 8-400 keV | fluence 30-400keV | α | β | $E_o$ keV | $E_p$ keV |
|---|---|---|---|---|---|---|---|
| GRB001102 † | 70° | 77.33 | 72.33 | 0.14 | -2.86 | 58 | 142 |
| GRB001225 | 37° | 165.12 | 143.63 | -1.15 | -1.90 | 277 | 235 |
| GRB010126 | 10° | 2.50 | 1.79 | -.92 | -3.46 | 90 | 97 |
| GRB010213 | 14° | .12 | .0091 | -1.50 | -10 | 12 | |
| GRB010225 | 23° | .39 | .18 | -1.27 | -10 | 37 | |
| GRB010326a | 60° | 15.10 | 14.05 | -0.13 | -3.12 | 81 | 151 |
| GRB010326b | 17° | .16 | .08 | -2.13 | -10 | 50 | 10 |
| GRB010612 | 13° | 4.78 | 4.23 | -0.90 | -10 | 189 | |
| GRB010613 | 38° | 26.58 | 19.18 | -1.15 | -10 | 80 | |
| GRB010629 | 28° | 5.06 | 3.12 | -1.41 | -3.50 | 83 | 49 |
| GRB010921 | 44° | 15.41 | 11.63 | -1.42 | -2.35 | 170 | 99 |
| GRB010923 | 60° | 5.78 | 4.08 | -0.57 | -3.63 | 80 | 114 |
| GRB010928 | 24° | 20.86 | 19.77 | -0.64 | -1.82 | 276 | 375 |
| GRB001115a | 30° − 150° | 24.41 | 23.90 | 0.53 | -1.55 | 55 | 139 |
| GRB001115b | | 11.07 | 9.76 | -0.91 | -1.42 | 105 | 114 |
| GRB010428 | 4° − 82° | .34 | .022 | -2.06 | -10 | 12 | 10 |
| GRB010609 | 48° − 195° | 5.24 | .15 | -3.14 | -10 | 11 | 10 |
| GRB010621** | | .84 | .024 | -2.31 | -10 | 10 | 10 |
| GRB010706 | | .11 | .097 | 0.29 | -10 | 58 | |
| GRB010827a | | .61 | .47 | -.79 | -2.44 | 56 | 68 |
| GRB010828 | 46° − 113° | .72 | .32 | -0.96 | -10 | 176 | |
| GRB010903 | 53° − 106° | 3.68 | 3.36 | 0.10 | -3.53 | 52 | 109 |
| GRB010917 | | 4.74 | 4.69 | 0.99 | -1.69 | 58 | 173 |

* The fluences are in units of $10^{-6} erg.cm^{-2}$
† This burst was in the limit of the field of view of FREGATE
** this events have been detected only by FREGATE

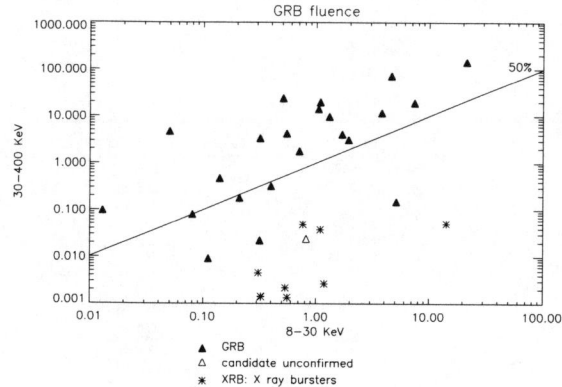

**FIGURE 4.** The fluence in 30-400 keV vs the fluence in 8-30 keV for 23 GRBs detected by FREGATE. The XRFs are placed under the line 50%.

## CONCLUSION

A preliminary study of 23 GRB candidates detected by FREGATE within its FOV shows the existence of a few soft events, XRFs (with $E_p < 15$ keV and most of the fluence below 30 keV). These XRFs can be compared to the detection of such soft events by GINGA and the WFC on Beppo SAX.

In its second year of observations, HETE-2 will provide fast localizations of several new XRFs, an approach which may shed a new light on the nature of these enigmatic sources.

## REFERENCES

1. Atteia et al, these proceedings.

2. Olive et al, these proceedings.
3. Band et al, Ap.J, 413:281-292 (1993).
4. J. J. Brainerd, G. Pendleton, R. Mallozzi, M. S. Briggs, R. D. Preece. *The BATSE Gamma-Ray Burst E-Peak Distribution:* unpublished manuscript available by request from Jim Brainerd at the e-mail address <brainjj@bellsouth.net>.
5. T. E. Strohmayer & E. E. Fenimore. *X-ray spectral characteristics of GINGA gamma ray bursts*, Ap.J, 500:873-887 (1998).
6. J.Heise, J. in't Zand, R. M. Kippen and P. M. Woods. *X-ray flashes and X-ray rich Gamma Ray Bursts*, Astro-ph/0111246 (2001).

# A Study of the Background in the HETE Equatorial Orbit

N. Butler[*], A. Dullighan[*], J-L. Atteia[†], N. Kawai[**], A. Yoshida[**] and T. Tamagawa[**]

[*]37-524 MIT CSR, Cambridge, MA, 02139
[†]CESR, BP 4346, 31028 Toulouse Cedex 4, France
[**]RIKEN, Wako, Saitama, 351-0198 Japan

**Abstract.** After a year of operation, we have become well acquainted with HETE's in-orbit background. In particular, as it is near solar maximum, we have observed through intense periods of geomagnetic disturbance. We describe the resulting deviations from our mean backgrounds. Aside from these periods, the background has been quite stable over this first year. We compare the averaged background in the WXM and Fregate instruments for the months of March and September, 2001.

## INTRODUCTION

The *HETE* satellite orbits at a mean altitude of 610 km, in a 4° band about the equator. The orbital ellipticity is 0.003. In this study, we examine count rate variations in two of *HETE*'s instruments, binned in geophysical coordinates. The Fregate instrument [1] is most sensitive to photons of energy between 8 – 400 keV, though photons can be detected out to about 1 MeV. We focus primarily on this instrument in the 8 – 40 keV and the 32 – 400 keV bands, bands "A" and "C" respectively. The WXM instrument [2] is sensitive to photon energies between 2 and 25 keV.

## DISCUSSION

### Longitude Variations (Mild Space Weather)

During May–July 2001, the WXM instrument and Band A of the Fregate instrument were dominated by emission from X-ray sources in the Galactic bulge. This is an annual occurrence due to *HETE*'s anti-solar pointing. We chose for comparison two months–March and September–away from this period of contamination. We exclude light curve regions where a burst is present or where the background has risen appreciably due to solar activity. Our selection excludes about 15% of the data.

In Figure 1, we plot the background in several energy bands as a function of Earth longitude. The similarity between the March and September lightcurves indicates that the *HETE* background is quite stable. The background is somewhat higher irrespective of longitude in March due to the Galactic bulge entering the edge of the instrumental fields of view. At low energies (2 – 40 keV) and at very high energies (> 400 keV), the dominant feature is due to solar protons in the South Atlantic Anomaly (SAA). (The SAA [centroid at $\sim 40°$ West longitude] has migrated somewhat Westward toward South America, since its naming.) Also present in Figure 1 is the count rate signature due to the electrons (and possibly positrons) trapped in Earth's radiation belt. This region is most pronounced in band C and appears much more extended (from $\sim 90°$ East longitude Eastward to $\sim 60°$ West longitude) than the SAA.

### Altitude and Latitude Variations (Mild Space Weather)

In Figures 2 and 3, data are plotted by longitude along with latitude or altitude. We observe counts rates which rise with increasing altitude. The maximum minus minimum rate ranges within 10% of the mean count rate in Fregate bands A and C. From the latitude plots, it appears that the electrons are clustered somewhat more to the North than the SAA protons.

### Longitude Variations (Intense Space Weather)

There were two periods–one at the beginning of the month and one toward the end in September–during

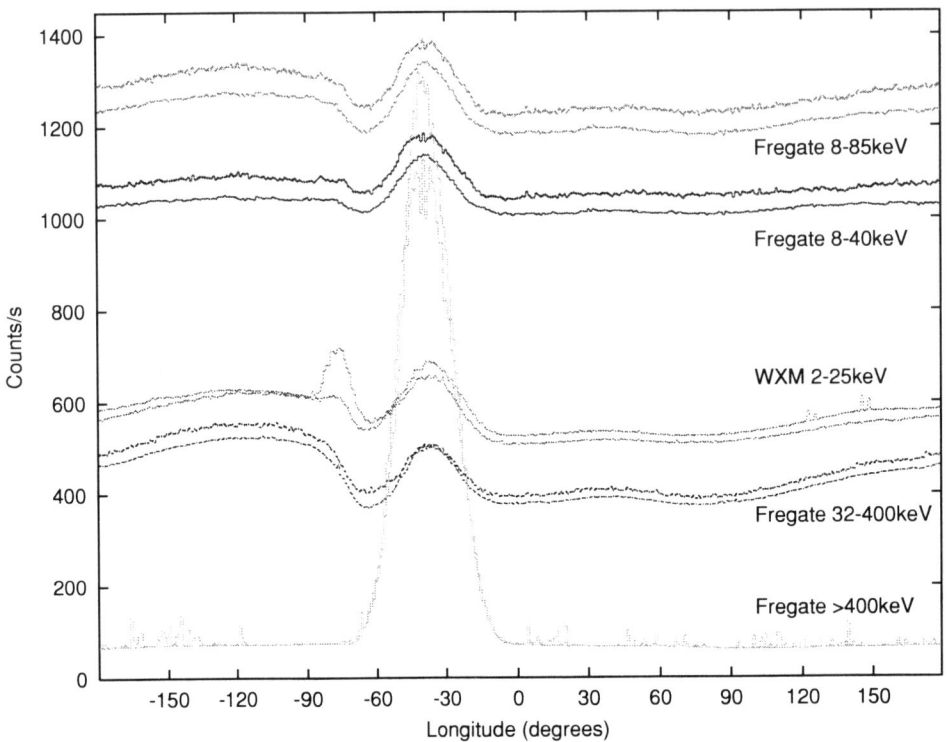

**FIGURE 1.** Count rates in March (upper curve in each pair) and September. Error bars are plotted; the instrumental backgrounds were well behaved over these periods.

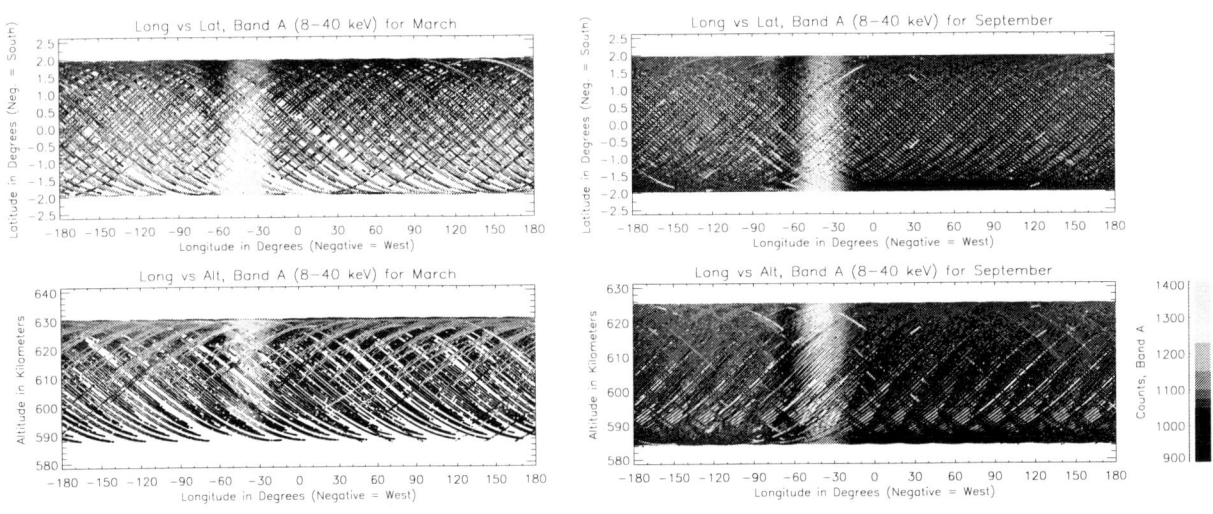

**FIGURE 2.** Count rates in March and September for Fregate Band A.

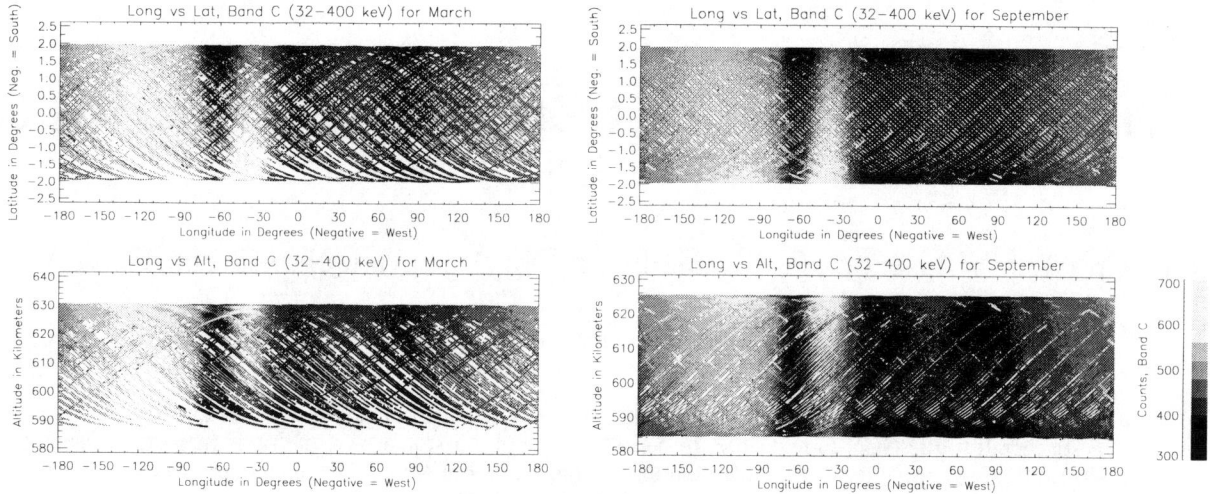

**FIGURE 3.** Count rates in March and September for Fregate Band C.

which solar flares initiated large increases in *HETE*'s orbit background. Figure 4 shows count rate deviations for Fregate on the 25th of September, over-plotted on the mean count rates in September. This activity occurred one day after an X2/2b class solar flare (and accompanying CME) at 10:38 UTC on the 24th of September. Increases are localized in two regions–over the SAA, and over the Eastern-most edge of the electron (see above) region. Operations are disrupted for several hours. Such episodes have occurred at a rate of $\sim 1$/month over the first year of the mission.

## ACKNOWLEDGMENTS

Special thanks to the *HETE* operations team for maintaining the science operations which made this study possible during even the most erratic episodes of geomagnetic and solar activity. Thanks to George Ricker for several discussions influencing this work.

## REFERENCES

1. Atteia, J.-L. et. al., "In-flight Performance and First Results from the Fregate Instrument on HETE", in *these proceedings.*, 2002.
2. Kawai, N. et. al., "The In-orbit Performance of the WXM Instrument on HETE", in *these proceedings.*, 2002.

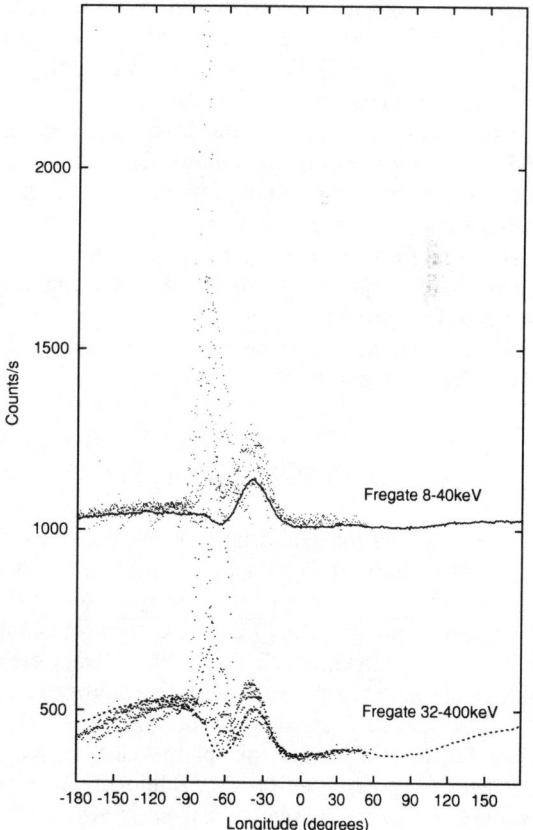

**FIGURE 4.** Background count-rate deviations (shown here for September, with the mean behavior over-plotted) occur in a narrow longitude band.

# Real-Time Message from the HETE Spacecraft via the HETE Burst Alert Network

G. B. Crew*, J. Villasenor*, G. Monnelly*, J. Doty*, R. Vanderspek* and G. Ricker*

*MIT Center for Space Research, Cambridge, MA, 02139 USA*

**Abstract.** The HETE Burst Alert Network [1] was designed to allow propagation of burst localizations from the HETE spacecraft to the GCN in real time. In periods where no burst is active, the same network is used to relay spacecraft housekeeping data to the Mission Operations Center at MIT. The Burst Alert Network thereby provides an invaluable view into the real-time operation of the HETE spacecraft. We describe the contents of real-time messages, both during and outside of burst activities.

## VHF OVERVIEW

Although designed to allow the rapid disemination of GRB Alerts, the Burst Alert Network also allows real-time monitoring of HETE state. The typical delay is the time required to transmit at 300 bits/sec.

Since the primary science instruments operate in the night portion of the orbit, the routine diagnostic message is different in the two portions of the orbit. During bursts, only the burst message is sent.

Note that the messages are tightly packed binary—we present here samples of their readable representations as produced by operations software.

At the end of this paper we present some of the details of the station hardware and software.

## MESSAGE CONTENT

All messages include a header which provides some basic information, followed by a variable length data portion. This is the same overall message format (Inter-Processor Protocol, or IPP) that is used throughout the HETE system. The header includes the message creation time on the satellite clock; this time is calibrated against the GPS (Global Positioning Satellite) epoch—seconds since Jan 6, 1980, the origin of the GPS Epoch, and ultimately expressed as UTC. Additional information available when the message received is also displayed for the benefit of the HETE operators and scientists. This includes: whether it is orbit day (62 min) or orbit night (35 min); the transit time (transmission + network delay); the receiving groundstation; and an assessment of message validity.

The data portion of the message is built up from several variable-length components, with the most important components broadcast first. Thus if the message is garbled in transmission, some portions are nevertheless usable (perhaps with some interpretation). The "priority" field of the message header is usurped for this purpose. In particular, fully-decoded, valid messages are marked with a 31, while variously suspect messages are marked with a "priority" greated than 50. Two of the stations (Cayenne and Marquesas) use a different hardware/software arrangement than is described here—for these stations the original priority (2) is presented, unchanged.

Software within Mission Operations at MIT (the `vhf_console`) receives the messages and displays them in a text representation for interpretation. However all of the automatic software processes the messages in their native binary.

### Satellite Diagnostics

The satellite housekeeping portion of the messages (25 bytes) is sent first, and includes reports from all major subsystems. We shall discuss some of the more important fields here.

**Communications** includes a comm-mode (CM) which is 0: off, 2: gps, 3: trying, 4: in contact; the NC field gives the time of the next primary groundstation contact in internal time units.

**Operating System** has a number of bits for the system configuration: notably the NODES field indicates which of the processing nodes are operating; and the LINKS field indicates which nodes can communicate; the number of kB held in mass memory (MM) in each of the four nodes. There is a total of 12 MB of memory on each of the nodes, which can hold data from several orbits.

- **Power** the batteries are charged using two power point trackers (PPTs) when HETE is in daylight, at night the batteries may discharge down to $\sim 80\%$; the bus voltage typically floats between 27 and 33 V; there are on/off flags for all the major subsystems, and a fault indicator.
- **ACS** has several modes; 3,5,7 are normally used in day, 2 (no stars) or 8 (OPT tracking) at night; the wheel speed is nominally 94 rad/s; `s_hat` points to the sun; `omega` is the drift rate.

## Instrument Diagnostics

Each science component has several bytes ($< 32$) for its routine diagnostic data. New messages are generated every 20 seconds or so when there is no active burst.

- **gamma** provides its observing mode (MODESURVEY or MODESURVEYOFF) and an internal burst counter; the remaining bytes cycle through various housekeeping for each of the 4 detectors on a slower timescale.
- **WXM** provides its observing mode; the SSI IRQs are software interrupts (real photons plus background) the nominal HV power supply values are 216 and 214; and the counting rates in the 4 detectors are given.
- **sxc** provides data on its observing program; the CCD temperatures are given; the best DC position would be a steady source; the event rate as counts / data frame. In survey mode it correlates its photons to provide a DC position which may correspond to a X-ray source.
- **opt** provides data on its observing program; the Star word is the number of stars visible in each of the four cameras (one hex digit for each of the two ACS cameras, canted at $\sim 40$ degrees off-axis, and the two boresight cameras; $f = 15$ or more); the real-time HETE attitude (RA, dec and roll) is displayed if the optical system has enough stars to match against its catalog.
- **TM** the trigger monitor presents the number of the next burst and instruments that are active; any old burst information is also displayed if it is recent.
- **GPS** this instrument can track up to 8 of the 32 satellites in the Global Positioning System Constellation: for each the ID, and C/N (carrier to noise ratio) is given, together with a status (1 = trying - 4 = acquired).

## Typical Daytime Message

```
GPS time 688158096.214 [20011026_190123], ORBIT DAY, TT=9.2,
   Maui [31]
SC:       M2 Tx 2-off CM-2 Free-55 Pri-3f
          NC = 0019 4EE4 8209 D3C2
          MS = 2B09 TIN70 RZP0 SEQ0 NODES = 3773 LINKS = FFFF
          MM  4596.0   3584.5   8866.5   7735.0 kB
          Bus = 32.635V ppt0 = 2.656A ppt1 = 2.628A
          Cur =  3.259A pscF = 0000  Charge = 93.93%
          SP.. RFon SAXon MWDon TCDon GPSon IPSon
          HTR... FGTon WXMon    SXC..  OPTon RBMon
          ACS mode7 s_hat (0.0010 -0.0015  1.0000)
          whl(94.22) omega(-7.14 -1.84 -6.25) '/s
gamma:    MODESURVEYHTOFF, burst #21
          U 2, msg 16; 0, 0 cts/5s; templ = 236
          U 3, msg 17; 0, 0 cts/5s; commande_HT = 285
WXM:      standard vhf diagnostics      mode is 0x0003  ON_PORTC
          SSI IRQs: 0
          voltage -- State: 0x80   HVA:   0    HVB:    0
          LAST_CMD: 0x5000
          COUNTS -- XA:    0  XB:    0  YA:    0  YB:    0
sxc:      idle
opt:      idle
gps-status: 500493.562500 s gdop: 3.00 Val: (ok) 00 00 17 00
gps-SatId:       1     8    11    20    22    27    31     3
gps-State:       1     4     4     4     4     4     4     4
gps-C/N:      0.00 32.50 41.75 42.75 36.25 36.75 38.00 35.25
gps-hk: vco 6.945 agc 3.346 ref 1.230 temp 37.1 C (!ETX)
TM:       active, next burst ID 1824
          gamma down, wxm down, sxc down, opt down
                                              Repeated 1 times
                                              Repeated 2 times
```

## Typical Nighttime Message

```
GPS time 687813490.135 [20011022_191757], ORBIT NIGHT, TT=12.7,
   Bangalore [31]
SC:       M0 Tx 2-off CM-0 Free-55 Pri-3f
          NC = 0019 4E94 0976 9BAC
          MS = 2B09 TIN70 RZP0 SEQ0 NODES = 3773 LINKS = FFFF
          MM  2070.0   1583.0   8312.5   5618.5 kB
          Bus = 29.313V ppt0 = -0.008A ppt1 = -0.003A
          Cur = -2.163A pscF = 0000  Charge = 92.34%
          SP.. RFon SAXon MWDon TCDon GPSon IPSon
          HTR... FGTon WXMonHV  SXCon OPTon RBMon
          ACS mode8 s_hat (0.0000  0.0000  1.0000)
          whl(94.39) omega(-1.26 -0.93  2.09) '/s
gamma:    MODESURVEY, burst #13
          U 2, msg 3948; 1793, 477 cts/5s; -15V = 423
          U 3, msg 3949; 1556, 481 cts/5s; +28V = 0
WXM:      standard vhf diagnostics      mode is 0x0007  ON_PORTC_HV
          SSI IRQs: 6240
          voltage -- State: 0x8c   HVA: 216    HVB: 214
          LAST_CMD: 0x50d7
          COUNTS -- XA:  111  XB:  109  YA:   93  YB:  105
sxc:      running n36, next idle
          state:   OV=4,0 OB=3,8 AN=2,0 IM=0,0
          HK error flag = 0x0000
          FTEMPS = -17C, -49C
          Best DC position = 1194 1650 sig = 1.4 0.8
          Event rate = 17 cts/frame
          Latest sync ends (lower 8 bits):  165 165
opt:      running n53, next idle
          state:   OV=4,0 OB=3,8 AS=4,0 DR=3,2
          HK error flag = 0x0000
          CTEMPS = -40C, -40C
          Star word = 0xf6ff
          RA, dec, roll = 205.238 -10.388 162.587
          Latest sync ends (lower 8 bits):  235 235
TM:       active, next burst ID 1820, 1 OLD burst
          gamma UP, wxm UP, sxc UP, opt UP
OLD:      ID = 1819, XG triggered
          Trigger time = 687809245.928924 [20011022_180712.928]
          Running:      gamma NO,  wxm NO,  sxc NO,  opt NO
          Responded:    gamma NO,  wxm YES, sxc NO,  opt YES
          WXM flag 2, crit 27, S/N 15, duration 5120 ms, using WXM data
          WXM position (-0.5599 0.6361) goodness (3.8,2.8) virtue (8.0,5.0)
                                              Repeated 1 times
                                              Repeated 2 times
```

## Burst Information

The Alert message is short (< 32 bytes) to allow frequent retransmission and to increase the likelihood of its being delivered uncorrupted. It includes:

**ID** this is a unique trigger identifier

**Trigger Time** FREGATE or WXM trigger time

**Responded/Running** what science processing is active

**Trigger Criteria** details on which instrument triggered on which channels and at what timescale

**WXM Position** $\tan\theta_x$ and $\tan\theta_y$ with count and image SNR

**OPT Status** number of stars in each camera, HETE's attitude

The message must repeat for it to be considered real. When it does, an alert is forwarded to the GCN and the scientists' beepers are activated.

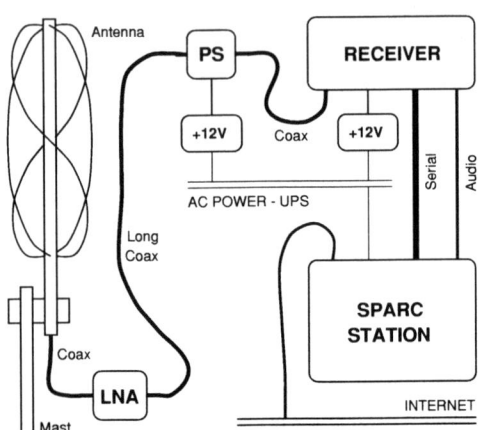

**FIGURE 1.** The burst alert station hardware consists of an quadifilar antenna, a receiver and a computer.

## Typical Burst Message

```
GPS time 677852530.887 [20010629_122157], ORBIT NIGHT, TT=9.2,
    Kwaj-PGS [31]

************ BEEP **************

burst!   in progress, ID = 1573, GAMMA triggered
         Trigger time = 677852480.522345 [20010629_122107.522]
         Running:        gamma YES, wxm YES, sxc  NO, opt YES
         Responded:      gamma YES, wxm YES, sxc  NO, opt NO
         TI=160ms, BE=30-400 keV, DC=2+3, pS=1, pG=1, pX=0, cts=49
         WXM flag 2, crit 15, S/N 6, duration 640 ms, using WXM data
         WXM detector mask 0x01, energy mask 0x0f
         WXM position (0.2536 0.2920) goodness (1.0,1.0)
         OPT bitmap is 0x000a, RA, dec, roll are 97.629 23.576 167.880
                                                        Repeated 1 times
SGS-Kwaj-PGS:  Burst 1573, GAMMA trigger at 677852480.5 on gamma data
                                                        Repeated 2 times
```

## STATION DESCRIPTION

The VHF message bits are encoded as a continuous-phase shift key (CPSK) 300 bit/sec broadcast at 137.962 MHz which is received in single side-band (SSB) mode by a commercial, computer-controlled radio (see Figure 1). (We use both the Drake R8 series as well as the AR 3030.) The antenna is typically a quadrifilar narrowly tuned in the VHF band. The received signal is presented to the computer as a 2 kHz bandwidth audio signal as shown in the accompanying Figure 2.

This signal is sampled at 8 kHz and processed first to determine if a signal is present, and if so, where within the audio band it may be found. In practice, it is easiest to pick out the two spikes (the frequencies of the 0 and 1 bits) using a maximum-entropy (MEM) spectral representation. In the course of a typical pass, the doppler shift is 6 kHz from one horizon to the other—the software must constantly tune the receiver to maintain the signal within the mid-range ($\sim$1.5 kHz) of the audio

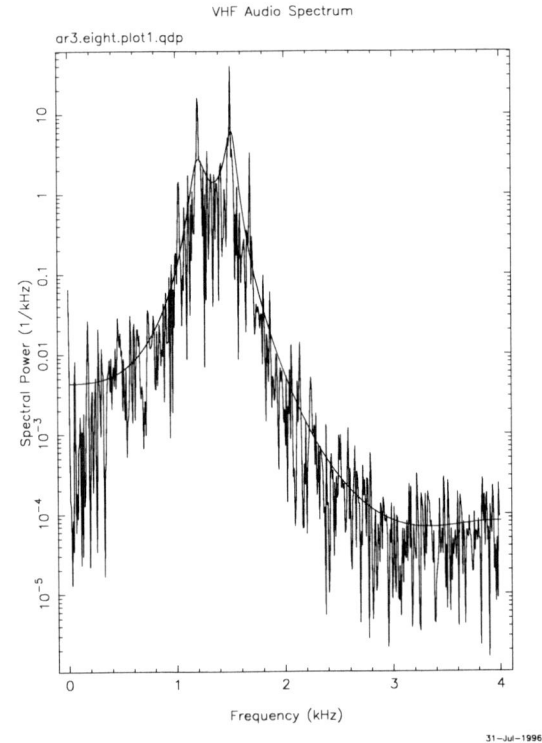

**FIGURE 2.** The VHF audio signal. The ragged component is the FFT spectrum, the smooth, double-peaked line is MEM spectral representation.

band. (The typical computer audio port is optimized for speaking, rather than general applications.)

The software then filters the audio samples to obtain a bandwidth limited, decimated, complex baseband signal from which the continuous signal phase relative the assumed carrier may be obtained. It is then a simple matter to read off the bits, descramble them, and assemble complete messages. (The signal is trivially scrambled to ensure that the 0 and 1 bits broadcast have approximately equal frequencies.)

The message is preceded and followed by special bytes that, in the presence of a damaged signal, make it possible to find the start and end of each message. When a complete message has been decoded, it is forwarded over the internet to the burst alert software running at MIT. This consists of four components:

- the CNS (which is responsible for routing messages, between ground processes)
- the VHF CONSOLE which converts the binary message into a human-readable form,
- the GCN CONSOLE which converts burst alerts into GCN Notice binary data,
- the DUTY SCIENTIST who has responsibility for the validity of the alerts.

There are three principle limitations of the system:

- RF noise at the groundstations which at best may flip some of the bits and at worst make the signal undetectable.
- Power at the remote groundstations.
- Network stability to the remote groundstations.

The overall performance of the system may best be demonstrated through the end-to-end timing of the delivery of the burst alert messages. The starting point is the time the burst photons appear in the data, and the end point is the time the alert is delivered to the GCN.

The histograms in the Figures 3 and 4 (one data set—two time scales) present the delay in seconds between the burst trigger time and the delivery of the alert message to the GCN. This sample includes only bursts preceding the Woods Hole meeting that occurred when access to the GCN was enabled (190 alerts total—the GCN is sometimes "down" due to network issues, and sometimes it is disabled at MIT for operational reasons).

The minimum delay is the sum of:

- the time to detect the burst ($\sim 5$ s)
- twice the transmission time ($2 \times \sim 2$ s)
- network delay to MIT ($\sim 1$ s)

Longer delays result when there is appreciable RF interference, or when stations are offline (e.g. due to power outages).

Overall, the system is performing as designed. The biggest improvement would come from more reliable network and power at some of the remote stations; however, we do not have the resources to affect this appreciably.

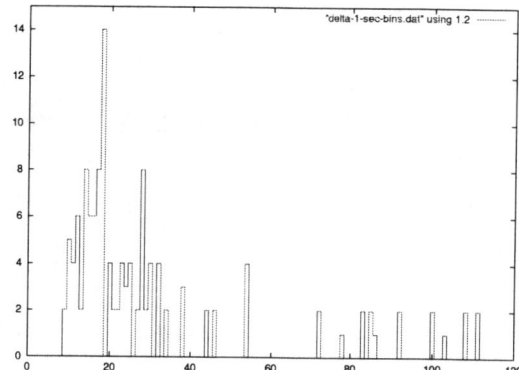

**FIGURE 3.** Histogram of end-to-end burst timing—bursts while in contact.

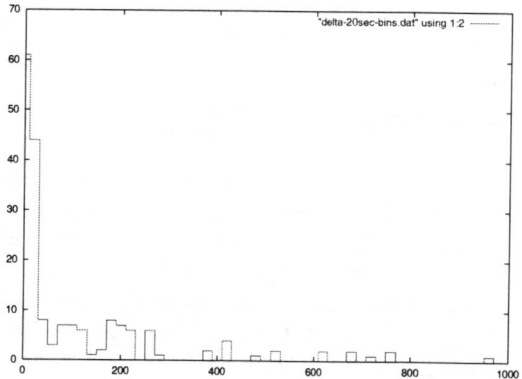

**FIGURE 4.** Histogram of end-to-end burst timeing—late bursts.

## ACKNOWLEDGMENTS

The VHF system has had many hands on it. We'd specifically like to acknowledge the efforts of Greg Huffman, who built the flight transmitter; Tim Jones and Joe Kendall who wrote the HETE-1 flight VHF software on which the HETE-2 version is based; and Latif Lokman, who wrote the first implementation of the computer-based audio deconvolution software.

Finally, we are grateful for the efforts of Michel Boer and his colleagues who have provided us with two stations of an original design which provide additional coverage.

## REFERENCES

1. Villasenor, *et al.* 2002, this proceedings.

# HETE Mission Operations

G. B. Crew*, R. Vanderspek*, J. Doty*, J. Villasenor*, G. Monnelly*, N. Butler*, G. Prighozin* and G. Ricker*

*MIT Center for Space Research, Cambridge, MA, 02139 USA*

**Abstract.** HETE Mission Operations are unique in that a small team is responsible for full-time operation of the HETE spacecraft, three primary ground stations, and fourteen Burst Alert stations. These requirements have led to the development of a suite of operations procedures and software which attempt to maximize operations efficiency while minimizing the amount of human interaction required. HETE Mission Operations are reviewed here.

## OPERATIONS PHILOSOPHY

HETE is a low-cost mission; the satellite was designed to be managed by a small operations team. In particular, the satellite itself has to be robust and relatively simple to operate. In addition, a high degree of automation of the routine operations was imperative. Moreover, HETE is "walk-away safe", and in fact it is indeed rare for the satellite to require urgent operator attention on matters of health and safety.

One salient feature of the HETE mission is that (essentially) the same software was used for both ground testing and flight operations. A second is that HETE is almost continuously in contact with the ground through the network of secondary (burst-alert) stations. This makes it possible to monitor HETE almost continuously.

HETE was launched from Kwajalein, in the Republic of the Marshall Islands at 5:38 UT on Oct 9, 2000. The first contact was made at Kwajalein (Figure 1) during the first orbit. During the first few months, student operators from the MIT Department of Aeronautics and Astronautics were employed to assist with daily operations. In practice, they spent large amounts of time watching the satellite. They got very bored; then they graduated. Since then, the satellite has been managed continuously by the half-dozen duty scientists.

At present, a single Duty Scientist (DS) handles most of the routine satellite operations for a one-week shift. The operation software alerts the DS to situations requiring attention through a beeper or email. In addition all members of the team check in on HETE from time to time and alert the DS to situations that the automated software canot catch. In practice, the DS and the other members of the team are largely free to pursue the HETE gamma-ray burst science.

**FIGURE 1.** Kwajalein PGS. The October 2000 launch team (N. Butler, G. Crew, J. Villasenor and B. Dill) pose with ceremonial coconut, for scale. The large dome atop the container protects the S-band antenna from the weather; the "water-heater" VHF antenna is mounted on the mast to the right.

## GROUNDSTATIONS

Communications with HETE are made with a network of two flavors of groundstations. The burst alert stations (also known as secondary groundstations), although designed to receive and relay notification of GRB alerts, perform the important operational task of providing us with almost contionous monitoring of HETE status. See the three other papers [1, 2, 3] for details on the burst alert stations and the GCN. This means that the DS receives almost continuous information on HETE's state between contacts with the primary groundstations (which can deliver uplink as well as receive the downlink data).

**FIGURE 2.** Singapore: The radome is visible atop the CRISP building at the right.

The primary groundstations operate unattended at remote equatorial sites. Two of the three primary groundstations (Singapore and Kwajalein) contain identical components that were built and deployed for the original HETE-1 mission. The Singapore station (Figure 2) was moved from its original deployment in Miyazaki, Japan to its present location at the National University of Singapore. The Kwajalein station was packed into a commercial shipping container and sent to Kwajalein in the fall of 1999 to support the first launch attempt in January 2000. After arrival, the radome (top left) and VHF antenna were installed on the roof of the container and local power and telephone connections were made. The third station (Cayenne, French Guyana) was first modified from an existing station (Toulouse, France) to track HETE-1, and later relocated to French Guyana for HETE-2. This station uses a 3 m dish antenna (Figure 3) and is operated at lower power than the other two stations. When it is windy, the antenna is automatically parked.

## CONTACT MANAGEMENT

A schedule of groundstation contacts with HETE is automatically determined from the orbital elements and power considerations. HETE-2 is essentially the same satellite as HETE-1 (lost in November, 1996 due to a rocket failure), which was designed for an inclined orbit where at most a single (primary) groundstation would be available for contact each orbit. (The S-band radio consumes considerable power of the power budget.) HETE-2 was placed in an equatorial orbit, and so in principle all three of the groundstations are available to make contact. However, the battery subsystem was designed around the requirement of the single contact per orbit; thus there is insufficient power margin to expect to use all three of the groundstations on every orbit. On the other hand, the solar panels and power system are performing better than expected—daytime contacts are easily powered directly from the solar panels. On most orbits it is possible to make two (or in some cases) three contacts. Finally, there are occasional network outages or maintenance activities which may prevent us from using a station.

Once the contact schedule has been determined, a suite of scripts converts the current science program into a set of scheduled commands which are delivered to each groundstation prior to contact. For example, the loss of the optical blocking filters on the SXC instrument precludes the use of that instrument for the two weeks surrounding full moon; the instrument is turned off for that interval. Conversely, the GPS instrument is granted extra operations time during that period. Thus, the FREGATE, WXM and SXC instruments are operated for the half-orbit when the earth is not in the field-of-view. The GPS instrument is activated in the day-time portion of the orbit in order to determine HETE's position. The amount of observing time for each instrument is adjusted subject to constraints of overall downlink capacity and power.

## DOWNLINK PROCESSING

Following a contact the downlink is automatically retrieved from the groundstation and processed. This includes analyzing some of the satellite housekeeping to anticipate undesirable situations (*e.g.* low power) and

**FIGURE 3.** Cayenne PGS Antenna.

alert the DS. All science data is processed and placed where team members can examine the data. In addition several robots search through the FREGATE data looking for bursts that may not have triggered an on-board burst alert.

The GPS data is reduced to provide accurate position information for the IPN as well as orbital elements for planning future contacts. When the mission was planned, it was thought that NORAD two-line elements would be routinely available to locate HETE. Since launch, however, we have learned that this is not always the case.

## BURST HANDLING

In the typical case, HETE detects the burst and relays the burst time, position and triggering information to the ground in real-time using the VHF-band radio. The alert is forwarded automatically to the GCN if the burst is likely to be valid. Since HETE's primary science are the Gamma Ray Bursts, several team members are beeped to respond to the downlink burst data (which typically arrives < an hour later), discuss the burst, assess its validity and take followup actions as desirable. Space weather plays an important role in false triggers—many of the false triggers have been due to solar flares or particle events. To some extent these can be reduced by monitoring Kp and other magnetospheric indices and turning off the GCN at disturbed times.

## PRIMARY GROUNDSTATION HARDWARE AND SOFTWARE

A schematic of the RF hardware is shown above together with the rack-mounted RF stack electronics (RFSE—everything but the antenna, motor, diplexor and computer—see Figure 4). From the top down: G2E (the interface to the computer), Receiver, Positioner and Bit-Sync, Transmitter, GPS (not used—ntp is good enough), Power Amplifier. This S-band system supports data rates of 31.25 kbps uplink and 250 kbps downlink.

The groundstation software is modular—several programs control the dish positioning (sattrack), link hardware, the packet protocol, pass management and data exchange with MIT. Normally the groundstation is operated in a store-and-forward fashion, but real-time commanding is available. For each pass, a file of uplink data is delivered to the groundstation, which automatically drives the antenna, makes contact, delivers the uplink and collects the downlink into priority-split files for return to MIT.

**FIGURE 4.** RF hardware. With the exception of the dish and computer, all components fit in a single rack-mount container.

## ACKNOWLEDGMENTS

The deployment and operation of three groundstations world-wide would not have been possible without the efforts of a largely unsung cadre of friends.

The Kwajalein PGS was sponsored by MIT with local support from Raytheon contractors: Robert Miller, Rickey Huber, Lewis Smith.

The Cayenne PGS was sponsored by SUPAERO, Centre Aeronautique et Spatial: Christian Colongo, Eric Metral, Jacques Lamaison, Etienne Perrin; with local support from Institut des Etudes Superieures de Guyane: Oukaour Amrane, Henri Clergeot.

The Singapore PGS was sponsored by RIKEN: N. Kawai and A. Yoshida; with local support from the National University of Singapore: Choong Weng Mak, Lim Geok Quee, Lee Kok Wah, Leong Keong Kwoh.

In addition, the flawless performance of the corresponding systems on the flight is a credit to the engineers who built them. We'd like to mention folks here: Bob Dill, Jim Francis, Greg Huffman, Tye Brady, Frank Larosa, Jerry Roberts, Fred Miller, David Breslau.

## REFERENCES

1. Crew, G., *et al.* 2002, these proceedings.
2. Villasenor, J., *et al.* 2002, these proceedings.
3. Vanderspek, R., *et al.* 2002, these proceedings.

# Sensitivity of the FREGATE Experiment

J-P. Dezalay*, J-F. Olive*, J-L. Atteia*, J-P. Lestrade*, C. Barraud*, N. Butler[†], G. Crew[†], J. Doty[†], K. Hurley**, G. Ricker[†] and R. Vanderspek[†]

*C.E.S.R., 9 Avenue du Colonel Roche, 31028 Toulouse Cedex 4, FRANCE
[†]M.I.T. Center for Space Research, 70 Vassar St., Cambridge, MA 02139
**UC Berkeley Space Sciences Laboratory, Berkeley, CA 94720-7450

**Abstract.** During the first year of operation, the FREGATE instrument onboard HETE-2 has detected a large number of Gamma-Ray Bursts in the 6-400 keV energy range. In this paper, the sensitivity of FREGATE is computed in various energy ranges using several parameters such as the observed background, the angle of arrival of the photons on the detector, and the trigger criteria. From the current list of GRB candidates and with the duty cycle of FREGATE, we estimate the instrument burst detection rate within the field of view. The sensitivity and the burst rate are in good agreement with preflight simulations and with the results of previous experiments.

## INTRODUCTION

The FREGATE instrument onboard the HETE-2 satellite is designed to detect gamma-ray transients in the 6-400 keV energy range (a detailed description of the experiment characteristics is given in these proceedings [1]). The overall performance of a GRB detector is measured by its burst detection rate (in bursts per year for a full sky coverage) and by the minimum flux from the source required to trigger the experiment, i.e., the sensitivity. In this paper, we have calculated these two quantities to see how they compare to preflight simulations. Moreover, these parameters are important to compare FREGATE's results with those of previous experiments.

## THE FREGATE BURST LIST

Between the start of operations on October 16, 2000 and September 30, 2001, FREGATE recorded 35 events among which 32 (see table 1) were confirmed by other instruments (including the WXM). For each event in the current FREGATE GRB list the table 2 summarizes the conditions of the detection of the event. A burst is considered to be in the FREGATE's field of view (FOV), if at least a part of the event flux hits the detector without passing through the collimator. In this case, the angle of arrival of the photons with respect to the detector axis is less than 70° [1].

When the burst direction is not known from either WXM or IPN data, the count rate in the 6 - 80 keV energy range is checked. If the event is not visible in this energy range, it is because the event is outside the FOV and the photons have been absorbed by the shield. Following this procedure, we found 27 bursts inside the FREGATE FOV. From October 16, 2000 through September 30, 2001, the experiment had 116.7 days of observation (corrected for daytime) equivalent to an average duty cycle of 33.5%.

Assuming a 100% duty cycle, FREGATE would detect 81 bursts per year within its FOV. Therefore, the FREGATE burst rate for full sky coverage can be estimated to be $\approx 250$ GRB yr$^{-1}$ in the 6 - 400 keV energy range. At the end of this first year of operation, the duty cycle increased to 41.5% and remained stable [1]. At this value the expected number of events in FREGATE is 43-44 per year of which 33-34 are in the FOV.

Among the 35 events in our sample, only 5 exhibit a duration $T_{90} <2$ s. For this first year of operation, the observed ratio is therefore less than 15%. This proportion of short bursts is low when compared with the 25% observed by BATSE [5] or PHEBUS [2] but is in line with that observed by the BeppoSax WFC [3].

**TABLE 1.** Statistics on FREGATE burst list

| | |
|---|---|
| GRB candidates | 35 |
| Confirmed GRBs | 32 |
| Events in FOV | 27 |
| Triggered events | 21 |
| Localized events | 16* |
| Short events ($T_{90} < 2$ s) | 5 |
| Burst rate in FOV (GRB / yr) ($4\pi$ sr) | 250 |

* with WXM or IPN [4]

**TABLE 2.** FREGATE GRB list

| Event | Time (s) | Trig. number | FOV | Confirmed by | Location |
|---|---|---|---|---|---|
| GRB001102 | 57236 | No trigger | Yes | KONUS, ULYSSES, NEAR | IPN |
| GRB001105 | 59133 | No trigger | No | KONUS, ULYSSES, NEAR | IPN (GCN 874) |
| GRB001106 | 54915 | No trigger | Yes | KONUS, ULYSSES, NEAR | IPN |
| GRB001115a | 46146 | No trigger | Yes | KONUS | |
| GRB001115b | 50810 | No trigger | Yes | KONUS | |
| GRB001225 | 25759 | No trigger | Yes | SAX, ULYSSES, NEAR | IPN |
| GRB010126 | 33162 | # 1487 | Yes | RXTE, ULYSSES, NEAR, KONUS | IPN (GCN 921, 922, & 928) |
| GRB010213 | 45332 | No trigger | Yes | WXM | WXM (GCN # 934) |
| GRB010225 | 60733 | # 1491 | Yes | WXM | |
| GRB010308 | 56337 | No trigger | No | KONUS, ULYSSES, SAX | IPN |
| GRB010326 | 11701 | # 1495 | Yes | KONUS, ULYSSES, SAX | IPN (GCN 1014, 1015, & 1030) |
| GRB010326 | 30792 | # 1496 | Yes | KONUS, WXM | WXM (GCN 1018) |
| GRB010428 | 36291 | No trigger | Yes | ULYSSES | |
| GRB010607 | 53720 | Fregate only | Yes | ULYSSES, MARS-O, KONUS | IPN |
| GRB010609 | 20368 | No trigger | Yes | ULYSSES | |
| GRB010612 | 9194 | # 1546 | Yes | SAX, ULYSSES, KONUS, WXM | WXM,IPN (GCN 1065, 1066) |
| GRB010613 | 27235 | # 1547 | Yes | KONUS, ULYSSES, WXM | WXM,IPN (GCN 1067) |
| GRB010619 | 55038 | No trigger | No | SAX | |
| GRB010621 | 40382 | No trigger | Yes | Unconfirmed | |
| GRB010628 | 4204 | # 1569 | No | KONUS, ULYSSES | |
| GRB010629 | 44468 | # 1573 | Yes | KONUS, ULYSSES, SAX, WXM | WXM,IPN (GCN 1075, 1076) |
| GRB010706 | 20567 | No trigger | Yes | KONUS | |
| GRB010722 | 14512 | No trigger | Yes | KONUS | |
| GRB010726 | 5490 | # 1611 | No | KONUS, ULYSSES, MARS-O | IPN |
| GRB010801 | 66638 | # 1669 | No | ULYSSES, SAX, KONUS | |
| GRB010827a | 38910 | # 1723 | Yes | KONUS | |
| GRB010827b | 84560 | # 1724 | Yes | ULYSSES | |
| GRB010828 | 15505 | # 1729 | Yes | ULYSSES, KONUS | |
| GRB010903 | 84488 | # 1748 | Yes | ULYSSES | |
| GRB010917 | 56924 | # 1756 | Yes | Unconfirmed | |
| GRB010921 | 18950 | # 1761 | Yes | ULYSSES, SAX, WXM | WXM,IPN (GCN 1096, 1097) |
| GRB010923 | 33870 | # 1764 | Yes | SAX, ULYSSES, KONUS | IPN (GCN 1102, 1104) |
| GRB010928a | 26179 | # 1769 | No | Unconfirmed | |
| GRB010928b | 60826 | # 1770 | Yes | WXM, DMSP (GCN 1103) | |
| GRB010929 | 1358 | # 1771 | No | KONUS, ULYSSES | |

# FREGATE TRIGGER SENSITIVITY

The minimum flux necessary to trigger FREGATE depends on the spectral and temporal properties of the burst as well as on the background level and the characteristics of the instrument such as the trigger timescales and the detector response. These parameters may vary significantly from burst to burst, or from the direction of arrival, or the seasonally-changing background in the FREGATE FOV.

*Count rate needed to trigger.* The triggering procedure of FREGATE requires that counts recorded on one of the four trigger timescales i.e., 20, 164, 1310, and 5242 ms exceed the background level fluctuations by 4.5 to 6 σ in at least two detectors (see [1]). The background level is measured in two trigger energy ranges, i.e., LE = 6 - 80 keV and HE = 30 - 400 keV. During normal operation, the background level in the LE band increases by a factor of 2 or more when the Galactic Center is in the FREGATE field of view. To take into account the background variability we calculate the instrument sensitivity with the Galactic Center in the field of view (hereafter GC background) and without it (hereafter AC background). The variation of the time profile in the trigger windows is not accounted for and the burst is assumed to last longer than the trigger window.

*Burst counts.* To model the GRB spectrum, we chose a Band law with typical values for the parameters ($\alpha = -1, \beta = -2, E_0 = 150$ keV ). This burst spectrum is then convolved with the detector response matrix to determine the predicted count rate for the instrument in the trigger energy ranges as a function of angle.

*Trigger sensitivity.* The sensitivity of FREGATE is the minimum burst flux for which the resultant burst counts exceed the counts necessary to trigger. Figure 1 summarizes this study and displays the minimum fluxes necessary to trigger FREGATE for different background

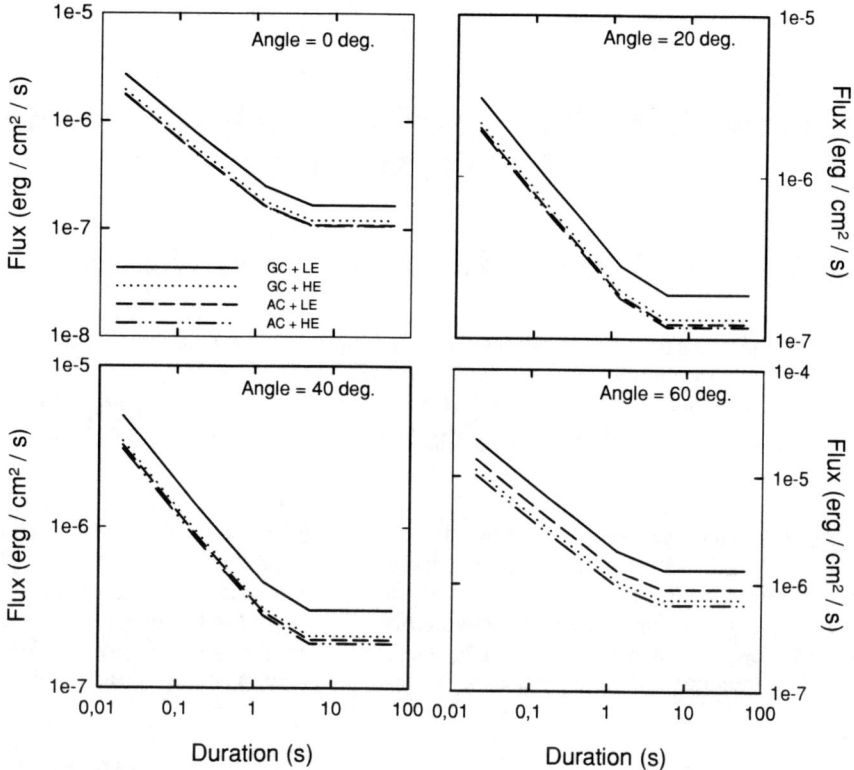

**FIGURE 1.** Sensitivity of FREGATE in the 6 - 400 keV for the two cases of background levels (with the Galactic Center (GC) in the FOV or outside AC) and the two cases of energy ranges (LE = 6 - 80 keV and HE = 30 - 400 keV), as a function of the duration of the trigger timescale.

conditions and several angles of arrival (with respect to the detector axis) as a function of the duration of the trigger window. The average sensitivity of FREGATE in the 6 - 400 keV range is on the order of $10^{-7}$ erg/cm$^2$/s for low angles and the longest triggering timescales. This result is in good agreement with preflight simulations.

*Comparison with BATSE.* In order to compare with BATSE we have also calculated the minimum flux to trigger FREGATE in the energy range 50 - 300 keV. For an average angle of 20° and the 1310-ms trigger criterion the sensitivities are

- $2.0 \times 10^{-7}$ erg/cm$^2$/s (1.1 phot/cm$^2$/s) for GC + HE
- $1.8 \times 10^{-7}$ erg/cm$^2$/s (1.0 phot/cm$^2$/s) for AC + HE

Assuming a BATSE burst detection rate of 666 GRBs yr$^{-1}$ full sky, the number of events in the 4B Catalog [5] having a peak flux in the 50 - 300 keV energy range on the 1024 ms timescale greater than 1.1 photons/cm$^2$/s is compatible with a burst detection rate of $\approx 270$ GRBs yr$^{-1}$ full sky. This rough comparison indicates that our result is in good agreement with BATSE data.

## CONCLUSION

In this paper, we show that the GRB sensitivity of FREGATE is approximately $10^{-7}$ erg/cm$^2$/s in the 6 - 400 keV energy range and the burst detection rate, when corrected for the duty cycle and the limited field of view, is approximately 250 yr$^{-1}$. These results are in good agreement with preflight simulations and with the results of past experiments. The FREGATE sensitivity, over the predicted lifetime of HETE2, will yield several hundred bursts allowing us to improve the statitics on short bursts and on X-ray rich events.

## REFERENCES

1. Atteia, J-L., et al., 2002, these proceedings.
2. Dezalay, J-P., et al., 1996, ApJ, 471, L27.
3. Gandolfi, G., et al., 2002, these proceedings.
4. Hurley, H., et al., 2002, these proceedings.
5. Paciesas, W.S., et al., 1999, ApJS, 122, 465.

# Localization of GRBs by Bayesian Analysis of Data from the HETE WXM

Carlo Graziani[*], Donald Q. Lamb[*] and The HETE Science Team[†]

[*]*Department of Astronomy & Astrophysics, University of Chicago, 5640 South Ellis Avenue, Chicago, IL 60637*
[†]*An international collaboration of institutions including MIT, LANL, U. Chicago, U.C. Berkeley, U.C. Santa Cruz (USA), CESR, CNES, Sup'Aero (France), RIKEN, NASDA (Japan), TESRE (Italy), INPE (Brazil), TIFR (India)*

**Abstract.** We describe a new method of transient point source localization for coded-aperture X-ray detectors that we have applied to data from the HETE Wide-Field X-Ray Monitor (WXM). The method is based upon the calculation of the likelihood function and its interpretation as a probability density for the transient source location by an application of Bayes' Theorem. The method gives a point estimate of the source location by finding the maximum of this probability density, and credible regions for the source location by choosing suitable contours of constant probability density. We describe the application of this method to data from the WXM, and give examples of GRB localizations which illustrate the results that can be obtained using this method.

## INTRODUCTION

The HETE Wide-Field X-Ray Monitor (WXM) is composed of two crossed, one-dimensional coded-aperture cameras. GRB locations are inferred from position histograms in the X and Y cameras, using the known mask pattern and detector response.

Several processing methods for coded-aperture data have been proposed, including cross correlation [1], least-squares fitting [2], and Maximum Entropy [3, 4]. Skinner and Nottingham [5] have described a maximum-likelihood fitting technique.

In [6] we presented a Bayesian scheme for analyzing coded-aperture data from such transient events. The method is based on the calculation of the joint likelihood function for two stretches of data: the stretch covering the transient event itself, and a stretch before and/or afterwards, which provides information about the background. We interpret the likelihood thus obtained as a posterior probability density for the transient event location by an application of Bayes' Theorem.

We have implemented the method in the HETE data-processing pipeline, where it is routinely used to obtain GRB locations. Here we present the implementation details, and some of the results that have been obtained to date.

## IMPLEMENTATION

When high-resolution WXM data from a HETE trigger are received on the ground, background and burst time intervals are automatically determined by SNR maximization. The software then essentially compares the background-to-burst change in the WXM position histogram data with a Monte Carlo point source model, using the method described in [6] to calculate the posterior probability density at each interrogated source location.

Computation of the Monte Carlo point source model (or "template") is somewhat time-consuming, so that it is not possible to pre-compute templates in a dense grid spanning all possible locations in the field-of-view. We have therefore adopted a two-stage approach:

- *Coarse Location* — coarsely spaced pre-computed templates are compared to the data to produce a location accurate to about 1/3°.
- *Fine Location* — a freshly-generated template at the coarse location is shifted around the immediate neighborhood of that location, to map the posterior probability distribution.

### Coarse Location

The WXM is composed of two 1-dimensional coded-aperture cameras that resolve locations in perpendicular directions (X and Y). We can use 1-dimensional arrays of

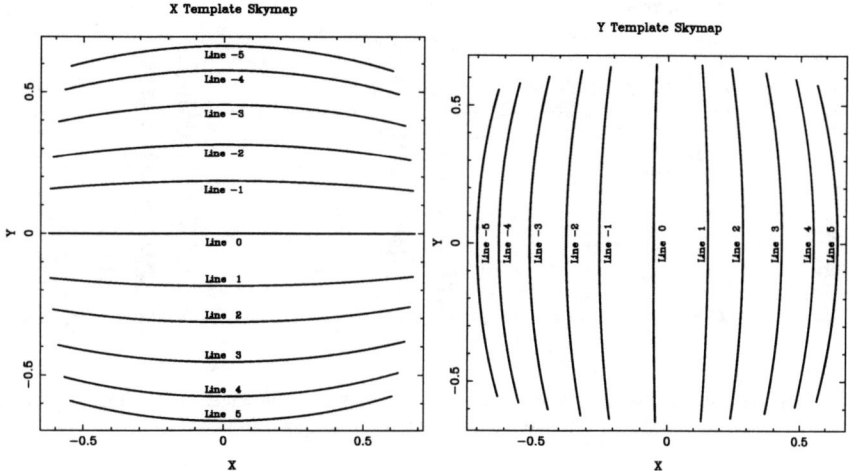

**FIGURE 1.** Lines of Templates for coarse location. The plots are equal-area projections, with an opening half-angle of about 40 degrees.

sky locations with pre-computed templates in order to get coarse X and Y locations. Given such an array, the code simply computes the posterior density at each location and plots it as a function of template number, reporting the highest value.

One such array that we occasionally use has 22 lines of templates — 11 each in X and in Y. The lines are chosen so as to expose select combinations of adjacent WXM detector wires (the detector in each WXM camera has six wires, some of which might not be illuminated by a point source at a certain location). This allows the X-detector to supply some Y information, and vice-versa. Figure 1 shows a skymap of this template configuration.

Another template array is a simple cross - two lines of sky locations, one in the X direction and one in the Y direction, intersecting at the center of the FOV. This array uses less information, but can be run much faster, and is usually adequate for the purpose of obtaining a coarse localization.

In both cases the spacing between adjacent templates on a line is about 18', which is the fundamental resolution element of WXM. This is thus also the accuracy of a coarse location obtained in this fashion.

Figure 2 shows examples of the output from the coarse location procedure, for GRB010326B and GRB010629. Each set of two panels shows the log-likelihood profile as a function of template number, for the X and Y template lines. The two panels on the left show the result of the analysis of GRB010326B performed using the full 22-line template array. The two panels on the right show the result of the analysis of GRB010629 performed using the cross template array. In both cases the best-fit X and Y angles are marked by a vertical line.

## Fine Location

The posterior density function allows us to go beyond merely obtaining a point estimate of the location — we can also use it to infer *contours* of prescribed probability content around the best-fit location. That is, we can use it to get error boxes.

We proceed by calculating one very bright Monte-Carlo template at the best-fit coarse location. We fit that template with an empirical model constructed by smearing and transforming the coded mask pattern. This transformation is designed to give a simplified account of physical processes such as scattering and detector penetration. An example of such a fit is shown in Figure 3, which was produced in the analysis of GRB010629.

We then calculate the posterior density on a $20 \times 20$ grid of points in the neighborhood of the coarse location, using at each point a template calculated by smearing the mask pattern with the best-fit empirical parameters determined by the fit of the model to the MonteCarlo template.

We issue not only a refined best-fit location, but also constant-density contours containing 68.3%, 95.5%, and 99.7% (statistical) probability of including the correct location. Figure 4 shows the location contours produced in this way for GRB010629.

For the purposes of quick communication, and to simplify follow-up observations, we also produce "circularized" error-boxes by calculating the radius of a circles about the best-fit location that just contain 68.3%, 95.5% and 99.7% (statistical) probability of including the correct location. Since these are not constant probability contours, they necessarily subtend slightly larger solid angles than the iso-density contours. The difference is usually not very large.

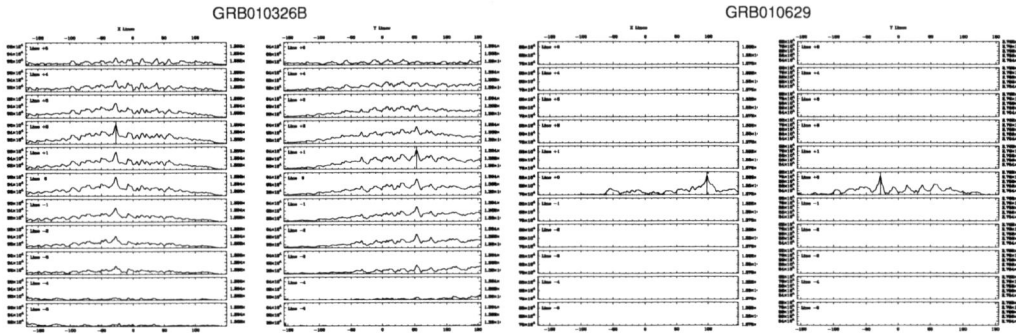

**FIGURE 2.** Coarse location log-likelihood profiles. Each set of two panels shows X and Y localizations. Left: GRB010326B, 22-line array of templates. Right: GRB010629, "Cross" configuration of templates.

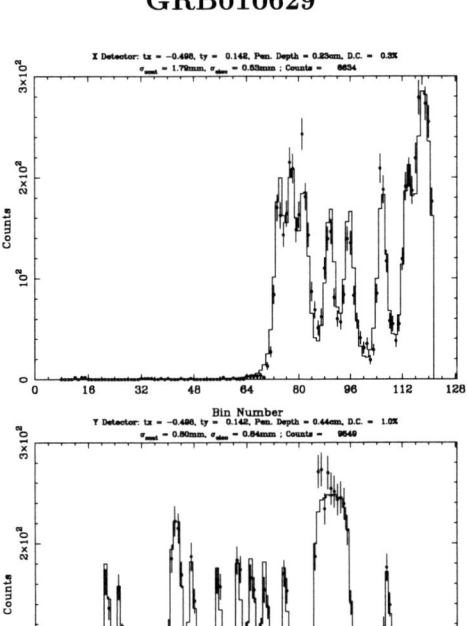

**FIGURE 3.** Fit of empirical model to Monte Carlo simulation. The assumed direction of the source is the coarse location of GRB010629.

The statistical probabilities inferred here must be supplemented by the systematic location error deduced from our calibration studies.

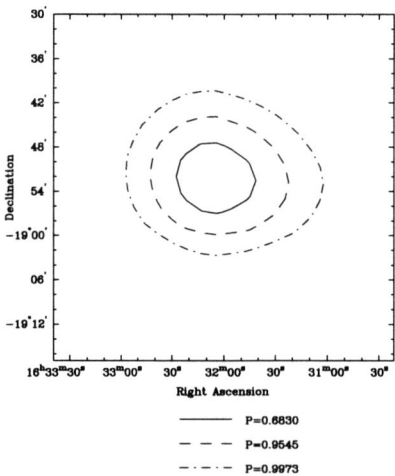

**FIGURE 4.** Skymap showing location probability contours obtained by the fine location of GRB010629.

# REFERENCES

1. Fenimore E. E., *Appl. Opt.* **17**, 3562 (1978).
2. Doty J. P. "The All Sky Monitor for the X-ray Timing Explorer," in *X-ray Instrumentation in Astronomy II*, edited by L. Golub, SPIE Conference Proceedings 982, Bellingham, Washington, 1988, p. 164.
3. Sims M., Turner M.J.L., and Willingale R., *Space Sci. Instrum.* **5**, 109 (1980).
4. Willingale R., Sims M., and Turner M.J.L., *Nucl. Instrum. Methods Phys. Res.* **221**, 60 (1984).
5. Skinner G. K., and Nottingham M. R., *Nucl. Instrum. Methods Phys. Res.* **A333**, 540 (1993).
6. Graziani, C., Lamb, D. Q., and Slawinski, R. "Determination of X-Ray Transient Source Positions by Bayesian Analysis of Coded Aperture Data," in *All-Sky X-Ray Observations in the Next Decade*, edited by M. Matsuoka and N. Kawai, RIKEN, Wako, Japan, 1997, pp. 303-308.

# A "Spiffy" Trigger for Gamma-Ray Bursts

Carlo Graziani

*Department of Astronomy & Astrophysics, University of Chicago, 5640 South Ellis Avenue, Chicago, IL 60637*

**Abstract.**
The traditional design of trigger algorithms for GRB experiments requires the specification of the background and burst samples in terms of acquisition times that are of fixed duration and of fixed elapsed time from each other. One such set of acquisition times is required for each characteristic timescale of GRB variation that one desires to detect. One then slides each set through the trigger data searching for samples that maximize the signal-to-noise of the background-subtracted burst sample.

Here we describe a new triggering approach in which the times at which the background and burst samples are acquired are allowed to vary dynamically. Two background samples bracket a burst sample. The background and burst durations and elapsed time between them are allowed to be free parameters, which are maximized using the downhill simplex method. This produces great flexibility in the timescales that are available for detecting GRBs.

## INTRODUCTION

The search for untriggered GRBs is has been an active field of research at least since the public release of BATSE data. Most recently, Kommers et al. [1] and Stern et al. [2] have described searches of BATSE data directed at revealing GRBs that occurred without being detected by the BATSE on-board triggers, either for operational reasons or because their spectral or temporal morphologies were poor fits to the on-board trigger criteria.

Naturally, the same considerations apply to GRB detection by HETE. The HETE mission deploys an unprecedentedly varied set of trigger criteria — the FREGATE DSP trigger [3] uses four timescales and operates in two energy bands, while the WXM XG trigger [4, 5] is typically configured to apply thirty or so criteria, some on WXM data and others on FREGATE data. Nevertheless the variety of GRB morphologies, and operational considerations, can result in GRBs that are not detected in flight. It is important to develop a strategy to mine HETE survey data for such untriggered GRBs.

Typical ground searches for untriggered bursts use detection methods that largely mirror on-board trigger algorithms [1, 2]. Background and burst samples are specified in terms of acquisition times that are of fixed duration and of fixed elapsed time from each other. Each GRB timescale — risetime or duration — is probed by a different fixed choice of these parameters. Each such fixed set of time windows is then swept through the time series being probed for transient events, searching for samples that maximize the signal-to-noise of the background-subtracted burst sample.

This scheme has the disadvantage of being rather inflexible about the the timescales that are probed. This inflexibility is especially troublesome when seeking weak signals, for which inaptly chosen burst or background samples may lead to a signal dilution that prevents detection.

In this work we describe an alternative approach that has been quite successful in identifying extremely weak events. In this approach, the background and burst samples are treated as free parameters, which are varied using the downhill simplex method of Nelder & Mead [6] to maximize the signal-to-noise ratio of the background-subtracted burst sample.

## IMPLEMENTATION

The operation of the code, `spiffy-trigger`, is illustrated in Figure 1.

The trigger operates as a simplified "bracket trigger" [4, 5], in that the background is estimated using samples before and after the burst sample. The background is assumed constant, so no interpolation (linear or otherwise) is performed to obtain the background rate during the burst sample.

This restriction is not an essential feature of the method, but rather merely a simplification. The two background intervals are restricted to remain equidistant from the burst sample, so as to prevent the maximization procedure from exploiting a monotonic increase or de-

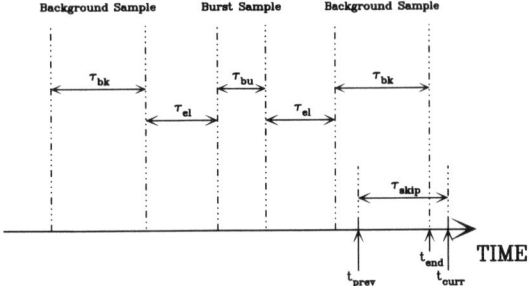

**FIGURE 1.** Illustration of the definition of the time parameters that are varied to search for transient events.

crease in the background rate to estimate an erroneously low background, by driving one of the background samples to a region of lower background without driving the other to a region of higher background.

The code operates on a time-series of integer counts. It advances a trigger window of fixed duration through the time series by steps of size $\tau_{skip}$. It sets up a burst sample interval, of duration $\tau_{bu}$, bracketed at an elapsed time $\tau_{el}$ by two background sample intervals of duration $\tau_{bk}$, the second of which ends at time $t_{end}$. It calculates the SNR for the burst sample, assuming a background rate calculated by a weighted average of the count rates in the two background intervals.

The SNR is computed as follows: Assume for the sake of generality that the two background accumulation times may differ, so that we accumulate $n_{bk1}$ counts in the first background during an accumulation time $\tau_{bk1}$, and $n_{bk2}$ counts in the later background accumulation time $\tau_{bk2}$. Denoting the estimated background counts during the burst sample by $\mu_{bu}$, and assuming the Gaussian approximation to the Poisson distribution, it is a straightforward exercise in Gaussian estimation to show that

$$\mu_{bu} = \frac{\tau_{bk1} + \tau_{bk2}}{\tau_{bu}} \Sigma^2, \quad (1)$$

$$\Sigma^2 = \tau_{bu}^2 \left( \frac{\tau_{bk1}^2}{n_{bk1}} + \frac{\tau_{bk2}^2}{n_{bk2}} \right)^{-1}, \quad (2)$$

where $\Sigma^2$ is the variance in the estimate $\mu_{bu}$.

Denote by $n_{bu}$ the counts that we accumulate during the burst sample. Then the net signal in the burst sample is $s = n_{bu} - \mu_{bu}$. The variance in $s$ is the sum of $\Sigma^2$ and the variance in $n_{bu}$. Triggering is essentially hypothesis testing, with the null hypothesis consisting of the assumption that the count rate in the burst sample is the same as what is estimated using the background samples. Thus the appropriate choice for the variance of $n_{bu}$ is "model variance", that is $\sigma_{bu}^2 = \mu_{bu}$. Thus the SNR of the burst sample is

$$\text{SNR} = \frac{n_{bu} - \mu_{bu}}{\left( \mu_{bu} + \Sigma^2 \right)^{1/2}}. \quad (3)$$

This is the quantity that `spiffy-trigger` endeavors to maximize.

The code uses the simplex method to vary the four parameters $t_{end}$, $\tau_{bk}$, $\tau_{el}$, and $\tau_{bu}$, which are viewed by the simplex minimization routine as continuous parameters. A very lax convergence criterion is imposed — the absolute variation of the SNR must be less than 0.1 across the simplex — because in triggering there is no point in determining the SNR to great accuracy, and because we don't want to spend many CPU cycles chasing noise.

The parameter $t_{end}$ is constrained to be later than the end of the trigger window in the previous invocation. Consequently, the arrangement of burst and background samples "accordions out" backwards in time from the current time, without repeating choices of intervals made during previous iterations.

When there is no transient event in the data, the simplex will typically not wander very far from its initial configuration. On the other hand if there is a transient event, and the initial simplex includes a vertex corresponding to a configuration in which the burst sample even partially includes the event, the simplex will rapidly climb the SNR slope, dynamically adjusting its timescales until the event is well-bracketed.

Since the simplex does not wander far if it doesn't find much at the outset, it is important to ensure that $\tau_{skip}$ is not so large that a short event may "fall between the cracks" — that is, fail to have any of its constituent time samples included in a burst sample probed by the initial simplex. It is therefore a good idea to ensure that at simplex initialization, $\tau_{bu} > \tau_{skip}$ for at least one of the simplex vertices. This ensures that every data sample passes through the burst sample of at least one initial simplex parameter vertex.

Constraints on the time parameters are imposed by making the SNR function return a large negative value when the constraints are violated. The previously-discussed constraint on the parameter $t_{end}$ is enforced in this way. The code also uses this parameter-constraint mechanism to prevent intervals from encroaching upon each other, to ensure that $\tau_{bk}$, $\tau_{el}$, and $\tau_{bu}$ remain positive-valued, and to keep all intervals inside the current trigger window.

Other useful constraints that it is good practice to enforce are a minimum value for $\tau_{el}$ (so that burst and background samples are well-separated), a minimum duration for $\tau_{bk}$ (so as to minimize the risk of the background nestling into a low fluctuation), and a maximum

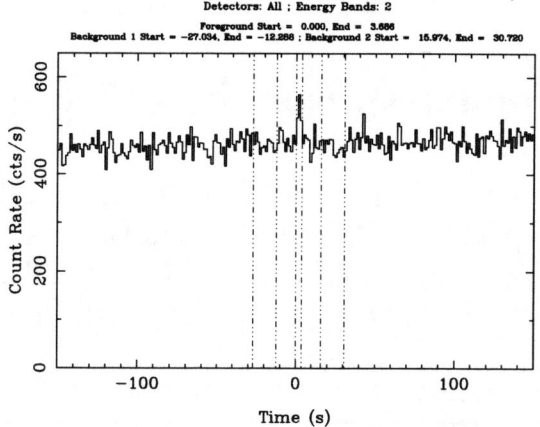

**FIGURE 2.** GRB detected by the trigger robot on 19 June 2001 at 15:17:01 UTC. The SNR of this event was 6.1, the duration was 3.7 s. This event was also detected by BeppoSax.

duration $\tau_{bu}$ (so as to minimize the risk of triggering on very long duration trends in the background).

## DEPLOYMENT

`spiffy-trigger` is currently used in three different contexts within the HETE project:

- The Chicago ground location pipeline [7] uses `spiffy-trigger` to identify the burst sample time with maximal signal-to-noise in the WXM data. This sample is used throughout the subsequent location analysis.
- A robot script that runs after every downlink uses `spiffy-trigger` to search for untriggered bursts in FREGATE band C (40-300 keV) 1.3s resolution survey data. During normal HETE operation, it tends to see about 1 possible GRB per week, above and beyond detecting all triggers picked up in flight that are sufficiently hard, and long (or short but bright) to register at this timescale and in this energy band. Figure 2 shows an example of such an event, which was confirmed by BeppoSax.
- The general untriggered burst search described by Butler & Doty [8] uses `spiffy-trigger` in parallel to Butler & Doty's wavelet trigger, and runs on all survey data products. GRB011212 was in fact detected on the ground in this pipeline, by both the wavelet algorithm and by `spiffy-trigger`.

## CONCLUSIONS

The `spiffy-trigger` algorithm can probe a wide spectrum of burst timescales. It is still possible that initialization with a very short $\tau_{bu}$ might miss a very long, slow-rising event, or that a very long initial $\tau_{bu}$ might cause the SNR of a weak, short event to be too diluted to register before convergence is reached. However, careful choice of the range of $\tau_{bu}$ spanned by the initial simplex can address this issue to a large extent.

In any event, the algorithm may be re-run with radically different initial values of $\tau_{bu}$. For example, re-running the algorithm three times, with $\tau_{bu}$ set initially to 0.1s, 3s, and 100s — with suitably chosen initial simplices — one may probe a range of timescales that would probably require hundreds of criteria for a traditional trigger algorithm to examine.

In principle, there is no reason the `spiffy-trigger` algorithm could not be deployed in flight in a future mission. The floating-point operations that it performs are not particularly expensive, particularly for modern space computing hardware. While more complex than a traditional trigger, it is not vastly more so, and its complexity is offset by its great flexibility, configurability, and dynamic range of burst timescales to which it is sensitive.

## REFERENCES

1. Kommers, J. M., Lewin, W. H. G., Kouveliotou, C., van Paradijs, J., Pendleton, G. N., Meegan, C. A., and Fishman, G. J., *ApJ* **134**, 385 (2001)
2. Stern, B. E., Tikhomirova, Y., Kompaneets, D., Svensson, R., & Poutanen, J., *ApJ*, **563**, 80, 2001
3. Atteia, J-L., Boer, M., Cotin, F., Couteret, J., Dezalay, J-P., Ehanno, M., Evrard, J., Lagrange, D., Niel, M., Olive, J-F., Rouaix, G., Souleille, P., Vedrenne, G., Hurley, K., Ricker, G., Vanderspek, R., Crew, G. Doty, J. and Butler, N. "In-Flight Performance and First Results of FREGATE", these proceedings, (2002)
4. Fenimore, E. E., and Galassi, M. "The HETE Triggering Algorithm", in *Gamma-Ray Bursts in the Afterglow Era*, edited by E. Costa, F. Frontera, and J. Hjorth, Springer, Berlin, 2001, pp. 393-395.
5. Tavenner, T., Fenimore, E., Galassi, M., Vanderspek, R., Preger, B., Graziani, C., Lamb, D., Kawai, N., Yoshida, A., Shirasaki, Y., and Tamagawa, T. "The Effectiveness of the HETE-2 Triggering Algorithm", these proceedings, (2002)
6. Nelder, J. A. and Mead, R. "A Simplex Method for Function Minimization" *Comput. J.* **7**, 308-313, (1965)
7. Graziani, C., and Lamb, D.Q., "Localization of GRBs by Bayesian Analysis of Data from the HETE WXM", these proceedings (2002)
8. Butler, N. and Doty, J. "Using Wavelets to Detect HETE Untriggered Bursts", these proceedings, (2002)

# FREGATE observation of a strong burst from SGR1900+14

J-F. Olive[*], K. Hurley[†], J-P. Dezalay[*], J-L. Atteia[*], C. Barraud[*], N. Butler[**], G. B. Crew[**], J. Doty[**], G. Ricker[**], R. Vanderspek[**], D. Q. Lamb[‡], N. Kawai[§], A. Yoshida[‖], Y. Shirasaki[††], T. Sakamoto[‡‡], T. Tamagawa[§§], K. Torii[§§], M. Matsuoka[¶¶], E. E. Fenimore[‡‡], M. Galassi[‡‡], T. Tavenner[‡‡], T. Q. Donaghy[‡] and C. Graziani[‡]

[*]*Centre d'Etude Spatiale des Rayonnements, CNRS/UPS, 31028 Toulouse Cedex 04, France*
[†]*UC Berkeley Space Science Laboratory, Berkeley, CA 94720-7450*
[**]*Massachusetts Institute of Technology, Center for Space Research, Cambridge, MA, US*
[‡]*Department of Astronomy and Astrophysics, University of Chicago,
5640 South Ellis Avenue, Chicago, IL 60637.*
[§]*Department of Physics, Tokyo Institute of Technology, 2-12-1 Ookayama, Meguro-ku, Tokyo 152-8551, Japan.*
[‖]*Department of Physics, Aoyama Gakuin University, Chitosedai 6-16-1 Setagaya-ku, Tokyo 157-8572, Japan*
[††]*National Astronomical Observatory, Osawa 2-21-1, Mitaka, Tokyo 181-8588 Japan.*
[‡‡]*Los Alamos National Laboratory, P.O. Box 1663, Los Alamos, NM, 87545.*
[§§]*RIKEN (Institute of Physical and Chemical Research), 2-1 Hirosawa, Wako, Saitama 351-0198, Japan.*
[¶¶]*Tsukuba Space Center, National Space Development Agency of Japan, Tsukuba, Ibaraki, 305-8505, Japan.*

**Abstract.**
After a long period of quiescence, the soft gamma repeater SGR1900+14 was suddenly reactivated on April 2001. On July 2$^{nd}$ 2001, a bright flare emitted by this source triggerred the WXM and FREGATE instruments onboard the HETE-2 satellite. Unlike typical short ($\sim 0.1$ s) and spiky SGRs recurrent bursts, this event features a 4.1 s long main peak, with a sharp rise ($\sim 50$ ms) and a slower cutoff ($\sim 250$ ms). This main peak is followed by a $\sim 2$ sec decreasing tail. We found no evidence of any precursor or any extended 'afterglow' tail to this burst. We present the preliminary spectral fits of the total emission of this flare as observed by the FREGATE instrument between 7 and 150 keV. The best fit is obtained with a model consisting of two blackbody components of temperatures 4.15 keV and 10.4 keV. A thermal bremsstrahlung can not be fitted to this spectrum. We compare these features and the burst energetics with the other strong or giant flares from SGR1900+14.

## INTRODUCTION

With only four (maybe five) objects known, the Soft Gamma Repeaters (SGRs, see [7] for a review) are rare sources. They undergo repeated, unpredictable periods of intense activity during which they emit few hundreds of brief ($\sim 0.1$ s) and intense ($\sim 10^3 - 10^4\ L_{Edd}$) bursts of soft $\gamma$-rays. Besides these **classic** bursts, very rarely, SGRs emit **giant** flares[1], which are extremely energetic (typically $\sim 10^{44}$ erg) and much longer events, lasting for several minutes. The SGR active phases usually last only from a few weeks to a few months. They are separated by long periods (from years to decades) of quiescence during which the SGRs are detected as persistent soft X-ray sources associated with supernova remnants.

Among the few Soft Gamma Repeaters, SGR1900+14 is a kind of prototype. This source was first detected in 1979 [20] when it burst 3 times in 2 days. Its activity resumed in 1992 [14] and afterwards in May 1998 [15], [8] and April 2001 [5], [10], [11]. During the quiescence state, ASCA and RXTE observations revealed a low luminosity ($\sim 3 \times 10^{34}$ ergs s$^{-1}$) soft X-ray source with a periodicity of 5.16 s and a high value of period derivative[2] ($\sim 6 \times 10^{-11}$ s s$^{-1}$, [9], [17], [27]). SGR1900+14 lies just outside G42.8+0.6, a $10^4$-year-old galactic su-

---

[1] So far, only two such giant flares have been detected : the famous events of March 5 1979 from SGR0526-66 and August 27 1998 from SGR1900+14.

[2] A similar spin period and rapid spindown is also measured for SGR 1806-20 [18]

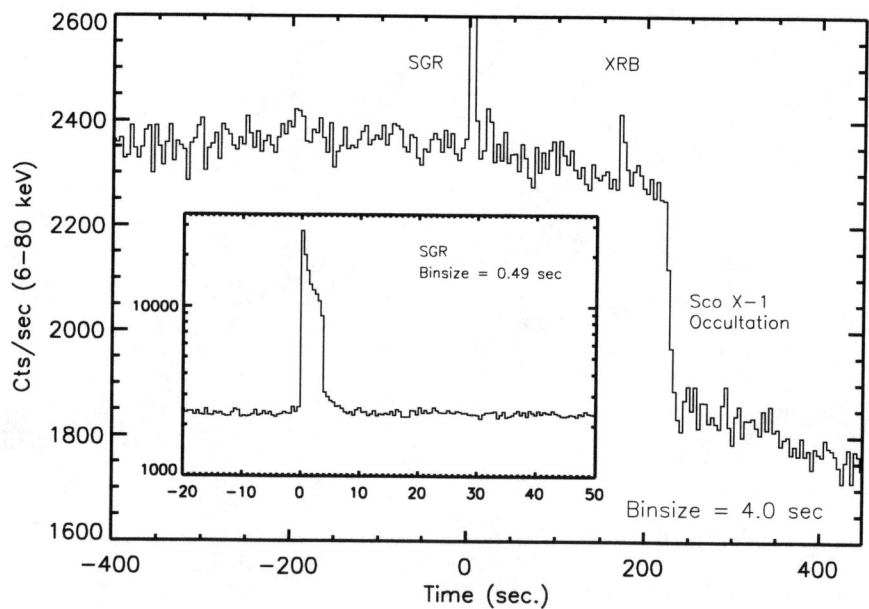

**FIGURE 1.** The FREGATE 6–80 keV time history centered on the trigger time. The peak is cut off on the linear plot.

pernova remnant. It is plausible that the SGR is a young neutron star, born in this supernova. If the spindown of the neutron star is due to magnetic braking, assuming purely dipole radiation, the inferred magnetic field is found to be $B \sim 8 \times 10^{14}$ G. All these indices support the hypothesis formulated in 1992 that SGR are strongly magnetized young neutron stars (i.e. *magnetars* [2]). In a magnetar, the energy of the bursts are drawn from the magnetic field energy which dominates all other sources of energy including the neutron star rotation [25].

The 2001 reactivation period of SGR1900+14 started with a burst detected by *Ulysses* on April 17 [11] and a strong $\sim 40$ s flare detected by *BeppoSax* the day after [5], [10]. In June-July 2001, SGR1900+14 was very active and entered the antisolar field-of-view of the HETE instruments. Many short **classic** bursts were detected at that time and are reported elsewhere. At 03:34:06.53 UTC on 2 July, the HETE-2 FREGATE and WXM instruments detected and localized a strong burst lasting 4.1 s from SGR1900+14 [24]. In this paper, we report on the preliminary timing and spectral analysis of this strong burst as observed with FREGATE. The WXM data on this burst are also presented in these proceedings [26].

## The FREGATE observations

*Large scale timing analysis*

The four identical units of X/$\gamma$–ray detectors onboard HETE-2, named FREGATE, are sensitive to photons between 6 and 400 keV (see [1] for a full description of the experiment and operating modes). Using the continuous time history data (resolution 0.16 s in 4 energy bands) we have searched indices for the SGR activity before (i.e. precursors) and after (i.e. extended tail or 'afterglows') the main peak for which the instruments triggered. For that purpose, we have built light curves in various energy bands and with various time resolutions. We didn't find any such features. As an example, Figure 1 shows the FREGATE 6–80 keV time history centered on the trigger

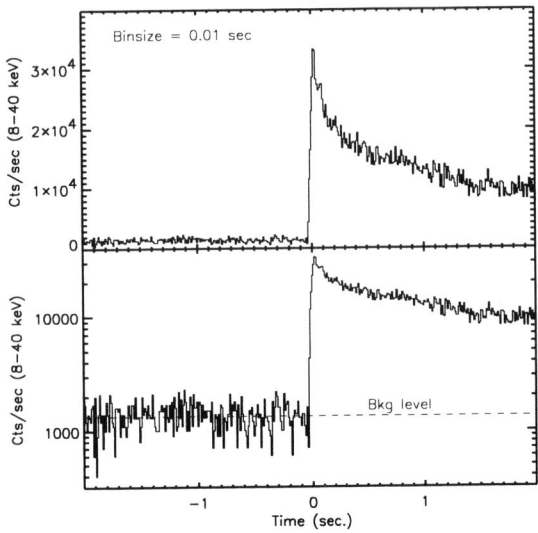 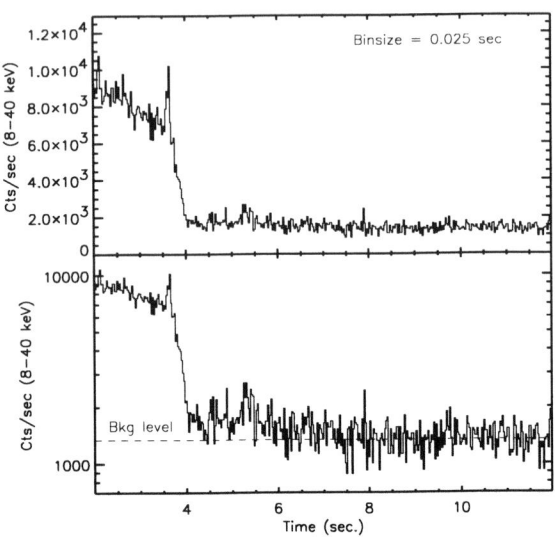

**FIGURE 2.** The SGR flare light curves in 8-40 keV energy in a linear scale (top) and logarithmical scale (bottom)

time. On the large scale plot (more than 10 minutes with 4 s resolution), we see that the count rate goes back to the background level immediatly after the 4.1 s flare (labelled 'SGR' on the plot). Three other features to note: 170 s after the trigger the very soft (i.e. no photons above 30 keV) and short ($\sim$ 5 s) excess labelled 'XRB' on the plot is probably due to a galactic X-ray burster. Unfortunatly this burst was outside the WXM field-of-view and was not localized. About four minutes after the trigger, the large step visible in this light curve is due to an Earth occultation of Sco X-1. Finally, about 20 s after the SGR peak, the slight increase in the count rate (marginally significant) is possibly due to the variability of Sco X-1 or to a background fluctuation.

*The SGR flare*

The inset in Figure 1 shows a zoom with a logarithmic scale on the peak of the burst, with 0.49 s time resolution. With the exception of the main peak, the only noticeable feature is a short $\sim 2 - 3$ s tail rapidly decreasing to the background level after the peak.

A zoom on the first part of the burst is shown in Figure 2 (left). This light curve has been constructed with the 'Burst Data' of FREGATE (256k photons tagged in time with a resolution of 6.4 $\mu$s and in 256 energy channels for each detector). The dashed line represents the linear interpolation of the background level taken for times less than 2 s before and greater than 10 s after the trigger time. Several important features can be seen in this plot. First, no short precursors are detected during the few seconds before the main peak. The rise of the burst is quasi-linear in time and is fully resolved ($\sim$ 50 ms). This sharp rise is followed by a short ($\sim$ 20 ms) spike occuring at the peak of the flare. Then the intensity starts to decrease in a complex way. A zoom on the last part of the burst is shown in Figure 2 (right). The main peak terminates with a second short spike (lasting $\sim$ 100 ms) immediately followed by a rapid decrease of the intensity lasting for $\sim$ 0.35 s. After the main peak, a short spiky tail lasting for $\sim$ 2 s is visible.

## The spectrum and energetics of the flare

As a first spectral analysis, we have built the spectrum of the whole main peak (4.05 s duration) and a background spectrum for 10 s of data before the burst using the 'Burst Data' of FREGATE. The deconvolution matrices and the energy-to-channel relations for that date have been computed using the method validated with the Crab observations of FREGATE [21]. A 2 % systematic error has been added to the statistical errors to account for the calibration uncertainties affecting this high-level spectrum. We have tried with XSPEC to fit the 7-150 keV FREGATE spectrum with several simple spectral models widely used for such bursts.

During these trials, the spectrum readily appeared to feature two main characteristics : (I) it is strongly curved below 20 keV, (II) it extends up to 150 keV. None of the single component models (thermal bremsstrahlung OTTB, powerlaw PL, blackbody BB, broken powerlaw, powerlaw with an exponential cutoff, etc...) provides an acceptable fit over the whole FREGATE range. As an example, we show in Figure 3 (right) the unfolded spectrum derived with an OTTB model for energies above 25 keV. If the fit is marginally acceptable ($kT_{br}$ = 24.8 keV, $\chi^2$ = 1.49 for 47 dof), the model totally fails to reproduce the low energy part of the spectrum below 15 keV and the situation is even worse with the other simple models.

Among the composite models that we have tried (OTTB+BB, BB+PL, etc..) only the sum of two blackbody components (2BB) produces a good fit ($\chi^2$ = 1.127 for 66 dof) over the full energy range (see Figure 3, left). The temperatures we derived are $kT_1$ = 4.15 ± 0.1 keV and $kT_2$ = 10.4 ± 0.4 keV.

The flux of the burst was computed by extrapolating and integrating the fitted 2BB model in the energy band > 25 keV. We have computed the luminosity and the total energy for both spectral components and the total emission assuming an isotropic emission at a source distance of 10 kpc. All these results are reported in Table 1 and compared to those of 3 major bursts from the source in the same energy range.

## DISCUSSION AND CONCLUSIONS

On August 27, 1998 *Konus-Wind*, *Ulysses*, and *BeppoSAX* detected a **giant** flare from SGR1900+14 [8], [3], [19]. Its luminosity of $1.41 \times 10^{41}$ ergs s$^{-1}$ (above 25 keV) and exceptional duration ($\sim$ 6 min) made this event the most intense burst ever detected at Earth, thousands of times more energetic than the other bursts of this source. The pulsations at 5.16 s were clearly detected during the burst.

Only two days later (August 29), a strong bright burst was detected simultaneously by BATSE and RXTE (named, after [13], the **unusual** burst). This event exhibits a 3.5 s burst peak preceded by complex ($\sim$ 1 s) precursor and followed by a long ($\sim 10^3$ s) tail modulated at the 5.16 s pulsation period. Its luminosity of $6.43 \times 10^{40}$ ergs s$^{-1}$ is a factor of only 2 less than the giant August 27 burst. Nevertheless, due to its shorter duration, the total energy carried by this burst is much less by a factor $\sim$ 200.

On April 18, 2001 *BeppoSAX* was triggered by an intense X-ray burst from SGR1900+14 [5], [10]. The event, also detected by *Ulysses*, lasted $\sim$ 40 s and was modulated with the 5.16 s period. Unfortunately, no spectral data are available for this burst. Nevertheless, assuming an optically thin thermal bremsstrahlung spectral model with $kT \sim$30 keV [4] the inferred 25 − 100 keV fluence is $\sim 6.5 \times 10^{-6}$ ergs s$^{-1}$ cm$^{-2}$ ($\sim 7.8 \times 10^{40}$ ergs s$^{-1}$ at 10 kpc) that is similar to the previous burst. Nevertheless, considering its energetics and unusual duration, this event has been qualified as an **intermediate** burst.

The morphology of the July 2$^{nd}$, 2001 flare as observed with FREGATE resembles the August 29, 1998 **unusual** flare. Its duration is similar and the total energy carried by the burst above 25 keV is a factor of only 3–4 less. It is not surprising that we did not detect any precursor or pulsed afterglow at the level reported with *RXTE* [13]. If we scale down these features by a factor of $4 \times S_{HETE}/S_{RXTE} \sim 160$, they are undetectable against the somewhat large FREGATE background due to its extended field-of-view.

The main difference between the two bursts resides in the energy spectrum of the main peak. The **unusual** burst was so bright that the *RXTE*-PCA detectors was saturated during the majority of the peak. Nevertheless, the burst rise and burst falloff spectra could be fitted with a classical OTTB model[3] with temperatures of 17.2 ± 2 keV and 15.4 ± 2.5 keV respectively. The BATSE spectrum of this burst could also be fitted above 25 keV by an OTTB model with kT = 20.6 ± 0.3 keV. Both analyses combined suggest a nonvarying spectrum for the *unusual* burst [13]. The spectrum presented here can also be fitted by an OTTB model **above 25 keV** but it is much more curved than this model at lower energies. There is no doubt that such a feature would have been detected with *RXTE* if present for the unusual burst. Tentatively we have tried to fit our spectrum with a composite model consisting of two blackbodies with temperatures $kT_1$ = 4.15 ± 0.1 keV and $kT_2$ = 10.4 ± 0.4 keV. The equivalent radii for an isotropic emission are 25 and 3.5 km respectively (at 10 kpc). These values are suggestive of emission regions close to the stellar surface but our modelling is probably too crude to draw any definite conclusion. Many effects, such as the anisotropy of the heat flow through an ultramagnetized neutron star envelope, the reprocessing by a light element atmosphere and the general relativity correction can modify a thermal spectrum near a magnetar surface leading to different values of temperatures and radii [23].

A detailed discussion of the FREGATE observations of this burst in the framework of the magnetar model is in preparation and will be published in the near future [22].

## REFERENCES

1. Atteia J-L. *et al.* 2002, these proceedings
2. Duncan, R. C. & Thompson, C. 1992, ApJ, 392, L9
3. Feroci M. *et al.* 1999, Apj, 515, L9
4. Guidorzi C. *et al.* 2001, IAU Circular No 7611

---

[3] Alternatively a PL+BB model also gives a good fit with $kT \sim$ 2.4 − 2.5 keV and $\gamma \sim$ 1.2 − 1.6.

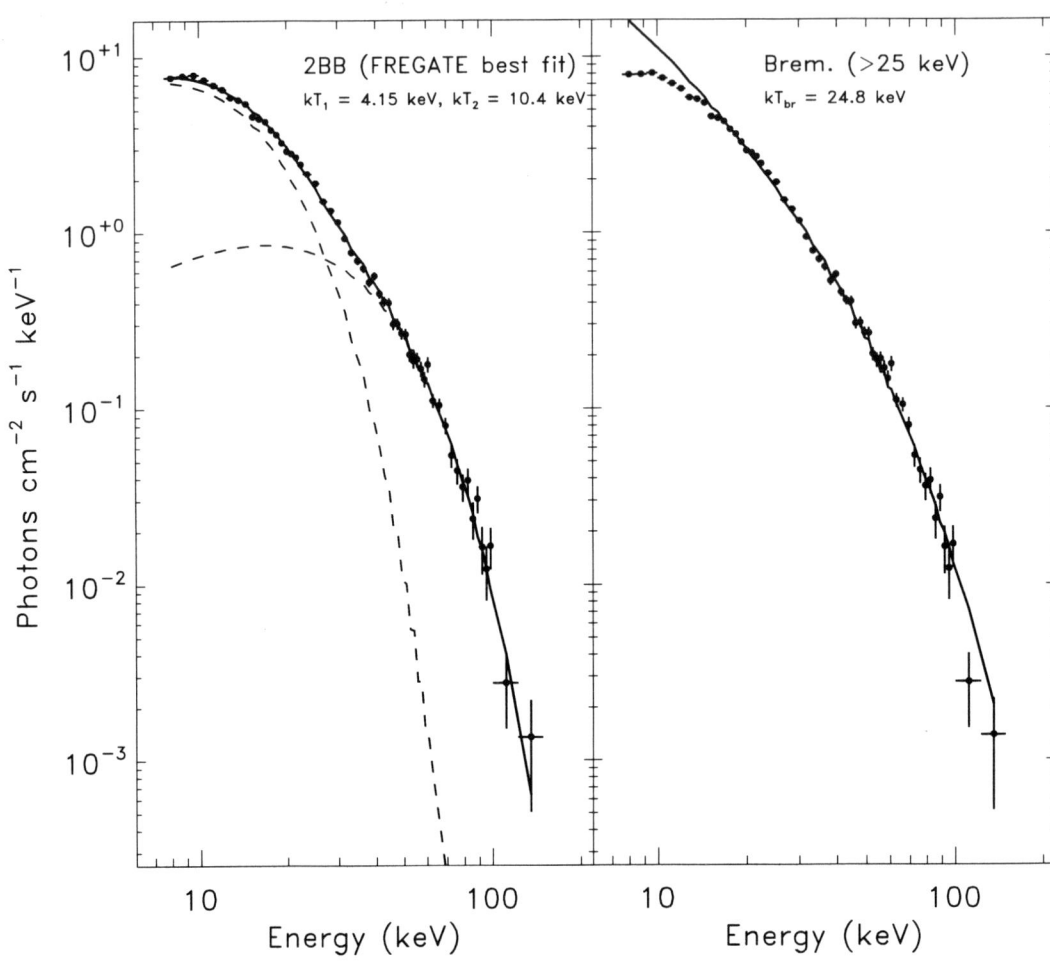

**FIGURE 3.** The unfolded total emission spectrum of the July 2 2001 burst from SGR1900+14 seen by FREGATE. Right : a thermal bremsstrahlung above 25 keV can roughly reproduce the data but totally fails below ∼ 15 keV. Left : using the best fit composite model with 2 blackbody in the extended range 7–150 keV

**TABLE 1.** Summary for the energetics of the SGR1900+14 July 2$^{nd}$ flare using the best spectral model (a sum of two blackbodies) in the range > 25 keV. For the calculations of the luminosity and total energy $E_{tot}$ we assume a source distance of 10 kpc. These parameters are compared to those of 3 major bursts from the source in the same energy range

| Burst | Component | Flux $10^{-6}$ ergs s$^{-1}$ cm$^{-2}$ | Luminosity $10^{40}$ ergs s$^{-1}$ | $E_{tot}$ $10^{40}$ erg |
|---|---|---|---|---|
| Jul. 2$^{nd}$ 2001 flare | 4.15 keV BB | 0.30 | 0.36 | 1.45 |
|  | 10.4 keV BB | 1.10 | 1.33 | 5.40 |
|  | 4.05 s total burst | 1.40 | 1.69 | 6.85 |
| August 27, 1998 (giant) | ∼ 370 s burst |  | 14.1 | 5200 |
| August 29, 1998 (unusual) | ∼ 3.5 s burst |  | 6.43 | 22.5 |
| April 18, 2001 (intermediate) | ∼ 40 s burst |  | 7.8 | 313 |

5. Guidorzi C. *et al.* 2001, GCN 1041
6. Hurley, K. *et al.* 1999, ApJ, 510, L111
7. Hurley K. 2001, AIP Conf. Proc., 599, 160, Bologna.
8. Hurley, K. *et al.* 1999a, ApJ, 510, L107
9. Hurley, K. *et al.* 1999b, ApJ, 510, L111
10. Hurley K. *et al.* 2001, GCN 1043
11. Hurley K. *et al.* 2001, GCN 1045
12. Hurley, K. *et al.* 1999b, Nature 397, 41
13. Ibrahim A. I. *et al.* 2001, Apj, 558, 237
14. Kouveliotou, C. *et al.* 1993, Nature, 362, 728
15. Kouveliotou, C. *et al.* 1998, IAU Circular no. 6929, June 3, 1998
16. Kouveliotou, C. *et al.* 1998, Nature, 393, 235
17. Kouveliotou *et al.* 1999, ApJ, 510, L115
18. Kouveliotou, C. *et al.* 1998b, Nature, 393, 235
19. Mazets, E.P. *et al.* 1999, Astron. Lett., 25, 635
20. Mazets, E.P. *et al.* 979, Soviet Astron. Lett., 5(No.6),343
21. Olive J.F. *et al.* 2002, these proceedings
22. Olive J.F. *et al.* 2002, in preparation
23. Perna R. *et al.* 2001, ApJ, 557, 18
24. Ricker G. *et al.* 2001, GCN 1078
25. Thompson, C. and Duncan, R. C. 1995, MNRAS, 275, 255
26. Torii K. *et al.* 2002, these proceedings
27. Woods, P.M. *et al.* Kouveliotou, C., 1999, ApJ, 524, L55

# In-Flight Verification of the FREGATE Spectral Response

J-F. Olive*, J-P. Dezalay*, J-L. Atteia*, C. Barraud*, N. Butler[†], G. B. Crew[†], J. Doty[†], G. Ricker[†] and R. Vanderspek[†]

*Centre d'Etude Spatiale des Rayonnements, CNRS/UPS, 31028 Toulouse Cedex 04, France
[†]Massachusetts Institute of Technology, Center for Space Research, Cambridge, MA, US

**Abstract.**
We present the first results of the in-flight validation of the spectral response of the FREGATE X/γ detectors on–board the HETE–2 satellite. This validation uses the Crab pulsar and nebula as reference spectra.

## INTRODUCTION

The four identical FREGATE X/γ–ray detectors onboard HETE-2, are sensitive to photons between 6 and 400 keV (see [1] for a full description of the experiment and operating modes). In this short paper, we describe briefly the Monte Carlo simulations of the detectors, the ground calibrations, and the validation of the in-flight performances using the Crab pulsar and nebula as standards. We compare the spectral parameters obtained with FREGATE with those reported by other instruments in similar energy ranges.

## SIMULATIONS AND CALIBRATIONS

In parallel with the construction of the detectors, the energy response matrix of FREGATE was calculated from extensive Monte Carlo simulations of the detector using CERN's GEANT package. More than 50 regions of different materials were included in the simulation with special care for the regions in the detector field-of-view (e.g. the graded shield and the beryllium window). The parameters (position, angles, energy, etc.) of the incoming photons are generated within a separate code including several options (point-like radioactive source at a finite distance with several photons and branching ratios, parallel flux for celestial sources, etc.). Once the photon characteristics are generated, the tracking code is the same in all cases. The Monte-Carlo simulations were checked before launch with the help of ground calibrations using a large set of radioactive sources (9 sources with photons energies spanning from 8 to 1300 keV) and angles of incidence from on-axis to 60° in 5° steps. We have first determined realistic gains and resolutions for our detectors. Then, the simulated spectra (in a point-like configuration) have been folded with the detector response functions and normalized knowing the source activity for the given energy, the duration of the calibration run and the solid angle presented by the detector to the photon flux. Finally, we compared the simulated spectra with those from the calibrations. For all angles and energies, we found that the relative differences in the full energy peak and in the diffusion continuum were less than 10 %, which is on the order of the uncertainties of the source activity (see Figure 1 for few examples of comparison). Since we concentrate in this paper on the validation of the in-flight performances, we will not discuss these comparisons any further.

## IN-FLIGHT CALIBRATIONS

Since launch, the gains of each detector have been continuously monitored using on-board calibration $^{133}$Ba sources (two lines at 81 and 356 keV). The positions of these lines show a weekly orbital variation of a few percent, and a long-term drift on a ~ 100 day scale periodically corrected by adjusting the high voltage commands. No significant variation of the relative line width has been found. These in-flight calibrations are used to find the correct channel-to-energy relations during the HETE mission. FREGATE generates two types of scientific data with good spectral resolution: permanently the 'Spectral Data' (**SD**: four 128 channel energy spectra every 5 or 10 seconds); and when a trigger occurs : the 'Burst Data' (**BD**: 256k photons tagged in time with a resolution of 6.4 $\mu$s in time and in 256 channels in energy). The process of spectral deconvolution for both data types has been tested during the in-flight verification phase, using the Crab pulsar (for the BD) and nebula (for the SD). The counting rates in the 6–200 keV range obtained from the

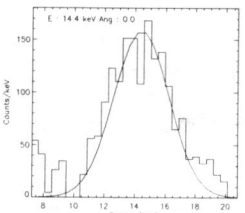 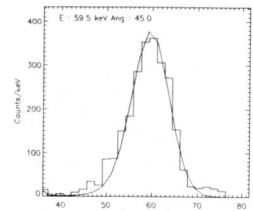 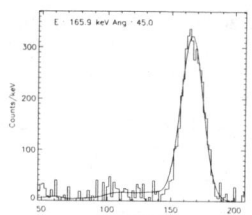 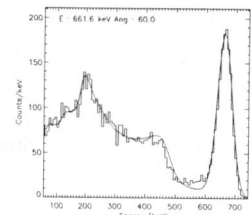

**FIGURE 1.** Comparison of experimental spectra of several radioactive sources from ground calibrations (histograms) and the corresponding simulated spectra (solid lines). From left to right : $^{60}$Co (14 keV, 0°), $^{241}$Am (59 keV, 45°), $^{139}$Ce (166 keV, 45°) and $^{137}$Cs (662 keV, 60°). The simulations are normalized using the source activity for the given energy, the duration of the calibration run, and the solid angle of the detector seen by the source. In other words, it is an absolute comparison.

four detectors have been rebinned in the same histogram to improve the statistical significance. We have checked that fitting the four spectra simultaneously gives the same spectral parameters within the error bars. As usual for the Crab, we used a single power law model :

$$\frac{dN}{dE} = A_{30} \, E_{30}^{-\gamma} \; \text{ph cm}^{-2} \, \text{s}^{-1} \, \text{keV}^{-1} \qquad (1)$$

where $E_{30}$ is defined as $E_{30} = E/30$ keV

### The Crab pulsar

The data consist of 10 artificial burst triggers recorded on 01/07/2001, when the Crab was about 23° off axis. Considering that the triggers are separated by only a few minutes, the photon times were not barycenter corrected. We have constructed the folded light-curves (with a 33 phase bin epoch folding) for periods around the expected Crab period ([3]). For each light-curve, we have computed the $\chi^2_{red}$ value. For the maximal value of this statistical parameter ($\chi^2_{red} = 10.1$) we have found the best pulsation period ($\nu$ =29.832951 Hz). This value is $\sim 10^{-3}$ Hz smaller than the extrapolated value from the ephemeris ($\nu_0$ =29.834086 Hz) which is compatible with a Doppler shift due to the earth's motion in the solar system and the satellite motion in its orbit. The corresponding phasogram (Figure 2) consists of two peaks of similar intensity, separated by $\sim 0.4$ in phase and connected by an interpulse. Even if the peaks are a little broad (due to the lack of barycentric corrections), this phasogram looks very similar to the ones reported at X and $\gamma$ energies. The off-pulse level and spectrum were obtained for phases outside the pulsed emission (dashed level, Figure 2).

We obtained a good fit with the spectral model described above ($\chi^2_{red} = 1.1$ for 86 dof). The rebinned unfolded spectrum can be seen in Figure 4. The spectral parameters are reported in Table 1. The spectral index ($1.87 \pm 0.13$) is fully consistent with those reported in similar energy ranges. The amplitude ($A_{30}$) is about 10 %

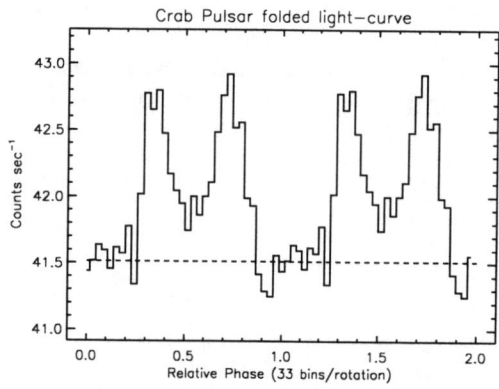

**FIGURE 2.** The Crab pulsar light-curve folded at the best pulsation period. Two periods are shown.

lower than the canonical value (although roughly consistent within the error bars). This can be explained because, in absence of timing corrections, the FREGATE light curve is smoothed and our definition of the total pulsed interval (0.6 in phase) is greater than the one usually used ($\sim 0.5$-0.55).

### The Crab Nebula

The Crab was periodically occulted by the Earth in the FREGATE field-of-view with an inclination angle less than 10° on December, 2000. The occultation dates (Crab rise and Crab set) were calculated. From the SD, we have extracted sets of 80 successive spectra centered on the occultation dates. With a screening procedure, the data polluted with solar flares or high background were eliminated, leading to 90 'good' occultations. Then we have added all these 90 occultations (spectrum by spectrum), to get an 'averaged' set of 80 successive spectra centered on the Crab steps (a Crab set is time-reversed before

**TABLE 1.** Summary for Crab pulsar and Crab nebula spectral parameters (FREGATE results and [5]). The spectral index is represented by $\gamma$. The parameter $A_{30}$ (intensity at 30 keV) is in units of $10^{-3}$ ph cm$^{-2}$ s$^{-1}$ keV$^{-1}$. While given by the authors at a different energy, this amplitude has been recalculated using the quoted best-fit slope.

| Reference | Energy range | Pulsar (phase averaged) $A_{30}$ | $\gamma$ | Nebula (total emission) $A_{30}$ | $\gamma$ |
|---|---|---|---|---|---|
| FREGATE | 6-200 keV | **0.89 ± 0.13** | **1.87 ± 0.13** | **7.23 ± 0.2** | **2.16 ± 0.03** |
| [5] | 15-130 keV | 1.04 | 2.06 ± 0.3 | 7.05 | 2.18 ± 0.04 |
| [2] | 15-180 keV | – | – | 7.46 | 2.06 ± 0.01 |
| [4] | 20-200 keV | 1.06 | 1.92 ± 0.09 | 7.48 | 1.94 ± 0.02 |
| [6] | 20-250 keV | 1.16 | 1.96 ± 0.05 | 7.18 | 2.19 ± 0.02 |

adding). Next, the light-curve for each of the 128 energy channels has been fitted with a polynomial function of fourth degree to account for the background orbital variation plus a centered step to account for the Crab contribution in the channel (see an example in Figure 3). Thus, the Crab spectrum is built channel-by-channel for each of the four FREGATE detectors.

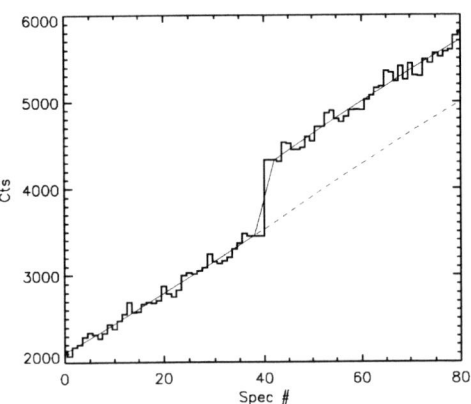

**FIGURE 3.** Example of light curve obtained by folding the Spectral Data of FREGATE with respect to the Crab occultation times. The amplitude of the step represents the number of counts due to the Crab nebula.

Again, we obtained a good fit with a power law model ($\chi^2_{red} = 1.19$ for 84 dof, $\gamma = 2.16 \pm 0.03$) with spectral parameters fully consistent with those reported in similar energy ranges (see the unfolded spectrum in Figure 4 and the parameters in Table 1).

## CONCLUSION

For both the Crab pulsar and nebula spectra, the spectral parameters derived with FREGATE are fully consistent with the canonical values. This demonstrates our capability to perform a detailed spectral analysis of sources within the FREGATE field-of-view, even for low statistics spectra such as the pulsed spectrum presented here.

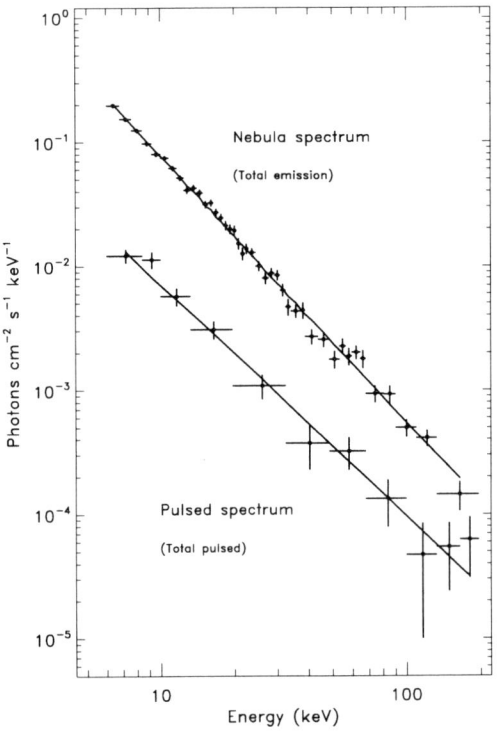

**FIGURE 4.** Unfolded spectrum of the total nebula emission and the phase averaged Crab pulsed emission

## REFERENCES

1. Atteia J-L. et al., 2002, these proceedings.
2. Jung G.V., 1989, *Ap. J.*, 338, 972.
3. Jodrell Bank Pulsar Timing Results Monthly Ephemeris // (http://www.jb.man.ac.uk/pulsar/crab.html)
4. Hasinger G. et al., 1984, Adv. Space Res., 3, No 10-12, 63.
5. Ubertini P. et al., 1994, *Ap. J.*, 421, 269.
6. Strickman M. et al., 1979, *Ap. J.*, 230, L15.

# In-Flight Characterization of the HETE Soft X-Ray CCD Cameras

G. Prigozhin*, J Villasenor*, R. Vanderspek*, G. Jernigan[†], J. Doty*, G. Crew* and G. Ricker*

*Center for Space Research, Massachusetts Institute of Technology
[†]Space Sciences Laboratory, University of California, Berkeley

**Abstract.** We have developed a set of software tools that allow to monitor the performance of the flight X-ray CCD cameras as soon as data arrive at MIT. An emission line at 5.9 keV from the on-board Fe-55 radioactive calibration source is clearly visible in the spectra and provides the means to measure the gain and the noise for each observation in each of the 4 CCD chips in operation. Both parameters can change with time, depending on the phase of the moon and the amount of light leaking into the system. Time vs. position scatter plots were found to be an extremely powerful tool in understanding of the device performance. They illustrate the evolution of the light leaks produced by the dark Earth at the beginning and the end of each orbit. With a bright X-ray source in the field of view the shadow of the mask projected on the surface of the CCD clearly shows the motions of the spacecraft.

## INTRODUCTION

HETE-2 is equipped with two orthogonally oriented CCD cameras which allow celestial X-ray sources to be localized with high precision (approximately 30″). A detailed description of the cameras and burst localization algorithms can be found in several papers presented at this conference [1, 2]. In this paper we describe monitoring of the routine operation of the Soft X-ray CCD cameras (SXC) while spacecraft is orbiting the Earth and sends down non-burst data produced by the instrument.

SXC is in operation only during orbit night, when no light from the Sun can compromise its performance. Because of that a natural chunk of data is an event list corresponding to one orbit night. The data telemetered down from the spacecraft contain only "events", or pulse height of the signal above the bias.

## IN FLIGHT CCD PERFORMANCE

One of the most informative ways to monitor CCD performance is to look at a scatter plot of pulse height as a function of column number (a plot of one orbit worth of data for the 4 CCD chips is shown on Fig. 1). Each dot in such a plot corresponds to an event above the threshold. It clearly shows $Mn\ K_\alpha$ line in all 4 CCD sectors produced by the in-flight calibration source (radioactive

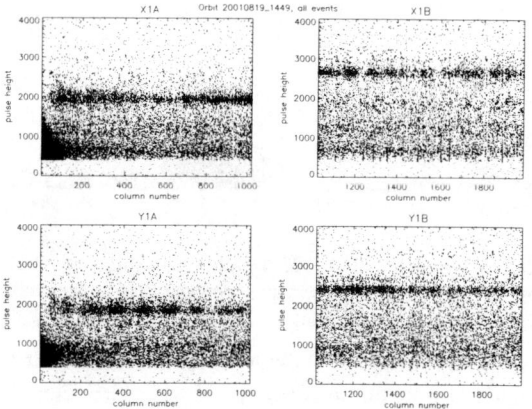

**FIGURE 1.** Scatter plots for 4 different CCD chips.

$Fe^{55}$). The lumpiness of the characteristic line photon distribution across the columns is caused by the shadow mask between the source and the CCDs. The dark area at low column numbers in sectors X1A and Y1A is due to light leaks at the beginning and at the end of each orbit. These light leaks are obvious in the corresponding light curves (see Fig. 2) for this particular orbit. The light contamination can only be seen when some part of the Earth surface is above the top plane of the SXC and is likely to be caused by city lights.

The plots on Fig. 3 show typical examples of the

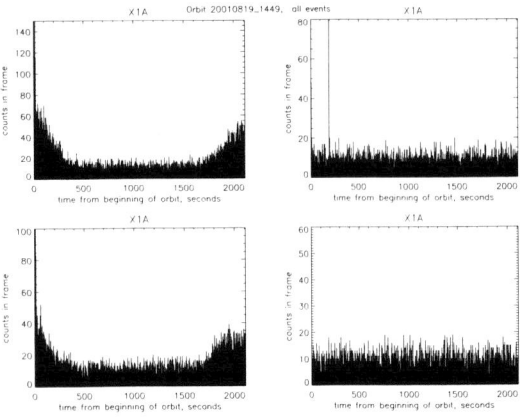

**FIGURE 2.** Lightcurves for one orbit night in 4 different CCD chips.

spectra in each of the 4 CCD sectors. Sharp cutoff at

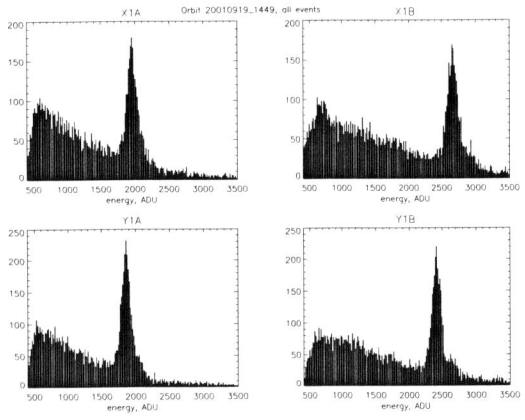

**FIGURE 3.** Spectra of the X-ray photons collected by 4 different CCD chips during the same orbit night as in previous figures.

low energies is caused by the beryllium filter and by the drop of CCD quantum efficiency below 1 keV due to absorption in polysilicon gates.

Strong $Mn$ line from the calibration source provides a reference point for measuring gain and energy resolution for every detector. The unresolved $Mn\,K_\beta$ line is also visible in the spectra. Measurements of the $Mn\,K_\alpha$ line are being done regularly for every set of data corresponding to one orbit night. The plots on Fig. 4 show the location of the line centroid and on Fig. 5 the width (FWHM) of the line as a function of time for 3 months of operation for each of the CCD chips. The gaps in the data indicate the full Moon intervals when the SXC is turned off. Line width near the time of the full Moon increases significanly due to light leaks and associated increase of noise.

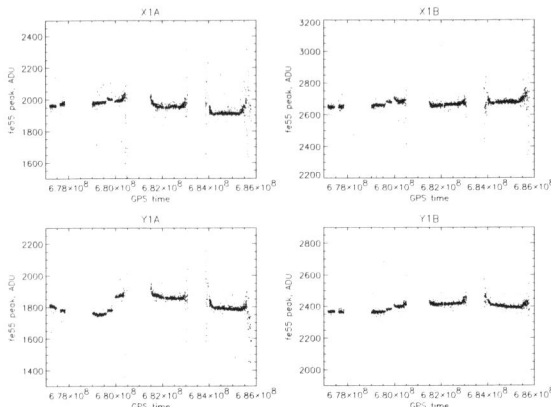

**FIGURE 4.** Location of $Mn\,K_\alpha$ line centroid as a function of time for 3 months of operation.

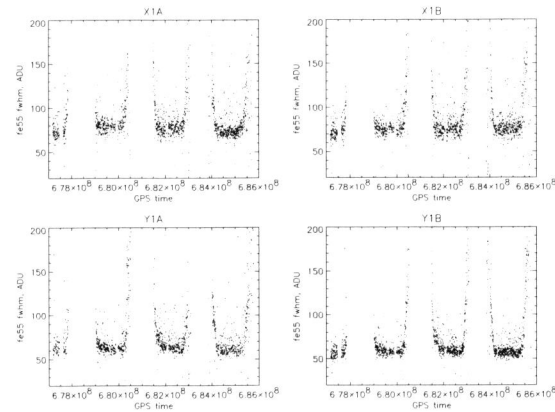

**FIGURE 5.** FWHM of $Mn\,K_\alpha$ line as a function of time for 3 months of operation.

## POSITION PLOTS

Extremely helpful in understanding of the status of the CCD instrument turned out to be the scatter plots showing time of an event vs. position of an event. An example of such a plot on Fig. 6 (again, each dot corresponds to an event above threshold) for the same orbit 20010819_1449 as the plots above on Figures 1, 2, 3 illustrates the evolution of the light leak during orbit night. Data points from CCD sectors X1A and X1B are combined into one plot for camera X1, sectors Y1A and Y1B are glued together into Y1 plot. Light contaminates smaller and smaller number of columns as satellite goes deeper into the Earth's shadow. Near the orbit dawn both cameras start seeing increasing amount of light again.

Several vertical dark streaks correspond to "warm" columns with higher generation rate of electrons. Disappearance of some of them closer to the orbit end results from the spacecraft cooling during the orbit night.

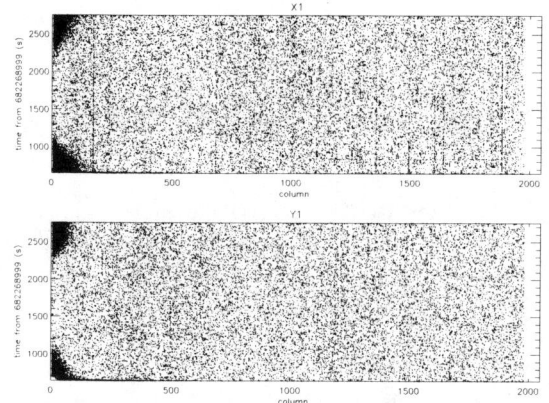

**FIGURE 6.** Time vs. position scatter plot for orbit 20010819_1449, showing light contamination at the beginning and at the end of the orbit night.

Another example of such a plot with a strong X-ray source in the field of view clearly illustrates the motion of the spacecraft (see Fig. 7). The pattern formed on the

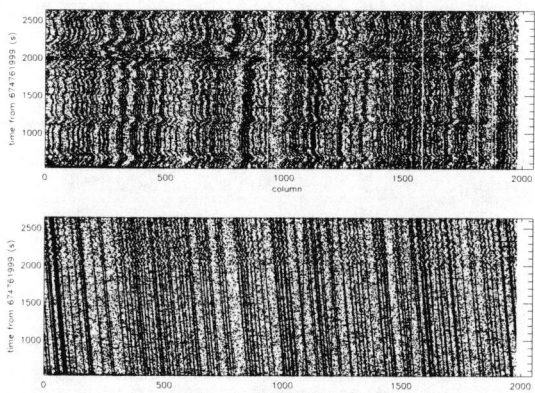

**FIGURE 7.** Time vs. position scatter plot with strong X-ray source (*Sco* X-1) in the field of view shows shadow mask pattern on the CCD drifting due to spacecraft movements.

surface of the CCD by the X-rays passing through the shadow mask shifts with time as the spacecraft changes its orientation relative to the strong source. The plot below corresponds to the period when *Sco* X-1 was in the field of view of both cameras. Since the columns of the CCD of the Y camera (and the slits of the shadow mask) are oriented perpendicular to the spacecraft spin wheel, the motions seen by this camera (lower plot) are slow and steady. The X camera, oriented perpendicular to the Y, sees a lot of fast movements caused by the actions of the Attitude Control System. Near the end of the orbit Y camera sees the nutation of the satellite.

## CONCLUSION

The plots shown above are produced for each orbit immediately after the data arrive from the primary ground stations to the MIT and are archived and displayed on the spacecraft's Web site. This allows to quickly evaluate the current status of the SXC cameras and helps to monitor the spacecraft health.

## REFERENCES

1. J. Villasenor et al., *These proceedings*.
2. G. Monnelly et al., *These proceedings*.

# X-Ray Bursts Observed by the HETE-2 Satellite

T. Sakamoto*, D. Takahashi†, N. Kawai*, A. Yoshida†, Y. Shirasaki**, T. Tamagawa‡, K. Torii‡, M. Matsuoka**, E. Fenimore§, M. Galassi§, D. Q. Lamb¶, C. Graziani¶ and HETE-2 team

*Graduate school of Science and Engineering, Tokyo Institute of Technology
†Department of Physics, Aoyama Gakuin University
**National Space Development Agency of Japan (NASDA)
‡Cosmic Radiation Laboratory, The Institute of Physical and Chemical Research (RIKEN)
§Los Alamos National Laboratory
¶Department of Astronomy & Astrophysics, University of Chicago

**Abstract.** In the period that the Galactic center region was in the field of view of HETE-2 satellite from May to September 2001, it detected bursts from well-known X-ray bursters (XRBs). More than 130 events were localized with the Wide-Field X-ray Monitor (WXM) on HETE-2 in these four months to the XRBs including X 1728-34, SAX J1750-29, Aql X-1, NGC 6624, and GS 1826-238. Localization accuracy is better than 18 arc-minutes for most of the events. In this paper, we summarize the XRBs detected by HETE-2 and show the light curves of the several events.

## INTRODUCTION

The Galactic center was observed by HETE-2 satellite from May to September 2001. In this region, there are many Low Mass X-ray binaries (LMXB) which cause the bursting activities known as X-ray bursts (XRBs). HETE-2 detected more than 135 XRB events during this period. These events are important for verifying the localization capability of Wide-Field X-ray Monitor (WXM) [1] on board in HETE-2 which is the primary instrument for determining the position of celestial events. These data are also of interest for studying the activity of X-ray bursters. In this paper, we summarize the XRBs detected by HETE-2, localization errors and display some light curves of several events.

## RESULTS

We localized all the potential XRB events using the RIKEN localization algorithm [2] used for ground analysis of GRBs. The localized celestial positions are compared with the XTE ASM catalog. Our sample includes 53 triggered events and 82 untriggered events.

We list the XRB candidate in table 1. Two events could not be associated with known X-ray sources; they may be extragalactic X-ray rich GRBs.

As we can see in figure 1, the most of the XRBs events are localized within 0.2 degrees of known XRB sources. Note that events with variety of statistics are sampled in these plots.

**TABLE 1.** The summary of the number of the XRBs detected by HETE-2.

| XBRs | Number of events |
|---|---|
| GS 1826-238 | 38 |
| X1728-34(GX354-0) | 30 |
| Aql X-1 | 17 |
| NGC 6624 | 11 |
| SAXJ1750-29 | 11 |
| X1916-053 | 9 |
| X1812-121 | 5 |
| SL1735-269 | 4 |
| NGC 6652 | 3 |
| X1724-307 | 2 |
| X1702-429 | 1 |
| M 15 | 1 |
| GCX-1 | 1 |
| unidentified | 2 |
| Total | 135 |

## REFERENCES

1. Kawai, N., et al., 2002, these proceedings.
2. Shirasaki, Y., et al., 2002, these proceedings.

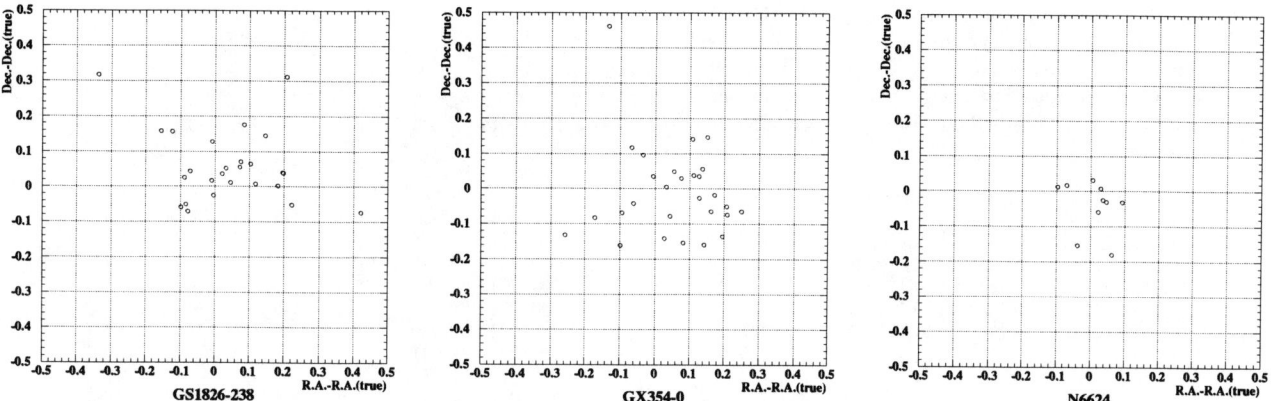

**FIGURE 1.** Localization errors are plotted for several XRB sources. The horizontal (vertical) axis is the difference in R.A. (Dec.) between WXM localization and the catalog position (XTE ASM catalog). The center (0, 0) corresponds to a location consistent with the catalog. The plots are produced only for the events within 0.5 degrees of the sources.

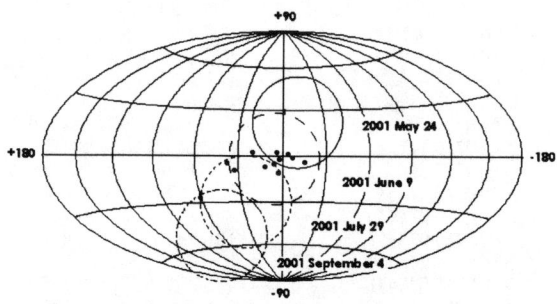

**FIGURE 2.** The filled circles are XRBs detected by HETE. The open circles are the field of view of WXM at the date described near the circle.

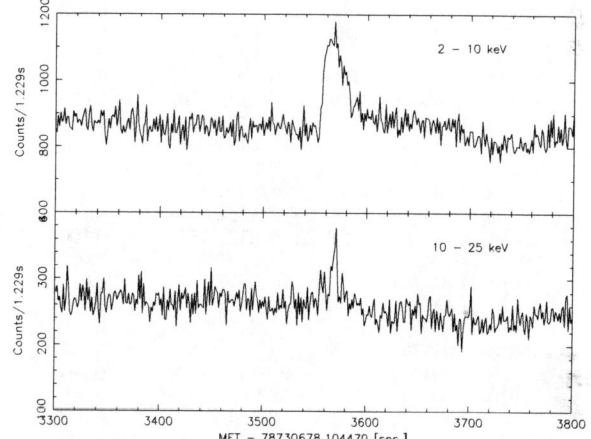

**FIGURE 4.** The energy resolved light curve of Aql X-1 (2001 July 26, 2:24:30 UT).

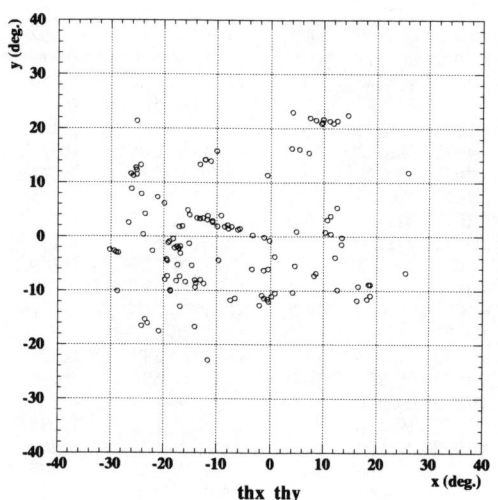

**FIGURE 3.** All the XRB events detected by HETE-2 are plotted in the detector coordinate.

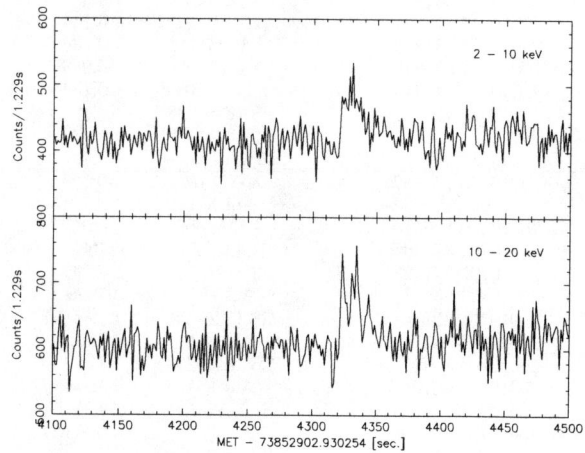

**FIGURE 5.** The energy resolved light curve of X1812-121 (2001 May 30, 16:06:02 UT).

**TABLE 2.** The list of XRBs detected by HETE-2 WXM. Localized celestial positions (J2000) are described for unidentified events. Note that "S" at the $\Delta\theta$ implies that the anti-solar position is used to determine the aspect of the spacecraft. This method has a systematic error of a few degrees.

| date (UT) | ID | Counterpart | $\Delta\theta$ |
|---|---|---|---|
| 20010524 230954 | - | SAXJ1750-29 | 0.337 |
| 20010526 010924 | - | X1724-307 | 0.299 |
| 20010527 222305 | 1529 | X1728-34 | 0.173 |
| 20010528 042834 | 1530 | X1728-34 | 0.183 |
| 20010529 143808 | 1531 | X1728-34 | 0.143 |
| 20010530 160602 | - | X1812-121 | 0.073 |
| 20010603 134639 | 1539 | X1728-34 | 0.215 |
| 20010608 082606 | - | X1702-429 | 0.771 |
| 20010608 210500 | 1540 | X1728-34 | 0.070 |
| 20010609 050648 | - | SAXJ1750-29 | 0.142 |
| 20010609 115346 | 1542 | X1728-34 | 0.469 |
| 20010609 212920 | - | SAXJ1750-29 | 0.260 |
| 20010610 213953 | - | X1728-34 | 0.169 |
| 20010610 232757 | 1543 | SAXJ1750-29 | 0.216 |
| 20010613 122410 | 1548 | X1728-34 | 0.530 |
| 20010614 013630 | 1549 | X1728-34 | 0.195 |
| 20010614 044749 | 1550 | SL1735-269 | 0.049 |
| 20010616 045155 | - | X1728-34 | 0.095 |
| 20010616 144053 | 1554 | SAXJ1750-29 | 0.358 |
| 20010616 163608 | 1556 | X1728-34 | 0.053 |
| 20010617 015433 | 1557 | GC X-1 | 0.091 |
| 20010617 034808 | * | X1728-34 | 0.111 |
| 20010617 035232 | - | SL1735-269 | 0.027 |
| 20010617 100957 | - | GS 1826-238 | 0.084 |
| 20010618 004815 | - | X1728-34 | 0.121 |
| 20010618 182100 | 1559 | X1728-34 | 0.100 |
| 20010618 194507 | - | GS 1826-238 | 0.091 |
| 20010621 075306 | - | X1728-34 | 0.035 |
| 20010621 141300 | - | SAXJ1750-29 | 0.024 |
| 20010621 170559 | - | X1728-34 | 0.099 |
| 20010622 013701 | - | X1728-34 | 0.252 |
| 20010622 191403 | - | X1728-34 | 0.194 |
| 20010623 080235 | - | SAXJ1750-29 | 0.040 |
| 20010623 173601 | - | GS 1826-238 | 0.067 |
| 20010624 161555 | - | X1728-34 | 0.225 |
| 20010624 210546 | - | SAXJ1750-29 | 0.020 |
| 20010625 002458 | - | X1728-34 | 0.033 |
| 20010627 021121 | - | X1728-34 | 0.108 |
| 20010627 181955 | - | (247.73, -7.60) | NA |
| 20010627 231707 | - | X1728-34 | 0.070 |
| 20010628 215231 | - | X1812-121 | 0.151 |
| 20010628 234222 | - | X1728-34 | 0.184 |
| 20010629 073848 | - | X1728-34 | 0.089 |
| 20010629 123457 | - | X1724-307 | 0.134 |
| 20010629 152841 | - | GS 1826-238 | 0.191 |
| 20010629 234432 | - | GS 1826-238 | 0.123 |
| 20010630 141818 | - | SAXJ1750-29 | 0.167 |
| 20010630 142447 | - | X1728-34 | 0.144 |
| 20010630 190054 | - | Aql X-1 | 0.151 |
| 20010630 190154 | - | X1728-34 | 0.062 |
| 20010701 224309 | - | (299.43, -27.91) | NA |
| 20010702 211128 | - | GS 1826-238 | 0.043 |
| 20010703 094732 | - | Aql X-1 | 3.797 (S) |
| 20010705 051341 | - | GS 1826-238 | 1.397 (S) |
| 20010705 082551 | - | Aql X-1 | 3.306 (S) |
| 20010705 134105 | - | GS 1826-238 | 1.195 (S) |
| 20010706 101045 | - | Aql X-1 | 3.164 (S) |
| 20010706 120414 | - | SL1735-269 | 2.155 (S) |
| 20010707 040805 | - | Aql X-1 | 4.024 (S) |
| 20010707 115707 | - | GS 1826-238 | 1.660 (S) |
| 20010707 123300 | - | SAXJ1750-29 | 1.803 (S) |
| 20010707 201643 | - | GS 1826-238 | 1.763 (S) |
| 20010708 170203 | - | GS 1826-238 | 1.379 (S) |
| 20010709 073217 | - | Aql X-1 | 3.241 (S) |
| 20010709 092707 | - | GS 1826-238 | 2.143 (S) |
| 20010710 105609 | - | GS 1826-238 | 0.179 |

| date (UT) | ID * | Counterpart | $\Delta\theta$ † |
|---|---|---|---|
| 20010710 191120 | 1579 | Aql X-1 | 0.037 |
| 20010710 192553 | - | GS 1826-238 | 0.106 |
| 20010711 075348 | - | GS 1826-238 | 0.396 |
| 20010711 131120 | 1580 | Aql X-1 | 0.092 |
| 20010711 160955 | - | GS 1826-238 | 0.283 |
| 20010711 194620 | 1581 | X1812-121 | 0.206 |
| 20010712 003030 | 1583 | GS 1826-238 | 0.206 |
| 20010712 013718 | * | SL1735-269 | 0.206 |
| 20010712 034945 | 1584 | X1916-053 | 0.023 |
| 20010712 111802 | * | Aql X-1 | 0.221 |
| 20010712 130047 | - | GS 1826-238 | 0.185 |
| 20010713 040130 | * | NGC 6624 | 0.057 |
| 20010713 052931 | * | GS 1826-238 | 0.192 |
| 20010713 230051 | - | X1728-34 | 0.125 |
| 20010714 055533 | - | GS 1826-238 | 0.617 |
| 20010714 070930 | 1585 | X1916-053 | 0.168 |
| 20010714 101850 | 1589 | X1728-34 | 0.153 |
| 20010714 150400 | 1590 | SAXJ1750-29 | 0.099 |
| 20010714 151000 | 1591 | NGC 6624 | 0.193 |
| 20010714 181530 | 1592 | GS 1826-238 | 0.442 |
| 20010714 182420 | 1593 | NGC 6624 | 0.025 |
| 20010714 201330 | 1594 | X1916-053 | 0.131 |
| 20010714 232430 | 1595 | X1728-34 | 0.175 |
| 20010715 005800 | 1596 | NGC 6624 | 0.161 |
| 20010715 040200 | 1597 | NGC 6624 | 0.069 |
| 20010715 055220 | - | X1728-34 | 0.171 |
| 20010715 072000 | 1598 | NGC 6624 | 1.182 |
| 20010715 104021 | - | GS 1826-238 | 0.092 |
| 20010715 121925 | - | NGC 6652 | 0.052 |
| 20010715 214202 | - | NGC 6624 | 0.033 |
| 20010716 010630 | 1599 | NGC 6624 | 0.092 |
| 20010716 011930 | 1600 | Aql X-1 | 0.127 |
| 20010716 071412 | - | GS 1826-238 | 0.821 |
| 20010716 220600 | 1605 | NGC 6624 | 0.085 |
| 20010717 073531 | - | GS 1826-238 | 0.198 |
| 20010717 154327 | - | GS 1826-238 | 0.086 |
| 20010717 155700 | 1606 | NGC 6624 | 0.061 |
| 20010717 235830 | - | GS 1826-238 | 0.021 |
| 20010719 080432 | - | NGC 6624 | 0.035 |
| 20010719 123858 | - | NGC 6652 | 0.215 |
| 20010719 125700 | - | X1916-053 | 0.063 |
| 20010719 131139 | - | GS 1826-238 | 0.042 |
| 20010720 094938 | - | X1916-053 | 0.063 |
| 20010720 100646 | - | GS 1826-238 | 0.100 |
| 20010720 142842 | - | GS 1826-238 | 0.104 |
| 20010721 065200 | 1607 | X1916-053 | 0.062 |
| 20010721 081058 | - | GS 1826-238 | 0.076 |
| 20010721 164610 | - | GS 1826-238 | 0.120 |
| 20010722 035250 | 1609 | X1916-053 | 0.143 |
| 20010722 054800 | - | X1812-121 | 0.267 |
| 20010722 182242 | - | GS 1826-238 | 0.029 |
| 20010724 170121 | - | NGC 6652 | 0.181 |
| 20010725 170130 | 1610 | X1812-121 | 0.135 |
| 20010726 024530 | 1612 | Aql X-1 | 0.138 |
| 20010727 042830 | 1613 | X1916-053 | 0.141 |
| 20010727 043112 | 1613 | GS 1826-238 | 0.209 |
| 20010727 123550 | 1618 | GS 1826-238 | 0.112 |
| 20010728 001631 | 1620 | GS 1826-238 | 0.167 |
| 20010728 155904 | 1628 | GS 1826-238 | 0.365 |
| 20010729 034708 | 1629 | GS 1826-238 | 0.775 |
| 20010729 051920 | 1631 | X1916-053 | 0.090 |
| 20010729 095319 | 1632 | Aql X-1 | 0.272 |
| 20010729 193158 | 1638 | Aql X-1 | 0.086 |
| 20010730 032501 | 1643 | Aql X-1 | 0.268 |
| 20010730 113104 | 1647 | Aql X-1 | 0.207 |
| 20010730 113655 | 1648 | GS 1826-238 | 0.128 |
| 20010730 200534 | 1653 | Aql X-1 | 0.264 |
| 20010731 051713 | 1658 | Aql X-1 | 0.133 |
| 20010904 042930 | 1749 | M 15 | 2.919 (S) |

* whether or not triggered by HETE ("*" or ID number: triggered, "-": untriggered)
† difference of the localized position in degree

# The Effectiveness of the HETE-2 Triggering Algorithm

Tanya Tavenner*, Ed Fenimore*, Mark Galassi*, Roland Vanderspek[†], Barbara Preger**, Carlo Graziani[‡], Don Lamb[‡], Nobuyuki Kawai[§], Atsumasa Yoshida[§], Yuji Shirasaki[§] and Toru Tamagawa[§]

*Los Alamos National Laboratory, Los Alamos, NM USA*
[†]*Massachussetts Institute of Technology, Center for Space Research, Cambridge, MA USA*
**Consiglio Nazionale delle Ricerche, Istituto di Astrofisica Spaziale, Frascati (Roma) ITALY*
[‡]*Department of Astronomy and Astrophysics, University of Chicago, Chicago, IL USA*
[§]*Institute for Chemistry and Physics (RIKEN), Wako, Saitama JAPAN*

**Abstract.** We determine the most effective trigger criteria for detecting GRBs on HETE-2. Our simulations include a full Monte Carlo tracking of the photons through the instrument, as well as randomly selected BATSE bursts. The HETE-2 flight triggering algorithm runs about thirty different triggers simultaneously. These criteria range from the traditional trigger style that uses one background which is located before the moment of interest (the potential burst), to most of the current HETE-2 criteria which include two backgrounds which bracket the moment of interest. The dual background style allows the software to be more effective when dealing with trends in the HETE-2 backgrounds. As a result, we have been able to set the threshold of the triggers to less than half of what has been used in previous experiments such as BATSE. Our simulations show that bracketed triggers give HETE-2 almost three times as many triggered GRBs and 23% more correct localizations than traditional trigger criteria run on the same gamma ray bursts. We have also shown that our new trigger set on HETE-2 finds 5% more gamma-ray bursts and produces 11% more correct on-board localizations than the previous set we ran on the satellite.

## TRIGGERING IN GAMMA RAY BURST MISSIONS

Triggers aboard satellites such at the High Energy Transient Explorer (HETE-2) serve two important functions. First, they tell the satellite when a sudden source of gamma-rays or x-rays has appeared so that it can start sending high resolution data to the ground. Second, they tell the on-board localization algorithm which time samples to use to find the location of the potential GRB.

Gamma-ray bursts (GRBs) are unpredictable, and satellite telemetry band-passes are not large enough to send every photon to the ground in real time. This means that it is important to have an on-board method for telling the satellite when to switch operation modes, so that it can capture the event with the maximum time and energy resolution. This is done by searching for a "significant" increase in the photon count rate over a background count rate. Most previous experiments (Vela, PVO, ISEE-3, Ginga, and BATSE) searched for such increases over a few time scales (candidate foreground regions) that ranged from 0.064 sec to 4 sec. Backgrounds were estimated by taking an average of the count rate from a period assumed to be well before the burst (which was usually 16 sec to 30 sec before the foreground region).

Triggers searching for a sudden increase in counts have to overcome two major obstacles. The first is that the background follows Poisson statistics, which can cause false triggers. It is also common for the background to have trends caused by solar or earth related effects such as solar flares reflected off the atmosphere, or the South Atlantic Anomaly (SAA) (see Figure 1). Trends are much harder to overcome than statistics. Given a rising trend and a background set thirty seconds before the foreground, the trigger could easily find a "significant" increase in photon counts, giving rise to a false trigger as in Figure 1. Both obstacles can be overcome by setting a high enough threshold value on the triggers, at the expense of sensitivity.

Significance is usually determined by calculating how many standard deviations separate the foreground count rate from the background count rate assuming Poisson statistics. The sigma ($\sigma$) level (or threshold) was usually never set below $\sim 11\sigma$ in previous experiments, therefore statistical fluctuations were never a critical factor in false

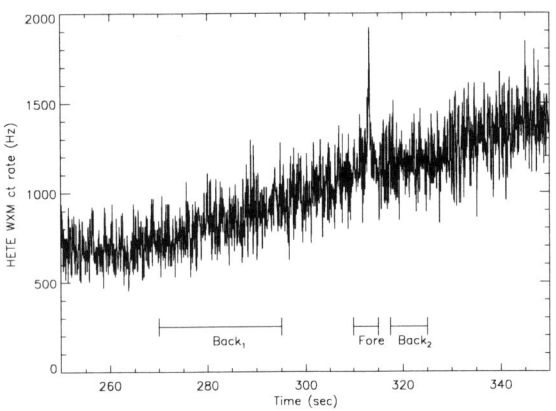

**FIGURE 1.** A triggering scenario where single background, traditional trigger criteria might easily trigger on the background trend before they even encounter the burst. Traditional on-board triggering systems use only a single background period (e.g., "Back$_1$") before the candidate trigger sample (i.e., "Fore"). The trend can show a significant increase in counts in the foreground region when only one background is used, leading to a false trigger. In HETE-2, we use two background regions which bracket the foreground region. We can then remove most of the effects of trends by making a linear interpolation of what the background should be in the foreground region. Many triggers run simultaneously on the satellite, covering a wide temporal and energy parameter space.

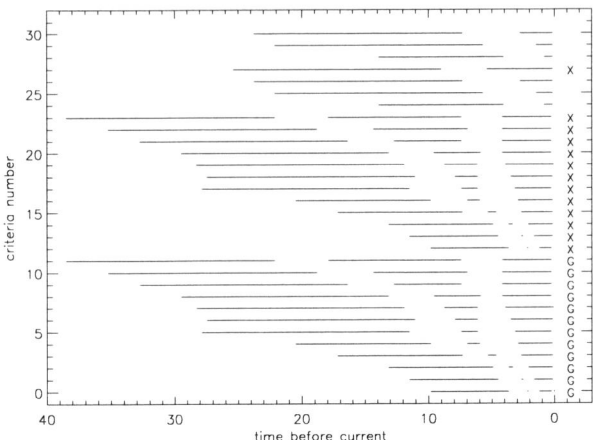

**FIGURE 2.** Short Trigger Set: A visual representation of the background and foreground regions of the "short" trigger set. The short, medium and long trigger sets refer to the length of the second background. (ie: The first background in trigger criteria 11 starts about 39 seconds before the trigger was run and ends at 22 seconds. The foreground runs from about 18 seconds to about 7 seconds, and the second background starts at 5 seconds and runs until the current time when the trigger is run.) This is the trigger criteria set that ran on HETE-2 from about June 2001 until early December 2001. Criteria 0 - 11 are gamma-ray triggers that the WXM (x-ray instrument) runs on FREGATE (gamma-ray instrument) data. Criteria 12 - 23 are identical to 0 - 11, but are applied to the WXM's x-ray data. Criterias 24 - 30 are traditional style triggers run on x-ray data. Because they do not have second backgrounds they are able to trigger more quickly on bursts. However they are also more likely to produce false triggers on trends, and thus require a higher threshold.

triggers. Such a high threshold was required to overcome the trends in the background. Even with that threshold most experiments (e.g., PVO, Ginga, ISEE-3) still had a false trigger rate of 90%. BATSE was able to achieve a 50% false trigger rate because crude on-board locating allowed sources that appeared to be inside the satellite (ie, particle events) to be rejected.

HETE-2 is able to use a threshold very close to the statistical limit because it uses two backgrounds which bracket the foreground instead of one background set before the foreground (see Figure 2). The first background on HETE-2 is still set up to be approximately thirty seconds before the foreground, but the second background is one to three seconds after the foreground. The second background is always shorter than the first background so that HETE-2 has a rapid response to GRBs. Bracketing triggers allow a linear interpolation of what the background should be in the foreground region. This allows HETE-2 to remove most of the effects of trends, and use a much smaller threshold than previous experiments.

## HOW TRIGGERS AFFECT LOCALIZATIONS ON HETE-2

HETE-2's gamma-ray instrument, FREGATE, searches for the first sign of a gamma ray burst (GRB). However, while it is very good at finding them, it cannot localize the bursts. The Wide-field X-ray Moniter (WXM) uses coded apertures to localize the x-rays from the burst. Many triggers are run concurrently on the data from FREGATE and the WXM, with foregrounds ranging from 160 ms to 13 seconds. As photons arrive, each section of data is tested as a potential foreground region on different time scales. Only when the excess goes above a threshold (which is between 4.5 $\sigma$ and 5.5 $\sigma$ depending on the foreground duration) does the satellite tell the ground that a trigger has occurred. The initial trigger usually comes from FREGATE. Additional triggers are then run on the WXM data in order to find the best possible foreground and background regions for localization. The SNR of the foreground region compared to the interpolated background is called the trigger score. Each time a new trigger beats the previous trigger's score, it produces a new localization. The localization code uses the foreground region and the first background region to produce its on-board localization. Typically, three to nine additional triggers

**TABLE 1.** Differences in Trigger Criteria

|  | Traditional | Short | Medium | Long |
|---|---|---|---|---|
| Total number that triggered | 224 | 638 | 673 | 722 |
| Number that localized in x and y | 176 | 216 | 240 | 256 |
| Number that localized in x | 207 | 266 | 272 | 287 |
| Number that localized in y | 184 | 268 | 277 | 290 |

and localizations are produced per event. (S/N)$^2$ of x plus the (S/N)$^2$ of y is called the LOC_SNR. The best localization reported to ground is typically the one with the highest LOC_SNR.

Given the coupling between triggers and localizations, modifying the trigger criteria not only affects the total number of triggered events, it also affects the number of correct on-board localizations reported to ground.

## THE SIMULATIONS

We now turn to several important questions. How bracketing triggers compare to traditional triggers in finding events. How having a background inside the burst effects the trigger's ability to find the burst. And how different backgrounds and spacings (components of the trigger criteria) affect the number of bursts both found and correctly localized. We addressed these questions by simulating 2000 bursts and running several different sets of trigger criteria on them. Our simulations include a full Monte Carlo tracking of the photons through the instrument, as well as randomly selected BATSE events placed at a random locations within the field of view of the instrument.

Our first two sets of trigger criteria are single background, traditional criteria and "short" bracketed criteria. (see Figure 2). Because we did not simulate trends with the bursts, by comparing these two types we are able to see how many triggers HETE-2 gains with the lower thresholds of bracketed triggers. As Table 1 shows, we have 184% more triggers (GRBs that our codes found) with the short bracketed triggers than we have with the traditional triggers. We also get 23% more correct on-board localizations with the bracketed triggers.

One of the problems with using two backgrounds is that the second background might actually be inside the burst. The burst itself would then be treated as a trend. This means that it would subtract some of the burst's counts from the foreground region (see Figure 3). This gives the foreground region less significance than it would have had with a traditional trigger. We address this issue by creating a set of "long" trigger criteria whose bracketing backgrounds are of equal length and are set far away from the foreground such that there is little chance that either one could be in the burst (see Figure 4).

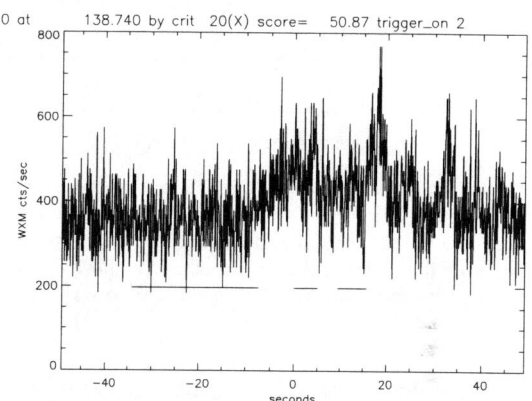

**FIGURE 3.** An example of the second background being inside the burst. The triggering code interpolates between the two backgrounds and finds a background count rate for the foreground region. While this is excellent for removing background trends, in a case like this it actually lowers the effective significance of the burst because it thinks that the burst itself is a trend.

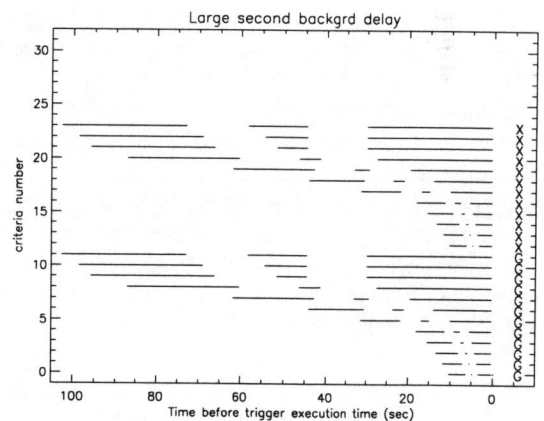

**FIGURE 4.** Long Trigger Set: Ideal case with no background trends and no time constraints. By comparing this to flight we are able to see how many bursts and localizations we miss by having one or both backgrounds in the burst.

Because we have a background in the burst, we find (by comparing "long" and "short" in Table 1) that we lose 11% of the bursts that it is possible to trigger on, and 19% of the on-board localizations. On the surface this long set appears to be the best possible approach, but is not appropriate for the HETE-2 flight software because

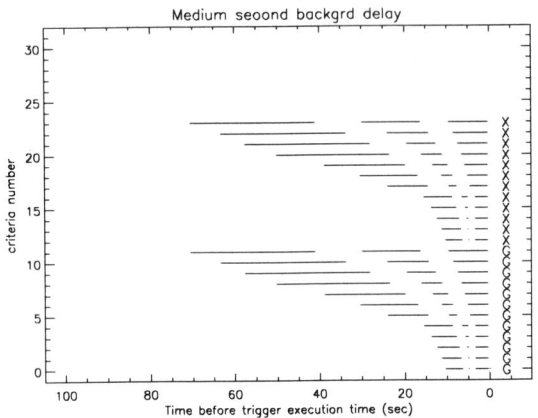

**FIGURE 5.** Medium Trigger Set: The new set of flight criteria on HETE-2. Slightly longer backgrounds and slightly longer spaces between the backgrounds and foreground than the short set (former flight set). These criteria find 5% more GRB's and produces 11% more localizations than the short set.

it does not deal effectively with trends, and it does not detect bursts quickly. We are able to use this set of trigger criteria for our comparisons because our simulations are of the ideal case and have no trends in them.

Our last question deals with how to make the trigger criteria better so that the satellite is able to find and localize more bursts. We have shown that longer background and larger spacing work well in the ideal case. So now we look at what we call the "medium trigger set" which has slightly longer backgrounds and slightly longer spaces between backgrounds than the short set (see Figure 5). Table 1 shows that with no trends the medium set triggers on 5% more GRBs and gets 11% more localizations than the short set. In simulations using HETE-2 flight data to define orbital variations of the background (such as the SAA), the medium set was shown to be less likely to produce false triggers than the short set. For all of these reasons the on-board HETE-2 flight, trigger criteria set was changed from the short set to the medium set in early December of 2001.

results, HETE-2 switched to the medium trigger set in mid-December 2001.

## CONCLUSIONS

The use of bracketing triggers allows HETE-2 to set a lower threshold than previous experiments, and yields nearly three times as many triggered events in our simulations than the traditional triggers. It also gives 23% more correct localizations than traditional triggers. While we do lose some GRBs to having our second background in the burst, our new flight criteria are a significant improvement over the old criteria. The new criteria (medium set) find 5% more bursts and 11% more localizations than the old (short) criteria set. Based on these

# What HETE Sends to the GCN

R. Vanderspek*, G. B. Crew*, S. Barthelmy[†] and the HETE Science Team**

*MIT Center for Space Research, Cambridge, MA, 02139 USA
[†]NASA Goddard Space Flight Center, Greenbelt, MD, USA
**An international collaboration of institutions including MIT, LANL, U. Chicago, U.C. Berkeley, U.C. Santa Cruz (USA), CESR, CNES, Sup'Aero (France), RIKEN, NASDA (Japan), TESRE (Italy), INPE (Brazil), TIFR (India)

**Abstract.** Information about bursts detected by the HETE spacecraft is propagated in real time to the Mission Operations Center at MIT via the Burst Alert Network ([4], [11]). Upon receipt at MIT, pertinent burst information are reformatted and automatically sent to the GCN for distribution to the observing community; if, after ground analysis, more or better information become available, it is sent to the GCN manually by a HETE Duty Scientist. We review the contents of these messages and the procedures used to distribute them. We also summarize the results of the HETE User's Group Meeting, which took place at the Woods Hole GRB Conference, in which ideas were exchanged between the observing community and the HETE and GCN teams.

## INTRODUCTION

The HETE satellite has the ability to localize a GRB on board and then transmit the coordinates of the burst to the ground, all within seconds of the detection of the burst: once received on the ground, burst data are sent to the GRB Coordinates Network (GCN; [2]) at GSFC, which immediately forwards them to the GRB observing community. This feature makes possible prompt ground-based observations of the GRB and its afterglow, in some cases *even while the burst is still in progress*.

The flow of decisions and information from the time of the detection of the GRB on board to the time the GCN Notice is received by the observer is complicated, and it has been noted that the contents of the GCN Notices have, at times, been confusing. In this paper, we walk the reader through the on-board and ground-based decision tree which leads to the generation of the GCN Notice. The contents of this paper and any updates to the interface will be available at the HETE web site (http://space.mit.edu/HETE) for the duration of the HETE mission.

While various features of the HETE flight and ground hardware and software will be mentioned here, the reader should refer to other papers in these proceedings for more detailed information: on Fregate, see [1]; on the WXM, its triggering algorithms, and error estimations, see [7], [10], [5], [6], [9]; on the VHF system on board and the Burst Alert Network, see [11], [4].

## THE BURST ALERT NETWORK AND THE TRIGGER MONITOR

The key players in the distribution of burst messages, other than the science instruments themselves, are the Burst Alert Network and on-board Trigger Monitor process.

Besides its primary S-band antenna for high-bandwidth data downlink, the HETE satellite is equipped with a low-power, low-bandwidth (300 baud), omnidirectional VHF transmitter which is used to broadcast burst information during and immediately after a burst trigger (and basic spacecraft and instrument housekeeping when there is no burst in progress). The burst messages are received by the nearest station(s) in the *Burst Alert Network*. The Burst Alert Network (BAN; [11]) is a set of low-cost, receive-only ground stations distributed along the equator. Each of these stations is capable of receiving and forwarding to MIT the VHF message broadcast by HETE: details, including information about transmission time and delays, are in [4].

The *Trigger Monitor* is an on-board software process which acts as the central clearinghouse for burst information on board. Upon receipt of a trigger message from one of the instruments, the Trigger Monitor immediately notifies the other instruments on board. As burst data products are produced by the other instruments, they are sent to the Trigger Monitor.

The Trigger Monitor is responsible for condensing the

burst data products it receives into VHF burst messages. These messages are made as short as possible, as the time to transmit the VHF message is part of the overall response time of a ground-based instrument to a detected GRB.

## FROM GRB TO GCN NOTICE

This section describes the steps taken by flight and ground software from the detection and evaluation of a GRB to the generation of a GCN Notice.

### Burst Detection on HETE

Two trigger algorithms are used on board HETE to search for bursts in the Fregate and WXM data. One operates in the digital signal processor (DSP) which controls the Fregate instrument, while the other runs in the transputer processor common to both the Fregate and WXM instruments.

*DSP trigger.* The Fregate flight software searches for enhancements of the Fregate counting rate on four timescales (20 ms, 160 ms, 1.3 s, and 5.2 s). The Fregate flight software defines a trigger as a significant ($> \sim 5\sigma$) enhancement is seen in at least two of the four Fregate detectors in either the 8-80 keV band or the 30-400 keV band [1].

*Transputer trigger.* The flight transputer software searches both WXM and Fregate time-history data for enhancements on timescales ranging from 80 ms to over 10s. This method uses estimates of the gamma-ray or X-ray background rate measured before and after the trigger period, and so is less sensitive to trends in the background than the DSP trigger [10].

When either of these processes detects a trigger, it sends a message to the Trigger Monitor. The Trigger Monitor immediately distributes a message giving details of the trigger to the three science instruments and the optical star camera system: these instruments then switch into a burst processing mode, described below. The Trigger Monitor also sends the initial burst message to the ground via the VHF transmitter.

### Burst Processing on HETE

The focus of on-board burst processing is the collection of high-resolution raw data from each instrument for ground analysis and the calculation of the location of the burst and the orientation of the spacecraft.

*Fregate.* Once a trigger is detected, the Fregate instrument collects 65000 time-tagged photons detected before, during, and after the burst for later downlink. If the DSP trigger detected the burst, the flight software compares the background level ca. 100s after the trigger to the pre-trigger background level: if it is clear that there is a strong rising trend in the background level, the Trigger Monitor is notified that the trigger is "invalid" (see below).

*WXM.* Regardless of the source of the trigger, the focus of the WXM flight software after a trigger is the determination of the position of the source in the WXM field-of-view. To achieve this, the WXM data from the time near the trigger are re-examined on many timescales to find the time period with the highest S/N ratio. As soon as a period is found with a better S/N ratio than the one previously reported, that period is reported to the imaging process, described in more detail below. As more time periods are examined, periods with better S/N may be found: each is reported to the imaging process, which analyzes each in turn.

*WXM imaging.* The WXM imaging algorithm ([5]) compares the background-subtracted pattern of the X-rays from the GRB with a set of pre-calculated pattern "templates" for a range of incident angles. In each WXM module, the template which has the highest correlation with the detected pattern is considered best: an "image SNR", which describes the peak signal-to-noise in the cross-correlation of the background-subtracted image with the mask, describes the quality of the correlation. If the image SNR exceeds a threshold in both modules *and* exceeds the image SNR of any previous correlation calculated for this burst (necessarily done with a different time period), the burst localization (the angles of the best templates), the image SNRs, and the lightcurve SNRs, which describe the signal-to-noise of the excess over background in the time period selected, are reported to the Trigger Monitor.

*SXC.* The Soft X-ray Camera on HETE ([12], [8]) operates in a mode where raw data are continuously collected and sent to the ground: when a burst occurs, the SXC does not change its mode of operation, other than to increase the priority of the data messages to the ground, to insure timely reception of the burst data.

*SXC imaging.* The SXC imaging process, discussed in [8], provides a continuous stream of "best positions" during normal operations. These "best" positions may reveal the location of a steady source or a transient source, if there is one in the field-of-view; the S/N associated with the position allows one to distinguish between real sources and noise in the cross-correlation map. The SXC "best position" in X and Y and the associated S/N are reported to the Trigger Monitor. If a better position, based on S/N, is measured while the burst is in progress, that position is sent to the Trigger Monitor. During burst analyses, the SXC "best position" within 1° of the most recent WXM position is also reported. At the time of this writing, SXC positions are not distributed to the GCN in real time.

*Optical star cameras.* When in operation, the optical star cameras on HETE continuously calculate the spacecraft aspect. Because of the effect of scattered light, the optical cameras cannot run near the orbit terminators, so the typical overlap of star camera and WXM/Fregate operations is about 90%. When a trigger is announced by the Trigger Monitor, the optical camera system sends to the Trigger Monitor its best estimate of the spacecraft aspect at the time of the trigger to the Trigger Monitor; if no aspect is available, a message saying "no aspect available" is sent to the Trigger Monitor.

The process of on-board data collection and analysis continues until each declares its analysis complete. Fregate processing continues until the validation measure has been made and the time-tagged photons have been stored in on-board memory: this typically takes about four minutes. WXM processing continues until the imaging analysis cycle is complete: this can take from 30 seconds to five minutes, depending on the duration and complexity of the GRB. The SXC and star camera systems do not change their modes of operation during a burst, so they play no role in determining the end of burst processing.

## The VHF Burst Message

The VHF burst message, sent to MIT via the BAN, varies in length and content depending on the actions of the flight processors. This message always contains:

- Burst trigger time
- Burst ID
- Energy band of trigger
- Timescale of trigger
- Triggering instrument
- A measure of the brightness of the trigger

As processing on board continues, the message can also include

- WXM burst location and significance
- SXC burst location and significance
- Optical aspect
- Burst invalidation (DSP triggers only)

The VHF burst message is continuously broadcast: when new new burst data are received from the science instruments, the message is changed, but then continuously repeated.

Because the BAN coverage is not 100%, the Operations Center at MIT may not receive all the VHF messages sent by the spacecraft; the VHF messages are therefore written in a manner where each message contains all the information needed to understand the state of on-board burst processing at the time of transmission.

## Real-Time GCN Notices

There are currently three types of real-time GCN messages: *S/C_Alert*, *S/C_Update*, and *S/C_Last*.

*S/C_Alert.* The S/C_Alert message is sent out upon receipt of the first valid VHF message at MIT. It contains, at a minimum, the burst time, number, energy range, and timescale.

*S/C_Update.* The S/C_Update message is sent if there is new information about the burst analysis occurring on board: almost invariably, this message contains a localization calculated by the WXM flight software. S/C_Update messages can be sent as early as six seconds after the S/C_Alert message, and there can be multiple S/C_Update messages for a single burst.

*S/C_Last.* The S/C_Last message is sent when the on-board burst analysis is complete. The S/C_Last message contains the best information about the burst at the time of transmission. The S/C_Last message is typically sent 4–5 minutes after the onset of the burst.

If the VHF messages with the original burst data are missed, it is possible that the S/C_Alert message contains a burst localization, and no S/C_Update message with a localization, which would otherwise have been issued, is distributed. If all of the VHF messages sent *during* the burst analysis are missed and only the summary message is received, messages of type S/C_Alert and S/C_Last, each with the full burst information, is distributed.

*Criteria for Sending to GCN*

As a rule, S/C_Alert and S/C_Last messages are distributed for all bursts detected by HETE in the 8–80 or 30–400 keV band (except in situations described in the next section). However, since many triggers detected on board result in a WXM localization, strict criteria on the quality of the localization must be met before the coordinates are distributed to the GCN. These criteria, which are limits on the image SNR and lightcurve SNR in the WXM, do not restrict the distribution of localizations based on error box size: localizations that meet these criteria should have 90% errors of ∼10' radius, while those that do not meet these criteria could be incorrect by tens of degrees.

Thus, S/C_Update messages with burst localizations are only sent if the localizations are significant enough. At the time of this writing, only WXM positions are distributed in real time, and the threshold for distribution of a WXM position is that the image SNR and the lightcurve SNR in both X and Y modules must be at least 3.0. Those localizations that meet these criteria should be valid 90-95% of the time. This restriction will be lessened once the image SNR and lightcurve SNR are included in the GCN message (see below).

*Restrictions on GCN messages*

During the course of normal operations, there are times when the flow of data from the satellite to the GCN is restricted. The most common examples are

**Geomagnetic storms** During times of high geomagnetic activity or after the passage of a Coronal Mass Ejection (CME), the particle density in the region over western South America increases dramatically, and the probability of having a false trigger due to particles is high. During these periods, the HETE operations team may restrict distribution of GCN messages until the particle activity subsides or that region of the Earth is in orbit day.

**Galactic bulge sources** Because HETE points in the anti-solar direction, the galactic bulge is in the fields-of-view of the instruments from April to August of each year. Because of the high rate of XRBs from sources in this region, GCN distribution of bursts detected in the 2-25 band may be suppressed.

**Trigger tests** During tests of triggering thresholds, no GCN messages are distributed.

**GCN interface down** The interface between the HETE operations center at MIT and the GCN center at GSFC does go down from time to time, in which case prompt distribution of GCN messages is not possible.

# GCN NOTICES AFTER GROUND ANALYSIS

Once HETE passes over a primary ground station, the full load of burst data will be downlinked to the ground, and detailed analyses can begin. As ground analyses proceed, new and/or improved results will become available: these can be manually sent to the GCN as GCN Notices for immediate distribution by the HETE Duty Scientist. Information about bursts that are detected by ground processes or information that does not clearly fit in a standard HETE-related GCN Notice is distributed by GCN Circular.

Ground analysis of a burst begins as soon as the full burst data reach MIT after a Primary Ground Station contact (from a few minutes to over an hour after the burst, depending on where in its orbit HETE was at the time of the burst). Automated software performs standard analyses of the downlinked data, and the HETE Duty Scientist is notified to make the final decisions. A followup GCN Notice, of type **HETE_Gnd_Analysis**, will be distributed under the following circumstances:

- There was no position calculated on board, but ground analyses find a significant position.
- There was a position calculated on board and ground analyses can improve the coordinates and/or reduce the error box size.
- There was a position calculated on board, but there is actually no significant position in the data.

In general, if there is a position in a ground analysis GCN Notice, it should be considered accurate.

# INTERPRETING HETE GCN NOTICES

The key question in the minds of users of HETE GCN messages is "can I believe this position?". We offer the following guidelines in the interpretation of GCN messages.

*"Is this localization good?"*

The quality of localizations distributed in real time by the flight software depends on the image and lightcurve SNR in both WXM modules. The current restriction that all four values exceeding 3.0 is meant to ensure that the localization has a ∼90% probability of being correct; a lower limit of 2.5 on each of the four values reduces that probability to ∼50%. Once the image and lightcurve SNR are distributed in the GCN Notices, the user can determine the correct threshold to use: a plot of the "good" range of values will be posted on the HETE web page.

A caveat to these limits is that particle events can have an extremely good lightcurve SNR and a believeable image SNR. The ratio of these two values typically exceeds 5 for particle events, whereas the ratio is less than 3 for detected bursts. If the ratio is quite high, it would be wise to check the longitude of the spacecraft at the time of the trigger to see if it is near South America; the HETE web page will also indicate whether the trigger was due to particles.

*"Burst declared invalid"*

If the Fregate flight software detects an steep rise in the background count rate near the time of the event, it will declare the trigger "invalid". This declaration is propagated to the GCN Notice in the words "burst declared invalid", "invalidity bit is true", and/or "definitely not a GRB". This happens only rarely, but it has traditionally been a good measure of whether a burst was due to a rising background. If the localization of such a burst is significant, it could, in principle, still be a valid trigger, but this has not yet happened in the HETE mission.

*Localizations after ground analysis*

Localizations calculated after ground analysis of flight data should be considered accurate. On the occasions where the ground analysis was incorrect, the operations team had been asked to analyze data collected in a non-standard way (*e.g.*, no optical aspect at the time of the trigger, localization only in one WXM module) and errors were made in real-time software development. Most of these issues have been addressed, so we do not anticipate future errors in ground analysis localizations.

*Error radii*

All localization errors quoted in GCN Notices and Circulars are 90% probability errors. These errors are generated from measurements of statistical and systematic errors in the WXM and star camera data. These errors will become smaller as the quality of the astrometric calibration of both instruments improves.

Error regions calculated by the WXM are typically ellipses, but with aspect ratios that rarely exceed 2:1. The GCN Notice will quote a circular error region which has a 90% probability of containing the burst location.

*Multiple positions for a single burst*

Because of the nature of the WXM triggering and imaging algorithms, described above, it is possible for multiple localizations to be distributed for a given burst. Earlier in the mission, when there were no restrictions on the image SNR of a position calculated on board, there were instances where 2–4 positions were distributed, each several degrees away from the others. With the implementation of the restrictions that only "better" localizations can be distributed and only positions with image SNR $>$ 3.0, the instances of multiple or invalid localizations have diminished to near zero.

## FUTURE MODIFICATIONS

The following modifications to the HETE GCN message distribution were discussed during the HETE User's Group Meeting at Woods Hole and will have been implemented by the time of publication of this proceedings.

**Spacecraft Longitude** The longitude of the spacecraft will be included in the GCN message, to allow observers to judge whether the event might be particle-induced.

**WXM image and lightcurve SNR** The WXM image and lightcurve SNR will be included in the GCN message, and individual users will be able to specify the thresholds above which they receive a GCN message (true, Scott?).

**Burst summaries to the web** We have already implemented a means of putting a short description of the trigger type to the HETE bursts web page (http://space.mit.edu/HETE/Bursts): with this, the user can immediately see the longitude of the trigger and any comments that a HETE Duty Scientist has made. This feature will be enhanced with lightcurve plots and further trigger information.

## ACKNOWLEDGMENTS

The HETE team would like to thank the participants in the HETE User's Group Meeting at the Woods Hole Conference for their participation, understanding, and support.

## REFERENCES

1. Atteia, J-L., et al. 2002, these proceedings.

2. "GRB Coordinates Network (GCN): A Status Report"; S.D.Barthelmy, T.L.Cline, P.Butterworth; AIP "Gamma 2001 Workshop"; vol 587, p213, 2001
3. Butler, N., et al. 2002, these proceedings.
4. Crew, G. B., et al. 2002, these proceedings.
5. Fenimore, E. E., et al. 2002, these proceedings.
6. Graziani, C., et al. 2002, these proceedings.
7. Kawai, N., et al. 2002, these proceedings.
8. Monnelly, G., et al. 2002, these proceedings.
9. Shirasaki, Y., et al. 2002, these proceedings.
10. Tavenner, T., et al. 2002, these proceedings.
11. Villasenor, J. N., et al. 2002, these proceedings.
12. Villasenor, J. N., et al. 2002, these proceedings.

# First Year Operations of the HETE Burst Alert Network

J. Villasenor*, G. Crew*, G.Monnelly*, J.Doty*, R.Foster*, G. Ricker*, R. Vanderspek*, N.Kawai[†], A.Yoshida[†], M. Boer**, K.Hurley[‡], R.Manchanda[§], G.Pizzichini[¶], Joao Braga[∥] and G.Azzibrouck[††]

*MIT Center for Space Research, Cambridge, MA, 02139 USA
[†]Institute for Chemistry and Physics, Wako, Saitama, Japan
**Centre d'etude Spatiale des Rayonnements, Tolouse, France
[‡]University of California, Berkeley Space Sciences Laboratory, Berkeley, CA, USA
[§]Tata Institute for Fundamental Research, Mumbai, India
[¶]Consiglio Nazionale delle Ricerche, IASF, Bologna section, Bologna, Italy
[∥]Instituto Nacional de Pesquisas Espaciais, Sao Jose dos Campos, Brazil
[††]Universite de Masuku, Gabon

**Abstract.** The BAN (Burst Alert Network) is comprised by an equatorial belt of VHF receiving stations dedicated to relaying via the internet HETE burst alerts and spacecraft status reports to the MIT Operations Center. Messages created by the satellite are sent to the GCN within seconds. Each of the 14 BAN stations consists of low cost (total is less than $3000) components: a simple antenna, a preamplifier, a PC, and a receiver. HETE broadcasts a repeated VHF message stream to minimize the data corruption due to low signal level and varying local noise conditions. This low technology approach has proven to be successful for HETE operations. Real time coverage of the satellite has steadily increased from the time of the launch in 2000 to a value near 80% in 2002.

## INTRODUCTION

The Burst Alert Network (BAN) is a unique approach to real time monitoring of low earth orbiting satellites. The central idea is to establish low cost stations, situated close to the ground track of the satellite and accessible via the internet, which relay data worldwide just a few seconds after receipt.

A traditional satellite-to-satellite link using the TDRSS can absorb a substantial amount of the operations budget for a small mission[1], and adds system complexity and operational constraints to the satellite (e.g., a higher spacecraft power budget to accommodate the continuous S-band link). A few factors allow our alternative solution.

The burst alert message from HETE contains a small number of bytes, since on-board processing summarizes the GRB burst data. Transmission at lower frequencies is then possible and desirable [2]; in particular, VHF components are relatively cheap and commercial equipment are readily available. Multiple ground stations can then be established at a minimal cost. HETE's orbit limits the visibility of each ground station to 13 minutes on each 100 minute orbit, but an equatorial array of stations with overlapping horizons provides nearly continuous coverage and further improves reliability.

Finally, the explosive worldwide growth of the internet in recent years has allowed us to tap suitable remote sites for inclusion into the BAN. Local ISPs now provide connectivity to some of these stations which previously would have been considered inaccessible.

## STATION DEPLOYMENT

Since the launch of HETE, we have activated and continue to operate all of the 12 original, pre-launch stations. Two additional stations were added to augment the BAN in noisy locations. Table 1 lists the BAN stations. Except for a small gap in the Pacific (where there is no land mass), 99% of the HETE orbital path is potentially covered by a station.

---

[1] A typical NASA station costs > $50,000 a year to operate
[2] Fewer bits allow a lower transmission rate, and in turn a lower frequency to obtain an acceptable bit error rate

**TABLE 1.** BAN station locations and average horizon to horizon contact times for one orbit.

| Location | Longitude (deg) | Latitude (deg) | contact (min) |
|---|---|---|---|
| Malindi | 40.19 | -3.0 | 13.6 |
| Bangalore | 77.38 | 12.58 | 11.6 |
| Singapore | 103.83 | 1.33 | 13.7 |
| Palau | 134.5 | 7.33 | 13.6 |
| Kwajalein | 167.72 | 8.72 | 12.8 |
| Kiritimati | -157.1 | 1.9 | 13.6 |
| Maui | -156.26 | 20.71 | 6.6 |
| Galapagos | -90.28 | -0.67 | 13.7 |
| Marquesas | -139.0 | -9.78 | 12.6 |
| Cayenne | -52.35 | 4.95 | 13.4 |
| Natal | -35.21 | -5.84 | 13.3 |
| Ascension | -15 | -7 | 12.8 |
| Gabon | -1.66 | 13.61 | 13.6 |

**TABLE 2.** Ground station hardware

| Component | Varieties | Cost Range |
|---|---|---|
| Antenna | RHCP half turn quadrifilar RHCP full turn quadrifilar linear Yagi | $500-$1000 |
| Receiver | custom Drake R8A, Drake R8B AR3030a, AR3030b | $1000-$1200 |
| Pre-amp | Lunar SSB Electronics | $200-$500 |
| Computer | E-machines w/ Unix PCs w/ Windows | $500-$1000 |

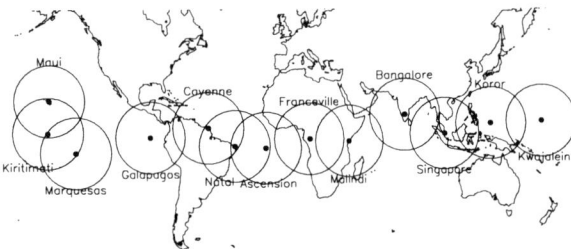

**FIGURE 1.** Locations of all the HETE ground stations

**FIGURE 2.** Components of a complete ground station.

All stations are automated and connected to the internet; most are remotely monitored and maintained. The VHF messages are forwarded to the MIT Control Center (MCC) for further processing.

## VHF HARDWARE

The HETE spacecraft broadcasts messages at a frequency of 137.96 MHz, at 300 bps CPFSK (h=1), and a transmitted power of 1 W. The VHF antenna is a simple 2.5 foot whip extending perpendicularly from one of the four solar panels.

Each ground station consists of a fixed antenna, a preamplifier, a receiver, and a low-cost computer running either Unix or Windows (Table 2, Fig. 2).

The signal demodulation and $\pm 3$ kHz Doppler correction is carried out by hardware in the Marquesas and Cayenne stations, using custom built receivers developed by our partners at CESR and CNES (Toulouse). Most stations however, apply a software solution developed at MIT: a commercial receiver is used as a tuneable mixer controlled by the PC. The audio output from the receiver is digitized and demodulated by the computer. The center frequency and Doppler corrections are tracked using the maximum entropy method, and a serial command is then sent to retune the receiver to the correct band [1].

Both methods perform satisfactorily. The software approach is novel because it allows the possibility of reception at other bands, and can be remotely reconfigured. Moreover, the widespread use of VHF receivers by amateur ham operators ensures the commercial availability and low cost of these receivers.

Various types of antennas were field tested and deployed. The full turn, circularly polarized quadrifilars (Kwajalein, Galapagos) provide better reception at lower elevations, which extends the temporal coverage. The cheaper and more compact half turn quadrifilars are sensitive above 20 degrees, although reception occasionally occurs below that. A linear 2 element Yagi antenna array also proved to be very successful when used at a high latitude location (Maui).

## RESPONSE AND PERFORMANCE

The growing real time coverage of the BAS is shown in Fig 3 and Fig 4. The stations were activated on different dates, as local problems were debugged. The most significant factors affecting performance are:

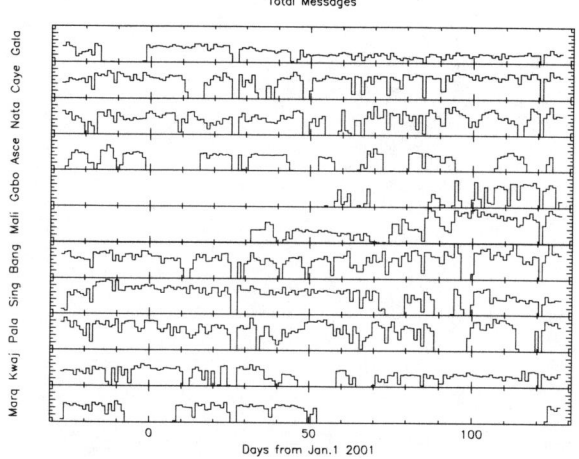

**FIGURE 3.** Daily message tally coming from the different stations as a function of time. The occasional periods of non-activity for some stations reflects the difficulty of maintaining and correcting problems by remote intervention at those isolated sites.

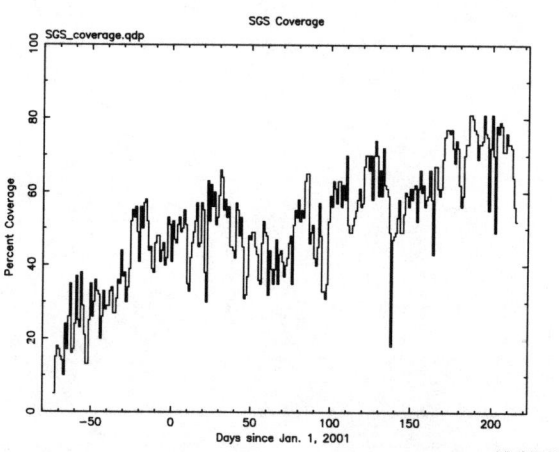

**FIGURE 4.** Increasing SGS coverage with time. The coverage is the actual time that HETE was in contact with one or more stations.

*Satellite–to–station geometry*: For unfavorable spacecraft roll angles, the body of the spacecraft can block the direct line of sight of the whip antenna to the receiving station. Diurnal variation in reception is observed in all stations.

*Varying reception due to local conditions*: The general solution to reducing local noise is to place the antenna suitably far from interfering sources; however, this could not be optimized prior to the launch, and the placement was often established based on proximity to the internet port. Satisfactory performance was nevertheless obtained in participating stations which are even telecommunications facilities. General ionospheric conditions plays only a minor role in reception.

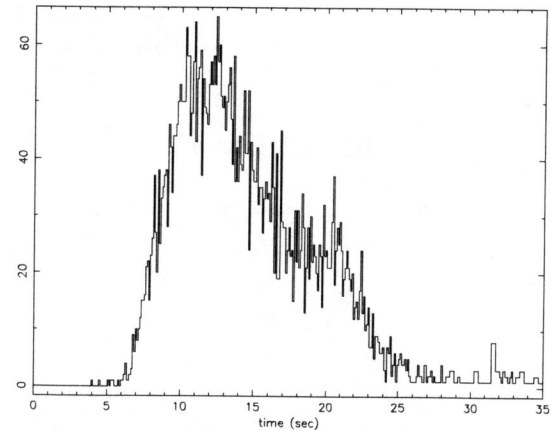

**FIGURE 5.** Histogram of the transit travel times from message creation on board the spacecraft to reception at the MIT Control Center.

*Internet availability*: Continuous internet access is available on all stations except Ascension, Kiritimati, and Marquesas, which dial-up and connect just before each pass. Prolonged periods of outage were mostly due to the lack of internet access, due to various factors including power outage and problems with the local ISP. Internet breakdown has been by far the likeliest cause for any downtime. There is significant station-to-station interconnectivity variation.

A typical VHF message is displayed on an operations console 10 seconds after it is created on the spacecraft (Fig. 5). This number factors in message transmission time (which at 300 bps can last 3-4 seconds for normal diagnostics, shorter for burst alerts), and internet delays.

## ACKNOWLEDGMENTS

The BAN was established and is currently maintained with the assistance of NASA, CNES, USGS, IRIS, Kwajalein Missile Range (KMR), Charles Darwin Research Foundation (CDRF), Redsat, Italian Space Agency (ASI), National University of Singapore (NUS) and the Two Boats School in Ascension (TBS). The BAN owes its existence from the contributions of the following people:

Mike Doucette, Kenton Philips, Charlie Sievers, Fred Miller, Bill Mayer, and Ed Boughan (MIT); John Derr and Rhett Butler (USGS); Robert Miller, Jerry Landess, and Mike Hart (KMR); Francois Lassere, Patrick Gelard, Frederick Lemagner, Henri Clergeot, Etienne Perrin, Eric Metral and Amrane Oukaour (CNES), Mauricio Patino (Redsat); Joe da Silva, Leonardo Vivar and Heidi Snell (CDRF); Parag Shah and Joe da Silva (TIFR); June Sim and Roy Drinkwater (TBS); Alex Nichols and Gary Peterson (TRW); Choong Weng Mak, Lim Geok Quee,

Lee Kok Wah, Leong Keong Kwoh (NUS); Ken Torii (RIKEN); Ennio Morelli and Fulvio Gianotti (ASI).

## REFERENCES

1. Crew, G., *et al.*, these proceedings.

# Soft Gamma-Ray Repeaters as Observed by the HETE-2 WXM

K. Torii*, A. Yoshida[†], N. Kawai**, Y. Shirasaki[‡], T. Tamagawa*, T. Sakamoto**, M. Matsuoka[§], E. Fenimore[¶], M. Galassi[¶], J.L. Atteia[∥], K. Hurley[††], N. Butler[‡‡], D.Q. Lamb[§§], C. Graziani[§§], T. Donaghy[§§], R. Vanderspek[‡‡], G. Ricker[‡‡] and HETE-2 Science Team

*RIKEN, Wako, Saitama 351-0198, Japan*
[†]*Aoyama Gakuin University, Shibuya-ku, Tokyo 150-8366, Japan*
**Tokyo Institute of Technology, Meguro-ku, Tokyo 152-8551, Japan, and RIKEN*
[‡]*JST/NASDA, Tsukuba, Ibaraki 305-8505, Japan*
[§]*NASDA, Tsukuba, Ibaraki 305-8505, Japan*
[¶]*Los Alamos National Laboratory, Los Alamos, NM 87545, USA*
[∥]*CESR, Toulouse Cedex 4, France*
[††]*University of California, Berkeley, Berkeley, CA 94720-7450, USA*
[‡‡]*MIT Center for Space Research, MIT, Cambridge, MA 02139, USA*
[§§]*Department of Astronomy & Astrophysics, University of Chicago, Chicago, IL 60637, USA*

**Abstract.** The Wide Field X-ray Monitor (WXM) onboard HETE-2 spacecraft observes the rectangular sky region of about 1.6 str centered at the anti-solar direction. SGR1806-20 and SGR1900+14 were monitored with the WXM instrument for about 60 days between June and August 2001 while they were within the field of view.

We have firmly detected 2 and 4 bursts from SGR1806-20 and SGR1900+14, respectively. A burst from SGR1900+14 occurred at 2001 July 2, 03:34 UT had a high fluence of $\sim 8 \times 10^{-6}$ ergs cm$^{-2}$ (6 – 40 keV) and broad-band burst spectrum was obtained by WXM and FREGATE (GCN Circular 1078). Here we present preliminary analyses of these bursts, and other $\sim$20 possible bursts from the SGRs detected by HETE-2.

## INTRODUCTION

The gamma-ray burst monitor satellite *HETE-2* observes the sky region of anti-solar direction. Although the main observational targets of the satellite are classical GRBs, its instruments are capable of monitoring two SGRs, SGR 1806-20 and SGR 1900+14 for about 60 days when the field of view crosses Galactic plane.

The soft gamma-ray repeaters are a class of objects which show occasional outbursts of intense soft gamma-rays. By now, four objects, SGR 0526-66, SGR 1627-41, SGR 1806-20, and SGR 1900+14 are known. One source, SGR 0526-66 is in the Large Magellanic Clouds and the others are in the Galaxy.

The most interesting aspect of the SGRs is that they are considered to have a super-strong magnetic field of $\sim 10^{15}$G. Two pieces of observational evidence support this idea. One is the large spin-down rates ($\dot{P}$) of their spin periods. If we apply the equation of spin-down via the magnetic dipole radiation, the corresponding dipole field on the surface of the (neutron) star becomes $\sim 10^{15}$G. The other evidence is that they show outbursts of peak luminosity far exceeding the Eddington luminosity. The gravity force is not strong enough to bound the emitting plasma of such a high luminosity and the strong magnetic field is considered to bound the plasma in the giant outbursts.

## HETE-2 OBSERVATIONS

In June – August, 2001, the field of view of *HETE-2* came across the Galactic plane where the two SGRs, 1806-20 and 1900+14 were monitored.

We have surveyed the triggered and untriggered bursts from SGRs (Table 1). Gamma Channel B (6 – 80 keV) data were analyzed by using wavelet based analysis tool. The events with signal to noise ratio of larger than 5.5 and the duration of less than 2 s were picked up.

Relatively strong bursts give hardware triggers to the

**TABLE 1.** The list of bursts.

| Burst date/time (UT) | Burst ID | Note |
|---|---|---|
| 20010608_045835.20 | | |
| 20010612_023313.97 | 1546 | |
| 20010618_194249.78 | 1560 | SGR1806−20 |
| 20010623_044321.59 | | |
| 20010623_155453.31 | 1566 | SGR1806−20 |
| 20010623_221345.48 | | |
| 20010627_232418.36 | 1568 | SGR1900+14 |
| 20010628_011040.00 | 1569 | |
| 20010628_035839.04 | 1570 | |
| 20010628_150636.79 | 1571 | SGR1900+14 |
| 20010701_055556.74 | 1574 | |
| 20010701_063629.77 | | |
| 20010702_001412.50 | 1575 | |
| 20010702_033409.35 | 1576 | SGR1900+14 |
| 20010702_093116.43 | 1577 | |
| 20010703_065507.58 | 1578 | SGR1900+14 |
| 20010703_114539.07 | | |
| 20010703_161034.12 | | |
| 20010703_205527.72 | | |
| 20010703_212300.38 | | |
| 20010704_100300.59 | | |
| 20010704_181657.27 | | |
| 20010708_185117.30 | | |
| 20010712_050420.23 | | |
| 20010716_122431.23 | 1603 | |
| 20010728_093740.46 | | |
| 20010730_081757.95 | | |

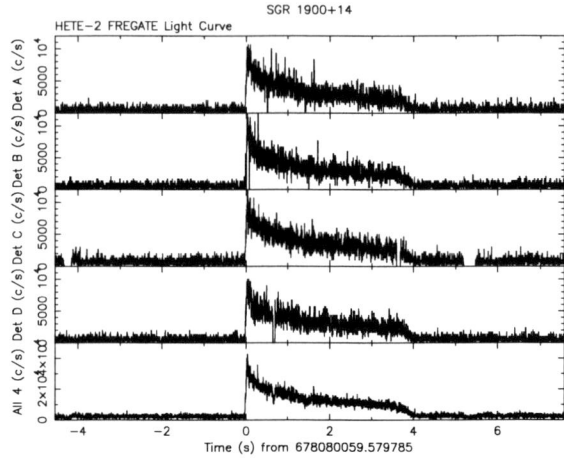

**FIGURE 1.** FREGATE light curves for the BID 1576. Four curves from different FREGATE detectors are shown.

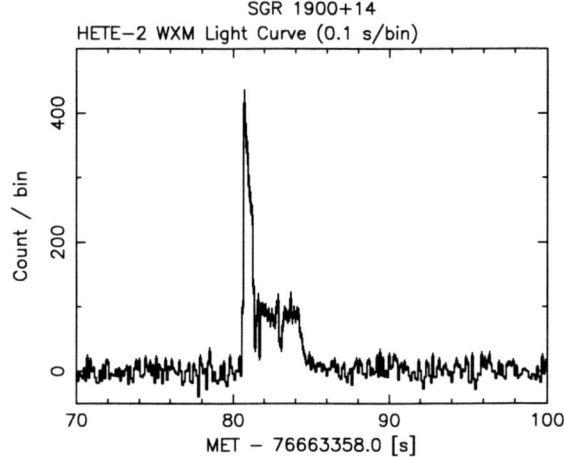

**FIGURE 2.** WXM light curve (total energy range and total wires) for the BID 1576.

*HETE* FREGATE instruments. Burst IDs 1560, 1566, 1568, 1571, 1576, and 1578 belong to this class of bursts.

The WXM light curves were created for each burst summarized in the table. If the burst is significantly detected by the WXM, then localizations were made. As a result, two bursts at 20010618_194249.78 and 20010623_155453.31 were localized to SGR1806-20, and four bursts at 20010627_232418.36, 20010628_150636.79, 20010702_033409.35, and 20010703_065507.58 were localized to SGR1900+14.

## THE STRONG BURST ON 2001 JULY 2

The burst from SGR1900+14 on 2001 July 2 [1] had the highest fluence in the Table 1. Figure 1 shows the high resolution FREGATE light curve and figure 2 shows the 0.1-s resolution light curve. The burst lasting ∼4 s is clearly seen in both instruments.

We find sharp dips in the WXM light curve, which are not seen in the FREGATE data. These dips are due to saturation in DSP processing. We have therefore assumed that the light curve for WXM should be the same as that for the FREGATE Channel A (6–40 keV) and estimated the dead time effect every 0.5 s.

We have thus extracted the burst spectrum as shown in figure 3 where the dead time effect is approximately taken into account.

We have compared the burst spectrum with that of the Crab nebula. We have calculated the spectral ratio between the burst and the Crab nebula. Then, we have fitted the ratio with a power-law function to derive approximate values of the power-law indices. Figure 4 shows the time resolved values of the power law index derived by fitting the spectral ratios SGR/Crab. Gradual steepening (softening) of the spectrum is seen.

## Summary

We have presented timing and spectral data for the SGRs. Detailed spectral analyses and combined analyses of the WXM and FREGATE spectra will be presented in

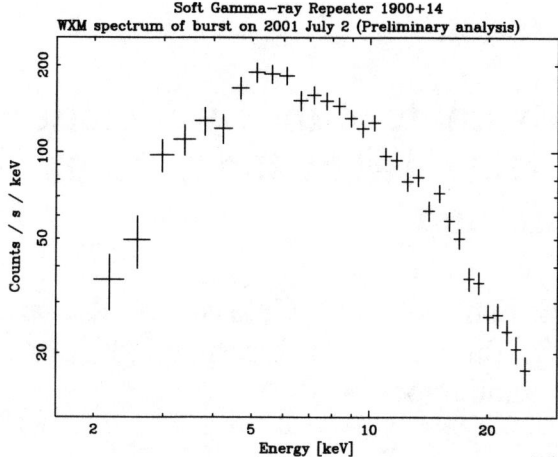

**FIGURE 3.** WXM spectrum for the BID 1576.

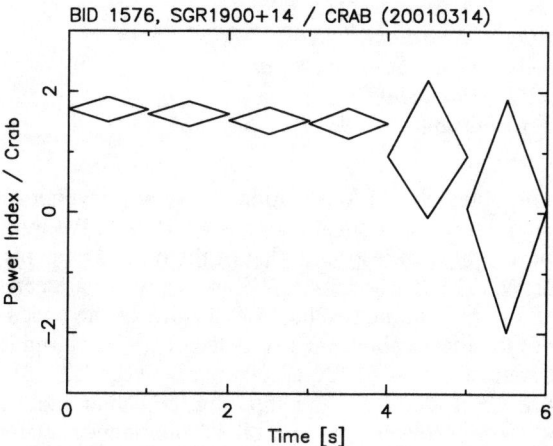

**FIGURE 4.** Spectral ratio of the SGR 1900+14 / Crab nebula.

a future article.

# REFERENCES

1. Ricker, G., et al., 2001, GCN Circulars, 1078

# Astrometric Calibration and Estimate of the Systematic Error in WXM Localizations Obtained by the Chicago Bayesian Method

C. Graziani[*], Y. Shirasaki[†], T. Donaghy[**], E. Fenimore[‡], M. Galassi[‡], N. Kawai[§], D.Q. Lamb[**], T. Sakamoto[§], D. Takahashi[¶], T. Tamagawa[||], T. Tavenner[‡], K. Torii[||], A. Yoshida[¶] and R. Vanderspek[††]

[*]*Department of Astronomy & Astrophysics, University of Chicago, 5640 South Ellis Avenue, Chicago, IL 60637*
[†]*JST/NASDA, Tsukuba, Ibaraki 305-8505, Japan*
[**]*Department of Astronomy & Astrophysics, University of Chicago, Chicago, IL 60637, USA*
[‡]*Los Alamos National Laboratory, Los Alamos, NM 87545, USA*
[§]*Department of Physics, Tokyo Institute of Technology, Meguro-ku, Tokyo 152-8551, Japan*
[¶]*Aoyama Gakuin University, Shibuya-ku, Tokyo 150-8366, Japan*
[||]*RIKEN, Wako, Saitama 351-0198, Japan*
[††]*MIT Center for Space Research, MIT, Cambridge, MA 02139, USA*

**Abstract.** WXM gives GRB localizations in instrument coordinates. WXM localizations must be converted to celestial coordinates using spacecraft aspect information obtained by the optical cameras on HETE. We must therefore accurately determine the alignment of the WXM boresight with respect to that of the optical cameras, in order to accurately determine the celestial coordinates of WXM burst locations. We use a seven-parameter model that treats as free parameters the three Euler angles of a pure rotation, two horizontal shifts of the coded-aperture masks with respect to the detectors, and the heights of the masks above the two detectors. We determine the alignment by fitting the model to a set of 252 WXM localizations of Sco X-1 obtained between 23 April and 28 June 2001. We estimate the systematic error in WXM GRB locations by comparing the actual and the calculated locations of Sco X-1. We find that the systematic error corresponding to a 68.3% confidence region is 1.7′, and the systematic error corresponding to a 90% confidence region is 2.4′. We find that this astrometric solution also provides a satisfactory fit to an independent sample of SGR and XRB events. These results are consistent with the astrometric calibration and the systematic error in WXM localizations derived independently using the RIKEN localization method.

## INTRODUCTION

The spacecraft placement of WXM is offset from that of the optical cameras (the OPT subsystem). The coordinate frames of WXM on of the optical cameras are only aligned to within machining tolerances and thermal shrinkage effects. These misalignments are of the order of $0.4°$, and must be corrected so that HETE can provide WXM locations accurate to $5' - 15'$. Any systematic effects inherent in the localization procedure compound the misalignment effects, and must also be corrected.

The HETE team employs two independent WXM localization pipelines, which use different location algorithms. The existence of two independent WXM localization pipelines has been invaluable as a check of the correctness of the design and implementation of each approach. We describe here the study of astrometric correction of the Chicago (Bayesian) location algorithm [1]. The calibration of the RIKEN location algorithm is described elsewhere in these proceedings.[2]

We have performed this astrometric calibration on-orbit, using sources whose locations are accurately known — Sco X-1, several X-Ray Burst (XRB) sources, and two Soft-Gamma Repeaters (SGRs).

## THE DATA

The calibration was performed using "RAW" data of Sco X-1 obtained in the course of the daily health check observations made during the period 2001 April 23 and 2001 June 28. There are 252 Sco X-1 locations deter-

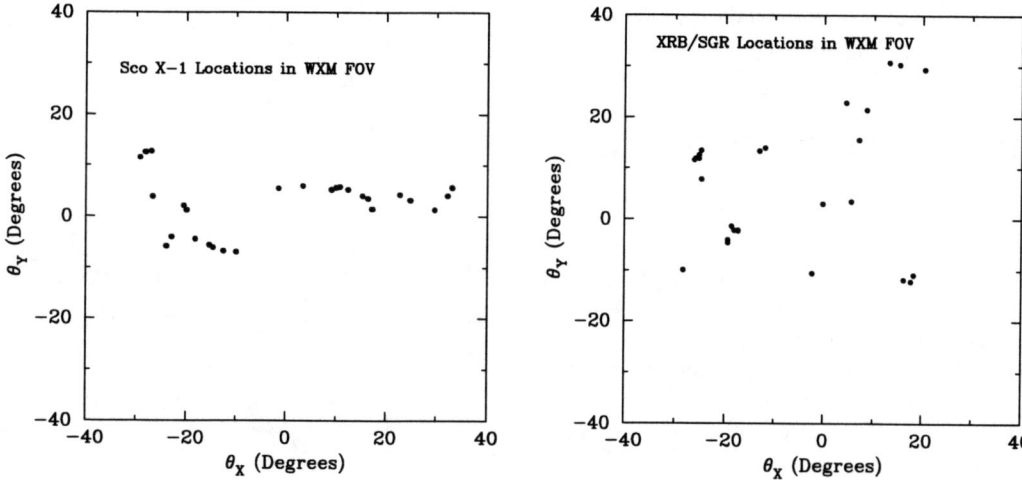

**FIGURE 1.** Location of the calibration sources in the WXM FOV. Left Panel: Location of Sco X-1 during each of the 27 health check sessions used for astrometric calibration. Right panel: Locations of 28 XRB and SGR sources.

mined using 27 sessions of 180s each. The statistical errors for these locations are estimated from the scatter in the locations independently for each session. The location of Sco X-1 in the WXM FOV for each of these 27 sessions is shown in the left panel of Figure (1).

We also derived the locations of 23 XRBs from known XRB sources, of three SGR1900+14 bursts, and of two SGR1806-20 bursts. We used these locations as an independent check on the quality of the fit to the Sco X-1 data. The statistical errors for these locations are produced directly by the Bayesian location process. Their locations in the WXM FOV are shown in the right panel of Figure (1).

## THE ASTROMETRIC MODEL

The transformation from WXM to OPT coordinates is assumed to have the following form:

$$\tan\left(\theta_X^{(True)}\right) = S_X \tan\left(\theta_X^{(Meas)}\right) + d_X, \qquad (1)$$

$$\tan\left(\theta_Y^{(True)}\right) = S_Y \tan\left(\theta_Y^{(Meas)}\right) + d_Y, \qquad (2)$$

$$\vec{n}^{(Opt)} = \mathbf{M}(\varepsilon_1, \varepsilon_2, \varepsilon_3) \cdot \vec{n}^{(WXM)}\left(\theta_X^{(True)}, \theta_Y^{(True)}\right). \qquad (3)$$

The projection angles $\theta_X^{(Meas)}$ and $\theta_Y^{(Meas)}$ are the angles that result from the Bayesian location analysis. The model corrects these angles to produce "true" projection angles $\theta_X^{(True)}$ and $\theta_Y^{(True)}$. The unit vector $\vec{n}^{(WXM)}$ is the direction vector to the source defined by the projection angles $\theta_X^{(True)}$, $\theta_Y^{(True)}$. The unit vector $\vec{n}^{(Opt)}$ is the direction vector to the source in the OPT frame.

$\mathbf{M}$ is a rotation matrix. $\varepsilon_1$, $\varepsilon_2$, and $\varepsilon_3$ are Euler angles. $S_X$ and $S_Y$ are scale parameters that can represent unknown changes in the height of the coded aperture masks. $d_X$ and $d_Y$ are shifts that can represent unknown mis-alignments of the masks with respect to the detectors. The model thus has 7 free parameters.

## SCO X-1 CALIBRATION RESULTS

With no astrometric correction, the $\chi^2$ of the fit is $2.9 \times 10^5$ for 504 DOF, and the RMS deviation (computed versus actual locations) is 0.39°.

The deviations from the true locations are shown in the left panel of Figure (2), together with the estimated location errors, which are typically in the 1'-2' range. The misalignment of the WXM and OPT frames is clearly manifested in the Figure.

The best-fit astrometric correction results in a $\chi^2 = 1517$ for 497 DOF. The fit gives an RMS deviation of 2.0'. The deviations from the true locations are shown in the right panel of Figure (2).

While clearly a distinct improvement over the null correction fit, the quality of this fit is rather poor — the Q-value for $\chi^2 = 1517$ from the $\chi^2$ distribution with 497 DOF is $\sim 10^{-103}$. This excess $\chi^2$ is attributable to systematic error. The source of the error is presumably a compounding of the inadequacy of the astrometric model with the limitations of the location analysis.

We estimate the magnitude of the systematic error as follows: we add (in quadrature) a systematic error to the statistical error so as to bring the $\chi^2$/DOF down to about 1. In this way we find that the systematic error corresponding to a 68.3% confidence region is 1.7', and the systematic error corresponding to a 90% confidence region is 2.4′. These results are consistent with those found using the RIKEN location algorithm.[2]

## AN INDEPENDENT CHECK: THE SGR/XRB SAMPLE

When the astrometric model that best fits the Sco X-1 data is applied to the XRB/SGR bursts and the derived locations are compared with the known locations of the sources, the result is $\chi^2 = 39.6$ for 56 DOF (P=0.05). The RMS deviation is 7.0'. The Sco X-1 astrometric solution thus provides a satisfactory fit to this independent sample of event locations. No systematic error is needed in this fit, because these sources are less bright — and thus less accurately located — than Sco X-1. The typical statistical error for this sample is about 6'.

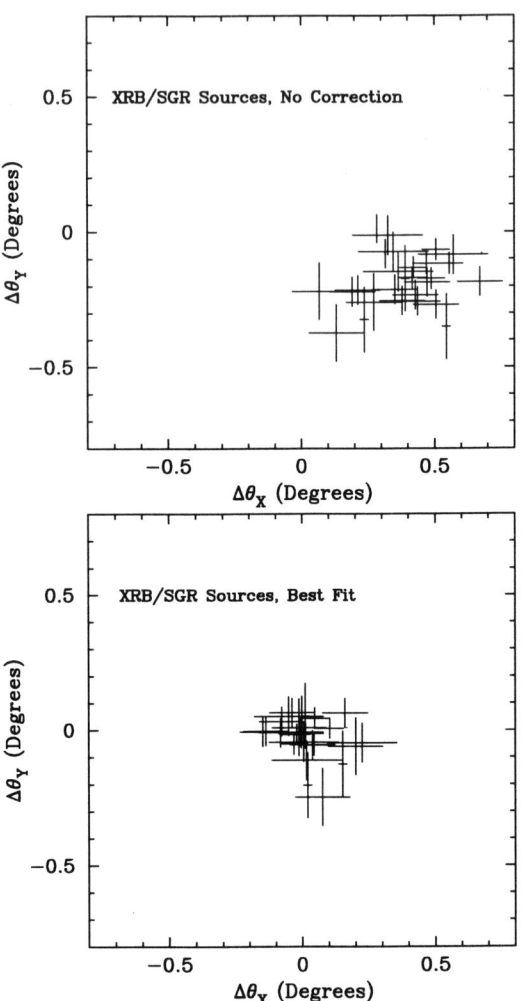

**FIGURE 3.** Deviations of derived locations of XRB/SGR sample from true source location. Top: Assuming no astrometric correction. Bottom: Assuming the astrometric correction that best fits the Sco X-1 data.

## REFERENCES

1. Graziani, C., and Lamb, D. Q., "Localization of GRBs by Bayesian Analysis of Data from the HETE WXM", these proceedings, 2002.
2. Shirasaki, Y., et al.,"Astrometric Calibration and Estimate of the Systematic Error in HETE WXM Localizations Obtained by the RIKEN Cross-Correlation Method," these proceedings, 2002.

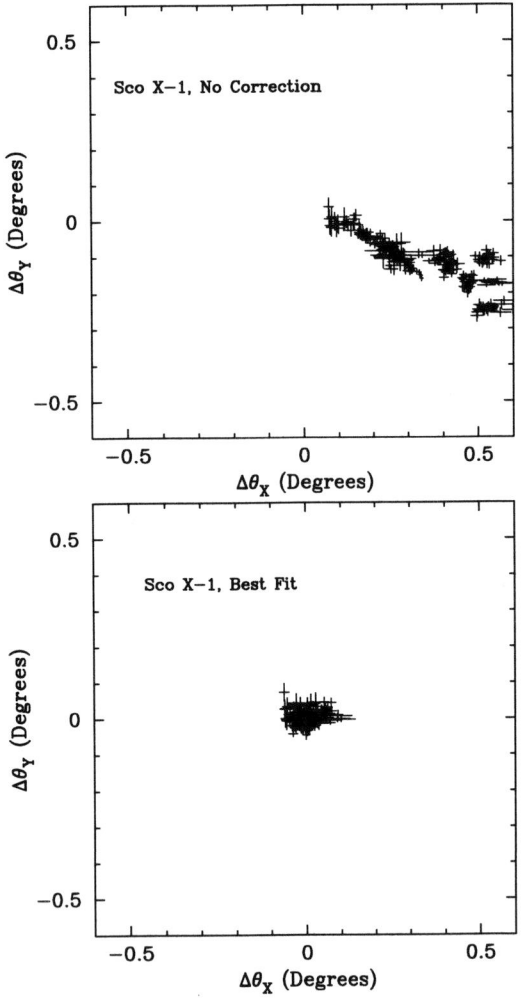

**FIGURE 2.** Deviations of derived locations of Sco X-1 from true source location. Top: Assuming no astrometric correction. Bottom: Assuming the astrometric correction that best fits the Sco X-1 data.

# Astrometric Calibration and Estimate of the Systematic Error in HETE WXM Localizations Obtained by the RIKEN Cross Correlation Method

Y. Shirasaki*, N. Kawai†, A. Yoshida**, M. Matsuoka‡, T. Tamagawa§, K. Torii§, T. Sakamoto†, E. Fenimore¶, M. Galassi¶, D. Lamb‖, C. Graziani‖, R. Vanderspek††
and HETE-2 Science Team

*JST/NASDA, Tsukuba, Ibaraki 304-8505, Japan
†Tokyo Institute of Technology, Meguro-ku, Tokyo 152-8551, Japan
**Aoyama Gakuin University, Shibuya-ku, Tokyo 150-8366, Japan
‡NASDA, Tsukuba, Ibaraki 304-8505, Japan
§RIKEN, Wako, Saitama 351-0198, Japan
¶Los Alamos National Laboratory, Los Alamos, NM 87545, USA
‖University of Chicago, Chicago, IL 60637, USA
††Massachusetts Institute of Technology, Cambridge, MA 02139-4307, USA

**Abstract.** Wide field X-ray Monitor (WXM) of HETE-2 is designed to localized the GRBs which occurred in its wide field of view ($60° \times 60°$) with $10'$ accuracy. In order to maintain such a good accuracy, we have been monitoring the positional response of proportional counter and correcting the change at regular intervals. We also have determined the difference of the alignments between WXM and the optical camera system, which is crucially important for determining the coordinate of the GRBs. The localization accuracy is estimated to be $\sim 3'$ for 1 Crab burst with 20 sec duration including the systematic error of $2.2'$.

## INTRODUCTION

Wide field X-ray Monitor (WXM) plays the most important roll in the GRB localization by HETE-2 satellite. WXM consists of two identical units of one-dimensional position sensitive X-ray detectors. They are placed in orthogonal directions to each other for measuring the X and Y directions independently. One unit consists of a one-dimensional coded mask and two 1-D position-sensitive proportional counter (PSPCs) placed 187 mm below the mask. The location of the GRB is determined by measuring a set of two shift distances of the mask pattern in the X and Y directions. Each PSPC has three anode wires made of a carbon fiber with 10 $\mu$m diameter in its upper cells and four veto wires in its lower cells, and filled with 1.4 atm Xenon gas (3% $CO_2$ quench gas). More details are described in [1].

During the period from December of 2000 to July of 2001, we have performed calibration of the detector using steady X-ray sources, Crab nebular and Sco X-1, and also several transient sources. The results were compared with the independent analysis with the Chicago method [2], and were used to establish the scale and alignment of the WXM coordinate system with respect to that of the spacecraft.

## CALIBRATION OF PC

In order to calibrate the positional response of the PSPCs, we have been monitoring the charge division ratio, which is a measure of X-ray incident position, using X-rays of the $Fe^{55}$ isotopes attached at the side walls. The isotopes irradiate the fixed positions of each anode at $x = \pm 40$ mm. From this data we can inspect the linearity of the charge division ratio to the physical position at any time. This is crucially important for accurate localization, for an example, at least 0.5 mm of positional accuracy is needed to achieve $10'$ directional accuracy considering 187 mm of separation distance between the coded mask and the PSPCs.

The zero point adjustment is also carried out for all the anode wires using the data of a bright steady source Sco X-1. The determination accuracy of the PSPC coordinate on the ground calibration was $\sim 0.2$ mm due to the lim-

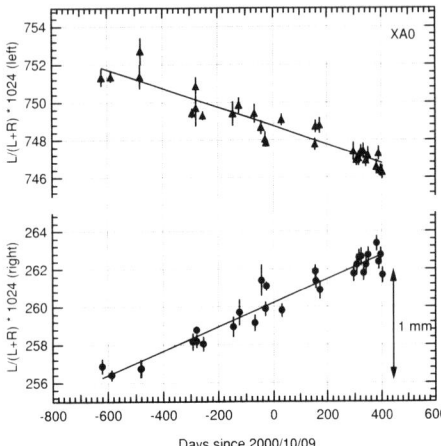

**FIGURE 1.** Monitoring of the positional calibration. This is an example for the case of XA0 anode. The rate of the change is about 0.5% per year. This change is corrected every half year or more frequently.

ited experimental condition, and also the precise alignment between the coded mask and the PSPC were not measured on the ground. This calibration has been performed by measuring the systematic differences among the Sco X-1 directions calculated independently for each anode. The estimated differences are less than $0.07°$, and the zero point adjustments of no larger than 0.23 mm are applied to all the anode so that the differences are canceled.

## RIKEN LOCALIZATION METHOD

The RIKEN localization method is based on the variance weighted cross correlation method which searches the direction where the simulated position histogram best matches the observed one. The procedure consists of two parts. First we determine a coarse location $(\theta_{x,0}, \theta_{y,0})$ using the model histograms simulated for the directions spaced by $0.2°$ along $\theta_x = 0$ and $\theta_y = 0$ axes. Next the fine location is determined using the models simulated along the $\theta_x = \theta_{x,0}$ and $\theta_y = \theta_{y,0}$ axes by every $0.05°$ step.

In the model calculation, we take into account the following effects: (1) the slant penetration effect, (2) positional resolution $\sigma(E,x)$ as a function of energy and X-ray absorption position, (3) obscuration function, $f(x)$, by the PC body structure, the side wall, and the coded mask. The model histogram for X-rays of energy $E$ is expressed as

$$P(x,E) = \varepsilon(E) \cdot \frac{1}{\sqrt{2\pi}\sigma(E,x')} \cdot \frac{1}{\lambda_x(E)} \int_{x_{min}}^{x_{max}} dx'' f(x'')$$

$$\times \int_{x''}^{x''+d\cos(\theta_x)} dx' e^{-\frac{(x-x')^2}{2\sigma^2(E,x')}} e^{-\frac{(x'-x'')}{\lambda_x(E)}} \quad (1)$$

where $\varepsilon$ is the transmission coefficient of Be window with $100\,\mu m$ thickness and thermal shield with $7.62\,\mu m$ thickness, $\lambda_x$ is the projection mean free path onto the wire direction, $x_{min}$ and $x_{max}$ give the usable range of anode wire which is from -57 mm to +57 mm, $d$ is the depth of an anode cell. Considering the source spectrum of $F(E)$, the spectrum weighted model histogram is written as,

$$M(x) = \int F(E)P(x,E) \quad (2)$$

We usually assume $E^{-1.5}$ spectrum for the burst analysis, and for the purpose of WXM aspect calibration we use an appropriate spectrum type for each X-ray source.

The cross correlation score $C/\sigma_C$ is calculated as:

$$C(j) = \psi_i [n \sum_i M_{ij} S_i - \sum_i S_i] \quad (3)$$

$$\sigma_{C(j)}^2 = \psi_i^2 [n^2 \sum_i M_{ij}^2 \sigma_{S_i}^2 + \sum_i \sigma_{S_i}^2] \quad (4)$$

where subscript $i$ and $j$ represent position bin and directional bin, respectively, $\psi$ is the normalization constant, $S$ is a background subtracted observed histogram, $\sigma_S$ is the standard deviation of $S$. The score is evaluated at every $0.05°$ and the maximum value is searched with $0.01°$ step by applying interpolation.

## WXM ASPECT CALIBRATION AND SYSTEMATIC ERROR ESTIMATE

To derive the burst location in the celestial coordinate from the location in the WXM coordinate, we need to estimate the alignment of WXM with respect to the optical cameras which are used for determining the spacecraft attitude. For this purpose, the WXM aspect calibration is performed using the data of the Crab Nebula, Sco X-1, SGR 1900+14, Aql X-1 and GS1826-238. The data type of the Crab Nebula and Sco X-1 is RAW data type and that of SGR and XRBs is TAG data type. The raw data is the most primitive data format containing the pulse height values measured at both the ends of anode wire, so it is possible to converted to the physical position using the best calibration parameters at any time. On the other hand, the TAG is the data type containing a physical position determined by on-board software with 1 mm resolution, so it does not have enough resolution to be corrected by the refined parameters. The zero point correction described in the previous section was not applied to the TAG data used in this analysis, so the model histograms are calculated considering the zero point offset. The locations of each source in the WXM FOV are

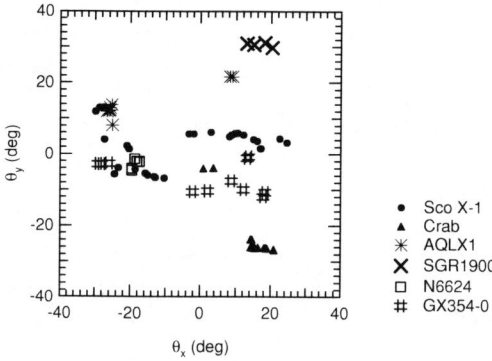

**FIGURE 2.** The location of the X-ray sources used for WXM aspect calibration. The coordinate is defined by the optical coordinate.

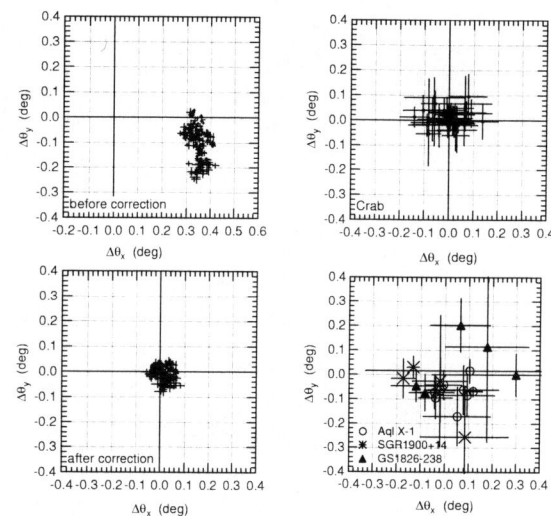

**FIGURE 3.** Localization error distribution of Sco X-1 for null aspect correction (left top) and aspect correction (left bottom) and of Crab (right top) and transient sources (right bottom).

plotted in FIGURE 2. The conversion from the WXM coordinate to the optical camera coordinate is assumed to be a pure rotational conversion around three axes, i.e. $V_{opt} = R_x(\alpha) \cdot R_y(\beta) \cdot R_z(\gamma) \cdot V_{wxm}$, where $V_{wxm}$ and $V_{opt}$ are unit vectors of the source location in the WXM and the optical camera coordinate, $R_x(\alpha)$ represents rotation around $x$ axis by $\alpha$ degree, and so on. The three Euler angles are derived by the $\chi^2$ minimization method. We obtained the best Euler angles as $\alpha = 0.070°$, $\beta = 0.340°$ and $\gamma = 0.108°$.

In FIGURE 3, the distribution of localization errors of Sco X-1 is shown for the cases of null aspect correction and of the best aspect correction. The R.M.S. deviations for these distributions are $21.6'$ and $2.4'$, respectively. The same plot for Crab, SGR and XRBs are shown in FIGURE 3, the R.M.S. deviations are $2.7'$ for Crab and $6.3'$ for SGR and XRBs.

The reduced $\chi^2$ for the fitting is 4.29 for 250 d.o.f., so we expect a non-negligible systematic error in the localized position. The systematic error is estimated so that the reduced $\chi^2$ defined by equation (5) becomes unity.

$$\chi^2_{tot} = \frac{(\theta_x^{loc} - \theta_x^{src})^2}{\sigma^2_{x,stat} + \sigma^2_{sys}} + \frac{(\theta_y^{loc} - \theta_y^{src})^2}{\sigma^2_{y,stat} + \sigma^2_{sys}} \qquad (5)$$

In this way, we obtain the systematic error of $1.56' \pm 0.3'$ for each $x$ and $y$ directions, so $2.2'$ is an expected systematic error of total direction.

## CONCLUSION

We have performed in-flight calibration of WXM to provide GRB localization in good accuracy. The estimated systematic error, which is in part caused by an attitude estimation error, is $2.2'$. Typical localization accuracy for a 1 Crab burst with 20 sec duration is $6.3'$ in R.M.S. deviation including the systematic error.

## REFERENCES

1. Shirasaki Y, e. a., "Performance of the wide-field x-ray monitor on board the High-Energy", in *Proc. SPIE Vol. 4012, p. 166-177, X-Ray Optics, Instruments, and Mission III*, 2000, vol. 4012, pp. 166–177.
2. Glaziani C, e. a., "Astrometric Calibration and Estimate of the Systematic Error in HETE WXM Localizations Obtained By the Chicago Bayesian method ", in *These proceedings*, 2002.

# II. GLOBAL PROPERTIES OF GAMMA-RAY BURSTS

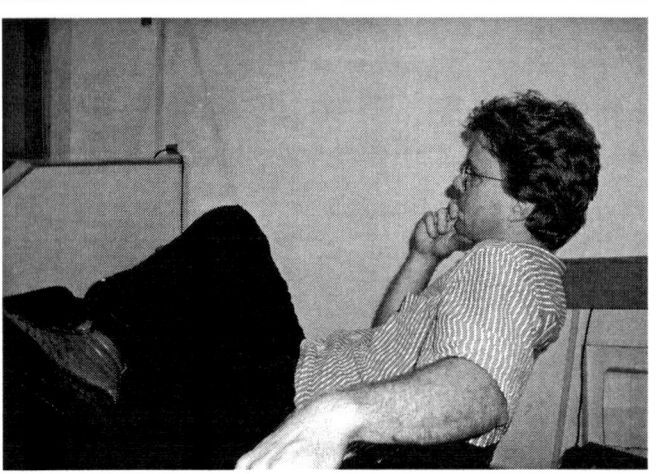

# Spectral Properties of GRBs Observed with BeppoSAX

E. Costa[*] and F. Frontera[†]

[*]*Istituto di Astrofisica Spaziale e Fisica Cosmica del CNR, via del Fosso del Cavaliere 100, 00133-I Roma, Italy*
[†]*Physics Department, University of Ferrara, Via Paradiso 12 , 44100-I Ferrara, Italy and Istituto di Astrofisica Spaziale e Fisica Cosmica del CNR via P.Gobetti 101, 40129-I Bologna, Italy*

**Abstract.** The *BeppoSAX* mission has not only significantly improved the Gamma-Ray Burst science through the discovery of the afterglows but is also providing important data on the prompt events. The Gamma-Ray Burst Monitor is building a catalogue of GRBs that, due to a recently achieved, coarse but suitable positioning capabilities, can usefully integrate and extend the BATSE catalogues. We show the good relative calibration of the two experiments. Wide Field Cameras are providing a sample of about 40 GRBs, in an important band so far relatively ill covered. Moreover the combined use of the two instruments is providing a unique information on the spectral evolution of the GRBs. We show some recent spectral results based on these data and discuss the impact on our knowledge of GRB physics.

## INTRODUCTION

The *BeppoSAX* satellite has opened a new era in GRB science through the discovery of the afterglow [10] but it has also been providing highly competitive results on the prompt emission of the events, the *classic* GRB. After the crisis following the loss of the last gyroscope the new giro-less pointing mode has been activated and commissioned. At the time of this Conference the payload is working nominally at the same level since May 1997.

The Gamma-Ray Burst Monitor [12, 11](40–700 keV), in spite of some limitations, deriving from its primary role of anticoincidence detector for the Phoswich Detector System (PDS), is one of the largest GRB detector arrays ever flown, and benefits of the *BeppoSAX* equatorial orbit (3.9° inclination, circular starting from 600 km) that gives a very low instrument background and a small background modulation (less than 15% ). Moreover this background is slowly decreasing through the orbit decay.

The *BeppoSAX* Wide Field Cameras (2-30 keV) [20] are the first instrument performing a long term monitoring of 5% of the sky with 5 arcminutes positioning capability, millisecond timing and the moderate but suitable energy resolution of wire proportional counters.

The joint analysis of GRBM and WFC provides the best instrument combination, not only for the identification and localization of GRBs, but also for the study of the time resolved GRB spectra in a band never covered by an instrumentation in a single satellite.

We will report on some of the latest results obtained with the *BeppoSAX* GRBM and WFCs.

## GRBM ALONE

### Directional Capabilities

The *BeppoSAX* GRBM is made of 4 detection units of CsI(Na) equipped with an electronics providing continuously, for each unit, 1 s ratemeters in two energy bands (40–700 keV and >100 keV), spectra integrated on 128 s in 240 energy channels and, in the case of a GRB trigger, 40–700 keV ratemeters for about 100 s, with high time resolution (down to 0.5 ms). The counting rate from the 4 units depends on the exposed area to the direction of the burst. At low energies this is proportional to the pure projection minus the fraction shadowed by other shields. At higher energies, this is less straightforward because, with increasing transparency of the detectors, volume effects increase their role [29]. Further complications come from the relevant backscattering of the Earth atmosphere. By taking into account all these effects the direction of the burst can be derived from the comparison of the GRB counting rate in different detectors. This concept, very successfully applied by BATSE, can in principle work with GRBM data as well. But the GRBM is in the center of the *BeppoSAX* payload and the distribution of materials outside and inside the 4 detection units is very irregular. In order to achieve at least a coarse information on the burst direction, which is needed to derive the source count flux and express it in physical units, the PDS/GRBM team has implemented a very extended Monte Carlo code with a very detailed description of the materials of SAX satellite and their position with respect to different detectors [9]. The code is adjusted in order to arrive to describe correctly the data of the on–ground calibrations (after subtraction of local backscattering effects). As a result we have now a reasonably good description of the response of each GRBM detector at any energy and angle. By this response matrix we can derive nowadays a confidence region in the sky map for the large majority of bursts [17, 25]. In Fig. 1 we give two examples of GRB localization by this method. The statistical error is of the order of $10°$ and a further error of $10°$ is due to systematic effects. In Fig. 2 we show a case in which the GRBM has allowed the determination of the GRB position [28] from two possible solutions given by the IPN (K. Hurley, private communication).

*Spectral Capabilities*

The spectral capability of the GRBM detection units 1 and 3, which have their axes parallel to those of the WFCs, was tested soon after the *BeppoSAX* launch and is well known for GRB directions within the FOV of the WFCs. To deconvolve the GRB spectra detected by other detection units or by the units 1 and 2 in the case of directions outside the WFC FOV, we use the same matrix used for positioning GRBs. With a positioning capability of the order of $20°$ the uncertainty in the spectral normalization is at most 20%. For GRBs with a good statistics we can use the 240 channel spectra. In Fig. 3 we show the pulse height spectrum of GRB991216 as detected by GRBM. The measured count spectrum is compared with that expected using the best fit parameters obtained using the BATSE data for this GRB [21] convolved with the GRBM matrix for the GRB direction. Also the K-edge from Lead present in the nearby *BeppoSAX* instrument (HPGSPC) is present both in the data and in the matrix. The agreement is impressive, especially if we consider that the BATSE spectrum is used without any free parameter! This also shows that the relative calibration of GRBM and BATSE is excellent and data from the two missions can be combined for data compilations. This can be important for comparative studies in relatively small samples such as GRB with known redshift.

*Data distribution*

Following the status of knowledge of the instrument a systematic study of the GRBM data archive has been started. By combining together bursts triggered onboard and on–ground validated with bursts triggered off-line using the 1 s ratemeters we have a burst rate of 0.7 day$^{-1}$. After the on-ground validation essential data of each detected burst are included in an alert message which is distributed to the scientific community that has made a request [17]. The on-ground validation is automatical. The time delay from the GRB onset to the alert message ranges from 3 to 6 hrs, depending on the relative occurrence time of the GRBs with respect to visibility time of the satellite from the ground station in Malindi.

# SOME SPECTRAL RESULTS FROM WIDE FIELD CAMERAS

At the time of this conference WFCs had detected 43 GRBs and localized with a typical error radius of 3 arcminutes. Thirty–two of them were followed up with *BeppoSAX* Narrow Field Instruments which discovered an afterglow source in at least 80% of the cases (in practice all GRBs pointed within a reasonably short time window). Some important results on the GRBs are the subject of different talks at this Conference [16, 19, 32]. Here we want to review a few spectral data, obtained with Wide Field Cameras and GRBM, that seem to deviate from the standard synchrotron model, likely as a consequence of local absorption and/or Compton scattering. We think these data can enlighten us either on the fireball itself or on the close environment. We recall that a small but significant set of data on Fe lines has been collected in afterglow spectra [30, 34, 5, 31]. These can be interpreted either as fluorescence lines or as recombination features from the heavily photoionized circumburst medium (CBM). The geometry of the CBM material which is responsible of this emission is still subject of discussion (e.g., [33, 27]). In spite that we have no special reason to assume an isotropic and homogeneous distribution of this medium, some track of the ionization process should be also left in the spectral data of the prompt emission. We have found evidence of this ionization in a few cases, as it will be discussed below.

*Transient absorption in GRB990729*

From the study of the combined analysis of time-resolved spectra [13] of GRBM and WFC it is apparent that a simple thin synchrotron cannot explain the low energy part of the photon spectra, below the peak energy $E_p$ of the $EF(E)$ distribution, which is in several cases described by a power law with a slope steeper than the maximum synchrotron spectral slope (-2/3). Beside this effect, likely connected to Comptonization in a highly dense plasma, we find that in some spectra a low–energy exponential cut-off is consistent with a photoelectric absorption with a hydrogen column density $N_H$. In some cases this looks variable with time. This is a likely ev-

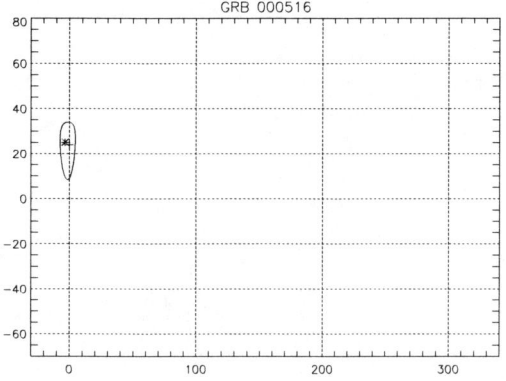
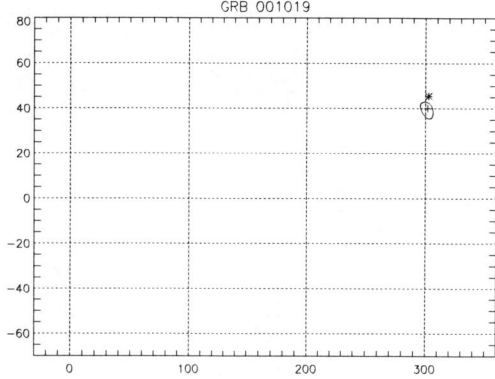

**FIGURE 1.** Two GRBs localized by GRBM. The asterisk denotes the IPN position, while the *plus* indicates the GRBM centroid. The confidence level is only statistics (90%). A systematic error of about 10 deg has to be added. The axes are in local coordinates.

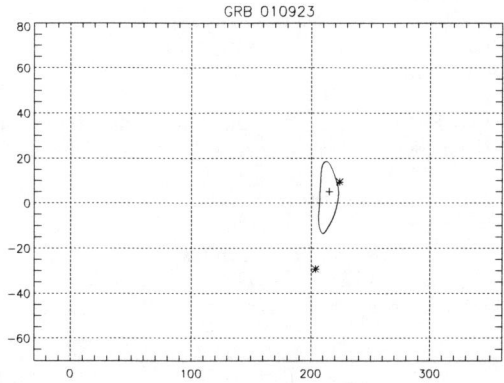

**FIGURE 2.** GRB010923, a GRB localized by GRBM. The *asterisks* denote the two possible IPN positions obtained with Ulysses/Konus/GRBM (Hurley, private communication), while the *plus* indicates the GRBM centroid [28]. The confidence level is only statistical (90%). In this case the coarse positioning by the *BeppoSAX* GRBM is sufficient to solve the ambiguity between the two candidate boxes from IPN. The axes are in local coordinates.

idence for photoionization of the GRB environment by the prompt burst radiation (e.g. [7]).

A decreasing $N_H$ was observed in the spectral evolution of GRB980329, from $\sim 2 \times 10^{23}$ cm$^{-2}$ to $\sim 1 \times 10^{22}$ cm$^{-2}$ [13].

A $N_H$ decrease from $4 \times 10^{22}$ cm$^{-2}$ (GRB onset) to a 3σ upper limit of $1 \times 10^{22}$ cm$^{-2}$ (much higher than the Galactic $N_H = 1.6 \times 10^{20}$ cm$^{-2}$) has been observed from GRB010222 [35].

Lazzati and Perna [24] show that the evolution of $N_H$ is connected with the density and size of the absorbing medium. From the $N_H$ behavior of GRB980329, Lazzati and Perna [24] demonstrate that GRB980329 is associated to overdense regions in molecular clouds, with properties similar to those of star formation globules (Bok globules).

*Transient Absorption in GRB990705*

A transient absorption feature at $3.8 \pm 0.3$ keV has been discovered in the X-ray spectrum of the prompt emission of GRB990705 [1]. The edge is present in the first 13 s of the burst and fades in the later part of the prompt emission (see Fig. 4 and Table 1).

If interpreted as an Fe K absorption edge due to a shell of neutral material around the GRB location, we derive a redshift of $0.86 \pm 0.17$, which is in agreement with the optical redshift ($z_{opt} = 0.84$) of the associated host galaxy [3]. The derived Iron relative abundance ($\sim 75$) should imply the existence of an iron-rich environment along the line of sight to the GRB, which is typical of supernova explosions. The short duration of the feature could be consequence of the photoionization of this environment by the GRB photons [7].

The drawback of this interpretation is the large mass of Fe required (tens of solar masses), unless the absorbing Iron is clumped (covering factor of a few percent) and a clump is by chance along the line of sight [8].

An alternative explanation [23] is that the feature is an absorption line due to resonant scattering of the GRB photons on H-like Iron (transition 1s-2p, $E_{rest} = 6.927$ keV [22]). The line broadening should derive from the outflow velocity dispersion (up to $\sim 0.1c$). The Fe mass required is $\sim 0.16 M_\odot$, while the Fe relative abundance $A_{Fe} \sim 10$.

Independently of the specific model, the data point to the presence of a iron-rich circumburst environment, typical of a young supernova remnant.

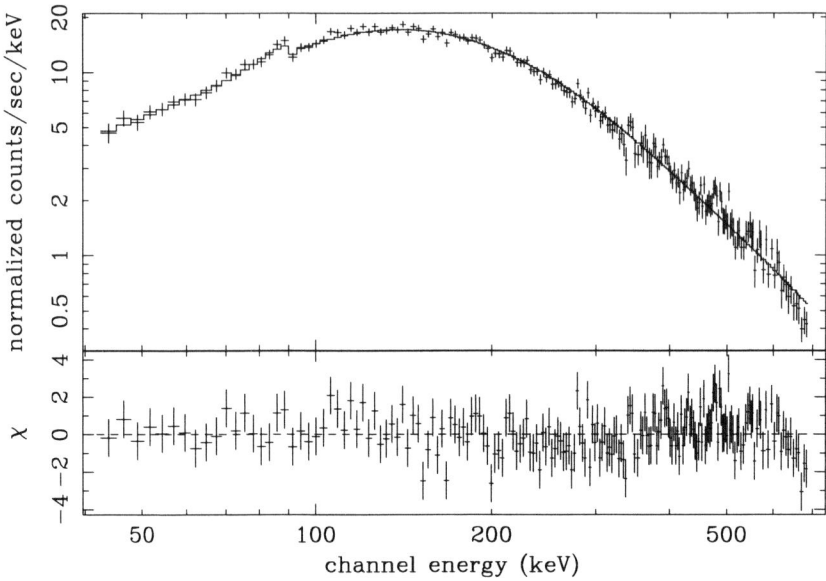

**FIGURE 3.** Spectral deconvolution of GRB991216 observed with the GRBM unit 4, using as input parameters those found by Jimenez et al. [21] with BATSE data. The good fit shows the reliability of the GRBM response matrix. From Montanari et al. [25]

**TABLE 1.** Transient absorption feature in GRB990705. From [14].

|   | $t_1$ - $t_2$ (Time (s) from onset) | $N_H$ ($1 \times 10^{22}$ cm−2) | $\tau_{Fe}$ | Photon Index $\Gamma$ |
|---|---|---|---|---|
| A | 0-6 | 1.23±.43 | 1.52±0.53 | 1.08±.03 |
| B | 6-13 | 1.32±.18 | 1.63±0.22 | 1.09±0.02 |
| C | 13-20 | 0.55±.18 | 0.68±0.32 | 1.25±0.02 |
| D | 20-28 | 0.24±.21 | 0.29±0.26 | 1.29±0.02 |

### Transient Emission Feature in GRB990712: first evidence a GRB photosphere?

The very X-ray rich GRB990712 at redshift $z = 0.433$ shows a transient feature [15]. The feature is present in an intermediate part of the burst (see Fig. 5). It lasts 2 s only but it is statistically significant. The feature is definitely larger than the detector resolution and cannot be fitted with a narrow emission line. If fitted with a broad line the center is at $4.4 \pm 0.8$ keV (corresponding to $6.3 \pm 1.1$ keV in the local frame) and the width $3.3 \pm 1.6$ keV.

If interpreted as a blackbody component we have a best fit value of $kT_{bb} = 1.3 \pm 0.3$ keV (corresponding in the GRB frame to $kT_{bb} = 1.9 \pm 0.4$ keV). The blackbody luminosity is $(2.5 \pm 0.9) \times 10^{49}$ ergs s$^{-1}$.

The interpretation of the line in terms of an Fe recombination line would be consistent with the measured centroid energy when the redshift is taken into account, but it appears unlikely: it would imply the presence of at least $10 M_\odot$ of matter, at radii of a few light seconds, with none of this obstructing the line of sight (since we do not see any edge).

Alternatively we can interpret the feature as the thermal emission from the fireball photosphere, at the time when the fireball expansion becomes optically thin. We stress that no evidence of such a photosphere, theoretically predicted [26], was found so far.

According to Meszaros and Rees [26] the following equations hold

$$L_{phot} = L_{52} \left(\frac{\eta}{\eta_\star}\right)^{8/3} \quad (1)$$

$$\Theta_{phot} = \Theta_0 \left(\frac{\eta}{\eta_\star}\right)^{8/3} \quad (2)$$

where $\eta$ is the flow Lorentz factor, $\eta_\star \approx 1000(L_{52}\mu_1^{-1} Y \Gamma_0)^{1/4}$ and $\Theta_0 = 1236(L_{52}^{0.5}\mu_1^{-1}\Gamma_0^{-1})^{1/2}$ keV, $L_{52}$ being the fireball wind luminosity in units of $10^{52}$ erg s$^{-1}$, $Y$ being the number of electrons per baryon ($\approx 1$), and $\mu_1$ being the mass, in units of $10 M_\odot$, of the rotating black hole, from 6 times the gravitational radius of which the fireball is assumed to start its expansion.

From the measured values of $L_{ph}$ ($= 2.5 \times 10^{49}$ erg s$^{-1}$) and $\Theta_{ph}$ ($= 1.86$ keV after correction for the

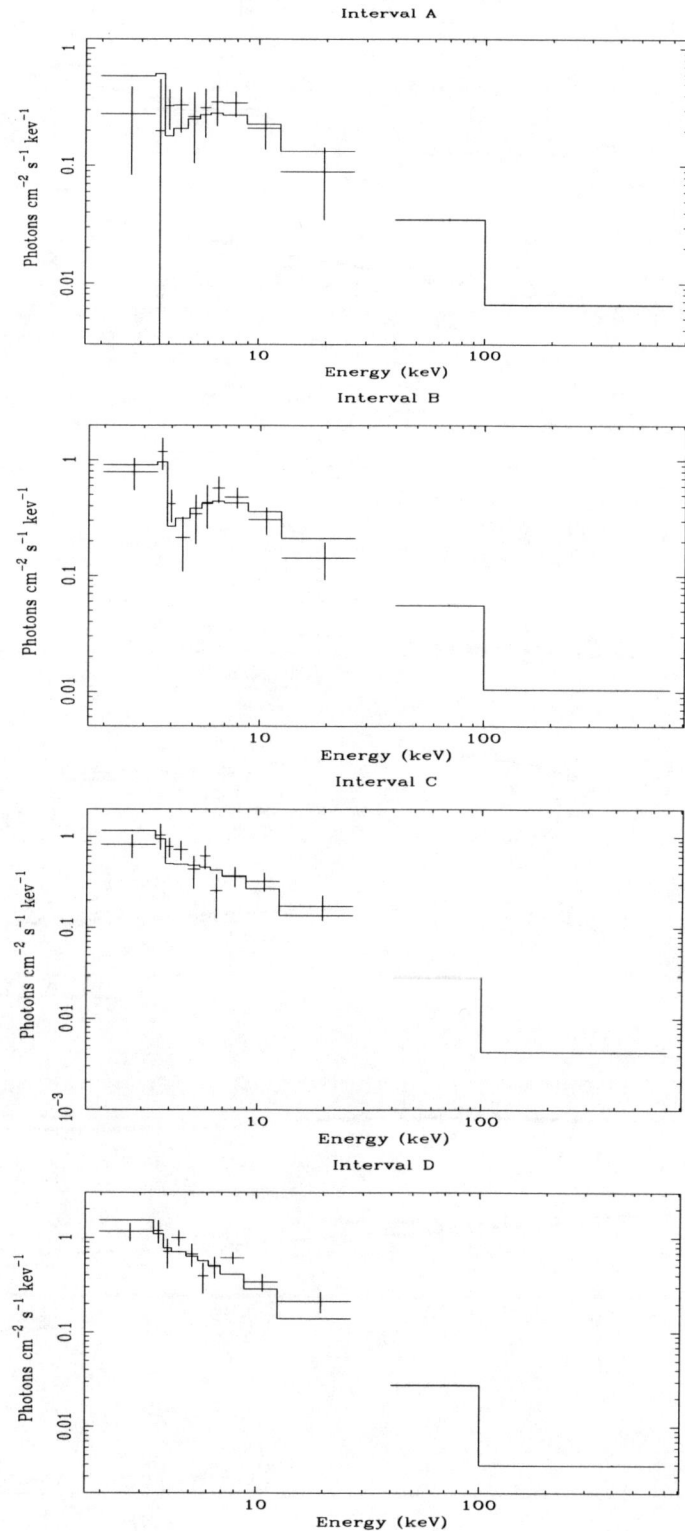

**FIGURE 4.** Spectral evolution of GRB990705 in the first 4 of the 7 intervals in which the GRB time profile was subdivided. From [14].

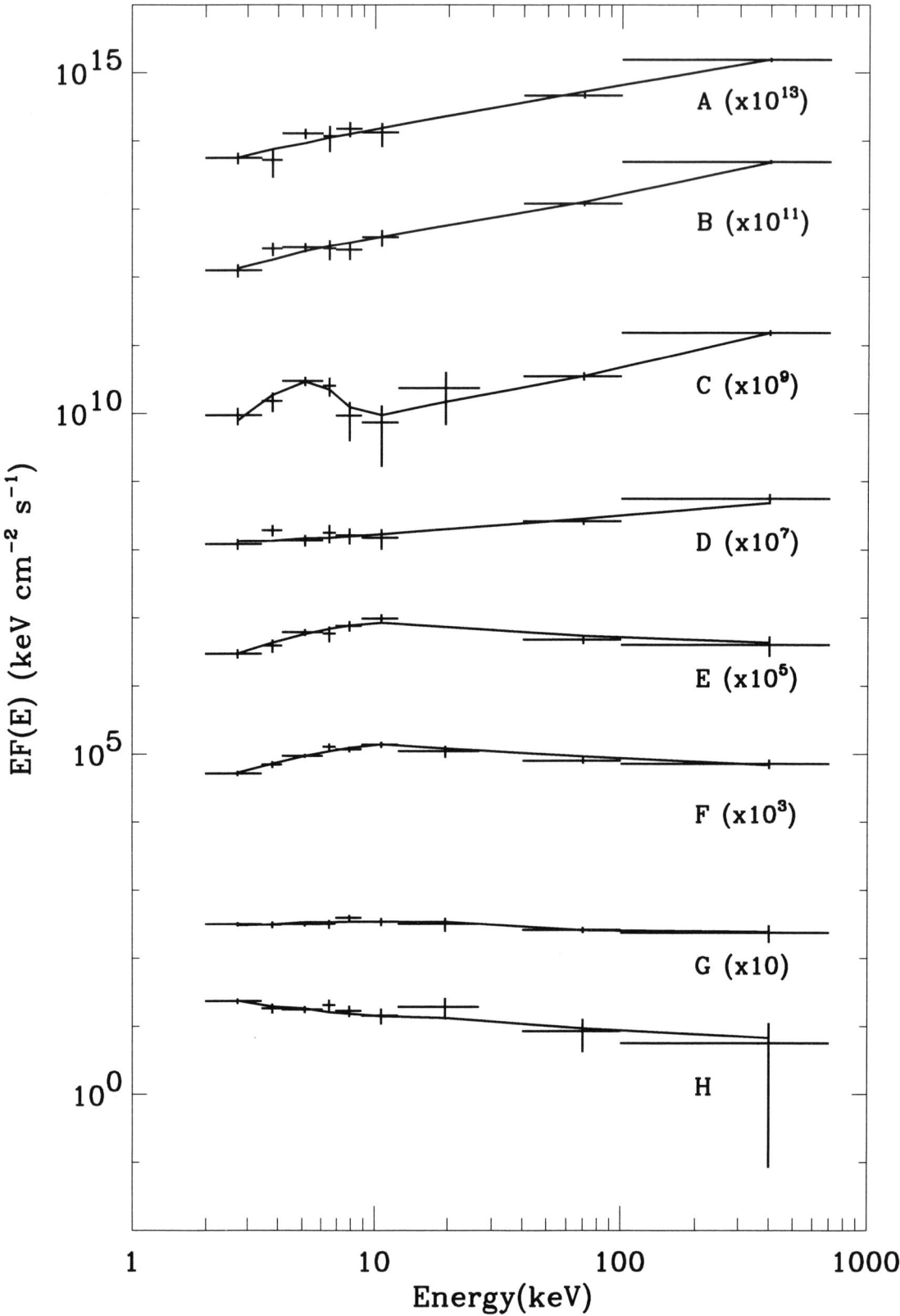

**FIGURE 5.** Spectral evolution of GRB990712. From [15]

redshift) we find $\eta = 100$ and $L_{52} = 2$ (assuming unit values for Y, $\Gamma_0$ and $\mu_1$). The most important implication of our result is that the luminosity of the fireball wind is about 100 times higher than the 2-700 keV luminosity ($\simeq 2 \times 10^{50}$ erg/s). As a consequence the efficiency of conversion of wind energy in electromagnetic energy is only 1%.

## *BEPPOSAX* SPECTRA AND COSMOLOGY

To conclude we discuss how spectral data from *BeppoSAX* can impact on cosmological studies. BeppoSAX has identified the new class of short transient events, similar to GRBs but with significantly softer spectrum (X-ray Flashes [18]). They are discussed in detail in another communication to this same conference [19]. We want to stress the point that they could include GRBs at moderate distance with low peak energy but, as well, GRBs with very high redshift, possibly exceeding the present best score of GRB000131 at $z = 4.50$.

The afterglow data from GRB000214 have been proposed as the first determination of a redshift from X-ray data only [5]. But also GRB990705 has allowed for such determination [1] that, furthermore, has been confirmed by an optical red-shift [4]. This BeppoSAX measurement demonstrates the possibility that with a detector with a better spectral resolution, especially at lower energies, the red-shift of a GRB, even at very high red-shift can be measured from the Fe K edge. GRBs could become the deepest tracers of the early universe.

Another contribution of GRB prompt spectra to Cosmology comes from collective studies of spectral properties of GRBs with known red-shift [2]. Combining WFC and GRBM data the spectra are fitted with a Band law [6]. The agreement is very good and the peak luminosity and the bolometric energy in the frame of the burst can be computed. After a discussion on the selection effects Amati et al. [2] arrive to the following results:

- the energy released by the burst (assuming isotropic emission) $E_{rad}$ is correlated with $1+z$;
- the peak energy $E_p$ of the EF(E) sectrum is correlated with $E_{rad}$;
- the slope of the low energy side of the Band law $\alpha$ is correlated with $1+z$ as well.

Probably the later correlation is a consequence of the other mentioned correlations and of a correlation between $\alpha$ and $E_p$ also found which is interpreted as consequence of the $\alpha$ dependence on the spectral curvature near $E_p$. The sample is relatively meagre (12 bursts) but the data seem to suggest that an evolutionary effect is present and can be determined from broad band spectra of bursts of known redshift.

## ACKNOWLEDGMENTS

Beppo-SAX is a major mission of the Italian Space Agency (ASI) with a contribution of the Dutch Space Agency (NIVR). We wish to thank Enrico Montanari for having allowed us the publication of figures taken from his paper in preparation.

## REFERENCES

1. Amati, L. et al. *Science*, **290**, 953 (2000)
2. Amati, L. et al. *A&A*, in press (2002)
3. Andersen, M.I. et al., *Gamma-Ray Bursts in the Afterglow Era*, E. Costa, F. Frontera, J. Hjorth eds. (Springer Publ.), p. 133 (2001)
4. Andersen, M.I. et al., in preparation (2002)
5. Antonelli, A.L. *ApJ*, **545**, L39 (2000)
6. Band, D. et al. *ApJ*, **413**, 281 (1993)
7. Boettcher, M. et al. *A&A*, **343**, 111 (1999)
8. Boettcher, M. et al. *ApJ*, **567**, 441 (2002)
9. Calura, F. et al., in: *AIP Conf. Proc.s*, **526**, 721 (2000)
10. Costa, E. et al. *Nature*, **387**, 783-785 (1997)
11. Costa, E. et al. *Adv. Space Res*, **22**, 1129 (1998)
12. Frontera F. et al. *A&A Suppl.*, **122**, 357 (1997)
13. Frontera F. et al. *ApJS*, **127**, 59 (2000)
14. Frontera F. et al., in: *Gamma-Ray Bursts in The Afterglow Era*, E. Costa, F. Frontera, J. Hjorth eds. (Springer Publ.), p. 106 (2001)
15. Frontera F. et al. *ApJ*, **550**, L47 (2001)
16. Gandolfi, G.G. et al., *these Proc.s* (2002)
17. Guidorzi C. et al. in: *Gamma-Ray Bursts in The Afterglow Era*, E. Costa, F. Frontera, J. Hjorth eds. (Springer Publ.), p. 43 (2001)
18. Heise, J. et al., in : *Gamma-Ray Bursts in The Afterglow Era*, E. Costa, F. Frontera, J. Hjorth eds. (Springer Publ.), p. 16 (2001)
19. Heise, J. et al. *these Proc.s* (2002)
20. Jager, R. et al. , *A&A Suppl.*, **125**, 557 (1997)
21. Jimenez, R., Band, D. & Piran, T., *ApJ*, **561**, 171 (2001)
22. Kato, T. , *ApJS*, **30**, 397 (1976)
23. Lazzati, D. et al. , *ApJ*, **556**, 471 (2001)
24. Lazzati, D., Perna, R. *MNRAS*, **330**, 383 (2002)
25. Montanari, E. et al., in preparation (2002)
26. Mészáros, P. & Rees, M. *ApJ*, **530**, 292 (2000)
27. Mészáros, P. & Rees, M. *ApJ*, **556**, L37 (2001)
28. Montanari, E., Guidorzi, C., Frontera, F., & Amati, L. *GCN Circ.* **1104** (2001)
29. Pamini, M. et al. *Nuovo Cimento*, **13C**, 337 (1990)
30. Piro, L. et al. *ApJ*, **514**, L73 (1999)
31. Piro, L. et al. *Science*, **290**, 955 (2000)
32. Piro, L. *these Proc.s* (2002)
33. Vietri, M et al. *ApJ*, **550**, L43 (2001)
34. Yoshida, A. et al. *A&A Suppl.*, **138**, 433 (1999)
35. in 't Zand, J.J.M. et al. *ApJ*, **559**, 710 (2001)

# The Use of $E_{pk}$ Decay Rates to Find Standard Candles in GRBs

Dan Kocevski* and Edison Liang*

*Department of Physics and Astronomy, Rice University, Houston, Texas, 77005

**Abstract.** We have examined a possible correlation between the decay rate of $E_{pk}$ and the luminosity distance for 15 gamma-ray bursts (GRB), in an attempt to determine if the decay constant ($\Phi_o$) of the primary pulse within a burst may be used as a distance indicator. We show from physical first principles that the invariant quantity $\Phi_o d^2$ should be a constant if the luminosity of GRBs is indeed a standard candle. We find tentative evidence that this quantity may be constant for clean separable FRED pulses, whereas chaotic pulses with unresolved structure do not yield the expected correlation. This is believed to be due to an inevitable overestimating of $\Phi_o$ that occurs when trying to measure the $E_{pk}$ decay rates of overlapping pulses. This method has the potential of yielding distances and hence luminosities of clean GRBs with separable pulses and may be related to the underlying physics of the lag-luminosity correlation.

## INTRODUCTION

It has been known for some time that the break in the gamma-ray burst spectra evolves to lower energies with time, so that the average energy of the arriving photons becomes softer as the burst proceeds. Liang and Kargatis [4] quantified this trend as an exponential decay of the peak power energy $E_{pk}$ as a function of photon fluence.

$$E_{pk} = E_o e^{-\Phi/\Phi_o} \quad (1)$$

Where $E_o$ is the max of $E_{pk}$, $\Phi(t)$ is the photon fluence integrated from the start of the pulse, and $\Phi_o$ is the decay constant. Here $E_{pk}$ is defined as the maximum of the $\nu F_\nu$ spectra, where $\nu$ is the photon energy and $F_\nu$ is the specific energy flux. This result is open to various interpretations, the simplest being that the decay of $E_{pk}$ is governed by radiative cooling. This type of exponential decay of the break energy with photon fluence is expected if the average energy of the emitted photons is directly proportional to the average emitting particle energy such as in thermal bremsstrahlung or multiple Compton scatterings. This interpretation, however, is not unique. In any case, it is seen empirically that the decay constant $\Phi_o$ is directly related to the cooling of the peak energy of a GRB pulse.

It can be shown from physical first principles that the decay of the $\nu F_\nu$ spectral peak is directly related to the absolute luminosity of a GRB under certain physical conditions. In the case of pure radiative cooling, we know that the rate of change of energy of the source is

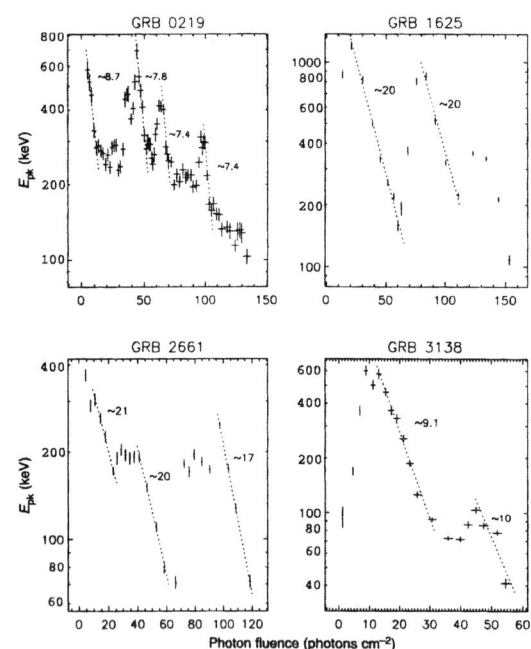

**FIGURE 1.** Examples of $E_{pk}$ Decay. Taken from [4], ©Nature 1996.

directly proportional to the luminosity emitted.

$$\frac{-dE(t)}{dt} \propto L \quad (2)$$

There would be additional terms if energy was being released via mechanisms other than radiation. Now, if we assume that the peak in the $\nu F_\nu$ spectra is a characteristic energy that is related to the average energy $\langle E \rangle$ of the emitting particles, then we can set $\langle E \rangle = E_{pk}$ and use the Liang-Kargatis relation to obtain:

$$\frac{-dE_{pk}}{dt} = \frac{F_E}{\Phi_o} = \frac{L}{\Omega \Phi_o d^2} \quad (3)$$

Where $\Omega$ is the beaming solid angle. Therefore the spectral evolution of $E_{pk}$ is directly related to the absolute luminosity of a GRB divided by the quantity $\Phi_o d^2$, which is a frame invariant.

It has been suggested [7], [5] that the spectral lag observed in GRBs may be related to the evolution of the peak energy through the various BATSE channels. Kocevski and Liang [5] quantify this by showing a linear relationship between the spectral lag and $\Phi_o$ over flux for a sample of GRBs. This interpretation of the lag observed in GRB is consistent with equation 3 if the quantity $\Phi_o d^2$ is approximately a constant for the peak pulse within a GRB. While there is no current justification for such a constant, it is conceivable that some limiting mechanism, analogous to an Eddington limit, may operate to make the energy release rate a maximum value among different GRBs. This would lead the proportionality constant $\Phi_o d^2$ between the energy release rate and the luminosity to approach a constant value among different bursts, and may in itself be used to find the distances to GRBs. Therefore, a measure of the rate of spectral evolution in a GRB may potentially yield the burst's absolute luminosity per solid angle.

## DATA ANALYSIS

To test the idea that the quantity $\Phi_o d^2$ of the peak pulse is a constant among different GRBs, we set forth to measure the decay constant $\Phi_o$ for all BATSE bursts with independently measured redshift. This amounts to 12 GRBs with known distances, of which 3 were dropped due to low signal to noise ratios which prevented reliable spectral fits. We also included in our sample 5 bright bursts with separable FRED (Fast Rise Exponential Decay) profiles which tend to have high S/N ratios and long clean decays allowing for very reliable $\Phi_o$ measurements. Unfortunately, the redshifts for these five bursts are not known, but can be calculated using the well known Lag-Luminosity correlation proposed by Norris et. al [6]. These calculated redshifts were obtained from [7] where the authors estimated the distances and luminosities to 112 GRBs using a combination of the lag-luminosity and variability-luminosity correlations [2].

To perform the data analysis we obtained the High Energy Resolution (HER) and when available the Medium Energy Resolution (MER) BATSE data for our entire

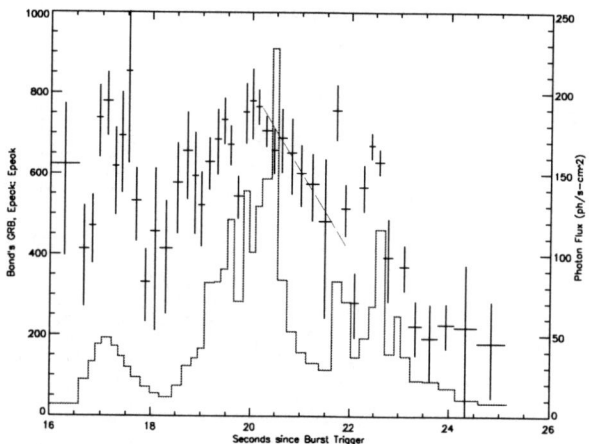

**FIGURE 2.** $E_{pk}$ evolution plotted over the burst's light curve for GRB 991216 (BATSE Trigger 7906). The linear fit yields $\Phi_o$.

sample and performed time-resolved spectral fits via $\chi^2$-minimization with the empirical Band model [1]. For bursts that had both HER and MER data available, we tended to use the MER data due to its higher time resolution, 16ms as opposed to 64ms for the HER data. The higher time resolution is required when trying to measure the decay of $E_{pk}$ for bursts that have highly chaotic and overlapping pulses. An example of the time-resolved spectral fits that were preformed is shown in Figures 2 and 3, where the log of $E_{pk}$ vs. photon fluence is plotted over the burst's light curve which is shown in photons cm$^{-2}$ s$^{-1}$. In Figure 2 the peak energy of the primary pulse in 991216 can be seen to decay monotonically on the semi-log plot, and a linear fit to this trend directly yields $\Phi_o$. The same is true in Figure 3 for GRB 990712 but in this instance the burst is a FRED and it can immediately be seen that the decay is much better sampled, giving a more robust determination of the decay constant.

The largest source of error in the measurement of the decay constants originates from the chaotic and overlapping nature of 8 out of the 15 GRB in our sample. In bursts that have highly overlapping pulses, it becomes very difficult to obtain a sufficient number of data points during the decay phase of the pulse in order to measure $\Phi_o$ before the rise portion of another pulse begins. This generally leads to an overestimating of $\Phi_o$ due to the unresolved pulses down to the 16-64ms level. For example, a GRB consisting of three highly overlapping FRED pulses, each separated only by 30ms, observed with 64ms HER data would appear to be a single pulse with unresolvable structure. The spectral components of all three pulses would add to produce an $E_{pk}$ decay that occurs over a much longer timescale then any one particular pulse, resulting in an overestimate of the decay constant for that burst. Due to this effect, the most reli-

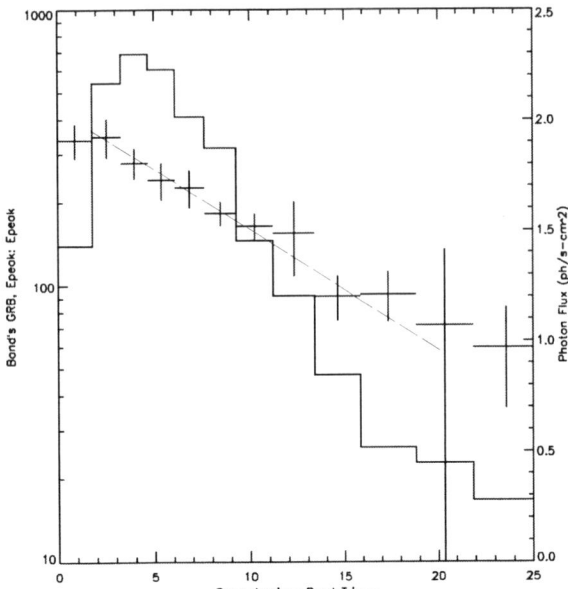

**FIGURE 3.** $E_{pk}$ evolution for GRB 7648 (990712). The decay constant is much more well defined for the FRED bursts.

able $\Phi_o$ measurements came from the bursts with clean and separable FRED pulses, whereas the results received from the more chaotic bursts are less dependable.

## RESULTS

The quantity $\Phi_o d^2$ along with individual $\Phi_o$ and z values for our entire sample are shown in Table 1. The first 9 GRBs have known redshifts of which 8 exhibit overlapping structure, the last 6 bursts are characterized by the FRED profile. It can be seen that for the more chaotic bursts the $\Phi_o d^2$ values span several orders of magnitude and hence do not appear to be constant. This is in contrast with the FRED results which cluster about a unitless value of $2.0 \times 10^{57}$. Additionally, it can be seen from Figure 4 that the bursts with high variability and structure all have $\Phi_o d^2$ values that are higher then those of the FRED bursts. We believe that this is due to the overestimating that was discussed in the last section, namely that the overlapping nature of these bursts leads to an inevitable overestimating of the decay constant. If this overestimating does indeed occur, then it is expected that all the complex bursts would have apparent $\Phi_o d^2$ values that are above their true value. In Figure 4 we've plotted the $\Phi_o$ values for 6 FRED events (1 with known distance and 5 with calculated distance) vs. $d^{-2}$. From this plot we can see that there is a linear relationship between $\Phi_o$ and Distance$^{-2}$, which is expected if the quantity $\Phi_o d^2$ is a constant. A linear fit to this data gives the mean value for $\Phi_o d^2$ of $1.99 \times 10^{57} \pm 0.15$. All of the non-FRED events

**TABLE 1.** $\Phi_o d^2$ values for our entire sample

|        | *     | $\Phi_o$ [†]    | $\Phi_o d^2$ (×10$^{57}$) |
|--------|-------|-----------------|--------------------------|
| 970508 | 0.825 | 1.07 ± 0.16     | 0.776 ± 4.95             |
| 970828 | 0.958 | 43.6 ± 10.9     | 4.61 ± 0.17              |
| 971214 | 3.42  | 12.59 ± 1.17    | 5.10 ± 1.08              |
| 990123 | 0.966 | 16.24 ± 2.43    | 6.14 ± 0.92              |
| 990506 | 1.6   | 76.62 ± 14.2    | 118.0 ± 0.22             |
| 990510 | 1.3   | 55.63 ± 13.7    | 15.3 ± 0.38              |
| 990703 | 1.619 | 5.53 ± 5.53     | 8.76 ± 2.22              |
| 991216 | 1.02  | 95.53 ± 9.13    | 48.5 ± 0.46              |
| 000131 | 4.5   | 73.53 ± 7.5     | 8.82 ± 3.00              |
| 910807 | 0.41  | 37.68 ± 2.91    | 2.16 ± 0.17              |
| 911016 | 0.56  | 18.40 ± 1.0     | 3.78 ± 0.21              |
| 911031 | 0.66  | 11.92 ± 0.95    | 2.06 ± 0.17              |
| 930612 | 0.25  | 106.9 ± 22.69   | 2.01 ± 0.43              |
| 950624 | 0.56  | 15.79 ± 1.58    | 1.83 ± 0.183             |

* Redshift obtained from [3] and [7]
[†] Decay constant in units of photons s$^{-1}$ cm$^{-2}$

have values that are above this mean. This results in two noteworthy relationships:

$$d = 4.43 \times 10^{28} (\Phi_o)^{-\frac{1}{2}} \quad (4)$$

$$L = \frac{1.99 \times 10^{57} Flux}{\Phi_o} \quad (5)$$

Where both $\Phi_o$ and Flux are measured from the peak pulse within the GRB.

It must be noted that all of these results were obtained with only long, bright GRBs. We have not examined any bursts with durations shorter then a few seconds, so we can say nothing about whether these relationships hold for short GRBS. Also, these results are given per steradian, due to the lack of information about the beaming angle of the events. Furthermore, the associated error in the calculated distance from Schaefer et. al. was assumed to be 10%.

We wish to expand this study by looking at other bursts that contain separable FRED pulses to study whether this trend holds for a wider sample of GRBs. Eventually, in order to test our theory that overlapping pulses lead to overestimates of $\Phi_o$ measurements we will need to performed spectral deconvolution techniques to try to separate out spectral components due to different pulses within a burst. Once this is done, we can measure the decay constant for each pulse individually and examine the resulting $\Phi_o d^2$ values. Only after this can it be certain that this distance indicator holds for all GRB and not simply the FRED events.

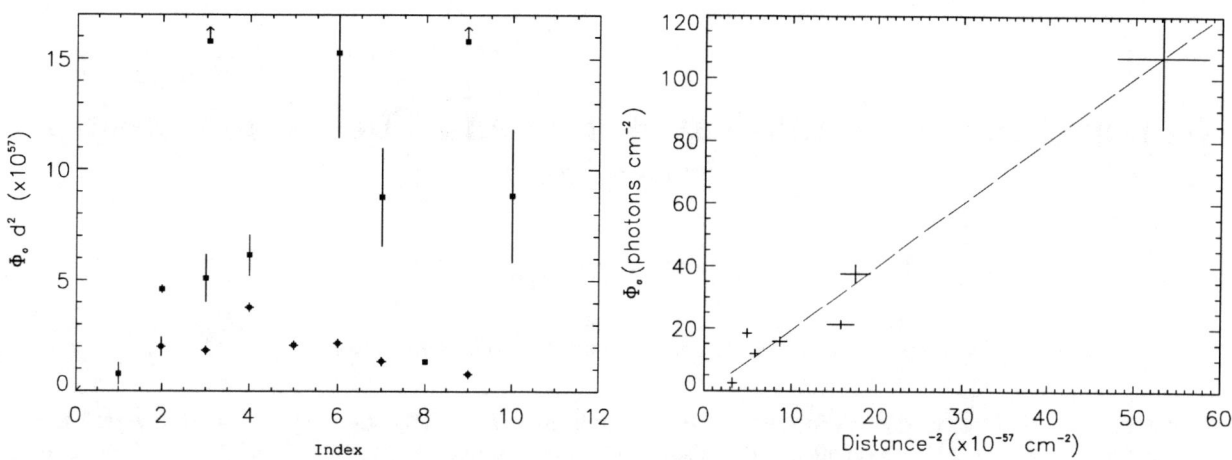

**FIGURE 4.** (Left) $\Phi_o d^2$ for our entire sample. The GRBs with known redshift are represented by squares whereas the bursts with calculated z are shown with circles. (Right) A plot of $\Phi_o$ vs. distance$^{-2}$. The linear fit to the data gives the mean $\Phi_o d^2$ value. The FRED bursts all have values lower then the non-FRED bursts, including the one FRED with known redshift. Two of the data points for the non-FRED subset fall beyond the plot's range.

## ACKNOWLEDGMENTS

We thank Brad Scheafer, Felix Ryde, and Rob Preece, for their thoughtful suggestions and advice. This work was supported by NASA grant NAG 53824 and the NASA GSRP fellowship program.

## REFERENCES

1. Band, D., et al. 1993. ApJ, 428, 21
2. Fenimore, E.E. et. al. 2000, ApJ submitted astro-ph/0004176
3. Greiner, J. 2001, http://www.aip.de/ jcg/grbgen.html
4. Liang, E., & Kargatis, V. 1996. Nature, 381, 49
5. Kocevski, D. & Liang, E. submitted to ApJ
6. Norris, J. P., et al. 2000. ApJ, 534, 248
7. Scheafer, B. et. al. 2001. ApJ, 563, 123

# Kinematics of Gamma-Ray Bursts and Their Relationship to Afterglows

Jay D. Salmonson

*Lawrence Livermore National Laboratory, P.O. Box 808, Livermore, CA 94551*

**Abstract.** A strong correlation is reported between gamma-ray burst (GRB) pulse lags and afterglow jet-break times for the set of bursts (seven) with known redshifts, luminosities, pulse lags, and jet-break times. This may be a valuable clue toward understanding the connection between the burst and afterglow phases of these events. The relation is roughly linear (i.e. doubling the pulse lag in turn doubles the jet break time) and thus implies a simple relationship between these quantities. We suggest that this correlation is due to variation among bursts of emitter Doppler factor. Specifically, an increased speed or decreased angle of velocity, with respect to the observed line-of-site, of burst ejecta will result in shorter perceived pulse lags in GRBs as well as quicker evolution of the external shock of the afterglow to the time when the jet becomes obvious, i.e. the jet-break time. Thus this observed variation among GRBs may result from a perspective effect due to different observer angles of a morphologically homogeneous populations of GRBs.

Also, a conjecture is made that peak luminosities not only vary inversely with burst timescale, but also are directly proportional to the spectral break energy. If true, this could provide important information for explaining the source of this break.

## INTRODUCTION

Only recently, with the discovery of afterglows and in turn, redshifts for a handful of gamma-ray bursts, has there been progress in trend spotting within the seemingly chaotic variety of gamma-ray burst shapes and sizes. Norris et al. [1] discovered an anti-correlation between the isotropic peak gamma-ray luminosity, $L_{pk}$, of GRBs and the pulse lag, $\Delta t$. This lag is the time delay of the arrival of a burst pulse in the BATSE detector low energy channels compared to its arrival in the high energy channels. Similarly Fenimore and Ramirez-Ruiz [2] and also Reichart et al. [3] have shown that a measure of the variability of GRB lightcurves correlates with this peak luminosity. Most recently Frail et al. [4] have shown that the isotropic gamma-ray energy, $E_{iso}$, is anti-correlated with the jet-break time, $\tau_j$. The jet-break time is when the afterglow lightcurve changes (typically seen as a break) its decay rate, which is thought to be a manifestation of the finite opening angle of the jet.

As demonstrated in Salmonson and Galama [5] these correlations are closely related and are likely manifestations of the same physical effect. As discussed in the next section, we find an unexpectedly tight relationship between spectral lags and jet-break times. Thus we argue that transitivity suggests that $L_{pk}$, $E_{iso}$, $\Delta t$ and $\tau_j$ are all interrelated by power-laws. In Salmonson [6, 7] it was argued that the lag-luminosity relationship, $L_{pk}$ vs. $\Delta t$, derives from kinematics: the variation in velocity of the relativistic ejecta with respect to the observer. In particular, the Doppler factor, dependent upon the speed and angle of the emitter with respect to the observer, will increase observed luminosity and decrease observed timescales. In Salmonson and Galama [5] we argue that all of these relationships originate from kinematic variations among bursts.

## DISCOVERY OF A CORRELATION BETWEEN PULSE LAGS AND JET-BREAK TIMES

In Salmonson and Galama [5] we compare the two burst timescales: the redshift corrected jet-break time, $\tau_j$, and the redshift corrected lags, $\Delta t$. We assembled a complete sample of seven bursts for which there are data for $\Delta t$, $\tau_j$ and redshift $z$ (GRB 971214 has only a lower limit for $\tau_j$, so was not used in fits, but is shown in the figures). Using the CCF31 0.1 lags, $\Delta t_{(CCF31\ 0.1)}$, determined by cross-correlating pulses in BATSE channels 1 & 3 down to 0.1 of the peak luminosity [1], a good fit results:

$$\tau_j = \frac{t_j}{1+z} = 28^{+18}_{-11} \left( \frac{\Delta t_{(CCF31\ 0.1)}}{1\ \text{sec}} \right)^{0.89 \pm 0.12} \text{days} \quad (1)$$

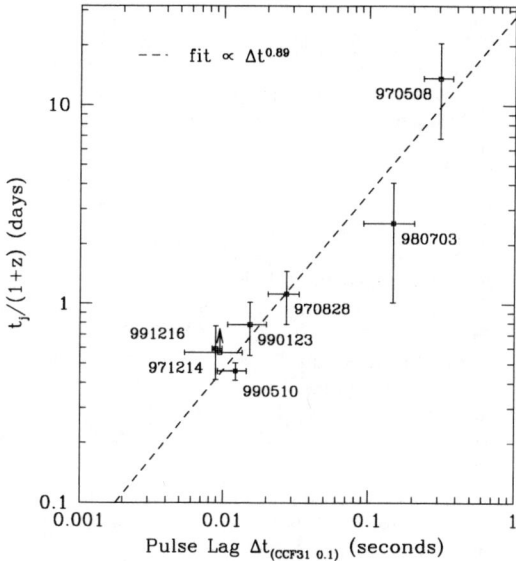

**FIGURE 1.** Plot from [5] of redshift-corrected burst pulse lags, $\Delta t$, observed between BATSE channels 1 and 3, versus observed jet-break times, corrected for redshift, $\tau_j \equiv t_j/(1+z)$. Jet break times $t_j$ are from Frail et al. [4] and pulse lags are from Norris et al. [1]. The fit, given by Eqn. (1), does not include GRB 971214 which only has a lower limit on the jet-break time.

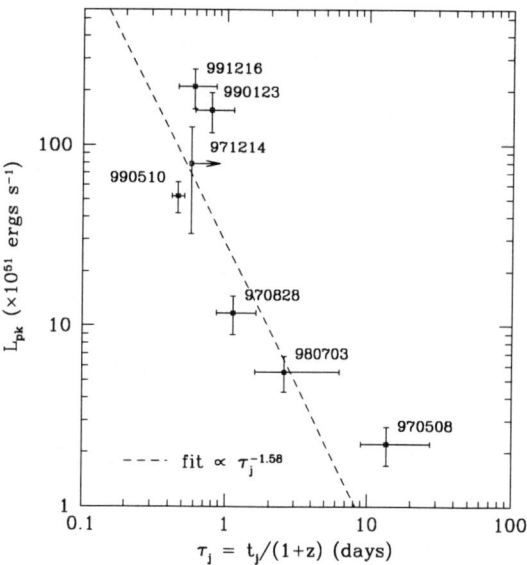

**FIGURE 2.** Plot from [5] of redshift-corrected burst peak luminosities $L_{pk}$, versus redshift-corrected observed jet-break times $\tau_j \equiv t_j/(1+z)$. Jet break times $t_j$ are from Frail et al. [4] and luminosities are calculated from Jimenez et al. [12]. Because GRB 971214 only has a lower limit on the jet-break time, it is not included in the fit (given by Eqn. 2).

(shown in Fig. 1) with a reduced chi-squared $\chi_r^2 = 4.7/4$ and a respectable goodness-of-fit $Q = 0.31$ [8].

The existence of such a close relationship between one timescale associated with the GRB itself, and another timescale solely deriving from the afterglow is surprising. The standard GRB paradigm [9] says that the GRB derives from internal shocks in an uneven relativistic wind, while the afterglow comes from a shock sweeping into the ISM, obeying simple self-similar scaling laws and thus not depending on initial conditions imposed by the GRB. In Salmonson and Galama [5] we discuss three possible models to explain the relationship of Eqn. (1).

## CONJECTURE: LUMINOSITY IS CORRELATED TO BREAK ENERGY

An enduring mystery in GRBs is the relative constance [10] of the observed break energy, $E_0$, of GRB spectra, represented by a broken power-law "Band function" [11]. This mystery is doubly troubling in light of the several relationships described earlier in this paper. How can $E_0$ be constant within a factor of about three while timescales, luminosities, and energies vary over almost two orders of magnitude? Light might be shed on this issue with the observation that there appears to be a correlation between $L_{pk}\tau_j^\beta$ and $E_0$, where $\beta = 1.58$ is the index for the $L_{pk}$ vs. $\tau_j$ power-law relationship (Eqn. 2) found by Salmonson and Galama [5].

This correlation is demonstrated by comparing the fit $L_{pk}$ vs. $\tau_j$ with that of $L_{pk}/E_0/(1+z)$ vs. $\tau_j$. As in Salmonson and Galama [5] we find

$$L_{pk} = 28^{+6}_{-5} \times 10^{51} \left(\frac{\tau_j}{1 \text{ days}}\right)^{-1.58 \pm 0.23} \text{ ergs s}^{-1} \quad (2)$$

(shown in Fig. 2) with $\chi_r^2 = 29/4$ and $Q \sim 10^{-6}$. While the correlation is plainly apparent, the fit is poor, suggesting this relationship is not consistent with a simple power-law.

Now in order to compare with Eqn. (2), I fit $L_{pk}/E_0/(1+z)$ vs. $\tau_j$ and find

$$\frac{L_{pk}}{E_0(1+z)} = 4.9 \pm 8 \left(\frac{\tau_j}{1 \text{ days}}\right)^{-1.43 \pm 0.19} \times 10^{49} \text{ergs s}^{-1} \text{keV}^{-1} \quad (3)$$

(shown in Fig. 3) with $\chi_r^2 = 9.9/4$ and $Q = 0.04$. The fit is substantially improved. For the sake of demonstration, the errors for $L_{pk}$ from Eqn. (2) are used in this fit. Realistically one should also factor in the errors in determining $E_0$.

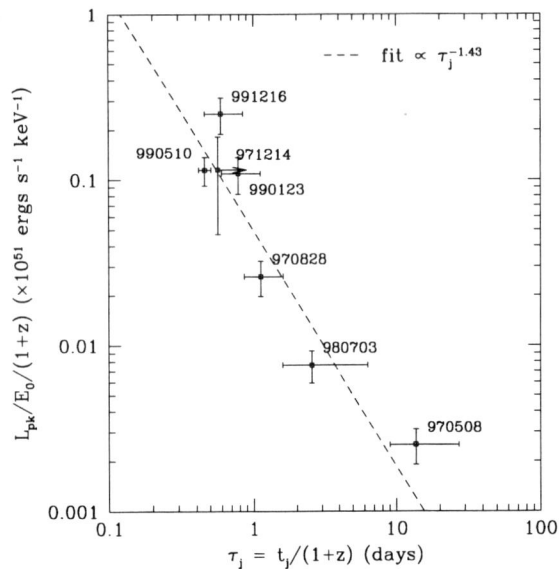

**FIGURE 3.** Plot of redshift-corrected burst peak luminosites $L_{pk}$ divided by redshift corrected spectral break energies $(1+z)E_0$, versus redshift-corrected observed jet-break times $\tau_j \equiv t_j/(1+z)$. The fit is given by Eqn. (3). Spectral break energies, $E_0$, are from Jimenez et al. [12]. See the caption of Fig. 2 for details.

This improvement in the fit of Eqn. (3) over that of Eqn. (2) leads one to hypothesize the existence of an additional dependence on $E_0$ in the $L_{pk}$ vs. $\tau_j$ relationship. Thus I propose

$$L_{pk} \propto E_0^\alpha \tau_j^{-\beta} \quad (4)$$

where $\alpha$ and $\beta$ are both roughly unity. This dependence might be indicative of a mechanism for the spectral breaking independent of the physical mechanism behind the relations described previously. One intriguing possibility is that it may indicate a filtering of the gamma-rays that is responsible for the spectral break. As such, the more effective the filter, the lower the energy, $E_0$, at which the spectrum is broken *and* the more attenuated is the photon flux at all energies. Such a filtering mechanism would require both of these effects to account for the dependence suggested in Eqn. 4. Future observations will be necessary to confirm this conjecture, and future work will elaborate on its physical cause.

This work was performed under the auspices of the U.S. Department of Energy by University of California Lawrence Livermore National Laboratory under contract W-7405-ENG-48.

## REFERENCES

1. Norris, J. P., Marani, G. F., and Bonnell, J. T. (**2000**). "Connection between energy-dependent lags and peak luminosity in gamma-ray bursts," ApJ, **534**, 248.
2. Fenimore, E. E., and Ramirez-Ruiz, E. (**2000**). "Redshifts for 224 batse gamma-ray bursts determined by variability and the cosmological consequences," astro-ph/0004176, submitted to ApJ.
3. Reichart, D. E., Lamb, D. Q., Fenimore, E. E., Ramirez-Ruiz, E., Cline, T. L., and Hurley, K. (**2000**). "A possible cepheid-like luminosity estimator for the long gamma-rayb bursts," ApJ, **552**, 57.
4. Frail, D. A., Kulkarni, S. R., Sari, R., Djorgovski, S. G., Bloom, J. S., Galama, T. J., Reichart, D. E., Berger, E., Harrison, F. A., Price, P. A., Yost, S. A., Diercks, A., Goodrich, R. W., and Chaffee, F. (**2001**). "Beaming in Gamma-Ray Bursts: Evidence for a Standard Energy Reservoir," ApJ, **562**, L55–L58.
5. Salmonson, J. D., and Galama, T. J. (**2001**). "Discovery of a tight correlation between pulse lag/luminosity and jet-break times: a connection between gamma-ray burst and afterglow properties." astro-ph/0112298, submitted to ApJ.
6. Salmonson, J. D. (**2000**). "On the kinematic origin of the luminosity-pulse lag relationship in gamma-ray bursts," ApJ, **544**, L115–L117.
7. Salmonson, J. D. (**2001**). "On the kinematics of grb 980425 and its association with sn 1998bw," ApJ, **546**, L29–L31.
8. Press, W. H., Flannery, B. P., Teukolsky, S. A., and Vetterling, W. T. (**1988**). *Numerical Recipes in C: The Art of Scientific Computing* (Cambridge University Press).
9. Piran, T. (**2000**). "Gamma-ray bursts - a puzzle being resolved." Phys. Rep., **333**, 529–553.
10. Mallozzi, R. S., Paciesas, W. S., Pendleton, G. N., Briggs, M. S., Preece, R. D., Meegan, C. A., and Fishman, G. J. (**1995**). "The nu f nu peak energy distributions of gamma-ray bursts observed by batse," ApJ, **454**, 597+.
11. Band, D., Matteson, J., Ford, L., Schaefer, B., Palmer, D., Teegarden, B., Cline, T., Briggs, M., Paciesas, W., Pendleton, G., Fishman, G., Kouveliotou, C., Meegan, C., Wilson, R., and Lestrade, P. (**1993**). "Batse observations of gamma-ray burst spectra. i - spectral diversity," ApJ, **413**, 281–292.
12. Jimenez, R., Band, D., and Piran, T. (**2001**). "Energetics of gamma ray bursts," ApJ, **561**, 171.

# An Observational Evidence for the Difference Between the Short and Long Gamma-Ray Bursts

L.G. Balázs*, P. Mészáros[†], Z. Bagoly**, I. Horváth[‡] and A. Mészáros[§]

*Konkoly Observatory, Budapest, Box 67, H-1525, Hungary*
[†]*Dept. of Astronomy, Pennsylvania State University, 525 Davey Lab. University Park, PA 16802, USA*
**Lab. for Information Technology, Eötvös University, Budapest, Pázmány Péter sétány 1/A, H-1518, Hungary*
[‡]*Dept. of Physics, Bolyai Military University, BJKMF, Budapest, Box 12, H-1456, Hungary*
[§]*Astron. Inst. of the Charles University, 180 00 Prague 8, V Holešovičkách 2, Czech Republic*

**Abstract.**
The intrinsic fluence and duration distributions of gamma-ray bursts are well represented by log-normal distributions. This allows a bivariate log-normal distribution fit to be made to the BATSE short and long bursts separately. A statistically significant difference between the long and short groups is found. We argue that the effect is probably real. Applying the Cramér's theorem these results lead to some predictions for models of long and short bursts.

## INTRODUCTION

The simplest grouping of gamma-ray bursts (GRBs) is given by their well-known bimodal duration distribution. This divides bursts into long ($T_{90} > 2$ s) and short ($T_{90} < 2$ s) duration groups [6]. The bursts measured with the BATSE instrument on the Compton Gamma-Ray Observatory are usually characterized by 9 observational quantities, i.e. 2 durations, 4 fluences and 3 peak fluxes [7]. In [1] we used the principal components analysis (PCA) technique to show that these 9 quantities can be reduced to only two significant independent variables, or principal components (PCs). The observational fact, that the dominant principal component consists mainly of the durations and the fluences, may be of consequence for the physical modeling of the burst mechanism.

In this paper we investigate the nature of this principal component decomposition, and in particular, we analyze quantitatively the relationship between the fluences and durations implied by the first PC. We analyze the distribution of the observed fluences and durations of the long and the short bursts, and we present arguments indicating that the intrinsic durations and fluences are well represented by log-normal distributions. The implied bivariate log-normal distribution represents an ellipsoid in these two variables, whose major axis inclinations are statistically different for the long and the short bursts.

Our GRB sample is selected from the Current BATSE Gamma-Ray Burst Catalog according to two criteria, namely, that they have both measured $T_{90}$ durations and fluences (for the definition of these quantities see [7], henceforth referred to as the Catalog). The Catalog in its final version lists 2041 bursts for which a value of $T_{90}$ is given. The fluences are given in four different energy channels, $F_1, F_2, F_3, F_4$, whose energy bands correspond to [25,50] keV, [50,100] keV, [100,300] keV and $> 300$ keV. The "total" fluence is defined as $F_{tot} = F_1 + F_2 + F_3 + F_4$, and we restrict our sample to include only those GRBs which have $F_i > 0$ values in at least the channels $F_1, F_2, F_3$. Concerning the fourth channel, whose energy band is $> 300$ keV, if we had required $F_4 > 0$ as well, this would have reduced the number of eligible GRBs by $\simeq 20\%$. Hence, we decided to accept also the bursts with $F_4 = 0$, rather than deleting them from the sample. Using therefore these two cuts, we are left with $N = 1929$ GRBs, all of which have defined $T_{90}$ and $F_{tot}$, as well as peak fluxes $P_{256}$. This is the sample that we study.

## FITTING THE LOGARITHMIC FLUENCES AND DURATIONS BY THE SUPERPOSITION OF TWO BIVARIATE DISTRIBUTIONS

We assume here that the distributions of the variables $T_{90}$ and $F_{tot}$, for both the short and long groups, can well be approximated by log-normals. As it was already noted in a previous contribution [4], this is an acceptable assumption. In this case it is possible to fit *simultane-*

**TABLE 1.** The best fit parameters of the sum of two bivariate log-normal distributions for $x = \log T_{90}$ and $y = \log F_{tot}$ for the sample with $N = 1929$.

| | | | |
|---|---|---|---|
| $a_{x1}$ | -0.08 | $a_{x2}$ | 1.54 |
| $a_{y1}$ | -6.22 | $a_{y2}$ | -5.29 |
| $\sigma'_{x1}$ | 0.73 | $\sigma'_{x2}$ | 0.67 |
| $\sigma'_{y1}$ | 0.46 | $\sigma'_{y2}$ | 0.37 |
| $\tan\alpha_1$ | 0.91 | $\tan\alpha_2$ | 2.29 |
| W | 0.32 | | |

*ously* the values of $\log F_{tot}$ and $\log T_{90}$ by a single two-dimensional (bivariate) normal distribution. This distribution has five parameters (two means $a_x$, $a_y$, two dispersions $\sigma_x$, $\sigma_y$, and the correlation coefficient $r$, where $x = \log T_{90}$, $y = \log F_{tot}$). An equivalent set of parameters consists of taking the same two means with two other dispersions $\sigma'_x$ $\sigma'_y$, and (instead of the correlation coefficient) the angle $\alpha$ between the axis $\log T_{90}$ and the semi-major axis of the "dispersion ellipse". (In the case of bivariate normal distributions, the constant probability curves define ellipses with well-defined axis directions). In this case $\alpha$ and the correlation coefficient are related unambiguously through analytical formulas [5]. If the data are well fitted by this bivariate normal distribution, then the distributions of each of the variables by themselves must also be univariate normal distributions (the marginal distributions are also normal).

A crucial point in this analysis is that, when the $r$-correlation coefficient differs from zero, then the semi-major axis of the dispersion ellipse represents a linear relationship between $\log T_{90}$ and $\log F_{tot}$, with a slope of $m = \tan\alpha$. This linear relationship between the logarithmic variables implies a power law relation of form $F_{tot} = (T_{90})^m$ between the fluence and the duration, where $m$ may be different for the two groups.

Fitting the data with the superposition of two bivariate log-normal distributions can be done by a standard search for 11 parameters with $N = 1929$ measured points. (Both log-normal distributions have five parameters; the eleventh parameter defines the weight of the first log-normal distribution.) We will use $\tan\alpha$ as the fifth parameter for both partial distributions ("terms"). Figure 1. shows the values of $x = \log T_{90}$ and $y = \log F_{tot}$ for the $N = 1929$ GRBs. Each GRB defines a point in the $x,y$ plane with coordinates $x_i, y_i$ ($i = 1,2,...,N$). The theoretical curve is a sum of two normal distributions. The normalization constant of the first [second] term is $NW$ [$N(1-W)$], where $W$ is the weight ($0 \leq W \leq 1$). For the first (second) term the parameters are $a_{x1}, a_{y1}, \sigma_{x1}, \sigma_{y1}, \alpha_1$ ($a_{x2}, a_{y2}, \sigma_{x2}, \sigma_{y2}, \alpha_2$).

We obtain the best fit to the 11 parameters through a maximum likelihood (ML) estimation. The results are collected at Table 1.

# DISCUSSION AND CONCLUSION

We have presented evidence indicating that there is a power law relationship between the logarithmic fluences and $\log T_{90}$ of the GRBs in the Current BATSE Catalog, based on a maximum likelihood estimation of the parameters of the bivariate distribution of these measured quantities.

An intriguing corollary of these results is that the exponents in the power law dependence between fluence and duration differ significantly for the two groups of short ($T_{90} < 2$ s) and long ($T_{90} > 2$ s) bursts.

These two results may have an interesting impact on the models of GRBs.

As it was already discussed in [4], the application of the mathematical Cramér's theorem [3, 8] ensures that there is also the same power law relationship between the total emitted energies and the intrinsic durations. Because the exponents are different for the short and long subgroups, these subgroups should also be generated by different scenarios.

This results, together with the conclusion that for the short (long) bursts the total released energy is proportional to the intrinsic duration (to the square of intrinsic duration) were already announced by [4].

Nevertheless, a care is needed yet. According to the Bayes theorem [8], the probability density $P(F_{tot}, T_{90})$ of fluence and $T_{90}$ is given by

$$P(F_{tot}, T_{90}) = \int_0^\infty P(F_{tot}, T_{90}|P)G(P)dP,$$

where $P$ denotes the peak-flux (either on 64 ms, or 256 ms or 1024 ms trigger), $G(P)$ is the probability density of $P$, and $P(F_{tot}, T_{90}|P)$ is the so-called kernel. (The meaning of $P(F_{tot}, T_{90})$ is straightforward: $NP(F_{tot}, T_{90})dF_{tot}dT_{90}$ defines the number of GRBs in the intervals $[F_{tot}, (F_{tot} + dF_{tot})]$ and $[T_{90}, (T_{90} + dT_{90})]$, respectively, where $N$ is the number of GRBs.) Unfortunately, $G(P)$ is well biased by instrumental effects, and - to have the real intrinsic biasfree relation between the fluence and duration - the kernel should be known. Of course, in principle, due to the biases in $G(P)$, it is not sure that the the observed relation between the fluence and $T_{90}$ and the relation coming from the kernel are identical.

We proceeded a new estimation of these phenomena by different approximations of kernel, and - it seems - *our previous proclamations further hold.* The details of our studies will be published elsewhere [2].

# ACKNOWLEDGMENTS

This research was supported in part through OTKA grants T024027 (L.G.B.), F029461 (I.H.) and T034549;

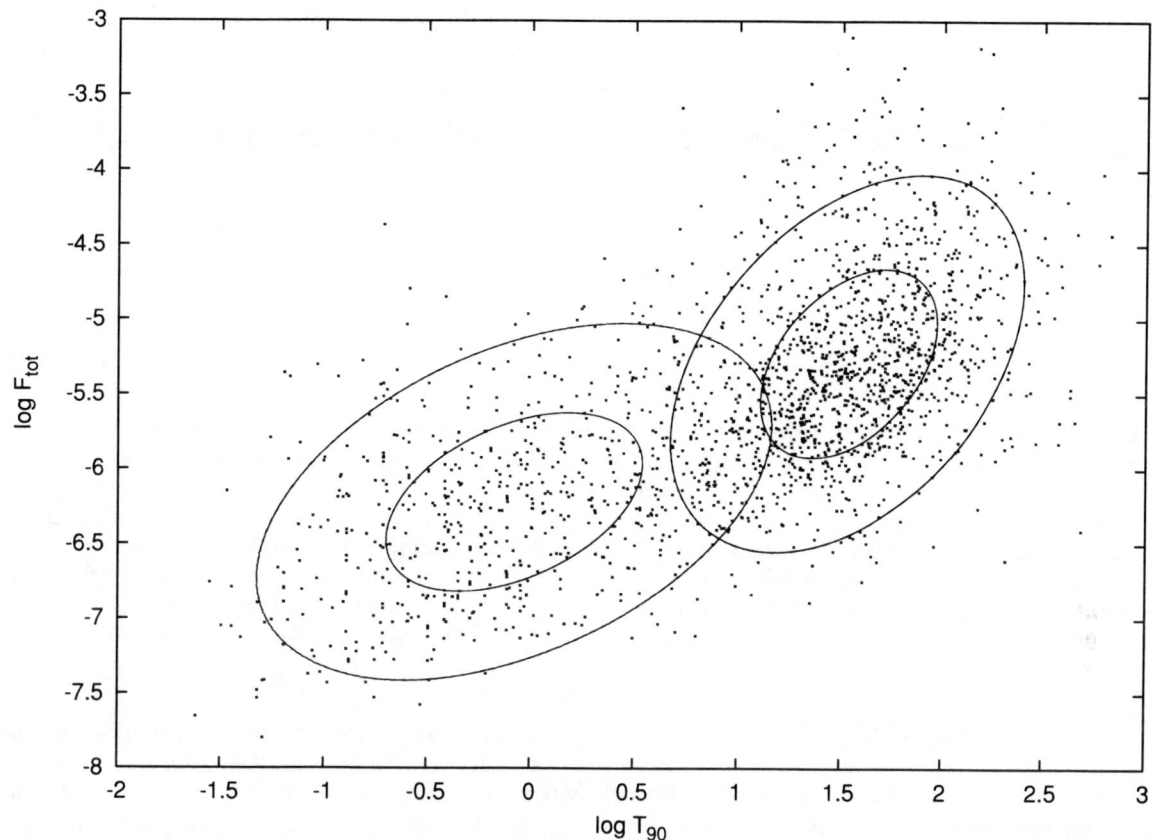

**FIGURE 1.** The best fit of two bivariate log-normal distributions for the whole BATSE sample (1929 GRBs). The ellipses give the 1σ and 2σ probabilities.

NASA grant NAG5-2857, Guggenheim Foundation and Sackler Foundation (P.M.); and Czech Research Grant J13/98: 113200004 (A.M.).

## REFERENCES

1. Bagoly, Z., Mészáros, A., Horváth, I., Balázs, L.G., and Mészáros, P., *ApJ*, **498**, 342, 1996.
2. Balázs, L.G., Mészáros, P., Bagoly, Z., Horváth, I., Mészáros, A., in preparation, 2002.
3. Cramér, H., *Random variables and probability distributions, Cambridge Tracts in Mathematics and Mathematical Physics, No.36*, Cambridge University Press, Cambridge, 1937.
4. Horváth, I., Balázs, L.G., Mészáros, P., Bagoly, Z., Mészáros, A., in *Proceedings of Second Gamma-Ray Bursts in the Afterglow Era*, Rome, Italy, 17-20 October 2000, eds. E.Costa et al., Springer, p.66
5. Kendall, M. and Stuart, A., *The Advanced Theory of Statistics*, Griffin, London, 1976.
6. Kouveliotou, C., et al., *ApJ*, **413**, L101. 1993.
7. Meegan, C.A., et al., Current BATSE Gamma-Ray Burst Catalog, (2000) URL http://gammaray.msfc.nasa.gov/batse/grb/catalog/current.
8. Rényi, A. *Wahrscheinlichtkeitsrechnung*, VEB Deutscher Verlag der Wissenschaften, Berlin, 1962

# The Statistical Determination of the Burst Energy Distribution

David L. Band

*GLAST SSC, Code 661, NASA/GSFC, Greenbelt, MD 20771*

**Abstract.** The shape of the distribution of the apparent energy radiated by a gamma-ray burst must be determined by the statistical analysis of well-defined burst samples. To convert the fluence into an energy requires the distance to the burst. A spectroscopic redshift is preferred, but the redshift can be estimated from the host galaxy's magnitude. The observed burst is drawn from the portion of the distribution above the detection threshold, and consequently this threshold must be known to recover the distribution. The detection threshold is the most restrictive of the thresholds of the different observations that characterize a burst (e.g., detecting the burst, localizing the afterglow, determining the redshift of the host galaxy). Current burst samples either have unknown detection thresholds, or rely on the currently uncertain luminosity-variability or luminosity-lag correlations.

## INTRODUCTION

What is the distribution of the apparent energy radiated by a gamma-ray burst? The apparent energy is the total gamma-ray energy the burst would have emitted if the observed flux were radiated isotropically. This energy is a fundamental physical burst characteristic. If bursts are beamed then the true energy is the apparent energy times the beaming fraction (see [4]). Why use the energy and not the peak photon flux or some other intensity measure? The choice of the fundamental measure of burst intensity is a theoretical prejudice. The peak photon flux is often used because observational studies present its distribution: detectors trigger on the peak count rate, which is roughly proportional to the peak flux, giving a relatively clean detection threshold. Therefore the preference for peak flux is based on instrumental, not physical, grounds. In the internal shock model, the peak flux is an accident of the collision of different shells. While also dependent on the burst-to-burst details of colliding shells, the total apparent energy has a better chance of being a fundamental burst property.

## METHOD

I am interested in determining the shape of the distribution function of the apparent energy; the normalization of this function—proportional to the spatial density of bursts—is not a goal of these studies. I consider each burst as a separate sample drawn from the observable distribution function; consequently the data space for each burst can be considered separately. For a given burst this space consists of all values of the apparent energy. The efficiency of detecting the burst in this data space is quite simple: unity above the detection threshold, and zero below. The burst sample need only be chosen so that there is no correlation between the bursts' intrinsic apparent energy and the criteria by which the bursts were selected. Thus selection effects such as a bias against small, faint host galaxies or the difficulties in determining spectroscopic redshifts in certain redshift ranges (e.g., as strong spectral lines redshift in and out of a spectrometer's bandpass) are irrelevant as long as these effects do not select for a different burst population. However, if the bursts in small galaxies are intrinsically different from bursts in large galaxies, then the bias against small, faint host galaxies invalidates the assumption about the burst sample. Although in my studies thus far I have not searched for redshift evolution in the shape of the distribution function, such evolution can easily be considered by giving the distribution function's parameters a redshift dependence. Non-uniform redshift sampling of the burst population will not bias the results as long as this sampling is not correlated with intrinsic burst populations. The data space in which bursts are detectable is restricted to higher energy values for larger redshifts, which the methodology accommodates.

Studies which attempt to determine both the shape and normalization of the distribution function, such as Donaghy et al.[2] and Lloyd-Ronning et al.,[6] must consider each burst as having been drawn from a more complex distribution function. Each burst now populates a two dimensional redshift-apparent energy data space. The efficiency with which bursts were detectable in this data space is complicated by the detection threshold which varied with the type of burst (bursts are generally detected by peak flux, not fluence) and over the period the burst sample was collected. The burst sample must be constructed with much greater care so that the detection efficiency over the data space can be determined.

I have been using parametric models of the apparent energy distribution function. I construct the likelihood function as the product of obtaining the observed energy fluence given the assumed distribution function. The parameters of a given function are calculated by maximizing the likelihood function with respect to the function's parameters. The ratio of maximum likelihoods in the frequentist formulation of statistics, or the odds ratio in the Bayesian, compares different functional forms.

The probability of obtaining the observed energy fluence is calculated not from the total distribution function but from the portion of the distribution function which is above the detection threshold; of course, the distribution function must be therefore be renormalized. The detection threshold must be known considering all processes which affected the inclusion of the burst in our sample.

If we derive the apparent energy from the burst itself (e.g., using the proposed variability-luminosity[3, 8] or lag-luminosity[7] correlations to derive the luminosity from which the energy can be scaled from observed quantities), then all we need is the threshold for detecting the burst. This is usually determined and reported for gamma-ray detectors such as BATSE.

But if we need other observations—such as optical observations to determine the redshift—then the detection threshold is the most restrictive criterion. For the typical Beppo-SAX afterglow case:

1. Beppo-SAX must detect and localize the gamma rays
2. A decision must be made for a Beppo-SAX search for an X-ray afterglow
3. Beppo-SAX must detect and localize a fading X-ray source
4. A decision must be made to mount an optical or radio campaign
5. The optical or radio campaign must detect and localize a fading source
6. A constant, extended optical source—the host galaxy—must underlie the well localized afterglow
7. A decision must be made to observe the optical afterglow or the host galaxy spectroscopically
8. Spectral lines useful for a redshift determination must be detected

Steps 1, 3, 5, 6, and 8 introduce obvious detection thresholds. Steps 2, 4, and 7 introduce less obvious detection thresholds: an observation may not have been attempted unless the burst was sufficiently bright. Note that steps 3, 5, 6, and 8 may introduce selection effects because of correlations between the observed quantity and the burst energy: bursts in galaxies (step 6) may be brighter than bursts outside galaxies (if they exist); bursts deep within dust clouds which hide the optical afterglow may be intrinsically brighter.

We need the apparent energy E and its threshold value $E_T$ for each burst in a dataset. This is usually derived from the energy fluence, its threshold value, the redshift and a cosmological model—here $(H_0, \Omega_B, \Lambda) = (65, 0.3, 0.7)$. If the proposed variability-luminosity[3, 8] and lag-luminosity[7] correlations are valid, then the energies are the luminosity scaled by the ratio of the fluence to the peak flux.

Burst energies over the same photon energy range should be compared. If the fluences are over the photon energy range which includes most of the burst energy, the fluence can be considered bolometric, and k-corrections are not necessary. The 20–2000 keV range is nearly bolometric. Where possible, I integrate the spectral fits to a burst over the burst duration and the desired energy range. For the BATSE bursts with spectral data this results in a fluence that is somewhat smaller (by a factor of $\sim$1–2) than the BATSE catalog values.[5]

The fluence is almost never the quantity which triggers an observation. Therefore I calculate the threshold fluence by scaling from the quantity which does trigger an observation, usually the peak flux.

A spectroscopic redshift is the ideal. However, if such observations are absent, the redshift can be estimated from other data. For example, a bright host galaxy is more likely to be nearby than a faint one. Thus a redshift probability distribution can be calculated from the magnitude of the host galaxy. Fairly accurate redshifts can be estimated from a galaxy's colors, but unfortunately multicolor host galaxy magnitudes are rarely reported. The observed distributions of galaxy magnitude vs. redshift (e.g., from the Hubble Deep Field) can be converted into a redshift distribution given a galaxy magnitude. However, galaxy surveys weight each observed galaxy equally, and we do not expect intrinsically dim nearby galaxies and intrinsically bright distant galaxies with the same apparent magnitudes to have the same burst rate. Thus the survey results need to be weighted, which requires a model of burst occurrence.

How do we choose the weighting? Likelihoods are calculated using different host galaxy models for a burst sample with both spectroscopic redshifts and galaxy magnitudes, and the model with the highest likelihood value is favored. Jimenez et al.[5] considered four models using a sample of 10 bursts; these models with the resulting likelihood values $L$ are:

1. Unit weighting ($L = 10^{-3}$)—all galaxies have equal probabilities of hosting a burst
2. Current epoch luminosity weighting ($L = 3 \times 10^{-4}$)—both k- and e-corrections (color and evolution) are applied, which translates the observed luminosity into the luminosity the galaxy would currently have in the fiducial band; this is approximately equivalent to weighting by mass
3. Burst epoch luminosity weighting ($L = 10^{-2}$)—only k-corrections (color) are applied, which translates the observed luminosity into the galaxy's luminosity in the fiducial band; this is weighting by the luminosity at the time of the burst
4. Star formation weighting ($L = 10^{-4}$)—model 2 is then weighted by the cosmic star formation rate as a function of redshift

Thus the third model is favored, and was used.[5]

## RESULTS

I consider four burst samples, the first from Jimenez et al.[5] and the other three from Band:[1]

1. The Jimenez et al.[5] sample—8 BATSE bursts with spectroscopic redshifts and 4 with redshifts estimated from the host galaxy magnitude. The fluence threshold was calculated by scaling from $C_{max}/C_{min}$ (i.e., from the peak flux). This sample was used to calculate not only the apparent gamma-ray energy, but also the peak photon flux and the afterglow energy.
2. BATSE bursts with spectroscopic redshifts—9 bursts with spectroscopic redshifts and fluences calculated from integrating spectral fits. The fluence threshold was calculated by scaling from $C_{max}/C_{min}$.
3. The Frail et al.[4] sample—17 bursts; basically the second sample with an additional 8 bursts with fluences from other burst detectors. A fluence threshold of $10^{-6}$ erg cm$^{-2}$ was assumed for the non-BATSE bursts.
4. The Fenimore and Ramirez-Ruiz[3] sample—220 BATSE bursts with catalog fluences and redshifts derived from the variability-luminosity correlation. The threshold is scaled from the sample peak flux threshold.

The first sample was only fit with a lognormal distribution while the other three samples were fit with both lognormal and simple power law distributions.

The lognormal and power law distributions are both acceptable distributions for samples 2-4. For samples 2 and 3 the frequentist Likelihood Ratio Test mildly favors the power law distribution but mildly favors the lognormal distribution for sample 4. The Bayesian Odds Ratio favors neither for samples 2 and 3, and strongly favors the lognormal distribution for sample 4.

The best fit parameter values differ significantly for the different samples. For example, the lognormal distribution's central energy $E_0$ is an order of magnitude larger for samples 1-3 (using spectroscopic redshifts) than for sample 4 (using the variability-luminosity correlation).

These discrepancies may result from imperfect samples. The detection thresholds for samples 1-3 are unknown and probably underestimated. The validity of the variability-luminosity correlation is still open to debate.

Finally, the shape of the likelihood contours for the lognormal distribution does not rule out broader distributions centered at lower energies than the best fit value.

Definitive results await a larger, better-defined sample.

## ACKNOWLEDGMENTS

I thank R. Jimenez and T. Piran, with whom I collaborated on the early part of this research. I also thank C. Graziani who improved the underlying statistics. Most of this work was performed under the auspices of the U.S. Department of Energy by the Los Alamos National Laboratory under Contract No. W-7405-Eng-36.

## REFERENCES

1. Band, D. *ApJ*, 563, 2001 in press [astro-ph/0105259].
2. Donaghy, T., Lamb, D. Q., and Reichart, D. "Determining the GRB (Redshift, Luminosity)-Distribution Using Burst Variability," in these proceedings, 2002.
3. Fenimore, E., and Ramirez-Ruiz, E. 2002. *ApJ*, submitted [astro-ph/0004176].
4. Frail, D. A., et al. *ApJL*, 562:L55-58, 2001 [astro-ph/0102282].
5. Jimenez, R., Band, D., and Piran, T. 2001. *ApJ*, 561:171-177, 2001 [astro-ph/0103258].
6. Lloyd-Ronning, N. M., Fryer, C. L., and Ramirez-Ruiz, E. "Cosmological Aspects of Gamma-Ray Bursts: Luminosity Evolution and an Estimate of the Star Formation Rate at High Redshifts," in these proceedings, 2002.
7. Norris, J., Marani, L., and Bonnel, J. *ApJ*, 534:248-257, 2000 [astro-ph/9903233].
8. Reichart, D. E., et al. *ApJ*, 552:57-71, 2001 [astro-ph/0103258].

# Seven-year Konus GRB Catalog

T. L. Cline[*], S. D. Barthelmy[*], P. S. Butterworth[*], T. B. Sheets[*], R. I. Aptekar[†], D. D. Frederiks[†], S. V. Golenetskii[†], V. N. Il'inskii[†], E. P. Mazets[†] and V. D. Pal'shin[†]

[*]NASA/GSFC, Greenbelt, MD USA
[†]Ioffe Physico-Technical Institute, St. Petersburg, Russia

**Abstract.** A complete catalog of all Konus gamma ray burst (GRB) event profiles, collected from GGS-Wind launch in November, 1994 to the present, is under development and will be processed in CD form. All Konus GRB and soft gamma repeater (SGR) event data are presently available on the GCN Web site[1], with over 850 GRB and over 75 SGR event profiles logged. The location of the Wind spacecraft, always outside the magnetosphere, provides a steady background which may facilitate precise GRB phenomenological studies.

## SUMMARY

The Konus experiment has been active on board the GGS-Wind spacecraft for over 7 years. The location of the Wind spacecraft is always within a few light-seconds distance from the Earth, varying at times from a nearby cislunar trajectory to the location of the first Langrangian point. The detector array of two large, unshielded gamma ray sensors on the opposite faces of the spacecraft, has a nearly omnidirectional celestial view that is essentially free of occultation by the Earth.

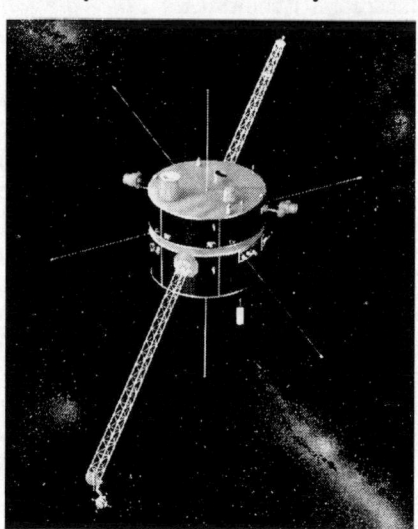

**FIGURE 1.** The GGS-Wind Spacecraft

Konus is a major element in the interplanetary GRB Network (IPN), and became its most sensitive near-Earth vertex after the Earth-orbiting Compton-GRO mission was terminated in June 2000. Many observations of GRB counterparts were enabled in 1999-2000 by the Ulysses-Konus-NEAR network, including one with the greatest redshift, at z = 4.5. Konus has also made basic contributions to studies of other transients, including soft gamma repeaters (SGRs), the giant SGR flare, and the bursting pulsar. Only the Compton-GRO BATSE provided a more extensive GRB catalog than Konus can, that has enabled a great variety of GRB studies. However, Konus has the advantage over orbiting instruments, such as BATSE, of a steady interplanetary background, unaffected by passages through the trapped radiation. The fact of this constant background may be of particular advantage in enabling some comparative studies of GRB event characteristics.

We are therefore preparing a catalog in CD form of Konus GRB histories and spectra in order to make this database available to the community. This compilation, for consistency, will consist of triggered events only. In addition to time histories, certain spectral information for each event, not presently exhibited on the Web site, will be included. NASA plans to continue GGS-Wind data recovery at least through 2002.

---

[1] http://gcn.gsfc.nasa.gov/gcn/konus_grbs.html

# Comments on Anisotropic Distributions of Faint BATSE GRBs

Jon Hakkila*, Timothy W. Giblin*, William S. Paciesas[†], Richard J. Roiger**, David J. Haglin** and Charles A. Meegan[‡]

*Department of Physics and Astronomy, College of Charleston, Charleston, SC*
[†]*Department of Physics and Astronomy, University of Alabama in Huntsville, AL*
**Department of Computer and Information Sciences, Minnesota State University, Mankato, MN*
[‡]*NASA NSSTC, Huntsville, AL*

**Abstract.** We study the angular anisotropies of faint intermediate and extremely short BATSE bursts in CGRO spacecraft coordinates. Our goal is to determine whether or not biases in detector effeciency can account for the observed anisotropies. We conclude that the faint intermediate burst anisotropy is not statistically meaningful, and that some of the very shortest bursts might be transient events other than gamma-ray bursts.

## INTRODUCTION

In the past few years, claims have been made that specific gamma-ray bursts are anistropically distributed on the celestial sphere. Anisotropic angular distributions might indicate that these bursts are Galactic in origin, that they are associated with specific classes of astronomical objects, or that some sources repeat. The claims involve intermediate bursts [1] and the very shortest bursts [2].

We suspect that anisotropies in the faint burst angular distributions might be due to statistical variations or might be the result of instrumental and sampling errors. We have demonstrated [3] that instrumental and sampling biases are likely responsible for creating an intermediate burst class from selected long bursts. Faint bursts are particularly susceptible to instrumental and sampling biases, since (1) their properties can be incorrectly measured in systematic ways, and (2) faint bursts with specific temporal and/or spectral characteristics can trigger at the expense of others.

## ANISOTROPIC INTERMEDIATE BURST DISTRIBUTION

We study the intermediate burst angular distribution in CGRO spacecraft coordinates. Meszaros et al. [1] have found an angular anisotropy of faint intermediate bursts in the celestial coordinates, based on the expected number of bursts found in discrete cells. Their intermediate bursts have 2 sec < T90 < 10 sec and 256 > 0.65 photons cm$^{-2}$ sec$^{-1}$. The authors reject the null hypothesis (angular isotropy in this statistic) for the intermediate subclass at the 96.4% confidence level. A greater anisotropy is found for the faint intermediate bursts (2 sec < T90 < 10 sec and 0.65 photons cm$^{-2}$ sec$^{-1}$ < p256 < 2 photons cm$^{-2}$ sec$^{-1}$; the authors reject the null hypothesis for these bursts at the 99.3% confidence level.

We examine angular anisotropy for these subsets in CGRO (spacecraft) coordinates. *Since the satellite has been rotated to look at new objects roughly twice a month over the duration of the CGRO mission, there is no direct correlation between a burst's CGRO coordinates and its celestial coordinates.* If the angular distribution of a suspected burst subclass is anisotropic in CGRO coordinates, then the anisotropy might be due to instrumental effects rather than intrinsic ones. Our measure of isotropy in spacecraft coordinates is based on the distribution of angles from each burst to the BATSE detector face most nearly pointing in its direction. The burst brightness measured in a BATSE detector is roughly proportional to the cosine of the angle from the detector normal, and two or more BATSE detectors must observe a burst above a preset threshold in order that the burst trigger. Thus, as demonstrated in Figure 1, no angular anisotropy should be detected in the distribution of bright bursts (solid line), whereas very faint bursts might be observed at angles between detector faces (dotted line).

The intermediate duration burst distribution (dotted line) is compared to a normalized random distribution. A $\chi^2$ test rejects the null hypothesis (isotropy) at the 97.3% confidence level. Using the same rejection criteria as

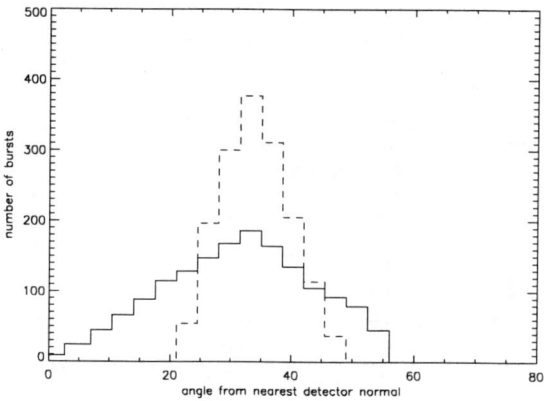

**FIGURE 1.** Expected distributions of bright (solid line) and faint (dotted line) bursts in GRO coordinates, based on the angle of each burst from the nearest detector normal.

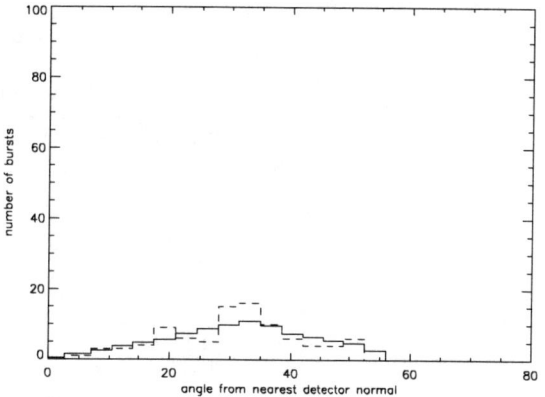

**FIGURE 2.** Meszaros et al. [1] faint intermediate burst angular distribution in GRO coordinates.

Meszaros et al. [1] (95% confidence level), we might be tempted to argue that the anisotropy is real and is instrumental in nature. However, when we compare the means of the two distributions using a Wilcoxon rank order test (a shift of the mean might indicate a systematic error in the detector performance, as opposed to a random error), we find that they are similar (the null hypothesis is only rejected at the 86.0% confidence level). The intermediate bursts are distributed slightly anisotropically in CGRO coordinates, but the anisotropy appears to be due to random rather than systematic variations.

In Figure 2 we compare the distribution of faint intermediate burst nearest angles (dotted line) to a normalized random burst distribution (solid line). A $\chi^2$ test only rejects isotropy at the 52.7% confidence level while the Wilcoxon rank order test only rejects it at the 83.6% confidence level. This distribution is thus isotropic in spacecraft coordinates; any angular isotropy in celestial coordinates does not result from instrumental biases.

Surprisingly, the distribution of bright intermediate values (which Meszaros et al. [1] found to be isotropically distributed) is anisotropic in spacecraft coordinates, as demonstrated in Figure 3 (dotted line, compared to a solid line normalized random burst distribution). A $\chi^2$ test rejects the null hypothesis at the 99.9% confidence level while the Wilcoxon rank order test rejects it at the 95.1% confidence level. There is no physical reason why bright intermediate bursts should be located abnormally relative to BATSE detector faces whereas faint ones should be located normally; the instrumental biases are much more likely to influence faint bursts than bright ones. Furthermore, the observed angular distribution is inconsistent with any form of the distribution expected from known instrumental biases. If we believe that this anisotropy is due to chance, then we must likewise be-

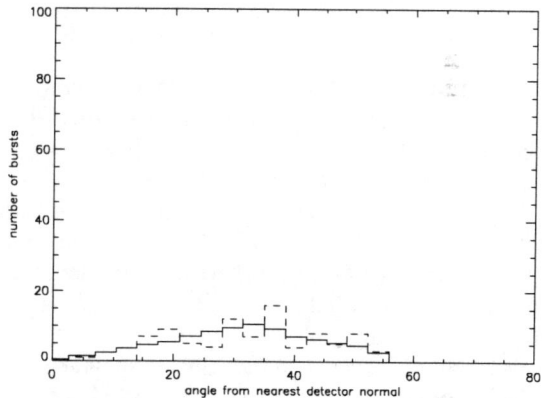

**FIGURE 3.** Meszaros et al. [1] bright intermediate burst angular distribution in GRO coordinates.

lieve that the celestial anisotropy of faint intermediate bursts seen by Meszaros et al. [1] is due to chance.

## VERY SHORT BURST DISTRIBUTION

Cline *et al.* [2] find the very shortest BATSE GRBs (e.g. those with T90 < 100ms) to be distributed anisotropically. We suspect that any angular anisotropy could be due to instrumental and/or sampling error, since these bursts are by definition faint. The Cline et al. [2] count-in-cells examination of 42 bursts finds the null hypothesis of isotropy to be rejected at the 99.9984% confidence level; they attribute this to a galactic population of evaporating primordial black holes. There are 46 bursts with T90 < 100 ms listed in the BATSE Current Catalog.

We find no evidence of an anisotropy in spacecraft coordinates. We note that two short GRBs (BATSE triggers

**TABLE 1.** Thirteen short, soft BATSE triggers. The trigger locations appear to be anisotropically distributed, and many are consistent with common locations or with other known objects. However, it is noted that large localization errors detract somewhat from the apparent anisotropy.

| BATSE trigger | RA (°) | DEC (°) | error (°) | hr321 | ch4 fluence | location is consistent with |
|---|---|---|---|---|---|---|
| 2132 | 42.11  | 84.41  | 9.51  | 2.71 | 0       | source1 |
| 2159 | 318.99 | 71.46  | 18.54 | 2.25 | 0       | source1 |
| 2463 | 59.24  | 14.29  | 18.13 | 1.05 | 7.3e-8  | source2 |
| 2464 | 59.39  | 46.28  | 15.29 | 0.51 | 0       | source2 |
| 3910 | 74.53  | 52.18  | 6.90  | 0.99 | 8.4e-9  | source2 |
| 5620 | 63.09  | 38.22  | 6.6   | 1.66 | 2.9e-08 | source2 |
| 2757 | 340.57 | -39.28 | 9.00  | 0.02 | 2.9e-6  | source3 |
| 5536 | 346.48 | -30.66 | 5.35  | 1.90 | 0       | source3 |
| 3384 | 56.64  | -32.75 | 7.71  | 6.17 | 0       | source4 |
| 3799 | 57.86  | -34.95 | 9.32  | 2.69 | 0       | source4 |
| 5458 | 251.01 | -20.48 | 10.79 | 0.69 | 0       | 3 SGRs |
| 5992 | 178.14 | -42.35 | 11.77 | 1.25 | 0       | ? |
| 6645 | 0.47   | -40.22 | 4.85  | 2.32 | 1.3e-7  | ? |

2463 and 2464) have been considered for reclassification as "unknown" sources partly on the basis of their soft spectra; their HR321 hardness ratios ($\langle HR321 \rangle = 0.8$) are considerably softer than the average for short bursts ($\langle HR321 \rangle = 3.3$), and both have essentially no emission at energies greater than 300 keV. In fact, examination reveals that thirteen of 46 bursts have no high energy emission or are very soft (HR321 $\leq$ 2.5). The localization errors on all of these faint bursts are large, which has not been taken into account in the Cline et al. anaysis, and probably dilutes that study's significance. However, we note that four of the thirteen triggers have angular positions consistent with one common source, six have positions consistent with three other common sources, and one has a position consistent with any of three known soft gamma repeaters (SGRs). *We do not necessarily believe that a small number of sources is responsible for these short, soft bursts; repetition and soft spectra are not typical short burst characteristics. Instead, we consider it possible that some of these short soft events are not gamma-ray bursts.*

## CONCLUSIONS

We suspect that the marginally anisotropic faint intermediate angular distribution does not represent an anisotropic source distribution; we believe instead that the distribution results from statistical variations caused by small-number statistics.

Very short BATSE bursts might be anisotropically distributed, but the significance of the anisotropy could be overestimated due to binning and lack of incorporating localization errors into the analysis. On the basis of spectral characteristics, it is possible that some of very short bursts are actually other types of transient events misclassified as gamma-ray bursts. We are examining these hypotheses in greater detail.

## ACKNOWLEDGMENTS

We would like to gratefully acknowledge NASA support under grant NRA-98-OSS-03 (the Applied Information Systems Research Program).

## REFERENCES

1. Meszaros, A., Bagoly, Z., Horvath, I., Balazs, L. G., and Vavrek, R., *ApJ*, **539**, 98–101 (2000).
2. Cline, D. B., Matthey, C., and Otwinowski, S., "Non-Isotropic Angular Distribution for Very Short-Time Gamma-Ray Bursts?", in *Proceedings of the Fifth Huntsville Gamma-Ray Burst Symposium*, edited by M. Kippen, R. S. Mallozzi, and G. J. Fishman, AIP Conference Proceedings 526, American Institute of Physics, New York, 2000, pp. 97–101.
3. Hakkila, J., Haglin, D. J., Pendleton, G. N., Mallozzi, R. S., Meegan, C. A., and Roiger, R. J., *ApJ*, **538**, 165–180 (2000).

# The Internal Luminosity Functions of BATSE 5B Gamma-Ray Bursts

Jon Hakkila[*], Timothy W. Giblin[*], Thomas M. Freismuth[*], Kevin C. Young[*], Amanda J. Sprague[*], Andrew D. Stallworth[*], David J. Haglin[†], Richard J. Roiger[†] and William S. Paciesas[**]

[*]*Department of Physics and Astronomy, College of Charleston, Charleston, SC*
[†]*Department of Computer and Information Sciences, Minnesota State University, Mankato, MN*
[**]*Department of Physics and Astronomy, University of Alabama in Huntsville, AL*

**Abstract.** The Internal Luminosity Function (ILF) is the differential distribution of luminosity measured within a gamma-ray burst (GRB). Most GRBs are found to have pseudo power-law ILFs; the properties of the ILF power-law index $\alpha$ has been examined by Horack and Hakkila for a sample of 50 bright GRBs (ApJ 479, 371). We measure $\alpha$ values for 348 BATSE GRBs, then correlate these with other measured GRB properties.

## INTRODUCTION

The internal luminosity function $\psi(L)$ (or ILF) [1] is the distribution of luminosity within a gamma-ray burst. The quantity $\psi(L)dL$ represents the fraction of time during which a burst's luminosity lies between $L$ and $L+dL$. A practical method for calculating $\psi(L)$ from BATSE data is to use 64 ms counts in the 50-300 keV spectral range; this approach avoids the inclusion of a spectral model, a detector response matrix, and a spectral deconvolution of each 64 ms burst interval.

For our calculation of the ILF we have chosen to use the concatenated BATSE 64-ms data type found at HEASARC's Compton Gamma-Ray Observatory Science Support Center. A few minor modifications have been made to the background-subtracted data for each burst. First, a constant background level is identified from (assumed) Poisson variations in the BATSE energy channels and burst time histories are adjusted to these levels. Second, Monte Carlo models of Poisson variations are used to noisify time intervals with poor (e.g. 1024 ms) resolution data. Third, a distribution function is constructed by binning count rates relative to an assumed minimum (for quality control purposes, we examine three different values of the minimum: $1\sigma$, $2\sigma$, and $3\sigma$ above the constant background rate). Fourth, expected Poisson background rates are subtracted from each bin so that only "true" counts remain. Finally, the ILF is normalized from the requirement that

$$\sum_i \psi(L)_i \Delta L_i = 1 \qquad (1)$$

The resulting $\psi(L)$ distribution represents the fraction of the total detection time during which the burst was observed to have a luminosity between $L$ and $L+dL$.

The function $\psi(L)$ has previously been measured for the 50 brightest bursts in the BATSE 3B Catalog [1]. The ILF is found to take on a quasi power-law form for most bursts such that

$$\psi(L) \propto L^\alpha \qquad (2)$$

with power-law indices $\alpha \approx -3/2$. A steep power-law index ($\alpha < -3/2$) indicates that the burst spends more time emitting low intensity emission than high-intensity emission. A flat power-law index ($\alpha > -3/2$) indicates that the burst spends a relatively large amount of time emitting high-intensity photons.

A probable anti-correlation has been found between $\alpha$ and T90 duration [1] which indicates that longer bursts spend more time than shorter bursts emitting relatively low-intensity flux. A correlation has also been identified between $\alpha$ and hardness ratio HR32 which indicates that harder bursts spend more time than softer bursts emitting relatively low-intensity flux.

## ANALYSIS

Our current sample contains 348 bursts with measured $\alpha$ values and their errors $\sigma_\alpha < 0.2$. The sample is biased towards long, bright bursts, since some bursts are too faint or too short for $\alpha$ to have their IFL values measured.

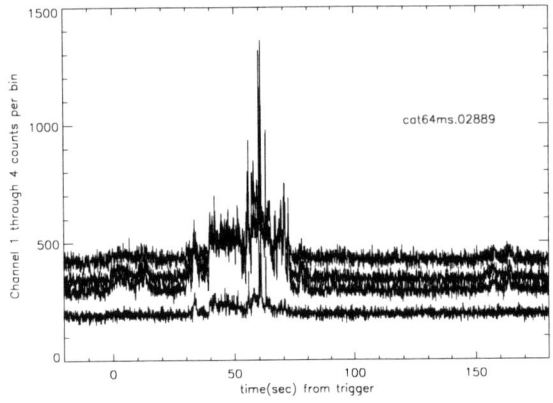

**FIGURE 1.** Time history of BATSE trigger 2889. All four BATSE energy channels are plotted; low energy channel 1 is the noisiest and high energy channel 4 is the quietest.

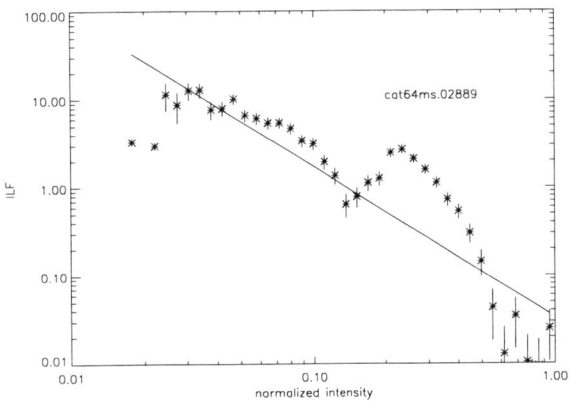

**FIGURE 2.** Best-fit ILF power-law index $\alpha$ of channels $2+3$ for BATSE trigger 2889.

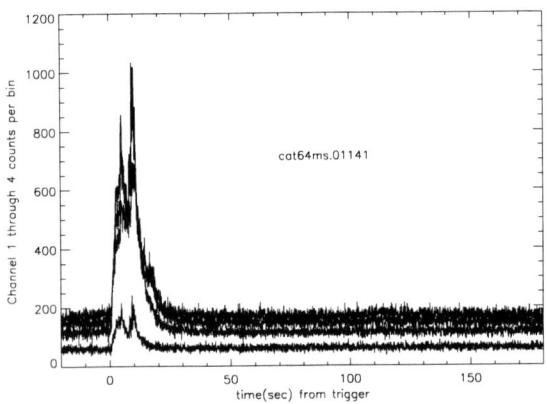

**FIGURE 3.** Time history of BATSE trigger 1141. All four BATSE energy channels are plotted; low energy channel 1 is the noisiest and high energy channel 4 is the quietest.

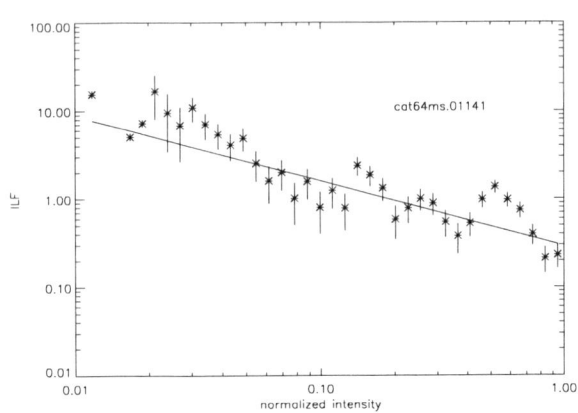

**FIGURE 4.** Best-fit ILF power-law index $\alpha$ of channels $2+3$ for BATSE trigger 1141.

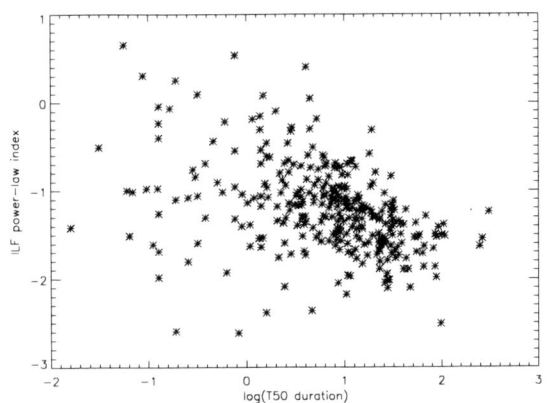

**FIGURE 5.** $\alpha$ vs. log(T50 duration).

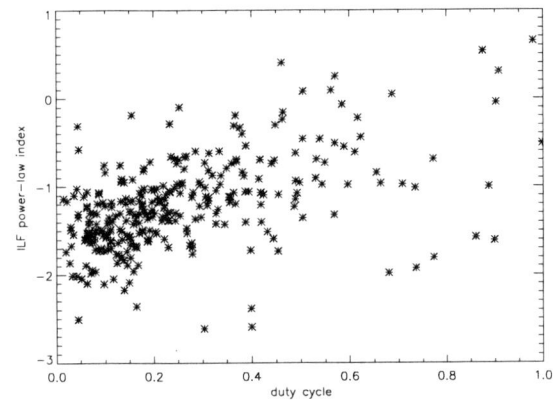

**FIGURE 6.** $\alpha$ vs. duty cycle.

**TABLE 1.** Attributes compared to $\alpha$.

| Measured attribute | Probability that attribute is uncorrelated with $\alpha$ |
|---|---|
| log(P1024) | $p = 1.8 \times 10^{-1}$ |
| log(fluence) | $p = 6.2 \times 10^{-3}$ |
| log(T90) | $p = 1.1 \times 10^{-16}$ |
| log(T50) | $p = 4.6 \times 10^{-17}$ |
| log(HR32) | $p = 4.1 \times 10^{-2}$ |
| log(duty cycle) | $p = 1.2 \times 10^{-21}$ |

We are currently analyzing BATSE TTE data in order to obtain $\psi(L)$ properties for short and faint bursts.

The ILF is demonstrated for two bursts with extreme best-fit $\alpha$ values, BATSE triggers 2889 and 1141. Trigger 2889 (Figure 1) has a steep power-law index ($\alpha = -1.8$, Figure 2) indicating that the low-intensity flux is well-sampled with respect to the high-intensity flux. Trigger 1141 (Figure 3) has a flat power-law index ($\alpha = -0.8$, Figure 4) indicating that the low-intensity flux is sparsely-sampled with respect to the high-intensity flux. *Based on inspection of a large number of BATSE GRBs, we note what appears to be a relationship between $\alpha$ and burst morphology. Simple bursts such as FREDs have flat power-law indices while complex bursts with large pulse intensity variations have steep power-law indices.*

We have compared $\psi(L)$ properties (primarily $\alpha$) to other measured BATSE data attributes: $\alpha$ to log(1024 ms peak flux), log(50-300 keV fluence), log(T90 duration), log(T50 duration), log(HR32 hardness ratio), and duty cycle. Duty cycle [2] measures the persistence of GRB emission relative to the burst's peak emission.

All comparisons between $\alpha$ and other parameters are examined (Table 1) with a Spearman rank correlation test in order to determine the probability that the two parameters are uncorrelated. We verify the anti-correlation between $\alpha$ and duration (as measured using both log(T90) and log(T50); see Figure 5) found previously [1]. Only weak evidence exists for correlations between $\alpha$ and log(fluence) and $\alpha$ and log(HR32), which contradicts our earlier results. No correlation is found between $\alpha$ and log(1024 ms peak flux). An extremely strong correlation is found between $\alpha$ and duty cycle (Figure 6).

## DISCUSSION

The anti-correlation of $\alpha$ with duration is quite strong. We want to make sure that this effect is real, rather than a reflection of some instrumental and/or sampling bias. We have considered several instrumental scenarios that could cause this relation. One possible scenario is that shorter bursts are simply time-compressed versions of longer bursts. Given enough time compression, counts could pile up in 64 ms bins. This would cause most of the emission to be measured as high-intensity emission, such that low-intensity emission would be undersampled. The measured value of $\alpha$ calculated from 64 ms data would be smaller for short bursts than it would be for long bursts. *However, shorter bursts do not appear to be time-compressed versions of longer bursts, as shorter bursts typically have fewer pulses and shorter interpulse intervals than long bursts (e.g. [3] ). Thus, the anti-correlation appears related to real burst physics.*

What does the $\alpha$ vs. duration anti-correlation tell us about gamma-ray burst physics? Both $\alpha$ and duty cycle indicate burst persistence. The $\alpha$ vs. duration correlation therefore implies that shorter bursts are more persistent emitters than longer bursts. We hypothesize that more energy is stored in the few pulses of shorter bursts than is stored in the many pulses of longer ones. Fewer, more energetic shocks could be emitted by the central engine in short bursts than in longer bursts. Conversely, an external medium that more persistently dissipates shock energy could be responsible for the more frequent, less-persistent pulses found in longer bursts.

## CONCLUSIONS

We have measured the ILF for 348 BATSE gamma-ray bursts as part of an ongoing project. We verify the anti-correlation between best-fit ILF power-law index $\alpha$ and duration found previously [1], and we find a new correlation of $\alpha$ with duty cycle [2]. The anti-correlation of $\alpha$ with duration appears to be real as opposed to being instrumental in nature. We conclude from this preliminary analysis that shorter bursts are more persistent emitters than longer bursts, which may indicate that shorter bursts are more efficient emitters than longer bursts.

## ACKNOWLEDGMENTS

This project is supported by NSF grant AST-0098499 and NASA grant NRA-98-OSS-03.

## REFERENCES

1. Horack, J. M., and Hakkila, J., *The Astrophysical Journal*, **479**, 371+ (1997).
2. Hakkila, J., Preece, R. D., and Pendleton, G. N., "A Simple BATSE Measure of GRB Duty Cycle", in *AIP Conf. Proc. 526: Gamma-ray Bursts, 5th Huntsville Symposium*, 2000, pp. 83+.
3. Fenimore, E. E., *The Astrophysical Journal*, **518**, 375–379 (1999).

# The AMANDA Search for High Energy Neutrinos from Gamma-Ray Bursts

R. Hardtke* for the AMANDA Collaboration

*Dept. of Physics, University of Wisconsin, Madison, WI 53706*

**Abstract.** If GRBs accelerate protons as well as electrons, they may be the source of the highest energy cosmic rays. Detection of neutrinos from GRBs would confirm hadronic acceleration. AMANDA uses the Antarctic icecap as a Cherenkov medium for detecting such high energy neutrinos. We searched data recorded during 1997 for neutrinos coincident with northern hemisphere GRBs detected by BATSE. BATSE provides the time and location information that reduces the background of atmospheric neutrinos from which the high energy neutrinos must be separated. Quality cuts reduce the number of misreconstructed atmospheric muon tracks, resulting in a nearly background-free search. The current search result, based on one year of data, is consistent with no signal and we place an upper limit on the neutrino flux from GRBs.

## INTRODUCTION

Gamma-ray bursts (GRBs) are short, intense, and randomly distributed eruptions of high energy photons. The likely mechanism for achieving such high energies is the conversion to radiation of the kinetic energy of ultrarelativistic electrons and protons that have been accelerated in a relativistically expanding fireball.

Gamma-rays are produced primarily by the synchrotron radiation of accelerated electrons. Neutrinos are the decay products of pions produced when accelerated protons interact with the intense radiation field of the burst: $p + \gamma \to \pi^+ \to \mu^+ + \nu_\mu \to e^+ + \nu_e + \bar{\nu}_\mu + \nu_\mu$.

GRB neutrino searches are scientifically important for many reasons [1]. First, they are a direct probe of the fireball model of GRBs. Second, they may unveil the source of the highest energy cosmic rays. Third, the relative timing of photons and neutrinos over cosmological distances will allow unrivaled tests of special relativity. Fourth, the fact that photons and neutrinos should suffer the same time delay traveling through the gravitational field of our galaxy will lead to better tests of the weak equivalence principle.

## DETECTOR

In 1997, the Antarctic Muon and Neutrino Detector Array (AMANDA) consisted of 10 strings and 302 optical modules (OMs) deployed in the icecap near the geographic South Pole. An OM consists of a photomultiplier tube housed in a protective glass sphere. See Fig. 1.

Relativistic muons produce Cherenkov light that is detected by the OMs. The muon path is reconstructed using the timing and topology of the OM detections. In this analysis, AMANDA uses the earth as a filter, searching for upgoing neutrino-induced muons from the northern hemisphere while rejecting downgoing cosmic ray induced muons from the southern hemisphere. See [2] for details.

We searched data recorded by AMANDA during the Antarctic winter of 1997 for high energy upgoing neutrinos coincident with northern hemisphere GRBs detected by the Burst and Transient Satellite Experiment (BATSE) aboard the Compton Gamma-Ray Observatory. BATSE provides the time and location of the bursts, thereby reducing the background of downgoing muons from which upgoing neutrinos must be separated. Analysis of 1998-2001 data, partially recorded by the completed AMANDA-II detector of 677 OMs, is in progress.

## MONTE CARLO

The neutrino flux can be calculated as a function of the relative ratio of protons and electrons in the fireball. It can be normalized by the assumption that GRBs are the source of the observed highest energy cosmic ray flux [3]. In this case, energy should be approximately equally transferred to electrons and protons in the fireball [4, 5]. The rest of the calculation follows established particle physics techniques.

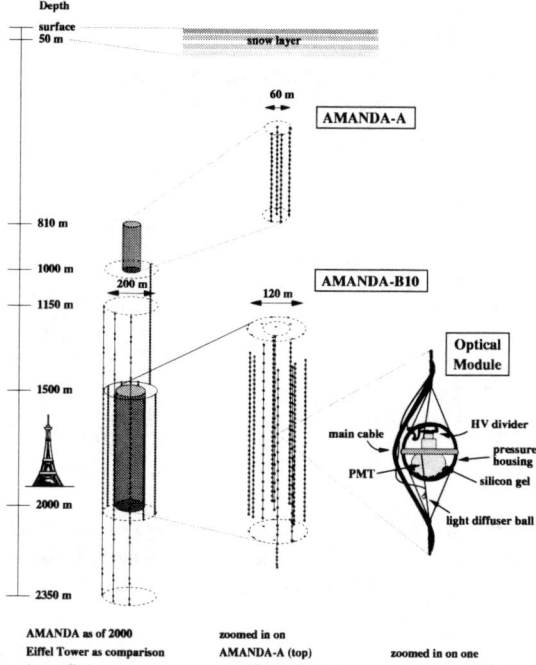

**FIGURE 1.** AMANDA detector.

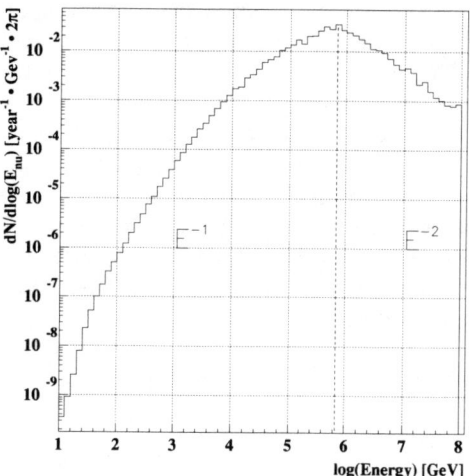

**FIGURE 2.** Energy spectrum of GRB neutrinos at AMANDA trigger level for $\Gamma = 300$.

The expected neutrino event rate in AMANDA has been determined from a full Monte Carlo simulation of the GRB signal and detector. GRB neutrinos are generated following a broken power law energy spectrum [6]:

$$\frac{d\phi}{dE_\nu} = \frac{A}{E_B E} \text{ for } E < E_B; \quad \frac{d\phi}{dE_\nu} = \frac{A}{E^2} \text{ for } E > E_B \quad (1)$$

where $E_B$ is the energy of the break in the spectrum. Its value depends on the boost factor of the fireball. The normalization constant A is determined by assuming GRBs are the source of the highest energy cosmic rays.

We re-calculated the expected neutrino event rates following the method of [7] and [8], taking into account burst-to-burst fluctuations in energy and distance and the absorption of PeV neutrinos in the Earth.

We draw attention to the fact that neutrino measurements are sensitive to the Lorentz boost factor, $\Gamma$, of the expanding fireball [7]. $\Gamma$ has been only indirectly determined by observations. Models show that GRBs would be rendered optically thick if $\Gamma \ll 100$. The efficiency for producing pions in the p-$\gamma$ fireball collisions varies as $\Gamma^{-4}$ [6] and the neutrino energy varies as $\Gamma^2$. See [9] for an alternative GRB model.

Highly transparent sources with large $\Gamma$ will preferentially emit photons. However, a moderately reduced value of $\Gamma$ will create a prolific neutrino source due to a more opaque beam dump with higher photon density.

An important consequence of fluctuations is that the signal is dominated by a few neutrino-bright bursts, which greatly simplifies their detection. Although we expect much less than one neutrino event in AMANDA from an average GRB, a burst with favorable characteristics would produce multiple events in the detector.

In the Monte Carlo, muons produced by neutrinos near the detector are tracked [10] until they reach the OMs. The simulation includes the muon track, its emission spectrum of Cherenkov photons, their propagation in the ice, the OM detection, the pulse transfer from the OM to the surface data acquisition system, and the event trigger. The Monte Carlo events were filtered and reconstructed in the same way as the data.

The GRB spectrum at trigger level is shown in Fig. 2. The result represents the convolution of the neutrino flux of Eq. (1) and the probability of conversion of the neutrinos to muons near the detector. For a boost factor of $\Gamma = 300$, the neutrino energy peaks near 700 TeV.

## ANALYSIS

In 1997, AMANDA recorded muon events at ~70Hz. Using the geometry of the OMs involved in each event, an initial line fit for the muon track was calculated. Events with a line fit originating north of declination $\delta \geq -40°$ underwent full reconstruction methods [2].

Emission of high-energy neutrinos produced in the internal shocks of GRBs should coincide with the $\gamma$-ray emission [6]. Neutrinos that may be emitted hours or days later [11] or before the GRB [12] are not the focus of this search. To ensure that our search window included the period of initial $\gamma$-ray emission, we used the earlier of the BATSE trigger time or T90 start time and then subtracted 1 second.

We estimated the background for each angular bin near a given $(\theta, \phi)$, where $\theta$ and $\phi$ are local zenith and azimuthal coordinates. The background is dominated by down-going cosmic ray muons that are misreconstructed as upgoing tracks. Data recorded an hour before and after a GRB, but excluding the [-1,+5] minutes around the GRB trigger, was used to estimate the background.

The expected number of background events is calculated by measuring the ratio of events observed in the search bin to the total number of events observed in the background data:

$$\varepsilon(\theta, \phi) = \frac{N_{\text{searchbin}}^{\text{offtime}}}{N_{\text{allbins}}^{\text{offtime}}},$$

where $N_{\text{searchbin}}^{\text{offtime}}$ is the number of events observed in the direction of a GRB during a non-GRB time window (i.e., when no signal is expected) and $N_{\text{allbins}}^{\text{offtime}}$ is the number of all events observed in the sky ($\delta \geq -40°$) during the same non-GRB time window.

We then calculate the expected number of background events in our search bin during a GRB as:

$$<n_{\text{bg}}> = \varepsilon(\theta, \phi) \times N_{\text{allbins}}^{\text{ontime}},$$

where $N_{\text{allbins}}^{\text{ontime}}$ is the number of all observed events in the sky during the search window.

Thus our background estimates are robust to overall changes of the trigger rate. Deadtime and downtime are completely accounted for in this way. The stability of the efficiency $\varepsilon(\theta, \phi)$ was studied for each burst and from burst to burst. A total of 78 GRBs occurred in the northern hemisphere and met data quality criteria [13].

Using the simulation, we designed a search that accounts for the zenith-dependent angular resolution of the AMANDA, the low number of signal events expected, and the number of background events in the search bin. We found that a single cut on the number of "direct" hits gives the best search sensitivity. Direct hits have a small time delay relative to the calculated arrival time of Cherenkov photons from the fitted track. Well-reconstructed tracks have many of these hits, which are delayed less than 75 nsec by scattering in the ice. The number of direct hits and the search bin size $\psi$ were optimized for the duration and zenith position of each GRB.

## RESULTS AND OUTLOOK

The analysis results in a virtually background-free search with 0 or 1 event detected during each GRB. Because the search did not yield statistically significant bursts, we derive a combined upper limit using the 78 searches. In Fig. 3, the neutrino event upper limit is shown for $\Gamma = 300$.

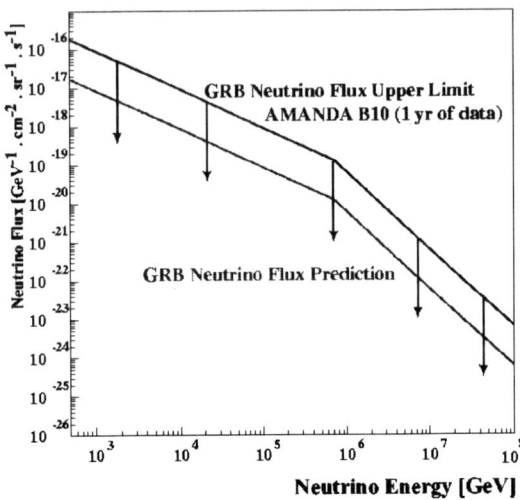

**FIGURE 3.** AMANDA GRB neutrino upper limit. The limit is within a factor of ten of the Alvarez-Muniz, Halzen and Hooper flux [7] after analysis of one year of data.

We are currently searching for neutrinos coincident with GRBs from 1998 until BATSE was turned off in May, 2000. This data will increase the trials of the current analysis by a factor of $\sim 3$.

The 19-string AMANDA-II detector was completed in 2000 [14]. Despite the loss of BATSE soon after completion, this enlarged detector will continue to search for neutrinos from GRBs detected by other satellites. The effective area of AMANDA-II for GRB neutrino searches is approximtaely 60,000m$^2$.

## REFERENCES

1. Halzen, F., *Proc. of WIN99* (1999).
2. Andres, E., et al., *Nature*, **410**, 441–443 (2001).
3. Waxman, E., *Ap. J.*, **452**, L1 (1995).
4. Waxman, E., *Phys. Rev. Lett.*, **75**, 386 (1995).
5. Vietri, M., *Ap. J.*, **453**, 883 (1995).
6. Waxman, E., and Bahcall, J., *Phys. Rev. Lett.*, **78**, 2292 (1997).
7. Alvarez-Muniz, J., Halzen, F., and Hooper, D., *Phys. Rev. D*, **62**, 093015 (2000).
8. Halzen, F., and Hooper, D., *Ap. J. Lett.*, **527**, L93 (1999).
9. Dar, A., and DeRujula, A., *astro-ph/0012227 and astro-ph/0105094* (2001).
10. Lipari, P., and Stanev, T., *Phys. Rev. D*, **44**, 11 (1991).
11. Dermer, C., *astro-ph/0005440* (2000).
12. Meszaros, P., and Waxman, E., *Phys. Rev. Lett.*, **87**, 171102 (2001).
13. Bay, R., Ph.D. thesis, University of California-Berkeley (2000), available at astro-ph/0008255.
14. Barwick, S., *Proc. of the 27th ICRC* (2001).

# A Gamma-Ray Burst Bibliography, 1973-2001

K. Hurley

*UC Berkeley Space Sciences Laboratory, Berkeley, CA 94720-7450*

**Abstract.**
On the average, 1.5 new publications on cosmic gamma-ray bursts enter the literature every day. The total number now exceeds 5300. I describe here a relatively complete bibliography which is on the web, and which can be made available electronically in various formats.

## INTRODUCTION

I have been tracking the gamma-ray burst literature for about the past twenty-one years, keeping the authors, titles, references, and key subject words in a machine-readable file. The present version updates previous ones reported in 1993 [1], 1995 [2], 1997 [3] and 1999 [4]. In its current form, this information is in a Microsoft Word 97 "doc" format. My purpose in doing this was first, to be able to retrieve rapidly any articles on a given topic, and second, to be able to cut and paste references into manuscripts in preparation. The following journals have been scanned on a more or less regular basis starting with the 1973 issues:

Advances in Physics
Annals of Physics
Astronomical Journal
Astronomische Nachrichten
Astronomy and Astrophysics (including Supplement Series)
Astronomy and Astrophysics Review
Astronomy Letters (formerly Soviet Astronomy Letters)
Astronomy Reports (formerly Soviet Astronomy)
Astrophysical Journal (letters, main journal, and supplements)
Astrophysical Letters and Communications
Astrophysics and Space Science
ESA Bulletin
ESA Journal
IAU Circulars
IEEE Transactions on Nuclear Science
Journal of Astrophysics and Astronomy
Monthly Notices of the Royal Astronomical Society
Nature
Nuclear Instruments and Methods in Physics Research Section A
Observatory
Physical Review (main journal A and letters)
Proceedings of the Astronomical Society of Australia
Publications of the Astronomical Society of Japan
Publications of the Astronomical Society of the Pacific
Reports on Progress in Physics
Science
Scientific American
Sky & Telescope

In addition, the following journals either have been scanned, but less regularly in the past, or in some cases, are no longer being scanned:

Acta Astronomica
Annals of Geophysics
Astrofizika
Astroparticle Physics
Bulletin of the American Astronomical Society
Bulletin of the American Physical Society
Bulletin of the Astronomical Society of India
Chinese Astronomy and Astrophysics
Chinese Physics Letters
Cosmic Research
Journal of Atmospheric and Terrestrial Physics
Journal of the British Interplanetary Society
Journal of the Royal Astronomical Society of Canada
New Astronomy
Progress in Theoretical Physics
Solar Physics
Soviet Physics

The above lists are not exhaustive. For example, where theses or internal reports have come to my attention, I have included them, too. To be included, an article had to have something to do with GRB or SGR theory, observation, or instrumentation, or be closely related to one of these topics (e.g., merging neutron stars, AXPs, high-z supernovae, etc.), and must have been published. With only a few exceptions, preprints or internal reports which were never published have not been included.

## ORGANIZATION OF THE BIBLIOGRAPHY

The overall organization is chronological by year. Within a given year, articles published in journals are listed first, in alphabetical order by first author. Then come theses and conference proceedings articles. The latter are listed in the order in which they appear in the proceedings. The entries are numbered consecutively, so that paper copies which are kept on file can be retrieved quickly. However, to avoid having to renumber this entire file when a new article is added, numbers are skipped at the end of each year and reserved for later inclusion. The complete author list follows, as it appears in the journal, along with the title, journal, volume number, page number, and year. A line containing key words follows this. These are generally not the same key words as the ones listed in the journal, nor are they taken from the title or any particular list. Rather, they are meant to reflect the true content of the article, and provide a list of machine-searchable topics. In general, however, key words have not been included for conference proceedings articles. An example of an entry is the following:

5163. Guetta, D., Spada, M., and Waxman, E., On the Neutrino Flux from Gamma-Ray Bursts, Ap. J. 559, 101, 2001
Key Words: p-gamma interactions, photomeson production, 10^14 eV neutrinos

## A FEW INTERESTING STATISTICS

The number of articles published each year since 1973 is shown in figure 1. Starting with one article per month in 1973, it began to exceed one per day in 1994, and reached over 1.5 per day in 2000, enough, in principle, to base an entire journal on. Several milestones are indicated as the probable causes of

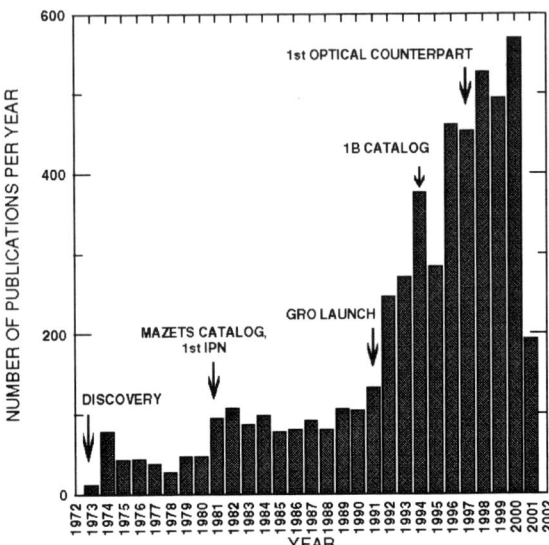

**FIGURE 1.** The number of publications by year.

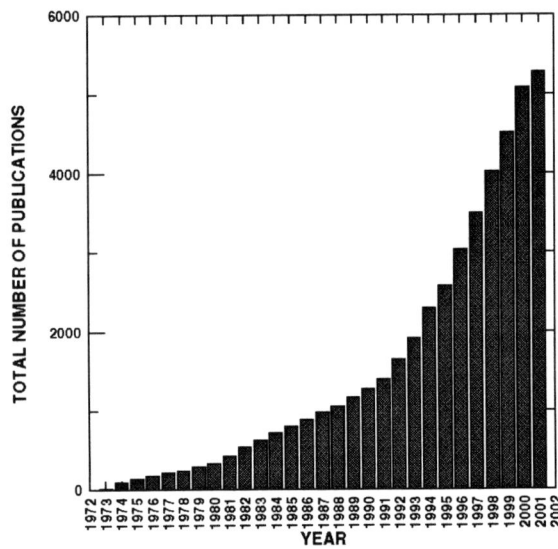

**FIGURE 2.** The cumulative number of publications by year.

sudden increases in the number of publications per year. Note that there are still about as many papers published as there are gamma-ray bursts observed. The cumulative total is shown in Figure 2. The cutoff date is mid-2001. At any given time, there may be about 100 articles waiting to be entered into the file, so the completeness, including an estimate of the number of articles which were missed for any reason, is about 98%.

The volume of the literature (it would take about 600 pages simply to print out the bibliography) has necessitated the development of a program which can search for and extract particular titles. I have written such a program in Microsoft Word Basic (a variant of the BASIC programming language). It allows one to extract all titles between two dates whose entries contain a particular key phrase, key word, or author, and write them to a separate file.

## AVAILABILITY

A web version of this bibliography may be found at ssl.berkeley.edu/ipn3/index.html. However, although the bibliography is updated on an approximately daily basis, the most up-to-date version is usually not at the website. It is available in plain ASCII, "doc", and "rich text format" (rtf) format files, which can be sent to anyone interested, as can the Word Basic program. Please contact me at khurley@sunspot.ssl.berkeley.edu to request copies, and indicate your preference for the format. I would appreciate it if users would communicate errors and omissions to me.

## ACKNOWLEDGMENTS

This work was carried out under JPL Contract 958056.

## REFERENCES

1. Hurley, K., "A Gamma-Ray Burst Bibliography, 1973-1993", in *Gamma-Ray Bursts, Second Workshop*, edited by G. Fishman, J. Brainerd, and K. Hurley, AIP Conference Proceedings 307, American Institute of Physics, New York, 1994, pp. 726–729.
2. Hurley, K., "A Gamma-Ray Burst Bibliography, 1973-1995", in *Gamma-Ray Bursts, Third Huntsville Symposium*, edited by C. Kouveliotou, M. Briggs, and J. Fishman, AIP Conference Proceedings 384, American Institute of Physics, New York, 1996, pp. 985–989.
3. Hurley, K., "A Gamma-Ray Burst Bibliography, 1973-1997", in *Gamma-Ray Bursts, Fourth Huntsville Symposium*, edited by C. Meegan, R. Preece, and T. Koshut, AIP Conference Proceedings 428, American Institute of Physics, New York, 1998, pp. 87–91.
4. Hurley, K., "A Gamma-Ray Burst Bibliography, 1973-1999", in *Gamma-Ray Bursts, Fifth Huntsville Symposium*, edited by R. M. Kippen, R. Mallozzi, and J. Fishman, AIP Conference Proceedings 526, American Institute of Physics, New York, 2000, pp. 3–7.

# Explaining GRB Lag with Spectral Evolution

Dan Kocevski* and Edison Liang*

*Department of Physics and Astronomy, Rice University, Houston, Texas, 77005*

**Abstract.** The spectral lag observed in Gamma-ray bursts (GRBs) has been shown by Norris et al. to be correlated to the absolute luminosity of a GRB. Despite the apparent importance of this GRB property, there has yet to be a full explanation of its origin. We put forth that the lag is directly due to the evolution of the GRB spectra. In particular, as the energy at which the GRB's $\nu F_\nu$ spectra is a maximum ($E_{pk}$) decays through the four BATSE channels, the photon flux peak in each individual channel will inevitably be offset producing what we measure as lag. We test this hypothesis by measuring the $E_{pk}$ decay constant $\Phi_o$ for a sample of clean single peaked bursts with known lag. We find a linear correlation between $\Phi_o$ and lag, demonstrating a direct relationship between lag and the decay of $E_{pk}$.

## INTRODUCTION

Gamma-ray burst spectra have a well known property of evolving as the burst proceeds. This evolution is characterized by two distinct features: an overall softening of the GRB spectra with time and a delay in the arrival of low energy photons. Although GRBs show remarkable variety in most of their properties, such as duration and light curve structure, the evolution of the GRB spectra appears to be a universe trend that is observed in a large number of bursts. Cheng et. al. (1995) was the first to quantify the observed delay between the low and high energy photon by using the cross-correlation technique to measure the amount of time delay between the high and low energy BATSE light curve peaks. For the purpose of trigging on burst events the BATSE detector is typically subdivided into four broad energy channels from 25 keV to above 300 keV, each channel producing a different light curve for a particular GRB event. The cross-correlation function can be used to look for variable components between these similar light curves, yielding correlation coefficients between such properties as the temporal offset of the peak pulse. Cheng et. al. found that almost all of the bursts they examined showed a delay in the 25-50 keV photon arrival times, which they hypothesized was contributed to scattering near in the environment surrounding the GRB. Norris et. al. (2000) later used a similar approach of using a CCF method to measure the lag between the BATSE channel 3 (100-300 keV) and channel 1 (25-50 keV) light curves for all GRBs with independently measured redshift. What they found was a anticorrelation between the delay of the low energy photon arrival times and the absolute luminosity of the GRB, yielding one of the first distance indicators that could be obtained from the gamma-ray data alone. They found that bursts with high $z$ exhibited little or no lag, whereas closer bursts exhibited the largest time delay. This rules out the possibility that the lag could be due simply to relativistic effect such as time dilation which would be expected to work in the opposite fashion. If the lag was due to environmental effects, such as scattering near the source, then the lag-luminosity correlation could only be explained by an evolution of the source environment with distance.

We put fourth another explanation for the spectral lag observed in GRBs, one that was first proposed by B. Scheafer (2001), that the evolution of the GRB spectra toward lower energies is directly related to the delay in the low energy light curves. It has been known for some time that the GRB spectra tends to soften with time (Golenetski et. al. 1983, Norris et. al. 1986), but Liang and Kargatis 1996 were the first to quantify the evolution as a exponential decay as a function of photon fluence.

$$E_{pk} = E_o e^{-\Phi/\Phi_o} \quad (1)$$

Where $E_{pk}$ is the max of the $\nu F_\nu$ spectra, $\Phi(t)$ is the photon fluence integrated from the start of the burst, and $\Phi_o$ is the decay constant. In other words, the average energy of the arriving photons becomes softer as the burst progresses. This result is open to various interpretations, the simplest being that the decay of $E_{pk}$ is governed by radiative cooling, although this interpretation is not unique. We believe that as the peak in the $\nu F_\nu$ spectra decays through the various BATSE channels, the time to peak in the individual light curves will correlate to the hardness of $E_{pk}$. One simple way to test this hypothesis is to com-

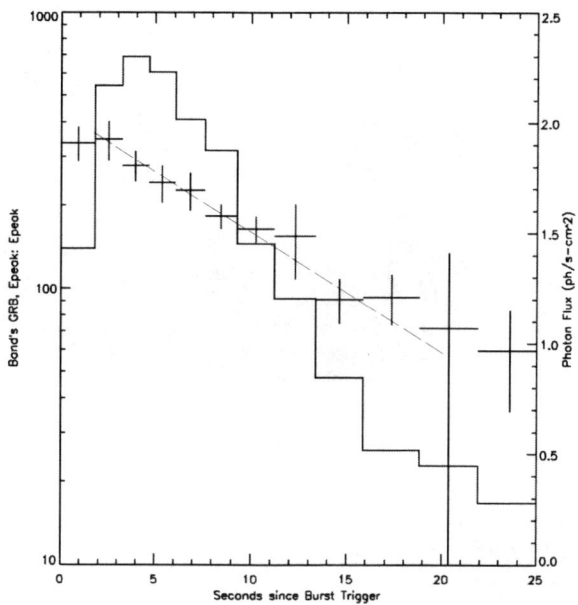

**FIGURE 1.** $E_{pk}$ evolution for GRB 7648 (990712). The decay constant is much more well defined for the FRED bursts.

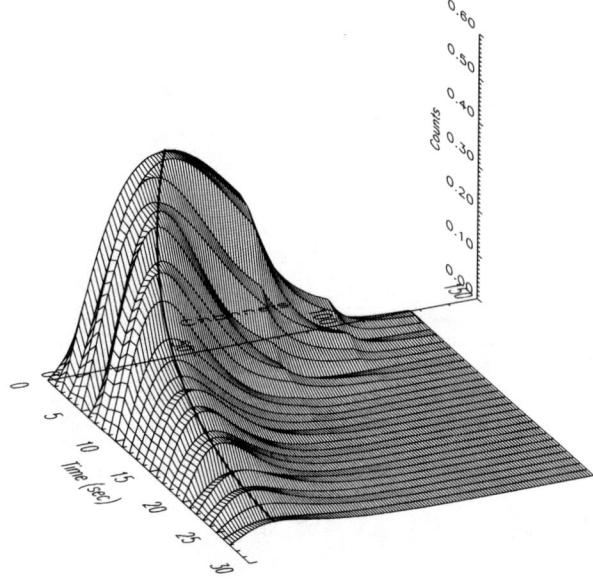

**FIGURE 2.** A 3 dimensional spectral plot of GRB 911016 (907) showing the hard to soft evolution of the peak energy. The solid line traces the decay of $E_{pk}$.

pare the timescales of GRB spectral decay to the burst's spectral lag. It would be expected that the bursts that have the longest decays, and hence smallest $\Phi_o$, would have the largest lag.

## DATA ANALYSIS

In order to test the possible correlation between $E_{pk}$ decay rates and spectral lag we obtained the High Energy Resolution (HER) BATSE data for a sample of 6 GRBs with known lag and performed time-resolved spectral fits via $\chi^2$-minimization with the empirical Band model (Band et. al. 1993). These 6 bursts were choosen for two reason: first because they are characterized by clean seperable FRED (fast rise exponential decay) pulses which tend to give reliable $\Phi_o$ measurements and secondly because their lags had been previously measured by Scheafer et. al. 2001, where the authors measure the lag and variability and hence the luminosity for 112 GRBs. An example of the time-resolved spectral fits that were preformed is shown in Figure 2, where the log of $E_{pk}$ vs. photon fluence is plotted over the burst's light curve which is shown in photons $cm^{-2} s^{-1}$. The peak energy in 990712 can be seen to decay monotonically on the semi-log plot, and a linear fit to this trend directly yields $\Phi_o$.

## RESULTS

Figure 2 shows a 3 dimensional time resolved spectral plot for GRB 951213, with time on the x-axis, BATSE LAD spectroscopic channel number on the y-axis and counts on the z-axis. Each slice in the y-plane represents a Band model fit to BATSE high energy resolution data, whereas a slice in the x-plane reproduces the GRB light curve. The evolution of the peak energy, which is represented by the solid line, can clearly be seen to begin at high energies and decay down near the BATSE detector threshold. The individual light curves that are used to measure the lag, typically those of channels 1 and 3, can be thought of as being produced by summing all of the photons between that particular channel's energy range. For the channel 1 light curve this corresponds to 25-50 keV, which in Figure 2 falls between in the first 3 bins of the y-axis. If a slice in the x plane was taken in the middle of these three spectroscopic channels the resulting light curve would be seen to peak very late in the burst. If the same was done for the 100-300 keV energy range which occurs in about the middle of the y-axis, then the resulting light curve peak would be very early in the burst history. We believe that this is the fundamental factor that contributes to the production of GRB lag. If the decay in the above case was very short, then the time delay between the channel 1 and 3 light curve peaks would be relatively small, but if the decay of $E_{pk}$ took several tens of seconds, then it's expected that the low energy light curve would peak at a very late time. Note

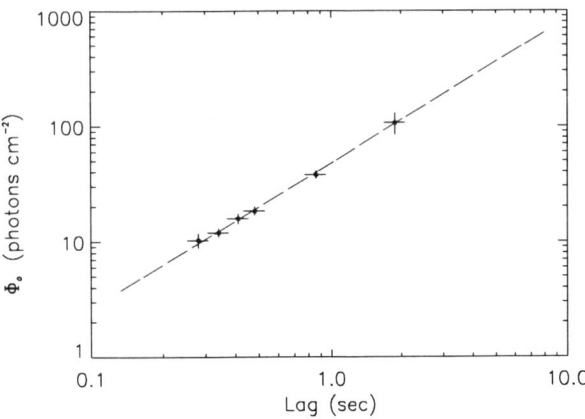

**FIGURE 3.** A plot of $\Phi_o$ vs. the GRB spectral lag for our entire sample. Note that the bursts with the fastest decay (lower $\Phi_o$) have the shortest lags.

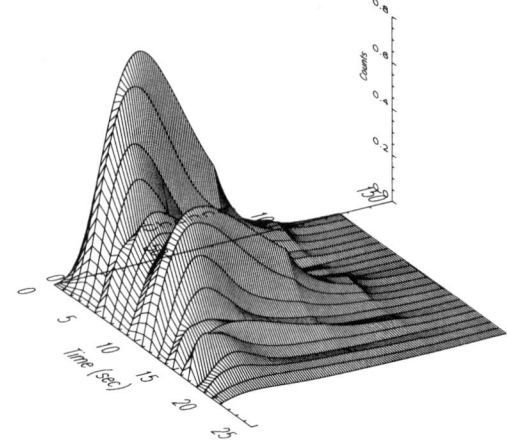

**FIGURE 4.** A 3 dimensional spectral plot for GRB 910807 (647). The two individual peaks both exhibit a hard to soft evolution of $E_{pk}$.

that it is not necessary for $E_{pk}$ to decay through all four channels, even if the peak energy of the burst were below 100 keV, the peak in the channel 3 light curve would still occur very near the onset of the burst due to the geometry of the spectra.

Figure 3 shows a plot with the resulting $\Phi_o$ measurement vs. the spectral lag for our entire sample of bursts. A general trend can be seen that bursts with large $\Phi_o$ values result in longer lags, supporting the notion that longer $E_{pk}$ decay timescales lead to larger lags. A power law fit to the $\Phi_o$ vs. Lag reveals a nearly linear correlation of the two parameters with an index of $1.178 \pm 0.058$.

## DISCUSSION

The results shown in Figure 3 reveal that the spectral lag measured in GRB is directly connected to the evolution of the burst spectra. There are several other important factors that must be considered in this evolution other then $E_{pk}$ decay that could affect the resulting measure of lag. One important factor is the evolution of the spectral indices above and below the break energy. Crider et. al. 1999 showed that the low energy power law index ($\alpha$ in the Band model) of the GRB photon spectra evolves to lower values in 58% of the pulses they studied. This steepening of the photon spectral index to negative values is a result of an increased flux of low energy photons later in the burst. This effect is in addition to the decay of $E_{pk}$ which already tends to push the peak photon flux to lower energies. The evolution of $\alpha$ would also have the effect of creating a peak in the low energy light curve later in the burst history. Therefore we would still expect to observe a lag between the 25-50 keV and 100-300 keV light curves in a hypothetical scenario where only the $\alpha$ parameter evolves and $E_{pk}$ remains constant. These two effects are both expected to contribute to the production of the spectral lag, although determining the amount of contribution that each one makes has been left for future work.

The results shown in Figure 3 inescapably opens up a number of ideas about the nature of the well known lag-luminosity correlation. If the correlation between $E_{pk}$ decay and lag is confirmed, then the absolute luminosity of a GRB is directly related to the rate of spectra evolution. Several ideas have been proposed to explain the lag-luminosity correlation (Salmonson 2000, Ioka et. al. 2001, Scheafer et. al. 2001) but any such explanation would have to address this relationship between luminosity and spectral evolution.

## ACKNOWLEDGMENTS

We thank Brad Scheafer, Felix Ryde, and Rob Preece, for their thoughtful suggestions and advise. This work was supported by NASA grant NAG 53824 and the NASA GSRP fellowship program.

## REFERENCES

Band, D., et al. 1993. ApJ, 428, 21
Crider et. al. 1997, ApJ, 479, L39
Ioka et. al. 2001 ApJ 554L, 163I
Liang, E., & Kargatis, V. 1996. Nature, 381, 49
Norris, J. P., et al. 2000. ApJ, 534, 248
Salmonson, J. 2000. ApJ, 544L, 115S
Scheafer, B. et. al. in preparation, astro-ph/0101461
Scheafer, B. et. al. in preperation astro-ph/0101462

# Observational appearance of relativistic outflows in the presence of neutron component

A.A. Belyanin*, E.V. Derishev[†], V.V. Kocharovsky** and Vl.V. Kocharovsky*

*Dept. of Physics, Texas A&M University, College Station, TX 77843-4242*
[†]*MPI für Kernphysik, Saupfercheckweg 1, D-69117 Heidelberg, Germany*
**Institute of Applied Physics, Russian Academy of Science 46 Ulyanov st., 603950 Nizhny Novgorod, Russia*

**Abstract.** The presence of free neutrons in relativistic outflows from compact ultra-luminous objects alters their dynamical and radiative properties. The neutron-proton decoupling at the acceleration phase produces energetic photons and neutrinos. In the magnetized outflows (e.g., MHD-driven jets in hypernova-like models), the decoupling is rather likely to convert the major part of outflow's energy into neutrino emission. At the time of external shock formation, the interplay between primary and secondary (started by protons from neutron decay) shocks causes a reach variety of phenomena. These include appearance of one or more secondary peaks on a GRB or afterglow light-curve, pre-burst acceleration of ambient gas, bimodal distribution of GRB durations, etc.

## FORMATION OF THE NEUTRON COMPONENT

For a nucleon density in the wind base $10^4 - 10^6$ g/cm$^3$ typical for GRBs, the dissociation temperature is about 0.7 MeV for helium and is even less for other nuclei. In fact, the temperature $T_0$ in the source is about an order of magnitude higher and amounts to 3–10 MeV. Ultrarelativistic plasma flows in GRBs are formed during either catastrophic events in neutron-star evolution or the collapse of central parts of supermassive stars [1–5]. Therefore, the matter in the base of the fireball is initially enriched by neutrons, and dissociation of the nuclei composing this matter makes the ratio of number densities of neutrons and protons amount to $n_n/n_p \gtrsim 1$.

The weak-interaction processes $p + \bar{\nu}_e \to n + e^+$ and $p + e^- \to n + \nu_e$ also lead to creation of free neutrons in GRB sources. Due to the temperature decrease in the course of fireball acceleration, the time scale for weak-interaction processes increases sharply. Quite soon, at a temperature exceeding $(m_n - m_p)c^2$, the density ratio becomes independent of the temperature behavior and remains equal to $\eta \equiv n_n/n_p \approx 1$.

When the plasma temperature falls to about 70 keV, the recombination of protons and neutrons becomes possible. The process comprises two stages. First, a proton and a neutron form a deuteron via the reaction $p + n \to d + \gamma$ and then the reaction chain splits in different channels leading to the synthesis of helium according to the resulting reaction $3d \to {}^4\text{He} + p + n$. The recombination rate is limited mainly by the first reaction and is negligible if $\tau_r = 2\langle\sigma v\rangle_d n_p/3 < c/(R_r \Gamma_r)$, where $\langle\sigma v\rangle_d \approx 5 \cdot 10^{-20}$ cm$^3$/s is the constant of the deuterium-synthesis rate, $R_r$ the distance from the source at which recombination becomes possible, $\Gamma_r$ the Lorentz factor at this distance. To calculate the recombination rate of protons and neutrons, we use the continuity equation that determines the dependence of the proton number density $n_p$ on the radius:

$$n_p = \frac{L}{4\pi R^2 (1+\eta)\Gamma\Gamma_\ell m c^3}, \quad (1)$$

where $m$ is the nucleon mass and $\Gamma_\ell = L/\dot{M}c^2$ the limiting Lorentz factor in the absence of decoupling. The recombination becomes significant only if the following condition is met:

$$\tau_r \approx \left(\frac{R_0}{1\text{ km}}\right)\left(\frac{T_0}{1.5\text{ MeV}}\right)\frac{1}{(1+\eta)\Gamma_\ell} > 1. \quad (2)$$

The radius $R_0$ in the fireball base is assumed to be $3R_g$, where $R_g$ is the Schwarzschild radius of the central object.

## NEUTRON DECOUPLING

At the base of the fireball neutrons collide with protons frequently enough to be advected by the bulk flow. This means that the time between collisions of neutrons with other nucleons is less than the acceleration time scale $R/(c\Gamma)$, i.e.,

$$\sigma_{\rm np} v n_{\rm p} R > c\Gamma \quad \Rightarrow \quad L \gtrsim 12(1+\eta)\Gamma_\ell \frac{\sigma_{\rm T}}{\sigma_{\rm np}} L_{\rm edd}. \quad (3)$$

The expression on the right-hand side is calculated at the base of the fireball where $\Gamma \approx 1$ and $R \approx 3R_g$. Also, we take into account that the relative velocity of protons and neutrons and the cross-section of their collisions at the threshold of decoupling are given by $v \sim c/2$ and $\sigma_{\rm np} \approx 6 \cdot 10^{-26}$ cm$^2$, respectively. In the obtained formula, the $4\pi$ luminosity is related to the Eddington limit $L_{\rm edd} = 2\pi m c^3 R_g/\sigma_{\rm T} \approx 1.3 \cdot 10^{38}(M/M_\odot)$ erg/s, where $\sigma_{\rm T} \approx 6.6 \cdot 10^{-25}$ cm$^2$ is the Thomson cross-section.

In the course of acceleration and expansion of the wind, the decrease in the neutron-proton collision rate is much more rapid than the increase in the acceleration timescale. As a result, upon reaching certain distance, neutrons move almost freely. If the fireball acceleration continues beyond this distance, the Lorentz factor of the proton component exceeds that of the neutron component, i.e., decoupling takes place. The criterion of the decoupling is identical to Eq. (3) calculated at the distance $R_s \sim \Gamma_\ell R_0$ from the source, where the fireball acceleration is terminated. This condition can be presented in a more illustrative form:

$$\Gamma_\ell \gtrsim \Gamma_* = T_0/T_*, \quad T_* \approx 5 \text{ keV}. \quad (4)$$

For a given luminosity, the distance at which the decoupling takes place decreases with decreasing $\dot{M}$. Therefore, the terminal Lorentz factor $\Gamma_{\rm n}$ of the neutron component as a function of $\Gamma_\ell$ reaches the maximum at $\Gamma_\ell \approx \Gamma_*$ and then decreases as $\Gamma_\ell^{-1/3}$. The maximum value is approximately equal to $\Gamma_*$. The limiting Lorentz factor $\Gamma_{\rm p}$ of the proton component increases monotonically with increasing $\Gamma_\ell$ and, if neutrino losses are negligible (see below), can be calculated using the energy conservation law: $\Gamma_{\rm p} = \Gamma_\ell + \eta(\Gamma_\ell - \Gamma_{\rm n})$. The value of $\Gamma_{\rm n}$ can be approximately calculated using the formula

$$\Gamma_{\rm n} \approx \frac{\Gamma_{\rm p} \Gamma_*^{4/3}}{\left(\Gamma_*^4 + 2.37 \Gamma_{\rm p}^4\right)^{1/3}}. \quad (5)$$

The value of the decoupling parameter $\Gamma_{\rm p}/\Gamma_{\rm n}$ ranges typically from 1 to 10.

## ELECTROMAGNETIC CASCADE

The decoupling causes pion production in inelastic collisions between protons and neutrons [7]. Decay of both charged and neutral pions leads to injection of electrons and positrons with energies $\gtrsim 35$ MeV, since the background radiation is very opaque for photons above $\varepsilon_t \sim 3$ MeV. These electrons and positrons scatter off background photons, initiating electromagnetic cascade.

Location of the effective photosphere for energetic quanta from pion decay having $\varepsilon_i \gg \varepsilon_t$ is defined by absorption on the soft cascade photons. Setting absorption depth for energetic photons to unity, we obtain a fraction of the total fireball energy carried away by these photons:

$$\delta E_{\gamma\pi} \simeq \frac{2\varepsilon_i \sigma_{\pi 0} \Gamma_p n_p \varepsilon_t}{m \sigma_T \Gamma_n n_e m_e} \left(\frac{\varepsilon_t}{\varepsilon_i}\right)^{3/5} \sim 10^{-3} \left(\frac{\Gamma_p}{\Gamma_*}\right)^{7/30}. \quad (6)$$

Most of the cascade output is concentrated near $\varepsilon_t$ since for photons of that energy the photosphere location is closest to the central source. When absorbed, each quantum from pion decay produces $\sim 10$ photons with energy around $\varepsilon_t$. After correction for different photospheris radii at $\varepsilon_t$ and $\varepsilon_i$, we obtain that photon fluence at the maximum of cascade spectrum is related to the fluence of unprocessed quanta in the following way:

$$F_{\rm cas} \simeq N_t \frac{\Gamma_p}{2\Gamma_n} \left(\frac{\varepsilon_i}{\varepsilon_t}\right)^{3/5} F_{\gamma\pi} \sim 15 \left(\frac{\Gamma_p}{\Gamma_*}\right)^{4/3} F_{\gamma\pi}. \quad (7)$$

## NEUTRINO LOSSES

Neutrino emission is interesting mostly due to the fact that it can consume a large fraction of the fireball energy. The neutrino losses increase if the pion momenta are changed partially or completely before the decay. This is possible, e.g., due to interaction with the magnetic field. The decay time of a pion, equal to $2.6 \cdot 10^{-8}$ s, exceeds the inverse gyrofrequency if the magnetic field is greater than 600 G. The energy density of such a field is many orders of magnitude smaller than the thermal energy density of the plasma in the decoupling region. Since the requirements to the uniformity of the magnetic field are not stringent in this case, we suppose that this effect is always present. Let us describe the effect of the magnetic field by the parameter $\theta$, the angle between the field lines and radius. This parameter varies from 0 to $\pi/2$ for different models of GRB sources and jets.

If $\Gamma \gg \Gamma_n$, all the three types of pions are created with equal probabilities, and their total energy in the center-of-mass frame of the colliding nucleons is approximately equal to $mc^2\sqrt{\Gamma/\Gamma_n}$. Taking into account the effect of the magnetic field, we find that neutrinos in the laboratory frame carry away the energy

$$\epsilon_\nu \approx mc^2 \Gamma^2 \Gamma_n^{-1} \frac{\left(1 - \sqrt{1 - 4\Gamma_n/\Gamma}\cos^2\theta\right)}{4}.$$

The total energy of the emitted neutrinos per proton is equal to $E_\nu \approx \int_{R_d}^{\infty} \epsilon_\nu \sigma_c n_n \frac{dR}{\Gamma}$, where $\sigma_c \approx 2 \cdot 10^{-26}$ cm$^2$ is the cross-section of inelastic proton-neutron collisions. We divide the integration region into three parts: (i) from the decoupling radius $R_d$ to the point where $\sin\theta = \sqrt{2\Gamma_n/\Gamma}$; (ii) from the previous point to the saturation radius $R_s = \Gamma_p R_d/\Gamma_d$; and (iii) from $R_s$ to infinity. With accuracy sufficient for an estimate, we adopt $n_n = \eta n_p \Gamma^2/(2\Gamma_n^2)$, where $\Gamma = (R/R_d)\Gamma_d$ in regions (i) and (ii), and $\Gamma = \Gamma_p$ in region (iii). If $\sin\theta < \sqrt{2\Gamma_n/\Gamma_p}$, then the formula $\epsilon_\nu \approx mc^2\Gamma/2$ is valid in any region. In the opposite case, $\epsilon_\nu \approx (mc^2\Gamma^2\Gamma_n^{-1}\sin^2\theta)/4$ in regions (ii) and (iii).

The result of integration divided by the terminal proton energy $\Gamma_p mc^2$ shows how large are the neutrino losses compared to the power of the fireball:

$$\delta E_\nu \approx 0.15\left(\frac{\Gamma_d}{\Gamma_n}\right)^3 \eta\tau_n \times \begin{array}{l} \frac{\Gamma_n}{\Gamma_p}\left[\ln\frac{\Gamma_p}{\Gamma_n} + 0.7\right]; \\ \sin^2\theta + \frac{\Gamma_n}{\Gamma_p}\left[\ln\frac{2}{\sin^2\theta} - 0.7\right] \end{array}. \quad (8)$$

Here the first line is for $\sin\theta < \sqrt{2\Gamma_n/\Gamma_p}$ and the second – for $\sin\theta > \sqrt{2\Gamma_n/\Gamma_p}$; $\tau_n \approx 4$ is the optical depth of the decoupling region for neutrons and the factor $(\Gamma_d/\Gamma_n)^3$ is equal to about 2.5. Because of the adopted approximations, Eq. (8) is only applicable if $\Gamma_p/\Gamma_n \gtrsim 4$. Note that if $\eta \approx 1$ and the magnetic field is perpendicular to the radius, then over 50% of the source power is lost due to neutrino emission. The energy of a neutrino for a typical source with $\Gamma_p \sim 10^3$ and $\Gamma_p/\Gamma_n \sim 4$ is about 30 GeV in the limiting case of a small angle $\theta$ and about $200\sin^2\theta$ GeV in the opposite case.

# CLASSIFICATION OF LIGHTCURVES

Below we assume that a GRB source generates a fireball with comparable fluxes of protons (ions) and neutrons having either equal or different Lorentz factors, $\Gamma_p$ and $\Gamma_n$. We also assume that the fireball ejection lasts for a time period much shorter than a GRB duration.

The plasma component of a fireball pushes the surrounding medium and forms a shock at the interface, while neutrons propagate freely until they decay into charged particles. There are two independent alternatives: (i) the neutron flow may decouple from the proton one or may not, and (ii) the lifetime of a free neutron, $t_n$, either exceeds or is smaller than the deceleration time of the proton shock, $t_p$. The above two alternatives produce four combinations; each one gives rise to a distinct type of lightcurves [8].

*First case:* $t_n < t_p$, no decoupling. It is the simplest case, which is practically indistinguishable from the standard GRB model considering a shock formed by one-component fireball (e.g., [9]). The resulting type I lightcurve has a single peak, and its rise time and decay time are comparable with the total duration of burst.

*Second case:* $t_n < t_p$, neutron flow decouples. The proton shock moves ahead of slower neutrons, gradually decelerating. Neutrons decay before the time their decay products catch up to the primary shock. Two shells collide at a radius defined by the equation $\int_0^{R_c} \beta\, dR = \beta_n R$, where $\beta$ is (decreasing with radius) velocity of the proton shock, $\beta_n$ is (constant) velocity of neutrons. At the radius $R_c$ the Lorentz factors of protons and neutrons satisfy the relation $\Gamma_n^2/\Gamma^2 \simeq 7$ (radiative deceleration) or $\Gamma_n^2/\Gamma^2 \simeq 4$) (adiabatic deceleration). The faster inner shell catches up to the GRB envelope, thus forming a secondary shock. The bulk Lorentz factor of GRB envelope is boosted to significantly higher values, that causes another peak on the lightcurve. The second peak may appear at the afterglow stage if the ratio $\Gamma_p/\Gamma_n$ is large.

*Third case:* $t_n > t_p$, no decoupling. Chargeless neutrons overtake and surpass a shock initiated by protons and then some of the neutrons decay producing secondary protons. The latter push interstellar gas ahead and hence prolong the deceleration of the primary shock till all neutrons have decayed. Decay products do not carry frozen-in magnetic field, so that the expansion of the secondary shock (which is in front of the primary one in this case) starts in the adiabatic regime and continues in the radiative regime when magnetic field increases. The final result of proton-neutron interplay in type III GRBs is a single-peaked lightcurve with slow rise followed by relatively sharp outburst, which occurs when the radiative regime of the shock deceleration establishes.

*Fourth case:* $t_n > t_p$, neutron flow decouples. Type IV GRBs generate the most sophisticated lightcurves, which may be considered as a hybrid of types II and III. At the beginning, the primary shock moves faster than neutrons. Neutrons surpass the primary shock when it slows down. Those of them that have already decayed by this time boost the GRB envelope and produce the second pulse, while others pass through the shock creating the geometry characteristic for type III bursts. After that point type IV GRBs follow the scenario for the third case, which predicts one more peak on the lightcurve when secondary protons form a radiative shock.

## REFERENCES

1. S. E. Woosley, 1993, ApJ 405, 273
2. B. Paczyński, 1998, ApJ 494, L45
3. E. V. Derishev, V. V. Kocharovsky, and Vl. V. Kocharovsky, 1998, Radiophys. Quantum Electron. 41, 7
4. E. V. Derishev, V. V. Kocharovsky, and Vl. V. Kocharovsky, 1999, JETP Lett. 70, 652
5. B. Paczyński, 1990, ApJ 363, 218
6. S. L. Shapiro and S. A. Teukolsky, *Black Holes, White Dwarfs, and Neutron Stars*, John Wiley and Sons, New York (1985).
7. E.V. Derishev, V.V. Kocharovsky, and Vl.V. Kocharovsky, 1999, ApJ 521, 640
8. Derishev E.V., Kocharovsky V.V., Kocharovsky Vl.V., 1999, A&A 345, L51
9. Mészáros, P., Laguna, P., and Rees, M.J., 1993, ApJ 415, 181

# The Results of Statistical Tests of the Angular Distribution of Gamma-Ray Bursts

R. Vavrek[*], L.G. Balázs[†], A. Mészáros[**], I. Horváth[‡] and Z. Bagoly[§]

[*]*Max-Planck-Institut für Astronomie, 17 Königstuhl, D-69117 Heidelberg, Germany*
[†]*Konkoly Observatory, Budapest, Box 67, H-1525, Hungary*
[**]*Astron. Inst. of the Charles University, 180 00 Prague 8, V Holešovičkách 2, Czech Republic*
[‡]*Dept. of Physics, Bolyai Military University, BJKMF, Budapest, Box 12, H-1456, Hungary*
[§]*Lab. for Information Technology, Eötvös University, Budapest, Pázmány Péter sétány 1/A, H-1518, Hungary*

**Abstract.**
The spherical variants of multiscale methods - Voronoi tesselation, minimal spanning tree, and multifractal analysis - are used to test the angular distributions of three subgroups of gamma-ray bursts (GRBs) collected in BATSE Gamma-Ray Burst Catalog. They verify the isotropy of the sky distribution of gamma-ray bursts. The short - in some tests also the intermediate - subclass exhibit anisotropic distribution. The distribution of long subclass seems to be isotropic.

## INTRODUCTION

In the last years the authors provided [1, 2, 6, 7] several different tests verifying the intrinsic isotropy in the angular sky-distribution of gamma-ray bursts (hereafter GRBs) collected at BATSE Catalog [5]. Shortly summarizing the results of these studies one may conclude: A. The short subgroup ($2\,s > T_{90}$) and the remaining GRBs are distributed differently. B. The long subgroup ($T_{90} > 10\,s$) seems to be distributed isotropically; C. The intermediate subgroup ($10\,s > T_{90} > 2\,s$) is distributed anisotropically on the $(96-97)\%$ confidence level; D. For the short subgroup ($2\,s > T_{90}$) the assumption of isotropy is rejected on the 92% confidence level. (About the definition of subclasses see Horváth [3]; $T_{90}$ is the duration of a GRB, during which time the 90% of radiation is received [5].)

Recently, independently by different tests, a Russian group [4] confirmed these results with one essential difference: for the intermediate a much bigger - namely 99.89% confidence level - of anisotropy is claimed.

In this article the results of new tests are collected. Some results of these calculations were already presented [9]; this paper completes the calculations.

The methods of Voroni tesselation (VT), minimal spanning tree (MST) and multifractal analysis (MFR) were already shortly described in [9]; therefore, here mainly the results are collected and shortly discussed.

## SAMPLES

We conducted tests of randomnesses for five different subgroups. The choice of these subgroups is motivated by authors' earlier studies. The first motivation comes from the separation of GRBs into short, intermediate and long subgroups, respectively. These subgroups are defined by the conditions $2\,s > T_{90}$; $2\,s < T_{90} < 10\,s$; $T_{90} > 10\,s$ (see [3, 6, 7] and references therein for more details). In addition, because the introduction of the intermediate subgroup is not unambigous yet (see again, e.g., Horváth [3]), we will study two "long" subgroups: with $T_{90} > 2\,s$ ("Long1") and with $T_{90} > 10\,s$ ("Long2"). The second motivation comes from [7]. In this paper, surprisingly, the dim part of intermediate subgroup defined by the condition $0.65 < P_{256} < 2$, where $P_{256}$ is the peak-flux of a burst on the 256 ms trigger in units photons/(cm$^2$s) [5], shows an evident anisotropic distribution. Because in any earlier studies of authors [1, 2, 6, 7] the short subgroup showed a "suspicious" anisotropy at a small but remarkable $\simeq 90\%$ confidence level (this level is is not enough to reject the assumption of isotropy, but is clearly remarkable), it is a possibility that such defined dim part of short subgroup will show - similarly to the intermediate subgroup - again a more clear anisotropy. In any case, the study of such dim part of short subgroup may be useful. Add still that, similarly to [7], we again always have taken $0.65 < P_{256}$ in order to avoid the faintest objects, which can cause instrumental biases [7]. This truncation in peak-flux avoids the problems with changing thresh-

**TABLE 1.** Results of VT, MST and MFR analyses. "No" means that the rejection on the > 90% confidence level does not occur; results obtained by $\chi^2$ test are denoted by *; Kolmogorov-Smirnonov(KS) statistics is referred by +; if in one of the bins we detect a large difference with respect to the simulations, then there is no sign; if one applies more tests, only the highest obtained confidence level is indicated; "dist." means "distribution"; "RF" means "round factor"; $A$ ($P$) is the area (length of the boundary) of a Voronoi cell; $N_v$ is the number of its vertices; $\alpha_i$ are the inner angles ($i = 1, 2, ..., N_v$; "chord" is the name of boundary curve; $L_{MST}$ is the length of minimal spanning tree.

|  |  | Long1/2 | Short1 | Short2 | Intermediate |
|---|---|---|---|---|---|
| RF average | $4\pi A/P$ | No | No | $> 98\%^+$ | No |
| RF homogeneity | $1 - \frac{\sigma(RF_{av})}{RF_{av}}$ | No | No | No | No |
| AD factor | $1 - \left(1 - \frac{\sigma(A)}{\langle A \rangle}\right)^{-1}$ | No | No | No | No |
| Cell area dist. | $A$ | No | No | No | $> 99.2\%$ |
| Cell vertex (edge) dist. | $N_v$ | No | No | No | No |
| Cell chords dist. | $C$ | No | No | No | No |
| Inner angle dist. | $\alpha_i$ | No | $> 99.5\%^*$ | No | No |
| Shape factor dist. | $A/P^2$ | No | $> 99.9\%$ | $> 90\%$ | $> 90\%$ |
| Modal factor dist. | $\sigma(\alpha_i)/N_v$ | No | $> 98\%$ | No | No |
| MST variance dist. | $\sigma(L_{MST})$ | No | No | No | No |
| MST average dist. | $L_{MST}$ | No | $> 90\%$ | No | $> 92\%$ |
| MST angle dist. | $\alpha_{MST}$ | No | $> 96\%$ | No | No |
| MFR spectra | $f(\alpha)$ | No | $> 99.9\%$ | $> 96.0\%$ | $> 99.9\%$ |

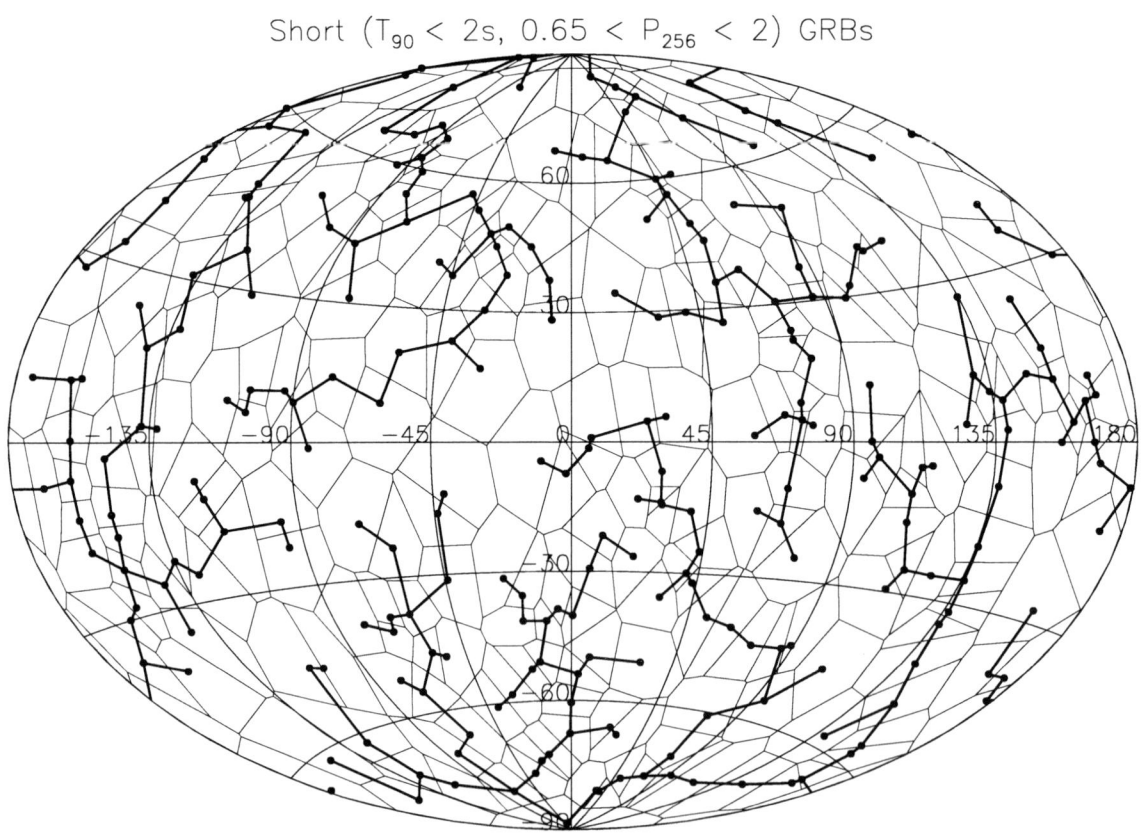

**FIGURE 1.** The Voronoi diagram and the minimal spanning tree (MST) for the GRBs of Short1 sample in equatorial coordinates and in Aitof projection.

old, biases in durations of faint GRBs, etc.

All this means that 5 different samples of GRBs were tested by three different statistical tests (VT, MST, MFR spectra): Sample "Short1" with $T_{90} < 2$ and $0.65 < P_{256} < 2$ (261 GRBs); Sample "Short2" with $T_{90} < 2$ and $0.65 < P_{256}$ (406 GRBs); Sample "Intermediate" with $2 < T_{90} < 10$ and $0.65 < P_{256}$ (253 GRBs); Sample "Long1" with $T_{90} > 2$ and $0.65 < P_{256} < 2$ (676 GRBs); Sample "Long2" with $T_{90} > 10$ and $0.65 < P_{256}$ (966 GRBs), where $T_{90}$ is in seconds and $P_{256}$ in photons/(cm$^2$s). In addition, in the case of VT and MST methods, respectively, different measures were used to obtain the confidence levels.

## RESULTS AND CONCLUSIONS

The results of tests are collected in Table 1. From this it follows:

1. The null hypothesis of the randomness of long subgroup holds. There is no difference between the Long1 and Long2 samples. This is in accordance with the earlier results [1, 2, 6, 7, 4].

2. The randomness in the intermediate subclass is rejected on the $> 95\%$ confidence level. This is again in accordance with the earlier results.

3. The short subgroup seems to be distributed also anisotropically (or at least non-randomly), and the $> 95\%$ confidence level seems to be also reached. This is a new result compared with the earlier results. In addition, the dimmer part of short GRBs seems to distributed more anisotropically; this is a similar to the behavior of intermediate subgroup.

All these results are highly remarkable.

The first important result concerns the intermediate subgroup. Both the earlier paper of authors and also of the Russian group [4] found here an anisotropy. This is supported also here. This is a remarkable new phenomenon, because the character, origin and other behaviors of this subgroup give a challenge in the topic of GRBs.

The second important result concerns the short subgroup. It was already believed by earlier tests that this subgroup hardly can be distributed randomly. But only these studies confirmed this behaviour. In fact, the structures going across the whole sky in the distribution of short GRBs are clearly seen both from the Voronoi diagram and from MST (see Fig.1). This structures represent further challange in the topic

One must add still the following. Recently the direct confirmations of high redshifts of GRBs (up to redshift 4.5) from afterglows is confirmed for the long GRBs only [8]. The concrete redshifts of short and intermediate GRBs are unknown yet. This fact, together with the found non-randomnesses for these subclasses, clearly show that further attention should be paid for these subgroups.

## ACKNOWLEDGMENTS

This research was supported in part through OTKA grants F029461 (I.H.), T034549, and by Czech Research Grant J13/98: 113200004 (A.M.).

## REFERENCES

1. Balázs, L.G., Mészáros, A. & Horváth, I., *A&A*, **339**, 1, 1998.
2. Balázs, L.G., Mészáros, A., Horváth, I. & Vavrek, R., *A&A Suppl.*, **138**, 417, 1999.
3. Horváth, I., *ApJ*, **508**, 757, 1998.
4. Litvin, V.F., Matveev, S.A., Mamedov, S.V. & Orlov, V.V., *Pis'ma v Astron. Zhurnal*, **27**, 489, 2001.
5. Meegan, C.A., et al., Current BATSE Gamma-Ray Burst Catalog, (2000) URL http://gammaray.msfc.nasa.gov/batse/grb/catalog/current.
6. Mészáros, A., Bagoly, Z. & Vavrek, R., *A&A*, **354**, 1, 2000a.
7. Mészáros, A., Bagoly, Z., Horváth, I., Balázs, L.G. & Vavrek, R., *ApJ*, **539**, 98, 2000b.
8. Mészáros, P., *Science*, **291**, 79, 2001.
9. Vavrek, R., Balázs, L.G., Mészáros, A., Horváth, I., & Bagoly, Z., in *Proceedings of Second Roma Workshop on GRBs*, ed. N. Masetti, Springer, in press.

# The Dynamics of Magnetized Outflows in GRBs

Nektarios Vlahakis and Arieh Königl

*Department of Astronomy & Astrophysics and Enrico Fermi Institute, University of Chicago, 5640 S. Ellis Ave., Chicago, IL 60637,* `email: vlahakis@jets.uchicago.edu, arieh@jets.uchicago.edu`

**Abstract.** Using relativistic, axisymmetric, ideal MHD, we examine the outflow from a debris disk around a newly formed stellar-mass black hole, taking into account the baryonic matter, the electron-positron/photon fluid, and the large-scale electromagnetic field. We clarify the relationship between the thermal (fireball) and magnetic (Poynting flux) acceleration mechanisms, identify the parameter regimes where qualitatively different behaviors are expected, and demonstrate that the observationally inferred properties of the GRB outflows can be attributed to magnetic driving. We show that the Lorentz force can convert up to 50% of the initial total energy into kinetic energy of a collimated flow of baryons. This energy, in turn, may be converted into radiation by internal shocks. We examine how baryon loading and magnetic collimation affect the structure of the flow.

## INTRODUCTION

According to currently accepted GRB formation scenarios, the burst is powered by the extraction of rotational energy from the central black hole or neutron star, or, alternatively, from the debris disk left behind when the mass near the origin collapses into a black hole. In either case, strong ($\gtrsim 10^{14}$ G) magnetic fields provide the most plausible means of extracting the energy on the burst time scale. The magnetic energy might be dissipated near the origin in a series of flares, giving rise to a "magnetic" fireball [10, 7]. Alternatively, the magnetic field may have a large-scale, ordered component that could help guide and collimate the outflow, and, if it is strong enough, also contribute to its acceleration (e.g., [15, 8, 3]). Collimation is, in fact, consistent with the observational indications for GRB jets (e.g., [13, 11, 1]). Furthermore, if the outflow is largely Poynting flux-dominated, the implied lower radiative luminosity near the origin could alleviate the baryon contamination problem. Magnetic fields have thus come to be regarded as the favored means of driving GRB outflows.

## THE MHD DESCRIPTION

We assume that the outflow originates in a debris disk around a stellar-mass black hole (e.g., [8, 2]) and concentrate on the case of a magnetically driven, axial jet that has a sufficiently high baryon loading to insure that it is matter dominated when it becomes optically thin, and in which a significant fraction of the Poynting flux is eventually converted into kinetic energy. As discussed by [14], such configurations can be validly described by the MHD approximation, and, furthermore, they are not expected to dissipate a substantial amount of magnetic energy along the way. Correspondingly, we adopt the equations of relativistic, ideal MHD to describe the flow. We anticipate that, near the origin, the optical depth is large enough to ensure local thermodynamic equilibrium. We therefore assume that the gas (consisting of baryons with their neutralizing electrons as well as of photons and pairs) evolves adiabatically with a polytropic index of 4/3. Assuming a quasi-steady poloidal magnetic flux function $A$ and changing variables from $(A,\ell,t)$ to $(A,\ell,s=ct-\ell)$, with $\ell$ the arclength along a poloidal fieldline, it can be shown that all terms with derivatives with respect to $s$ are negligible when the flow is highly relativistic. As is elaborated on in [18], the equations are then effectively time independent and the motion can be described as a frozen pulse whose internal profile is specified through the variable $s$ (see also [12]).

Assuming also axisymmetry, the full set of MHD equations can be partially integrated to yield five field-line constants [17]. Two integrals remain to be performed, involving the Bernoulli and transfield force-balance equations. The solutions derived in [18] are obtained under the most general ansatz for radial self-similarity [in spherical coordinates $(r,\theta,\phi)$], in which the shape $r(\theta,A)$ of the poloidal field lines is given as a product of a function of $A$ times a function of $\theta$: $r = \mathcal{F}_1(A)\mathcal{F}_2(\theta)$.

We approximate the outflow as a pair of shells that move in opposite directions from the debris

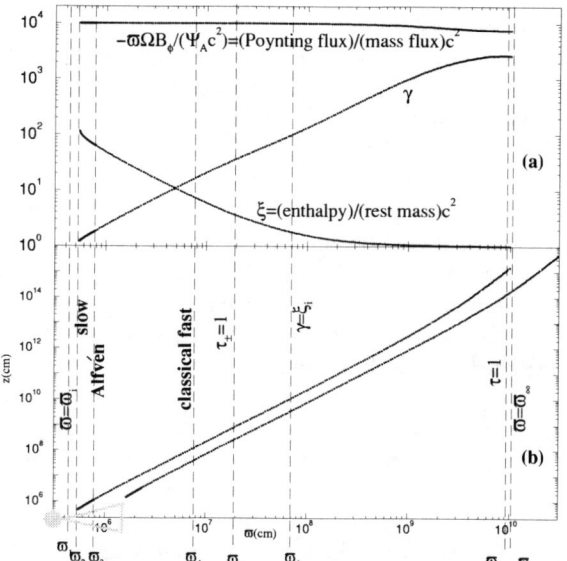

**FIGURE 1.** (*a*) The Lorentz factor $\gamma$, the ratio $\xi$ of the enthalpy to the rest energy, and the ratio of the Poynting flux to the rest-energy flux (*top* curve) are shown as functions of $\varpi$, the distance from the axis of rotation, along the innermost field line. (*b*) The meridional projections of the innermost and outermost field lines are shown on a logarithmic scale, along with a sketch of the black-hole/debris-disk system. The fieldlines have parabolic shape $z \propto \varpi^2$ for $\varpi \lesssim 10^9$ cm and become asymptotically cylindrical. The vertical lines mark the positions of the various transition points along the innermost field line.

3. $\varpi_3 = c/\Omega$: Alfvén ($\approx$ light) surface.
4. $\varpi_4 = (c/\Omega)\sqrt{\mu/\xi_i}$: classical fast-magnetosonic surface. The bulk of the magnetic acceleration occurs downstream from this point through the "magnetic nozzle" mechanism [5]. In essence, this term is a shorthand for the fact that an MHD flow can continue to accelerate until it crosses the modified fast-magnetosonic singular surface. This effect is not purely relativistic: it is manifested in an exact solution of the nonrelativistic MHD equations where all the singular surfaces (including the modified-fast one) are crossed [16], and there, too, most of the acceleration occurs downstream of the classical fast-magnetosonic surface (but *upstream* of the modified-fast surface).
5. $\varpi_5 = 25\left(\frac{kT_i}{m_ec^2}\right)\varpi_i$: The opacity due to pairs $\tau_\pm = 1$.
6. $\varpi_6 = \xi_i\varpi_i$: end of thermal acceleration, $\gamma \approx \xi_i$. The entire initial thermal energy of the photons and pairs has by this point been converted into baryon kinetic energy. Magnetic acceleration effectively starts here.
7. $\varpi_7 = 5 \times 10^3 (\Delta\varpi)_{i,6}^{1/2} \rho_{0i,2}^{1/2} \varpi_i$: $\tau = \tau_b = 1$, the flow becomes optically thin to Compton scattering.
8. $\varpi_8 = \varpi_\infty \approx (\mu/2)\varpi_i$: cylindrical flow regime. Near-equipartition between the Poynting and the baryon kinetic-energy fluxes is attained, with $\gamma = \gamma_\infty \approx \mu/2$.

In figures 1 and 2 we present an exact self-similar solution of the relativistic MHD equations. Figure 1 demonstrates the validity of the scaling $\gamma \propto \varpi$ over several decades in $\varpi$ (see [17]) and the separation of the thermal ($\gamma < \xi_i$) and magnetic ($\gamma > \xi_i$) acceleration regimes. It also manifests the significant collimation from an initial opening half-angle $\approx 24°$ to a very nearly cylindrical geometry (attained on scales $\gtrsim 10^{14}$ cm from the origin). It is, furthermore, seen that approximate equipartition ($\gamma \approx -\varpi\Omega B_\phi/\Psi_A c^2$) holds during the final phase of the flow. Figure 2 demonstrates that the thermal pressure is everywhere smaller than the comoving magnetic pressure.

The requirement that the flow be optically thin to Compton scattering in its final (cylindrical) stage, corresponds to an upper bound on the baryon loading: $\rho_{0i,2} \lesssim (\mu/10^4)^2 (\Delta\varpi)_{i,6}^{-1}$.[1]

disk, each having an initial meridional cross section $(\Delta z)_i \times (\Delta\varpi)_i$, with $(\Delta\varpi)_i = 10^6 (\Delta\varpi)_{i,6}$ cm comparable to $\bar{\varpi}_i = 10^6 \bar{\varpi}_{i,6}$ cm (the mean radius of the debris disk) and $(\Delta z)_i \approx c\Delta t = 3 \times 10^{11} \Delta t_1$ cm (where $\Delta t = 10\Delta t_1$ s is the total duration of the burst). The total baryonic mass of outflowing matter is $M_b = 2 \times 10^{-7} \gamma_i \rho_{0i,2} \bar{\varpi}_{i,6} (\Delta\varpi)_{i,6} \Delta t_1 M_\odot$, where $\rho_{0i} = 10^2 \rho_{0i,2}$ g cm$^{-3}$ is the initial rest-mass density. If $\mu c^2$ is the total energy-to-mass flux ratio, the total energy is $\mathcal{E}_i = \mu M_b c^2 = 4 \times 10^{51} (\mu/10^4) \gamma_i \rho_{0i,2} \bar{\varpi}_{i,6} (\Delta\varpi)_{i,6} \Delta t_1$ ergs, and initially it resides predominantly in the electromagnetic field; the initial thermal energy (associated with the enthalpy of the photons and pairs) is $\xi_i M_b c^2 = (\xi_i/\mu)\mathcal{E}_i$, where $\xi_i = 400(kT_i/m_ec^2)^4(\rho_{0i,2})^{-1}$ (we approximate the initial radiation field by a blackbody distribution and the $e^\pm$ pairs by a Maxwellian distribution).

## STRUCTURE OF THE FLOW

As the flow moves outward it passes through the following points (see Fig. 1):
1. $\varpi_1 = \varpi_i$: origin of the outflow, $\gamma_i \approx 1$.
2. $\varpi_2 = \varpi_i\sqrt{\frac{3}{2}}$: slow-magnetosonic surface, $V_p \approx \frac{c}{\sqrt{3}}$.

## APPLICATION TO GRBs

To demonstrate how the model outflows could account for observed GRBs, we adopt a $\gamma$-ray fluence of $\sim$

---

[1] We note, however, that the increase of $\gamma$ with $\varpi$ is slower than linear as the cylindrical regime is approached (see Fig. 1), which implies that $\varpi_\infty$ is larger than $(\mu/2)\varpi_i$ and the actual upper bound on the initial baryon mass density is larger than the analytic estimate (typically by a factor of a few).

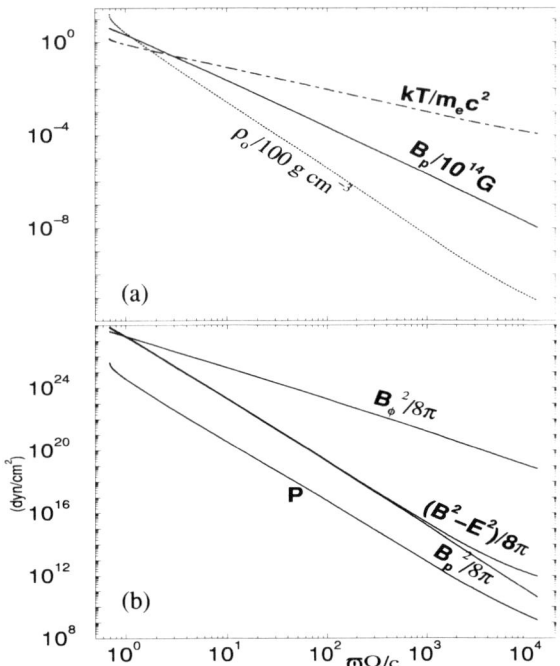

**FIGURE 2.** (a) The temperature $T \propto \varpi^{-1}$, poloidal magnetic field $B_p \propto \varpi^{-2}$, and baryonic rest-mass density $\rho_0 \propto \varpi^{-3}$. (b) Toroidal and poloidal magnetic pressure, comoving magnetic pressure $(B^2 - E^2)/8\pi$ and total (lepton+radiation) pressure $P$.

$10^{-4}$ ergs cm$^{-2}$ and a distance $\sim 3$ Gpc, which imply an isotropic equivalent energy of $\sim 10^{53}$ ergs. If the $\gamma$-ray emitting material is in fact confined to a pair of cones of opening half-angle $\sim 3°$, then the actual radiated energy is $\sim 1.5 \times 10^{50}$ ergs. The emission is believed to be powered by the kinetic energy of the relativistic outflow, and the most common interpretation is in terms of internal shocks produced by the collision of overtaking shells [12]. Assuming that the kinetic-to-radiative energy conversion efficiency is $\sim 10\%$ and that the magnetic stresses transfer $\sim 1/2$ of the initial energy into baryon kinetic energy, we infer $\mathcal{E}_i \approx 3 \times 10^{51}$ ergs. If the energy is deposited in the form of a Poynting flux over $\Delta t \approx 10$ s, then the field at the origin must be $\sim (\mathcal{E}_i c / \bar{\varpi}_i^3 (\Delta\varpi)_i \Delta t \Omega^2)^{1/2} \approx 3 \times 10^{14}$ G for $\bar{\varpi}_i \approx (\Delta\varpi)_i \approx 10^6$ cm and $\Omega \approx 10^4$ s$^{-1}$ (representative of a debris disk that extends to the last stable orbit of a rotating solar-mass black hole).

The variability properties of the observed GRBs have been interpreted in terms of $N = 100 N_2$ distinct shells that are ejected with slightly different Lorentz factors and whose subsequent collisions give rise to the "internal" shocks responsible for the $\gamma$-ray emission (e.g., [12]). Adopting $\gamma \propto \varpi \propto \sqrt{z}$, one may integrate the equation of motion for each shell and show that two neighboring shells starting with $\Delta\gamma_i \sim 1$ will not collide for as long as they move in the flow acceleration zone. Only after the shells reach the constant-velocity, cylindrical flow regime, will the two shells collide (at a height $z_f \approx \gamma_\infty^2 (\Delta z)_i / N = 7.5 \times 10^{16} (\mu/10^4)^2 \Delta t_1 / N_2$ cm, assuming an initial separation $\sim (\Delta z)_i / N$). A lower bound on the Lorentz factor in this region can be obtained from the requirement that the optical depth to photon-photon pair production be less than 1 [6]. The solution exhibited in Figure 1 demonstrates that magnetically accelerated flows can attain the requisite high values of $\gamma$.

Real GRB outflows evidently have a finite opening half-angle (estimated to be at least $5°$ on average; [9]) and thus are not accurately represented by our asymptotically cylindrical solutions. This discrepancy should not, however, affect our basic conclusions about the robustness of the magnetic acceleration mechanism for these flows. It remains a challenge for future work to find exact solutions with conical asymptotics. We note that, although the outflowing "shells" conceivably do not fill the entire solid angle into which they are ejected, the internal shock mechanism requires that multiple shells be ejected along any given direction (e.g., [4]). The guiding property of a magnetic field in an MHD-driven outflow provides a natural physical basis for this picture.

In conclusion, we have shown that ordered magnetic fields can transform up to $\sim 50\%$ of the energy deposited by the central source of a GRB into kinetic energy of a collimated flow of baryons with $\gamma \sim 10^3$.

## REFERENCES

1. Frail D. A., et al., *ApJL*, **562**, L55 (2001)
2. Fryer, C. L., Woosley, S. E., & Hartmann, D. H., *ApJ*, **526**, 152 (1999)
3. Katz, J. I., *ApJ*, **490**, 633 (1997)
4. Kumar, P, & Piran, T., *ApJ*, **535**, 152 (2000)
5. Li, Z.-Y., Chiueh, T., & Begelman, M.C., *ApJ*, **394**, 459 (1992)
6. Lithwick, Y., & Sari, R., *ApJ*, **555**, 540 (2001)
7. Mészáros, P., Laguna, P., & Rees, M. J., *ApJ*, **415**, 181 (1993)
8. Mészáros, P., & Rees, M. J., *ApJL*, **482**, L29 (1997)
9. Mészáros, P., Rees, M. J., & Wijers, R., *NewA*, **4**, 303 (1999)
10. Narayan, R., Paczyński, B., & Piran, T., *ApJL*, **395**, L83 (1992)
11. Panaitescu, A., & Kumar, P., *ApJ*, **554**, 667 (2001)
12. Piran, T., *Physics Reports*, **314**, 575 (1999)
13. Sari, R., Piran, T., & Halpern, J. P., *ApJL*, **519**, L17 (1999)
14. Spruit, H. C., Daigne, F., & Drenkhahn, G., *A&A*, **369**, 694 (2001)
15. Thompson, C., *MNRAS*, **270**, 480 (1994)
16. Vlahakis, N., Tsinganos, K., Sauty, C., & Trussoni, E., *MNRAS*, **318**, 417 (2000)
17. Vlahakis, N., & Königl, A., *ApJL*, **563**, L129 (2001)
18. Vlahakis, N., & Königl, A. in preparation

# Hard X-Ray Afterglows of Short GRBs

Davide Lazzati[*], Enrico Ramirez-Ruiz[*] and Gabriele Ghisellini[†]

[*]*Institute of Astronomy, Madingley Road CB3 0HA Cambridge, U.K.*
[†]*Osservatorio Astronomico di Brera, via Bianchi 46, 23807 Merate (LC), Italy*

**Abstract.** We report the discovery of a transient and fading hard X-ray emission in the BATSE lightcurves of a sample of short γ-ray bursts. We have summed each of the four channel BATSE light curves of 76 short bursts to uncover the average overall temporal and spectral evolution of a possible transient signal following the prompt flux. We found an excess emission peaking $\sim 30$ s after the prompt one, detectable for $\approx 100$ s. The soft power-law spectrum and the time-evolution of this transient signal suggest that it is produced by the deceleration of a relativistic expanding source, as predicted by the afterglow model.

## INTRODUCTION

Since their discovery, γ-ray bursts (GRBs) have been known predominantly as brief, intense flashes of high-energy radiation, despite intensive searches for transient signals at other wavelengths. Fortunately, the rapid follow-up of *Beppo*SAX [1] positions, combined with ground-based observations, has led to the detection of fading emission in X-rays [2], optical [3] and radio [4] wavelengths. These afterglows in turn enabled the measurement of redshifts [5], firmly establishing that GRBs are the most luminous known events in the Universe and involve the highest source expansion velocities.

The detection of afterglows that follow systematically long bursts has been a major breakthrough in GRB science. Unfortunately no observation of this kind was possible for short bursts. Our physical understanding of their properties was therefore put in abeyance, waiting for a new satellite better suited for their prompt localization.

In this paper we show that afterglow emission characterizes also the class of short bursts. In a comparative analysis of the BATSE lightcurves of 76 short bursts we detect a hard X-ray fading signal following the prompt emission with a delay of $\sim 30$ s. The spectral and temporal behavior of this emission is consistent with the one produced by a decelerating blast wave, providing a direct confirmation of relativistic source expansion. For a more detailed discussion, see Lazzati, Ramirez-Ruiz and Ghisellini [6].

## DATA ANALYSIS

The detection of slowly variable emission in BATSE lightcurves is a non trivial issue, since BATSE is a non-imaging instrument and background subtraction can not be easily performed. We selected from the BATSE GRB catalog a sample of short duration ($T_{90} \leq 1$ s), high signal-to-noise ratio, GRB lightcurves with continuous data from $\sim 120$ s before the trigger to $\sim 230$ s afterwards. We aligned all the lightcurves to a common time reference in which the burst (binned to a time resolution of 64 ms) peaked at $t = 0$ and we binned the lightcurves in time by a factor 250, giving a time resolution of 16.0 s. The time bin $[-8 < t < 8$ s$]$ containing the prompt emission was removed and the remaining background modelled with a $4^{\text{th}}$ degree polynomial. The bursts in which this fit yielded a reduced $\chi^2$ larger than 2 in at least one of the four channels were discarded. Note that we did not subtract this best fit background curve from the data. This procedure was used only to reject lightcurves with very rapid and unpredictable background fluctuations and is based on the assumption that the excess burst or afterglow emission is not detectable in a single lightcurve. This procedure yielded a final sample of 76 lightcurves, characterized by an average duration $\langle T_{90} \rangle = 0.44$ s and fluence $\langle \mathcal{F} \rangle = 2.6 \times 10^{-6}$ erg cm$^{-2}$.

To search for excess emission following the prompt burst, we added the selected binned lightcurves in the four channels independently. The resulting lightcurves are shown in the upper panels of Fig. 1 by the solid points. Error bars are computed by propagating the Poisson uncertainties of the individual lightcurves.

The lightcurves in the third and fourth channels can be successfully fitted with polynomials. The third (110–

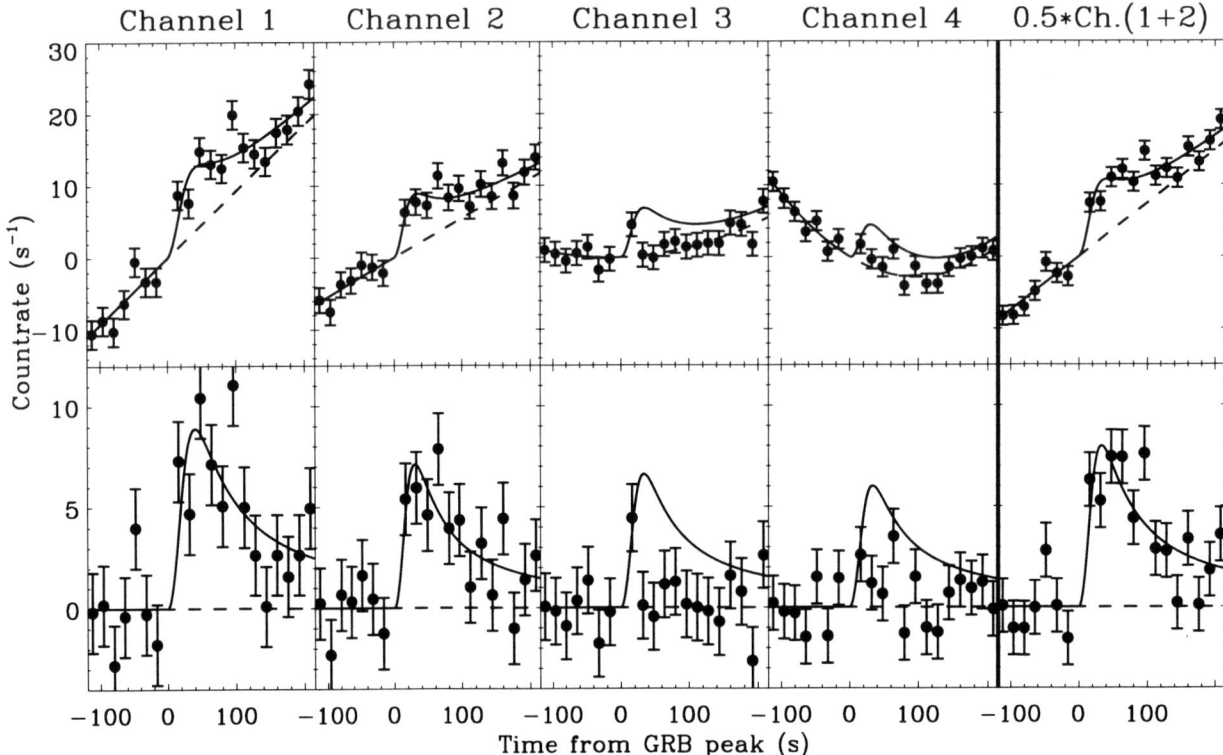

**FIGURE 1.** Overall lightcurves in the 4 BATSE channels (from left to right) of the sample of short bursts (see text). The rightmost panels show the average signal in the first and second channels. The time interval of the burst emission has been excluded. The upper panels show the lightcurves without background subtraction (a constant has been subtracted in all panels for viewing purposes in order to have zero counts at $t = 0$). The solid line is the best fit background plus afterglow model (in the channel 3 and 4 panels the $3\sigma$ upper limit afterglow is shown). The dashed line shows the background contribution in all channels. The lower panels show the same data and fit after background subtraction.

325 keV) and fourth (> 325 keV) channel lightcurves can be fitted with a quadratic model, yielding $\chi^2/\text{d.o.f.} = 17/18$ and $\chi^2/\text{d.o.f.} = 18.5/17$ respectively. In the first two channels, a polynomial model alone does not give a good description of the data. In the first (25–60 keV) channel, a cubic fit yields $\chi^2/\text{d.o.f.} = 42/16$, while in the second (60–110 keV) we obtain $\chi^2/\text{d.o.f.} = 26/16$. A more accurate modelling of the first two channels lightcurves can be achieved by allowing for an afterglow emission following the prompt burst. We model the afterglow lightcurve with a smoothly joined broken power-law function:

$$L_A(t) = \frac{3 L_A}{\left(\frac{t_A}{t}\right)^2 + \frac{2t}{t_A}}; \quad t > 0 \quad (1)$$

which rises as $t^2$ up to a maximum $L_A$ that is reached at time $t_A$ and then decays as $t^{-1}$. Adding this afterglow component to the fit, we obtain $\chi^2/\text{d.o.f.} = 28.5/16$ and $14.7/16$ in the first and second channels, respectively. The $\chi^2$ variation, according to the F-test, is significant to the $\sim 3.5\sigma$ level in both channels. The fact that the fit in the first channel is only marginally acceptable should

**TABLE 1.** Fit results. Quoted errors at 90% levels, upper limits at $3\sigma$ level.

| | E (keV) | $L_A$ (cts s$^{-1}$) | $t_A$ (s) |
|---|---|---|---|
| Channel #1 | [25-60] | $8.9 \pm 2.6$ | $40 \pm 16$ |
| Channel #2 | [60-110] | $7.1 \pm 2.4$ | $30^{+16}_{-10}$ |
| Channel #1+2 | [25-110] | $16 \pm 3.5$ | $33.5^{+24}_{-15}$ |
| Channel #3 | [110-325] | < 6.6 | 33.5 (fixed) |
| Channel #4 | > 325 | < 6.0 | 33.5 (fixed) |

not surprise. This is because the excess is due to many afterglow components peaking at different times, and has therefore a more "symmetric" shape than Eq. 1. A fully acceptable fit can be obtained with a different shape of the excess, but we used the afterglow function for simplicity. By adding together the first two channels, the afterglow component is significant at the $4.2\sigma$ level. The results of the fit are reported in Tab. 1.

## DISCUSSION

In order to understand whether the excess is residual prompt burst emission or afterglow emission, we computed its four channel spectrum. To convert BATSE count rates to fluxes, we computed an average response matrix for our burst sample by averaging the matrices of single bursts obtained from the `discsc_bfits` and `discsc_drm` datasets. The resulting spectrum is shown in Fig. 2. The dark points show the spectrum at $t = 30$ s (the peak of the afterglow in the second BATSE channel), which is consistent with a single power-law $F(\nu) \propto \nu^{-1}$ (black dashed line). Grey points show the time integrated spectrum (fluences has been measured with a growth curve technique), consistent with a steeper power-law $F(\nu) \propto \nu^{-1.5}$ (grey dashed line). These power-law spectra are much softer than any observed burst spectrum (independent of their duration). This spectral diversity, together with the fact that a single power-law does not fit the data, suggest that the emission is not due to a tail of burst emission but more likely to an early hard X-ray afterglow. This also confirms earlier predictions that the mechanism responsible for the afterglow emission is different from that of the prompt radiation.

An interesting comparison can be made with the early afterglow of long GRBs. Connaughton [7] finds that on average the BATSE countrate is $\sim 150$ cts s$^{-1}$ at $t = 50$ s after the main event. In our short GRB sample, the countrate at the same time is $\sim 15$ cts s$^{-1}$. Since the luminosity of the average early X-ray afterglow is representative of the total isotropic energy of the fireball, we can conclude that the isotropic equivalent energy of the short bursts is on average ten times smaller than that of the long ones (or that their true energy is the same, but the jet opening angle is three times larger). Indeed, the $\gamma$-ray fluence of long bursts is on average ten times larger than that of short bursts.

In the case of short bursts, the analysis of the afterglow emission is made easier by the lack of superposition with the prompt burst flux. For this reason the time and luminosity of the afterglow peak can be directly measured while in long bursts it had to be inferred from the shape of the decay law at longer times. In our case, however, the lightcurve in Fig. 1 is the result of the sum of many afterglow lightcurves, with different peak times and luminosities. For a given isotropic equivalent energy $E$, afterglows peaking earlier (with larger $\Gamma$) are expected to be brighter and should dominate the composite lightcurve. On the other hand, for a given Lorentz factor $\Gamma$, afterglow peaking earlier (with lower $E$) are dimmer. The fact that the $\sim 35$ s timescale is preserved, suggests that there are only few very energetic bursts with a large bulk Lorentz factor.

We must also remain aware of other possibilities. For instance, we may be wrong in assuming that the cen-

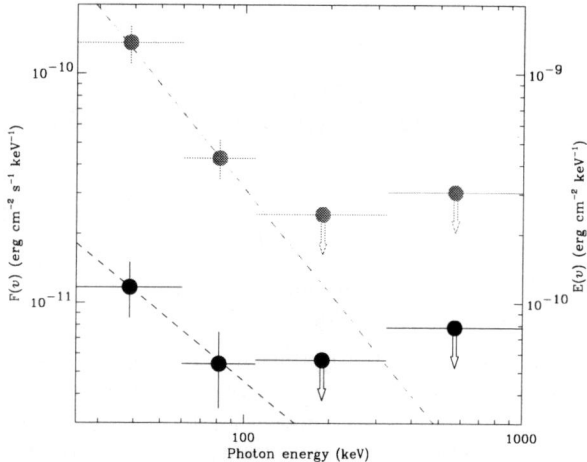

**FIGURE 2.** Spectrum of the peak afterglow emission. Black dots (and left vertical axis) show the spectrum at $t = 30$ s. Gray dots (right vertical axis) show the time integrated spectrum as obtained from the four BATSE channel counts. Error bars are for 90% uncertainties, while arrows are 3$\sigma$ upper limits.

tral object goes dormant after producing the initial explosion. A sudden burst followed by a slowly decaying energy input could arise if the newly formed black hole slowly swallows the orbiting torus around it or if the central object becomes a rapidly-spinning pulsar rather than a black hole. This luminosity may dominate the continuum afterglow at early times before the blast wave decelerates. Under this interpretation, the hard X-ray transient following the prompt emission could be attributed to the central object itself rather than to a standard decelerating blast wave. Contrary to what is observed, this emission should smoothly decay after the main episode, unless this energy is converted into a relativistic outflow which is in turn converted to radiation at a larger radius.

## REFERENCES

1. Boella, G., Butler, R. C., Perola, G. C., Piro, L., Scarsi, L., and Bleeker, J. A. M., *A&AS*, **122**, 299–307 (1997).
2. Costa, E. e. a., *Nature*, **387**, 783–785 (1997).
3. van Paradijs, J. e. a., *Nature*, **386**, 686–689 (1997).
4. Frail, D. A., Kulkarni, S. R., Nicastro, S. R., Feroci, M., and Taylor, G. B., *Nature*, **389**, 261–263 (1997).
5. Metzger, M. R., Djorgovski, S. G., Kulkarni, S. R., Steidel, C. C., Adelberger, K. L., Frail, D. A., Costa, E., and Frontera, F., *Nature*, **387**, 878–880 (1997).
6. Lazzati, D., Ramirez-Ruiz, E., and Ghisellini, G., *A&A*, **379**, L39–L43 (2001).
7. Connaughton, V., *ApJ in press (astro-ph/0111564)* (2001).

# Constraints on the Bulk Lorentz Factor of GRB 990123

Alicia M. Soderberg* and Enrico Ramirez-Ruiz*

*Institute of Astronomy, Madingley Road, Cambridge CB3 0HA, ENGLAND*

**Abstract.** GRB 990123 was a long, complex gamma-ray burst accompanied by an extremely bright optical flash. We present the collective constraints on the bulk Lorentz factor for this burst based on estimates from burst kinematics, synchrotron spectral decay, prompt radio flash observations, and prompt emission pulse width. Combination of these constraints leads to an average bulk Lorentz factor for GRB 990123 of $\Gamma_0 = 1000 \pm 100$ which implies a baryon loading of $M_{jet} = 8^{+17}_{-2} \times 10^{-8} M_\odot$. We find these constraints to be consistent with the speculation that the optical light is emission from the reverse shock component of the external shock. In addition, we find the implied value of $M_{jet}$ to be in accordance with theoretical estimates: the baryonic loading is sufficiently small to allow acceleration of the outflow to $\Gamma > 100$.

## INTRODUCTION

The discovery of a prompt and extremely bright optical flash in GRB 990123 [1], implying an apparent peak (isotropic) optical luminosity of $5 \times 10^{49}$ erg s$^{-1}$, has lead to widespread speculation that the observed radiation arose from the reverse shock component of the burst. The reverse shock propagates into the adiabatically cooled particles of the coasting ejecta, decelerating the shell particles and shocking the shell material with an amount of internal energy comparable to that of the material shocked by the forward shock. The typical temperature in the reverse-shocked fluid is, however, considerably lower than that of the forward-shocked fluid. Consequently, the typical frequency of the synchrotron emission from the reverse shock peaks at lower energy. It is believed to account for the bright prompt optical emission from GRB 990123. The reverse shock emission stops once the entire shell has been shocked and the reverse shock reaches the inner edge of the fluid. Unlike the continuous forward shock, the hydrodynamic evolution of the reverse-shocked ejecta is more fragile. As we will demonstrate, the temperature of the reverse-shocked fluid is expected to be mildly-relativistic for GRB 990123 and thus the evolution of the ejecta deviates from the Blandford & McKee (BM) [2] solution that determines the late profile of the decelerating shell and the external medium. In general, the determination of $\Gamma_0$ for optical flashes, would play an important role in discriminating between *cold* and *hot* shell evolution. Moreover, the strong dependence of the peak time of the optical flash on the bulk Lorentz factor $\Gamma$ provides a way to estimate this elusive parameter. We discuss how different constraints on $\Gamma_0$ for GRB 990123 are consistent with optical emission from the reverse shock. We assume $H_0 = 65$ km s$^{-1}$ Mpc$^{-1}$, $\Omega_M = 0.3$, and $\Omega_\Lambda = 0.7$.

## THE ROLE OF $\Gamma$

Relativistic source expansion plays a crucial role in virtually all current GRB models [3, 4]. The Lorentz factor is not, however, well determined by observations. The lack of apparent photon-photon attenuation up to $\approx 0.1$ GeV implies only a lower limit $\Gamma \approx 30$ [5], while the observed pulse width evolution in the gamma-ray phase eliminates scenarios in which $\Gamma >> 10^3$ [6, 7]. The initial Lorentz factor is set by the baryon loading, that is $m_0 c^2$, where $m_0$ is the mass of the expanding ejecta. This energy must be converted to radiation in an optically-thin region, as the observed bursts are non-thermal. The radius of transparency of the ejecta is

$$R_\tau = \left(\frac{\sigma_T E}{4\pi m_p c^2 \Gamma_0}\right)^{1/2} \approx 10^{12} - 10^{13} \text{cm}, \quad (1)$$

where $E$ is the isotropic equivalent energy generated by the central site. The highly variable $\gamma$-ray light curves can be understood in terms of internal shocks produced by velocity variations within the relativistic outflow [8]. In an unsteady outflow, if $\Gamma$ were to vary by a factor of $\approx 2$ on a timescale $\delta T$, then internal shocks would develop at a distance $R_i \approx \Gamma^2 c \delta T \geq R_\tau$. This is followed by the development of a blast wave expanding into the external medium, and a reverse shock moving back into the ejecta. The inertia of the swept-up external matter decelerates the shell ejecta significantly by the time it reaches the deceleration radius [9],

$$R_d = \left(\frac{E}{n_0 m_p c^2 \Gamma_0^2}\right)^{1/3} \approx 10^{16} - 10^{17} \text{cm}. \quad (2)$$

Given a certain external baryon density $n_0$, the initial Lorentz factor then strongly determines where both internal and external shocks develop. Changes in $\Gamma_0$ will modify the dynamics of the shock deceleration and the manifestations of the afterglow emission.

## REVERSE SHOCK EMISSION AND $\Gamma_0$

It has been predicted that the reverse shock produces a prompt optical flash brighter than 15th magnitude with reasonable energy requirements of no more than a few $10^{53}$ erg emitted isotropically [10, 11]. The forward shock emission is continuous, but the reverse shock terminates once the shock has crossed the shell and the cooling frequency has dropped below the observed range. The reverse shock contains, at the time it crosses the shell, an amount of energy comparable to that in the forward one. However, its effective temperature is significantly lower (typically by a factor of $\Gamma$). Using the shock jump conditions and assuming that the electrons and the magnetic field acquire a fraction of the equipartition energy $\varepsilon_e$ and $\varepsilon_B$ respectively, one can describe the hydrodynamic and magnetic conditions behind the shock.

The reverse shock synchrotron spectrum is determined by the ordering of three break frequencies, the self-absorption frequency $\nu_a$, the cooling frequency $\nu_c$ and the characteristic synchrotron frequency $\nu_m$, which are easily calculated by comparing them to those of the forward shock [10, 11, 12]. The equality of energy density across the contact discontinuity suggests that the magnetic fields in both regions are of comparable strength.

Assuming that the forward and reverse shocks both move with a similar Lorentz factor, the reverse shock synchrotron frequency is given by

$$\nu_m = 5.840 \times 10^{13} \varepsilon_{e,-1}^2 \varepsilon_{B,-2}^{1/2} n_{0,0}^{1/2} \Gamma_{0,2}^2 (1+z)^{-1/4} \text{Hz}, \quad (3)$$

while the cooling frequency $\nu_c$ is equal to that of the forward shock. Here we adopt the convention $Q = 10^x Q_x$ for expressing the physical parameters, using cgs units. The spectral power $F_{\nu_m}$ at the characteristic synchrotron frequency is

$$F_{\nu_m} = 4.17 D_{28}^{-2} \varepsilon_{B,-2}^{1/2} E_{53} n_{0,0}^{1/2} \Gamma_{0,2} (1+z)^{3/8} \text{Jy}. \quad (4)$$

The distribution of the injected electrons is assumed to be a power law of index $-p$, above a minimum Lorentz factor $\gamma_i$. For an adiabatic blast wave, the corresponding spectral flux at a given frequency above $\nu_m$ is $F_\nu \approx F_{\nu_m}(\nu/\nu_m)^{-(p-1)/2}$, while below $\nu_m$ is characterised by a synchrotron tail with $F_\nu \approx F_{\nu_m}(\nu/\nu_m)^{1/3}$. Similar relations to those found for a radiative forward shock hold for the reverse shock [13].

Unlike the synchrotron spectrum, the afterglow light curve at a fixed frequency strongly depends on the hydrodynamics of the relativistic shell, which determines the temporal evolution of the break frequencies $\nu_m$ and $\nu_c$. The forward shock is always highly relativistic and thus is successfully described using the relativistic generalisation of theory of supernova remnants. In contrast, the reverse shock can be mildly relativistic. In this regime, the shocked shell is unable to heat the ejecta to sufficiently high temperatures and its evolution deviates from the BM solution [14]. Shells satisfying

$$\xi \approx 0.01 E_{52}^{1/6} \Delta_{11}^{-1/2} \Gamma_{0,2}^{-4/3} n_{0,1}^{-1/6} > 1, \quad (5)$$

are likely to have a Newtonian reverse shock[1], otherwise the reverse shock is relativistic and it considerably decelerates the ejecta. The width of the shell, $\Delta$, can be inferred directly from the observed burst duration by $\Delta = cT_{\text{dur}}/(1+z)$ assuming the shell does not undergo significant spreading [3].

If $\xi > 1$, then the reverse shock is in the sub-relativistic temperature regime for which there are no known analytical solutions. In order to constrain the evolution of $\Gamma$ in this regime it is common to assume $\Gamma \propto R^{-g}$ where $3/2 \leq g \leq 7/2$ [10, 14]. For an adiabatic expansion, $\Gamma \propto T^{-g/(1+2g)}$ and so $\nu_m \propto T^{-3(8+5g)/7(1+2g)}$ and $F_{\nu_m} \propto T^{-(12+11g)/7(1+2g)}$. The spectral flux at a given frequency expected from the reverse shock gas drops then as $T^{-2(2+3g)/7(1+2g)}$ below $\nu_m$ and $T^{-(7+24p+15pg)/14(1+2g)}$ above. For a typical spectral index $p = 2.5$, the flux decay index varies in a relatively narrow range ($\approx 0.4$) between limiting values of $g$.

## CONSTRAINTS ON THE $\Gamma_0$ OF GRB 990123

Despite ongoing observational attempts, the optical flash associated with GRB 990123 remains the only event of its kind detected to date. Observations of this optical flash appear to be in good agreement with early predictions for the reverse shock emission [15]. As a result, numerous studies have been done on this event in which reverse shock theory has been applied to burst observations in order to constrain physical parameters and burst

---

[1] It should be remarked that equations (3) and (4) refer to this reverse shock regime. Equations for the relativistic case are relatively similar, the biggest discrepancy being that the peak flux is inversely proportional to $\Gamma$ (see equations (7)-(9) of [13]).

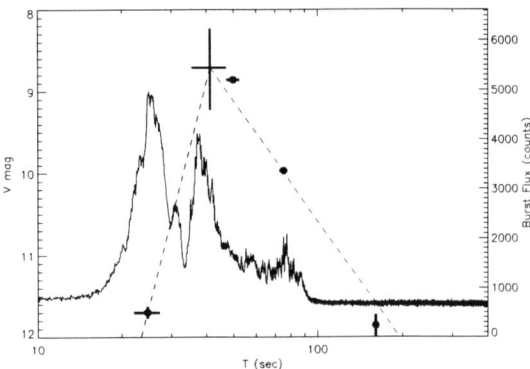

**FIGURE 1.** BATSE and ROTSE light curves for GRB 990123 as a function of time from the BATSE trigger. Dashed lines represent theoretical predictions for the rise $\propto T^{3p-3/2}$ and decay $\propto T^{-(21+73p)/96}$ of an adiabatic reverse shock light curve, assuming the shell is thin and cooling slowly. We predict that the optical flash peaked $\approx 41 \pm 6$ seconds after the trigger.

properties, including $\Gamma_0$. Current estimates on the bulk Lorentz factor for GRB 990123 stretch over nearly an order of magnitude, with values ranging from $\approx 200$ [11] to $\approx 1200$ [16]. It should be noted, however, that these estimates were made before accurate burst parameters for GRB 990123 were known, and consequently they include approximations and parameters from other GRB afterglows.

By fitting multi-frequency afterglow light curves, physical parameters for 8 GRBs have recently been reported [17]. Best fit values presented for GRB 990123 are $E_{j,50} = 1.5^{+3.3}_{-0.4}$ (initial jet energy), $\theta_0 = 2.1^{+0.1}_{-0.9}$, $n_{0,-3} = 1.9^{+0.5}_{-1.5}$, $\varepsilon_{e,-2} = 13^{+1}_{-4}$, $\varepsilon_{B,-4} = 7.4^{+23}_{-5.9}$, and $p = 2.28^{+0.05}_{-0.03}$ with a rough estimate for the bulk Lorentz factor of $\Gamma_0 = 1400 \pm 700$ [18]. Using these physical parameters, we present a comprehensive examination of the constraints on $\Gamma_0$ and report a best-fit value based on an analysis of these constraints.

*Time of Peak Flux.* Observational estimates for the time of peak flux enabled a measurement of the initial Lorentz factor with reasonable accuracy using the physical parameters specific to GRB 990123. Assuming the optical flash was the result of the reverse shock, the initial bulk Lorentz factor

$$\Gamma_0 = 237 \, E_{52}^{1/8} \, n_{0,0}^{-1/8} \, T_{\text{peak},1}^{-3/8} \, (1+z)^{3/8} \quad (6)$$

where $T_{\text{peak}}$ is the time of peak flux in the observer frame. Light curves for the optical flash and $\gamma$-ray emission are shown in Figure 1. Dashed lines represent theoretical predictions for the rise $\propto T^{3p-3/2}$ and decay $\propto T^{-(21+73p)/96}$ of an adiabatic reverse shock light curve, assuming the shell is thin, i.e. $\Delta < (E/(2n_0 m_p c^2 \Gamma_0^8))^{1/3}$,

and cooling slowly [13]. Using the recently reported physical parameters, one finds GRB 990123 to have a marginal thickness, as predicted by [13]. The observed rise time is, however, in good agreement with that of a thin shell $\approx T^{5.5}$ (in contrast with $\approx T^{1/2}$ for a thick shell). The shape of the light curve is determined by the time evolution of the three spectral break frequencies, which in turn depend on the hydrodynamical evolution of the fireball. In the case of GRB 990123, the typical synchrotron frequency $\nu_m = 1.5 \times 10^{14}$ Hz is well below the cooling frequency $\nu_c = 1.0 \times 10^{19}$ Hz, and therefore places the burst in a regime with a flux decay governed by the relation $F_\nu \propto T^{-(21+73p)/96}$. This implies a decay of $\approx T^{-2}$ for the optical flash of GRB 990123. Applying these light curve predictions to the prompt optical data and taking observational uncertainties as well as burst parameter uncertainties into account, we predict that the optical flash peaked $T_{\text{peak}} \approx 41 \pm 6$ s after the GRB started. Substitution for $T_{\text{peak}}$ in equation (6) gives $\Gamma_0 = 770 \pm 50$.

*Synchrotron Spectral Decay.* Observations of the optical peak brightness enable further accurate constraints on the value of $\Gamma_0$. The synchrotron spectrum from relativistic electrons comprises four power-law segments, separated by three critical frequencies. The prompt optical flash in GRB 990123 is observed at a frequency that falls well below the cooling frequency, but above the typical synchrotron frequency: $\nu_a < \nu_m < \nu_{\text{obs}} < \nu_c$. The synchrotron spectrum for this spectral segment is given by $F_{\text{obs}} = F_{\nu_m}(\nu_{\text{obs}}/\nu_m)^{(p-1)/2}(1+z)^{1/2-p/8}$ where $\nu_{\text{obs}}$ is taken to be the ROTSE optical frequency. Assuming the optical peak flux $F_{\text{obs}}$ observed in GRB 990123 is radiation arising from the reverse shock, we find $\Gamma_0 = 1800^{+600}_{-500}$.

*Radio Flare Observations.* In addition to the optical flash associated with GRB 991023, a short ($\sim 30$ hr.) radio flare was also observed [19]. Since the emission did not grow stronger with time thereby demonstrating the properties of a typical radio afterglow, it was subsequently proposed that both the prompt optical emission and the radio flare arose from the reverse shock [11]. As the reverse shock cools the emission shifts to lower frequencies which rapidly approach the observational radio band. The received flux peaks when $\nu_m$ crosses the band while decreasing mildly relativistically as $\nu_m \propto T^{-3(8+5g)/7(1+2g)}$. At $t < 1.2$ days after the initial GRB trigger, the radio flare from 990123 was observed to be increasing thereby implying $\nu_m > \nu_{\text{obs}}$. Using eqn (3) and the evolution of $\nu_m$, we compute the $\Gamma_0$ which places the peak frequency in the radio band at $t \sim 1.2$ days. We find $\Gamma_0 = 2000^{+400}_{-200}$.

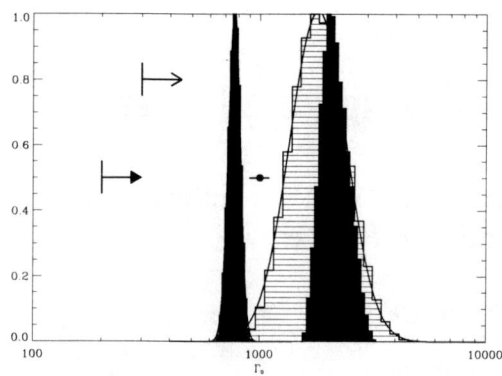

**FIGURE 2.** Collective constraints on $\Gamma_0$ for GRB 990123. These include estimates from the burst kinematics (narrowest distribution), synchrotron spectral decay (widest distribution), radio flash observations (medium width distribution), prompt emission pulse width (filled arrow), and jet modelling (unfilled arrow). We find a best fit value of $\Gamma_0 = 1000 \pm 100$.

*Prompt Emission Pulse Width.* Although observations of a reverse shock induced optical peak enable fairly accurate calculations of the bulk Lorentz factor, it remains possible to obtain information on $\Gamma_0$ in situations when an optical flash has not been detected. Consider an internal shock which produces an instantaneous burst of isotropic $\gamma$-ray emission at a time, $t$, and radius, $R_i$, in the frame of the central engine. The kinematics of colliding shells implies that although photons are emitted simultaneously, the curvature of the emitting shell spreads the arrival time of the emission over a period of $\Delta T_p$, thereby producing the observed width in individual pulses. The delay in arrival time between on-axis photons and those at $\theta \approx \Gamma^{-1}$ is a function of the radius of emission and the Lorentz factor according to: $\Delta T_p/(1+z) = R_i/(2c\Gamma^2)$ where $\Gamma = \Gamma_0$ in the early phase of the expansion [20]. In order to allow photons to escape, $R_i$ must be larger than the radius of transparency $R_\tau$. This imposes a lower limit on the initial bulk Lorentz factor such that: $\Gamma_0 > (R_\tau/2c\Delta T_p)^{1/2}(1+z)^{1/2}$. We determine $\Delta T_p$ for GRB 990123 by measuring the average pulse width through autocorrelation methods based on those described in [21] and find $\Delta T_p \approx 0.45$ s. Using the burst parameters to estimate $R_\tau$ from equation (1) and applying the inequality relation defined above, we find a lower limit of $\Gamma_0 > 200$. Figure 2 displays the collective constraints on $\Gamma_0$ for GRB 990123. An additional lower limit of $\Gamma_0 > 300$ constraint from afterglow modelling is also included [22]. The combination of these constraints leads to an average bulk Lorentz factor for GRB 990123 of $\Gamma_0 = 1000 \pm 100$ which implies $R_\tau \approx 1.3 \times 10^{14}$ and a baryon loading of $M_{jet} = 8^{+17}_{-2} \times 10^{-8} M_\odot$.

## CONCLUSIONS

We have shown that the collective constraints for the bulk Lorentz factor of GRB 991023 are compatible with current reverse shock theory. In addition, our best fit value of $\Gamma_0 \approx 1000$ provides confirmation of the ultra-relativistic nature of GRBs. The implied values for $R_\tau$ and $M_{jet}$ are in accordance with GRB theory: the radius of transparency is within theoretical estimates and the baryonic loading is sufficiently small to allow acceleration of the outflow to $\Gamma > 100$. As we have discussed, the bulk Lorentz factor plays a crucial role throughout all stages of GRB evolution, and therefore the ability to constrain $\Gamma_0$ provides clues on the nature of gamma-ray bursts.

We thank A. Blain, D. Lazzati, M. J. Rees and E. Rossi for helpful conversations. AMS was supported by the NSF GRFP. ER-R thanks CONACYT, SEP and the ORS for support.

## REFERENCES

1. Akerlof, C. W. et al., *Nature*, **398**, 400, (1999).
2. Blandford, R. D., McKee C. F., *Phys. Fluids*, **19**, 1130, (1976).
3. Piran, T., *Physics Reports*, **314**, 575, (1999).
4. Mészáros, P., *Science*, **291**, 79, (2001).
5. Mészáros P., Laguna P., Rees M. J., *ApJ*, **415**, 181, (1993).
6. Lazzati, D., Ghisellini G., Celotti, A., *MNRAS*, **309**, L13, (2001).
7. Ramirez-Ruiz, E., Fenimore E. E., *ApJ*, **539**, 712, (2000).
8. Rees, M. J., Mészáros P., *ApJ*, **430**, L93, (1994).
9. Mészáros, P., Rees M. J., *ApJ*, **405**, 278, (1993).
10. Mészáros, P., Rees M. J., *MNRAS*, **306**, L39, (1999).
11. Sari, R., Piran T., *ApJ*, **517**, L109, (1999a).
12. Panaitescu, A., Kumar P., *ApJ*, **543**, 66, (2000).
13. Kobayashi, S., *ApJ*, **545**, 807, (2000).
14. Kobayashi, S., Sari R., *ApJ*, **542**, 819, (2000).
15. Sari, R., Piran T., *ApJ*, **520**, 641, (1999b).
16. Wang, X. Y., Dai Z. G., Lu T., *MNRAS*, **319**, 1159, (2000).
17. Panaitescu, A., Kumar P., *ApJ*, **560**, L49, (2001b).
18. Panaitescu, A., Kumar P., *ApJ*, **554**, 667, (2001a).
19. Kulkarni, S. P. et al., *ApJ*, **522**, L97, (1999).
20. Fenimore, E. E., Madras C. D., Nayakshin S., *ApJ*, **473**, 998, (1996).
21. Fenimore, E. E., Ramirez-Ruiz E., Wu B., *ApJ*, **518**, L73, (1999).
22. Panaitescu, A., Kumar P., *ApJ submitted (astro-ph/0109124)*, (2002).

# BATSE 5B Sky Exposure and Trigger Efficiency

Jon Hakkila*, Geoffrey N. Pendleton[†], Charles A. Meegan**, Michael S. Briggs[‡], R. Marc Kippen[§] and Robert D. Preece[‡]

*Department of Physics and Astronomy, College of Charleston, Charleston, SC*
[†]*COLSA Corporation, Huntsville, AL*
**NASA NSSTC, Huntsville, AL*
[‡]*Department of Physics and Astronomy, University of Alabama in Huntsville, AL*
[§]*Space and Remote Sensing Sciences, LANL, Los Alamos, NM*

**Abstract.** Sky Exposure and Trigger Efficiency are presented for the BATSE 5B burst catalog. Previously published BATSE sky exposure has been improved by an updated CGRO orbital model. Trigger efficiency is presented for the long soft class of bursts as well as for the short hard class of bursts. This demonstrates that BATSE is significantly more sensitive to hard bursts, implying that (1) short bursts are being oversampled relative to long bursts based on spectral characteristics, and (2) there is little evidence for a class of undetected hard bursts.

## INTRODUCTION

The BATSE 5B catalog (Briggs *et al.* 2002, in preparation) summarizes triggered gamma-ray burst observations made by the Burst And Transient Source Experiment on NASA's Compton Gamma-Ray Observatory prior to its June 4, 2000 de-orbit. Many important uses to which the catalog are put (*e.g.*, angular and intensity distributions, population studies) require knowledge of the instrumental response and sensitivity to bursts that were not detected.

No gamma-ray burst experiment is capable of detecting all gamma-ray bursts. We have chosen to summarize BATSE's trigger limitations in terms of *sky exposure*, which refers to the experiment's angular sensitivity to gamma-ray burst detection, and *trigger efficiency*, which refers to the flux-dependent instrumental sensitivity.

## SKY EXPOSURE

Although BATSE's eight detectors allowed for a $4\pi$ steradian view of the sky, several factors (primarily earth blockage in CGRO's low-earth orbit) contribute to decreasing the angular sensitivity. The procedure used to calculate sky exposure has been described previously ([1] [2]) and has been used to calculate exposure for prior BATSE gamma-ray burst catalog releases. It has been assumed that the sky exposure is primarily a function of declination. Earth blockage introduces the main anisotropy; over many orbits anisotropies in right ascension become insignificant compared to anisotropies in declination.

In this study we extend the sky exposure analysis approach to the final BATSE 5B catalog. Some minor modifications have been made to the pre-existing sky exposure software. The primary improvement is that an improved CGRO orbital model has now been added to the code. This model more accurately accounts for times when CGRO's orbit became highly eccentric due to the satellite being reboosted. The improved software has been used to re-evaluate sky exposure for previously-published catalgs. Although there have been minor changes to previously published values, corrections are typically less than 5% different than the previously-published values. It should be noted that some time intervals have been excluded from the exposure calculations based on time periods when triggers were disabled or had peculiar configurations; these time intervals are TJD 9918 to TJD 9922 (due to phosphorescent events), TJD 10062 to TJD 10069 (due to specific detector triggers for the bursting pulsar), and TJD 10416 to TJD 10418 (due to a flight software crash). It has been assumed that observations started on TJD 8362 and ended on TJD 11691.

The sky exposure is plotted in Figure 1. The properties of the anisotropic sky coverage are summarized in Table 2. It should be noted that the exposure is better away from the direction of the earth (toward the equatorial poles), in the northern hemisphere (away from the South Atlantic Anomaly, over which the experiment was turned off), and near solar minimum (when the experi-

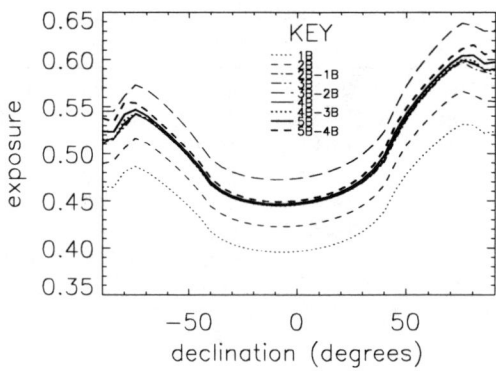

**FIGURE 1.** BATSE 5B Sky Exposure

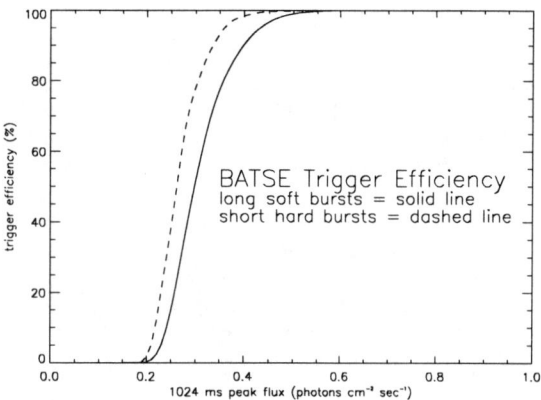

**FIGURE 2.** An Improved BATSE Trigger Efficiency Model for Long Soft and Short Hard Bursts

ment was triggering on solar flares and magnetospheric events rather than on gamma-ray bursts). The exposure never reaches a value of 1.0 because trigger readouts and high-voltage offtimes effect trigger efficiency in complex ways and are thus treated as deadtime.

## TRIGGER EFFICIENCY

BATSE's trigger efficiency is the experiment's flux-dependent sensitivity to gamma-ray bursts. The procedure used to calculate the trigger efficiency has been described previously ([3] [2]); the efficiency as calculated describes the probability of exceeding the 1024 ms threshold, in the nominal 50-300 keV trigger energy range and at the nominal level of $5.5\sigma$ as a function of peak flux in at least two detectors. The efficiency calculation procedure includes atmospheric scattering corrections (which increase the trigger efficiency near the detector threshold) and is based on a counts background obtained during a complete CGRO orbital precession cycle. Previous calculations have assumed that all bursts have specific (but canonical) spectral forms, and have demonstrated the strong dependence of efficiency on burst spectral characteristics.

We have generalized the trigger efficiency calculation by finding the joint efficiency probabilities produced by 52 bright bursts with different spectra; this estimate of BATSE's overall efficiency is indicated in Figure 2 (solid line). As expected, the composite efficiency spans a larger flux range than the efficiencies found for individual bursts.

Trigger efficiency has also been identified for the class of short, hard bursts based on 30 spectra; this efficiency curve is plotted in Figure 2 (dotted line). BATSE is significantly more sensitive to these bursts than it is to longer, softer bursts. This is because many hard inci-

dent photons are detected as softer photons, as found from the off-diagonal detector response matrices [4]. BATSE's increased sensitivity to hard bursts strongly supports claims that no unknown class of hard gamma-ray bursts exists brighter than BATSE's detection threshold (e.g. [5]). The lower threshold to hard bursts also indicates that more hard bursts are being detected than would be if the threshold was the same for hard and soft bursts.

BATSE's trigger efficiency for the faintest bursts has an angular dependence. Since two detectors are required for a trigger, BATSE's sensitivity is poorest for faint bursts triggering directly over a detector face. If enough faint bursts could be observed, and if the orientation of CGRO remained constant, then a symmetric "pumpkin diagram" would be observed [6]. The strength of this anisotropy depends on burst spectral characteristics and temporal morphology, but naively becomes important when the total counts (used for localizing a burst) are less than 2.5 times the minimum detection threshold (count rate times trigger timescale) in the trigger energy channels (used for triggering a burst). The trigger efficiency rapidly becomes anisotropic as one looks at extremely faint bursts. We decided to determine the significance of this anisotropy by rotating the "pumpkin diagram" efficiency through all actual CGRO pointings, then inspecting the output for obvious anisotropies in celestial coordinates. We found that the large-scale components of these anisotropies (corresponding to dipole and quadrupole moments) were washed out by the many CGRO pointings. However, we also found that certain sky areas have indeed been sampled more effectively than others, primarily for the faintest bursts. However, we note that these anisotropies are only apparent for a small number of extremely faint bursts, and that spectral and temporal burst structure (as well as atmospheric

**TABLE 1.** BATSE Exposure Statistics

| Statistic | 1B | 2B | 2B-1B | 3B | 3B-2B | 4B | 4B-3B | 5B | 5B-4B |
|---|---|---|---|---|---|---|---|---|---|
| Avg exposure | 0.4289 | 0.4571 | 0.4819 | 0.4813 | 0.5113 | 0.4817 | 0.4824 | 0.4846 | 0.4886 |
| Time ($10^6$ s) | 27.9 | 59.9 | 31.9 | 108.2 | 48.3 | 168.7 | 60.5 | 287.4 | 118.7 |
| Expos. ($10^6$ s) | 12.0 | 27.4 | 15.4 | 52.1 | 24.7 | 81.3 | 29.2 | 139.3 | 60.0 |
| Bursts | 260 | 585 | 325 | 1122 | 537 | 1637 | 515 | 2702 | 1065 |
| $\langle \cos\theta \rangle$ | -0.00762 | -0.00776 | -0.00787 | -0.00836 | -0.00903 | -0.00847 | -0.00867 | -0.00865 | -0.00890 |
| $\langle \sin^2 b \rangle - 1/3$ | -0.00450 | -0.00442 | -0.00436 | -0.00442 | -0.00442 | -0.00444 | -0.00448 | -0.00454 | -0.00466 |
| $\langle \sin\delta \rangle$ | 0.01576 | 0.01605 | 0.01627 | 0.01729 | 0.01867 | 0.01752 | 0.01792 | 0.01788 | 0.01841 |
| $\langle \sin^2\delta \rangle - 1/3$ | 0.02468 | 0.02424 | 0.02390 | 0.02423 | 0.02422 | 0.02435 | 0.02458 | 0.02486 | 0.02557 |

scattering effects) have been neglected in our analysis. The latter effects should further dilute the suspected trigger efficiency anisotropies. It should also be noted these same extremely faint bursts typically have large localization errors, which makes it very difficult to associate their locations with angular anisotropies in the trigger efficiency. As a result, we believe that the trigger efficiency can be treated as being generally isotropic for the triggered sample of BATSE bursts.

## CONCLUSIONS

BATSE's sky exposure has been obtained for the entire time span of the CGRO mission. The exposure has been observed to decrease with increased solar activity, indicating that many solar flares and particle events caused deadtime that affected BATSE's ability to trigger on gamma-ray bursts. The exposure is better toward the celestial poles (due to earth blockage of CGRO's sky), and is better in the northern sky than in the southern sky (as a result of the satellite high voltage being turned off whil in the South Atlantic Anomaly).

Unlike sky exposure, BATSE's trigger efficiency cannot be calculated for the entire mission because of gaps in the data stream due to failure of the onboard flight recorders two years into the mission. Thus, BATSE's trigger efficiency has been obtained based on background count rates found during one complete orbital precession cycle. The experiment is more efficient in detecting hard bursts than soft ones, indicating that a missing population of hard bursts brighter than BATSE's detection threshold is unlikely.

## ACKNOWLEDGMENTS

Support from the entire BATSE team over the CGRO mission is acknowledged. The last phase of this project was funded under NASA cooperative agreement NCC 8-200.

## REFERENCES

1. Hakkila, J., Meegan, C. A., Pendleton, G. N., Henze, W., McCollough, M. L., Kommers, J. M., and Briggs, M. S., "BATSE Sky Exposure", in *Proceedings of the Fourth Huntsville Gamma-Ray Burst Symposium*, edited by C. A. Meegan, R. D. Preece, and T. M. Koshut, AIP Conference Proceedings 428, American Institute of Physics, New York, 1998, pp. 144–148.
2. Paciesas, W. S., Meegan, C. A., Pendleton, G. N., Briggs, M. S., Kouveliotou, C., Koshut, T. M., Lestrade, J. P., McCollough, M. L., Brainerd, J. J., Hakkila, J., Henze, W., Preece, R. D., Connaughton, V., Kippen, R. M., Mallozzi, R. S., Fishman, G. J., Richardson, G. A., and Sahi, M., *ApJS*, **122**, 465–496 (1999).
3. Pendleton, G. N., Hakkila, J., and Meegan, C. A., "The BATSE Trigger Efficiency as a Function of Intensity and Energy Range", in *Proceedings of the Fourth Huntsville Gamma-Ray Burst Symposium*, edited by C. A. Meegan, R. D. Preece, and T. M. Koshut, AIP Conference Proceedings 428, American Institute of Physics, New York, 1998, pp. 899–903.
4. Pendleton, G. N., Paciesas, W. S., Mallozzi, R. S., Koshut, T. M., Fishman, G. J., Meegan, C. A., Wilson, R. B., Horack, J. M., and Lestrade, J. P., *Nuclear Instruments and Methods in Physics Research Section A*, **364**, 567–577 (1995).
5. Harris, M. J., and Share, G. H., *ApJ*, **494**, 724–733 (1998).
6. Brock, M. N., Meegan, C. A., Fishman, G. J., Wilson, R. B., Paciesas, W. S., and Pendleton, G. N., "BATSE's sky sensitivity map", in *Gamma-Ray Bursts*, 1992, pp. 399–403.

# A Comparison of Unsupervised Classifiers on BATSE Catalog Data

Jon Hakkila*, Richard J. Roiger†, David J. Haglin†, Timothy W. Giblin* and William S. Paciesas**

*Department of Physics and Astronomy, College of Charleston, Charleston, SC*
†*Department of Computer and Information Sciences, Minnesota State University, Mankato, MN*
**Department of Physics and Astronomy, University of Alabama in Huntsville, AL*

**Abstract.** We classify BATSE gamma-ray bursts using unsupervised clustering algorithms in order to compare classification with statistical clustering techniques. BATSE bursts detected with homogeneous trigger criteria and measured with a limited attribute set (duration, hardness, and fluence) are classified using four unsupervised algorithms (the concept hierarchy classifier ESX, the EM algorithm, the Kmeans algorithm, and a kohonen neural network). The classifiers prefer three-class solutions to two-class and four-class solutions. When forced to find two classes, the classifiers *do not* find the traditional long and short classes; many short soft events are placed in a class with the short hard bursts. When three classes are found, the classifiers clearly identify the short bursts, but place far more members in an intermediate duration soft class than have been found using statistical clustering techniques. It appears that the boundary between short faint and long bright bursts is more important to the classifiers than is the boundary between short hard and long soft bursts. *We conclude that the boundary between short faint and long hard bursts is the result of data bias and poor attribute selection.* We recommend that future gamma-ray burst classification avoid using extrinsic parameters such as fluence, and should instead concentrate on intrinsic properties such as spectral, temporal, and (when available) luminosity characteristics. Future classification should also be wary of correlated attributes (such as fluence and duration), as these bias classification results.

## INTRODUCTION

Successful separation of gamma-ray bursts into meaningful classes can potentially provide tremendous insight into the physics responsible for them. However, gamma-ray burst classification is made difficult by the large range of burst spectral and temporal behaviors, the variation of spectral behaviors during a burst, and the many instrumental and sampling biases that obscure measurement of these behaviors. Gamma-ray burst properties therefore overlap, and classification techniques are needed that can identify clusters of bursts with similar properties, despite these overlaps.

To this end, statistical clustering techniques have been applied to gamma-ray burst classification [1]. These techniques strongly support the existence of two previously-known gamma-ray burst classes (e.g. [2]) and identify a third class. The characteristics, or *attributes*, that appear to most easily delineate these classes are *fluence* (time integrated flux in the 50 to 300 keV spectral range) *HR321 spectral hardness* (the 50-300 keV fluence divided by the 25 to 50 keV fluence), and *T90 duration* (the time interval during which 90% of the fluence is recorded). Statistical clustering techniques find three gamma-ray burst classes: (1) long, bright bursts of intermediate hardness, (2) short, faint, hard bursts, and (3) soft bursts of intermediate duration and fluence. The properties of these classes are summarized in Table 1.

Machine learning pattern recognition algorithms are alternative approaches to statistical clustering techniques. These automated classifiers typically operate in either of two modes (some can operate in both modes); *supervised* (in which the classifier is trained with known classification instances) and *unsupervised* (in which classification occurs without training examples).

**TABLE 1.** Statistical clustering technique classes.

| Attributes | Class 1 Long | Class 2 Short | Class 3 Intermediate |
|---|---|---|---|
| T90 | long | short | intermediate |
| Fluence | bright | faint | intermediate |
| Hardness | intermediate | hard | soft |

In a previous application of supervised classification [3] to gamma-ray burst data, we have demonstrated that the intermediate class does not necessarily represent a separate source population. Instead, instrumental and sampling biases can cause some long bursts to take on intermediate burst characteristics. Due to the hardness intensity correlation, faint long bursts are typically softer than bright long bursts; this has been attributed to a cosmological redshift. Additionally, certain faint long bursts can have their fluences and durations systematically measured as shorter and fainter than they would if they were brighter; we have named this bias the *fluence duration bias* [3]. Simply put, fluences and durations of faint long bursts can be underestimated due to the unrecognizability of low signal-to-noise burst pulses.

## COMPARISON OF UNSUPERVISED CLASSIFIERS

The purpose of our GRB ToolSHED project [4] is to provide an online suite of data mining tools and a gamma-ray burst database within which scientists can search for data clustering and correlations. Our tool has a variety of supervised and unsupervised classifiers.

Using more than one classifier aids data analysis *because there is no correct way of classifying a dataset*. Each classifier operates under a different set of assumptions concerning relationships between data attributes. Some classifiers are designed to work with nominal data while others are not, some employ Bayesian while others employ frequentist statistics, some assume *a priori* distributions of attribute values while others do not. The results of any classifier can change if the size and makeup of the data set is altered. Data errors can influence the results, since these are not yet rigorously treated by any multidimensional classification approach.

We present here a systematic comparison of several unsupervised algorithms; this approach allows us to assess the individual learning behaviors of these classifiers collectively (each classifier already optimizes its behavior individually). The comparison is meant to contrast performance while demonstrating that the differences in these algorithms still allow them to find similar burst classes. For this analysis, we apply the concept hierarchy classifier ESX [5] [6], a kohonen neural network, the unsupervised Kmeans algorithm, and the unsupervised EM algorithm to BATSE 4B data. We compare the results to a smaller dataset classified using statistical clustering techniques [1].

In order to avoid instrumental biases, we have limited our database to bursts detected with a homogeneous set of BATSE trigger criteria. Bursts included are non-overwriting and non-overwritten BATSE 5B bursts triggering on energy channels 2+3 with the trigger threshold set at $5.5\sigma$ above background.

The base ten logarithmic values of fluence, HR321 hardness ratio, and T90 duration have been identified as significant classification attributes. In order to aid our systematic study, we require all four classifiers to use only these three attributes.

As with all statistical formulations, validity of the result is based on some predetermined significance. When allowed to find an optimum number of classes based on a default significance, the classifiers typically recover three to four burst classes. In addition to a preferred number of classes, we forced all four classifiers to recover two, three, and four classes.

### Three Class Results

The properties of three recovered classes are demonstrated in Table 2. At first glance the classes appear to be similar to the three classes found by statistical clustering techniques. However, closer inspection indicates that the intermediate classes found by unsupervised classifiers have far more members than the intermediate class found by statistical clustering techniques; the intermediate class has so many members that it has become the dominant class.

### Two Class Results

In order to explain why the intermediate class dominates the distribution, we examine how members of this large intermediate class are treated when classifiers are forced to recover only two classes. The results are remarkably consistent: all four classifiers fail to delineate the traditionally-accepted short and long burst classes. Each classifier places a large number of soft intermediate bursts in with the hard short burst class (see Figure 1). This is surprising, since the traditional long and short burst classes are clearly separated in the hardness vs. duration parameter space. The reason that this boundary is not well-defined is that a sharper one exists in the fluence vs. duration parameter space (Figure 2); *the delineation between short faint bursts and long bright bursts is more significant than the delineation between short hard bursts and long soft bursts.*

### Analysis

Fluence appears to play a more important role in gamma-ray burst classification than it should. First of all, fluence is clearly an extrinsic attribute as opposed

**TABLE 2.** Mean properties of three classes found by unsupervised classification

| Class | Property | ESX | kohonen | EM | Kmeans |
|---|---|---|---|---|---|
| short | No. bursts | 194 | 239 | 144 | 173 |
| | log(fluence) | -6.71 | -6.63 | -6.76 | -6.72 |
| | log(T90) | 0.02 | 0.13 | -0.23 | -0.12 |
| | log(HR321) | 0.44 | 0.39 | 0.58 | 0.53 |
| intermediate | No. bursts | 354 | 334 | 232 | 352 |
| | log(fluence) | -5.95 | -5.94 | -6.28 | -6.07 |
| | log(T90) | 1.38 | 1.51 | 1.02 | 1.28 |
| | log(HR321) | 0.06 | 0.07 | 0.04 | 0.06 |
| long | No. bursts | 250 | 225 | 422 | 273 |
| | log(fluence) | -5.19 | -5.06 | -5.39 | -5.16 |
| | log(T90) | 1.71 | 1.71 | 1.70 | 1.78 |
| | log(HR321) | 0.31 | 0.31 | 0.21 | 0.26 |

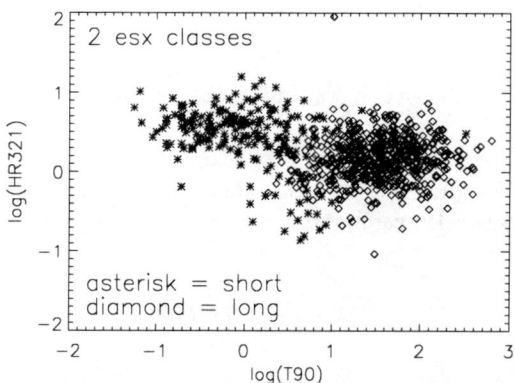

**FIGURE 1.** Hardness and duration properties of two burst classes found using unsupervised classifiers.

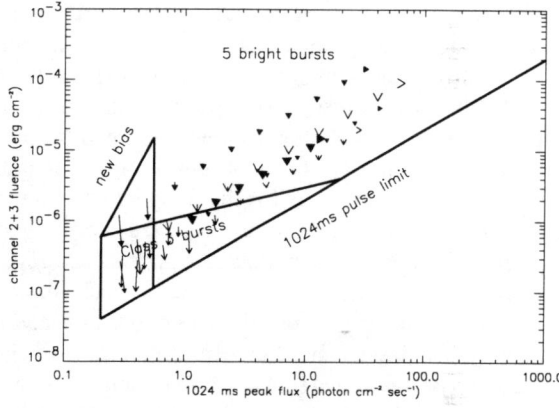

**FIGURE 3.** Region of the fluence vs. peak flux diagram occupied by Intermediate (Class 3) bursts. Also demonstrated is how the fluence duration bias can cause some bursts to have their fluences and durations arbitrarily shortened (assuming they had fainter peak fluxes than their measured values).

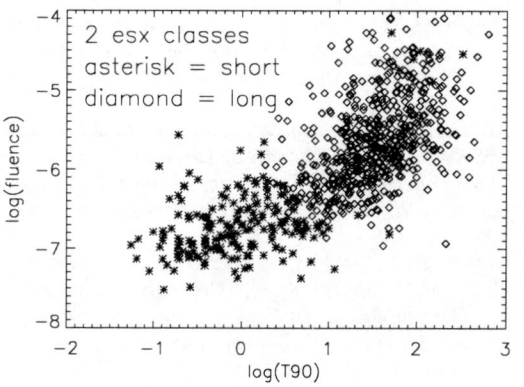

**FIGURE 2.** Fluence and duration properties of two burst classes found using unsupervised classifiers.

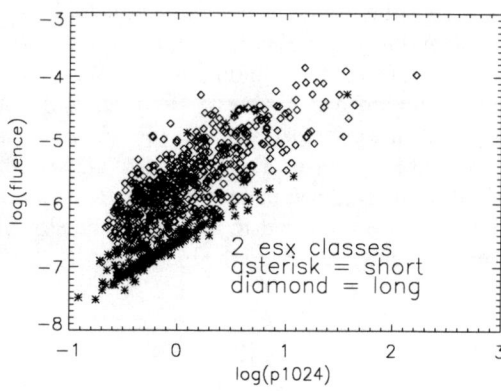

**FIGURE 4.** Fluence and peak flux properties of two burst classes found using unsupervised classifiers.

to hardness and duration; there is no reason why fluence clustering should relate to any physical differences between burst classes. Secondly, fluence correlates with duration, as a longer burst has more temporal intervals in which to place flux. As a result of this correlation, clustering in the duration attribute can also cause clustering in the fluence attribute, and the use of fluence as a classification attribute magnifies the relative clustering importance of duration relative to hardness. The break between short faint and long bright bursts therefore appears to be in part a result of attribute selection.

What is the physical mechanism that produces the duration vs. fluence clustering? Rephrasing the question, why are some traditional long bursts noticeably shorter and fainter than others? *We hypothesize that the fluence duration bias is responsible for making some long bursts appear to be shorter and fainter than they otherwise would be; these would include the intermediate bursts identified previously by statistical clustering techniques.*

If the fluence duration bias is responsible for the duration vs. fluence clustering, then we would expect the additional short faint bursts to occupy the same general region of fluence vs. peak flux parameter space as the intermediates identified by statistical clustering techniques. We have shown previously [3] that long bursts suffering from the fluence duration bias are expected to occupy this region of the fluence vs. peak flux diagram (see Figure 3). Figure 4 demonstrates the fluence vs. peak flux distributions of long bright and short faint bursts found by unsupervised classification. Indeed, the short faint class is made up of traditional short hard bursts mixed with intermediates. *This supports our hypothesis, and provides strong and self-consistent evidence that the fluence duration bias plays a significant role in altering gamma-ray burst properties.*

We are now able to address why the unsupervised classifiers place so many bursts in the intermediate class when three recovered classes are allowed. By allowing three classes, the classifiers are free to recover the traditional short hard class. All four classifiers do a fairly good job at doing this. The delineation of the short class removes the necessity of finding a common divider for fluence, duration and hardness. Hardness and fluence now become the principal attributes for delineating intermediate and long bursts. The correlation between hardness and intensity allows many soft bursts to be placed in the intermediate class, and thus expands the size of the intermediate class relative to the long class.

## CONCLUSIONS

Caution should be excercised when applying unsupervised classifiers to scientific data. The classifiers help the user to find weakly-defined clusters, but they do not interpret the mechanisms responsible for them.

We have demonstrated that fluence is an improper classification attribute. Fluence is an extrinsic attribute. Furthermore, fluence correlates with the duration attribute because the concept of duration is included in the fluence definition. Fluence should be replaced in future gamma-ray burst classification with more appropriate intrinsic attributes. This supports our previous conclusions [3], as well as those of Paciesas *et al.* (this conference), that spectral parameters are useful gamma-ray burst classification attributes.

We have also demonstrated additional evidence that the fluence-duration bias alters gamma-ray burst properties. When forced to recover two gamma-ray burst classes, unsupervised classifiers use fluence and duration rather than hardness and duration to define the classes. The fluence and duration boundary used to delineate the classes is consistent with previous models of the fluence duration bias.

## ACKNOWLEDGMENTS

We would like to gratefully acknowledge NASA support under grant NRA-98-OSS-03 (the Applied Information Systems Research Program).

## REFERENCES

1. Mukherjee, S., Feigelson, E. D., Babu, G. J., Murtagh, F., Fraley, C., and Raftery, A., *ApJ*, **508**, 314–328 (1998).
2. Kouveliotou, C., Meegan, C. A., Fishman, G. J., Bhat, N. P., Briggs, M. S., Koshut, T. M., Paciesas, W. S., and Pendleton, G. N., *ApJ*, **413**, L101–L104 (1993).
3. Hakkila, J., Haglin, D. J., Pendleton, G. N., Mallozzi, R. S., Meegan, C. A., and Roiger, R. J., *ApJ*, **538**, 165–180 (2000).
4. Haglin, D. J., Roiger, R. J., Hakkila, J., Pendleton, G. N., and Mallozzi, R., "A GRB Tool Shed", in *AIP Conf. Proc. 526: Gamma-ray Bursts, 5th Huntsville Symposium*, 2000, pp. 877+.
5. Roiger, R. J., Geatz, M. W., Haglin, D. J., and Hakkila, J., "ESX - A Tool for Knowledge Discovery", in *Proceedings of the Federal Data Mining Symposium & Exposition '99*, edited by W. T. Price, AFCEA International, Fairfax, VA, 1999, pp. 109–120.
6. Roiger, R. J., Hakkila, J., Haglin, D. J., Pendleton, G. N., and Mallozzi, R. S., "Unsupervised Induction and Gamma-Ray Burst Classification", in *Proceedings of the Fifth Huntsville Gamma-Ray Burst Symposium*, edited by M. Kippen, R. S. Mallozzi, and G. J. Fishman, AIP Conference Proceedings 526, American Institute of Physics, New York, 2000, pp. 38–42.

# III. GAMMA-RAY BURST PROGENITORS AND PROMPT EMISSION

## Sources and Central Engines

## Prompt Emission

# The Central Engines of Gamma-Ray Bursts

S. E. Woosley*, Weiqun Zhang* and A. Heger[†]

*Department of Astronomy and Astrophysics, University of California, Santa Cruz, CA 95064, USA*
[†]*Enrico Fermi Institute, University of Chicago, Chicago, IL 60637, USA*

**Abstract.**
Leading models for the "central engine" of long, soft gamma-ray bursts (GRBs) are briefly reviewed with emphasis on the collapsar model. Growing evidence supports the hypothesis that GRBs are a supernova-like phenomenon occurring in star forming regions, differing from ordinary supernovae in that a large fraction of their energy is concentrated in highly relativistic jets. The possible progenitors and physics of such explosions are discussed and the important role of the interaction of the emerging relativistic jet with the collapsing star is emphasized. This interaction may be responsible for most of the time structure seen in long, soft GRBs. What we have called "GRBs" may actually be a diverse set of phenomena with a key parameter being the angle at which the burst is observed. GRB 980425/SN 1988bw and the recently discovered hard x-ray flashes may be examples of this diversity.

## INTRODUCTION

Recent years have seen a welcome decline in estimates of the energy that must be provided by the central engine of GRBs - from almost $10^{54}$ erg to $\sim 10^{52}$ erg (including the supernova that accompanies the GRB). While retaining their title as the *brightest* explosions in the modern universe, current estimates of kinetic and neutrino energies have demoted GRBs to merely comparable to supernovae. This decrease has come about mostly because the beaming long indicated by the theoretical models has been verified experimentally.

While the energy requirements on the model have become less problematic, other demands have arisen. The successful model must not only deliver a few times $10^{51}$ erg in highly relativistic ejecta, it must collimate those ejecta into a narrow outflow with typical width 0.1 radian. Adherents of the internal shock model further need the Lorentz factor to vary rapidly so that the efficiency for making gamma-rays is not small. Diverse light curves should have a natural explanation and the events must occur in star forming regions. At least occasionally, GRBs should give be accompanied by Type I supernova, by which we mean a supernova, without hydrogen, powered at peak light by the decay of radioactive $^{56}$Ni and $^{56}$Co.

One model that shows promise in satisfying these constraints is the *collapsar* model. Collapsars occur naturally in star forming regions, make bright Type I supernovae, offer a natural mechanism for narrow jet collimation, and can explain the diverse light curves of long GRBs. They may also explain why GRBs have a nearly constant total energy. The model also has its difficulties. The large requisite angular momenta are difficult to achieve when current estimates of magnetic torques and mass loss rates are included and the model also offers no clear route to making short hard GRBs. However, unlike other models, the collapsar model makes many testable predictions.

These predictions (and postdictions) include: a) association of GRBs with massive stars and star formation; b) an increasing fraction of GRBs at low metallicity (high redshift); c) collimation of the outflow into a narrow jet with typical opening angle 0.1 radian; d) an asymmetric Type I supernova accompanying every GRB; e) an event rate of about 1% that of ordinary supernovae (we only see a few tenths of a percent of these as GRBs); f) a distribution of circum-source matter consistent with the wind of a Wolf-Rayet star at the end of its life; g) possible large production of $^{56}$Ni by the disk of the accreting black hole; h) continuing energetic outflows for a long time after the GRB; i) diverse phenomena seen at different polar angles; and j) a jet whose temporal structure (and hence a GRB whose light curve) is relatively insensitive to the central engine, but set by the interaction of the jet with the overlying star.

Not all of these predictions are unique to the collapsar model. Any model that produces the same relativistic outflow in the middle of a massive helium star will give similar results. In the end, details like the duration of the event, the $^{56}$Ni production, and other specific properties of the supernova will help to distinguish the operation of an engine that, except for faint signals in neutrinos and gravity waves, is not directly observable.

## BASIC COLLAPSARS

Generically a collapsar is a rotating massive star, devoid of hydrogen envelope, whose central regions collapse to a black hole surrounded by an accretion disk [49, 22]. Accretion of at least a solar mass through this disk produces outflows that are further collimated by passage through the stellar mantle. These flows attain high Lorentz factor as they emerge from the stellar surface and, after traversing many stellar radii, produce a GRB and its afterglows by internal and external shocks respectively.

There are actually several ways to make a collapsar and each is likely to have different observational characteristics.

- A standard (Type I) collapsar is one where the black hole forms promptly in a helium core of approximately 15 to 40 $M_\odot$. There never is a successful outgoing shock after the iron core first collapses. A massive, hot proto-neutron star briefly forms and radiates neutrinos, but the neutrino flux is inadequate to halt the accretion. For iron cores near 1.9 solar masses, as can occur in massive metal-deficient stars [50], and soft equations of state, eventual collapse to a black hole is assured. For most other equations of state, collapse to a black hole is also certain if the iron core accretes an additional half-solar mass or so without launching an outgoing shock. Such an occurrence seems very likely in helium cores of high mass because of the rapid accretion that characterizes the first second after core collapse [12]. Such high mass helium cores are more frequently realized in systems with low metallicity because low metallicity inhibits mass loss. The mass threshold for forming a black hole promptly is not known with certainty, but based upon two dimensional models with crude neutrino physics is thought to be around 10 solar masses of helium. A lower limit is the 6 solar mass helium core of SN 1987A (a main sequence star of $\sim 20$ $M_\odot$) which certainly did not form a black hole promptly [8] because we saw the neutrino signal continue for $\sim 10$ s.
- A variation on this theme is the "Type II collapsar" wherein the black hole forms after some delay - typically a minute to an hour, owing to the fallback of material that initially moves outwards, but fails to achieve escape velocity [23]. The time scale for such an event is set by the interval between the first outgoing shock and the GRB event. Such an occurrence is again favored by massive helium cores and might have occurred in SN 1987A [7]. The binding energy rises rapidly above main sequence masses of 25 $M_\odot$ in metal-poor stars [50]. Delayed black hole formation of this sort seems unavoidable above some threshold mass that is probably smaller than that required for Type I collapsars. There is nucleosynthetic evidence in low metallicity stars in our own Galaxy for a progenitor population where most of the iron failed to escape the star [10].

  Type II collapsars should be common events, probably more frequent than Type I. They are also capable of producing powerful jets that might make gamma-ray bursts. Unfortunately their time scale may be, on the average, too long for the typical long, soft bursts. If the GRB-producing jet is launched within the first 100 s or so of the initial supernova shock, it still emerges from the star before the supernova shock has gotten to the surface, i.e., when the star is still dense enough to provide collimation. Their accretion disks are also not hot enough to be neutrino dominated and this may affect the accretion efficiency [27].

- A third variety of collapsar occurs for extremely massive metal-deficient stars (above $\sim 300$ $M_\odot$) that may have existed in the early universe [1, 14]. For non-rotating stars with helium core masses above 137 $M_\odot$ (main sequence mass 280 $M_\odot$), it is known that a black hole forms after the pair instability is encountered [17]. It is widely suspected that such massive stars existed in abundance in the first generation after the Big Bang at red shifts $\sim 5 - 20$. For *rotating* stars the mass limit for black hole formation will be raised. The black hole that forms here, about 100 $M_\odot$, is more massive, than the several $M_\odot$ characteristic of Type I and II collapsars, but the accretion rate is also much higher, $\sim 10$ $M_\odot$ s$^{-1}$, and the energy released may also be much greater. The time scale in the lab frame for this accretion is of order 20 s or so, not so different from GRBs. However, one must dilate both the spectrum and the time scale according to the red shift. The spectrum even in the lab frame is unknown, but if it were similar to nearby GRBs, one might expect long, hard x-ray flashes rather than classical GRBs.

For Type I and II collapsars it is also essential that the star loses its hydrogen envelope before death. No jet can penetrate the envelope in less than the light crossing time which is typically 100 s for a blue supergiant and 1000 s for a red one. After running into 1/Γ of its rest mass, a ballistic jet loses its energy.

## PROGENITORS

As Heger & Woosley show elsewhere in these proceedings, a bare helium star born (e.g., from a merger) with equatorial rotation 10% of Keplerian and low metallicity can retain enough angular momentum to form a centrifugally supported disk around a central black hole of $\sim 3\,M_\odot$. Without magnetic fields, the angular momentum is sufficient to form a Kerr black hole and support most of the star in an accretion disk. However, when an approximate treatment of angular momentum transport by magnetic fields is included [39] along with mass loss, the resulting rotation become too low to form centrifugally supported disks in the inner part of the core. Even though our knowledge of magnetic torques inside evolved massive stars is still quite uncertain, this is a concern for the collapsar model.

The mass loss rate of Wolf-Rayet stars (WR-stars) during helium burning (i.e., most of their lifetime) is observed to be large, but has an uncertain dependence on the metallicity. A value of $\sim 10^{-5}\,M_\odot\,y^{-1}$ should be typical for collapsar progenitors (Heger & Woosley, this volume) implying a number density (of *helium* nuclei) outside the star $\sim 10^3 r_{16}^{-2}\,cm^{-3}$, large compared with what is inferred from afterglows [28]. This wind is clumpy and has an uncertain angular distribution. Perhaps the polar region has a lower mass loss.

Because the mass loss carries away both mass and angular momentum, it is detrimental to collapsar production. To the extent that mass loss is suppressed in such stars by low metallicity, one may expect the fraction of stars that become GRBs to increase with redshift. Lower mass loss rates for red supergiant stars with low metallicity also raises the maximum mass of helium core that can result from the evolution of single stars. For solar metallicity this number is about $12\,M_\odot$ [50]. For 1/4 solar metallicity, the number is already considerably higher.

No observations constrain the mass loss rate of WR-stars during the post-helium burning phases (100 - 1000 years), nor have the necessary stability analyses been carried out to see if such stars are stable. The loss rate could be quite high (or conceivably, even lower if these stages *are* stable). Unlike supergiants, the surface of a WR-star remains in sonic communication with the central core up to the last few minutes of core collapse. Strong acoustic waves from pulsationally unstable oxygen and silicon burning stages could, in principle, expel large quantities of material out to $\sim 10^{15}\,cm$ (the escape speed times the duration of oxygen burning). Alternatively, it is known that helium cores between about 40 and $65\,M_\odot$ encounter the pulsational pair instability [17]. Up to several solar masses can be ejected with supernova-like energies from days to years prior to the collapse of the core to a black hole. These are probably too massive to be common GRBs, but could occur occasionally, especially at low metallicity.

## COLLAPSARS - A STANDARD BOMB?

Frail et al. [11] have suggested that GRBs are a standard bomb in the sense that, within an order of magnitude, they all have the same total kinetic energy, $\sim 3 \times 10^{51}$ erg. This can be understood in the collapsar model as an approximate constancy of the total accreted mass and black hole mass. Collapsars of Type I have an accretion rate of $\sim 0.1\,M_\odot\,s^{-1}$ for about 20 s into a black hole initially of about $3\,M_\odot$. Slower accretion rates can continue for a longer time, but the total energy in all cases is limited by the fact that the jet explodes the star shutting off its own accretion. The gravitational binding energy outside the iron core of a $15\,M_\odot$ helium star is $\sim 2 \times 10^{51}$ erg (contributing in part to the difficulty of exploding these stars by ordinary means). Some of this falls into the black hole, but the jet needs to deposit at least $10^{51}$ erg in the star simply to unbind it and shut off the accretion. Much more energetic jets will not be produced because the mass in the disk is limited and the explosion shuts off the flow.

## THE JET-STAR INTERACTION

The propagation of relativistic jets through the massive, partially collapsed progenitors of collapsars has been considered in refs. [2, 51] and the reader is referred to those papers for extensive discussions (see also Zhang, Woosley, & MacFadyen, this volume). Regardless of the initial Lorentz factor and opening angle, after a few seconds the jet inside the star is characterized by two shocks, one at the leading "head" moving subrelativistically, and another deeper in where the initial outflow runs into material piled up behind the leading shock. Some of the material it runs into has also slowed due to interaction with the jet walls. Only the initial outflow deep within the star remembers the properties of the central engine. In calculations this outflow is taken either to be born highly relativistic ($\Gamma \sim 100$) or to have such large internal energy that it becomes highly relativistic before going very far.

Between the two shocks is material with moderate Lorentz factor ($\Gamma \sim 10$) and large internal energy per baryon, roughly $10\,m_o c^2$. It is this material that, after exiting the star, makes most of the prompt burst. The large Lorentz factors that characterize GRBs ($\gtrsim 100$) are developed outside the star as expansion converts internal energy back into highly relativistic motion. That is conversion of internal energy in the moving frame give $\Gamma \sim 10$ which translates into $\Gamma \sim 200$ in the laboratory frame. Considerable lateral expansion of the jet also occurs after exiting the star. This is also true of the mildly relativistic cocoon of matter surrounding the exiting jet (see also [25]).

## GRB Light Curves

Within this context, all short time scale variability of the central engine itself is washed out by the first shock. Variations of energy input where the jet is born, do not manifest themselves in the GRB light curve.

Still, GRB light curves are known to be diverse and complex. Where does the time structure originate? We believe that it comes from the jet-star interaction. Mixture of nearly stationary matter into the jet by the (relativistic) Kelvin-Helmholtz instability can load the jet with baryons and slow it down. A particularly large wavelength instability can even temporarily block the flow giving bursts that seem to turn on and off multiple times.

Modulating the jet in this fashion has two effects. First, it can give rise to complex light curves even for a constant power input at the base. Second, it can provide the variable Lorentz factor necessary for the internal shock model.

## Luminosity-Variability-Angle Correlations

The angle with which the jet emerges increases with time as the star is blown aside. An observer situated at a relatively large angle may not even see the GRB until it has been operating at smaller polar angles for some time. If the jet power or its efficiency for conversion to gamma-rays declines with time, it will appear less luminous. This effect will lead to an "arrow-head" structure for the distribution of relativistic material in the jet. Salmonson & Galama [36] have discussed how such a structure leads to increased break times (and thus larger inferred opening angles) for observers off axis.

Further, if the central engine provides constant power, the energy per unit area in the jet that emerges will be larger for more focused jets. The focusing is assumed to be dependent on the structure of the progenitor star. Narrower cylindrical jets will have a larger ratio of surface to volume and will experience more Kelvin-Helmholtz instability, thus making them more variable.

Putting these effects together, one expects that jets with larger opening angles will make less luminous, less variable GRBs. This is apparently the case [35, 11].

## SUPERNOVAE

One of the earliest predictions of the collapsar model was that each GRB should be accompanied by a supernova-like display. The idea that a black hole formation in a rotating helium star would make a supernova of some sort was discussed by Bodenheimer & Woosley [6]. Woosley [49] extended the idea to GRB production. In what many regard as a gross understatement, these early models were characterized as "failed supernovae" because the usual mechanism for producing an outgoing shock and a successful explosion (neutron star formation and neutrino emission) was assumed to be inoperable. Calculations showing that the passage of a relativistic jet through the star not only leads to collimation of the jet, but explosion of the star were carried out by MacFadyen, Woosley, & Heger [22, 23] and Khokhlov et al. [20]. The kinetic energy energy available to the explosion is approximately the work the jet does prior to breaking out of the star [51]. For a jet power of $\sim 5 \times 10^{50}$ erg s$^{-1}$ (both jets) and a traversal time 5 - 10 s, this gives $\sim 3 \times 10^{51}$ erg, comparable to but somewhat greater than the kinetic energy of an ordinary supernova. For a typical 10 s GRB this gives jet energies - after break out - also of about $3 \times 10^{51}$ erg per jet, similar to what is inferred from afterglows and a kinetic energy conversion of 20%. Of course the supernova is initially grossly asymmetric and one might infer a much more energetic explosion viewing the supernova along the jet axis.

Lacking a hydrogen envelope, the supernova will be Type Ib or Ic with an optical luminosity given entirely by the yield of $^{56}$Ni. It is not generally appreciated how poorly determined this yield is in most GRB models. In ordinary (spherically symmetric) supernovae the iron-group yield (mostly $^{56}$Ni) is set by the amount of ejected material that experiences explosion temperatures in excess of $5 \times 10^9$ K. This in turn is given by the strength of the explosion and the density structure at the edge of the collapsing iron core. In Type I collapsars however, the material that would have become $^{56}$Ni falls into the black hole. In Type II collapsars some $^{56}$Ni is ejected, but not as much as in an explosion that left a neutron star. Depending on the fallback mass, most of the $^{56}$Ni reimplodes. The jet itself subtends a small solid angle and carries a small, albeit very energetic mass. It cannot propagate outwards until the mass flux inwards at the pole has declined, i.e., the density has gone down. This makes it hard for the jet itself to synthesize much $^{56}$Ni. How then is the supernova visible?

We think that the $^{56}$Ni may be made not by the jet, but by the disk wind ([22, 27] and MacFadyen, this volume). In the parlance of Narayan et al., it could be that at late times (after $\sim$10 s), a neutrino-dominated accretion disk (NDAF) switches to a convection dominated accretion disk (CDAF) with a large fraction of the mass flow being ejected. On the other hand, MacFadyen and Woosley found considerable mass outflow even from NDAFs. We postulate that a certain fraction of the accreting matter - composed initially of nucleons or iron group elements - is ejected at high velocity (~0.1 c) by the accretion disk. If the energy of the burst is given by the amount of material accreted and the outflow fraction is constant, the brightness of the supernova could correlate with the energy of the GRB. This simple relation is could be complicated by uncertain efficiency factors for converting disk energy into jet energy and mass loss from the disk. If the accretion rate is very high and the disk viscosity quite low, the density in the disk may be so high that electron capture must be considered.

In any case, it is quite possible to get a highly variable amount of $^{56}$Ni, and therefore a variable luminosity for the peak of the supernova. SN 1998bw may not be a standard candle. Observationally, besides SN 1998bw, evidence for supernova-like light curves, color, and time history has been found in at least three GRBs: GRB 011121 [5]; GRB 980326 [4], and GRB 970228 [34, 15]. What we would all like to see is the spectrum of a putative supernova accompanying a cosmologically distant GRB.

## ALTERNATE MODELS

### Merging Neutron Stars

The principal alternative model to the collapsar remains the merging neutron star pair or neutron star - black hole pair discussed elsewhere in this proceedings. These have the admirable properties of being associated with events that are known to occur in nature and have sufficient angular momentum to form an accretion disk around the black hole after the merger. An energy of $3 \times 10^{51}$ erg in relativistic ejecta is more challenging for these models than some others, but easily within reach of those employing magnetohydrodynamics (MHD) to extract black hole rotational energy or disk binding energy. However, even though a few of these might happen in star-forming regions, the vast majority are expected to occur outside [13]. It may also be difficult for merging neutron stars to collimate their outflows within 0.1 radians, at least in those versions where neutrino transport produces the jet. Given the difficulty the collapsar model has in making short hard bursts, we continue to associate compact mergers with this subclass [13].

Other popular models for the long, soft GRBs associated with massive star death involve either the delayed production of a black hole or the prompt production of a very magnetic, rapidly rotating neutron star.

### "Supranovae"

It has been suggested by Vietri& Stella [41, 42] and others that GRBs may result from the delayed implosion of rapidly rotating neutron stars to black holes. The neutron star is "supramassive" in the sense that without rotation, it would collapse, but with rotation, collapse is delayed until angular momentum is lost. The momentum can be lost by gravitational radiation and by magnetic field torques. Vietri and Stella assume that the usual pulsar formula holds and, for a field of $10^{12}$ gauss, a delay of order years (depending on the field radius and mass) is expected. When the centrifugal support becomes sufficiently weak, the star experiences a period of runaway deformation and gravitational radiation before collapsing into a black hole. It is assumed that ~0.1 M$_\odot$ is left behind in a disk which accretes and powers the burst in a manner analogous to the merging neutron star model.

The model has several advantages. It, as well as the collapsar model that it in some ways resembles, predicts an association of GRBs with massive stars and supernovae. Moreover it produces a large amount of material enriched in heavy elements located sufficiently far from the GRB as not to obscure it. The irradiation of this material by the burst or afterglow can produce x-ray emission lines as have been reported in several bursts [30, 31, 33].

However, the supranova model also has some difficulties (see also [24]). First, it may take fine tuning to produce a GRB days to years after the neutron star is born. Shapiro [38] has shown that neutron stars requiring differential rotation for their support will collapse in only a few minutes. The requirement of rigid rotation reduces the range of masses that can be supported by rotation to, at most, ~20% above the non-rotating limit [38, 37]. Small changes in the angular momentum and mass cause large variations in the delay between the supernova and GRB. This is because the pulsar radiation formula employed depends on $\omega^{-4}$ and the critical angular momentum where collapse ensues depends on the excess mass above the non-rotating limit. That the combination would have been just right in a (randomly selected) event like GRB 011211 [33] to give a delay of a few days is highly constraining.

Second, the supernova had best happen years and not days before the GRB (in conflict with Reeves et al). Supernovae are optically thick to gamma-rays until well after their optical peak, that is, at least a month even for Type I. Nor can any relativistic jet penetrate an object whose light crossing time is well in excess of the duration of the central engine (~$10^{12}$ cm). Supranovae that are younger might make lines, but they don't make GRBs. Third, the very success of the collapsar model in producing a collimated jet with opening angle near 5 degrees must be held in its favor. This collimation, as well as the time structure in the GRB light curve require a high pressure stellar mantle to be present when the black hole launches its jet [51]. Finally, the timing of supernovae seen in conjunction with GRBs demands a simultaneous explosion. The optical maximum of a Type I supernova of any subclass occurs a few weeks after explosion. This time modulated by the redshift is consistent with SN 1998bw/GRB 980425 and with supernovae seen in the tails of the optical afterglows of several other GRBs.

The collapsar gets around these restrictions by producing a jet that exits the star while the central engine is still on and making a supernova nearly simultaneously. Lines might be energized by a continuation of the same jet at late times (see "Post-Burst Phenomenology"). In the event that it proves necessary to eject appreciable matter just prior to the GRB, one may want to consider pulsationally driven mass loss (see "Progenitors").

## "Magnetar Model"

Another model, championed most recently by Wheeler et al [43], is the "super-magnetar" model (see also [40]). As usual, the iron core of a massive star collapses to a neutron star. For whatever reasons, unusually high angular momentum perhaps, the neutron star acquires at birth an extremely powerful magnetic field, $10^{15}$ - $10^{17}$ gauss. If the neutron star additionally rotates with a period of a ms or so, up to $10^{52}$ erg in rotational energy can be extracted on a GRB time scale by a variation of the pulsar mechanism. This model has the attractive features of being associated with massive stars, making a supernova as well as a GRB, and utilizing an object, the magnetar, that is implicated in other phenomena - soft gamma-ray repeaters and anomalous x-ray pulsars. It has the unattractive feature of invoking the magnetar fully formed in the middle of a star in the process of collapsing without consideration of the effects of neutrinos or rapid accretion. The star does not have time to develop a deformed geometry or disk that might help to collimate jets. To break the symmetry, Wheeler et al invoke the operation of a prior LeBlanc-Wilson [21] jet to "weaken" the confinement of the radiation bubble along the rotational axis. Numerical models to give substance to this scenario are needed (though see [44]).

## GRBS - A UNIFIED MODEL

According to the "Unified Model" for active galactic nuclei (e.g., [3]), one sees a variety of phenomena depending upon the angle at which the source is viewed. These range from tremendously luminous blazars, thought to be jets seen on axis, to narrow line radio galaxies and Type 2 Seyferts thought to be similar sources seen edge on. Given that an accreting black hole and relativistic jet may be involved in both, it is natural to seek analogies with GRBs.

In the equatorial plane of a collapsar, probably little more is seen than an extraordinary supernova that, were it close enough to observe, would be an exceptionally bright radio source. Roughly 1% of all supernovae might be of this variety - perhaps a larger fraction at high redshift. Given that we have seen ~1000 relatively nearby supernovae, it would not be surprising to find a few in the cataloged sample. They would be of Type Ib/c and perhaps extraordinarily energetic. SN 1998bw could be a prototype, but without the high velocities that come from observing the event at high latitude. There are indications that a few of these may have been seen. Besides SN 1998bw there are SN 1997ef and 1997ey [26], and perhaps SN 2002ap. These supernovae, all of Type I, are characterized by a large inferred kinetic energy (at least for the equivalent isotropic explosion) and a variable, but occasionally large mass of $^{56}$Ni. Very high velocity intermediate mass elements were also seen in SN 1998bw [29]. Completing the connection from the other end, there are also an increasing number of GRBs which show evidence for supernova-like activity in the tail of their optical afterglow, most recently in GRB 011211 [5].

Perhaps the most interesting phenomena are those at intermediate angles. The models clearly show, and nature generally demands that the edges of jets are not discontinuous surfaces. Moving off axis, one expects and calculates a smooth decline in the Lorentz factor and energy of relativistic ejecta. These low energy wings with moderate Lorentz factor come about in three ways [51]. First, the jet that breaks out still has a lot of internal energy. Expansion of this material in the comoving frame leads to a broadening of the jet. Some of the material is even *decelerated* by expansion pushing back towards the origin. As a result a small amount of material with low energy ends up moving with intermediate Lorentz factors - say 10 - 30 and at angles up to several times that of the main GRB-producing jet. Second, as the star explodes from around the jet, the emerging beam opens up. At late times the outflow continues (see "Post-burst Phenomenology"), but with decreased power. Third, the jet is surrounded by a hot mildly relativistic cocoon. This material has low energy, but can expand to large angles. As a consequence of the spreading of the jet, a large region of the sky, much larger than that which sees the main GRB, will see a hard transient with less power, lower Lorentz factor, and perhaps coming from an external shock instead of internal ones.

We have speculated for some time now [47, 48, 45, 46] that these ordinary bursts seen off axis might appear as hard x-ray transients of one sort or another. We have identified them with GRB 980425 and with the class of hard x-ray flashes reported by Heise et al. [18]. We do not say that these are ordinary high $\Gamma$ jets seen just beyond a sharp edge [19]. The events are made by matter moving towards us.

In the particular case of GRB 980425, it is important to know if the optical afterglow of a *normal* GRB that was *not* directed at us would have had an observable effect on the supernova light curve. After all, the optical afterglow is not nearly so beamed as the GRB and might be visible at lower latitude. However, Granot et al. [16] show that the afterglow would be invisible at the time the supernova was studied provided that the polar angle to our line of sight is greater than about 3 or 4 times that of the main GRB. This does raise the interesting possibility though that some future event might show the supernova and afterglow more nearly balanced in a "soft" relatively faint GRB.

## POST-BURST PHENOMENOLOGY

After the main burst is over, accretion continues at a decaying rate. The lateral shock launched by the jet starts at the pole and wraps around the star, but does not reach into the origin at the equator (one may envision an angle-dependent "mass cut"). Consequently, some reservoir remains to be accreted at late time. This accretion occurs at a rate given by the viscosity of the residual disk and the free fall time of material farther out not ejected in the supernova. MacFadyen, Woosley, & Heger [23] and Chevalier [9] estimate the accretion rate from fall back to be $\sim 3 \times 10^{-6} t_4^{5/3}$ $M_\odot$ s$^{-1}$. Here $t_4$ is the elapsed time since core collapse in units of $10^4$ s. Given the slow rate, the disk that forms is not neutrino dominated and there may be considerable high velocity flow from its surface (MacFadyen, this volume; [22, 27]). The outflow will still be jet-like in nature since the equatorial plane is blocked by the disk and its energy will be $\sim 5 \times 10^{46} t_4^{-5/3} \varepsilon_{01}$ erg s$^{-1}$ where $\varepsilon_{01}$ is the efficiency for converting rest mass into measured in percent. This is comparable to the energy in x-ray afterglows and might be important for producing the emission lines reported in some bursts [32, 24] and for providing an extended tail of hard emission in the GRB itself.

As a consequence of this continuing outflow, the polar regions of the supernova made by the GRB remain evacuated and the photosphere of the object resembles an ellipse seen along its major axis but with conical sections removed along the axis. An observer can see deeper into the explosion than they could have without the operation of the jet's "afterburner".

## CONCLUSIONS

The collapsar model is able to explain many of the observed characteristics of GRBs. Here we have explored some of its predictions (enumerated in the "Introduction"). Probably the greatest challenges facing the model today are not the large energy associated with GRBs, or even the relativistic collimated flow. They are an understanding of how the necessary angular momentum comes about in the precollapse star - presumably by the special circumstances that make GRBs rare compared with supernovae - and of how accretion energy in the disk is transformed into jets. The former is a problem we share with competing models for GRBs like the supranova and millisecond magnetar models; the latter is also a long standing obstacle in understanding AGNs. There is hope that numerical simulation might address both in a few years.

We have described a "unified theory of GRBs" in which diverse phenomena are expected depending upon the angle at which a standard model for the explosion is viewed. In this theory the Lorentz factor and the energy of relativistic ejecta vary both with polar angle and with time. This paradigm is similar to the unified theory of AGNs. Some of the phenomena it predicts, like hard x-ray flashes are just now being discovered. Others like the long duration gamma-ray transients expected from Type II and III collapsars may await discovery.

In the collapsar model, the light curves of GRBs reflect more the interaction of the jet with the star as it emerges than time variability of the central engine itself. We have mentioned how correlations in break times, luminosity, and opening angle might come about and discussed some of the special properties of the accompanying supernova.

In the near future we hope to carry out the next steps in realistic collapsar simulation - special relativistic studies (in three dimensions) of jet propagation inside the star and longer time scale calculations of the supernova it produces.

## ACKNOWLEDGMENTS

This work has been supported by the HETE-2 grant (MIT-SC-292701), the NASA Theory Program (NAG5-8128), and by the Scientific Discovery Through Advanced Computing (SciDAC) program of the DOE (DE-FC02-01ER41176). Alex Heger was partly supported by the Alexander von Humboldt Society (FLF-1065004). We are grateful for helpful conversations with and calculations by Andrew MacFadyen regarding the nature of collapsars.

# REFERENCES

1. Abel, T., Bryan, G. L., & Norman, M. L. 2002, Science, 295, 93
2. Aloy, M. A., Müller, E., Ibanez, J. M., Martí, J. M., & MacFadyen, A. 2000, ApJL, 531, 119
3. Antonucci, R. 1993, ARAA, 31, 473
4. Bloom, J. S., Kulkarni, S. R., Djorgovski, S. G., Eichelberger, A. C., Cote, P., Blakeslee, J. P., Odewahn, S. C., Harrison, F. A., et al. 1999, Nature, 401, 453
5. Bloom, J. S., Kulkarni, S. R., Price, P. A., Reichart, D., Galama, T. J., Schmidt, B. P., Frail, D. A., Berger, E. 2002, ApJ, in press, astroph-0203391
6. Bodenheimer, P., & Woosley, S. E. 1983, ApJ, 269, 281.
7. Brown, G. E. & Bethe, H. A. 1994, ApJ, 423, 659
8. Burrows, A. 1988, ApJ, 334, 891
9. Chevalier, R. A. 1989 ApJ, 346, 847
10. Depagne, E., Hill, V., Spite, M., Plez, B., Beers, T. C., Barbuy, B., Cayrel, R., Anderson, J. et al. 2002, ApJ, in press
11. Frail, D. A., Kulkarni, S. R., Sari, R., Djorgovski, S. G., Bloom, J. S., Galama, T. J., Reichart, D. E., Berger, E., et al. 2001, ApJL, 562, 55.
12. Fryer, C. L. 1999, ApJ, 522, 413
13. Fryer, C. L., Woosley, S. E., Hartmann, D. H. 1999, ApJ, 526, 152
14. Fryer, C. L., Woosley, S. E., & Heger, A. 2001, ApJ, 550, 372
15. Galama, T. J., Tanvir, N., Vreeswijk, P. M., Wijers, R. A. M. J., Groot, P. J., Rol, E., van Paradijs, J., Kouveliotou, C. et al. 2000, ApJ, 536, 185
16. Granot, J., Panaitescu, A., Kumar, P., & Woosley, S. E. 2002, ApJL, in press, astroph-0201322
17. Heger, A., & Woosley, S. E. 2002, ApJ, 567, 532
18. Heise, J., in't Zand, J., Kippen, R. M., Woods, P. M. 2001, GRBs in the Afterglow Era, eds. Costa, Frontera, & Hjorh, ESO Astrophysics Symposia, (Springer), 16
19. Ioka, K., & Nakamura, T. 2001, ApJL, 554, 163
20. Khokhlov, A. M., Höflich, P. A., Oran, E. S., Wheeler, J. C., Wang, L., Chtchelkanova, A. Yu. 1999, ApJL, 524, 107
21. LeBlanc, J. M., & Wilson, J. R. 1970, ApJ, 161, 541
22. MacFadyen, A., Woosley, S. E. 1999 ApJ, 524, 262
23. MacFadyen, A., Woosley, S. E., & Heger, A. 2001, ApJ, 550, 410
24. McLaughlin, G. C., Wijers, R. A. M. J., & Brown, G. E., & Bethe, H. A. 2002, 567, 454
25. Meszaros, P., & Rees, M. J. 2001, ApJL, 556, 37
26. Nakamura, T., Umeda, H., Iwamoto, K., Nomoto, K., Hashimoto, M., Hix, W. R., & Thielemann, F-K. 2001, ApJ, 555, 880.
27. Narayan, R., Piran, T., & Kumar, P. 2001, ApJ, 557, 949
28. Panaitescu, A., & Kumar, P. 2001, ApJL, 560, 49
29. Patat, F., Cappellaro, E., Danziger, J., Mazzali, P. A., Sollerman, J., Augusteijn, T., Brewer, J. Doublier et al. 2001, ApJ, 555, 900.
30. Piro, L., Costa, E., Feroci, M., Frontera, F., Amati, L., dal Fiume, D., Antonelli, L. A., et al. 1999, ApJL, 514, L73.
31. Piro, L., Garmire, G., Garcia, M., Stratta, G., Costa, E., Feroci, M., Mészáros, P., Vietri, M., et al. 2000, Science, 290, 955
32. Rees M. J., & Meszaros, P. 2000, ApJL, 545, 73
33. Reeves, J. N., Watson, D., Osborne, J. P., Pounds, K. A., O'Brien, P. T., Short, A. D. T., Turner, M. J. L., Watson, M. G., et al, 2002, Nature, 416, 512.
34. Reichart, D. 1999, ApJL, 521, 111
35. Reichart, D. E., Lamb, D. Q., Fenimore, E. E., Ramirez-Ruiz, E., Cline, T. L., & Hurley, K. 2001, ApJ, 552, 57
36. Salmonson, J. D., & Galama, T. 2002, ApJ, 569, 682
37. Salgado, M., Bonazzola, S., Gourgoulhon, E., & Haensel, P. 1994, A&AS, 108, 455
38. Shapiro, S. L. 2000, ApJ, 544, 397
39. Spruit, H.C. 2001, accepted by A&A; astro-ph/0108207
40. Usov, V. 1992, Nature, 357, 472
41. Vietri, M., & Stella, L. 1998, ApJL, 507, L45
42. Vietri, M., & Stella, L. 1999, ApJL, 527, L43
43. Wheeler, J. C., Yi, I., Höflich, P., & Wang, L. 2000, ApJ, 537, 810
44. Wheeler, J. C., Meier, D. L., & Wilson, J. R. 2002, ApJ, 568, 807
45. Woosley, S. E. 2000, GRBs, 5th Huntsville Symposium, eds. Kippen, Mallozzi, & Fishman, AIP, Vol 526, 555
46. Woosley, S. E. 2001, GRBs in the Afterglow Era, eds. Costa, Frontera, & Hjorh, ESO Astrophysics Symposia, (Springer), 555
47. Woosley, S. E., Eastman, R. G., Schmidt, B. P. 1999, ApJ, 516, 788
48. Woosley, S. E. & MacFadyen, A. I. 1999, A&AS, 138, 499
49. Woosley, S. E. 1993, ApJ, 405, 273.
50. Woosley, S. E., Heger, A., & Weaver, T. A. 2002, RMP, in press.
51. Zhang, W., Woosley, S. E., & MacFadyen, A. 2002, ApJ, in preparation

# Merging Neutron Star – Black Hole Binaries

## M. Ruffert[*] and H.-Th. Janka[†]

[*]*University of Edinburgh, Edinburgh EH93JZ, Scotland, U.K.*
[†]*MPI für Astrophysik, Postfach 1317, 85741 Garching, Germany*

**Abstract.** We report on ongoing investigations of the merger model for the central engine of gamma-ray bursts and show first results of a new set of simulations for neutron star black hole binaries. The masses of the neutron star and black hole are the same in all cases, but we change the numerical resolution and the gravitational potential of the black hole by using descriptions that follow [12] and [1]. Since our code is basically Newtonian, this procedure is supposed to mimic the general relativistic effects that are associated with the existence of an innermost stable orbit and the variation of this orbit and of the horizon radius with the spin and the growing mass of the accreting black hole. The neutron star, which initially is in orbit about the black hole, is tidally disrupted but in several cases the remnant manages to temporarily move to a more eccentric orbit. A second and final episode of mass transfer follows, producing the standard outcome of a high-density torus around the black hole. Our simulations still need to be analysed with respect to the neutrino-antineutrino annihilation luminosities, but we expect similar values as in the double neutron star models.

## INTRODUCTION

Although most recent observations and measurements of gamma-ray bursts favour the collapsar model as central engine, these detections are all instances of the "long" type of gamma-ray burst (GRB). Much less is known about the "short" GRBs, except that their duration and hardness are distinct from those of long GRBs [6]. While the collapsar model works well for the long GRBs it does have difficulty in generating short burst durations (cf. [18]. The model involving the merger of compact objects, i.e. neutron stars and black holes, not only roughly matches the observed occurrence rates, but the natural (dynamic) timescales (ms) are of the order of the variability timescale of short GRBs.

We are now extending our previous investigations to look at the merging of a binary consisting of a neutron star and a black hole (NS-BH). In previous work of merging double neutron star binaries (NS-NS) we showed that neutrino-antineutrino annihilation could provide the luminosity necessary to power a short GRB, if that neutrino luminosity is sustained for the typical durations of short GRBs, i.e. 0.1–1 sec. An asymmetric merger involving a black hole and a neutron star promises a priori to yield a more massive torus around the black hole, since the black hole is already present and no mass needs to be 'lost' to first form the black hole. On the other hand, the deeper potential well present from the outset might facilitate more mass to be accreted anyway. Numerous effects of similar nature, e.g. deeper potential well allows more energetic neutrino emission, etc., act to both increase as well as decrease the clean energy finally available for the fireball.

Our main initial aim when performing these computations is to compare the NS-BH results — torus mass, accretion rates, neutrino emission, annihilation luminosities — to the results produced by NS-NS mergers. The NS-BH results will depend on additional parameters specific to these models, e.g. mass and spin of black hole, and these will have to be varied, too, to assess their influence.

The conference talk and consequently this proceedings contribution can only report on preliminary results, since the calculations are still in progress. Thus results are not available yet for all models of interest and only a few models are completed.

## NUMERICAL PROCEDURES AND INITIAL CONDITIONS

### Numerical Procedures

Here we will only summarise the computational procedures that we used. Details of the hydrodynamic method as well as the neutrino relevant algorithms can be found in [16,17]. The nested grid was described by [14] and [13]. The implementation of the black hole was explained in [8,9,15].

The hydrodynamical simulations were done with a code based on the Piecewise Parabolic Method (PPM) developed by [7]. The code is basically Newtonian, but contains the terms necessary to describe gravitational-wave emission and the corresponding back-reaction on the hydrodynamical flow [5]. The gravitational potential and other quantities that follow from a Poisson equation are obtained from fast Fourier transforms. Other spatial derivatives are evaluated as standard centred differences on the grid.

In order to describe the thermodynamics of the neutron star matter, we use the EoS of [10] for a compressibility modulus of bulk nuclear matter of $K = 180\,\text{MeV}$ in tabular form. Use of a physical EoS instead of a simple ideal gas law implies that the adiabatic index is a function of space, i.e. of density, temperature and composition in the star, as well as of time, i.e. the value of the adiabatic index of the bulk of the matter changes when the neutron star strips some of its mass.

Energy loss and changes of the electron abundance due to the emission of neutrinos and antineutrinos are taken into account by an elaborate "neutrino leakage scheme". The energy source terms contain the production of all types of neutrino pairs by thermal processes and additionally of electron neutrinos and antineutrinos by lepton captures onto baryons. The latter reactions act as sources or sinks of lepton number, too, and are included as source terms in a continuity equation for the electron lepton number. Matter is rendered optically thick to neutrinos due to the main opacity producing reactions, which are neutrino-nucleon scattering and absorption of electron-type neutrinos onto nucleons.

The presented simulations were done on multiply nested and refined grids. With an only modest increase in CPU time, the nested grids allow one to simulate a substantially larger computational volume while at the same time they permit a higher local spatial resolution of the merging stars. The former is important to follow the fate of matter that is flung out to distances far away from the merger site either to become unbound or to eventually fall back. The latter is necessary to adequately resolve the strong shock fronts and steep discontinuities of the plasma flow that develop during the collision. The procedures used here are based on the algorithms that can be found in [2,3,4]. Each grid had $64^3$ or $128^3$ zones, the size of the smallest zone was $\Delta x = \Delta y = \Delta z = 1.27$ or 0.64 km, respectively. The zone sizes of the next coarser grid levels were doubled to cover a volume of 648 km side length. We used 4 nested grids.

The black hole was modeled as a vacuum sphere on the grids. Any gaseous matter that flowed into that region was taken out of the grids and the mass, velocity and angular momentum of the black hole changed by the appropriate amounts. The potential of the black hole was approximated either by a purely Newtonian dependence (i.e. $1/r$), or a Paczyński-Wiita type potential (i.e. $1/(r - r_s)$, cf. [12] which models the effect of the last stable orbit, or by its extension by [1] which mimics the effect of a rotating (Kerr) black hole on the location of the innermost stable circular orbit. For rotating black holes, the radii of both the event horizon ($r_h$) as well as of the last stable circular orbit ($r_o$) depend on the angular momentum via the Kerr parameter. The pseudo-potential diverges at the horizon. Therefore the radius of the vacuum sphere was chosen to be the arithmetic average $(r_h + r_o)/2$.

**TABLE 1.** Models that have been or will be computed, with their characterising parameters. $N$ is the number of grid zones per dimension in the orbital plane, $L$ the size of the largest grid, $a_0$ the initial centre to centre distance of the neutron star from the black hole. "BH spin" indicates the magnitude of the Kerr parameter used in the phenomenological model by [1] to mimic rotating black holes.

| Model  | $N$ | Potential | $L$ km | BH spin | $a_0$ km |
|--------|-----|-----------|--------|---------|----------|
| Newt   | 64  | Newton    | 648    | 0       | 50.      |
| Newt2  | 128 | Newton    | 648    | 0       | 50.      |
| PaWi   | 64  | PaWi      | 648    | 0       | 58.      |
| PaWi2  | 128 | PaWi      | 648    | 0       | 58.      |
| Ker6   | 64  | Kerr      | 648    | +0.6    | 55.      |
| Ker6/2 | 128 | Kerr      | 648    | +0.6    | 55.      |
| Ker9   | 64  | Kerr      | 648    | +0.9    | 54.      |
| Ker9/2 | 128 | Kerr      | 648    | +0.9    | 54.      |
| Ret9   | 64  | Kerr      | 648    | -0.9    | 63.      |
| Ret9/2 | 128 | Kerr      | 648    | -0.9    | 63.      |

With these new models we have not yet performed the post-processing step, to evaluate the neutrino-antineutrino ($\nu\bar{\nu}$) annihilation in the surroundings of the merged stars. This can only be performed after the hydrodynamical evolution is complete and will allow us to construct maps showing the local energy deposition rates per unit volume when a quasi-steady state has been reached. Spatial integration finally yields the total rate of energy deposition outside the neutrino emitting high-density regions.

## Initial conditions

We considered neutron stars with a baryonic mass of about 1.63 $M_\odot$ and the black hole mass was initially chosen to be 2.5 $M_\odot$. The black hole mass obviously increases during the simulation as it absorbs matter from the neutron star. With the equation-of-state that we used the neutron star has a radius of approximately 15 km.

The distributions of density $\rho$ and electron fraction $Y_e \equiv n_e/n_b$ (with $n_e$ being the number density of elec-

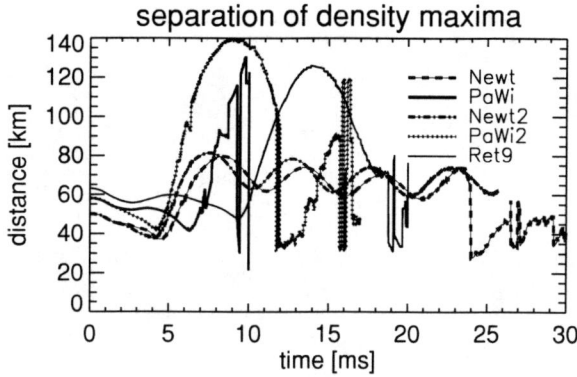

**FIGURE 1.** Distance of centre of neutron star from centre of black hole as function of time for several models.

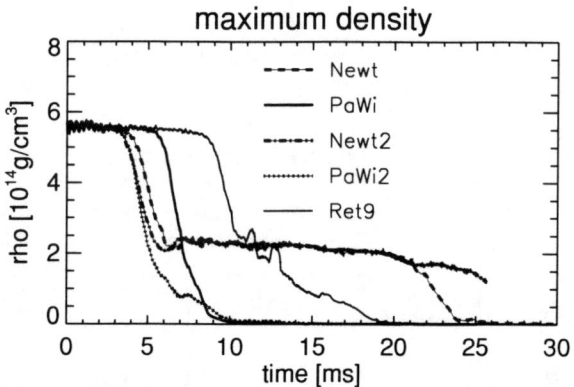

**FIGURE 2.** Value of maximum density within the computational volume as function of time for several models.

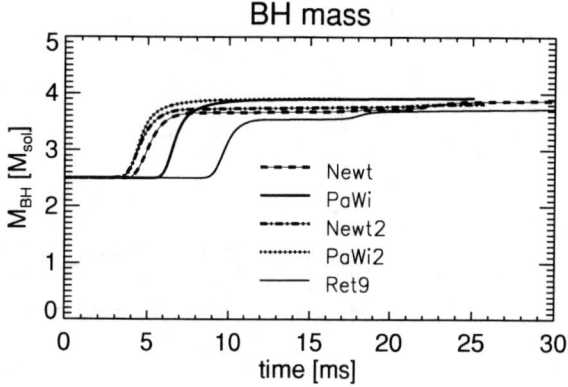

**FIGURE 3.** Black hole mass as function of time for several models.

trons minus that of positrons, and $n_b$ the baryon number density) were taken from a one-dimensional model of a cold, deleptonized (neutrino-less) neutron star in hydrostatic equilibrium and were the same as in [16]. The initial central temperature was set to a value of typically a few MeV, the temperature profile decreasing towards the surface such that the thermal energy was much smaller than the degeneracy energy of the matter.

We prescribed the initial orbital velocities of the coalescing neutron star and black hole according to the motions of point masses, as computed from the quadrupole formula. The tangential components of the velocities of the centres of the compact objects correspond to Kepler velocities on circular orbits, while the radial velocity components reflect the emission of gravitational waves leading to the inspiral of the orbiting bodies. No additional spin was added, so the neutron star can be considered as "irrotational".

Table 1 gives a list of ongoing merger simulations. "PaWi" refers to a Paczyński-Wiita type potential [12] while "Kerr" indicates the use of the Artemova type potential [1]. The Table also lists the initial centre-to-centre distance between black hole and neutron star. Since we do not relax the neutron stars at the start the simulations the initial distances have to be chosen sufficiently large that most of the induced oscillations have been damped away by numerical viscosity before the dynamical phase of the merging is reached.

## RESULTS

Figs. 1 and 2 show distinct episodic phases of the merging process. First, the orbits shrink slowly while the neutron star retains its mass and density structure (constant maximum density except for small-amplitude oscillations). A rapid mass transfer from the neutron star to the black hole follows and is shut off again within 1–2 ms (compare also Figs. 3, 4 and 5); accretion rates of nearly $10^3$ $M_\odot$/s are reached during this time. The neutron star is then flung into an eccentric orbit which will at first take it away from the black hole but eventually they meet again (similar behaviour has been reported by [11]). Note that the neutron star in models Newt and Newt2 orbits several times before the second and final plunge into the black hole, and during that time its maximum density is around $2 \cdot 10^{14}$ g/cm$^3$, and its mass is about 0.25 $M_\odot$. Note also the rough similarities between the results of the low-resolution model Newt and the high-resolution model Newt2. On the contrary, models PaWi and PaWi2 do not seem to agree that well, indicating the need for even higher numerical resolution.

Although most of the gaseous matter is accreted by the black hole, some becomes unbound and escapes. Fig. 6 displays logarithmically the amount of unbound mass as function of time, typically several percent of a solar mass by the end of the simulation. Together with mass the black hole accretes angular momentum, too. This is

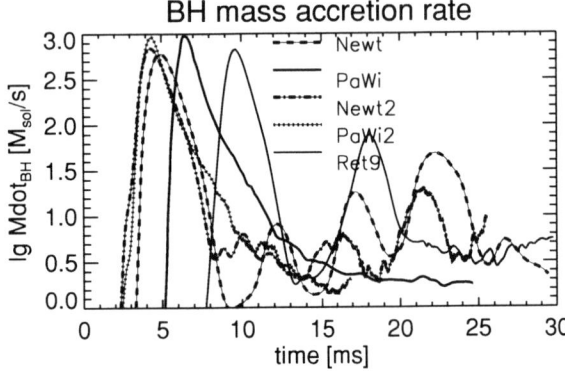

**FIGURE 4.** Black hole mass accretion rate as function of time for several models.

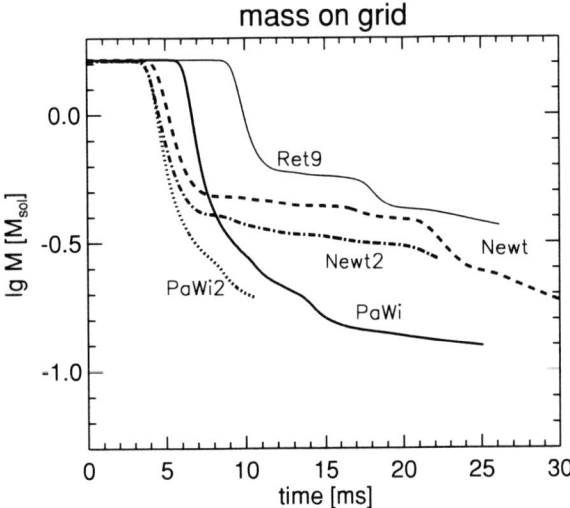

**FIGURE 5.** Mass of gas within the computational volume as function of time for several models.

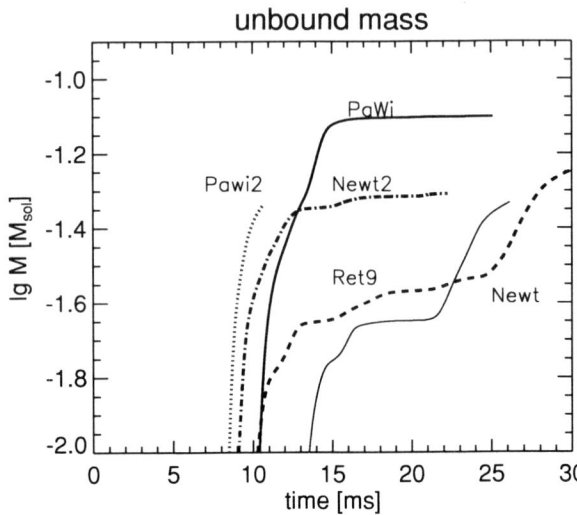

**FIGURE 6.** Mass of gas that is unbound when it leaves the computational volume for several models.

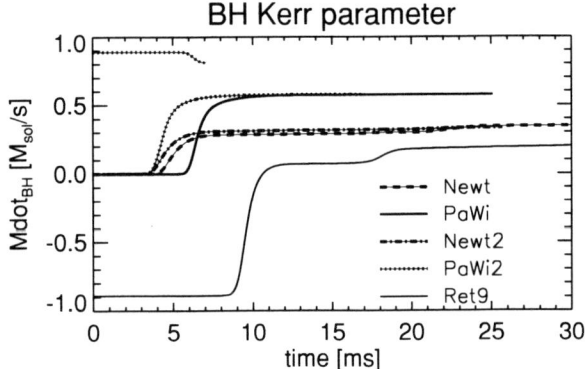

**FIGURE 7.** Black hole rotational (Kerr) parameter as function of time for several models. The topmost line shows model Ker9.

reflected in the time dependence of the rotational (Kerr) parameter shown in Fig. 7. In all models except model Ker9 the value of their rotational parameter increases. For these models the accreted matter spins up the black hole in the direction of the rotation of the disk (for the black hole with the retrograde spin in model Ret9 this means a spin-down). Only the black hole in model Ker9 is initially spinning so rapidly, that the accreted matter has less specific angular momentum. Note that this regime is highly relativistic and general relativistic effects are expected to be strongest here.

The neutrino luminosities (Fig. 8) are roughly similar to what has been reported previously [8,15]: they rise up to several $10^{53}$ erg/s. Given a similar geometry (torus around black hole) we anticipate similar neutrino-antineutrino annihilation luminosities and finally available clean energies for the fireball.

The gravitational wave luminosity is shown as function of time plotted logarithmically in Fig. 9 and linearly in Fig. 10. These results are to be accepted with caution, since our code is Newtonian. While the maximum luminosity is fairly similar, around $3-7 \cdot 10^{55}$ erg/s, the subsequent evolution depends sensitively on whether a remnant neutron star remains in orbit or not. If so, during the period before the remnant gets tidally disrupted, a residual gravitational wave luminosity is present but at a reduced level of approximately two orders of magnitude below the peak value. The time-integrated energy carried away by gravitational waves is shown in Fig. 11.

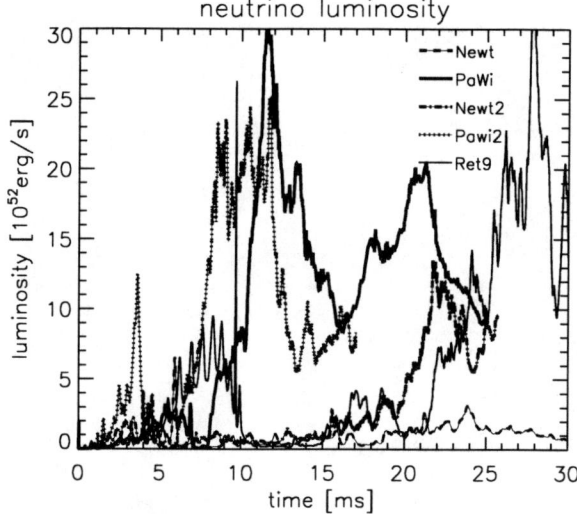

**FIGURE 8.** Total neutrino luminosity as function of time for several models.

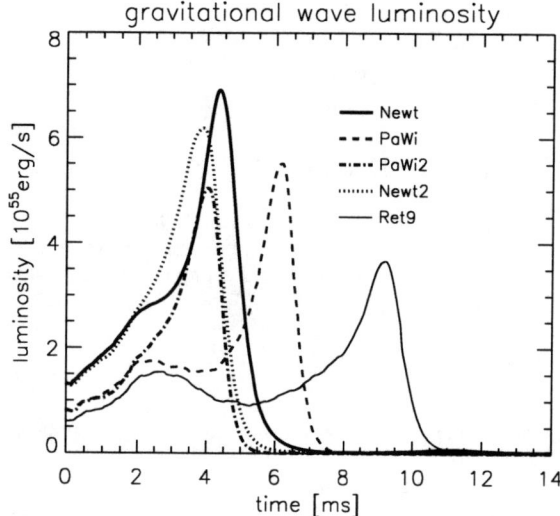

**FIGURE 10.** Gravitational wave luminosity as function of time for several models.

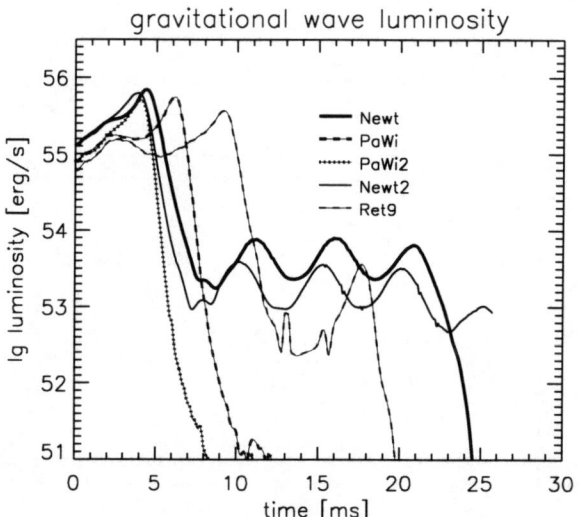

**FIGURE 9.** Gravitational wave luminosity as function of time for several models.

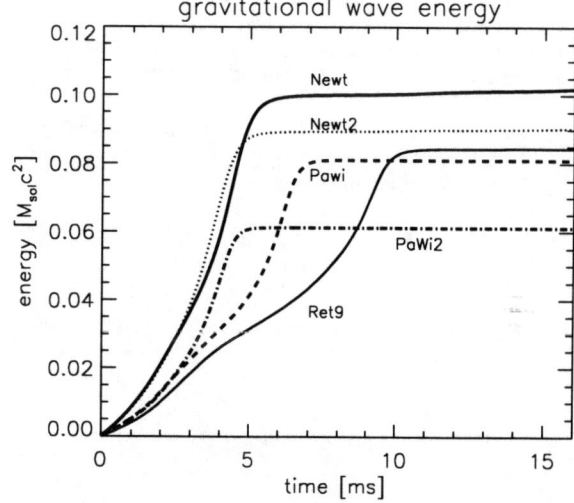

**FIGURE 11.** Integrated emitted gravitational wave energy as function of time for several models.

## CONCLUSIONS

We show the first results of ongoing simulations investigating the mergers of neutron star-black hole binaries with the gravitational potential of the black hole being generalized to account for the effects of the existence of an innermost stable circular orbit. The emission of neutrinos is included to ascertain the viability of this model as a possible central engine of gamma-ray bursts. In some cases a low-mass neutron star remnant seems to be left over after a first episode of mass transfer. This remnant eventually gets tidally disrupted, too. Some of the quantities we display are the maximum density, the mass and mass accretion rate of the black hole, and the neutrino and gravitational wave luminosities.

## ACKNOWLEDGMENTS

MR gratefully acknowledges support by a PPARC Advanced Fellowship and HTJ by the Sonderforschungsbereich 375 on "Astroparticle Physics" of the Deutsche Forschungsgemeinschaft. Most of the computations reported here were performed using the UK Astrophysical

Fluids Facility (UKAFF).

# REFERENCES

1. Artemova I.V., Björnsson G., Novikov I.D., 1996, ApJ 461, 565.
2. Berger M.J. 1987, J. Numer. Anal., 24, 967.
3. Berger M.J., & Colella P. 1989, J. Comput. Phys., 82, 64.
4. Berger M.J., & Oliger J. 1984, J. Comput. Phys., 53, 484.
5. Blanchet L., Damour T., & Schäfer G. 1990, MNRAS, 242, 289.
6. Briggs M.E., 2002, these proceedings.
7. Colella P., & Woodward P.R. 1984, J. Comput. Phys., 54, 174.
8. Janka H.-T., Eberl T., Ruffert M., & Fryer C.L. 1999, ApJ, 527, L39.
9. Eberl T. 1998, Diploma Thesis, Technical University Munich.
10. Lattimer J.M., & Swesty, F.D. 1991, Nucl. Phys., A535, 331
11. Lee W.H. 2000, MNRAS 318, 606.
12. Paczyński B., & Wiita P.J. 1980, A&A, 88, 23.
13. Ruffert M. 1992, A&A, 265, 82.
14. Ruffert M., & Janka H.-T. 1998, A&A, 338, 535.
15. Ruffert M., & Janka H.-T. 1999, A&A, 344, 573.
16. Ruffert M., Janka H.-T., & Schäfer G. 1996, A&A, 311, 532.
17. Ruffert M., Janka H.-T., Takahashi K., & Schäfer G. 1997, A&A, 319, 122.
18. Woosley S., 2002, these proceedings.

# Comparing GRB Progenitors

## C. L. Fryer

*T-6, MS B288, Los Alamos National Labs, Los Alamos, NM 87545*

**Abstract.** Here I review 6 of the best-known black-hole accretion disk (BHAD) gamma-ray burst (GRB) progenitors, highlighting the differences between each progenitor.

## GAMMA-RAY BURST PROGENITORS

Most of the proposed GRB mechanisms invoke an energy source fueled by accretion through a disk onto a black hole. The accretion energy is converted into a fireball jet through either neutrino annihilation or some magnetic field mechanism. Although most models boil down to this black-hole accretion disk (BHAD) mechanism, a number of BHAD GRB progenitors have been proposed [1,2].

Table 1 lists 6 of the best-known progenitors of BHAD GRBs and various properties that are associated with them. None of these properties are set in stone, and the wily theorist will, no doubt, be able to adapt their favorite model to keep up with observations. Nevertheless, understanding the basic features of each progenitor can tell us what obstacles each theorist must overcome with his/her favorite progenitor as the data increases.

## Disk Structure

One of the major differences between the BHAD GRB progenitors is the structure of the disks. For a neutrino driven engine to work, the density of the accretion disk (and hence accretion rate of that disk) must be extremely high. Although it is not clear that neutrino annihilation can work for any of the progenitors, we can rule out this engine for both the WD/BH and supranova progenitors. But the magnetically driven engine, in part because it has not been calculated in detail, can potentially work on all BHAD progenitors.

The duration of these bursts can also be estimated from their disk structure. Because of the low angular momentum in the disks produced by DNS and BH/NS mergers, their disks are very compact and accrete very rapidly ($< 1$ s [3]). The supranova model has even less angular momentum and its disks will be very short-lived. Although with proper manipulation, one can extend the duration of a gamma-ray burst beyond that of the actual disk accretion phase, most scientists believe that these 3 models will only produce short duration bursts. On the other hand, it is hard to make the accretion phase of the disks produced in Collapsars, WD/BH mergers, and He-mergers short enough to explain short-duration GRBs (again, models to make short-duration bursts already exist.)

Beaming also depends upon the disk structure and the environment immediately surrounding the disk. Collapsars and He-mergers have stellar envelopes which help to collimate the outburst [4]. The stellar envelope is ultimately blown off the black hole, producing a supernova like explosion. Binary merger GRBs, as well as the supranova model, do not have this collimating envelope and, unless magnetic fields collimate the outflow, these bursts would have weak beaming.

## Environment

Although all of the progenitors can trace their origins back to massive star formation regions (SFR), some of the progenitors (DNS, BH/NS and WD/BH) have high enough velocities and long enough delays after they receive those velocities that they will certainly leave the SFR within which they formed. A fraction of DNS and BH/NS GRB progenitors have high systemic velocities and long merger times and would not only travel beyond their SFR, but beyond the confines of their host galaxy before merging to produce GRBs (17-39% in a Milky Way massed galaxy [1]). A recent paper has instead argued that DNS GRBs would occur within the host galaxy unless the galaxy were very small (some GRB host galaxies have masses much less than that of the Milky Way [5]).

The descrepancy between the results in [1,5] is caused by a new mechanism to form DNS systems [6]. They form many tight C/O star binaries which then produce

**TABLE 1.** Gamma-Ray Burst Progenitors

| BHAD Prog.* | DNS | BH/NS | WD/BH | Collapsar | He-Merger | Supranova |
|---|---|---|---|---|---|---|
| Neutrino Engine | ? | ? | no | ? | ? | no |
| Magnetic Engine | ? | ? | ? | ? | ? | ? |
| Duration | Short | Short | Long | Long | Long | Long |
| SN Assoc. | no | no | no | just after | just after[†] | before |
| Beaming | Weak? | Weak? | None | Strong | Strong | ? |
| Environment | ISM, none | ISM, none | ISM | Winds | Winds, ISM? | ISM |
| Distribution ($> 1$ Mpc) | 17-39% | 17-39% | in Gal. | in SFR | in SFR | in SFR |
| $<z>/<z_{SFR}>$ | 0.5-0.8 | 0.5-0.8 | $<1$ | $>1$ | $\gtrsim 1$ | $\sim 1$ |
| Iron Lines | no | no | no | ? | ? | ? |
| Binary Prog. | yes | yes | yes | mostly yes | yes | no |
| GWs | Strong | Strong | Weak | Weak | Weak | Weak |

* DNS, BH/NS, and WD/BH refer to GRBs produced by the merger of binaries consisting of, respectively, two neutron stars, a neutron star and a black hole, and a white dwarf and a black hole (or neutron star). Collapsar refers to GRBs produced through the collapse of massive stars, He-Merger refers to GRBs produced after a helium star mergers with either a black hole or neutron star, and supranovae are objects produced by the collapse of a massive, rapidly-spinning, neutron star. Most of these are reviewed in [1,2]

[†] Some bursts will also have a supernova days to months before.

tight DNS binaries which merge very rapidly. Because these binaries merge rapidly, they have no time to escape the host. What this scenario requires is that the system first form helium star binaries which go through a common envelope phase to remove the helium layer, leaving behind two tight C/O stars. In [1], most of the helium star binary systems which went through a common envelope phase actually merged to form a single, rapidly rotating helium star. Discussions with N. Langer and A. Heger have convinced us that the evolutionary path used by [1] is correct for most of these systems and very few, if any, of these short-period C/O binaries actually form (so Table 1 uses these results). However, the study of interacting binary systems is still in its infancy, and my current intuition could be wrong.

The environment surrounding BHAD GRBs also varies with progenitor. Some DNS and BH/NS GRBs will occur outside of their host galaxy and the medium in which the fireball expands will be extremely diffuse. Collapsars, Supranovae, and He-mergers will primariy occur in dense star forming regions. In some cases, particularly with collapsars whose progenitors are very massive stars ($>40M_\odot$), the fireball will blast through a windswept medium.

Supranovae may occur within a year of the supernova explosion that produced the rapidly spinning neutron star whose collapse powers the GRB. For this GRB progenitor, the fireball will first travel through the supernova shocked ejecta. This feature allows a natural explanation for the observed iron lines. In particular, although iron emission lines can be explained by a number of models, it is difficult to explain the observations of iron absorption lines with both the He-merger and the Collapsar model [7]. Although this may lead one to favor the Supranova model, remember that iron lines have only been seen in long-duration gamma-ray bursts, and it requires a bit of magic to argue that supranovae can be long-duration outbursts. It might be easier to believe that either the iron absorption line was noise or some sort of fluke or that we have not yet come up with the correct explanation of such lines.

## Gravitational Waves and Redshift Distribution

As a number of gravitational wave (GW) detectors come on line, a new possible emission source may be detected in GRBs. DNS and BH/NS mergers are the strongest known astrophysical sources for detectors like LIGO and they are being studied in great detail by the general relativity community. GRB progenitors involving stellar collapse exhibit much weaker GW signals [8] and, unless we are extremely lucky with a nearby burst, the upgraded LIGO will not detect GRBs from these progenitors. One could argue that if the stellar core fragments prior to forming a black hole or if the disk fragments around the black hole, a strong signal can be produced, but such fragmentation is highly unlikely. Hence, only the DNS and BH/NS mergers will be strong sources for the currently proposed gravitational wave detectors.

What about redshift distribution of GRB progenitors and the possibility of using GRBs to probe the cosmological star formation history? We can not assume that the star formation rate is simply proportional to the GRB rate. Recall that some DNS and BH/NS GRBs have long merger times which would mean that at high redshifts, there would be a noticeable shift between the star for-

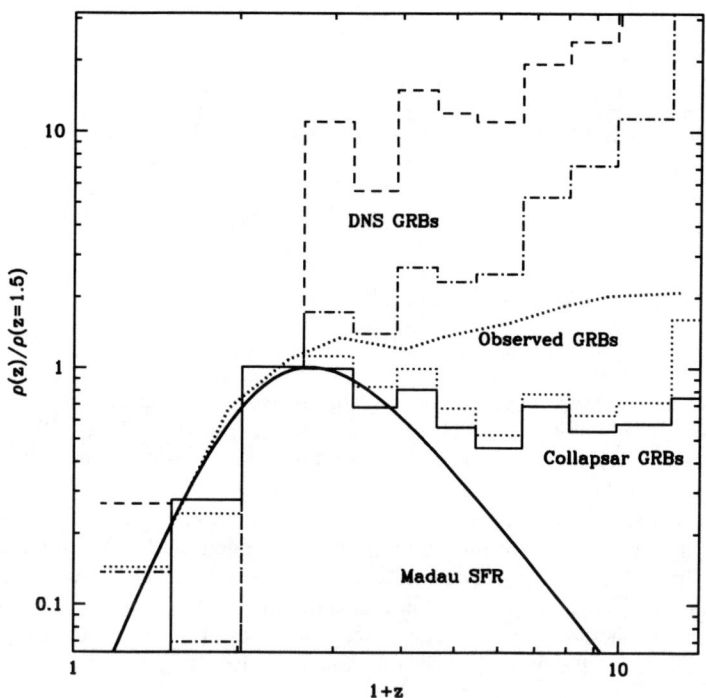

**FIGURE 1.** Estimates of the star formation rate based on various progenitor models and using a best-fit GRB distribution (thick dotted line). The solid and dotted lines correspond to the collapsar model, with and without the effects of a flattening IMF respectively. This provides a lower-limit for the slope of the SFR (with the best-fit data). The dashed and dot-dashed lines correspond, respectively, to the same models (with and without IMF flattening) for the compact merger models. Results are from [9].

mation time and the GRB outburst. Hence, if DNS and BH/NS binaries produced all GRBs, the star formation rate at high redshifts would be higher than what we would simply get by assuming the star formation rate were proportional to the GRB rate (Fig. 1). At the other extreme, since Collapsars are produced only by very massive stars with low mass-loss, they are likely to be more common (per solar mass formed into stars) at high redshift than at low redshift. Hence, the star formation rate at high redshifts would be lower than what we expect by assuming the star formation rate were proportional to the GRB rate (Fig. 1). Even if we don't know the exact GRB progenitor, if we have a large enough redshift sample of GRBs, we can already constrain the data [9]. But if we knew the true GRB progenitor, GRBs could prove the most powerful probe of the star formation history of the universe.

## REFERENCES

1. Fryer, C. L., Woosley, S. E., & Hartmann, D. H. 1999, ApJ, 526, 152.
2. Vietri, M., & Stella, L. 1998, ApJ, 492, L59
3. Ruffert, M., Janka, H.-T., Takahashi, K., Schaefer, G. 1997, A&A, 319, 122.
4. Aloy, M. A., Müller, E., Ibanez, J. M., Marti, J. M., & MacFadyen, A. 2000, ApJ, 531, L119.
5. Belczynski, K., Bulik, T., & Rudak, B. 2001, astro-ph/0112122.
6. Belczynski, K., & Kalogera, V. 2001, ApJ, 550, L183.
7. Böttcher, M., Fryer, C. L., & Dermer, C. D. 2001, astro-ph/0110625.
8. Fryer, C. L., Holz, D. E., Hughes, S. A. 2001, astro-ph/0106113.
9. Lloyd-Ronning, N.M., Fryer, C.L., Ramirez-Ruiz, E. 2001, astro-ph/0108200.

# Collapsar Disks and Winds

A. I. MacFadyen

*Caltech, MC 130-33, Pasadena, CA 91125, USA*

**Abstract.**
Winds blown from collapsar accretion disks may produce observable stellar explosions independent of any GRB-(and afterglow)-producing jets which may be simultaneously produced. The production of winds is controlled by the accretion disk physics, in particular, the nature of disk cooling via neutrino emission and photo-disintegration of heavy nuclei. These temperature-dependent processes depend on the stellar angular momentum via the depth of the gravitational potential at the Kepler radius where the disk forms. Wind-driven stellar explosions which do not make a GRB (or only a faint one) may occur and constitute a new class of supernova explosion. SN1998bw and 1997ef may be examples. A key feature of collapsar winds is that they are capable of producing the radioactive $^{56}$Ni necessary to power a supernova light curve. It is possible to make a GRB in a star without significant production of $^{56}$Ni. Such a star would not make an observable supernova and no such component would be expected in the light curve of the optical afterglow.

## INTRODUCTION

Simulations of the collapse of massive rotating stars indicate the viability of rotating stellar collapse (collapsars) as the central engine for the long duration gamma ray bursts[1][2].

Collapsars form dense accretion disks ($\rho \gtrsim 10^9$ gm cm$^{-3}$) which are extremely optically thick to photons ($\tau_\gamma \sim 10^{19}$). As the stellar gas spirals through the disk, photons are trapped and accrete with the gas. This is in distinction from "thin" accretion disks in which photons are assumed to escape to infinity carrying away the locally dissipated energy. Since photons are trapped, viscous dissipation of orbital energy increases the disk entropy, pressure gradients are important for the force balance and the disk is "thick." Such non-radiating accretion flows are capable of ejecting gas away from the black hole [3, e.g.]. Accretion in these disks is inefficient with significant fractions of the gas supplied at large radii being ejected from the system.

An important feature of collapsar disks, is the realization at sufficiently high accretion rates ($\dot{M} \sim 0.1 M_\odot s^{-1}$) of temperatures ($T \sim 10^{10}$ K) and densities ($\rho \sim 10^9$ gm cm$^{-3}$) at which the loss of thermal energy to neutrino emission and photodisintegration of heavy nuclei allows for accretion with a range of efficiency.

## WINDS

MacFadyen & Woosley 1999 showed that collapsar disks eject comparable amounts of material in a wind as is accreted by the central black hole. The fraction of accreted gas depends on the efficiency of neutrino cooling at removing entropy from the accreting gas. The remainder is ejected from the black hole as an outflowing wind. Recent semi-analytic work [4] has mapped the parameter space of inefficient neutrino-cooled accretion in agreement with detailed calculations of [1] for limited parameters.

Of particular interest in the case of collapsars is the chemical composition of the wind. Collapsar disks are hot enough to completely photodisintegrate heavy nuclei to free nucleons (neutrons and protons). Recent simulations [5] and [1] show expulsion of free nucleon gas in the wind. This is important for two reasons: 1) energetics: free nucleons combining to iron group nuclei (e.g. Nickel-56) release 8 MeV/nucleon or $1.5 \times 10^{52}$ erg per solar mass of recombined material. 2) observability: this ejected material provides a long term energy supply to the explosion (through radioactive decay of $^{56}$Ni) enabling the gas to shine on time scales of months.

## LIGHT CURVE

Models of the light curve of the energetic and peculiar Type Ibc supernova SN1998bw require large quantities of $^{56}$Ni (M($^{56}$Ni) $\sim 0.5 M_\odot$) [6][7]. Conventional models require large explosion energies to produce sufficient nickel and fit the light curve. In addition abnormally high expansion velocities were inferred from the unusual spectrum indicating a large explosion energy ($\sim 10^{52}$ erg). Several groups have also interpreted deviations from power law decay of GRB optical transients as supernovae light curve components and have matched them with appropriately shifted 1998bw light curves [8, e.g.].

Since supernovae are invoked to interpret these observations it is important to note that a stellar explosion (e.g., jets piercing a star) is not necessarily a supernova. Supernovae, as an observable phenomenon, require a persistent source of energy input to power a light curve for long times (weeks to months). It is necessary to make $^{56}$Ni in the explosion so that radioactive decay (to Cobalt to Iron) injects energy into the gas so that it can shine. Lacking a persistent source of energy input, a stellar explosion would be unobservable via electromagnetic radiation. Explosion energy released in the optically thick star would simply be converted to expansion kinetic energy with little or no light emitted.

In conventional core collapse supernovae some nickel is thought to be produced via explosive nuclear burning behind the explosion shock. However, current models for these "delayed" supernova explosions have trouble producing the $10^{51}$ erg for a normal supernova (in fact, some current models fail to get any explosion at all!) and are unlikely to be capable of producing the higher energies required for 1998bw.

## ANGULAR MOMENTUM

As we have seen, neutrino cooling and photodisintegration of heavy nuclei are crucial for allowing gas to accrete efficiently. The neutrino cooling depends sensitively on temperature (e.g., $Q_\nu^- \propto T^6$ for neutrino losses due to pair capture on free nucleons) and therefore on the radius where the disk forms. This radius is, in turn, dependent on the angular momentum of the accreting gas with the disk first forming at the Kepler radius $R_{kep} \equiv j^2/GM = 2.5 \times 10^7 j_{17}^2 M_3^{-1}$ cm, where $j_{17}$ is the specific angular momentum of the accreting gas in units of $10^{17}$ cm$^2$ s$^{-1}$ and $M_3$ is the mass of the central black hole in units of three solar masses. The virial temperature for gas falling to its Kepler radius $T_{vir} = GMm_p/3k_B R_{kep} = 3.3 \times 10^{10} M_3^2 j_{17}^{-2}$ K, where $m_p$ is the proton mass and $k_B$ is the Boltzmann constant. In terms of gravitational radii, $R_G \equiv GM/c^2$, this temperature is $T_{vir} = m_p c^2 / 3k_B r = 1.8 \times 10^{12} r^{-1}$ K (assuming a Newtonian potential), where $r \equiv R/R_G$. We see that gas with $j_{17} \approx 1$ is heated to above $10^{10}$ K so that it is fully photodisintegrated to free neutrons and protons from it's original composition of Silicon, Oxygen and Helium. This means that capture of electron-positron pairs onto the free neutrons and protons serves as an efficient neutrino emission process which cools the gas and helps it to accrete efficiently. Gas with $j_{17} \gtrsim 2.6$, however, heats to less then $5 \times 10^9$ K. At these lower temperatures the heavy nuclei fail to photodisintegrate and pair capture neutrino cooling is suppressed. This gas is therefor poorly cooled and subject to being driven from the disk.

It is worth noting that gas with $1 \lesssim j_{17} \lesssim 2.6$ is partially photodisintegrated. Photodisintegration acts as a loss for thermal energy for the gas and thus is effectively a cooling process, robbing about $10^{19}$ erg of thermal energy from every gram of photodisintegrated nuclei. This process helps the gas to accrete and provides free nucleons which enhance the neutrino cooling.

The above discussion assumed $M_3 = 1$ though the scaling with $M_3$ is apparent.

## THE AFTERBURNER

Interestingly, energy lost to photodisintegration becomes available again if the gas is ejected from the disk and begins to reassemble.

Effectively, gravitational energy is temporarily stored in the freeing of nucleons from the heavy nuclei. These nucleons are volatile in the sense that the have a huge nuclear energy source if they manage to escape the energetic photons trapped in the (optically thick) accretion disk. Accretion physics may provide the nucleons with opportunity to escape the disk's nuclei-disintegrating photon bath. Once free they can quickly recover nuclear binding energy by reassembling into iron group elements. This process can be explosive since the nuclei may recombine in seconds compared to millions of years it took them to assemble (burn) the first time around during the slower pre-explosion nuclear burning stages. In fact the reassembly, plus the kinetic luminosity of the disk wind, may power extremely energetic explosions. SN1998bw may be an example.

## GRB WITH SUPERNOVA

The collapsar model relies on rapid accretion into the central black hole to power relativistic jets which pierce the star and make a GRB and afterglow via internal and external shocks. It is notable that for an interesting range of angular momentum the accretion of the star simultaneously feeds the black hole rapidly and powers a wind [1].

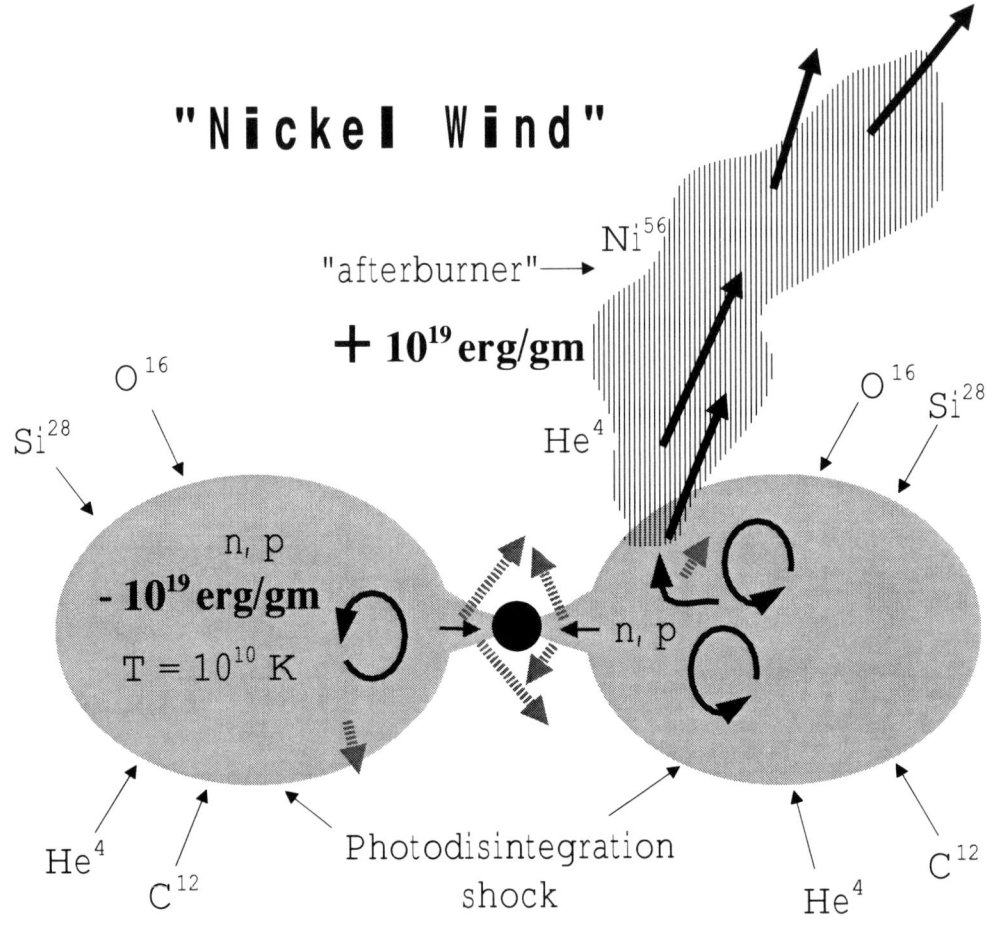

**FIGURE 1.** Cross section of a collapsar accretion disk feeding a stellar mass black hole (center). The disk is embedded in a collapsing star that is falling onto the disk at rates above $0.1 M_\odot$ s$^{-1}$. A wind (the striped region in the upper right) is blown from the collapsar disk at speeds of up to $\sim 40,000$ km s$^{-1}$. The wind is composed of free neutrons and protons which can recombine to iron group elements injecting $1.5 \times 10^{51}$ erg per 0.1 $M_\odot$ of reassembled nucleons. $^{56}$Ni in the wind can power a long term "supernova" light curve via radioactive decay of nickel and cobalt. The black solid arrows indicate the velocity of the gas flow while the thick dashed lines represent neutrino emission. The wind is shown only in the upper right quadrant for clarity but is in reality present in all four quadrants.

There are three interesting regimes determined by the angular momentum present in the collapsing star:

The following values of angular momentum correspond to important transition radii in the accretion flow:

$j_{isco}$ angular momentum of the innermost stable circular orbit. This is the minimum angular momentum needed to form a disk around a black hole.

$j_\nu$ angular momentum of gas that falls deep enough in the gravitational potential to photodisintegrate the heavy nuclei to free nucleons activating pair capture neutrino emission as an efficient coolant.

$j_\gamma$ angular momentum of gas that falls deep enough to cool partially. Some gas accretes and some is expelled in a wind. The relative amount depends on the exact value of $j$

1. $j_{isco} < j < j_\nu$ - efficient neutrino cooling allows rapid accretion into black hole with plenty of power potentially going into jets with little or no outflows expected. This kind of star would not be expected to produce a bright supernova since little or no $^{56}$Ni is expected to be present in the exploding star. A possible caveat is that there is some nickel production via explosion burning in the lateral jet shock but the temperature is low in this region and not much mass is involved.

2. $j_\nu < j < j_\gamma$ - some gas accretes and some is ejected in a wind rates can be comparable depending on j. This can make both a GRB and a "supernova"

3. $j > j_\gamma$ Gas doesn't cool efficiently so doesn't feed the hole rapidly. Not good for making an accretion powered GRB. Outflows may results with some recombination nickel possible if some gas is heated above $5 \times 10^9$ K by a combination of virialization and viscous dissipation. Of interest here is explosive burning of centrifugally supported oxygen.

A less interesting regime is $j < j_{lso}$ for which the gas falls directly to the innermost stable circular orbit without forming an accretion disk.

Note that electromagnetic extraction of black hole spin energy is a possible source of jet energy even for "slowly" accreting black holes. Convective motions may even be favorable for building up large magnetic fields needed to extract the hole spin energy.

## VISCOSITY

The above scenario assumes significant viscosity in the disk gas corresponding to a Shakura-Sunyaev alpha viscosity parameter $\gtrsim 0.1$. The temperature and density of the disk wind and hence the nucleosynthesis depend on the disk viscosity. Observations of of the supernova powered by a collapsar wind may help to constrain the viscosity of the collapsar accretion disk.

## WORK IN PROGRESS

Detailed hydrodynamical simulations of the disks and winds described here using the FLASH code are currently underway.

## REFERENCES

1. MacFadyen, A. I., and Woosley, S. E., ApJ, **524**, 262–289 (1999).
2. MacFadyen, A. I., Woosley, S. E., and Heger, A., ApJ, **550**, 410–425 (2001).
3. Hawley, J. F., Balbus, S. A., and Stone, J. M., ApJ, **554**, L49–L52 (2001).
4. Narayan, R., Piran, T., and Kumar, P., ApJ, **557**, 949–957 (2001).
5. MacFadyen, A. I., and Woosley, S. E., ApJ, *in preparation* (2002).
6. Iwamoto, K., Mazzali, P. A., Nomoto, K., Umeda, H., Nakamura, T., Patat, F., Danziger, I. J., Young, T. R., Suzuki, T., Shigeyama, T., Augusteijn, T., Doublier, V., Gonzalez, J.-F., Boehnhardt, H., Brewer, J., Hainaut, O. R., Lidman, C., Leibundgut, B., Cappellaro, E., Turatto, M., Galama, T. J., Vreeswijk, P. M., Kouveliotou, C., van Paradijs, J., Pian, E., Palazzi, E., and Frontera, F., Nature, **395**, 672–674 (1998).
7. Woosley, S. E., Eastman, R. G., and Schmidt, B. P., ApJ, **516**, 788–796 (1999).
8. Bloom, J. S., Kulkarni, S. R., Djorgovski, S. G., Eichelberger, A. C., Cote, P., Blakeslee, J. P., Odewahn, S. C., Harrison, F. A., Frail, D. A., Filippenko, A. V., Leonard, D. C., Riess, A. G., Spinrad, H., Stern, D., Bunker, A., Dey, A., Grossan, B., Perlmutter, S., Knop, R. A., Hook, I. M., and Feroci, M., Nature, **401**, 453–456 (1999).

# Erupting Fireballs, Nozzles and Precursors

Enrico Ramirez-Ruiz*, Andrew I. MacFadyen† and Davide Lazzati*

*Institute of Astronomy, Madingley Road, Cambridge CB3 0HA, England
†Department of Astronomy and Astrophysics, University of California, Santa Cruz, CA 95064, USA

**Abstract.** Recent observations suggest that long-duration γ-ray bursts and their afterglows are produced by highly relativistic jets emitted in core-collapse explosions. As the jet makes its way out of the stellar mantle, a bow shock runs ahead and a strong thermal precursor is produced as the shock breaks out. Such erupting fireballs produce a very bright γ-ray precursor as they interact with the thermal break-out emission. The detection of such precursors would offer the possibility of diagnosing not only the radius of the stellar progenitor and the initial Lorentz factor of the collimated fireball, but also the density of the external environment.

## INTRODUCTION

A generic scheme for a cosmological γ-ray burst (GRB) model has emerged in the last few years (see [1] for a review). According to this scheme the observed γ-rays are emitted when a relativistic energy flow is converted to radiation. Possible forms of the energy flow are kinetic energy of relativistic particles or electromagnetic Poynting flux [2]. This energy must be converted to radiation in an optically thin region, as the observed bursts are not thermal. The ultimate energy source of this relativistic outflow is the gravitational energy release associated with temporary mass accretion onto a black hole, which may result from the collapse of a massive rotating star [3] In this scenario, the γ-rays are thought to be produced in shocks occurring after the relativistic jet has broken free from the stellar envelope, whose density is reduced along the rotation axis due to an early phase of accretion [4, 5]. A strong terminal wave breaking out of the envelope is expected to produce a transient thermal emission that should appear as a precursor signal prior to the observed GRB [6, 7, 8].

In this paper we explore the interaction of such erupting fireballs with the shock break-out emission. We show that a substantial fraction of the fireball energy can be converted into a collimated, bright γ-ray precursor via the Compton drag process [9, 10]. We suggest that detailed observation of this prompt emission provides a potential tool for diagnosing the radius of the stellar progenitor and the initial Lorentz factor of the fireball (see [11]). It also provides a means for probing the external environment surrounding the stellar progenitor and the radial distance at which the observed γ-rays are produced. We assume $H_0 = 65$ kms$^{-1}$Mpc$^{-1}$, $\Omega_m = 0.3$, $\Omega_\Lambda = 0.7$.

## ERUPTING FIREBALLS

Numerical simulations of rotating helium stars in which iron core collapse does not produce a successful traditional neutrino-powered explosion [4, 12, 5] have identified a range of stellar progenitors and initial conditions in which a jet would not be able to break free from its stellar cocoon. This is expected in a large fraction of cases where the stellar envelope is too thick, for example in stars with small radiative mass-loss. A highly relativistic jet is likely to escape if the star loses its hydrogen envelope before collapsing and if the jet produced by the accretion maintains its energy for longer than it takes the jet to reach the surface of the star (MW99). Otherwise, acceleration of the explosion debris to a sufficiently high Lorentz factor ($> 10^2$ [13]) is unlikely.

A collimated fireball propagating inside a funnel cavity would be stopped by the envelope when its momentum flux is insufficient to accelerate the impacted stellar mantle to a speed comparable to its own. The jet would then be stalled at a distance $\approx \sqrt{L_j/(\Omega \rho_{\rm env} v_j^3)}$ [14], where $L_j$ is the total luminosity of the jet, $\Omega$ its collimation solid angle and $\rho_{\rm env}$ the density along the rotation axis of the star. At this distance, the relativistic jet is abruptly decelerated to $\Gamma \approx 1$. In order to break through the star, the energy injected into the envelope, $E_j = L_j \Delta t_j$, should be enough to unbind the impacted envelope material. Thus, inside a rotating massive star whose core has collapsed (leading to a central black hole), gas fall-back would drive for a time $\Delta t_j = r_*/\bar{v}_{\rm h} \approx 10^3 r_{*,13}$ s a slowly advancing standoff shock inside the envelope (see Figure 1; here we adopt the convention $Q = 10^x Q_x$, using cgs units), where the average speed of the jet head $\bar{v}_{\rm h}$ is about $c/2$. For small opening angles $\theta \ll 1$, the thickness

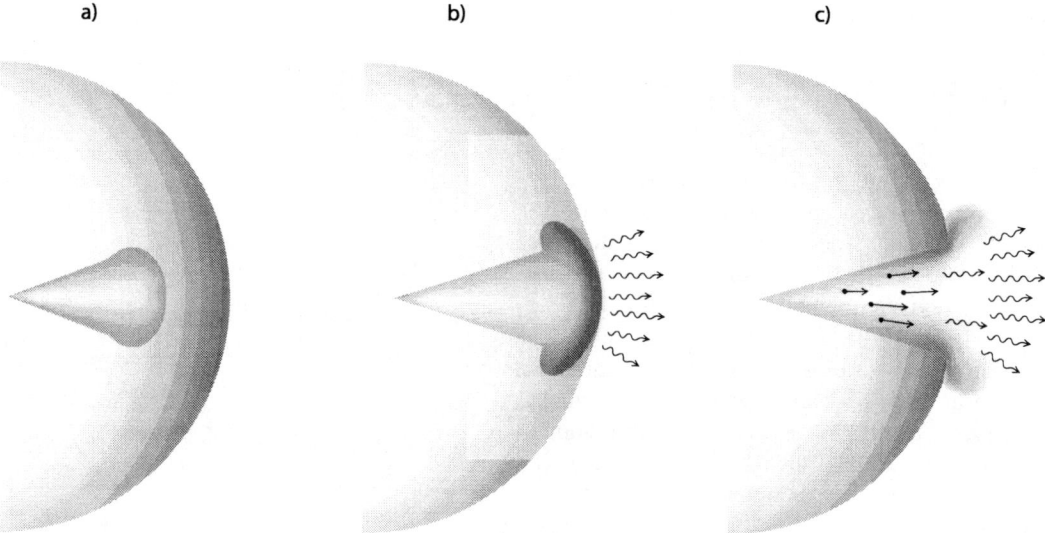

**FIGURE 1.** Diagram illustrating the propagation of the jet through the stellar mantle. Initially, the jet is unable to move the envelope material to a speed comparable to its own and thus is abruptly decelerated. As the jet propagates a bow shock runs ahead of it (a). The bow shock of the jet will both heat material and cause it to expand sideways. A strong thermal precursor is produced as the shock breaks through the stellar surface and exposes the hot shocked material (b). The fireball escapes the stellar envelope and interacts with very dense soft photon emission (c), converting the fireball bulk energy into radiation with a remarkably high efficiency.

of the shocked plasma shell is $\Delta \approx \sqrt{3}\theta r_*/8$, where $r_*$ is the stellar radius. The bow shock of the jet will heat the shocked plasma, converting a large fraction of its internal energy into radiation (Figure 1b). The high opacity of the plasma shell would cause the radiation to thermalize with a black body emission at a temperature [8]

$$T_{\rm p} \approx 0.3 \left[ \frac{E_{j,51}}{r_{*,13}^2 \Delta t_{j,2} \theta_{-1}^2} \right]^{1/4} \text{keV}. \quad (1)$$

This emission would not be appreciably beamed. Once the strong shock wave breaks the stellar surface, the temperature of the shocked plasma decreases roughly as a power-law $T(t) \approx T_{\rm p}(1 + 3ct/\Delta)^{-1/3}$. These analytical estimates are consistent with numerical results [5].

A relativistic collimated fireball may result if the engine operates for a sufficiently long time to allow the standoff shock (i.e. the jet head) to break out of the surface of the star. High Lorentz factors ($\Gamma > 100$) can be achieved if the jet continues to be powered after jet break out. The column of stellar material pushed ahead of the jet ($0.1 - 1 M_\odot$) escapes the star and expands sideways, leaving a decreasing amount of material ahead of the jet. This allows the hot, low-density jet plasma with moderately relativistic bulk velocity to accelerate to Lorentz factors determined by the energy loading per baryon in the jet. The jet remains physically beamed due to relativistic effects. The shocked plasma temperature at the time at which the fireball crosses the surface ($t \approx \Delta/c$) is $T_{\rm s} \approx T_{\rm p}/4^{1/3}$. Thus, the erupting fireball escapes the stellar envelope while interacting with very dense soft photon emission with typical energy $\Theta_{\rm s} = kT_{\rm s}/(m_e c^2)$ (Figure 1c). A fraction $\approx \min(1, \tau_j)$ of the photons are scattered by the inverse Compton effect to energies $\approx 2\Gamma_0^2 \Theta_{\rm s}$, where $\tau_j$ is the Thomson optical depth of the collimated fireball and we assume that a constant $\Gamma_0$ has been reached at the stellar surface. The radius of transparency of the fireball is

$$r_\tau \approx 2 \times 10^{14} \theta_{-1}^{-1} \Gamma_{0,2}^{-1/2} E_{j,51}^{1/2} \text{ cm}. \quad (2)$$

Due to relativistic aberration, the scattered photons propagate in a narrow $1/\Gamma$ beam. The net amount of energy $E_{\rm ic}$ extracted by the Compton drag process is $\approx \Gamma^2 \min(1, \tau_j) \Delta \pi (\theta r)^2 a T_{\rm s}^4$, where $\Delta \pi (\theta r)^2$ is the volume filled by the soft photon radiation. The Compton drag process can be very efficient in extracting energy from the collimated fireball

$$\xi = \frac{E_{\rm ic}}{E_j} \approx \theta_{-1} r_{*,13} \Delta t_{j,2}^{-1} \Gamma_{0,2}^2 \qquad r < r_\tau \quad (3)$$

even for jets erupting from stars with small radii $r_* < 10^{12}$ cm. Note that the Compton drag process limits the maximum speed of expansion so that $\xi < 1$. When the fireball becomes transparent, the amount of scattered photons is correspondingly reduced, and the process becomes less efficient. Each seed photon is boosted by $\approx \Gamma^2$

in frequency, yielding a spectrum peaking at

$$h\nu \approx 2\Gamma_0^2(3kT_s) \approx 11\Gamma_{0,2}^2 \frac{E_{j,51}^{1/4}}{r_{*,13}^{1/2}\Delta t_{j,2}^{1/4}\theta_{-1}^{1/2}}(1+z)^{-1} \text{MeV}. \quad (4)$$

The observed variability time scale is related to the typical size $\Delta$ of the shocked plasma region containing the thermal photon field, its curvature $r_*$, and the mean free path $\lambda$ of a photon inside the fireball. The observed timescale is hence given by $\max[\lambda/c, r_*/(c\Gamma_0^2), \Delta/(c\Gamma_0^2)]$, where $\lambda \sim 10^{10}(r_{*,13}\theta_{-1})^2\Gamma_{0,2}t_{j,2}E_{j,51}^{-1}$ cm. For large Lorentz factors, the duration is dominated by the mean free path term, while for slow fireballs the curvature is more important. Taking into account the conservation of the number of photons and the increase by a factor $2\Gamma_0^2$ of the photon energy, the peak luminosity of the boosted component is

$$L_{ic} \approx 2^{-5/3}\Gamma_0^2 \frac{\delta t_b}{\delta t_{ic}}\frac{4\pi}{\Omega}L_b \quad (5)$$

where subscripts ic and b refer to the boosted and break out quantities, respectively. When the boosted duration $\delta t_{ic}$ is dominated by curvature effects, we have $L_{ic} \approx 0.3\theta\Gamma_0^4 L_b$. The above observational signatures would be present even if $\Gamma_0$ is low, as expected for stars in which the jet either fails to maintain sufficient collimation or loses a significant amount of energy before breaking out of the star.

## PRECURSOR EMISSION

The prompt thermal signal emerging from shock break-out would precede by $\approx \Delta/c \approx 6\,\theta_{-1}r_{*,13}$ seconds the prompt emission produced through the Compton drag process. In most cases, and especially if we consider the BATSE [20-600] keV spectral window, the break out emission is too soft to be detected (see equation 1). For compact star progenitors ($r_* < 10^{12}$ cm), however, the break out emission could be detectable with instruments like Ginga and the *Beppo*SAX wide field cameras. The boosted component, on the other hand, could be hard and may be difficult to disentangle from the internal shock emission. In fact, this up-scatter emission should appear as a transient signal $(r_\gamma - r_*)/(2c\Gamma_0^2)$ seconds prior to the main burst, where $r_\gamma = \max(r_\tau, r_{int})$ is the radius at which the $\gamma$-rays are produced and $r_{int}$ is the radius of internal shocks. As shown by equation 2, the radius of transparency of the fireball is likely to happen at some distance from the stellar surface and thus the time delay between the up scatter emission and the burst is given by $r_\tau/(2c\Gamma_0^2) \approx 0.3\,\theta_{-1}^{-1}E_{j,51}^{1/2}\Gamma_{0,2}^{-5/2}$ s. If some bursts are characterized by low-intermediate Lorentz factors, precursors with typical photon energies $h\nu \sim 100\Gamma_{0,1}^2$ keV

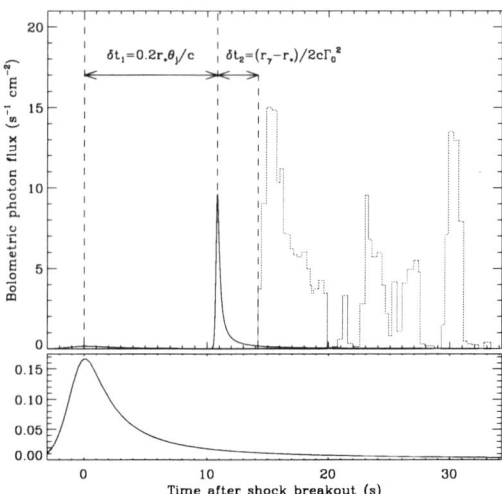

**FIGURE 2.** Lightcurve of the break-out, Compton-drag and internal shock components of a GRB. The lightcurve is computed to match the observation of GRB900126: $r_* = 8 \times 10^{11}$ cm, $\theta = 0.9$, $E_j = 3 \times 10^{51}$ erg, $\Delta t_j = 20$ s, $\Gamma_0 = 8$ and for a redshift $z = 1$. The main panel shows all the three components of the lightcurve while the lower panel shows a close up of the break out component

should have been observed by BATSE $\sim 100\Gamma_{0,1}^{-5/2}$ s before the main event. Such precursors may have indeed been observed [15].

## GRB 900126: A BROAD RELATIVISTIC JET?

On 1990 January 26, the Ginga experiment discovered X-ray emission in the 2–10 keV energy range $\sim 10$ s before the onset of a $\gamma$-ray event [16]. The $\gamma$-ray signal shows two distinct peaks, separated by $\sim 6$ s, both of which have rise times of $\approx 1.5$ s. The peak energy values in the $\gamma$-ray emission vary from $\approx 120$ keV in the first peak to $\approx 80$ keV in the second. The spectrum of the precursor X-ray emission can be described by that of a black body with a temperature $kT = 1.58 \pm 0.26$ keV and a flux $F \sim 2.5 \times 10^{-9}$ erg cm$^{-2}$ s$^{-1}$ [16]. The peak luminosity of the first spike is $\sim 9.5 \times 10^{-6}$ erg cm$^{-2}$ s$^{-1}$, i.e. $\sim 2600$ times brighter than the thermal precursor.

Knowing the flux and observed temperature of the thermal precursor, it is possible to estimate the radius $r_s$ of the emitting surface as a function of redshift. In the redshift range $0.5 < z < 10$, we obtain $10^{11} < r_s < 3 \times 10^{11}$ cm. In the framework of the shock break out, the radius of the emitting surface is given by $r_s \approx r_*\theta$ and the thermal precursor should precede the boosted emission by $0.2r_*\theta(1+z)/c \approx 2(1+z)$ s. The first peak of the γ–ray emission is indeed observed several seconds after the thermal precursor. If this second peak is interpreted as the Compton boosted emission described above, we can derive the Lorentz factor from the peak frequency of the second pulse and the temperature of the precursor, yielding $\Gamma_0 \approx 8$. The expected ratio of luminosity between the thermal precursor and the boosted pulse would then be $\approx 2500\theta$, to be compared with the measured value of $\approx 2600$. This would consequently imply that the fireball of GRB900126 was only moderately beamed and the radius of the progenitor star was $r_* \approx 10^{11}$ cm. The expected shock break out temperature is (see equation 1) $\approx 1(E_{j,51}/\Delta t_{j,2})^{1/4}$ keV, so that a comparison with the observed $T \approx 1.6(1+z)$ keV implies a total energy of few $\approx 10^{52}$ erg. The fluence of the burst was $\mathcal{F} = 4 \times 10^{-5}$ erg cm$^{-2}$ which, for a $z = 1$ burst with a mild ($\theta \approx 1$) beaming, corresponds to $E_j = 5 \times 10^{52}$ (Figure 2). On the contrary, if the softening between the first and second γ-ray peaks is caused by the same mechanism that is responsible for both of these emissions (i.e. internal shocks), the lack of detection of the Compton drag transient in the 1.5–400 keV band would imply that $\Gamma_0 \geq 20$.

## DISCUSSION

Many massive stars produce supernovae when forming neutron stars in spherically symmetric explosions, but some may fail neutrino energy deposition, forming a black hole in the center of the star and possibly a GRB [4]. One expects various outcomes ranging from GRBs with large energies and durations, to asymmetric, energetic supernovae with weak GRBs. The prompt transients produced by the Compton drag of the shock breakout emission would provide a natural test to distinguish between these different stellar explosions. The detection of these prompt multi-wavelength signatures would be a test of the collapsar model; and the precise measurement of the time delay between emissions would constrain the dimensions of the stellar progenitor, the Lorentz factor of the fireball and the radius of the burst emitting region ($r_{\rm int}$ or $r_\tau$).

This very hard prompt emission would propagate ahead of the collimated fireball loading the external medium with $e^\pm$ pairs. In most early discussions [17, 18, 19, 20], the concern has been raised that $e^\pm$ pair loading in low-density environments is rather inefficient, converting only a few percent of the bulk motion energy into $e^\pm$ pairs. We have shown here that the Compton drag mechanism can be an effective catalyst for converting bulk motion energy into γ-rays close to the stellar surface. Numerous $e^\pm$ pairs can then be produced as some of the photons in the beam are backscattered and interact with other incoming photons. The process discussed here suggests that the $e^\pm$ pairs can play a substantial role in both the dynamics of the fireball and the nature of the afterglow emission, as they are produced well before the fireball becomes optically thin. This suggests that if GRBs are the outcome of the collapse of massive stars involving a relativistic fireball jet, the time structure, dynamics and efficiency of the prompt and afterglow emissions may have a more complex dynamic than the standard models suggest.

## REFERENCES

1. Mészáros, P., *Science*, **291**, 79 (2001).
2. Rees, M. J., *A&AS*, **138**, 491 (1999).
3. Woosley, S. E, *ApJ*, **405**, 273 (1993).
4. MacFadyen, A. I., Woosley, S. E., *ApJ*, **524**, 262 (1999).
5. MacFadyen, A. I., Woosley, S. E., Heger, A., *ApJ*, **550**, 410 (2001).
6. Colgate, S. A., *ApJ*, **163**, 221 (1974).
7. Chevalier, R. A., *ApJ*, **259**, 302 (1982).
8. Mészáros, P., Waxman, E., *PhRvL*, **87**, 1102 (2001).
9. Lazzati, D., Ghisellini, G., Celotti, A., Rees, M. J., *ApJ*, **529**, L17 (2000).
10. Ghisellini, G., Lazzati, D., Celotti, A., Rees, M. J., *MNRAS*, **316**, L45 (2001).
11. Ramirez-Ruiz, E., MacFadyen, A. I., Lazzati, D., *MNRAS*, **331**, 197 (2002).
12. Aloy, M. A. et al., *ApJ*, **531**, L119 (2000).
13. Mészáros, P., Rees, M. J., *ApJ*, **476**, 232 (1997).
14. Wheeler, J. G., Yi, I., Hoflich, P., Wang, L., *ApJ*, **537**, 810 (2000).
15. Koshut, T. M. et al., *ApJ*, **452**, 145 (1995).
16. Murakami, T. et al., *Nature*, **350**, 592 (1991).
17. Thompson, C., Madau, P., *ApJ*, **538**, 105 (2000).
18. Madau, P., Blandford, R., Rees, M. J., *ApJ*, **541**, 712 (2000).
19. Mészáros, P., Ramirez-Ruiz, E., Rees, M. J., *ApJ*, **554**, 660 (2001).
20. Beloborodov, A. M., *ApJ*, **565**, 808 (2002).

# Gamma-Ray Bursts: The Tip of the Iceberg?

Maurice H.P.M. van Putten

*MIT 2-378, 77 Massachusetts Avenue, Cambridge, MA 02139*

**Abstract.** The spin-energy $E_{rot}$ of a Kerr black hole surrounded by a torus may power emissions in multiple windows. The recently determined true GRB-energies $E_\gamma = 3 - 5 \times 10^{50}$erg indicate a minor fraction $E_j/E_{rot} \simeq 0.1\%$ in baryon poor output, here considered as jets along open magnetic flux-tubes extending from the horizon to infinity. A major fraction $E_{gw}/E_{rot} \simeq 5\%$ is expected in gravitational radiation from the torus. A LIGO/VIRGO detection of $\alpha = 2\pi \int f dE_{gw}$ in excess of the neutron star limit $\alpha^* \simeq 0.005$ promises a calorimetric test for Kerr black holes. A sample of LIGO/VIRGO detections should obey the distribution of de-redshifted GRB-durations.

## INTRODUCTION

Black hole-torus systems may represent high-energy astrophysical transient sources. They feature the prospect of multi-window emissions powered by the spin energy $E_{rot}$ of a Kerr black hole. This could take the form of outflows along the axis of rotation accompanied by emissions from the torus in various channels: gravitational radiation, Poynting flux-dominated and baryonic winds and, when sufficiently hot, neutrino emissions. Ultimately, these systems may provide definitive tests for Kerr black holes as objects in Nature – the most compact energy reservoirs in angular momentum.

Rotating black holes where discovered by Kerr as exact solutions to general relativity [1]. The specific angular momentum of their radiation is at least twice that of the black hole, which suggests that Kerr black holes may be luminous under appropriate conditions. Identifying Kerr black holes will require observational evidence for their defining properties (see [34]): a compact horizon surface in common with non-rotating Schwarzschild black holes; frame-dragging of space-time is described by an angular velocity $-\beta$ of zero angular momentum observers; a compact energy reservoir of energy $E_{rot} = 2M \sin^2(\lambda/4)$, where $a/M = \sin \lambda$ denotes the ratio of specific angular momentum $a$ to the black hole mass $M$. In an extreme Kerr black hole, about half of the rotational energy corresponds to the top ten percent of the angular velocity $\Omega_H = \tan(\lambda/2)/2M$.

Black hole-torus systems harboring Kerr black holes are leading candidates as the inner engine of gamma-ray bursts (see [9] for a review). GRBs are characteristically non-thermal in the range of a few hundred keV with a bi-modal distribution in durations of short bursts around 0.3s and long bursts around 30s [2]. Black hole plus disk or torus systems may represent failed-supernovae [10] hypernovae [11] or black hole-neutron star coalescence [12], where the former is intimately connected to star forming regions [11, 13]. With GRBs remnants potentially found in Soft X-ray transients [14] GRO J1655-40 [15] and V4641Sgr [16], the putative black hole assumes the observed mass-range $3 - 14M_\odot$ (Fig. 1). We recently identified long/short bursts with rapidly/slowly spinning black holes in a state of suspended-/hyperaccretion [17]. A mean de-redshifted duration on the order of tens of seconds corresponds to the lifetime of rapid spin in suspended accretion in the presence of superstrong magnetic fields.

Here, we focus on baryon poor jets along the axis of rotation along with gravitational radiation from the torus. As proposed input to GRBs, the former will represent a minor fraction of $E_{rot}$ as inferred from the recently determined true GRB-energies [18, 19]. The latter is expected to be a major fraction of the output which could be representative for $E_{rot}$. We thus expect LIGO/VIRGO to detect a distribution of durations in gravitational waves which corresponds to the presently observed redshift-corrected distribution of GRB-durations.

## MULTI-WINDOW EMISSIONS

A Kerr black hole is expected to be luminous over *all* horizon angles, in response to a generally uniform magnetic flux. Thus, we expect emissions along its axis of rotation as well as into the surrounding matter. A detailed analysis is based on the topology of the magnetic field, partly by equivalence to pulsar magnetospheres and the formation of open flux-tubes.

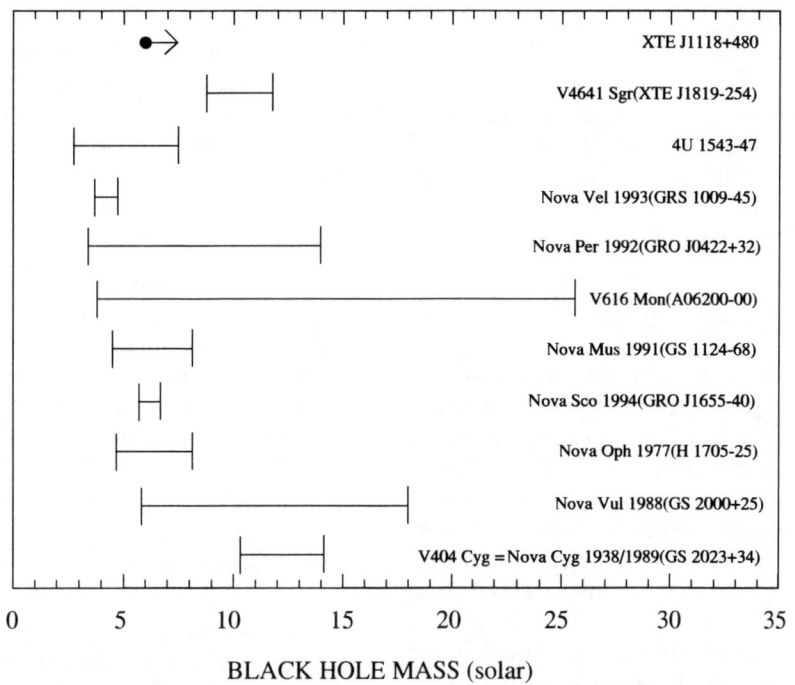

**FIGURE 1.** Shown is the distribution of black hole masses in X-ray novae. The top four are XTE J118+408 [20] V4641 Sgr [16], 4U 1543-47 [21] and Nova Vel 1993 [22]; the lower seven are from [23]. This mass distribution manisfests a certain diversity in black hole masses of about $3-14M_\odot$. [Reprinted from van Putten, *Physics Reports*, 345 ©2001 Elsevier B.V.]

In its lowest energy state, a rotating black hole surrounded by a torus magnetosphere develops an equilibrium magnetic moment [9]

$$\mu_H \simeq aBJ_H, \quad (1)$$

where $B$ denotes the mean of the poloidal magnetic field in the surrounding torus magnetosphere. It corresponds to the Wald equilibrium charge [27] consistent with the no-hair theorem [28, 9]. It carries over to a largely force-free magnetosphere around a black hole [29] and, in scaling, is analogous to the equilibrium charge on a neutron star [30]. The black hole hereby maintains essentially a maximal and uniform horizon flux at arbitrary rotation rates.

## Baryon-poor jets

A rapidly rotating black hole may support an open flux-tube supported by Eqn.(1). These are endowed with slip/slip- and ingoing/outgoing-boundary conditions on the horizon/infinity. The charge-density about the axis of rotation of the black hole satisfies $\rho = -(\Omega+\beta)B/2\pi$ (see [9] for references), where $\Omega$ denotes the angular velocity of the open flux-tube; a sign-change from positive in a lower section attached to the black hole to negative in the semi-infinite section above occurs at some height above the black hole when $0 < \Omega < \Omega_H$. This permits a continuous current along the open flux-tube from infinity into the hole, with outflow to infinity and inflow into the black hole. Open magnetic flux-tubes are a remarkable natural phenomenon, perhaps most dramatically demonstrated by solar activity. Magnetic mediated outflows in general, therefore, require the creation of open magnetic flux-tubes which formally extend to infinity. Open magnetic flux-tubes may be created from a torus magnetosphere around a black hole, by a change in topology [9]. This change in topology represents a transient fast magnetosonic wave, which might be excited as a nonlinear feature to strong Alfvén waves or by superradiance within the torus magnetosphere. It produces a coaxial structure of two flux-tubes, an inner tube supported by the equilibrium magnetic moment of the black hole and an outer tube supported by the torus. The flux in the inner/outer tube is $\pm 2\pi A_\phi$ in terms of the vector potential $A_\phi$. The outer flux-tube endowed with no-slip/slip boundary conditions on the surface of the torus/infinity.

The inner flux-tube forms an powerful artery for the spin-energy of the black hole. In asymptotically charge-separation equilibrium, it assumes an angular velocity $\Omega_+$ on the horizon and $\Omega_-$ at larger distances by the slip-slip boundary conditions. These lower and upper sections are separated by differential rotation. For a net flux $2\pi A_\phi$, the ingoing boundary conditions produce a current $I_+ = (\Omega_H - \Omega_+)A_\phi$ in the small angle approximation [31, 9]; likewise, the current at infinity is $I_- = \Omega_- A_\phi$ for ultrarelativistic winds [9]. This leaves a Faraday-induced potential $V = (\Omega_+ - \Omega_-)A_\phi$ along the differentially rotating inner tube. Global current closure is over the outer flux-tube, which corotates with the angular velocity $\Omega_T$ of the torus by its no-slip boundary conditions. By current continuity, differential rotation hereby creates a baryon-poor jet with luminosity [9]

$$L_j = \frac{1}{2}\Omega_T(\Omega_H - 2\Omega_T)A_\phi^2. \qquad (2)$$

These low-σ outflows may represent the baryon-poor input to cosmological gamma-ray bursts (GRBs). When the low-σ outflow is hidden from view by an intervening medium, the observed spectrum will be different from black body radiation. The observed spectrum - but less so the characteristic energies – depends on the subsequent evolution of this medium and its interaction with the interstellar medium. In particular, the non-thermal GRB emissions may hereby be the testimonial to a hypernova progenior wind, associated with its rapid rotation and consistent with the recently discovered iron line emissions [32]. This introduces shocks and synchrotron radiation. The observed emissions, therefore, should be non-thermal with approximately similar characteristic energies as emitted from the gap. The detailed structure of the emissions further depends on the interaction of this accelerated wind mass with the interstellar medium [4].

The recently inferred true GBR-energies of $3-5 \times 10^{50}$erg [18, 19] imply $E_j/E_{rot} \simeq 0.01\%/\epsilon$ in baryon poor outflows. Here $\epsilon \simeq 0.15\%$ denotes the radiation efficiency, assuming $E_{rot} \simeq M/3$ with $M = 7M_\odot$. As a specific application of (2), this indicates an open flux-tube with half-opening angle $\theta_H \simeq 35^o$ [33]. The half-opening angles on the celestial sphere will satisfy $\theta_j \leq \theta_H$. The duration of the burst and the observed half-opening angles $\theta_j$ may be uncorrelated, upon collimation by external parameters such as hydrostatic static pressure in a GBR-precursor winds or the interstellar medium, or positively correlated, upon collimation by winds [24] coming off the torus.

## Gravitational radiation

A surrounding torus receives energy and angular momentum by equivalence in poloidal topology to pulsar magnetospheres: the inner face of a torus around a black hole receives energy and angular momentum, as does a pulsar when infinity wraps around it. The outer face looses angular momentum and energy to infinity, as does a pulsar in flat space-time. These equivalences becomes apparent when working in a frame of references fixed to the horizon of the black hole and, respectively, infinity (Mach's principle). When the black hole spins rapidly, it develops a state of suspended accretion for the duration of rapid spin of the black hole. The high incidence of the black hole-luminosity onto the inner face indicates that the emissions from the torus may be luminous. We thus find an output

$$E_{gw} = 1 - 2\%M \qquad (3)$$

in gravitational radiation – about two orders of magnitude higher than the inferred true GRB-energies.

The major fraction $E_{gw}/E_{rot} \simeq 5\%$ emitted at twice the Keplerian frequency, i.e., $f = 1 - 2\text{kHz}$, promises black hole-torus systems to be viable sources for the upcoming broadband gravitational wave observatories LIGO [25] and VIRGO [26]. Thus, black hole-torus systems may have a compactness parameter [34]

$$\alpha = 2\pi \int_0^{E_{gw}} f dE > \alpha^* \simeq 0.005 \qquad (4)$$

in excess of the limit for rapidly spinning neutron stars. This provides for the first time a *calorimetric* compactness test for Kerr black holes. The proposed association to GRBs predicts that a future sample of LIGO/VIRGO detections will satisfy a distribution of durations which obeys the distribution of redshift corrected GRB-durations $T/(1+z)$ (Fig. 2). The displayed spread in $T/(1+z)$ is consistent with the narrow mass range of $3-14M_\odot$ in SXTs. Indeed, we expect a positive correlation between $T/(1+z) \propto M^2$ and $E_{gw} \propto M$ (as well as $E_j$ and $E_\gamma$) in view of $E_{rot} \propto M$ and a black hole-to-torus coupling $\propto M^{-1}$ for a universal ratio of poloidal magnetic field energy-to-kinetic energy in the torus [35].

## SUMMARY

We have described a prospect for multi-window emissions from Kerr black holes powered by their rotational energy. Surrounded by a torus, a Kerr black hole is luminous over *all* horizon angles in its lowest energy state. These systems are long-lived in a state of suspended accretion, which operates by equivalence in poloidal topology to pulsar magnetospheres. Quite generally, the powerful competing torques acting on the torus introduce turbulent shear flow, which may stimulate the formation of a quadrupole moment in its mass distribution. We expect a minor fraction in baryon-poor jets from a differentially rotating tube along the axis of rotation and a major fraction in gravitational radiation from the torus. (Further output is expected in torus winds and, when sufficiently hot, neutrino emissions.) These black hole-torus systems are predicted to be powerful LIGO/VIRGO-sources of gravitational radiation, permitting for the first time a calorimetric compactness test for Kerr black holes. A

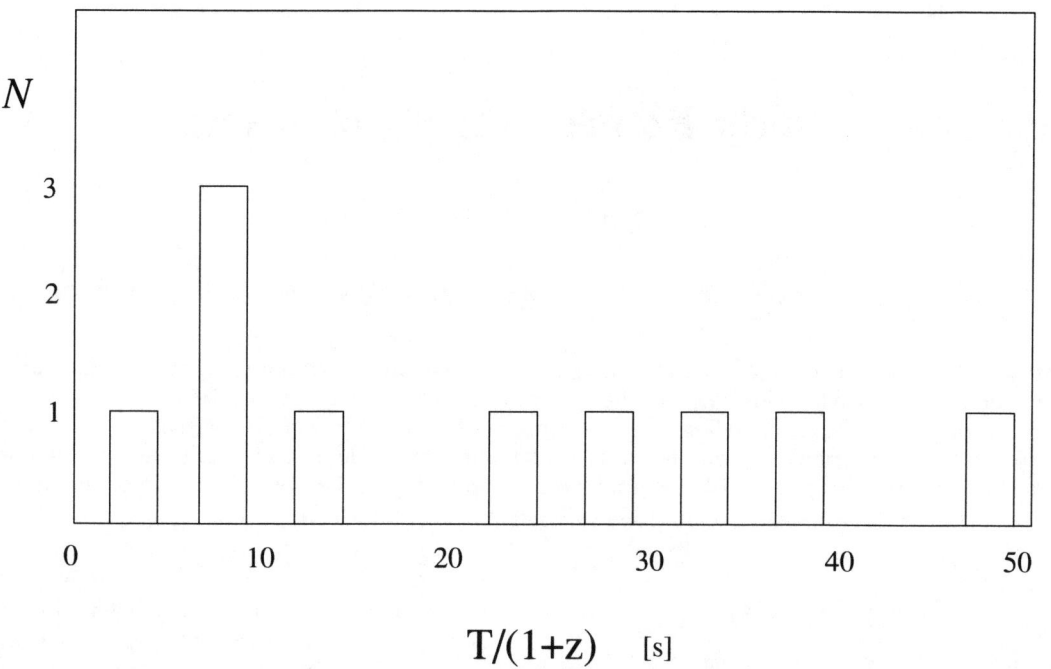

**FIGURE 2.** Shown is the distribution of redshift corrected durations, obtained from 10 GRBs with individually determined redshifts from their afterglow emissions (GRB000926,GRB000418,GRB000301c,GRB990510,GRB990123,GRB980613, GRB980425,GRB971214,GRB970508,GRB970228). This distribution represents the life-time of the inner engine. In the black hole-torus model, the proposed gravitational radiation from the torus is simultaneous with the baryon-poor output powering the GRB. LIGO/VIRGO detections of these emissions (from cosmologically nearby sources) are expected to obey a similar distribution.

sample of LIGO/VIRGO detections is predicted to obey the distribution of redshift corrected GRB-durations.

## ACKNOWLEDGMENTS

This work is partially supported by NASA Grant 5-7012, and MIT C.E. Reed Award and a NATO Collaborative Linkage Grant. The author thanks E. Costa and A. Levinson for stimulating discussions.

## REFERENCES

1. Kerr, R.P., 1963, Phys. Rev. Lett., 11, 237
2. Kouveliotou, C., et al., 1993, ApJ, 13, L101
3. Rees, M.J., & Meszaros, 1992, Mon. Not. R. Astron. Soc., 152, 258, 41P; ibid., 1994, Astroph. J., 430, L93
4. Piran, T., 1999, Phys. Rep. 314, 575; ibid., 2000, 333, 529
5. Schmidt, M., 1999, Astron. Astroph. Suppl. Ser. 138, 409
6. Meegan, C.A., et al., 1992, Nature, 355, 143
7. Unruh, W.G., 1974, Phys. Rev. D., 10, 3194
8. Hawking, S.W., 1975, Commun. Math. Phys., 43, 199
9. van Putten, M.H.P.M., 2001, Phys. Rep. 345, 1; in Proc. 2nd KIAS Workshop High Energy Emission around Black Holes, KIAS, Korea, to appear; astro-ph/0109429
10. Woosley, S.E., 1993, Astroph. J., 405, 273
11. Paczynski, B.P., 1998, Astroph. J., 494, L45
12. Paczynski, B.P., 1991, Acta Astron., 41m, 257
13. Bloom, J.S., Kulkarni, S.R., Djorgovski, S.G., 2000, Astro-ph/0010176
14. Brown, G.E., et al., NewA, 5, 191
15. Israelian, G., et al., 1999, Nature, 401, 142
16. Orosz, J.A., et al., 2001, 555, 489
17. van Putten, M.H.P.M., & Ostriker, E.C., 2001, ApJ, 552, L31
18. Frail, D.A., et al., ApJ, 562, L55
19. Piran, T., et al., ApJ, 560, L167
20. McClintock, J.E., et al., 2001, ApJ, 551, L147
21. Orosz, J.A., et al., 1998, ApJ, 499, 375
22. Filippenko, A.V., et al., 1999, Pub. ASP 111 (792), 969
23. Bailyn, C.D., et al., 1998, ApJ, 499, 367
24. Levinson, A., & Eichler, D., 2000, Phys. Rev. Lett., 85, 236
25. Abramovici, A., et al., 1992, Science, 256, 325
26. Bradaschia, C., et al., 1992, Phys. Lett. A., 163, 15
27. Wald, R.M., 1974, Phys. Rev. D., 10, 1680
28. Carter, B., 1968, Phys. Rev., 174, 1559
29. Lee, C.-H., Lee, C.H., & van Putten, M.H.P.M., 2001, MNRAS, 324, 731
30. Cohen, J.M., Kegeles, L.S., & Rosenblum, A., 1975, ApJ, 201, 783
31. Punsly, B., & Coroniti, F., 1990, ApJ, 350, 518
32. Piro, L., et al., 2001, Science, 290, 955
33. van Putten, M.H.P.M., & Levinson A., 2001, ApJ, 555, L41
34. van Putten, M.H.P.M., 2001, ApJ, 562, L51
35. Coward, D., & van Putten, M.H.P.M., in preparation

# On the Progenitors of Collapsars

A. Heger* and S. E. Woosley*

*Department of Astronomy and Astrophysics, University of California, Santa Cruz, CA 95064, USA

**Abstract.** We study the evolution of stars that may be the progenitors of common (long-soft) GRBs. Bare rotating helium stars, presumed to have lost their envelopes due to winds or companions, are followed from central helium ignition to iron core collapse. Including realistic estimates of angular momentum transport [1] by non-magnetic processes and mass loss, one is still able to create a collapsed object at the end with sufficient angular momentum to form a centrifugally supported disk, i.e., to drive a collapsar engine. However, inclusion of current estimates of magnetic torques [2] results in too little angular momentum for collapsars.

## INTRODUCTION

One of the most promising models for the "long variety" of gamma-ray bursts (GRBs) is the so-called *collapsar* model [3]. It assumes that a sufficiently massive stellar core collapses into a black hole and the infalling outer layers form a disk around it. Energy dissipated in the disk or the rotation of the black hole itself is assumed to power a jet of high Lorentz factor ($\Gamma \sim 200$) that escapes from the engine along the polar axis to large distance ($\sim 10^{15}$ cm) and powers a GRB by interaction with the circumstellar medium or by internal shocks.

The traversal time for the relativistic jet through the hydrogen envelopes of typical massive stars is hundreds to thousands of seconds. Thus, at the time of the GRB, bare helium stars, which have radii of only a few light seconds (about a solar radius), are required if the lifetime of the engine and the GRB are not to be short compared to the time it takes the jet to drill through the star.

Two essential ingredients for the collapsar model are a sufficiently massive core to form a black hole and a sufficient rotation rate at the time of collapse to allow the formation of a disk. The question we address here is: *What can be expected for the rotation rates of massive stellar cores when they collapse?*

## PROGENITOR MODELS

We calculate the evolution of bare helium cores with an initial mass of $15\,M_\odot$. The stars are assumed to rotate rigidly initially with two different surface rotation rates corresponding to $10\,\%$ and $30\,\%$ of a Keplerian orbit. The former could be either the result of a massive single star ($\sim 40\,M_\odot$) that has lost its envelope early during helium burning or a close binary that lost its envelope to a companion. The latter might require a binary merger. The evolution of the helium core and its rotation is followed as described by Heger et al. [1] using fine surface zoning.

Two different evolutionary paths are considered. The first neglects mass loss, possibly corresponding to WR stars of very low metallicity, while in the second, it is taken into account. We use the WR mass loss rate given by Wellstein & Langer [4, equation 1], reduced by a factor 3 to account for effect of "clumping" [5]. An initial stellar metallicity of 1/10 solar is assumed along with a WR mass loss rate that scales as the square root of metallicity [6], reducing the mass loss rate by an additional factor of 3.

## RESULTS

Figures 1 - 4 give the results for collapsar progenitors that follow the evolution of the angular momentum in the stellar interior till the onset of iron core collapse.

### Rotation Profile and Disk Formation

To decide whether a centrifugally supported accretion disk can form around a central black hole (which we assume either forms promptly or by fallback), we compare the angular momentum calculated as a function of interior mass to that a test particle would require at the last stable orbit around a Schwarzschild or Kerr black hole (Figures 2 and 3). Though mass loss significantly reduces the angular momentum at core collapse, enough remains in the equatorial regions of the star to form a centrifugally supported disk.

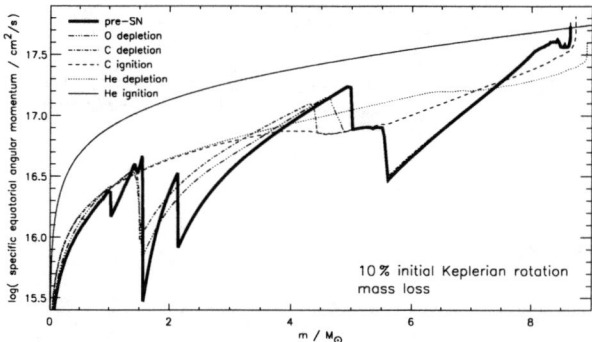

**FIGURE 1.** Angular momentum in the equatorial plane as a function of the interior mass coordinate, $m$, at different evolutionary stages.

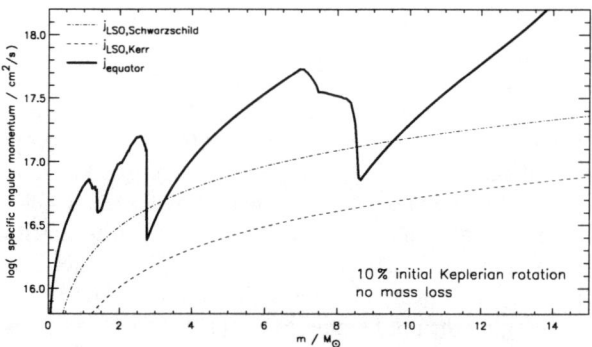

**FIGURE 2.** Equatorial angular momentum (*thick black line*) of a 15 $M_\odot$ helium core of initially 10 % Keplerian rotation as a function of the interior mass coordinate, $m$. The *dashed-dotted line* shows the specific angular momentum required for a test particle at the last stable orbit around a Schwarzschild black hole of mass equal to the mass coordinate, $m$. The *dashed line* shows the same for a extreme Kerr black hole (spin parameter $a = 1$).

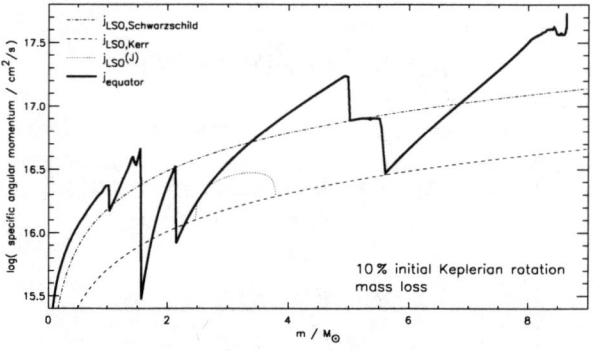

**FIGURE 3.** Same as Figure 2, but mass loss due to stellar winds is included. The *dotted line* shows the specific angular momentum a test particle requires at the last stable orbit around a black hole that has formed with the mass and integrated angular momentum below the given mass coordinate. Where the dotted line is missing, sufficient angular momentum is available to form a Kerr black hole.

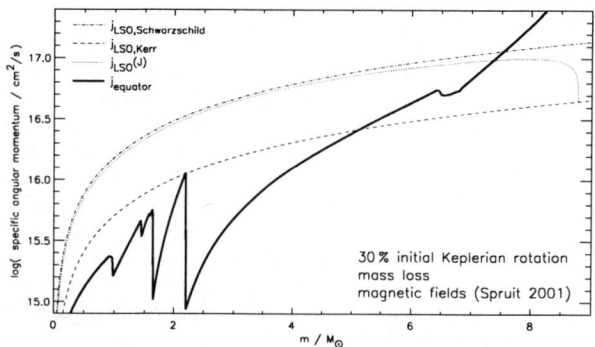

**FIGURE 4.** Same as Figure 3, but an initial rotation rate of 30 % Keplerian rotation was assumed and angular momentum transport by magnetic fields according to a prescription by Spruit [2] was included.

## Magnetic Fields

We also calculated models including magnetic torques that might result from a dynamo process recently described by Spruit [2]. They lead to considerable spin-down of the core, especially when combined with wind mass loss (Table 1 and Figure 4).

For comparison, preliminary calculations [7] of rotating stellar models of 15, 20, and 25 $M_\odot$ stars with hydrogen envelopes and initial surface rotation velocities of $200\,\mathrm{km\,s^{-1}}$, using the same prescription for torques, resulted in pulsar birth rotation periods of 7.65, 5.50, and 3.99 ms. Without magnetic fields, including all other forms of angular momentum transport [1], the same calculations previously gave rotation periods of $\sim 0.2$ ms in these same stars. Here it is assumed that the innermost 1.6 $M_\odot$ of the collapsing star produces a rigidly rotating neutron star of 1.45 $M_\odot$ gravitational mass, while conserving the total angular momentum contained in the precollapse model.

Calculations that used the effects of magnetic fields as described by Spruit & Phinney [8, not shown here] always resulted in much too low core rotation for the collapsar model of GRBs.

## Binary Interaction

In a sufficiently close binary system the envelope can be tidally locked to the orbit of the star. Depending on separation and mass ratio, this rotation can reach several 10 % of Keplerian. We find that maintaining 10 % Keplerian surface rotation can be sufficient for collapsars. If employing the dynamo process by Spruit [2], however, the resulting presupernova rotation is too slow even for a system with 20 % Keplerian surface rotation.

To simulate a merger of a binary system after or at

**TABLE 1.** Presupernova properties of initially 15 M$_\odot$ helium core models. The first three columns define the initial model and physics employed (magnetic fields according to Spruit [2], amount of rotation, mass loss by winds). Next we give the period a pulsar would have if it formed in this star, then the non-dimensional spin parameter, $a_{BH}$, a black hole would acquire, if all the angular momentum below the mass coordinate indicated were to go into the black hole of that mass (formal values in excess of 1 are shown in brackets solely to give a measure for the angular momentum available in the model), and in the last column we show the mass ranges in which the equatorial mass could form a centrifugally supported accretion disk around a central compact object.

| magnetic fields | rotation (% Keplerian) | mass loss | pulsar period (ms) | $a_{BH}$ (2 M$_\odot$) | $a_{BH}$ (2.5 M$_\odot$) | $a_{BH}$ (3 M$_\odot$) | $a_{BH}$ (4 M$_\odot$) | mass coordinate ranges for which an equatorial disk could form (M$_\odot$) |
|---|---|---|---|---|---|---|---|---|
| – | 10 | – | 0.09 | (2.5) | (3.4) | (3.8) | (3.7) | 0 – 15 |
| – | 10 | yes | 0.23 | (1.1) | 0.90 | 0.98 | (1.2) | 0 – 2.5, 2.7 – 8.6 |
| – | 30 | – | 0.06 | (4.1) | (4.8) | (5.6) | (5.9) | 0 – 15 |
| – | 30 | yes | 0.18 | (1.7) | (1.3) | (1.3) | (1.7) | 0 – 8.4 |
| yes | 10 | – | 0.45 | 0.58 | 0.76 | 0.80 | 0.88 | 2.4 – 2.8, 5.7 – 15 |
| yes | 10 | yes | 3.4 | 0.08 | 0.09 | 0.11 | 0.10 | 8.5 – 8.8 |
| yes | 30 | – | 0.26 | (1.2) | (1.4) | (1.4) | (1.5) | 0 – 15 |
| yes | 30 | yes | 1.9 | 0.16 | 0.14 | 0.16 | 0.17 | 7.4 – 8.8 |

the end of central helium burning (Case C) an additional model was calculated assuming rigid rotation with 50 % Keplerian surface rotation at core helium depletion and including the dynamo process of Spruit [2]. Again the magnetic stress kept the star in rigid rotation till carbon burning, removing too much angular momentum for the collapsar model to work. *Without* magnetic fields, as shown above, even single stars may already retain sufficient angular momentum for collapsar progenitors.

## Decoupling of Core Rotation

*At what evolution stage does the core rotation need to decouple from the surface?* Figure 5 shoes that this needs to happen before central carbon burning – assuming angular momentum is locally conserved from this time on and no further transport occurs. In the model star of Figure 5 the decoupling would need to happen no later than when a central density of $\sim 10{,}000\,\mathrm{g\,cm^{-3}}$ is reached.

## CONCLUSIONS

A bare helium star of low metallicity can retain enough angular momentum to form a centrifugally supported disk around a central black hole of $\sim 3\,\mathrm{M}_\odot$, as required by the collapsar model for GRBs. Without magnetic

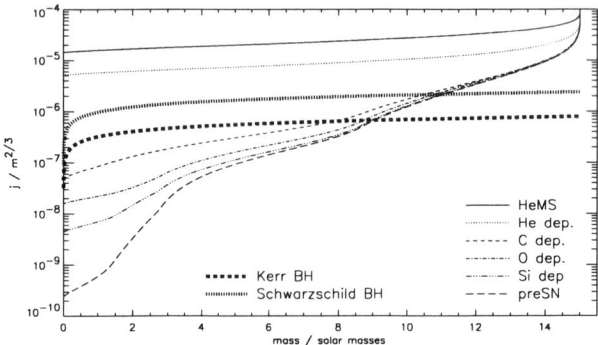

**FIGURE 5.** Angle averaged specific angular momentum for a rigidly rotating 15 M$_\odot$ star with Keplerian surface rotation at different evolution stages. The specific angular momentum has been scaled by $m^{-2/3}$ to remove the singularity at $m = 0$ from the plot. *Thick lines* show the specific angular momentum required for the last stable orbit around a black hole.

fields, the angular momentum is sufficient to form a Kerr black hole and support most or all of the star in an accretion disk. However, if we include an approximate treatment of angular momentum transport by magnetic fields, the resulting spin rates become too low to form centrifugally supported disks in the inner part of the core. Even a binary helium star merger at the end of central helium burning might not be able to avoid this fate. Mass loss can lead to an additional significant spin-down of the core, especially if magnetic fields couple it effectively to the envelope. Even in case of Keplerian surface rotation, the core rotation needs to decouple before carbon ignition in order to make a Kerr black hole. The dynamo process recently proposed by Spruit [2] seems too efficient to form collapsar progenitors from single stars or helium star mergers. This is even more so for the magnetic field modeling suggested by Spruit & Phinney [8].

## ACKNOWLEDGMENTS

We thank Henk Spruit for a preview of his work and many helpful discussions. This work has been supported by the NSF (AST-9731569), NASA (NAG5-8128), the DOE (B347885), and the AvH (FLF-1065004).

## REFERENCES

1. Heger, A., Langer, N., Woosley, S.E. 2000, ApJ, 528, 368
2. Spruit, H.C. 2002, A&A, 381, 923
3. Woosley, S.E. 1993, ApJ, 405, 273
4. Wellstein, S., Langer, N. 1999, A&A, 350, 148
5. Hamann, W.-R., Koesterke, L. 1998, A&A, 335, 1003
6. Van Beveren, D. 2001, astro-ph/0201110
7. Heger, A., Woosley, S.E., Spruit, H.C. 2002, ApJ, in prep.
8. Spruit, H.C., Phinney, E.S. 1998, Nature, 393, 139

# Gamma-Ray Bursts from Accreting Black Holes

William H. Lee* and Enrico Ramirez–Ruiz[†]

*Instituto de Astronomía, UNAM, Apdo. Postal 70-264, Cd. Universitaria, México D.F. 04510 MEXICO
[†]Institute of Astronomy, Madingley Road, Cambridge, CB3 0HA, U.K.

**Abstract.** We show the hydrodynamical evolution of massive accretion disks around black holes, formed when a neutron star is disrupted by a black hole in a binary system. The time dependence is followed in two dimensions assuming azimuthal symmetry for 0.2 seconds, using an ideal gas equation of state. The disk evolves because of the transport of angular momentum due to viscosity. We estimate the energy released in neutrinos, as well as through magnetic–dominated mechanisms, and find it can be as high as $E_\nu \approx 10^{52}$ erg and $E_{BZ} \approx 10^{51}$ erg respectively, in approximately 0.1–0.2 seconds. Of the former, only a small fraction (a few per cent) would potentially be capable of producing a burst through $\nu\bar{\nu}$ annihilation, while in principle the latter could do so almost entirely. Thus these systems could in principle account for the energetics of short gamma ray bursts.

## INTRODUCTION

Black hole (BH)–neutron star, or double neutron star (NS) binaries (such as PSR1913+16) will coalesce due to angular momentum losses to gravitational radiation. These mergers are currently viewed as promising candidates for the progenitors of gamma-ray bursts [1, 2], at least for the subclass of short bursts with durations typically ranging between $10^{-2}$ and 2 s. This is much longer than the typical dynamical timescale $\approx$ ms and so it requires that the *central engine* evolve into a configuration that is stable enough to survive the violent merging, while keeping enough binding energy to power the burst. Event rates for binary coalescence of compact binaries are consistent with the observed GRB rate [3].

The formation of a BH with a debris torus around it could power a burst through the release of a fraction of its gravitational energy $E_G \simeq 10^{52}$erg [4] into a relativistic fireball. Possible forms of this outflow are kinetic energy of relativistic particles generated by $\nu\bar{\nu}$ annihilation or an electromagnetic Poynting flux. The intrinsic dynamical timescale of the system could account for the rapid variability observed in burst light curves. Strong magnetic fields anchored in the dense matter surrounding the BH could produce the large amplitude variations in the energy release [5, 6]. The main requirement is that the torus itself should survive at least as long as the characteristic burst duration. A differentially rotating neutron torus is a natural site for the onset of a dynamo process that winds up the magnetic field to the required intensity [7].

We show here the evolution of realistic disks resulting from dynamical coalescence calculations, on timescales comparable to the durations of short GRBs. No assumptions are made a priori regarding the structure of the disk.

## DYNAMICAL EVOLUTION OF A MASSIVE ACCRETION DISK

The hydrodynamical evolution of coalescing black hole–neutron star binaries has been considered in detail [8, herafter L01] using an ideal gas equation of state for a variety of initial configurations. We will use here the results of the runs labeled C50 and C31 in that paper, with initial mass ratios $q_b = 0.5$ and $q_b = 0.31$ respectively (both runs used an adiabatic index $\Gamma = 2$).

To follow the dynamical evolution of the accretion disks on a sufficiently long timescale, we have mapped the output from the three–dimensional simulations to two dimensions by assuming azimuthal symmetry (without assuming reflection symmetry with respect to the $z = 0$ plane). We use a 2D Smooth Particle Hydrodynamics (SPH) code, adapted from the version used for the 3D calculations. The equation of state is $P = \rho u(\Gamma - 1)$, where $u$ is the internal energy per unit mass, and the fluid is given the azimuthal velocity $v_\phi$ required to maintain centrifugal equilibrium. We show simulations for $\Gamma = 2$ and $\Gamma = 4/3$. The self–gravity of the disk is neglected, and the black hole (at the origin), produces a Newtonian point–mass potential $\Phi_{BH} = -GM_{BH}/r$. Accretion is modeled by an absorbing boundary at the Schwarzschild radius $r_{Sch} = 2GM_{BH}/c^2$. Only the inner regions of the accretion disk, with $0 < |z| < 200$ km and $0 < r < 400$ km are initially mapped.

We have introduced into the equations of motion the terms arising from the viscous stress tensor $t_{\alpha\beta}$, with all components included, using the formalism developed for SPH in [9], adapted to cylindrical coordinates with azimuthal symmetry $(r,z)$. The magnitude of the viscosity is fixed with an $\alpha$–prescription, with $\eta = \alpha\rho c_s^2/\Omega_k$, where $\eta$ is the dynamical viscosity coefficient, $c_s = (\Gamma P/\rho)^{1/2}$ is the adiabatic sound speed and $\Omega_k$ is the Keplerian angular velocity. The energy dissipation per unit mass due to the viscous processes is $Tds/dt = \eta\sigma_{\alpha\beta}\sigma_{\alpha\beta}/2\rho + (Tds/dt)_{art}$, where $s$ is the entropy per unit mass and $\sigma_{\alpha\beta}$ is the shear tensor. The last term is due to the artificial viscosity included in SPH (we use the form given in [10] to minimize its effect on the evolution of the disk). We assume that all the energy disspiated in the flow is radiated away in neutrinos. The dissipation rate given above allows us to calculate a luminosity, which we hereafter refer to as a neutrino luminosity $L_\nu$. All the dynamical simulations followed the disk evolution for $t = 0.2$ seconds.

The choice of different parameters shown in Table 1 allows us to explore the effect of numerical resolution on the results, as well as that of the magnitude of the viscosity, the compressibility of the gas and the initial mass of the black hole. For a lower black hole mass $M_{\rm BH}$ the accretion disk is initially more massive, and the spin angular momentum of the black hole is larger. Thus we expect this type of system to potentially be able to provide and release a greater amount of energy.

Initially the disk flattens slightly, on a vertical free–fall timescale $t_{ff} \approx \sqrt{r^3/GM_{\rm BH}} \approx 1-2$ ms. Thereafter the evolution is more gradual, and a characteristic circulation pattern in the disk is established (see Figure 1). Viscosity transports angular momentum outwards, the outer boundary of the disk moves to larger radii, and the disk becomes slowly thinner. In the equatorial plane, the gas flows clearly outwards, while it is on the surface of the disk and very close to the black hole ($r \approx 50$ km) in the equator that accretion takes place. There are large–scale eddies along the boundary that divides inflow from outflow all along the surface of the disk. This pattern is observed in all our simulations, and is essentially stationary once it becomes established.

There are minor differences between the dynamical runs, the most important one being the value of $\alpha$, which fixes the magnitude of the viscosity. For $\alpha = 0.01$, the disk spreads radially to a lesser extent, and more importantly, the accretion rate is initially much lower than in the rest of the runs (by about an order of magnitude). By $t = 0.1$ s, it is still a factor of three lower in run G than in run E (where $\alpha = 0.1$). Hence for a lower viscosity, the disk retains a larger amount of mass for a longer time, spreading its energy release over a longer period (see Table 2).

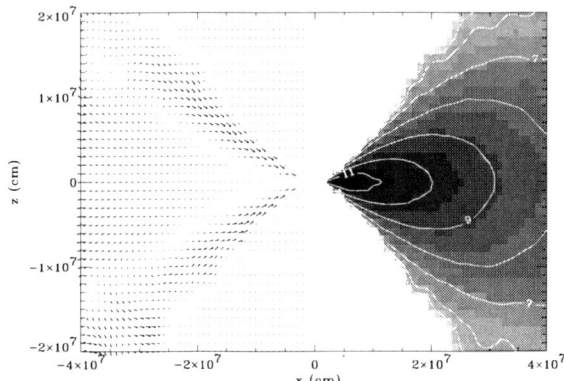

**FIGURE 1.** Logarithmic density contours (in g cm$^{-3}$) and velocity field for run E at $t = 0.11$s. The longest arrows correspond to $v = 2.9 \times 10^8$cm s$^{-1}$.

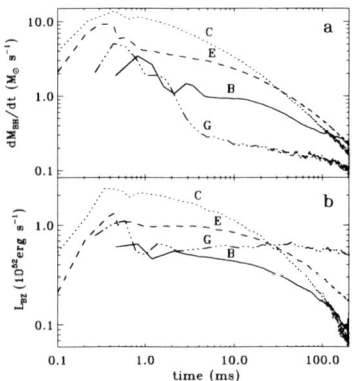

**FIGURE 2.** Accretion rate (top) onto the black hole and Blandford–Znajek luminosities (bottom) for several runs.

The two main forms of energy release from the disk we consider are i) neutrino emission and ii) MHD flow, either through the Blandford–Znajek effect or by means of a Poynting flux. In the first case, we make a rough estimate for $L_\nu$ and $E_\nu$ from the dissipated energy because of viscosity. For the magnetic–dominated case, we assume that the field strength is directly related to the internal energy density in the gas. In the inner regions of the disk (at $r = 20$ km in the equatorial plane), $\rho c_s^2$ is of order $10^{30}$ erg cm$^{-3}$, and even larger for the cases with low viscosity. An estimate for the Blandford–Znajek luminosity (see Figure 2b) is computed as $L_{BZ} \approx 10^{50} a^2 (M_{\rm BH}/3\ M_\odot)(B/10^{15}\ {\rm Gauss})^2$ where $a \simeq 0.3$ is the Kerr parameter of the black hole and the magnitude of the magnetic field is computed using $B^2/8\pi = \rho c_s^2$ (this gives a field $B \approx 10^{16}$ G). For a larger viscosity, the gas in the disk drains into the black hole on a shorter

**TABLE 1.** Initial parameters for the accretion disks

| Run | $q_b$ | $\Gamma$ | $\alpha$ | $\frac{M_{BH}}{M_\odot}$ | $\frac{M_{disk}}{M_\odot}$ | N* |
|---|---|---|---|---|---|---|
| A | 0.31 | 2.0 | 0.1 | 5.57 | 0.266 | 21,609 |
| B | 0.31 | 4/3 | 0.1 | 5.57 | 0.266 | 21,609 |
| C | 0.50 | 2.0 | 0.1 | 3.85 | 0.308 | 19,772 |
| D | 0.50 | 4/3 | 0.1 | 3.85 | 0.308 | 19,772 |
| E | 0.50 | 4/3 | 0.1 | 3.85 | 0.308 | 79,489 |
| F | 0.50 | 4/3 | 0.01 | 3.85 | 0.308 | 19,772 |
| G | 0.50 | 4/3 | 0.01 | 3.85 | 0.308 | 39,658 |

* Number of SPH particles used

**TABLE 2.** Luminositites and energies during the dynamical evolution

| Run* | $L_\nu$ $10^{52}\frac{erg}{s}$ | $E_\nu$ $10^{52}$ erg | $L_{BZ}$ $10^{51}\frac{erg}{s}$ | $E_{BZ}$ $10^{50}$ erg |
|---|---|---|---|---|
| A | 3.25 | 1.38 | 1.00 | 2.94 |
| B | 3.61 | 1.08 | 1.75 | 4.05 |
| C | 4.00 | 1.79 | 2.00 | 6.47 |
| D | 5.00 | 1.52 | 2.50 | 6.25 |
| E | 4.65 | 1.57 | 2.95 | 7.41 |
| F | 1.86 | 0.44 | 6.00 | 11.7 |
| G | 1.84 | 0.44 | 5.95 | 11.5 |

* $E_\nu$ and $E_{BZ}$ are given at $t = 0.2$ s. $L_\nu$ and $L_{BZ}$ are given at $t = 0.1$ s

timescale, and thus the drop in $L_{BZ}$ is much faster than for a low value of $\alpha$ (in run G, $L_{BZ} \approx 5 \times 10^{51}$ erg/s is practically constant).

Dynamical instabilities could make the disk lifetime shorter than the viscous timescale. In the present calculations we can determine if the disks are unstable with respect the Toomre criterion. We find the Toomre parameter $Q = c_s \kappa / \pi G \Sigma$ to be larger than unity for all cases. The profile of $Q$ in the disk is such that there is a minimum $Q_{min}$ at a radius $r_Q$. The minimum occurs at $t = 0$, with $(r_Q, Q_{min}) \approx (50 \text{ km}, 5)$. As the disk evolves, both parameters increase (since the density in the disk drops), and by $t = 0.2$, typically $(r_Q, Q_{min}) \approx (150 \text{ km}, 10)$. Additionally, the disks are stable with respect to the runaway radial instability [11] because their distribution of specific angular momentum is far from being constant, with $j \propto r^{0.4-0.45}$ (see L01).

## DISCUSSION

Here we have only made an estimate for the total neutrino luminosity $L_\nu$, and not for the *annihilation* luminosity $L_{\nu\bar{\nu}}$, which is would determine if a relativistic fireball could be launched or not. The calculation of $L_{\nu\bar{\nu}}$ requires the use of a more realistic equation of state, which we will explore in future work. For now we note that, regardless of the efficiency of energy conversion from neutrino luminosity to annihilation luminosity (which could be quite low, on the order of 1 per cent, see [12]), it appears that the time–dependence of $L_\nu$ is such that neutrinos could only be responsible for a very short energy release ($L_\nu \propto t^{-5/4}$, which means that $L_{\nu\bar{\nu}} \propto L_\nu^2 \propto t^{-5/2}$), and thus would be unable to power a burst lasting several tenths of a second or more. This does not mean that it would have a negligible impact on the structure of the burst itself.

The evolution of the accretion disks such as the ones treated here should be studied with time–dependent models, since the system is clearly not in a steady state, even from its inception. For example, if one assumes a steady state $\alpha$–type solution for the disk structure, the circulation pattern seen in Figure 1 is not taken into account, and in fact it lengthens the lifetime of the disk, by moving matter to larger radii continously.

## ACKNOWLEDGMENTS

We thank M. Rees and W. Kluźniak for many helpful convsersations. WHL thanks the IoA for its hospitality. Financial support for this work was provided in part by CONACyT (27987E), PAPIIT (IN-110600) and the Royal Society.

## REFERENCES

1. Eichler, D., Livio, M., Piran, T., and Schramm, D. N., *Nature*, **340**, 126 (1989).
2. Kluźniak, W., and Lee, W. H., *ApJ*, **494**, L53 (1998).
3. Kalogera, V., Narayan, R., Spergel, D. N., and Taylor, J. H., *ApJ*, **556**, 340 (2001).
4. Rees, M. J., *A& AS*, **138**, 491 (1999).
5. Usov, V. V., *Nature*, **357**, 472 (1992).
6. Mészáros, P., and Rees, M. J., *ApJ*, **482**, L29 (1997).
7. Kluźniak, W., and Ruderman, M., *ApJ*, **505**, L113 (1998).
8. Lee, W. H., *MNRAS*, **328**, 583 (2001).
9. Flebbe, O., Münzel, S., Herold, H., Riffert, H., and Ruder, H., *ApJ*, **431**, 754 (1994).
10. Balsara, D., *J. Comp. Phys.*, **121**, 357 (1995).
11. Abramowicz, M. A., Calvani, M., and Nobili, L., *Nature*, **302**, 597 (1983).
12. Popham, R., Woosley, S. E., and Fryer, C., *ApJ*, **518**, 356 (1999).

# Neutron Star Binaries as Central Engines of GRBs

## S. Rosswog

*Dep. Phys. and Astronomy, University of Leicester, LE1 7RH Leicester, UK*

**Abstract.** We describe the results of high resolution, hydrodynamic calculations of neutron star mergers. The model makes use of a new, nuclear equation of state, accounts for multi-flavour neutrino emission and solves the equations of hydrodynamics using the smoothed particle hydrodynamics method with more than $10^6$ particles. The merger leaves behind a strongly differentially rotating central object of $\sim 2.5$ $M_\odot$ together with a distribution of hot debris material. For the most realistic case of initial neutron star spins, no sign of a collapse to a black hole can be seen. We argue that the differential rotation stabilizes the central object for $\sim 10^2$ s and leads to superstrong magnetic fields. We find the neutrino emission from the hot debris around the freshly-formed, supermassive neutron star to be substantially lower than predicted previously. Therefore the annihilation of neutrino anti-neutrino pairs will have difficulties powering very energetic bursts ($\gg 10^{49}$ erg).

## INTRODUCTION

There is growing observational evidence that the subclass of long Gamma Ray Bursts (GRBs) is related to star forming regions (e.g. [4]). While this connection is immediately evident for the short-lived progenitors of collapsars, the question of whether neutron star binaries merge close to star forming regions or not is not a settled one. Bloom et al. (2002), for example, argue against compact object mergers as central engines of (at least the subclass of the long) GRBs. Belczynski et al. (2001), however, claim to have identified new formation channels that lead to classes of very tight, short-lived neutron star binaries. They find typical inspiral times of the order $10^6$ years and therefore neutron star binaries would merge very close to their birth places, even if equipped with high systemic velocities due to kicks in asymmetric supernova explosions.

Recent afterglow observations of long bursts suggest that they are beamed and require $E_\gamma \sim 5 \cdot 10^{50}$ erg [8]. The energy requirements for the subclass of short bursts are so far essentially unconstrained. Due to their gravitational binding energy of several times $10^{53}$ erg, neutron star binaries certainly do possess the energy reservoirs necessary to power a (long) burst, however, how to transform the available energy into gamma rays is still far from being clear. Among the suggested mechanisms are magnetic energy extraction processes (e.g. [6, 11, 22]) and the annihilation of neutrino-antineutrino pairs emitted from the hot neutron star debris during the coalescence [7].

## HIGH RESOLUTION SIMULATIONS OF THE MERGER EVENT

To study the possible role of neutron star coalescences for either long or short bursts we have performed detailed high-resolution simulations of the merger event [15]. To solve the equations of fluid dynamics we apply the smoothed particle hydrodynamics method (SPH) together with a largely improved artificial viscosity tensor [14]. We treat the self-gravity of the neutron star fluid in a Newtonian way, but we add the forces emerging from the emission of gravitational waves to drive the system towards coalescence. The microphysical properties of the hot and dense neutron star matter are described using an equation of state (EOS) that is based on the tables of Shen et al. [19, 20]. We have extended the EOS to the low density regime via a gas consisting of neutrons, protons, alpha particles, electron-positron pairs and photons. This new EOS covers the whole relevant parameter space in density, temperature and electron fraction. To allow for cooling and compositional changes, we have implemented a detailed neutrino treatment that accounts for three neutrino flavours ($\nu_e, \bar{\nu}_e$, and $\nu_x$, which is collectively used for the four heavy-lepton neutrinos) and takes the relevant emission processes (lepton capture on nucleons, electron-positron pair annihilation and plasmon decay) into account. The opacities are calculated from the absorption processes of the electron-type neutrinos on nucleons and scattering off nucleons and nuclei, the latter process becoming the dominant opacity source even for moderate mass fractions of heavy nuclei. The calculations are performed efficiently on shared-memory

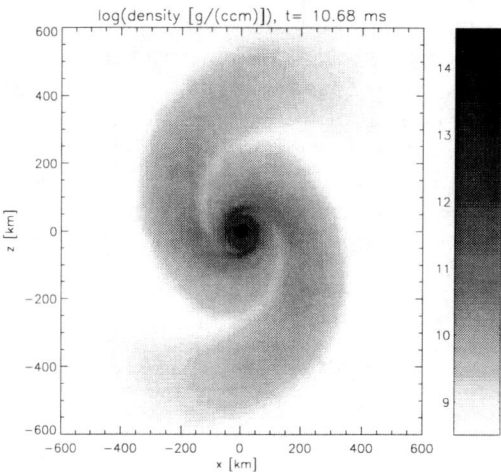

**FIGURE 1.** Coalescence of a corotating binary system of 1.4 $M_\odot$ per star. Colour-coded is the matter density in the orbital plane. More than $10^6$ SPH-particles were used for this calculation.

parallel computers with more than $10^6$ SPH-particles. For a more detailed description of the input physics and the computational methods we refer to Rosswog et al. [15, 16].

## RESULTS

In this latest set of calculations we explore systems with two initial spin configurations (for details see [15]): tidally locked, corotating systems for the ease of constructing initial equilibrium models and systems without initial neutron star spins. The latter ones are the most relevant spin configurations since the intrinsic neutron star viscosity is too low to lead to a tidal locking during the short phase where the binary components undergo tidal interaction [3, 10]. We find central objects of 2.3 to 2.6 $M_\odot$ which are strongly differentially rotating. For the most realistic, irrotational case the maximum density does not even reach the initial density of a single, cold non-rotating neutron star. Apart from the thermal pressure and the differential rotation there may be further effects that, at least temporarily, stabilize the merger remnant against collapse to a black hole: the presence of non-leptonic, negative charges together with trapped neutrinos [13], for example, can substantially increase the maximum possible mass. One also expects magnetic seed fields to be amplified in the differentially rotating remnant to enormous field strengths ($\sim 10^{17}$ G) [6]. Fields of this strength can substantially modify the structure of the central object and provide additional support against collapse [5]. Due to our ignorance of the high-density equation of state it cannot be ruled out that the end product of the coalescence is a stable supermassive neutron star of $\sim 2.8$ $M_\odot$. It seems, however, more likely that the central object is only temporarily stabilized and once the stabilizing effects weaken (e.g. neutrinos have diffused out after $\sim 10$ s, magnetic braking has damped out differential rotation) collapse to a black hole will set in. The time scale until collapse is difficult to determine, since it depends sensitively on poorly known physics and on the specific system parameters. We expect the most important effect to come from the differential rotation [12, 1]. Assuming the dominant effect to come from magnetic dipole radiation (the viscous time scale is estimated to be $\sim 10^9$ s [18]), this time is given by

$$\tau_c \sim \frac{18c^3 M}{5B^2 R^4 \omega^2}$$
$$\sim 10^2 s \left(\frac{M}{2.5 M_\odot}\right) \left(\frac{10^{16} G}{B}\right)^2 \times \left(\frac{15 km}{R}\right)^4 \left(\frac{3000 s^{-1}}{\omega}\right)^2,$$

where $M, B$ and $R$ are mass, magnetic field and radius of the central object. Such an object, an at least temporarily stabilized, supermassive neutron star with enormous magnetic field strength, is at the heart of many suggested GRB models (e.g. [6, 22, 9]). Kluzniak and Ruderman, for example, estimate that the magnetic field becomes buoyant at $\sim 10^{17}$ G, floats up and breaks through the surface of the remnant as a sub-burst. This process, winding up the field to buoyancy, subsequent floating up and sub-burst, would continue until the energy stored in differential rotation is used up (or collapse sets in), $\sim 100$ s. Therefore even long bursts, with substructures on millisecond time scales set by the motion of the magnetized fluid, could result from a neutron star merger. Once the differential rotation has been damped out, the system could still continue as a gamma ray burster using the energy stored in rigid rotation [21] in case it remains stable, or, in case of collapse, one would be left with the 'classic' GRB-engine, a black hole plus a debris torus.

The annihilation of neutrino anti-neutrino pairs into electron-positron pairs above the poles of the merger remnant has been suggested [7] as a process to produce a fireball from thermal energy stored in the disk. We find that the main neutrino-emitting region is the inner, shock- and shear-heated region of the torus around the central object. The prevailing densities lie between $10^{10}$ and $10^{12}$ gcm$^{-3}$ and temperatures are $\sim 2$-$3$ MeV. In our simulations we find typical neutrino luminosities of $\sim 10^{53}$ erg/s with mean energies of $\sim 10, \sim 15$ and $\sim 25$ MeV (see Fig. 2) for an initially corotating system, and slightly higher values for irrotational systems [16]. The dominant emission process is the capture of positrons onto neutrons which occurs in the hot, neutron-rich inner regions of the debris torus. The found luminosities

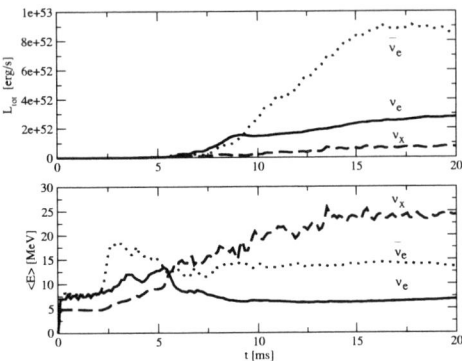

**FIGURE 2.** Total neutrino luminosities and mean energies (in MeV) for the $\nu_e$, $\bar{\nu}_e$ and $\nu_x$ emitted in the coalescence of a corotating neutron star binary system of twice 1.4 $M_\odot$. The abscissa gives the time in ms.

are lower than those found previously by Ruffert and Janka [17] by roughly a factor of 3 to 4. The annihilation efficiency is currently being investigated in detail, but the preliminary results indicate that it will be difficult to power Gamma Ray Bursts largely in excess of $\sim 10^{49}$ ergs.

## SUMMARY

We have performed high resolution calculations of neutron star coalescences with more than $10^6$ SPH-particles, using a new, nuclear equation of state and accounting for the effects of cooling and compositional changes from neutrino reactions by a detailed multi-flavour neutrino treatment. For the most realistic case with negligible neutron star spins, we do not see any sign of a collapse to a black hole and therefore argue that the outcome of the coalescence is a (at least temporarily) stabilized, supermassive, hot, differentially rotating object with huge magnetic fields. Such an object could produce a GRB in various ways, e.g. via a relativistic electron-positron wind [22] or via the so-called DROCO-mechanism (differentially rotating compact object) suggested by Kluzniak and Ruderman (1998) where the field inside the differentially rotating central object becomes locally wound up until it becomes buoyant at $\sim 10^{17}$ G, floats up and breaks through the surface as a sub-burst. This self-limited process would continue until the differential rotational energy is used up and would therefore continue for several seconds with substructures given by the fluid instabilities on a millisecond time scale.

We have further analyzed the neutrino signal expected from the event. The neutrino emission is dominated by anti-neutrinos produced in positron captures, the total luminosities lie around $\sim 10^{53}$ erg/s and are substantially lower than those found in previous investigations. Preliminary analysis of the neutrino anti-neutrino pair annihilation efficiency suggests that it is difficult to power energetic bursts with energies $\gg 10^{49}$ ergs via this mechanism. Whether this energy is enough to power a short GRB will have to be clarified by future afterglow observations.

## ACKNOWLEDGMENTS

The calculations reported above were performed on the United Kingdom Astrophysical Fluids Facility (UKAFF) and on the supercomputer of the Math Modelling Center of the University of Leicester.

## REFERENCES

1. Baumgarte T., Shapiro S., Shibata M., 2000, ApJ, 528, L29
2. Belczynski, K., Bulik, T. and Rudak, B., astro-ph/0112122(2001)
3. Bildsten,L. and Cutler,C., ApJ, 400, 175 (1992)
4. Bloom, J.S, Kulkarni, S.R. and Djorgoski, S.G., AJ, 123, 1111 (2002)
5. Cardall, C, et al. ApJ, 554, 322 (2001)
6. Duncan, R. and Thompson, C., ApJ, 392, L9 (1992)
7. Eichler.D. et al., Nature, 340, 126 (1989)
8. Frail, D. et al., ApJ, 562, L55 (2001)
9. Kluzniak, W. and Ruderman, M., ApJ, 505, L113 (1998)
10. Kochanek, C.S., ApJ, 398, 234 (1992)
11. Narayan, R., Paczynski, B. and Piran, T., ApJ, 395, L83 (1992)
12. Ostriker J., Bodenheimer P., 1968, ApJ, 151, 1089
13. Prakash, M. et al., Phys. Rev D52, 661 (1995)
14. Rosswog, S., Davies, M.B., Thielemann, F.-K. and Piran, T., A&A 360, 171 (2000)
15. Rosswog, S. and Davies, M.B., MNRAS in press (2002)
16. Rosswog, S. et al., in preparation (2002)
17. Ruffert, M. and Janka, T., A&A, 380, 544 (2001)
18. Shapiro, S.L., ApJ, 544, 397 (2000)
19. Shen, H. et al., Nucl. Phys. A637, 435 (1998)
20. Shen, H. et al., Prog. Theor. Phys. 100, 1013 (1998)
21. Usov, V.V., Nature, 357, 472 (1992)
22. Usov, V.V., MNRAS, 267, 1035 (1994)

# A Model for Short Gamma-Ray Bursts: Heated Neutron Stars in Close Binary Systems

Jay D. Salmonson* and James R. Wilson*

*Lawrence Livermore National Laboratory, P.O. Box 808, Livermore, CA 94551*

**Abstract.** In this paper we present a model for the short (< second) population of gamma-ray bursts (GRBs). In this model heated neutron stars in a close binary system near their last stable orbit emit neutrinos at large luminosities ($\sim 10^{53}$ ergs/sec). A fraction of these neutrinos will annihilate to form an $e^+e^-$ pair plasma wind which will, in turn, expand and recombine to photons which make the gamma-ray burst. We study neutrino annihilation and show that a substantial fraction ($\sim 1/2$) of energy deposited comes from inter-star neutrinos, where each member of the neutrino pair originates from each neutron star. Thus, in addition to the annihilation of neutrinos blowing off of a single star, we have a new source of baryon free energy that is deposited between the stars. To model the $e^+e^-$ pair plasma wind between stars, we do three-dimensional relativistic numerical hydrodynamic calculations.

Preliminary results are also presented of new, fully general relativistic calculations of gravitationally attracting stars falling from infinity with no angular momentum. These simulations exhibit a compression effect.

## INTRODUCTION

In Salmonson et al. [1] a model was presented for the production of a GRB in a close neutron star binary system, near its last stable orbit. In that model the stars undergo compression due to non-linear general relativistic effects. Vortices and shocks within the stars will convert this compressional energy into thermal energy which will be radiated from the stars in neutrinos. These neutrinos emerging from the neutron stars will partially recombine via $\nu\nu \to e^+e^-$, an effect that is substantially augmented (up to 30 times) by bending of neutrino paths by strong gravitational fields [2]. Thus an $e^+e^-$ pair plasma fireball emerges from the neutron stars and expands relativistically. A key parameter studied in Salmonson et al. [1] was the entropy per baryon, $s$, of the plasma, representing the amount of baryons entrained in the $e^+e^-$ fireball. It was found through 1D relativistic hydrodynamic simulations that if $s \sim 10^8$, then prompt gamma-ray emission would result from the eventual recombination of $e^+e^- \to \gamma\gamma$. If the entropy is as low as $s \sim 10^6$, effectively all of the energy would be transferred as kinetic energy to the baryons, thus resulting in gamma-ray emission from an external shock as the relativistic baryons sweep into the interstellar medium. Each scenario was studied in detail in Salmonson et al. [1].

In the current work we employ three enhancements. The first is from recent refined simulations of the compression by Wilson & Mathews which show timescales $\sim 1$ second, thus this model is best suited to describe the short class of GRBs (< second). Second, we use new calculations by Salmonson and Wilson [3] of the $\nu\nu \to e^+e^-$ between the neutrons stars where each neutrino of the annihilation pair originates from each star. This effect is found to be of about the same importance as that of the previously considered annihilation from individual netron stars [2]. Thus we have a new source of baryon-free $e^+e^-$ plasma between the neutron stars. Third, we implement fully relativistic 3D hydrodynamics to model the plasma expansion in the complex, rotating inter-star environment.

## THE MODEL

In this model we estimate that 10% of the binding energy of a neutron star ($\sim 10^{53}$ ergs) is converted to thermal energy via compression, vortices and shocks. This $10^{52}$ ergs of energy is released as a monotonically increasing luminosity of neutrinos over a timescale of order $\sim 1/10$ second as found by calculations of J. R. Wilson & G. J. Mathews. This neutrino luminosity $\sim 10^{53}$ ergs sec$^{-1}$ annihilates into an $e^+e^-$ pair plasma. Annihilation of high neutrino luminosities in the strong gravitational field of the neutron stars can have high efficiencies; near unity [1]. About half of the pair plasma energy is deposited uniformly around the neutron stars due to single star neutrino annihilation [2] (Figure 1) and the

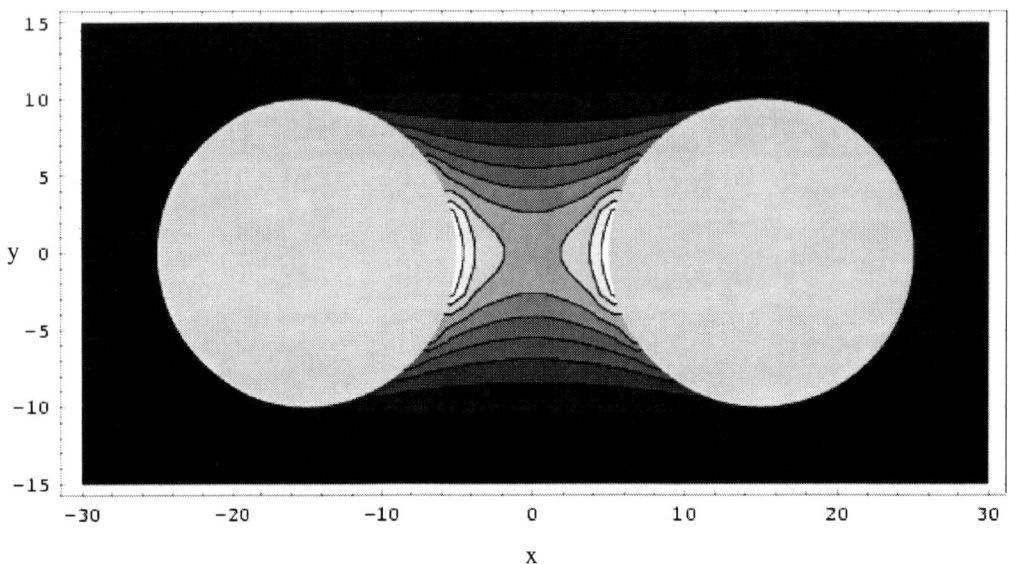

**FIGURE 1.** A contour plot of interstar neutrino annihilation energy depostion rates on a slice through a neutron star binary system [3], with stellar radii 10 km and separation 30 km. Lighter colors correspond to higher levels of energy deposition where black includes zero.

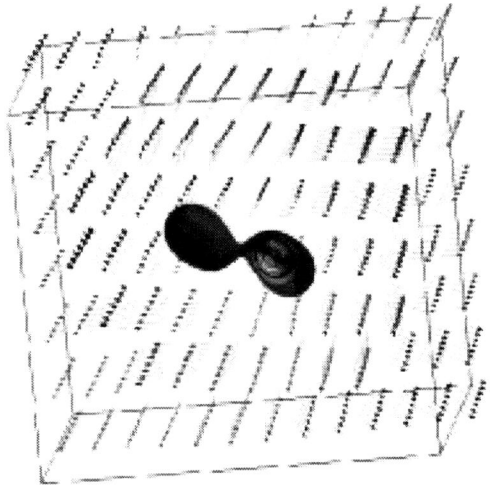

**FIGURE 2.** Three dimensional relativistic simulation of two 10 km radius neutron star separated by 30 km emitting $10^{53}$ ergs/sec of energy in $e^+e^-$ pair plasma and about 1 % equivalent mass in baryons. The contour map, with right star cutaway, is of baryon density. The vector field is the 3-velocity of expanding plasma. This problem settles down to a static flow after about one orbit with period $\sim 1/300$ second.

other half is deposited between the stars due to interstar neutrino annihilation [3].

This plasma deposition morphology then becomes input for the 3D relativistic hydrodynamic code, which calculates the expansion of the plasma (Figure 2). These simulations show a plasma of very high entropy expanding out along the plane of symmetry between the neutron stars. In the regions around the stars lower entropy plasma is formed because a baryon wind is blown from the stars.

Thus this model predicts a variety of bursts. Viewed along the axis of rotation, a prompt quasi-thermal burst of duration $\sim 1/10$ second will result from the annihilation of fireball pairs. Because of the dearth of baryons left over to sweep into the interstellar medium, we do not predict the existence of an afterglow. This agrees with preliminary searches of the data archives for short-burst afterglows, which appear to be missing [4].

Viewed far from the axis of rotation a very different burst results. The lower entropy means that there will not be a prompt burst from pair annihilation in the fireball. However, there will be a baryon wind sweeping into the interstellar medium. Thus we expect a burst that decays into an afterglow as a power-law. This behavior will be made chaotic and complex by the rapid rotation of the binary system.

## THE COMPRESSION EFFECT IN STRONGLY GRAVITATING SYSTEMS

In Wilson et al. [5] it was reported that neutron star binaries near their last stable orbit undergo a compression due to non-linear general relativistic (GR) effects. This effect could be strong enough to crush the stars to black holes

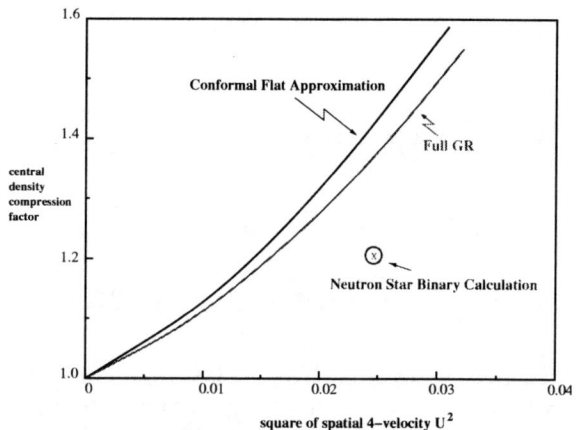

**FIGURE 3.** A schematic diagram of the increase in central density as a function of the spatial 4-velocity of two close neutron stars observed in three different calculations. The point marked with an 'X' shows the magnitude of the compression calculated by Mathews and Wilson [6] in neutron star binary calculations at their last stable orbit. The curves show two calculations for two neutron stars falling toward eachother from infinity, with no angular momentum. One curve is done in full general relativity (GR), made possible by the axisymmetry of this problem, and the other with the conformal flat approximation (CFA) used in the neutron star binary calculations [6]. One can see that all the CFA does a good job of reproducing the solution of the exact, full GR calculations. Also, we see that compression is observed for both rotating and linear systems.

and perhaps release binding energy as thermal neutrinos in the process. Since that report, this effect has remained controversial.

Wilson recently has done a similar calculation of two stars gravitationally falling together without angular momentum. The calculations were done both in full GR and using the "conformal flat approximation" (CFA), often cited by critics to be the spurious source of the compression effect. Preliminary results, schematically shown in Figure 3, demonstrate a compression effect for both calculations. Thus, not only is the CFA not the source of the compression, but perhaps the compression effect, here demonstrated in a non-rotating system, is more general than previously thought.

This work was performed under the auspices of the U.S. Department of Energy by University of California Lawrence Livermore National Laboratory under contract W-7405-ENG-48.

## REFERENCES

1. Salmonson, J. D., Wilson, J. R., and Mathews, G. J. (**2001**). "Gamma-ray bursts via the neutrino emission from heated neutron stars," ApJ, **553**, 471–487.
2. Salmonson, J. D., and Wilson, J. R. (**1999**). "General relativistic augmentation of neutrino pair annihilation energy deposition near neutron stars," ApJ, **517**, 859–865.
3. Salmonson, J. D., and Wilson, J. R. (**2001**). "Neutrino annihilation between binary neutron stars," ApJ, **561**, 950.
4. Gandolfi, G. e. a. (**2001**). "Bepposax results on short gamma ray bursts," To appear in the proceedings of Gamma-Ray Burst and Afterglow Astronomy 2001.
5. Wilson, J. R., Mathews, G. J., and Marronetti, P. (**1996**). "Relativistic numerical model for close neutron-star binaries." Phys. Rev. D, **54**, 1317–1331.
6. Mathews, G. J., and Wilson, J. R. (**2000**). "Revised relativistic hydrodynamical model for neutron-star binaries." Phys. Rev. D, **61**, 127304.

# Relativistic Jets in Collapsars

Weiqun Zhang*, S. E. Woosley* and A. I. MacFadyen*

*Department of Astronomy and Astrophysics, University of California, Santa Cruz, CA 95064, USA

**Abstract.** We examine the propagation of two-dimensional relativistic jets through the stellar progenitor in the collapsar model for gamma-ray bursts. In agreement with previous studies, we find that the jet is collimated by its passage. Moreover, interaction of the jet with the star causes mixing that sporadically decelerates the jet, leading to a highly variable Lorentz factor. The jet that finally emerges has a moderate Lorentz factor, but a very large internal energy loading. In a second series of calculations we follow the emergence of such enegy-loaded jets from the star. For the initial conditions chosen, conversion of the remaining internal energy gives a terminal Lorentz factor of approximately 150. Implications of our calculations for GRB light curves, the luminosity-variability relation, and the GRB-supernova association are discussed.

## INTRODUCTION

Growing observational evidence supports the contention that at least the long-soft gamma-ray bursts (GRBs) are the consequence of massive stars being exploded by relativistic jets. The "collapsar" model [1, 2] is one of the most promising models. Here we examine in some detail the propagation of relativistic jets along the rotational axis of a spinning, collapsing star (see also Aloy et al. [3]). We are particularly interested in the interaction of the jet with the star - how it is focused and modulated during its passage to the stellar surface - and in the evolution of the jet immediately after it breaks free.

Our simulations begin with a $15 M_\odot$ helium star derived from a $40 M_\odot$ main sequence star [4] whose iron core has collapsed to a black hole [2]. Because of the difficulty of carrying too large a dynamic range on the computational grid, we break up our study [5] into two stages: the propagation of the jet a) inside the star, and b) outside, in the stellar wind, after breakout. The results from the first series of calculations provide the initial conditions for the second, and show that the emerging jet still contains a lot of internal energy. The conversion of this remaining energy determines the final kinetic energy and Lorentz factor of the jet as well as its opening angle.

## JETS INSIDE STARS

We presume that a highly relativistic jet has already formed inside the inner boundary of our first computational grid at $2 \times 10^8$ cm and study its evolution only after it enters the grid. In particular, an axisymmetric jet is injected radially through the inner boundary with an opening angle, $\theta_0$, here taken to be a free parameter. The initial jet is additionally defined by its total energy flux, $\dot{E}$, initial Lorentz factor, $\Gamma_0$, and the ratio of its kinetic energy to total energy, $f_0$. Two models were followed: (A) $\dot{E} = 10^{51}$ erg s$^{-1}$ (counting both axes - up and down), $\theta_0 = 20°$, $\Gamma_0 = 50$, $f_0 = 0.33$; and (B) $\dot{E} = 10^{51}$ erg s$^{-1}$, $\theta_0 = 5°$, $\Gamma_0 = 50$, $f_0 = 0.33$.

After a short time the jet in both models consists of a supersonic beam, a cocoon composed of shocked jet material, shocked medium gas, a terminal bow shock, a working surface, and backflows. In both models, the jet is narrowly collimated and its beam, very thin.

Along the polar axis, the beam divides into two regions: an unshocked region with characteristics sensitive to the initial conditions, and a shocked region. In the unshocked region, the internal energy of the initial beam is converted by expansion into kinetic energy. At the head of the unshocked region, most of this kinetic energy is converted back into internal energy by shock dissipation. The Lorentz factor rises in the unshocked region because the internal energy is converted into kinetic energy, then drops sharply again when the beam is shocked (Fig. 1). In the shocked region, the jet dynamics imprints time structure on the Lorentz factor. Because the jet has a larger initial opening angle, the jet material in Model A flowing through the first shock has less energy per area. The average Lorentz factor in the shocked region in Model A is thus lower than in Model B. Along the polar axis, the shocked beam is still highly supersonic and is thus shocked once again before it encounters the contact discontinuity between the jet and the star. Also, jet mate-

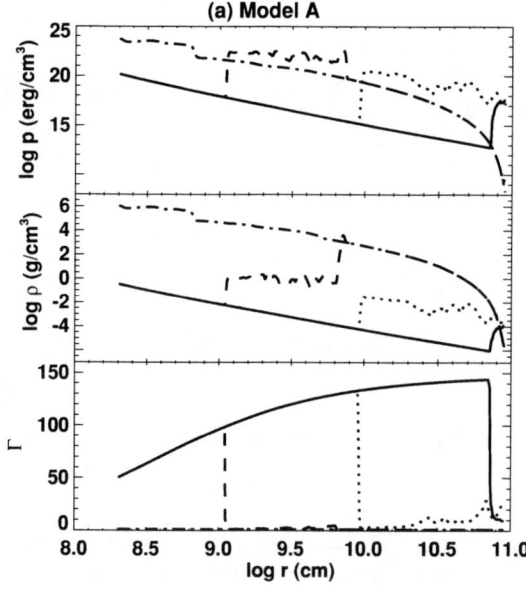 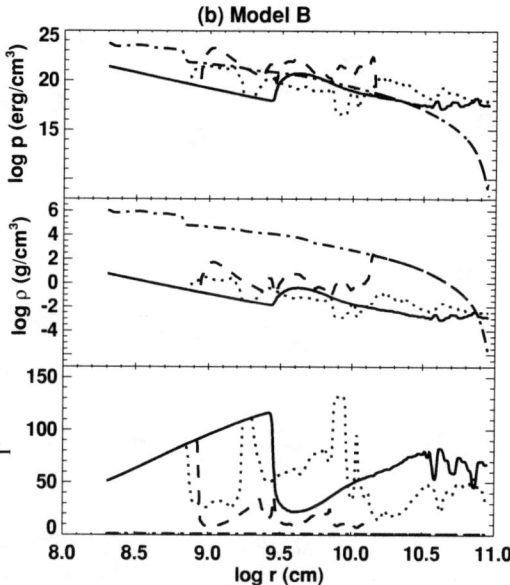

**FIGURE 1.** Pressure, density, and Lorentz factor vs. radius along the polar axis for: (a) Model A (*left*), and (b) Model B (*right*). Different lines are for different times: (a) $t = 0.0$ s (*dash dot line*), $t = 2.0$ s (*dashed line*), $t = 8.0$ s (*dotted line*), and $t = 20.0$ s (*solid line*); (b) $t = 0.0$ s (*dash dot line*), $t = 1.0$ s (*dashed line*), $t = 4.0$ s (*dotted line*), and $t = 10.0$ s (*solid line*).

rial at the head of the beam flows backwards along the beam and forms a cocoon. The jet beam and cocoon are thinner in Model B and more likely to be affected by the Kelvin-Helmholtz instability and by oblique shocks moving towards the axis. Mixing between the jet and the cocoon imposes time structure on the Lorentz factor. Consequently, the jet in Model B is more highly variable.

As the jets pass through the star, they are narrowly collimated by external pressure. Eventually though, the jets break free of the star with small opening angles of $\sim 4°$, and $3°$ respectively, for Models A and B. The Lorentz factor of the jets at breakout are $\sim 10$, and 50, for Models A and B. However, both still contain a lot of internal energy, $\sim 80$-$95\%$ and $\sim 50$-$75\%$ of the total energy for Models A and B, respectively. After conversion of this internal energy to kinetic, the jets attain final Lorentz factors $\sim 100$-$200$ in both cases.

## JETS IN THE STELLAR WIND

After it breaks out the star, a jet loaded with internal energy will inevitably experience lateral expansion. To study the fate of the jet and its role in making GRBs, we carried out a second series of simulations that followed jet propagation in the stellar wind. In this study a constant jet with opening angle $5°$ is injected into the stellar wind at the outer edge of the star. The jet is left on for 10 seconds in a radial direction and then gradually shut off (linearly with time). Again we considered two models:

(A2) $\dot{E} = 10^{51}\,\mathrm{erg\,s^{-1}}$, $\theta_0 = 5°$, $\Gamma_0 = 10$, $f_0 = 0.0667$;
(B2) $\dot{E} = 10^{51}\,\mathrm{erg\,s^{-1}}$, $\theta_0 = 5°$, $\Gamma_0 = 50$, $f_0 = 0.33$. These two Models A2 and B2 should approximately reflect the behavior of our earlier Models A and B after they break out.

In both cases, the Lorentz factor increases to over 100 as the jet propagates in the stellar wind, but there are large differences between Models A2 and B2. Although the initial opening angle for both models is $5°$, the opening angle at the head of the jet at $t = 35$ s is $15°$ and $5°$ respectively for Models A2 and B2 (Fig. 2). In Model A2, the jet initially has a large internal energy loading (93% of the total energy) and Lorentz factor 10, so the lateral expansion causes a $15°$ opening angle at $t = 35$ s. In Model B2, the internal energy loading is low (67% of the total energy) and the initial Lorentz factor is 50, so the lateral expansion is small. This series of simulations shows that both the Lorentz factor and internal energy loading when the jet breaks out of the star will decide its final opening angle.

When the jet is shut down gradually after 10 seconds, its Lorentz factor is assumed to decrease. As a result, jet experiences more lateral expansion at late times. Although the details depend on the precise recipe used, a jet characterized by a narrow head and wide wings of low energy material may occur in many cases. These wings are mildly relativistic (Fig. 2).

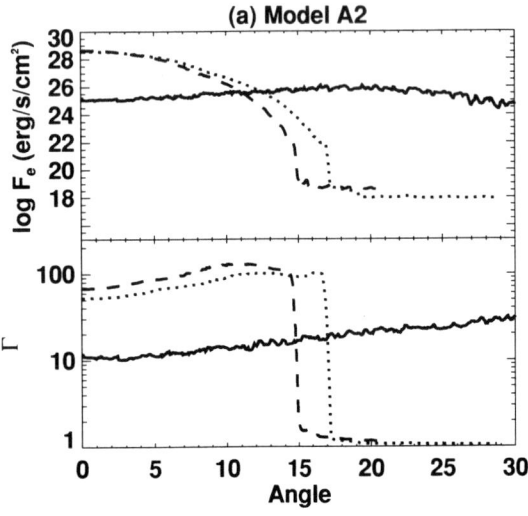
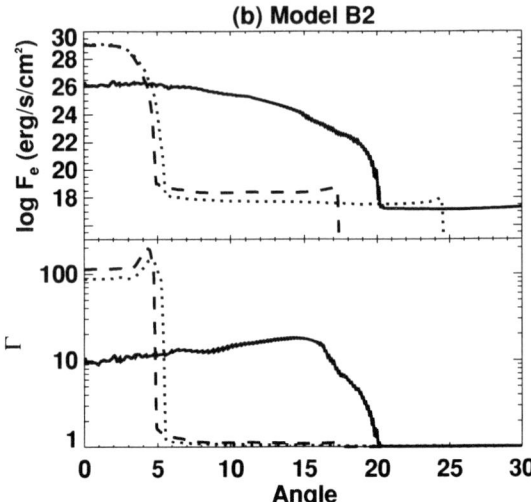

**FIGURE 2.** Pressure, density, and Lorentz factor vs. radius along the polar axis for: (a) model A (*left*), and (b) model B (*right*). Different lines are for different times: (a) $t = 0.0$ s (*dash dot line*), $t = 2.0$ s (*dashed line*), $t = 8.0$ s (*dotted line*), and $t = 20.0$ s (*solid line*); (b) $t = 0.0$ s (*dash dot line*), $t = 1.0$ s (*dashed line*), $t = 4.0$ s (*dotted line*), and $t = 10.0$ s (*solid line*).

## CONCLUSIONS

Our special relativistic calculations show, insensitive to the initial opening angle, that a jet originating near the center of a collapsing massive star will emerge with an opening angle ∼5 degrees. The final opening angle depends on the internal energy loading and Lorentz factor at breakout. This has implications for the observed GRB energy and event rate.

We find that the mixing of the jet with the surrounding star by the Kelvin-Helmholtz instability acts to modulate the beam appreciably. Even for a constant beam energy and angle at the base, what emerges at the top has a highly variable time structure and opening angle. *It is these variations, which have nothing to do with the central engine, that determine the emission efficiency and light curve of a GRB.* If a large amount of matter mixes in, the burst may even turn off for a while.

Our calculations also demonstrate a correlation between opening angle and variability. A narrower jet experiences more instabilities and therefore the GRB is more variable. For a given total energy input at the bottom, a narrower jet also carries more energy per solid angle. Perhaps this explains the observed correlation between luminosity and variability [6].

As the jet angle opens with time, because the star is exploding, and as the jet energy declines with time, because the black hole is accreting at a decreased rate, the energy per solid angle may decline precipitously. Material emerging with lower Lorentz factor but still high internal energy loading can expand and become visible at larger angles. These effects might explain the "anomalous" GRB 980425. Indeed, *all* long GRBs might be accompanied by "wings" visible over a larger range of angles. Such weak GRBs viewed far from the opening angle of the jet early on might dominate a distance-limited sample of GRBs.

## ACKNOWLEDGMENTS

This research has been supported by NASA (NAG5-8128 and MIT-292701) and the DOE ASCI Program (B347885).

## REFERENCES

1. Woosley, S.E. 1993, ApJ, 405, 273
2. MacFadyen, A.I., & Woosley, S. E. 1999, ApJ, 524, 262
3. Aloy, M.A., Müller, E., Ibáñez, J.M[a]., Martí, J.M[a]., & MacFadyen, A. 2000, ApJL, 531, L119
4. Heger, A., & Woosley, S.E. in preparation
5. Zhang, W., Woosley, S.E., & MacFadyen, A.I. 2002, ApJ, in preparation
6. Reichart, D. E., Lamb, D. Q., Fenimore, E. E., Ramirez-Ruiz, E., Cline, T. L., & Hurley, K. 2001, ApJ, 552, 57

# X-Ray Flashes and X-Ray Counterparts of Gamm-Ray Bursts

## John Heise

*Space Research Organization Netherlands, Sorbonnelaan 2, 3584CA Utrecht, Netherlands*

**Abstract.**
The brightest transients in the sky, Supernovae and Gamma-Ray Bursts, are associated with the collapse of cores of massive stars. They shine in the optical and in the gamma-ray sky. On theoretical grounds one would expect to see similar events in the x-ray and ultra-violet sky. Here we summarize recent observational evidence demonstrating the existence of X-ray bursts, termed X-ray Flashes (XRFs). We argue that they are most likely very large explosions on a cosmological distant scale, similar to Gamma-Ray Bursts (GRBs). They may either be highly redshifted GRBs, GRBs viewed at a large angle or another geometrical effect. Or they may be a new type of cosmic explosion expanding mildly relativistically, much larger than the initial expansion of supernova remnants, but less than the extreme relativistic cosmic fireballs of Gamma-Ray Bursts.

The ratio of the energy contained in the X-ray part (2-10 keV) of Gamma-Ray Bursts to the γ-range (50-350 keV) varies widely. GRBs for which this fraction is typically more than half, are referred to as X-ray Rich GRBs. The new class of x-ray transients, the X-ray Flashes, show a further extension of this ratio to above 1, where most of the energy is contained in the x-ray range. They have properties similar to the X-ray counterpart of GRBs, but do not trigger Gamma-Ray Burst experiments since they are not detected in the gamma-ray range above typically 100 keV. X-ray Flashes nevertheless appear to be related to Gamma-Ray Bursts.

Observationally X-ray Flashes are found as a subset of Fast X-ray Transients, which have been seen by almost all x-ray satellites. Their nature has remained unclear in most cases. It was generally assumed that the origin of FXTs is a mixture of detector artifacts and several types of astronomical events, including coronal emission from late type stars. Using the Wide Field camera on board BeppoSAX we have made a systematic study of FXTs and identify at least two types on the basis of their duration. The subset with a typical duration of order an hour contain identifications with active coronal sources. The other subset which last typically of the order of minutes, are the X-ray Flashes.

A recent X-ray Flash, XRF011030 or SAX J2043.6+7717 is shown to be associated with an afterglow in the radio and in the X-ray range and thus demonstrates the relation with a cosmic explosive event, which may very well be a new type of cosmic explosion associated with stellar collapse.

## INTRODUCTION

The brightest transients in the sky, supernovae (SNe) and gamma-ray bursts (GRBs), mark the death of stars. SNe shine in the optical sky and are powered by the collapse of cores of massive stars to neutron stars. Recent observational advances have established that GRBs are also cataclysmic events with explosive energies similar to those of SNe. An explosion is primarily described by the energy, $E_0$, and the mass of the exploding debris (or "ejecta", $M_B$); together, these parameters determine the initial speed (or equivalently the Lorentz factor, $\Gamma_0$) of the ejecta. The duration and the appearance (i.e. the electromagnetic band) of the transient (as well as that of its decay or "afterglow") depends strongly on $\Gamma_0$. SNe with $M_B \sim 1 M_\odot$ (or $\Gamma_0 \sim 1.006$) understandably shine in the optical sky, aided at early times by energy from radioactive decay. GRBs have $\Gamma_0 \sim 100$ (and inferred $M_B \sim 10^{-5} M_\odot$) which explains why the GRBs shine predominantly in the γ-ray sky. However, there is no convincing model explaining the clustering of $M_B$ for GRBs. It has long been realized (e.g. [1]) that, allowing for a range of $M_B$ one would expect explosive events which shine in the X-ray sky and the Ultra-violet sky. Separately, should GRBs occur at very high redshifts ($z \sim 5 - 10$) then such GRBs would manifest as long duration X-ray bursts. Thus for theoretical reasons X-ray bursts are expected on two grounds.

Here we summarize recent observational evidence demonstrating the existence of X-ray bursts, termed X-ray Flashes (XRF). Like GRBs, the XRFs are isotropically distributed on the sky with a rate of comparable to that of GRBs, about 1 burst per day over the entire sky. We first discuss the X-ray counterparts of GRBs, then

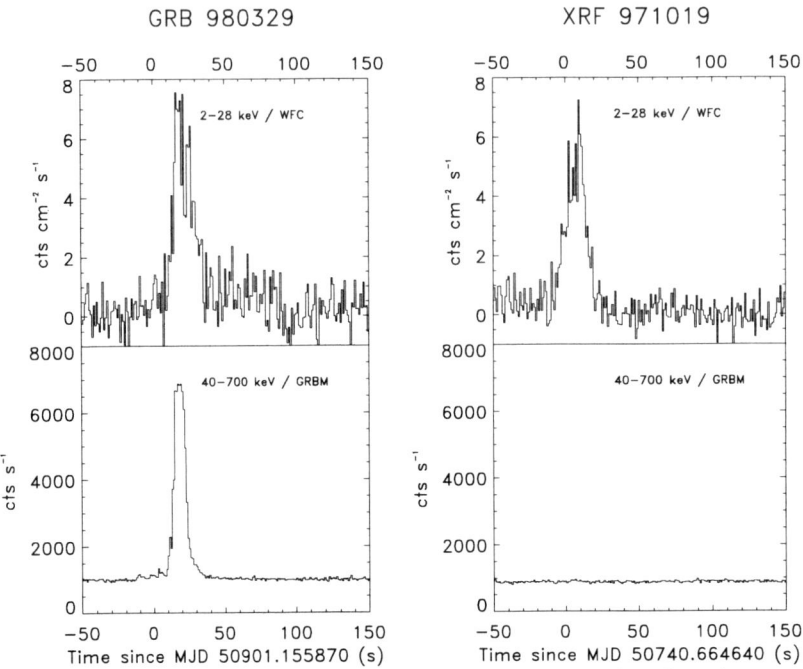

**FIGURE 1.** A GRBs (left) and X-Ray Flash (right) in the X-ray and γ-ray range plotted on the same scale demonstrates the absence of γ-rays of X-ray Flashes

show the recent evidence for at least two types of Fast X-ray Transients, one of which are the X-ray Flashes. The discovery of afterglows in the radio and x-ray range of a recent XRF demonstrates the link between XRFs and explosive events.

## X-RAY COUNTERPARTS OF GRBS

X-ray (1-8 keV) counterparts of GRBs were first detected in 1973 and 1974. Strohmayer et al. [2a] summarize the results observed with the *Ginga* satellite in the range 2-400 keV. Out of 120 GRBs in the operational period of 4.5 years between 1987 March and 1991 October, 22 events were studied. The average flux ratio of the X-ray energy (2-10 keV) to the gamma-ray energy (50-300 keV) is 0.24 with a wide distribution from 0.01 to more than unity. Photon spectra are well described by a low-energy slope, a bend energy, and a high-energy slope. The distribution of the bend energy extends to below 10 keV (see also Kippen et al., this volume).

The Wide Field Cameras (WFCs) on board Bep-poSAX (2-25 keV) combine a large field of view ($40^o \times 40^o$, full width to zero response) with a resolving power of 5 arcmin and allow for a fast position determination. Since the launch in 1996 about 50 X-ray counterparts of GRBs have been localized and studied. Typically error circles around the localization are between 2 and 3 arcmin with 99% confidence levels. These positions have been established after an average delay of 4 to 5 hours and are quickly disseminated. The error regions fall within the field of view of most optical, radio and x-ray telescopes and have triggered the discovery of afterglows in these wavelength bands. The average deviation between the WFC position and the optical transients found is within the error circle radius and consistent with statistics.

The average rate of GRB prompt counterparts observations is about 9 per year. The X-ray counterparts can be very bright with peak intensities that range between $10^{-8}$ and $10^{-7}$ erg/s/cm$^2$. The average spectra are characterized by a power law shape, with photon indices between 0.5 and 3 (see Heise et al. 2000 [17], Fig. 2). The T90 durations (the time interval from the time that 5% of the fluence is accumulated to 95%) range between 10 and 200 sec (see Heise et al. 2000 [17], Fig. 3).

## FAST X-RAY TRANSIENTS

Fast X-ray Transients (FXTs) generally are loosely defined as new temporary x-ray sources in the sky which disappear on a time scale of less than a few hours and which are not related to known persistent x-ray sources. The term has always been used as a trash can for all sorts of ill understood flares, bursts, flashes and related phenomena and may even include (in a few cases) x-ray detector artifacts. The origin of some of the events has been associated with RS CVns, nearby dMe stars, superflares from pre-main sequence stars and occasionally with the x-ray counterpart of gamma-ray bursts (GRBs).

Almost every X-ray satellite has seen such transient

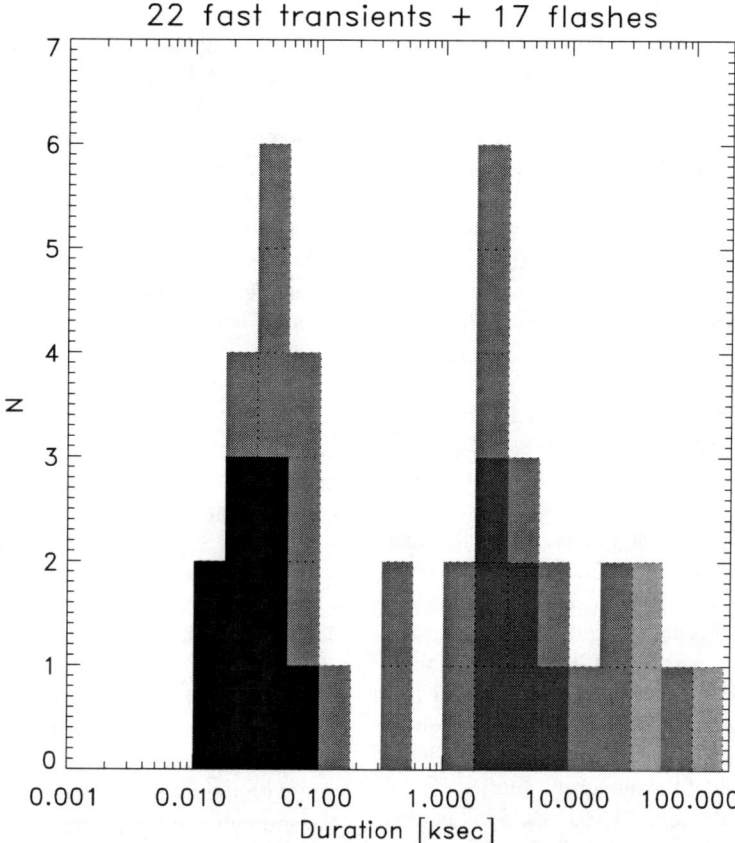

**FIGURE 2.** Histogram of the durations of Fast X-ray Transients observed with the Wide Field Cameras on BeppoSAX, showing a bimodal distribution. The long duration events ($\sim 1$ hour) contain identifications with coronas of active stars. All other long duration events are identifiable in the sense that a known X-ray source is present in the error box. The short duration events ($\sim$ min) are the X-ray Flashes (XRFs) discussed in the text. XRFs seen in the lowest BATSE channels are shaded darker in the black and white print of this figure.

phenomena in their field of view. The first one was detected[3] by the UHURU satellite in 1972. In the Ariel V sky survey [4] 27 events were observed in 5.5 years, obtained with a time resolution of one satellite orbit ($\sim 100$ min). About 20% of the sources are identified with active coronae in RS CVn systems and these are the longest ones, lasting of the order of an hour. Pye & McHardy[4] conclude that all are consistent with as yet unknown coronal sources. The frequency is estimated as one FXT every $\sim 3$ days above $4 \times 10^{-10}$ erg/s/cm$^2$ (2-10 keV).

Ambruster *et al.* observed[5] 10 FXTs with the A1 instrument on HEAO-1 (0.5-20 keV) above $\sim 7 \times 10^{-11}$ erg/s/cm$^2$, of which 4 are identified with coronal systems: 3 flare stars and 1 RS CVn systen. The duration is $> 10$ s and $< 1.5$ hr. In the A2 instrument (2-60 keV), 5 more FXTs have been detected[6]. Connors *et al.* suggest[6] that most of the events are hard coronal flares from dMe-dKe stars, with a flare rate of $2 \times 10^4$ per year above $\sim 10^{-10}$ erg s$^{-1}$cm$^{-2}$.

FXTs have also been seen[7] as serendipitous x-ray flashes in the Imaging Proportional Counter (IPC) onboard the Einstein Observatory. The 42 sources found are much shorter (up to 10 s) and weaker (limiting sensitivity of $10^{-11}$ erg s$^{-1}$cm$^{-2}$ in the 0.2-3.5 keV band). Grindlay suggested[8] that some FXTs might be the early x-ray afterglow of $\gamma$-ray bursts. A search[9] for such x-ray afterglows in the ROSAT all-sky survey leads to 22 candidates in the energy range 0.1-2.4 keV. The authors believe in a flare star origin in many, if not all, cases.

## Two types of Fast Transients

We observe Fast X-ray Transients, using the BeppoSAX-WFCs, in the x-ray range 2-25 keV. The total operation period of more than 5 years with the WFC has produced a homogeneous database with sufficient positional accuracy to search for identifications, and sufficient time coverage to study the light curves. A total of about 40 sources have been detected. They are seen at positions including high galactic latitude. The observed sky distribution is consistent with being isotropic.

There appears to be two classes of Fast X-ray Transients with different outburst time scales, as shown in a histogram of durations in Fig. 2. The bimodal distribution has a peak around a few minutes (17 sources) and a peak around an hour (22 sources). A subset of the short ones are X-ray counterparts of GRBs. Heise et al. called[17] the remaining FXTs of a short duration X-ray flashes. Their spectra are powerlaw like (which distinguish the flashes from type I x-ray bursts). The powerlaw photon index ranges between very soft spectra with photon index 3 to hard spectra with index 1.2. This powerlaw does not extend to the $\gamma$-ray range (see below), but a break occurs to a steeper spectrum, typically between 30 to 50 keV.

The long duration FXTs last between $2 \times 10^3$ and $2 \times 10^5$ s. In the latter class 9 out of 22 sources have been identified with galactic coronal sources (6 flare stars and 3 RS CVn variables). All other long duration sources are identifiable: they contain a known X-ray source in their error box. Since also all early Ariel V and HEAO 1 fast transient identifications with coronal sources pertain to long duration transients ($\sim 1-2$ hours), we consider it probable that the identity of the remaining sources on the longer time scales all are coronal sources. This is also the time scale observed in dedicated studies of coronal sources.

## Properties of X-ray flashes

X-ray flashes thus are bright x-ray sources with peak fluxes in the range $10^{-8}$ to $10^{-7}$ erg s$^{-1}$cm$^{-2}$, the same range as the x-ray counterparts of classical GRBs observed in the WFCs. The observed frequency of flashes is about one half of the frequency of GRBs, so they are quite common with a frequency of 1 per 2 days. How do we distinguish flashes from GRBs? The BATSE energy range starts at 25 keV, which is lower than the GRBM, which starts at 40 keV. BATSE detections of these soft events are therefore likely. 17 GRBs observed in the WFCs are also detected by BATSE and there is BATSE data during the times of 10 flashes. 9 flashes out of these 10 flashes that were potentially observable by BATSE, are actually detected[18] in either the lowest or the lowest two BATSE energy channels, resp. 25-50 keV and 50-100 keV. These events did not trigger the BATSE instrument. So we have a sample of both GRBs and flashes for which we can compare the x-ray fluence versus the $\gamma$-ray fluence.

Flashes are clearly distinct from GRBs in a plot of x-ray flux or fluence versus $\gamma$-ray flux or fluence. Gamma ray Bursts that are X-ray rich, such as GRB980326, GRB981226[19], and GRB990705[20] have a $\gamma$-ray flux on the low side of the distribution but still above the x-ray flashes. They may be considered as bridging cases between GRBs and XRFs.

The ratio of peak flux and fluence of XRFs in the X-ray range and gamma range extends the range observed in normal GRBs by a large factor. Peak flux ratios of XRFs extend up to a factor of 100, and fluence ratios extend up to a factor of 20.

## XRF011030 (SAX J2043.6+7717)

On 2001 October 30 we detected a new FXT, using the Wide Field Camera unit 1 on board the BeppoSAX satellite. The source is designated XRF011030, with the prefix for X-ray flash, and after localization it was called SAX J2043.6+7717. The observations started about 100 s before the first activity from this source was detected. The x-ray 2-28 keV light curve is shown in Fig. 3. It shows significant structure and the total duration is at least 1200 s. The source has a total fluence of $9 \times 10^{-7}$ erg cm$^{-2}$. The spectrum fits a powerlaw with a photon index of $-1.9 \pm 0.1$. A black body fit is inconsistent with the data at 99.97% confidence. The position of the source from the prompt x-ray observations has been disseminated[10, 11, 12] to the scientific community. This resulted in the discovery[13] of a new transient radio source on 2001 Nov 8.80 UT, 10 days after the flash. The new source was detected with high significance with a flux density of $181 \pm 18\mu$Jy. This new source is located near the center of the WFC error circle at (epoch 2000) R.A.=20:43:32.3, Dec.=+77:17:18.9 ($\pm 1$ arcsec in both coordinates). There is no visible optical counterpart[14] (limiting magnitude of R=22.9 magnitude, 5$\sigma$) to the radio source nor was a bright optical or IR transient seen following the detection of the X-ray flash. Harrison *et al.* detect[15] an X-ray source within 1.2" of the radio transient. The X-ray spectrum over 0.2-5 keV is consistent with a power law of photon index $\sim 1.45$, with interstellar absorption levels of $N_H \sim 2 \times 10^{21}$ cm$^{-2}$, consistent with the galactic value along the line of sight. The unabsorbed 2-10 keV flux is $2.4 \times 10^{-13}$ erg cm$^{-2}$ s$^{-1}$. They associate the x-ray source with the variable radio object.

## DISCUSSION

What could SAX J2043.6+7717 be? By several accounts, SAX J2043.6+7717 does not appear to be a traditional GRB. First, no simultaneous signal was detected in the Gamma-Ray Burst Monitor GRBM on the same platform or any other currently active gamma-ray detector. The

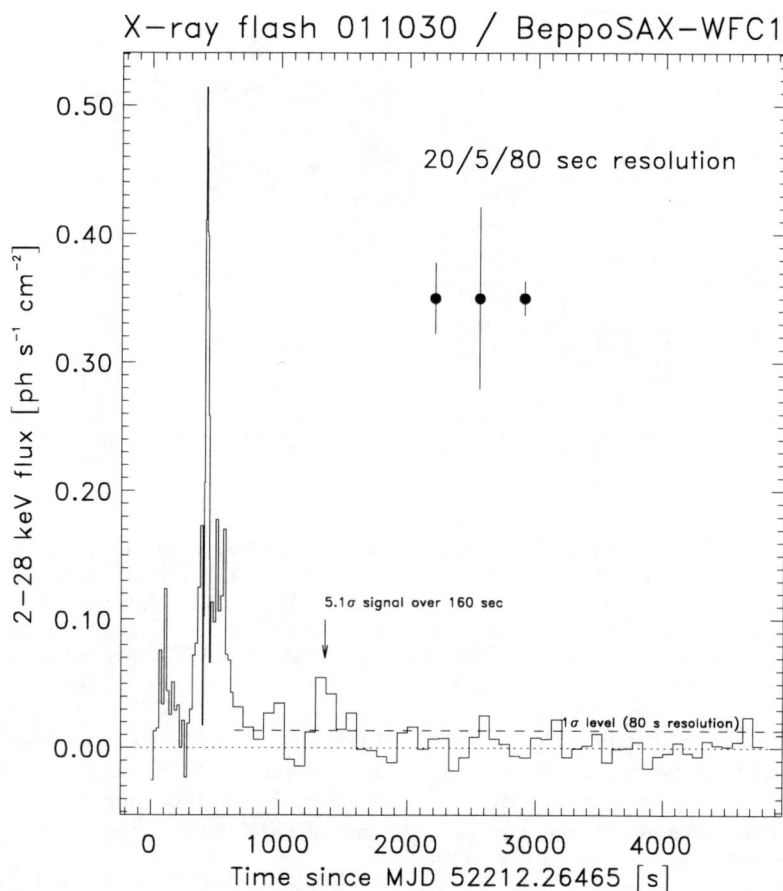

**FIGURE 3.** The lightcurve in the range 2-28 keV of XRF011030, the longest X-ray Flash observed so far. It has a structure reminiscent to the lightcurve of long GRBs. The light curve has been binned in 20 s bins during the beginning, in 5 s bins during the peak and in 80 s bins in the remainder in order to show the structure where sufficient signal-to-noise is available. Typical error bars in each bins is given in the figure.

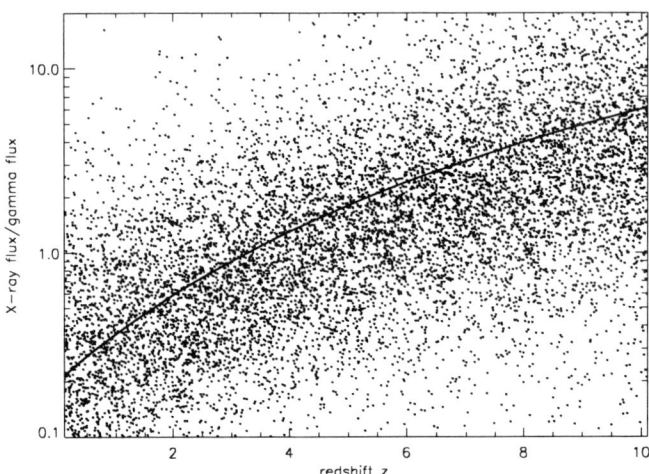

**FIGURE 4.** The ratio of the X-ray (2-10 keV) flux over the γ-ray flux (35-350 keV) for simulated GRB spectra, plotted as a function of redshift. Each dot represents a GRB with Band spectral parameters ($\alpha, E_{break}, \beta$) drawn from the observed distribution as measured by BATSE in the γ-range and extrapolated into the X-ray range. The drawn line represents the average GRB spectrum. The figure shows that indeed on average a GRB observed at large redshift becomes an XRF, but the spread is large. The reverse is not true: XRFs do not imply an object at large redshift. Both nearby XRFs and distant GRBs may be expected.

upper limit in the γ-range implies that the peak energy of the transient must be lower than ≈ 40 keV. In contrast, GRBs are distinguished by peaking in the 50-500 keV range. Second, GRBs with such a long duration are rare, e.g. the longest T90 duration listed in the 4th BATSE catalog[16] out of 1234 bursts is 674 s (trigger 3458) and in general have strong subpulses. In contrast SAX J2043.6+7717 lasted more than a thousand seconds.

The second possibility, especially given the galactic latitude of SAX J2043.6+7717 (b=+20.7 degrees), may argue for this source to be a galactic transient. However, the X-ray spectrum rules out SAX J2043.6+7717 being a type I X-ray burst. This then leaves us with the possibility that SAX J2043.6+7717 is a Fast X-ray Transient.

The position of SAX J2043.6+7717 in the histogram of durations indicates that it is an x-ray flash. More importantly, an identification with a galactic stellar object like an RS CVn, a nearby dMe star, etc., which have relatively bright quiescent optical/IR counterparts, is ruled out because of the absence of a visible optical counterpart[14] to the radio source.

We conclude that most likely SAX J2043.6+7717 is a member of the newly recognized class of X-ray flashes[17]. The new bright radio source 9 days after the event indicates that we are dealing with an explosive phenomenon.

Given that we know little about such events (with XRF011030 being the first event localized to arcsecond accuracy) it is not appropriate to call such non-triggered transients as "X-ray rich, gamma-ray poor" GRBs (or other equally oxymoronic names). Recognizing that astronomical research is largely based on empiricism, we suggest that events such as the one on XRF011030 characterized by the following criteria be termed as "X-ray flashes" (XRF):

1. Strong non-thermal emission in the X-ray (2-10 keV) band; this criterion distinguishes them from type I bursts);

2. Weak in the traditional GRB gamma-ray band, 50-350 keV; this explains why typical XRFs do not trigger gamma-ray burst monitors in general. We define the ratio $R = F_{2-10}/F_{50-350}$ with $F_{2-10}$ the peak X-ray flux in 2-10 keV range and $F_{50-350}$ the peak gamma-ray flux in 50-350 keV range. Both energy ranges are choosen such that they are part of most X-ray and γ-ray experiments. When $R > 1$ we define the object as an XRF, otherwise a GRB. In these terms, GRBs in the range $0.5 < R < 1$ could be called X-ray rich GRBs. The ratio R averaged over 23 GRBs observed with GINGA is 0.25;

3. Durations less than a few thousand seconds; this differentiates XRFs from the other fast X-ray transients which have durations of several hours (many of which are stars with intense coronal activity and some binaries containing a compact object such as SAX J1819-255);

4. Display no strong quiescent optical/IR counterpart; this criterion distinguishes XRFs from stars with strong coronal activity (e.g., strong flares from pre-main sequence stars, dMe stars, RS CVn stars, and Be stars etc);

## XRFs as redshifted GRBs

As to the origin of the x-ray flashes, two possibilities comes to mind. Firstly, X-ray flashes could be related to the same type of physical event as GRBs, but observed differently because of different observing orientation or a different distance. For example XRFs could be GRBs at large cosmological redshift $z > 5$, when gamma rays would be shifted into the x-ray range and the typical spectral break energy at 100 keV shows at 16 keV. A typical average spectrum of a GRB (spectral slope $\alpha = -1$ before the break, a break energy of 100 keV and a spectral slope $\beta = -2.25$ after the break), would have a ratio of x-ray over $\gamma$-ray fluence of around a few percent (see Fig. 4). This ratio becomes larger than one at large redshift: a typical $\gamma$-burst becomes an x-ray flash. Note, however, that the spread in spectral properties of GRBs is large. Extrapolations of these spectra into the x-ray range would predict a large variation in the ratio of x-ray to $\gamma$-ray fluence. These intrinsic variations are much larger than the average increase with redshift.

A time dilation in the duration of classical GRBs is not observed. The intrinsic distribution of durations might dominate the effect caused by cosmological redshift. Moreover, if the average decay in time of the prompt emission has a powerlaw shape (such as is the case for the afterglow emission which in some cases smoothly joins the later stages of the prompt emission), no time dilation is expected from the part with a powerlaw shape.

If XRFs are on average at a much larger redshift than GRBs do we see the effect of time dilation in XRFs? Initially we thought (Heise et al. [17], Fig. 2) that there was no such effect and that the average duration for an XRF is similar to the average duration of the X-ray counterparts of GRBs. However, observational selection effects in finding long duration XRFs may be large. The long duration of XRF011030, e.g., is in part determined by a weak feature seen after 1200 s, which we may have missed in other flashes. This triggered us to renew the search for longer duration XRFs. This program is ongoing, but preliminary results indicate several newly found XRFs with a duration of several hundred seconds, the longest one even being 2500 s. The end result will be that, indeed, the average duration of XRFs is longer than the equivalent value for GRBs. Note in this respect also that XRF011030 itself is longer than GRBs in the BATSE catalogue.

The geometrical relation between XRFs and GRBs might be another than based on distance. It might be based on viewing angle. GRBs are strongly beamed. It is possible that XRFs are those explosive events which we do not see head-on, but at a certain angle, where the relativistic boosting is determined by the transverse component of the expansion and a lower Lorentz factor leads to x-ray energies rather than gamma-ray energies.

## XRFs as a new type of cosmic explosion

The second possibility to explain x-ray flashes draws upon their similarity with GRBs. The statistical properties of X-ray flashes display many of the properties of GRBs, except that they are not seen at $\gamma$-rays and do not trigger $\gamma$-ray burst instruments. X-ray flashes therefore may show an extension of the physical circumstances which lead to relativistic expansion and the formation of gamma-ray bursts, the scenario called the cosmic fireball. Interestingly, Katz[1], in 1994, while discussing the cosmic fireball scenario with extreme relativistic motion (large bulk Lorentz factors $\Gamma$), remarks that one should look for x-ray bursts and uv-bursts, because such bursts are expected at mildly relativistic expansion at lower bulk Lorentz factor. $\Gamma$ depends on the ratio of rest mass energy (the baryon load) to the total energy of the burst. A low baryon load leads to high Lorentz factors and probably high energy gamma-ray bursts. A high baryon load (also called a "dirty fireball") leads to a smaller Lorentz factor and presumably a softer gamma-ray burst: possibly the x-ray flash.

In many progenitor models for GRBs, the so called collapsar models, the gamma burst is produced by the final collapse of an accretion torus around a recently formed collapsed object (black hole). In many cases the stellar debris around the birthgrounds of gamma bursts prevents one from observing any prompt high energy radiation at all, making GRBs a rare phenomenon as compared to supernovae. In these models the explosions that do break out of the envelope material are accelerated to high Lorenz factors and can be observed as GRBs. In many cases it can be expected that the envelope material is too massive, the explosion is stopped and does not produce an observable effect in high energy x- and $\gamma$ radiation. It seems plausible that x-ray flashes bridge these two extremes in stellar collapses: the stellar collapses in a relatively clean direct circumburst environment that are observed as normal GRBs and the stellar collapses that go unobserved in prompt x- and $\gamma$ radiation.

## REFERENCES

1. Katz, J., 1994, *Astrophys. J.* **422**, 248
2a. T.E. Strohmayer, E.E. Fenimore, T. Murakami, A. Yoshida, Ap.J. **500**, 873 (1999)
3. Forman, W., Jones, C., Cominsky, L., et al. *Astrophys. J. Supp. Series* **38**, 357 (1978)
4. Pye, J. P., & McHardy, I. M. The Ariel V sky survey of fast-transient X-ray sources. *Mon. Not. R. astr. Soc.* **205**, 875 (1983)
5. Ambruster, C.W., & Wood, K.S. The HEAO A-1 all-sky survey of Fast X-ray Transients. *Astrophys. J.* **311**, 258 (1986)
6. Connors, A., Serlemitsos, P. J. & Swank, J. H. Fast

Transients: A search in X-rays for short flares, bursts, and related phenomena. *Astrophys. J.* **303**, 769 (1986)
7. Gotthelf, E. V., Hamilton, T. T., & D.J. Helfand, D. J. The Einstein observatory detection of faint X-ray flashes. *Astrophys. J.* **466** 779 (1996)
8. Grindlay, J. E. 1999, *Astrophys. J.* **510**, 710
9. Greiner, J., Hartmann, D. H., Voges, W., et al. Search for GRB X-ray afterglows in the ROSAT all-sky survey. *Astr. Astrophys.* **353**, 998 (2000)
10. GCN notices 1118, 1119 (2001)
11. GCN notice 1123 (2001)
12. GCN notice 1138 (2001)
13. Taylor, G. B., frail, D. A., Kulkarni, S. W., GCN notice 1136 (2001)
14. Bloom, J. S., Gil de Paz, A., Harrison, F. A., & Kulkarni, S. R. GCN notice 1137 (2001)
15. Harrison, F. A., Yost, S., Fox, D., Heise, J., Kulkarni, S. R., Price, P. A., and Berger, E., GCN notice 1143 (2001)
16. Paciesas et al. 4th BATSE catalog *Astrophys. J. Sup. Series* **122**, 465 (2000)
17. Heise, J., in 't Zand, J., Kippen, R. M., & Woods, P. M. 2000, Second Rome Workshop "Gamma Ray Bursts in the afterglow era", October 2000
18. Kippen, R.M., Woods, P. M., Heise, J., in 't Zand, J., Preece, R., & Briggs, M. S. in Second Rome Workshop "Gamma Ray Bursts in the afterglow era", Rome, october 2000
19. Frontera, F., Antonelli, L. A., Amati, L., et al., *Astr. Astrophys.* submitted (astro-ph/0002527)
20. Feroci, M., Antonelli, L. A., Soffitta, P., in 't Zand, J. J. M., Amati, L., Costa, E., Piro, L., Frontera,, F., Pian, E., Heise, J., Nicastro, L. GRB 990704: the most X-ray rich BeppoSAX gamma-ray burst. *Astr. Astrophys.* submitted (astro-ph/0108414)
23. Kippen, R.M., Woods, P. M. et al. 2001 HETE2-conference "Gamma Ray Bursts in the afterglow era", Woods Hole, November 2001
24. Strohmayer, T. E., Fenimore, E. E., Murakami, T., Yoshida, A. 1998 *Astrophys. J.* **500**, 873–887 (1998)
25. Wheaton, W. A., et al. 1973, *Astrophys. J. Lett.* **185**, L57

# Analytic Modeling of GRB Spectral and Temporal Evolution

Edison Liang* and Dan Kocevski*

*Department of Physics and Astronomy, Rice University, Houston, Texas, 77005

**Abstract.** We model the temporal profiles of BATSE GRB pulses with an analytic form based on first principles which give excellent fits to most FRED-like pulses. We've used these models to extract the asymptotic rise and decay power law indices for a sample of FRED pulses. Utilizing these results we can test a variety of currently popular emission models in the source comoving frame. Some tentative conclusions are listed.

## ANALYTIC MODEL FOR PULSE PROFILE

We model the time-profile of individual GRB pulses with an analytic form derived from first principles. Our starting point is the Liang-Kargatis [7] spectral decay law which is followed by most hard-to-soft pulses:

$$dE_p/dt = -L/A_o \qquad (1)$$

where $E_p$ is the spectral break energy, L is the intrinsic energy luminosity and $A_o$ is a constant for each pulse. At the same time, a correlation between L and $E_p$ was found for the decay of most pulses: $L \propto E_p^k$ where k has a broad distribution centered around 1.8 ([4], [5], [8]). But since L is anti-correlated with $E_p$ for most pulses during the rise phase, we therefore propose a simple, but flexible parametric relation between L and $E_p$ which would allow for both rise and decay of the pulse: $L \propto t^r \cdot E_p^s$ where r and s are indices to be fixed by fitting the pulse profile. Physically, this form allows for a positive correlation between L and $E_p$ due to the intrinsic emission mechanism but also an additional explicit dependence of L on time t, say due to environmental factors such as the evolution of the magnetic field or optical depth while the $E_p$ decay may be due to cooling.

When we combine the above two equations we have a simple differential equation which can be integrated to obtain the normalized time profiles of L and $E_p$:

$$L(t) = t^r/(1+r.t^{1+r}/d)^{(d+r)/(1+r)} \qquad (2)$$

$$E_p(t) = (1+r.t^{1+r}/d)^{(1-d)/(1+r)} \qquad (3)$$

where $d = (s+r)/(s-1)$. Sample profiles based on these equations are illustrated in Fig.1.

We see that as t becomes large the above equations become $L\ t^{-d}, E_p\ t^{-d+1}$ asymptotically, while for t ap-

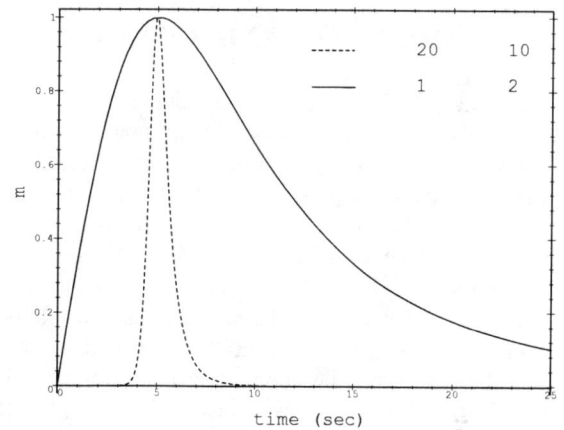

**FIGURE 1.** Analytic pulse profile for parameters r=20,d=10 and r=1,d=2

proaching zero $L\ t^r, E_p$ *constant*. Hence in the asymptotic decay L and $E_p$ are correlated in time and the above k-index is simply related to d via: $d = k/(k-1)$. From Fig.1 we see that indeed this theoretical time profile resembles the typical FRED pulse shape, especially the characteristic asymmetry between rise and decay (fast rise slow decay). We also note that while the r and d values controls the underlying asymptotic behavior of the pulse shape where it is often dominated by background noise or contaminated by neighboring pulses, in fitting the observed pulse profiles r and d are mainly constrained by the middle section of the pulse (e.g. how sharply peaked and asymmetric the pulse is) where the signal to noise is high. Hence this analytic approach allows us to estimate the asymptotic rise and decay rate of a pulse profile even where the pulse is no longer visible.

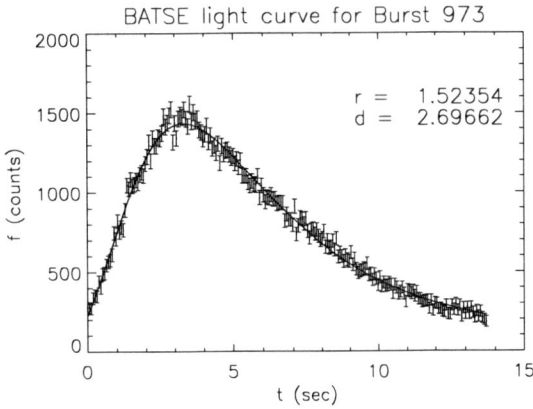

**FIGURE 2.** Fit of analytic profile to BATSE trigger 973. Best fit values are listed

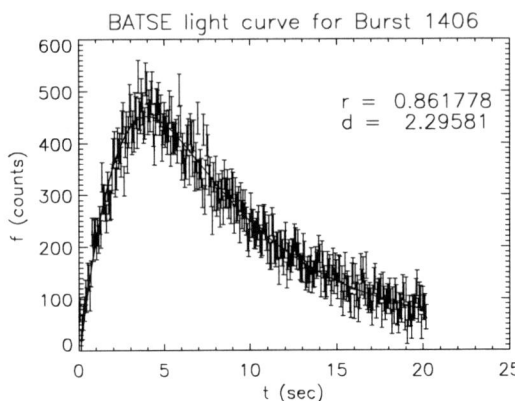

**FIGURE 3.** Fit of analytic profile to BATSE trigger 1406

## MODEL FITS AND RESULTS

Fig.2 and Fig.3 give fits of this analytic model to sample FRED pulses. We have systematically fitted two dozen cleanly separable pulses with high signal to noise. The results for the r distribution are plotted in Fig.4 and for the d distribution plotted in Fig.5. We see that r is cluster around 1 with a broad tail, which may be due to uncertainly caused by the lack of data points during the rise phase for many bursts. d has a moderate distribution centered around 2.3. This implies that the corresponding k distribution should be centered around 2.3/1.3 = 1.8 which is indeed consistent with the results of [4], [4], [5] and [8], even though the two results are obtained using totally different and independent methods.

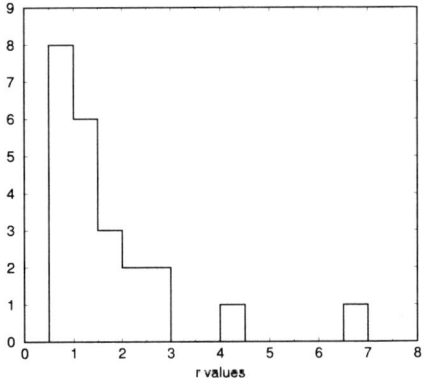

**FIGURE 4.** Distribution of r values for FRED pulses

## IMPLICATIONS FOR EMISSION MECHANISMS

The above results are obtained in the detector/observer frame. However to apply these results to constrain the intrinsic emission mechanisms we need to translate these results to the comoving emitter frame. What is the relation of the observed decay rate to the intrinsic decay rate in the source comoving frame? Fenimore et al [2], [3] have computed the effect of relativistic kinematics and angular spread due to the spherical curvature of the emitting shell on the observed time profile. They concluded that for intrinsic power-law spectral emissions, if angular spread dominates then the time profile decay index d should be correlated with the spectral index, due to off-axis relative redshifting of the composite spectrum. We have examined this in our FRED data and concluded that there is absolutely no correlation between d and the spectral indices.. This implies that the observed decay profile is intrinsic to the source. In this case there are two regimes, depending on whether the observed profile is diluted by angular spread or not: one is that the observed photon number flux decay index is the same as the decay index in the source comoving frame. The other is that the photon number decay index in the source comoving frame equals the observed index plus 2. For the FRED pulses we have analyzed our result is consistent with the photon number flux decaying according to $1/t$, which is consistent with most of the pulses analyzed by Ryde [8]. In this case photon flux in the source comoving frame may be decaying either as $1/t'$ or $1/t'^3$. where t' is proper time in the comoving frame. However, Ryde [8] has also found a second subpopulation of pulses with a photon number decay index centered around 3, which would suggest that in this subpopulation $1/t'^3$ is likely the photon decay rate in the source comoving frame. Hence in the comoving frame we have 2 possibilities allowing for dilution due to angular spread: photon number flux could be decaying either as $1/t'$ or $1/t'^3$.

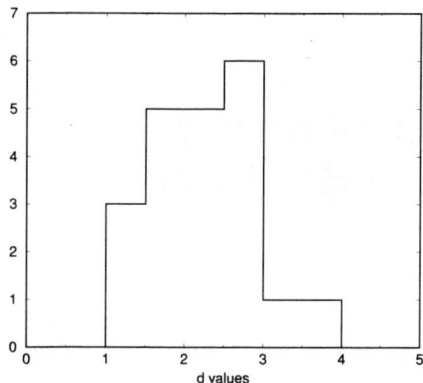

**FIGURE 5.** Distribution of d values for FRED pulses

We have analyzed many popular emission models under both scenarios in conjunction with the above (r, d) distributions. These models include optically thin synchrotron, self-absorbed synchrotron, small-pitch-angle synchrotron [1], synchrotron-self- Compton, external Compton, and synchrotron saturated-Compton. So far none of these models seems to be completely consistent with the entire BATSE database if we assume that cooling dominates $E_p$ evolution. It may be that GRBs emit by a variety of mechanisms or cooling does not dominate $E_p$ evolution. For example evolution of the magnetic field or the pitch angle could dominate $E_p$ evolution in models in which cooling is balanced by nonthermal energization. Equally likely, the true underlying emission mechanism is none of the above. It remains to be seen which turns out to be the correct scenario. In any case we conclude that systematic analysis of BATSE spectral evolution data remain the most powerful tool in the study of GRB emission mechanisms.

## ACKNOWLEDGMENTS

This work was partially supported by NASA grants NAG 5-7980 and NGT 8-52899.

## REFERENCES

1. Epstein, R. 1973, ApJ 183, 593
2. Fenimore, E., et. al. 1996, ApJ, 473, 998
3. Fenimore, E. &Sumner, M.C. 1998, ApJ
4. Golenetskii, S. V. et al. 1983, Nature, 306, 451
5. Kargatis, V. E., et. al. 1994, ApJ, 422, 260
6. Kocevski, D. & Liang, E. P. 2001, AIP Conf. Proc. No. 586, 623, ed. Wheeler, J. C & Martel, H.
7. Liang, E. & Kargatis, V. 1996, Nature, 381, 49
8. Ryde, F. 2000, Stockholm University Ph.D. Thesis

# Observations of the Highest Energy Gamma Rays from Gamma-Ray Bursts

Brenda L. Dingus

*Physics Department, University of Wisconsin, Madison, WI 53711*
*dingus@physics.wisc.edu*

**Abstract.** EGRET has extended the highest energy observations of gamma-ray bursts to GeV gamma rays. Such high energies imply the fireball that is radiating the gamma rays has a bulk Lorentz factor of several hundred. However, EGRET only detected a few gamma-ray bursts. GLAST will likely detect several hundred bursts and extend the maximum energy to a few 100 GeV. Meanwhile new ground based detectors with sensitivity to TeV gamma rays from gamma-ray bursts are beginning operation.

Prior to the launch of the Compton Gamma-Ray Observatory, the spectra of GRBs were known to extend beyond 10 MeV in several bursts. [1] Already these measurements ruled out thermal emission. However, EGRET has extended the maximum energy emission from GRBs to 20 GeV, and even detected longer duration emission than measured at MeV energies. GLAST will observe many more gamma-ray bursts, and should detect many more gamma rays in each burst. However, even before the GLAST launch in 2006, several new TeV ground based observatories with the ability to observe gamma-ray bursts are now becoming operational, and one burst has been claimed to have TeV emission.

## IMPORTANCE OF HIGH-ENERGY OBSERVATIONS

Several models predict GeV to TeV gamma rays on varying timescales due to varying emission mechanisms. [2, 3, 4] While the BATSE spectra are well fit by a low-energy power law breaking to a steeper high-energy power law, these models point to the possibility of another component of the emission at higher energies. For example, Figure 1 shows the BATSE observations as explained by synchrotron emission with higher energy gamma-rays produced by inverse Compton. In this model of external shock emission, the spectral energy distribution is sensitive to the initial Doppler boost factors as seen in Figure 1.

Not only are these high-energy gamma rays difficult to produce, but also they are easily attenuated because they have sufficient energy to pair produce with ambient light. A large Doppler boost factor or a very collimated beam of gamma rays at all energies is required in order to avoid the opacity of high-energy gamma rays to pair production with high density of lower energy gamma-rays in the source. Calculations of Lithwick & Sari [5] show that bulk Lorentz factors of a few hundred are required to explain the observations of high-energy gamma rays.

The GeV to TeV gamma rays from GRBs are also useful because of the expected attenuation by the extragalactic infrared to optical background light encountered during the transit of these gamma rays to us. [6, 7] Measurements of high-energy gamma rays probe these extragalactic photons whose density is unmeasured, but interesting in order to constrain the galaxy and star formation of the early universe. Unfortunately, the expected level of extragalactic background light is such that sources at $z \sim > 0.3$ are not observable by ground-based TeV observatories. While the typical redshift observed from GRBs is about one, the number of nearby GRBs is poorly constrained. Also, other classes of GRB sources may exist and have a different cosmological distance distribution. Due to the theoretical uncertainties and the important ramifications, high-energy gamma-ray observations are clearly needed.

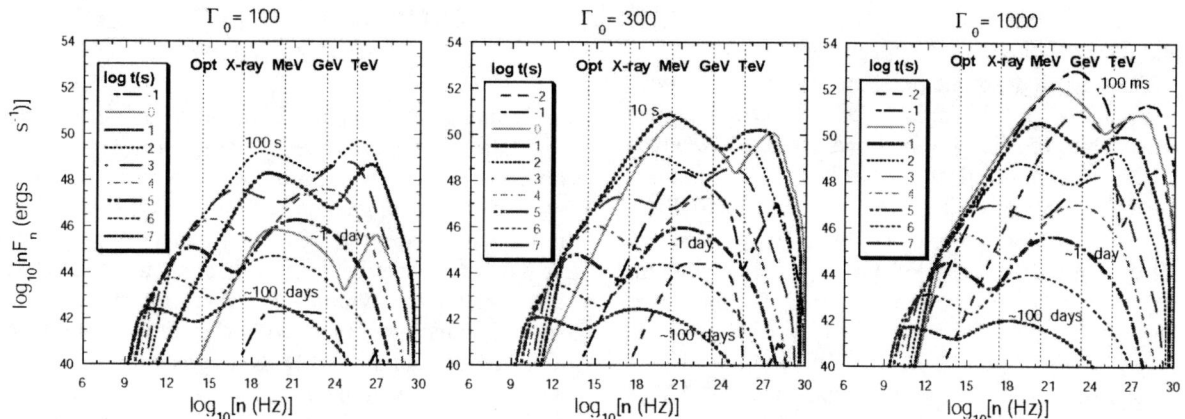

**FIGURE 1.** The spectral evolution of a GRB produced by the shock created when a single blast wave of different initial bulk Lorentz factors $\Gamma_0$ interacts with an external medium [2]. As in active galactic nuclei, two peaks are expected with the lower energy due to synchrotron emission and the higher energy due to inverse Compton scattering.

## EGRET GRB OBSERVATIONS

EGRET imaged individual gamma rays with energies above 30 MeV, recording their time, direction, and energy. Figure 2 gives the EGRET results for the ~100 GRBs detected by BATSE and within EGRET's field of view. Four bright bursts were detected by EGRET in this standard mode, and a separate paper has been published on each one [8].

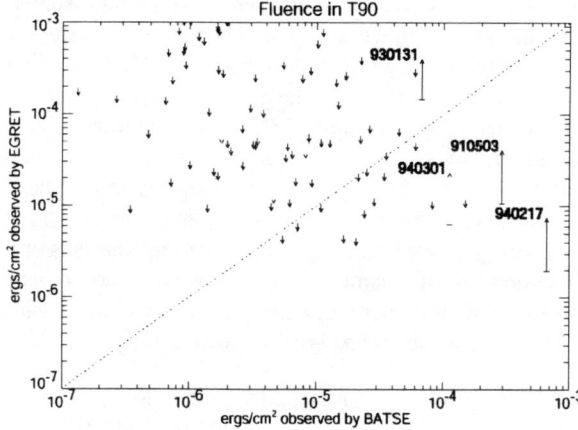

**FIGURE 2.** Fluence (or the upper limit of the fluence) above 30 MeV as observed by EGRET is plotted vs the fluence observed by BATSE in its 4 energy channels for the BATSE-defined time interval of T90. The dates of the 4 bursts detected by EGRET are given. The fluences of these bursts are only lower limits because of the large deadtime (110 msec per trigger) of EGRET. The size of these error bars reflects the 90% statistical confidence interval of the fluence.

The average spectrum of the 4 bright bursts detected by EGRET contains only 45 gamma rays observed within T90, the time interval of 90% of the flux as defined by BATSE. The average spectrum is hard with a differential photon spectral index of 1.95 ± 0.25, and there is no indication of a spectral break up to 10 GeV; however, only 4 gamma rays above 1 GeV are detected. Because these are the brightest BATSE bursts, possibly all GRBs have high energy emission during T90, but EGRET did not have the sensitivity to detect weaker bursts.

Limited statistics makes the duration of the highest energy emission difficult to determine. In the burst of GRB940217[8], an 18±4 GeV γ-ray was detected 4500 seconds after the end of the BATSE detected emission. The probability of a γ-ray > 18 GeV occurring in 5400 seconds, the time interval during which EGRET was observing this direction, and with a position consistent with the GRB is only $5 \times 10^{-6}$. Also in this burst, 18 events > 30 MeV are detected with a background of 4.7 (chance probability of $2 \times 10^{-6}$) in the time interval from the end of the burst until the end of the observation 5400 seconds later. Two of the other 3 bursts mentioned in the previous paragraph have a marginal ~3 σ excess in longer time intervals of 100 to a few 1000 seconds. These 4 bursts are among the 5 brightest in EGRET's field of view as determined in BATSE's energy range. A few other of the brightest bursts have marginal excesses of 3-4 σ in similar time intervals [9] but are not detected during the BATSE T90 time. These observations would be consistent with the model of Figure 1 which predicts that the higher energy bump initially at TeV energies will degrade in energy and move to GeV energies at later times.

## GLAST EXPECTATIONS

GLAST will detect many more GRBs than EGRET did, and many more gamma rays per burst. GLAST's increased sensitivity is a combination of four factors:

- GLAST's effective area is ~ 6 times that of EGRET and does not decrease at high energies;
- EGRET has a deadtime per gamma ray of ~110 ms, comparable to GRB pulse widths at lower energies where time profiles are well characterized, whereas GLAST will not be deadtime limited;
- GLAST's field of view is ~ 4 times that of EGRET;
- GLAST will be slewed continuously to keep the Earth out of its field of view, which results in approximately twice the observing time over EGRET.

The results of a detailed simulation of GLAST's ability to detect GRBs and measure their direction are shown in Figure 3. The simulation uses the GRB duration and flux distributions based on the measurements in the BATSE energy range and uses the EGRET TASC measurements from 1-10 MeV to determine high energy power law spectra to extrapolate to the energy range of GLAST. With these assumptions, GLAST is predicted to detect more than 200 bursts per year with at least 50 of those bursts having over 100 γ-rays above 100 MeV-- sufficient to measure power-law spectral indices with a typical error of 0.1. GLAST will also have a gamma-ray burst monitor to measure the burst spectra from 10 keV to 30 MeV[10].

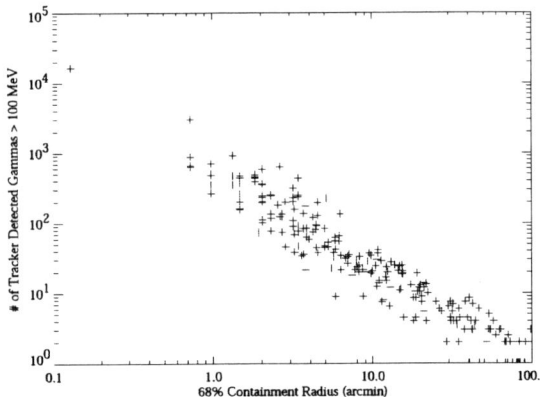

**FIGURE 3.** Simulation of the capabilities of GLAST to detect high-energy gamma rays and to determine the source direction for the ~200 bursts which GLAST will detect each year. [11]

GLAST will not launch until 2006; however, AGILE [12] will launch 2-3 years sooner. While the geometrical area of AGILE is not much bigger than EGRET's, the sensitivity to GRBs will be much improved. The field of view is much larger and the instrument scans the sky in a manner similar to GLAST. Plus the deadtime will not be an issue. Also prior to GLAST, will be AMS on the space station, and AMS may have the capability to detect γ-rays above a few GeV [13].

## GROUND-BASED TEV GRB OBSERVATIONS

Imaging atmospheric Cherenkov telescopes have revolutionized ground-based gamma-ray astronomical observations of *known* sources. However, due to the short duration of gamma-ray bursts and their relative infrequent occurrence of approximately once per day, gamma-ray bursts have been difficult to observe with this technique. Whipple Observatory has slewed to several burst positions within a few minutes, but detected no emission [14]. New observatories will be capable of slewing even faster. Also, new satellites, such as SWIFT and HETE, will send better prompt localizations, so that ground-based observers don't have to spend time searching for the correct position within a region larger than their field of view.

However, large field of view detectors with high duty factors and low energy thresholds are needed to observe the prompt emission, since the time delay of the notification plus the telescope slew time is of order the duration of the gamma-ray burst. Nearly 100% live time and > 1 sr field of view are possible if the particles in the air shower are detected. However, previous extensive air shower detectors have only had low energy thresholds of 10-100 TeV. Lower energy showers have fewer particles and reach shower maximum at higher altitude. Therefore, in order to have a low energy threshold below 1 TeV, the particle detectors must cover a larger fraction of the area (previous arrays covered < 1%) and should be placed at mountain altitudes. The sensitivity of air shower arrays is not yet as good as imaging atmospheric Cherenkov telescopes, because angular resolution is still the principal technique for rejecting the isotropic background of cosmic rays. However, preliminary work is being done on gamma-hadron separation using information from extensive air showers [15].

EAS γ is an extensive air shower array in Tibet located at an altitude of 4300 m, which is an overburden of 600 g/cm$^2$. Extensive air showers of ~ 3 TeV were detected with an angular resolution of ~ 1 degree. Both the Crab and the blazar Mrk501 were significantly detected [16]. Another observatory, named ARGO-YBJ, is also being built at the same Tibet site. The detector is made of gas resistive plate chambers and covers the ground completely over an area of ~ 6000m$^2$. When the construction is finished, the detector will have a very low energy threshold of a few 100 GeV[17].

Milagro uses a covered 80 m x 60 m x 8 m pond of water located at 2650 m altitude near Los Alamos, New Mexico. Relativistic particles in the air shower emit Cherenkov light in the water, resulting in detection of over 50% of the particles incident on the pond. Milagrito, a prototype for Milagro operated for just over a year and detected Mrk501 [18]. Milagrito data was searched for evidence of emission from 54 satellite detected GRBs with one potential counterpart, GRB 970417a, found. The probability that this burst was not a statistical fluctuation of the background is $2.9 \times 10^{-5}$. Since 54 bursts were searched, the probability that this observation is due to a random fluctuation is $1.5 \times 10^{-3}$.[19] The implied fluence of this burst at TeV energies (if it is not due to a random fluctuation of the background) is much greater than the extrapolation of the BATSE fluence would predict as shown in Figure 4.

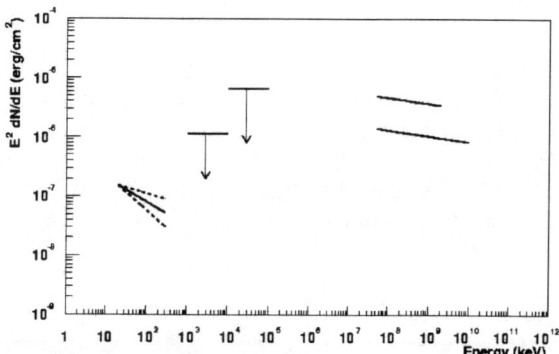

**FIGURE 4.** Spectral energy distribution for GRB970417a observed by Milagrito. A fit to the BATSE spectrum [20] is shown at low energies, the upper limits from the EGRET TASC at a few MeV, and at high energies 2 alternative spectra with different maximum energies, which are consistent with the Milagrito observation, are given [19]. TeV gamma-rays are attenuated by the extragalactic background light, therefore the lowest (highest) fluence spectrum requires the burst be nearer than z=0.03 (0.1) which corresponds to an isotropic luminosity of $>10^{49}$ ($5 \times 10^{51}$) ergs.

Milagro began operation in December 1999 and has detected the Crab and blazar Mrk 421. [21] Upper limits on the 24 bursts in Milagro's field of view were presented at this conference, and one of these bursts, GRB010921, has a redshift of 0.45. [22] Milagro is undergoing upgrades to be completed in 2002 that will lower the energy threshold as well as improve background rejection, angular resolution and energy resolution.In conclusion, EGRET has detected only a few GRBs, but has shown that there is valuable information to be gained by looking for higher energy emission. New space based detectors will be designed to observe GRBs , and new ground based detectors are becoming operational that can observe GRBs. There may even be classes of GRBs that will only be detected at these high energies.

## ACKNOWLEDGMENTS

The author acknowledges support from NASA, NSF and the University of Wisconsin.

## REFERENCES

1. Share, G.H. et al. *Adv. Space Res* **6** 15 (1986).
2. Dermer, C. D. & Chiang, J., *GeV to TeV Gamma-Ray Astrophysics*, eds. Dingus, B.L., Salamon, M. H., & Kieda, D.B., AIP Conf. Proc. **515**, New York: American Institute of Physics 225 (1999)
3. Zhang, Bing & Meszaros, Peter *Ap J* **559** 110 (2001).
4. Pilla, R.P. & Loeb, A. *Ap J Lett* **494** 167 (1998).
5. Lithwick, Yoram & Sari, Re'em *ApJ* **555** 540 (2001).
6. Primack J. et al. *High Energy Gamma-Ray Symposium*, eds. Aharonian, F. & Volk, H. J., AIP Conf. Proc. **558**, New York: American Istitute of Physics 463 (2001)
7. Salamon, M. H. & Stecker, F. W. *Ap J* **493** 547 (1998).
8. GRB910503 Schneid, E.J. et al. *A & A* **255** L13 (1993); GRB930131 Sommers, M. et al. *Ap J Lett* **422** L63 (1993); GRB 940217 Hurley, K. et al. *Nature* **372** 652 (1994); GRB940301 Schneid, E.J. et al, *Ap J* **453** 95 (1994)
9. Dingus, B.L., Catelli, J.R., & Schneid, E.J. *25th Cosmic Ray Conference*, eds. Potgieter, M. S., Raubenheimer, B.C., & vander Walt, D.J. Potchefstroom: Potchefstroomse Universiteit **Vol 3**, 29 (1997)
10. Bonnell, J., Norris, J., Dingus, B., & Scargle, J. *4th Huntsville GRB Symposium*, eds. Meegan, C.A., Preece, R.D., & Koshut, T.M., AIP Conf. Proc. **428**, New York: American Institute of Physics 884 (1998)
11. Meegan, C. et al in these proceedings (2002)
12. Tavani, M. et al *A & A Supp* **138**, 569 (1999)
13. Battiston, R., Biasini, M., Fiandrini, E., Petrakis, J., Salamon, M. H. *Astropart. Ph.* **13**, 51 (2000)
14. Connaughton, V. et al *Ap J* **479**, 859 (1997)
15. Sinnis, C. et al. *27th Cosmic Ray Conference*, Hamburg, Germany, Copernicus Gessellschaft 2579 (2001)
16. Amenomori, M. et al *Ap J* **532**, 202 (2000)
17. Bacci, C. et al, *A & A Supp* **138**, 597 (1999)
18. Atkins, R. et al , *Ap J Lett* , **525**, 25L (1999)
19. Atkins, R. et al., *Ap J Lett*, **533**, 119L (2000)
20. Kaneko, Y. & Preece, R. private communication.
21. Sullivan, G. et al *27th Cosmic Ray Conference*, Hamburg, Germany, Copernicus Gessellschaft 2773 (2001), also astro-ph/0110513
22. McEnery, J. et al GGN 1162 (2001), also in these proceedings (2002)

# Spectral Characteristics of X-Ray Flashes Compared to Gamma-Ray Bursts

R. M. Kippen[*†], P. M. Woods[**†], J. Heise[‡], J. J. M. in 't Zand[§‡], M. S. Briggs[*†] and R. D. Preece[*†]

[*]*University of Alabama in Huntsville, Huntsville, AL 35899, USA*
[†]*National Space Science & Technology Center, 320 Sparkman Dr., Huntsville, AL 35805, USA*
[**]*Universities Space Research Association, Huntsville, AL 35806, USA*
[‡]*SRON National Institute for Space Research, Sorbonnelaan 2, 3584 CA Utrecht, The Netherlands*
[§]*Astronomical Institute, Utrecht University, P.O. Box 80 000, 3508 TA Utrecht, The Netherlands*

**Abstract.** X-ray flashes (XRFs) are a new type of fast transient source observed with the *Beppo*SAX Wide Field Cameras (WFC) at a rate of about four per year. Apart from their large fraction of 2–26 keV X-rays, the bulk properties of these events are similar to those of classical gamma-ray bursts (GRBs). By investigating the wideband spectra of ten events detected in common with WFC and BATSE, we explore the possibility that XRFs are a low-energy branch of the GRB population. We find that XRF spectra are similar to those of GRBs, and that their low peak energies could be an extension of known GRB properties.

## INTRODUCTION

In the study of prompt emission from gamma-ray bursts (GRBs) one is overwhelmed by a large amount of observational data with comparatively little physical understanding. In this realm, one of the keys to better understanding lies in the identification of new, or extreme behavior that differs from that of the bulk ensemble. One such enlightening characteristic would be the identification of the lowest energy bursts, which could help to indicate the limiting form of the radiation emission process.

Most observed GRBs have energy spectra that exhibit $\nu \mathcal{F}_\nu$ peak power at energies ($E_{\text{peak}}$) in the range $\sim$50–300 keV. For example, the distribution of time-averaged $E_{\text{peak}}$ for 156 bright bursts measured with *Compton*-BATSE [1] is approximately log-normal, with a centroid of $\sim$175 keV and a width of $\sim$0.5 decade (FWHM). However, in apparent contradiction to the BATSE results, several bursts with significant X-ray ($\sim$2–10 keV) emission have been observed. In particular, the spectra of several bright bursts measured with *Ginga* suggest that either the distribution of $E_{\text{peak}}$ extends below 10 keV, or there is a second (X-ray) spectral component in some bursts [2]. The later hypothesis is supported by the fact that 15% of all BATSE GRBs analyzed show a low-energy excess above the standard continuum model [3].

The most recent observational clue related to the question of low-energy bursts is the discovery [4, 5] of several unknown "X-ray Flashes" (XRFs; also referred to as "Fast X-ray Transients") using the *Beppo*SAX Wide Field Cameras (WFC). These events are distinguished from Galactic transient sources by their isotropic spatial distribution and short ($\sim$10–100 s) durations. Furthermore, they are distinguished from GRBs based on their non-detection above 40 keV with the *Beppo*SAX GRB Monitor — implying larger X/$\gamma$ ratios than GRBs. In fact, the distribution of X/$\gamma$ ratio overlaps considerably with that of GRBs. In other respects, such as duration, temporal structure, spectrum (X-ray) and spectral evolution, XRFs exhibit properties that are qualitatively similar to the X-ray properties of GRBs. This similarity led to the suggestion that the XRFs are in fact "X-ray rich" gamma-ray bursts [4, 5].

To investigate the nature of the WFC X-ray flashes, and their true relation to GRBs, untriggered BATSE data were searched, and nine of ten observable flash sources were found to have significant flux above 25 keV (extending in most cases to >100 keV). These data were used to directly compare the *gamma-ray* properties of XRFs to those of the numerous BATSE GRBs [6]. This comparison showed that the flashes are similar in most respects to the long-duration class of GRBs, except that they are significantly softer (based on gamma-ray hardness ratios) and weaker (based on gamma-ray peak flux or fluence) than most long GRBs. In addition, the XRFs appear to be consistent with a low-intensity extrapola-

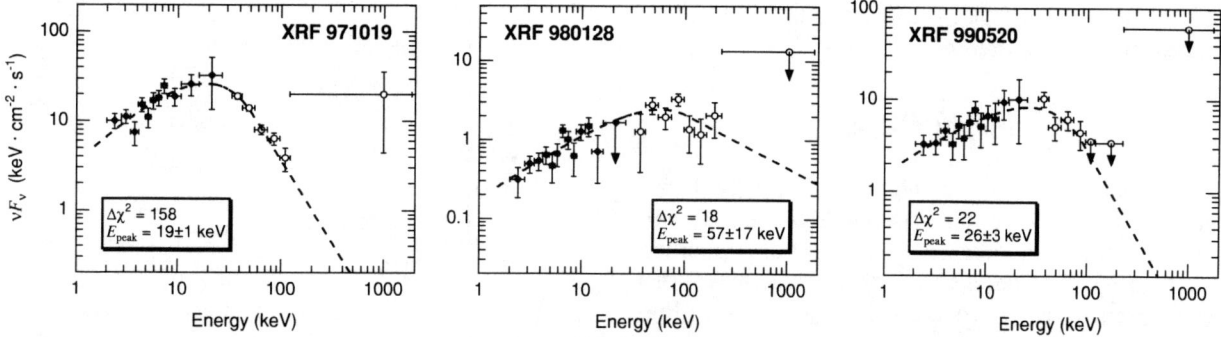

**FIGURE 1.** Model-dependent deconvolution of spectral data from WFC (*solid diamonds*) and BATSE (*open circles*) for three X-ray flashes. The best-fit Band GRB function is shown as dashed lines. Also indicated are the change in chi-squared ($\Delta\chi^2$) from a single power law to the Band function and the best-fit values of $E_{peak}$ with $1\sigma$ errors.

tion of the GRB hardness-intensity (H-I) correlation — suggesting that the flashes could indeed represent a low-energy (or X-ray rich) extension of the GRB population. If XRFs are X-ray rich GRBs, they represent a major fraction (~30%) of the full GRB population.

In this paper, we attempt to further quantify the comparison of X-ray flashes to GRBs by studying wide-band spectral properties. Only by combining WFC X-ray data with BATSE gamma-ray data can spectra with $E_{peak}$ below the gamma-ray regime be parameterized.

## JOINT SPECTRAL ANALYSIS

In the 3.8 years when BATSE and WFC were operating nearly simultaneously, a total of 36 GRBs and 17 XRFs were detected with WFC. Due to Earth occultation and data outages, only 18 of the GRB and 10 XRF sources were observable with BATSE. For these events, we have the unique opportunity to study the wide-band energy spectra from 2 keV to $\gtrsim 1$ MeV (depending on brightness). Note that one of the ten XRF sources is not a significant detection in the BATSE data. This event is nonetheless included in the joint spectral analysis because the BATSE data do constrain the wide-band spectrum.

Joint WFC/BATSE spectral analysis has been successfully applied to some of the GRBs. A standard $\chi^2$ fitting technique is used to compare model spectra, folded through the appropriate instrument response functions, to the observed (background-subtracted) counting rates. For example, the time-averaged spectrum of GRB 990510 was found to be well described over 3 decades in energy by the now-standard Band GRB spectral form with $E_{peak} \approx 143$ keV [7].

Here, the same analysis tools are applied to time-integrated data from each of the XRF sources. The WFC data are from source images obtained by correlating with the WFC coded mask. The approximate energy range is 2–26 keV. The gamma-ray data are from the BATSE Large Area Detectors (LADs), with 16 energy channels covering the energy range ~25 keV to 1.8 MeV. Accumulation time intervals are based on the entire duration of significant emission in the WFC data. The spectrum of each flash was compared to three models: a simple power law (two free parameters), the "Comptonized" model (three free parameters that describe a power law with a high-energy exponential cutoff), and Band's GRB model (four free parameters describing two smoothly connected power laws) [see, e.g., 1]. The Comptonized (COMP) and Band models describe a curved spectrum that constrains $E_{peak}$.

All of the flashes are adequately described by either the COMP or Band models, which have average $\chi^2/\nu$ values of 0.93 and 0.95, respectively. In contrast, the power-law model can be rejected in most cases, with an average $\chi^2/\nu = 1.57$. Based on the change in $\chi^2$ between the single power law and the other models, eight of the ten XRFs show significant evidence ($\Delta\chi^2 > 4$) for a curved spectrum resembling that of GRBs. Examples of spectra for the three brightest XRFs are shown in Figure 1. For these events the curvature is very significant and the Band spectral parameters are well-constrained. For the weaker events the parameters are not as well constrained, particularly the high-energy spectral index in the Band function. This is typical of weak GRBs, where the COMP model is often used in place of the Band function due to low S/N ratio at high energies.

## COMPARISON TO GRBS

When comparing spectral parameters between XRFs and GRBs it is important to consider biases due to the way the parameters are measured and how the samples are selected. Unfortunately, we do not have large samples of GRBs measured and selected in exactly the same manner as the XRFs. Hence, we must make the comparisons with the data that are available. The resulting biases cloud the comparison, as discussed below.

## Bright GRBs

The sample of 156 bright, high-fluence BATSE bursts analyzed by Preece et al. [1] represents the best current knowledge of detailed GRB spectral properties. The high signal-to-noise ratio in this sample means that time-averaged parameters are well constrained. In Figure 2, the distributions of $E_{peak}$ and low-energy power-law index $\alpha$ for these bursts are compared to those of the jointly-fit XRFs. The bright-burst distributions are represented by best-fit log-normal functions, which provide good descriptions of the data. Also included are BATSE-only spectral parameters for the 18 WFC-selected GRBs observed with BATSE. These bursts were fit using BATSE LAD data in the energy range ~25 keV to 1.8 MeV. In all cases, the spectral model is the Band GRB function. Similar results were obtained using the Comptonized model.

To quantify the comparisons, the K-S test was used to evaluate statistical differences between the WFC-selected samples and the log-normal functions for the bright BATSE GRBs. The results are that XRFs have significantly lower values of $E_{peak}$ than the bright BATSE GRBs, with K-S probability $P_{KS} = 1.5 \times 10^{-8}$, while the WFC-selected GRBs have $E_{peak}$'s that are consistent with those of the bright BATSE bursts. This latter agreement is not surprising since most of the WFC GRBs are rather bright. It does, however, indicate that (at least for bright bursts) biases due to different sample selections are small. Biases due to the fact that the XRFs were fit over a different energy range (and using different instrument data) than the GRBs remain unquantified.

The $\alpha$ distributions of WFC XRFs and WFC-selected GRBs are each statistically consistent with that of bright BATSE bursts, with $P_{KS} = 0.11$ and 0.32, respectively. Detailed comparison of the high-energy power-law index $\beta$ is problematic because it is not well constrained for many of the XRFs. However, the mean value for the four brightest flashes is $\langle \beta \rangle \approx -2.5 \pm 0.5$, which is consistent with the average of $-2.1$ obtained for bright GRBs. Thus, excluding the potential measurement biases mentioned above, it appears that the main outstanding spectral difference between GRBs and XRFs is that XRFs have lower $E_{peak}$ values, on average.

## Dim GRBs

Another significant bias that must be considered is the fact that weak bursts have different spectral properties than bright bursts [see, e.g., 8, 9]. In particular, there is

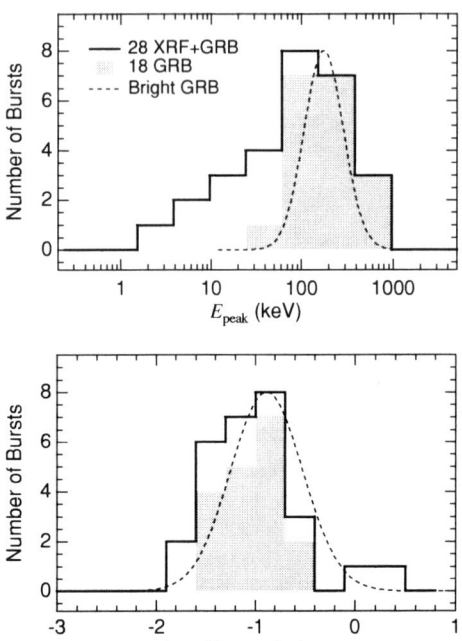

**FIGURE 2.** Jointly fit time-averaged spectral parameters of XRFs compared to BATSE-only parameters of GRBs. Dashed curves are log-normal fits to parameters from 156 bright GRBs [1], normalized to the peaks of the XRF+GRB histograms.

a strong correlation between $E_{peak}$ and peak GRB flux (measured in the 50–300 keV energy range). Thus, if XRFs are related to GRBs, we expect them to have lower $E_{peak}$, on average.

For a detailed comparison with weak (and bright) GRBs, we use the spectral catalog of Mallozzi et al. [10], which contains model fit parameters for 1,275 BATSE bursts (obtained with LAD data). Although the catalog contains spectral parameters for several spectral models, we employ the COMP model, which is more robust for fitting weak bursts with few high-energy counts. In practice, the spectral parameters derived from COMP and Band model fits are similar [10].

Figure 3 shows distributions of $E_{peak}$ versus duration ($T_{50}$), power-law index ($\alpha$) and peak flux ($P_{1024}$, evaluated on the 1.024 s timescale in the energy range 50–300 keV) for the XRFs and BATSE GRBs. Since the XRFs have (gamma-ray) durations comparable with long GRBs, the Mallozzi et al. catalog has been selected for long bursts with $T_{50} > 1$ s and that have standard peak fluxes and fluences available. This results in a sample of 802 long GRBs that are used for comparison in the two right-most plots in Figure 3.

The plot of $E_{peak}$ versus peak flux is particularly informative since it shows how the XRFs compare with the long GRB H-I relation. For this sample of GRBs, the correlation is very significant, with a Spearman rank or-

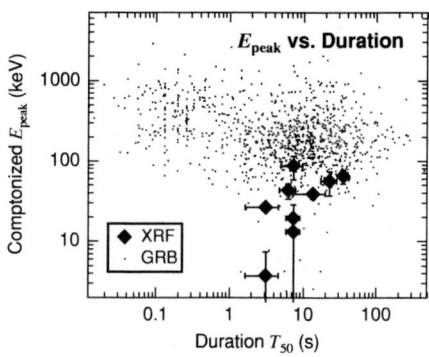 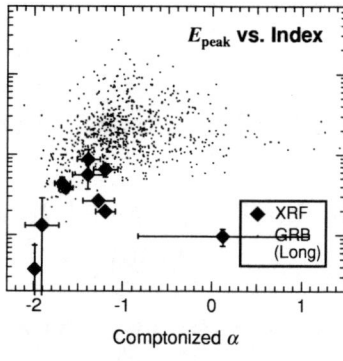 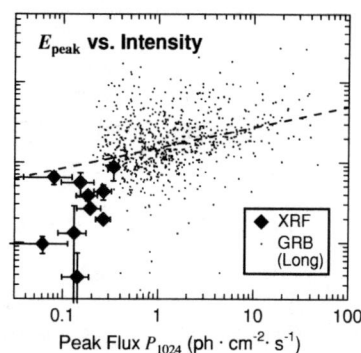

**FIGURE 3.** Jointly fit time-averaged spectral parameters of XRFs compared to BATSE-only [10] parameters of GRBs. Measurement errors for the GRBs are excluded from the plots for clarity. The dashed line at right indicates the best-fit power law to the GRB distribution.

der probability of $5.5 \times 10^{-16}$. At the extremes of the distribution are the bright GRBs corresponding to the Preece et al. sample, with $P_{1024} > 10$ ph·cm$^{-2}$·s$^{-1}$ and $\langle E_{peak} \rangle \sim 350$ keV, and the weakest GRBs near the BATSE trigger threshold with $\langle E_{peak} \rangle \sim 100$ keV.

Statistically, the XRFs are inconsistent even with the weakest 5% of GRBs ($P_{1024} < 0.3$), with a two-distribution K-S probability of $P_{KS} \approx 10^{-5}$. However, the average XRF flux of $P_{1024} = 0.16$ is nearly factor of two below that of the weakest BATSE bursts.

Assuming the H-I correlation extrapolates unchanged below the BATSE threshold with a simple power law (see Figure 3), the expected mean and (log-normal) standard deviation of $E_{peak}$ at the average XRF flux level are 94 keV and 0.25 decades, respectively. The XRFs are marginally inconsistent with this extrapolation, with $P_{KS} \approx 10^{-4}$.

## CONCLUSION

We have shown that X-ray flashes have significantly curved time-averaged energy spectra that are quantitatively similar to those of gamma-ray bursts. The main difference is that XRFs have peak power at significantly lower energies than most GRBs. Combined with our previous work, this lends further support to the idea that XRFs could be a low-energy extension of the GRB population.

We also find that, statistically speaking, XRFs are marginally inconsistent with a power-law extrapolation of the GRB hardness-intensity distribution. However, given the considerable systematic uncertainties in this extrapolation, the marginal statistical inconsistency is not a strong argument to rule out a GRB/XRF association. It is also important to note the two key biases in this comparison: Firstly, the observed XRFs are purely X-ray selected, so they probably have lower $E_{peak}$ than average. Secondly, the XRF spectra were fit using X-ray and gamma-ray data, whereas the GRBs used only gamma-ray data, so $E_{peak}$ for weak GRBs could be overestimated. With these considerations, it is not unreasonable to conclude that XRFs are indeed low-energy, or X-ray rich GRBs.

Final conclusion of this problem will undoubtedly require more *wide-band* XRF measurements. Continued XRF observations with HETE-II and *Beppo*SAX may be sufficient, but it will be difficult to compare these data to known GRBs since they lack high-sensitivity gamma-ray measurements.

## REFERENCES

1. Preece, R. D., Briggs, M. S., Mallozzi, R. S., et al., *ApJSS* **126**, 19 (2000).
2. Strohmayer, T. E., Fenimore, E. E., Murakami, T., & Yoshida, A., *ApJ* **500**, 873 (1998).
3. Preece, R. D., Briggs, M. S., Pendleton, G. N., et al., *ApJ* **473**, 310 (1996).
4. Heise, J., in't Zand, J., Kippen, R. M., & Woods, P. M., in *Proc. 2nd Rome Workshop: Gamma-Ray Bursts in the Afterglow Era (Oct. 2000)*, eds. N. Masetti et al., in press (astro-ph/0111246).
5. Heise, J., et al., these proceedings.
6. Kippen, R. M., Woods, P. M., Heise, J., et al., in *Proc. 2nd Rome Workshop: Gamma-Ray Bursts in the Aferglow Era (Oct. 2000)*, eds. N. Masetti et al., in press (astro-ph/0102277).
7. Briggs, M. S., Preece, R. D., van Paradijs, J., et al., in *Gamma-Ray Bursts: 5th Huntsville Symp.*, eds. R. M. Kippen et al., AIP Conf. Proc. **526**, New York, 2000, p. 125.
8. Nemiroff, R. J., Norris, J. P., Bonnell, J. T., et al., *ApJ* **435**, L133 (1994).
9. Mallozzi, R. S., Paciesas, W. S., Pendleton, G. N., et al., *ApJ* **454**, 597 (1995).
10. Mallozzi, R. S., Pendleton, G. N., Paciesas, W. S., et al., in *Gamma-Ray Bursts: 4th Huntsville Symp.*, eds. C. Meegan et al., AIP Conf. Proc. **428**, New York, 1998, p. 273.

# Spectral Properties of Short Gamma-Ray Bursts

William S. Paciesas*, Michael S. Briggs*, Robert D. Preece* and Robert S. Mallozzi[†]

*University of Alabama in Huntsville and National Space Science and Technology Center, Huntsville, AL
[†]deceased

**Abstract.**
It is well known that short GRBs have harder spectra than long GRBs, but the spectral differences between these two modes of the duration distribution have not been examined in detail. Using a database of standard model fits to GRBs in the BATSE 4B catalog, we compare the distributions of spectral parameters of short and long bursts. We also investigate the duration dependence of the same parameters within each mode. We find a consistent pattern of duration dependence: the mean values of all parameters differ significantly between the two duration classes, with short bursts being consistently harder. However, the same parameters show only weak dependence on duration within each class. We discuss the implications of these results for our understanding of GRBs.

## INTRODUCTION

The division of GRBs into subclasses is an area of intense current interest. Although the exact nature and number of subclasses is debatable, there is general agreement that short duration GRBs are a distinct subclass [1, 2, 3, 4, 5, 6]. Recently, additional evidence has come from a study by Norris et al. [7], who found no measurable energy-dependent pulse lag in the time histories of short events. This is in contrast to long GRBs, which clearly show such a lag [8], even for short sub-pulses. Moreover, in bursts with measured redshifts (which thus far are all long events), the energy-dependent pulse lag appears to be anti-correlated with burst luminosity [8]. Thus, if the mechanism producing the lag works for short bursts, they must be intrinsically more luminous than long bursts, and therefore more distant. Alternatively, a different mechanism may operate in short events. Either way, the evidence seems to support separate classification of short and long GRBs.

In the currently favored fireball model, the prompt burst emission is thought to be optically thin synchrotron or synchrotron self-Compton emission from internal shocks [9, 10], as external shocks are unable to produce the observed temporal structure [11, 12]. Detailed studies of the spectra of a number of bright GRBs, including both long and short events, have shown good consistency with the synchrotron shock model [13, 14, 15, 16, 17]. However, more comprehensive analyses have uncovered problems with this interpretation [18, 19, 20]. In particular, some GRB spectra are harder at low energies than the synchrotron limit [18, 19]. These conclusions are based mostly on spectroscopy of long bursts, but the problem may be most acute for short bursts because their spectra are on average harder.

Recent work [21, 22] has characterized the range of spectral behavior in bright, long bursts in some detail, but the spectral properties of the class of short bursts have only been characterized using hardness ratios. Phebus data showed that short GRBs are harder than long ones [1] (confirmed by numerous subsequent analyses of BATSE data), but detailed study of the spectral differences between short and long bursts has not been done. In particular, the consistency of short burst spectra with the synchrotron shock model predictions has not been properly tested. It is clear that a better characterization of the spectral differences between short and long bursts is warranted. For the foreseeable future, the BATSE data base will provide the best sample of bursts for this purpose.

## DATA ANALYSIS

The BATSE *CONT* datatype is derived from the large area detectors and is independent of the BATSE trigger. However, the *CONT* data have 16-channel energy resolution and 2 s time resolution, so they are not optimal for the analysis of the spectra of short events because the 2 s integration degrades the signal-to-noise ratio. Nevertheless, a database of *CONT* fits was conveniently available [23], so we used these data to investigate spectral differences between short and long GRBs. Preliminary results from our study have been published elsewhere [24].

The *CONT* fit database contains spectral fits for ~1200 GRBs from the BATSE 4B catalog [25]. Fit results for two spectra per burst (peak flux interval and total fluence interval) are available, generally from four different spectral models. For a given event, fit results may not be available for all models due to poor statistics and/or lack of fit convergence.

We extracted spectral parameters for all GRBs in the *CONT* database for three of the models (Comptonized, broken power laws, and the Band GRB function). The BAND function [26] (a smoothly broken power-law) is conventionally used to fit GRB spectra, but it is more likely to encounter convergence problems when the signal-to-noise ratio is low. The Comptonized model (power-law with exponential high-energy cut-off) has one fewer parameter than the other two models, and thus the database contains more successful fits with this model than the others.

For short events, there is little difference between the peak flux and fluence intervals because of the 2 s *CONT* time resolution, whereas long GRBs typically have harder spectra at the time of peak flux. Thus, differences in hardness between short and long events should be most pronounced in the fluence spectra, and this is indeed what our analysis indicates.

For simplicity in the present analysis, we ignored the errors on the fit parameters, thus giving each fit result equal weight. However, we only included fit results where all parameters were allowed to vary.

Figure 1 shows the parameter distributions for the Comptonized model fits to peak flux intervals separately for short and long bursts. The short bursts are significantly harder in both spectral index and cut-off energy. This is confirmed quantitatively using the Wilcoxson Rank-Sum test (IDL function RS_TEST), where the probability that the spectral index distributions have the same mean is less than $10^{-8}$. The probability that the cut-off energy distributions have the same mean is also less than $10^{-8}$.

Similarly, Figure 2 shows the parameter distributions for the Band GRB function fits to peak flux intervals. In this case, the short bursts are significantly harder in all three parameters; RS test significances are $3.1 \times 10^{-5}$, $3.0 \times 10^{-7}$, and $1.2 \times 10^{-6}$ for the low-energy index $\alpha$, peak energy $E_p$, and high-energy index $\beta$, respectively.

The problem of fit convergence is most severe for the Band function, and thus the corresponding sample size is smallest. However, this function is of most interest for comparison with other GRB spectral analysis, so in the remainder of this paper we confine the discussion to the Band function fit results, and specifically to the peak

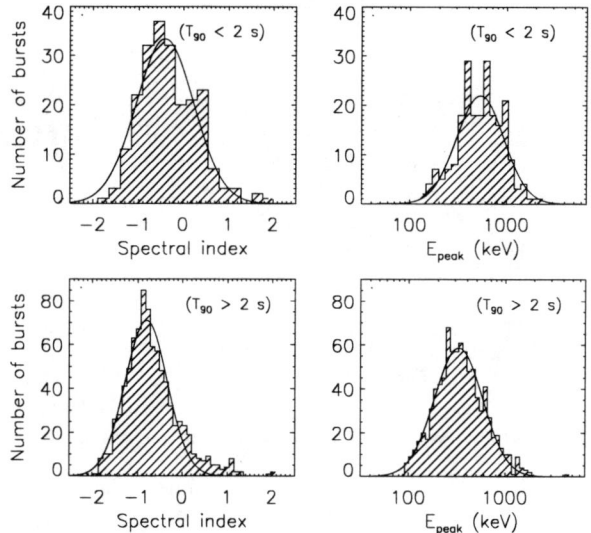

**FIGURE 1.** Frequency distributions of spectral parameters for short (upper panels) and long (lower panels) GRBs, from fits to a Comptonized model (power-law with exponential high-energy cut-off). The left panels show the spectral index and the right panels show the cut-off energy. Solid curves show the best-fitting Gaussians.

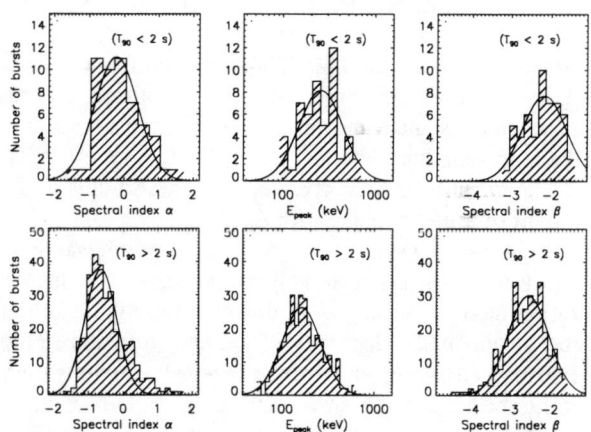

**FIGURE 2.** Similar to Figure 1, but from fits to the Band GRB function. The left panels show the low-energy spectral index $\alpha$, the middle panels show the peak energy $E_p$, and the right panels show the high-energy spectral index $\beta$.

flux spectra. We binned the results for each spectral fit parameter according to the burst duration, and computed the means and standard deviations of the data in each duration bin.

Figure 3 shows the individual fit parameters as a function of duration as well as the binned results. Within a given duration interval, the thin vertical bars show the standard deviation of the data, and the thick vertical bars show the error in the mean. Although the distributions

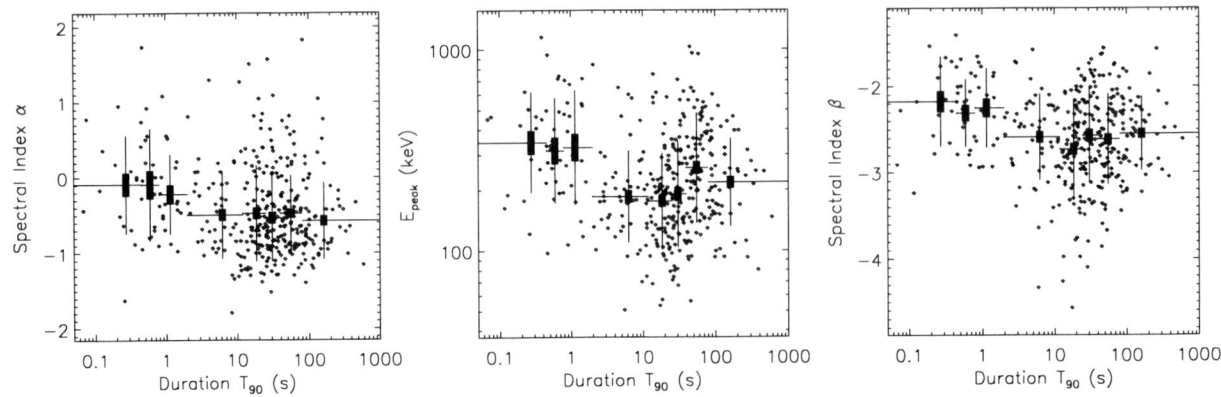

**FIGURE 3.** Spectral parameters as a function of burst duration from fits to the Band GRB function. The left panel shows $\alpha$, the center panel $E_p$, and the right panel $\beta$. Dots are un-binned data for individual bursts. Thin vertical bars show the standard deviation of the parameter distribution within a duration bin and thick vertical bars show the error in the mean of each distribution.

are broad, the trend of their mean values in all cases shows the largest discontinuity at a duration $T_{90} \sim 2$ s, the conventional boundary between short and long duration modes. Furthermore, within each mode the mean values are relatively constant. One way to quantify this uses the Kruskal-Wallis H-test (IDL function KW_TEST), which tests the hypothesis that three or more sample populations have the same mean. For the short mode, the KW test significances are 0.80, 0.94, and 0.68 for $\alpha$, $E_p$ and $\beta$, respectively, consistent with the hypothesis that the mean is constant within the mode. For the long mode, the corresponding significances are 0.83, $1.4 \times 10^{-3}$, and 0.44, indicating that the constant hypothesis holds for the means of $\alpha$ and $\beta$, but not for $E_p$.

Figure 4 shows the 64-ms peak flux dependence of the spectral parameters separately for the short and long duration modes. For clarity in this case only the errors in the means of the distributions are shown in the figure. For long GRBs, $E_p$ shows the previously known softening trend with decreasing peak flux [27]. However, the short GRBs do not show a similar trend, $E_p$ remaining relatively constant as peak flux decreases. In the highest peak flux bin, the short GRBs have roughly the same $E_p$ as long GRBs, but for weaker fluxes the short GRBs always have larger $E_p$ on average.

The spectral indices of long GRBs show relatively little dependence on peak flux, with perhaps a slight hardening trend as peak flux decreases, whereas a more pronounced hardening trend is present for short events. Again, the agreement in average parameter values between short and long modes is closest for the brightest events. The spectral indices of all but the brightest short GRBs are harder than long GRBs of the same peak flux.

## DISCUSSION

The general problem that low-energy spectral indices of many GRBs are harder than the synchrotron limit ($\alpha < -2/3$) is particularly acute for short GRBs. Figures 1 and 2 show that the majority of short GRBs violate the limit for both spectral models. For the Comptonized model, the short bursts have $\langle \alpha \rangle = -0.41 \pm 0.04$ and for the Band function model, $\langle \alpha \rangle = -0.24 \pm 0.09$. There has been speculation that the hardness of the low-energy spectra is due at least partly to an instrumental effect [28], but further work failed to support this [29]. On the other hand, the data may be consistent with a suitably modified synchrotron model. Synchrotron self-absorption can improve the model fit for some bursts [28], but this would require rather high magnetic fields and particle densities. Alternatively, the pitch-angle distribution of the accelerated electrons may be highly anisotropic [28, 29], in which case small pitch-angle scattering can produce harder low-energy spectra [30]. However, this mechanism still requires that $\alpha \leq 0$, whereas a significant fraction ($\sim 30$–$40\%$) of the short GRBs have $\alpha > 0$. As Figure 4 shows, the short GRBs with $\alpha > 0$ tend to have low peak flux, which may indicate that instrumental effects play a significant role. If the small pitch-angle scattering hypothesis is correct, then the data imply that the electron pitch-angle distribution in short GRBs is more anisotropic than in long GRBs.

The center panel of Figure 3 shows that the energy spectra of short and long GRBs have different characteristic break energies that otherwise depend only weakly on duration. Unlike the distributions of power-law spectral indices, the observed $E_p$ distribution is a convolution of the intrinsic distribution with the (unknown) distribution of cosmological redshifts. The intrinsic distribution is itself determined by at least two parameters, one re-

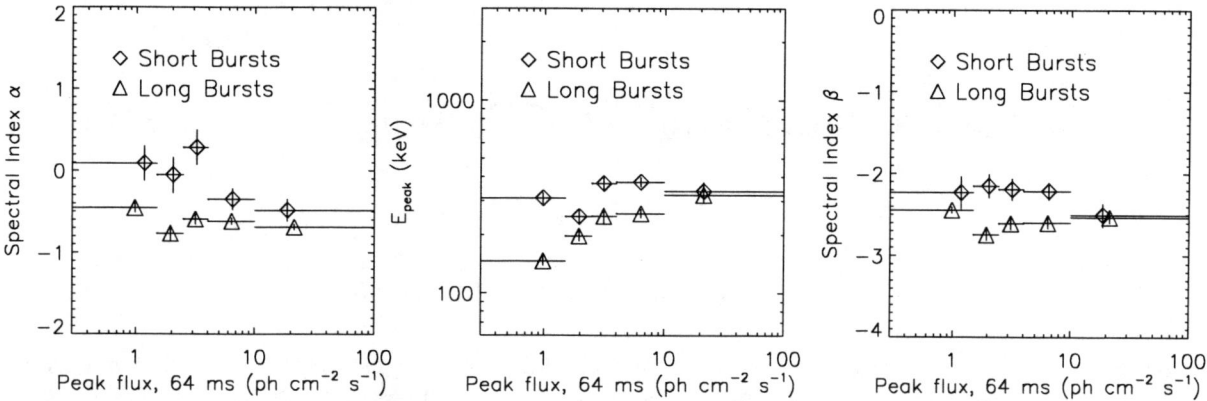

**FIGURE 4.** Band function spectral parameters as a function of 64-ms peak flux for short (diamonds) and long (triangles) GRBs. Vertical bars show the errors in the mean of the distribution in each peak flux bin.

flecting the basic emission mechanism and the other reflecting the bulk Lorentz factor of the emitting particles. Given this, the observed BATSE distribution is surprisingly narrow, and the extent to which this is an instrumental artifact has been the subject of some debate, the summary of which is beyond the scope of this paper. We contend that even if instrumental effects are important, these do not vary significantly versus duration for the peak-flux spectra of the bursts in our sample. This is supported by the relative constancy of $\langle E_p \rangle$ within each duration mode. Thus, the discontinuity in $E_p$ between the short and long duration modes is likely to be a real effect. This could represent a difference in the typical redshift between the two GRB classes, but since the spectral indices also show similar trends with duration, it is likely that at least some of the difference in $E_p$ is intrinsic to the sources.

## SUMMARY

Spectral parameters of short GRBs are generally harder than those of long GRBs, independent of the spectral model. Within each duration mode, the dependence of the spectral parameters on duration is weak. The difference is most likely due to different physical emission conditions in the two source classes.

## REFERENCES

1. Dezalay, J.-P., et al., "Short Cosmic Events: A Subset of Classical GRBs?", in *Gamma-Ray Bursts: Huntsville, AL 1991*, edited by W. S. Paciesas and G. J. Fishman, AIP, New York, 1992, pp. 304–309.
2. Kouveliotou, C., et al., *ApJ*, **413**, L101 (1993).
3. Belli, B. M., *Ap&SS*, **231**, 43 (1995).
4. Mukherjee, S., et al., *ApJ*, **508**, 314 (1998).
5. Horváth, I., *ApJ*, **508**, 757 (1998).
6. Hakkila, J., et al., *ApJ*, **538**, 165–180 (2000).
7. Norris, J. P., Scargle, J. D., and Bonnell, J. T., "Short Gamma-Ray Bursts Are Different", in *Gamma-Ray Bursts in the Afterglow Era*, edited by E. Costa, F. Frontera, and J. Hjorth, Springer, Berlin, 2001, pp. 40–42.
8. Norris, J. P., Marani, G. F., and Bonnell, J. T., *ApJ*, **534**, 248 (2000).
9. Katz, J., *ApJ*, **432**, L107 (1994).
10. Tavani, M., *AP&SS*, **231**, 181 (1995).
11. Fenimore, E. E., et al., *ApJ*, **473**, 998 (1996).
12. Sari, R., and Piran, T., *ApJ*, **485**, 270 (1997).
13. Tavani, M., *Phys. Rev. Lett.*, **76**, 3478 (1996).
14. Tavani, M., *ApJ*, **466**, 768 (1996).
15. Cohen, E., et al., *ApJ*, **488**, 330 (1997).
16. Schaefer, B. E., et al., *ApJ*, **492**, 696–702 (1998).
17. Bromm, V., and Schaefer, B. E., *ApJ*, **520**, 661–665 (1999).
18. Crider, A., et al., *ApJ*, **479**, L39 (1997).
19. Preece, R. D., et al., *ApJ*, **506**, L23 (1998).
20. Preece, R. D., et al. (2002), in preparation.
21. Preece, R. D., et al., *ApJ*, **496**, 849 (1998).
22. Preece, R. D., et al., *ApJS*, **126**, 19 (2000).
23. Mallozzi, R. S., et al., "Gamma-Ray Burst Spectra and the Hardness-Intensity Correlation", in *Gamma-Ray Bursts: 4th Huntsville Symposium*, edited by C. A. Meegan, R. D. Preece, and T. M. Koshut, AIP, New York, 1998, pp. 273–277.
24. Paciesas, W. S., et al., "Spectral Properties of Short Gamma-Ray Bursts", in *Gamma-Ray Bursts in the Afterglow Era*, edited by E. Costa, F. Frontera, and J. Hjorth, Springer, Berlin, 2001, pp. 13–15.
25. Paciesas, W. S., et al., *ApJS*, **122**, 465–495 (1999).
26. Band, D. L., et al., *ApJ*, **413**, 281 (1993).
27. Mallozzi, R. S., et al., *ApJ*, **454**, 597–603 (1993).
28. Lloyd, N. M., and Petrosian, V., *ApJ*, **543**, 722 (2000).
29. Lloyd-Ronning, N. M., and Petrosian, V., *ApJ*, **565**, 182–194 (2002).
30. Epstein, R., *ApJ*, **183**, 593 (1973).

# Towards an Understanding of Prompt GRB Emission

## Nicole M. Lloyd-Ronning

*Canadian Institute for Theoretical Astrophysics*

**Abstract.** We discuss the prompt emission of Gamma-Ray Bursts in different spectral energy bands. First, we suggest that a three-part synchrotron emission model [1, 2] is a good description of the $\sim 20$ keV - 1 MeV gamma-ray emission of GRBs. We show that this model provides excellent fits to the data and naturally explains the observed global correlations between spectral parameters. In particular, we show there exists a negative correlation between between the peak of the $\nu F_\nu$ spectrum, $E_p$, and the low energy photon index $\alpha$ for bursts with $-2/3 < \alpha < 0$, and suggest that this correlation is due to the mechanism responsible for producing $\alpha$'s above the value of $-2/3$ - namely, a decreasing mean pitch angle of the electrons. We then discuss the physical origin of the increasing number of GRBs that are observed to peak in the X-ray energy band ($\sim 5 - 40$ keV). Although either a cosmological (i.e. high redshift) or intrinsic interpretation for the low values of $E_p$ is viable at this point, the data appear to suggest that intrinsic effects are playing the dominant role. Finally, we briefly comment on the prompt GRB optical emission ($\sim$ eV) and very high energy emission ($> 10$ MeV), and how these spectral bands may be used to place additional constraints on the physics of gamma-ray bursts.

## INTRODUCTION

Just as broadband observations were (and are) necessary to establish the general external shock paradigm of the GRB afterglow, so are broadband data necessary to fully understand the physics of the prompt GRB emission. The prompt ($\sim$ first one hundred seconds or so) emission gives us a unique opportunity to explore the physics of internal - and possibly external - shocks, address fundamental issues associated with the behavior of relativistic plasmas, particle acceleration and turbulence, and possibly even gain insight into the GRB progenitor itself. In this paper, I will present some recent advancements in our understanding of GRB prompt emission, and discuss important remaining questions associated with the initial phases of a burst.

## PROMPT GAMMA-RAY EMISSION

Most GRB spectra in the energy range ($\sim 20$ keV - $\sim$ MeV) are well described by a so-called Band spectrum [3]. This model is essentially a smoothly broken power with a low energy photon spectral index $\alpha$, a high energy photon index $\beta$, and a break energy $E_p$. There have been some attempts to explain or interpret the global properties of these spectral parameters in terms of a physical model [4, 5, 6], with inconclusive results. In particular, it was thought from very early on that the radiation mechanism responsible for GRBs (in this energy range) is synchrotron emission [7]. However, problems with this model - particularly with the behavior of the low energy $\gamma$-ray spectrum - were quickly brought to light. First, it was noted [5] that a number of bursts have a low energy spectral index $\alpha$ that falls above the so-called "line-of-death" value of $-2/3$, the asymptotic limit of an instantaneous (i.e. non-cooling) optically thin synchrotron spectrum from an isotropic, power-law electron distribution with some minimum cutoff energy. Even more fatal for the standard synchrotron picture, it was pointed out [6] that if particles are injected at once and then left to radiate, the synchrotron cooling times are very short (much less than the detector integration time). In this case, all of the particles cool and the electron spectrum becomes a soft power law (of index $-2$) at low energies. As a result, the asymptotic (upper) limit of the low energy photon spectrum is $\alpha = -3/2$. Nearly *all* GRBs have observed values of $\alpha$ that fall above this limit! However, as we will show below, when some of the simplifying assumptions in these models are modified, synchrotron emission in fact does a very good job of explaining the observed data.

### A Three-part Synchrotron Model

Our model (described in more detail in [1] and [2]) modifies the usual simple picture of optically thin syn-

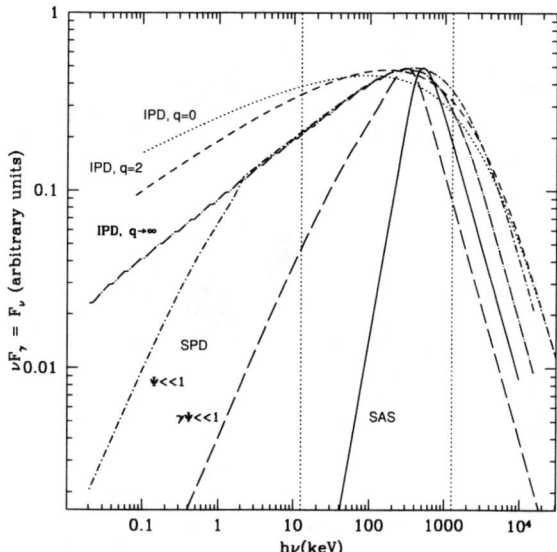

**FIGURE 1.** Synchrotron spectra in the various emission regimes (see text for explanation). The dotted vertical lines represent the approximate width of the BATSE spectral window.

chrotron emission from a power law distribution of electrons with a sharp low energy cutoff, by accounting for: 1) the possibility of a smooth cutoff to the low energy electron distribution, 2) radiation from an anisotropic electron distribution with a small mean pitch angle, 3) synchrotron self-absorption, and 4) the important instrumental effect in which the value of the fitted parameter $\alpha$ decreases as $E_p$ approaches the lower edge of the BATSE window. We have envisioned a realistic scenario in which particle acceleration and synchrotron losses occur continually and simultaneously behind each internal shock, with the characteristic acceleration time shorter than the loss time; this means that synchrotron loss effects are only evident in the particle distribution spectrum at energies much larger than what is relevant for our discussion here (at energies where the inequality is reversed and the loss time becomes shorter than the acceleration time). The emission in this model can be characterized by three distinct regimes.

1) **IPD.** This is the familiar optically thin synchrotron emission from a power law electron energy spectrum, with an isotropic pitch angle distribution. In contrast to most analyses, however, here we consider an electron distribution with a smooth low energy cutoff: $N(\gamma) \propto \frac{(\gamma/\gamma_m)^q}{1+(\gamma/\gamma_m)^{p+q}}$. Note that for high energies ($\gamma > \gamma_m$), the spectrum goes as $\gamma^{-p}$, while for low energies ($\gamma < \gamma_m$), the spectrum goes as $\gamma^q$. Hence, $q$ denotes the steepness of the electron low energy cutoff (note that an actual "cutoff", in the sense that $N(\gamma) \to 0$ as $\gamma \to 0$, requires $q > 0$). The asymptotic behavior of the synchrotron (photon number) spectrum for $q > -1/3$ is:

$$F_\gamma = \begin{cases} \nu^{-2/3} & \nu \ll \nu_m = \frac{2}{3}\nu_B \sin\Psi \gamma_m^2 \\ \nu^{-(p+1)/2} & \nu \gg \nu_m \end{cases} \quad (1)$$

where $F_\gamma$ is the photon flux, $\Psi$ is the electron pitch angle, and $\nu_B = \frac{eB}{m_e c}$ where $B$ is the magnetic field. Note that the peak of the $\nu F_\nu$ spectrum will occur at $E_p \propto \nu_m \propto B\sin\Psi\gamma_m^2$, and that the aymptotic low energy index below this break is $\alpha = -2/3$.

In this regime, we expect a positive correlation between $E_p$ and $\alpha$ due to instrumental effects alone. If $E_p$ is close to the edge of the BATSE window, the low energy photon index may not yet have reached its asymptotic value and a smaller (or softer) value of $\alpha$ (relative to the asymptotic value) will be determined. A smooth cutoff to the electron energy distribution will exacerbate this effect because for a smoother cutoff (or a smaller value of $q$), the asymptote is reached at lower energies relative to $E_p$. Note that a dispersion in the smoothness of the low energy cutoff will tend to wash this correlation out to some degree, as seen in Figure 4 of [1]. For the cases of small pitch angle radiation and the self-absorbed spectrum (see below), this instrumental effect will be weaker because the low energy asymptotes are reached more quickly (i.e. at energies closer to $E_p$) than for the isotropic optically thin case (see Figure 1).

2) **SPD.** This type of synchrotron spectrum results from optically thin synchrotron emission by electrons with a mean pitch angle $\Psi \ll 1$; the analysis of synchrotron radiation in this regime was first done by [8]. For low density, high magnetic field plasmas expected in GRBs, the Alfvén phase velocity is greater than the speed of light and (therefore) the speed of the relativistic particles under consideration here. In this case, the pitch angle diffusion rate of the electrons interacting with plasma turbulence is smaller than the acceleration rate; consequently, the accelerated electrons could maintain a highly anisotropic distribution as required in the small pitch angle model. The shape of this spectrum depends on just how small the pitch angle is. For $\Psi \ll 1$, but $\Psi\gamma_m \sim 1$, we have:

$$F_\gamma = \begin{cases} \nu^0 & \nu \ll \nu_s = \frac{2}{3}\nu_B/(\gamma_m\Psi^2) \\ \nu^{-2/3} & \nu_s \ll \nu \ll \nu_m \\ \nu^{-(p+1)/2} & \nu_m \gg \nu \end{cases} \quad (2)$$

There are two breaks in this spectrum - one at $\nu_m$ and one at $\nu_s$. Because the Band spectrum can only accommodate one break, spectral fits to this model will put the parameter $E_p$ at one or the other of these two breaks, but most likely at $\nu_m$ because for $p > 5/3$ (or for high energy photon index $\beta < -4/3$ which is the case for most bursts), the break across $\nu_m$ is more pronounced than across $\nu_s$. In this case, the low energy photon index $\alpha$ will fall somewhere between $-2/3$ and 0.

However, as the pitch angle $\Psi$ decreases such that $\Psi \ll 1/\gamma_m$, then the $\nu^{-2/3}$ portion of the spectrum disappears, and only the $\nu^0$ portion is left. In this case we have:

$$F_\gamma = \begin{cases} \nu^0 & \nu \ll \nu_s = \frac{4}{3}\nu_B\gamma_m \\ \nu^{-(p+1)/2} & \nu_s \gg \nu, \end{cases} \quad (3)$$

where $E_p \propto B\gamma_m$ (see [8] for a more detailed description of the behavior of the spectrum in this regime).

Here, we expect evidence of a <u>negative correlation</u> between $E_p$ and $\alpha$ as we transition from the IPD to the SPD regime, i.e. for $-2/3 < \alpha < 0$. In this case, the pitch angle decreases so that $E_p \propto \sin\Psi$ decreases, if all other physical parameters ($B$ and $\gamma_m$) remain constant. In addition, as we go from the small pitch angle regime, $\Psi\gamma_m \sim 1$ ($\Psi \ll 1$), to the very small pitch angle regime, to $\Psi\gamma_m \ll 1$, the $\nu^{-2/3}$ portion of the spectrum disappears, and we are left with only the $\nu^0$ portion. In other words, as the mean of the pitch angle distribution decreases to very small values, $E_p$ decreases and the value of $\alpha$ increases from $-2/3$ to 0. This negative correlation will compete with the positive instrumental correlation mentioned above.

3) **SAS.** If the magnetic field and density are such that the medium becomes optically thick to the synchrotron photons with frequency $\nu < \nu_a$, then, for $\nu_a < \nu_m$, we have the following spectrum:

$$F_\gamma = \begin{cases} \nu^1 & \nu \ll \nu_a, \\ \nu^{-2/3} & \nu_a \ll \nu \ll \nu_m, \\ \nu^{-(p+1)/2} & \nu_m \gg \nu \end{cases} \quad (4)$$

In that case, $E_p \propto \nu_a \sim 10(nl)^{3/5}B^{2/5}\gamma_m^{-8/5}\Gamma^{9/5}$ Hz, where $l$ and $n$ are the path length and particle density in the co-moving frame, and we have assumed an electron energy distribution index $p = 2$. For $\nu_a > \nu_m$ we just have one break at $\nu_a$ with a low energy photon index of $\alpha = 3/2$ (in both the isotropic and small pitch angle cases). The possibility of self-absorption in GRBs is a controversial issue. We have shown [1] that there are bursts for which a self-absorbed spectrum is a better fit than an optically thin one. We also found that in these cases, the absorption frequency tends to be near the lower edge of the BATSE window. In addition to this, Strohmeyer et al. [9] found that a number of bursts observed by GINGA with $E_p$'s in the range 2 to 100 keV have steep ($\alpha \sim 1$) low energy spectral indices consistent with a self-absorbed spectrum. This raises interesting questions about the physics of the ambient plasma, because self-absorption in a GRB requires fairly large ($\sim 10^8 G$) magnetic fields and particle densities ($\sim 10^8 cm^{-3}$). The physical processes required to

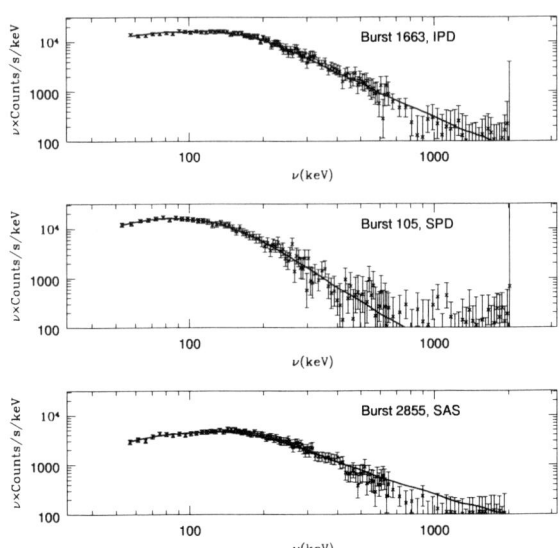

**FIGURE 2.** Three time resolved spectra fit to our generalized synchrotron model. The best fit emission regime corresponds to what is predicted by the low energy Band parameter $\alpha$.

achieve these conditions will need to be theoretically established if the data prove self-absorption to be a viable model.

## How the Data Stand Up

**Spectral Fits:** We fit each of three representative GRB spectra shown in Figure 2 to all 3 emission scenarios and then evaluate the fits based on their values of a reduced $\chi^2$. Each fit is taken at a time during the burst spectral evolution when the $\alpha$ parameter corresponded to the respective model. For example, in the top panel - burst 1663 - the spectrum is from a time when $\alpha \sim -2/3$, while in the middle panel - burst 105 - the spectrum is from a time in the profile when $\alpha = 0$. Similarly, for the bottom panel, this spectrum corresponds to a time when $\alpha = 1$. The reduced $\chi^2$ are 0.34, 0.33, and 0.50 for the top, middle and bottom panels respectively. In general the best model turns out to correspond to the emission regime suggested by Band's $\alpha$ values, which confirms our proposed method of physically interpreting Band fits based on the bursts' low energy photon index (for example, an IPD fit to the spectrum of burst 105 gave a $\chi^2 > 1$ compared to the $\chi^2 = 0.33$ for an SPD fit).

**Global $\alpha$ Distribution:** As discussed above, the low energy photon index $\alpha$ is the best parameter for distinguishing between the various synchrotron regimes. Figure 3 shows a histogram of $\alpha$ (taken from 2,026 time resolved spectra with Band spectral fits, from [11]) with

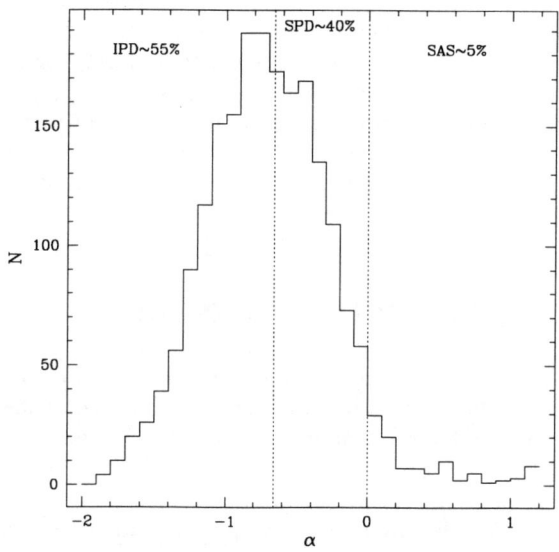

**FIGURE 3.** Histogram of the low energy spectral index $\alpha$, with each emission regime delineated. The data are from Preece et al., 1999.

each regime clearly marked. Note that there are a significant number of spectra in the SPD regime.

Although the error bars on $\alpha$ can make some difference in the numbers of spectra in each regime, this does not affect the qualitative nature of our conclusions below (see [2] for a more detailed discussion on how the error on $\alpha$ affects this distribution). We now discuss the correlations present in the data and their consistency with what we expect in the context of the three synchrotron emission scenarios.

**Observed Correlations** Figure 4 shows the binned average correlation between $E_p$ and $\alpha$ present in $\sim 2000$ time resolved spectra from [11]. We have sorted $\alpha$ in ascending order and binned the data every 100 points (the horizontal error bars indicate the size of the bins). We then computed the average $E_p$ for these 100 points. The most intriguing result is that the correlation appears to be positive for $\alpha < -0.7$ and negative for $\alpha > -0.7$. Performing a Kendell's $\tau$ test on all of the (unbinned) data, we find a 9σ *positive correlation between $\alpha$ and $E_p$ in the IPD regime*. To account for both the error in $\alpha$ and $E_p$, we have performed this test on all permutations of correlations between the lower and upper values of $\alpha$ (from the 1σ error bars) with the lower and upper values of $E_p$. In addition we have averaged the value of the correlation statistic $\tau$ from 100 sets of data, in which - for each data point - $\alpha$ and $E_p$ are drawn from Gaussian distributions with means equal to the parameter values given in the catalog and standard deviations corresponding to the error bars. In all cases, we find a highly significant ($> 6\sigma$)

correlation. The positive correlation between $\alpha$ and $E_p$ in the IPD regime can be simply understood by the instrumental effect discussed in the previous section.

On the other hand, we find a 4σ *negative correlation between $\alpha$ and $E_p$ in the SPD regime*. Again, to account for the error in both $\alpha$ and $E_p$, we have performed this test on all permutations of correlations between the lower and upper values of $\alpha$ (from the 1σ error bars) with the lower and upper values of $E_p$, and have also averaged the $\tau$ value from 100 sets of data drawn from distributions based on the existing data, according to the prescription described in the above paragraph. In all of these cases, we find a significant ($> 3\sigma$) negative correlation. As mentioned above, this type of correlation is natural in the small pitch angle regime, as a result of a decreasing average pitch angle in the electron distribution.

The observed trends are consistent with what is expected from our model in each emission scenario, and tell us something important about the role various effects play in the correlations. For example, the dashed line in Figure 4 shows how $E_p$ should change as a function of $\alpha$ in the BATSE spectral window, if only the mean of the electron pitch angle $\Psi$ changes (all other parameters such as $B$ and $\gamma_m$ remaining constant). The fact that the observed correlation is weaker could be due to a number of different physical effects.[1] Of course, we expect that the correlation will be washed out to some degree by dispersion in the intrinsic values of $\Psi$ and $\gamma_m$, as well as variation in the magnetic field from burst to burst. It is also possible that the minimum electron Lorentz factor or the magnetic field of the electrons *increases* as we transition to a physical regime in which electrons are accelerated primarily along the magnetic field lines, which would in turn cause a more gradual decrease of $E_p$ with $\Psi$ (or $\alpha$). This may be a very plausible explanation - there may exist physical situations which require either a higher magnetic field or characteristic electron Lorentz factor, in which it is very efficient to accelerate along the magnetic field lines. On the other hand, the electrons' Lorentz factors could be only mildly relativistic (instead of $\sim 100$ as we assumed for the dashed line in Figure 4), which would lead to a smaller relative decrease ($\sim 1/\gamma_e$) in $E_p$ as a function of pitch angle. Finally, it is of course possible that this model is incorrect and an alternative explanation is needed to accommodate bursts above the $\alpha = -2/3$ line of death. Still, it is encouraging that the global distributions and observed trends are accomodated well by our scenario.

---

[1] The positive instrumental correlation discussed in the previous section will also play a small role in reducing the strength of the negative correlation.

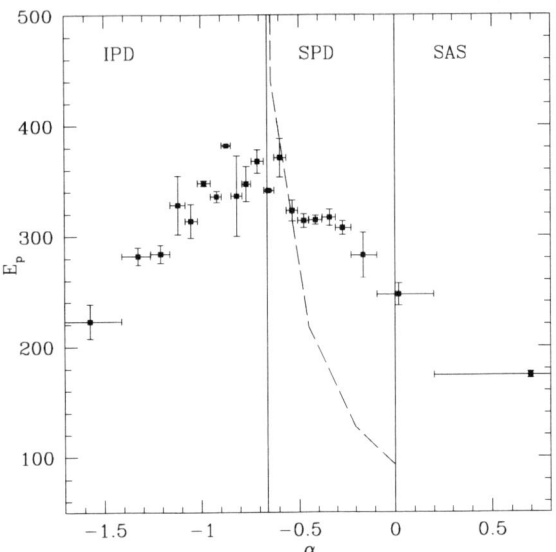

**FIGURE 4.** Observed $E_p$ vs. $\alpha$, for $\sim 2000$ time resolved spectra. Note the change in the sign of the correlation as $\alpha$ transitions from the IPD to SPD regime.

## Particle Acceleration and Remaining Questions

Our results bring to light the fact that particle acceleration in GRBs is a quite poorly understood problem. Usually, it is assumed that the radiating particles in GRBs are accelerated via repeated scatterings across the (internal) shocks. This mechanism, however, predicts several features in the electron distribution not borne out by the data. First, it has been shown [12] that these repeated crossings of the shock result in a power law particle distribution with a well defined index, $p = -2.23$, which would give a high energy synchrotron photon index $\beta$ of $-1.62$ (or $-2.12$ for the "cooling" spectrum, e.g. [13]). Although this is consistent with some afterglows, this is certainly is not true for many bursts in the prompt phase. In our synchrotron models above, the high energy photon index $\beta = -(p+1)/2$, where $p$ is the high energy index of the emitting particle distribution. The parameter $\beta$ can vary by a factor of 4 (or more!) throughout a single burst (see, e.g., [11]), reflecting a huge variation (from 1 to 9) in the parameter $p$ of the underlying particle distribution - this is well beyond the statistical limits placed on $p$ by shock acceleration simulations.[2] In ad-

dition, shock acceleration predicts an *isotropic* distribution of electrons. Our work suggests that in a large fraction of GRBs, the particle acceleration is not isotropic but along the magnetic field lines. Thus, there are many crucial open questions related to the physics of particle acceleration in relativistic plasmas, and it is clear that a complete investigation of this phenomenon in the context of GRBs is necessary.

## PROMPT X-RAY EMISSION

Although most GRBs emit the bulk of their prompt radiation in the soft gamma-ray energy band (as their names imply), with the availability of recent X-ray observations of the GRB prompt emission from GINGA [9], Beppo-SAX [15, 16], untriggered low energy BATSE data [17], and HETE-II [18], an increasing number of GRBs with spectra that peak in the X-ray energy range ($\sim 1 - 40$ keV) have been discovered. Current estimates [17, 15], suggest that $\sim 30\%$ of all bursts may have $E_p$ values below $\sim 40$ keV. There has been considerable speculation over the possibility that these low $E_p$ spectra may be a result of redshift effects - e.g. that these GRBs occur at extremely high redshifts ($(1+z) > 5$), and therefore their observed spectra are shifted into the X-ray band. However, as we show below, intrinsic properties of the GRB could also very well produce low peak energies. Below we discuss the role of both cosmological and intrinsic effects in producing X-ray rich GRBs.

### Cosmological Effects

It is straightforward to ask: if we take a burst with a "typical" flux and peak energy at a redshift of 1 and we redshift it until its peak energy falls in the X-ray band, *would the burst be detectable (by current instruments)?*. Consider a burst at redshift $z$ with specific luminosity $L(v_o)$ in the cosmological rest frame of the source. The observed flux per unit frequency is

$$f_v \propto \frac{L(v(1+z))}{d\Omega d_{metric}^2 (1+z)} \quad (5)$$

where $v = v_o/(1+z)$, $d\Omega$ is the geometric solid angle into which the GRB outflow is confined, $d_{metric}$ is the metric distance defined as

$$d_{metric} = \int_0^z (c/H_o) \frac{dz}{\sqrt{\Omega_\Lambda + \Omega_m(1+z)^3}}. \quad (6)$$

For our purposes, we assume a simple standard synchrotron spectrum (in the IPD regime). A burst at a redshift of 1 with $E_p = 200$ and a flux of $10^{-6} erg/cm^2/s$

---

[2] Preliminary results indicate that, instead, our adopted picture of stochastic acceleration can produce the necessary electron distributions needed to explain the observed photon spectra in at least the IPD synchrotron emission scenario.

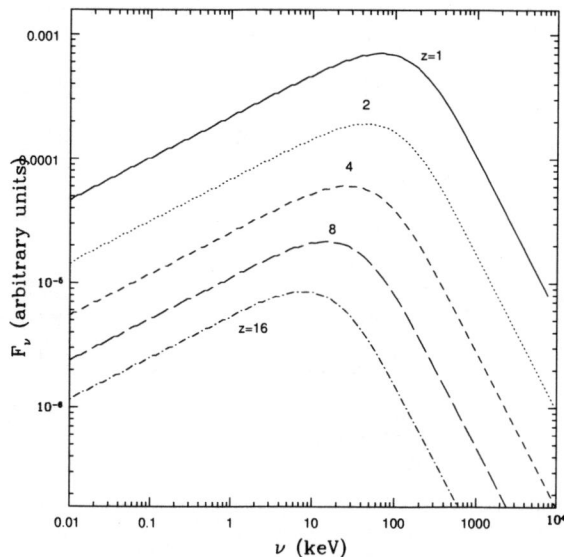

**FIGURE 5.** A sample $F_\nu$ spectrum as a function of redshift.

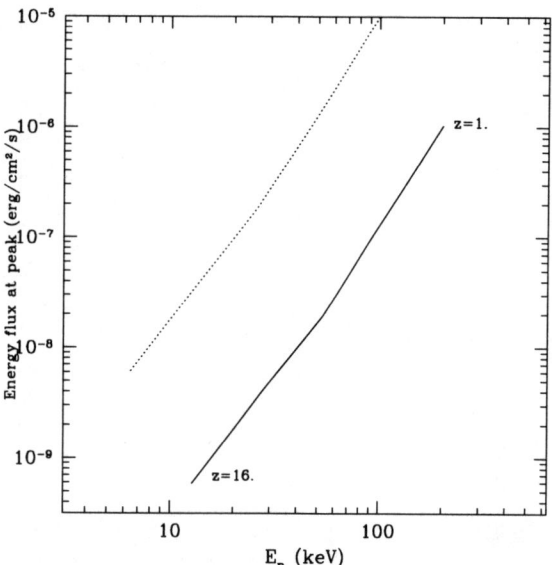

**FIGURE 6.** Energy flux at the peak of the $F_\nu$ spectrum vs. $E_p$ for changing redshifts of the burst, given two initial "starting" values of $E_p$ and flux at $z=1$.

would have an observed flux of about $5 \times 10^{-9}$ when $E_p$ is redshifted to 30 keV. On the other hand, for a burst with $E_p = 100$ keV and a flux of $10^{-5} erg/cm^2/s$ at $z=1$, it is easily detectable (flux = $10^{-8}$ erg/cm$^2$/s) out to a redshift of 10, where $E_p = 10$ keV. We note that the Heise BeppoSAX sample [15] has fluxes between $10^{-7}$ and $10^{-8}$ erg/cm$^2$/s, while the fluxes from Kippen's BATSE sample [17, 23] ranged from $10^{-7}$ to $5 \times 10^{-9}$ erg/cm$^2$/s. Unfortunately, because of the broad GRB luminosity function and intrinsic $E_p$ distribution (i.e. because there are no standard intrinsic luminosity and $E_p$ values for all GRBs) all we can say at this point is that it is *possible* that these low $E_p$ bursts are very high redshift bursts. We do note that the highest fluxes in both data samples require the bursts to be on either the high end of the intrinsic luminosity distribution or the low end of the intrinsic $E_p$ distribution in order to be consistent with the redshift interpretation. Nonetheless, we can neither rule out nor strongly favor the high redshift interpretation of these bursts.

*Luminosity Evolution*

There have been some suggestions that GRB luminosity function evolves with redshift, in the sense that bursts at high redshift tend to have intrinsically higher luminosity [19]. *This statement is based on those bursts with redshifts and luminosities from the luminosity-variability relation [20, 21], and so caution should be exercised until this relation is more definititively shown to provide valid redshifts.* However, if we do adopt the redshifts and luminosities from the L-V relation, then there is significant evidence that GRBs are brighter at higher redshifts. This effect allows for an even greater probability that very high redshift bursts are detectable by current instruments. For example, [19] find the average GRB luminosity $L \sim (1+z)^{1.4\pm0.5}$. Therefore, the observed flux decreases less rapidly as a function of redshift, than if there were no luminosity evolution. [3]

## Intrinsic Effects

The analysis above considered how a particular burst behaves as it is moved out in redshift space, for its particular set of observed properties. However, as has been suggested repeatedly and confirmed by bursts with measured redshifts, GRB intrinsic properties vary substantially from burst to burst. In some cases, the intrinsic properties can act in such a way as to mimic redshift effects.

For example, again assuming a standard synchrotron spectrum (which we do to be concrete, but which is not necessary), we can investigate how the $\nu F_\nu$ flux changes

---

[3] However, we mention one important caveat: If whatever causes the GRB luminosity function to evolve also causes $E_p$ to evolve, then there would be a decreased dependence of $E_p$ on redshift and one would have to place the burst at even higher redshifts to shift $E_p$ into the X-ray band.

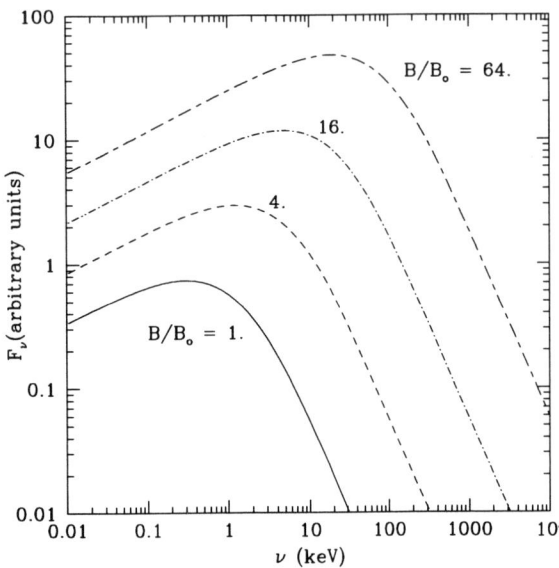

**FIGURE 7.** A synchrotron $F_\nu$ spectrum as a function of magnetic field.

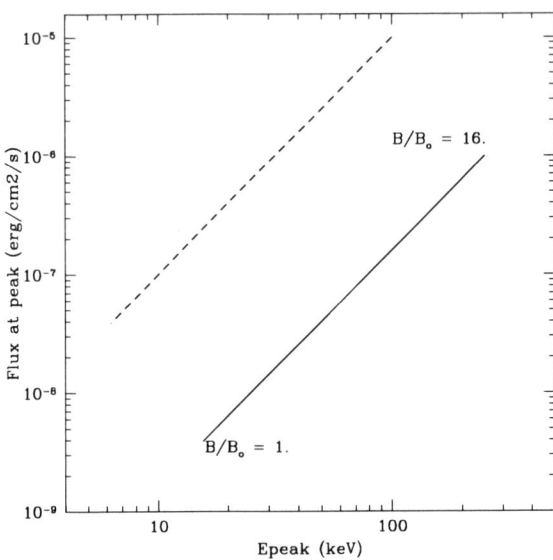

**FIGURE 8.** Same as Figure 6, but for a changing magnetic field.

as a function of various physical parameters. We take the magnetic field as an example. Figure 7 shows how the spectrum shifts with decreasing magnetic field from the γ-ray to X-ray regime. The flux at the peak is approximately proportional to $B^2$ while the peak energy goes as $B$. Figure 8 shows the peak flux as a function of $E_p$. Here we see that changing the magnetic field could also very reasonably produce X-ray bursts. Other possibilities include decreasing the bulk Lorentz factor, or changing the electron spectrum (such as the density or minimum energy of the electrons).

There are in fact several pieces of evidence suggesting that intrinsic effects are playing the dominant role in producing the low $E_p$ bursts:

1. One clue comes from the fact that the Kippen et al. bursts [17, 23], fall along an extension of the well known observed hardness-intensity correlation [24] in BATSE GRBs (see Figure 3 of [23]). Lloyd, Petrosian, & Mallozzi [22] found that this correlation cannot be produced by redshift effects alone, given any reasonable (i.e. not a δ-function) GRB luminosity function. That is, even for a fairly narrow intrinsic peak energy distribution, the observed hardness-intensity correlation cannot be due to cosmological expansion (e.g. lower flux and lower $E_p$ do not necessarily imply higher redshifts because the broadness of the GRB luminosity function washes out this effect). In fact, they found that a correlation between the intrinsic peak energy and the emitted energy of the GRB would naturally reproduce the observed correlation, and that such a relation between peak energy and total energy/luminosity is natural in (but not limited to) a synchrotron emission model.

2. The observed durations of the XRBs are not unusually long or smooth. At the highest redshifts (e.g. $1 + z \sim 10$), the duration and pulse width are dilated by a factor of 10. Although there exists no standard in GRB temporal properties, one might expect that if these bursts were at very high redshifts, their time profiles would be particularly long or smooth - this does not seem to be the case. Heise et al. [15] also show that the durations of the bursts in his sample are completely consistent with the global population and not unusually long.

3. The Kippen and Heise sample of low $E_p$ bursts are based on the *average* (over the duration of the burst) spectral properties of the burst. However, there have been some time resolved spectral analyses in which observations of X-ray rich *pulses* within a burst are reported (e.g. [25, 16]). For at least these bursts, the X-ray "richness" of the pulses *must* therefore be coming entirely from intrinsic effects.

## PROMPT OPTICAL AND VERY HIGH ENERGY EMISSION

**Optical** There are several mechanisms in a GRB that may produce prompt emission at optical wavelengths. And in fact in one burst - GRB 990123 - such emission was seen as a bright optical flash ($m_v \sim 9$) during the prompt phase [26]. The most common interpretation of this optical flash is that it resulted from the blast wave reverse shock propagating into the dense GRB ejecta.

Because the temperature in this region is smaller relative to the shocked ambient medium by a factor of $\sim \Gamma^2$, the emission from this interaction peaks in the optical energy band. Certain conditions (sensitive to $\Gamma$) are required to actually produce such a flash (see [27, 28] for discussion); if this interpretation is correct, it offers the possibility of constraining - among other things - the bulk Lorentz factor and baryonic content of the GRB outflow.

An alternative interpretation of this bright optical flash involves the interaction of the GRB radiation front [29, 30, 31] with the ambient medium. In this scenario, the γ-ray photons travelling ahead of the blast wave are side-scattered by the ambient medium. These photons then pair produce with the forward streaming photons, and pair enrich the medium. The available energy must be distributed among these pairs, and therefore the peak of the initial afterglow emission is lower in energy (relative to the standard picture without pairs) by a factor of $(m_e/m_p)^2$. As a result, the very early GRB afterglow (possibly coincident with the prompt emission) is brightest in the optical. In order for this mechanism to work efficiently, the medium is constrained to a particular density profile (see [29] for more details). Hence, if this mechanism succeeds, the prompt optical GRB emission affords us the opportunity to constrain the surrounding GRB environment.

**Very High Energy** If synchrotron emission is indeed responsible for the prompt GRB emission in the low energy gamma-ray band (see above), then we expect an inverse-Compton component from this emission. The inverse Compton component is boosted by a factor of $\Gamma^2$ and therefore peaks in the $\sim 10$ GeV range. There is also the possibility of observing very high (i.e. TeV) emission from a high energy proton component [32], as well as other high energy phenomena (such as neutrinos; see [33] for a discussion). A number of upcoming instruments (GLAST, Agile, Milagro, etc) will be sensitive in the range from 10's of MeV to TeV energies and have the potential of shedding much light on this relatively unexplored but important energy range of of GRB emission.

## CONCLUSIONS

We have discussed the prompt emission of GRBs, focusing particularly on the gamma-ray and X-ray emission properties. We find that a generalized synchrotron emission model does a good job of explaining the behavior of the prompt spectra in the $\sim 20$ keV $-1$ MeV range. Moreover, we find that although the increasing number of low $E_p$ ($< 40 keV$) GRBs may be consistent with a high redshift interpretation, the data appear to suggest that in fact variations in the bursts' intrinsic properties (e.g. magnetic field, electron energy density, etc.) are probably responsible for producing these "X-ray rich" bursts. We emphasize the importance of broadband (from eV to TeV) observations of prompt emission in order to gain a complete understanding of the physics of Gamma-Ray Bursts.

## ACKNOWLEDGMENTS

I would like to thank the organizers for an interesting and stimulating conference. I would also like to thank Vahe' Petrosian, with whom much of the work on the prompt gamma-ray emission was done.

## REFERENCES

1. Lloyd, N.M. & Petrosian, V. 2000, *ApJ*, 543, 722
2. Lloyd-Ronning, N.M. & Petrosian, V. 2001, *ApJ*, in press
3. Band, D., et al. 1993, *ApJ*, 413, 281
4. Tavani, M. 1996, *ApJ*, 466, 768
5. Preece, R.D., et al. 1996, *ApJ*, 473, 310
6. Ghissellini, G., Celotti, A., Lazzati, D. 2000, *MNRAS* 313, L1
7. Katz, J. 1994, *ApJ*, 432, L107
8. Epstein, R.I. 1973, *ApJ*, 183, 593
9. Strohmeyer, T.E., et al. 1998, *ApJ*, 500, 873
10. Dung, R. & Petrosian, V. 1994, *ApJ*, 421, 550
11. Preece, R.D., et al. 1999, *ApJS*, 126, 19
12. Kirk, J.G. et al., 2000, *ApJ*, 542, 235
13. Sari, R., Piran, T., & Narayan, R. 1998, *ApJ*, 497, L17
14. Piran, T. 1999, Physics Reports, 314, 575
15. Heise, J. et al. 2001, these proceedings.
16. Frontera, F., et al. 2000, *ApJS*, 127, 59
17. Kippen, R. M., et al. 2001, these proceedings
18. Barraud, C., et al. 2001, these proceedings
19. Lloyd-Ronning, N. M., Fryer, C. L., & Ramirez-Ruiz, E. 2001, *ApJ*, submitted (astro-ph/0108200)
20. Fenimore, E. E., & Ramirez-Ruiz, E. 2001, *ApJ*, submitted (astro-ph/0004176)
21. Reichart, D. E., et al. 2001, *ApJ*, 552, 57
22. Lloyd, N.M. et al., 2000, *ApJ*, 534, 227
23. Kippen, M. et al., 2001 in "Gamma-Ray Bursts in the Afterglow Era", Rome, in press.
24. Mallozzi, R.S. et al. 1996, in Gamma-Ray Bursts, AIP Conf. Proc. 384, eds. C. Kouveliotou, M.F. Briggs, G.J. Fishman (New York: AIP), 204.
25. Smith, D.A., et al. 2001, *ApJ*, submitted
26. Akerlof, C., et al., 1999 Nature 398, 400
27. Sari, R. & Piran, T. 1999, *ApJ*, 517, L109
28. Soderberg, A. & Ramirez-Ruiz, E., these proceedings
29. Beloborodov, A.M., 2001, *ApJ*, in press
30. Thompson, C. & Madau, P. 2000, *ApJ*, 538, 105
31. Meszaros, P., Ramirez-Ruiz, E., Rees, M.J. 2001, *ApJ*, 554, 660
32. Totani, T., 1999, *MNRAS*, 307, L41
33. Waxman, E. 2000, *ApJS*, 127, 519

# Luminosity and Variability of Collimated Gamma-Ray Bursts

Shiho Kobayashi*, Felix Ryde† and Andrew MacFadyen**

*Department of Earth and Space Science, Osaka University, Toyonaka, 560
†Center for Space Science and Astrophysics, Stanford University, Stanford, CA 94305, USA
**Astronomy Department, University of California, Santa Cruz, CA 95064, USA

**Abstract.** Within the framework of the internal shock model, we study luminosity and variability in gamma-ray bursts from collimated fireballs. In particular we pay attention to the role of photosphere due to $e^{\pm}$ pairs produced by internal shock synchrotron photons. It is shown that the observed Cepheid-like relationship between luminosity and variability can be interpreted as a correlation between opening angle of the fireball jet and mass included at the explosion with a standard energy output. We also show that such a correlation can be a natural consequence of the collapsar model. Using a multiple-shell model, we numerically calculate the temporal profiles of gamma-ray bursts. Highly collimated jets, in which the typical Lorentz factors are higher than in wider jets, can produce more variable temporal profiles due to smaller angular spreading time scales at the photosphere radius. Our simulations reproduce the observed correlation.

## INTRODUCTION

The temporal profiles of GRBs are often very variable and each profile looks very different. Quantitative measures of the variability have been suggested allowing for a systematic study of their morphology. Several studies have explored the possibility that quantities directly measurable in GRB light curves could be related to the luminosities of the bursts. Stern, Poutanen & Svensson [18] concluded that there is an intrinsic correlation between luminosity and the complexity of GRBs. Fenimore & Ramirez-Ruiz [2] found that the luminosities of seven bursts with known redshifts are correlated with the variabilities. Based on this work, Reichart et al. [15] also reported a possible Cepheid-like luminosity estimator for long bursts.

Recently, Frail et al. [3] reported that the gamma-ray energy releases, corrected for geometry, are narrowly clustered around $10^{51}$erg, and suggested that the wide variation in fluences and luminosities of GRBs is due entirely to a distribution of the opening angles. If this is the case, the Cepheid-like relation implies a correlation between the opening angle and the variability: variable bursts are radiated from highly collimated jets. Such a correlation is actually found in the observations [8].

As a result of relativistic beaming, an observer can see only a limited portion of the ejecta. There should be no observable distinction between a spherical ejecta and a conical ejecta until the ejecta has slowed down in the afterglow phase. Therefore, the correlation between variability and opening angle should be attributed to the properties of the central engine. In this paper, we will explore the relations among variability, luminosity and jet opening angle in the framework of the internal shock model. We show that the variability-opening angle relation and the Cepheid-like relation can be interpreted as a correlation between the opening angle and the mass included at the explosion.

## CENTRAL ENGINE: COLLAPSAR

The model currently favored for long bursts is the collapsar model in which GRBs are caused by relativistic jets expelled along the rotation axes of collapsing massive stars. The jets are powered by a black hole with a surrounding accreting torus. Energy from the accretion is pumped into jets via electrodynamic processes or by neutrino annihilation. In addition spin energy of the rotating black hole may be tapped by magnetic fields anchored in the accretion disk.

According to Frail et al. [3] and Piran et al. [14], the wide variation in the fluences of GRBs originates from the differences in opening angles of the jets. A wide jet radiates gamma-ray photons into a large solid angle, resulting in a dim burst. Though we do not know yet what physical mechanism results in the wide variation of the opening angles, it is likely that a wide jet involves a large mass $M$ at the explosion, and that it results in a flow with a lower Lorentz factor $\Gamma \sim E/M \propto \theta^{-2}$.

MacFadyen, Woosley & Heger [10] showed that jets of equivalent energy injected into a stellar envelope can be focussed by the pressure of the star. The degree of focusing is a function of the assumed entropy of the jet which was parametrized by $E_{int}/E_{tot}$. "Colder" jets were squeezed into tightly collimated flow while "hotter" jets of the same energy expanded sideways (See Fig 10 of [10]). The result was that "hotter" jets are broader and sweep up a larger mass of stellar material along the rotation axis of the star. While the calculations in [10] were non-relativistic, the results should hold in special relativity since the asymptotic Lorentz factor is an inverse function of the swept up mass. Current fully relativistic numerical simulations indicate the the results do hold [19].

MacFadyen et al. [10] also experimented with the effect of the collapsar density structure on jet collimation and found a large difference in jet opening angle for two density structures corresponding to low and high disk mass. The difference was due to the assumed viscosity parameter, but may also result from initial angular momentum and density structure of the progenitor stars. Calculations of the collapsar density structures and their effect on relativistic jet opening angle are currently underway.

# INTERNAL SHOCKS AND THE TEMPORAL STRUCTURE

Internal shocks arise in a relativistic wind with a nonuniform velocity when the fast moving flow catches up with the slower one. The wind can be modeled by a succession of relativistic shells [7]. A collision of two shells is the elementary process, and produces a single pulse of gamma-ray. Three time scales, the cooling time, the hydrodynamic time and the angular spreading time, are relevant to the pulse width. With the relevant parameters the cooling time is negligible compared to the other two time scales [17]. Let $d$ and $D$ be the width and separation of the shells. Then the hydrodynamic time scale $\sim d/c$ and the angular spreading time scale $\sim D/c$ determine the rise and the decay time of the pulse, respectively. Since most observed pulses rise more quickly than they decay, the pulse width is mainly determined by the angular spreading time $\sim D/c$ [12, 16].

The whole light curve of a GRB is given by the superposition of the resulting pulses from each collision. Since all shells are moving towards us with almost the speed of light, it is possible to show that we observe pulses arising from collision between shells mostly according to their positions inside the wind, i.e., according to the time when those shells were emitted by the central engine [7, 11]. The relative positions of the thin shells inside the wind are also determined by the separations. Then, the variability time scales in a temporal profile reflect well the shell separations at the central engine, provided that we can observe all collisions.

However, it has recently been shown that the Thomson optical depth due to $e^{\pm}$ pairs produced by synchrotron photons plays an important role in the internal shock model [1, 4]. In order to obtain high radiative efficiency and the characteristic clustering of spectral break energies of GRBs in the range 0.1-1 MeV, the collision radii are required to be similar to the photosphere radius. Since collisions producing narrow pulses occur at small radii $\propto D$, the photosphere might obscure these, leaving only the wider pulses visible. This will make the temporal profile smooth. In a wider jet, the typical Lorentz factor $\Gamma$ is smaller, a larger fraction of the collisions occur at small radii $\propto \Gamma^2$ below the photosphere, therefore, the smoothing effect is expected to be stronger.

## The Multiple-Shell Model

We here discuss the smoothing effect by using a multiple-shell model. We represent the irregular wind by $N$ relativistic shells in a manner similar to that in [7]. Because of the relativistic beaming effect, we can study the emission from a jet by using the spherical shells. We assign an index $i(i = 1, N)$ to each shell according to the order of the emission from the central engine. Each shell is characterized by four variables: Lorentz factor $\Gamma_i$, mass $m_i$, width $d_i$ and the distance to the outer neighbor shell $D_i$. We assume that the Lorentz factors and the separations are distributed uniformly in logarithmic spaces; between $\Gamma_{min}$ and $\Gamma_{max}$ and between $D_{min}$ and $D_{max}$, respectively. The width is assumed to be a constant value $d = D_{min}$. Since the internal shock process is very efficient only when the central engine produces shells with comparable masses [9], we consider this equal mass case. The mass is determined by the isotropic explosion energy $E_{iso}$ as $m = E_{iso}/\Sigma\Gamma_i c^2$. Assuming the correlation between the Lorentz factor and the mass involved in the jet: $\Gamma_{max} = a \times \theta^{-2}$, the isotropic explosion energy itself is related to the geometrically corrected explosion energy, $E_{iso} = 2a^{-1}\Gamma_{max}E$.

Consider a two shell collision. A rapid shell with $\Gamma_r$ catches up to a slower one with $\Gamma_s$ and the two merge to temporarily form a single shell. Using conservation of energy and momentum we calculate the Lorentz factor of the merged shell to be $\Gamma_m \sim \sqrt{\Gamma_r \Gamma_s}$. The internal energy of the merged shell is the difference of kinetic energy before and after the collision, $E_{int} \sim mc^2(\Gamma_r + \Gamma_s - 2\Gamma_m)$. A fraction $\varepsilon_e$ of this internal energy goes into the electrons, and it is radiated via synchrotron emission. If $\varepsilon_e < 1$, the merger stays hot after the emission. As a result, the merger will spread, transforming the remaining internal energy back to kinetic energy [9].

The synchrotron photons can be scattered by electrons within the merger. The Thomson optical depth is increased significantly when taking into account $e^{\pm}$ pairs produced by internal shocks [1, 4]. The typical energy of the synchrotron emission, in the shell frame, is well below the pair production threshold. However, since the photon spectrum extends to high energy as a power law with spectral index $\sim -1$, there exists an equal energy of photons per logarithmic energy intervals, and there may exist a large number of photons beyond the threshold. The pairs produced by these photons could contribute significantly to the Thomson optical depth. The photosphere radius $R_{\pm}$, where the Thomson optical depth becomes unity, can be estimated by assuming that a significant fraction, $\sim 1/2$, of the radiative energy $\varepsilon_e E_{int}$ is converted to pairs. Since the number density of the pairs is given by $n_{\pm} \sim \varepsilon_e E_{int}/8\pi m_e c^2 R^2 \Gamma_m^2 d$, the photosphere radius is

$$R_{\pm} \sim \left(\frac{\sigma_T}{4\pi m_e c^2} \frac{\varepsilon_e E_{iso}}{N\Gamma_{max}}\right)^{1/2} \left(\frac{\Gamma_r}{\Gamma_s}\right)^{1/4} \quad (1)$$

If a collision happens below the photosphere, the whole internal energy produced by the collision is converted to kinetic energy again via the shell spreading. A simplified description of this process is to assume that the two shells reflect with the same relative velocity.

Numerous collisions happen during the evolution of the multiple-shells. Each collision produces a pulse. However, the main pulses are produced by collisions between the fastest shells and the slowest shells at $R \sim \Gamma_{min}^2 D_i$, for which the photosphere radii take almost the same value. Then, we can define two characteristic collision radii, $\Gamma_{min}^2 D_{min}$ and $\Gamma_{min}^2 D_{max}$, and a typical photosphere radius $R_{\pm}$. If $R_{\pm} > \Gamma_{min}^2 D_{max}$, the first collisions for most shells occur below the photosphere, but the shells reflect each other with the same velocity. Then, it is still possible to produce bright pulses when the shells propagate beyond the photosphere and collide into other shells. However, if an enormous number of collisions and reflections happen below the photosphere, the shells are ordered with increasing values of the Lorentz factors and no collision happens any more. Especially, in order to produce bright pulses, the shells need to come out from the photosphere before the slowest shells are sorted out at the tail of the wind with a width $\Delta \sim ND_{max}/\log(D_{max}/D_{min})$. Using this sorting radius $\sim \Gamma_{min}^2 \Delta$ and the minimum collision radius $\Gamma_{min}^2 D_{min}$, GRBs are classifies into the following three cases.

(1) Narrow Jet Case: $\Gamma_{min} > (R_{\pm}/D_{min})^{1/2}$. Since all collisions occur above the photosphere, the variability should be less dependent on $\Gamma_{min}$ and only on the distribution of the separations at the central engine.

(2) Intermediate Jet Case: $(R_{\pm}/\Delta)^{1/2} < \Gamma_{min} < (R_{\pm}/D_{min})^{1/2}$. Some collisions occur above the photosphere to produce bright pulses wider than $\sim R_{\pm}/c\Gamma_{min}^2$. The variability of the temporal profile should highly depend on $\Gamma_{min}$.

(3) Wide Jet Case: $\Gamma_{min} < (R_{\pm}/\Delta)^{1/2}$. The shells are almost ordered with increasing values of the Lorentz factors within the photosphere, only minor collisions happen. The resulting bursts are very dim and it could be difficult to detect.

## Numerical Results

It is observationally found that the isotropic peak luminosities $L$ are correlated with their variability measures. We now show that such a correlation exists in our numerical simulations also. A numerical temporal profile depends on the specific realizations of random Lorentz factors and random separations that are assigned to each shell, as well as on the model parameters. We have assumed $N = 100$, $\Gamma_{max}/\Gamma_{min} = 10$, $D_{min}/c = 1$ msec, $D_{max}/c = 1$ sec and $\varepsilon_e = 0.1$.

The gamma-ray energy released from a GRB is narrowly clustered around $10^{51}$ erg [3], and the conversion efficiency from the explosion energy into the gamma-rays is about 10% [4, 9]. Then, the standard explosion energy can be estimated as $\sim 10^{52}$ erg. The observed wide jets with $\theta \sim 0.2 - 0.3$ also should have ultra-relativistic Lorentz factors to be optically thin to high energy photons, assuming $\Gamma_{max} = 100 \, (\theta/0.2)^{-2}$, the isotropic explosion energy is given by $E_{iso} \sim 10^{54} \, (\Gamma_{min}/20)$ erg.

For a given $\Gamma_{min}$, we calculate the temporal profiles for 100 realizations, and evaluate the mean isotropic peak luminosity and the mean variability measure. Here the peak luminosity is calculated with 1 sec resolution. In Figure 1, the solid line shows the mean values, while the dashed lines depict the 1$\sigma$ error. The numerical results reasonably fit the observational data. Since the numerical variability is calculated with $\Delta t = 64$ msec bins to compare with the BATSE data, the mean variability reaches an asymptotic value around $\Gamma_{min} \sim (R_{\pm}/c\Delta t)^{1/2} \sim 200$. The breaks around $V \sim 0.3$ in Figure 1 are due to this low time resolution. If $\Gamma_{min}$ is small, the slowest shells are sorted out at the tail before the shells reaches the photosphere, so the luminosity rapidly decreases for $V < 0.05$ in Figure 1. In range of $0.05 < V < 0.3$, power law fits to numerical results give $L \propto V^{2.2}$. Our numerical results also give a power law similar to ones reported by [2, 15].

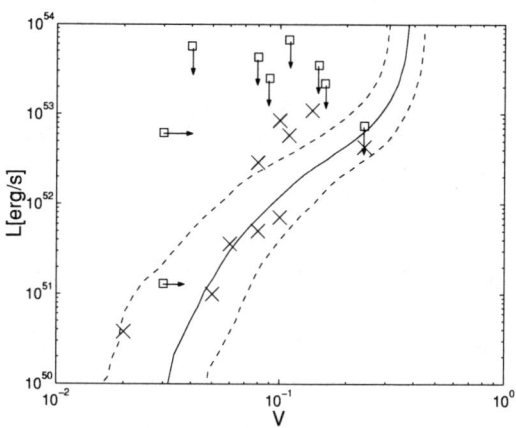

**FIGURE 1.** Variability vs. Peak Luminosity with 1 σ error bars of 100 random simulations. Observational data: measurements(crosses) and upper or lower limits (squares).

## CONCLUSIONS

We have shown that there exists a correlation between the jet opening angle θ and the gamma-ray light curve variability $V$. Though the correlation is based on only seven events at present and needs to be further confirmed with more events, it is naturally expected if the luminosity $L$ is correlated with the variability [2, 15], and if GRBs have a standard energy output [3, 14]. This correlation might give us a way to measure the opening angle for a long burst directly from the GRB light curve.

We have shown that the $θ - V$ relation, or equivalently, due to the constancy of burst energy, the $L - V$ relation can be interpreted as the correlation between the opening angle of a fireball jet and the included mass at the explosion. We also show that such a correlation can be a natural consequence of the collapsar model. Using a multiple-shell model, we numerically calculate the temporal profile and estimate the luminosity and the variability. Our simulations reproduce the observed correlation.

If the typical Lorentz factor is small enough, the shells are ordered within the photosphere, only minor collisions happen. The resulting dim burst is smooth and has a soft spectrum. GRB980425 and recently reported Fast X-ray Transients [5, 6] might be classified into this "Wide Jet Case".

Norris, Marani & Bonnell [13] found that the isotropic peak luminosity is inversely proportional to the spectral lag which is defined as the time lag of the peak luminosity between different energy bands. A detailed study of the peak lag requires a discussion of the spectrum of the internal shock emission and it is beyond the scope of this paper. However, if the time lag is also determined by the angular spreading time scale, which is the key time scale in our model, the relation can be derived qualitatively. In our model the isotropic peak luminosity is scaled by the opening angle $L ∝ θ^{-2}$, while the angular spreading time is related to the opening angle as $t_{ang} ∝ Γ^{-2} ∝ θ^4$. Then we obtain $L ∝ t_{ang}^{-1/2}$.

## REFERENCES

1. Asano,K., & Kobayashi, S. 2002, in preparation.
2. Fenimore,E., & Ramirez-Ruiz, E. 2000, Submitted to ApJ, (astro-ph/0004176).
3. Frail, D. A., et al. 2001, ApJ, 562, L55.
4. Guetta,D., Spada,M., & Waxman E. 2001, ApJ, 557,399.
5. Heise, J., et al. 2001, Proc. Second Rome Workshop.
6. Kippen, R. M., et al. 2001, Proc. Second Rome Workshop, astro-ph/0102277.
7. Kobayashi,S., Piran,T. & Sari,R. 1997, ApJ, 490, 92.
8. Kobayashi,S., Ryde, R. & MacFadyen, A. 2002, ApJ in press.
9. Kobayashi,S., & Sari,R. 2001, ApJ, 542, 819.
10. MacFadyen,A., Woosley,S. & Heger, A. 2001, ApJ, 550, 410.
11. Nakar,E., & Piran,T. 2001, submitted to MNRSA (astro-ph/0103210).
12. Norris, J.P. et al. 1996, ApJ, 459, 393.
13. Norris, J.P., Marani,G.F., & Bonnell,J.T. 2000, ApJ, 534, 248.
14. Piran,T., et al. 2001, ApJL in press (astro-ph/0108033).
15. Reichart,D. E., et al. 2001, ApJ, 552, 57.
16. Ryde,F., & Petrosian,V. 2001, in preparation.
17. Sari,R., Narayan, R., & Piran,T. 1996, ApJL, 473, 204.
18. Stern,B., Poutanen,J. & Svensson,R. 1999, ApJ, 510, 312.
19. Zhang,W, Woosley, S., & MacFadyen, A. 2002, in preparation.

# Determining Bolometric Corrections for BATSE Burst Observations

Luis Borgonovo*, Felix Ryde†, Miguel de Val Borro* and Roland Svensson*

*Stockholm Observatory, SCFAB, SE-106 91 Stockholm, Sweden
†Center for Space Science and Astrophysics, Stanford University, Stanford, CA 94305, USA

**Abstract.** We compare the energy and count fluxes obtained by integrating over the finite bandwidth of BATSE with a measure proportional to the bolometric energy flux, the φ-measure, introduced by Borgonovo & Ryde. We do this on a sample of 74 bright, long, and smooth pulses from 55 GRBs. The correction factors show a fairly constant behavior over the whole sample, when the signal-to-noise-ratio is high enough. We present the averaged spectral bolometric correction for the sample, which can be used to correct flux data.

## INTRODUCTION

The gamma-ray burst (GRB) spectrum is often peaked in the $EF_E$ representation (where $F_E$ is the energy flux) at some energy $E_{pk}$ in the γ-ray band [1]. The limited spectral coverage of the used detector might affect the assigned measure of the total energy-integrated flux. To obtain a true bolometric flux a correction is needed. This is most important for the photon flux and less affects the energy flux as the energy spectrum is often peaked within the detector energy window.

To circumvent this issue Borgonovo & Ryde [2] proposed to use the value of $EF_E$ at $E_{pk}$ as a representation of the energy flux, since it is proportional to the bolometric flux. We denote this quantity by φ. It was shown that the power-law hardness-intensity correlations (HIC) that were studied for pulse decays were better if the φ value was used instead of integrating the energy flux over the BATSE band. This could indicate that the latter method suffers from effects which are not part of the correlation. The φ-method thus provides an efficient way of studying detailed features in the observed HICs.

This measure is limited to the cases where the peak actually exists within the studied band $[E_{min}, E_{max}]$. Representing the spectrum by the Band et al. model [1], this means that the low and high energy power law indices must have $\alpha > -2$ and $\beta < -2$. In the most common case where $E_{max} > (\alpha - \beta)E_{pk}/(\alpha+2)$, the proportionality between φ and $F = \int F_E dE$ becomes

$$\frac{F}{\varphi} \equiv k(\alpha, \beta, y_{min}, y_{max}) = \frac{e^{(\alpha+2)}}{(\alpha+2)^{\alpha+2}}$$

$$\times \left[ \Gamma(\alpha+2)\{P(\alpha+2, \alpha-\beta) - P(\alpha+2, y_{min})\} \right.$$
$$\left. + \frac{(\alpha-\beta)^{\alpha-\beta} y_{max}^{\beta+2} - (\alpha-\beta)^{\alpha+2}}{(\beta+2)e^{\alpha-\beta}} \right], \quad (1)$$

where $y_{min} = (\alpha+2)E_{min}/E_{pk}$, and $y_{max} = (\alpha+2)E_{max}/E_{pk}$. $\Gamma(\alpha+2)$ and $P(\alpha+2, y)$ are the gamma function and the incomplete gamma function, respectively (see, e.g., Press et al. [3]). In the case that α and β do not have a strong dependence on $E_{pk}$, the only $E_{pk}$ dependence in $k(\alpha, \beta, y_{min}, y_{max})$ is in $y_{min}$ and $y_{max}$. In particular, when the integration is chosen over the whole energy range from 0 to ∞, there will be no dependence at all. The φ-value can be interpreted as the integral over the whole energy range or, at least, over a range for which $y_{min} \ll 1$ and $y_{max} \gg 1$, and should, under some circumstances, be a better representation of the bolometric flux for the study of the hardness-intensity correlation.

## BOLOMETRIC CORRECTION TO THE ENERGY FLUX

For this work, we use a sample of 74 pulses taken from 55 GRBs, observed by the Burst and Transient Source Experiment (BATSE) on the *Compton Gamma-Ray Observatory*. The burst data have a time resolution of multiples of 64 ms and the count flux is obtained by adding up the four energy channels from the Large Area Detectors (LADs) of BATSE. We search within bright bursts (with peak fluxes larger than 2 photons s$^{-1}$cm$^{-2}$) for cases

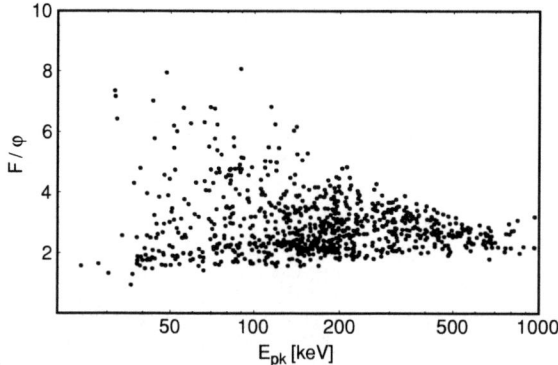

**FIGURE 1.** Ratio $F/\varphi$ versus $E_{pk}$ for 846 time-resolved spectra from 55 GRBs. The integration of the energy flux $F$ was made over the range 20 – 2000 keV.

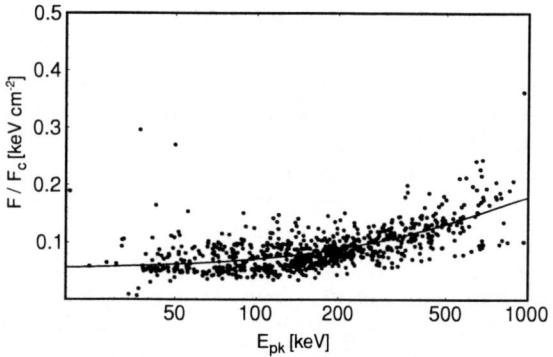

**FIGURE 2.** Ratio of the instantaneous energy flux to the count flux versus the corresponding $E_{pk}$ (in keV), for the same sample of time-resolved spectra as in Fig. 1. A best-fit second-order polynomial $0.05235 + 1.924 \times 10^{-4} E_{pk} - 6.624 \times 10^{-8} E_{pk}^2$ is also shown. The dispersion parameter is 0.1516.

containing long, smooth pulse structures with a general "fast rise-slow decay". The light curve for each burst in the sample was rebinned to achieve a signal-to-noise-ratio ($S/N$) of at least 30. All selected pulses have at least 4 time-bins at the given $S/N$. The spectrum for each time-bin was fitted with the Band et al. [1] model, allowing a deconvolution to find the energy spectrum and the peak energy. The energy spectrum was integrated over the energy band of the BATSE detector (20 – 2000 keV), for the strongest illuminated LAD, to find the instantaneous energy flux.

In Figure 1, the spectral bolometric correction for 846 time-resolved spectra from the 55 GRBs is shown. Average values can be used to correct the energy flux data when $E_{pk} = E_{pk}(t)$ is known. The data show a greater dispersion for smaller $E_{pk}$. This is most likely a bias introduced by the fact that most of the spectra correspond to the long decay phase of pulses, during which the intensity is positively correlated with the hardness. Therefore, the signal-to-noise ratio is usually much higher at the peak of the pulses when the spectra are *harder*, i.e., for higher $E_{pk}$, and decreases to the chosen limit ($S/N = 30$) as the pulse decays.

## BOLOMETRIC CORRECTION TO THE COUNT OBSERVATIONS

We now redo the above analysis on the count flux $F_c$ instead. The count data have not been deconvolved and for proper physical interpretation the effective correction must be estimated and understood.

Figure 2 shows the ratio of the instantaneous energy flux $F$ (in keV s$^{-1}$cm$^{-2}$) to the corresponding count flux $F_c$ (in counts s$^{-1}$) for the same time-resolved spectra as in Figure 1. The data were fitted with a second-order polynomial which was found to be $0.05235 + 1.924 \times$

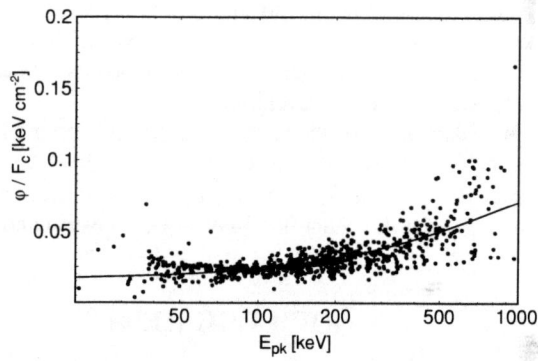

**FIGURE 3.** Ratio of the $\varphi$-measure to the count flux $F_c$ versus $E_{pk}$. A second-order polynomial fit to the data $0.01597 + 8.320 \times 10^{-5} E_{pk} - 2.829 \times 10^{-8} E_{pk}^2$ is also shown. The dispersion parameter is 0.07605.

$10^{-4} E_{pk} - 6.624 \times 10^{-8} E_{pk}^2$ (with $E_{pk}$ given in keV). The dispersion was measured as the mean of the squared ratios between the fit residuals and the fit expected values. The $F/F_c$ fit has a dispersion of 0.15.

In Figure 3, the ratio of the $\varphi$-measure (in keV s$^{-1}$cm$^{-2}$) to the count flux (in counts s$^{-1}$) over the same sample of spectra is shown. Again the data were fitted with a second-order polynomial given by $0.01597 + 8.320 \times 10^{-5} E_{pk} - 2.829 \times 10^{-8} E_{pk}^2$ (with $E_{pk}$ in keV). This adjusts the data with higher accuracy for small $E_{pk}$ values. The calculated dispersion is 0.08, significantly smaller than in the $F/F_c$ case, shown in Figure 2.

# DISCUSSION

To interpret the BATSE spectral data, one must have in mind that due to the GRB spectra being broad compared to the observed energy window, one must apply a bolometric correction factor which is unknown. To circumvent this problem we propose the use of the φ measure that is also proportional to the bolometric flux.

The flux $F$ is found from integrating the deconvolved spectrum. This spectrum is model-dependent as the deconvolution is based on the model spectrum. This could introduce additional scatter into the determination. A second problem, when one aims at studying single pulses in the light curve, arises from the fact that the observed spectra may contain contributions from other pulses which even could be unresolved. Furthermore, additional soft components [4] could also affect the measured flux value and thus weakening the correlations. This makes the φ-method better to use.

The $F/\varphi$ ratio shows an approximately constant behavior over the studied sample, with a larger dispersion for smaller $E_{pk}$. This fact can be explained as a bias introduced by the HIC correlation.

We also analyze the correction for the count flux. The $\varphi/F_c$ ratio is shown to have a substantially smaller dispersion than the $F/F_c$ case. Therefore, the effective correction for the count flux may be better estimated.

# ACKNOWLEDGMENTS

This research made use of data obtained through the HEASARC Online Service provided by NASA/GSFC. We are also grateful to GROSSC, STINT, NOTSA, and the A. E. W. Smitts Fund at Stockholm University for support.

# REFERENCES

1. Band, D., et al., *ApJ* **413**, 281 (1993).
2. Borgonovo, L., and Ryde, F., *ApJ* **548**, 770 (2001).
3. Press, W. H., Teukolsky, S. A., Vetterling, W. T., and Flannery, B. P., *Numerical Recipes in Fortran* 2nd Ed., Cambridge Univ. Press, Cambridge, 1992.
4. Preece, R. D., Briggs, M. S., Pendleton, G. N., et al., *ApJ* **473**, 310 (1996).

# BATSE-EGRET Combined Spectral Fits

M.M. González*, Y. Kaneko†, R. Preece† and B.L. Dingus*

*1150 University Av., Madison WI 53706
†National Space Science and Technology Center, 320 Sparkman Dr, Huntsville, AL 35805

**Abstract.** Combining BATSE and EGRET data yields spectral fits over the broadest energy range for the prompt phase of gamma-ray bursts. The spectra from the BATSE data have previously been reported; however, the addition of higher energy EGRET data further constrains the values of the peak energy and high-energy spectral index. The EGRET data from the TASC (Total Absorption Shower Counter) begin at 1 MeV and for brighter bursts extend beyond 10 MeV. These data are combined using the separate response matrices of each detector and a common spectral fitting algorithm, RMFIT, which has been extensively used with BATSE data. The spectrum from GRB910503 is presented.

## INTRODUCTION

Observations at all energy ranges play a crucial role in understanding GRBs, especially during the prompt phase of the burst. BATSE observations have yielded spectra in the energy range of 0.02 to few MeV at these times. For bursts with high $E_{peak}$ of $\sim$ 1MeV, determination of $E_{peak}$ and the spectral index above $E_{peak}$ are poorly measured.

The EGRET-TASC calorimeter [1] is larger (6400cm$^2$) and thicker ($8r.l = 20$cm) than BATSE with its LAD detectors (each of area 1964cm$^2$ and 1.3cm thick). Therefore, the TASC offers observations at higher energies that will help to restrict $E_{peak}$ and the high energy spectral index, $\beta$, as well as deviations from the Band function [2].

The goal of this work is to combine BATSE and EGRET-TASC observations to give a better spectral fit over larger energy intervals. We show the spectral fit for one burst, GRB910503, BATSE trigger number 143, which was a bright burst.

## EGRET-TASC

EGRET-TASC is a 76x76x20cm self-triggered NaI(Tl) calorimeter. Its burst/solar mode records gamma-ray burst and solar flares in the energy range of 1-200 MeV. A solar spectrum is accumulated and recorded every 32s. Four burst spectra are also accumulated for 1,2,4,16s sequential time intervals upon receiving a prompt BATSE trigger. The solar spectrum recorded before and after the burst allows background subtraction.

The complete CGRO mass model and the EGS-4 Monte Carlo code are used to determine its response for a given direction and energy.

TASC is sensitive in all directions but does not have position information and depends on the BATSE position determination. However, because of its large effective area (Fig.1), TASC becomes a good detection system of bright bursts from any direction at 1-200MeV energies and complements BATSE observations of bright bursts.

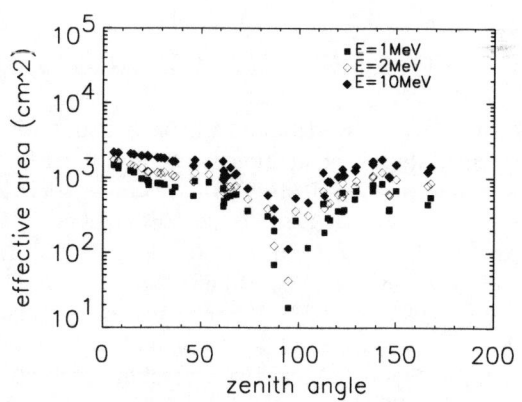

**FIGURE 1.** EGRET-TASC effective area dependence on the angle from the z-axis of EGRET for three different energies.

## BATSE SPECTRAL FITTING

The BATSE spectral analysis software RMFIT was used to obtain spectral fits combining BATSE and TASC data. RMFIT is written as object-oriented code within IDL, with the spectral fitting engine written as a FORTRAN dynamically-linked library. Some of its features are:

- Display of high-energy astrophysical data as time sequences of spectra.
- Interactive selection of regions of interest for spectral fitting.
- Interactive selection of regions for background subtraction. Both numerical and visual cues aid in the evaluation of the goodness of the background model.
- A fast fitting engine that implements model, rather than data, variances to allow for more accurate fitting of data that have both high- and low-counts statistics in their energy spectra.
- User selection from a number of common additive and multiplicative spectral models, including user-defined functions.
- Several ways to analyze time-resolved spectral fit parameters, as well as storing the time-resolved spectral model in FITS format.
- Cross-platform compatibility through IDL.
- Joint fitting of several sets of data from different detectors, instruments, and/or spacecraft. XSPEC data and responses can easily be converted into BATSE FITS (BFITS) format. Any normalization mis-match between different sets of data can be resolved via a multiplicative factor as an extra model parameter.

## SPECTRA FITTING OF GRB910503

The EGRET-TASC data and response matrix were reformatted to be compatible with the BATSE spectral fitting program RMFIT. The resulting spectral fitting combining GRB910503 data from EGRET-TASC and the brightest detector for this burst, BATSE-LAD6 in the time intervals 7:04:14.688-7:04:21.688 UT and 7:04:14.634-7:04:21.738 UT, and the energy intervals 1.5 – 160.0MeV and 32.3 – 1945.9keV respectively, is shown in Figures 2 and 3. A resulting normalization factor for EGRET-TASC data in the fit to adjust for the variable and large deadtime ($\sim$ 70%) of EGRET-TASC can be seen in Figure 3.

Since the time resolution of BATSE-LAD6 (HERB) data is 128 ms, the time interval and binning were selected to match as close as possible the the first 7 seconds in the EGRET-TASC's accumulated time intervals, where most of the emission at BATSE energies was observed as seen in Figure 4.

Table 1 shows results for $E_{peak}$, $\alpha$, and $\beta$ fitting values. For consistency, a fitting of only EGRET-TASC data was done with this program and the $\beta$ value obtained

**TABLE 1.** Resulting $E_{peak}$, $\alpha$ and $\beta$ values from fitting BATSE, EGRET data separately and jointly for GRB910503.

| | 910503 | | |
|---|---|---|---|
| | $E_{peak}$ | $\alpha$ | $\beta$ |
| BATSE | 708± 18 | -0.91± 0.01 | -2.59± 0.11 |
| EGRET | — | — | -2.21± 0.03 |
| joint | 662± 14 | -0.89± 0.01 | -2.29± 0.02 |

agrees with that reported previously by Schneid [3]. As it is observed, the values of $\alpha$ are $\beta$ are mainly determine by BATSE-LAD6 and EGRET-TASC data respectively. However, the value of $E_{peak}$ shifts and is better constrained using the combined fit even though it is out of the EGRET-TASC energy range.

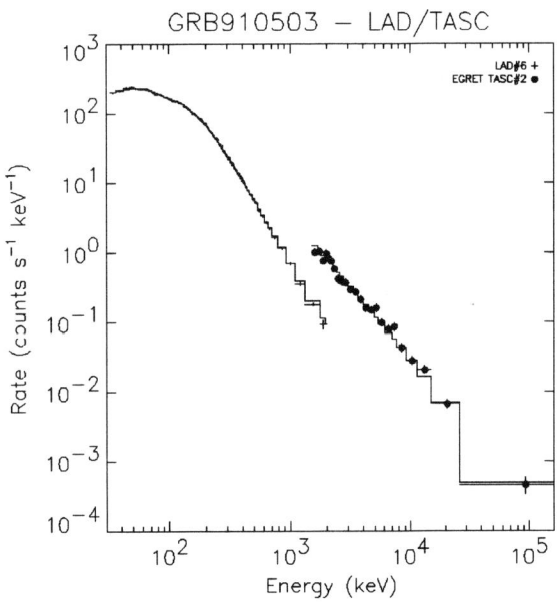

**FIGURE 2.** Preliminary results of the rate fit (solid line) for burst 910503 combining BATSE-LAD6 (crosses) and EGRET-TASC (circles) observations.

## CONCLUSION AND FUTURE PLANS

A preliminary combined BATSE and EGRET-TASC spectrum fit has been done for GRB910503. The combined fit results in a different $E_{peak}$ and $\beta$ with smaller uncertainties. New analysis tools have been developed. The TASC's data for these bursts, as well as their response matrices, will be written in the same format as BATSE, i.e. bfits, and will be submitted to the GRO SSC for public release.

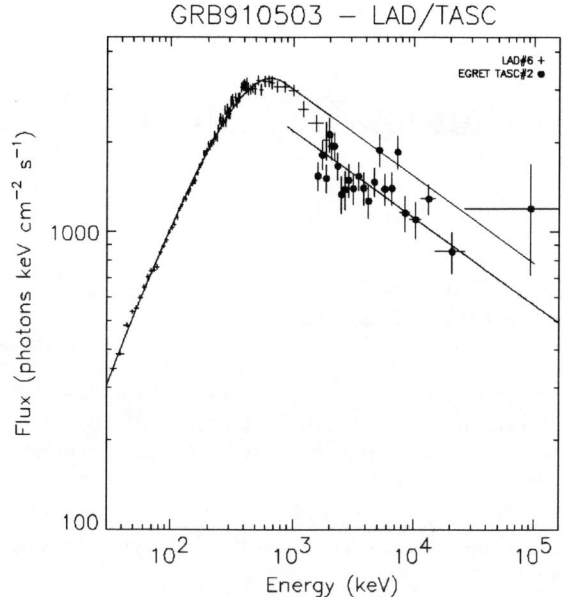

**FIGURE 3.** Preliminary flux fit (large solid line) for burst 910503 combining BATSE-LAD6 (crosses) and EGRET-TASC (circles) observations. The normalization factor for EGRET-TASC data can be seen and is a free parameter in the fit to adjust for the variable and large deadtime ($\sim 70\%$) of the EGRET-TASC.

Fitting for different spectral models will also be explored with these broad-band spectra.

The EGRET-TASC observed a few bursts per year [4]. The same combined spectra fitting will be done for these bursts leading to improved estimates of $E_{peak}$ and $\beta$ and better understanding of the high $E$ spectra of GRBs.

## ACKNOWLEDGMENTS

This work was supported by NASA grant NAG5-9712.

## REFERENCES

1. D.J.Thompson and et al, *ApJ. Sup.*, **86**, 629 (1993).
2. D.Band and et al, *ApJ*, **413**, 281 (1993).
3. E.J.Schneid and et al, *A&A*, **255**, L13–L16 (1992).
4. J.Catelli, B.Dingus and E.Schneid, "EGRET Observations of Bursts at MeV Energies", in *Gamma-Ray Bursts 4th Huntsville Symposium*, AIP Conference Proceedings 428, 1997, p. 309.

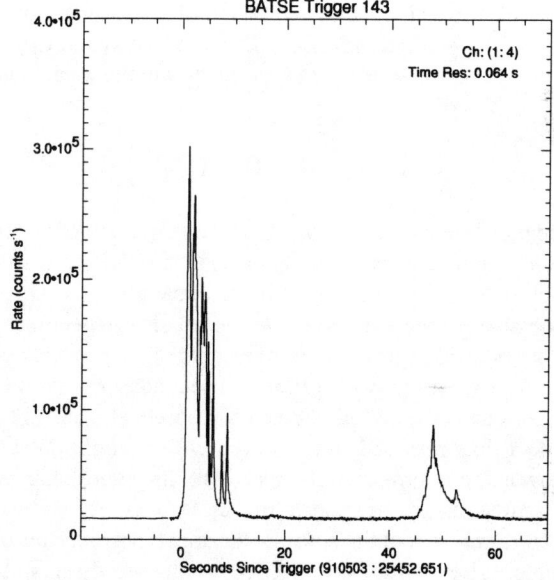

**FIGURE 4.** Light curve observed by BATSE for GRB910503. Spectral fit presented was for the time interval 1-8s after the trigger time.

# Spectral Analysis of Bright Gamma-Ray Bursts

G. Ghirlanda[*], A. Celotti[*] and G. Ghisellini[†]

[*]SISSA/ISAS, Trieste - Via Beirut 2, 34014 Trieste, Italy
[†]Osservatorio Astronomico di Brera, via Bianchi 46, Merate (LC), Italy

**Abstract.** We present the time integrated and time resolved spectral analysis of a sample of bright bursts selected from the BATSE archive. We fitted four different spectral models to the time integrated and time resolved spectra of the flux pulses. We point out that the found (marginal) differences in the parameter distributions can be ascribed to the different spectral shape of the employed models and that a smoothly curved model best fits the observed spectra. We characterize the spectral shape of bright bursts and compare the low energy slope of the fitted spectra with the prediction $N(E) \propto E^{-2/3}$ of the synchrotron theory, finding that this limit is violated in a considerable number of time resolved spectra around the peaks, both during the rise and decay phase.

## INTRODUCTION

The nature and emission mechanisms responsible for the prompt emission of Gamma–ray bursts (GRB) are still unclear. In order to identify the physical process(es) responsible for the emission it would be ideally necessary to study spectra resolved on the shortest time–scales of variability, typically of a few milliseconds, observed in bright bursts [3] and predicted on theoretical basis [11]. In fact time resolved spectra, even within single pulses, show a strong time evolution, and are, in general, harder than time integrated spectra [6], [2].

Here we present the study of the spectral properties of single pulses within bright GRBs, compare the results of the spectral analysis of the time average spectrum with the time resolved spectra of the very same burst in order to quantify systematic differences and examine any spectral 'violation' (with respect to the predicted slope in the case of e.g. synchrotron emission) for the entire burst evolution. A more detailed analysis is given in [7].

## DATA ANALYSIS

The main characteristics of the BATSE detectors have been described by [3]. We selected bursts with a peak flux, on the 64 ms time-scale, higher than 20 phot/cm$^2$sec. The data used were mainly the HERB: time sequence of 128 channel spectra with a minimum integration time of 0.128 s.

Within the selected bursts each peak was analyzed separately considering the spectrum time-integrated over the duration of the peak and the sequence of time resolved spectra comprised within the same peak. We fitted the background subtracted spectra with 4 spectral models: the BAND one (Band et al. [1]) which consists of 2 power laws joined smoothly by an exponential roll–over, the Broken Power Law (BPLW hereafter) which has a sharp break between the two power law segments, the COMP model which comprises a low energy power law ending–up in an exponential cutoff, and the Synchrotron Shock Model (SSM) [12] based on optically thin synchrotron emission from relativistic particles (and for the first time fitted to the time resolved spectra).

## RESULTS

From a statistical point of view, the comparison of the spectral fits obtained with the 4 models shows that in terms of the reduced $\chi^2$, the BAND and COMP models can better represent the time resolved spectra of bright bursts, but some counter examples exist showing that in general within a single pulse more than one time resolved spectrum can be fitted by different spectral models.

### The parameters distributions

The distributions of the *low energy power law spectral index* $\alpha$ (Fig.1), for the BAND and COMP model, are similar, like in the case of the time integrated spectra, and both have a mode of $-0.85 \pm 0.1$, also consistent with the BPLW average value $-1.15 \pm 0.1$. Note that qualitatively the extension of the $\alpha$ distribution of the BPLW model (*solid line* in Fig.1) towards lower values

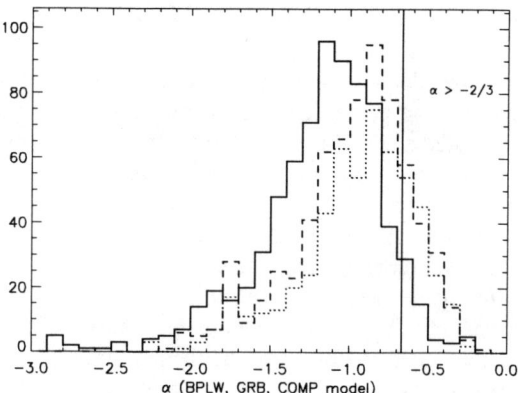

**FIGURE 1.** Low energy power law spectral index ($\alpha$) distributions derived from the time resolved spectral analysis. *Solid line:* BPLW model, *dotted line:* BAND model, *dashed line:* COMP model. The vertical line represents the synchrotron limit ($\alpha = -2/3$) for the low energy spectral shape.

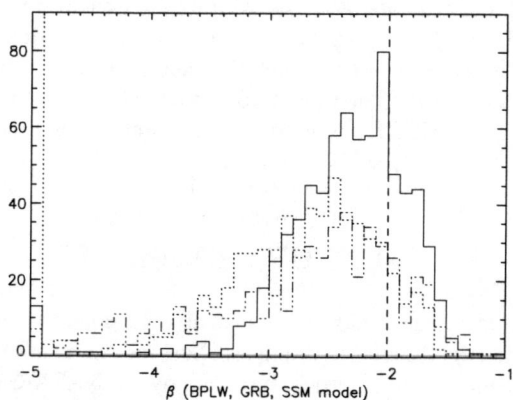

**FIGURE 2.** High energy power law spectral index ($\beta$) distributions for the time resolved spectra. *Solid line:* BPLW model; *dotted line:* BAND model; *dot–dashed line:* SSM model. Also shown (bin with $\beta = -5$) the time resolved spectra with undetermined high energy spectral index for the BAND model.

could be attributed to the fact that at low energies this model (which has a sharp break) tends to under–estimate the hardness of the spectrum compared to a smoothly curved model.

The average low energy spectral slope obtained from the time resolved spectra is harder than that obtained with the time integrated pulse spectra for all the three models (BAND, BPLW, COMP). This is a consequence of time integration (i.e. hardness averaging) of the spectral evolution (which can be also very dramatic) over the entire rise and decay phase of the pulse.

The *high energy spectral index* $\beta$ distributions are reported in Fig.2. The average value is $-2.45 \pm 0.1$ and

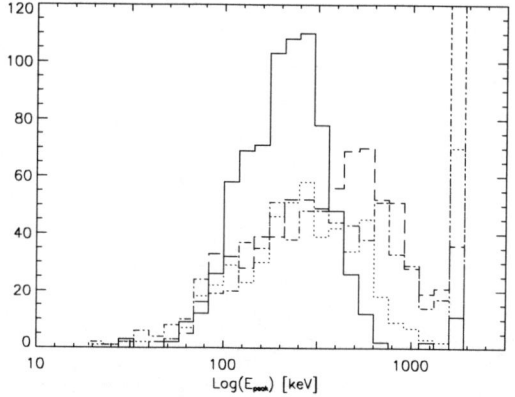

**FIGURE 3.** $E_{peak}$. Peak energy distribution for the 4 spectral models. *solid line:* BPLW model, *dotted line:* BAND model, *dashed line:* COMP model, *dot–dashed line:* SSM model. Spectra with undetermined peak energy (i.e. the high energy threshold 1800 keV assumed as lower limit) are reported in the last bin.

$-2.05 \pm 0.1$ for the BAND and BPLW model respectively, the former being harder than that obtained from the pulse average spectrum. The SSM average $\beta$ is -2.17 which is consistent with that found from the average pulse spectral analysis.

The most important spectral parameter obtained in fitting the spectrum with these models is $E_{peak}$ corresponding to the peak of the $EF_E$ spectrum, and thus to the energy where most of the power is released. $E_{peak}$ is coincident with the break energy $E_0$ for the BPLW and COMP and is equal to $(\alpha + 2)E_0$ for the BAND model. This characteristic energy can be obviously calculated only for those spectra (BPLW and BAND model) with $\beta < -2$ and its distribution is presented in Fig.3. The average is $E_{peak} = 280^{+72}_{-57}$ keV for the BAND model, consistent, within the errors, with the BPLW most probable value of $211^{+25}_{-22}$ keV. The COMP model, instead, gives a highly asymmetric peak energy distribution with a mode of $595^{+104}_{-88}$ keV because the lack of an high energy power law component tends to over–estimate the energy corresponding to the start of the exponential cutoff. The SSM model has an average $E_{peak} \sim 316^{+64}_{-52}$ keV with a wide distribution.

## The Synchrotron limit violation

A strong prediction of the optically thin synchrotron model is that the asymptotic low energy photon slope $\alpha$ should be lower than or equal to $-2/3$ [8].

We find that 13.7% of the time, resolved spectra fitted with the BAND model are inconsistent with $\alpha \leq -2/3$ at $2\sigma$. A similar percentage of spectra violating the

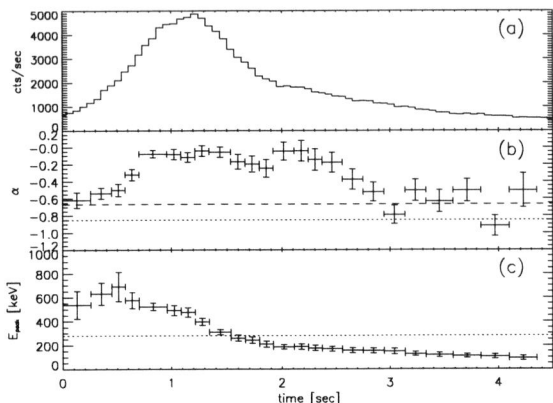

**FIGURE 4.** Trigger 2083. Spectral evolution of the BAND model fitted to the time resolved spectra. Light curve on the 64 ms time-scale (panel a); low energy spectral index and (*dashed*) synchrotron shock model limit $\alpha = -2/3$ (b); peak energy (c). For reference the average values of $\alpha$ and $E_{peak}$ obtained from the time resolved spectra (*dotted line*) and the synchrotron model limit are reported (*dashed line*).

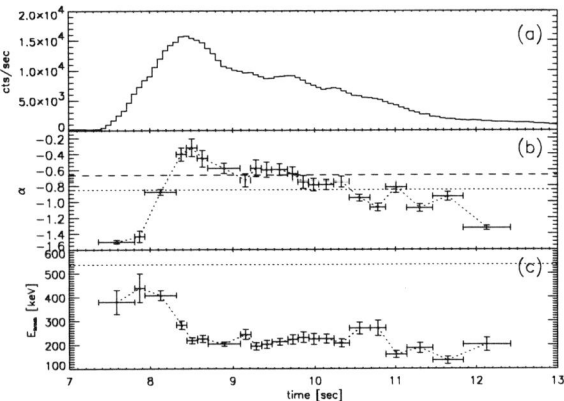

**FIGURE 5.** Trigger 5614. Spectral evolution of the COMP model fitted to the time resolved spectra. Light curve on the 64 ms time-scale (panel a); low energy spectral index and (*dashed*) synchrotron shock model limit $\alpha = -2/3$ (b); peak energy (c). For reference the average values of $\alpha$ and $E_{peak}$ obtained from the time resolved spectra (*dotted line*) and the synchrotron model limit (*dashed line*) are reported.

$\alpha$ limit is found for the COMP model (∼11.7%, of course mostly for the same spectra). Moreover ∼ 21% of the time resolved peak spectra violate the synchrotron limit, indicating that this violation happens during the peak phase and not preferentially before or after it: in Fig.4 we show the spectral evolution in the case of GRB921207 (fitted with the BAND model) and another example is reported in Fig.5.

## CONCLUSIONS

We considered a sample of bright bursts detected by BATSE and performed a uniform analysis for the time integrated and the time resolved (typically 128 ms) spectra with four different models adopted and proposed in the literature.

We find that even at this time resolution no model can better represent the data and different spectra require different shapes, re-confirming the erratic behavior of bursts and also possibly indicating that time resolution on time-scales comparable with the variability one are needed. The parameter distributions are consistent with the results reported by [10] although the average spectral shape (both from the time resolved and time integrated spectra) is harder because we selected only bright bursts and restricted the spectral analysis to the pulse phase.

A considerable number of spectra are characterized by extremely hard low energy components with spectral index greater than −2/3 (i.e. the limit predicted by synchrotron theory [8]). The $\alpha$ limit violation, also found by [5] and [2], is evident around the peak both during the rise and decay phase, and this could indicate that at some stages of the burst evolution (possibly near the peak of emission itself) alternative radiative processes, other than synchrotron, can be dominant.

## ACKNOWLEDGMENTS

This research has made use of data obtained through the High Energy Astrophysics Science Archive Research Center Online Service, provided by the NASA/Goddard Space Flight Center. We are grateful to D. Band for useful discussions on the BATSE data analysis. We thank Marco Tavani for having provided his code of the SSM model. Giancarlo Ghirlanda and AC acknowledge the Italian MUIR for financial support.

## REFERENCES

1. Band D. et al., *ApJ* **413**, 281-292 (1993)
2. Crider A. et al., *ApJ* **479**, L39-L42 (1997)
3. Fishman G. J. & Meegan C. A., *ARAA* **33**, 415-458 (1995)
4. Ford L. et al., *ApJ* **439**, 307-321 (1995)
5. Frontera F. et al., *ApJ* **127**, 59-78 (2000)
6. Liang E. P. & Kargatis V. E., *Nature* **318**, 495 (1996)
7. Ghirlanda G., Celotti A. & Ghisellini G., 2001, A&A submitted
8. Katz J.I., *ApJ* **432**, L107 (1994)
9. Preece R. D. et al., *ApJ* **496**, 849-862 (1998)
10. Preece R. D. et al., *ApJ SS* **126**, 19-36 (2000)
11. Rees M. J. and Meszaros P., *ApJ* **430**, L93 (1994)
12. Tavani M., *ApJ* **466**, 768-778 (1996)

# Extended Power-Law Decays in BATSE Gamma-Ray Bursts: Signatures of External Shocks?

Timothy W. Giblin*, Valerie Connaughton†, Jan van Paradijs**, Robert D. Preece†, Michael S. Briggs†, Chryssa Kouveliotou‡, Ralph A. M. J. Wijers§ and Gerald J. Fishman¶

*Department of Physics and Astronomy, The College of Charleston, Charleston, SC
†Department of Physics, University of Alabama in Huntsville, Huntsville, AL
**Deceased
‡Universities Space Research Association, Huntsville, AL
§Department of Physics and Astronomy, SUNY, Stony Brook, NY
¶NASA, National Space Science and Technology Center, Huntsville, AL

**Abstract.**
We present a temporal and spectral analysis of a subset of BATSE GRBs with smooth extended emission tails to search for signatures of the "early high-energy" afterglow, i.e., the onset of the afterglow emission that initially peaks in gamma-rays and subsequently evolves into X-Ray, optical, and radio emission according to the internal/external shock model. From our sample of 40 GRBs we find that the temporal decays are best described with a power-law $\sim t^\beta$, rather than an exponential, with a mean index $\langle \beta \rangle \approx -2$. Spectral analysis shows that $\sim 20\%$ of these events are consistent with a fast-cooling synchrotron spectrum for an adiabatic blast wave; three of which are consistent with the blast wave evolution of a jet, with $F_\nu \sim t^{-p}$. This behavior suggests that, in some cases, the emission may originate from a narrow jet, possibly consisting of "nuggets" whose angular size are less than $1/\Gamma$, where $\Gamma$ is the bulk Lorentz factor.

## INTRODUCTION

The observed GRB afterglow spectrum is well-described by synchrotron emission that arises from the interaction of a relativistic blast wave with bulk Lorentz factor $\Gamma \sim 10^{2-3}$ with the ambient medium [1], [2]. The often highly variable phase of the GRB is attributed to internal shocks within the flow and may reflect the activity of the progenitor. In some cases, internal and external shock emissions will overlap in time, while in other cases the afterglow emission may be delayed with respect to the GRB (e.g., [3], [4], [5]).

Evidence for both cases has surfaced in the combined GRBM/WFC/NFI data from BeppoSAX. The X-ray afterglow emission may be delayed in time from the prompt GRB (e.g., GRB970228, [6]) or may begin during the GRB emission (e.g., GRB980519, [7]). In the latter case, it is not clear if the X-ray afterglow is a separate underlying emission component or a continuation of the GRB itself. Evidence for overlapping shock emission has also been found in GRB990123 [8], GRB980923 [9], and GRB920723 [10].

We expect some fraction of BATSE GRBs to show a signature of the external shock emission characterized by a soft gamma-ray tail component that decays as a power-law in the late time histories, either disconnected or possibly superposed upon the variable gamma-ray emission. Our investigation focuses on the combined temporal and spectral behavior of a sample of 40 BATSE GRBs that exhibit these smooth decays during the later phase of their time histories. Many of these events fall into a category of bursts traditionally referred to as "FREDs" (Fast Rise, Exponential-like Decay), bursts with rapid rise times and a smooth extended decay. We examine the temporal behavior and spectral characteristics of the decay emission for the events in our sample and compare their spectra with the model synchrotron spectrum.

## DATA ANALYSIS

Our dataset was collected by performing a visual scan of all BATSE DISCSC time histories in the 25-2000 keV range for smooth extended decay features. Our search resulted in a sample of 40 bursts, 17 with a FRED-like profile and 23 that exhibit a period of variability followed by a smooth decay. Source count rates were obtained by subtracting the background model rates for each DISCSC channel. Background models were generated by modeling pre- and post-burst background intervals with a polynomial. This method was adequate for bursts with durations less than $\sim 200$ seconds. For longer bursts the long term variations in the background can inhibit knowledge of when the tail emission drops below the background. We therefore applied an orbital background subtraction method [11] to events with durations that exceed 200 seconds.

## Temporal Analysis

We modeled the smooth decay of the background subtracted source count rates (25-300 keV) with a power-law function $\sim (t-t_0)^\beta$. Fit intervals $[\tau_1, \tau_2]$ were selected in a systematic manner. For FRED-like bursts, $\tau_1$ was chosen as the time of half-width at half-maximum intensity of the burst. For bursts with variability, $\tau_1$ was defined as the bin following the end time of the variable emission. The fit interval end time, $\tau_2$, was defined as the time when the amplitude of the tail count rates (computed from a 16 bin moving average) first falls within 1-$\sigma$ of the background model. The fitted model parameter values were insensitive to arbitrarily larger $\tau_2$.

A series of Monte Carlo simulations revealed a bias in the distribution of fitted slopes hinged on the correlation of $t_0$ and $\beta$. This bias results when the value of $\tau_1$ is too far out in the tail of the power-law where the curvature of the power-law decay is undersampled, resulting in a broad $\chi^2$ minimum and asymmetric uncertainties in the fitted parameter values. Examples of three burst decays from our sample are displayed in Figure 1. The mean index for our sample is $\langle \beta \rangle = -2.03 \pm 0.51$. An exponential decay model gave poor fits to the data.

## Spectral Analysis

We modeled the time-integrated photon spectrum *during the decay* with a smoothly broken power-law or a broken power-law using the WINGSPAN software package and the BATSE 16 channel CONT data. The entire burst emission was selected for FREDs. The slopes of the spectral energy flux, $F_\nu$, are easily found from the low- and high-energy photon indices through the relations: $\alpha = \alpha_{low} + 1$ and $\alpha' = \alpha_{high} + 1$.

Acceptable fits were obtained for 20 bursts (a few spectra were better fit with a single power-law; details will be presented in the forthcoming ApJ paper). Com-

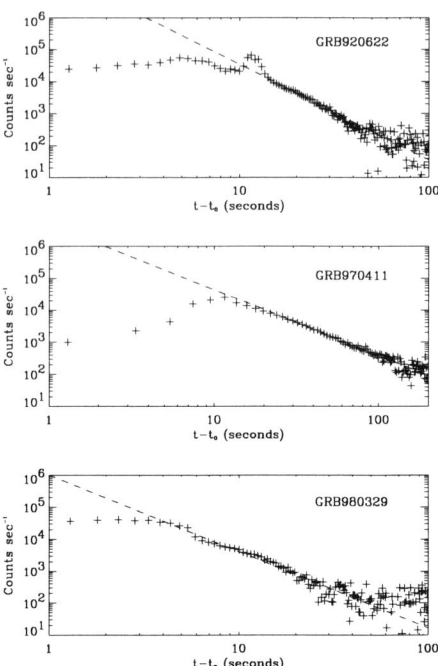

**FIGURE 1.** Logarithmic time histories of three events in the 25-300 keV range. The dashed line is the best-fit power-law model for each burst.

parison of the fitted spectral parameters with the expected high-energy synchrotron spectrum [12], [13] reveals 8 high-energy afterglow candidates based on the following constraints: (i) in the fast-cooling mode, the spectral slope below the high-energy break ($\nu_m$) is *always* $-1/2$ for radiative or adiabatic evolution, and (ii) a value of $p$ derived from the fitted spectral slope consistent with values derived from afterglows observed at X-ray, optical, and radio wavelengths ($2.0 \leq p \leq 2.5$). None of the events are consistent with $\alpha' - \alpha = 1/2$, the change in slope across $\nu_c$ expected for slow-cooling.

## Temporal Index vs. Spectral Index

Figure 2 shows a plot of temporal vs. spectral index for the 20 events with spectral fit paramters. For comparison with synchrotron afterglow models, we plot the possible linear relationships between $\alpha'$ and $\beta$. This plot should be interpreted with a certain degree of caution, since the expected values of $\alpha'$ are somewhat restricted by the possible range of $p$ values predicted by Fermi acceleration models (e.g., [14]). Most interestingly, however, five events are consistent with the $\beta = -p$ relation for adiabatic jet evolution.

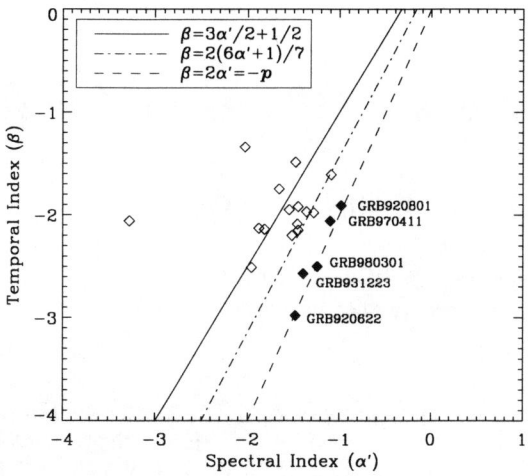

**FIGURE 2.** High-energy spectral index vs. temporal index for the 20 events with spectral fits. Also plotted are the linear relationships expected from the evolution of a fast-cooling adiabatic spherical blast wave (*solid line*), fast-cooling radiative spherical blast wave (*dash-dot line*), and a jet (*dashed line*). The five bursts labeled on the plot (*fill diamonds*) are within 1$\sigma$ of the $\beta = 2\alpha'$ line.

## DISCUSSION

A significant fraction of bursts in our sample show deviations from the synchrotron model. From a catalog of BATSE GRB spectra, the low -and high-energy spectral indices follow well-defined distributions [15]. For example, the distribution of low-energy power-law indices given in Figure 7 of Preece et al. (2000) peaks near unity. Roughly 200 of the 5500 spectra in the distribution are consistent with the expected value of $\alpha_{low}$ below $\nu_m$, or about 4%. If we adopt the hypothesis that GRB spectra are not synchrotron spectra, then on average 4% of the time we expect to measure parameters consistent with the synchrotron spectrum purely by chance. Thus we can expect one event from our sample to have $\alpha_{low} = -1.5$ by chance. Our total of 8 candidates obviously exceeds this limit, suggesting that these decays are sources of synchrotron emission.

The five bursts labeled in Figure 2 are candidates for jet outflows. In general, the jet-break in the light curve may occur at early times after the initial shock, as in the case of GRB980519, where evidence exists for an early break [16]. Since the break time in the observer's frame goes as $t_b \propto \theta_c^2$ [17], a very early break requires a very small $\theta_c$. If $\theta_c < \theta_b$ initially, then the slope is steep from the start. This implies one of two possiblities: (1) very narrow emission spots, or "nuggets", within a narrow collimation angle, or (2) a very small value of $\Gamma$ such that $\Gamma^{-1} > \theta_c$. The second option is not likely since the observed emission is in the keV to MeV range and $\nu_m \propto \Gamma^4$, requiring a high Lorentz factor.

Three of the five events (GRB920622, GRB980301, and GRB970411) are strong candidates because they belong to the group of 8 high-energy afterglow candidates selected on the basis of their spectral properties. Spectra of the other two events (GRB920801 and GRB931223) are only marginally consistent with synchrotron emission from an external shock.

## CONCLUSION

We have found evidence for external shock emission in BATSE GRBs in the form of smooth decay emission that is best described with a power-law function rather than an exponential, consistent with the results of Ryde and Svensson (2001) who studied the decay phase of a sample of GRB pulses with a broad range of durations. Spectral analysis indicate that 20% of the events in our sample are consistent with synchrotron emission expected from an external shock. Interestingly, three of these events have decay rates consistent with that expected from the evolution of a jet. In most cases, the emission is coincident in time with the GRB. Complete details of this analysis can be found in a forthcoming ApJ paper.

## ACKNOWLEDGMENTS

Tim Giblin, Valerie Connaughton, and Ralph Wijers acknowledge support from NAG5-11017.

## REFERENCES

1. Mészáros, P., and Rees, M. J., 1997, , 476, 232
2. Galama, T., J., et al., 1998, , 500, L97
3. Sari, R., and Piran, T., 1999, , 520, 641
4. Mészáros, P., and Rees, M. J., 1999, , 306, L39
5. Vietri, M., 2000, Astroparticle Physics, 14, 211
6. Costa, E., 2000, in Gamma-Ray Bursts: 5th Huntsville Symposium, eds. Kippen, R. M., Mallozzi, R. M. (AIP Conference Proceedings: New York), pp. 365.
7. in't Zand, J. J. M., et al. 1999, , 516, L57
8. Galama, T., J., et al., 1999, , 398, 394
9. Giblin, T., W., et al., 1999, , 524, L47
10. Burenin, R. A., et al. *A & A* **344**, L53 (1999).
11. Connaughton, V., 2001, , submitted.
12. Sari, R., Piran, T., and Narayan, R., 1998, , 497, L17
13. Granot, J., and Sari, R., 2001, these proceedings.
14. Gallant, Y. A., et al., 1999, A&AS, 138, 549
15. Preece, R. D., et al., 2000, , 126, 19
16. Sari, R., et al., 1999, , 519, L17
17. Rhoads, J. E., 1999, , 525, 737
18. Ryde, F. and Svensson, R., 2001, Gamma-Ray Bursts in the Afterglow Era: 2nd Workshop, in press.

# RXTE All-Sky Monitor Observations of GRB Light Curves

Alan M. Levine[*], Hale Bradt[*], Ron Remillard[*] and Donald A. Smith[†]

[*]*M.I.T. Center for Space Research, 70 Vassar St., Cambridge, MA 02139*
[†]*Univ. of Michigan, Department of Physics, 500 E. University Ave., Ann Arbor, MI, 48109*

**Abstract.**
X-ray light curves for 4 gamma-ray bursts detected with the All-Sky Monitor on RXTE are presented. For two of the bursts, we also present simultaneous higher energy light curves obtained by BATSE and the *BeppoSAX* GRBM. We discuss the important features of the 4 sets of light curves and their shortcomings. We also briefly discuss future observations of X-ray light curves and the problems they could address.

## INTRODUCTION

X-ray observations of gamma-ray bursts (GRBs) yield positions, spectra, and light curves. We are all now keenly aware that the positions derived from the X-ray observations are essential for the critical task of searching for optical or radio counterparts and also for searching for afterglows with sensitive narrow-field X-ray instruments. The X-ray spectral results have helped to establish the power law nature of the GRB spectra which supports the interpretation that the emission is produced by synchrotron radiation [see 1]. The importance to date, however, of the study of X-ray light curves (over time intervals of seconds or minutes immediate to the strong burst phase) is less clear. An abbreviated summary of the history of X-ray observations of GRBs may be found in Smith et al. [2].

From the study of the X-ray light curves a few things have been learned. One, the range of peak fluxes and durations that can be expected from GRBs has been reasonably well established. The X-ray flux can be quite strong, i.e., more than 1 Crab, and the duration of a GRB in X-rays can be long, e.g., over 100 s. Two, multiband light curves show the general shape and variability of the spectrum. While the changes in the source spectrum inferred from observed differences in light curves for different bands can be dramatic, it is also evident that the differences from one band to the next generally follow smooth trends, rather than being wildly chaotic. This is consistent with the results of X-ray spectral studies which have found the spectrum to be crudely describable by power laws which may show breaks or high energy cutoffs [e.g., 3, 4]. Three, the spectral variability often takes the form of softening through the bursts and also through the individual peaks that make up a burst. This softening trend is evident in the tendency of the pulses that make up GRBs to have longer durations at lower photon energies [3, 4, 5]. Four, GRBs may be reasonably strong in X-rays at times when the gamma-ray flux is very weak.

Although not designed for the purpose, the All-Sky Monitor (ASM; [6]) on the *Rossi X-ray Timing Explorer (RXTE)* has yielded positions and light curves from the X-ray emission of 26 events which have been confirmed to be genuine GRBs by detections in gamma-rays by instruments on other spacecraft [2, 7, 8]. X-ray light curves for 15 of these GRBs have been presented and discussed in Smith et al. [2]. Herein we show a few light curves not presented in that article, and we point out how they support the general conclusions summarized above. We also use them to illustrate the shortcomings of contemporary measurements of the X-ray light curves of GRBs. Further information on ASM studies of GRBs, including light curves, may be found at http://xte.mit.edu/asmgrblc.

## SOME GRB LIGHT CURVES

Figure 1 shows light curves for GRB 960228, the earliest GRB known to have been detected with the ASM, having occurred only two weeks after the ASM began its regular monitoring routine. However, it was not found until years after the event, when the ASM data were being reanalyzed. This event was observed simultaneously in two of the Scanning Shadow Cameras (SSCs) of the ASM. Exposed fractional areas were 0.26 and 0.33 in SSCs 1 and 2 and correspond to effective areas of 7 and 9 cm$^2$, respectively, during the dwell (90 s exposure) displayed in Fig. 1. This is the brightest event observed with

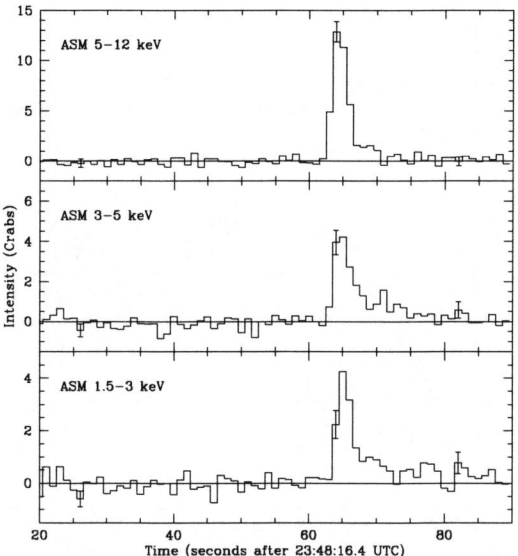

**FIGURE 1.** ASM light curves for GRB 960228. These light curves are background-subtracted and flux calibrated, and represent the average intensities determined from both SSCs 1 and 2 during a single 90 s dwell. Error estimates are based solely on counting statistics. The source intensity is given in Crabs where 1 Crab has a flux of $1.0 \times 10^{-8}$, $0.68 \times 10^{-8}$, and $1.1 \times 10^{-8}$ ergs cm$^{-2}$ s$^{-1}$ in the 1.5-3, 3-5 and 5-12 keV spectral bands, respectively. The absolute calibration of intensities determined from ASM data is uncertain by roughly 20%.

the ASM to date, as judged by the peak intensities measured in time bins of 1 to 10 s, so that even with small effective areas, the counting statistics yield good signal to noise ratios in the 3 individual energy channels with 1 s time resolution for the stronger portions of the burst. The event appears to be relatively extended in duration at lower energies, but much larger counting statistics would be needed to reach more definitive conclusions about the spectral variations. Data may be available to produce higher energy light curves since this GRB was detected by the Konus instrument on *Wind* [see Table 2 in 8], but we have not yet attempted to obtain them.

Figure 2 shows light curves for GRB 960610, another gamma-ray burst that was found in the ASM data long after the event. This GRB was also observed with BATSE, so the ASM light curves can be compared with BATSE counting rates from 3 higher energy bands. This event was typical in terms of the peak fluxes of ASM-detected GRBs; it is not particularly strong. The data show that the event lasted longer at lower energies. This latter characteristic is typical of GRBs, but this event may be somewhat unusual in that the initial rise appears to occur earlier at lower photon energies. The 5-12 keV ASM light curve is substantially negative for about a 15 second interval $\sim$ 30 s before the reference time. This suggests that

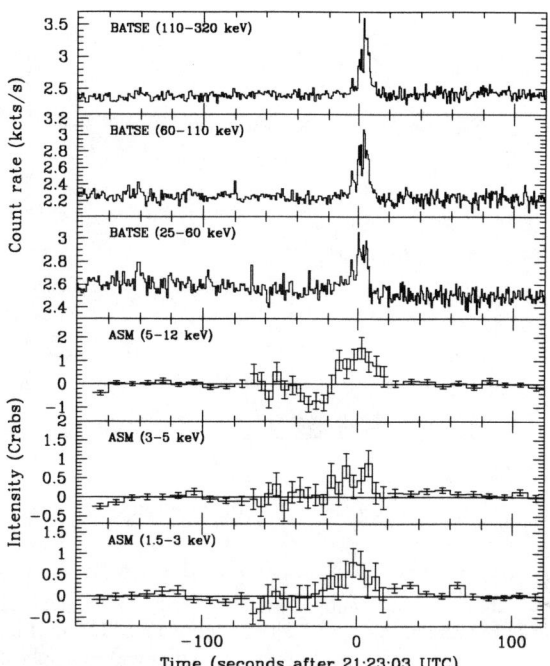

**FIGURE 2.** Light curves for GRB 960610 from ASM and BATSE data. The ASM light curves are background-subtracted and flux calibrated. The BATSE light curves are shown as raw count rates. Typical error estimates based on counting statistics are shown. The ASM data were obtained from SSC 2 (first dwell) and SSC 1 (second and third dwells).

there may be a problem with the background subtraction in this one band, but we have found no other indications of possible background subtraction problems, and there do not appear to be problems with the background in the other spectral bands. Indeed, only one other source listed in the ASM catalog was in the field of view, i.e., PKS0219-164, and in the 5-12 keV band it either was not detected or was very weak, as expected. The BATSE light curves show finer temporal structure than do the ASM light curves, because of the limitations of the counting statistics of the ASM data. Finally, one may notice the short gaps in the ASM data that occur between dwells.

GRB 010126 was a fairly strong event detected by the Konus instrument on *Wind* as well as by instruments on the *NEAR* and *Ulysses* spacecraft [8]. Light curves from the ASM are shown in Figure 3. The background estimate for the 1.5-3 keV band appears to be high by about 0.4 Crab. Differences in the temporal profiles beyond those that can be reasonably attributed to counting statistics are apparent. These differences are unusual in that there may be more structure, i.e., a deeper dip between peaks around 10 seconds after the reference time, in the lower energy channels.

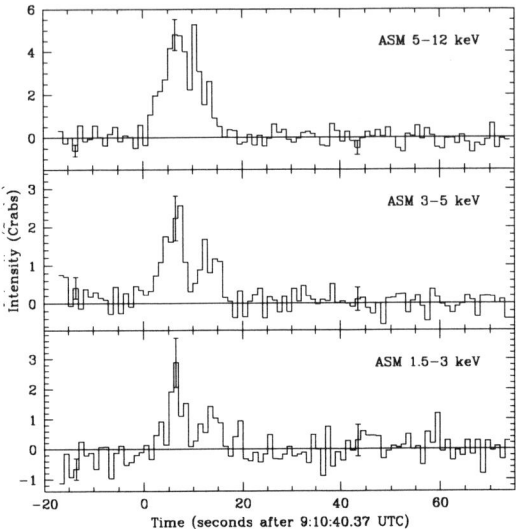

**FIGURE 3.** ASM light curves for GRB 010126. These light curves are background-subtracted and flux calibrated, and represent the average intensities determined from SSC 2 during a single 90 s dwell. Error estimates ($\pm 1\sigma$) are based solely on counting statistics.

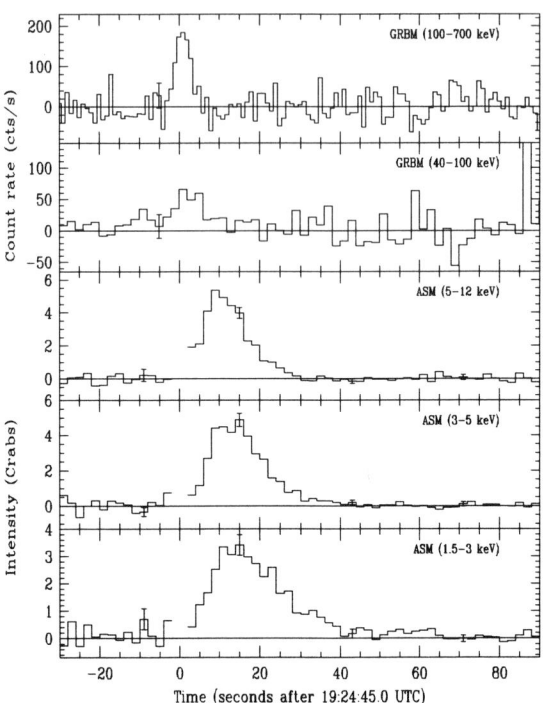

**FIGURE 4.** Light curves for GRB 010728 from ASM and *BeppoSAX* GRBM data. The ASM light curves are background-subtracted and flux calibrated. The GRBM light curves are shown as raw count rates. Typical error estimates based on counting statistics are shown. The ASM data were obtained from SSC 1 (first and second dwell) and SSC 2 (second and third dwells). The data from the third dwell are not shown since they show no evidence of any detected flux.

ASM and *BeppoSAX* GRBM light curves for GRB010728 are shown in Figure 4. If we have correctly interpreted the times of the data from both instruments, the results in Figure 4 indicate that the event began markedly earlier at higher energies. The good statistics from this relatively strong event allow one to compare the profiles for the three spectral bands from the ASM with one another, and to thereby notice that the event begins later and decays more slowly in the lowest energy band. In the 1.5-3 keV band, the tail of the event is perhaps still noticeable 60 seconds after the reference time. The trend of beginning later at lower energies is evident both upon comparison of the GRBM data with the ASM data and upon intercomparison of the data in the three ASM spectral bands. The gap between dwells in the ASM data happens at the particularly inopportune time of the peak in the GRBM data.

The spectral coverage of this GRB includes both gamma-ray and X-ray bands. Nonetheless, there is a big hole in the spectral coverage between 12 and 40 keV. Data from a few energy channels in that hole could have been rather useful in illuminating the strong trend in start time of the burst and in determining whether the trend is real or an artifact. The 12-40 keV band would also be important in analysis of the spectrum, e.g., looking for the peak energy in a $\nu F_\nu$ vs. $\nu$ plot.

## PROSPECTS

The observations which yield either X-ray or gamma-ray light curves have been made by a number of instruments, all of which have necessarily been wide-field monitors of some type because of the unpredictability of the appearance of GRBs with respect to both time and direction. Thus all of the instruments have produced data with high backgrounds and nearly all have also had relatively small effective areas. They thereby yielded light curves with rather limited counting statistics, a problem that was exacerbated when the data were subdivided into multiple spectral channels. The BATSE instrument was a notable exception, in that it was a large instrument with a large effective area, so that even with its high background, it produced light curves with good statistical properties. For wide-field instruments, insult is often added to injury when multiple X-ray sources are present in the field of view, making the non-GRB background variable or difficult to determine. Many of the instruments to date have not covered the extremely wide energy range that is desirable, or perhaps even necessary, for many studies. Comparison of our Figures 1 and 3 with Figures 2 and 4 illustrates this visually for the light curves, but full spectral coverage is even more important for getting accurate measurements of spectra. Another problem particularly plagues the ASM; that is that the ASM often produces light curves which do not fully cover each detected GRB event from beginning to end.

The X-ray light curves from the ASM and other instruments leave no doubt in our minds that better quality observations in the future will reveal important and possibly unexpected facts about the nature of the sources and the events. Future studies with advanced instrumentation will give us a better idea of whether the X-ray light curves are really much smoother that their gamma-ray counterparts at short time scales of 1 second or less. Searches for precursors will be both more sensitive and more complete in both temporal and spectral domains. Studies of time lags among different spectral bands will likely lead to interesting conclusions in analogy to the studies of time lags in the gamma-ray regime [e.g., 9]. Improved searches for the transition, if any, from burst to afterglow require much higher signal to noise in the tails of the bursts. Few if any of the short hard category of GRBs [10, 11] have been detected in X-rays, let alone localized and studied in detail [4, 12]. Bigger more sensitive instruments are needed to bring them into the sphere of real astronomy.

Better X-ray instruments will also lead to advances in spectral analysis of bursts, including studies of the locations of spectral breaks and their variations, unambiguous separation of the spectral components of events, searches for low energy absorption by circumsource or interstellar matter, and, of course, searches for discrete spectral features.

All of these considerations suggest to us that continued observation and study of X-ray light curves with the instruments and data available at present will yield additional interesting, if not definitive, results on GRBs, and will also help to provide a solid foundation for guiding the astrophysics community in the process of designing an advanced generation of X-ray instruments.

## ACKNOWLEDGMENTS

We are very grateful to Marco Feroci and the *BeppoSAX*/GRBM Team for kindly providing the *BeppoSAX* GRBM data for GRB 010728. D. A. Smith is supported by NSF Fellowship 00-136.

## REFERENCES

1. Piran, T., *Physics Reports*, **314**, 575 (1999) (astro-ph/9810256)
2. Smith, D. A., et al., submitted to *ApJ.*, (2002) (astro-ph/0103357)
3. Strohmayer, T.E., Fenimore, E. E., Murakami, T., Yoshida, A., *ApJ.*, **500**, 873 (1998)
4. Frontera, F., et al., *ApJS.*, **127**, 59 (2000)
5. Sazonov, S. Y., Sunyaev, R. A., Terekhov, O. V., Lund, N., Brandt, S., & Castro-Tirado, A. J., *A&AS.*, **129**, 1 (1998)
6. Levine, A. M., Bradt, H., Cui, W., Jernigan, J. G., Morgan, E. H., Remillard, R., Shirey, R. E., & Smith, D. A., *ApJ.(Letters)*, **469**, L33 (1996)
7. Smith, D. A., et al., *ApJ.*, **526**, 683 (1999)
8. Bradt, H., et al., in *Proceedings Of The Joint CNR/ESO Meeting, "Gamma-Ray Bursts in the Afterglow Era: 2nd Workshop"*, Rome, Italy, October 17-20, 2000, eds. F. Frontera, E. Costa, J. Hjorth, ESO Astrophysics Symposia" series, Springer Verlag, 2001 (astro-ph/0108004)
9. Norris, J. P., Marani, G. F., & Bonnell, J. T., *ApJ.*, **534**, 248 (2000)
10. Kouveliotou, C., et al., *ApJ.(Letters)*, **413**, L101 (1993)
11. Kouveliotou, C., *ApJS.*, **92**, 637 (1994)
12. Gandolfi, G., et al., in *Proc. 5th Huntsville Gamma-Ray Burst Symp.*, ed. R. M. Kippen, R. S. Mallozzi, & G. J. Fishman, AIP, New York, 2000, p. 23 (astro-ph/0001011)

# Temporal Properties of Short and Long Gamma-Ray Bursts

S. McBreen*, F. Quilligan*, B. McBreen*, L. Hanlon* and D. Watson[†]

*Department of Experimental Physics, University College Dublin, Dublin 4, Ireland*
[†]*X-Ray Astronomy Group, Department of Physics and Astronomy, Leicester University, Leicester LE1 7RH, UK*

**Abstract.** A temporal analysis was performed on a sample of 100 bright short GRBs with $T_{90}<2$ s from the BATSE Current Catalog along with a similar analysis on 319 long bright GRBs with $T_{90}>2$ s from the same catalog. The short GRBs were denoised using a median filter and the long GRBs were denoised using a wavelet method. Both samples were subjected to an automated pulse selection algorithm to objectively determine the effects of neighbouring pulses. The rise times, fall times, FWHM, pulse amplitudes and areas were measured and their frequency distributions are presented. The time intervals between pulses were also measured. The frequency distributions of the pulse properties were found to be similar and consistent with lognormal distributions for both the short and long GRBs. The time intervals between the pulses and the pulse amplitudes of neighbouring pulses were found to be correlated with each other. The same emission mechanism can account for the two sub-classes of GRBs.

## INTRODUCTION

It has been recognised that GRBs may occur in two sub-classes based on spectral hardness and duration with $T_{90}>2$ s and $T_{90}<2$ s [2, 6, 7]. The bimodal distribution can be fit by two Gaussian distributions to the logarithmic durations [3]. A variety of statistical methods have been applied to the temporal properties of the long GRBs with $T_{90}>2$ s. It is important to compare the temporal properties of the long and short GRBs to determine the similarities and differences between the two classes in an objective way. Detailed temporal analyses have been performed on a large sample of short and long bright GRBs. The results from the long sample [8] can be used as templates for comparison with a similar analysis of short GRBs.

## DATA ANALYSIS

The sample of 100 short GRBs was selected from the Time Tagged Event data at 5 ms from the BATSE Current Catalog. The four energy channels were combined to maximise the signal to noise ratio. The sample of 319 long GRBs was selected from the "discsc" 64ms data also from the BATSE Current Catalog. The energy channels were combined as for the short GRBs. Bursts from both samples were background subtracted by selecting a pre- and/or post-burst section.

A median filter was used to denoise the short GRBs [4] and a wavelet method [8] was used to denoise the long GRBs. The same pulse selection method was applied to each sample of denoised GRBs. The pulses selected had a threshold of 5 $\sigma$ above background ($\tau_\sigma \geq 5$) and were isolated from adjacent pulses by at least 50% ($\tau_i \geq 50\%$). A value of $\tau_i \geq 50\%$ implies that the two minima on either side of the pulse maximum must be at or below half the maximum value. A total of 313 pulses were selected from the sample of short GRBs with $\tau_\sigma \geq 5$ and 181 of these had $\tau_i \geq 50\%$. A total of 3358 pulses with $\tau_\sigma \geq 5$ were selected from the sample of long GRBs, 1575 of which had $\tau_i \geq 50\%$.

## RESULTS

The distributions of rise times ($t_r$), fall times ($t_f$) and full width at half maxima (FWHM) for the isolated pulses ($\tau_i \geq 50\%$) are presented in Fig. 1. The distribution of time intervals between the pulses ($\Delta T$) with $\tau_\sigma \geq 5$ is also given in Fig. 1. The distributions of pulse amplitudes and areas are given in Figs. 2 and 3 for the isolated pulses observed by two BATSE large area detectors. The median values of the distributions are presented in Table 1. The Spearman Rank Order correlation coefficients $\rho$ along with associated probabilities for the time intervals separated by N pulses are given in Table 2. The $\Delta T$ values are normalised by $T_{90}$ for each burst and show that there is a high degree of correlation over many intervals for the long GRBs. The Spearman Rank Order correlation coeffcents for the pulse amplitudes with N are listed in Table 3 for the short and long bursts.

**TABLE 1.** The median values of the pulse properties in GRBs. All pulses with $\tau_\sigma \geq 5$ were used for the time intervals, not just isolated pulses as for the pulse properties.

| Pulse Property | Short GRBs | Long GRBs |
|---|---|---|
| Rise Time (sec) | 0.035 | 0.64 |
| Fall Time (sec) | 0.056 | 1.10 |
| FWHM (sec) | 0.045 | 0.58 |
| Time Interval (sec) | 0.095 | 1.34 |
| Pulse Area (count rate) | $1.5 \times 10^5$ | $1.5 \times 10^5$ |
| Pulse Amplitude (count rate) | $1.0 \times 10^4$ | $8 \times 10^3$ |

**TABLE 2.** Spearman Rank Order correlation coefficients $\rho$ for time intervals between pulses. The value of N indicates the number of pulses between the correlated time intervals. The two values for $\rho$ and the probability are for unnormalised/normalised time intervals. The values are normalised by $T_{90}$. The first two lines refer to the short GRBs and the remaining lines refer to the long GRBs.

| N | Number of Intervals | $\rho$ | Probability |
|---|---|---|---|
| 1 | 140 | 0.42/0.30 | $1.7 \times 10^{-12}/3.4 \times 10^{-4}$ |
| 2 | 84 | 0.48/0.32 | $4.9 \times 10^{-6}/3.0 \times 10^{-3}$ |
| 1 | 2751 | 0.42/0.56 | $< 10^{-48}$ |
| 2 | 2499 | 0.34/0.48 | $< 10^{-48}$ |
| 5 | 1929 | 0.24/0.37 | $5 \times 10^{-26}/ < 10^{-48}$ |
| 10 | 1395 | 0.20/0.29 | $3 \times 10^{-13}/6 \times 10^{-27}$ |
| 15 | 890 | 0.16/0.25 | $3 \times 10^{-6}/4 \times 10^{-14}$ |

**TABLE 3.** Spearman Rank Order correlation coefficients $\rho$ for the pulse amplitudes of neighbouring pulses (N=1) and pulses separated by N pulses. The two values for $\rho$ and the probability are for unnormalised/normalised pulse amplitudes. The later are normalised by the maximum pulse amplitude in the burst. The first two lines refer to the short GRBs and the remaining lines refer to the long GRBs. The maximum peak was removed for the short GRBs in the normalised sample.

| N | Number of Pulses | $\rho$ | Probability |
|---|---|---|---|
| 1 | 213/107 | 0.39/0.24 | $5 \times 10^{-9}/1.1 \times 10^{-2}$ |
| 2 | 140/84 | 0.13/−0.09 | 0.13/0.42 |
| 1 | 3039 | 0.72/0.57 | $< 10^{-48}$ |
| 3 | 2499 | 0.55/0.32 | $< 10^{-48}$ |
| 5 | 2098 | 0.52/0.24 | $< 10^{-48}/3 \times 10^{-29}$ |
| 7 | 1777 | 0.48/0.15 | $< 10^{-48}/6 \times 10^{-11}$ |

## DISCUSSION

There are remarkable similarities between the statistical properties of the two sub-classes of GRBs. The distributions of the $t_r$, $t_f$, FWHM, pulse amplitude, pulse area

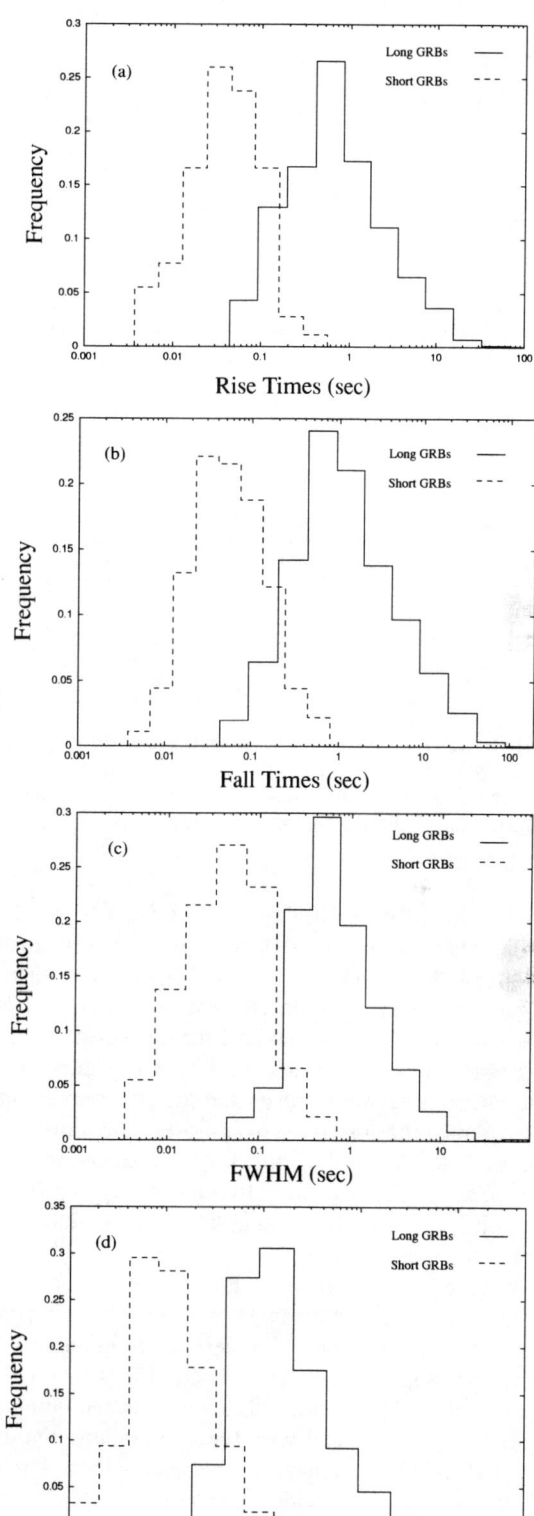

**FIGURE 1.** Normalised distributions of $t_r$, $t_f$, FWHM, and $\Delta T$ (a-d).

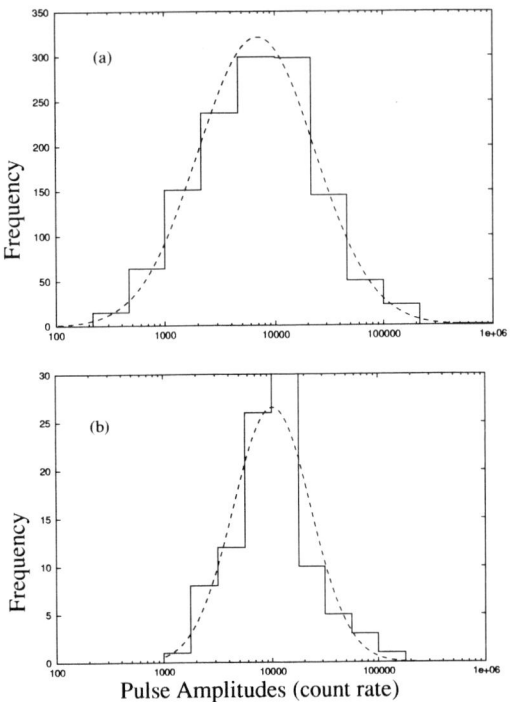

**FIGURE 2.** Pulse Amplitude distributions for the long GRBs (a) and the short GRBs (b). The dashed lines indicate lognormal fits to the data. The resolution of the long sample is 64 ms and the resolution of the short sample is 5 ms.

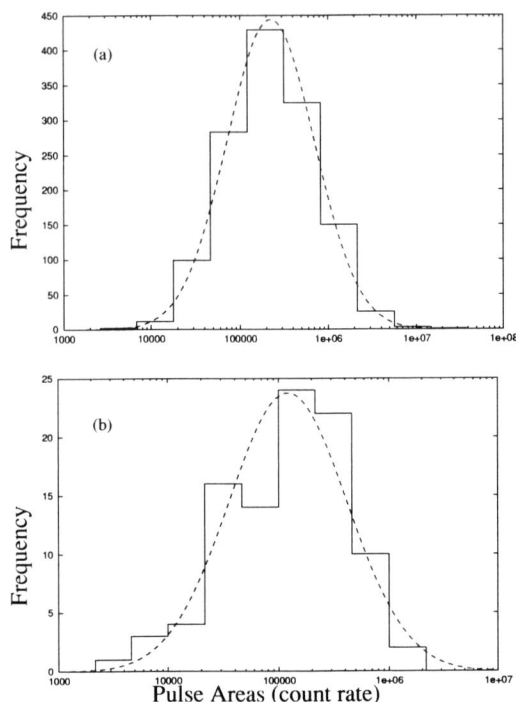

**FIGURE 3.** Pulse Area distributions for the long GRBs (a) and the short GRBs (b). The dashed lines indicate lognormal fits to the data.

and $\Delta T$ for GRBs with $T_{90}<2$s and $T_{90}>2$s are very similar and both are well described by lognormal distributions [4, 8]. In long GRBs with $T_{90}>2$s the values of $\Delta T$ are not random but consistent with a lognormal distribution with a Pareto-Levy tail for a small number of long time intervals in excess of 15 s. The values of the time intervals between pulses and the pulse amplitudes were found to be correlated over most of the long GRBs (Tables 2 and 3). In the short GRBs adjacent and subsequent time intervals and pulse amplitudes were found to be correlated at a lower significance level due to the smaller number of pulses.

The clear conclusion is that the same emission mechanism can account for the two types of GRBs. This conclusion is in agreement with a very different analysis of the temporal structure of short GRBs [5]. The external shock model [1] has serious difficulties in accounting for GRBs with $T_{90}<2$s and with the non-random distribution of correlated time intervals between pulses. The results presented here provide considerable support for the internal shock model [9]. The internal shock model can account for the results obtained for long and short GRBs provided the cause of the pulses and the correlated values of $\Delta T$ can be attributed to the central engine.

## CONCLUSIONS

Samples of short and long bright GRBs have been denoised and analysed by an automatic pulse selection algorithm. The results show that in both cases the distribution of the properties of isolated pulses and time intervals between all pulses are similar and compatible with lognormal distributions. The same mechanism seems to be responsible for both long and short GRBs and may be attributable to the internal shock model.

## REFERENCES

1. Dermer, C. D. & Mitman, K. E. 1999, ApJ, 513, L5
2. Kouveliotou, C., Meegan, C. A., Fishman, G. J. et al. 1993, ApJ, 413, L101
3. McBreen, B., Hurley, K. J., Long, R., and Metcalfe, L. 1994, MNRAS, 271, 662
4. McBreen, S., Quilligan, F., McBreen, B., Hanlon, L., and Watson, D. 2001, A&AL, In press
5. Nakar, P. & Piran, T. 2001, [astro-ph/0103192]
6. Norris, J. P., Scargle, J. D., and Bonnell, J. T. 2000, [astro-ph/0105108]
7. Paciesas, W. S., Preece, R. D., Briggs, M. S., and Malozzi, R. S. 2001, [astro-ph/0109053]
8. Quilligan, F., McBreen, B., Hanlon, L., et al. 2002, A&A, 385, 377
9. Rees, M. J. and Mészáros, P. 1994, ApJ, 430, L93

# Low-Energy Study of Gamma-Ray Bursts using Two BATSE Spectroscopy Detectors

Michael J. Pangia* and Robert D. Preece[†]

*Department of Chemistry & Physics, Georgia College & State University, Milledgeville, GA 31061, USA
[†]Department of Physics, University of Alabama at Huntsville, Huntsville, AL 35899, USA

**Abstract.** We have analyzed data from the Burst and Transient Experiment (BATSE) Spectroscopy Detectors (SDs), on board NASA's Compton Gamma-Ray Observatory. Preece et al. (1996) have previously identified the low-energy ($\sim 10$ keV) characteristics of gamma-ray bursts (GRBs) and found 14% of 86 bright bursts showed low-energy excesses greater than $5\sigma$ above the extrapolated spectral model. The current work extends the previous study by analyzing GRBs for which two SDs meet the selection criteria. Essentially, we required that the GRB be bright, and that adequate low-energy coverage existed from two SDs, not just one. A sequence of spectral fits were performed on a series of independent spectra making up the time history of each burst, with an additional low-energy component tracking the excess. We studied correlations between the low-energy spectral fit parameters and the dynamic GRB characteristics. Five GRBs met the selection criteria, of which three exhibited a correlation between the low-energy amplitude parameter and the photon flux during the majority of the event. No correlation was apparent for the remaining two GRBs.

## INTRODUCTION

Using BATSE data from the Large Area (LADs) and Spectroscopy Detectors (SDs), Preece et al. [1] studied 86 bright Gamma-Ray Bursts (GRBs) in search of a low-energy ($\sim 10$ keV) component of the GRBs spectra. They found that 14% of the GRBs had a definite low-energy component, referred to as a low-energy excess. Their basic method was to fit the energy spectrum of each GRB to a suitable model and rank the discrepancy between the model fit and the data at low-energy. The criterion for identifying a low-energy excess was $5\sigma$ above the model fit. The use of SD data was essential to their study, and the present one as well, because it extends down to the low-energy range when operating in a high-gain mode. The purpose of this work is to study the dynamic behavior of the low-energy component looking for correlations between the fit parameters characterizing the low-energy component and the photon flux.

## GRB SELECTION

The criteria for selecting the GRBs are: 1) the GRB must be bright (total fluence $> 2 \times 10^{-5}$ ergs/cm$^2$ and a low-energy excess greater than about $3\sigma$), 2) the two SDs registering the most activity for the GRB must both be in a Hi-Gain mode ($4\times$ or $8\times$), and 3) the azimuthal angles from the SDs to the source must be less than 70°. Of these criteria, the second one was the most restrictive. The third criterion was used to eliminate the possibility that the field of view of the SDs was obscured by the LADs or other equipment onboard the Compton Gamma-Ray Observatory. A review of the entire set of 2704 BATSE-triggered GRBs yielded only five that met all the criteria. These are GRB 920517 (Trig 1609), GRB 930706 (Trig 2431), GRB 940228 (Trig 2852), GRB 940526 (Trig 2992), and GRB 000101 (Trig 7929).

## PROCEDURES

BATSE data was acquired and recorded in a number of formats. The properties of the data types vary in resolution and coverage in both time and energy. A table of the different types and their properties is given in Ref. 2. Of particular interest are the continuous sampling mode (2.048 s resolution) and the burst-triggered higher temporal resolution mode with the finest resolution used in our study of 0.128 s. The continuous type used is the low-energy channel of the DISCSP data type, and the burst-triggered type is SHERB. The data type SHER is the corresponding background data that is needed to determine the uncertainties in the data and the model, and to determine the signal-to-noise ratio.

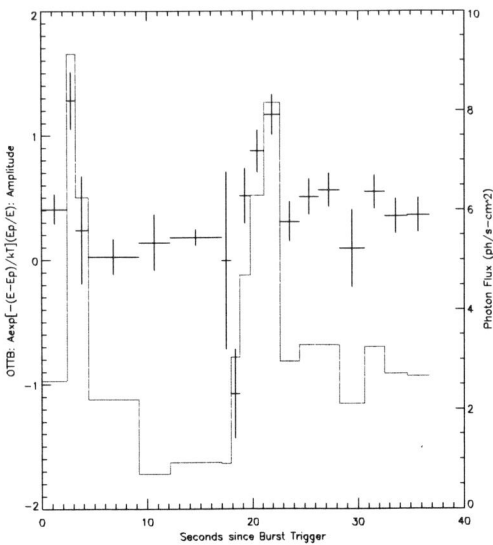

**FIGURE 1.** Plot of the OTTB A-parameter and photon flux for Trig 2852 using Procedure 1.

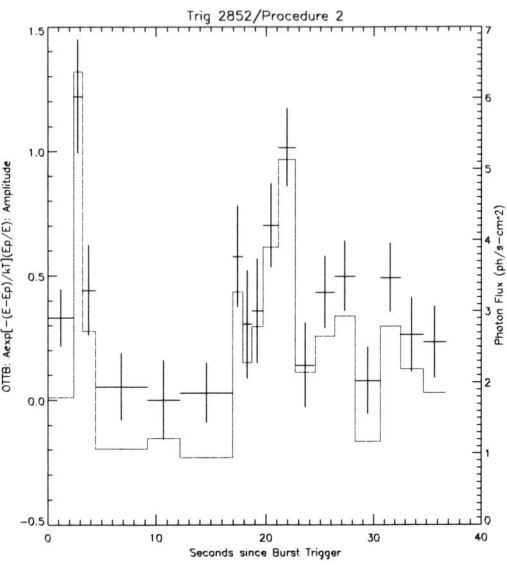

**FIGURE 3.** Plot of the OTTB A-parameter and photon flux for Trig 2852 using Procedure 2.

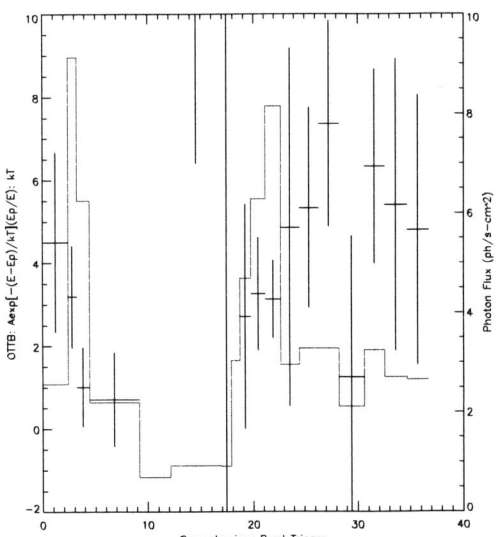

**FIGURE 2.** Plot of the OTTB T-parameter and photon flux for Trig 2852 using Procedure 1.

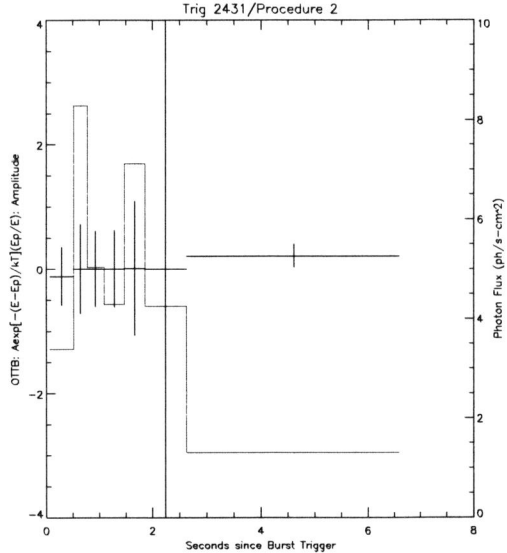

**FIGURE 4.** Plot of the OTTB A-parameter and photon flux for Trig 2431 using Procedure 2.

The procedure developed by Preece et al. [1] to study the low-energy aspects of GRBs with BATSE data is to fit the data to a representative spectral function using the program RMFIT. For the present study, two slightly different procedures were used, with Procedure 2 yielding better correlation. For each procedure, the total spectral model is a linear sum of two functions, representing the low-energy and main components of the spectrum. The low-energy function is the optically thin thermal bremsstrahlung (OTTB) model, which has the following form

$$f = A\,exp(-E/kT)\,exp(E_P/kT)\,(E/E_P)^{-1} \quad (1)$$

where A is the amplitude, T is the temperature, and $E_P$ is a 'pivot' energy (constrained to 10 keV) that defines the normalization. The main part of the spectrum is fit by the Band GRB model (two smoothly-joined power-laws), or, if that does not converge, a model known as the Comptonized Spectral Model (see Ref. 2 for the formulae).

In combining the different data, a signal-to-noise level criterion of 15 was applied to the DISCSP data of the SD with the second brightest signal to improve on the accuracy of the fit. This determines the time intervals into which each data set is rebinned. Then, the fitting is actually done in steps. For Procedure 1 the steps are: A) fit the spectrum of the entire duration of the GRB using only the main component function, B) again fit the entire duration of the GRB, but with the OTTB function added in and the fit parameters determined from Step A used as the initial values for the main component, and C) fit the spectrum for each time bin separately using the fit parameters determined in the Step B as initial values. Steps A & B are the same for both Procedures 1 & 2. Step C of Procedure 2 differs from Procedure 1 in that the OTTB T-parameter is fixed to the value determined in the Step B, in which case, in regard to the OTTB function, only its amplitude A-parameter is being fit for each time bin.

## RESULTS AND CONCLUSIONS

Correlations between the OTTB parameters and the photon flux were investigated using Procedures 1 & 2 described in the previous section. Figure 1 is a combined plot of the OTTB A-parameter (indicated by crosses) obtained using Procedure 1 and the photon flux (indicated by the continuously drawn dashed line) as a function of the time since the start of Trig. 2852. In this and the following figures, the vertical extent of the crosses gives the uncertainty in the fit parameter, and its horizontal extent matches the time interval. Figure 2 compares the OTTB T-parameter with photon flux of Trig 2852 in the same fashion as Fig. 1. It is evident from Fig. 1 that the OTTB A-parameter is correlated with the photon flux only during the peaks in activity, whereas a correlation for Fig. 2 does not appear to exist. Trig 2993 was the only other GRB that showed a correlation using Procedure 1, again for only the A-parameter.

Also evident from Fig. 2 is that the fractional uncertainties are considerable for the OTTB T-parameter, which, because of the functional form of Equation 1, may be masking a correlation not only for the OTTB T-parameter, but for the A-parameter as well. The implication is that, although low-energy coverage from two SDs is a minimum for fitting a two-parameter model (Equation 1 with $E_P$ fixed), the fit parameters are not being constrained to an acceptable level of accuracy. This motivates using Procedure 2, which fixes the OTTB T-parameter and studies just the A-parameter. Figure 3 shows the result of Procedure 2 for Trig 2852. Notice the correlation now exists for the entire duration. Procedure 2 also yields good correlation for the majority of the duration of Trigs 1609 & 2993. To show that Procedure 2, does not universally result in a good correlation, Figure 4 is the result for Trig 2431. Neither is a correlation apparent for Trig 7929.

In closing, the results of this study complement the study of Preece et al. [1] by having investigated the dynamic behavior of the low-energy excess for the five BATSE-triggered GRBs that our selection criteria. Even with the use of two SDs covering low energy, the fits could not be sufficiently constrained (Procedure 1), but, nonetheless, a correlation was evident in two cases. With the additional constraint of fixing the OTTB T-parameter (Procedure 2), a good correlation was found for three of the GRBs. We feel that this study supports the idea that a low-energy component simultaneously exists with the main component of, at least, a bright GRB.

## ACKNOWLEDGMENTS

Funding for one of the authors (MJP) was provided through a 2001 NASA/ASEE Summer Faculty Fellowship overseen by the University of Alabama at Huntsville.

## REFERENCES

1. Preece, R. D., Briggs, M. S., Pendleton, G. N., Paciesas, W. S., Matteson, J. L., Band, D. L., Skelton, R. T., and Meegan, C. A., *The Astrophysical Journal*, **473**, 310-321 (1996).
2. Preece, R. D., Briggs, M. S., Mallozi, R. S., Pendleton, G. N., Paciesas, W. S., and Band, D. L., *The Astrophysical Journal Supplement Series*, **126**, 19-36 (2000).

# Analytical Descriptions of Gamma-Ray Burst Pulses

Felix Ryde[*,†], Dan Kocevski[†] and Edison Liang[†]

[*]*Center for Space Science and Astrophysics, Stanford University, Stanford, CA 94305*
[†]*Department of Physics and Astronomy, Rice University, Houston, TX 77005*

**Abstract.** We present a discussion on the shape of a fundamental pulse in a GRB light curve. As has been previously noted by Ryde et al. the shape is dictated by the spectral behavior during the pulse. We review this analysis and expand it to also include the rise phase. We discuss two alternative analytical descriptions; the proposal already made by Kocevski & Liang and a variation of this.

## INTRODUCTION

A gamma-ray burst (GRB) light curve often consists of several individual, asymmetric pulses with a fast rise and a slower decay. Each pulse is assumed to be associated with a separate emission episode caused by shocks from collisions in an episodic relativistic outflow. A complex GRB is a superposition of several such episodes. The complete spectral and temporal evolution of a pulse can be characterized by the energy flux, $F_E(E,t)$. We focus on the following three main observables, the peak of the energy spectrum, $E_p(t)$, the instantaneous energy flux, $F(t)$, and the derived quantity, the energy fluence $\mathcal{E}(t) = \int^t F dt'$. The relations between these three observables are given by two empirical correlations.

1. The hardness-intensity correlation (HIC; Golenetskii et al. [1]), chosen here as $E_p(F)$, which relates the flux of the source to the 'hardness' of the spectrum (here represented by $E = E_p$).
2. The Hardness-Fluence Correlation, $E_{pk}(\mathcal{E})$ (HFC; Liang & Kargatis, [2]), represented here as $E_p(\mathcal{E})$, relates the hardness to the time running integral of the flux, i.e. the fluence.

As first shown by Ryde & Svensson [3] these two empirical correlations dictate the shape of the pulse light curve. First, we review their results which were specifically for the decay phase of a pulse and thereafter we expand their analysis to also include the rise phase. We motivate and describe two possible analytical descriptions within the internal shock model of GRBs. In this model there are three main time scales relevant for a pulse: the shell crossing time, the angular spreading time, and the cooling time. As discussed by Ryde & Petrosian [4], [5] the first two time scales are comparable and dominate over the third. In this paper we focus on a situation in which the shell crossing time is dominant and determines the pulse widths and shape.

## DECAY PHASE OF A PULSE

For the decay phase of a pulse the most common behavior of the HIC is

$$F = F_0(E_p/E_{p,0})^\gamma, \qquad (1)$$

where $E_{p,0}$ and $F_0$ are the initial values of the peak energy and the energy flux at the beginning of the decay phase in each pulse and $\gamma$ is the power law index. The original study [1] found the power law index to vary between $1.5 - 1.7$ over the whole GRB. Moreover, Borgonovo & Ryde [6] studied a sample of 82 GRB pulse decays and found them to be consistent with a power law HIC in, at least, 57% of the cases and for these found $\gamma = 2.0 \pm 0.7$.

The HFC represents the observation that the rate of change in the hardness is proportional to the luminosity of the radiating medium (or, equivalently, to the energy density);

$$\dot{E} = -\frac{F}{\Phi_0}. \qquad (2)$$

Using the energy flux, $F$, this corresponds to a linear relation between the $E_p$ and the time integrated energy flux, the energy fluence

$$E_p(t) = E_{p,0} - \frac{\mathcal{E}(t)}{\Phi_0} \qquad (3)$$

where $\Phi_0$ is the decay constant. This behavior is most often seen over the entire pulse.

We now follow Ryde & Svensson [3] to find an analytical description of the energy flux decay. Note that in the original description they used the photon flux, and we will here reformulate their results in terms of the energy flux instead, which makes a physical interpretation easier. The combination of the two empirical relations gives the following differential equation governing the spectral evolution

$$\dot{E} = -\frac{F_0}{\Phi_0 E_0^\gamma} E^\gamma \qquad (4)$$

The solution gives that the energy flux during a decay phase of a pulse follows

$$F(t) = \begin{cases} F_0 \left(1 + \frac{(\gamma-1)t}{T}\right)^{-\gamma/(\gamma-1)} & \text{if } \gamma \neq 1 \\ F_0 e^{-t/T} & \text{if } \gamma = 1, \end{cases} \qquad (5)$$

and the corresponding peak energy follows

$$E_p(t) = \begin{cases} E_{p,0} \left(1 + \frac{(\gamma-1)t}{T}\right)^{-1/(\gamma-1)} & \text{if } \gamma \neq 1 \\ E_{p,0} e^{-t/T} & \text{if } \gamma = 1, \end{cases} \qquad (6)$$

where $T \equiv \Phi_0 E_{p,0}/F_0$. Note that this gives the possibility to measure $\Phi_0$ directly from the light curve and $E_{p,0}$. Introducing $d \equiv \gamma/(\gamma - 1)$ ($d$ as in the asymptotic *decay* of the energy flux) we describe the peak and the energy flux decays as

$$F(t) = F_0 \left(1 + \frac{t}{T(d-1)}\right)^{-d} \qquad (7)$$

$$E_p(t) = E_{p,0} \left(1 + \frac{t}{T(d-1)}\right)^{1-d} \qquad (8)$$

Note that this description is for $\gamma \neq 1$.

## INCLUSION OF THE RISE PHASE

The evolution of $E_p$ is still from hard to soft during the main part of the rise phase of a pulse. The energy flux and the $E_p$ are thus anti-correlated during the rise phase.

In most physical models both the peak of the energy spectrum and the luminosity are proportional to the random Lorentz factor of the shocked electrons to some power:

$$E_p(t) \propto \Gamma_r^a(t) \cdot g(t), \qquad (9)$$

and

$$F_E(t) \propto \Gamma_r^b(t) \cdot h(t). \qquad (10)$$

The functions $g(t)$ and $h(t)$ parameterize the unknown time dependencies on particle densities, optical depth, magnetic field and kinematics, etc. The correlation between hardness and the energy flux can thus be described as

$$F_E(t) \propto E_p^\gamma \cdot f(t) \qquad (11)$$

with $\gamma = b/a$ and $f(t) = h(t)g(t)^{-\gamma}$.

During the decay phase of a pulse the power law relation dominates the HIC which has been been manifested in previous studies. However, the HIC during the rise phase will be dominated by the unknown function $f(t)$.

We will now assume two simple representations of $f(t)$ and recapitulate the analytical discussion in the previous section to find what such a description will lead to for the shape of the whole pulse.

## Scenario 1

As a first step we assume a simple prescription (see also Kocevski & Liang [7], [8]) of the function

$$f(t) \propto t^r \qquad (12)$$

where the power law $r$ represents the *rise*. We have to introduce two new parameters, the energy flux and the time of the peak of the light curve, $F_m$ and $t_m$:

$$F = F_m \left(\frac{E_p}{E_m}\right)^\gamma \left(\frac{t}{t_m}\right)^r \qquad (13)$$

which combined with the HFC now gives the differential equation

$$\dot{E}_p = -\frac{F_m}{\Phi_0 E_m^\gamma t_m^r} E_p^\gamma t^r \qquad (14)$$

Note that $E_p(t = 0) = E_0$ and $E_p(t = t_m) = E_m$ and $F(t = t_m) = F_m$ and that $\tau_m \equiv F_m/E_m \Phi_0$. Now, we define the exponent $d$ (for finite $r$) to describe the asymptotic power law behavior of the light curve, $F(t) \to t^{-d}$ as $t \to \infty$ which gives $\gamma = (d+r)/(d-1)$. We now define the peak of the pulse to be at $t = t_m$ by letting $dF/dt(t = t_m) = 0$ and get the condition that for $t_m \neq 0$, $\tau_m = t_m(d+r)/[(d-1)r]$. We then finally arrive at the pulse shape and the peak energy evolution as

$$\frac{F}{F_m}(t) = \left(\frac{t}{t_m}\right)^r \left(\frac{d}{d+r} + \frac{r}{d+r}\left(\frac{t}{t_m}\right)^{r+1}\right)^{-\frac{r+d}{r+1}} \qquad (15)$$

$$\frac{E_p}{E_m}(t) = \left(\frac{d}{d+r} + \frac{r}{d+r}\left(\frac{t}{t_m}\right)^{r+1}\right)^{-\frac{d-1}{r+1}} \qquad (16)$$

Equation (15) describes the GRB pulse shape by two parameters apart from the position of the peak: $F_m, t_m, r, d$. Direct measurement of an observed light curve gives $F_m, t_m, r, d$ which gives $\tau_m$, which in its turn, gives the product $E_m \cdot \Phi_0$. As $E_m$ can be measured from the $E_p$ decay, $\Phi_0$ can be deduced.

As the description in the previous section deals only with the decay phase, the HIC power law index, $\gamma$, will correspond to both the implicit power law behavior, $E^\gamma$ as well as the explicit behavior, $t^r$ in this scenario. Here $\gamma$ denotes the power law in the second description.

The two descriptions coincide for $r \to 0$ and for $E_{p,0} \to E_m$ which gives

$$F = \frac{F_m}{\left(1 + \frac{t}{(d-1)\tau_m}\right)^d} \quad (17)$$

and thus $T = \tau_m$ by identification. Note that $t_m \to 0$ as $r \to 0$. It is also readily seen that the asymptotic behaviors as $t \to \infty$ are the same.

## Scenario 2

It is reasonable that the rise phase is connected to some transient process, e.g., an initial decrease in optical depth, increase in the number of energized particles, or the merging (or crossing) of the shells. After this initial phase, the decay behavior as described by equations (7) and (8) should emerge.

We therefore try the following prescription

$$f(t) \propto 1 - e^{-t/\tau_r} \quad (18)$$

where the time constant $\tau_r$ now represents the *r*ise phase. This corresponds to $h(t) \propto g(t)^\gamma$ as $t \gg \tau_r$. We therefore have

$$F = F_m \left(\frac{E_p}{E_m}\right)^\gamma \frac{1 - e^{-t/\tau_r}}{1 - e^{-t_m/\tau_r}} \quad (19)$$

which gives the differential equation

$$\dot{E}_p = -\frac{F_m}{\Phi_0 E_m^\gamma (1 - exp(-t_m/\tau_r))} E_p^\gamma (1 - e^{-t/\tau_r}) \quad (20)$$

In the same manner as before we define the decay index $d$ by requiring $F(t) \to t^{-d}$ as $t \to \infty$. This gives $\gamma = d/(d-1)$. Solving the differential equation (20) and combining with equation (19) we find

$$F(t) = \frac{A_0(1 - e^{-t/\tau_r})}{(1 + (t + \tau_r e^{-t/\tau_r})/\tau_d)^d} \quad (21)$$

$$E_p(t) = \frac{A_1}{(1 + (t + \tau_r e^{-t/\tau_r})/\tau_d)^{(d-1)}} \quad (22)$$

where $A_0$ and $A_1$ are analytical functions of $d, \tau_d, \tau_r, \Phi_0, F_m, E_m$ and $t_m$. There is no analytical expression for the peak time $t_m$, but it can be solved for numerically if the other parameters are known.

Equation (21) describes the pulse by four parameters: $A_0, \tau_r, \tau_d$, and $d$. This description coincides with the decay function (Eqs. 7 and 8) for $\tau_r \to 0$ or $t \to \infty$ with $\tau_d \to (d-1)T$ (and $E_m, F_m \to E_{p,0}, F_0$). Note that in this description there are two explicit time constants $\tau_r$ and $\tau_d$, while in equation (15) the decay constant is an expression of $r$ and $t_m$. For a sharp rise phase the explicit time dependence will disappear. This is, however, not the case for equation (15).

## DISCUSSION

The main difference in these new descriptions is that we now have introduced an explicit time dependence of the energy flux, apart from the implicit time dependence, through the peak energy, which enables us to give an analytical description of the rise phase as well.

A general problem in dealing with *CGRO* BATSE pulses is that the rise phases are short (FREDs). Furthermore, many pulses have some contamination from previous pulses. Another complication to bear in mind is that the description above assumes a monotonically decaying $E_p$. Often the $E_p$ rises quickly, but a few time bins are inevitably lost and in some cases all the rise phase is lost and only the transition phase can be used.

All of these issues lead to the result that the $r$ and $\tau_r$ parameters are occasionally difficult to constrain. To avoid this problem, one can include more of the rise phase even though the peak energy is observed to rise. A motivation for this is that the peak energy of the studied pulse can very well be monotonic and the observed lower value of the peak energy is due to previous pulses contaminating the spectrum, leading the fit to the lower values. A detailed comparative analysis using these three prescriptions is given in Ryde et al. [9].

## ACKNOWLEDGMENTS

We are grateful to Dr. Vahé Petrosian for valuable discussions. F.R. acknowledges financial support from the Swedish Foundation for International Cooperation in Research and Higher Education (STINT) and the Ludovisi Boncompagni, née Bildt, foundation.

## REFERENCES

1. Golenetskii, S. V., Mazets, E. P., Aptekar, R. L., and Ilyinskii, V. N., *Nature* **306**, 451 (1983).
2. Liang, E. P., and Kargatis, V. E. *Nature*, **381**, 495 (1996).
3. Ryde, F., and Svensson, R., *ApJ*, **529**, L13 (2000).
4. Ryde, F., and Petrosian, V. *ApJ*, submitted (2001).
5. Ryde, F., and Petrosian, V., these proceedings (2002).
6. Borgonovo, L., and Ryde, F., *ApJ*, **548**, 770 (2001).
7. Kocevski, D., and Liang, E., *ApJ*, to be submitted.
8. Kocevski, D., and Liang, E., these proceedings.
9. Ryde, F., Kocevski, D., and Liang, E., *ApJ*, in preparation.

# The Expected Thermal Precursors of Gamma-Ray Bursts in the Internal Shock Model

Frédéric Daigne[*] and Robert Mochkovitch[†]

[*]CEA/DSM/DAPNIA, Service d'Astrophysique, C.E. Saclay, 91191 Gif sur Yvette Cedex, France
[†]Institut d'Astrophysique de Paris, 98 bis bd. Arago 75014 Paris, France

**Abstract.** The prompt emission of GRBs probably comes from a highly relativistic wind which converts its kinetic energy into radiation via the formation of shocks within the wind itself. Such "internal shocks" can occur if the wind is generated with a highly non uniform distribution of the Lorentz factor. We estimate the expected thermal emission of such a relativistic wind when it becomes transparent. We compare this emission to the emission produced by internal shocks. In most cases we predict a rather bright thermal emission that could easily be detected. This favors wind acceleration mechanisms where the energy is initially injected in the form of magnetic rather than internal energy. Such scenarios can produce thermal X-ray precursors comparable to those observed by GINGA and WATCH/GRANAT.

## THE PHOTOSPHERE OF A HIGHLY RELATIVISTIC WIND

The nature of the source responsible for the energy release leading to a GRB is not discussed here. We suppose that a relativistic wind carrying the energy has emerged with an average Lorentz factor $\bar{\Gamma} \geq 100$ and that the acceleration is complete at $r = r_{\text{acc}}$, where the energy injection rate is $\dot{E}(t)$ and the initial Lorentz factor is $\Gamma(t)$, for $t = 0$ to $t = t_{\text{w}}$. At $r = r_{\text{acc}}$, the wind is still optically thick. Photons emitted by a given layer must successively cross all the previously emitted layers before escaping from the relativistic wind. The optical depth $\tau(r)$ and the corresponding photosphere radius $r_{\text{ph}}$ are approximatively given by

$$\tau(r) \simeq \frac{\kappa \dot{E}}{8\pi c^3 \Gamma^3 r} \text{ and } r_{\text{ph}} \simeq \frac{\kappa \dot{E}}{8\pi c^3 \Gamma^3}, \quad (1)$$

where $\kappa$ is the Thomson opacity. Fig. 2 shows $r_{\text{ph}}$ for a wind lasting $t_{\text{w}} = 10$ s with $\dot{E} = 10^{52}$ erg/s and an initial Lorentz factor which is plotted in Fig. 1.

## SPECTRUM AND TIME PROFILE OF THE PHOTOSPHERIC EMISSION

In the framework of the fireball model (see e.g. Piran [1]), the luminosity and temperature of the photosphere

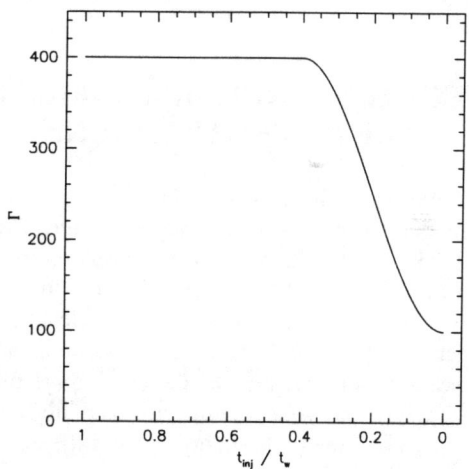

**FIGURE 1.** The initial Lorentz factor $\Gamma(t)$ used in our example as a function of the injection time $t_{\text{inj}}/t_{\text{w}}$.

are given by:

$$L_{\text{ph}} \simeq \dot{E}\left(\frac{r_{\text{ph}}}{r_{\text{acc}}}\right)^{-2/3},$$

$$kT_{\text{ph}} \simeq 1.3 \text{ MeV } \dot{E}_{52}^{1/4} \mu_1^{-1/2} \left(\frac{r_{\text{ph}}}{r_{\text{acc}}}\right)^{-2/3}, \quad (2)$$

where the mass of the central black hole is $\mu_1 10 \text{ M}_\odot$. Fig. 2 shows $L_{\text{ph}}$ and $kT_{\text{ph}}$ in our example. The spectrum $dn_{\text{ph}}(E)/dE$ and the count rate $C_{\text{ph}}^{12}$ in any energy

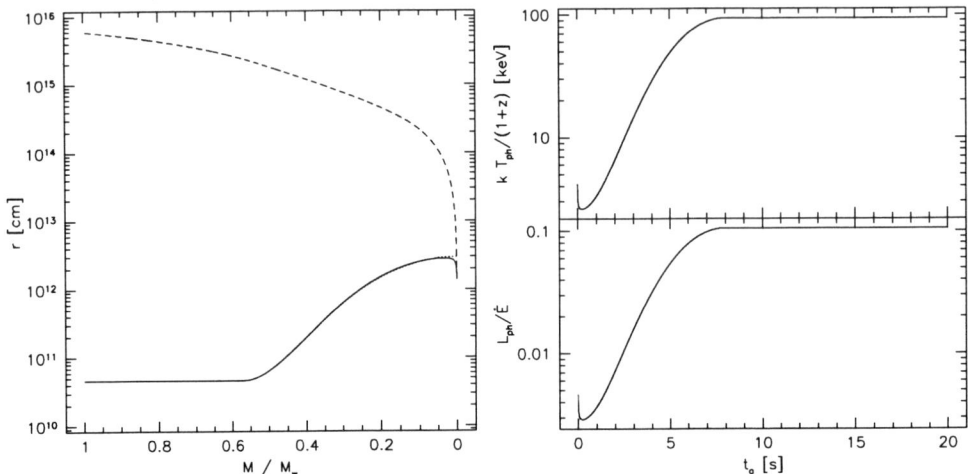

**FIGURE 2.** *Left:* The photosphere radius $r_{ph}$ (solid line) as a function of the mass coordinate $M/M_w$ in the wind ($M_w$ is the total mass). The dashed line shows the radius at which the photons emitted at the photosphere of each layer escape from the wind. *Right:* The photospheric temperature and luminosity as a function of the arrival time $t_a$ of photons. A redshift of $z=1$ is assumed.

band $[E_1; E_2]$ are computed assuming a Planck distribution for the photons. The result for our example is plotted in Fig. 3.

## SPECTRUM AND TIME PROFILE OF THE INTERNAL SHOCKS

The internal shock luminosity is $L_{IS} = f_\gamma \dot{E}$, where the efficiency $f_\gamma$ is the product of 3 terms : (i) the efficiency $f_d$ of the kinetic to internal energy conversion in shocks within the wind; (ii) the fraction $\alpha_e$ of the internal energy which is injected into non-thermal electrons; (iii) the fraction $f_{rad}$ of the electron energy which is radiated. The spectrum is represented by the usual GRB-function [2], with four parameters : the low and high energy slopes $\alpha$ and $\beta$, the peak energy $E_p$ and the amplitude (related to $L_{IS}$). This allows to compute the spectrum $dn_{IS}(E)/dE$ and the count rate $C_{IS}^{12}$. The result for our example is plotted in Fig. 3. The internal shocks have been simulated using the model developed in Daigne and Mochkovitch [3]. We have adopted $\alpha = -1.0$, $\beta = -2.25$ and $E_p = 200$ keV which are the typical values observed by Preece et al. [4].

Notice how the photospheric thermal emission is dominant even in the $\gamma$-ray range, in contradiction with the observations. In the next section, we show that this feature is related to the moderate efficiency of internal shocks and we propose a possible solution which decreases the photospheric emission and leaves a dominant non-thermal contribution from internal shocks.

## DISCUSSION

The ratio $R_{12} = C_{ph}^{12}/C_{IS}^{12}$ of the photospheric over internal shock count rates is given by [5] :

$$R_{12} = 1.6 \, f_{\gamma \, 0.1}^{-1} \dot{E}_{52}^{-1/4} \mu_1^{1/2} \frac{(1+z)E_p}{200 \text{ keV}}$$
$$\times \frac{\mathcal{I}_{Band}}{\mathcal{I}_{Planck}} \frac{\int_{x_1}^{x_2} \frac{x^2 \, dx}{\exp x - 1}}{\int_{x_1}^{x_2} f_{Band}(x) dx} \, . \quad (3)$$

As in our example, this ratio is usually greater than 1 in the X- and $\gamma$-ray bands so that the thermal emission dominates. This is in disagreement with the observations showing a non-thermal spectrum. Eq. 3 suggests 2 solutions :

**(i) Increase of $f_\gamma$** : however, even with extreme values of $\alpha_e \simeq 1$ and $f_{rad} \simeq 1$, $f_\gamma$ will always be limited by the dissipation efficiency which cannot exceed $f_d \simeq 0.1 - 0.4$. This is insufficient to avoid a well detectable thermal emission.

**(ii) Decrease of $L_{ph}$ and $kT_{ph}$** : the only remaining solution to recover non-thermal bursts is to assume a less hot and luminous photosphere. We define $\lambda$ as the fraction of the energy released by the source which is initially in the form of internal energy (the standard fireball model corresponds to $\lambda = 1$). Then the photospheric luminosity and temperature are scaled respectively by $\lambda$ and $\lambda^{3/4}$ so that the ratio $R_{12}$ given by Eq. 3 is now multiplied by $\lambda^{3/4}$. Small values of $\lambda$ can be expected, if, for instance, a large fraction of the energy released by the source is initially in magnetic form. Fig. 4 shows the new spectrum and profiles obtained for our example when $\lambda = 0.01$ is adopted. The photospheric emission is

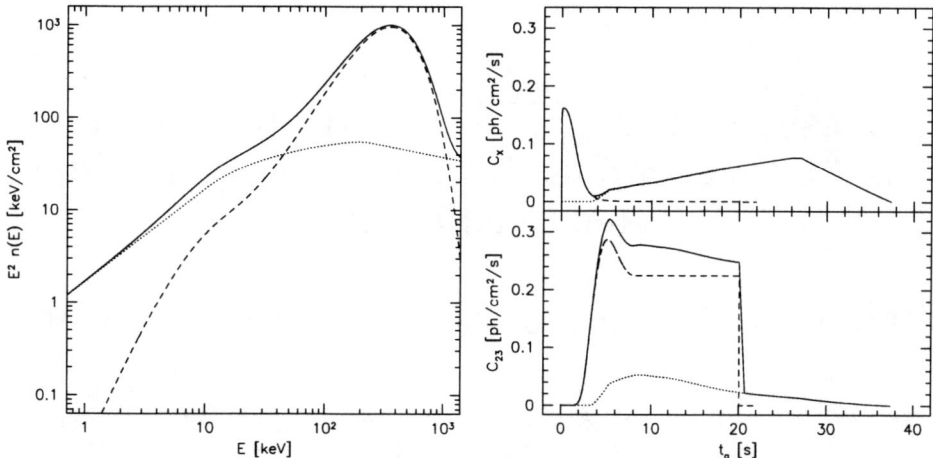

**FIGURE 3.** *Left:* $E^2 n(E)$ is plotted as a function of the photon energy in keV (thick line). The spectrum is dominated by the photospheric thermal emission (dashed line). The non-thermal emission from the internal shocks is also plotted (thin line). *Right:* The count rate of the total (thick line), photospheric (dashed line) and internal shocks (thin line) emission are plotted as a function of the arrival time $t_a$. *Top:* X-ray profile (3.5–8.5 keV); *Bottom:* $\gamma$-ray profile (50–300 keV). A redshift of $z = 1$ is assumed.

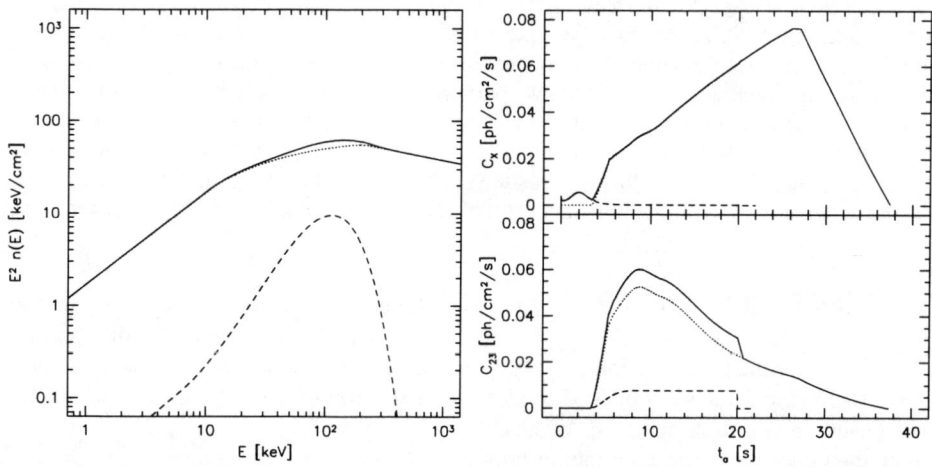

**FIGURE 4.** Same as in Fig. 3 when the photosphere is less hot and luminous ($\lambda = 0.01$).

now only detectable in the X-ray band at the very beginning of the burst. This could correspond to the X-ray precursors observed by GINGA [6] and WATCH/GRANAT [7] in several bursts. This last point, as well as the other results presented in this contribution, will be discussed in more details in a forthcoming paper [5].

## ACKNOWLEDGMENTS

F.D. acknowledges financial support from a postdoctoral fellowship from the French Spatial Agency (CNES).

## REFERENCES

1. Piran, T., *Physics Reports*, **314**, 575 (1999).
2. Band, D., Matteson, J., Ford, L., Schaefer, B., Palmer, D., Teegarden, B., Cline, T., Briggs, M., Paciesas, W., Pendleton, G., Fishman, G., Kouveliotou, C., Meegan, C., Wilson, R., and Lestrade, P., *ApJ*, **413**, 281 (1993).
3. Daigne, F., and Mochkovitch, R., *MNRAS*, **296**, 275 (1998).
4. Preece, R. D., Briggs, M. S., Mallozzi, R. S., Pendleton, G. N., Paciesas, W. S., and Band, D. L., *ApJS*, **126**, 19 (2000).
5. Daigne, F., and Mochkovitch, R., *to be submitted to MNRAS* (2001).
6. Murakami, T., Inoue, H., Nishimura, J., van Paradijs, J., and Fenimore, E. E., *Nature*, **350**, 592 (1991).
7. Sazonov, S. Y., Sunyaev, R. A., Terekhov, O. V., Lund, N., Brandt, S., and Castro-Tirado, A. J., *A&ASS*, **129**, 1 (1998).

# The Self-Consistent keV to TeV Spectra of Gamma-Ray Bursts Produced by the Synchrotron-Self-Compton Emission in Relativistic Shocks

E.V. Derishev[*], V.V. Kocharovsky[†], Vl.V. Kocharovsky[**] and P. Mészáros[‡]

[*]*MPI für Kernphysik, Saupfercheckweg 1, D-69117 Heidelberg, Germany*
[†]*Institute of Applied Physics, Russian Academy of Science 46 Ulyanov st., 603950 Nizhny Novgorod, Russia*
[**]*Dept. of Physics, Texas A&M University, College Station, TX 77843-4242*
[‡]*Dept. of Astronomy and Astrophysics, Pennsylvania State University, University Park, PA 16803*

**Abstract.**
The spectra of Gamma-Ray Burst (GRB) emission have been investigated theoretically by a number of authors (e.g., [1, 2, 3]) for different versions of a general fireball-shock scenario [4, 5]. However, the existing models give rather coarse predictions and suffer from many uncertainties. Here we present a refined analytical model of synchrotron-self-Compton (SSC) emission. We take an arbitrary injection spectrum of relativistic particles and let them cool self-consistently, taking into account both synchrotron and inverse Compton (IC) losses and making corrections for the Klein-Nishina cut-off. When two-photon absorption is negligible, the problem is reduced to the single integral equation for the ratio of synchrotron to total losses as a function of particle's energy. The spectra of synchrotron and IC components, as well as the electron distribution, can be derived from this function in a regular way. The low-energy portion of the composite synchrotron spectrum in the fast cooling regime may have spectral index varying from 1/2 to 1, and the spectrum of the IC component is nearly flat around its maximum.

## THE EQUATION FOR SSC COOLING

The simplest yet self-consistent description of GRB emission processes within the framework of SSC model is the following. Given an injection spectrum of electrons, we calculate their cooling distribution taking into account both synchrotron and IC losses, which can be calculated provided the synchrotron spectrum is known. The IC spectrum is obtained by means of perturbation theory. The subsequent Comptonization cascades are negligible in the GRB emitting regions [6].

Let us introduce a function $\eta(\gamma)$, which is the ratio of synchrotron luminosity $\mathcal{L}_s$ to the total luminosity $\mathcal{L}_s + \mathcal{L}_{ic}$ (including Compton losses $\mathcal{L}_{ic}$) for an electron with the Lorentz factor $\gamma$, so that

$$1/\eta = 1 + \mathcal{L}_{ic}/\mathcal{L}_s. \quad (1)$$

By definition, $0 < \eta < 1$. The synchrotron and IC luminosity of a single electron are

$$\mathcal{L}_s = \frac{4}{3}\sigma_T \gamma^2 c \frac{B^2}{8\pi}$$
$$\mathcal{L}_{ic} = \frac{4}{3}\gamma^2 c \int_0^\infty \sigma(\omega) w_{sy,\omega} d\omega, \quad (2)$$

where $\omega$ is the photon frequency and $w_{sy,\omega}$ the energy density of the synchrotron radiation per unit frequency interval. Below we assume that synchrotron radiation of an electron is monochromatic with the frequency $\omega(\gamma) = 0.5\gamma^2 \frac{eB}{m_e c}$ and $\sigma(\omega) = \sigma_T \Theta(\omega_* - \omega)$, where $\Theta$ is the step function and $\omega_* = m_e c^2/(\hbar\gamma)$.

The integral in Eq. (2) is then reduced to the product of $\sigma_T$ and the energy density of synchrotron radiation emitted by all electrons with a Lorentz factor less than $\gamma_*$, where $\gamma_*$ satisfies the equation $\omega(\gamma_*) = \omega_*$. If the electron distribution over their Lorentz factors is stationary, it can be found as

$$N_e(\gamma) = \frac{S(\gamma)}{\dot{\gamma}} = \frac{S(\gamma) m_e c^2}{\mathcal{L}_s + \mathcal{L}_{ic}} \propto \frac{\eta(\gamma) S(\gamma)}{\gamma^2}, \quad (3)$$

where $S(\gamma)$ is the electron injection rate integrated over $\gamma' \geq \gamma$. We represent this function as follows: $S(\gamma) = S_e p(\gamma); p(1) = 1$, so that $S_e$ is the total electron injection rate. Now we can find the energy density of synchrotron radiation below the Klein-Nishina cut-off (for an electron with the Lorentz factor $\gamma$):

$$w_{sy}(\omega < \omega_*) = \frac{\tau S_e m_e c^2}{V} \int_1^{\gamma_*} p(\gamma') \eta(\gamma') d\gamma'. \quad (4)$$

Here $\tau$ is the variability timescale and $V$ the volume of the emitting region. Since the electrons are injected with the average energy $\bar{\gamma} m_e c^2$ ($\bar{\gamma} = \gamma_i$ for monoenergetic injection) and almost all this energy is eventually radiated, we can substitute $\tau S_e m_e c^2$ with $E_r/\bar{\gamma}$, where $E_r \sim L_{\text{peak}} \tau$ is the energy radiated in one pulse (it includes the whole lightcurve in the case of external shock model).

For convenience, we introduce a new variable $x \equiv \gamma/\gamma_0$, where $\gamma_0$ satisfies the relation $\gamma_*(\gamma_0) = \gamma_0$, i.e.,

$$\gamma_0 = \left(\frac{2 m_e^2 c^3}{\hbar e B}\right)^{1/3}. \quad (5)$$

For electrons with the Lorentz factor $\gamma_0$ the Klein-Nishina cut-off is at their own synchrotron frequency. In GRBs $1 \ll \gamma_0 < \gamma_p$, where $\gamma_p$ is the Lorentz factor of electrons responsible for the emission at the peak of the spectrum. Finally, we obtain from relation (1) the following equation for $\eta(x)$:

$$\frac{1}{\eta(x)} = 1 + \mathcal{K} \int_0^{1/\sqrt{x}} p(x') \eta(x') \, dx' \quad (6)$$

For $x \ll x_p^{-2}$ one has $\eta = (1 + \tau_{ic})^{-1}$, where $\tau_{ic} \simeq 4/3 \gamma_p^2 \sigma_T n_e L$ and $L$ is the size of the emitting region. The Compton dominance parameter $\mathcal{K}$ is defined as

$$\mathcal{K} \equiv \frac{E_r \gamma_0}{W_m \bar{\gamma}}. \quad (7)$$

Here $\varepsilon_p$ is photon energy at the maximum of synchrotron spectrum (in the comoving frame), $W_m = (B^2/8\pi)V$ the total energy of magnetic field in the emitting region. The weighted average of $\eta$, $\bar{\eta} = \int p \eta \, d\gamma/\bar{\gamma}$, shows what fraction of the total energy radiated by GRB concentrates in the synchrotron (sub-MeV) spectral domain. In GRBs, the parameter $\mathcal{K}$ is within $0.1 \lesssim \mathcal{K} \lesssim 100$, but in a typical burst it is of the order of unity, no matter whether it is produced by internal or external shocks.

## ASYMPTOTIC SOLUTIONS

Let us consider the most illustrative case of monoenergetic injection at $x_i = x_p$. In the absence of analytical solution, it is still possible to use Eq. (6) to set lower and upper limits on $\eta(x)$:

$$\eta(x) > \frac{1}{1 + K/\sqrt{x}}, \quad (8)$$

which is based on the inequality $\eta < 1$, and

$$\eta(x) < \left[1 + \frac{K}{\sqrt{x}} - \frac{2K^2}{x^{1/4}} + 2K^3 \ln\left(1 + \frac{1}{K x^{1/4}}\right)\right]^{-1}, \quad (9)$$

which utilizes Eq. (8). This procedure may be repeated again, starting from Eq. (9) as an upper limit, and so on.

Equations (8) and (9) also give a good approximation of the function $\eta(x)$ for certain values of $x$. Let us introduce the reference point $x_{1/2}$ so that $\eta(x_{1/2}) = 1/2$. Then Eq. (8) represents the asymptote for $x \ll 1/x_{1/2}^2$ and Eq. (9) gives the asymptote for $x \gg x_{1/2}^4$. The latter may be reduced to

$$\eta(x) \simeq \left[1 + \frac{2}{3 x^{3/4}}\right]^{-1} \quad (10)$$

for $x^{1/4} \gg 1/\mathcal{K}$. Both asymptotes join more or less smoothly at $x_{1/2}$ if $x_{1/2} < 1$, so that Eqs. (8) and (9) give nearly complete knowledge about the function $\eta(x)$. Otherwise, there is an intermediate region where both approximations are unsatisfactory.

An important characteristic of GRBs is the relaxation parameter. It is equal to the ratio of GRB variability timescale $\tau$ to the synchrotron cooling time for an electron with the Lorentz factor $\gamma_p$:

$$R = \frac{\Gamma \tau}{m_e c} \gamma_p \sigma_T \frac{B^2}{6\pi}. \quad (11)$$

This parameter determines the length of quasistationary part of the electron distribution and the time it takes to settle the distribution.

Equation (6) makes sense only if the electron distribution can be considered stationary in its most important part – at $x \sim x_p$ – which is responsible for the bulk synchrotron emission. For a group of electrons with a particular $x$, this requires the cooling timescale to be less than GRB variability timescale $\tau$ as well as the inverse Compton cooling rate to be settled down. Since it is the synchrotron emission of electrons with $x \sim 1/\sqrt{x_p}$ that determines the IC losses of the electrons with $x \sim x_p$, the quasistationary part of the distribution (if present) must extend from $x_p$ down to at least $1/\sqrt{x_p}$. Large enough relaxation parameter ($\mathcal{R} > x_p^{3/2}$) guarantees stationary state for $1/\sqrt{x_p} \lesssim x \lesssim x_p$, although this part of the electron distribution function may come to equilibrium much faster due to possible dominance of IC losses at the lower end of the interval.

## THE SPECTRA

The GRB sub-MeV spectrum is just blueshifted synchrotron spectrum, and is actually given by $\eta(x)$. Indeed, the luminosity (in the shock comoving frame) of electrons with the Lorentz factors between $\gamma$ and $\gamma + d\gamma$ is $dI = \eta(x) S(x) m_e c^2 d\gamma$ and $d\gamma = (d\gamma/d\omega) d\omega$. As a result,

$$I_\omega \propto \frac{\eta(x) S(x)}{\sqrt{\omega}}, \quad x = \left(\frac{2 \hbar^2}{m_e c^3 e B}\right)^{1/6} \sqrt{\omega}. \quad (12)$$

In particular, Eq. (12) gives the following power-law asymptotes:

$$\omega I_\omega \propto \omega^{3/4} \quad (13)$$

for $1/x_{max}^2 < x \ll \min[x_{1/2}, 1/x_{1/2}^2]$, and

$$\omega I_\omega \propto \omega^{1/2} \quad (14)$$

for $x < 1/x_{max}^2$ and $x \gg \max[x_{1/2}, x_{1/2}^4]$ if $\mathcal{K} \sim 1$.

The self-consistent synchrotron spectrum extends down to $\omega_r = \omega_p(x_r/x_p)^2$, $\omega_p = \varepsilon_p/\hbar$. Below this frequency $I_\omega$ reproduces the synchrotron spectrum of a single particle because the electron distribution is cut at $x = x_r$, which is the root of the equation $x_p\eta = x\mathcal{R}$ and marks the point where the cooling timescale becomes equal to the GRB variability timescale. It is supposed that $x_r < 1/\sqrt{x_p}$, otherwise the function $\eta$ should be derived anew in the non-stationary case.

The spectrum of comptonized radiation can be represented analytically as a convolution of synchrotron spectrum with the same spectrum but taken as a function of re-scaled frequency. However, some qualitative conclusions about the main features in the spectrum of IC radiation, such as location of the maximum and its shape, can be made on the basis of asymptote expression (10) even without analyzing the exact expression. We take a rather crude approximation assuming that the most energetic electrons produce monochromatic comptonized radiation with photon energy $\varepsilon \simeq \gamma m_e c^2/2$, that leads to a minor error as far as the resulting IC spectrum is softer than the spectrum of those synchrotron photons which undergo Comptonization. So, the hardest part of IC spectrum is well described by the following equation:

$$\varepsilon I_\varepsilon \propto (1-\eta)x \propto \varepsilon^{1/4}, \quad (15)$$

where $x = x_p\varepsilon/\varepsilon_{ic}$, $\varepsilon_{ic} = \gamma_p m_e c^2$, and Eq. (10) is used to substitute $(1-\eta)$.

## DISCUSSION

The presented theory describes two parts of the GRB spectrum – sub-keV to MeV radiation attributed to the synchrotron mechanism and ultra-energetic GeV to TeV radiation produced by the inverse Compton mechanism. In the soft part of the spectrum there is a break at the frequency (here and below in observer's frame) $\Gamma\omega_r \sim \Gamma\omega_p/\mathcal{R}^2$ where the low-energy asymptote of the synchrotron spectrum of a single particle replaces the stationary spectrum generated by a steady-state ensemble of electrons. The expression for stationary spectrum are applicable between $\Gamma\omega_r$ and $\Gamma\omega_p$.

Very broad maximum of the IC spectrum in the absence of self-absorption is expected at the energy $\Gamma\varepsilon_{ic}/2$.

The actual value is sensitive to GRB parameters which are not observed directly, so that $\Gamma\varepsilon_{ic}/2$ falls in the range 10 GeV – 100 TeV (the lower and upper bounds correspond to the internal shock model and the external shock model, respectively). The IC spectrum may be cut at an energy less than $\Gamma\varepsilon_{ic}/2$ if there is two-photon absorption internal to the source or due to the interaction with the infrared background radiation.

A significant result of our analysis is that the cooling distribution of electrons produces a broad range of spectral indices. The low energy portion of the synchrotron spectrum can be as steep as $I_\omega = const$. Still steeper spectra may be produced if there is additional to the self-consistent synchrotron emission source of soft photons.

We generalize Eq. (6) to take into account the non-synchrotron sources of soft radiation. The procedure is straightforward and yields an additional term in sub-integral function which now takes the form $q(x') + p(x')\eta(x')$. The contribution of external sources is

$$q(x) = \frac{I_\omega^{(ex)}(x)}{\gamma_0 S_e m_e c^2} \frac{d\omega}{dx}, \quad (16)$$

where $I_\omega^{(ex)}$ is their luminosity per unit frequency interval (in the fireball comoving frame). If, for example, the photospheric emission dominates GRB spectrum at low frequencies, then the ratio of IC losses to synchrotron ones will become more strongly dependent on the Lorentz factor for those electrons which have the Klein-Nishina cut-off in the Rayleigh-Jeans part of the photospheric spectrum. Being imprinted in the cooling distribution, such a dependence may give rise to a very steep portion in the synchrotron spectrum provided the low-energy asymptote in the spectrum of a single particle allows that (see, e.g., [7]). Numerical solutions show that the effect of photospheric emission on the IC losses causes significant steepening of low-energy synchrotron spectra.

Steep spectra with $\alpha > 1$ may be obtained in alternative ways (e.g., [8]), but this leads to low radiative efficiency of GRBs and raises concerns about the visibility of emission from fireball photosphere against the nonthermal component.

## REFERENCES

1. Sari, R., Piran, T., 1997, MN RAS 287, 110.
2. Ghisellini, G., Celotti, A., 1999, ApJ 511, L93.
3. Dermer, C.D., Chiang, J., Mitman, K.E., 2000, ApJ 537, 785.
4. Mészáros, P., Rees, M. J., 1992, MNRAS 257, P29
5. Mészáros, P., Rees, M. J., 1994, MNRAS 269, L41
6. E.V. Derishev, V.V. Kocharovsky, and Vl.V. Kocharovsky, 2001, A&A 372, 1071.
7. Medvedev, M.V., 2000, ApJ 540, 704.
8. Panaitescu, A., Mészáros, P., 2000, ApJ 544, L17

# Inverse Compton Model of Gamma-Ray Burst Spectra

E. Liang, M. Boettcher and D. Kocevski

*Rice University, Houston, TX 77005-1892*

**Abstract.** We discuss sample model spectra of GRBs generated by inverse Comptonization of soft photon sources using our state-of-the-art Monte Carlo code. Results using both soft blackbody sources and self-absorbed synchrotron sources are presented. Applications to BATSE GRB spectra with soft excesses are discussed.

## INVERSE COMPTON SPECTRA

Most GRB spectra in the BATSE energy range are well fit by the Band et al [1] empirical form with low energy power law index $\alpha$, high energy power law index $\beta$ and break energy $E_p$ corresponding to the peak energy of the $\nu F_\nu$ distribution. Many emission models have been proposed over the years to explain this broken power law shape, including optically thin synchrotron, self-absorbed synchrotron, small-pitch-angle synchrotron, cold Compton attenuation, saturated Comptonization, Fe-absorption, plasma absorption, free-free absorption etc. So far none of these models is completely consistent with BATSE spectra and spectral evolution data. Here we explore another mechanism: (unsaturated) inverse Comptonization of a soft photon source. This model is motivated by the fact that ~15% of BATSE spectra [2] show soft x-ray excess, in which the soft x-ray flux lies above the extrapolation of the Band et al power law from the hard x-rays, suggesting an additional soft component. In addition, the detection of strong simultaneous optical emission and soft-x-ray excess during GRB 990123 hints at a soft component extending from the x-rays to the optical. This soft component could then be naturally upscattered to the BATSE range in an emission region of significant Thomson depth, similar to the currently popular models of blazars. Hence it is reasonable to study the prospects of modeling the BATSE spectra as the Compton upscattering of a soft photon source and investigate the consequences of this model.

Using a state-of-the-art Monte Carlo code we have systematically modeled the inverse Compton spectra, using both a soft blackbody source (BBC) as well as a self-absorbed synchrotron source (SSC). The inverse-Comptonizing electrons are assumed to have a power law distribution with minimum Lorentz factor $\gamma_{min}$, maximum Lorentz factor $\gamma_{max}$, and power law index p.

## MONTE CARLO SIMULATIONS

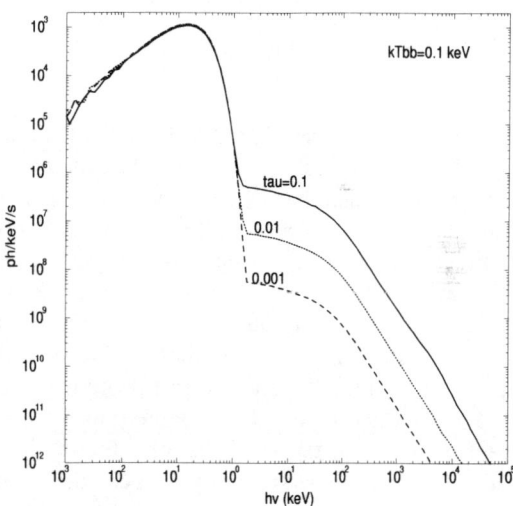

**FIGURE 1.** Inverse Compton spectra of a 0.1 keV blackbody source for sample Thomson depths. Here $\gamma_{min}$ =10 and $\gamma_{max}$=infinity. Electron power law index p=2.5.

Figure 1 shows sample Monte Carlo simulation outputs of inverse Comptonization of a $kT_{bb}$ = 0.1 keV blackbody source by relativistic electrons of various Thomson depths. In these blackbody Comptonization or BBC runs p=2.5, $\gamma_{min}$ = 10 and $\gamma_{max}$ = infinity. It is evident that the gamma-ray spectral break $E_p$ is given by $\gamma_{min}^2 \cdot kT_{bb}$, as expected. More importantly, the photon number index below the break, which would correspond to the Band et al $\alpha$ index, is only slightly < 0. In fact, we have examined the asymptotic behavior of α as $\gamma_{min}$ becomes large: it asymptotes towards 0 in the limit of large $\gamma_{min}$. This is shown in Fig. 2. Thus the low energy power law index of the inverse Compton

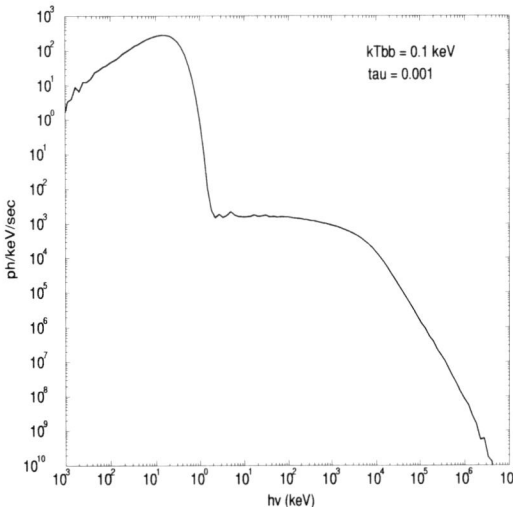

**FIGURE 2.** Same as Fig.1 except $\gamma_{min}$ =300. In this case the slope below $E_p$ approaches 0.

component is harder than that of optically thin synchrotron (α = -2/3), so that it is capable of fitting many more hard GRB spectra whose α's lie above the so-called "line of death" for optically thin synchrotron models [3, 4, 5]. However we have explored the entire BBC parameter space but found no case in which α can be > 0. We also find that the index α is insensitive to the slope of the soft photon source below its peak. Figure 3 shows the same results as Fig. 1 but in terms of the $\nu F_\nu$ distribution. Note the remarkable similarity of such spectra to those of blazars.

Figure 4 shows sample results in which the soft photon source is provided by the self-emitted synchrotron radiation of the same electron population (i.e. synchrotron self-Compton, or SSC model). Here the synchrotron spectrum is cut off below a certain frequency due to self-absorption. In these cases the Compton peak is related to the synchrotron peak by a factor of $\gamma_{min}^2$ as expected. For a steep electron power law the upscattered spectrum resembles those of the blackbody case, with α approaching 0. However, as we make the electron power law index harder and harder (lower p) or lower the

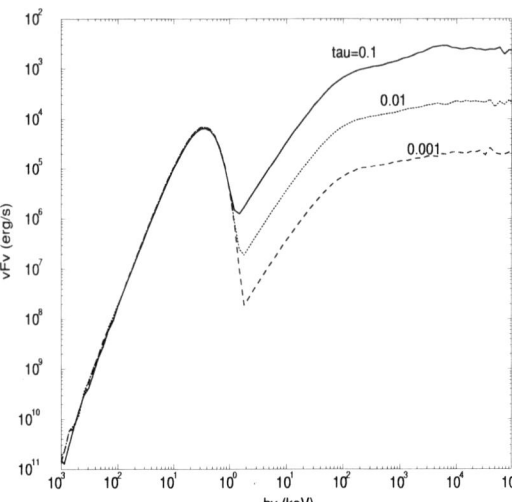

**FIGURE 3.** $\nu F_\nu$ plot of Fig.1. Note the resemblence to typical blazar spectra.

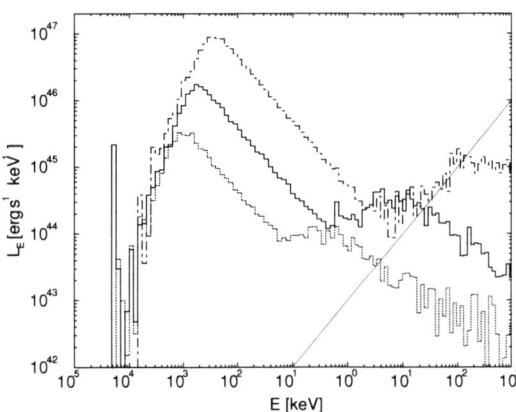

**FIGURE 4.** Sample SSC energy spectra with self-absorbed synchrotron sources. Emission parameters are listed at the top. Solid diagonal line corresponds to energy slope of +1 which corresponds to α =0

Thomson depth, the Compton peak becomes less and less prominent and the effective Band index α decreases. Figure 5 shows that as p drops to 2.5 and $\gamma_{min}$ =10, the entire spectrum approaches a single power law, with the inverse Compton peak basically buried under the superposed synchrotron and Compton power laws. Hence if the BATSE spectral data is to be successfully modeled by inverse Comptonization, we need a fairly large minimum $\gamma_{min}$ to separate the synchrotron and Compton peaks, and a steep electron power law index, say p≥ 3.

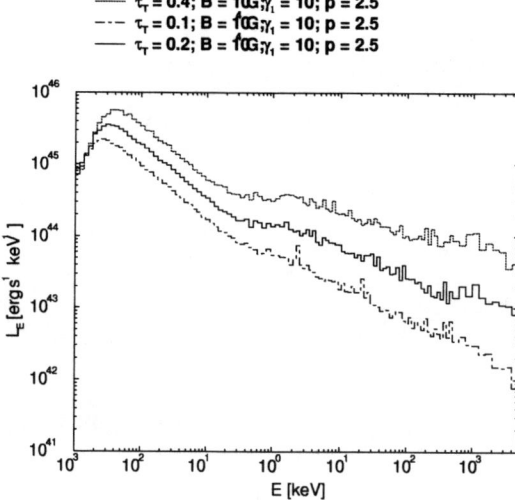

**FIGURE 5.** SSC spectra showing the gradual disappearance of $E_p$ as the Thomson depth is decreased, for sufficiently low p and $\gamma_{min}$.

## COMPARISON WITH BATSE DATA

Figure 6 shows one of the time-resolved $\nu F_\nu$ spectrum of GRB991216, showing evidence of soft x-ray excess. While the exact amplitude and slope of the soft x-ray excess remains uncertain due to detector response uncertainties at the lowest energies, there is little doubt that a significant fraction of BATSE bursts [2] show hints of soft x-ray excesses below ~30 keV. A casual comparison of Fig.6 with Fig.3 shows remarkable similarity. However, to use such data to constrain or confront the SSC or BBC models we will have to wait for future broadband experiments such as SWIFT, which can hopefully capture or at least tightly constrain the peak of the low energy x-ray/optical component. Only by coordinated studies of the two spectral components and their variability, as it has been done for blazars, can we hope to pin down the physical parameters and distinguish models.

## IMPLICATIONS FOR SPECTRAL EVOLUTION

The BBC and SSC models predict different spectral evolution behaviors. But in both cases a broadband spectrum can provide unique

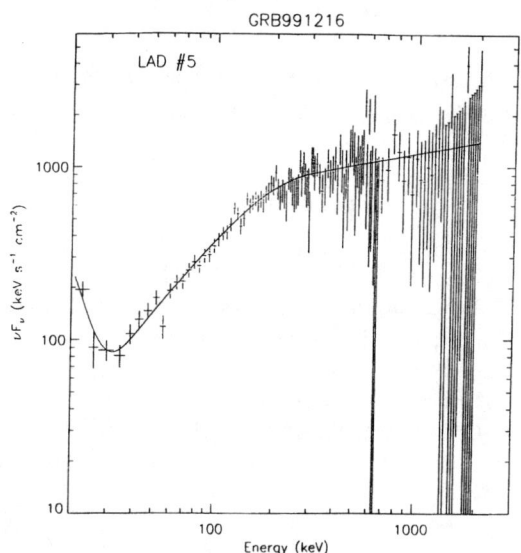

**FIGURE 6.** Time resolved $\nu F_\nu$ spectrum of GRB991216 showing an apparent soft x-ray excess. Note the similarity of this spectrum with Fig.3.

diagnostics of the emission parameters, provided that we can measure both spectral components in detail. Assuming that the soft photon source and the Comptonizing electrons are comoving, for BBC to first order we have:

$$E_p \sim kT_{bb} \cdot \gamma_{min}^2 \text{ and } L_c \sim L_{bb} \cdot \tau_T \cdot \gamma_{min}^2$$

where $L_{bb}$ is the blackbody luminosity, $\tau_T$ is the Thomson depth of the Comptonizing region and $L_c$ is the luminosity of the Compton component. So if we can measure $E_p$, $kT_{bb}$, $L_c$ and $L_{bb}$ from the spectrum, we can estimate independently $\gamma_{min}$ and $\tau_T$ and study how they evolve with time. We can then test if the evolution of $\gamma_{min}$ is consistent with Compton cooling [6].

On the other hand, for SSC we have:

$$E_p \sim E_a \gamma_{min}^2 \text{ and } L_c \sim L_{sa} \cdot \tau_T \cdot \gamma_{min}^2$$

where the synchrotron self-absorption energy $E_a \sim \tau_T^{2/(p+4)} \cdot \gamma_{min}^{(2p-2)/(p+4)} \cdot B^{(p+2)/(p+4)}$ and the self-absorbed synchrotron luminosity $L_{sa} \sim N \cdot \tau_T^{(1-p)/(p+4)} \gamma_{min}^{(5p-5)/(p+4)} B^{(2p+3)/(p+4)}$. Here N is the total number of synchrotron emitting leptons. In this case if we can measure $E_p$, $E_a$, $L_c$ and $L_{sa}$, we can uniquely determine the four emission parameters N, $\tau_T$, $\gamma_{min}$, B, and study their time evolution. We can then test if the evolution of $\gamma_{min}$ is consistent with SSC cooling [6]. Hence in principle detailed analysis of GRB spectral evolution may be used to test the BBC and SSC emission models.

## ACKNOWLEDGMENTS

This work was partially supported by NASA grants NAG5-7980, NAG 5-9223 and NGT 8-52899.

## REFERENCES

1. Band, D. et al., ApJ 413, 281 (1993).
2. Preece, R. et al., AIP Conf. Proc. 526, 175, ed. Kippen, R. M. et al. (AIP, NY, 2000).
3. Preece, R. et al., ApJ 496, 849 (1998).
4. Crider, A. et al., ApJ 479, L39 (1997).
5. Crider, A. et al., ApJ 519, 206 (1999).
6. Liang et al., contribution in this volume (2002).

# The Physics of Pulses in GRBs

Frédéric Daigne* and Robert Mochkovitch[†]

*CEA/DSM/DAPNIA, Service d'Astrophysique, C.E. Saclay, 91191 Gif sur Yvette Cedex, France
[†]Institut d'Astrophysique de Paris, 98 bis bd. Arago 75014 Paris, France

**Abstract.** The temporal and spectral properties of GRB pulses have been summarized in the "pulse paradigm" proposed by Norris et al [1]. Long pulses are asymmetric and reach their maximum earlier in higher energy bands while shorter pulses tend to be more symmetric with negligible time lags between energy channels. The decay phase can be fitted by an exponential (leading to the so-called FRED shape for the pulses) but in many cases the count rate is also well represented by a $1/(1+t/\tau)$ law. We have checked how the pulse paradigm is reproduced by the internal shock scenario using a simple phenomenological model to test different assumptions relative to the emission process during shocks. We show that the internal shock model can account for most of the burst temporal and spectral properties if some specific conditions are satisfied by the emission process.

## TEMPORAL AND SPECTRAL EVOLUTION DURING PULSE DECAY

During the decay phase a substantial fraction of GRB pulses satisfies both the hardness-intensity correlation (HIC) [2] and the hardness-photon fluence correlation (HFC) [3]. Combining the HIC and the HFC, Ryde & Svensson [4] have shown that the photon flux and the peak energy follow simple power laws

$$N(t) = \frac{N_0}{1+t/\tau} \quad \text{and} \quad E_p(t) = \frac{E_{p,0}}{(1+t/\tau)^\delta} \quad (1)$$

It is also possible to relate $N(t)$ and $E_p(t)$ to the energy flux $F_E(t)$

$$F_E(t) = N(t)E_p(t)\frac{\phi_0}{\varphi(E_p)} = \frac{N_0 E_{p,0}}{(1+t/\tau)^{1+\delta}}\frac{\phi_0}{\varphi(E_p)} \quad (2)$$

with

$$\varphi(E_p) = \int_{E_1/E_p}^{E_2/E_p} \mathcal{B}(x)dx \quad \text{and} \quad \phi_0 = \int_0^\infty x\mathcal{B}(x)dx \quad (3)$$

$\mathcal{B}(x)$ being the Band function. We have used Eq.(2) to obtain the evolution of $F_E$ for 14 pulses belonging to the sample studied by Ryde & Svensson [4] and which satisfy both the HIC and the HFC during the decay phase. To compute $\varphi(E_p)$ we adopted standard low and high energy indices $\alpha = -1$ and $\beta = -2.5$ in the Band function. We find that $F_E$ also follows a power law

$$F_E(t) \propto \frac{1}{(1+t/\tau)^\varepsilon} \quad (4)$$

the value of $\varepsilon$ depending only weakly on the choice made for $\alpha$ and $\beta$.

Since $\varphi(E_p)$ is a decreasing function of time during pulse decay there is a shift between the distributions of $\varepsilon$ and $1+\delta$ for the 14 pulses ($\langle\varepsilon\rangle = 1.5$ while $\langle 1+\delta\rangle = 1.81$). But the most striking difference is that $\varepsilon$ is much less scattered than $(1+\delta)$ with a mean dispersion $\sigma_\varepsilon = 0.17$ instead of $\sigma_\delta = 0.28$.

We then tried to develop a pulse model which could be able to reproduce the temporal and spectral behavior illustrated by Eqs (1) and (4).

## A SIMPLE PULSE MODEL

We consider a relativistic wind where a slow shell of mass $M_0$ and Lorentz factor $\Gamma_0$ decelerates a more rapid part of the flow characterized by a constant mass flux $\dot{M}$ (in the source frame) and Lorentz factor $\Gamma_1 > \Gamma_0$. We do not solve the true hydrodynamical problem but rather approximate the flow evolution by assuming that fast material is "accreted" by the slow shell at the rate

$$\frac{dM}{dt} = \dot{M}(1-\gamma^2) \quad (5)$$

where $t$ is the observer time and $\gamma = \Gamma/\Gamma_1$ ($\Gamma$ and $M$ being the current Lorentz factor and mass of the slow shell). Due to the accretion of fast moving material, the Lorentz factor of the slow shell increases as

$$\frac{d\gamma}{dM} = \frac{1-\gamma^2}{2M} \quad (6)$$

which can be integrated to give

$$\mu = \left(\frac{1+\gamma}{1-\gamma}\right) / \left(\frac{1+\gamma_0}{1-\gamma_0}\right) \quad (7)$$

where $\mu = M/M_0$ and $\gamma_0 = \Gamma_0/\Gamma_1$. Introducing $t_0 = M_0/\dot{M}$ and $\tau = t/t_0$, Eqs (5 – 7) yield

$$\frac{d\gamma}{d\tau} = Q(1-\gamma^2)(1-\gamma)^2 \quad (8)$$

with $Q = \frac{1}{2}\left(\frac{1+\gamma_0}{1-\gamma_0}\right)$. Equation (8) has the analytical solution

$$\tau = [F(\gamma) - F(\gamma_0)]/Q \quad (9)$$

where the function $F(\gamma)$ is given by

$$F(\gamma) = \frac{1}{8}\text{Log}\left(\frac{1+\gamma}{1-\gamma}\right) + \frac{1}{4(1-\gamma)} + \frac{1}{4(1-\gamma)^2} \quad (10)$$

When $\tau \gtrsim 2$, the solution for $\gamma(\tau)$ is well approximated by

$$\gamma(\tau) \simeq 1 - \frac{1}{2\sqrt{Q\tau}} \quad (11)$$

Once $\gamma(\tau)$ is known it is possible to calculate the dissipated power

$$\dot{E}(\tau) = \frac{\dot{M}\Gamma_1 c^2}{2}(1-\gamma^2)(1-\gamma)^2 \quad (12)$$

At large $\tau$, it behaves as $\tau^{-3/2}$ since

$$\dot{E}(\tau) \propto (1-\gamma)^3(1+\gamma) \propto \tau^{-3/2} \quad (13)$$

for $\tau \gtrsim 2$. The dissipated power given by Eq.(12) is however different from what the observer will see since the energy released at a given time $t$ is spread over an interval $\Delta t$ corresponding to the difference in arrival time for photons emitted from a shell of radius $r$ moving at a Lorentz factor $\Gamma$

$$\Delta t = \frac{r}{2c\Gamma^2} \quad (14)$$

The solution for $\dot{E}$ including angular spreading has been obtained numerically. It differs from the analytical expression (12) at early times but preserves the power law decay at late times. We believe that the $-3/2$ slope, which is equal to the average value of $\varepsilon$ obtained in the last section suggests that GRB pulses result from the deceleration of fast moving material by comparatively slower layers in relativistic outflows. In our approach, internal shocks take place in a continuous medium and not among a system of discrete thin shells. The hydrodynamical timescale is then larger than the angular spreading time and controls the pulse shape in the decay phase.

# EMISSION PROCESSES AND SPECTRAL EVOLUTION

We now use the analytical model to follow the spectral evolution during pulse decay. If the dissipated energy is radiated by the synchrotron process the peak energy $E_p$ is

$$E_p = E_{\text{syn}} \propto B\Gamma_e^2\Gamma \quad (15)$$

where $B$ is the magnetic field and $\Gamma_e$ the characteristic electron Lorentz factor behind the shock. With classical equipartition assumptions $B$ and $\Gamma_e$ can be expressed as

$$B = (8\pi\alpha_B n\varepsilon)^{1/2} \quad \text{and} \quad \Gamma_e = \frac{\alpha_e}{\zeta}\frac{m_p}{m_e}\varepsilon \quad (16)$$

where $n$ is the baryon density and $\varepsilon c^2$ the dissipated energy per unit mass (in the comoving frame); $\alpha_B$ and $\alpha_e$ are the equipartition parameters and $\zeta$ is the fraction of electrons which is accelerated. Finally,

$$E_{\text{syn}} \propto n^{1/2}\varepsilon^{5/2}\Gamma \quad (17)$$

where the density $n$ is proportional to $r^{-2}$ ($r$ being the shock radius $r \sim \Gamma^2 ct$) and $\varepsilon$ is given by

$$\varepsilon = \frac{(1-\gamma)^2}{2\gamma} \quad (18)$$

This leads to the following expression for $E_{\text{syn}}$

$$E_{\text{syn}} \propto \frac{(1-\gamma)^5}{\gamma^{7/2}t} \quad (19)$$

which behaves as a power law ($E_p \propto t^{-7/2}$) when $(1-\gamma) \sim t^{-1/2}$. This is much steeper than the observed spectral evolution since the slope $\delta$ of the HIC is always smaller than 1.5 in our pulse sample. Instead of using Eq.(17) we therefore parametrize the peak energy with the more general expression

$$E_p \propto n^x \varepsilon^y \Gamma \propto \frac{(1-\gamma)^{2y}}{\gamma^{4x+y-1}t^{2x}} \quad (20)$$

which becomes $E_p \propto \frac{1}{t^{2x+y}}$ at late times. The exponents $x$ and $y$ can be different from their standard synchrotron values $1/2$ and $5/2$ if the equipartition parameters $\alpha_B$, $\alpha_e$ or $\zeta$ vary with $n$ or $\varepsilon$. For example, Daigne & Mochkovitch [5] adopted a fraction $\zeta$ of accelerated electrons proportional to $\varepsilon$ so that $\Gamma_e$ remains constant which leads to $x = y = 0.5$ and $E_p \propto t^{-3/2}$. The observed values of $\delta$ however require even more drastic changes in the assumptions concerning the emission process (which must satisfy $2x+y \lesssim 1$) and we have therefore considered below the case $x = y = 0.25$ ($\delta \simeq 2x+y = 0.75$).

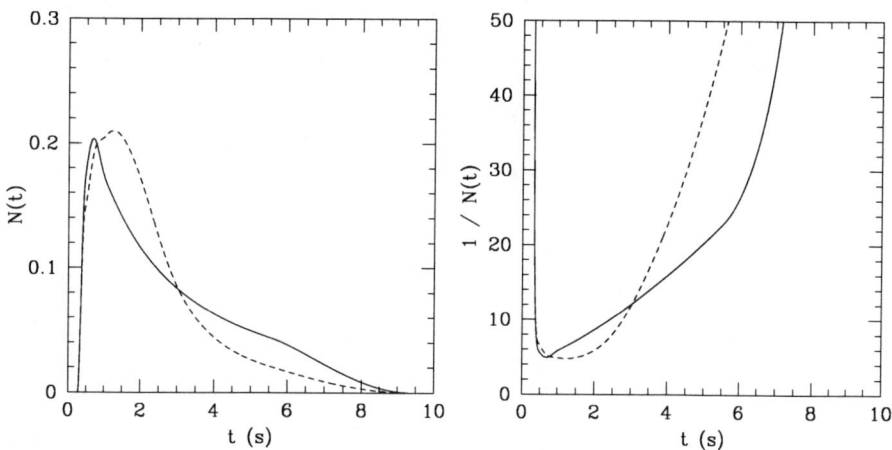

**FIGURE 1.** Pulse profiles produced by a relativistic outflow decelerated by a slower shell (see text for the other model parameters). dashed line: case (1) with $x = y = 0.5$ in Eq.(20); full line: case (2) with $x = y = 0.25$. The left panel shows the photon flux $N(t)$ and the right panel $1/N(t)$.

## TEMPORAL PROFILES

Temporal profiles can be obtained from Eqs (1), (12) and (20) of our model. We have represented in Fig.1 two pulses obtained with $\Gamma_0 = 100$, $\Gamma_1 = 400$, $\dot{M}\Gamma_1 c^2 = 3\,10^{52}$ erg.s$^{-1}$, $t_0 = 0.8$ s and two choices for the exponents $x$ and $y$: $x = y = 0.5$ (case 1) and $x = y = 0.25$ (case 2). It can be seen that, for the same initial distribution of the Lorentz factor, cases (1) and (2) lead to quite different pulse profiles. The rise time appears to be much shorter in case (2) than in case (1). A short rise time implies that (i) the energy is initially released at a high rate but also that (ii) it is mostly radiated in the spectral range of the detector. With the adopted distribution of the Lorentz factor the dissipated power $\dot{E}$ is maximum at $t = 0$ (see Eq.(12)) and condition (i) is satisfied. However, in case (1) where $E_p \propto n^{1/2}$ the value of the peak energy lies beyond the BATSE range during the early evolution when the density at the shock radius is high. In case (2), the increase of $E_p$ at high density remains moderate (since $E_p \propto n^{1/4}$) and condition (ii) is also satisfied, leading to a short rise time.

Another remarkable difference between cases (1) and (2) is illustrated in the right panel of Fig.1 where $1/N(t)$ has been represented as a function of time. Case (2) exhibits an extensive linear portion lasting about 5 seconds after maximum which corresponds to a $1/t$ decay phase while no such trend is visible in case (1). Reducing the values of $x$ and $y$ so that $2x + y \lesssim 1$ therefore improves the temporal and spectral evolution of our synthetic pulses. In case (2), the decay phase is well decribed by Eq.(1) (with $\delta = 0.75$) and then follows both the HIC and the HFC. We checked that similar results can be obtained with other choices of $x$ and $y$ as long as $2x + y \lesssim 1$.

## CONCLUSION

We have developed a simple model where GRB pulses are produced when a rapid part of a relativistic outflow is decelerated by a comparatively slower shell. We obtain an analytical solution for the dissipated power $\dot{E}$ showing that during pulse decay $\dot{E} \propto t^{-3/2}$ in agreement with the average behavior of 14 BATSE pulses which satisfy both the HIC and the HFC. To compute the spectral evolution of our synthetic pulses we parametrize the peak energy as $E_p \propto n^x \varepsilon^y \Gamma$. At late times, we get $E_p \propto t^{-(2x+y)}$ which constrains $x$ and $y$ since in observed bursts $E_p \propto t^{-\delta}$ with $\delta \lesssim 1$. The synchrotron process with standard equipartition assumptions corresponds to $x = 0.5$ and $y = 2.5$ ($2x + y = 3.5$) and gives a much too steep spectral evolution. One has to suppose that the equipartion parameters vary with $n$ and $\varepsilon$ to reduce $x$ and $y$. We considered the case $x = y = 0.25$ and the resulting pulses then follow both the HIC and the HFC during the decay phase. It therefore appears that relaxing the usual equipartition assumptions to decrease $x$ and $y$ below their usual values 0.5 and 2.5 strongly improves the temporal and spectral evolution of synthetic bursts.

## REFERENCES

1. Norris, J. P., Nemiroff, R. J., Bonnell, J. T., Scargle, J. D., Kouveliotou, C., Paciesas, W. S., Meegan, C. A., and Fishman, G. J., , **459**, 393+ (1996).
2. Golenetskii, S. V., Mazets, E. P., Aptekar, R. L., and Ilinskii, V. N., , **306**, 451–453 (1983).
3. Liang, E., and Kargatis, V., , **381**, 49–51 (1996).
4. Ryde, F., and Svensson, R., , **529**, L13–L16 (2000).
5. Daigne, F., and Mochkovitch, R., , **296**, 275–286 (1998).

# IV. AFTERGLOWS OF GAMMA-RAY BURSTS

## Afterglow Theory

## Radio, Infrared, and Optical Observations

## X-Ray Afterglows

# Properties of Gamma-Ray Burst Jets Obtained from Afterglow Modelling

A. Panaitescu* and P. Kumar[†]

*Dept. of Astrophysical Sciences, Princeton University, Princeton, NJ 08544*
[†]*Institute for Advanced Study, Olden Lane, Princeton, NJ 08540*

**Abstract.**
We model the broadband emission and determine the basic characteristics of eight GRB afterglows whose features allow the determination of the burst collimation. The resulting post-GRB jet energies are consistent with a universal value of $3 \times 10^{50}$ erg. The jet initial apertures are in the 2-14 degree range, half of the jets being narrower than 3 degrees. We find that homogeneous circumburst media are consistent with the emission of all these afterglows while, with a couple of exceptions, wind-like density profiles are not. The external densities we find are typically in the 0.1-50 particles per cc range. For all eight afterglows the total energy in shock-accelerated electrons is close to equipartition with protons. However the slope of the power-law electron distribution is not universal, varying between 1.4 and 2.8. From the jet parameters in the afterglow phase we infer initial bulk Lorentz factors between 70 and 300. Our results on the jet energy, opening and Lorentz factor provide constraints on theoretical models of GRB jets.

## INTRODUCTION

Today, the leading model for the GRB progenitor focuses on the black hole-debris torus system resulting either when a massive star undergoes core collapse ([1],[2]) or when the components of a binary system containing one or two compact objects merge ([3],[4]). The outflow is ejected along the rotation axis of the black hole, where the baryonic contamination is the lowest. The kinetic energy of the resulting jet can be converted into radiation either after its dissipation in *internal shocks* if the flow is unsteady ([5],[6]), or in the *external shock* caused by the interaction with the circumburst medium ([7]). The latter is also the mechanism at work in the ensuing afterglow, whose emission is detected sometimes up to more than one hundred days.

The afterglow flux at a given frequency and time depends on the jet dynamics, the local density of the external medium, and the particle acceleration and magnetic field generation in relativistic shocks. Their properties can be constrained by modelling the observed broadband (radio to X-ray) afterglow emission, providing valuable clues in identifying the type of GRB progenitor. Highly collimated ejecta interacting with a medium whose density decreases as $r^{-2}$ (i.e. a pre-ejected wind) are the signatures of a collapsing massive star. A more tenuous and homogeneous circumburst medium and, perhaps, less collimated ejecta point toward the merging of two compact objects.

Due to the highly relativistic motion of the GRB source, during the γ-ray phase and the early afterglow we receive emission from only a small region of the jet surface, around the line of sight toward its center. The GRB/afterglow emission at these stages depend on the jet energy per solid angle, thus early observations do not allow us to determine the outflow's collimation and its energy. As the jet is decelerated by the sweeping-up of the circumburst medium, the angular spread of the region whose radiation is relativistically beamed toward the observer increases. The observer "sees" the jet edge when its Lorentz factor falls below the inverse of the jet aperture, which takes place at

$$t_j = 0.4 \, (z+1) \left(E_{0,50} n_0^{-1}\right)^{1/3} \theta_{0,-1}^2 \, \text{day} , \quad (1)$$

$z$ being the burst redshift, $E_{0,50}$ the initial jet energy measured in $10^{50}$ erg, $\theta_{0,-1}$ its initial half-opening measured in 0.1 radians, and $n_0$ the external medium density in $\text{cm}^{-3}$. The lack of shocked gas outside the jet boundaries yielding a steepening of the afterglow decay ([8]), a so-called "jet-break". This steepening is enhanced by the jet lateral spreading ([9]), which increases its area and the jet deceleration rate.

So far, there are eight well-observed GRB afterglows (980519, 990123, 990510, 991208, 991216, 000301c, 000926, and 010222) for which a break or a steep decay has been observed in their optical light-curves, allowing the determination of the jet initial aperture and energy.

## MODEL DESCRIPTION

The afterglow model used in this work is described in [10] – [12]. Analytical treatments of the dynamics of and radiation from relativistic fireballs can also be found in [9], and [13] – [19]. Below we summarize the main features of this model.

*Jet dynamics.* The dynamics of the GRB remnant is calculated by tracking its energy (taking into account radiative losses), mass and aperture (which increase as the jet expands and sweeps-up the surrounding gas). The jet spreads laterally at the speed of sound, which is calculated from the pressure and density within the shocked gas. The jet is considered uniform within its aperture and having sharp boundaries. The equations for the remnant dynamics are accurate in any relativistic regime.

*Emission of radiation.* The afterglow synchrotron/inverse Compton emission and the radiative losses are calculated from the electron distribution and the magnetic field in the shocked external medium. The observed light-curves are obtained by integrating the jet emission over its surface, taking into account the differential relativistic beaming and the spread in the photon arrival time across the jet surface. The observer is assumed to be located on the jet axis. As shown in [20], the effect of off-axis locations within the initial jet aperture on the resulting light-curves is quite small. In the shocked fluid comoving frame, the synchrotron spectrum between adjacent breaks is approximated as power-law. The breaks are at the self-absorption frequency, injection frequency $\nu_i$ corresponding to the minimum electron $\gamma_i$, and cooling frequency corresponding to the electron Lorentz factor for which the radiative (synchrotron + inverse Compton) cooling timescale equals the dynamical time.

*Parameters.* The model has *three* parameters that determine the jet dynamics: the initial jet energy $E_0$, initial half-angle $\theta_0$, and external particle density $n$ (or the normalization factor for a wind-like density profile, which is proportional to the ratio between the progenitor mass loss rate the wind speed) and *three* parameters pertaining to the microphysics of shocks: the fraction $\varepsilon_B$ of the post-shock energy density in magnetic fields, the fractional energy $\varepsilon_i$ in the injected electrons if they all had the same minimum Lorentz factor $\gamma_i$, and the power-law index $p$ of the shock-accelerated electron distribution $\mathcal{N}(\gamma) \propto \gamma^{-p}$

above $\gamma_i$. The parameter $p$ is constrained by both the afterglow power-law decay index $\alpha$ ($F_\nu \propto t^{-\alpha}$) and the optical spectral slope $\beta$ ($F_\nu \propto \nu^{-\beta}$). For $p \lesssim 2$ the total electron energy is divergent. For simplicity, in such cases we assume that above some $\gamma_* > \gamma_i$ the injected electron distribution becomes a steeper power-law of index $q > 2$. The $\gamma_*$ break is parameterized through the fractional energy $\varepsilon_e$ in the electrons with $\gamma_i < \gamma < \gamma_*$. The total number of model parameters is at most *nine*, including the $V$-band dust extinction in the host galaxy (assuming an SMC-like reddening curve), when it is required by the observed curvature of the optical spectrum or by consistency between the optical and $X$-ray afterglow emission.

*Light-curve breaks.* Because of its origin, the afterglow light-curve break associated with the ejecta collimation (i.e. the jet-break) is *achromatic*, occurring simultaneously at all frequencies (however, the magnitude of this break, i.e. the difference in the asymptotic light-curve indices $\alpha$, is frequency-dependent). Because the afterglow light-curve at a fixed frequency is determined by the continuous softening of the synchrotron/inverse Compton spectrum, a steepening of the afterglow fall-off may also occur when a *spectral* break crosses the observing band. As observations are typically made between $\sim 1/2$ day and $\lesssim 100$ days, it is quite likely to see the passage of the injection frequency $\nu_i$ through the radio domain. Furthermore, for $p \lesssim 2$, the synchrotron frequency $\nu_*$ corresponding to the high electron energy "cut-off" at $\gamma_*$ may cross the optical domain. The light-curve steepening produced by the passage of a spectral break is, evidently, highly *chromatic*.

## AFTERGLOW FITS

Figures 1-8 show the best fits to the broadband emission of eight afterglows, obtained with model described above, for a homogeneous external medium. In some cases (see captions) fits of comparable quality can be obtained with wind-like media. For clarity, these Figures show only the most constraining data, representing all the frequency domains were observations were carried out and providing long temporal monitoring.

A 5% error was assumed in the magnitude-to-flux conversion and Galactic extinction and the reported contributions of the host or contaminating galaxies were subtracted from the optical transient magnitudes. $X$-ray fluxes have been calculated from the reported band fluxes (2–10 keV, usually) and spectral slopes.

The radio fluctuations due to inhomogeneities in the intra-Galactic medium are calculated following the treatment given by [21] and [22], and are taken into account in the $\chi^2$ calculation. Where relevant, the amplitude of these

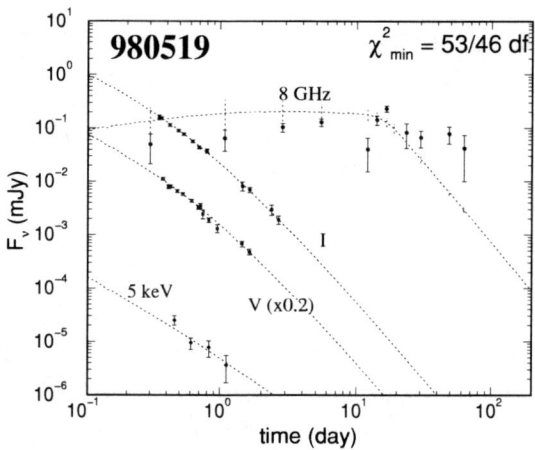

**FIGURE 1.** The decay of the optical emission of the afterglow 980519 exhibited a steepening at $\sim 0.5$ day, which can be well modeled with a jet-break. For this afterglow redshift was not measured. We assumed $z = 1$, typical for other GRBs. A jet interacting with a wind medium has $\chi^2_{min} = 73$.

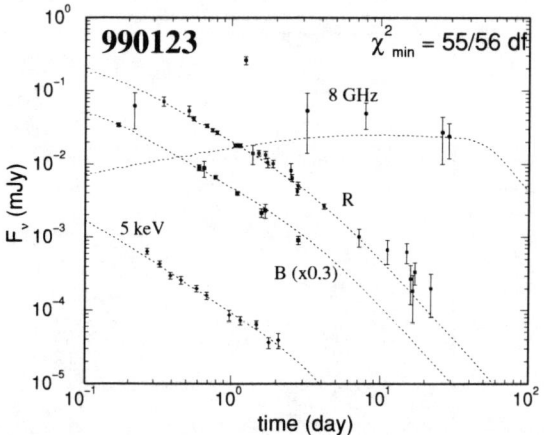

**FIGURE 2.** The optical light-curves of the afterglow 990123 show a break at $\sim 2$ days, consistent with a jet undergoing lateral expansion. The earliest two radio measurements exceed the jet model prediction, indicating a substantial contribution from *reverse* shock that swept-up the GRB ejecta. The best fit obtained with a wind medium has $\chi^2_{min} = 130$.

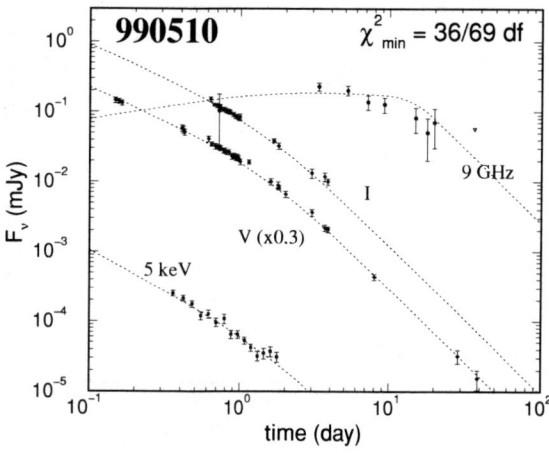

**FIGURE 3.** The fall-off of the optical light-curves of the afterglow 990510 steepened at 1-2 days and can be well accommodated with a jet-break. For a wind medium, $\chi^2_{min} = 128$.

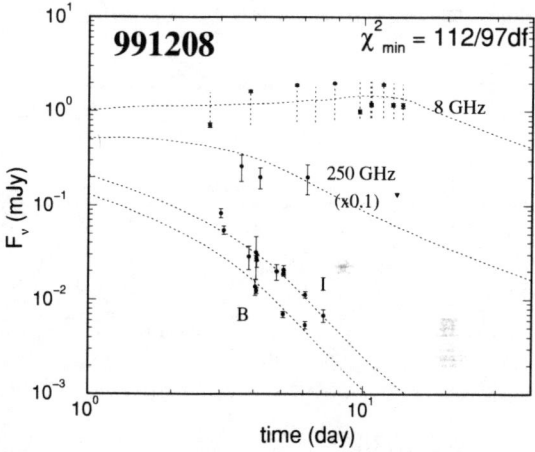

**FIGURE 4.** The steep decline of the optical emission of the afterglow 991208 and the almost constant radio emission suggest that the jet-break occurred at $\sim 1$ day. The shallow fall-off of the radio emission at 10-100 days (not shown here) requires $p \sim 1.5$, which brings the $\nu_*$ spectral break below the optical domain, explaining the steep fall-off exhibited by the optical emission. This afterglow has not been observed at X-rays. A wind medium also provides a good fit, with $\chi^2_{min} = 110$.

fluctuations (which quench, as the source angular size increases) is indicated in Figures 1-8 with dotted segments.

## JET PARAMETERS

The model parameters for the best fits shown in Figures 1-8 are presented in Figure 9, together with their 90% confidence levels estimated from the variation of $\chi^2$ around its minimum.

*Jet energy.* The most striking feature presented in Figure 9 is that the jet energies $E_0$ at the beginning of the afterglow phase span a relatively narrow range: $10^{50} - 4 \times 10^{50}$ erg, slightly smaller than the typical kinetic energy of a supernova. A small width of the jet energy distribution has been also inferred by [24] from the distribution of the X-ray luminosity at 0.5 days of 21 GRB afterglows.

A lower limit on the energy $E_\gamma$ lost by the jet during the GRB phase can be obtained from the observed 20

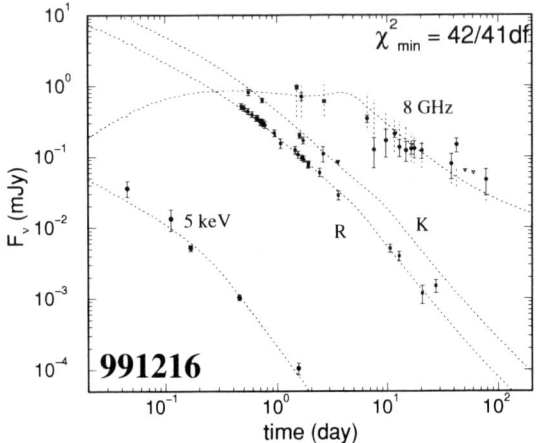

**FIGURE 5.** The nearly constant radio flux until several days and the steepening of the X-ray emission of the afterglow 991216 indicate the jet-break occurred at $\sim 1$ day. The injection frequency $\nu_i$ and the cut-off frequency $\nu_*$ cross the radio and optical domains, respectively, at several days. For a wind medium, the best fit has $\chi^2_{min} = 41$.

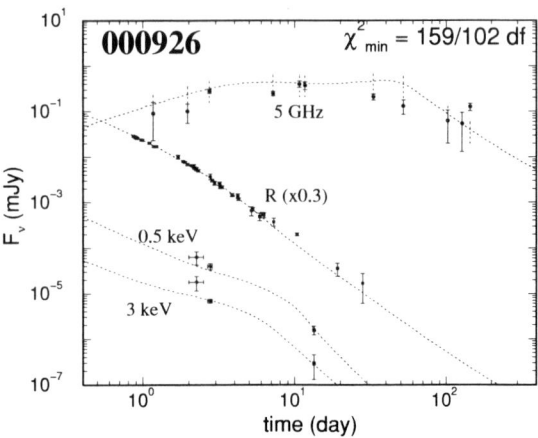

**FIGURE 7.** The steepening of the optical emission decay of the afterglow 000926 at $\sim 2$ days can be modeled with a jet-break. The X-ray emission of this afterglow lies well above the extrapolation of the synchrotron optical spectrum to higher frequencies, indicating a dominant inverse Compton emission in X-rays. For a wind medium, the best fit has $\chi^2_{min} = 270$.

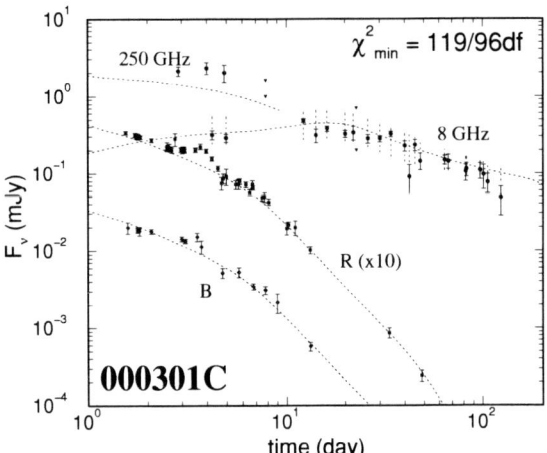

**FIGURE 6.** The strong light-curve steepening exhibited by the optical emission of the afterglow 000301 can be fit with a jet-break occurring at $\sim 2$ days and the passage of the $\nu_*$ frequency through the optical range at several days. Various mechanisms (gravitational microlensing, inhomogeneity in the surrounding medium, refreshed shock) can explain qualitatively the brightening seen at $\sim 4$ days in the optical emission. There are no X-ray measurements for this afterglow. For a wind-like medium $\chi^2_{min} = 140$.

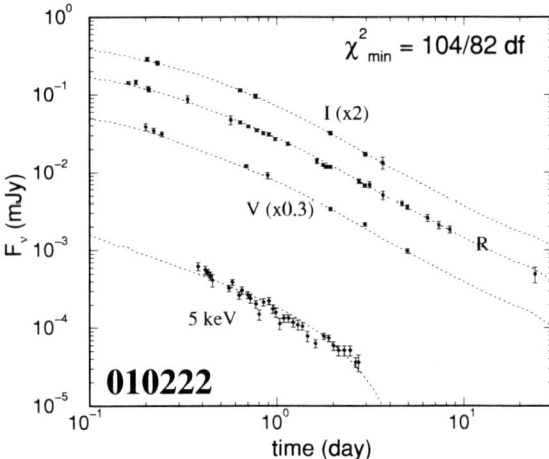

**FIGURE 8.** The decay of the optical afterglow 010222 steepens at $\sim 0.5$ days, and can be accommodated with a jet-break. For this afterglow radio measurements are not yet available. For a wind medium, $\chi^2_{min} = 130$

keV – 1 MeV fluences and the burst redshift, which give the isotropic $\gamma$-ray output $\mathcal{E}_\gamma$, if we make the assumption that the jet is uniform during the GRB phase as well, i.e. $E_\gamma = \mathcal{E}_\gamma \theta_0^2/2$. The resulting $E_\gamma$ span the range $5 \times 10^{49} \div 3 \times 10^{51}$, significantly wider than that of $E_0$, while the total jet energies $E_{jet} = E_\gamma + E_0$ (see Figure 10), stretch over a decade around $10^{51}$ ergs.

The above results imply that only a very small fraction ($\sim 0.1\%$) of the energy budget for GRB progenitors estimated by [23] is given to the highly relativistic ejecta that produce the GRB and the afterglow. Furthermore, the efficiency $\varepsilon_\gamma = E_\gamma/(E_\gamma + E_0)$ of the $\gamma$-ray mechanism are, with the exception of 980519, in the 50%–90% range, or even higher if a significant fraction of the GRB emission falls outside the 20-1000 keV range. However, the burst efficiencies inferred in this way may overestimate the true efficiency if the jet surface is rather inhomoge-

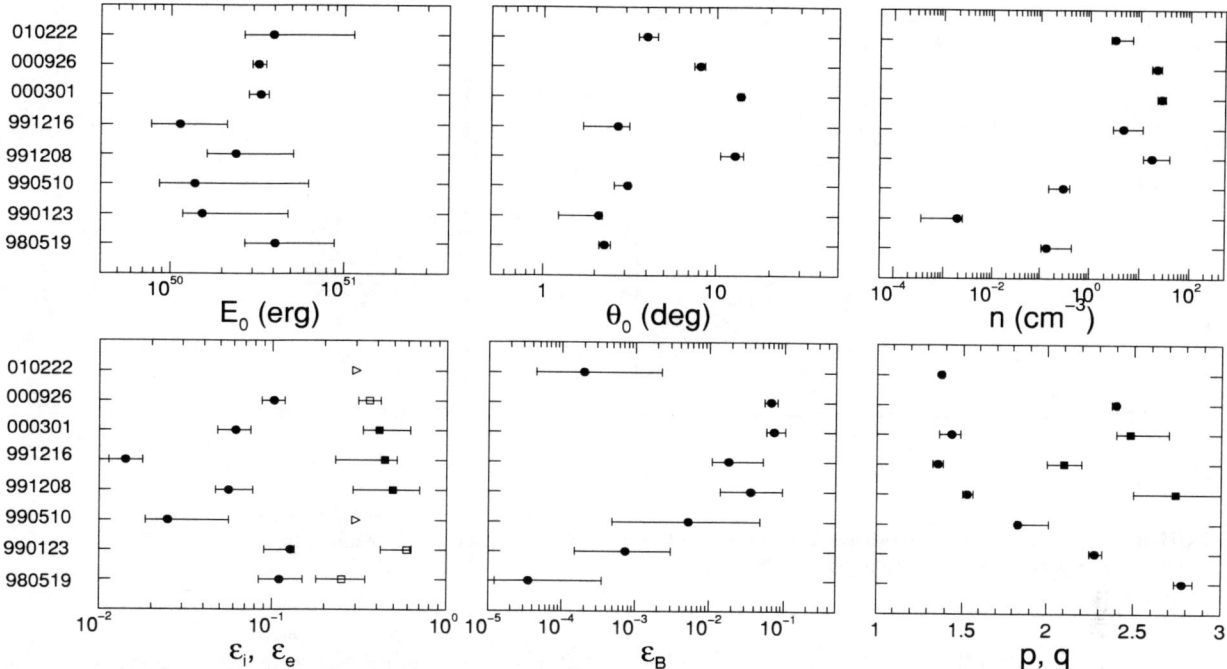

**FIGURE 9.** Parameters for the best fits shown in Figures 1-8: $E_0$ = jet energy after the GRB phase and beginning of the afterglow stage, $\theta_0$ = jet initial half-angle, $n$ = particle density of the circumburst medium, $\varepsilon_i$ = parameter for minimum energy of shock-accelerated electrons (shown with filled circles), $\varepsilon_e$ ($> \varepsilon_i$) = fraction of internal energy stored in electrons with Lorentz factor up to the $\gamma_*$-break (filled squares, triangles for lower limits) or total electron energy ($\frac{p-1}{p-2}\varepsilon_i$) assuming that the electron distribution extends to infinity (open squares), $\varepsilon_B$ = fractional energy in magnetic field, $p$ = index of injected power-law electron distribution above $\gamma_i$ (filled circles), $q$ ($> p$)= index of the electron distribution above $\gamma_*$ (open squares).

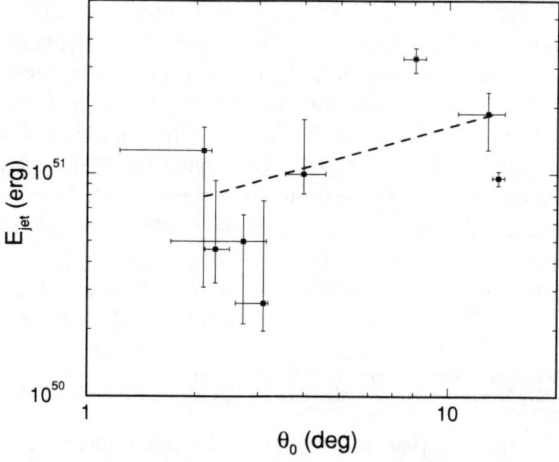

**FIGURE 10.** Total jet energy (i.e. including the jet $\gamma$-ray output) versus the initial jet half-angle. The linear fit in log-log space is $E_{jet} = 5.6 \times 10^{50} \theta_0^{0.46}$.

neous during the GRB phase, so that an observer is biased toward detecting those jets for which the small patch visible during the GRB phase is a bright spot on the jet surface, i.e. a region of higher energy per solid angle.

*Jet initial opening.* As illustrated in Figure 9, the initial jet aperture spans about one decade, varying from $2°$ to $14°$. The distribution of $\theta_0$ in other GRBs could be broader, as narrow jets with early jet-break times (eq. 1) could fall below detection until first observations ($\sim 1/2$ days), while wide jets have jet-break times occurring too late, when the afterglow light is dominated by the host galaxy.

Figure 10 suggests that wider jets are more energetic, the linear correlation coefficient of $E_{jet}$ and $\theta_0$ being $0.45 \pm 0.07$. Given the small number of afterglows included in this work, the correlation is not statistically significant.

*Jet initial Lorentz factor and mass.* The afterglow emission is very weakly dependent on the jet Lorentz factor $\Gamma_0$ at the beginning of the afterglow phase, as $\Gamma_0$ determines the evolution of the radiative losses in the early afterglow. Thus $\Gamma_0$ cannot be directly constrained through afterglow modelling. However, the inferred jet parameters can be used to determine the jet Lorentz

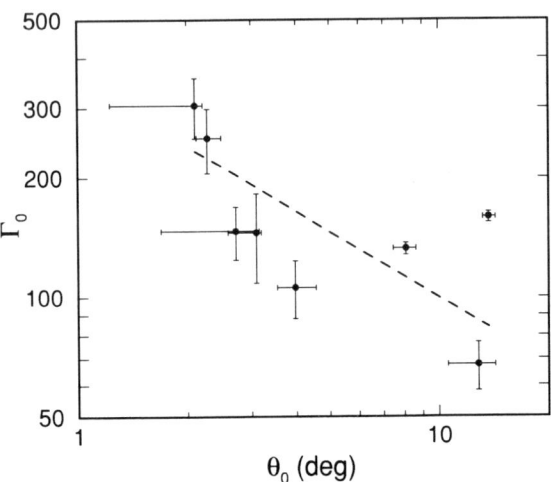

**FIGURE 11.** Jet Lorentz factor at the beginning of the afterglow phase versus the jet opening. The linear fit in log-log space is $\Gamma_0 = 350\ \theta_0^{-0.55}$.

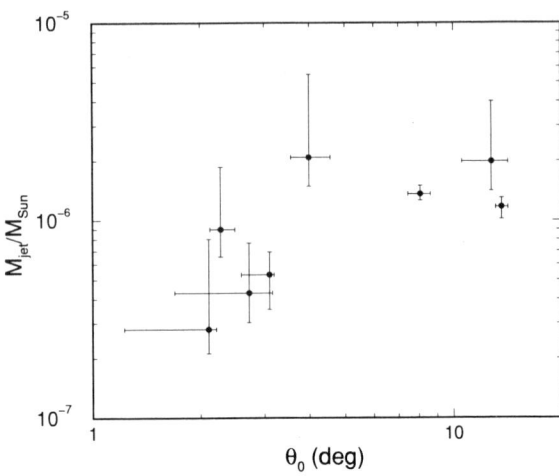

**FIGURE 12.** Jet mass (inferred from its energy and Lorentz factor) versus the jet's initial half-angle.

factor during the afterglow phase:

$$\Gamma \simeq 400 \left(\frac{E_{0,50}}{\theta_{0,-1}^2 n_0}\right)^{1/8} \left(\frac{t}{1+z}\right)^{-3/8}, \qquad (2)$$

where the usual notation $X = 10^n X_n$ was used and $t$ is measured in seconds. Thus $\Gamma_0$ could be calculated if the jet deceleration timescale $t_0$ were known. Here we assume that the observed GRB duration is a good measure of $t_0$ (note from eq. 2 that $\Gamma$ has a moderate dependence on $t$, thus the error due to this assumption is likely not too large).

The resulting initial jet Lorentz factors are shown in Figure 11, and range between 70 and 300. The linear correlation coefficient between $\Gamma_0$ and the initial jet aperture is $-0.54 \pm 0.04$, corresponding to a 15% chance of obtaining by chance this correlation in the null hypothesis. That wider jets have lower bulk Lorentz factors may explain the GRB pulse lag-time anticorrelation with the burst peak luminosity found by [25] and [26] if, during the GRB phase, the peak luminosity is correlated with the jet energy per solid angle and the pulse duration is anticorrelated with the jet Lorentz factor. Both these assumptions seem natural in the *internal shocks* model.

From the jet energy $E_0$ at the beginning of the afterglow phase one can calculate the jet mass $M_{jet} = E_0/(c^2\Gamma_0)$. The resulting values (shown in Figure 12) are within a factor 10 around $10^{-6} M_\odot$, and appear correlated with the jet initial half-angle (linear correlation coefficient is $0.56 \pm 0.26$). This is a natural expectation in most models of GRB progenitors, as the baryonic content increases at larger angles from the progenitor's rotation axis, along which the relativistic jet is expelled.

*External Medium.* The afterglow modelling has shown that a wind-like external medium is compatible with the emission of the afterglows 991208 and 991216 and can marginally accommodate the afterglows 000301c and 010222, but cannot explain the broadband emission of 980519, 990123, 990510 and 000926. Therefore our findings do not fully support entirely the existence of pre-ejected winds around GRB sources, as it would be expected if the GRB progenitors were massive (Wolf-Rayet) stars ([19]).

The particle density shown in Figure 9 for homogeneous media range from values typical for the interstellar medium (980519, 990510) to those of diffuse interstellar hydrogen clouds (991208, 991216, 000301c, 000926, 010222). In one case (990123) we find an external density below $10^{-2}$ cm$^{-3}$, characteristic of a hot component of the interstellar medium or a galactic halo. These values are 2–5 orders of magnitude smaller than those implied by the $N_H$ column densities inferred by [27] from the soft X-ray absorption of 980703, 990123, 990510, and 980519, suggesting that either the GRB is not embedded in the absorbing medium or that the gas in the vicinity of the GRB was evacuated.

*Microphysical parameters.* As illustrated in Figure 9, the afterglow modelling leads to the following properties for the acceleration of electrons at relativistic shocks: the total electron energy is close to equipartition with protons and the power-law distribution of electrons does not have an universal index $p$, ranging from $\sim 1.5$ to 3. In four of the afterglows analyzed here, a hard electron distribution is required by shallow fall-off of the radio and/or optical light-curve after the jet-break.

# CONCLUSIONS

There are two major results established by our modeling of GRB afterglows: 1) the jet energy at the beginning of the afterglow phase has a relatively narrow distribution (the full width being a factor $\lesssim 5$) and 2) the circumburst medium is more likely homogeneous than with the profile of a massive star's wind, having densities below those of dense molecular clouds were such stars reside. The former represents a useful constraint on models of GRB progenitors, the latter disfavors hypernovae as the origin of GRB jets.

There are, however, some caveats. These conclusions are based on a small sample of only eight afterglows, which may not be representative for the 40-50 afterglows detected January 2002. The eight afterglows were selected based on the existence of an optical light-curve break (or a steep decay) and sufficient data, covering more than one observational domain. The first criterion limits our analysis to jets whose initial opening is such that their jet-break times occur between $\sim 1/2$ day and $\sim 10$ days. It is thus possible that through this selection criterion we have limited the modelling only to the afterglows with a certain range of kinetic energies.

Furthermore, the jet parameters where "extracted" from observations with an afterglow model that is approximate. By far, the treatment of the jet surface undergoing lateral spreading as being uniform, i.e. the energy and mass per solid angle are the same in any direction within the jet, is the least reliable approximation. It may have little effect on the resulting jet energy, but it could be the reason for which, in our model, spreading jets interacting with wind-like media fail to produce the observed fast light-curve steepenings ([10]). Numerical simulations of the hydrodynamics of spreading jets ([28]) will, hopefully, allow a better estimation of the sharpness of light-curves breaks in this case. Nevertheless, there are other reasons for which wind-like media are not found consistent with the observed afterglow emission in some of the cases analyzed, independent of the effects of jet lateral spreading: rising radio fluxes at frequencies below the self-absorption frequency and too bright millimeter/sub-millimeter emission (see [11]), exceeding the observed fluxes.

We note that, as shown by [29], the interaction of the wind ejected by a massive Wolf-Rayet star with the surrounding medium could, in principle, form a quasi-homogeneous bubble on the length-scales of interest for afterglows ($\lesssim 1$ pc). The typical density of the bubble, $n \gtrsim 300$ cm$^{-3}$, is larger than found here, and could lead to high self-absorption frequencies, above the radio domain until few days, which in turn would lead to fast rising radio light-curves, inconsistent with the rather flat behavior observed in general.

Other potentially useful results in constraining the models of GRB progenitors, obtained by through afterglow modelling, are that the jet initial Lorentz factor ($\gtrsim 100$) at the beginning of the afterglow phase is anticorrelated with the jet initial opening, and that the latter is between 2 and 15 degrees. The former result has been obtained by approximating the jet deceleration timescale with the burst duration. The latter result implies that GRBs are highly beamed and have an occurrence rate that is $\sim 10^{-3}$ smaller than that of supernovae. The inferred jet initial apertures may be affected by our assumption of jet uniformity, if the afterglow brightness "reflects", indeed, the energy per solid angle of the ejecta moving toward the observer, as discussed in [30]. Then, for a jet with angular structure, the jet half-angles inferred here are in fact the angle between the jet axis and the direction toward the observer, providing an estimation of the jet angular scale, i.e. the scale over which the energy per solid angle changes significantly.

# REFERENCES

1. Woosley, S., *Astrophysical Journal*, **405**, 273-277 (1993)
2. Paczyński, B., *Astrophysical Journal*, **494**, L45-L48 (1998)
3. Narayan, R., Paczyński, B., and Piran, T., *Astrophysical Journal*, **395**, L83-L86 (1992)
4. Mészáros, P., and Rees, M.J., *Astrophysical Journal*, **397**, 570-575 (1992)
5. Rees, M.J., and Mészáros, P., *Astrophysical Journal*, **430**, L93-L96 (1994)
6. Piran, T., *Physics Reports*, **314**, 575-667 (1999)
7. Rees, M.J., and Mészáros, P., *Monthly Notices of the Royal Astronomical Society*, **258**, 41p-43p (1992)
8. Panaitescu, A., Mészáros, P., and Rees, M.J., *Astrophysical Journal*, **503**, 314-324 (1998)
9. Rhoads, J., *Astrophysical Journal*, **525**, 737-749 (1999)
10. Kumar, P., and Panaitescu, A., *Astrophysical Journal*, **541**, L9-112 (2000)
11. Panaitescu, A., and Kumar, P., *Astrophysical Journal*, **543**, 66-76 (2000)
12. Panaitescu, A., and Kumar, P., *Astrophysical Journal*, **554**, 667-678 (2001)
13. Mészáros, P., and Rees, M.J., *Astrophysical Journal*, **476**, 232-237 (1997)
14. Waxman, E., *Astrophysical Journal*, **485**, L5-L8 (1997)
15. Sari, R., Piran, T., and Narayan, R., *Astrophysical Journal*, **497**, L17-L20 (1998)
16. Granot, J., Piran, T., and Sari, R., *Astrophysical Journal*, **527**, 236-246 (1999)
17. Gruzinov, A., and Waxman, E., *Astrophysical Journal*, **511**, 852-861 (1999)
18. Wijers, R., and Galama, T., *Astrophysical Journal*, **523**, 177-186 (1999)
19. Chevalier, R., and Li, Z.-Y., *Astrophysical Journal*, **536**, 195-212 (2000)
20. Granot, J., Panaitescu, A., Kumar, P., Woosley, S., *Astrophysical Journal*, submitted (2002)
21. Goodman, J., *New Astronomy*, **2**, 449-460 (1997)

22. Walker, M., *Monthly Notices of the Royal Astronomical Society*, **294**, 307-311 (1998)
23. Mészáros, P., Rees, M.J., and Wijers, R., *New Astronomy*, **4**, 303-312 (1999)
24. Piran, T., Kumar, P., Panaitescu, A., and Piro, L., *Astrophysical Journal*, **560**, L167-L169 (2001)
25. Norris, J., Marani, G., and Bonnell, J., *Astrophysical Journal*, **534**, 248-257 (2000)
26. Salmonson, J., *Astrophysical Journal*, **544**, L115-L117 (2000)
27. Galama, T., and Wijers, R., *Astrophysical Journal*, **549**, L209-L212 (2001)
28. Granot. J, Miller, M., Piran, T., Suen, W., and Hughes, P., "Light Curves from an Expanding Relativistic Jet", in *Gamma-Ray Bursts in the Afterglow Era*, edited by E. Costa et al., Springer-Verlag, Berlin, 2001, pp. 312-314
29. Ramirez-Ruiz, E., Dray, L., Madau, P., and Tout, C., *Monthly Notices of the Royal Astronomical Society*, **327**, 829-840 (2001)
30. Rossi, E., Lazzati, D., and Rees, M.J., *Monthly Notices of the Royal Astronomical Society*, submitted (2002)

# GRB Remnants

Tsvi Piran and Shai Ayal

*Racah Institute for Physics, The Hebrew University, Jerusalem, Israel 91904*

**Abstract.**
The realization that GRBs are narrowly beamed implied that the actual rate of GRBs is much larger than the observed one. There are 500 unobserved GRBs for each observed one. The lack of a clear trigger makes it hard to detect these unobserved GRBs as orphan afterglows. At late time, hundreds or thousands of years after a GRB, we expect to observe a GRB remnant (GRBR). These remnants could be distinguished from the more frequent SNRs using their different morphology. While SNRs are spherical, GRBRs that arise from a highly collimated flow, are expected to be initially nonspehrical. We ask the question for how long can we identify a GRBR among the more common SNRs? Using SPH simulations we follow the evolution of a GRBR and calculate the image of the remnant produced by bremsstrahlung and by synchrotron emission. We find that the GRBR becomes spherical after $\sim 3000 \text{yr}(E_{51}/n)^{1/3}$ at $R \sim 12\text{pc}(E_{51}/n)^{1/3}$, where $E_{51}$ is the initial energy in units of $10^{51}$ erg and $n$ is the surrounding ISM number density in cm$^{-3}$. We expect $0.5(E_{51}/n)^{1/3}$ non-spherical GRBs per galaxy. Namely, we expect $\sim 20$ non spherical GRBRs with angular sizes $\sim \mu$arcsec within a distance of 10Mpc. These results are modified if there is an underlying spherical supernova. In this case the GRBR will remain spherical only for $\sim 150\text{yr}(E_{51}/n)^{1/3}$ and the number of non-spherical GRBRs is smaller by a factor of 10 and their size is smaller by a factor of 3.

## INTRODUCTION

A $\gamma$-ray burst (GRB) that originates within a galactic disk deposits $\sim 10^{51}$ergs into the ISM. This results in a blast wave whose initial phase produces the afterglow. The late phase of the blast wave evolution would result, as noted by Chevalier [1] in the context of supernova remnants (SNRs), in a cool expanding H I shell. The shell will remain distinct from its surrounding until it has slowed down to a velocity of $\approx 10 \text{km s}^{-1}$ [2], which should happen within $2.3 \cdot 10^6$ yr $E_{51}^{0.32}$ where $E_{51}$ is the initial energy in units of $10^{51}$ erg.

The observed rate of GRBs is one per $\sim 10^7$ yr per galaxy [3]. The implied GRB isotropic energy is of the order of $10^{53}$ergs. These estimates suggested that there are a few remnants per galaxy at any given time. As it was believed that the GRB explosions were much more energetic than SNs, Loeb and Perna [2] suggested that GRBRs would form HI supershells. This giant structures require much more energy than what a usual SN can supply.

However, the realization that GRBs are beamed [4, 5, 6, 7, 8] changed both estimate. First the rate of GRBs is much higher. Beamed GRBs illuminate only a fraction $f_b$ of the sky, thus their rate should be higher by a factor of $f_b^{-1}$. With $f_b \sim 0.002$ [9] we expect several thousand GRB remnants per galaxy. On the other hand the energy output of each GRB is much smaller [9, 10, 11]. Thus they cannot produce the giant HI shells.

How can we distinguish a GRBR from and SNR and for how long? Both GRBs and SNs deposit a comparable kinetic energy ($\sim 10^{51}$ erg) into the ISM. The energy injection in a GRB is in a form two narrow relativistic beams containing $\sim 10^{-5} M_\odot$. A SN deposits this energy spherically with $\sim 10 M_\odot$. In both cases the expected late evolution is similar. At this late stage both remnants are in the Sedov [12] regime where all the kinetic energy is in the ejecta and all the mass is in the surrounding ISM. A key distinguishing feature unique to GRB remnants could be their beamed nature. We expect that the beamed emission would lead to a distinct double shell morphology at intermediate times. The late time behavior of the GRB remnant is expected to be spherical. To establish how many H I shells are GRB remnants we need to find out the expected morphology of GRB remnants and how long they stay non-spherical and distinguishable from SNRs. Establishing how many of the H I shells are GRB remnants would make it possible to directly estimate the local rate of GRBs, determine $\varepsilon$, the efficiency of converting the explosion energy into $\gamma$-rays, and the beaming factor $f_b$ [2].

We model the intermediate evolution of a beamed

GRB by two blobs of dense material moving into the ISM in opposite directions. We follow numerically the hydrodynamic evolution [13]. We find that the morphology depends on the dimensionless ratio between the accumulated mass and the initial kinetic energy, $\mu \equiv Mc^2/E_0$. When $\mu \sim 2.1 \times 10^5$ (at $t \sim 3000\text{yr}(E_{51}/n)^{1/3}$ and $R \sim 12\text{pc}(E_{51}/n)^{1/3}$), the remnant becomes spherical and indistinguishable from a SNR.

An additional complication arises if the GRB is accompanied by a supernova, as suggested in the Collapsar model [14, 15, 16]. The supernova produces an underlying massive spherical Newtonian shell that propagates outwards. At $\mu \sim 3000$ corresponding to $t \sim 150\text{yr}(E_{51}/n)^{1/3}$ and $R$ $4\text{pc}(E_{51}/n)^{1/3}$ this shell will catch the non-spherical GRBR and the system will quickly become spherical. In this case the number of non-spherical GRBRs is smaller by a factor of 10 and their size is smaller by a factor of 3.

## THE NUMERICAL SIMULATIONS

### The Model

A GRB occurs when a compact 'inner engine' ejects two ultra-relativistic beams. Internal collisions within these beams leads to the GRB (See a schematic description in Fig. 1.). Later external shocks caused by collisions with circumstellar matter produce the afterglow. The matter slows down during this interaction and its bulk Lorenz factor $\Gamma$, decreases. The ejecta stays collimated only until $\Gamma$ drops below $\sim 1/\theta_0$, at approximately $2.9\,\text{hr}(E_{51}/n_1)^{1/3}(\theta_0/0.1)^{8/3}$ after the GRB [18, 4] where $\theta_0$ is the initial angular width. At this time the matter starts expanding sideways causing, for an adiabatic evolution, an exponential slowing down [18]. The ejecta continues to expand sideways at an almost constant radial distance from the source $R_0 \sim 0.3\,\text{pc}\,E_{51}^{1/3}n^{-1/3}$ until it becomes non-relativistic. At this stage, we begin our simulation.

Without a detailed numerical modeling of the relativistic phase of the ejecta we have only an approximate description of the initial conditions. We expect the angular width of the ejecta to be $\sim 1$ rad and we are constrained by the energy conservation:

$$R_0 \sim 0.3\,\text{pc}\,E_{51}^{1/3}n^{-1/3}(v_0/c)^{-2/3}. \quad (1)$$

Our initial conditions comprise two identical blobs moving at $v_0 \sim c/3$ in opposing directions into the ISM. Both the blobs and the ISM are modeled by a cold $\gamma = 5/3$ ideal gas. The blobs are are denser than the ISM.

Luckily the intermediate and late evolution of the ejecta are insensitive to the initial conditions. Already

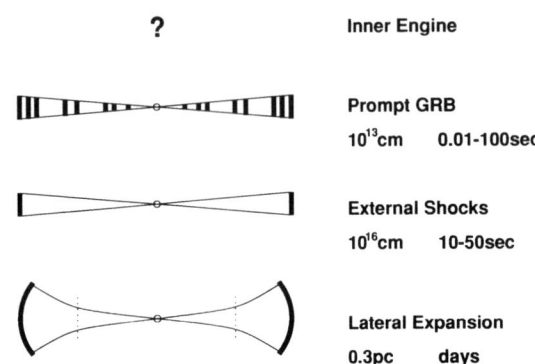

**FIGURE 1.** A schematic evolution of a GRB during its relativitistic phase. From top to bottom: (a) An inner engine accelerates relativistic jets. (b) Collisions within the jets produce the observed GRB. (c) External shocks produce the afterglow. (d) The jets expand sideways as they slow down.

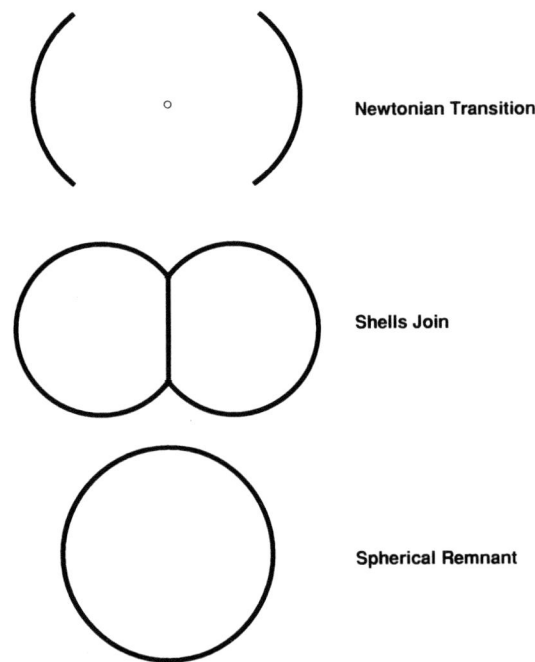

**FIGURE 2.** A schematic evolution of a GRBR. From top to bottom: (a) Initial conditions around the Newtonian transition. (b) Shells collision along the equatorial plane. (c) A late time spherical shell.

in the intermediate stage we are in the Sedov regime, the mass is dominated by the "external" ISM gas which washes out any variations in the initial conditions of the ejecta. Our numerical simulations [13] verified this expectation and different initial densities, angular widths and shapes of the blobs led to essentialy similar late time configurations.

Our code is based on the Newtonian version of the smooth particle hydrodynamics (SPH) code introduced in [17]. The code was adapted for the specific problem at hand. We have also used the post Newtonian version of the code to take account of possible initial relativistic effects (with an initial blob velocity of $c/3$).

Once we choose the initial velocity. Equation (1) leaves us with the freedom of choosing two out of the three parameters $E_0$, $R_0$ and $n$, the initial energy, distance and ISM density respectively. In presenting the results we choose $E_0$ and $n$. To parameterize the evolution of the remnant we utilize the fact that mass scales linearly with initial energy and define the dimensionless parameter $\mu = Mc^2/E_0$ where $M$ is the accumulated shell mass. We define $M$ as all mass with density above $2n$. We show all subsequent results as functions of $\mu$. The simulations begin with $\mu \sim 1$. Conveniently $\mu$ scales linearly with time with: $t \sim 0.046\mu(E_{51}/n)^{-1/3}$ yr, as can be seen from Fig. 3.

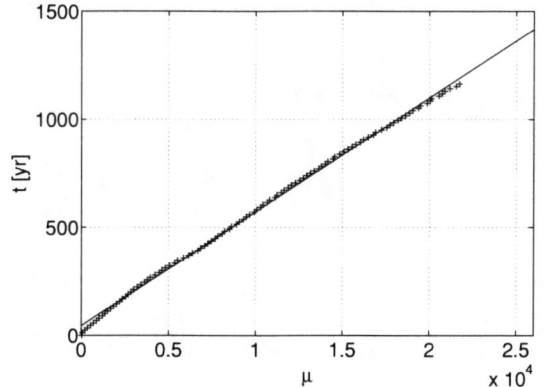

**FIGURE 3.** Time as a function of $\mu$. The linear relation between time and $\mu$ is $t \sim 0.046\mu(E_{51}/n)^{-1/3}$ yr.

## Results

As each blob collides with the ISM it produces a bow shock. This shock propagates also in the direction perpendicular to the blob's velocity. As the shocked blob material heats up it begins to expand backwards and a backwards going shock develops as well. The expected morphology of the remnant will therefore be of two expanding shells which will eventually join, producing yet another shock. At late times the shells merge and become a single spherical shell.

Fig. 2 depicts the expected schematic hydrodynamic behaviour after the Newtonain transition. This is indeed confirmed in the computation. Fig. 4, depicts the density contours along the evolution. We observe the expected evolution: from two individual blobs via a peanut shape configuration with a shock along the equator towards a more and more spherical configuration at late times.

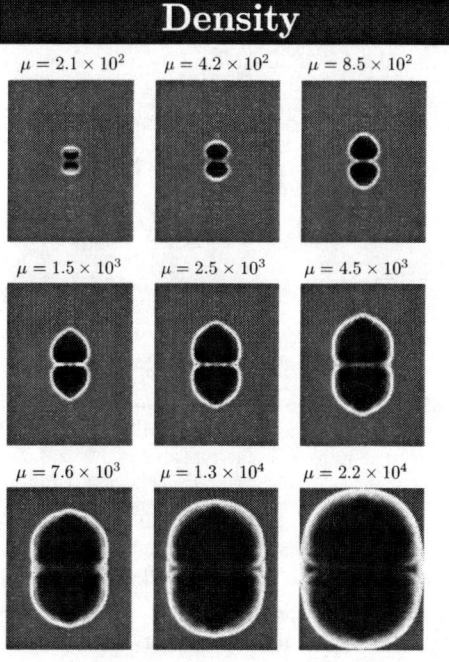

**FIGURE 4.** Equally spaced density contours ($\rho = 1.5n, 2n, \ldots, 3.5n$) at $\mu = 9.5 \times 10^2, 1.4 \times 10^3, 2.4 \times 10^3, 3.9 \times 10^3, 6.1 \times 10^3, 10^4, 1.8 \times 10^4, 3 \times 10^4, 5 \times 10^4$ (left to right, top to bottom)

The ratio $z_{\max}/r_{xy}$ can be approximated by a power law as shown in Fig 5. In our simulation this ratio is always between 1 and 2 so that the power law fit is very inaccurate. This ratio decreases in time as a power law with an exponent of $-0.15$. Extrapolating this power law we see that this ratio reaches a value of 1 at $\mu \sim 2.1 \times 10^5$. At this time the shock has a spherical shape with $z = r_{xy} \sim 15(E_{51}/n)^{1/3}$ pc. Even then the shock will not be completely spherically symmetric as there would still be a ring of shocks around the "equator" where the shells have collided.

Figures 6 and 7 depict the images of the remnant as a function of time and angles of inclination. We show images due to bremsstrahlung emission and synchrotron emission. The images are constructed assuming that all the gas is optically thin in the relevant frequencies. The bremsstrahlung luminosity (Fig. 6) was calculated assuming that the volume emissivity is proportional to $\rho^2 \varepsilon^{1/2}$ [19]. In calculating the synchrotron emissivity (Fig. 7) we assumed that both the magnetic field energy density and the number density of the relativistic electrons are proportional to the internal energy density of the gas with constant proportionality factors $\varepsilon_B$ and $\varepsilon_e$

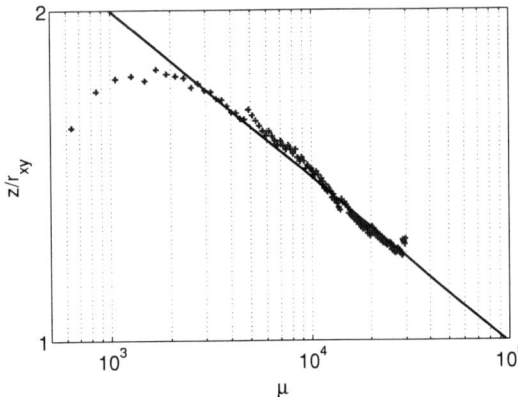

**FIGURE 5.** The ratio between the radius $r_{xy}$ of the shock and the $z$ position of the shock. The solid line is the best fit power law $\mu^{-0.15}$. The ratio will reach a value of 1 at $\mu \sim 2.1 \times 10^5$ yr

respectively. We further assume that the relativistic electron number density is a power law in energy. Under these assumptions the volume emissivity is proportional to $\rho^2 \varepsilon^2$ [e.g. 20]. In the late images there are two bright circles at the lines where the colliding blobs form a hot shocked region. In figures 8 and 9 we show the characteristic emission frequencies. For bremsstrahlung this is $kT/h$ where $T$ is the temperature of the gas. For synchrotron emission we assume $\varepsilon_B = 0.1$. The characteristic frequency in this case is the Larmor frequency $eB/m_e$ where $e$, $B$ and $m_e$ are the electron charge, the magnetic field and the electron mass respectively.

## DISCUSSION

The long time shape of a GRB remnant is insensitive to the exact initial morphology, angular width and density of the ejecta. Initially the remnant is highly non-spherical. It becomes spherical as time advances and the ratio between its height and radius approaches unity when

$$\mu \approx 2.1 \times 10^5 \, . \quad (2)$$

This corresponds to

$$t \sim 10^4 \, \text{yr} \, E_{51}^{1/3} n^{-1/3} \, , \quad (3)$$

and

$$R \sim 12 \text{pc} (E_{51}/n)^{1/3} \, . \quad (4)$$

After this time it will be difficult to distinguish a GRB remnant from a SNR on the basis of its morphology alone.

Using as the observed GRB rate $R_{GRB} = 10^{-7} \text{yr}^{-1} \text{gal}^{-1}$ [3] the expected number of non-

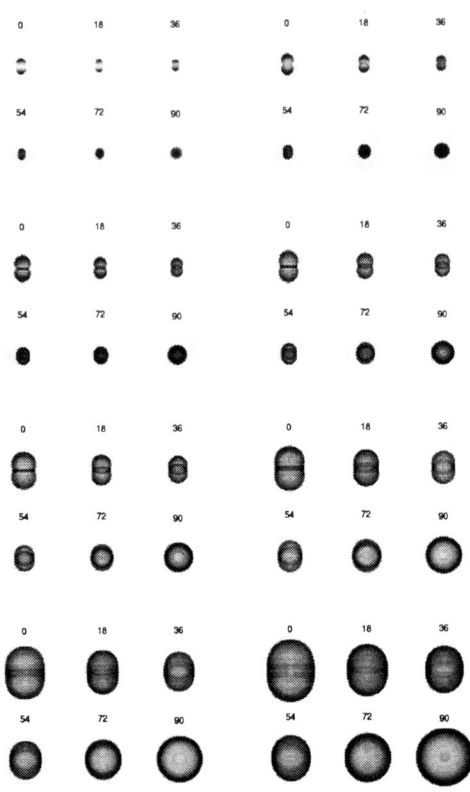

**FIGURE 6.** Images of the remnant, bremsstrahlung emission. The number above each image is the angle of inclination in degrees. The images are shown at the same $\mu$ as the last 8 panels of figure 4.

spherical GRBRs per galaxy is:

$$0.5 \, \text{yr}^{-1} \left(\frac{f_b}{500}\right)^{-1} \left(\frac{R_{GRB}}{10^{-7}}\right) E_{51}^{1/3} n^{-1/3} \, . \quad (5)$$

This value depends of course critically on the typical beaming factor, $f_b$. It should be compared with the expectation of 100 similar aged ($10^4$ yrs) SNRs per galaxy. We would expect 20 non spherical GRBRs up to a distance of 10 Mpc. The angular sizes of these GRBRs would be around a $\mu$arcsec.

### Implications to DEM L 316

DEM L 316 [21] in the LMC looks like two colliding bubbles (see Fig. 10). It is thought to result from a collision between two SNRs. This requires, of course, an unlikely coincidence in the timing and the location of the two SNes. An interesting possibility is that DEM L 316 is a GRBR. Does this fit our model? DEM L 316 is far from spherical and has a distinct double shell morphology, most similar to our results at $\mu \sim 10^4$

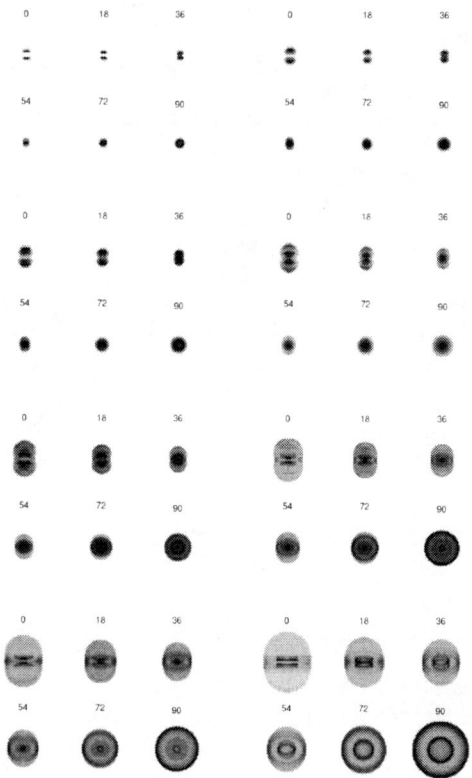

**FIGURE 7.** Images of the remnant, synchrotron emission. The $\mu$ are the same as the last 8 panels in figure 6

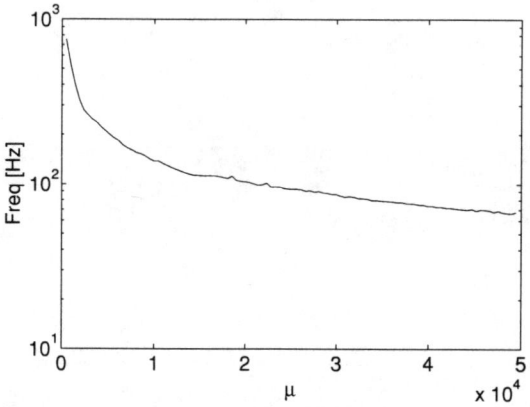

**FIGURE 8.** The characteristic synchrotron frequency as a function of $\mu$.

(see Fig. 4). The $\mu$ ratio measured for DEM L 316 is $\approx 7 \times 10^5$. However, according to our results this is far after the spherical transition. A GRB remnant would already be spherical at this stage. This discrepancy rules out the possibility of fitting DEM L 316 with our model for a GRB remnant.

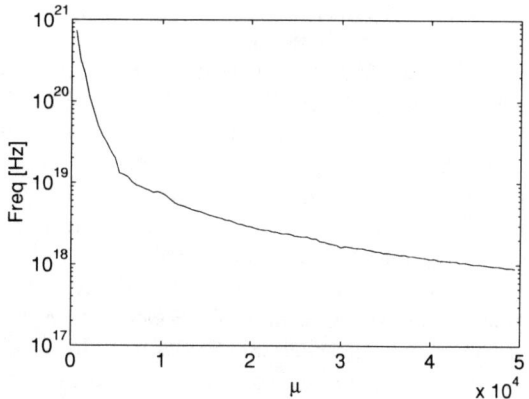

**FIGURE 9.** The characteristic bremsstrahlung frequency as a function of $\mu$.

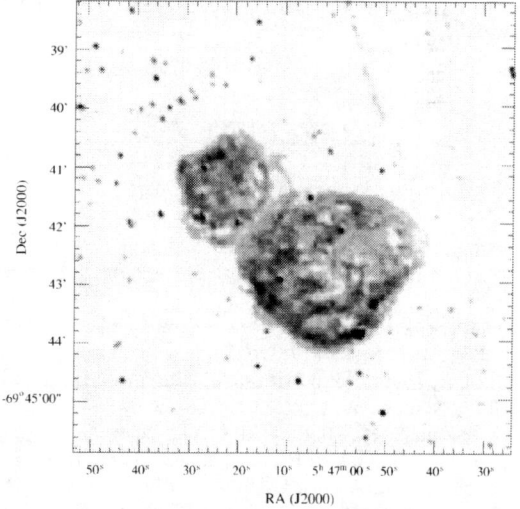

**FIGURE 10.** DEM L 316 in the LMC (from [21]).

## An Underlying Supernova

Our model should be modified if the GRB relativistic beams are accompanied by an underlying spherical supernova, as would be expected in the Collapsar model [14, 15, 16]. In this case a spherical shell of $\sim 10 m_\odot$, the supernova ejecta, will accompany the GRB beams. This ejecta propagates at a much lower, Newtonain velocity, with initial values of $\sim 10^4$km/sec. However, it will not slow down while the outer GRB ejecta is piling up the external matter and is slowed down. Eventually it will catch the GRB ejecta. It is clear that at this stage the SN shell would tend to make the GRBR bow shock more spherical. To determine when this will happen one needs another set of numerical simulations. These are in progress now. However, we can attempt to estimate when the slower SN ejecta will catch up the slowing down

GRBR remnant. Assuming that the SN ejecta does not slow down (as the GRBR ejecta clears the surrounding ISM matter) we find that this will happen at: $\mu \approx 3000$, namely at $t \approx 150\text{yr}(E_{51}/n)^{1/3}$ and $R \approx 4\text{pc}(E_{51}/n)^{1/3}$. This happens around the time that the two shells collide on the equator. We expect to see an enhanced emission due to this collision and then the system will become quickly spherical. The expected number of non spherical GRBRs and their corresponding sizes would be smaller by a factor of 10 then the values estimated earlier for the simple evolution of the beamed GRB ejecta. Thus we expect one or two non spherical GRBRs with distances up to 20 Mpc and their sizes would be around 0.1 $\mu$arcsec.

## ACKNOWLEDGMENTS

This research was supported by a grant from the US-ISRAEL BSF.

## REFERENCES

1. Chevalier, R. A. 1974, Ap. J., 188, 501
2. Loeb, A. & Perna, R. 1998, Ap. J. Lett., 503, L35
3. Schmidt, M. 1999, Ap. J. Lett., 523, L117
4. Sari, R., Piran, T., & Halpern, J. P. 1999, Ap. J. Lett., 519, L17
5. Halpern, J. P., Kemp, J., Piran, T., & Bershady, M. A. 1999, Ap. J. Lett., 517, L105
6. Sari, R. 1999, in Proc. of the 5th Huntsville Gamma-Ray Burst Symposium
7. Kulkarni, S. R., Djorgovski, S. G., Odewahn, S. C., Bloom, J. S., Gal, R. R., Koresko, C. D., Harrison, F. A., Lubin, L. M., Armus, L., Sari, R., Illingworth, G. D., Kelson, D. D., Magee, D. K., van Dokkum, P. G., Frail, D. A., Mulchaey, J. S., Malkan, M. A., McClean, I. S., Teplitz, H. I., Koerner, D., Kirkpatrick, D., Kobayashi, N., Yadigaroglu, I. ., Halpern, J., Piran, T., Goodrich, R. W., Chaffee, F. H., Feroci, M., & Costa, E. 1999, Nature, 398, 389
8. Harrison, F. A., Bloom, J. S., Frail, D. A., Sari, R., Kulkarni, S. R., Djorgovski, S. G., Axelrod, T., Mould, J., Schmidt, B. P., Wieringa, M. H., Wark, R. M., Subrahmanyan, R., McConnell, D., McCarthy, P. J., Schaefer, B. E., McMahon, R. G., Markze, R. O., Firth, E., Soffitta, P., & Amati, L. 1999, Ap. J. Lett., 523, L121
9. Frail, D. A. and Kulkarni, S. R. and Sari, R. and Djorgovski, S. G. and Bloom, J. S. and Galama, T. J. and Reichart, D. E. and Berger, E. and Harrison, F. A. and Price, P. A. and Yost, S. A. and Diercks, A. and Goodrich, R. W. and Chaffee, F., 2001. Ap. J. Lett., 562, L55
10. Piran, T. and Kumar, P. and Panaitescu, A. and Piro, L., 2001, Ap. J. Lett., 560, L167.
11. Panaitescu, A. and Kumar, P., 2001, Ap. J. Lett., 560, L49
12. Sedov, L. I. 1959, Similarity and Dimensional Methods in Mechanics (New York: Academic Press)
13. Ayal, S. & Piran, T., 2001, Ap. J., 555, 23
14. S. E. Woosley, Ap. J., **405**, 273 (1993)
15. B. Paczynski, Ap. J. Lett., **494**, L45 (1998).
16. A. I. MacFadyen and S. E. Woosley, Ap. J., **524**, 262 1999, Ap. J., **524**, 262
17. Ayal, S. and Piran, T. and Oechslin, R. and Davies, M. B. and Rosswog, S., 2001, Ap. J., 550, 846
18. Rhoads, J. E. 1997, Ap. J. Lett., 487, L1
19. Lang, K. R. 1980, Astrophysical Formulae (Springer-Verlag)
20. Shu, F. H. 1991, The Physics of Astrophysics, Vol. 1 (University Science Books)
21. Williams, R. M., Chu, Y. H., Dickel, J. R., Beyer, R., Petre, R., Smith, R. C., & Milne, D. K. 1997, Ap. J., 480, 618

# Optical and X-Ray Afterglows in the Cannonball Model

## A. De Rújula

*Theory Division, CERN, CH-1211 Geneva 23, Switzerland*

**Abstract.**
The Cannonball Model is based on the hypothesis that GRBs and their afterglows are made in supernova explosions by relativistic ejecta similar to the ones observed in quasars and microquasars. Its predictions are simple, and analytical in fair approximations. The model describes well the properties of the γ-rays of GRBs. It gives a very simple and extremely successful description of the optical and X-ray afterglows of *all* GRBs of known redshift. The only problem the model has, so far, is that it is contrary to staunch orthodox beliefs.

## INTRODUCTION

The idea that GRBs are due to collimated emissions is not recent. In the case of GRBs from quasars it was discussed by Brainerd [1]; in the case of a funnel in an explosion, by Meszaros & Rees [2]. In what is no doubt the relevant case: jets in stellar gravitational collapses, the idea has been developed over the years by Dar and collaborators [3] to [11]. Now we know that long-duration GRBs are cosmological, originate in galaxies, are associated with supernovae (SNe) and have energies that would be ridiculously large for a stellar spherical explosion (a fireball). GRBs must be "jetted".

In the currently dominating scenarios, the ejecta that beget a GRB and its afterglow (AG) are thrown off in a uniform cone with an opening angle $\theta_j(0)$. This cone expands sideways: the angle $\theta_j(t)$ subtended by the ejecta *increases* with time, delineating a *firetrumpet*, as in Fig. (1). Relativistic jets are ubiquitous in astrophysics. The ejecta of these real jets, as seen from their emission point up to the point where they eventually stop and expand, generally subtend angles that *decrease* with time: just the opposite of the assumed behaviour of firetrumpets. In the analysis of the observed jets, e.g. [12, 13, 14], it is the fixed angle of observation —and not the angle subtended by the ejecta— that plays a key role.

The Cannonball (CB) model is based on the contention that GRBs and their AGs are made by relativistically jetted balls of ordinary matter which, by a mechanism [11] that I will outline, stop expanding soon after their emission. The CB idea gives a good description of the properties of the γ-rays in a GRB, that we modelled in simple approximations [8]. It gives an excellent and complete description of optical and X-ray afterglows, which we have modelled in full detail, as I outline here.

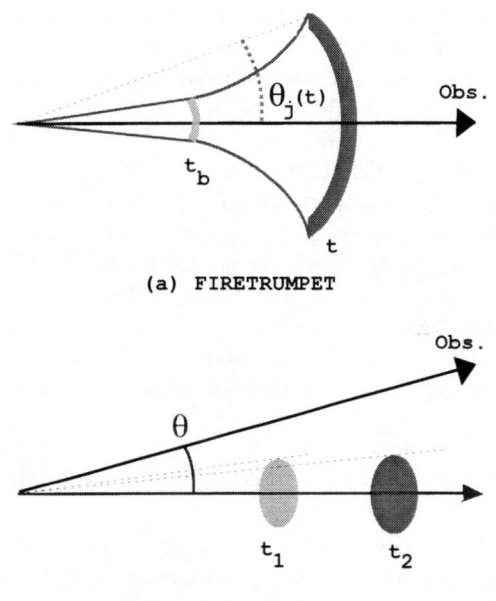

**FIGURE 1.** (a) A firecone or, more properly, a *firetrumpet*. In these scenarios the cone expands conically for a distance, after which the jet angle $\theta_j$ widens faster as its firefront travels. (b) Cannonballs (shown here, somewhat pedantically, a bit Lorentz-contracted) subtend decreasing angles as they travel. The only relevant angle in the CB model is the observer's viewing angle $\theta$.

## THE GRB AND ITS ENGINE

We assume that in core-collapse SN events a tiny fraction of the parent star's material, external to the newly-born compact object, falls back in a time of the order of very roughly one day [15, 7]. Given the considerable specific angular momentum of stars, it settles into an accretion disk around the compact object. The subsequent sudden episodes of catastrophic accretion —occurring with a chaotic time sequence that we cannot predict— result in the emission of CBs, lasting till the reservoir of accreting matter is exhausted. The emitted CBs initially expand in their rest system at a speed $\beta_T c$, of the order of the speed of sound in a relativistic plasma ($\beta_T = 1/\sqrt{3}$). These considerations are illustrated in Fig. (2).

From this point onwards, the CB model relies on processes whose outcome can be approximately worked out in an explicit manner. The collision of the CB with the SN shell heats the CB (which is not transparent at this point to $\gamma$'s from $\pi^0$ decays) to a surface temperature that, by the time the CB reaches the transparent outskirts of the SN shell, is $\sim 150$ eV, further decreasing as the CB travels [8]. The resulting quasi-thermal CB surface radiation, Doppler-shifted in energy and forward-collimated by the CB's fast motion, gives rise to an individual pulse in a GRB [8]. The GRB light curve is an ensemble of such pulses, often overlapping one another. The energies of the individual GRB $\gamma$-rays, as well as their typical total fluences, require CB Lorentz factors $\gamma$ of $O(10^3)$.

The CB model also explains the "Fe lines" seen in some X-ray AGs, as boosted hydrogen-recombination lines [9]. Their properties require $\gamma \sim 10^3$ and a baryonic number per cannonball $N_{CB} \sim 6 \times 10^{50}$. Even in a GRB with very many significant pulses, the total mass of a jet of CBs would be comparable to that of the Earth: peanuts, by stellar standards.

The rest of this note is based on [11], and concentrates on afterglows, for which the CB theory is very simple.

## OPTICAL AFTERGLOWS: THEORY

When an expanding CB, in a matter of (observer's) seconds, becomes transparent to its enclosed radiation, it loses its internal radiation pressure. If it has been expanding at a speed comparable to that of relativistic sound, should it not inertially continue to do so? No! We assume CBs to enclose a magnetic field maze, as the observed ejections from quasars and microquasars do. The interstellar medium (ISM) the CBs traverse has been previously partially ionized by the forward-beamed GRB radiation. The neutral ISM fraction is efficiently ionized by Coulomb interactions as it enters the CB. In analogy to processes occurring in quasar and microquasar ejections, the ionized ISM particles are multiply scattered, in a "collisionless" way, by the CBs' turbulent magnetic fields. In the rest system of the CB the ISM swept-up nuclei are isotropically re-emitted, exerting an inwards

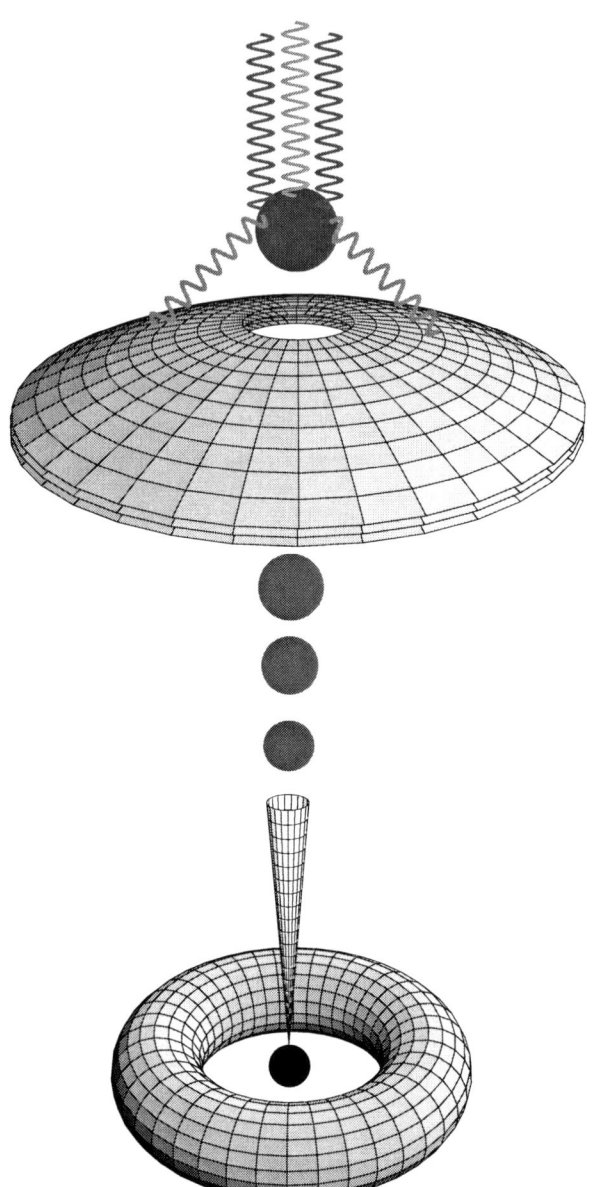

**FIGURE 2.** An "artist's view" (not to scale) of the CB model of GRBs and their AGs. A core-collapse SN results in a compact object and a fast-rotating torus of non-ejected fallen-back material. Matter (not shown) catastrophically accreting into the central object produces a narrowly collimated beam of CBs, of which only some of the "northern" ones are depicted. As these CBs pierce the SN shell, not precisely on the same spot, they heat and re-emit photons, that are Lorentz-boosted and collimated by the CBs' relativistic motion.

force on the CB's surface. This allows one to compute *explicitly* the CB's radius as a function of time [11]. The radius, for typical parameters, reaches a constant value $R_{max} \sim 10^{14}$ cm in minutes of observer's time.

The ISM nuclei (mainly protons) that a CB scatters

also decelerate its flight: its Lorentz factor, $\gamma(t)$, is calculable. Travelling at a large $\gamma$ and viewed at a small angle $\theta$, the CB's emissions are strongly relativistically aberrant: in minutes of observer's time, the CBs are parsecs away from their source. For a constant CB radius and an approximately constant ISM density, $\gamma(t)$ has an explicit analytical expression [11]. Typically $\gamma = \gamma(0)/2$ at a distance of order 1 pc from the source.

The ISM electrons entering a CB bounce off its enclosed magnetic domains, lose energy effectively by synchrotron radiation, and acquire a predictable power-law energy spectrum, $dn_e/dE \propto E^{-3.2}$, which implies a given distribution of the radiated photons: $\nu dn_\gamma/d\nu \propto \nu^{-1.1}$ [16]. The emitted energy rate, in a CB's rest system, is equal to the rate at which the ISM electrons bring energy into the CB[1]. The AG fluence in the CB model is of the form:

$$F_\nu = f \nu^{-\alpha} [\gamma(t)]^{2\alpha} \left[ \frac{2\gamma(t)}{1+\gamma(t)^2 \theta^2} \right]^{3+\alpha}, \quad (1)$$

with $\alpha \sim 1.1$ and $f$ an *explicit* normalization proportional to the ISM electron density, $f \propto n_e$.

We assume that all long-duration GRBs are associated with SNe and, as a bold ansatz, we take these SNe to have the fluence of SN1998bw, properly transported in time and frequency to the GRB's redshift [17]. An observed AG's fluence is then the sum of this SN, the background galaxy emission, and the CBs' contribution, Eq. (1). Remarkably, this very simple theory very successfully describes the optical AGs, *at all times*, of *every* GRB of known redshift.

## X-RAY AFTERGLOWS: THEORY

In the CB model, the X-ray emission by a GRB is more complex than its optical emission. During the GRB the emitted light at all energies is mainly of thermal origin (although it does not have a thermal spectrum) and, in a fixed energy interval, it decreases exponentially with time [8]. A few seconds after the last GRB pulse (the last CB), this pseudothermal emission becomes a subdominant effect. For the next few hours, the evolution of a CB is interestingly complicated. In particular, its originally ionized material should recombine into hydrogen and emit Lyman-$\alpha$ lines that are seen Doppler-boosted to keV energies [9]. Later, the CBs settle down to a much simpler phase, which typically lasts for months, till the CBs finally stop moving relativistically.

The X-ray AG is initially dominated by thermal bremsstrahlung (TB), which has a harder spectrum than synchrotron emission. This period of TB-dominance begins at a time $t_{trans}$, a few seconds after the end of the GRB, when the last CB becomes transparent to its enclosed radiation. A few minutes later, both the X-ray and optical AGs are dominated by synchrotron radiation, and their shapes are *achromatic*, as in Eq. (1).

The TB X-ray fluence decreases with time as $R^{-3}T^{1/2}$, with $R$ the CB's radius, still increasing linearly with time at an early stage. Depending on whether a CB's cooling soon after $t_{trans}$ is dominated by TB, or by adiabatic losses, the X-ray fluence is $\propto t^{-5}$ or $\propto t^{-4}$, the first behaviour being expected for our typical CB parameters.

The previous considerations justify a very simple description of the X-ray light curves:

$$F_X(t) \simeq f_X(t_{trans}) \left[ \frac{t_{trans}}{t} \right]^5 + F_{sync}(t), \quad (2)$$

where $t$ is the observer's time since the ejection of the (last) CB, and $F_{sync}(t)$ is the synchrotron fluence in the X-band, i.e. Eq. (1) integrated in the relevant energy interval. The normalization $f_X$ is, once again, explicit in the CB model. Equation (2) provides an excellent description, *at all times,* of *all* the X-ray AGs of GRBs of known $z$.

## OPTICAL AFTERGLOWS: RESULTS

GRBs have varied numbers, $n_{CB}$, of gamma-ray pulses, or CBs, which may have different initial Lorentz factors $\gamma_0$ and baryon numbers $N_{CB}$. This and other complications are eased by the fact that the AG light curve is the sum of temporally unresolved individual CB afterglows: we can characterize, as in Eq. (1), the AG with the parameters of one single CB, whose values represent a weighted average. The parameters to be fit are $f$, $\theta$ and $\alpha$ in Eq. (1), as well as two parameters entering the expression for $\gamma(t)$: $\gamma_0$ and the deceleration parameter $x_\infty \equiv N_{CB}/(\pi R_{max}^2 n_p)$, with $n_p$ the ISM proton number density light-centuries away from the GRB's progenitor.

The fits to the CB model are generally excellent. An example, GRB 970228, is shown in Fig. (3). Our only bad fit, that to GRB 000301c, is shown in Fig. (4). The occasional misfit is to be expected: the AG fluences are proportional to $n_e$, which is not constant for kpc distances, even in the halo of galaxies. The fitting procedure —which attributes to the errors a counterfactual purely-statistical origin— results in tiny 1 $\sigma$ spreads for the parameters, and in excellent confidence levels, on which one **should not** place excessive confidence.

The spectral slope in the optical-to-X-ray interval, $\alpha$, is the only parameter for which we have no reason to expect a range of different values. It is extremely satisfactory that the fitted values of $\alpha$ (extracted from the AG's

---

[1] The kinetic energy of a CB is mainly lost to the ISM protons it scatters; only a fraction $\sim m_e/m_p$ is re-emitted by electrons, as the AG.

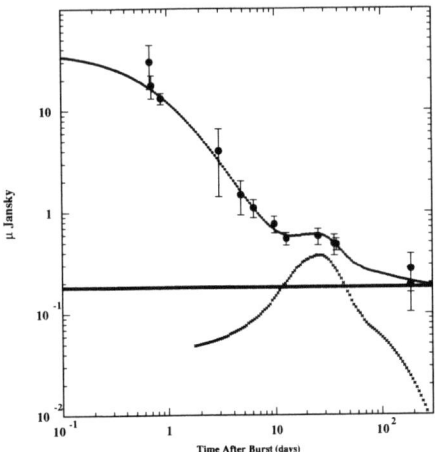

**FIGURE 3.** Comparison between the fitted R-band afterglow (upper curves) and the observations, not corrected for extinction, for GRB 970228, at $z = 0.695$, without subtraction of the host galaxy's contribution (the straight line). The contribution from a 1998bw-like supernova placed at the GRB's redshift, corrected for extinction, is indicated by a line of crosses. The SN bump is clearly discernible.

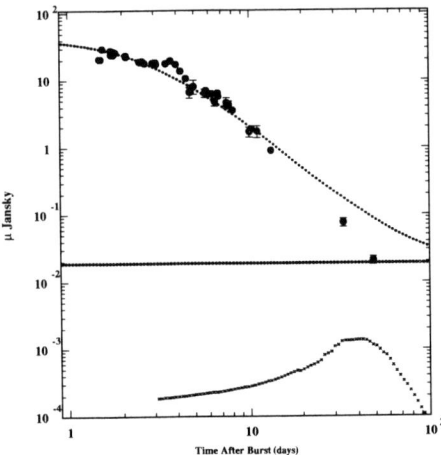

**FIGURE 4.** Comparison between the fitted R-band afterglow and the observations, not corrected for extinction, for GRB 000301c, at $z = 2.040$, without subtraction of the host galaxy's contribution (the straight line). The contribution from a 1998bw-like supernova placed at the GRB's redshift is too weak to be observable.

temporal shape at fixed ν) are, within errors, compatible with *all* of the GRBs having a universal behaviour with the theoretically predicted value: $\alpha \approx 1.1$. Most spectral measurements agree with this result and, for the ones that do not (and some of the ones that do) there is always a good reason not to worry: poorly-understood absorption. The distribution of $\gamma_0$ values agrees snugly with our expectation from the properties of the GRB: $\gamma_0 \sim 10^3$, and it is surprisingly narrow: $\Delta\gamma_0/\gamma_0 \sim 0.2$. The values of $x_\infty$ should be fairly spread: they depend on $n_p$ close to the progenitor (that determines $R_{max}$) and $n_e$ in the region where the AG is emitted. Indeed, the $x_\infty$ values range for an order of magnitude above and below a typical prediction. The overall normalization $f$ of the optical afterglows is, with the other parameters fixed, $f \propto n_{CB} N_{CB}$. The results from optical AG fits range for an order of magnitude above and below the prediction for a single dominant CB, $n_{CB} = 1$, and our typical expected $N_{CB}$. This must be partially due to absorption uncertainties, for the X-ray fits result in 1/3 as much spread.

To summarize, the distributions of *all* parameters are in extremely good agreement with the expectations of the CB model and, if anything, they are astonishingly close to what they would be for "standard candle" GRBs.

### GRB 970508: a case of gravitational lensing

The AG of GRB 970508 has a most peculiar shape. An attempt to fit it with Eq. (1) is shown in Fig. (5): it is a miserable failure. But suppose the light from the CBs of this GRB is gravitationally lensed by a star or a binary, of mass $\sim 2 M_\odot$, placed roughly half-way to their position (the probability for something like this to happen is a few per-cent). The lensed AG, whose CB model parameters are entirely conventional, is shown in Fig. (6): the fit is fantastically good. Comparing Figs. (5) and (6), notice how time-asymmetric the amplification is (the time scale is logarithmic!). This is because, as the lensing occurs, the CBs are — in a specific predicted fashion— slowing down from an initial superluminal speed $v_\perp \sim 500$ c. Seing a result like this, one knows *in one's bones* that one is on the right track!

### GRB 990123 and its mother's wind

For this GRB, there are good optical data starting exceptionally early: 22.18 seconds after its detected beginning [18]. The AG rises abruptly to a second point at $t = 47.38$ s, and decreases thereafter. We can very simply describe this AG from this point onwards, as shown in Fig. (7). From the fitted values of z, $\gamma_0$ and $\theta$, we conclude that at that time the CBs are a mere $x = 0.46$ pc away from the progenitor star. This is precisely where the density profile ought to be $n \propto r^{-2}$, induced by the parent-star's wind and ejecta. This early, the CB's deceleration is negligible: an $r^{-2}$ density profile implies an optical AG declining as $t^{-2}$. The shape and normalization of the early AG are precisely the expected ones (the

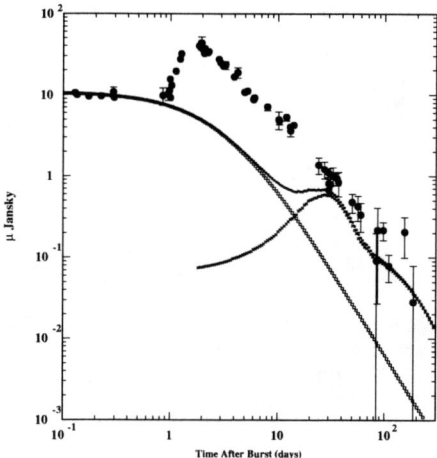

**FIGURE 5.** Comparisons between a fitted R-band afterglow (upper curves) and the observations, not corrected for extinction, for GRB 970508, at $z = 0.835$. The galaxy has been subtracted. The fit is a total disaster.

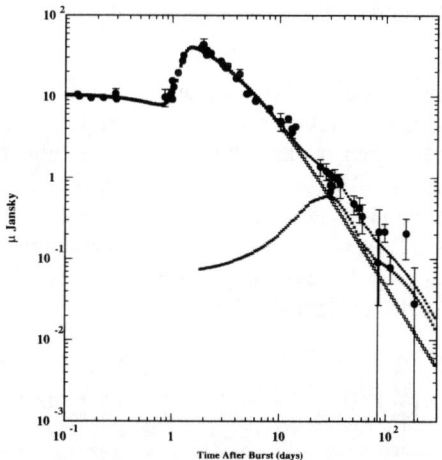

**FIGURE 6.** The R-band AG of GRB 970508, fit with the additional effect of gravitational lensing by a $\sim 2M_\odot$ intervening object. This time the fit is excellent. A SN1998bw-like contribution is necessary.

inferred local density is $0.54\,\text{cm}^{-3}$ at $x \simeq 0.46$ pc). The CB model describes the full history of an optical AG.

## X-RAY AFTERGLOWS: RESULTS

Once again, I can only show a subset of our results. The X-ray AG of GRB 010222, and its CB-model's description, are shown in Fig. (8). The early decline

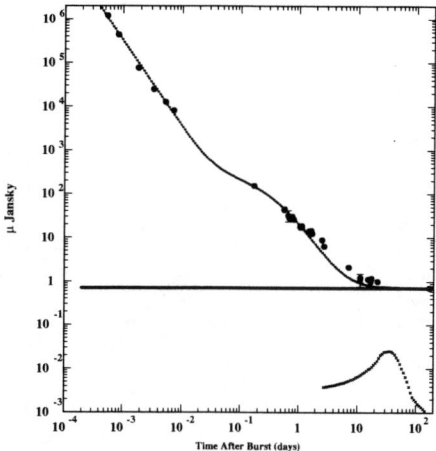

**FIGURE 7.** Comparisons between the fitted R-band afterglow and the observations, not corrected for extinction, for GRB 990123, at $z = 1.600$, without subtraction of the host galaxy. The data on this AG begins a very short time after the GRB. The starting $t^{-2}$ behaviour is that expected if the CBs are moving trough a density profile, $n \propto r^{-2}$, induced by the parent-star's pre-SN wind and ejections.

is dominated by thermal bremsstrahlung and has the expected $t^{-5}$ decline. The late AG is achromatic: the shape of the late X-ray fluence is that obtained with the parameters resulting from the fit to the optical AG. The normalizations at early and late times are in the expected range. The approximation of a constant ISM density in the normalization of $F_{sync}(t)$ in Eq. (2) should be inappropriate between $\sim 2 \times 10^{-3}$ and $\sim 0.2$ days. There are no data in that domain except for GRB 991216 and perhaps 970508, which suggest an initial density variation $\propto 1/r^2$, resulting in an observed $\sim t^{-2}$ decline, as in the optical AG of GRB 990123, shown in Fig. (7).

Our worst but most significant fit to an X-ray AG is that to GRB 980425, shown in Fig. (9). Unlike the observers [19] we assume the AG was produced by the CBs and *not* by conventional ejecta of the associated SN 1998bw.

## Even GRB 980425 is "normal"

The optical AG of this close-by GRB is dominated by SN1998bw, and we have extracted its AG parameters ($\theta$, $\gamma_0$, etc.) from the X-ray fit of Fig. (9). The fitted viewing angle is $\theta \sim 8.3$ mrad; $\gamma_0 \sim 750$, etc. are normal. If the CBs of GRB 980425 had been viewed from a typical viewing angle, $\theta \leq 1/\gamma_0$, the equivalent isotropic energy would have been in the range of all other GRBs. If for this case the ISM density and CB radius were the

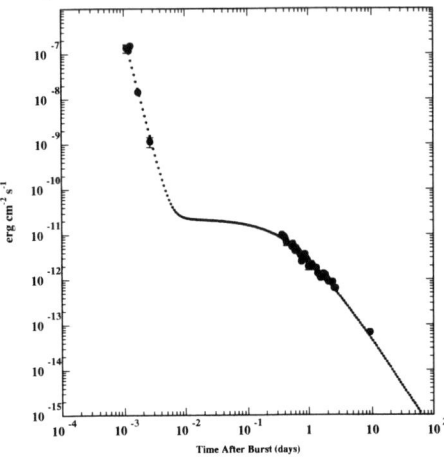

**FIGURE 8.** The early-time and late-time X-ray AG of GRB 010222 in the 2–10 keV, fitted with a constant density along the CB trajectory. The early decline is $\propto t^{-5}$; the late behaviour is achromatic: "parallel" to the optical AG curve.

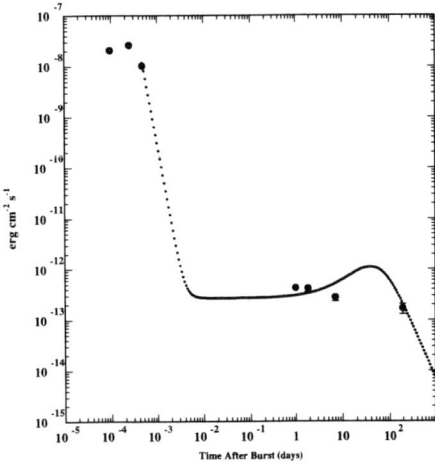

**FIGURE 9.** CB-model fit to the X-ray afterglow of the SN1998bw/GRB 980425 pair. The flatish domain we call "plateau". It is so extensive because $\theta\gamma_0 \gg 1$ and it takes time to reach the maximum at $\theta\gamma(t) \sim 1$.

same as for other GRBs, the predicted intensity of the X-ray "plateau" is $\sim 4 \times 10^{-13}$ erg cm$^{-2}$ s$^{-1}$, as observed [20]. The "normal" GRB energy and X-ray AG fluence strongly support the association of SN1998bw with (a not exceptional) GRB 980425.

The parameters of the X-ray AG can be used to predict the magnitude and shape of the optical AG of the blended SN 1998bw/GRB 980425 system, see Fig. (10). The CBs' contribution dominates at late time and is in perfect

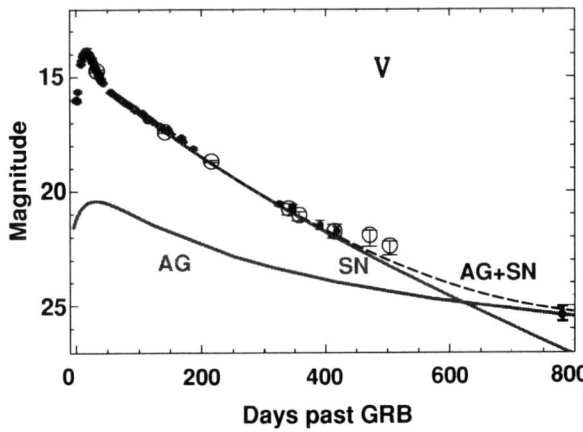

**FIGURE 10.** The V-band light curve of SN1998bw/GRB 980425. The blue "SN" curve is a fit to the SN [21], dominated after day $\sim 40$ by $^{56}$Co decay. The red "AG" curve is our prediction for the CB-induced AG component, as given by Eq. (1), with the parameters determined from the X-ray AG fit in the previous figure. The SN contribution dominates up to day $\sim 600$. The last point is an HST measurement at day 778, that precisely agrees with the (dashed) SN plus CB prediction for the total AG.

agreement with the HST observation [22] on day 778. At that time the SN and the CB (this is a single-pulse GRB) were far enough from each other, and close enough to us, to be resolvable! [7]. Alas, the rare occasion was missed.

Interpreted in the CB model, the fluence and soft spectrum of GRB980425, as well as its X-ray and optical AGs were "normal". It was simply much closer, and viewed at a much larger angle than other GRBs.

## Are GRBs associated with SNe?

The complete success of the CB model in describing optical AGs results in an excellent exposition [11] of the GRB/SN association. It is useful to discuss the issue in order of diminishing redshift. In the six more distant GRBs there is no evidence for *or against* a SN 1998bw-like component, two examples are given in Figs. (4) and (7). In the next five closer cases there is evidence, ranging from fairly weak to very strong; see Fig. (6) for GRB 970508, for which the light curve —in the CB model— clearly requires a SN component. The other four of these cases (000418, 980613, 991216 and 980703) are shown in Fig. (11). For the next three closer GRBs —991208 and 990712, shown in Fig. (11) and 970228, shown in Fig.(3)— a SN1998bw-like contribution is clearly required. Finally, GRB 980425, at $z = 0.0085$ is indeed associated to a SN [23]. The trend is clear. The closer a GRB is, the better the evidence for a SN. For the more distant GRBs, a SN could not be seen, even if it was

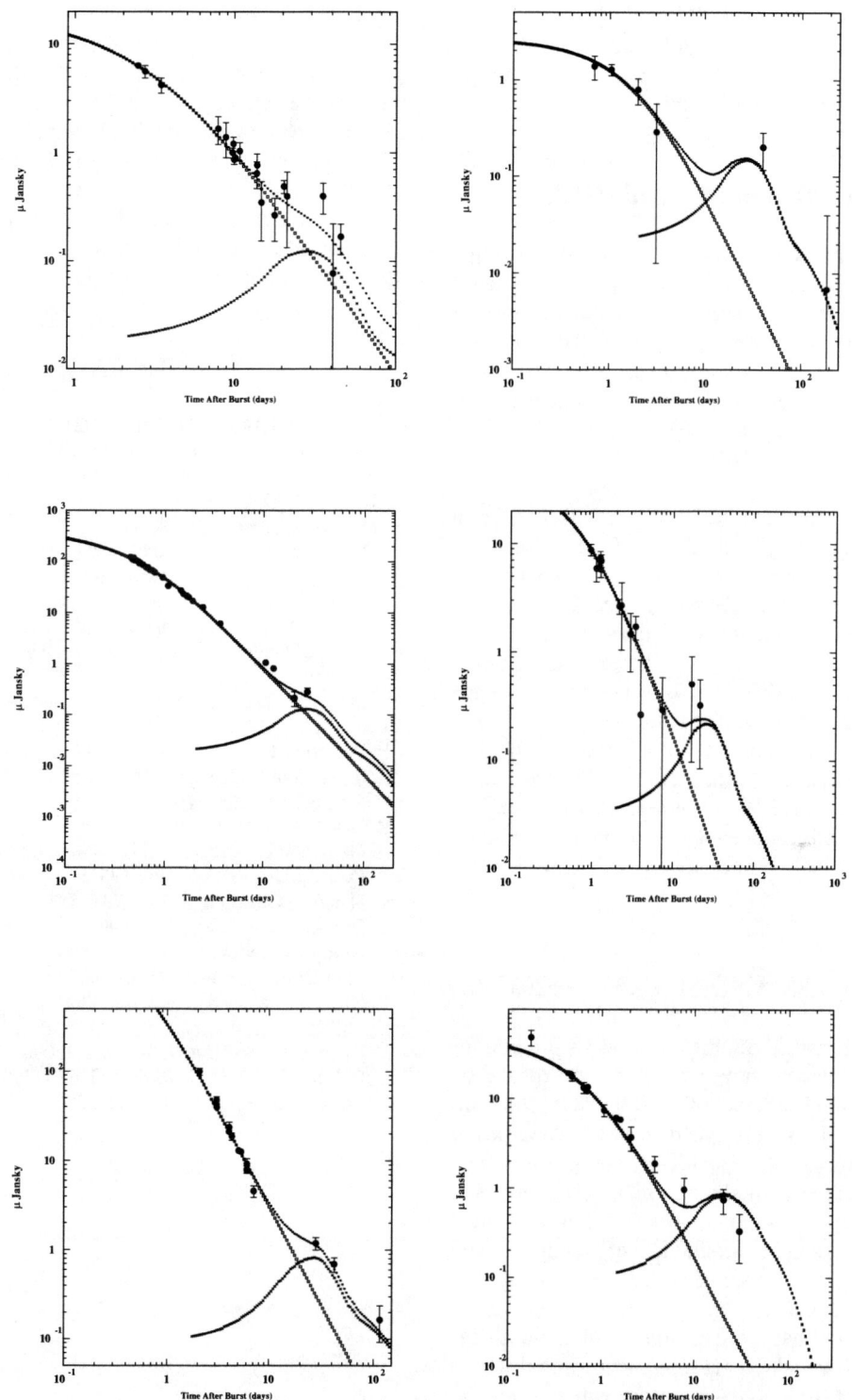

**FIGURE 11.** Galaxy-subtracted R-band AGs of some of the least distant GRBs, exhibiting the presence of a SN1998bw-like component. This SN appears to be a surprisingly standard candle. From left to right and top to botton: GRB 000418 (at $z = 1.119$), 980613 (1.096), 991216 (1.020), 980703 (0.966), 991208 (0.706), 990712 (0.434).

there. In all cases where the SN could be seen, it was seen; the evidence gaining in significance as the distance diminishes. The conclusion that all long-duration GRBs are associated with SNe is irresistible.

## What fraction of SNe emit GRBs?

The γ-ray fluence of a GRB [7] is $\propto [2\gamma_0/(1+\gamma_0^2\theta^2)]^3$. The θ dependence is the steepest parameter dependence of the CB model. It is reasonable to attribute the range of equivalent spherical energies mainly to the θ dependence (as if GRBs were otherwise standard candles). Excepting GRB 980425, the observed spread in equivalent energy then corresponds to a spread $\theta \approx 0$ to $\theta_{max} \approx 2.4/\gamma$. Thus the fraction of observable GRBs (with the current or past sensitivity) is $f(\gamma) = 2\pi\theta_{max}^2/(4\pi) \approx 2.84/\gamma^2$, with two jets of CBs per event. The rate of Type II/Ib/Ic SNe in the observable universe, multiplied by $f(10^3)$, coincides with the observed rate of GRBs, as if every SN emitted GRBs! But the observational errors are *very* large; distant SNe may be more frequent than in current estimates, if the star formation rate continues to increase above $z \sim 1$; the efficiency of GRB detection as a function of fluence has not been taken into account in this estimate. In spite of these uncertainties, the conclusion, in the CB model, is that *a very large fraction of SNe emit GRBs*. Then, why is SN1998bw peculiar? We have seen that in X-rays it is not: they were CB-induced. Other peculiarities should also be due to how close to the GRB "axis" this SN was observed.

## CONCLUSIONS AND SOCIAL AFFAIRS

We have not yet worked out the CB-model's predictions for radio afterglows. We have no CB-model explanation for the scintillation behaviour of GRB 970508 [24]. Other than that, the CB model explains well all properties of GRBs. In the case of optical and X-ray afterglows, which I have outlined, the CB-model's predictions are univocal (as opposed to multiple-choice), very explicit, analytical in fair approximations, quite simple, very complete, and extremely successful. I doubt that this statement applies to other models of GRBs.

I am not saying that the CB model will stand all future tests. But I would expect that —confronted with a simple and successful model— most scientists would, at least, say: *Hum!* and ask good questions. Not the case. Four of our papers on this subject, [7] to [10], have already been rejected by referees who found no single error, and/or stuck to bad numerical questions, even after being proved numerically (i.e. inarguably) wrong. This rejection statistics makes me feel that the GRB community has officially certified us... as crackpots. All this reflects, I suspect, the global rise of fundamentalism. Some people, almost literally in this case, still refuse to "look through the telescope". A GRB theorist, sitting on the first row in my talk, was kind enough to enact my social comments. Indeed, he rose in ire to exclaim: *I am glad that the referee system is working!* He then stuck to a bad numerical question. Most of the rest of the questions period was not this aggressive: there is still hope in science's eternal contest with faith.

## REFERENCES

1. Brainerd J.J., *ApJ*, **394**, 33L (1992).
2. Meszaros P., Rees M.J., *MNRAS*, **257**, 29P (1992).
3. Shaviv N.J., Dar A., *ApJ*, **447**, 863 (1995).
4. Dar A., 1997, astro-ph/9704187.
5. Dar A., *ApJ*, **500**, L93 (1998).
6. Dar A., Plaga R., *A&A*, **349**, 259 (1999).
7. Dar A., De Rújula, A., astro-ph/0008474, rejected by *A&A*.
8. Dar A., De Rújula, A., astro-ph/0012227, rejected by *A&A*.
9. Dar A., De Rújula, A., astro-ph/0102115, rejected by *A&A*.
10. Dar A., De Rújula, A., astro-ph/0105094, rejected by *Phys. Rev.*
11. Dado, S., Dar A., De Rújula, A., astro-ph/0107367, submitted to *A&A*.
12. Pearsons T.J., Zensus J.A., 1987, *Superluminal Radio Sources*, Cambridge Univ. Press 1987, pp. 1.
13. Mirabel I.F., Rodriguez L.F., *Nature*, **371**, 46 (1994); *ARA&A*, **37**, 409 (1999).
14. Ghisellini G., Celotti A., astro-ph/0103007.
15. De Rújula A., *Phys. Lett.*, **193**, 514 (1987).
16. Dar A., De Rújula, A., *MNRAS*, **323**, 391 (2001).
17. Dar A., *GCN Circ.*, 346 (1999).
18. Akerlof C., et al., *Nature*, **398**, 400 (1999).
19. Pian E., et al., *A&AS*, **138(3)**, 463 (1999).
20. Pian E., et al., *A&A*, **372**, 456 (2001).
21. Sollerman J., et al., astro-ph/0006406
22. Fynbo J.U., et al., *ApJ*, **542**, 89L (2000).
23. Galama T.J., et al., *Nature*, **395**, 670 (1998).
24. Taylor G.J., et al., *Nature*, **389**, 263 (1997).

# Afterglows with High Inferred Values of $\varepsilon_e$ and $\varepsilon_B$: A Clue to the Gamma-Ray Burst Environment?

Arieh Königl* and Jonathan Granot[†]

*Department of Astronomy & Astrophysics, University of Chicago, 5640 S. Ellis Ave., Chicago, IL 60637
[†]Institute for Advanced Study, Olden Lane, Princeton, NJ 08540
arieh@jets.uchicago.edu, granot@ias.edu

**Abstract.** Spectral modeling of GRB afterglows using synchrotron-radiation theory has implied high values for the fraction of the internal energy residing in relativistic electrons and positrons ($\varepsilon_e \gtrsim 0.1$), and in at least some cases also for the magnetic-to-internal energy ratio ($\varepsilon_B \sim 0.01 - 0.1$). These results are difficult to understand under the usual assumption that the afterglow-emitting shock propagates into a standard interstellar-medium or stellar-wind environment. We suggest that the high values of $\varepsilon_e$ and $\varepsilon_B$ can be naturally explained if the shock propagates, instead, into a pulsar-wind bubble. One possible scenario in which such an environment is produced is provided by the supranova model of GRB formation, wherein a supernova explosion leaves a rapidly rotating neutron star that emits intense pulsar-type radiation over a period of months to years before collapsing into a black hole and triggering the GRB. Guided by recent results on plerions, we construct a simple model of the bubble structure and show that it can reproduce the source parameter values inferred from the spectral fits.

## INTRODUCTION

Gamma-ray burst (GRB) sources are commonly interpreted in terms of nonthermally emitting shocks associated with relativistic (and possibly highly collimated) outflows from stellar-mass black holes or strongly magnetized and rapidly rotating neutron stars [e.g., 16, 13]. The prompt high-energy emission is thought to originate in the outflow itself, with the $\gamma$-rays attributed to internal shocks within the flow and with the associated optical "flash" and radio "flare" emission ascribed to the reverse shock that is driven into the outflowing material as it starts to be decelerated by the inertia of the swept-up ambient gas. By contrast, the longer-term, lower-energy afterglow emission is attributed to the forward shock that propagates into the ambient medium. The ambient gas is usually taken to be either the interstellar medium (ISM) of the host galaxy or a stellar wind from the GRB progenitor star.

It appears that most of the observed emission from GRBs and their afterglows represents synchrotron radiation. Source-energetics considerations imply that the emission efficiency must be high, and hence that the ratio $\varepsilon_e$ of the internal energy in relativistic electrons and positrons to the total internal energy density in the emission region is not $\ll 1$ and that the ratio $\varepsilon_B$ of the magnetic-to-internal energy densities is not $\ll \varepsilon_e$. In the case of afterglows, similar conclusions are obtained directly from spectral modeling, which indicates that $\varepsilon_e$ is typically $\gtrsim 0.1$ [e.g., 6, 15] and that $\varepsilon_B$ can be as high as $\sim 0.01 - 0.1$ in certain sources (e.g., GRB 970508; [26, 8, 4]).

If the shocked gas consists of protons and electrons, then only moderately high ($\lesssim 0.1$) values of $\varepsilon_e$ may be expected even under optimal circumstances [3]. For $\varepsilon_e$ to approach 1, it is probably necessary for the preshock gas to be composed primarily of $e^\pm$ pairs. A pair-dominated outflow is, in fact, a feature of certain GRB models. Furthermore, the radiative efficiency of the reverse shock (and possibly also of the forward shock during the early afterglow phase) could be enhanced through pair creation by the high-energy photons comprising the $\gamma$-ray pulse [23, 14]. There is, however, no natural way to account for large values of $\varepsilon_e$ during the later phases of afterglows in a typical ISM or stellar-wind environment.

It is in principle also possible to account for comparatively large values of $\varepsilon_B$ in internal and reverse shocks by appealing to shock compression of magnetized outflows [e.g., 22, 7]. However, in the case of afterglows in the standard scenario, the highest values of $\varepsilon_B$ that might be attained in this fashion (e.g., in a shock propagating into a magnetized wind from a progenitor star; see [2]) could at best account only for the low end of the actual range inferred in GRB afterglows ($\varepsilon_B \gtrsim 10^{-5}$; e.g., [15]). As an alternative to compressional amplification of a preshock field, various proposals have been advanced for generating strong magnetic fields in the shocks themselves, but so far none of the suggested mechanisms for which quantitative predictions have been obtained can account for a source like GRB 970508.

We propose that large values of $\varepsilon_e$ and $\varepsilon_B$ in afterglows are naturally accounted for if the outflow that gives rise to the $\gamma$-ray pulse expands into a pulsar-wind bubble (PWB). Such a bubble forms when the relativistic wind (consisting of relativistic particles and magnetic fields) that emanates from a pulsar shocks against the ambient gas and creates a "pulsar nebula." When a bubble of this type expands inside a supernova remnant (SNR), it gives rise to a "plerionic" SNR, of which the Crab and Vela remnants are prime examples. GRBs can arise inside PWBs under a number of plausible scenarios. Here we focus on the supranova model for the origin of GRBs [25], in which a rotationally supported "supramassive" neutron star (SMNS) forms by a superanova explosion that is triggered by the collapse of a massive star. In this picture, the SMNS loses angular momentum (and hence centrifugal support) through pulsar-type electromagnetic radiation until (on a time scale of months to years) it becomes unstable to collapse to a black hole, at which point a GRB outflow is produced.

## PULSAR-WIND BUBBLES IN YOUNG SUPERNOVA REMNANTS

In the supranova scenario, the SMNS loses rotational energy $\Delta E_{\rm rot} \approx 10^{53}$ ergs on the spindown time $t_{\rm sd} = \tau_{\rm sd}$ yr, which represents the time interval between the supernova explosion and the GRB event. It is thus natural to parameterize the pulsar wind luminosity by $\tau_{\rm sd}$: $L_w = \Delta E_{\rm rot}/t_{\rm sd} \approx 3 \times 10^{45} \Delta E_{53} \tau_{\rm sd}^{-1}$ ergs s$^{-1}$.[1] The wind luminosity consists of electromagnetic and particle contributions. The magnetic field is expected to be transverse to the flow direction, so the Poynting-to-particle energy flux ratio in the wind is given by $\sigma_w = B_w^2/4\pi\rho_w c^2$, where $B_w$ is the field amplitude and $\rho_w$ is the rest-mass density (both measured in the fluid frame). We have derived solutions for $\sigma_w$ in the range $10^{-3} - 1$ (see [11]), but here we only present results for $\sigma_w = 1$, a value indicated by recent X-ray observations of the Vela pulsar nebula [9]. We assume, for simplicity, that the composition is pure $e^\pm$ pairs. Furthermore, in view of recent model fits to the Crab nebula spectrum [1], we adopt $\gamma_w = 10^4$ as a typical Lorentz factor of SMNS outflows and assume that its magnitude does not change significantly on the time $t_{\rm sd}$.

As the powerful SMNS wind expands inside the SNR, it will compress the supernova ejecta into a thin shell and accelerate it [e.g., 18]. To within factors of order 1, the outer radius of the bubble at time $t_{\rm sd}$ can be approximated by $R_b \approx v_b t_{\rm sd} = 9.5 \times 10^{16} \beta_{b,-1} \tau_{\rm sd}$ cm, where $v_b = 0.1\beta_{b,-1}c$. To the extent that $v_b \propto (\Delta E_{\rm rot}/M_{\rm ej})^{1/2}$ (where $M_{\rm ej} \approx 10 M_\odot$ is the ejecta mass) has nearly the same value in all sources, the magnitude of $R_b$ can also be parameterized by $\tau_{\rm sd}$. We follow previous treatments of PWB structure [17, 10, 5] in our assumptions about the basic morphology of the bubble: we take it to be spherical, with an outer radius $R_b$, and assume that the pulsar wind propagates freely until it is shocked at a radius $R_s$. Our model differs, however, from previous treatments in that we take account of synchrotron radiation losses (which could be important during the early phase of the nebula) and that we allow for a departure from ideal MHD through the assumption of an "equipartition" upper bound on the electromagnetic-to-thermal pressure ratio in the shocked-wind bubble.[2] Given that the estimated internal acoustic speed is reasonably larger than $v_b$, we simplify the treatment by adopting a stationary-flow approximation within the PWB. This approximation entails a certain freedom in the choice of boundary conditions for the set of differential conservation equations (involving particle number, momentum, and energy) that we solve, but we have verified that the main results are not sensitive to which specific option is implemented.

For a given value of $\sigma_w$, the model PWBs depend on a single parameter, $a_1 \equiv \sigma_T L_w \gamma_w / 2 m_e c^3 R_b$, which measures the relative importance of radiative cooling within the bubble. Numerically, $1/a_1$ is of the order of the nominal radiative cooling time of the bubble in units of $R_b/c$, and hence the larger the value of $a_1$, the stronger the role that radiative cooling plays in determining the bubble structure.[3] Weakly cooling PWBs ($a_1 \lesssim 10^2$, or, for our fiducial values, $\tau_{\rm sd} \gtrsim 10$) have radial widths $\Delta R_b \equiv (R_b - R_s)$ that are of the order of $R_b$, whereas strongly cooling PWBs ($a \gg 10^2$, $\tau_{\rm sd} \ll 10$), have $\Delta R_b/R_b \ll 1$. Unless radiative cooling is very efficient, the interior of the bubble will have a "hot" equation of state (enthalpy

---

[1] The luminosity of the pulsar outflow can also be expressed in terms of the SMNS parameters using the magnetic dipole-radiation formula. This yields a characteristic spindown time (in years) $\tau_{\rm sd} \approx 6(\alpha/0.5)(M/2.5 M_\odot)^2 (R_c/15 {\rm km})^{-6} (\Omega/10^4 {\rm s}^{-1})^{-3} (B/10^{12} {\rm G})^{-2}$, where $M$, $B$, $R_c$, and $\Omega$ are the mass, polar surface magnetic field, circumferential radius, and (uniform) angular velocity of the SMNS, and $\alpha \equiv \Delta E_{\rm rot}/(GM^2\Omega/2c)$ is the portion of the rotational energy that needs to be lost before the SMNS becomes unstable to collapse. Given that the inferred surface magnetic fields of radio pulsars lie in the range $\sim 10^{12} - 10^{13}$ G, the value of $\tau_{\rm sd}$ could vary by roughly two orders of magnitude among different sources.

[2] In previous treatments it was usually assumed that ideal MHD remains applicable throughout the bubble. As is, however, well known [e.g., 17], such a model cannot describe a bubble with $v_b \ll c$ and $\sigma_w \approx 1$.

[3] Note that $a_1 \propto \tau_{\rm sd}^{-2}$ if $\beta_b$ and $\Delta E_{\rm rot}$ are approximately constant from source to source.

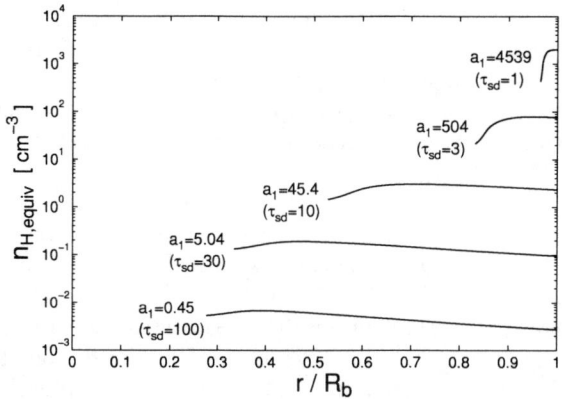

**FIGURE 1.** The effective hydrogen number density as a function of the normalized radius $r/R_b$ for $\sigma_w = 1$ and several values of the cooling parameter $a_1$. The dimensional scaling and the indicated pulsar spindown times $\tau_{sd}$ (in years) correspond to the fiducial parameter values.

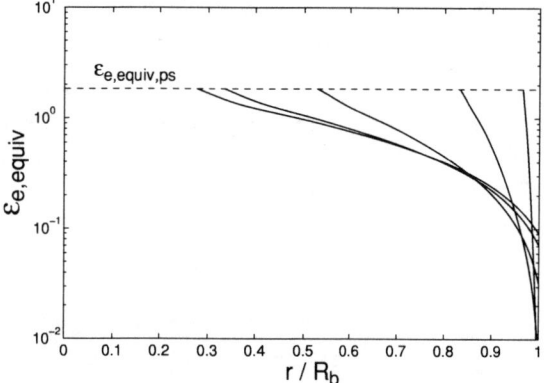

**FIGURE 2.** The equivalent electron energy fraction $\varepsilon_{e,\text{equiv}}$ as a function of the normalized radius for the PWB models depicted in Fig. 1. The *dashed* line shows the value of $\varepsilon_{e,\text{equiv}}$ immediately behind the wind shock.

density related to particle pressure $p$ by $w = 4p$). The sum of $w$ and $(B+E)^2/4\pi$ (where $B$ and $E$ are, respectively, the comoving magnetic and electric fields) represents the effective enthalpy density $w_{\text{eff}}$ that determines the compression in relativistic shocks that propagate within the bubble. In the "cold," weakly magnetized limit that is applicable to ISM and stellar-wind environments, $w_{\text{eff}} = w = \rho c^2$, where $\rho = n_H m_p$ is the rest-mass density of the medium. Shock models of GRB afterglows traditionally infer an ambient gas density based on this relation. This motivates us to define the quantity $n_{H,\text{equiv}} \equiv w_{\text{eff}}/m_p c^2$, which is plotted in Figure 1. Other properties of these solutions are discussed in [11].

## IMPLICATIONS TO GRB AFTERGLOWS

Pulsar wind-inflated bubbles provide an optimal environment for GRB afterglows since they are expected to contain a significant $e^\pm$ component and to be strongly magnetized. Thus they naturally yield high electron and magnetic energy fractions ($\varepsilon_e$ and $\varepsilon_B$) behind the propagating shock wave that gives rise to the afterglow emission. It must be noted, however, that the observationally inferred values of $\varepsilon_e$ and $\varepsilon_B$ are derived from spectral fits that are based on the standard model assumptions of a "cold" proton-electron preshock medium. We have therefore examined what would be the "equivalent" values (which we denote by $\varepsilon_{e,\text{equiv}}$ and $\varepsilon_{B,\text{equiv}}$, respectively) that one would derive if the afterglow-emitting shock propagated instead inside a PWB. For simplicity, we restricted attention to the three synchrotron-spectrum characteristics considered by [21], namely, the break frequencies $\nu_m$ and $\nu_c$ and the peak flux $F_{\nu,\text{max}}$, and we approximated the equivalent hydrogen number density inside the bubble as being roughly constant. Our models imply that $\varepsilon_{e,\text{equiv}}$ typically lies in the range $\sim 0.1 - 1$ (see Fig. 2) and that $\varepsilon_{B,\text{equiv}}$ varies from $\lesssim 1$ to $\lesssim 10^{-2}$ as $\sigma_w$ decreases from 1 to $10^{-3}$. Furthermore, we find that the standard expressions for a slow-cooling (adiabatic) shock that propagates into a uniform ambient medium remain approximately applicable (especially when the cooling inside the PWB is not too strong) if one replaces $n_H$, $\varepsilon_e$, and $\varepsilon_B$ by their "equivalent" counterparts.

The above considerations suggest that, as a rough check of the compatibility of the PWB model with observations, one can examine the consistency of the predicted values of $\varepsilon_{e,\text{equiv}}$, $\varepsilon_{B,\text{equiv}}$, and $n_{H,\text{equiv}}$ with the values of $\varepsilon_e$, $\varepsilon_B$, and $n_H$ that are inferred from the spectral data by using the standard ISM model. It is thus quite encouraging that our simple model reproduces the typical values inferred for $\varepsilon_e$ and can also account for comparatively high values of $\varepsilon_B$, which, as we noted above, have posed a challenge for the conventional interpretation. The derived values of $n_{H,\text{equiv}}$ are also consistent with the observationally inferred preshock particle densities: the typical range $\sim 0.1 - 50 \text{ cm}^{-3}$ estimated under the assumption of a uniform, "cold" ambient medium [e.g., 15] is reproduced by PWBs with $\tau_{sd}$ in the range $\sim 3 - 30$, whereas the less commonly derived low ($< 10^{-2} \text{ cm}^{-3}$) and high ($> 10^2 \text{ cm}^{-3}$) densities correspond to lower (respectively, higher) values of the cooling parameter $a_1$.

Another attractive feature of the PWB scenario is that it naturally gives rise to radial profiles of $n_{H,\text{equiv}}$ that, depending on the cooling parameter $a_1$ and the location within the bubble, may resemble a uniform medium (constant-$n_H$ ISM or interstellar cloud) or a stellar wind ($n_H \propto r^{-2}$, although strictly $n_{H,\text{equiv}}$ does not decline

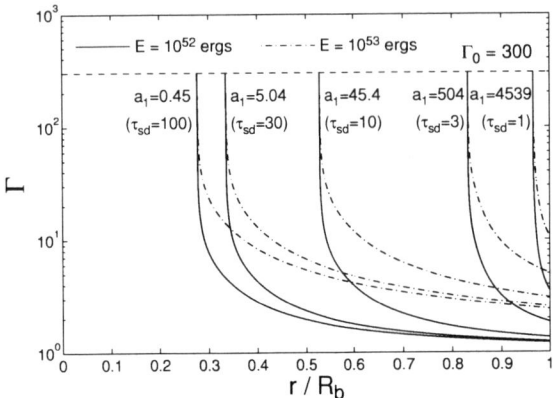

**FIGURE 3.** Lorentz factor of the shocked bubble material, plotted as a function of the normalized radius, for a spherical shock driven into a PWB by an outflowing mass of energy $E$ and initial Lorentz factor $\Gamma_0$. Results for two values of $E$ are shown for the bubble models of Figs. 1 and 2.

more steeply than $\sim r^{-1}$). Both types of behavior have, in fact, been inferred in afterglow sources. The unique aspect of the radial distribution of $n_{H,equiv}$ in this picture is that it spans a range of effective power-law indices $k$ that can vary from source to source, and, moreover, that the value of $k$ appropriate to any given afterglow is predicted to change with time as the afterglow-emitting shock propagates within the bubble. This leads to a more flexible modeling framework for the afterglow evolution and can naturally accommodate cases where a value of $k$ that is intermediate between those of a uniform ISM and a stellar wind could best fit the observations [see 12]. It also explains why afterglows associated with star-forming regions need not show evidence for a stellar-wind environment (as expected when the GRB progenitor is a massive star; in view of the results shown in Fig. 1, this model also makes it possible to understand how a source with such a progenitor could produce an afterglow with an implied value of $n_H$ that is much lower than the typical ambient density near massive stars). In addition, high values of $n_{H,equiv}$ in this picture are not subject to the objection that they might give rise to excess extinction.

The inferred radii of afterglow shocks typically lie between $\gtrsim 10^{17}$ cm and $\lesssim 10^{18}$ cm. These values are consistent with our estimate of $R_b$ for supranova–GRB time delays of $\lesssim 1$ yr to $\gtrsim 10$ yr. By solving the adiabatic shock-evolution equation, we have checked whether the afterglow-emitting gas will still be ralativistic (Lorentz factor $\Gamma \gg 1$) by the time the shock reaches the outer edge of the bubble at $R_b$. We found that, in all cases, the GRB outflow decelerates rapidly after entering the bubble (see Fig. 3). Only in the case of an energetic shock and a strongly cooling bubble is $\Gamma(R_b)$ appreciable (but even then it remains $\lesssim 10$). It is worth bearing in mind,

though, that if the outflow is collimated with a small opening half-angle $\theta_j$, then it will start to strongly decelerate due to lateral spreading when its Lorentz factor decreases to $\sim 1/\theta_j$ [19, 20], so that even the more energetic shocks could become nonrelativistic while they are still inside the PWB.

A GRB shock that reaches the supranova ejecta shell at $r = R_b$ with a Lorentz factor $> 1$ would be rapidly decelerated to subrelativistic speeds since the rest-mass energy of the shell [$\sim 2 \times 10^{55}(M_{ej}/10M_\odot)$ ergs] is in most cases much greater than the (equivalent isotropic) shock energy $E$. This is expected to lead to a discontinuous change in the shape and evolution of the afterglow spectrum, but the details remain to be calculated. The SNR shell could also manifest itself by imprinting X-ray features on the measured spectrum. Indeed, recent detections of such features in several GRB sources have been argued to provide strong support for the supranova scenario [e.g., 24]. The implications of these observations to the PWB afterglow model outlined in this contribution are discussed in [11].

This research was supported in part by NASA grant NAG 5-9063 (AK) and by NSF grant PHY-0070928 (JG).

## REFERENCES

1. Atoyan, A. M., *A&A*, **346**, L49 (1999)
2. Biermann, P. I., & Cassinelli, J. P., *A&A*, **277**, 691 (1993)
3. Bykov, A. M., & Mészáros, P., *ApJ*, **527**, 236 (1996)
4. Chevalier, R. A., & Li, Z.-Y., *ApJ*, **536**, 195 (2000)
5. Emmering, R. T., & Chevalier, R. A., *ApJ*, **321**, 334 (1987)
6. Freedman, D. L., & Waxman, E., *ApJ*, **547**, 922 (2001)
7. Granot, J., & Königl, A., *ApJ*, **560**, 145 (2001)
8. Granot, J., Piran, T. & Sari, R., *ApJ*, **527**, 236 (1999)
9. Helfand, D. J., Gotthelf, E. V., & Halpern, J. P., *ApJ*, **556**, 380 (2001)
10. Kennel, C. F., & Coroniti, F. V., *ApJ*, **283**, 694 (1984)
11. Königl, A., & Granot, J., *ApJ*, **574**, 000 (2002)
12. Livio, M., & Waxman, E., *ApJ*, **538**, 187 (2000)
13. Mészáros, P., *Science*, **291**, 79 (2001)
14. Mészáros, P., Ramirez-Ruiz, E., & Rees, M. J., *ApJ*, **554**, 660 (2001)
15. Panaitescu, A., & Kumar, P., *ApJ*, in press (2002)
16. Piran, T., *Phys. Rep.*, **314**, 575 (1999)
17. Rees, M. J., & Gunn, J. E., *MNRAS*, **167**, 1 (1974)
18. Reynolds, S. P., & Chevalier, R. A., *ApJ*, **278**, 630 (1984)
19. Rhoads, J. E., *ApJ*, **525**, 737 (1999)
20. Sari, R., Piran, T., & Halpern, J. P., *ApJ*, **519**, L17 (1999)
21. Sari, R., Piran, T. & Narayan, R., *ApJ*, **497**, L17 (1998)
22. Spruit, H.C., Daigne, F. & Drenkhahn, G., *A&A*, **369**, 694 (2001)
23. Thompson, C., & Madau, P., *ApJ*, **538**, 105 (2000)
24. Vietri, M., Ghisellini, G., Lazzati, D., Fiore, F., & Stella, L., *ApJ*, **550**, L43 (2001)
25. Vietri, M., & Stella, L., *ApJ*, **507**, L45 (1998)
26. Wijers, R. A. M. J., & Galama, T. J., *ApJ*, **523**, 177 (1999)

# Gamma-Ray Burst Afterglows: Two Post-Standard Effects

Bing Zhang* and Peter Mészáros*

*Pennsylvania State University, 525 Davey Lab, University Park*

**Abstract.** We discuss two "post-standard" effects in gamma-ray burst (GRB) afterglows. (1) Though synchrotron emission of the relativistic electrons is the canonical interpretation to the GRB afterglow broadband emission, some high energy spectral components also play an important role under certain conditions. It is found that the proton synchrotron emission and some other hadron-related emission components are usually not important in most circumstances, in contrast with some of the previous rough estimates. The synchrotron self-inverse Compton (IC) emission of the electrons, on the other hand, is found to be important in a large shock parameter space, and an extended (hours) IC-origin GeV afterglow for a typical ($z = 1$) GRB is predicted, which is readily detectable by the future mission, GLAST, if the shock parameters as inferred from the present afterglow fitting are typical among most long bursts. (2) In some types of the gamma-ray burst central engines, a significant energy input into the fireballs may in principle occur after the afterglow is set up, either in the form of some kinetic-energy-dominated shells or a Poynting-flux-dominated flow. Both cases are modeled, with distinct injection signatures that may bring us the information about the nature of the injection flow, or even the nature of the central engine.

## INTRODUCTION

A gamma-ray burst (GRB) afterglow sets up when the GRB fireball starts to decelerate after collecting a sufficient mass of the ambient medium. Though initially there is a short-term reverse shock crossing the injective shell, in long terms the emission mainly comes from the relativistic electrons continuously accelerated from the external forward shock that propagates into the medium. In the simplest "standard" model (e.g. [1]), five assumptions are made, i.e., (1) the fireball is isotropic; (2) the ambient density is constant; (3) the fireball is in the relativistic regime; (4) synchrotron radiation (SR) of the electrons is the main emission mechanism; and (5) the central engine injects the energy essentially impulsively. There are already plenty of discussions about some more realistic effects by discarding the first three hypotheses, e.g., the jet effect, the wind effect, and the transition from the relativistic regime to the Newtonian regime. Here we discuss the "post-standard" effects concerning the last two assumptions.

## HIGH ENERGY SPECTRAL COMPONENTS

The simplest mechanism proven to be quite adequate to interpret the broadband afterglow data from a dozen of GRBs is the synchrotron emission of the relativistic electrons (e.g. [2]), at least in the low-energy regime. In the high energy regime, some other high energy spectral components should exist and may, under certain conditions, become the dominant radiation mechanisms. These include the synchrotron self-inverse Compton (IC) component of the electrons, synchrotron emission of the protons, as well as some other hadron-related emission components due to photon-meson interactions (e.g., synchrotron emission from the positrons produced by $\pi^+$ decay and the gamma-rays produced directly from $\pi^0$ decay, [3]). Previously, these components were usually treated singly, without a clear manifest about their significance with respect to the main electron synchrotron component in various bands or about their relative importance with respect to each other. We have performed a coherent study about these high energy spectral components as well as their afterglow signatures[4].

In a toy model, electron-SR component is modeled by a four-segment broken power law[2]. To access the relative importance of the high energy components, the proton-SR and the electron-IC components can be also modeled by two four-segment broken power laws. The ratios between a relevant parameter (e.g. the peak flux density, the injection frequency, the cooling frequency, and so on) of either of these two components with respect to that of the electron-SR component can be readily obtained, which are dependent on several unknown free parameters, among which the equipartition parameters of the electrons, $\varepsilon_e$, and of the magnetic field, $\varepsilon_B$, in

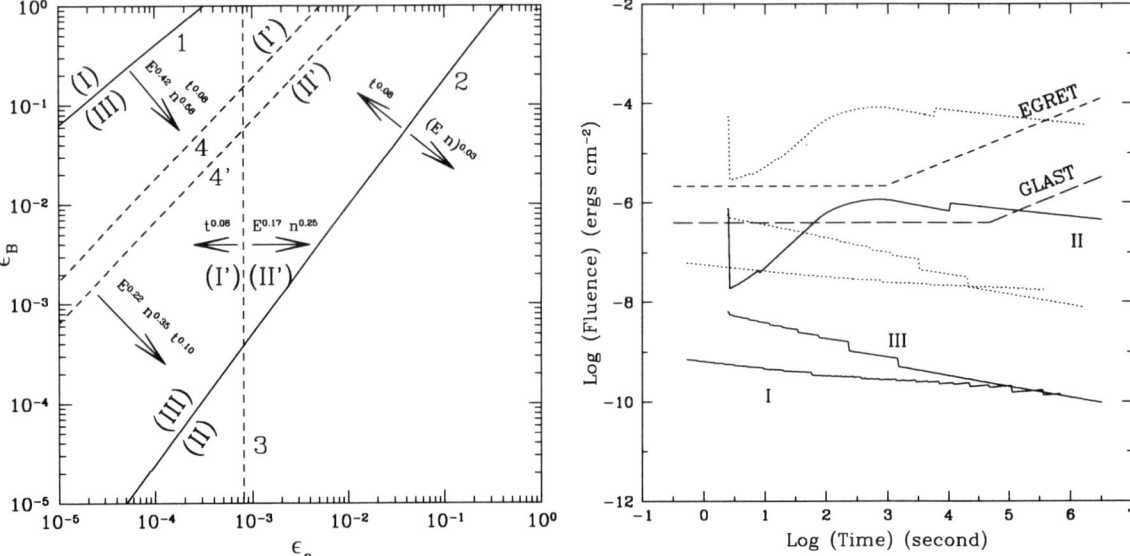

**FIGURE 1.** (a) Regions in the $\varepsilon_e - \varepsilon_B$ parameter space in which various emission components dominate the spectrum. Below the electron-SR cutoff: I. proton-SR dominated regime; II: electron-IC dominated regime; III: electron-SR dominated regime. Above the electron-SR cutoff: I'. hadron-component dominated regime; II'. electron-IC dominated regime. (b) GeV afterglows for the bursts in various regimes. Solid curves are for $z = 1$, and dotted curves are for $z = 0.1$. The detection thresholds for GLAST and EGRET are also plotted (from [4]).

the shock-heated region are of great interests. The broadband spectra can be calculated within a wide $\varepsilon_e - \varepsilon_B$ parameter regime with the contributions from different spectral components specified and compared. The results are summarized in Fig.1a. Below the electron-SR cutoff, three parameter space regions are identified: (I) The proton-SR shows up in a regime with a very small $\varepsilon_e$ and a large $\varepsilon_B$. The parameter space for this regime is small, and the absolute flux level is low, since one gets a prominent proton component by suppressing the electron-SR flux level (through a small $\varepsilon_e$). Some previous rough estimates (e.g. [5] [3] [6]) seem to have overestimated the flux level of this component. (II) The electron-IC shows up in a regime with a relative large $\varepsilon_e/\varepsilon_B$ ratio. The parameter space for this regime is quite substantial, which include the regime where IC cooling is important. The flux level of this component is also high. (III) The electron-SR dominates in a regime with a relative small $\varepsilon_e/\varepsilon_B$ ratio (but not too small to make the proton-SR to show up). The parameter space is also large. Above the electron-SR cutoff, the parameter space is separated in two regimes, (I') a small regime where the hadron-related components dominate, and (II') a large regime where the electron-IC dominates. Overall, electron-IC is the most promising high energy emission component in GRB afterglows.

## GEV AFTERGLOWS

The electron-IC component forms a second bump in the afterglow broadband spectrum. At a fixed frequency this bump may sweep into the band at a certain time. In X-rays, for a moderate dense medium, the IC component will show up in about a day ([7] [8] [4]). This can naturally account for the X-ray late-time bump observed in GRB 000926 ([4] [9]). In the GeV band, the IC-sweeping time is earlier, e.g., in hours. For a typical regime II burst, such a burst will be detectable by GLAST (with at least 5-photon detection) at a typical cosmological distance, e.g., $z = 1$ (Fig.1b). For a nearby regime II burst (e.g. $z = 0.1$), the GeV afterglow level is above the EGRET threshold (Fig.1b), and some of such bursts should have been detected (e.g. the case of GRB 940217, [10]).

In the future Swift-GLAST era, GeV afterglow detections will bring profound implications for the GRB studies. Here are some examples.

1. Due to IC-cooling, two possible fitting solutions for the GRB fireball and shock parameters exist for a same set of the observables (which correspond to our regimes II and III, respectively), if one only use the low-energy afterglow data ([8]). Whether or not a significant GeV afterglow is detected will break the degeneracy.

2. To determine the parameter regime in which a burst lies, one can determine unambiguously the values $\varepsilon_B$, and present a test to the hypothesis that GRBs are the sources

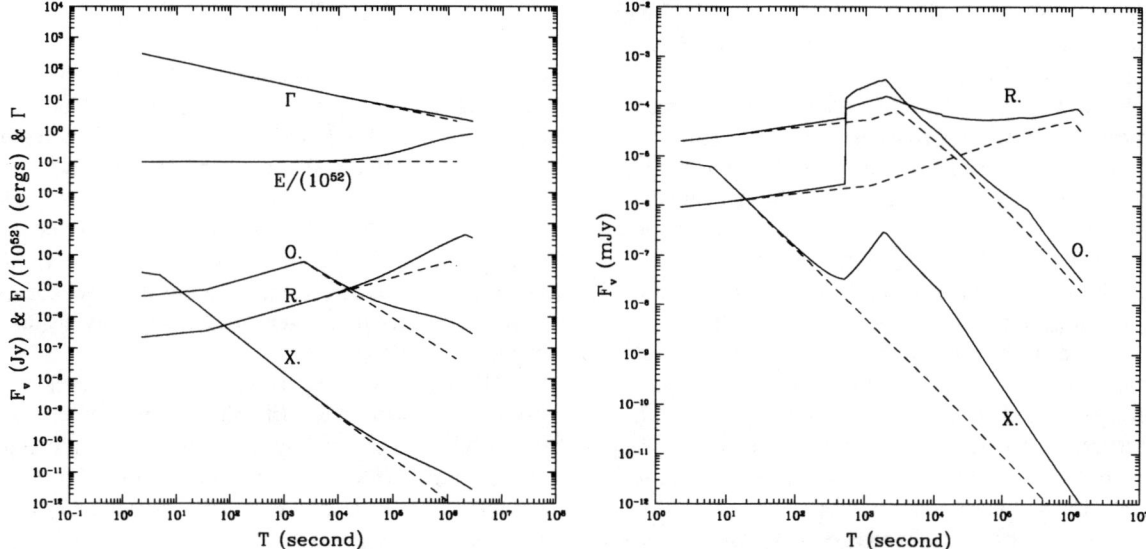

**FIGURE 2.** Two sets of the afterglow broadband lightcurves with injection signatures. X: X-ray; O: optical; R: radio. Solid and dashed curves are for the cases with or without the injection component. (a) The case for a continuous injection from a Poynting-flux-dominated flow; (b) The case for a violent collision between two shells in the afterglow phase (from [12]).

of the UHECRs.

3. Closely monitoring the broadband afterglows may unveil whether the shock parameters evolve with time.

At even higher energies (e.g. TeV), the IC-sweeping time is even earlier. The intrinsic photon-photon opacity has a weak dependence on time ([4]). Thus it is optimistic to detect TeV early afterglow emission from the nearby bursts by the experiments like Milagro. At early times, signals from the reverse shock as well as the internal shocks may be also important, and a closer modeling is necessary.

## FIREBALLS WITH CONTINUOUS ENERGY INJECTIONS

It is well-known that the GRB central engines are rather erratic. This is manifested by the spiky, irregular GRB lightcurves. The duration of the bursts can vary in a rather wide range which is intrinsic to the central engine. On the other hand, the afterglow onset time is mainly determined by the medium density. Since these two times are independent with each other, it is likely that under certain circumstances an additional injection of energy into the fireball will occur in the afterglow phase. In certain central engines such as a millisecond magnetar or a long-term black hole - torus system, a continuous energy injection with a lower luminosity than in the prompt phase can in principle go on for a long period of time, the accumulative effect of which will also bring noticeable effects on the GRB blastwave dynamics as well as the lightcurves in various bands. We have performed a systematic study on these issues ([11] [12]).

We classify the injection cases into two distinct categories. The crucial criterion is whether an injection excites more shocks, and hence, more emission sites[12]. (1) In the first case, such additional emission sites, if any, are believed to be unimportant, so that one can only calculate the synchrotron emission (or IC) of the electrons at the forward external shock propagating into the ambient medium. A continuous Poynting-flux-dominated wind, such as that associated with the millisecond magnetars, is likely of this kind. Even if the injection is dominated by the kinetic energy of the baryons, the injection is analogous to the Poynting wind case if the injective shell is not energetic and fast enough. (2) In the second case, a matter shell with a large enough energy runs into the forward shell with a large enough relative velocity. In such a case, a violent collision occurs, and two new strong shocks form between the impulsive and the injective shells. Together with the original forward shock, there are totally three shocks which accelerate particles to relativistic energies. Therefore there are three emission sites, and the lightcurves are expected to be more complicated.

## INJECTION SIGNATURES

We have calculated the broadband afterglow lightcurves with an injection component for both the Poynting-flux-dominated case and the violent collision case. The examples for the both cases are shown in Fig.2. Here is a summary of our findings.

1. In both cases, the modifications to the lightcurves are through different approaches. For the Poynting wind case, the lightcurve signatures are introduced through the change of the blastwave dynamics, as the total energy in the injection component exceeds the impulsively injected energy. For the collision case, the signatures are introduced through the sudden excitation of two additional emission sites due to the collision.

2. As a result, the injection signature in the Poynting wind case should be very smooth, while that in the collision case should be rather abrupt, as indicated in Fig.2. In both cases, a post-burst injection energy $E_{inj} \sim E_{imp}$ is generally required to cause a noticeable signature in the afterglow lightcurves. For a violent collision, a faster rear shell can partially compensate for a lower rear shell total energy.

3. In both cases, the final Lorentz factor of the blastwave, as well as the emission flux level, are boosted up with respect to the case without post-injection.

4. A prominent characteristic of the post-injection process is the conspiratorial variation of the flux in all bands. For the Poynting wind case, the signature is achromatic, with a similar smooth shape in all the bands. For the collision case, the rising time is essentially the same, but the shape of the bump varies considerably in different bands. Thus it is really essential to monitor a signature in various bands throughout a time of interest in order to discriminate the injection mechanism from some other models to interpret the afterglow bumps (e.g. [13][14][15]).

5. For a strong collision to occur (Fig.2b), the relative Lorentz factor between the two colliding shells has to exceed a certain value defined by the energy ratio between the two colliding shells. The condition is more demanding than usually believed (e.g. [16]). We expect that a violent injection most likely arises from a late injection of a high entropy shell from a long-live central engine.

A long standing goal of GRB research is to unveil the nature of the central engine. The progress is made slowly mainly because the engine is masked by the internal and external shocks where the observed emissions come from. Injection signatures provide a possibility for us to "reach" the central engine. Future broadband detections of some injection signatures accompanied by some more detailed modeling may tell us about the nature of GRB energy flow (e.g. Poynting wind or a matter shell), and even the nature of the central engine (e.g. a long-live black hole - torus system or a millisecond magnetar). Since injections likely occur in early afterglow phase, future broadband GRB detector, Swift, will play an essential role in identifying such features.

# CONCLUSIONS

The "standard" GRB afterglow model was proven to be successful in general, while more and more data indicate that such a model ought to be improved to include more post-standard effects. Extensive efforts have been made in identifying a jet, wind, or dense ambient medium associated with a GRB. We have discussed here the possible implications of two other post-standard effects. Some predictions are made in anticipating the new discoveries in the Swift-GLAST era. Following conclusions are reached.

1. High energy afterglow (e.g. GeV) observations are essential to constrain external shock physics. If the shock parameters as inferred by the present data are typical among long GRBs, an extended GeV afterglow (hours) is detectable by GLAST for a burst located at a typical cosmological distance ($z = 1$).

2. A variety of injection signatures can be used to infer the nature of the GRB energy flow as well as the nature of the central engine. Such a progress is expected to be made in the future Swift era, or even in the HETE-2 era.

# ACKNOWLEDGMENTS

This work is supported by NASA NAG5-9192 and NAG5-9153.

# REFERENCES

1. Mészáros, P., and Rees, M. J., *ApJ*, **476**, 232–237 (1997).
2. Sari, R., Piran, T., and Narayan, R., *ApJ*, **497**, L17–L20 (1998).
3. Böttcher, M., and Dermer, C. D., *ApJ*, **499**, L131–L134 (1998).
4. Zhang, B., and Mészáros, P., *ApJ*, **559**, 110–122 (2001).
5. Vietri, M., *Phys. Rev. Lett.*, **78**, 4328–4331 (1997).
6. Totani, T., *ApJ*, **502**, L13–L16 (1998).
7. Panaitescu, A., and Kumar, P., *ApJ*, **543**, 66–76 (2000).
8. Sari, R., and Esin, A. A., *ApJ*, **548**, 787–799 (2001).
9. Harrison, F. A., and et al., *ApJ*, **559**, 123–130 (2001).
10. Hurley, K., and et al., *Nature*, **372**, 652–654 (1994).
11. Zhang, B., and Mészáros, P., *ApJ*, **552**, L35–L38 (2001).
12. Zhang, B., and Mészáros, P., *ApJ*, **566**, in press (astro-ph/0108402) (2002).
13. Bloom, J. S., and et al., *Nature*, **401**, 453–456 (1999).
14. Esin, A. A., and Blandford, R., *ApJ*, **534**, L151–L154 (2000).
15. Garnavich, P., Loeb, A., and Stanek, K., *ApJ*, **544**, L11–L14 (2000).
16. Kumar, P., and Piran, T., *ApJ*, **532**, 286–293 (2000).

# Afterglow Lightcurves, Viewing Angle and the Jet Structure of Gamma-Ray Bursts

Elena Rossi*, Davide Lazzati* and Martin J. Rees*

*Institute of Astronomy, Madingley Road CB3 0HA Cambridge, U.K.

**Abstract.** Gamma ray bursts are often modelled as jet-like outflows directed towards the observer; the cone angle of the jet is then commonly inferred from the time at which there is a steepening in the power-law decay of the afterglow. We consider an alternative model in which the jet has a beam pattern where the luminosity per unit solid angle (and perhaps also the initial Lorentz factor) decreases smoothly away from the axis, rather than having a well-defined cone angle within which the flow is uniform. We show that the break in the afterglow light curve then occurs at a time that depends on the viewing angle. Instead of implying a range of intrinsically different jets – some very narrow, and others with similar power spread over a wider cone – the data on afterglow breaks could be consistent with a standardized jet, viewed from different angles. We discuss the implication of this model for the luminosity function and compare with data. We also discuss some predictions.

## INTRODUCTION

In the model which assumes a massive star as the progenitor of GRB, the flux of energy released has to expand in the irregular cavity of the stellar envelope. In this scenario is likely to create a jet that collects more baryons in the outer parts then near the symmetry axis. So even if the jet has been created homogeneous, the transfer of mass from the borders of the star results in the final Lorentz factor, $\Gamma$, and the energy per unit solid angle, $\varepsilon$, both depending on the angle $\theta$ made with its symmetry axis. We considered here inhomogeneous GRBs jets with a standard total energy, opening angle and local energy distribution, $\varepsilon \propto \theta^{-2}$. We show that this jet structure can reproduce the observed correlation between isotropic energy and break-time recently found [1, 2]. We also discuss a more general case in which $\varepsilon \propto \theta^{\alpha_\varepsilon}$ and we show that $1.5 \leq \alpha_\varepsilon \leq 2.2$ is still consistent with data.

## DYNAMICS AND BREAK TIME

We suppose that all long GRBs have jets with a standard beam profile for $0 < \theta < \theta_j$. We consider a relativistic outflow where both the bulk Lorentz factor and the energy per unit solid angle depend as power laws on the angular distance from the center $\theta$

$$\varepsilon = \begin{cases} \varepsilon_c & 0 \leq \theta \leq \theta_c \\ \varepsilon_c \left(\frac{\theta}{\theta_c}\right)^{-2} & \theta_c \leq \theta \leq \theta_j \end{cases} \quad (1)$$

and

$$\Gamma = \begin{cases} \Gamma_c & 0 \leq \theta \leq \theta_c \\ \Gamma_c \left(\frac{\theta}{\theta_c}\right)^{-\alpha_\Gamma}, \quad \alpha_\Gamma > 0 & \theta_c \leq \theta \leq \theta_j. \end{cases} \quad (2)$$

A lower limit for $\theta_c$ is $\theta_c > 1/\Gamma_{max} \sim 10^{-3}$ degrees, where $\Gamma_{max} \sim 10^5$ is the maximum value to which the fireball can be accelerated [3]. The power law index of $\Gamma$, $\alpha_\Gamma$, is not important for the dynamics of the fireball and the computation of the light curve as long as $\Gamma(t=0, \theta) \equiv \Gamma_0(\theta) > \theta^{-1}$ and $\Gamma_0(\theta) \gg 1, \forall \theta$. Consider an observer at an angle $\theta_o < \theta_j$ with respect to the axis of the jet. He measures an isotropic equivalent energy $E_{iso} = 4\pi\varepsilon(\theta_o)$ from the $\gamma$-ray fluence. If also the afterglow emission is dominated by the component pointing the earth, he will infer $\theta_o$ as the half-opening angle of the jet by means of the break time in the light curve, $t_b$ [4] $\theta_j \propto t_b^{3/8}(E_{iso}/n)^{-1/8}$, where $n$ is the external medium density. Since the total energy inferred from all viewing angles is $E_{tot} \simeq 2\pi\theta^2\varepsilon$ =const, the observer will derive the same conclusions obtained by [1, 2] of GRBs as fireballs with very different apertures but a standard energy reservoir injected in. We evaluate how the contributions of the other components add to the light curve from the zone with $\theta \sim \theta_o$. The result is that when the cone with $\theta_o$ has just started spreading the part of the blast generated by the core of the jet becomes visible to the observer, and its contribution to the light curve is not dominant; the regions with $\theta \gg \theta_o$ spread at later time but they never manage to have an energy per unit solid angle greater than $\varepsilon_o$. Therefore the regions with $\theta \ll \theta_o$,

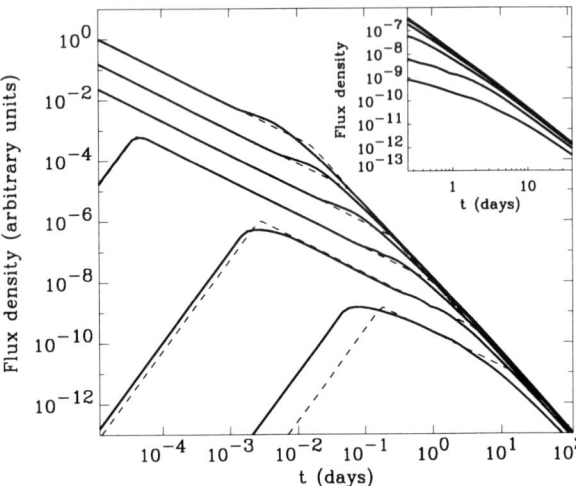

**FIGURE 1.** The light curves of an inhomogeneous jet observed from different angles. From the top we show $\theta_o = 0.5, 1, 2, 4, 8, 16°$. The break time is related only to the observer angle: $t_b \propto \theta_o^2$. The dashed line is the on-axis light curve of an homogeneous jet with an opening angle $2\theta_o$ and an energy per unit solid angle $\varepsilon(\theta_o)$. The blow up is the time range between 4 hours and 1 month, where most of the optical observations are performed. Comparing the solid and the dashed lines for a fixed $\theta_o$, it is apparent that we can hardly distinguish the two models by fitting the afterglow data. To compute the light curve we have assumed that the observer is on the jet axis of the cones with $\theta \geq \theta_o$. This predicts a transition between the two power-law branches sharper than what we expect from the exact integration. The slight flattening of the lightcurves just before the break is likely to be due to the numerical approximations we adopted.

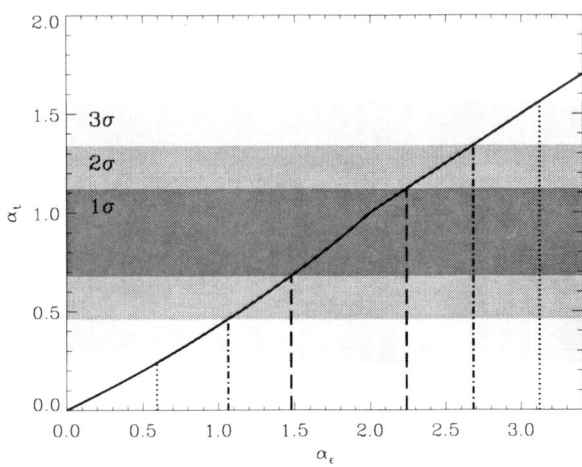

**FIGURE 2.** The relation between the indices $\alpha_t$ ($E_{iso} \propto t_b^{-\alpha_t}$) and $\alpha_\varepsilon$ ($\varepsilon \propto \theta^{-\alpha_\varepsilon}$) from an inhomogeneous jet. The shaded regions show the best fit to F01 data $\alpha_t = 0.9 \pm 0.22$.

(or those with $\theta$ substantially above $\theta_o$), don't determine the overall shape of the light curve, because they are hidden by the dominant emission from the component along the line of sight. Consequently the energy per unit solid angle that is measured modeling the afterglow is $\varepsilon_o$ and the time break depends only on the viewing angle $\theta_o$ (see Fig. 1).

Since we find that the break time is related to the viewing angle according to $t_b \propto \theta_o^2$ we obtain

$$E_{iso} \propto \theta_o^{-2} \propto t_b^{-1}, \quad (3)$$

consistent with [1] data but with a different interpretation: the observed distribution of $E_{iso}$ and its relation with $t_b$ is due to an inhomogeneous jet and the possibility to view it from different angles. An other difference is in the computation of the total energy. In our model it can be calculated from Eq. 1

$$E_{Total} = 2\pi\varepsilon_c\theta_c^2\left(1 + 2\ln\frac{\theta_j}{\theta_c}\right) = E_{jet}\left(1 + 2\ln\frac{\theta_j}{\theta_c}\right) \quad (4)$$

and compared to $E_{jet} = 2\pi\theta_o\varepsilon_o = 2\pi\varepsilon_c\theta_c^2$, the total energy inferred from observation, $E_{jet} \leq E_{Total}$. To give an example, for a fireball with $\theta_c = 1°$ and $\theta_j = 20°$, we have $E_{Total}/E_{jet} \simeq 6$, i.e. the true energy of the fireball can be one order of magnitude larger than what inferred with the models of [1, 2].

Under our assumptions, each $\gamma$-ray luminosity corresponds to a particular viewing angle. From the probability to see a jet at a certain off-axis angle we can derive the expected luminosity function

$$P(y) \propto 10^{-y}, \quad (5)$$

where $y = \log\varepsilon$.

Since there are only a small number of GRBs with observed redshift, the comparison of our predicted luminosity function with data is far from being definitive. Recently Bloom et al. (2001) [5] published an histogram of the bolometric k-corrected prompt energies for a flux limited sample of 17 GRBs and Schmidt (2001) [6] derived a luminosity function without using any redshift but assuming how the comoving GRBs rate varies with redshift. In these case a power-law luminosity function was derived, but flatter than the $n(L) \propto L^{-2}$ predicted in the simplest version of our model. On the other hand when we consider luminosity functions not based on assumed burst rate evolution but derived by measuring the burst distance scale through the variability-luminosity relation [7] the comparison is more favorable to our prediction. Given the somewhat contradictory observational results discussed above, we conclude that more accurate spectral and fluence measures and a larger sample of bursts are needed for a proper comparison.

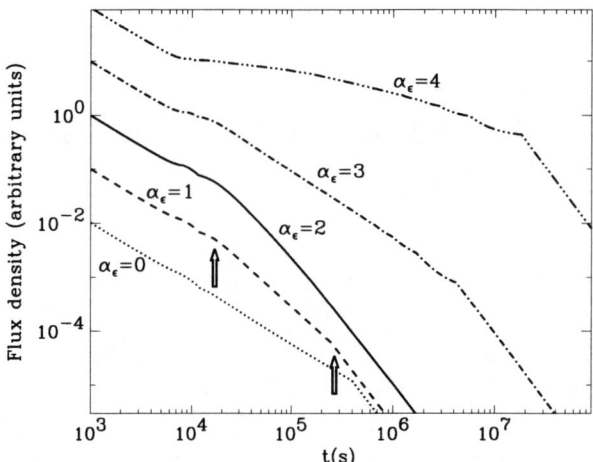

**FIGURE 3.** The lightcurves of inhomogeneous jets with different indices $\alpha_\varepsilon$ ($\varepsilon \propto \theta^{-\alpha_\varepsilon}$). The lightcurves have the same $\varepsilon_o$ but they are plotted shifted by factors of 10 for clarity (they would be indistinguishable at small time). The arrows highlight the location of the two breaks for $0 < \alpha < 2$.

## GENERAL CASE: $\theta \propto \theta^{\alpha_\varepsilon}$

In this paper we concentrated for simplicity on a beam profile $\varepsilon \propto \theta^{-2}$ which is consistent observational results but it is interesting to briefly discuss other power-law relations $\varepsilon \propto \theta^{-\alpha_\varepsilon}$ (see Fig. 3). A decay flatter then 2 would cause two breaks in the light curve: the first due to the cone pointing the observer when $\Gamma(\theta_o) \sim \theta_o^{-1}$ and the second, at later times, when the observer sees the edge of the jet and $\Gamma(\theta_j) \sim \theta_j^{-1}$. The power law index after the first break is flatter then $t^{-p}$ because the cones with $\theta_o < \theta \leq \theta_j$ enter the line of sight with $\varepsilon \geq \varepsilon_o$ and substantially modify the light curve shape. With a steeper decay in the distribution of $\varepsilon$ the time break and the emission after that break would be dominated by the jet along the axis rather then by the very much weaker part directed to the observer. The jet break would then be preceded by a prominent flattening in the lightcurve, especially for $\alpha_\varepsilon > 3$, difficult to reconcile with observations. we derive the relation between the index $\alpha_\varepsilon$ and $\alpha_t$ (where $E_{iso} \propto t_b^{\alpha_t}$). We obtain $\alpha_t = \alpha_\varepsilon/2$ for $\alpha_\varepsilon \geq 2$ and $\alpha_t = 3\alpha_\varepsilon/(8-\alpha_\varepsilon)$ for $\alpha_\varepsilon < 2$, when the first break is considered. In Fig. 2 we plot this relation overlaid on the interval in $\alpha_t$ allowed by observations (we used F01 data). We derive $1.5 \leq \alpha_\varepsilon \leq 2.2$, at the $1\sigma$ level. Further $\gamma$-ray and afterglow observations will allow to constrain this parameter much better in the future.

## PREDICTIONS OF THE MODEL

Besides the luminosity function discussed in § 5, there are several ways in which this model can be proved or disproved. First, we have shown that the real total energy of the fireball can easily be an order of magnitude larger than what estimated [1, 2]. In this case, after the fireball has slowed down to sub-relativistic speed, radio calorimetry [8] should allow us to detect the excess energy. In addition, in this model we naturally predict that the more luminous part of the fireball have higher Lorentz factors. This may help explaining the detected luminosity-variability correlation [7] due to relativistic time contraction ([9, 10]). Another constrain is given by polarization. Since fireball anisotropy is a basic ingredient of this model, inducing polarization [11]. The time evolution of the polarized fraction and of the position angle are however different from a uniform jet ([12]). Finally, the properties of the bursts should not depend on the location of the progenitor in the host galaxy, and therefore this model can accommodate the marginal detection of an $E_{\rm iso}$-offset relation ([13]) only if a distribution of $\theta_j$ is considered.

## REFERENCES

1. Frail, D. A. e. a., *ApJ*, **562**, L55–L58 (2001).
2. Panaitescu, A., and Kumar, P., *ApJ subm. (astro-ph/0109124)* (2001).
3. Piran, T., *Phys. Rep.*, **314**, 575+ (1999).
4. Sari, R., Piran, T., and Halpern, J. P., *ApJ*, **519**, L17–L20 (1999).
5. Bloom, J. S., Frail, D. A., and Sari, R., *AJ*, **121**, 2879–2888 (2001).
6. Schmidt, M., *ApJ*, **552**, 36–41 (2001).
7. Fenimore, E. E., and Ramirez-Ruiz, E., *ApJ subm. (astro-ph/0004176)* (2001).
8. Frail, D. A., Waxman, E., and Kulkarni, S. R., *ApJ*, **537**, 191–204 (2000).
9. Kobayashi, S., *this volume* (2001).
10. Ramirez-Ruiz, E., and Lloyd-Ronning, N. M., *New Astronomy*, **7**, 197–210 (2002).
11. Ghisellini, G., and Lazzati, D., *MNRAS*, **309**, L7–L11 (1999).
12. Rossi, E., Lazzati, D., Salmonson, J., and Ghisellini, G., *in prep.* (2002).
13. Ramirez-Ruiz, E., Lazzati, D., and Blain, A. W., *ApJ*, **565**, L9–L12 (2002).

# On the Color Indices and Absolute Brightnesses of the Optical Afterglows of GRB

V. Šimon*, R. Hudec*, N. Masetti† and G. Pizzichini†

*Astronomical Institute, Academy of Sciences of the Czech Republic, 25165 Ondřejov, Czech Republic
†Istituto Tecnologie e Studio delle Radiazioni Extraterrestri, CNR, via Gobetti 101, 40129 Bologna, Italy

**Abstract.** The study of the color indices and luminosities of 17 optical afterglows (OAs) of GRBs, including the most recent one, GRB010222, showed that the color variations during the decline of OAs are quite small during $t - T_0 < 10$ days and allow a comparison among them, even for the less densely sampled OAs. The colors in the observer frame concentrate at $(V - R)_0 = 0.40 \pm 0.13$, $(R - I)_0 = 0.46 \pm 0.18$, $(B - V)_0 = 0.47 \pm 0.17$, except for GRB000131 and GRB980425. However, large scatter is observed in $(U - B)_0$. The color evolution of the OAs is negligible although their brightness declines by several magnitudes during $t - T_0 < 10$ days. Such a strong concentration of the color indices also suggests that the intrinsic reddening (inside their host galaxies) must be quite similar and relatively small for all these events. The absolute brightness of OAs in the observer frame lies within $M_{R_0} = -26.5$ to $-22.2$ for $(t - T_0)_{rest} = 0.25$ days. The general decline rate of the OA sample considered here seems to be independent of the absolute optical brightness of the OA, measured at some $t - T_0$ identical for all OAs, and the light curves of all events are almost parallel, when corrected for the redshift-induced time dilation.

## INTRODUCTION

The color indices of the optical afterglows (OAs) of GRBs are a powerful and innovative approach to the study of such events. They might help us to:

- search for common properties of the afterglows
- understand the related physical processes
- find out whether an optical event is related to a GRB even without available γ-ray detection by using the color indices of the OAs
- search for relations among colors, luminosities and the decay rates of the OAs (if the redshift $z$ is known)
- constrain the properties of the local interstellar medium of the GRBs.

## COLLECTION AND ANALYSIS OF THE DATA

This analysis makes use of the data published in the GCN circulars archive [2], in Jochen Greiner's Web page [5] and in the journals. Because of space limitations, the reader is referred to these web pages for full bibliographic references on each OA.

At present, suitable *multicolor* photometry is available for 17 OAs and for the supernova SN1998bw which is a possible counterpart of GRB980425. However, this photometry often comprises unorganized observations. The light curves of the OAs in the individual passbands were therefore plotted and critically examined. Fortunately, mostly they were found to be free of complicated rapid changes on the time scale of hours to a few days. A meaningful interpolation between the neighbouring measurements was thus possible, at least in the early parts of the light curves ($t - T_0 < 10$ days, where $T_0$ refers to the time of the GRB). It enabled us to obtain at least one color index for each OA. The typical standard deviations of the indices, derived from the errors quoted in the original literary sources, lie within 0.04–0.3 mag. The indices were corrected for the Galactic reddening, computed from the maps by [10]. The light contribution of the host galaxies was quite small for $t - T_0 < 10$ days and could be neglected. More details can be found in [11].

### Color diagrams

The color indices $(R - I)_0$, plotted versus $t - T_0$ in the observer frame, were found to occupy a narrow belt with negligible evolution for the first 10 days (Figure 1a). Very similar results were found also for $(V - R)_0$ and

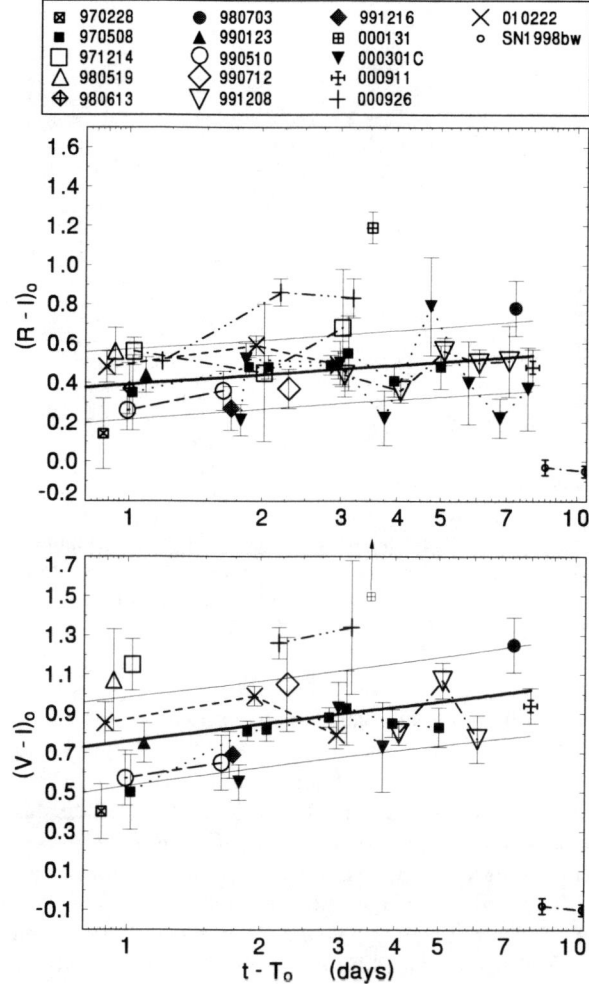

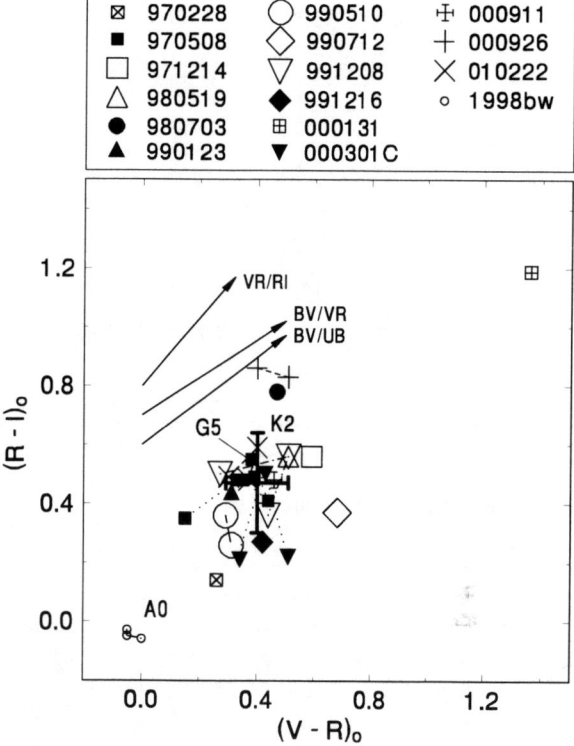

**FIGURE 1.** Temporal evolution of the color indices of the respective OAs during $t - T_0 < 10$ days. Colors of all OAs correspond to their final decline branch, except for the first point of GRB970508. The solid line represents the fit to the ensemble of data, with the exception of GRB000131 and GRB980425 (SN1998bw). Notice that the color evolution is very small, when compared with the standard deviation of the fit (thin lines).

**FIGURE 2.** $V - R$ vs. $R - I$ diagram of the respective OAs for $t - T_0 < 10$ days. Notice the strong concentration of most OAs in a well defined position in the diagram. The centroid of the whole ensemble of OAs, except for GRB000131 and GRB980425 (SN1998bw), including the standard deviations, is marked by the large cross. The representative reddening paths for $E_{B-V} = 0.5$ are shown. Positions of the main-sequence stars are included for comparison.

$(B - V)_0$. Notice that the fit to the data (thick solid line) in Figure 1a has a small non-zero slope but it still lies within the standard deviation (thin line). In order to check the color evolution, an additional color index $(V - I)_0$ which relates measurements over a broader range of wavelengths was calculated (Figure 1b). It can be seen that the reddening trend looks stronger for $(V - I)_0$ but also the scatter is larger. Notice also that the reddening trend is not very clear when evolution of a given OA is traced. This speaks in favour of just a small evolution of colors of the OAs during the first 10 days. The common color-color diagrams could therefore be built even if the observations of the various OAs came from different $t - T_0$'s. $V - R$ vs. $R - I$ diagram is shown in Figure 2. $B - V$ vs. $V - R$ diagram looks very similar. On the contrary, $U - B$ vs. $B - V$ diagram displays an appreciable scatter, but mostly just in the $U - B$ direction.

The representative reddening paths for $E_{B-V} = 0.5$ are shown in Figure 2. Because the observer in a given passband will detect radiation at progressively shorter wavelengths with increasing $z$, the reddening paths, appropriate for $U - B$, $B - V$, $V - R$ and $R - I$, are included in Figure 2. It can be seen that both the lengths and directions of the vectors are similar for all reddening paths.

The observed passbands of the OAs differ from those in the rest frame because of the effects introduced by the redshift $z$. The OAs considered here have $z = 0.43 - 4.5$, with typical $z \sim 1$. The color indices of all OAs with known $z$ were plotted versus $z$. An example is shown in Figure 3. This plot confirms that any dependence of the color on $z$ is quite weak and within $1\sigma$ for $(R - I)_0$. Very similar results are obtained also for $(V - R)_0$ and

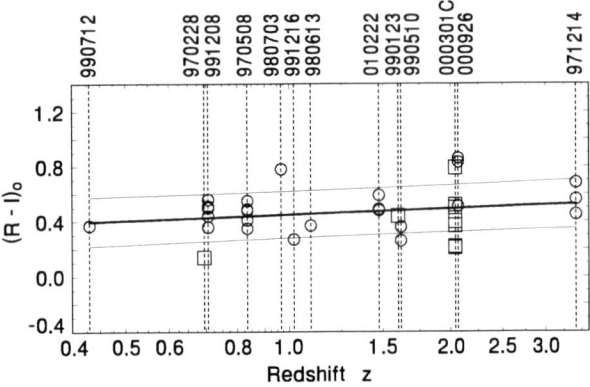

**FIGURE 3.** The color index $(R-I)_0$ of all OAs with known $z < 3.5$ as a function of $z$. Any dependence of the color on $z$ is quite weak and within $1\sigma$. Very similar results were obtained also for $(V-R)_0$ and $(B-V)_0$.

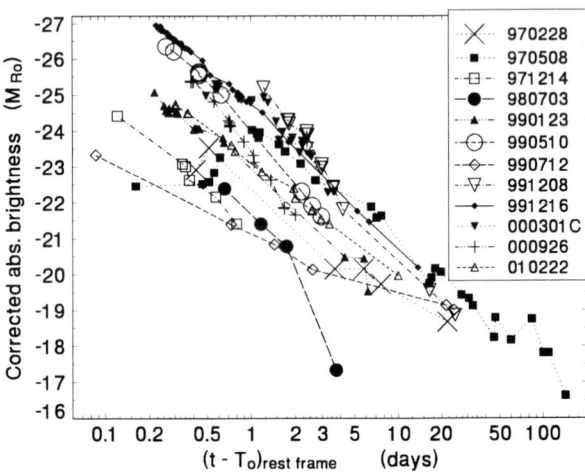

**FIGURE 4.** The absolute brightness of the OAs in the $R$-band as a function of time. Absolute brightness of the OA was corrected for the Galactic reddening and the light contribution of the host galaxy. The time intervals are in the rest frame. Zeroth order $k$-correction was applied to the observed $R$ magnitudes of the respective OAs. See text for details.

$(B-V)_0$, at least for the OAs with $z < 3.5$ (i.e. except GRB000131, located at $z = 4.5$). The only exception may be $(U-B)_0$ where a large scatter is apparent but this index could be determined just for a few OAs.

## Absolute brightnesses of the OAs

The absolute brightness in the $R$-band ($M_{R_0}$; Figure 4) of each OA was calculated from the redshift $z$ according to Eq.15.3.25 of [13]. $H_0 = 60$ km s$^{-1}$ Mpc$^{-1}$ was used. $M_{R_0}$ has then been corrected assuming OA power law spectra with $F_\nu \propto \nu^{-\beta}$, where $F_\nu$ is the flux and $\beta \sim 1$ is the spectral index (see below). We took into account which rest-frame band is corresponding to the $R$ band as seen in the observer frame. These corrections were $\sim -0.3$ mag when the rest-frame wavelength of the OAs was $\lambda_{\text{rest}} = 1400 - 2800$ Å, $\sim 0.2$ when $\lambda_{\text{rest}} = 2800 - 3750$ Å and $\sim 0.9$ when $\lambda_{\text{rest}} = 3750 - 4800$ Å. We can consider this as a sort of 'zeroth order' $k$-correction. $M_{R_0}$ was corrected for the contribution of the host galaxy in this case because the light curves in Figure 4 extend far beyond $t - T_0 < 10$ days and the light of the host cannot be neglected in the late phases of the OAs. The time intervals $t - T_0$ in the observer frame were transformed into the rest frame times $(t - T_0)_{\text{rest}}$.

## CONCLUSIONS

We have shown that the typical color indices of the OAs inside the time interval $0.14 < t - T_0 < 10$ days are $(V-R)_0 = 0.40 \pm 0.13$, $(R-I)_0 = 0.46 \pm 0.18$, $(B-V)_0 = 0.47 \pm 0.17$. Their changes are negligible although the brightness of all OAs declines by several magnitudes during this time interval. This fact implies that the shape of the spectrum of the OA does not change significantly while the luminosity of the OAs decreases by a large amount. On the other hand, the $(U-B)_0$ index displays a large scatter. The position of the OAs in the $U-B$ vs. $B-V$ diagram is often quite different from other sources.

The strong concentration of most OAs in $V-R$ vs. $R-I$ diagram and $B-V$ vs. $V-R$ diagram, despite the smearing introduced by the shifts of the passbands due to various values of $z$, implies that the spectral shape of the OAs is very smooth, with no bumps or strong lines, between the observed $B$ to $I$ passbands. This represents an interval between about 2000 and 5600 Å in the rest frame for the OAs with known $z$. The average $(R-I)_0$, $(V-R)_0$ and $(B-V)_0$ colors of the OAs are consistent with a power-law shaped optical spectral distribution with $\beta \sim 1$. This is in accordance with the theoretical treatment of the GRB afterglow 'fireball' emission model [9].

The fact that the spectra of most OAs are similar although their luminosity in a given $t - T_0$ appears to be different (Figure 1b) is most likely due to the following reason: the spectral shape in the 'fireball' model [9] does not depend on the input energy, while the luminosity of the afterglow at a particular epoch does depend on it. So, the higher the GRB input energy is, the brighter the OA is; this would also suggest that GRBs and their afterglows *are not* standard candles.

The strong concentration of most color indices also points to the fact that *the intrinsic reddening (that is inside their host galaxies) must be quite similar and relatively small for all these events.* The outlying position of GRB000131 can be explained by the presence of the Lyman α spectral break in the optical due to high redshift, and not by intrinsic reddening. The argument for the small reddening of the OAs is that in the case of a large reddening it would be quite unlikely to obtain such similar values of absorption in all cases. These GRBs are therefore unlikely to come directly from the inner (densest) parts of the star-forming regions (but they may lie on the earth-watching side of a star-forming region). Alternatively, the density and the dust abundance of the local interstellar medium might be substantially reduced by the intense high-energy radiation of the GRB trigger, as modeled by [12]. The study of colors of the OAs therefore provides us with direct information about the environment of GRB.

We are however aware that in some cases the OAs appear to have steeper optical spectra (and thus redder colors), as, for example, GRB980329 [8], GRB990705 [7], GRB000131 [1], GRB000418 [6], GRB000630 [4]. These GRBs show β ∼2, or higher. This is most likely due to strong local absorption in the burst environment (as in the cases of GRB990705 and GRB000418) or very high redshift $z$ (GRB000131), or possibly both (GRB980329; [3]). Thus, our results suggest that the sample of OAs considered here is very little affected by reddening effects induced by both strong local absorption or high redshift. This means that here we are dealing with GRB afterglows which are at $z < 4$ and not deep inside dense dust clouds in their host galaxies. At present, the available data enable the determination of the color indices of the redder OAs according to the criteria from Section 2 only for the OA of GRB000131. Nevertheless, the strong concentration of the color indices of the OAs in $V - R$ vs. $R - I$ (Figure 2) and $B - V$ vs. $V - R$ diagrams allows one to infer that there may not be smooth transition between the events considered in our analysis and these OAs with steep optical spectra.

The absolute brightness of the OAs, corrected for the host galaxy, lies within $M_{R_0} = -26.5$ to $-22.2$ for $(t - T_0)_{rest} = 0.25$ days. This spread of $M_{R_0}$ is not significantly influenced by the shifts of λ, caused by the different redshifts of OAs. The general decline rate of most OA considered here seems to be independent of the absolute optical brightness of the OA as measured at some fixed $t - T_0$ for all OAs, and the light curves of all events are almost parallel, when corrected for the redshift-induced time dilation.

Our study is an example of how the procedure known from variable star studies can be applied in high-energy astrophysics. The results presented here confirm the importance of color information in the strategy of the OA searches and analyses. Color-color information plus fading profile of OA appear to be powerful tools to confirm OAs in optical studies. They can yield a valuable phys-ical conclusion regarding the model, the origin, and the position of GRBs. This procedure is also suitable for a study of the orphan OAs. In general, it is important to provide optical observations of the OAs in a well organized way. If possible, the observations should be carried out with standard filters, and various passbands should be taken immediately, i.e. during the same night.

## OPEN QUESTIONS

Why do we observe such low optical extinctions in contrast to the X-rays with column densities of the order of $10^{22} - 10^{23}$ cm$^{-2}$?

Are the prompt X-ray and/or optical/UV flashes indeed so intense that they destroy the absorbing dust completely up to the large distances?

Are GRB indeed in dense environment as the X-ray data suggest?

## ACKNOWLEDGMENTS

This study was supported by the project ES002 by the Ministry of Education and Youth of the Czech Republic and the grant 205/99/0145 of the Grant Agency of the Czech Republic. We also acknowledge the CNR-AVČR Joint Research Program No. 3 (1998/2000).

## REFERENCES

1. Andersen, M., Hjorth, J., Pedersen, H., et al., 2000, A&A, 364, L54
2. Barthelmy, S., http://gcn.gsfc.nasa.gov/gcn/gcn3_archive.html
3. Fruchter, A.S., 1999, ApJ, 512, L1
4. Fynbo J.U., Jensen, B.L., Gorosabel, J., et al., 2000, A&A, 369, 373
5. Greiner, J., http://www.aip.de/People/JGreiner/grbgen.html
6. Klose, S., Stecklum, B., Masetti, N. et al., 2000, ApJ, 545, 271
7. Masetti, N., Palazzi, E., Pian, E., et al., 2000, A&A, 354, 473
8. Palazzi, E., Pian, E., Masetti, N., et al., 1998, A&A, 336, L95
9. Sari, R., Piran, T., Narayan, R., 1998, ApJ, 497, L17
10. D.J. Schlegel, D.P., Finkbeiner, M. Davis, 1998, ApJ, 500, 525
11. Šimon, V., Hudec, R., Pizzichini, G., Masetti, N., 2001, A&A, 377, 450
12. Waxman, E., Draine, B.T., 2000, ApJ, 537, 796
13. Weinberg, S., 1972, Gravitation and cosmology: principles and applications of the general theory of relativity, (Wilby, New York), p. 485

# SCUBA Sub-Millimeter Observations of Gamma-Ray Bursters

I. A. Smith[*], R. P. J. Tilanus[†], R. A. M. J. Wijers[**], N. Tanvir[‡],
P. Vreeswijk[§], E. Rol[§] and C. Kouveliotou[¶]

[*]*Department of Physics and Astronomy, Rice University, MS-108, 6100 South Main, Houston, TX 77005 USA*
[†]*Joint Astronomy Centre, 660 N. Aohoku Place, Hilo, HI 96720 USA*
[**]*Department of Physics and Astronomy, SUNY, Stony Brook, NY 11794-3800 USA*
[‡]*Department of Physical Sciences, University of Hertfordshire, College Lane, Hatfield, Herts AL10 9AB, UK*
[§]*Astronomical Institute 'Anton Pannekoek', University of Amsterdam and Center for High-Energy Astrophysics, Kruislaan 403, 1098 SJ Amsterdam, The Netherlands*
[¶]*NASA Marshall Space Flight Center, SD-50, NSSTC, 320 Sparkman Drive, Huntsville, AL 35805 USA, and Universities Space Research Association*

**Abstract.** We summarize our ongoing program of sub-millimeter observations of gamma-ray bursts (GRBs) using the Sub-millimetre Common-User Bolometer Array (SCUBA) on the James Clerk Maxwell Telescope (JCMT). Sub-millimeter observations of the early afterglows are of interest because this is where the emission peaks in some bursts in the days to weeks following the burst. Of increasing interest is to look for underlying quiescent sub-millimeter sources that may be dusty star-forming host galaxies.

## INTRODUCTION

The study of gamma-ray bursters (GRBs) has made a significant leap forward through the discovery of transient afterglow counterparts and quiescent host galaxies. These have shown that the bursts are due to enormous explosions taking place throughout the early Universe.

During and after the burst, the observed multiwavelength emission can come from several components. The "prompt" emission comes from the initial explosion. A reverse shock can give an optical and/or radio flash. The later "afterglow" emission comes from the expanding fireball as it sweeps up the surrounding medium. Finally, the "quiescent" constant emission comes from any underlying host galaxy.

Since 1997, we have been performing sub-millimeter observations of GRBs using the Sub-millimetre Common-User Bolometer Array (SCUBA) on the James Clerk Maxwell Telescope (JCMT) on Mauna Kea, Hawaii. Our program aims to study both the transient emission due to the explosion, as well as any quiescent emission from the host galaxy.

The detailed SCUBA results for the first eight bursts we observed (GRB 970508, 971214, 980326, 980329, 980519, 980703, 981220, and 981226) are described in [1]. GRB 990123 is discussed in [2]. Observations of GRB 990520 were made in mediocre weather [3]. Details of our SCUBA observations of GRB 991208, 991216, 000301C, 000630, 000911, and 000926 are contained in [4], where we also revisited the quiescent host galaxy implications of our earlier observations. The sub-millimeter emission from GRB 010222 that was detected by SCUBA appeared to have unusual properties [5, 6]; these observations will be reported elsewhere after a final sensitive SCUBA observation has been made to determine the contribution of any underlying host galaxy. Finally, a SCUBA survey of the host galaxies of dark GRBs is described in [7].

## IMPORTANCE OF SUB-MILLIMETER OBSERVATIONS

To understand GRBs and their environments, it is critical to have complete broad-band coverage when studying the prompt, afterglow, and host galaxy emissions. Observations in the sub-millimeter band can provide "clean" measures of the source intensity because: 1) they are unaffected by scintillation, which can cause serious fluctuations to the flux densities measured at longer radio wavelengths in some bursts; and 2) they are unaffected by extinction in the host galaxy and in our Milky Way, which can significantly diminish the optical flux.

## Afterglow Motivation

Both observations and theories show that the prompt and/or afterglow emission often peaks in the sub-millimeter in the hours to weeks following the burst.

The simplest picture is that the afterglow emission comes from synchrotron radiation of the energetic particles swept up by an expanding fireball; inverse Compton scattering may also be important in some cases. If these energetic electrons start with a simple power law energy distribution, one expects to see one or more breaks in the photon spectrum.

By carefully studying the fluxes at different wavelengths, it is possible to determine the evolving multiwavelength spectral shape, and thus the evolution of the break frequencies with time. The peak of the spectrum moves to longer wavelengths with time, generally maintaining a constant peak flux. For some bursts this takes weeks, while in others it is as short as days.

Different scenarios predict different evolutions for the spectral breaks. In principle, it is possible to determine whether the fireball expanded into a uniform density interstellar medium (as might be expected in the merger of binary compact objects), or into a radially decreasing density region (that would be expected for a prior stellar wind). It is also possible to study whether the fireball is beamed or isotropic, and the strength of the magnetic field in the afterglow emitting regions. For a beamed fireball, a brightening later due to a normal spherical supernova explosion may be seen.

However, the observed afterglow behaviors can be complex, and extracting the various features is not trivial and requires well-sampled, well-calibrated data spanning a broad range of wavelengths. The sub-millimeter detections or limits are therefore important for determining the breaks in the multiwavelength spectrum.

In the future, we hope to perform the initial sub-millimeter observations within the first day of the burst. It has been postulated that this could provide the strongest discriminator between different afterglow models [8, 9].

## Host Galaxy Motivation

If GRBs are due to the explosive deaths of high-mass stars, they are likely to be found in active star forming regions. Alternatively, if GRBs are due to the coalescence of two compact objects, the lifetime of the binary will likely be long, and the active star formation will have died down by the time the burst occurs; furthermore, the binary system may have obtained a substantial velocity if there was a supernova explosion with a large kick, moving the system far from its birthplace. GRBs would then be less likely to occur in star-forming regions. Thus there has been a great deal of interest recently in determining the star formation properties of the host galaxies.

At present, it is uncertain which type of galaxy has been responsible for the bulk of the star formation throughout the history of the Universe. There are numerous faint blue galaxies [10], but each one produces few stars. There are also ultraluminous infrared galaxies (ULIRGs) and starburst galaxies [11]; these are relatively rare, but they have very high star formation rates.

If GRBs are due to the explosive deaths of massive stars, they should trace the star formation history of the Universe. The GRBs can be detected to very high redshifts [12]. Thus studies of GRBs and the star formation rate (SFR) of their host galaxies could provide new insights into the question of which type of galaxy caused the bulk of the star formation, and whether this dominance has shifted over the history of the Universe.

There are three independent ways to determine the SFR of a host galaxy. Each method studies a different tracer of the recent star formation.

- Using optical lines. However, the inferred SFR can require a large (sometimes orders of magnitude) correction for the host galaxy extinction.
- Using the radio emission, which comes predominantly from (recent) supernovae and H II regions. This works well for relatively nearby galaxies.
- Using the sub-millimeter emission from warm and cold dust.

A significant advantage of the sub-millimeter method is that the sub-millimeter flux density that is observed is rather insensitive to the redshift of the source. Although the observed multiwavelength spectrum of a more distant galaxy has a lower overall amplitude, it is also redshifted to longer wavelengths, so that the peak in the spectrum from the dust emission is shifted closer to the sub-millimeter band [13]. The dust emission spectrum $S_\nu \propto \nu^\alpha$ is steeply rising towards higher frequencies, with $\alpha \sim 3 - 4$ based on observations and simple dust models (e.g. [14]). This means that it is possible to detect dusty star-forming galaxies out to high redshifts. Depending on the temperature of the dust, for very large $z \gtrsim 7$, the peak of the dust emitting spectrum may be redshifted to a wavelength $\sim 850 \mu m$, and $\alpha$ will be much flatter.

SCUBA has recently discovered several dusty star-forming galaxies out to high redshifts [15, 16, 17, 18], and it appears that the star formation rate does not drop rapidly beyond $z \sim 1$ [19]. However, there is still some uncertainty; for example, there may be a substantial contribution to the energy output of the sub-millimeter-bright galaxies from active galactic nuclei. If GRBs are regularly found to be associated with extremely luminous dusty galaxies, this could provide independent evidence that these are sub-millimeter bright because of their prodigious star formation.

If there is a connection between GRBs and dust-enshrouded star formation in luminous galaxies, it has been suggested that $\gtrsim 20\%$ of GRB hosts should be brighter than 2 mJy at 850 $\mu$m [20]. However, the conversion between the observed sub-millimeter flux and the inferred SFR remains somewhat uncertain.

## SCUBA OBSERVING DETAILS

SCUBA uses two arrays of bolometers to simultaneously observe the same region of sky, $\sim 2.3'$ in diameter [21]. The arrays are currently optimized for operations at 450 and 850 $\mu$m.

Fully sampled maps of the 2.3′ region can be made by "jiggling" the array. However, for most cases, we have only been looking at the well-localized optical or radio transient coordinates by performing deep photometry observations using a single pixel of the arrays. The other bolometers in the arrays are used to perform a good sky noise subtraction.

A typical integration time of 2 hours gives an rms $\sim$ 1.5 mJy at 850 $\mu$m. However, the sensitivity depends significantly on the weather and the elevation of the source; since our Target of Opportunity (ToO) observations are done on short notice, sometimes these factors are less than ideal. The simultaneous 450 $\mu$m observations usually have too high an rms to be useful, unless the weather is extremely good.

The pointing accuracy of the JCMT is a few arcsec, and the pointing is checked several times during the night. The 850 $\mu$m bolometric pixel has a diffraction limited resolution of $14''$. Thus we can be sure that the target is always well centered in the bolometric pixel.

However, the large beam size combined with the large number of distant galaxies radiating strongly at this wavelength means that in any observation there is a non-negligible chance of detecting a quiescent sub-millimeter source that is completely unrelated to the GRB. The surface density of sub-millimeter galaxies is still somewhat uncertain [22, 23]. We currently estimate that the chance of detecting a random $\geq 4$ mJy source in any pointing is $\sim 1-3\%$, while the chance of detecting a random $\geq 1$ mJy source in any pointing is $> 10\%$. This chance of getting a false positive means that one must be cautious when claiming that a particular quiescent sub-millimeter source must be the host galaxy to the GRB. For individual cases, confirmation of the star formation rate is needed from observations at other wavelengths. Taking the sample of burst observations as a whole, if many more bursts are associated with quiescent sub-millimeter sources than is expected by chance, this would be good evidence that the majority of these are true associations.

Our ToO program (PI Smith) is designed to look for the afterglow emission of the bursts by making observations as soon as possible after the burst, and by separating observations by a few days to look for a variable source. However, by combining the data from all of our observations of a source we can also look for quiescent host galaxies: our best rms to date is 0.7 mJy for GRBs 990123 and 000301C.

Our companion project (PI Tanvir) performs sensitive observations under good observing conditions to look for quiescent host galaxies, long after any burst afterglow emission has faded away.

## RESULTS TO DATE

With the possible exception of GRBs 980329 and 010222, for all of the bursts that were observed by SCUBA between 1997 and 2001, any sub-millimeter emission is consistent with coming from the afterglow: we did not conclusively detect quiescent sub-millimeter counterparts to any of the other sources.

### Afterglow Results

In addition to helping to determine the breaks in the afterglow spectra between $10^{10}$ and $10^{14}$ Hz for several bursts, our SCUBA observations have provided the following highlights:

1) An interesting early result was the detection of a fading sub-millimeter counterpart to GRB 980329 [1]. The 850 $\mu$m SCUBA flux decayed rapidly with time. For a power law decay with the flux density $\propto t^{-m}$ where $t$ was the time since the burst, the best fit power law index was $m = 3.0$. However, $m$ was not tightly constrained: the 90% confidence interval was $m = 1.2$ to $m = 5.3$. A power law $S_\nu \propto \nu^{+0.9}$ gave a good fit to the joint VLA-OVRO-SCUBA afterglow spectrum. However, such a behavior has no good theoretical explanation.

2) For the short or intermediate duration (2 s in the $> 25$ keV energy range) GRB 000301C, there may have been a transient magnification of the 850 $\mu$m afterglow flux, consistent with an achromatic fluctuation seen at other wavelengths [24, 25]. These variations might be due to refreshing of the shock in the fireball, or they may be due to inhomogeneities in the ambient medium that the fireball is expanding into. The achromatic fluctuations have also been explained as a micro-lensing event due to a $0.5 M_\odot$ lens located half way to the burst [26];

but see also [27]. The peak magnification of ∼ 2 occurred 3.8 days after the burst.

3) For GRB 000926, our observation is consistent with Harrison et al. [28], who place the peak of the spectrum in the sub-millimeter a couple of days after the burst. The broad-band spectrum is best explained if inverse Compton scattering is important in the formation of the afterglow spectrum.

## Host Galaxy Results

Although based on a small sample, our observations to date tentatively indicate that there may be fewer bright quiescent sub-millimeter sources than might have been expected. Our sample spans a wide range of redshifts, from $z = 0.707$ to $3.418$. The inferred star formation rates ($M \geq 5 M_\odot$) are typically $\lesssim 300 \, M_\odot \, \mathrm{yr}^{-1}$. Our numerical estimates are based on Condon [30] and Carilli & Yun [31], and are subject to revision when the conversion of sub-millimeter flux density to SFR is better understood.

The SFR quoted here is the star formation rate for stars with masses $\geq 5 M_\odot$. Other methods of estimating the SFR often quote results using masses between 0.1 and 100 $M_\odot$. Our results will be smaller than these by a factor that depends on the initial mass function (IMF). For a Salpeter IMF, this factor would be ∼ 5. However, the situation may be significantly different for starburst galaxies, which may strongly favor the creation of high mass stars [29]; then the factor would be closer to 1.

In Smith et al. [1], we noted that the relatively bright sub-millimeter flux from GRB 980329 could be due in part to an underlying quiescent source. If true, this would be the first quiescent sub-millimeter host galaxy detected. If this ∼ 1 mJy quiescent source is a true association, we estimate a star formation rate ($M \geq 5 M_\odot$) in the host galaxy of ∼ 200 $M_\odot \, \mathrm{yr}^{-1}$. However, since the chance of detecting a random $\geq 1$ mJy source in any pointing is > 10%, we cannot rule out that this is an unrelated source in the JCMT beam.

## THE FUTURE

Observations of new bursts are continuing to produce surprises, and there is much left to learn about GRB afterglows and host galaxies. To obtain a complete picture of their nature will require the careful study of many bursts to expand our sample; in particular, we would like to have more high redshift sources in our sample. Sub-millimeter observations with a ∼ mJy sensitivity are a key component to the multi-wavelength coverage. To this end, our programs of ToO and quiescent host galaxy observations are ongoing.

## ACKNOWLEDGMENTS

The James Clerk Maxwell Telescope is operated by The Joint Astronomy Centre on behalf of the Particle Physics and Astronomy Research Council of the United Kingdom, the Netherlands Organisation for Scientific Research, and the National Research Council of Canada.

## REFERENCES

1. Smith, I. A., Tilanus, R. P. J., Van Paradijs, J., et al. 1999, A&A, 347, 92
2. Galama, T. J., Briggs, M. S., Wijers, R. A. M. J., et al. 1999, Nature, 398, 394
3. Smith, I. A., Van Paradijs, J., Tilanus, R. P. J., et al. 2000, in Gamma-Ray Bursts: 5th Huntsville Symposium (New York: AIP), 326
4. Smith, I. A., Tilanus, R. P. J., Wijers, R. A. M. J., et al. 2001, A&A, 380, 81
5. Fich, M., Phillips, R. R., Moriarty-Schieven, G., Tilanus, R. P. J., Frail, D. A., & Smith, I. 2001, GCN 971
6. Ivison, R. J., Jenner, C. E., Lundin, W. E., Tilanus, R. P. J., & Smith, I. A. 2001, GCN 1004
7. Barnard, V. E., Blain, A. W., Tanvir, N. R., Natarajan, P., & Smith, I. A. 2002, MNRAS, submitted
8. Panaitescu, A., & Kumar, P. 2000, ApJ, 543, 66
9. Livio, M., & Waxman, E. 2000, ApJ, 538, 187
10. Ellis, R. S. 1997, ARA&A, 35, 389
11. Sanders, D. B., & Mirabel, I. F. 1996, ARA&A, 34, 749
12. Lamb, D. Q., & Reichart, D. E. 2000, ApJ, 536, 1
13. Scott, D. 1998, Nature, 394, 219
14. Dwek, E., & Werner, M. W. 1981, ApJ, 248, 138
15. Smail, I., Ivison, R. J., & Blain, A. W. 1997, ApJ, 490, L5
16. Smail, I., Ivison, R. J., Blain, A. W., & Kneib, J.-P. 1998, ApJ, 507, L21
17. Hughes, D. H., Serjeant, S., Dunlop, J., et al. 1998, Nature, 394, 241
18. Barger, A. J., Cowie, L. L., Sanders, D. B., et al. 1998, Nature, 394, 248
19. Blain, A. W., Smail, I., Ivison, R. J., & Kneib, J.-P. 1999, MNRAS, 302, 632
20. Ramirez-Ruiz, E., Trentham, N., & Blain, A. W. 2002, MNRAS, 329, 465
21. Holland, W. S., Robson, E. I., Gear, W. K., et al. 1999, MNRAS, 303, 659
22. Blain, A. W., Ivison, R. J., & Smail, I. 1998, MNRAS, 296, L29
23. Barger, A. J., Cowie, L. L., & Sanders, D. B. 1999, ApJ, 518, L5
24. Masetti, N., Bartolini, C., Bernabei, S., et al. 2000, A&A, 359, L23
25. Berger, E., Sari, R., & Frail, D. A. 2000, ApJ, 545, 56
26. Garnavich, P. M., Loeb, A., & Stanek, K. Z. 2000, ApJ, 544, L11
27. Panaitescu, A. 2001, ApJ, 556, 1002
28. Harrison, F. A., Yost, S. A., Sari, R., et al. 2001, ApJ, 559, 123
29. Elmegreen, B. G., 1999, ApJ, 515, 323
30. Condon, J. J. 1992, ARAA, 30, 575
31. Carilli, C. L., & Yun, M. S. 1999, ApJ, 513, L13

# A Serendipitous Search for GRB Afterglows by Subaru/Suprime-Cam: A Test of GRB Beaming

T. Totani[*], S. Miyazaki[†], Y. Mizumoto[†], R. Ogasawara[†], T. Takada[†], N. Yasuda[†], M. Doi[**], W. Kawasaki[**], N. Kawai[‡], A. Yoshida[§] and Y. Urata[¶]

[*]*Princeton University Observatory, Peyton Hall, NJ08544-1001, USA*
[†]*NAO, Japan*
[**]*Univ. of Tokyo, Japan*
[‡]*TiTech, Japan*
[§]*Aoyama-Gakuin Univ., Japan*
[¶]*Tokyo Science Univ., Japan*

**Abstract.** Gamma-ray bursts (GRBs) are the most mysterious object in astronomy, and one of the key issues is whether they are isotropic explosions or strongly collimated jet-like ones. A direct consequence of the collimation (or beaming) is that we should be able to find many afterglows without observed prompt gamma-ray emission, since the collimation gets wider and wider at later stages. We describe a proposed serendipitous search for optical GRB afterglows as a test of GRB beaming, by the Suprime-Cam on the 8.2m Subaru telescope, which is a very unique facility for this test by its wide field of view and high sensitivity for faint objects. We will show that the detection probability of such afterglows is of order unity by this observation, if GRBs are beamed by a factor of $4\pi/\Delta\Omega > 100$, which is a typical beaming factor currently discussed. Therefore, positive detection is possible and it would have a strong impact on our understanding of GRBs. Even if undetected, it would set the strongest constraint on the GRB beaming ever obtained. We also show that discrimination of GRB afterglows from other transient objects is possible. The data taken by this proposal can also be used in other scientific purposes of the high-$z$ supernova search.

## INTRODUCTION

Gamma-ray bursts are now confirmed to be located at cosmological distances, and they are recognized as the most energetic explosion in the universe, while their origin still remains as a mystery. The redshift of one GRB 990123 was $z = 1.6$, and the total energy of this GRB emitted as gamma-rays is estimated by its energy flux and redshift as $E \sim 3 \times 10^{54}$ erg assuming that radiation is isotropic [4]. This energy is equivalent to a rest mass energy of $1.7 M_\odot c^2$ and it is almost impossible to explain by stellar mass scale explosions. This is why most researchers in this field now consider that GRBs are very likely strongly collimated, with a typical beaming factor of $4\pi/\Delta\Omega > 100$. It is then very important to test this hypothesis by observations, for better understanding of the nature of GRBs.

A direct prediction of such GRB beaming is that there should be a much greater number of GRB afterglows which do not have the prompt gamma-ray emission, than the number of prompt GRBs observable in the gamma-ray band. GRBs are believed to be produced by dissipation of ultra-relativistic outflow from the central engine with a Lorentz factor of $\Gamma \sim 100$–1000. Then the outflow is decelerated by interaction with interstellar matter, just like supernova remnants. After $\Gamma$ decreases down to $\Gamma \sim \theta_0$, where $\theta_0$ is the opening angle of the jet, the collimation of radiation starts to get wider and wider than that of outflow. Then the afterglow becomes observable from directions different from that of the original gamma-ray radiation.

Therefore a serendipitous search of GRB afterglows without information of prompt GRBs is a crucial test for the beaming of GRBs. Such searches have been performed in radio and x-ray bands [5, 11], but the radio search gave only a crude limit on the beaming factor ($b_0 \equiv 4\pi/\Delta\Omega_0 < 1.5 \times 10^3$ where $\Delta\Omega_0$ is the initial opening solid angle of the jet), and the x-ray search gave no significant constraints. It should also be noted that the radio search is not a "true search" but it was a constraint obtained simply by comparing the expected GRB afterglow counts to the radio source counts with an assumption that $\sim$3% of radio sources are variable. A true systematic serendipitous search for GRB afterglows as transient objects has not yet been performed in any wavelength. Here we propose to search optical afterglows serendipitously by the Subaru/Suprime-Cam, which enables us to search transient objects with unprecedented sensitivity and wide field of view. In fact, it will be shown that the combination of Subaru and S-Cam provides very unique facility for this search requiring both high sensitivity and wide field of view. We will show that the constraint obtained by this observation could reach as severe as $b_0 < 250$, which is the strongest ever obtained and close to the beaming factor typically discussed ($b_0 \sim 100$–$1000$), even in the case of a negative result. In other words, there is a reasonable chance of positive detection of GRB afterglows. It would prove the strong collimation of GRBs and hence have strong scientific impact. We will also show that the discrimination of afterglows from other transient objects is possible.

## EXPECTED DETECTION RATE OF BEAMED GRB AFTERGLOWS

The simplest estimate of the number of GRB afterglows in one snapshot observation is

$$N_{GRB} \sim R_{GRB} T_{GRB} (\Omega_{FOV}/4\pi)(\Delta\Omega(T_{GRB})/\Delta\Omega_0) , \quad (1)$$

where $R_{GRB}$ is the GRB rate in all sky, $T_{GRB}$ the time duration in which afterglows can be detected by a given sensitivity, $\Omega_{FOV}$ the field of view (FOV) of instruments, and $\Omega(T_{GRB})$ the opening angle of afterglow radiation at the time of $T_{GRB}$. It is suggested both theoretically and observationally that the relativistic outflow is decelerated to sub-relativistic speed ($\Gamma \sim 1$) about 1 month after the GRB (e.g., [9]). Therefore, we can approximate $\Omega(T_{GRB}) \sim 4\pi$ when $T_{GRB} > 1$ month. As we will show later, the Suprime-Cam can take $3 \times 3 = 9$ snapshots in one night with a sensitivity sufficient for the GRB afterglows at the time $T_{GRB} \sim 30$ days after the burst. Then, using typical parameters of GRBs and Suprime-Cam, we get:

$$N_{GRB} \sim 0.4 \left(\frac{R_{GRB}}{10^3 \text{yr}^{-1}(4\pi \text{ str})^{-1}}\right) \left(\frac{T_{GRB}}{30 \text{days}}\right) \quad (2)$$

$$\times \left(\frac{N_{FOV}}{9}\right)\left(\frac{\Omega_{FOV}}{27' \times 27'}\right)\left(\frac{b_0}{100}\right) , \quad (3)$$

or, we would have the sensitivity to the beaming of:

$$b_0 \sim 250 \left(\frac{R_{GRB}}{10^3 \text{yr}^{-1}(4\pi \text{ str})^{-1}}\right)^{-1} \left(\frac{T_{GRB}}{30 \text{days}}\right)^{-1} \quad (4)$$

$$\times \left(\frac{N_{FOV}}{9}\right)^{-1}\left(\frac{\Omega_{FOV}}{27' \times 27'}\right)^{-1}\left(\frac{N_{GRB}}{1}\right) , \quad (5)$$

where $N_{FOV}$ is the number of snapshots taken by the Suprime-Cam. Therefore it is statistically possible to detect an afterglow by this observation, if the beaming is stronger than $b_0 > 100$, as many GRB researchers now believe. Furthermore, continuation of this project in a long term would eventually detect (or constrain) the GRB beaming in the whole interesting range.

## OBSERVATION STRATEGY

Observation time of about 10 minutes results in a sensitivity limit of the S-Cam, $R \sim 25$ at S/N=5, and this magnitude corresponds to typical flux of GRB afterglows at about 30 days after the prompt burst (e.g., [1, 2]). Therefore, the sensitivity is enough to warrant the use of $T_{GRB} \sim 30$ days in the detection rate estimated above. On the other hand, this sensitivity is difficult to reach by smaller telescopes, and this point justifies the use of the Subaru/S-Cam for this purpose.

We should discuss how to discriminate GRB afterglows from other objects. A three-bands observation is known to be very powerful to discriminate the power-law spectrum of GRBs from stars and galaxies having curved spectra [6], even in one snapshot observation. Since we can remove steady sources by two-time observations, we expect that discrimination from stars (including variable stars) or galaxies would be an easy task. It is expected that there are about 15 supernovae per 1 deg$^2$ with the limiting magnitude of $R \sim 24.5$. They would also be discriminated by the color-color diagram, since their spectrum is basically thermal which is considerably different from GRB afterglows. In fact, there is a phase when a type Ia supernova shows a $B-V$ color of 0.0–0.5 which is similar to that of GRB afterglows, but at that time $V-I$ color is about $\sim -0.5$ for supernovae, which is very different from 0.5–1 for GRB afterglows (see, e.g., [7]). Finally, a part of quasars may have a similar position in the color-color diagram, since they also have power-law spectra. However, their variability amplitude is generally smaller than that of afterglows on a timescale of 10–30 days. The standard deviation in optical magnitude of quasars or active galactic nuclei is 0.10–0.15 mag, respectively, and variability greater than 0.3 mag is very rare [10]. On the other hand, flux decay by a factor of 2 or more is well expected for GRB afterglows on a time scale comparable with the time elapsed after the burst, with $\alpha \sim 1$–$2$. Therefore the amplitude of variability can be used to discriminate GRB afterglows from quasars whose number is expected to be 100–1000 deg$^{-2}$ down to $R \sim 25$. Furthermore, the data which we have already taken in this field can remove quasars because they are persistent on a much longer time scale than GRB afterglows, as mentioned above. Therefore, we conclude that afterglow discrimination from all known objects is possible.

Our observation will discover about 10 type Ia supernovae whose luminosities have not yet reached the light curve peak. These can potentially be used as supernova cosmology data which is important to determine the geometry of the universe, as is well known.

## PROJECT STATUS

Our pilot observation is scheduled on Jan. 12 and Feb. 14 of 2002, observing 9 FOVs of the S-Cam, with three passbands of R, B, i'. (In fact, the first observation is now performed when I am writing this article!) There is also a possibility that the search of faint transient objects such as supernovae and GRB afterglows is more systematically done in the future key projects of the Subaru telescope.

## REFERENCES

1. Fruchter, A.S. et al. 1999, ApJ, 516, 683
2. Harrison, F.A. et al. 1999, ApJ, 523, L121
3. Hartman, R.C. et al. 1999, ApJS, 123, 79
4. Kulkarni, S. et al. 1999, Nature, 398, 389
5. Perna, R. & Loeb, A. 1998, ApJ, 509, L85
6. Rhoads, J. 2001, ApJ in press, astro-ph/0008461
7. Riess, A.G., Press, W.H., & Kirshner, R.P. 1996, ApJ, 473, 88
8. Totani, T. & Kitayama, T. 2000, ApJ, 545, 572
9. Waxman, E., Kulkarni, S.R., & Frail, D.A. 1998, ApJ, 497, 288
10. Webb, W. & Malkan, M. 2000, ApJ, 540, 652
11. Woods, E. & Loeb, A. 1999, ApJ, 523, 187

# GRB Afterglows and Other Transients in the SDSS

Brian C. Lee[*], Daniel E. Vanden Berk[*], Don Lamb[†], Brian Wilhite[†], Daniel E. Reichart[**], John F. Beacom[*], Douglas L. Tucker[*], Brian Yanny[*], Kevork Abazajian[*], Jennifer Adelman[*], James Annis[*], Bing Chen[‡], Mike Harvanek[§], Arne Henden[¶], Kevin Hurley[∥], Zeljko Ivezic[††], Robert Kehoe[‡‡], Scot Kleinman[§], Richard Kron[*], Jurek Krzesinski[§], Dan Long[§], Timothy McKay[§§], Russet McMillan[§], Eric H. Neilsen[*], Peter R. Newman[§], Atsuko Nitta[§], Povilas Palunas[¶¶], Donald P. Schneider[***], Steph Snedden[§], James Wren[†††], Don York[†], John W. Briggs[‡‡‡], J. Brinkmann[§], Istvan Csabai[‡], Greg S. Hennessy[§§§], Stephen Kent[*], Robert Lupton[††], Heidi Jo Newberg[¶¶¶¶] and Chris Stoughton[*]

[*]*Experimental Astrophysics Group, Fermi National Accelerator Laboratory, P.O. Box 500, Batavia, IL 60510*
[†]*Department of Astronomy and Astrophysics, University of Chicago, 5640 South Ellis Avenue, Chicago, IL 60637*
[**]*Palomar Observatory, 105-24, California Institute of Technology, Pasadena, CA 91125*
[‡]*Department of Physics and Astronomy, Johns Hopkins University, 3701 San Martin Drive, Baltimore, MD 21218*
[§]*Apache Point Observatory, P.O. Box 59, Sunspot, NM 88349-0059*
[¶]*Universities Space Research Association / U. S. Naval Observatory, Flagstaff Station, P. O. Box 1149, Flagstaff, AZ 86002-1149*
[∥]*University of California at Berkeley, Space Science Laboratory, Grizzly Peak and Centennial Drive, Berkeley, CA 94720*
[††]*Princeton University Observatory, Peyton Hall, Princeton, NJ 08544-1001*
[‡‡]*Michigan State University, Department of Physics and Astronomy, East Lansing, Michigan 48824*
[§§]*University of Michigan, Department of Physics, 500 East University, Ann Arbor, MI 48109*
[¶¶]*Catholic University of America, NASA/Goddard Space Flight Center, Code 681, Greenbelt, MD 20771*
[***]*Astronomy and Astrophysics Department, Pennsylvania State University, 525 Davey Laboratory, University Park, PA 16802*
[†††]*Los Alamos National Laboratory, PO Box 1663, Los Alamos, NM 87545*
[‡‡‡]*Yerkes Observatory, University of Chicago, 373 West Geneva Street, Williams Bay, WI 53191*
[§§§]*U.S. Naval Observatory, 3450 Massachusetts Ave., NW, Washington, DC 20392-5420*
[¶¶¶¶]*Physics Department, Rensselaer Polytechnic Institute, SC1C25, Troy, NY 12180*

**Abstract.** The Sloan Digital Sky Survey (SDSS) will image one quarter of the sky centered on the northern galactic cap and produce a 3-D map of galaxies and quasars found in the sample. An additional $225\,deg^2$ southern survey will be imaged repeatedly on varying timescales. Here we discuss both archival searches in the SDSS catalog (such as SDSS J24602.54+011318.8) and active searches with the SDSS instruments (such as for GRB 010222) for GRB afterglows and other transient objects.

## INTRODUCTION: THE SLOAN DIGITAL SKY SURVEY

The SDSS is a project to image $10{,}000\,deg^2$ of the Northern Galactic Cap in five different filters ($u,g,r,i,z$) to a depth of $r^* \sim 23$ and to perform followup spectroscopy of $10^6$ galaxies and $10^5$ quasars found in the photometry [13, 16]. The SDSS imaging survey is designed to be on an $AB_\nu$ system described in Fukugita et al. [2] where flat spectrum objects ($F_\nu \propto \nu^0$) have zero colors. The magnitudes in this poster are quoted on the preliminary $u^*, g^*, r^*, i^*, z^*$ system which may differ by at most a few percent from the system of Fukugita et al. [2]. The spectroscopic survey will map galaxies to a depth of $z \sim 0.2$, luminous red galaxies to $z \sim 0.5$, and quasars as distant as $z \sim 6.5$.

The survey uses two dedicated telescopes, the automated 0.5m "Photometric Telescope" (PT) which monitors extinction and provides calibration information, and the 2.5m survey telescope used for the main imaging and spectroscopic surveys, both of which are located at Apache Point Observatory (APO) in Sunspot, New Mexico. The PT is an f/8.8, 0.5m telescope equipped with *ugriz* filters on a filter wheel. The single SITe $2048 \times 2048$ CCD camera has a $41.5' \times 41.5'$ field of view. A normal exposure of 300 sec in $u$ and 120 sec each in $g$, $r$, $i$, and $z$ reaches limiting $10\sigma$ magnitudes of approximately 17.0, 16.0, 15.5, 15.0, and 15.0, although images as deep as $r^* \sim 19$ are regularly attained with longer exposures. Operation of the PT is fully automated (short of opening and closing the dome due to safety concerns) with targets for observation chosen from a queue based on assigned priorities and observability.

The 2.5m survey telescope is an f/5, $3°$ field of view telescope designed and constructed for the SDSS. The telescope has two interchangeable instruments, a 54 CCD imaging camera and a fiber-fed spectrograph capable of simultaneously observing 640 spectra. The imaging camera [3] includes an array of 30 $2048 \times 2048$ CCDs in six columns of five CCDs each, one CCD for each of the 5 filters. The camera operates in a drift scan mode, scanning the sky in great circles at sidereal rate. Astronomical objects are imaged sequentially in the order $r,i,u,z,g$ for 53.9 seconds in each filter. Because of the gaps between columns the telescope must observe a second such interleaved strip to make a complete stripe. The survey is designed to reach $u^*g^*r^*i^*z^*$ magnitude limits of 22.3, 23.3, 23.1, 22.3, and 20.8. Galaxies, quasars, and other targets are selected from this imaging data for the spectroscopic survey.

## ARCHIVAL SEARCHES IN THE SDSS CATALOG

Although the SDSS is not designed to search for GRB afterglows or variable objects of any sort, the multicolor observations make it possible to search for rare objects with non-stellar spectral energy distributions (SEDs) [14, 15, 13, 12, 7]. GRB afterglows are observed to have power law SEDs of the form $F_\nu \propto \nu^\beta$ with typical values of $\beta \sim -1$. Our initial studies have shown that such objects are rare but do exist in the SDSS catalog. Unfortunately reddened quasars have similar colors and can only be ruled out with temporal information. However several hundred $deg^2$ have already been scanned at least twice and the 225 $deg^2$ of the southern survey will be scanned dozens of times. Much work is currently under-

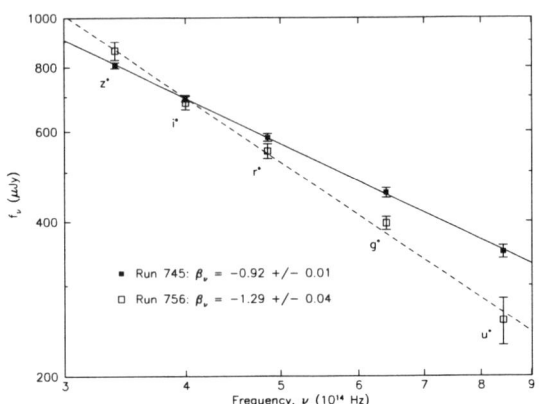

**FIGURE 1.** Spectral energy distribution of SDSS J124602.54+011318.8

way to find SNe and other variable objects in the southern survey within a day or less of observation.

Many objects which lie outside the stellar locus are automatically targeted for observation in the spectroscopic survey. In Vanden Berk et al. [15] we have used spectrophotometric magnitudes derived from these observations to provide temporal information and search for varying objects such as SNe. We have found one optical transient (OT) which shares many of its properties with GRB afterglows (see Vanden Berk et al. [14]). This transient was imaged at three separate epochs as part of the SDSS imaging survey and the spectrum was measured twice. A bright point source was observed in images separated by 2 days; all other SDSS and archival plate images show a galaxy 2.5 magnitudes fainter with no evidence of variability. ROTSE-I observations [5] from 50 days before to 50 days after show that the OT was not significantly brighter than the SDSS detection, indicating the SDSS detected the OT at its peak. The SED (Figure 1) is quite similar to that of GRB afterglows. While it is possible that this OT is a very unusual AGN, it seems much more likely that it is an "orphan afterglow" – a GRB afterglow without any observed $\gamma-ray$ flux [14].

## ACTIVE SEARCHES WITH THE SDSS TELESCOPES

In certain rare cases when conditions do not allow normal survey operations the SDSS observers have been able to manually follow up some GRB triggers, including GRB010222 with the 2.5m and 0.5m PT telescopes [9], and more recently GRB010921 [8] and GRB011019 with the PT alone. The 2.5m survey telescope observations of GRB010222 show the value of early multicolor observations. In addition to the measurement of the $g^*r^*i^*z^*$ spectral slope ($F_\nu \propto \nu^{-0.90\pm0.03}$ for GRB010222), the ob-

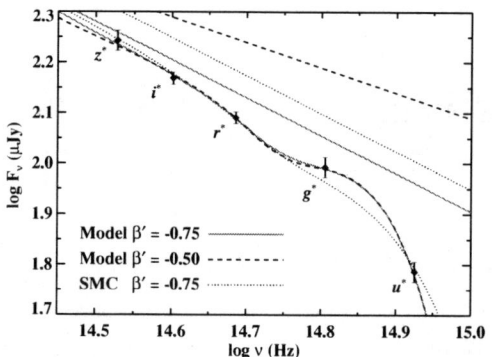

**FIGURE 2.** Extinction model fits to the 2.5m multiband observations.

served break to a steeper slope in $u^*$ (Figure 2) was not predicted or seen in spectra, and likely indicate extinction at the source, possibly in a star forming region modified by the GRB or its progenitor. GRB010921 was observed under partially cloudy conditions approximately 1 and 2 days after the HETE-I trigger, and was seen to fade significantly from the first night ($r \approx 19.6$) to the second. We have not yet detected a candidate object for GRB011019.

The above manual PT followup attempts have all suffered due to long delays between GRB detection and PT observation – the afterglow is often near the limiting magnitude of a 0.5m telescope by the time PT observations can be scheduled. With trivial modifications to the PT's scheduling code, GCN alert positions could be inserted in the target queue, resulting in observations starting on average 10 minutes after the trigger. Additional modifications could reduce the average response time considerably. Existing PT image processing code may allow for detection of transients within minutes, allowing early spectroscopic followup on non-SDSS telescopes such as the ARC 3.5m telescope at Apache Point. We are still in the early stages of studying whether a GRB followup mission could be funded and incorporated without any impact on the PT's primary mission as part of SDSS operations.

## CONCLUSIONS

Our searches for GRB afterglows and other transients in the SDSS have already made many valuable discoveries, including what may be the first orphan afterglow and a strong indication of dust extinction in an afterglow's environment. Additional detections of transients with properties similar to GRB afterglows can test the beaming hypothesis and provide the first measurements of the properties of so called "orphan afterglows." Early near simultaneous observations of GRB afterglows spanning the Sloan *ugriz* filters, especially in the bluer filters, are nearly non-existent, yet these observations (along with x-ray, near-IR, radio, and submillimeter observations) are needed to test the current ultra-relativistic external shock model of GRB afterglows, and to provide information about extinction of burst afterglows.

## ACKNOWLEDGMENTS

The Sloan Digital Sky Survey (SDSS) is a joint project of The University of Chicago, Fermilab, the Institute for Advanced Study, the Japan Participation Group, The Johns Hopkins University, the Max-Planck-Institute for Astronomy (MPIA), the Max-Planck-Institute for Astrophysics (MPA), New Mexico State University, Princeton University, the United States Naval Observatory, and the University of Washington. Apache Point Observatory, site of the SDSS telescopes, is operated by the Astrophysical Research Consortium (ARC).

Funding for the project has been provided by the Alfred P. Sloan Foundation, the SDSS member institutions, the National Aeronautics and Space Administration, the National Science Foundation, the U.S. Department of Energy, Monbusho, and the Max Planck Society. The SDSS Web site is http://www.sdss.org/.

## REFERENCES

1. Djorgovski, S. G. 2001, in Gamma-Ray Bursts in the Afterglow Era: 2nd Workshop, Proc. ESO Astroph. Symp., ed. N. Masetti et al., (Berlin: Springer Verlag), in press, (astro-ph/0107535)
2. Fukugita, M., Ichikawa, T., Gunn, J. E., Doi, M., Shimasaku, K., & Schneider, D. P. 1996, AJ, 111, 1748
3. Gunn, J. E. et al. 1998, AJ, 116, 3040
4. Henden, A. 2001a, GCN Circ. 961
5. Kehoe, R., et al. 2001, in Supernovae and Gamma-Ray Bursts: The Greatest Explosions since the Big Bang, ed. M. Livio, N. Panagia, & K. Sahu (Cambridge: Cambridge Univ. Press), 47
6. Kennicutt, R. C. 1998, ARA&A, 36, 189
7. Krisciunas, K., Margon, B., & Szkody, P. 1998, PASP, 110, 1342
8. Lamb, D. Q., et al. 2001, GCN Circ. 1125
9. Lee, B. C., et al. 2001, ApJ, 561, 183 (astro-ph/0104201)
10. McDowell, J., Kilgard, R., Garnavich, P. M., Stanek, K. Z., & Jha, S. 2001, GCN Circ. 963
11. Reichart, D. E. 2001, ApJ, 553, 235 (astro-ph/9912368)
12. Rhoads, J. E. 2001, ApJ, in press, (astro-ph/0008461)
13. Stoughton, C., et al. 2001, AJ, accepted
14. Vanden Berk, D. E., et al. 2001a, in preparation
15. Vanden Berk, D. E., et al. 2001b, in preparation
16. York, D. G. et al. 2000, AJ, 120, 1579

# Gamma-Ray Burst Optical Afterglow Observations at Nyrola Observatory

Arto Oksanen, Harri Hyvönen, Marko Moilanen

*Jyväskylän Sirius ry, Kyllikinkatu 1, FIN-40100 Jyväskylä, Finland*

**Abstract.** We present the results of observing five gamma-ray bursts (GRB000926, GRB010119, GRB010222, GRB000324 and GRB010412) at Nyrola Observatory, Finland. Two optical afterglows were successfully detected. Also the equipment and methods used for the observations are presented. Our example shows how dedicated amateur astronomers with modest equipment can make a significant contribution to the GRB research.

## INTRODUCTION

Nyrölä Observatory is an amateur observatory located in Finland. It is located in countryside near town Jyväskylä and is operated by a 200-member astronomy club. The equipment used in these observations consists of 16-inch Meade LX200 telescope, Santa Barbara Instrument Group ST7E CCD-imager and photometric B, V & R filters.

**FIGURE 1.** The 16-inch Meade LX200 telescope with SBIG ST7E CCD-imager of Nyrölä Observatory.

Our three-person GRB team has been involved with GRBs since 1999, when the observatory was joined to the GRB Coordinates Network. The observatory computer is receiving socket alerts and observers are notified via text messages to mobile phones (GSM/SMS) and email. Five GRB fields has been imaged so far: two showing optical afterglows and three negative observations. The observers have joined also to the AAVSO International GRB Network in 2000.

## OBSERVATIONS

### GRB 000926

The first successful observation was GRB 000926 after several unsuccessful observing attempts. This afterglow was observed after the optical transient coordinates were published on the GRB Coordinates Network (GCN 804) [1]. The total exposure time was 4080 seconds by co-adding 17 unfiltered four minute exposures. The limiting magnitude was over 20 magnitudes and the optical transient was identified and measured from the combined image by Hitoshi Yamaoka of Kyushu Univ., Japan [5]. See the final image in Fig. 2. This demonstrated that the detection of 20 magnitude targets was possible with our equipment. This was also the first ever GRB OT observed by an European amateur astronomers.

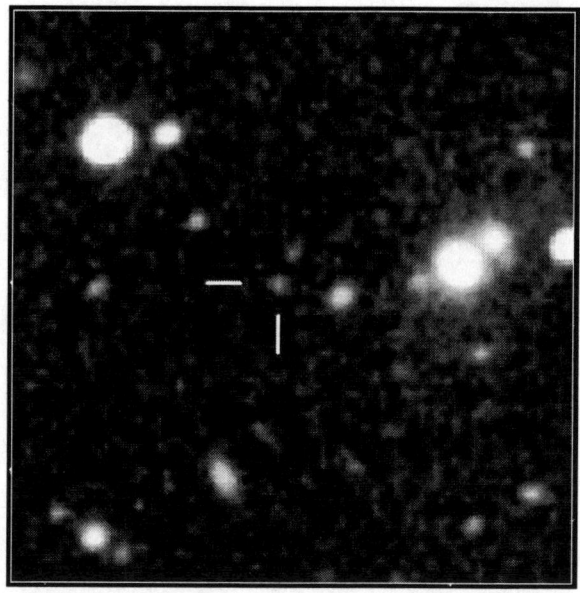

**FIGURE 2.** GRB 000926 optical afterglow imaged on September 28, 2000.

**FIGURE 3.** The optical counterpart of GRB010222 imaged on February 22, 2001.

*GRB 010119*

The second observation was of a short-hard burst. The complete error box of GRB 010119 was observed 42 hours after the burst. This time observation was done with a photometric Rc filter. The upper-limit for optical transient was determined R=19.5 magnitudes by the dimmest stars visible on the image. Observing results were published on GCN circular 920 [2]. This observation is also presented on a paper by K. Hurley et. al. *Afterglow upper limits for four short duration, hard spectrum gamma-ray bursts* [4].

*GRB 010222*

GRB 010222 was observed all night from 12 to 22 hours after the burst. Observations were performed with photometric filters and the fading behavior of the optical afterglow was detected in V and R bands. The light curve showing the observation is in Figure 4 and individual observations are listed on Table 1. The composite (BVR) image of the afterglow is presented on Figure 3. A GCN Circular was submitted by the observers [6].

*GRB 010324 and GRB 010412*

The last two observations were negative: GRB 010324 and GRB 010412. The upper limits for optical transients were C=18.8 and R=19.0 magnitudes respectably. The GRB010324 error box was observed by several observers of the International GRB Network operated by American Association of Variable Star Observers (AAVSO). The summary of the observations were published on GCN 1019 [3].

## CONCLUSIONS

Dedicated amateurs can give significant help in GRB research with modest 'off-the-shelf' equipment. Amateurs are often ready to observe these kind of transient objects, as they don't have pre-scheduled observations. The response time can be also very fast if the alerts are reaching the observer(s) already imaging other targets. By well-coordinated pro-am collaboration the amateur observations can be quickly forwarded by professional astronomers to the whole GRB community.

**TABLE 1.** All Nyrölä GRB OT observations. Magnitude reference is USNOA2.0 except for GRB010222 where Henden (GCN1025) sequence was used.

| Burst ID | Delay (hrs) | Filter | Magnitude |
|---|---|---|---|
| GRB000926 | 43 | unfiltered | 20 ± 0.5 |
| GRB010119 | 42 | R | > 19.5 |
| GRB010222 | 12.4 | B | 19.84 ± 0.31 |
| | 13.6 | R | 19.57 ± 0.13 |
| | 14.6 | V | 19.70 ± 0.13 |
| | 15.8 | R | 19.69 ± 0.15 |
| | 16.9 | V | 19.92 ± 0.14 |
| | 18.1 | R | 19.92 ± 0.17 |
| | 20.1 | V | 20.29 ± 0.21 |
| | 20.6 | R | 20.11 ± 0.31 |
| GRB010324 | 10.9 | unfiltered | > 18.8 |
| GRB010412 | 29 | R | > 19.0 |

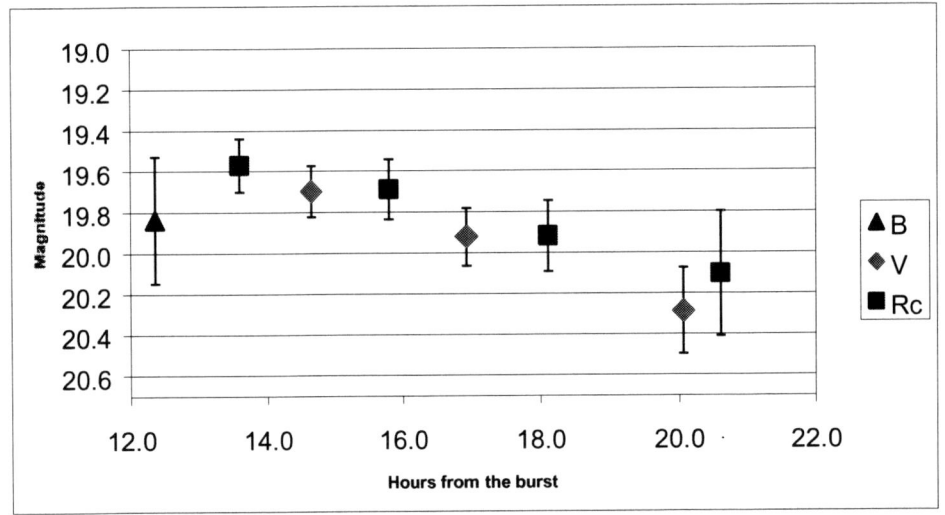

**FIGURE 4.** GRB 010222 optical observations of Nyrölä Observatory. The light curve shows fading of the optical transient in V and Rc bands during the observing run.

## ACKNOWLEDGMENTS

We want to thank Jyväskylän Sirius and Jenny and Antti Wihuri foundation for providing the observatory and its equipment, Gamma-ray Burst Coordinates Network and specially Dr. Scott Barthelmy for accepting us amateurs in, the American Association of Variable Star Observers for coordinating amateur GRB observations and Dr. Arne Henden for helping us with our observations and data reduction.

## REFERENCES

1. Fynbo, J., *GCN Circular 804*, 2000
2. Henden, A., *GCN Circular 920*, 2001
3. Henden, A., *GCN Circular 1019*, 2001
4. Hurley, K., et al, *ApJ (submitted)*, astro-ph/0107188, 2001
5. Kato, T., *GCN Circulars 813 & 817*, 2000
6. Oksanen A., *GCN Circular 990*, 2001

# The AAVSO International GRB Network

Aaron Price

*American Association of Variable Star Observers*
*Clinton B. Ford Astronomical Data and Research Center*
*25 Birch St., Cambridge, MA. 02138 USA*

**Abstract.** The AAVSO International GRB Network provides services to both amateurs and professionals to help detect GRB afterglows. The network leverages the unique abilities of amateur astronomers to offer global coverage to eliminate geographic and climatic restrictions to GRB alert reaction times. Additionally, public outreach is a critical component of the network and automated online chart making procedures have made it a useful tool for professionals. The financial support of NASA and the Curry Foundation is gratefully appreciated.

## INTRODUCTION

The AAVSO International GRB Network is a network of amateur and professional astronomers collaborating to image GRB afterglows. With over 150 members in 15 countries, the network offers global coverage and eliminates geographic and climate restrictions to GRB alert reaction times. To date, four GRB afterglows have been imaged by members of the network (GCNs 844, 969, and 990). In addition to afterglow detection, grassroots public outreach is an important part of network activity. The network was established at a joint AAVSO/NASA High Energy Astrophysics Workshop for Amateur Astronomers held in Huntsville, Alabama in April, 2000. A second follow up workshop is planned for July, 2002 in Hawaii.

## PROCEDURES

The network has developed a fully comprehensive set of procedures to follow when attempting to image a GRB afterglow. When a GRB alert is distributed via GCN, the first thing that happens is the automatic creation of finder charts and the notification of network members. The charts are placed on the AAVSO WWW site within a couple of minutes of an alert and include finder charts using DSS and the USNO PMM (usually POSS II) server. These charts include the drawing of the error pattern. In addition, photometry from USNOA2, GSC 2.2, and automatic minor planet searches of the field are placed online. This is a resource of which both amateurs and professionals make use. While the charts are being created, alerts are distributed to 97 network members. Of these 97, 31 receive their alerts via wireless devices (some provided by the AAVSO) and the rest receive them via e-mail.

## MEMBERSHIP

Network members are located in 15 countries scattered all across the globe. This gives the network the unique ability to cover just about any location in the sky all the time. The AAVSO has distributed CCDs to members in Finland, Hungary, Brazil, Australia, and New Zealand to help cover any gaps in our original coverage.

The network is primarily made up of advanced amateur astronomers who are very experienced and have good equipment. Most have telescopes of at least .25m (while many of the most active observers have access to larger telescopes through local clubs and other organizations) and use CCDs such as the SBIG ST-8E and ST-9 cameras. All have been trained in the proper use of standard filters and photometric reduction. Many members have years if not decades of observational experience and have contributed to other professional-amateur collaborations such as WEBT, CBA, and the AAVSO.

Even with this capability, the primary advantage amateurs have is an unlimited supply of enthusiasm. Amateurs often serve as the best educators and ambassadors to the public because of ties to local communities and their love of astronomy. The network has worked hard to increase public awareness of GRB research through a vast campaign of grassroots education and outreach. Members have given 54 talks in the United States and 7 more outside of the country and also have developed educational curricula for teachers available via the AAVSO WWW site and on CD-ROM. One such piece, a Power Point presentation on GRBs, has been requested by teachers 41 times and has been viewed online in HTML form over 5,000 times in six months.

The American Association of Variable Star Observers has been collecting observations of variable stars for over 90 years. Currently, the AAVSO International Database contains over 10 million observations of variable stars, a number increasing by at least 400,000 annually.

## ADVANTAGES FOR PROFESSIONALS

*Free Workers! Free Data!* These are mostly amateurs who are doing this for enjoyment so there is no need to request funds or fill out observing proposals. Also, false alarms are not discouraging since they do not use up valuable telescope time. All data collected is available to the public for free.

*Collaboration.* The AAVSO will gladly help you with any special requests you may have. AAVSO HQ has 90 years of experience in this area.

*Tools.* Automatic charts, photometry lists, and minor planet searches on GRB fields are available within minutes of a GCN at:
http://www.aavso.org/grb/archive

*Outreach.* Need to drum up support in the public? Members have given over 126 talks in 11 countries about GRBs. We also have enjoyed coverage by CNN, Space.com, *Sky & Telescope*, PBS.org, and many other publications.

## CONCLUSION

GRB afterglow hunting is an exciting area of pro-am collaboration. The AAVSO International GRB Network can help professionals by providing support resources, data, and increasing public awareness in the field of GRB research.

## ACKNOWLEDGEMENTS

The AAVSO and its members would like to thank the following people for their support and expertise: Scott Barthelmy, The Curry Foundation, Jerry Fishman, Peter Garnavich, Arne Henden, Kevin Hurley, Chryssa Kouveliotou, NASA, and George Ricker.

# Colour-Colour Diagram as a Tool for Prompt Search of GRB Afterglows; the Discovery of the GRB 001011 Optical/Near-Infrared Counterpart

J. Gorosabel[*], J.U. Fynbo[†], P. Møller[†], J. Hjorth[**], H. Pedersen[**], L. Christensen[**], B.L. Jensen[**], M.I. Andersen[‡], C. Wolf[§], J. Afonso[¶], M. A. Treyer[||], G. Mallén-Ornelas[††], A.J. Castro-Tirado[‡‡], A. Fruchter[§§], J. Greiner[¶¶], S. Klose[¶¶], C. Kouveliotou[***], N. Masetti[†††], E. Palazzi[†††], F. Frontera[†††], E. Pian[‡‡‡], N. Tanvir[§§§], P. M. Vreeswijk[¶¶¶], E. Rol[¶¶¶], I. Salamanca[¶¶¶], L. Kaper[¶¶¶], E. van den Heuvel[¶¶¶] and R.A.M.J. Wijers †

[*]*Danish Space Research Institute, Juliane Maries Vej 30, DK–2100 Copenhagen Ø, Denmark.*
[†]*European Southern Observatory, Karl–Schwarzschild–Straße 2, D–85748 Garching, Germany.*
[**]*Astronomical Observatory, University of Copenhagen, Juliane Maries Vej 30, DK–2100 Copenhagen Ø, Denmark.*
[‡]*Division of Astronomy, P.O. Box 3000, FIN-90014 University of Oulu, Finland.*
[§]*Max-Planck-Institut für Astronomie, Königstuhl 17, D-69117 Heidelberg, Germany.*
[¶]*Blackett Laboratory, Imperial College, Prince Consort Road, London 7 2BW, UK.*
[||]*Laboratoire d'Astronomie Spatiale, Traverse du Siphon, BP8, 13376 Marseille, France.*
[††]*Department of Astronomy, University of Toronto, 60 St. George Street, Toronto, ON, M5S 3H8, Canada.*
[‡‡]*Laboratorio de Astrofísica Espacial y Física Fundamental (LAEFF-INTA), P.O. Box 50727, E-28080, Madrid, and Instituto de Astrofísica de Andalucía (IAA-CSIC), P.O. Box 03004, E-18080 Granada, Spain.*
[§§]*Space Telescope Science Institute, 3700 San Martin Drive, Baltimore MD 21218, USA.*
[¶¶]*Astrophysikalisches Institut, Potsdam, and Thüringer Landessternwarte Tautenburg, Tautenburg, Germany.*
[***]*NASA MSFC, SD-50, Huntsville, AL 35812, USA.*
[†††]*Istituto Tecnologie e Studio Radiazioni Extraterrestri, CNR, Via Gobetti 101, 40129 Bologna, Italy.*
[‡‡‡]*Osservatorio Astronomico di Trieste, Via Tiepolo 11, I-34131 Trieste, Italy*
[§§§]*Department of Physical Sciences, University of Hertfordshire, College Lane, Hatfield, Herts AL10 9AB, UK.*
[¶¶¶]*University of Amsterdam, Kruislaan 403, 1098 SJ Amsterdam, The Netherlands.*
[†]*Department of Physics and Astronomy, State University of New York, Stony Brook, NY 11794-3800, USA.*

**Abstract.**
We present the results provided by an automatic colour-colour discrimination pipe-line developed to discern the different populations of objects present in GRB error boxes. We illustrate the technique by describing its application to GRB 001011, which enabled the discovery of its optical/near-infrared counterpart. The GRB 001011 afterglow is the first discovered with the assistance of colour-colour diagram techniques.

## INTRODUCTION

Since the first detection of an optical counterpart to a GRB [1] searches have been based either on comparing a single epoch image to Digital Sky Survey (DSS) images in order to search for new objects or by comparing images taken through the same filter at different epochs in order to find fading, transient objects.

It is clear that the identification process would benefit from alternative identification techniques. One such alternative is using colour-colour selection techniques similar to those used for quasar selection [2].

The main principle behind colour selection of optical afterglows (OAs) is that they have power-law spectral energy distributions, which can be distinguished from curved thermal spectra in colour-colour plots ([3, 4] and references therein).

We put in practice such principles by an automatic

software [5, 6]. This pipe-line allowed us to discover the GRB 001011 afterglow, using J, Ks and R-band data taken between 7.37 and 8.95 hours after the hours after the trigger with the NTT and the 1.54-m Danish telescopes.

We demonstrate that the colour selection techniques can be successfully applied to identify GRB afterglows, including the use of near-IR data. The following criteria is used to discriminate OAs from other objects in the GRB error box (see shaded area of Fig. 1 and 2):

$$R-Ks > 2.0, \quad J-Ks > 1.2$$

## WHY USE COLOUR-COLOUR DISCRIMINATION TECHNIQUES?

The colour-colour discrimination technique is an ideal complement to the traditional method aimed to find fading, transient objects.

- The method only requires three images, once they have been WCS calibrated.
- The technique is based on one-epoch images, allowing fast identifications and consequently prompt spectroscopy or/and polarimetry.
- Many OAs are fainter than the DSS limit at the time of the first optical follow-up observations. With 8-m class telescopes the technique can easily be used for bursts as faint as R=24. Hopefully it will facilitate the detection of afterglows from a larger fraction of well localised GRBs, that is around 30% according to the experience accumulated during the latest 3–4 years [7].
- In some cases the afterglow light curves show "plateaus" that could disguise their transient nature. The colour-colour pipe-line can identify these afterglows based only on one-epoch observations.
- As afterglow spectra ($F_\nu$) are fairly well described by pure power-laws, ($\nu^{-\beta}$) the method is independent of the redshift, at least for redshifts with a negligible Lyman-$\alpha$ blanketing along the line of sight. If the R-band is used to construct the colour-colour plot it is independent of the redshift for $z < 3.7$.
- GRB optical/infrared afterglow lightcurves are reasonably well described by achromatic power-law decays ($F_\nu$ decays as $t^{-\alpha}$, being $\alpha$ independent of frequency). Thus, the colours should remain approximately constant with time. Hence, the technique is valid at any epoch after the gamma-ray burst at least until the emission of the host galaxy becomes dominant, usually weeks after the gamma-ray event.

## APPLICATION OF THE TECHNIQUE: GRB 001011

The application of the colour-colour discrimination technique enabled us to detect the GRB 001011 near-IR/optical afterglow (see Fig. 1 and 2).

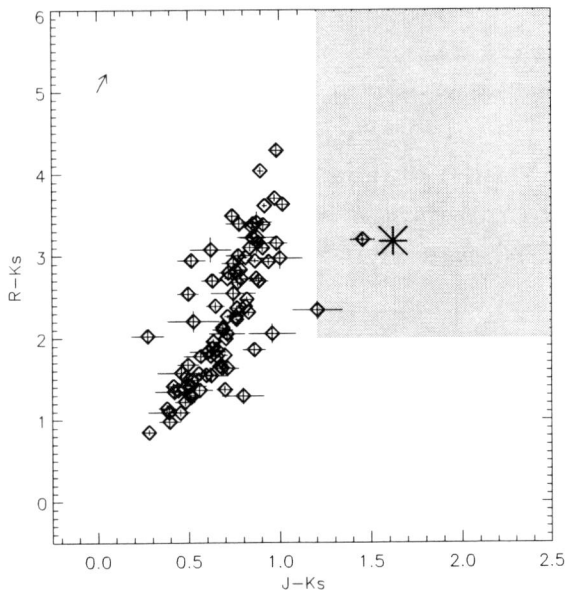

**FIGURE 1.** The figure shows an R-Ks versus J-Ks colour-colour diagram for the objects inside the GRB 001011 error box (open diamonds). The star represents the colours of the GRB 001011 optical/near-IR counterpart. The errors of the counterpart colours are smaller than the size of the star. The shaded region shows the OAs colour-colour space locus determined by our selection criteria R-Ks > 2.0, J-Ks >1.2. As can be seen, the colours of the GRB 001011 counterpart are consistent with the shaded area and inconsistent with almost all the other objects within the error circle. Only two non-transient objects in the error box have colours consistent with the shaded area. They are very likely quasars or compact galaxies.

The main properties of the afterglow are summarized below:

- The GRB 001011 afterglow evolution is consistent with an achromatic decay index of $\alpha = 1.33+/-0.11$.
- With no corrections for the intrinsic absorption, we derived a spectral index of $\beta = 1.25+/-0.05$. Thus, this value of $\beta$ has to be considered as an upper limit to the unextincted afterglow spectral index.
- The values of $\alpha$ and $\beta$ are consistent with a spherical afterglow model with an electron energy index $p = 2.5$. These values also would indicate that the cooling break, $\nu_c$, was located at frequencies lower than the R-band 8 hours after the gamma-ray event. This would make from GRB 001011 a very interesting system, since in most cases $\nu_c$ is higher than the optical frequencies at early times.

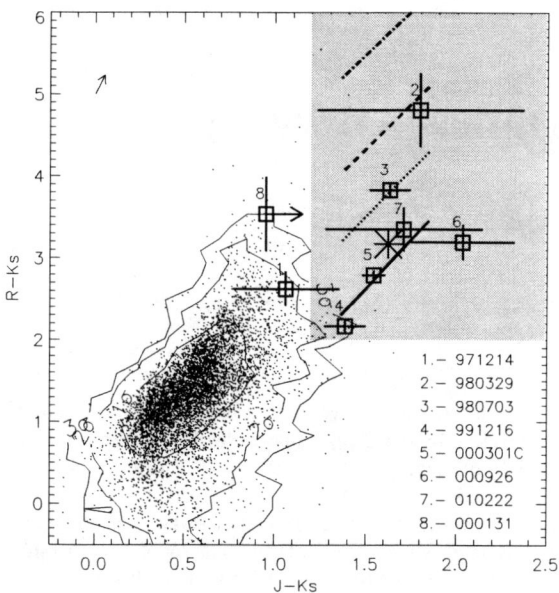
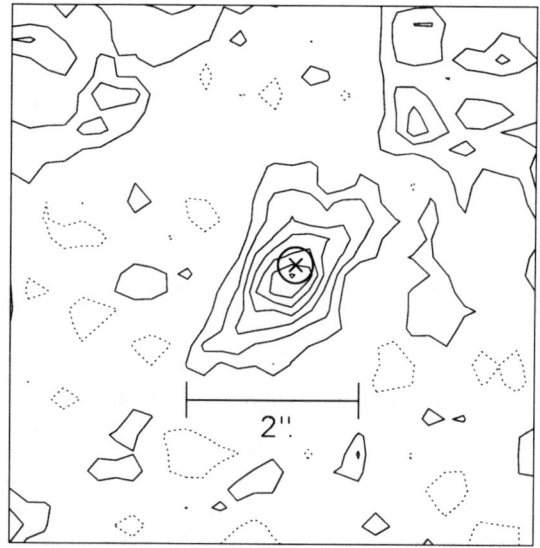

**FIGURE 2.** The figure shows an R-Ks vs J-Ks colour-colour diagram for the sources (dots) from the 2MASS and USNO catalogues (2MASS+USNO) found in a 10 x 10 degree window around GRB 001011. As it is shown, the counterpart (star) exhibits colours inconsistent at least at a 3σ level with the objects in the 2MASS+USNO catalogue, which mostly traces the stellar population. The seven open squares labeled from 1 to 7 represent the colours of afterglows with measured R-Ks and J-Ks colours to date. The additional square labeled with number 8 represents GRB 000131 which was not detected in the J-band [8]. As can be seen, at least seven of the eight open squares and the star are consistent with the shaded area. The solid straight line inside the shaded region represents the colour trace of a pure power-law SED with β ranging from 0.6 to 1.5. When Lyman-α blanketing is taken into account the pure power-law SED (straight solid line) is shifted upwards (three dashed lines). From the bottom to the top, the three dashed lines show the colours of a power-law SED when the Lyman-α blanketing is considered for an afterglow at $z$=5.0, 5.5 and 6.0, respectively.

- Images taken 7 months after the burst reveal an elongated object with R=25.38+/-0.25 fully consistent with the OA position, likely the host galaxy of GRB 001011 (see Fig. 3).

## ACKNOWLEDGMENTS

J. Gorosabel acknowledges support from the ESO visitors program and also the receipt of a Marie Curie Research Grant from the European Commission. We acknowledge the availability of the 2MASS and USNO catalogues. This work was supported by the Danish Natural Science Research Council (SNF). The observations presented in this paper were obtained under the ESO Large

**FIGURE 3.** The figure shows a contour plot of the co-added R-band images taken in May 2001 with the VLT. The total exposure time is 2400s with a mean seeing of 0.9". The cross shows the position of the optical counterpart, fully consistent with the object. The object is clearly elongated in the North-West direction. The image has been smoothed with a 3x3 pixel boxcar filter. North is to the top and East is to the left.

Program 165.H–0464.

## REFERENCES

1. van Paradijs, J., Groot, P., Galama, T., et al., *Nature*, **386**, 686–689 (1997).
2. Warren, S. J., Hewett, P. C., Irwin, M., and Osmer, P. S., *ApJS*, **76**, 1–22 (1991).
3. Rhoads, J. E., *ApJ*, **557**, 943–948 (2001).
4. Šimon, V., Hudec, R., Pizzichini, G., and Masetti, N., *A&A*, **377**, 450–461 (2001).
5. Gorosabel, J., Hjorth, J., Pedersen, H., GCN 849 (2001).
6. Gorosabel, J., Fynbo, J., Hjorth, J., et al., *A&A*,in press, (2001),URL http://xxx.lanl.gov/abs/astro-ph/0110007.
7. Fynbo, J., Jensen, B., Gorosabel, J., et al., *A&A*, **369**, 373–379 (2001).
8. Andersen, M., Hjorth, J., Pedersen, H., et al., *A&A*, **364**, L54–L61 (2000).

# Searches for Orphan Optical Afterglows

R. Hudec[a], V. Ríkal[b], J. Polcar[a,b], F. Hroch[a,b]

[a]*Astronomical Institute, Academy of Sciences of the Czech Republic, CZ-251 65 Ondřejov, Czech Republic*
[b]*Masaryk University, Brno, Czech Republic*

**Abstract.** The detection of orphan optical afterglows is expected to provide a very valuable contribution to the GRB analyses since the rate of such events may exceed the GRB rate, hence improved statistics may be expected. We describe and discuss a project based on the analyses of selected deep archival photographic plates to search and to study such events. We show that e.g. the UKSTU archive contains suitable sequences of deep (limiting magnitude 19-23) photographic plates with good sampling inside a two week period and taken in various filters. A suitable strategy for searches for orphan afterglows is presented and discussed, including methods for their verification and distinguishing from other highly variable astrophysical objects such as supernovae etc. The scientific aspects of such analyses are also addressed.

## INTRODUCTION

An orphan afterglow is an optical afterglow of a Gamma Ray Burst (GRB) without detectable Gamma ray emission (due to different beaming). These phenomena are predicted by theory but not yet confirmed by observation. The rate of Orphan Optical Afterglows (OOAs) may exceed the GRB rate, hence improved GRB statistics are expected with numerous consequences such as improved statistics of host galaxies, redshift distribution, cosmological conclusions, etc. The searches for OOAs assume independent optical searches.

## FEASIBILITY OF INDEPENDENT OPTICAL SEARCHES

The recent detection of optical afterglows and optical transients of gamma ray bursts allows to consider optical ground-based independent detection of these phenomena.

The recent results indicate that optical surveys achieving limiting magnitude better than 19...23 for stars and/or 10 for 1 min exposures may detect OAs and OTs of GRBs. This opens the possibility of independent optical searches. These searches must be of large field of view, i.e. CCD surveys and/or deep patrol plates are suitable.

## PROSPECTS OF OPTICAL SURVEYS FOR GRBS

The optical surveys may provide a larger sample (due to different beaming) and better localization accuracy (1 arcmin or better) than provide gamma ray satellite detectors. The larger sample of OAs and their host galaxies may be crucial for understanding the nature of GRBs as well as for related cosmological implications.

Further, the actual rate of OAs can place constraints on the afterglow appearance fraction and, perhaps, the initial beaming angle of GRB sources.

The UV flashes predicted by some theories such as Protheroe and Bednarek 1999 could be detected and studied. The corresponding delays regarding GRBs could serve to study the nature of the sources. This can be addressed only by surveys, not by follow-up devices since the flashes may preceed the GRBs [1]. The optical flashes preceding GRB expected by theory [2] can also be detected and analysed.

Not quite negligible is also the fact that the optical surveys are cost-effective.

# THE RATE OF OAs - ESTIMATED BEAMING

It is expected that sources emit jets from which the gamma ray emission is more beamed than the subsequent optical afterglow radiation due to the deceleration of the jet by the ambient gas and the corresponding decline in its relativistic beaming with time [3]. Because the shift to lower frequencies accompanies the shift to lower bulk Lorentz factor, the minimum solid angle into which the transient can radiate increases with time. A jet geometry hence implies a higher rate of OAs detections.

If bursts are highly collimated, the gamma rays will radiate into a small solid angle, the optical light into a larger one, and radio into a still larger one. If bursts emit isotropically, we do not expect OAs unaccompanied by GRBs. The ratio of transients detected hence allows the ratio of the mean solid angle into which transients radiate to be estimated.

# USING ASTRONOMICAL PLATES/SURVEYS

It seems to be feasible that both flaring (OTs) as well as fading (OAs) optical emission related to GRBs may be detected by optical sky patrols. Although the true rate of these triggers remains unknown, it is very probably that their rate is substantially below the background rate, hence good knowledge of all background triggers must be available as well as a reliable technique for their classification and elimination.

As background, we understand the false events not related to GRBs but with similar transient behavior. The background is represented mostly by the triggers from the following categories: SNe, AGN flares/brightening, stellar flares, variable stars, OTs of unknown nature and origin, as well as nonastrophysical triggers.

# OTs OF UNKNOWN NATURE AND ORIGIN

There are real OTs of unknown nature but of astrophysical origin detected both on emulsions and CCDs. Examples (real CCD detections):

- OT 970215: real CCD detection, V 13 mag, nothing down 20 mag on the position, amplitude more than 7 mag [4]

- OT 950806: real object: detected on 20 CCD frames, peak magnitude I 7.5, amplitude more than 10 mag, nothing down mag 21 48 hrs after detection [5].

- OT triggers found by SNe searches (Schmidt et al. 1999): mystery events found at a rate of about 0.15 $deg^{-2}$/per time scale (between 10 min and 3 days) lim mag R 23.5. In 3 cases, no host galaxy seen down mag R 24, in 1 case, host galaxy clearly visible. 2 events at a low galactic latitude and hence can be flare stars.

It should be noted that the SN searches reject events with timescales less than 3 min (as cosmic rays) hence they can detect OAs but not all OTs. Further, these searches are very limited so far (6 SN runs done, 2-6 sq. deg. per run, [6]), this means that the OAs may be among detected SNe.

# HOW TO DISTINGUISH OAS AND SNE (AND OTHER BACKGROUND EVENTS)

1. Light curve
2. Peak luminosity (only for objects with known redshift)
3. Color information: most of OAs have R-I = 0.46, V-R= 0.40, B-V=0.47 [7]

# THE PROJECT

Our project of searches for OOAs is based on the use of the UKSTU deep (lim mag 19-23) plates. We were able to localize there suitable plate series, with reasonable dense sampling over a time period of few tens of days, and taken in different filters. Two examples are given below. The column meaning of the plate lists below:

1- 2   Plate prefix
3- 7   Plate number
8      Plate suffix
9-11   Survey code (if relevant)
12-15 Non-survey code (T number)
16-20 Field number or other identification
21-25 RA (hhmmt) (t is tenth of a minute, e.g 7105 = 07h 10m 30s)
26-30 Dec (+/-ddmm)
31-36 Date of exposure (yymmdd)
37-40 LST of start of exposure (hhmm)
41-46 Emulsion
47-52 Filter
53-56 Exposure time (mmmt) (t is tenth of a minute, eg 700 = 70 minutes)
57-61 Plate grade
62-64 Prism code (for prism plates only)

**TABLE 1.** Position 1

| | | | | | | |
|---|---|---|---|---|---|---|
| OR17501 F | 1092 | 855 | 10400+0000 | 970405 0847 | 4415 | OG 590 600a |
| OR17504 F | 1092 | 855 | 10400+0000 | 970406 0921 | 4415 | OG 590 600a |
| OR17506 F | 1092 | 855 | 10400+0000 | 970407 0934 | 4415 | OG 590 600aI |
| OR17507 F | 1092 | 855 | 10400+0000 | 970408 0815 | 4415 | OG 590 750a |
| OR17510 F | 1092 | 855 | 10400+0000 | 970409 0949 | 4415 | OG 590 750a |
| I17518 | 1092 | 855 | 10400+0000 | 970412 0907 | IVN | RG 715 900bE |
| I17521 | 1092 | 855 | 10400+0000 | 970413 0856 | IVN | RG 715 900aE |
| I17523 | 1092 | 855 | 10400+0000 | 970414 0933 | IVN | RG 715 900a |
| I17528 | 1092 | 855 | 10400+0000 | 970415 0936 | IVN | RG 715 900aE |
| OR17536 F | 1092 | 855 | 10400+0000 | 970427 1027 | 4415 | OG 590 600a |
| I17541 | 1092 | 855 | 10400+0000 | 970428 0939 | IVN | RG 715 900a |
| OR17550 F | 1092 | 855 | 10400+0000 | 970430 1027 | 4415 | OG 590 600a |

**TABLE 2.** Position 2

| | | | | | |
|---|---|---|---|---|---|
| V 3235 | 97 5 | 1214400-25 | 007706 1113 | 51IIaD | GG 495 120 |
| V 3244 | 97 5 | 1214400-25 | 007706 1211 | 51IIaD | GG 495 120 |
| J 3245 | 97 5 | 1214400-25 | 007706 1212 | 16IIIaJ | GG 395 700 |
| V 3246 | 97 5 | 1214400-25 | 007706 1213 | 40IIaD | GG 495 120 |
| V 3251 | 97 5 | 1214400-25 | 007706 1215 | 35IIaD | GG 495 120 |
| V 3256 | 97 5 | 1214400-25 | 007706 1217 | 36IIaD | GG 495 120 |
| V 3261 | 97 5 | 1214400-25 | 007706 1219 | 24IIaD | GG 495 120 |
| V 3270 | 97 5 | 1214400-25 | 007706 1311 | 51IIaD | GG 495 120 |
| V 3272 | 97 5 | 1214400-25 | 007706 1313 | 35IIaD | GG 495 120 |
| V 3277 | 97 5 | 1214400-25 | 007706 1315 | 40IIaD | GG 495 120 |
| V 3282 | 97 5 | 1214400-25 | 007706 1317 | 27IIaD | GG 495 120 |
| V 3287 | 97 5 | 1214400-25 | 007706 1319 | 44IIaD | GG 495  41 |

## CONCLUSIONS

The recent results of OAs and OTs searches indicate that GRBs may be monitored and studied by observing their optical emission, i.e. independently of satellite projects. The rate of OAs may be higher than the rate of GRBs due to different beaming. It is feasible to use ground based optical devices to monitor OAs and OTs of GRBs. This opens a new observing window for GRBs, but these surveys must be of wide field and high sensitivity. There is however a background of false triggers (not related to GRBs but with similar transient behavior) with poorly known statistics (for faint magnitudes). The detected OAs and especially OTs (since they will be recorded only once due to their short duration in most cases) must be further studied in detail to eliminate them from background triggers.

## ACKNOWLEDGEMENT

The investigations of gamma-ray bursts and optical transients are supported by the project KONTAKT ES002 provided by the Ministry of Education and Youth of the Czech Republic and by the grant IAA3003206 provided by the Grant Agency of the Academy of Sciences of the Czech Republic. The investigation of plate defects has been supported by the Academic Link between the University of Westminster and Astronomical Institute Ondrejov provided by the British Council in Prague.

## REFERENCES

1. Protheroe R. J. and Bednarek W., 1999, astro-ph/9904279.
2. Paczynski B., 2001, Astro-ph/0108522.
3. Rhoads J. E. 1997, ApJ 487, L1.
4. Vidal-Saiz J. et al., IBVS Budapest 4324, 1996.
5. Toth I. et al., AA 315, 153, 1996.
6. Schmidt et al., 1999, private communication.
7. Šimon V. et al., AA 377, 2001, 450.
8. Anderson M. I. 1999 Nature, 398, 400.
9. Fenimore E. F., Epstein R. I. and Ho C. 1993, AAS 97, 59.
10. Halpern J. P., Kemp J., Piran T. and Bershady M. A. 1999, ApJ 517, L105.
11. Hjorth J. et al., 1999, GCN 403.
12. Kulkarni S. et al. 1999 Nature 398, 389.
13. Pain R. et al. 1996 ApJ 473, 356.
14. Reichart D. E., 1999, ApJ 521, 111.
15. Sari R., Piran T. and Halpern J. P. 1999 ApJ 519, L17.
16. Woods E. and Loeb A. 1995, ApJ 453, 583.
17. Woods E. and Loeb A. 1998 ApJ 508, 760.
18. Hudec R. 1993 Astroph. Letters and Communications 28, 359.
19. Park H. S. et al. 1997 ApJ 490, 99.

# Optical Follow-Up Observations of GRBs at the Kleť Observatory

J. Polcar[a,b], R. Hudec[a], M. Topinka[a], J. Tichá[c], M. Tichý[c], N. Masetti[d], G. Pizzichini[d]

[a] Astronomical Institute, Academy of Sciences of the Czech Republic, CZ-251 65 Ondřejov, Czech Republic
[b] Masaryk University Brno, Czech Republic
[c] Observatory České Budějovice, České Budějovice, Czech Republic
[d] TESRE CNR, Bologna, Italy

**Abstract.** The 57 cm telescope of the Kleť Observatory has been used for follow-up observations of selected GRBs. Limiting magnitudes as deep as mag 22 have been obtained. The results obtained are presented and discussed. We also address the capabilities of the current system for GRB analyses as well as further improvements including the newly developed 1.06 m aperture CCD telescope.

## INTRODUCTION

The 57 cm CCD telescope of the Kleť Observatory (belonging to the České Budějovice Observatory) has been used for optical follow-up observations of Gamma Ray Bursts (GRBs). The telescope is capable of covering the positions of GRBs detected by BeppoSAX and HETE satellites soon (less than 0.5 hour) after their detection down to limiting magnitude 21. In this paper, we present and discuss results obtained so far, with focus on procedures and methods used to search for optical transients and to determine the magnitudes of objects on the CCD images as well as the corresponding limiting magnitudes.

## LIST OF FOLLOW-UP OBSERVATIONS

The GRBs listed in the following Table have been covered by optical follow-up CCD images. For the actual list see http://www.klet.cz/GRB/.

**TABLE 1.** List of GRB follow up observations at the Kleť Observatory

| Object | Time interval | Time of GRB event | Time delay |
|---|---|---|---|
| GRB010623 (SGR1806-20) | 2001 Jun. 24, 23:10:01 to 23:25:17 UT | 2001 Jun. 23, 15:54:53 UT | 1 day 8 hrs 15 min |
| GRB010222 | 2001 Feb. 25, 03:15:21 to 03:24:04 UT | 2001 Feb. 22, 07:23:30 UT | 2 days 20 hrs 52 min |
| GRB010214 | 2001 Feb. 15, 01:38:38 to 01:52:17 UT | 2001 Feb. 14, 08:48:11 UT | 17 hrs 50 min 19 |
| GRB010213 | 2001 Feb. 15, 02:01:36 to 02:17:55 UT | 2001 Feb. 13, 12:35:35 UT | 1 day 14 hrs 26 min |
| GRB010126 | 2001 Jan. 27, 01:54:16 to 02:10:21 UT | 2001 Jan. 25, 09:10:40 UT | 1 day 17 hrs 43 min |
| GRB000630 | 2000 Jun. 30, 21:19:51 to 21:35:36 UT | 2000 Jun. 30, 00:30:53 UT | 21 hrs 49 min |
| GRB000620 | 2000 Jun. 20, 21:38:03 to 21:52:53 UT | 2000 Jun. 20, 05:33:38 UT | 17 hrs 04 min |
| GRB011030 | 2001 Oct. 31, 17:23:51 to 17:38:57 UT | 2001 Oct. 30, 06:28:02 UT | 1 day 10 hrs 56 min |

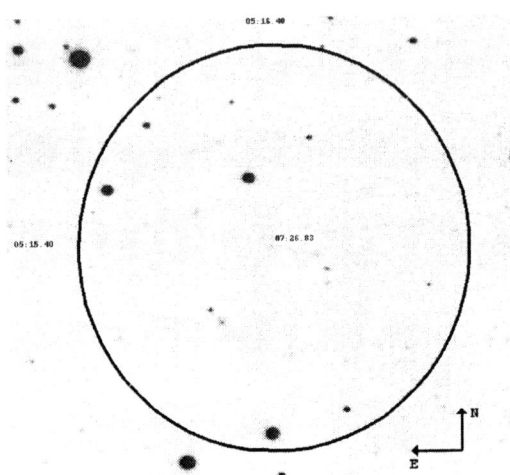

**FIGURE 1.** CCD image of GRB010222 with emphasized error box

**FIGURE 2.** Picture of error box of GRB010222 from DSS

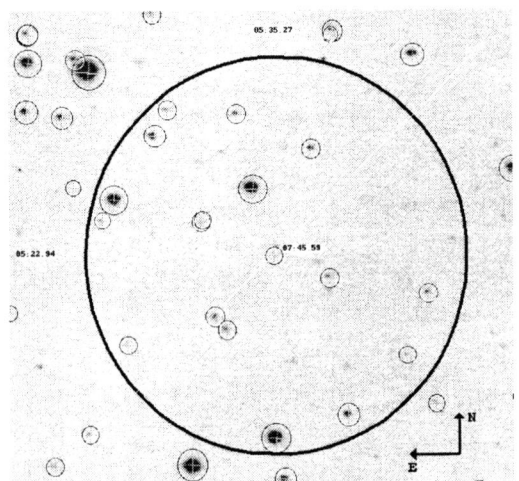

**FIGURE 3.** CCD image of GRB010222 with emphasized error box and USNO objects in circles

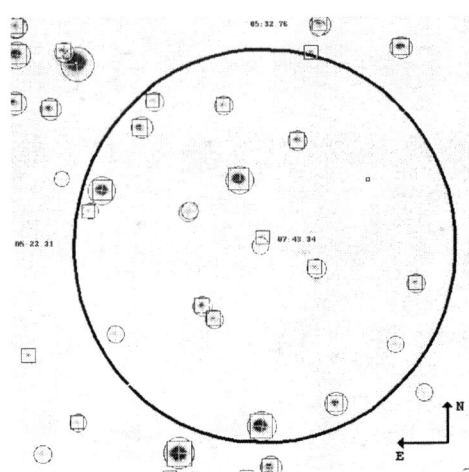

**FIGURE 4.** CCD image of GRB010222 with emphasized error box and USNO objects in circles. In squares there are objects detected by muniphot. Parameter TRESHOLD was set at 5.0.

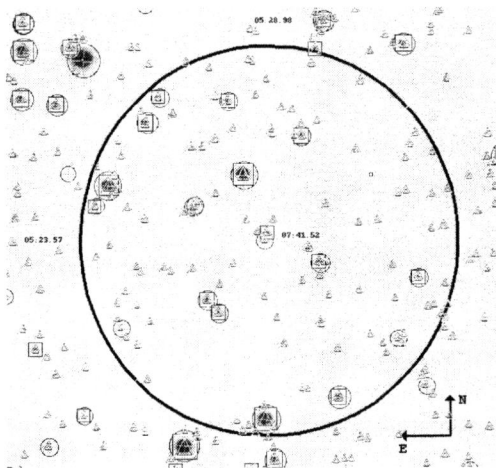

**FIGURE 5.** CCD image of GRB010222 with emphasized error box and USNO objects in circles. In squares and triangles there are objects detected by muniphot. In squares there are objects detected when TRESHOLD was set at 5.0, in triangles are objects detected when TRESHOLD was set at 1.0.

# EXAMPLES OF RESULTS

## GRB010222

Coordinates (J2000): alpha 14h52m12s, delta +43°01', error box 2.5'. Co-added six 60-sec images. Exposition: 2001 Feb 25, 03:15:21-03:24:04. Observers: Milos Tichy, Jana Tichá and Michal Kocer. Telescope: 0.57-m f/5.2 reflector. CCD camera: SBIG ST-8. Filter: No. Dark Frame: Yes. Flat Field: No

## Limiting magnitude estimation

Fitting parameters were:

| Final set of parameters | Asymptotic Standard Error |
|---|---|
| a = 1.0316 | +/- 0.02941   (2.85%) |
| b = -2.43342 | +/- 0.5096   (20.94%) |

Correlation matrix of the fit parameters:

|   | a | b |
|---|---|---|
| a | 1.000 | |
| b | -0.996 | 1.000 |

By measuring the objects at detection threshold level and by using the transformation mentioned above the limiting magnitude was estimated as 21. With respect of USNO catalogue magnitude error, the overall limiting magnitude error can be considered as 1.

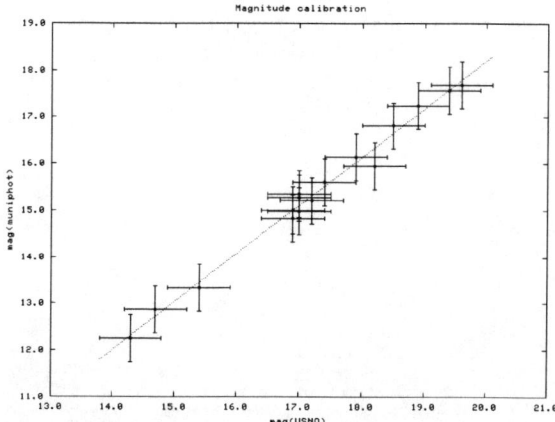

**FIGURE 6.** The magnitude calibration for GRB010222. The horizontal axis represents magnitude of objects from USNO catalog, vertical axis means magnitudes measured with muniphot.

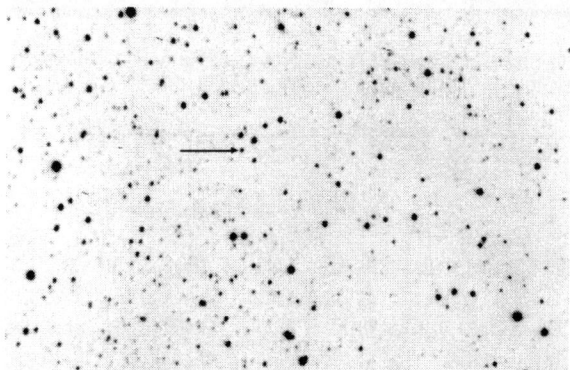

**FIGURE 7.** The optically variable object (probably new variable star) located nearly 1 arcmin from the Newton X-ray position of the SGR1806-20 detected on the Klet CCD image.

## GRB010213

Coordinates (J2000): alpha 10h 31m 36s, delta +05°31', error box 30'. Exposure: ten 60-sec exposures between 2001 Feb.15 02:01:36 - 02:17:55 UT. Dark Frame: Yes. Flat Field: No. Observer(s): Milos TICHY and Jana TICHA. Telescope: 0.57-m f/5.2 reflector of the Klet Observatory. CCD camera: SBIG ST-8. Filter: No.

## Limiting magnitude estimation

Fitting parameters were:

| Final set of parameters | Asymptotic Standard Error |
|---|---|
| a = 0.89802 | +/- 0.0472   (5.255%) |
| b = -0.431962 | +/- 0.7492   (173.4%) |

Correlation matrix of the fit parameters:

|   | a | b |
|---|---|---|
| a | 1.000 | |
| b | -0.994 | 1.000 |

By measuring the objects at detection threshold level and by using the transformation mentioned above the limiting magnitude was estimated as 21. With respect of USNO catalogue magnitude error, the overall limiting magnitude error can be considered as 1.

## CONCLUSIONS

The 57 cm CCD telescope at the Klet Observatory is reaching limiting magnitude of 22 well suited for optical GRB analyses. The recently developed 1.06 m CCD telescope will further increase the sensitivity of optical observations. The new telescope is expected to be put into operation in 2002. For the GRB observed, no optical afterglows have been detected within the limiting magnitudes 19 ... 22. One new variable star has been found 1 arcmin from the X-ray position of the SGR1806-20, exhibiting more than 1 mag brightening nearly 1 day after the gamma ray burst from this trigger. Methods to look for very faint triggers close to the magnitude limit have been exploited and tested.

## ACKNOWLEDGEMENTS

We acknowledge the support provided by the Grant Agency of the Czech Republic, Grant 205/99/0145.

## REFERENCES

1. http://www.klet.cz/~GRB
2. http://gcn.gsfc.nasa.gov/gcn

# Super-LOTIS and LOTIS for HETE2 GRB Triggers

H. S. Park*, G. G. Williams[†], E. Ables*, S. D. Barthelmy**, T. Cline**, N. Gehrels**, D. Hartmann[‡], K. Hurley[§], K. Lindsay[‡], R. Nemiroff[¶], W. Pereira[¶] and D. Perez-Ramirez[¶]

*Lawrence Livermore National Laboratory, Livermore, CA 94551*
[†]*Steward Observatory, Tucson, AZ 85721*
**NASA/Goddard Space Flight Center, Greenbelt, MD 20771*
[‡]*Dept. Of Physics and Astronomy, Clemson University, Clemson, SC 29634*
[§]*Space Science Laboratory, University of California, Berkeley, CA 94720*
[¶]*Dept. Of Physics, Michigan Technological University, Houghton, MI 49931*

**Abstract.** LOTIS (Livermore Optical Transient Imaging System) and Super-LOTIS are automatic telescope systems that search for prompt optical emission from gamma-ray bursts (GRBs). Both systems are capable of responding to the Gamma-ray burst Coordinate Network (GCN) triggers within seconds. These systems have been monitoring the GCN real-time data for automatic HETE2 GRB triggers since HETE2's launch. In this paper, we present the systems' capability and current status. We also present the result of the GRB010921 afterglow detection that was localized by HETE2 and the IPN.

## INTRODUCTION

Our understanding of GRBs was dramatically enhanced when the high-resolution X-ray detector on the Beppo/SAX satellite determined the position of a GRB with sufficient accuracy to enable a large telescope to observe a faint, fading afterglow days later. Optical and radio afterglows now have been observed for many GRBs during the last three years. These long-lasting but faint afterglows have been successfully explained in the fireball models as the result of the synchrotron interaction with surrounding material [1]. However, prompt counterparts associated with GRBs are rare. A few afterglows such as GRB970805, GRB980508, GRB981208 and GRB990123 show that the light curves at early times are different from the later time, smooth decaying light curves. Unlike the observed later-time afterglows, prompt optical measurements (or even stringent constraints on that optical emission) would provide information about the GRB progenitors. In order to detect many more prompt counterparts, we are currently operating two instruments, LOTIS and Super-LOTIS [2], connected via the Coordinate Distribution Network to current and future satellites including HETE2, GLAST, INTEGRAL, and Swift. With HETE2's real-time triggers we will measure optical flux as early as 30 sec after the burst at the sensitivity of $R \sim 17$ to 19.

**FIGURE 1.** Super-LOTIS at Kitt Peak National Observatory. The system is fully automated to respond to HETE2 real-time GRB triggers.

## SUPER-LOTIS

The Super-LOTIS telescope is a Boller & Chivens 0.6-m $f/3.5$ reflector. It has superb optical quality and is equipped with computer controllable drives that can point any part of the sky within 30~60 seconds. Its focal array is a Loral $2048 \times 2048$ pixel $15 \times 15$ $\mu$m/pixel CCD cooled to $-30°C$ with custom-built readout electronics. This focal plane array is placed at the primary focus of the mirror with a coma corrector yielding a $51' \times 51'$

field-of-view. Super-LOTIS began operation in October 2000 at Kitt Peak National Observatory in Arizona. It is housed in a building with a roll-off roof and is fully automated with a dedicated weather station, computer controlled electronic and mechanical hardware and interrupt driven on-line software. Recently, we installed a computer controllable filter wheel that will enable us to image in $V, R, I$ and $Clear$-bands.

We have responded to several HETE2 triggers after its automatic trigger distribution was enabled on May 31, 2001. In one case we responded to HETE Trigger 1759 in 105 second after the burst. The trigger was later determined not to be a GRB by the ground analysis, but we found that 66 seconds of the 105 second delay was due to the internet transmission time delay and 41 seconds was due to the time it took for the mount to move to this particular position. The internet delay was unusually long; and we have not understood what caused the delay for this particular trigger. Super-LOTIS can detect $R \sim 17$ objects with 30 sec integration times.

For late-time afterglows, we integrate longer and co-add the images enabling us to detect objects as dim as $R \sim 22$ objects. Many of the HETE2 GRB positions are available after the ground analysis. For these events, Super-LOTIS attempts to detect afterglows and has succeeded in detecting two afterglows (GRB010222 [2] and GRB010921 [3]). When it is not imaging GRB fields, Super-LOTIS systematically searches for other transient objects, such as novae and supernovae. It also monitors long- and short-period stellar variability.

## LOTIS

LOTIS was originally constructed to respond rapidly to real-time GRB triggers provided by BATSE which had a 1-$\sigma$ error of $2 \sim 10°$. LOTIS utilizes commercially available Canon $f/1.8$, 200 mm focal length lenses. With a $2048 \times 2048$ pixel $15 \times 15$ $\mu$m/pixel CCD array, each camera has an $8.8° \times 8.8°$ field-of-view. The mount can point to any part of the sky within 5 sec after receiving a trigger. LOTIS has 4 cameras, each viewing the same field but through different filters, $R, V$ and 2 $Clears$, for simultaneous color imaging of any GRB counterparts. Its sensitivity is $R \sim 13$ to 15 depending on the integration times, weather and filter types. LOTIS is located at a LLNL's test facility, 25 miles east of Livermore, California. The system operates every night, weather permitting, is fully automated and will respond to any HETE2 triggers.

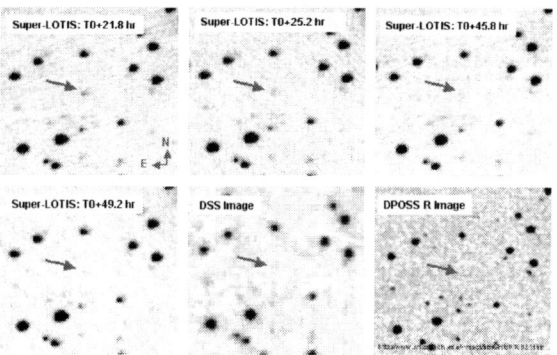

**FIGURE 2.** Super-LOTIS observation of GRB010921 afterglow. Each panel is a co-added image of twenty 50 sec exposures. The afterglow was detected at $m_{clear} = 19.4 \pm 0.2$. The DSS and DPOSS R image of the same area are also shown in the 5 and 6th panel.

## GRB 010921

On September 21.21934, 2001 UT the HETE2's FREGATE instrument detected a bright GRB (HETE Trigger 1761). GRB010921 had a duration of $\sim 12$ $s$ in the 8-85 keV band, a peak flux of $F_p > 3 \times 10^{-7}$erg cm$^{-2}$ s$^{-1}$ and a fluence of $S \sim 1 \times 10^{-6}$erg cm$^{-2}$ [4] [9]. This burst was also detected in the WXM X detector but not in the WXM Y detector resulting in a long narrow error box ($\sim 10° \times \sim 20'$). Ulysses and BeppoSAX detected this GRB producing an IPN annulus. The combined HETE2/IPN data produced an error box with an area of $\sim 310$ square arcmin and the coordinates were distributed 15 hours after the burst through a GCNC [5].

Even though there was no prompt trigger available from HETE2, LOTIS observed the GRB010921 afterglow area at September 21.255 UT (52 min after the burst) and at September 21.417 UT (4.75 hours after the burst) during a routine sky patrol. We searched for an optical transient (OT) in the reported GRB error box and found no transients brighter than $m_{clear} > 15.4 \pm 0.15$ and $V > 15.9 \pm 0.15$. No $R$-band data was acquired because the shutter for the $R$-band camera failed to open.

Super-LOTIS began imaging the GRB error box at September 22.128 (21.8 hours after the burst.) A total of twenty 50-second exposures were obtained during the first epoch. We re-observed the field at September 22.267, 23.128 and 23.269 that also consisted of twenty 50-second exposures. We searched for an OT by co-adding twenty images at each Epoch and detected the afterglow on the first two epochs at September 22.128 and 22.267 but no OT was seen on the images taken on September 23. The brightness of the afterglow during the first two epochs varied from $m_{clear} = 19.4 \pm 0.2$ to $m_{clear} = 19.9 \pm 0.2$. The 3-$\sigma$ limiting magnitude of the second night's image was $21.3 \pm 0.3$. Figure 1 shows the

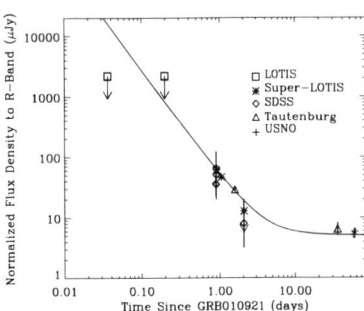

**FIGURE 3.** Light curve of the GRB010921 optical afterglow. The fluxes from different filters have been normalized to the $R$-band assuming a power-low spectrum of index $\beta = -2.3$ [6].

afterglow observed with Super-LOTIS.

Combining LOTIS and Super-LOTIS data with observations from the Sloan Digital Sky Survey Photometric Telescope and Tautenburg Schmidt telescope, we fit the data using a power law decay plus a constant host galaxy flux, $F = F_0(t-t_0)^{-\alpha} + F_{host}$. We find that the optical afterglow exhibited a power-law slope of $\alpha = 1.75 \pm 0.28$ using the measure host galaxy magnitude of $R = 21.93$. Other observations from different sites and different times [6] find the similar value which is a typical value for an optical afterglow prior to the jet break. This was predicted to take place $\sim 130$ days after the GRB based on the observed energetics [7] and was observed at $\sim 35$ days with HST [8]. From our data, we also find that this slope cannot be extended to early times ($t < 0.035$ day) [3] as seen in the light curve in Figure 2. The early time triggers are essential to resolve this kind of early time characteristics.

## LITE

We plan to upgrade the Super-LOTIS by installing an infrared camera. Measure GRB distance scales and recent theory of GRBs star-forming environment [10] suggest that the optical signal may be obscured compared to the infrared signals. Our planned LITE (Livermore Infrared Transient Experiment) will have a dichroic splitter enabling us to measure optical and near infrared-bands simultaneously. We will employ a $512 \times 512$ HgCdTe array with a filter wheel equipped with $J, H$ and $K$-bands.

## SUMMARY

We found that the early time GRB optical light curves are different from the late time light curves. Measurements of prompt optical and infrared counterparts of GRBs will provide important clues to our understanding of the GRB progenitors. With our current operating LOTIS and Super-LOTIS systems connected triggered in real-time by space-borne GRB detectors, we will measure many early-time optical and infrared fluxes associated with GRBs. Many more early time measurements such as GRB010921 will provide important clues for GRB progenitor physics.

## ACKNOWLEDGMENTS

This research is supported under NASA contract numbers S-03975G and S-57797F and under the auspices of the U.S. Department of Energy by University of California Lawrence Livermore National Laboratory under contract No. W-7405-Eng-48.

## REFERENCES

1. Wijers, R. A. M. J. et al., 1997, MNRAS *288*, L51.
2. Park, H. S. et al., AIP Conference Proceedings, *GAMMA 2001*. 2001, p.181.
3. Park, H. S. et al., 2002, Submitted to ApJ, astro-ph/0112397.
4. Ricker, G. et al., 2001, *GRB Circular Network* 1096.
5. Hurley, K. et al., 2001, *GRB Circular Network* 1097.
6. Price, P. A. et al., 2002, astro-ph/0201399.
7. Djorgovski, S. G. et al., 2001, *GRB Circular Network* 1108.
8. Price, P. A. et al., 2002, *GRB Circular Network* 1259.
9. Ricker, G. et al., 2002, astro-ph/0201461.
10. Lamb, D. et al., 2000, ApJ, *536*, 1.

# GRB 980329: Determining Density Without a Redshift

## S. A. Yost

*Caltech 220-47, Pasadena, CA 91125*

**Abstract.** We fit models to the late-time broadband dataset of gamma-ray burst (GRB) 980329. Despite being limited by sparse early optical data and no redshift measurement, we determine some parameters of the afterglow and its host robustly. The fireball expanded into a relatively dense medium with $n \sim 200\ cm^{-3}$. The host is far bluer than the afterglow, and is not compatible with high $z \sim 5$.

## INTRODUCTION

GRB 980329 was one of the earliest and brightest bursts detected by the BeppoSAX satellite. After fruitless optical searches, its afterglow was initially identified in the radio [1], which then led to the discovery of its faint R-band and relatively bright near-IR afterglow [2, 3, 4]. With evidence of a high column density nH in the X-ray [5], the very red nature of this burst's afterglow and the "dropout" of the R band flux has been postulated to be due to intrinsic host absorption by both Taylor[1] and Palazzi[2]. However, there is no direct redshift determination for this burst, despite observational efforts; Fruchter[6] proposed that the "dropout" could be due to absorption by the intergalactic medium's Lyman-alpha (Lyα) forest if the burst is at a redshift $z \geq 5$.

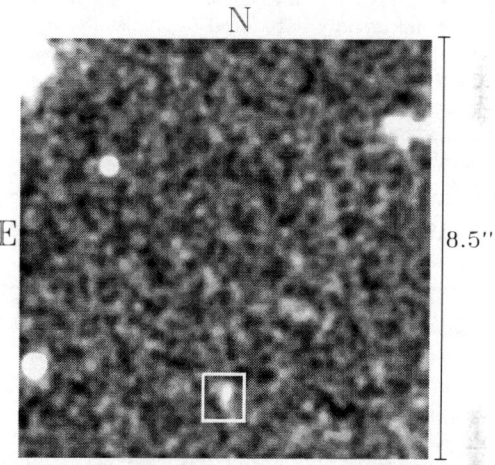

**FIGURE 1.** The host is well detected, indicated by the box in this HST longpass filter exposure. It is also well detected in late Keck eposures and has a colour $(R-I)_{host} = 0.2 \pm 0.3 << (R-I)_{OT} \sim 2$. Image courtesy of Joshua Bloom, Caltech.

## THE GRB REDSHIFT

The R band "dropout" and R-I=2 OT colour could be due to extinction in the host, or to Lyα flux suppression at $z \sim 5$. The latter would extinct the host flux similarly.

A good host spectrum was taken, but no lines were detected, leaving the redshift undetermined. However, the host is well detected at both R and I, (see Fig. 1) with a much bluer colour than the OT, rejecting Lyα as a significant effect on the R band and thus restricting $z \leq 4$.

## AFTERGLOW MODELLING

We fit late-time optical and radio points along with the early, published data to a broadband model of the afterglow emission (explained in some detail in Harrison [7]) and host galaxy. We assume several different fixed redshifts (z=1, 2, 3) in the modelling and find some robust physical parameters of the afterglow and its host. Results are presented in Table 1.

## CIRCUMBURST MEDIUM

The broadband model fits to a relatively high density, $n \gtrsim 200\ cm^{-3}$ (typical of a diffuse cloud), independent of the redshift assumed. The uncertainty in $n$ by changing $z$ is comparable to the modelling accuracy.

The density is chiefly determined by the high self-absorption frequency $v_a$, below which the spectrum $\propto v_a^2$. As can be seen in Figure 2 and 3, $v_a$ is certainly above 8.5 GHz at early times and comes out in the model near 100 GHz.

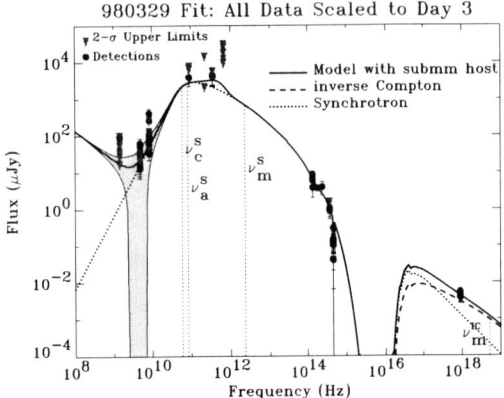

**FIGURE 2.** The model fit for z=2 with all the data used scaled to the common time of day 3. The model fit includes the model uncertainty introduced by interstellar scintillation, shown in pale grey. The host is included in the model in the radio and submm, but the optical data have the host component removed for clarity. The optical data are dereddened for Galactic effects and the model includes extinction in the host frame.

**TABLE 1.** Best-fit parameters from the modelling of GRB 980329

|  | z=1 | z=2 | z=3 |
|---|---|---|---|
| $F_{iso,t_{v_c=v_m}}$ ($10^{52}$ erg)* | 1.5 | 1.1 | 3.0 |
| $n(cm^{-3})$ | 370 | 250 | 230 |
| $p$ | 2.04 | 2.11 | 2.07 |
| $\epsilon_e$ (fraction of E) | 1.2 | 1.5 | 1.6 |
| $\epsilon_B$ (fraction of E) | 0.003 | 0.002 | 0.001 |
| $\theta_{jet}$ (rad)† | 1 | 1 | 1 |
| host $A(V)$ | 1.1 | 1.1 | 0.71 |
| host I ($\mu Jy$) | 0.085 | 0.086 | 0.086 |
| host H ($\mu Jy$) | 0.20 | 0.20 | 0.20 |
| host K ($\mu Jy$) | 0.61 | 0.63 | 0.63 |
| host 1.4 GHz ($\mu Jy$) | 19.9 | 16.4 | 15.9 |
| host 350 GHz ($\mu Jy$) | 1190 | 1160 | 1210 |
| $T_{eff}$(K)** | 4.7 | 4.5 | 4.5 |
| $\chi^2$ for 94 data pts | 69 | 65 | 65 |
| $t_{nonrel.}$ (days) | 25 | 48 | 81 |
| $t_{v_c=v_m}$ (days)‡ | 4.6 | 22 | 24 |
| $P(\alpha_{1.4}^{350})$§ | 0.54 | 0.52 | 0.21 |

* The isotropic-equivalent energy in the fireball at the time fast cooling ends
† Due to model uncertainties, jet angles $\geq 1$ rad are treated as isotropic
** Dust temperature T of the host, corrected for redshift, $T_{eff} = T/(1+z)$
‡ Time at which the cooling break becomes larger than the spectral break due to the peak in the input electron distribution and fast cooling ends
§ The probability of the fit's host 350-to-1.4 GHz spectral index at the assumed redshift, based upon the models of Carilli and Yun[8]

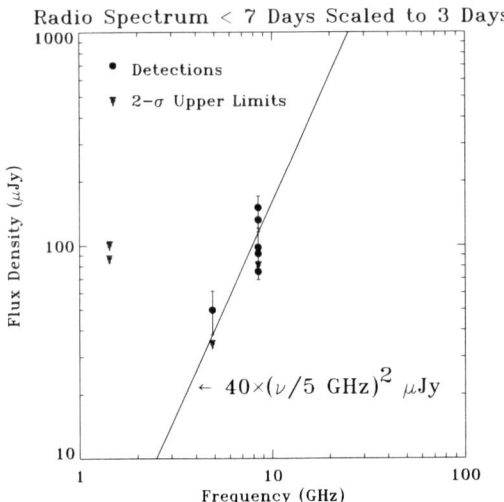

**FIGURE 3.** The first week's datapoints in the radio, scaled by the fit model to a common time of 3 days are shown, along with a line $\propto \nu^2$ overdrawn. Clearly at least the data below 8.5 GHz scales in this manner and is below the self-absorption frequency $\nu_a$.

From the equation for $\nu_a$ in Granot [9], we get for $n$ in $cm^{-3}$:

$$n \simeq 500 (\epsilon_e \frac{\nu_a}{10GHz} \frac{1+z}{3})^{5/3} (\frac{E}{10^{52}erg} \frac{\epsilon_B}{0.01})^{-1/3} \frac{K(p)}{K(2.2)} \quad (1)$$

$K(p) = (3p+2)(p-1)^{-8/3}(p-2)^{5/3}/(p+2)$.

A relatively high density is compatible with the high host extinction seen in the GRB afterglow.

## HOST EXTINCTION

Figure 4 shows a closeup of the optical spectrum at day 3, dereddened for the effects of the Galaxy, with the late time host flux removed. The I to K spectral slope is approximately -1.6, and the R "dropout" can be seen.

The synchrotron spectral slope is at its steepest $\nu^{-p/2}$; as a value of $p > 3$ isn't compatible with the broadband evolution, the extra steepening is modelled as extinction.

The I to K spectrum calls for a steep extinction law $A(\lambda)$, much like the SMC. We use the extinction curve parameterization of Reichart[10], with no 2175 bump or far-UV upturn, but a steep linear slope (c2) of $E(\lambda - V)/E(B-V)$ with frequency in $\mu m^{-1}$ of 3. We find the host contributes a significant extinction A(V)=1.

We cannot explain the extra R dropout of $\sim 2$ mag below the extinction curve with a standard extinction law. This presumed nonstandard extinction is beyond the scope of this effort.

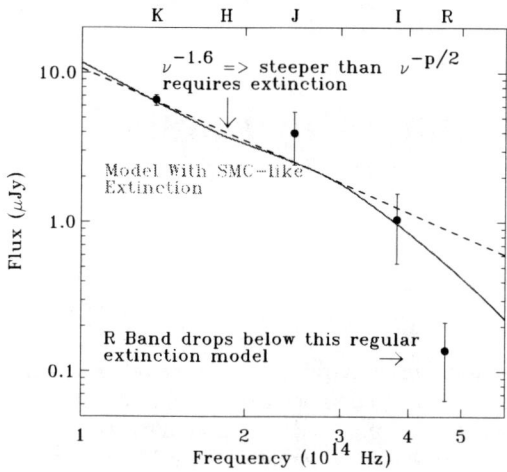

**FIGURE 4.** Earliest optical and near-IR data scaled by the z=2 fit model to day 3, and host-subtracted. The fit model with its steep extinction curve (see text) is overplotted (solid line). A powerlaw fit from I to K is shown with a dashed line, and the R band drop below both spectral curves is evident.

## CONCLUSIONS

We fit models to the late-time broadband dataset of gamma-ray burst (GRB) 980329 for various assumed redshifts. The host is far bluer than the afterglow, and is not compatible with high $z \sim 5$. We determined that the fireball expanded into a relatively dense medium with $n \sim 200 \, cm^{-3}$ independent of the redshift.

## REFERENCES

1. Taylor, G. B. i., *Astrophys. J. Letters*, **502**, 115–119 (1998).
2. Palazzi, E. i., *Astron. and Astrophys. Letters*, **336**, 95–99 (1998).
3. Gorosabel, J. i., *Astron. and Astrophys. Letters*, **347**, 31–34 (1999).
4. Reichart, D. E. i., *Astrophys. J.*, **517**, 692–699 (1999).
5. In't Zand, J. J. M. i., *Astrophys. J. Letters*, **505**, 119–122 (1998).
6. Fruchter, A. S., *Astrophys. J. Letters*, **512**, 1–4 (1999).
7. Harrison, F. A. i., *Astrophys. J.*, **559**, 123–130 (2001).
8. Carilli, C. L., and Yun, M. S., *Astrophys. J.*, **530**, 618–624 (2000).
9. Granot, J., Piran, T., and Sari, R., *Astrophys. J.*, **527**, 236–246 (1999).
10. Reichart, D. E., *Astrophys. J.*, **553**, 235–253 (2001).

# X-Ray Spectroscopy of Gamma-Ray Bursts

L. Piro

*Istituto Astrofisica Spaziale Fisica Cosmica, CNR, Roma, Italy*

**Abstract.** Observational evidence of iron absorption and emission lines in X-ray spectra of Gamma-Ray Bursts is quite compelling. I will briefly review the results, summarize different models and describe the connection with massive progenitors in star-forming regions implied by these results. This link is also supported by measurements of the X-ray absorbing gas in several GRB's, with column density consistent with that of Giant Molecular Clouds harbouring star-formation in our Galaxy, as well as by evidences gathered in other wavelengths. However, the volume density inferred by the fireball-jet model is much lower than typical of a GMC, and I will confront this with the alternative explanation of fireball expansion in a high dense medium, outlining the problems that both models have at present. Finally I will briefly summarize some results on dark GRB's, and describe the prospects of high resolution X-ray spectroscopy in getting closer to the central environment of GRB, and far in the Early Universe by using GRB as beacons to probe star and galaxy formation.

## INTRODUCTION

X-ray observations of GRB are playing a key role in several areas of the GRB research. Most of the fast and precise localizations of GRB have been obtained in X-rays by BeppoSAX, HETE-2 and Rossi XTE by combining a gamma-ray burst monitor trigger with the location accuracy provided by X-ray detectors.

X-ray spectroscopy of the prompt and afterglow emission is providing important clues as to the origin of GRB progenitors, the emission processes and the environment of different classes of GRB. The presence of emission and absorption Fe X-ray lines in some GRB is indicating massive progenitors in star forming regions. The gas in the local environment of the GRB or in the host galaxy affects also the X-ray spectral shape. In this regard, X-ray spectral measurements can provide information on the column density of the gas, its chemical composition and ionization stage.

The characterization of the environment is of particular importance to understand the origin of different classes of GRB's: optically bright and dark GRB's, X-ray rich GRB, and short GRB's. Some of the differences amongst two or more classes could be in fact the result of a different environment, rather that reflecting intrinsic differences in the central source. While this could not be the case of short GRB (thought to be produced by binary mergers rather than by massive progenitors [e.g. 1])[1], this sort of unification scenario of GRB could in principle account for other classes properties. For example, dark events could be associated to GRB whose optical light is effectively estinguished by dust either in the local environment or in the host galaxy [e.g. 2]. X-ray rich events could be associated with "dirty" fireballs [3], i.e. fireballs that, due to high baryon loading, expand with a much lower Lorentz factor, boosting the photons in the X-ray band rather than in the γ-ray range.

Another important piece of information that could play an important role in making up different observed properties is the distance. For example the properties of X-ray flashes and some dark GRB's can be explained if they lie at $z > 5 - 10$.

In this paper I will review the present picture and discuss future perspectives in assembling some of the missing elements, outlining, in particular, the role of X-ray spectroscopy.

---

[1] In this case the environment would make a difference in the *afterglow* properties because, lacking an external medium to make an external shock, the afterglow would be much fainter than in the case of long bursts

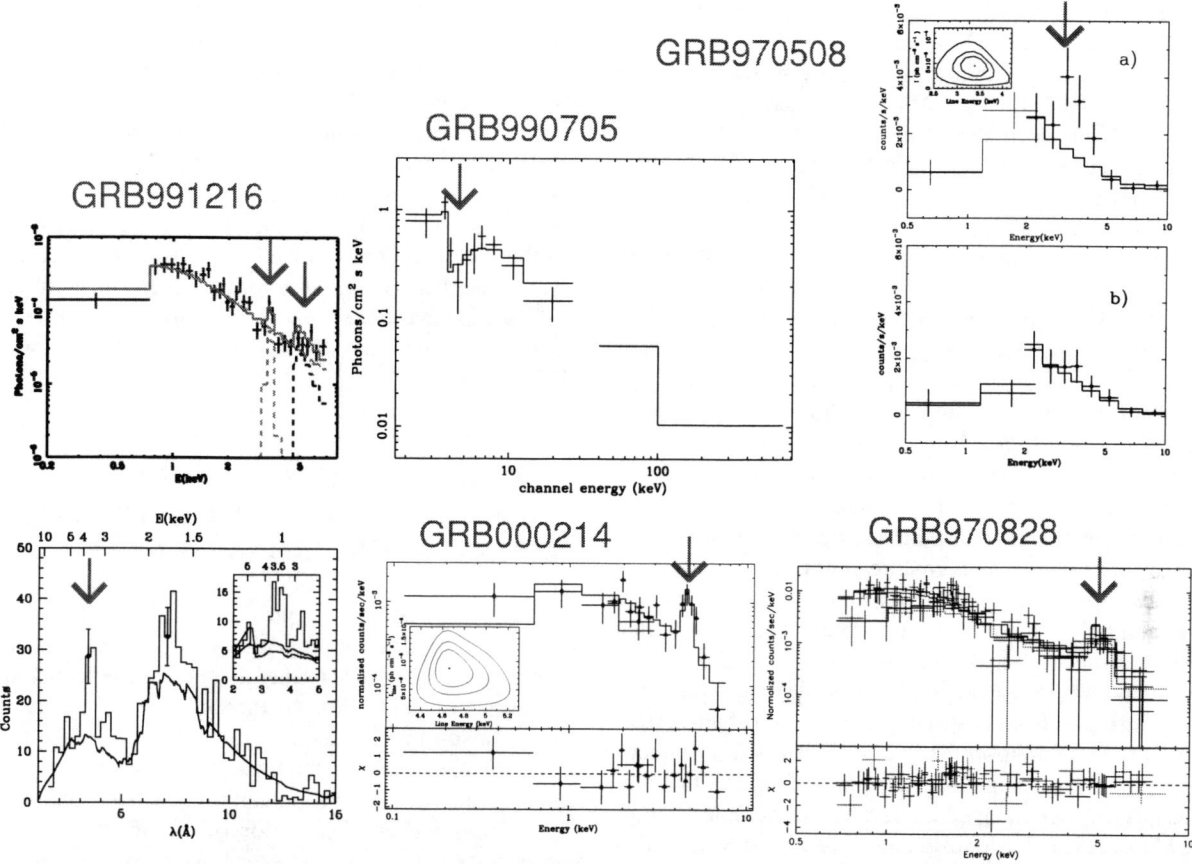

**FIGURE 1.** Iron features in GRB: GRB970508 [4]; GRB970828: [5]; GRB990214 [6]; GRB991216 (reprinted from [7], Copyright 2000, American Association for the Advancement of Science); GRB990705 (reprinted from [8], Copyright 2000, American Association for the Advancement of Science)

## X-RAY ABSORPTION AND EMISSION FEATURES

Iron features are ubiquitous in several classes of X-ray sources and have been used to probe the emission mechanisms and the close environment of these sources[9]. The observational evidence of iron features in the X-ray spectra of GRB is quite compelling. So far, there have been 5 detections of iron features in GRB (Fig.1), four in the afterglow phase (GRB970508 [4]; GRB970828: [5]; GRB990214 [6]; GRB991216 [7]), and one during the prompt phase (GRB990705 [8]), each at about 3 sigma level (with the exception of GRB991216, with a significance above 4 sigma). In three cases (GRB970508, GRB970828, GRB991216), the X-ray redshifts derived from the iron lines were consistent with those from optical spectroscopy. For the dark GRB000214, no optical redshift is available as yet. Finally, the case of GRB990705 is particularly important. No optical spectroscopy was available when the original paper was published[7]. The energy of the iron feature implied an X-ray redshift $z = 0.86 \pm 0.14$. This prediction is now confirmed by optical spectroscopy, that set the redshift of the host galaxy at $z = 0.8435$ [10]. These findings demonstrate that the measurement of the redshift by X-ray spectroscopy is more than a mere possibility, and that it can provide reliable results, a fact that is particularly important when optical spectroscopy is difficult or not possible at all.

Let us now summarize the observational picture and the ensuing theoretical implications. For the features in the three afterglows with an independent optical redshift, we can derive the line energy at the burst site and therefore the ionization stage of the medium. FeI to Fe XVII ions are characterized by a $K_\alpha$ line energy at $\approx 6.4$ keV, that rises to 6.7 keV for He-like ions and to 6.9 keV for H-like ions. The iron edge is located at 7.1 keV for neutral iron, rising slowly to 8.8 keV for He-like ions, and then to 9.3 keV for H-like ions.

In GRB991216 (z=1.02) the rest-frame line energy is $6.95 \pm 0.15$. There is also evidence of an additional emission feature at $9 \pm 1$keV, that is associated with a narrow H-like iron recombination edge *in emission* (also known as Radiative Recombination Continuum, RRC). In the case of GRB970828 (z=0.96), the emission feature at $9.3 \pm 0.5$ has been also attributed to the H-like recombination edge in emission. In the case of GRB970508, the line energy is $6.25 \pm 0.55$, i.e. consistent with an iron stage from neutral to H-like.

There is also marginal evidence of line variability on time scales of $\lesssim day$ in three afterglows (GRB970508, GRB970828 and GRB000214), while for GRB991216 the observation was too short (10 ksec). In the latter case a line broadening of $\approx 10\%$ is measured.

The first conclusion that can be drawn is that the line-emitting gas, *in the afterglow phase*, is highly ionized. The most obvious source of ionization is the GRB itself, that produces a copious flux of hard X-ray photons (*photoionization scenario*). In this setting, the medium has to be located outside the fireball region (i.e. at $R \gtrsim 10^{15}$cm: *distant reprocessor scenario*[e.g. 11, 7, 12]). The line variability is naturally expected from the light travel time between the GRB and the reprocessor.

In the early phase (i.e. on a time scale of $\approx 10$s), when the ionization front is still expanding, a substantial fraction of the medium in the line of sight is still to be ionized, thus producing X-ray absorption features, that will disappear when the medium is becomes completely ionized [e.g. 13, 14]. A transient iron absorption edge is then expected if the medium lies in the line of sight, exactly what has been observed in GRB990705. On longer time scales, ($\approx t_{rec}$, the recombination time scale), electrons recombine with ions, producing the line and the RRC [e.g. 15, 7], observed in the afterglow phase. This scenario requires the presence of a dense medium with high iron overabundance ejected before GRB, with a velocity (implied by the line width) of $\approx 0.1c$ and a mass of iron $\gtrsim 0.01$ $M_\odot$[e.g. 7]. The most straightforward scenario that emerges is that of a massive progenitor that ejects, before the GRB, a substantial fraction of its mass in a SN-like explosion, as in the case of the SupraNova model [16].

In the (*nearby reprocessor scenario* [17, 18]) ionizing photons would be produced by post-burst activity of the "remnant" of the central source after the GRB event. In this case the reprocessor, associated with the outer stellar envelopes of a massive progenitor, can be much nearer to the central source. Line variations are caused by the decaying radiation continuum of the remnant. While a massive progenitor is still assumed in this scenario, an highly iron-enriched medium (and therefore the associated SN event) is no more required, because of the higher densities and the greater reprocessing efficiency in producing a line with the requested luminosity ($L_{Fe} \approx 10^{44-45} erg s^{-1}$)[see 19, for a comparative discussion on the reprocessor scenario].

Finally, in the *shock heated* scenario, the gas is collisionally ionized by shock resulting from the interaction of GRB ejecta (either the fireball itself or the non-relativistic ejecta of SN-like explosion associated with the GRB) with the material pre-ejected before the GRB by massive progenitor systems [e.g. 20, 21, 22]. The presence of an iron recombination edge *in emission* is not straightforwardly explained in this scenario, because in the aforementioned models the plasma is expected to be near to thermal equilibrium. However Yonetoku et al. [23] argue that, under certain conditions, a plasma in a non-equilibrium state (NEI) can also produce a narrow RRC.

## STAR FORMING REGIONS, X-RAY ABSORPTION AND HIGH-DENSITY ENVIRONMENT

As discussed above, the presence of X-ray lines links long GRB to massive progenitors, and therefore to star-forming sites. There is further independent evidence supporting this connection. In the case of NS-NS coalescence, a substantial fraction of events should take place far from the center of the host galaxy, while the opposite applies to massive progenitors. Bloom et al. [24] have measured the distribution of the offsets of optical afterglows with respect to their host galaxy, finding that it is fully consistent with that of star-forming regions, while the delayed merging scenario can be ruled out at the $2 \times 10^{-3}$ level.

What are the implications of this scenario in terms of observable quantities? In a typical Giant Molecular Cloud harboring star formation in our Galaxy, densities are $n \approx 10^2 - 10^5$ cm$^{-3}$, the size is of order of 10 pc and the column density is $N_H \approx 5 - 10 \times 10^{21}$cm$^{-2}$ [25, 26].

Such values of column density can be measured by current X-ray satellites in relatively bright X-ray afterglows with $z \lesssim 3$ (note in fact that $N_{Hobs} \approx (1+z)^{-8/3} N_{Hrest}$ and the typical uncertainty on $N_H obs$ is $\approx 10^{20}$cm$^{-2}$). This is in fact the case. In the BeppoSAX sample at least three afterglows (GRB980703 [27, 28, 29]; GRB010222 [30, 28, 29] and GRB990123[28, 29]), show significant absorption, and two other some marginal evidence. $N_{Hrest}$ in the range $(2-20) 10^{21}$cm$^{-2}$. In addition, Chandra (and BeppoSAX) observations of two other afterglows give similar results (GRB000210:$N_{Hrest} = (5 \pm 1) \, 10^{21}$cm$^{-2}$ [31]; GRB000926: $N_{Hrest} = (4^{+3.5}_{-2.5}) \, 10^{21}$cm$^{-2}$[32]). These measurements of absorption provide further support to the scenario in which GRB are embedded in a star-forming GMC, in which the typical volume density is $n \approx 10^{2-5}$ cm$^{-3}$.

However, application of the standard fireball-jet model

to multi-wavelength data of afterglows leads to density estimates that are typically much lower that that expected in a GMC [e.g. 33]. There are two points that need to be stressed in this regard. First, the achromatic (i.e. energy-independent) break observed in the light curve of some afterglows is attributed to a collimated fireball. The break appears when the relativistic beaming angle $1/\Gamma$ becomes $\approx \theta$ (e.g. Rhoads 1997, Sari et al. 1999). The typical opening angle of the jet derived in these models is of $\approx 2-4°$ [e.g. 33, 34]. We have performed a systematic analysis of the X-ray spectra and light curves on a sample of BeppoSAX afterglows observed from few hours to about 2 days after the GRB [35] that shows that the fireball expansion *in the first two days* is consistent with a spherical expansion. This result implies that the average opening angle of the jet should be $\theta \gtrsim 10-20°$. As for a probable origin of the discrepancy we note that in [e.g. 33] the X-ray data are considered at just one energy, overlooking the fact that the X-ray window span almost two decades in energy, and therefore not including the *X-ray spectral* information.

For example, in the case of GRB010222, where a break in the light curves appears around 0.5 days [30, 36], the predictions of the standard jet model are not consistent with the X-ray temporal *and* spectral slopes [30]. On the contrary, the X-ray data are well described by a fireball undergoing a transition to a non-relativistic expansion (NRE, [37]). In fact, NRE can also produce an *achromatic break* in the afterglow light curves at $t_{NRE} \approx 3\frac{1+z}{2}(\frac{E_{53}}{n_6})^{1/3}$ days. This alternative explanation has been proposed in a few other cases (GB990123 [38], GRB000926 [32] - but see also [39], GRB010222 [30, 36]. A strong implication of this scenario is that, in those GRB's, the environment is composed of a dense medium ($n \approx 10^4 - 10^6 cm^{-3}$), i.e. typical of molecular clouds in star forming regions. Interestingly, in the case of GRB010222, a strong, constant sub-mm emission, has been attributed to enhanced star-forming activity ($\approx 600$ $M_\odot yr^{-1}$) in the galaxy hosting this burst[40]. Such high densities require a low magnetic field, $\lesssim 10^{-6}$ times the equipartition value, to keep the synchrotron self-absorption frequency in the observed region. Comparable values of the magnetic field have been also derived in other cases, like GRB971214 [41] and GRB990123 [42]. However, a first attempt to fit the broad-band data of GRB000926 with the NRE show that the radio data are not well described by the model [39].

In conclusion, at the present stage of analysis, both the NRE of a moderately-collimated fireball in a dense medium and the highly collimated jet scenario (in a low density medium) show some inconsistency with the data, that appears when the complete information from radio to X-rays is considered. More theoretical efforts are thus needed, in particular to reconcile the fireball model with an external medium typical of a star-forming region.

# RECENT FINDINGS AND REMAINING MYSTERIES

## GHOST and X-ray flashes

It is observationally well-established that about half of accurately localized gamma-ray bursts (GRBs) do not produce a detectable optical afterglow [43, 44], while most of them ($\approx 90\%$) have an X-ray afterglow [45]. Statistical studies have shown that the optical searches of these events, known variously as "dark GRBs", "failed optical afterglows" (FOA), or "gamma-ray bursts hiding an optical source-transient" (GHOST), have been carried out to magnitude limits fainter on average than the known sample of optical afterglows [46, 2]. Thus dark bursts appear to constitute a distinct class of events and are not the result of an inadequate search, but it is unclear whether this observational property derives from a single origin or it is a combination of different causes.

If the progenitors of long-duration GRBs are massive stars [47], as current evidence suggests (*e.g.*, [48, 7]), extinction of optical flux by dusty star-forming regions is likely to occur for a substantial fraction of events (*the obscuration scenario*[e.g. 2]). Another possibility is that dark GRBs are located at redshift $z \gtrsim 5$, with the optical flux being absorbed by the intervening Ly$\alpha$ forest clouds (*the intergalactic scenario* [e.g. 49]).

Dark bursts which can be localized to arcsecond accuracy, through a detection of either their X-ray or radio afterglow, are of particular interest. The first example is GRB 970828 for which prompt, deep searches down to R~24.5 failed to detect an optical afterglow [50, 51] despite it was localized within a region of only 10 arcsec radius by the ROSAT satellite [52]. Djorgovski et al. [53] recently showed how the detection of a short-lived radio transient for GRB 970828 allowed them to identify the probable host galaxy and to infer its properties (redshift, luminosity and morphology). In addition, they used estimates of the column density of absorbing gas from X-ray data, and lower limits on the rest frame extinction ($A_V > 3.8$) to quantify the amount of obscuration towards the GRB. The other best studied example is GRB000210 [31]. This burst had the highest $\gamma$-ray peak flux of any event localized by *BeppoSAX* as yet but it did not have a detected optical afterglow, despite prompt and deep searches down to $R_{lim} \approx 23.5$. *Chandra* observations allowed to localize the X-ray afterglow of GRB000210 to within $\approx 1 arcsec$ and a radio transient was detected with the VLA. The precise X-ray and radio positions allowed to identify the likely host galaxy of this burst, and to measure its redshift, $z = 0.846$. The proba-

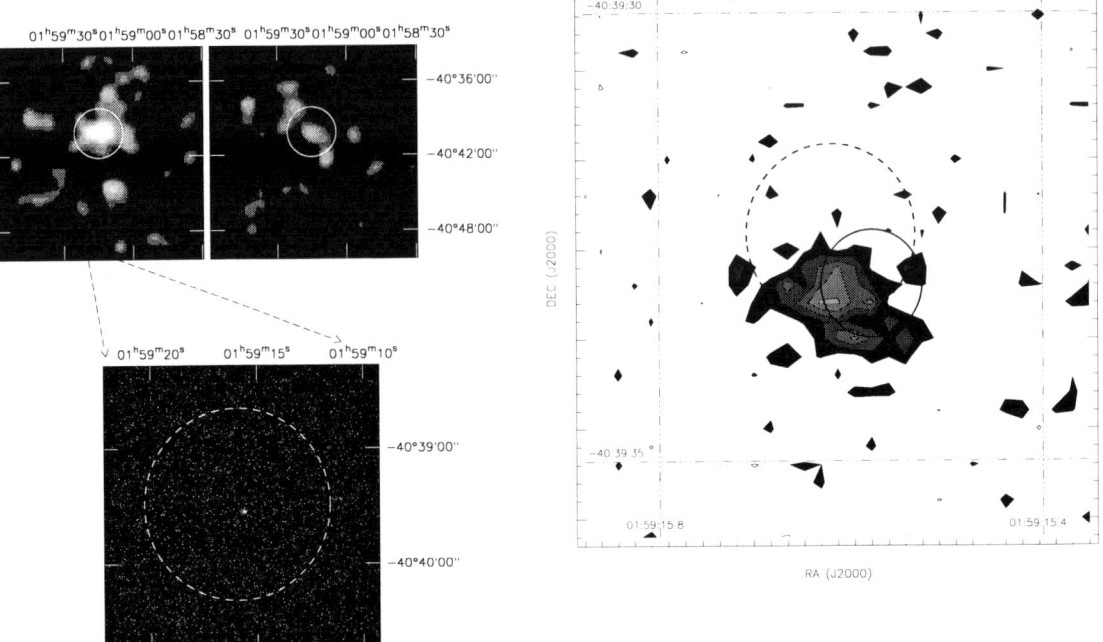

**FIGURE 2.** The dark GRB000210. The two images in the left upper panel show the decaying X-ray afterglow BeppoSAX MECS(1.6-10 keV) 8 hrs and 30 hrs after the GRB respectively. The circle is the WFC error box. The left lower panel is the Chandra ACIS-S image of the afterglow 21 hrs after the GRB. The dashed line is the BeppoSAX MECS error box. The right panel shows the optical image of the host galaxy taken with the VLT. The circles show the 90% error circles of the Chandra (continuous line) and radio (dashed line) afterglows. The redshift of the galaxy is z=0.846

bility that this galaxy is a field object is $\approx 10^{-2}$, but the chance that this happens in both the cases of GRB970828 & GRB000210 is negligible.

The X-ray spectrum of the afterglow of GRB000210 shows significant absorption in excess of the Galactic one corresponding, at the redshift of the galaxy, to $N_H = (5 \pm 1) \times 10^{21}$ cm$^{-2}$. The amount of dust needed to absorb the optical flux of this object is consistent with the above HI column density, given a dust-to-gas ratio similar to that of our Galaxy. If the absorption takes place in a GMC, a substantial fraction of the X-ray gas should be heavily ionized by the hard X-ray photons emitted by the GRB and its afterglow. This effect should change the absorption profile, in particular at lower energies, where the lighter elements are heavily ionized, and therefore become more transparent to the radiation. Nonetheless, the X-ray absorption observed in GRB000210 is consistent with a cold medium, an evidence that can be reconciled with the GMC scenario is the medium is condensed in high dense clouds ($n \gtrsim 10^9$ cm$^{-3}$ [31].

Given the extreme luminosity of GRBs and their probable association with massive stars, it is expected that some fraction of events will be located beyond $z > 5$ [54]. These would be probably classified as dark bursts because the UV light, which is strongly attenuated by absorption in the Ly$_\alpha$ forest, is redshifted into the optical band. We note that the four redshifts determined or suggested so far for dark GRBs ($z = 0.96$, GRB970828, [53]; $z = 1.3$, GRB990506, [55, 24]; $z \approx 0.47$, GRB000214, [6], GRB000210 [31]) are in the range of those measured for most bright optical afterglows, but whether this applies to the majority of these events is still to be assessed.

Particularly interesting in this respect is the case of the so-called X-ray flashes or X-ray rich GRBs discovered by *BeppoSAX*[56, 57]. No optical counterpart has been identified as yet in any of these bursts. The *intergalactic scenario* would naturally explain both the absence of optical afterglows and the high-energy spectrum, because the peak of the gamma-ray spectrum would be redshifted into the X-ray band. Alternatively, the paucity of gamma-ray emission can be explained by a "dirty" fireball [3]. In such a case the fireball will achieve a Lorentz factor much lower than previously considered, i.e. not high enough to boost the photons in the γ-ray range.

## Short GRB's

The distribution of GRB duration appears to be bimodal, with about 30% of events lasting less than 1 sec [58]. It is still unclear whether these events are intrinsically diverse from long bursts, i.e. if they are produced by different progenitors. So far, very little is known about these events, due to the lack of a counterpart. While the GRBM on board of BeppoSAX is detecting those events, no one has so far been observed in the X-ray range (and therefore precisely localized) by the WFC [59]. The probability that this is a chance fluctuation is getting interestingly low, suggesting a depleted X-ray emission compared to the class of long events [60]. If that is the case, the localization of afterglows of these events should rely only on experiments with good position accuracy in the γ-ray range, like the currently operating IPN or, in the future, SWIFT. If, instead, these events have an X-ray emission similar to long ones, we may hope to localize a few of them with BeppoSAX and HETE2.

## THE FUTURE OF X-RAY SPECTROSCOPY

X-ray spectroscopy of GRB is coming of age. Emission and absorption features as well as the properties of X-ray absorption are providing key information on the close environment of long GRB's, suggesting massive progenitors and a connection between these events and star-forming regions. The possibility of measuring the redshift directly from X-rays is of particular value for those classes of objects, like the majority of dark GRB, X-ray flashes and short GRB's, that still lack optical counterparts. The radiation intensity of GRB's is so high that they can be detectable out to much larger distances than those of the most luminous quasars or galaxies observed so far, and it is likely that high-z GRB are actually the constituents of one of those mysterious classes of events.

We should stress that, with present X-ray facilities (BeppoSAX, XTE, HETE2, Chandra, Newton), the progress in this field will follow "quantum" jumps. In fact, to get good-quality X-ray spectra, one should catch bright afterglows, but these should be not too many, because of the combination of ingredients needed (number of precise and fast locations, reaction time, number of TOO observations allocated to GRB programs). On the positive side, one can then hope that the next important discovery just lies behind the corner.

In the near future, we expect a further advancement in this area, when high quality X-ray CCD spectra of afterglows ($E/\Delta E \approx 50$) will be routinely available with SWIFT. Looking ahead in the future, high resolution X-ray spectroscopy ($E/\Delta E \gtrsim 1000$, like that provided by X-ray microcalorimeters) should open a new area of exploration ([61, 62]), that would bring us closer to the central engine of GRB, its environment and far in the Early Universe by using GRB as beacons to probe star and galaxy formation. We can foresee the possibility of resolving in detail line profiles, deriving information on the kinematics of the ejecta, or looking for narrow and faint emission lines, imprinted on the spectra by the low-velocity medium embedding the GRB. The measurement of the absorption edges produced by the ISM of the *host galaxy* will provide information on the chemical composition of galaxies in the Early Universe, thus opening the possibility to trace the metallicity history of the gas in the Universe and to probe the formation of the first stars [62, 63, 54].

## ACKNOWLEDGMENTS

BeppoSAX is a program of the Italian space agency (ASI) with participation of the Dutch space agency (NIVR)

## REFERENCES

1. Woosley, S. E., in *GRBs in the Afterglow Era*, edited by E. Costa, F. Frontera, and J. Hjorth, ESO-Springer, 2001, pp. 258–262.
2. Reichart, D. E., and Yost, S. A., (2001), apJ, in press; astro-ph/0107545.
3. Dermer, C. D., Chiang, J., and Boettcher, M., *ApJ*, **513**, 656–668 (1999).
4. Piro, L., et al., *A&A*, **331**, L41–L44 (1998).
5. Yoshida, A., Yonetoku, N. M., Murakami, T., Otani, C., Kawai, N., Ueda, Y., Shibata, R., and Uno, S., *ApJ*, **557**, L27–L30 (2001).
6. Antonelli, L. A., Piro, L., Vietri, et al.*ApJ*, **545**, L39–L42 (2000).
7. Piro, L., Garmire, G., Garcia, et al.*Science*, **290**, 955–958 (2000).
8. Amati, L., Frontera, F., Vietri, et al.*Science*, **290**, 953–955 (2000).
9. Piro, L., in *UV and X-ray spectroscopy of Laboratory and Astrophysical Plasmas*, edited by S. K. E. Silver, Cambridge Uni. Press, 1993, p. 448.
10. Andersen, M., et al. (2002), in preparation.
11. Lazzati, D., Campana, S., and Ghisellini, G., *MNRAS*, **304**, L31 (1999).
12. Weth, C., et al., *ApJ*, **534**, 581 (2000).
13. Perna, R., and Loeb, A., *ApJ*, **501**, 467– (1998).
14. Boettcher, M., Dermer, C. D., Crider, A. W., and Liang, E. . P., *A&A*, **343**, 111–119 (1999).
15. Paerels, F., Kuulkers, E., Heise, J., and Liedahl, D. A., *ApJ*, **535**, L25–L28 (2000).
16. Vietri, M., and Stella, L., *ApJ*, **527**, L43–L46 (1999).
17. Rees, M. J., and Mészáros, P., *ApJ*, **545**, L73–L75 (2000).
18. Mészáros, P., and Rees, M. J., *ApJ*, **556**, L37–L40 (2001).
19. Kallman, T. R., Mészáros, P., and Rees, M. J., Iron k lines from grb (2002), astro-ph/00110654.
20. Vietri, M., Perola, G. C., Piro, L., and Stella, L., *MNRAS*, **308**, L29 (1999).
21. Boettcher, M., *ApJ*, **539**, 102–110 (2000).
22. Boettcher, M., and Fryer, C. L., *ApJ*, **547**, 338–344 (2001).

23. Yonetoku, N. M., Murakami, T., Masai, K., Yoshida, A., Kawai, N., and Namiki, M., *ApJ*, **557**, L23–L26 (2001).
24. Bloom, J. S., Kulkarni, S. R., and Djorgovski, S. G.(2002), aJ, in press; astro-ph/0010176.
25. Balsara, D., Ward-Thompson, D., and Crutcher, R. M., *MNRAS*, **327**, 715–720 (2001).
26. Ward-Thompson, D., Scott, P. F., Hills, R. E., and Andre, P., *MNRAS*, **268**, 276–290 (1994).
27. Vreeswijk, P. M., *et al.ApJ*, **523**, 171–176 (1999).
28. Pasquale, M. D., et al. (2002), in preparation.
29. Stratta, G., et al. (2002), in preparation.
30. in' t Zand, J. . J. M., Kuiper, L., Amati*et al.ApJ*, **559**, 710–715 (2001).
31. Piro, L., Frail, D., Gorosabel, J., *et al.*(2002), apJ, submitted; astro/ph-0201282.
32. Piro, L., Garmire, G., Garcia, *et al.ApJ*, **558**, 442–447 (2001).
33. Panaitescu, A., and Kumar, P., *ApJ*, **554**, 667–677 (2001).
34. Frail, D. A., et al., *ApJ*, **562**, L55– (2001).
35. Stratta, G., Piro, L., Soffitta, P., et al., in *GRBs in the Afterglow Era*, edited by E. Costa, F. Frontera, and J. Hjorth, ESO-Springer, 2001, pp. 118–120.
36. Masetti, N., et al., *A&A*, **374**, 382 (2001).
37. Livio, M., and Waxman, E., *ApJ*, **538**, 187–191 (2000).
38. Dai, Z. G., and Lu, T., *ApJ*, **519**, L155–L158 (1999).
39. Harrison, F. A.,*et al.***559**, 523 (2001).
40. Frail, D. A., *et al.*(2001), submitted to ApJ; astro-ph/0108436.
41. Wijers, R. A. M. J., and Galama, T. J., *ApJ*, **523**, 177–186 (1999).
42. Galama, T. J., et al., *Nature*, **398**, 394–399 (1999).
43. Frail, D. A., Kulkarni, S. R., Wieringa, M. H., Taylor, G. B., Moriarty-Schieven, G. H., Shepherd, D. S., Wark, R. M., Subrahmanyan, R., McConnell, D., and Cunningham, S. J., , in *AIP Conf. Proc. 526: Gamma-ray Bursts, 5th Huntsville Symposium*, 2000, pp. 298–302.
44. Fynbo, J. U., *et al.A&A*, **369**, 373–379 (2001).
45. Piro, L., in *GRBs in the Afterglow Era*, edited by E. Costa, F. Frontera, and J. Hjorth, ESO-Springer, 2001, pp. 97–105.
46. Lazzati, D., Covino, S., and Ghisellini, G., (2002), mNRAS, in press, astro-ph/0011443.
47. Paczyński, B., *ApJ*, **494**, L45–L48 (1998).
48. Bloom, J. S., et al., *Nature*, **401**, 453–456 (1999).
49. Fruchter, A. S., *ApJ*, **512**, L1–L4 (1999).
50. Odewahn, S. C., *et al.IAU Circ*, **6735** (1997).
51. Groot, P. J.*et al.ApJ*, **493**, L27–+ (1998).
52. Greiner, J., Schwarz, R., Englhauser, J., Groot, P. J., and Galama, T. J., *IAU Circ*, **6757** (1997).
53. Djorgovski, S. G., , Frail, D. A., Kulkarni, S. R., Bloom, J. S., Odewahn, S. C., and Dierks, A., *ApJ*, **562**, 654 (2001).
54. Lamb, D. Q., and Reichart, D. E., *ApJ*, **536**, 1–18 (2000).
55. Taylor, G. B., Bloom, J. S., Frail, D. A., Kulkarni, S. R., Djorgovski, S. G., and Jacoby, B. A., *ApJ*, **537**, L17–L21 (2000).
56. Heise, J., in 't Zand, J., Kippen, M., and Woods, P., in *GRBs in the Afterglow Era*, edited by E. Costa, F. Frontera, and J. Hjorth, ESO-Springer, 2001, pp. 16–21.
57. Heise, J. (2002), this conference.
58. Kouveliotou, C., Meegan, C. A., Fishman, G. J., Bhat, N. P., Briggs, M. S., Koshut, T. M., Paciesas, W. S., and Pendleton, G. N., *ApJ*, **413**, 101–104 (1993).
59. Gandolfi, G., et al., in *AIP Conf. Proc. 526: Gamma-ray Bursts, 5th Huntsville Symposium*, edited by M. Kippen, R. Mallozzi, and G. J. Fishman, 2000, pp. 23–.
60. Gandolfi, G., et al. (2002), this conference.
61. Piro, L., et al.,in *GRBs in the Afterglow Era*, edited by E. Costa, F. Frontera, and J. Hjorth, ESO-Springer, 2001, p. 415.
62. Piro, L. (1999), probing the Early Universe with GRB and XEUS at www.ias.rm.cnr.it/sax/xeus.html.
63. Fiore, F., et al., *ApJ*, **544**, L7– (2000).

# BATSE Observations of GRB 991216

V. Connaughton*, T.W. Giblin[†], R.M. Kippen, R.D. Preece, M.S. Briggs* and C.A. Meegan**

*CSPAR-UAH, 5000 Technology Dr, Huntsville AL 35899
[†]Physics Dept., College of Charleston
**SD50, NASA-MSFC, AL 35812

**Abstract.** Observations of the afterglow of gamma-ray burst GRB 991216 have been reported at wavelengths from radio to the X-ray band. BATSE observations at energies above 20 keV show the gamma-ray emission to consist of 2 distinct impulsive episodes separated by 15 seconds and having a total duration of about 40 seconds, followed by a long tail which persists at least 1000 s beyond the trigger time. We present temporal and spectral characteristics of both the prompt and extended emission, and explore whether the tail might be the gamma-ray signature of the afterglow.

## INTRODUCTION

The afterglow associated with GRB 991216 (BATSE trigger 7906) was seen in follow-up observations at wavelengths from radio to X-ray as early as 1 hour after the GRB trigger [4]. The overall picture of the optical and X-ray afterglow is consistent with adiabatic evolution of collimated emission from a fireball expanding in a uniformly dense medium. There are, however, some deviations from the expectations of this scenario. In particular, the radio observations [5] which were made 1-2 days after the GRB trigger show the signal to be decaying at this early stage in contrast to most radio afterglows which peak later relative to the prompt GRB emission. A summary of the X-ray and optical observations is presented in Table 1.

## BATSE OBSERVATIONS OF GRB 991216

The top left plot of Figure 1 shows the complete count rate history from trigger time ($t_0$) to $t_0 + 600$s with the 3 other plots zooming in on distinct episodes of emission: at bottom left the triggering pulse (P1) which is a typical weak GRB lasting a few seconds; top right is a stronger, multi-peaked episode (P2) which is separated from the triggering pulse by a gap of 10s and lasts a further 20s or so; bottom right is a log time plot from $t_0 + 45$s to $t_0 + 1100$s showing a smoother decaying tail that is dim in comparison to the prior emission but can nevertheless be well-characterized because of its long duration.

## PROMPT EMISSION

The triggering pulse is a single-peaked event which is weak and lasts about 6s. It is not well-characterized spectrally by the Band function [1] often used to fit BATSE spectra, but this is not unusual for the weaker GRBs where the break in the spectrum is difficult to detect and the higher-energy power law ill-determined. In these cases the best fit is generally obtained with a single power law or Comptonized model which is a single power law with an exponential cut-off above an energy $E_p$. This pulse is consistent with a single power law photon index of $-1.8$ that does not appear to vary significantly during the emission.

It has been suggested [10] that this pulse, being softer than the subsequent emission, might be evidence for a photosphere of expanding ejecta and that it should therefore be a blackbody spectrum. An attempted fit of this type of spectrum implies a temperature of 40 keV but the blackbody fit is inferior to the power law fit and is especially divergent from blackbody expectation at energies above 50 keV.

After an apparent gap of 10s in the gamma-ray emission, a series of bright and spiky peaks is seen which appear to tail off after~ 20s. By itself, this episode is among the brightest and most fluent of BATSE bursts. The shape and evolution of the light-curve with energy and time is typical of many GRBs and the spectrum is well-fit by the Band function. For a spectrum integrated over the duration of P2 a low-energy power law index of $-1.15 \pm 0.01$ is seen below an $E_p$ of $390 \pm 18$ keV with the higher energy flux following a $-2.33 \pm 0.02$ power

**TABLE 1.** Temporal and spectral characteristics of afterglow observations of GRB 991216

|         | Time after trigger (hr) | Temporal decay index | Energy spectral index | Reference |
|---------|------------------------|---------------------|----------------------|-----------|
| X-ray   | 1      |       | $-1.8 \pm 0.3$ | Corbet et al. 1999    |
| X-ray   | 4 – 11 | -1.6  | -2.1           | Takeshima et al. 1999 |
| Optical | 11 – 19| -1.2  |                | Halpern et al. 2000   |
| Optical | 48+    | -1.53 |                | Halpern et al. 2000   |
| X-ray   | 37     |       | -2.1           | Piro et al. 1999      |

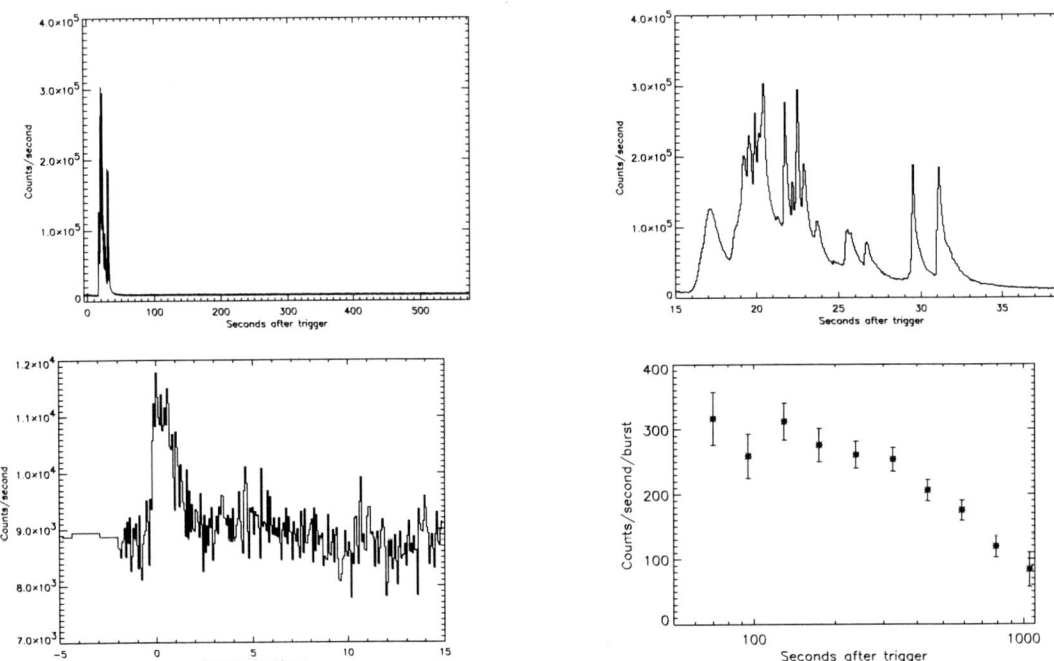

**FIGURE 1.** Count rate history from 20 keV–2 MeV for GRB 991216 from trigger time $t_0$ to $t_0 + 600$s (top left); the first pulse P1 (bottom left), the strong episode P2 (top right); and the background-subtracted tail in log-time (bottom right).

law. Beyond $t_0 + 32$s the fits to the Band function fail to converge although the flux is still high, suggesting that at this point the emission may be dominated by another component.

## EXTENDED EMISSION OF GRB 991216

The third emission period can be seen only after a careful treatment of the background rates and is not visible in the standard BATSE light-curves which use regions close to the triggered emission to fit and subtract background count rates. It is much weaker than the emission which precedes it and lasts at least as long as the variability time-scale of the rates registered by BATSE as the spacecraft moves through differing geomagnetic latitudes. For this tail period a background is obtained by averaging the rates from times on the day (15 orbits) before and after the emission period, when the spacecraft occupied a similar geomagnetic position. The rates shown in the bottom right plot of Figure 1 have these background rates subtracted. The signal persists at least until until $t_0 + 1100$s at which time the position becomes occulted to the spacecraft by the Earth, and may still be there after exiting the shadow of the Earth (at $t_0 + 4000$s) but the flux at this point is beyond the limit of the sensitivity of this technique.

The tail is bright enough that the occultation step is visible in the count rates of the background-subtracted light-curve – and occurs at the expected time. It can also be seen that after similar background subtraction, the light-curve from the period before the trigger at $t_0$ is flat.

Spectrally, the tail is best fit by a single power law with index $-1.8$ integrated over the duration of the tail. The period between P1 and P2 ($t_0 + 5$s $-t_0 + 15$s), which appears to be flat before background-subtraction, shows a flux of magnitude comparable to the average of the tail flux and has a $2.1 \pm 0.3$ photon index spectrum which is consistent with that of the tail.

Episode P2 exhibits the usual hardening of spectrum with increasing flux superimposed on a gradual softening of the spectrum as the burst progresses. In episode P1 the spectrum hardens until the peak of the emission and remains consistent with a photon index of $1.8 - 2.0$ with uncertainties that are larger than any evolution that might be present. In the tail, the photon index seems to oscillate around -1.8 rather than follow a trend. The RXTE-ASM X-ray afterglow spectrum of $-1.8 \pm 0.3$ at $t_0 + 3600$s is consistent with both the gamma-ray tail and with the later RXTE-PCA measurement of -2.1, but this PCA spectrum, however, is not consistent with the gamma-ray spectrum.

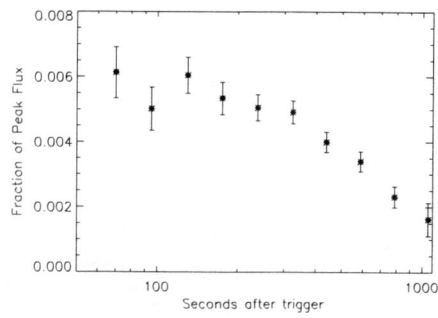

**FIGURE 2.** The summed-energy count rate history for the tail of GRB 991216 where counts are shown as a fraction of the peak count rate measured in the prompt emission in episode P2.

### Comparison with other gamma-ray tails

GRB 920723 [2] seen with GRANAT and the BATSE-detected GRB 980923 [6] were both observed to have a similar behavior to GRB 991216: bright, peaked episodes followed by long, smooth, dimmer tails with distinct spectral characteristics. In Figure 2 the count rates in the tail of GRB 991216 are given as a fraction of the peak count rates $P_F$ (measured during P2). During the early part of the tail a fraction of 6/1000 of the peak flux is observed, with this fraction decreasing by a factor of 4 just prior to occultation. This is comparable to the fraction in the tail of GRB 920723 (5/1000) but lower than GRB 980923 (3/100). It is consistent with the fraction of the peak flux seen in the tail of the summed background-subtracted signals of 200 medium (2s < duration < 30s) bursts at a time 40s after alignment of the GRB peaks [3].

Recent attempts to scale afterglow with prompt burst properties have focused on GRB fluence as a determining factor. Summing the counts in each episode suggests the tail contains 13% of the total fluence in counts between 20 keV and 2 MeV, comparable with the 12% tail fluence seen in the summed burst analysis [3], 20% in GRB 920723, and 7% in GRB 980923. It is likely that what is considered prompt emission overlaps with the beginning of the tail at some unknown time in each case.

### Temporal Characteristics

The flux in soft X-rays is known at $t_0 + 3600$s, and the soft X-ray spectrum is comparable to that in gamma rays 2000s earlier, so one can infer there is probably no cooling or synchrotron break moving through the X-ray–gamma-ray energy range during this time window. This is confirmed by the Chandra observations at $t_0 + 37$hr

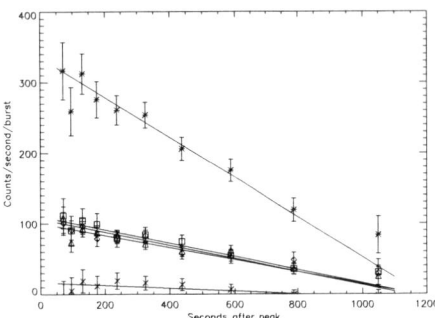

**FIGURE 3.** The count rates in the tail of GRB 991216 and the best fits over duration of the tail summed in energy (asterisks) and for individual channels (diamonds = 20 – 50 keV, triangles = 50 – 100 keV, squares = 100 – 300 keV). All are best fit by a straight line.

which imply that the cooling break frequency (which decays as $v \propto t^{-3/8}$) still lies above X-ray energies at this stage. One might, therefore, expect the gamma-ray emission to decay with a power-law in time similar to that measured by RXTE-PCA hours after the GRB trigger.

It can be seen, however, that this tail seems to decay linearly with time. Figure 3 shows the fits for summed energy and individual energy channels. All are best fit by a straight line - a power law is possible only with an un-physically large temporal offset.

The power law index of decay in the simple fireball expansion models for GRB afterglows represents at these early afterglow stages the decay of the emission flattened by relativistic beaming effects i.e. because we are seeing only an angle $1/\Gamma$ of the emission (where $\Gamma$ is the bulk Lorentz factor) and $\Gamma$ decreases with time, then as time progresses we see a larger fraction of the total emission. If $\Gamma$ cools more rapidly than the $t^{-3/8}$ expected in a slowly cooling adiabatic fireball, the decay will be flatter than expected. For a faster cooling fireball viable scenarios exist for a $t^{-3/7}$ decay, but ultimately the rate of change of $\Gamma$ is limited by the geometry of the fireball - since the radius must be increasing with time, a $t^{-1/2}$ is the fastest rate of change that can be observed, and this is not fast enough to produce this straight-line decay.

It is conceivable that the power law part of the gamma-ray decay starts later than the onset of the tail - a power law from $t_0 + 300$s, for example, can be well fit with a time decay index of $-1.1 \pm 0.2$, but a reason for such behavior is not evident. Another possibility is that the fireball is re-energized in the early stages of its expansion by some phenomenon such as that seen in the X-ray afterglow of GRB 970508 [9] more than a day after the prompt gamma-ray emission.

A decay of $t^{-0.6}$ was fit for the tail of 200 summed bursts [3] so that these shallow decays may be a feature of the early gamma-ray afterglow of GRBs.

## CONCLUSIONS

GRB 991216 is the first GRB with an afterglow-like gamma-ray tail for which a large body of afterglow observations exists at several wavelengths. The behavior and magnitude of the gamma-ray tail is consistent with other gamma-ray tails observed in individual bursts GRB 920723 and GRB 980923 and with the tail seen in the signal summed from hundreds of GRBs.

Superficially some parameters of the gamma-ray afterglow of GRB 991216 can be reconciled with observations at other energies - specifically the spectrum of the gamma-ray tail follows the correct form and is compatible with the ASM X-ray afterglow data.

There are problems with interpreting the temporal decay of this event, which is linear in time in the gamma-ray data instead of being compatible with the power-law decay measured in the X-ray observations. In addition to this inconsistency, the X-ray power-law decay is steeper than expected from the simplest flux/spectrum evolution model when considered in conjunction with the optical afterglow data. With a more detailed analysis and better understanding of the temporal behavior of the gamma-ray tail it may be possible to explore the association between the X-ray and gamma-ray extended components.

## ACKNOWLEDGMENTS

This work was supported by NASA grant NASA-11209.

## REFERENCES

1. Band, D. et al. (1993), *ApJ* **413**, 281.
2. Burenin, R.A. et al. (1999), *A&A Supp.* **138**, 443.
3. Connaughton, V. (2002), *ApJ* **567**, 1028.
4. Corbet, R. et al. (1999), GCN Circular 506.
5. Frail, D.A. et al. (2000), *ApJ* **538**, L129.
6. Giblin, T. et al. (1999), *ApJ* **524**, L47.
7. Halpern, J. et al. (2000), *ApJ* **543**, 697.
9. Piro, L. et al. (1998), *A&A* **331**, L41.
9. Piro, L. et al. (1999), GCN Circular 500.
10. Ruffini, R. et al. (2001), *ApJ* **551**, L113.
11. Takeshima, Y. et al. (1999), GCN Circular 462.

# A Radiative Recombination Edge in the X-ray Afterglow of GRB 970828, and Non-Equilibrium Ionization States

D. Yonetoku[*], T. Murakami[†], A. Yoshida[**], K. Masai[‡], N. Kawai[§] and M. Namiki[¶]

[*]*Department of Physics, Faculty of Science, Kanazawa University, Kakuma, Kanazawa, Ishikawa 920-1192*
[†]*ISAS, 3-1-1 Yoshinodai, Sagamihara, Kanagawa 229-8510, Japan*
[**]*Aoyama Gakuin University, Shibuya-ku, Tokyo 150-8366, Japan*
[‡]*Department of Physics, Tokyo Metropolitan University, 1-1 Minamiosawa, Hachioji, Tokyo 192-0397*
[§]*Department of Physics, Tokyo Institute of Technology, Meguro-ku, Tokyo 152-8551, Japan*
[¶]*RIKEN, Wako, Saitama 351-0198, Japan*

**Abstract.** We observed the X-ray afterglow of GRB 970828 with the Japanese X-ray satellite ASCA. The spectral data displayed a *hump* around $\sim 5$ keV. This *hump* structure is likely a recombination edge of iron in the vicinity of the source, taking into account the redshift $z = 0.9578$ found for the likely host galaxy of the associated radio flare. The Radiative Recombination Edge and Continuum model can interpret the spectrum as coming from a highly ionized plasma in a non-equilibrium ionization state.

## INTRODUCTION

We have already learned that GRBs occur at cosmological distances and are one of the most violent phenomena in the universe. However, the radiation mechanism and site of GRBs are still unknown; although the merging neutron star binary made in the baryon–clean environment was a popular scenario at the time when GRB afterglows were discovered, recent observations appear to suggest more "dirty" circumstances are expected in the hypernova [1], collapsar [2], or supranova scenarios [3].

The key observation to investigate the GRB environment is that of strong iron features reported in X-ray afterglows [4, 5, 6, 7]. Especially [6] report striking results from Chandra observations of GRB 991216. It clearly revealed both an iron K$\alpha$ emission line and a recombination edge interpreted to be from highly ionized iron.

GRB 970828 is one of the most important events so far because a spectral feature, possibly from highly ionized iron together with a large absorption column, was detected with the Japanese X-ray satellite ASCA. It had been interpreted as an iron emission line in previous paper, and a redshift of $z = 0.33$ was derived assuming He-like iron line [5].

However, a series of VLA observations detected a weak (4.5 $\sigma$) radio flare at 8.46 GHz at the period 3.5 days after the GRB inside the ROSAT error region. Subsequent optical observations with the Palomar 200-inch telescope revealed extended sources around the radio position [8]. It shows that the radio source is located between two bright optical peaks. A followup spectroscopic study by the Keck of the galaxy "A" revealed clear two emission lines which are interpreted as [O II] and [Ne III] lines. This gives a redshift of $z = 0.9578$ for the galaxy, which is likely the host of GRB 970828 [8].

The discrepancy in distance with ASCA and, moreover, the temporal detections of both forced us to doubt the reality of existence of the iron features. In this paper we report on properties of an X-ray afterglow of this interesting burst, GRB 970828, with the data obtained by ASCA and discuss a plasma state surrounding the GRB progenitor.

## SPECTRAL STRUCTURE OF THE X-RAY AFTERGLOW

An excess feature is found around 4.8 keV in the spectrum. An interpretation of this *hump* as an Fe K$\alpha$ line was presented in the previous paper [5]. However, an inferred redshift of $\sim 0.33$ did not match that of the host galaxy reported by [8]. Motivated by the detection of Radiative Recombination Edge and Continuum (RRC) with Chandra, we apply the the RRC structure for this feature as an alternative interpretation.

The RRC model is described as

$$\frac{dN}{dE} \propto \begin{cases} (kT_e)^{-3/2}\exp(-E/kT_e) & \text{if } E \geq 9.28 \text{ keV}; \\ 0 & \text{otherwise.} \end{cases}$$

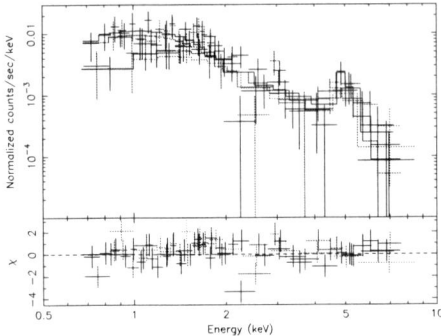

**FIGURE 1.** The X-ray spectrum of GRB 970828 fitted with the model of "absorbed power-law" with a Radiative Recombination edge and Continuum (RRC).

where $k$ is Boltzmann constant and $T_e$ is an electron temperature of the plasma, which can be determined from the width of the RRC structure.

Introducing three parameters for RRC, i.e., $E_{edge}$, $kT_e$ and normalization, we achieve a significant reduction of $\chi^2$, $\Delta\chi^2 = 13.2$. Hence a confidence level of the RRC model is found by an F-test to be 99.3 %. The best-fit gives the edge energy of $E_{edge} = 4.76^{+0.19}_{-0.25}$ keV, the electron temperature at the rest frame $kT_e = 0.8^{+1.0}_{-0.2}$ keV and the integrated RRC flux of $F_{RRC} = 1.7^{+6.4}_{-1.3} \times 10^{-5}$ photons cm$^{-2}$s$^{-1}$. The 90% error range of $E_{edge}$ above is consistent with the expected edge energy of H-like iron at 4.74 keV and He-like iron at 4.51 keV with a redshift of $z = 0.9578$. Thus, the spectral feature could be explained as a radiative recombination edge with a hard tail (continuum) above the edge energy in highly ionized plasma.

It should be noted that there is no redshifted iron emission line seen at an expected energy for such ionized plasma. We searched for a H-like iron emission line (6.97 keV at the rest frame) in the spectrum at "B". As shown in figure 1, we found no emission line around $6.97/(1+z) = 3.56$ keV with the upper limit of $F_{line} \leq 1.5 \times 10^{-6}$ photons cm$^{-2}$s$^{-1}$.

## DISCUSSION

### Spectral Simulation

We should note the fact that the iron features were found only in a small fraction (< 10 %) of X-ray afterglows, and sometimes the features were observed only during a certain interval. Most X-ray afterglows showed only upper limits. Especially, Yonetoku et al. (2000) [9] set an extremely low upper limit of almost 100 eV in EW for the bright GRB 990123 with ASCA. Therefore, there is a wide variety in the iron emission and the RRC intensity. In this paper, we try to explain these varieties of the iron features with the assumption that the line-emitting plasma state is in "non-equilibrium ionization (NEI) state" which has a low electron temperature as compared to the ionization degree.

We show spectra by numerical calculations in the NEI plasma state, not depending on the specific model. To explain the observed strong RRC without the K$\alpha$ line of iron, we calculate the emissivity using the NEI plasma radiation code [10, 11]. The code employs three mechanisms for the continuum, such as free-free emission, two-photon decay and radiative recombination. For line emissions, as well as excitation by electron-impact, fluorescence lines due to ionization and cascade lines due to recombination are taken into account. Radiation properties of a plasma were described by two parameters of the electron temperature ($T_e$) and the ionization degree represented in units of temperature ($T_z$), assuming a cosmic abundance. We study the emissivities in the range of $0.1 < kT_e < 10$ keV and $0.1 < kT_z < 100$ keV in every 0.1 keV step. We show representative spectra in figure 2.

The strong RRC compared with the K$\alpha$ line can be formed only in the regime of $T_e < T_z$, recombining plasma condition. We intend to find the condition of the plasma to account for the observed flux ratio ($F_{RRC}/F_{line}$), which is free from specific modeling with iron abundance, the emission measure, the geometry and so forth if the line and RRC are emitted from the same site. Thus, for a given $T_e$ and $T_z$ a priori, we carried out calculations of the emissivity ratio in the above wide range of $T_e - T_z$ space.

Figure 2-c and 2-d are the simulated plasma emissivities with a cosmic abundance to explain the ratio: $F_{RRC}/F_{line}$ of GRB 991216 and GRB 970828 respectively. The ratios of an integrated RRC flux to K$\alpha$ lines of the simulated ionization state are $F_{RRC}/F_{line} \sim 1$ and $\sim 4$ for figure 2-c and 2-d respectively, and the ratios are within the observational values in 90 % statistical error.

### The Reason for Strong RRC and Weak K$\alpha$ Line

To produce the strong RRC in quantum number $n = 1$, the iron must be almost fully ionized. The H-like K$\alpha$ line, which was observed with Chandra, dominates other ionization states at a temperature of $kT_z > 20$ keV. Above the temperature, Fe XXVII consists of more than 70 % of iron, thus we assume $kT_z > 20$ keV in the following discussion. In a condition of high electron temperature of $kT_e \sim kT_z > 20$ keV, i.e., equilibrium ionization, the capture rate of free electrons is small and the emissivity of the line and RRC also becomes small. Moreover, the free-free emission from high $T_e$ electrons dominates at

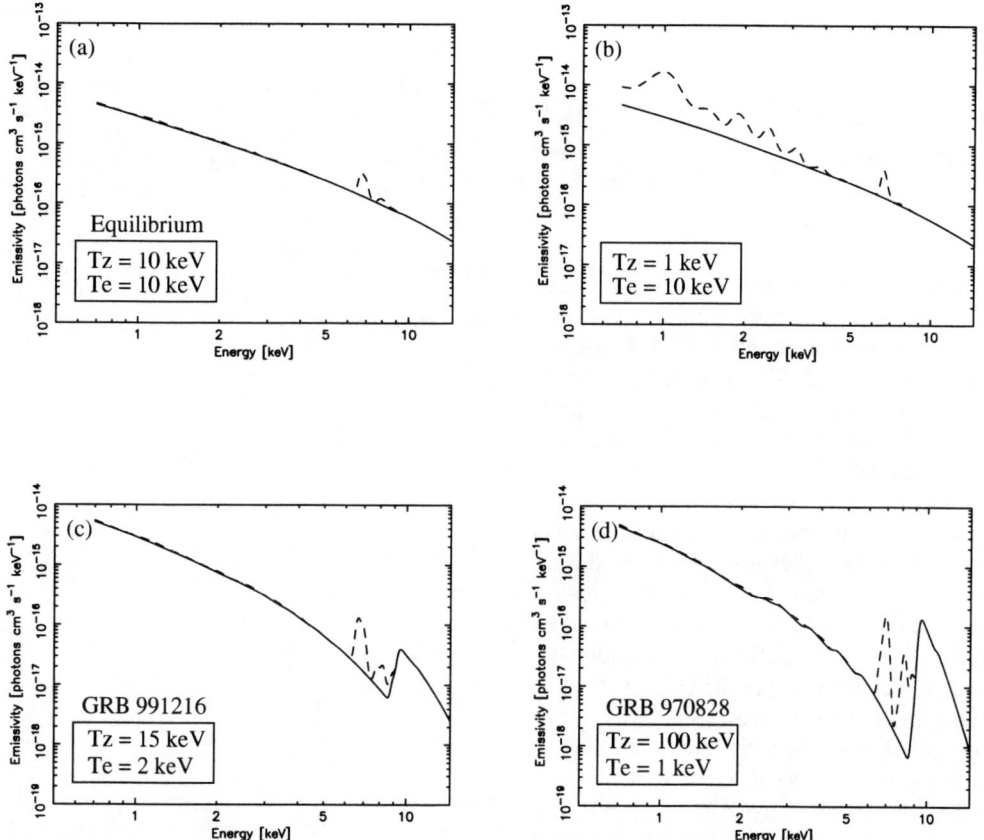

**FIGURE 2.** Simulated emissivities, convolved with the energy resolution of the ASCA-SIS for the cases; (a) $T_z = 10$ keV, $T_e = 10$ keV (equilibrium), (b) $T_z = 1$ keV, $T_e = 10$ keV, (c) $T_z = 15$ keV, $T_e = 2$ keV and (d) $T_z = 100$ keV, $T_e = 1$ keV. The solid lines represent emissivity of continuum and the dotted ones are for emission lines. The figure 1-c and 1-d are simulated for representing the cases of GRB 991216 and GRB970828 respectively only in view point of the observed ratios of $F_{\mathrm{RRC}}/F_{\mathrm{line}}$. We do not intend to reproduce the spectral shapes, which mostly consist of a non-thermal component.

the hard X-ray band. Therefore, the RRC may not be observed because it can be obscured by the free-free component.

The cross section of the electron capture into the $n$th quantum state can be expressed as

$$\sigma_n \propto \frac{1}{n^3}\left(\frac{3}{2}\frac{kT_e}{\varepsilon}+\frac{1}{n^2}\right)^{-1} \quad (1)$$

where $\varepsilon$ is an ionization energy of 9.28 keV for H-like iron [12]. Thus, the best condition to form the strong RRC would be the case of $kT_e \sim \varepsilon (\ll kT_z)$. In such a plasma state, $\sigma_n \propto n^{-3}$ and then most of free electrons recombine directly into the ground state ($n = 1$), compared to the $n \geq 2$ levels. However, free-free emission dominates the continuum.

With decreasing $kT_e$, the recombination rate increases, while free-free emission becomes suppressed. Recombination into $n \geq 2$ increases relatively and produces line emission. Thus, the K$\alpha$ line can be enhanced by cascades from $n \geq 3$ excited levels. This is the case for He-like K$\alpha$, but H-like K$\alpha$ (Ly$\alpha$) is little affected; a considerable fraction comes to direct transition to the ground state.

We summarize the above discussions in view point of the intensity of the RRC and K$\alpha$ line. The plasma state with a strong RRC but a weak K$\alpha$ line, which was observed with ASCA, is realized when $kT_e$ is slightly less than $\varepsilon$ but in high $kT_z$. We calculated the ratio of the RRC to K$\alpha$ lines ($F_{RRC}/F_{line}$) from the numerical emissivities. The condition of $F_{RRC}/F_{line} > 3.3$ of GRB 970828 can be explained by the case around $kT_e = 4$ keV and $kT_z = 100$ keV.

## CONCLUSION

1. Based on the redshift measured by the optical observations, we interpreted the spectral structure of GRB 970828 as a Radiative Recombination Edge and Continuum, and we solved the discrepancy of the redshift between X-ray and optical observations.

2. Although the RRC was identified, only the upper-limit of the iron line flux was settled ($F_{RRC}/F_{line} > 3.3$).

3. With numerical calculations, we pointed out the possibility of a "Non-Equilibrium Ionization State ($T_e < T_z$)" for the line- and RRC-forming region.

More detail discussions are given in [13, 14].

## REFERENCES

1. Paczyński, B. 1998 ApJ, 494, L45
2. Woosley, S. E. 1993 ApJ, 405, 273
3. Vietri M. et al. 1998, ApJ 507, L45
4. Piro, L. et al. 1999, ApJ, 514, L73
5. Yoshida, A. et al. 1999, A & A, supp. 138, 433
6. Piro, L. et al. 2000, Science, 290, 955
7. Antonelli, L. A et al. 2000, ApJ, 545, L39
8. Djorgovski, S. G. et al. 2001, ApJ, (submitted)
9. Yonetoku, D. et al. 2000, PASJ, 52, 509
10. Itoh, H. & Masai, K. 1989, MNRAS 236, 885
11. Masai, K. 1994, ApJ 437, 770
12. Nakayama, M. & Masai, K., 2001, A & A in press
13. Yonetoku, D. et al. 2001, ApJ, 557, L23
14. Yoshida, A. et al. 2001, ApJ, 557, L27
15. Murakami, T., et al. IAUC, No. 6732 (1997)

# The Prompt and Afterglow Emission of GRB 001109 Measured by BeppoSAX

L. Amati*, F. Frontera[†], J.M. Castro Cerón**, E. Costa[‡], M. Feroci[‡], G. Gandolfi[‡], P. Giommi[§], C. Guidorzi[¶], N. Masetti*, E. Montanari[¶], L. Piro[‡], P. Soffitta[‡] and J.J.M. in 't Zand[‖]

*Istituto di Astrofisica Spaziale e Fisica cosmica, CNR., Via Gobetti 101, I-40129 Bologna, Italy
[†]Dipartimento di Fisica, Universita' di Ferrara, Via Paradiso 12, I-44100,Ferrara, Italy
**Real Instituto y Observatorio de la Armada, Sección de Astronomía, 11.110 San Fernando-Naval (Cádiz) Spain
[‡]Istituto di Astrofisica Spaziale e Fisica cosmica, CNR., Via Fosso del Cavaliere, Roma, Italy
[§]ASI Science Data Center, via Galileo Galilei, I–00044 Frascati (RM), Italy
[¶]Dipartimento di Fisica, Universita' Ferrara, Via Paradiso 12, I-44100,Ferrara, Italy
[‖]Space Research Organization Netherlands, Sorbonnelaan 2, 3584 CA Utrecht, The Netherlands

**Abstract.** We present here preliminary results of BeppoSAX measurements of the prompt (2-700 keV) and afterglow (0.1 - 10 keV) emission of the 'dark' GRB 001109. The burst light curves show indication of pulse broadening with the decrease of the energy band and its average spectrum is well described by the standard Band function from X to gamma–rays. The X–ray afterglow emission shows a monotonic power–law decay and its spectrum can be satisfactorily fitted with a photo-electrically absorbed power–law. Both the decay index and the photon index are inside the range of common values found in GRBs afterglows. More detailed data analysis and interpretation is in progress.

## INTRODUCTION

GRB 001109 belongs to the intriguing class of 'dark GRBs', i.e. showing afterglow emission in the X-rays but not at optical or infrared wavelengths. In addition, it could be the closest among those GRBs for which a distance has been estimated. Indeed, the BeppoSAX WFC error box of this event (see next section) was pointed by several optical telescopes, leading to negative results in the optical and near–infrared energy bands [1, 2, 3]. In converse, a previously unknown radio source (VLA J1830+5518) was detected with the VLA [4]. This radio source had a 8.47 GHz flux density of 236±31 microJy 0.4 days after the GRB and its size was less than 0.4 arcsec. Subsequent observations at 4.8 GHz with the WSRT [5], performed 2.2, 4.3 and 10.3 days after the GRB, led to a non detection of radio sources in the error box. Assuming a spectral slope in the usual range (−1 to −0.5) for extra-galactic sources, Rol et al. [5] concluded that the source detected by [4] faded of a factor of ∼2 from the epoch of the VLA observations.
An extended object with constant flux, possibly a galaxy, was detected in a position coincident with that of the VLA radio source with near–infrared and optical observations [3, 6, 7]. The magnitudes of this object are ∼17, ∼18, ∼18.7, ∼20.1 and ∼20.7 in the K, H, J, I and R bands respectively. Combining the results of spectroscopic and astrometric measurements [7, 8] it turns out that, if the VLA radio source is indeed located in this galaxy and related to GRB 001109, the redshift of this event is 0.398 +/- 0.002 .

## BEPPOSAX OBSERVATIONS

GRB 001109 was detected simultaneously by the Gamma–Ray Burst Monitor (GRBM, 40–700 keV, [9]) and the Wide Field Cameras (WFC, 2–28 keV, [10]) aboard BeppoSAX on Nov. 09.3912 UT [11]. The WFC error box (2.5' radius) was pointed by the BeppoSAX Narrow Field Instruments (NFI, [12]) from Nov. 10.0762 (16.5 hr after the gamma-ray burst) to Nov. 10.9847 UT. A previously unknown source lying inside the error box was detected by the Low Energy Concentrator Spectrometer (LECS [13]) and the Medium Energy Concentrator Spectrometer (MECS [14]). The source, designated 1SAX J1830.1+5517, was located at R.A. = 18h30m07s.8, Decl. = +55o17'56" (equinox 2000.0; error radius 50") and had an average count rate of 0.00914±0.0007 count/s in the two MECS units, corresponding to a 2–10 keV flux of $(7.1\pm0.5)\times10^{-13}$ erg cm$^{-2}$ s$^{-1}$ [15]. The flux decreased by a factor of about 2 during the first 20000 s of the observation. A second BeppoSAX NFI observation of the field was performed from Nov. 13.9091 to Nov. 15.0077 UT, during which the average 2–10 keV flux faded to $(1.25\pm0.33)\times10^{-13}$ erg cm$^{-2}$ s$^{-1}$. Due to its position and fading behavior, 1SAX J1830.1+5517 was identified as the X-ray afterglow of GRB 001109.

## PRELIMINARY RESULTS

### Prompt emission

The event had a duration of $\sim 60$ s in both X– and gamma–rays; the 1 s peak flux and the total fluence were $(6.4\pm1.1)\times10^{-8}$ erg cm$^{-2}$ s$^{-1}$ and $(1.73\pm0.09)\times10^{-6}$ erg cm$^{-2}$ in the 2–28 keV energy band, and $(4.2\pm0.3)\times10^{-7}$ erg cm$^{-2}$ s$^{-1}$ and $(4.97\pm0.19)\times10^{-6}$ erg cm$^{-2}$ in the 40–700 keV energy band. As it can be seen from Fig. 1, the light curves in X (2–28 keV) and gamma (40–700 keV) rays are very similar, with indication of pulse broadening with decreasing energy. This is qualitatively in agreement with the predictions of standard optically thin synchrotron shock models for GRB emission. Also, the light curve structure, i.e. one main and distinct pulse followed by a 'train' of weaker pulses, is very similar to that of other GRBs simultaneously detected by WFC and GRBM, like GRB 970228 [16] and GRB 000214 [17]. As for this two events, the emission in the second half of the event is consistent with the hypothesis that it is generated by the same external shock producing the X–ray afterglow emission (see next section). In addition, also GRB 000214 is a 'dark' GRB, i.e. without a detected optical counterpart.

We fitted the WFC + GRBM time–integrated spectrum of GRB 001109 with the Band model [18] (Fig. 2). The best fit parameters values are: $\alpha = -1.15\pm0.11$, $\beta = -2.17\pm0.25$, $E_0 = 85\pm38$ keV and A = $0.34\pm0.12$ photons cm$^{-2}$ s$^{-1}$ keV$^{-1}$ (1$\sigma$ error, reduced chi–squared of 34.3/38). By assuming the possible redshift value of 0.398 (see section 1), a standard cosmological model, isotropic emission and by adopting the technique used by [19] to account for cosmological spectral redshift, from this fit we can derive a 'bolometric' (1-10000 keV) total radiated energy of $(4.0\pm0.2)\times10^{51}$ erg.

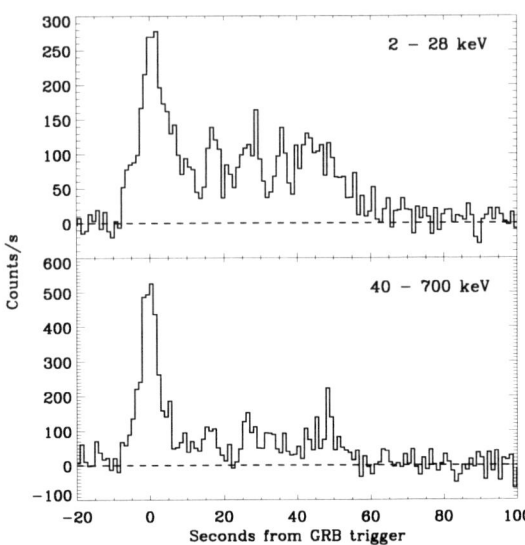

**FIGURE 1.** Top panel: WFC light curve (2–28 keV) with 1 s time resolution. Bottom panel: GRBM light curve (40–700 keV) with 1 s time resolution.

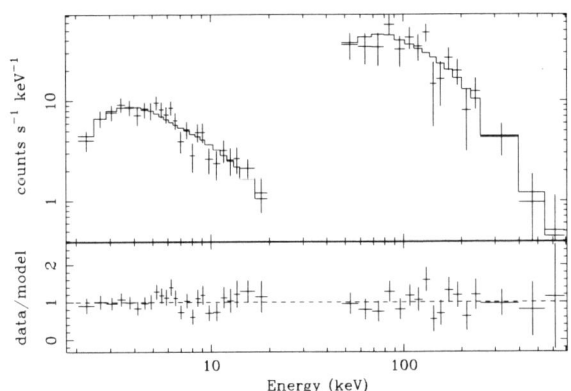

**FIGURE 2.** WFC + GRBM time–integrated spectrum of GRB 001109 fitted with the Band model [18].

### X–ray afterglow

In Fig. 3 we show the 2–10 keV light curve of the afterglow. The index of the best fit power–law is $1.27\pm0.25$ (1$\sigma$ error, reduced chi–squared of 2.4/6). Remarkably, the extrapolation of this law at the time of the prompt emission is consistent with the 2–10 keV average flux measured by the WFC during the second train of pulses. This indicates that the second part of the observed prompt emission could be generated by the same external shock originating the afterglow emission, as in the cases of GRB 970228 and GRB 000214 [16, 17].

The LECS + MECS X–ray afterglow spectrum averaged on TOO1 can be modeled with a simple photoelectrically absorbed power–law with best fit photon

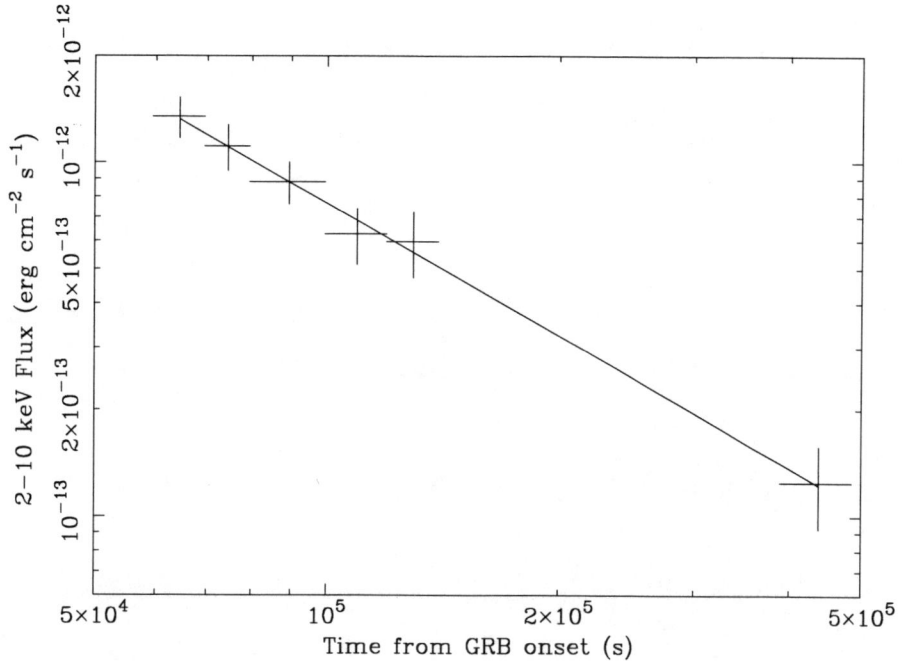

**FIGURE 3.** X–ray afterglow light curve in the 2–10 keV energy band fitted with a simple power–law

index $\Gamma = 2.4\pm0.3$ (1$\sigma$ error, reduced chi–squared of 17.8/18). Remarkably, we find a $N_H$ value is significantly above the galactic one in the direction of the GRB Also, the photon index is consistent with the high energy photon index measured during the last part of the prompt event, supporting the hypothesis of a common origin. The spectrum also shows an indication of an excess around 6.5–7 keV, which is not significant in the TOO1 averaged spectrum but we are investigating whether this feature could be more prominent by accumulating the spectrum over the first part of the observation. The high $N_H$ value is consistent with the non–detection of an optical / NIR counterpart. The presence of an emission feature could challenge the hypothesis of a very low redshift value resulting from spectroscopic measurements of the putative host galaxy.

## ACKNOWLEDGMENTS

This research was partly supported by the Italian Space Agency (ASI). We thank the teams of the BeppoSAX Operative Control Center and Scientific Data Center for their efficient and enthusiastic support to the GRB alert program.

## REFERENCES

1. Rumjantsev, V. et al. 2000, *GCN Circ.* **883** (2000).
2. Stanek, K.Z. et al. 2000, *GCN Circ.* **881** (2000).
3. Vreeswijk, P. et al. 2000, *GCN Circ.* **886** (2000).
4. Taylor, G.B. et al. 2000, *GCN Circ.* **880** (2000).
5. Rol, E. et al. 2000, *GCN Circ.* **889** (2000).
6. Greiner, J. et al. 2000, *GCN Circ.* **887** (2000).
7. Afanasiev, V. et al. 2001, *GCN Circ.* **1090** (2001).
8. Sokolov, V. et al. 2001, *GCN Circ.* **1092** (2001).
9. Costa, E., et al., *Adv.Sp.Res.* **22/7**, 1129 (1998).
10. Jager, R., et al., *Astron. Astrophys. Suppl. Ser.* **125**, 557 (1997).
11. Gandolfi, G. et al. 2000, *GCN Circ.* **878** (2000).
12. Boella, G., et al., *Astron. Astrophys. Suppl. Ser.* **122**, 299 (1997a).
13. Parmar, A.N., et al., *Astron. Astrophys. Suppl. Ser.* **122**, 309 (1997).
14. Boella, G., et al., *Astron. Astrophys. Suppl. Ser.* **122**, 327 (1997b).
15. Amati, L., et al., *IAU Circ.* **7519** (2001)
16. Frontera, F., et al., *Astrophys. J.* **493**, L67 (1998).
17. Antonelli, L.A., et al., *Astrophys. J.*, **545**, L39 (2000)
18. Band, D., et al., *Astrophys. J.* **413**, 281 (1993).
19. Amati, L., et al., *Astron. Astrophys.*, in press (2002)

# V. ASSOCIATIONS WITH GAMMA-RAY BURSTS: SUPERNOVAE, SOURCE ENVIRONMENTS, AND HOST GALAXIES

## Supernova Connection

## Source Environment

## Host Galaxies

# GRB 000911: Evidence for an Associated Supernova?

S. Covino*, D. Lazzati[†], G. Ghisellini**, D. Fugazza**, S. Campana**, P. Saracco**, P.A. Price[‡], E. Berger[‡], S. Kulkarni[‡], E. Ramirez–Ruiz[†], A. Cimatti[§], M. Della Valle[§], S. di Serego Alighieri[§], A. Celotti[¶], F. Haardt[‖], G.L. Israel[††] and L. Stella[††]

*Brera Astronomical Observatory, via E. Bianchi 46, 23807, Merate (LC), Italy.
[†]Institute of Astronomy, Madingley Road CB3 0HA Cambridge, U.K.
**Osservatorio Astronomico di Brera, via Bianchi 46, 23807 Merate (LC), Italy
[‡]Palomar Observatory, 105-24, California Institute of Technology, Pasadena, CA 91125, USA
[§]Osservatorio Astrofisico di Arcetri, Largo E. Fermi 5, 50125 Firenze, Italy
[¶]SISSA/ISAS, via Beirut 4, 34014 Trieste, Italy
[‖]Università dell'Insubria, via Lucini 3, 22100 Como, Italy
[††]Osservatorio Astronomico di Roma, via Frascati 33, 00040 Monteporzio Catone, Italy

**Abstract.** We present photometric and spectroscopic observations of the late afterglow of GRB 000911. We detect a moderately significant re–brightening in the $R$, $I$ and $J$ lightcurves, associated with a sizable reddening of the spectrum. This can be explained through the presence of an underlying supernova, outshining the afterglow $\sim 30$ days after the burst event.

## INTRODUCTION

An anomalous re–brightening was detected in the optical lightcurves of at least two gamma–ray bursts $\sim 30$ days after the burst event (GRB 980326: [1]; GRB 970228: [9], [5]). These have been tentatively interpreted as due to the simultaneous explosion of a supernova (SN) with a lightcurve similar to that of SN1998bw [4] which outshines the afterglow emission at the time of the SN peak.

Here we present the results of a simultaneous multi–filter observational campaign designed at detecting and studying the spectrum of the re–brightening component in the burst of September 11th, 2000. We observed the optical transient (OT) in five filters at three epochs. A low resolution spectrum was taken $\sim 36$ days after the burst explosion. A detailed analysis of these data is provided in [6].

## OBSERVATIONS AND DATA MODELING

We observed the OT associated with GRB 000911 with the ESO/VLT-Antu telescope (instruments, FORS1 and ISAAC), the Keck I telescope (instrument, LRIS) and the MSO 50-inch telescope. Furthermore, we obtained low resolution spectra with the FORS1 instrument (grism 150I, blocking filter OG590) on day 36, around the time of the expected peak SN emission. The resulting spectrum is shown by the filled circles in Fig. 2. The spectrum, despite its very low resolution, was determined to be considerably redder than any afterglow spectrum observed so far. A power–law fit yielded $F(v) \propto v^{-5.3\pm0.8}$ (1$\sigma$ error, $\chi^2 = 22$ for 21 degrees of freedom, hereafter d.o.f.).

We have modeled the data with a composite spectrum given by combining an external shock synchrotron component [7] plus a host galaxy. In addition, we examined the possible role of a supernova component.

The photometric data were dereddened for Galactic extinction and converted to flux densities. Extinction in the host galaxy was not modeled since no additional extinction was required by the data.

The lightcurve modelling was obtained as the sum of the above three contributions. First we considered an external shock afterglow component (hereafter ES) of the form:

$$F(v,t) = A_{ES} v^{-\alpha} t^{-2\alpha}. \quad (1)$$

This equation holds for a jet geometry after the break time [10]. Such a configuration is obtained by a broad-band fitting of the GRB 000911 afterglow [8]. The constant flux from a host galaxy was added in the five bands as a free parameter ($G_B$, $G_V$, $G_R$, $G_I$ and $G_J$).

First, an ES plus galaxy model was fitted to the data. The best fit gave a decay slope $\alpha = 0.724 \pm 0.006$ (temporal slope $\delta = 1.45 \pm 0.012$) with $\chi^2 = 24.4$ for 18

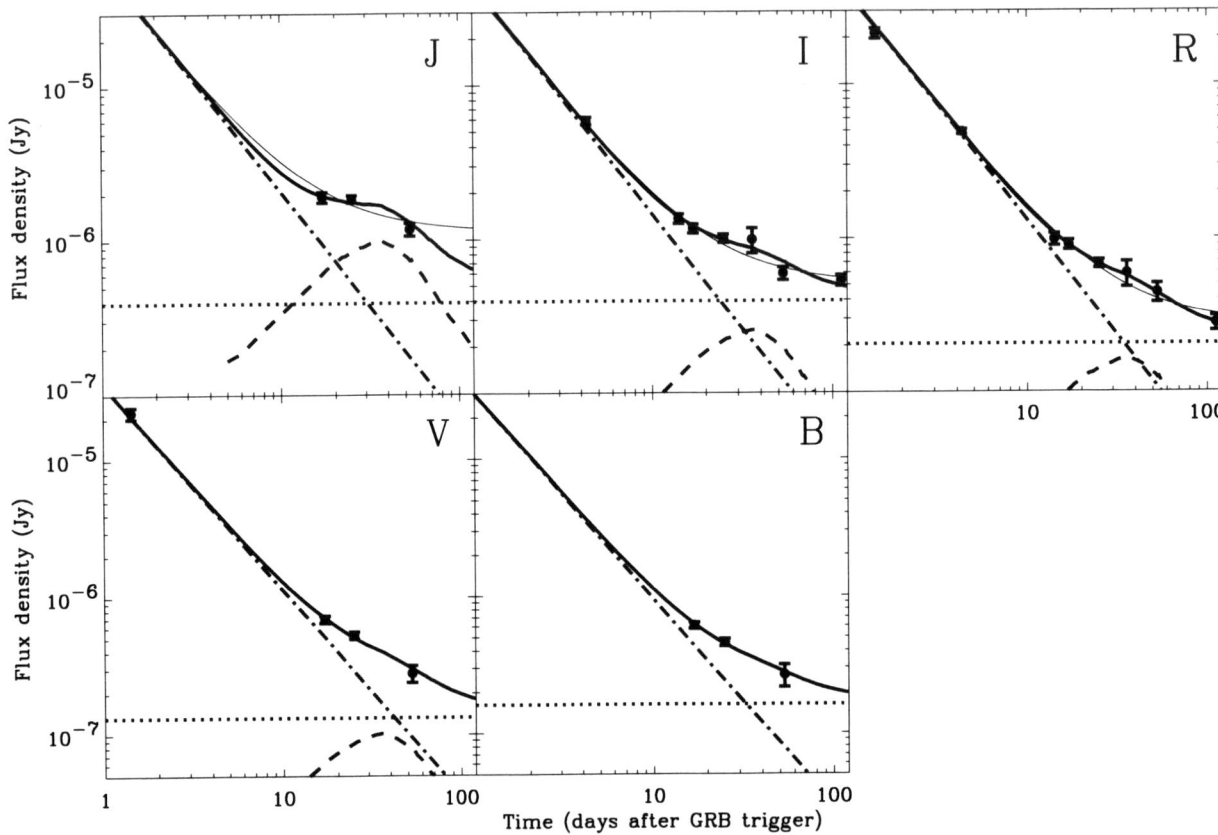

**FIGURE 1.** Lightcurves of the afterglow of GRB 000911. From top left to bottom right, the $J, I, R, V$ and $B$ lightcurves are plotted. The thick solid curves show the best fits obtained with our three component model. The dashed, dotted and dot–dashed lines show the SN, galaxy and ES components, respectively. The thin solid lines in the $J, I$ and $R$ panels indicate the best fit for a model comprising only the galaxy and ES (without SN). The thin line is indistinguishable from the thick solid line for the $V$ and $B$ filters.

d.o.f.. This is an acceptable fit, with chance probability of $P \sim 10\%$ to obtain a higher $\chi^2$ value. However, the fit can be improved by adding the SN component at the redshift $z = 1.06$ of the host galaxy [8]. The lightcurve of the supernova component was obtained by spline interpolation of the data of SN1998bw ($z=0.0085$; [4]). Cosmological parameters $H_0 = 65$ km s$^{-1}$ Mpc$^{-1}$ and $q_0 = 0.5$ were adopted to compute the flux as a function of redshift, and the time profile was stretched by a factor $1+z$. Following [1], the spectrum of the supernova was analytically extended in the rest frame ultraviolet assuming a power–law $F(\nu) \propto \nu^{-2.8}$ for $\lambda < 3600$ Å.

The addition of the supernova component changes slightly the spectral slope of the ES component ($\alpha = 0.748 \pm 0.006$) with $\chi^2 = 13.9$ ($P \sim 67\%$). An F test applied gives a statistical confidence of 99.8% (2.9$\sigma$) for the fit improvement. Alternatively, we allowed the redshift to vary. Interestingly, we obtain $Z_{SN} = 1.1$, in good agreement with the $z = 1.06$ measured spectroscopically by [8] for the host galaxy of GRB 000911.

The re–brightening component in the lightcurve is hence remarkably similar in luminosity, shape and color to the lightcurve of SN1998bw. As a final test, we allowed for a temporal shift $\Delta t$ between the SN and GRB explosions. Keeping all the other SN parameter fixed to the values of SN1998bw and the redshift $z = 1.06$, we obtain $\Delta t = 0^{+1.5}_{-7}$ days (1$\sigma$), showing that the SN explosion may anticipate the GRB but only by $\sim 1$ week (see, e.g., the Supranova model by [11]).

The best fit three component model (ES plus galaxy plus SN) is shown by the thick solid line in Fig. 1. The figure also shows the three individual components (ES, galaxy and SN).

To better constrain the models, we added photometric information to the spectrum. In the upper panel of Fig. 2, the spectrum (filled dots) is plotted together with the $J$ band measurement at day 25 (diamond). Note that the change in the $J$ magnitude from day 25 to day 36 is expected to be small (see the first panel of Fig. 1). In the lower panel, the best fit galaxy components obtained with the lightcurve fitting procedure (from left to right $G_B, G_V, G_R, G_I$ and $G_J$) are shown for the five photometric filters used (triangles). We modeled the ensemble of these data with a galaxy spectrum (templates from [2]) plus a power–law ES spectrum (parameters fixed to the best–fit values from the lightcurve) and a type Ic supernova spectrum a few days after the peak (SN 1987M, [3])

The two models were fitted as follows: the total template spectrum was fitted to the spectral data together with the $J$ band measurement; galaxy magnitudes derived from the template galaxy were fitted to the data in the lower panel. For the galaxy plus ES model, a formal best fit was obtained with a dust–enshrouded starburst galaxy template. The fit gave $\chi^2 = 58$ for 27 d.o.f. ($P \sim 0.05\%$). A better fit ($\chi^2 = 37$ for 27 d.o.f., $P \sim 9\%$) was obtained by adding a SN component, with a moderately dust–enshrouded starburst galaxy ($0.11 < E_{B-V} < 0.21$) as a template. The fit is shown in Fig. 2 overlaid on the data.

In order to obtain a single statistical indicator combining the photometric and spectral information, we finally fitted simultaneously all the available data with the appropriate galaxy template. This yielded $\chi^2 = 44$ (39 d.o.f.) and $\chi^2 = 69$ (40 d.o.f.) for the models with and without SN component, respectively. The $\chi^2$ decrease has a statistical significance of $4\sigma$, according to the F–test.

## CONCLUSIONS

We presented late time multifilter observations of the optical transient associated to GRB 000911. This set of observations was designed to detect and analyze the re-brightening associated with (some) GRB afterglows approximately one month after their explosion ([1]; [9]; [5]). In addition to photometric data, a low resolution spectrum was taken $\sim 36$ days after the burst explosion.

The lightcurve and spectrum were fitted with an external shock plus galaxy model, with the possible addition of a supernova, similar to SN1998bw. The addition of the SN component gives a better fit, with a statistical significance of $4\sigma$.

With the present data, a word of caution should be spent, since is not possible to unambiguously assess the presence of the SN. However, if future observations will allow us to better constraint the magnitude of the host galaxy and will confirm the presence of the rebrightening, we will be able to disentangle the SN component and to provide simultaneous multiband SEDs at the time of our 3 VLT observations. Such time resolved broad band SEDs will allow to better understand the spectral evolution of the bump in the lightcurve and hence to understand its physical origin.

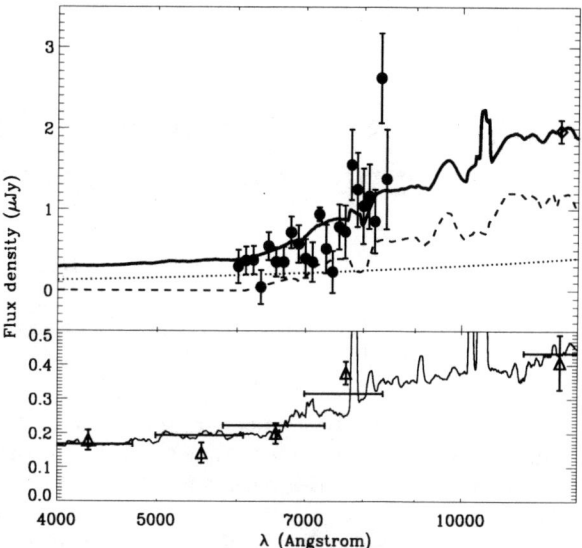

**FIGURE 2.** Spectrum of the OT of GRB 000911 observed 36 days after the burst explosion (upper panel, filled dots). The spectrum is modelled with a SN type Ic spectrum plus a background starburst galaxy and an ES component (see text). The solid line shows the total spectrum smoothed with a 110 Å boxcar filter, while the dashed line represents the supernova component (SN1987M, see text). The lower panel shows the template spectrum of the best fit galaxy model (thin solid line). Triangles are the galaxy photometric measurements as derived from the multiband fitting. The vertical position of the horizontal bars indicate the $BVRIJ$ filter fluxes derived from the galaxy template; their width is equal to the full width at half maximum of the filters.

## REFERENCES

1. Bloom, J. S. et al., 1999, Nat, 401, 453
2. Calzetti, D., Kinney, A.L. & Storchi–Bergmann, T., 1994, ApJ 429, 582
3. Filippenko, A. V., Porter, A. C. & Sargent, W. L. W., 1990, AJ, 100, 1575
4. Galama, T. J. et al., 1998, Nat, 395, 670
5. Galama, T. J. et al., 2000, ApJ, 536, 185
6. Lazzati D., Covino S., Ghisellini G. et al. 2001, A&A 378, 796
7. Meszaros, P. & Rees, M. J., 1997, ApJ, 476, 232
8. Price, P. A., Galama, T. J., Goodrich R. W. & Diercks A., 2000, GCN #796
9. Reichart, D. E., 1999, ApJ, 521, L111
10. Sari, R., Piran, T. & Halpern, J. P., 1999, ApJ, 519, L17
11. Vietri, M. & Stella, L., 1998, ApJ, 507, L45

# Signs and Consequences of a Supernova - Gamma-Ray Burst Association

Ralph A.M.J. Wijers

*Dept. of Physics and Astronomy, Stonybrook University, Stony Brook, NY 11794-3800, USA*

**Abstract.** The energy of gamma-ray bursts, and a variety of evidence suggesting their linkage with blue stars and star formation, has led to the strong suggestion of their connection with deaths or end products of massive stars. Most of this evidence is limited, however, to gamma-ray bursts of the long variety (duration greater than 2 s). For this class, recent evidence more specifically suggests that they are associated with a supernova-like phenomenon. Here this evidence is discussed, and some of its consequences explored. Finally, a specific model for the link between the two is discussed.

## INTRODUCTION

Since the discovery of the distance scale of gamma-ray bursts (GRBs) via their redshifts [1], which became possible due to the discovery of afterglows at X-ray to radio wavelengths [2, 3, 4], rapid developments led to an understanding of those afterglows. It is generally agreed that the expanding blastwave model [5, 6] is the correct explanation of the afterglow phenomenon. However, the origin of the burst itself and even more the nature of the original source of energy remain much less clear. The location of GRBs in distant, blue galaxies, the fact that their space density traces the star formation rate in the Universe fairly well (Fryer, this volume), the fact that they are associated with bright parts of their hosts (Fruchter, this volume), and the fact that they are often significantly extincted all speak to an association with massive stars. Mergers, however, remain a viable option at least for the short category of GRBs (Ruffert, this volume). In this work, I concentrate on the idea that core collapses of some types of massive star are the origin of long gamma-ray bursts (see also Woosley, this volume). For a more extensive review of these developments, see [7, 8].

## EVIDENCE FOR A SUPERNOVA-GRB ASSOCIATION

The first case of a supernova clearly associated with a GRB was supernova 1998bw (Fig. 1). It was found in the error box of GRB 980425 [9], in a galaxy at $z = 0.0085$, some 100 times closer than any other GRB. The associated supernova was of type Ic, and the radio loudest supernova ever. A possible interpretation of the extreme properties is that it was so energetic an explosion that formation of a black hole is energetically required [10], but asymmetry in the supernova itself may reduce the energy requirements to that available from the formation of a (possible ultra-magnetized) neutron star [11]. A similar controversy surrounds the nature of the shock: whereas it was initially claimed that the radio emission requires a relativistic shock [12], it was later shown that a fast but subrelativistic shock would do [13]. This may indicate that the phenomenon is continuous with other normal supernovae, in which evidence from hard X-ray emission and very wide lines sometimes indicates ejecta flowing out at up to tens of percents of the speed of light. The most problematic point, however, in generalizing this event to the origin of GRBs in general is its proximity. The modest gamma-ray flux of the event, combined with its low redshift, make the gamma-ray energy emitted (assuming isotropy) up to $10^5$ times lower than that of regular GRBs at $z \sim 1$. Therefore, it is not at all clear that the gamma-ray event associated with SN 1998bw is physically similar to 'normal' GRBs. To explain the discrepancy, it has been proposed that GRB 980425 was a normal, beamed GRB viewed far off axis (e.g., Woosley, this volume). Another interesting possibility is that the gamma rays are associated with shock break-out of a normal or somewhat energetic supernova, reviving a variety of the oldest GRB model [14] (see also [15]).

The discovery of the GRB 980425/SN 1998bw association inspired a closer look at some other anomalous afterglow events. GRB 980326 had a very rapidly fading afterglow [16]. When the fading behavior leveled off at $R \sim 25.5$, the natural assumption was that the host

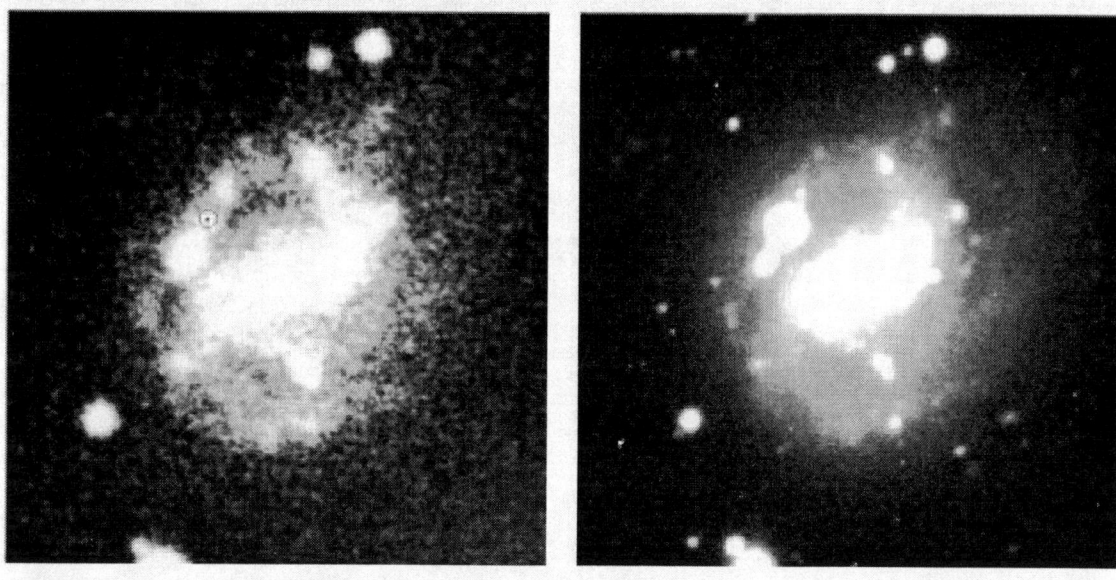

**FIGURE 1.** The supernova 1998bw associated with GRB 980425, before (left) and after (right) the GRB.

galaxy was beginning to contribute significantly to the light. However, two months later the source was undetected at $R > 27.3$, indicating that the light curve had a transient bump instead (Fig. 2). With SN 1998bw in mind, this was interpreted as a possible supernova event, this time in a GRB which was likely at a more usual redshift [17]. Unfortunately, the redshift of the GRB and/or host are unknown, and thus had to be treated as a free parameter. While the light curve agrees with the supernova hypothesis for $z \sim 1$, other explanations could not be excluded. Some natural suggestions have been infrared emission from dust heated by the burst [18] or a reflection echo of the afterglow light off nearby dust [19]; such models may be excluded by more detailed analysis [20].

A close investigation of the first-ever afterglow, of GRB 970228, then shed new light on an existing anomaly in this source. It had been noted early on that while the general light curve agrees very well with a relativistic blast wave model [21], the color of the afterglow at late times is much redder (contrary to those models) and the decay is not quite a pure power law [22]. In this case, the redshift of the host had been measured as $z = 0.695$ [23]. Fits of the data with a power-law decay with the light curve of SN 1998bw added in (scaled to $z = 0.695$) explain the color change quite well [24, 25] (Fig. 3).

There are a number of other claims of associations of less well localized GRBs with peculiar supernovae, or of subtle features in afterglow light curves that might be explained by a supernova contribution, but I consider the above ones be the entire body of evidence. The other claims owe their credibility, if any, to the above cases.

We therefore have one definite GRB-supernova association, which is however with a very special GRB of apparently extremely low energy, and two probable associations with normal GRBs, where either the lack of a redshift or the signal-to-noise prevent us from being too resolute in our inferences. However, the evidence is at worst very tantalizing and certainly justifies exploring the theoretical implications of a supernova connection. The definitive evidence will have to come from the absolute determination, from spectroscopy, that some of the light curve features are indeed supernovae. An issue that will remain unsettled for a rather long time, I fear, is how commonly long GRBs are associated with supernovae. Even if we establish with a few examples that the observed features are supernova signatures, there remains a problem in detecting these features in the first place: GRB 980425 was extremely nearby and there was no normal afterglow. In GRB 980326, either a more typical slow decline of the afterglow emission from the blast wave or a brighter host galaxy would have prevented detection of the supernova bump (all data for this burst are ground based). For GRB 970228, where the feature was barely detected, a somewhat faster decay conspired with late time HST observations to enable the detection: the host is brighter than the supernova, but the HST resolution allows us to separate it from the host. All the present detections therefore have some unusual feature that is not normally present, and even such highly luminous supernovae as SN 1998bw would not be detectable in the majority of afterglow data sets obtained thus far. In these circumstances, any statistical inference on the rate of GRB-

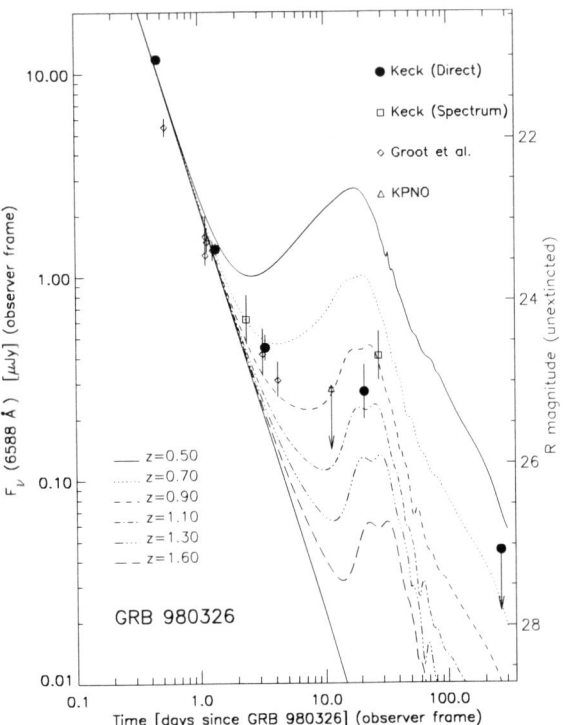

**FIGURE 2.** The $R$ band light curve of the afterglow of GRB 980326 with model fits that include a supernova like 1998bw at varying redshifts (from [17]; reprinted by permission from Nature copyright 1999 Macmillan Publishers Ltd.).

supernova associations is very hard to make.

## MODELS FOR THE CORE-COLLAPSE ORIGIN OF GRBS

Initially, the most popular model for GRBs at cosmological distances was a merger of two neutron stars, or a neutron star and a black hole [26, 27]. Among the reasons for this was the fact that an ultrarelativistic outflow was needed, and this was thought more likely to arise when there was not a large stellar envelope surrounding the explosion site, as is the case in core collapse. Nonetheless, Woosley suggested that one might use stellar core collapses when a black hole forms [28], and later Paczyński raised the possibility that the explosion could be so energetic in that case that even a spherical explosion might lead to ultrarelativistic ejecta [29]. It should be noted that since all these types of model are energetically quite close the energy issue is not a very discriminating one [30], but 'cleanness' of the fireball (i.e., little enough matter entrained in it) may well be. A specific model for formation of beamed GRBs from a rotating helium star collapsing to a black hole was worked out by MacFadyen and Woosley [31], which produces the required properties. Highly magnetized neutron stars may well be similarly successful as GRB central engines [32].

The collapsar model is discussed by Woosley in this volume, so here I concentrate on recent work to create a stellar evolution scenario that creates conditions conducive to the collapsar model. What is required is (i) a helium star, (ii) rapid rotation, and (iii) formation of a black hole. Condition (iii) implies that we need to seek progenitor stars in excess of $20-25M_\odot$. Condition (i) requires either extremely massive stars, which become Wolf-Rayet stars due to losing their hydrogen envelopes in a wind, or a primary star a in close binary, which loses its envelope to a companion when it evolves off the main sequence. Condition (ii) is harder to quantify, since the rotational evolution of stars is quite uncertain. However, if we choose the mass transfer option for condition (i) then we may have some help: angular momentum is lost from the core to the slow envelope during giant expansion, and so removing the envelope would halt the angular momentum loss from the core. Also, if we choose an initially wide binary and remove the envelope via spiral-in, this leads to some rotational spin-up of the core.

Brown and co-workers developed a model in which GRBs are the progenitors of galactic low-mass black hole binaries [33, 34], which are now visible as soft X-ray transients. In this model, a wide binary consisting of a massive O star primary and a low-mass companion is narrowed very much via spiral-in of the secondary when the primary becomes a red supergiant (in order to form a black hole, it must evolve undisturbed until then, i.e., not come into contact during the giant branch, see [34]). The spiral-in leads to a close binary with a rapidly rotating helium star orbited by the original secondary. When its core collapses it forms a rapidly rotating black hole. Assuming the presence of a magnetic field, spin energy will be extracted from the black hole electromagnetically, via the so-called Blandford-Znajek mechanism ([35], see Van Putten, this volume). Along the polar field lines, this powers two beamed outflows and a GRB. The field lines that connect to the equatorial torus of held-up stellar envelope and deposit energy into it thereby power the explosion of the entire envelope, i.e., the hypernova.

The first evidence for this scenario comes from a detailed study of GRO J1655−40 (Nova Sco 1994). It is peculiar among soft X-ray transients in having a very large space velocity. Then it was discovered that its companion has a very anomalous chemical composition [36]: it has a tenfold increase of medium-mass elements such as oxygen, silicon, and sulfur relative to the sun, but a solar abundance of iron. Within our model, we explain the abundances by material deposited onto the companion when the helium star exploded and formed a black hole.

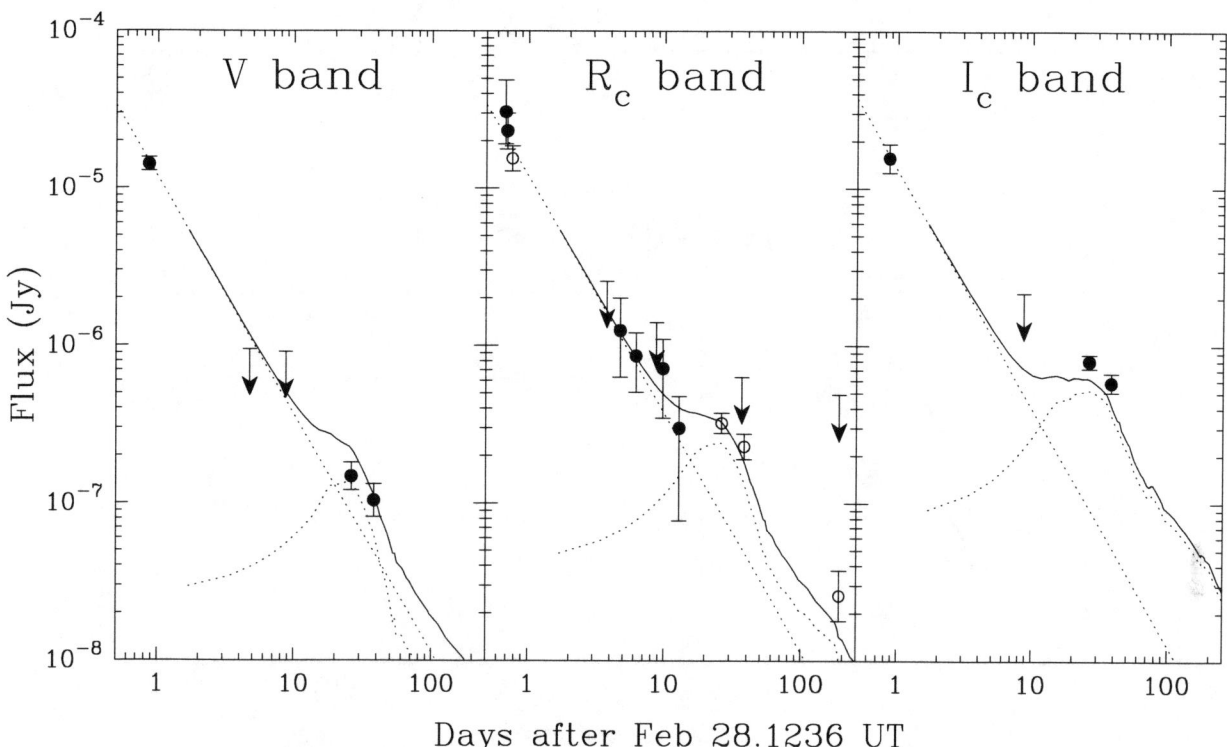

**FIGURE 3.** The *VRI* light curves of the afterglow of GRB 970228 with model fits (solid curves) that include a normal power-law component from a blast wave (dotted straight lines) plus a supernova like 1998bw (dotted curves) at the host redshift $z = 0.695$ [25].

The beamed nature of the explosion assured that the elements formed near the core, such as iron, are mostly ejected along the poles and thus do not find their way to the companion, whereas the elements of the torus do enrich it. At the same time, the ejection of about half the mass of the helium star causes the center of mass of the remaining binary to acquire the large observed space velocity. The abundance of sulfur also has particular relevance. Sulfur is normally produced in much smaller quantities than the alpha-nuclei oxygen and silicon in a supernova. But in 10–100 times more powerful explosions it can be produced almost as much as silicon [37]. We therefore infer that in the past history of GRO J1655−40, a black hole was formed in an asymmetric and unusually energetic explosion of a helium star. We associate this event with the occurrence of a gamma-ray burst.

While tantalizing, the evidence from Nova Sco 94 involves only one of many soft X-ray transients. It is somewhat special in that the companion is a giant star and is more massive than most donors. This means that its brightness and narrow lines make abundance analysis possible, contrary to most SXTs, and the donor mass makes the velocity induced by the mass ejection larger. Therefore, these precise tests of our model are unlikely to be repeated in many other SXTs. However, the model does make another prediction: if the helium star progenitor to the black hole spins rapidly enough, some of its outer layers will be centrifugally prevented from falling into the black hole immediately. Since the GRB mechanism soon starts, this matter will then be ejected before it can accrete onto the black hole. As a result, more rapidly rotating helium stars will leave lower-mass black hole remnants. Making the plausible hypothesis that the helium star spin period before the explosion equals the binary orbital period, we predict a correlation between orbital period and black hole mass in SXTs. This prediction, however, cannot be directly tested by observing the present SXTs, because post-explosion evolution and mass transfer modified both the orbital period and black hole mass. Much of the work is therefore to reconstruct the pre-explosion parameters of the systems via evolutionary calculations, which is only feasible for a subset of the systems. These, however, do agree with the predicted relation, as shown in Fig. 4.

Since the model predicts that the masses of the black hikes are centrifugally limited, much of the mass added to the black hole already has large angular momentum, giving the black hole a significant Kerr parameter. In Fig. 5 we show the expected Kerr parameter as a func-

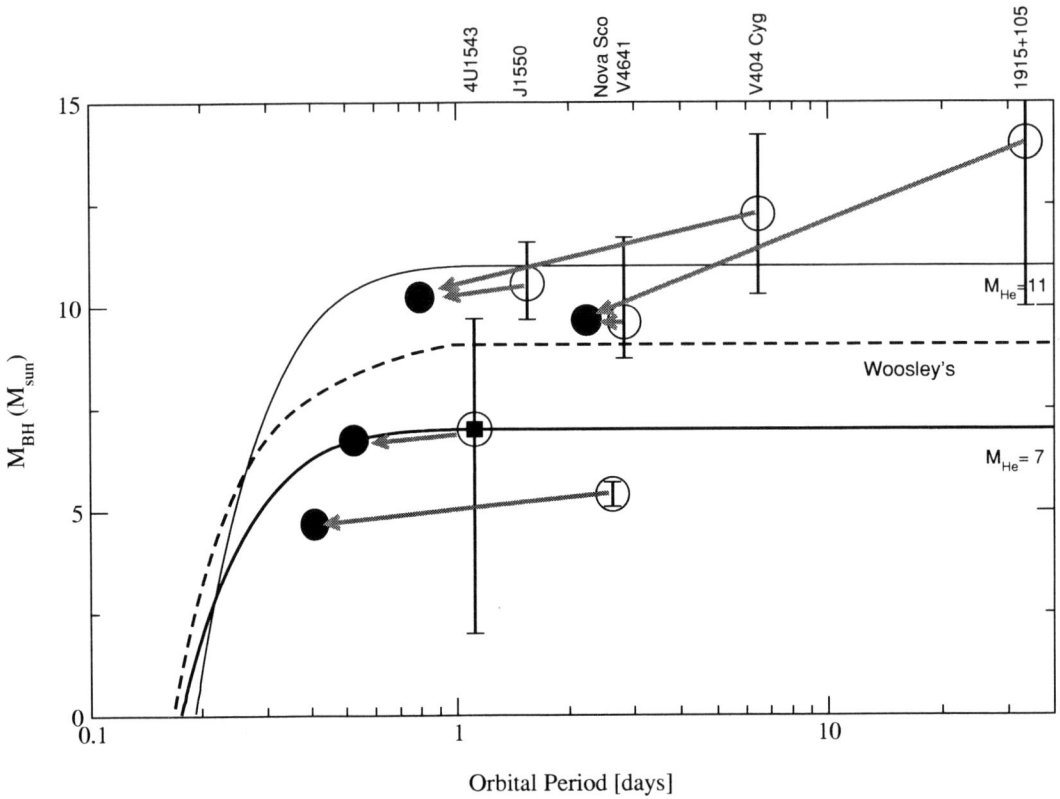

**FIGURE 4.** The correlation between orbital period and black hole mass of SXTs with evolved donors [34]. The correlation is much clearer for the reconstructed pre-explosion parameters (filled symbols) than for the present-day ones (open symbols). The model prediction for the relation depends on the initial helium stars mass. Solid curves are 7 and $11 M_\odot$ simple polytropes, and the dashed line an actual calculated model.

tion of pre-explosion orbital period. For periods less than about 0.5 days, the black holes appear to be good Blandford-Znajek power sources.

## CONCLUSION

The connection between supernovae and GRBs appears to be a very promising one, even though truly conclusive observational evidence still eludes us. Details such as the type of supernova that may cause a GRB and the fraction GRBs accompanied by supernovae are as yet quite unclear. At the same time, the theoretical pursuit of these ideas has led to a very fruitful exploration of stellar explosion mechanisms concomitant of black hole and magnetar formation. Some of the models developed for supernova-related GRBs have actually led to predictions for other astrophysical objects that may be related to GRBs, such as soft X-ray transients and asymmetric supernovae. Slowly, therefore, models are becom-

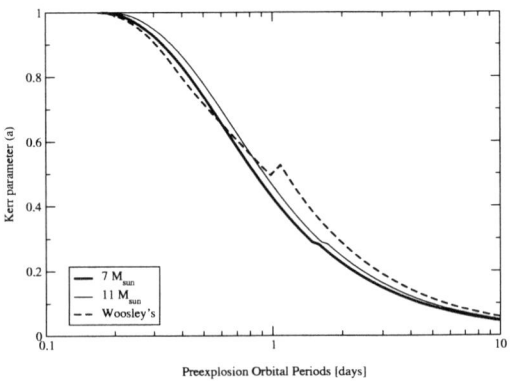

**FIGURE 5.** The predicted Kerr parameter of black holes formed in proto-SXTs [34]. Curves have the same meaning as in Fig. 4, and we see that short-period SXTs make quite good GRB power sources.

ing more constrained. Finally, it is a distinct possibility that GRB research may not only borrow from supernova studies, but also return something: the mechanism for GRBs is much more powerful in terms of explosive yield than normal supernovae. At the same time, the explosion mechanism in normal core-collapse supernovae is not yet well understood. In my view, if core collapses of stars are capable of yielding both normal supernovae and GRBs, then they must be capable of producing any level of energy between those extremes. Therefore, the complete zoo of stellar explosion phenomena may contain many more yet undiscovered types of object. Might it even be the case that the explosion mechanism for regular supernovae needs some re-thinking inspired by the GRB connection?

## ACKNOWLEDGMENTS

I am grateful to my colleagues in the GRACE and HST-GRB collaborations, without whom much of what I wrote would not be known.

## REFERENCES

1. Metzger, M. R., Cohen, J. L., Blakeslee, J. P., Kulkarni, S. R., Djorgovski, S. G., and Steidel, C. C., *IAU Circ.*, **6631** (1997).
2. Costa, E., Frontera, F., Heise, J., Feroci, M., in 't Zand, J., Fiore, F., Cinti, M. N., Dal Fiume, D., Nicastro, L., Orlandini, M., Palazzi, M., E. Rapisarda, Zavattini, G., Jager, R., Parmar, A., Owens, A., Molendi, S., Cusumano, G., Maccarone, M. C., Giarrusso, S., Coletta, A., Antonelli, L. A., Giommi, P., Muller, J. M., Piro, L., and Butler, R. C., *Nature*, **387**, 783–785 (1997).
3. van Paradijs, J., Groot, P. J., Galama, T., Kouveliotou, C., Strom, R. G., Telting, J., Rutten, R. G. M., Fishman, G. J., Meegan, C. A., Pettini, M., Tanvir, N., Bloom, J., Pedersen, H., Noerdgaard-Nielsen, H. U., Linden-Voernle, M., Melnick, J., van der Steene, G., Bremer, M., Naber, R., Heise, J., in 't Zand, J., Costa, E., Feroci, M., Piro, L., Frontera, F., Zavattini, G., Nicastro, L., Palazzi, E., Hanlon, L., and Bennett, K., *Nature*, **386**, 686–689 (1997).
4. Frail, D. A., Kulkarni, S. R., Nicastro, L., Feroci, M., and Taylor, G. B., *Nature*, **389**, 261–263 (1997).
5. Rees, M. J., and Mészáros, P., *Mon. Not. R. Astron. Soc.*, **258**, L41–L43 (1992).
6. Mészáros, P., and Rees, M. J., *Astrophys. J.*, **476**, 232–237 (1997).
7. van Paradijs, J., Kouveliotou, C., and Wijers, R. A. M. J., *Annu. Rev. Astron. Astrophys.*, **38**, 381–427 (2000).
8. Mészaros, P., *Annu. Rev. Astron. Astrophys.*, **40**, in press (2002).
9. Galama, T. J., Vreeswijk, P. M., van Paradijs, J., Kouveliotou, C., Augusteijn, T., Bohnhardt, H., Brewer, J. P., Doublier, V., Gonzalez, J. F., Leibundgut, B., Lidman, C., Hainaut, O. R., Patat, F., Heise, J., In't Zand, J., Hurley, K., Groot, P. J., Strom, R. G., Mazzali, P. A., Iwamoto, K., Nomoto, K., Umeda, H., Nakamura, T., Young, T. R., Suzuki, T., Shigeyama, T., Koshut, T., Kippen, M., Robinson, C., de Wildt, P., Wijers, R. A. M. J., Tanvir, N., Greiner, J., Pian, E., Palazzi, E., Frontera, F., Masetti, N., Nicastro, L., Feroci, M., Costa, E., Piro, L., Peterson, B. A., Tinney, C., Boyle, B., Cannon, R., Stathakis, R., Sadler, E., Begam, M. C., and Ianna, P., *Nature*, **395**, 670–672 (1998).
10. Iwamoto, K., Mazzali, P. A., Nomoto, K., Umeda, H., Nakamura, T., Patat, F., Danziger, I. J., Young, T. R., Suzuki, T., Shigeyama, T., Augusteijn, T., Doublier, V., Gonzalez, J. F., Boehnhardt, H., Brewer, J., Hainaut, O. R., Lidman, C., Leibundgut, B., Cappellaro, E., Turatto, M., Galama, T. J., Vreeswijk, P. M., Kouveliotou, C., van Paradijs, J., Pian, E., Palazzi, E., and Frontera, F., *Nature*, **395**, 672–674 (1998).
11. Höflich, P., Wheeler, J. C., and Wang, L., *Astrophys. J.*, **521**, 179–189 (1999).
12. Kulkarni, S. R., Frail, D. A., Wieringa, M. H., Ekers, R. D., Sadler, E. M., Wark, R. M., Higdon, J. L., Phinney, E. S., and Bloom, J. S., *Nature*, **395**, 663–669 (1998).
13. Waxman, E., and Loeb, A., *Astrophys. J.*, **515**, 721–725 (1999).
14. Colgate, S. A., *Can. J. Phys.*, **46**, S476–S480 (1968).
15. Tan, J. C., Matzner, C. D., and McKee, C. F., *Astrophys. J.*, **551**, 946–972 (2001).
16. Groot, P. J., Galama, T. J., Vreeswijk, P. M., Wijers, R. A. M. J., Pian, E., Palazzi, E., van Paradijs, J., Kouveliotou, C., in 't Zand, J. J. M., Heise, J., Robinson, C., Tanvir, N., Lidman, C., Tinney, C., Keane, M., Briggs, M., Hurley, K., Gonzalez, J. F., Hall, P., Smith, M. G., Covarrubias, R., Jonker, P., Casares, J., Frontera, F., Feroci, M., Piro, L., Costa, E., Smith, R., Jones, B., Windridge, D., Bland-Hawthorn, J., Veilleux, S., Garcia, M., Brown, W. R., Stanek, K. Z., Castro-Tirado, A. J., Gorosabel, J., Greiner, J., Jaeger, K., Bohm, A. B., and Fricke, K. J., *Astrophys. J.*, **502**, L123–L126 (1998).
17. Bloom, J. S., Kulkarni, S. R., Djorgovski, S. G., Eichelberger, A. C., Cote, P., Blakeslee, J. P., Odewahn, S. C., Harrison, F. A., Frail, D. A., Filippenko, A. V., Leonard, D. C., Riess, A. G., Spinrad, H., Stern, D., Bunker, A., Dey, A., Grossan, B., Perlmutter, S., Knop, R. A., Hook, I. M., and Feroci, M., *Nature*, **401**, 453–456 (1999).
18. Waxman, E., and Draine, B. T., *Astrophys. J.*, **537**, 796–802 (2000).
19. Esin, A. A., and Blandford, R. D., *Astrophys. J.*, **534**, L151–L154 (2000).
20. Reichart, D. E., *Astrophys. J.*, **554**, 643–659 (2001).
21. Wijers, R. A. M. J., Rees, M. J., and Mészáros, P., *Mon. Not. R. Astron. Soc.*, **288**, L51–L56 (1997).
22. Galama, T., Groot, P. J., van Paradijs, J., Kouveliotou, C., Robinson, C. R., Fishman, G. J., Meegan, C. A., Sahu, K. C., Livio, M., Petro, L., Macchetto, F. D., Heise, J., in 't Zand, J., Strom, R. G., Telting, J., Rutten, R. G. M., Pettini, M., Tanvir, N., and Bloom, J., *Nature*, **387**, 479–481 (1997).
23. Bloom, J. S., Djorgovski, S. G., and Kulkarni, S. R., *Astrophys. J.*, **554**, 678–683 (2000).
24. Reichart, D. E., *Astrophys. J.*, **521**, L111–L115 (1999).
25. Galama, T., Tanvir, N., Vreeswijk, P., Wijers, R., Groot, P., Rol, E., van Paradijs, J., Kouveliotou, C., Fruchter,

A., Masetti, N., Pedersen, H., Margon, B., Deutsch, E., Metzger, M., Armus, L., Klose, S., and Stecklum, B., *Astrophys. J.*, **536**, 185–194 (2000).
26. Eichler, D., Livio, M., Piran, T., and Schramm, D. N., *Nature*, **340**, 126–128 (1989).
27. Mochkovitch, R., Hernanz, M., Isern, J., and Martin, X., *Nature*, **361**, 236–238 (1993).
28. Woosley, S. E., *Astrophys. J.*, **405**, 273–277 (1993).
29. Paczyński, B., *Astrophys. J.*, **494**, L45–L48 (1998).
30. Mészáros, P., Rees, M. J., and Wijers, R. A. M. J., *New Astron.*, **4**, 303–312 (1999).
31. MacFadyen, A. I., and Woosley, S. E., *Astrophys. J.*, **524**, 262–289 (1999).
32. Wheeler, J. C., Meier, D. L., and Wilson, J. R., *Astrophys. J.*, **in press**, (astro–ph/0112020) (2002).
33. Brown, G. E., Lee, C.-H., Wijers, R. A. M. J., Lee, H.-K., Israelian, G., and Bethe, H. A., *New Astron.*, **5**, 191–210 (2000).
34. Lee, C. H., Brown, G. E., and Wijers, R. A. M. J., *Astrophys. J.*, **submitted**, (astro–ph/0109538) (2002).
35. Blandford, R. D., and Znajek, R. L., *Mon. Not. R. Astron. Soc.*, **179**, 433–456 (1977).
36. Israelian, G., Rebolo, R., Basri, G., Casares, J., and Martin, E. L., *Nature*, **401**, 142–144 (1999).
37. Nakamura, T., Umeda, H., Iwamoto, K., Nomoto, K., Hashimoto, M., Hix, W. R., and Thielemann, F., *Astrophys. J.*, **555**, 880–899 (2001).

# Evidence for Circumburst Extinction of Gamma-Ray Bursts with Dark Optical Afterglows and Evidence for a Molecular Cloud Origin of Gamma-Ray Bursts

Daniel E. Reichart[*] and Paul A. Price[*†]

[*]*Department of Astronomy, California Institute of Technology, Mail Code 105-24, 1201 East California Boulevard, Pasadena, CA 91125*
[†]*Research School of Astronomy and Astrophysics, Mount Stromlo Observatory, Cotter Road, Weston, ACT, 2611, Australia*

**Abstract.** First, we show that the gamma-ray bursts with dark optical afterglows (DOAs) cannot be explained by a failure to image deeply enough quickly enough, and argue that circumburst extinction is the most likely solution. If so, many DOAs will be "revived" with rapid follow up and NIR searches in the HETE-2 and Swift eras. Next, we consider the effects of dust sublimation and fragmentation, and show that DOAs occur in clouds of size $R \gtrsim 10 L_{49}^{1/2}$ pc and mass $M \gtrsim 3 \times 10^5 L_{49}$ M$_\odot$, where $L$ is the luminosity of the optical flash. Stability considerations show that such clouds cannot be diffuse, but must be molecular. Consequently, we compute the expected column density distribution of bursts that occur in Galactic-like molecular clouds, and show that the column density measurements from X-ray spectra of afterglows, DOAs and otherwise, satisfy this expectation in the source frame.

## EVIDENCE FOR CIRCUMBURST EXTINCTION OF BURSTS WITH DARK OPTICAL AFTERGLOWS

Optical afterglows have been detected for about 1/3 of the rapidly-, well-localized gamma-ray bursts (e.g., [1,2]). This data-rich subsample of the rapidly-, well-localized bursts has naturally been the focus of the vast majority of the field's attention and resources over the past five years, but in the end, it is a biased sample. The nature of the dark optical afterglows (DOAs) has only recently become a subject of greater interest, and as might be expected given that these are by definition data-poor events, contradictory initial findings: [1] find that $\gtrsim 3/4$ of the limiting magnitudes for the DOAs are brighter than the detected, but faint, optical afterglow of GRB 000630, and conclude that the DOAs are consistent with a failure to image deeply enough quickly enough. However, [2], using a smaller sample of DOAs, find that the distribution of limiting magnitudes for the DOAs, even if treated as detections, is significantly fainter than the distribution of magnitudes of the detected optical afterglows, and conclude that the DOAs cannot be explained by a failure to image deeply enough quickly enough.

In this section, we first resolve this issue of whether the DOAs can be explained by a failure to image deeply enough quickly enough by applying Bayesian inference, a statistical formalism in which limits can be treated as limits, instead of detections (e.g., [3,4]). Next, we argue that extinction by circumburst[1] dust is the most likely explanation for the vast majority of the DOAs. Finally, we consider the effects of dust sublimation and fragmentation [7,8,9], and place constraints on the sizes and masses of the circumburst clouds of DOAs.

### Evidence that DOAs are Fainter than Detected Optical Afterglows

The samples of [1] and [2] differ in the following ways: The sample of [1] is nearly complete – it contains limiting magnitudes for 95% of the DOAs preceding GRB 000630 – whereas the sample of [2] includes limiting magnitudes for BeppoSAX bursts only. Also, the sample of [2] relies less on GCN Reports, and more on published results. As BeppoSAX bursts appear to have been deeply imaged more often than bursts detected by other satellites, the different findings of these two papers

---

[1] By circumburst, we mean within the circumburst cloud, which is probably parsecs to tens of parsecs across ([5]; see also [6,7]).

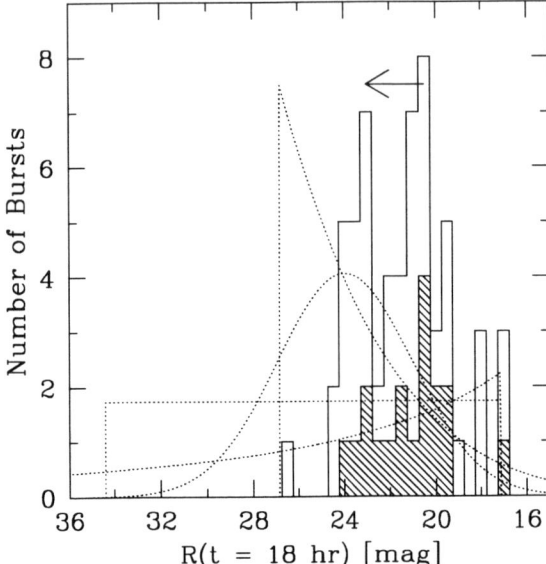

**FIGURE 1.** R-band magnitude distribution of the detected optical afterglows (hashed histogram), and limiting magnitude distribution of the DOAs (unhashed histogram, added to the hashed histogram), scaled to 18 hours after the burst. The dotted curves are the best-fit models (see text) to these data. Clearly, most bursts have afterglows fainter than R ≈ 24 mag 18 hours after the burst, independent of the assumed shape of the brightness distribution. From [10].

can be attributed to sample differences, and the practice of comparing limits to detections, which can turn such sample differences into sample biases, as we demonstrate below.

But first, we combine the samples of [1] and [2], keeping only the bursts through GRB 000630 to maintain the completeness of the [1] sample. As is done in these papers, we scale the data to a common time: We use a temporal index of −1.4, the median temporal index in Table 1 of [2], and we scale the data to 18 hours after the burst, the median observation time in our combined sample. In the event of multiple limiting magnitudes for a single burst, we adopt the most constraining. We plot the combined sample in Figure 1: The hashed histogram is a binning of the R-band magnitude distribution of the detected optical afterglows. The unhashed histogram, which is added to the hashed histogram, is a binning of the limiting R-band magnitude distribution of the DOAs.

Although these distributions appear to be similar, since one is a distribution of limits, they can be consistent only if the vast majority of the DOAs could have been detected had they only been imaged a little more deeply. Being this unlucky, consistently, is of course very improbable. We quantify this improbability with Bayesian inference: Let $n(R)$ be the normalized R-band magnitude distribution of all of the bursts. The likelihood function is then given by

$$\mathcal{L} = \prod_{i=1}^{N} \mathcal{L}_i, \quad (1)$$

where $N$ is the number of bursts in our combined sample, and

$$\mathcal{L}_i = \begin{array}{ll} n(R_i) & (R_i = \text{detection}) \\ \int_{R_i}^{\infty} n(R) dR & (R_i = \text{limit}) \end{array} \quad (2)$$

(e.g., [3,4]). We now consider a wide variety of two-parameter models for $n(R)$: a Gaussian, a boxcar, an increasing power law, and a decreasing power law. We have fitted these models to the data in Figure 1, we plot the best-fit models in Figure 1 (dotted curves), and we list the best-fit parameter values and 68% credible intervals in Table 1 of [10]. Also in Table 1 of [10], we list the fractions $f_{ALL}$ of all bursts, and $f_{DOA}$ of DOAs, that have afterglows fainter than R = 24 mag 18 hours after the burst, the magnitude scaled to this time of the faintest detected optical afterglow in our sample. The results are fairly independent of the assumed shape of the brightness distribution: On average, we find that ≈ $57^{+13}_{-11}$% of all bursts, or ≈ $82^{+22}_{-17}$% of DOAs, have afterglows that are fainter than R = 24 mag 18 hours after the burst. Consequently, the DOAs cannot be explained by a failure to image deeply enough quickly enough: As a whole, they are fainter than the detected optical afterglows.

## Evidence for Circumburst Extinction and Against Alternative Explanations

[11] find that the far-infrared luminosity-to-mass ratio for isolated and weakly interacting spiral galaxies is consistent with the average for Galactic molecular clouds, which suggests that most, if not all, of the star formation in such galaxies occurs in molecular clouds. Since bursts likely occur where massive stars form (see, e.g., [12] for a review of some of the evidence), and since the optical depths through such clouds are typically many magnitudes (e.g., [5]), circumburst extinction is a likely explanation for the vast majority of the DOAs.

[10] put circumburst extinction to a simple test: Considering only (1) strongly collimated bursts that burn completely through the optical depth of their circumburst clouds (e.g., [7,8,9]), and (2) weakly collimated bursts that do not, [10] find that circumburst extinction can explain snapshot optical vs. X-ray and optical vs. radio brightness distributions of all of the afterglows, DOAs and dark radio afterglows included.

We now address a number of alternative explanations:

**Galactic extinction:** [2] show that the distribution of Galactic column densities for the DOAs is consistent

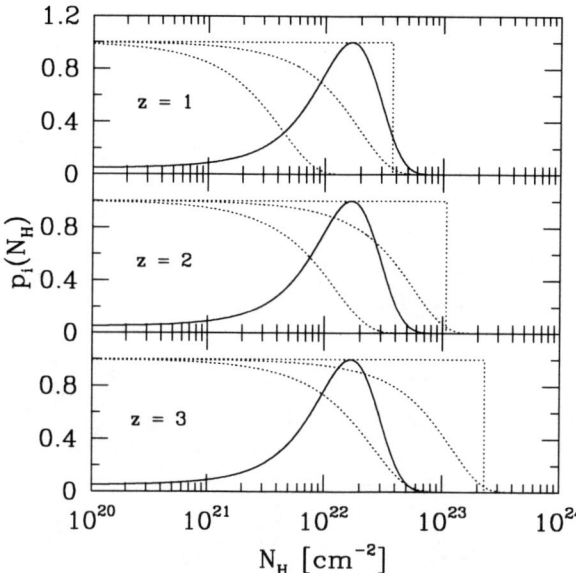

**FIGURE 2.** Source-frame column density probability distributions $p_i(N_H)$ for the four DOAs in Table 1 of [5]. The dotted curves mark bursts of unknown redshift, for which we adopt $z = 1$ (top panel), 2 (middle panel), and 3 (bottom panel). From [5].

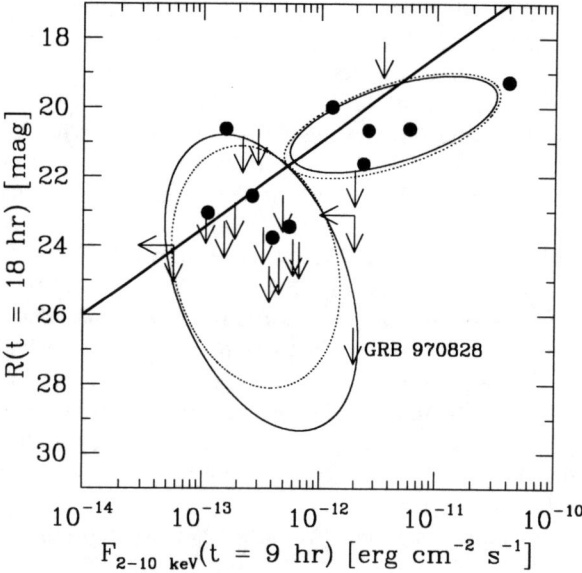

**FIGURE 3.** R-band magnitude or limiting magnitude, scaled to 18 hours after the burst, vs. 2 – 10 keV flux or limiting flux, scaled to 9 hours after the burst. The solid ellipse is from the best-fit model of §5 of [10] to all of the data, and the dotted ellipse is from the best-fit model to all of the data excluding GRB 970828. The ellipses should encompass $\approx 90\%$ of the data. Clearly, GRB 970828 does not dominate the fit, and most of the DOAs are X-ray faint, but given these X-ray fluxes, the DOAs are optically fainter than the expectation from changes in collimation angle and distance alone (solid line). From [10].

with the distribution of Galactic column densities for the bursts with detected optical afterglows, ruling this out as an explanation for the vast majority of the DOAs.

**Host galaxy extinction unrelated to the circumburst medium:** If burst host galaxies are like our galaxy, this cannot explain the vast majority of the DOAs for the very same reason: Most lines of sight through our galaxy do not result in a great deal of extinction.

However, the burst host galaxies might not be like our galaxy: [13] argue that the vast majority of the DOAs might be occurring in the nuclear regions of ultraluminous infrared galaxies (ULIRGs), which have column densities of $N_H \approx 10^{23} - 10^{24}$ cm$^{-2}$. We now test this idea. In Figure 2, we plot the source-frame column density probability distributions of the four DOAs in Table 1 of [5] (see below). Redshifts are not known for three of these bursts, so we adopt $z = 1$ (top panel), 2 (middle panel), and 3 (bottom panel) for these three bursts. Given the redshifts that have been measured for bursts to date, the probabilities that all three of these bursts have $z > 1$, 2, and 3 are 0.27, 0.013, and 0.0016, respectively. Even in the very unlikely event that all three of these bursts have $z \approx 3$, these four probability distributions are not consistent with the expected column density distribution for bursts that occur in the nuclear regions of ULIRGs, but are consistent with the expected column density distribution for bursts that occur in molecular clouds [5].

Another possibility is that burst host galaxies are very dusty. However, if this were the case, one would expect the X-ray bright afterglows to be DOAs as often as the X-ray dim afterglows, which is not the case (Figure 3). However, Figure 3 is naturally explained by circumburst extinction: The X-ray bright afterglows are probably strongly collimated, and consequently the burst and optical flash are more likely to burn through the optical depth of the circumburst cloud; the X-ray dim afterglows are probably weakly collimated, and consequently the burst and optical flash are less likely to do so [10].

**High redshift effects:** Lyman limit absorption in the source frame, absorption by the Lyα forest, absorption by excited molecular hydrogen in the circumburst medium [14], and source-frame extinction by the FUV component of the extinction curve (e.g., [3]) could all result in DOAs if at sufficiently high redshifts. However, [12] show that unless the burst history of the universe differs dramatically from the star-formation history of the universe, the redshift distribution of the bursts should be fairly narrowly peaked around $z \approx 2$, primarily because there is very little volume in the universe at low and high redshifts. [15] model the detection efficiency functions of BeppoSAX and IPN, and show that these satellites should detect even fewer bursts at high redshifts, pushing the expected typical redshift for bursts detected by these satellites down to the observed value of about one. However, based on their variability redshift estimates of 220 BATSE bursts, [16] (see also [17]) find that the burst history of the universe might differ dramatically from the star-formation history of the universe, with very many more bursts at high redshifts. However, using similar variability redshift estimates [4] for 907 BATSE bursts, [18] find that only $\approx 15\%$ of bursts above BATSE's detection threshold have $z > 5$ (and if the luminosity function is evolving, far fewer bursts below BATSE's detection threshold have $z > 5$). Consequently, the above high redshift effects probably affect $\lesssim 10\%$ of the bursts in our BeppoSAX- and IPN-dominated sample. Furthermore, [13] find that the variability redshift estimates for all of the DOAs for which high resolution BATSE light curves are available have $z < 5$.

A number of ad hoc hypotheses are also addressed in [10].

## Evidence for a Large, Massive Cloud Origin of DOAs

Neglecting for a moment extinction exterior to the circumburst cloud, we now consider the optical depth to a burst that is embedded a distance $r$ within its circumburst cloud, along the line of sight. Prior to sublimation and fragmentation [7,8,9], the column to the burst is optically thin, which we take to mean $\tau \lesssim 0.3$, if the hydrogen column density $N_H \lesssim 5 \times 10^{20}$ cm$^{-2}$, and optically thick, which we take to mean $\tau \gtrsim 3$, if $N_H \gtrsim 5 \times 10^{21}$ cm$^{-2}$. We mark column densities with dotted lines in Figure 4. Since sublimation and fragmentation burn through $\approx 10 L_{49}^{1/2}$ pc of optical depth [7,8,9], where $L = 10^{49} L_{49}$ erg s$^{-1}$ is the $1 - 7.5$ eV isotropic-equivalent peak luminosity of the optical flash, the post-sublimation/fragmentation column to the burst is optically thin if $r$ is less than this distance or $N_H \lesssim 5 \times 10^{20}$ cm$^{-2}$, and optically thick if $r$ is greater than this distance and $N_H \gtrsim 5 \times 10^{21}$ cm$^{-2}$. We show this in Figure 4 by plotting $\tau = 0.3$ and 3 for $L_{49} = 0.1$ (thin curves, left), 1 (thick curves), and 10 (thin curves, right; see [7] for details).

Consequently, modulo the value of $L$, bursts that occur to the lower left of the $\tau = 0.3$ curve have either relatively unextinguished optical afterglows, or optical afterglows that are extinguished by dust elsewhere in the host galaxy, or in our galaxy. Bursts that occur to the upper right of the $\tau = 3$ curve have highly extinguished afterglows. Since circumburst extinction appears to be responsible for most of the DOAs, and Galactic extinction and host galaxy extinction unrelated to the circumburst medium account for no more than perhaps a few of the

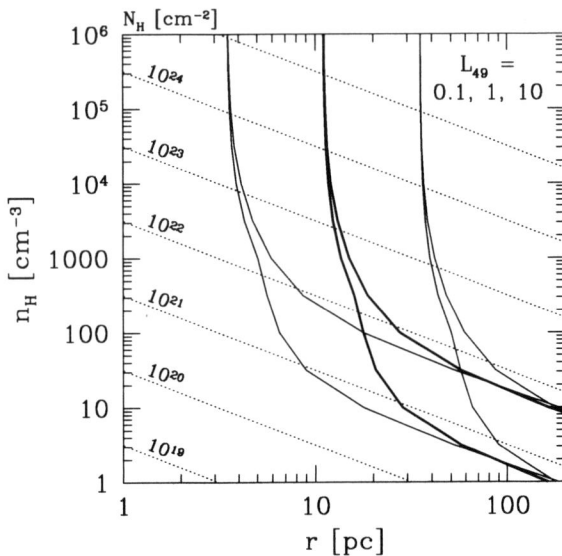

**FIGURE 4.** Post-sublimation/fragmentation optical depth $\tau$ to a burst that is embedded a distance $r$, along the line of sight, within a cloud of constant hydrogen density $n_H$. The three pairs of solid curves mark $\tau = 0.3$ (lower left) and 3 (upper right) for $L_{49} = 0.1$ (left), 1 (center), and 10 (right). For a given $L$, bursts that occur to the lower left of the $\tau = 0.3$ curve have afterglows that are relatively unextinguished by the circumburst cloud, and bursts that occur to the upper right of the $\tau = 3$ curve have highly extinguished afterglows. The dotted lines mark constant hydrogen column densities $N_H$. From [7].

DOAs detected to date, we find that most DOAs have $r \gtrsim 10 L_{49}^{1/2}$ pc and $N_H \gtrsim 5 \times 10^{21}$ cm$^{-2}$.

We now estimate the sizes and masses of the circumburst clouds of the DOAs. Taking the clouds to be spherical, and taking the bursts to be located at the cloud centers, both of which are reasonable approximations on average, we find that most DOAs occur in clouds of radius $R \gtrsim 10 L_{49}^{1/2}$ pc and mass $M \gtrsim 3 \times 10^5 L_{49}$ M$_\odot$ (Figure 4 of [7]). Clouds of this size and mass are typical of giant molecular clouds (e.g., [19]), and are active regions of star formation.

## EVIDENCE FOR A MOLECULAR CLOUD ORIGIN OF BURSTS

In this section, we first consider the equilibrium properties of clouds, and show that clouds of these sizes and masses, and hence the circumburst clouds of DOAs, cannot be diffuse, but rather are probably molecular, in which case they are active regions of star formation. Next, we show that Galactic-like molecular clouds have sufficiently high column densities to make bursts dark at optical wavelengths, and that sublimation of dust by the optical flash and fragmentation of dust by the burst and afterglow can leave a comparable number of bursts – bursts that are sufficiently near the earth-facing side of the cloud and/or in sufficiently small clouds – relatively unextinguished at optical wavelengths, which is in agreement with what is observed. Finally, we model and constrain the column density distribution of the bursts with detected optical afterglows, and show that these bursts are also consistent with a molecular cloud origin, and are not consistent with a diffuse cloud origin. Consequently, we find that all but perhaps a few bursts, dark or otherwise, probably occur in molecular clouds.

## Evidence for a Molecular Cloud Origin of DOAs: Cloud Equilibrium, Gravitational Collapse, and Fragmentation-Driven Bursts of Star Formation

We first consider the equilibrium properties of diffuse clouds: i.e., gas that is bound by pressure equilibrium with a warm or hot phase of the interstellar medium. Specifically, [20] considers the equilibrium properties of a uniformly magnetized, isothermal, non-rotating diffuse cloud, and introduces factors $c_1 = 0.53$ and $c_2 = 0.60$ that correct for the fact that in equilibrium the cloud will not be of uniform hydrogen density $n_H$, but will be centrally condensed, and will not be spherical with radius $R$, but will be moderately flattened along the direction of the magnetic field. [20] shows that clouds with $R > R_m$ are unstable to gravitational collapse, where

$$\frac{GM^2}{R_m^4} = \frac{25 p_m}{1 - (M_c/M)^{2/3}}, \qquad (3)$$

$M = 4\pi\rho R_m^3/3$ is the mass of the cloud, $\rho = \mu_H m_H n_H$, $\mu_H = 1.87$ is the mean molecular weight per hydrogen atom,

$$p_m = \frac{3.15 c_2 (kT/\mu)^4}{G^3 M^2 [1 - (M_c/M)^{2/3}]^3}, \qquad (4)$$

$T$ is the temperature of the cloud, $\mu = 1.44$ is the mean molecular weight,

$$M_c = \frac{0.0236(c_1 B)^3}{G^{3/2} \rho^2}, \qquad (5)$$

and $B$ is the strength of the magnetic field within the cloud. We plot $R_m(n_H)$ for $T = 10$ and $100$ K and $B = 3$ $\mu$G in Figure 5, and for $T = 30$ K and $B = 1$ and $10$ $\mu$G in Figure 2 of [5]: Most Galactic clouds have $10 < T < 100$ K, and $B = 3$ $\mu$G is typical of the Galactic interstellar medium.

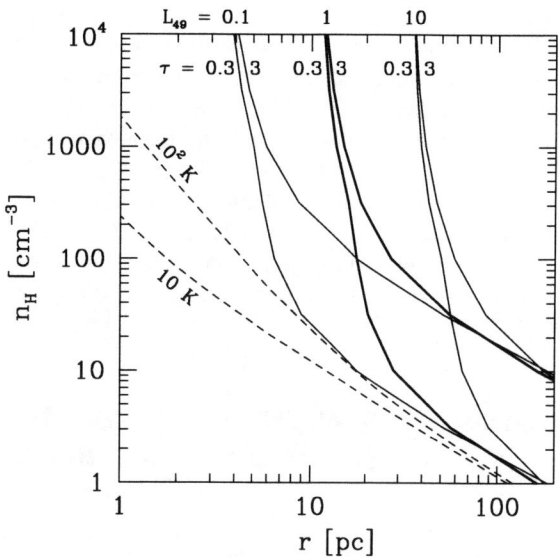

**FIGURE 5.** Post-sublimation/fragmentation optical depth $\tau$ to a burst that is embedded a distance $r$, along the line of sight, within a cloud of constant hydrogen density $n_H$, along the line of sight (solid curves, from Figure 4), and diffuse cloud stability criteria (dashed curves). The dashed curves mark the maximum radius for which a diffuse cloud is stable against gravitational collapse, for cloud temperatures $T = 10$ and $100$ K and cloud magnetic field strength $B = 3$ $\mu$G. From [5].

Also in Figures 5 and 2 of [5], we plot the curves of constant post-sublimation/fragmentation optical depth $\tau$ from Figure 4. Above, we argue that most of the DOAs occur to the upper right of the $\tau = 3$ curve, which implies relatively large sizes and high densities for their circumburst clouds. We show in Figures 5 and 2 of [5] that diffuse clouds of these sizes and densities are unlikely to be in equilibrium, or even near equilibrium, for typical temperatures and magnetic field strengths. Such clouds would be unstable to gravitational collapse, resulting in the collapse and fragmentation of the cloud until star formation re-establishes pressure equilibrium within the fragments, and the fragments are bound by self-gravity: This is precisely the structure of molecular clouds (e.g., [19]). Furthermore, molecular clouds regularly have sizes $R \gtrsim 10$ pc and masses $M \gtrsim 3 \times 10^5$ $M_\odot$, and are active regions of star formation. Consequently, DOAs probably occur in collapsing diffuse clouds/forming molecular clouds, and/or molecular clouds that re-established equilibrium some time ago. We confirm that molecular clouds have sufficiently high column densities to make a large fraction of bursts dark at optical wavelengths below.

[19] find the mass scale of the molecular cloud fragments to be on the order of a few solar masses, which

supports the idea that they are supported against further collapse and fragmentation by star formation. Since the lifetimes of massive stars are comparable to the free fall time of the progenitor cloud for these cloud masses, the first generation of massive star formation and supernovae should occur on this short timescale: $\sim 10^7 - 10^8$ yr. Consequently, DOAs are probably a byproduct of this burst of star formation if the molecular cloud formed recently, and/or the result of lingering or latter generation star formation if the molecular cloud formed some time ago.

## Consistency Check: The Column Density and Optical Depth Distributions of Bursts in Galactic-Like Molecular Clouds

We now consider the column density and optical depth properties of bursts in Galactic-like molecular clouds. First, we compute the mean column densities of the 273 molecular clouds in the sample of [19], which was constructed using data from the Massachusetts-Stony Brook CO Galactic Plane Survey. The mean column density of a cloud is given by $N_H = M/(\mu_H m_H A)$, where $M$ is the virial mass ([19] show that the clouds are in or near virial equilibrium), $A = 11.56(D\tan\sqrt{\sigma_l \sigma_b})^2$ is the effective cross section of the cloud [19], $D$ is the distance to the cloud, and $\sigma_l$ and $\sigma_b$ are the angular sizes of the cloud. We plot the distribution (dotted histogram) of mean column densities in Figure 6, and confirm the finding of [19] that $\mu_H N_H$ is narrowly peaked around $\approx 170$ $M_\odot$ pc$^{-2}$, independent of cloud mass and size.

Next, we correct this distribution for a number of effects. First, since bursts more likely trace cloud mass than cloud number, we weight these mean column densities by cloud mass, and renormalize the distribution (dashed histogram). Since molecular cloud column densities appear to be fairly independent of cloud mass, the distribution is not significantly changed. Also, we weight the $M < 7 \times 10^4$ $M_\odot$ clouds by an additional factor of $(M/7 \times 10^4 M_\odot)^{-3/2}$ to correct for an undercounting of low mass clouds on the far side of the Galaxy [19]. However, this affects the mass-weighted mean column density distribution negligibly.

Finally, we correct for geometrical effects: (1) bursts occur in their circumburst clouds, not behind them; and (2) molecular clouds are centrally condensed. First, we adopt the same density distribution for the clouds that [19] adopted to compute the virial masses: $\rho \propto r^{-1}$ within the effective radius of the cloud, and $\rho = 0$ beyond this radius. This density distribution results in a surface brightness profile that is similar to what is observed [19]. Next, we trace this density distribution with $10^4$ randomly-placed bursts for each cloud, and compute for

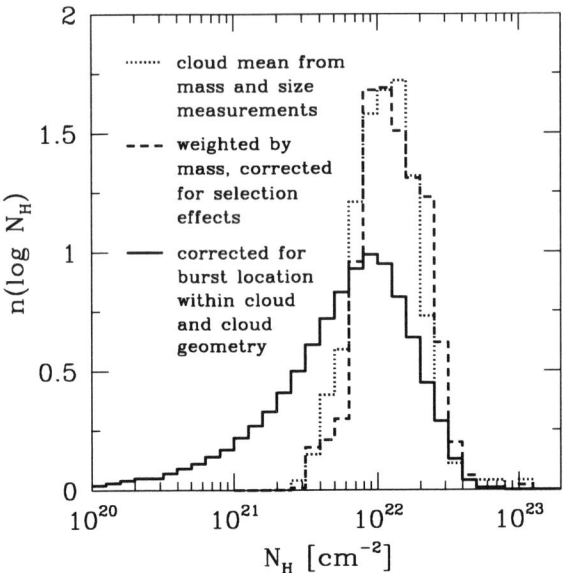

**FIGURE 6.** Expected column density distribution $n(\log N_H)$ (log-normalized to one) for bursts that occur in Galactic-like molecular clouds (solid histogram). The dotted histogram marks the mean column density distribution of the Galactic molecular clouds of [19], which we compute from their mass and size measurements. The dashed histogram marks the mass-weighted mean column density distribution, which we also correct for an undercounting of low mass clouds on the far side of the Galaxy (a negligible correction). The solid histogram marks this distribution after correcting for the fact that bursts occur in their circumburst clouds, not behind them, and for the fact that molecular clouds are centrally condensed. From [5].

each burst the distance and column density to the earth-facing side of the cloud. The effect of placing bursts in the clouds, as opposed to behind the clouds, is to lower the mean column densities on average by a factor of two, and sometimes (when bursts occur near the earth-facing side of the cloud) considerably more. However, this is partially offset by the central condensation of the clouds, which places more bursts in above average density (and column density) environments. We plot the final distribution – the expected column density distribution for bursts in Galactic-like molecular clouds – also in Figure 6 (solid histogram).

Lastly, we confirm that Galactic-like molecular clouds, the column densities of which appear to be fairly universal, at least within the Galaxy, have sufficiently high column densities to make a large fraction of bursts dark at optical wavelengths. To this end, we plot in Figure 7 the expected distribution (1 and 2 $\sigma$ dashed contours) of column densities and distances to the earth-facing side of the cloud for bursts in Galactic-like molecular clouds. Also in Figure 7, we plot the curves of constant post-sublimation/fragmentation op-

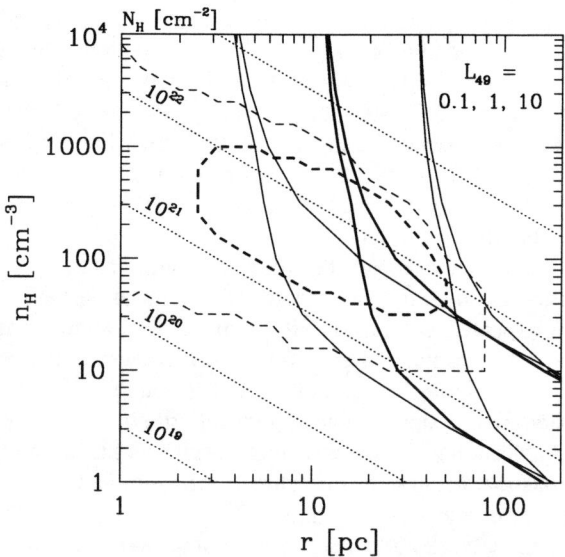

**FIGURE 7.** Expected distribution of column densities and distances to the earth-facing side of the cloud for bursts that occur in Galactic-like molecular clouds (dashed contours), and the curves of constant post-sublimation/fragmentation optical depth from Figures 4 and 5 (solid curves). The thick dashed contour contains 68.3% of the distribution, and the thin dashed contour contains 95.4% of the distribution. The dotted lines mark constant hydrogen column densities. From [5].

tical depth from Figures 4 and 5, and lines of constant column density. It appears that for low values of the isotropic-equivalent peak luminosity of the optical flash ($L_{49} \lesssim 0.1$), Galactic-like molecular clouds can make a large fraction of bursts dark at optical wavelengths. However, even for very low values of $L$, a reasonable fraction of bursts would be left relatively unextinguished at optical wavelengths. Since comparable fractions of DOAs and detected optical afterglows are observed (about 2/3 and 1/3, respectively; e.g., [1,2]), this suggests that the bursts with detected optical afterglows might also occur in molecular clouds. We confirm this by modeling and constraining the distribution of column densities, measured from absorption of the X-ray afterglow, of the bursts with detected optical afterglows below.

## Evidence for a Molecular Cloud Origin for Bursts with Detected Optical Afterglows: The Column Density Distribution

In this section, we model and constrain the column density distribution of the bursts with detected optical afterglows, and compare this distribution to the expected distribution for bursts in Galactic-like molecular clouds (solid histogram of Figure 6). To this end, we have reviewed the literature, and list in Table 1 of [5] observer-frame column densities and uncertainties that have been measured from the X-ray afterglows of 15 bursts, 11 of which have detected optical afterglows (this information is not yet available for the vast majority of the DOAs). Before modeling these data, we convert each measurement $N_{H,i}$ and upper and lower 1-$\sigma$ uncertainty $\sigma_{u,i}$ and $\sigma_{l,i}$ into probability distributions, first in the observer frame, and then in the source frame by correcting the observer-frame probability distribution for the Galactic column density along the line of sight, and the redshift of the burst if known. If $\sigma_{u,i} = \sigma_{l,i}$ (or $\sigma_{l,i} = N_{H,i}$), we take the observer-frame probability distribution to be a Gaussian of mean $N_{H,i}$ and standard deviation $\sigma_{u,i}$: $p_i(N_{H,obs}) = G(N_{H,obs}, N_{H,i}, \sigma_{u,i})$. Otherwise, we take the observer-frame probability distribution to be a skewed Gaussian: $p_i(N_{H,obs}) = G(N^s_{H,obs}, N^s_{H,i}, \sigma^s_i) dN^s_{H,obs}/dN_{H,obs}$, where $s$ is the skewness parameter, and $s$ and $\sigma_i$ are given by solving

$$\begin{aligned}\sigma^s_i &= (N_{H,i} + \sigma_{u,i})^s - N^s_{H,i} \\ &= N^s_{H,i} - (N_{H,i} - \sigma_{l,i})^s, \end{aligned} \qquad (6)$$

i.e., we take the observer-frame probability distribution to be a Gaussian in $N^s_{H,obs}$, which reduces to a Gaussian in $N_{H,obs}$ when $\sigma_{u,i} = \sigma_{l,i}$ ($s = 1$). The source-frame probability distribution $p_i(N_H)$ is then given by substituting $N_{H,obs} = N_{H,MW} + N_H(1+z)^{-2.6}$, where $N_{H,MW}$ is the Galactic column density along the line of sight (interpolated[2] from the maps of [21]), $N_H$ is the source-frame column density, and $(1+z)^{-2.6}$ scales $N_H$ to the observer frame (e.g., [22]). If multiple column density measurements are available for a burst, we average their source-frame probability distributions if the measurements were made from the same data, and we take the product of these distributions if the measurements were made from independent data (GRB 000926). We plot the source-frame probability distributions, peak-normalized to one, for the bursts with detected optical afterglows in the top panel of Figure 8. The dotted curves mark bursts of unknown redshift, in which case we fix $z = 0$: These distributions can be thought of as fuzzy lower limits (see below).

We model the column density distribution of the bursts with detected optical afterglows with a broken power law. We use a three-parameter function for flexibility, and a simple function (as opposed to, e.g., the more elegant skewed Gaussian) so the inner integral of Equation (8) can be evaluated analytically (otherwise,

---

[2] We use the $n_H$ FTOOL at http://heasarc.gsfc.nasa.gov/cgi-bin/Tools/w3nh/w3nh.pl.

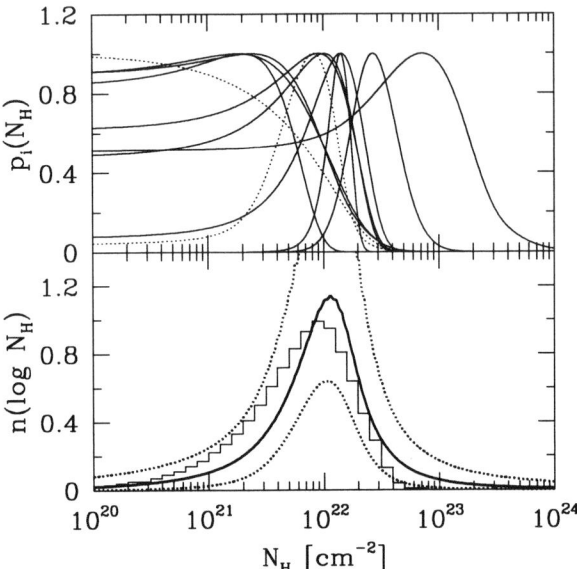

**FIGURE 8.** Top panel: Source-frame column density probability distributions $p_i(N_H)$ (peak-normalized to one) for the 11 bursts with detected optical afterglows in Table 1 of [5]. The dotted curves mark bursts of unknown redshift, for which we fix $z = 0$: These distributions can be thought of as fuzzy lower limits. Bottom panel: Likelihood-weighted average model distribution $n(\log N_H)$ (solid curve), and 1-σ uncertainty in this distribution (dotted curves). The solid histogram is the expected column density distribution for bursts that occur in Galactic-like molecular clouds from Figure 6. These distributions are consistent within the 1-σ uncertainties. From [5].

the error analysis would be computationally taxing). We note that the broken power law is log-normalized $[\int_{-\infty}^{\infty} n(\log N_H) d\log N_H = 1]$ to facilitate comparison of the fitted distribution to the solid histogram of Figure 6 (see below).

We fit the model $n(\log N_H)$ to the data $p_i(N_H)$ using Bayesian inference (e.g., [3]). The likelihood function is given by Equation (1), where $N = 11$ is the number of bursts with detected optical afterglows in our sample, and

$$\mathcal{L}_i = \begin{cases} \int_{-\infty}^{\infty} p_i(N_H) n(N_H) dN_H & (z_i \text{ known}) \\ \int_{-\infty}^{\infty} p_i(N_H') \left[ \int_{N_H'}^{\infty} n(N_H) dN_H \right] dN_H' & (z_i \text{ unknown}) \end{cases} \quad (7)$$

(e.g., [10]), or equivalently

$$\mathcal{L}_i = \begin{cases} \int_{-\infty}^{\infty} p_i(N_H) n(\log N_H) d\log N_H \\ \int_{0}^{\infty} p_i(N_H') \left[ \int_{\log N_H'}^{\infty} n(\log N_H) d\log N_H \right] dN_H' \end{cases}, \quad (8)$$

since $n(\log N_H) = 0$ for $N_H < 0$. We find that $\log(a/\text{cm}^{-2}) = 22.16^{+0.20}_{-0.33}$, $\arctan b = 0.92^{+0.38}_{-0.35}$ ($b \approx 1.31^{+2.25}_{-0.68}$), and $\arctan c = -1.46^{+0.53}_{-0.11}$ [$c \lesssim -1.33$ (1 σ)]. We plot the likelihood-weighted average model

distribution (solid curve), which is also log-normalized, and the 1-σ uncertainty in this distribution (dotted curves) in the bottom panel of Figure 8. We find that the column density distribution of the bursts with detected optical afterglows peaks around $10^{22}$ cm$^{-2}$, and spans a factor of a few to an order of magnitude or so. These findings verify and improve upon the earlier finding of [6] that these column densities span $10^{22} - 10^{23}$ cm$^{-2}$.

Lastly, we compare the fitted column density distribution of the bursts with detected optical afterglows to the expected column density distribution for bursts in Galactic-like molecular clouds (solid histogram, replotted from Figure 6), and find these distributions to be consistent within the 1-σ uncertainties. Furthermore, the fitted distribution is not consistent with the expectation for bursts that occur in diffuse clouds, the canonical column density of which is a few times $10^{20}$ cm$^{-2}$. Consequently, we find that all but perhaps a few bursts, dark and otherwise, probably occur in molecular clouds.

# REFERENCES

1. Fynbo, J. U., et al. 2001, A&A, 369, 373.
2. Lazzati, D., Covino, S., & Ghisellini, G. 2001, MNRAS, submitted (astro-ph/0011443).
3. Reichart, D. E. 2001, ApJ, 553, 235.
4. Reichart, D. E., et al. 2001, ApJ, 552, 57.
5. Reichart, D. E., & Price, P. A. 2002, ApJ, 565, 174.
6. Galama, T. J., & Wijers, R. A. M. J. 2001, ApJ, 549, L209.
7. Reichart, D. E. 2001, ApJ, submitted (astro-ph/0107546).
8. Waxman, E., & Draine, B. T. 2000, ApJ, 537, 796.
9. Fruchter, A. S., Krolik, J. H., & Rhoads, J. E. 2001, ApJ, 563, 597.
10. Reichart, D. E., & Yost, S. A. 2001, ApJ, submitted (astro-ph/0107545).
11. Mooney, T. J., & Solomon, P. M. 1988, ApJ, 334, L51.
12. Lamb, D. Q., & Reichart, D. E. 2000, ApJ, 536, 1.
13. Ramirez-Ruiz, E., Trentham, N., & Blain, A. W. 2002, MNRAS, 329, 465.
14. Draine, B. T. 2000, ApJ, 532, 273.
15. Weinberg, N., et al. 2001, in Gamma-Ray Bursts in the Afterglow Era: 2nd Rome Workshop, ESO Astrophysics Symposia XIX, eds. E. Costa, F. Frontera, and J. Hjorth (Berlin: Springer-Verlag), 252.
16. Fenimore, E. E., & Ramirez-Ruiz, E. 2001, ApJ, submitted (astro-ph/0004176).
17. Schaefer, D. E., Deng, M., & Band, D. L. 2001, ApJ, 563, 123L.
18. Reichart, D. E., & Lamb, D. Q. 2001, in Relativistic Astrophysics: 20th Texas Symposium, AIP Conference Proceedings 586, eds. J. C. Wheeler and H. Martel (Melville, New York: AIP), 599.
19. Solomon, P. M., et al. 1997, ApJ, 478, 144.
20. Spitzer, L. 1978, Physical Processes in the Interstellar Medium (New York: John Wiley).
21. Dickey, J. M., & Lockman, F. J., 1990, ARA&A, 28, 215.
22. Morrison, R., & McCammon, D. 1983, ApJ, 270, 119.

# X-Ray Spectroscopy of Gamma-Ray Bursts: the Path to the Progenitor

Davide Lazzati*, Rosalba Perna[†] and Gabriele Ghisellini**

*Institute of Astronomy, Madingley Road CB3 0HA Cambridge, U.K.
[†]Harvard Smithsonian Center for Astrophysics, 60 Garden street, Cambridge MA, 02138
**Osservatorio Astronomico di Brera, via Bianchi 46, 23807 Merate (LC), Italy

**Abstract.** Despite great observational and theoretical effort, the burst progenitor is still a mysterious object. It is generally accepted that one of the best ways to unveil its nature is the study of the properties of the close environment in which the explosion takes place. We discuss the potentiality and feasibility of time resolved X–ray spectroscopy, focusing on the prompt γ-ray phase. We show that the study of absorption features (or continuum absorption) can reveal the radial structure of the close environment, inaccessible with different techniques. We discuss the detection of absorption in the prompt and afterglow spectra of several bursts, showing how these are consistent with gamma-ray bursts taking place in dense regions. In particular, we show that the radius and density of the surrounding cloud can be measured through the evolution of the column density in the prompt burst phase. The derived cloud properties are similar to those of the star forming cocoons and globules within molecular clouds. We conclude that the burst are likely associated with the final evolutionary stages of massive stars.

## INTRODUCTION

It is widely believed that a good way to understand which is the progenitor of GRBs is by analyzing the properties of the interstellar medium that surrounds the explosion. This is because the fireball early self–similar evolution erases all the traces of its initial condition and hence any pre and during–explosion signature.

The three main classes of burst progenitor can be, in principle, easily distinguished by mean of the properties of their environment. If the burst are due to the merger event of binary coalescing systems of neutron stars [1] they are expected to take place in a uniform low density ($n \sim 0.1 - 10 \, \mathrm{cm}^{-3}$) intergalactic medium. This is due to the fact that the binary system has a long life ($\sim 10^9$ y) and a high proper motion ($v \sim 100 - 1000$ km/s) and can travel out of the original birth place before the merging event (but see Perna et al., this volume)

If the explosion of a GRB is coincident with the explosion of a massive rotating star (hypernovæ or collapsar, Woosley, this volume), it has to be surrounded primarily by the pre–explosion stellar wind. This wind will then impact on the molecular cloud in which the star was born, with a shock contact (terminal) discontinuity [2].

Finally, the bursts may be associated to supernova explosions but with some delay (see the supranova model [3]). In this case the burst should explode in an evacuated cavity, surrounded by a supernova shell and eventually by a molecular cloud medium.

These three radial profiles of the ambient media surrounding GRBs are sketched in the left panel of Fig. 1, where the solid line represents compact mergers and the dashed and dotted lines represent hypernovae and supranovae, respectively.

In principle the density profile can be traced by modelling the afterglow light curves and spectra. One must however be aware that most afterglow data are taken between half a day and several months after the burst explosion. As it is shown in the right panel of Fig. 1, in this period of time the fireball, no matter the progenitor model, runs through a uniform medium, the only difference being the normalization (probably the most uncertain of all the model parameters). The morphological difference of the left panel is then impossible to reconstruct with present day measurements.

There are several alternatives in order to measure the close environment density and structure. One is to be fast. In principle, if one can have (as we will have in the Swift era) detailed early time light curves, the whole radial density structure can be measured. However, it is likely that the emission mechanism of the afterglow gets more complicated as we approach the explosion site: reverse shock emission, late injection of energy from the inner engine and the superposition with the radiation from internal shocks will probably make the modeling of early time afterglows a delicate issue.

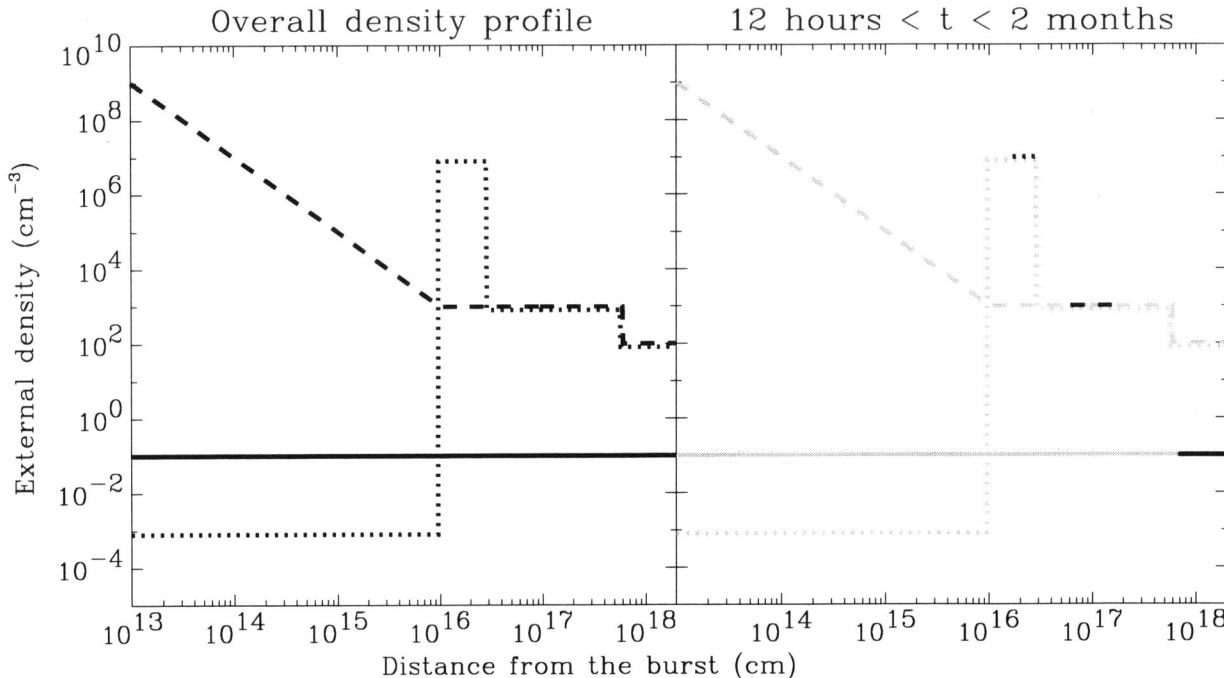

**FIGURE 1.** Radial density profiles for different GRB progenitors. the solid line represent compact mergers and the dashed and dotted lines represent hypernovae and supranovae, respectively. The left panel shows the pre-explosion setup, while in the right panel only the range of radii that the fireball travels between an observer time $t = 12$ hours and $t = 2$ months is highlighted.

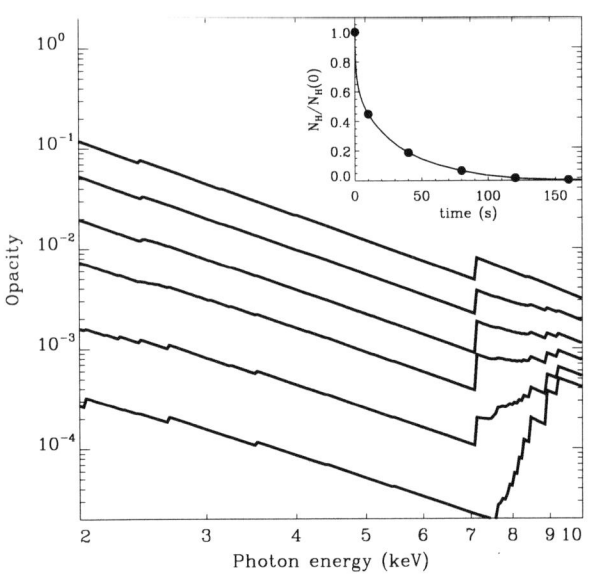

**FIGURE 2.** Opacity in the range [2-10] keV for a cloud with solar metallicity, $R = 3 \times 10^{18}$ cm and initial column density $N_H(0) = 3 \times 10^{21}$ cm$^{-2}$. In the main panel, from top to bottom, we plot the absorption at times $t = 0, 10, 40, 80, 120$ and 160 seconds. In the inset, the column density is shown as a function of time. Filled dots mark the column densities corresponding to the spectra plotted in the main panel.

An alternative is to look for echoes, i.e. photons that, initially emitted at large angles with respect to the line of sight, are scattered in the direction of the observer. Dust [4], Compton [5] and iron line [6] echoes have been proposed. Only iron lines have been securely observed, to date [7]. The modelling of echoes presents two difficulties: first, if GRB fireballs are highly collimated, there is little room for echoes. Secondly, it is difficult to disentangle photons scattered by a large angle at small distance from the burst site from photons scattered at small angles at a larger distance from the progenitor.

We here propose and analyze a method, based on prompt time-resolved X-ray spectroscopy of the burst photons, which is unbiased and rely on very well known physics. The propagation of the photons in the ambient medium will in fact imprint absorption features on the soft X-ray spectra. These features will become less deep as the ionization front expands, allowing us to measure the density and the radial profile of the surrounding material.

## COLUMN DENSITY EVOLUTION

There are in principle two ways of exploiting time resolved X-ray spectroscopy. In case of very good quality data, one can follow the opacity vs. time of a single, well

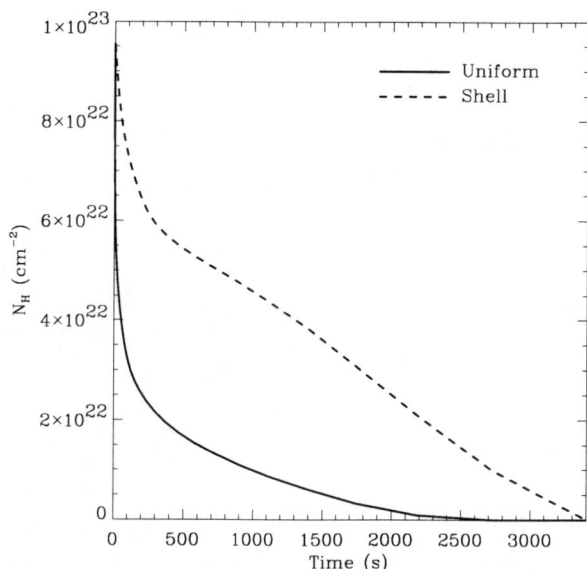

**FIGURE 3.** Evolution of the column density with time for a uniform (solid line) and shell (dashed line) environments. In both cases the initial column density is $N_H(0) = 10^{23}$ cm$^{-2}$ and the outer radius is $R = 1$ pc.

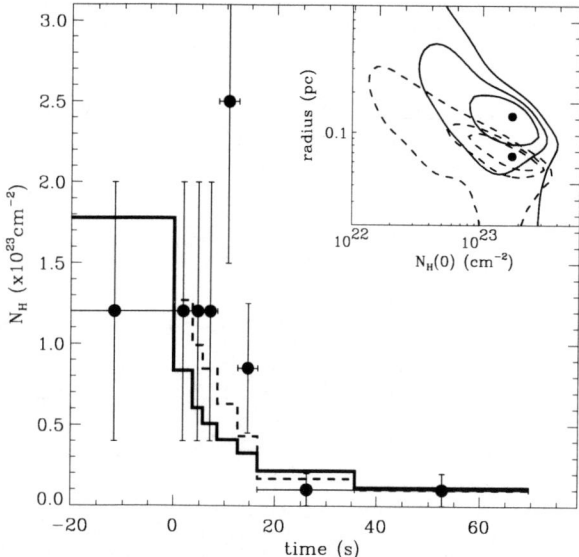

**FIGURE 4.** Evolution of the column density measured in the prompt X–ray spectrum of GRB 980329 [11]. A uniform and shell geometry for the absorbing column have been tested. The best fit models, integrated over the same time intervals of the data, are shown with a solid and a dashed line, for the uniform and shell geometries, respectively. The inset shows the 1$\sigma$, 90 per cent and 99 per cent confidence contours for the two fitted parameters. Again, solid contours refer to uniform density and dashed contours to a shell geometry.

isolated feature. An example is the iron $K_\alpha$ photoionization edge [8, 9] at 7.1 keV. This has the advantage of being completely model independent. On the other hand, the opacity of the iron edge for solar metallicity material is very small, and a Thomson thick cloud is necessary in order to observe a $\tau_{Fe} > 1$ feature.

At lower energies ([0.1-2] keV), the spectra are more crowded and it is very difficult to follow a single transition. However, the opacity is much larger, and even a $\tau_T < 0.01$ cloud can yield a very easily measurable signal. Usually the quantity of absorbing material is parametrized through the quantity $N_H$, i.e. the column of solar metallicity cold material that would absorb the same quantity of soft X–ray photons.

In the case of material surrounding GRBs, the assumption of solar metallicity can be reasonable, but the measured $N_H$ is much different from the real column density due to the progressive ionization the medium undergoes as the burst photons propagate through it. In order to estimate the amount of absorbing material as a function of time (which will show as $N_H$ in the spectra) we run many photoionization simulations (see [10] for more details). Fig. 2 shows the result of one of these simulations, in terms of the frequency–resolved time dependent opacity (main panel) and of the measured column density (inset). The advantage of the method is that different radial density profiles will give different time evolutions of the $N_H$ evaporation. For example, Fig. 3 shows the case of a uniform cloud and of a shell with the same initial column density. The shell material is more difficult to photoion-

ize and the column density evolution is then slower.

## Application to GRB data

Even though with limited spectral resolution and statistical quality, some time resolved column density measurement have been performed with real data. We show in Fig. 4 and Fig. 5 the cases of GRB 980329 [11], observed with *Beppo*SAX and GRB 780506 [12]. In both cases a fairly dense and compact region is derived as a best fit to the data. The radial profile could not be measured since in one case (GRB 980329) the error bars of the measurement were too large and in the other (GRB 780506) only one positive detection was made.

## DISCUSSION

We have shown that time resolved X–ray spectroscopy of the early phases of GRB emission can give us informations on the density and radial structure of the surrounding material. Given the capabilities of present day instrumentation, this can be effectively done in case of fairly dense and compact regions. If we impose that the

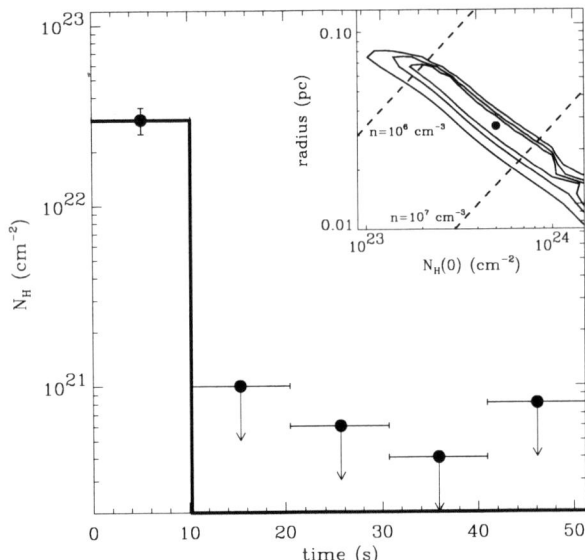
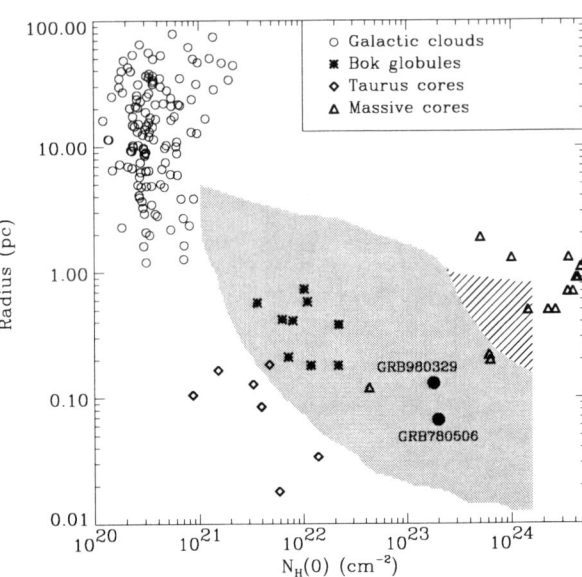

**FIGURE 5.** Same as Fig. 4 but for GRB 780506 [12]. The solid line shows the best fit model with $N_H(0) = 2 \times 10^{23}$ cm$^{-2}$ and $R = 2 \times 10^{17}$ cm integrated over the observed time intervals. The inset shows the 1$\sigma$, 90 per cent and 99 per cent contour levels for the two parameters and, with dashed lines, isodensity contours for $n = 10^6$ and $10^7$ cm$^{-3}$.

**FIGURE 6.** Comparison between the radius and column densities inferred and the properties of Galactic molecular clouds and some hierarchical structures embedded in them. The gray and line shaded areas are where variable column is observable with GRB measurements. Circles show radii and average column densities of a sample of Galactic molecular clouds. Asterisks refer to Bok globules, diamonds refer to dense cores in the Taurus molecular cloud and triangles refer to massive cloud cores. The two filled dots are the best fit values to the $N_H$ measurements of GRB 780506 and GRB 980329.

column density must not be negligible after 1 second of observation and that it must decrease by a factor of two after 100 seconds of GRB emission, we find that, for a uniform absorbing cloud, positive detections of $N_H$ variations should be performed if a GRB is surrounded by a cloud with size and column density marked with the gray shading in Fig. 6.

We compared these cloud properties with typical properties of molecular clouds and their overdense regions in our Galaxy. We find that if GRBs take place in random locations inside molecular clouds, their X–ray early spectra should show no sign of photoionization absorption, since all the material is ionized on a time scale of less than one second. On the other hand, massive stars are thought to be born inside overdense and compact regions within molecular clouds. If GRBs are associated with these regions, evolution of the X–ray absorbing column should be detectable. In particular, the variable absorption observed in the spectra of GRB 980329 and GRB 780506 can be explained if they were located in regions with properties close to those of Bok globules.

## REFERENCES

1. Eichler, D., Livio, M., Piran, T., and Schramm, D. N., *Nature*, **340**, 126–128 (1989).
2. Ramirez-Ruiz, E., Dray, L. M., Madau, P., and Tout, C. A., *MNRAS*, **327**, 829–840 (2001).
3. Vietri, M., and Stella, L., *ApJ*, **507**, L45–L48 (1998).
4. Esin, A. A., and Blandford, R., *ApJ*, **534**, L151–L154 (2000).
5. Madau, P., Blandford, R. D., and Rees, M. J., *ApJ*, **541**, 712–719 (2000).
6. Lazzati, D., Campana, S., and Ghisellini, G., *MNRAS*, **304**, L31–L35 (1999).
7. Piro, L. e. a., *Science*, **290**, 955–958 (2000).
8. Lazzati, D., Perna, R., and Ghisellini, G., *MNRAS*, **325**, L19–L23 (2001).
9. Lazzati, D., Ghisellini, G., Amati, L., Frontera, F., Vietri, M., and Stella, L., *ApJ*, **556**, 471–478 (2001).
10. Lazzati, D., and Perna, R., *MNRAS in press (astro-ph/0110486)* (2001).
11. Frontera, F. e. a., *ApJS*, **127**, 59–78 (2000).
12. Connors, A., and Hueter, G. J., *ApJ*, **501**, 307+ (1998).

# The Role of Dust in GRB Afterglows

## Donald Q. Lamb

*Department of Astronomy & Astrophysics, University of Chicago, 5640 South Ellis Avenue, Chicago, IL 60637*

**Abstract.** We show that the clumpy structure of star-forming regions can naturally explain the fact that 50-70% of GRB afterglows are optically "dark." We also show that dust echos from the GRB and its afterglow, produced by the clumpy structure of the star-forming region in which the GRB occurs, can lead to temporal variability and peaks in the NIR, optical, and UV lightcurves of GRB afterglows. We note that the detection of GRB "orphan" afterglows would provide strong evidence that the star-forming regions in which GRBs occur are clumpy.

## INTRODUCTION

There is increasing evidence that the long gamma-ray bursts (GRBs) are associated with galaxies undergoing copious star formation, and occur near or in the star-forming regions of these galaxies (see, e.g., [1] for a discussion of this evidence). Star-forming regions contain large amounts of dust that can extinguish the optical and UV light of GRB afterglows. Indeed, no optical afterglows have been detected for 60-70% of the long GRBs. Some of these failures may be due to the relatively large size of the GRB error box, or to a delay in observing the error box. Some may be because the GRB lies at a very high redshift, and the Lyman limit lies longward of the optical band [2,3]. However, the majority of the failures are most likely because the optical afterglow is faint or absent due to its extinction by dust in the host galaxy of the GRB [4].

We show that the clumpy structure of star-forming regions can naturally explain the statistics of optically "dark" GRB afterglows. We also show that dust echos from the GRB and its afterglow, produced by the clumpy structure of the star-forming region in which the GRB occurs, can lead to temporal variability and peaks in the NIR, optical, and UV lightcurves of GRB afterglows.

## STRUCTURE OF STAR-FORMING REGIONS

Star-forming regions are thought to be clumpy, with dense dust clouds embedded in a much less dense intercloud medium. A simple model of star-forming regions can therefore be characterized by three parameters: (1) the amount of dust equivalent to a radial optical depth $\tau_H$ of a homogeneous uniform density medium, the volume filling factor $f$ of the dust clumps, and the density contrast $k_1/k_2$ of the clumps relative to the interclump medium [5,6]. In this model, the dense dust clumps form connected structures, as is true in real star-forming regions.

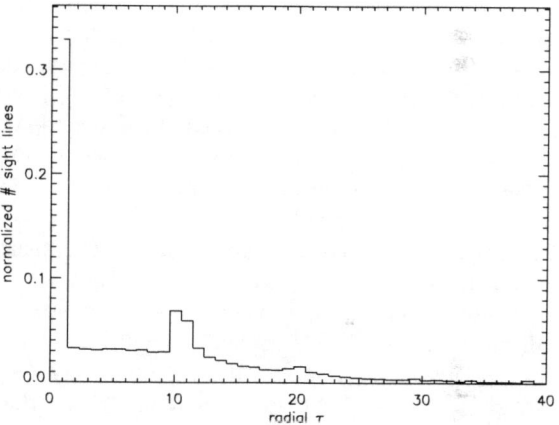

**FIGURE 1.** Distribution of optical depths, as experienced by random photons emitted isotropically by a central source in a clumpy medium with an amount of dust equivalent to a radial optical depth $\tau_H = 10$ in a homogeneous uniform density medium, a volume filling factor $f = 0.10$ of the clumps, and a density contrast $k_1/k_2 = 100$ between the clumps and the interclump medium. From [5].

Figure 1 shows the of a Monte Carlo calculation of the distribution of optical depths for random lines of sight (LOS) from the center of a clumpy star-forming region to an external observer. The parameters of this particular calculation are $\tau_H = 10$, $f = 0.10$, $k_1/k_2 = 100$, and the number $N^3$ of spatial bins is $20^3$ [5]. These parameter values lead to results that are consistent with observations of star-forming regions in the Milky Way.

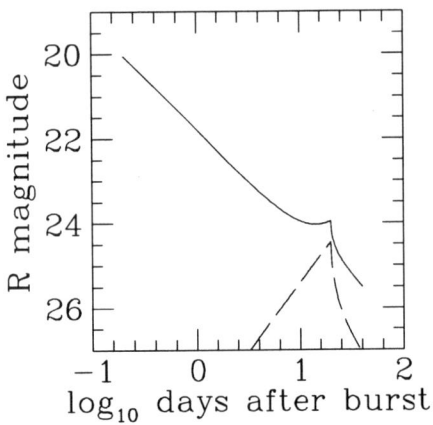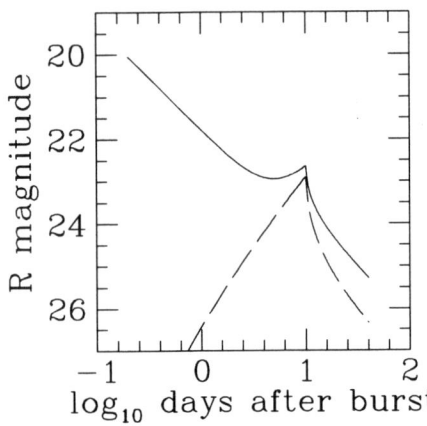

**FIGURE 2.** Contribution to the observed R-magnitude from the light forward-scattered by the dust cloud (dashed line), and the sum of this light and the light seen directly from the point-like source (solid curve) for two sets of parameters of the simple model described in the text. Left: $\delta t = 20$ days $A = 0.2$, $\varepsilon = 0.1$ days. Right: $\delta t = 10$ days $A = 0.5$, $\varepsilon = 0.1$ days.

Figure 1 shows that 35% of LOS have optical depths $\tau_{obs} = 1$ (the minimum value possible in this particular model), while the remaining 65% of LOS have $\tau_{obs} \gg 1$. The distribution of optical depths can be understood as follows. A substantial fraction of the photons emitted by the central source do not encounter a dense clump; these photons correspond to the LOS with $\tau_{obs} = 1$. The remaining photons emitted by the central source encounter one or more clumps and experience $\tau_{obs} \gg 1$. The distribution of $\tau_{obs}$ in this model is consistent with the statistics of GRB optical afterglows, 50-70% of which are optically "dark." It also implies that GRB "orphan" afterglows can exist, whereas models in which the star-forming region is approximately uniform do not.

## DUST ECHOS

Dust echos from the clumpy structure of the star-forming region can also produce variability and peaks $1^d - 30^d$ after the GRB in the NIR, optical, and UV lightcurves of GRB afterglows. As an illustrative example, we have calculated the optical light curve produced by the dust echo from a single clump at a substantial distance from the burst source but along the LOS from the burst source to the observer. The delay time $\delta t$ between the GRB and the echo is characterized by $\delta t = (1+z)R_{\perp,min}^2/cD$, where $D$ is the distance between the GRB and the dust cloud and $R_{\perp,min}$ is the minimum of the perpendicular extent of the dust cloud $R_{\perp,cloud}$, $\theta_{jet}(t)D$, and $\theta_{forward}D$. Here $\theta_{jet}(t)$ is the half opening angle of the afterglow jet and $\theta_{forward}$ is the half angular width ($\approx 10° - 20°$) of the forward scattering peak for scattering by dust grains.

The amplitude $A$ of the dust echo relative to the direct light from the afterglow is a function of the albedo $a$ of the dust, the scattering phase function $\Phi(\theta)$, and the optical depth $\tau_{clump}$ of the dust clump, all of which are wavelength dependent. For forward scattering, which we assume here, $\Phi(\theta) \approx 1$ [7]. Then in the limiting cases of small and large optical depths, $A \approx a\tau_{clump}$ and $A \approx a$, respectively. The prominence of the dust echo also depends on the time $\varepsilon$ after the GRB that the afterglow begins.

Figure 2 shows the dust echo from a single clump along the LOS from the burst source to the observer, assuming $\tau_{clump} < 1$. In the two cases shown, the rate of temporal decline of the GRB afterglow was taken to be a power law with $b = 1$. Most GRB afterglows decline more rapidly with time (i.e., $b = 1.3 - 2.25$). The dust echo is more prominent for afterglows that decline more rapidly, since the contrast between the direct light from the afterglow and the echo – which reflects the brightness of the afterglow at an earlier time – is then larger. Thus the examples we have shown are conservative.

Studies of the temporal variability of the NIR, optical, and UV lightcurves of the afterglows of GRBs may allow "reverberation mapping" of the structure of the star-forming regions in which GRBs occur.

## REFERENCES

1. Lamb, D. Q. 2000, Phys. Reports, **333**, 505
2. Fruchter, A. S. 1999, ApJ, **512**, L1
3. Lamb, D. Q. & Reichart, D. E. 2000, ApJ, **536**, 1
4. Reichart, D. E. & Yost, S. A. 2001, ApJ, submitted (astro-ph/0107545)
5. Witt, A. N. & Gordon, K. D. 1996, ApJ, **463**, 681
6. Witt, A. N. & Gordon, K. D. 2000, ApJ, **528**, 799
7. van de Hulst, H. C. 1957, Light Scattering by Small Particles (New York: Wiley), Chapter 9

# Observational Properties of Afterglows from Mergers of Compact Objects

Rosalba Perna* and Krzysztof Belczynski*

*Harvard-Smithsonian center for Astrophysics, 60 Garden Street, Cambridge, MA 01238, USA

**Abstract.** Data accumulated over the years have shown that the distribution of the GRB time duration is bimodal. While there is some evidence that long bursts are associated with star-forming regions, nothing is known regarding the class of short bursts. Their very short timescales are hard to explain with the collapse of a massive star, but would be naturally produced by the merger of two compact objects, such as two neutron stars (NS), or a neutron star and a black hole (BH). As for the case of long bursts, afterglow obervations for short bursts should help reveal their origin. By using the updated population synthesis code *StarTrack* ([1]), we simulate a cosmological population of merging NS-NS and NS-BH, and compute the distribution of their galactic off-sets, the density distribution of their environment, and, if indeed associated with GRBs, their expected afterglow characteristics.

## INGREDIENTS OF THE CALCULATION

### Population synthesis code

To represent the evolution of single stars, we employ the analytic formulae as in [2]; with these we are able to calculate the evolution of stars for Zero Age Main Sequence (ZAMS) masses $0.5-100 M_\odot$ and for metallicities: $Z = 0.0001 - 0.03$. Stellar evolution is followed from ZAMS through different evolutionary phases depending on the initial stellar mass: Main Sequence, Hertzsprung Gap, Red Giant Branch, Core Helium Burning, Asymptotic Giant Branch, and for stars stripped off their hydrogen-rich layers: Helium Main Sequence, Helium Giant Branch. We end the evolutionary calculations at the formation of a stellar remnant: a white dwarf, a neutron star or a black hole.

We use Monte Carlo techniques to model the evolutionary history and coalescence rates of binary GRB progenitors. The initial parameters of a given binary are drawn from assumed distributions. As we are interested only in NS-NS and NS-BH systems we evolve only massive binaries, with primaries more massive than 5 $M_\odot$. We generate a large number ($N \geq 10^6$) of primordial binaries and evolve them until formation of remnant system. During the evolution of every system we take into account the effects of wind mass-loss, asymmetric SN explosions, binary interactions (conservative/non-conservative mass transfers, common envelope episodes) on the binary orbit and the binary components. Once a binary consists of two compact remnants, we calculate its merger lifetime, the time until the components merge due to gravitational radiation and associated orbital decay.

Our NS-NS population includes the new subpopulation with very short lifetimes ($\sim 1$ Myr) identified by [1], which dominates (81%) the total population. This population merges well within the host galaxies.

### Galaxy model

We describe the potential of a spiral galaxy as the sum of three components: a bulge, a disk and a dark matter halo. The potential of the disk and the bulge is modeled according to [3]: $\Phi_{b,d}(R,\eta) = GM_{d,b}/\sqrt{R^2 + (a_{b,d} + \sqrt{z^2 + c_{b,d}^2})^2}$ where $a_{b,d}$ and $c_{b,d}$ are parameters (which depend on whether one considers the bulge or the disk), $M_{d,b}$ is the mass either of the bulge or the disk, $R = \sqrt{x^2 + y^2}$ is the coordinate in the plane of the disk, and $\eta$ is the coordinate in the plane perpendicular to the disk.

The dark matter halo is described by an isothermal sphere with core radius $r_0$. For a galaxy of mass $M$, the parameters $a_{b,d}$, $c_{b,d}$ and $r_0$ are scaled on the Milky Way values under the assumption of constant surface brightness.

The density distribution of the three components is determined according to the above equations.

At a redshift $z$, the probability that a merger occurs in a galaxy of mass $M$ is given by $P_{\text{merg}}(M,z)dM = AP_{\text{gal}}(M,z)MdM$, where $A$ is a normalization factor and the probability $P_{\text{gal}}(M,z)$ of finding a galaxy of mass $M$ at redshift $z$ is computed using the Press-Schecter formalism. To each galaxy, a random inclination with respect to the line of sight to the observer is assigned. We consider galaxy masses in the range $\{10^8 - 10^{11} M\odot\}$.

Figure 1 shows the probability distribution for the locations of merger events at various redshifts and for three values of the mass of the host galaxy. Both classes of NS-NS and NS-BH binaries are considered.

## GRB afterglow

We adopt the model as in [4], where the flux is described by a broken power law at frequencies $\nu_m$ (synchrotron frequency) and $\nu_c$ (cooling frequency), and the peak value of the spectral distribution, $F_{\nu,\max} = 110\, n^{1/2} \xi_B^{1/2} E_{52} d_{28}^{-2} (1+z)$ mJy, is obtained when the observing frequency $\nu = \nu_m$ in the slow cooling regime, and when $\nu = \nu_c$ in the fast cooling regime.

## RESULTS

The results of the simulations for a randomly generated sample of 10,000 merger events (for each of the two progenitor types NS-NS and NS-BH) are displayed in Figures 2-4. In particular, Fig. 2 shows the distribution of physical offsets $\theta = d_{\text{proj}}/d_A$, where $d_A$ is the angular diameter distance and $d_{\text{proj}}$ is the distance of the events from the center of their hosts, projected into the plane perpendicular to the observer. Given their rather short lifetime, NS-NS events occur rather close to the galactic centers. The typical ambient densities in which the events occur are shown in Figure 3. NS-NS mergers probe typical ISM densites, while a substantial fraction of NS-BH events occurs in a very low-density environment. Finally, the last two figures show, for both the case of NS-NS mergers (Fig. 4), and NS-BH mergers (Fig. 5), the integrated afterglow flux in the 2-10 keV energy band at the observation times $t_{\text{obs}} =$ 1hr, 3hr, 12hr after the burst. Given the detection thresholds of the current X-ray instruments *SAX* and *CXO*, and the upcoming *Swift* (in the range of a few $\times \{10^{-15} - 10^{-14}\}$ erg/cm$^2$/s), a sizeable fraction of events should be detectable if observed within the first few hours.

## SUMMARY

We have studied the properties that a population of GRB events due to mergers of compact objects should have with special emphasis on the related afterglows. By using

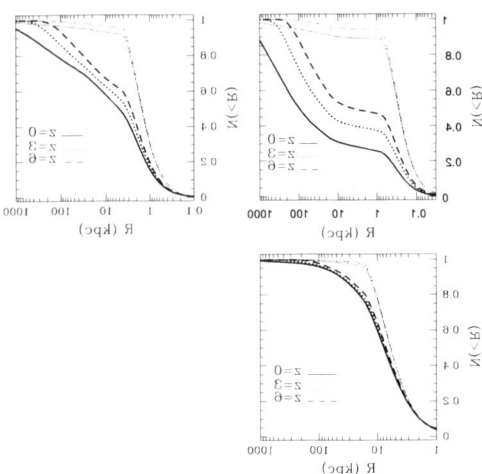

**FIGURE 1.** Distribution of distances of NS-NS (red lines) and NS-BH (blue lines) mergers from the host galaxy center for a face-on galaxy at different redshifts (obtained with the population synthesis code *StarTrack*). The mass of the galaxy is $M = 10^8 M_\odot$ (top left panel), $M = 6 \times 10^9 M_\odot$ (top right panel), and $M = 10^{11} M_\odot$ (bottom panel). The NS-NS population is dominated by the subpopulation [1] of short-lived objects. Note the evolution with redshift of the probability distribution for the distances. It is particularly important to take this into account for long-lived binaries in small galaxies.

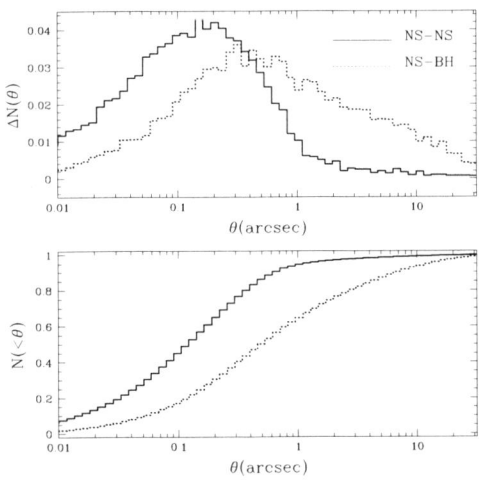

**FIGURE 2.** Distribution of offsets from the centers of the host galaxies for the two populations of NS-NS mergers and NS-BH mergers.

a Monte Carlo type approach, our simulations take into account the mass distribution of the host galaxies as a function of redshift, as well as the redshift evolution of the probability distributions for the location of the mergers within galaxies of various mass (cfr. Fig. 1). This

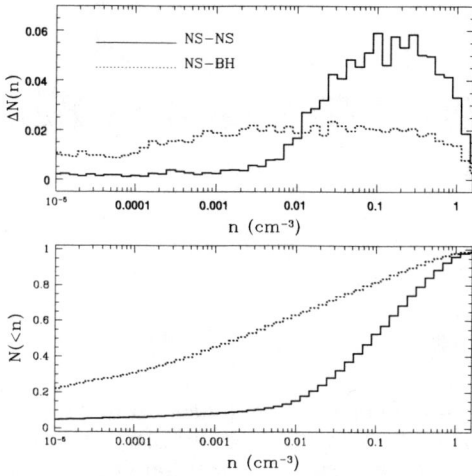

**FIGURE 3.** Density distribution for the environment in which the two populations of NS-NS mergers and NS-BH mergers are expected to occur.

last effect needs to be taken into account especially for binaries whose merger time can be comparable with the Hubble time for a sizeable fraction of them (such as the NS-BH population).

Our population of double neutron star binaries includes the new groups of short-lived binaries [1], which dominate the merger rates. Therefore our results regarding this population differ from previous studies on the same subject, and the derived distributions trace rather closely the star forming regions in the disk. The densities in which they occur are typical ISM densities; hence their afterglows, even though dimmer due to the smaller energy released, should still be observable with current X-ray instruments for a large fraction of them. Afterglows produced as a result of NS-BH mergers are even dimmer, but most of them should still be detectable if observed within the first few hours.

## REFERENCES

1. Belczynski, T., Bulik, and Kalogera, V., *APJ* (2001).
2. Hurley, J. R., Pols, O. R., and Tout, C. A., *MNRAS*, **315**, 543–569 (2000).
3. Miyamoto, M., and Nagai, R., *PASJ*, **27**, 533–543 (1975).
4. Sari, R., Piran, T., and Narayan, R., *APJL*, **497**, L17 (1998).

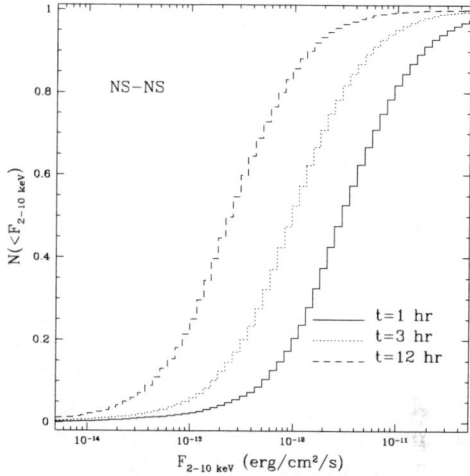

**FIGURE 4.** Distribution of the integrated afterglow flux in the 2-10 keV energy band at several observation times after the burst for the populations of NS-NS mergers. At early times, a considerable fraction of events should be detectable with current X-ray instruments.

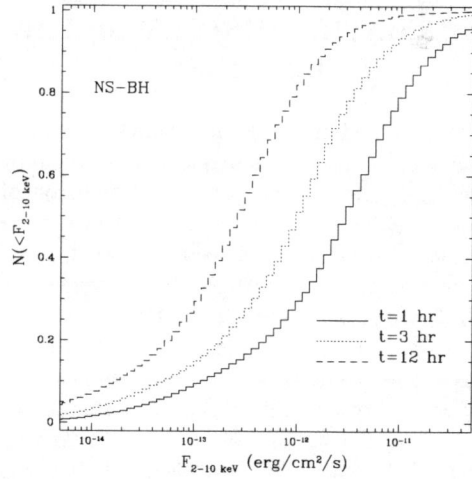

**FIGURE 5.** Same as in Fig.4, but for the population of NS-BH mergers.

# Radio, Sub-mm, and X-Ray Studies of Gamma-Ray Burst Host Galaxies

### E. Berger

*Palomar Observatory, California Institute of Technology 105-24, Pasadena, CA 91125*

**Abstract.** The study of gamma-ray burst (GRB) host galaxies in the radio, sub-mm, and X-ray wavelength regimes began only recently, in contrast to optical studies. This is mainly due to the long timescale on which the radio afterglow emission decays, and to the intrinsic faintness of radio emission from star-forming galaxies at $z \sim 1$, as well as source confusion in sub-mm observations; X-ray observations of GRB hosts have simply not been attempted yet. Despite these difficulties, we have recently made the first detections of radio and sub-mm emission from the host galaxies of GRB 980703 and GRB 010222, respectively, using the VLA and the SCUBA instrument on JCMT. In both cases we find that the inferred star formation rates ($\sim 500$ $M_\odot$) and bolometric luminosities (few $\times 10^{12}$ $L_\odot$) indicate that these galaxies are possibly analogous to the local population of Ultra-Luminous Infrared Galaxies (ULIRGs) undergoing a starburst. However, there is a modest probability that the observed emission is due to AGN activity rather than star formation, thus requiring observations with Chandra or XMM. The sample of GRB hosts offers a number of unique advantages to the broader question of the evolution of galaxies and star formation from high redshift to the present time since: (i) GRBs trace massive stars, (ii) are detectable to high redshifts, and (iii) have immense dust penetrating power. Therefore, radio/sub-mm/X-ray observations of GRB hosts can potentially provide crucial information both on the nature of the GRB host galaxies, and on the history of star formation.

## BACKGROUND: THE STAR FORMATION HISTORY OF THE UNIVERSE

The formation and evolution of galaxies from high redshift to the current time is a major focus of modern cosmology. This involves mapping the conversion of gas into stars, and the associated buildup of heavy elements. The former is succinctly parameterized by using the light from massive stars as a surrogate for the star-formation rate (SFR; see [1]) and the latter has focused on studies of the intergalactic medium (e.g. [2]) and damped-Ly$\alpha$ systems (e.g. [3]). GRBs can contribute to both these important areas through multi-wavelength studies of their host galaxies, and through absorption spectroscopy of their optical afterglows. Here I address the first issue.

Adelberger & Steidel (2000)[4] provide a balanced review of the various diagnostics to measure the evolution of SFR: optical (rest-frame UV), far-infrared (FIR), sub-millimeter, and decimeter radio measurements. Each of these techniques has its distinct strengths *as well as* weaknesses. For example, radio measurements offer superb astrometry but the current VLA sensitivity is only able to identify the tip of the star-formation iceberg [5]. Optical surveys offer the highest sensitivity but are vulnerable to dust extinction and may well miss galaxies forming stars at the most prodigious rate. The FIR and sub-mm approach has maximal sensitivity to dusty galaxies, but it lacks astrometric precision thereby creating a non-trivial bottleneck of requiring detection at other wavelengths; in particular detection by the VLA.

The principal question is the following: Given that each of these techniques provides a restricted view of the cosmic SFR history, and that the optical/UV technique has provided by far the largest sample, can we conclude that optical/UV studies have more or less accounted for the bulk of the cosmic star formation?

Adelberger & Steidel (2000) seem to think so; a number of other authors, especially those using long wavelength techniques (e.g. [6, 7]) are of the opinion that this issue is not settled. In the near term, we have reached a stalemate since sub-mm surveys have reached the confusion limit. SIRTF, through its IRAC survey, and ALMA, with its unprecedented $\mu$Jy sensitivity will certainly contribute to this critical sub-field of modern astronomy. Nonetheless, since the bulk of the stellar energy is effectively radiated in the FIR band, all these techniques require significant extrapolation to measure the true power radiated by galaxies. In the distant future, one can envis-

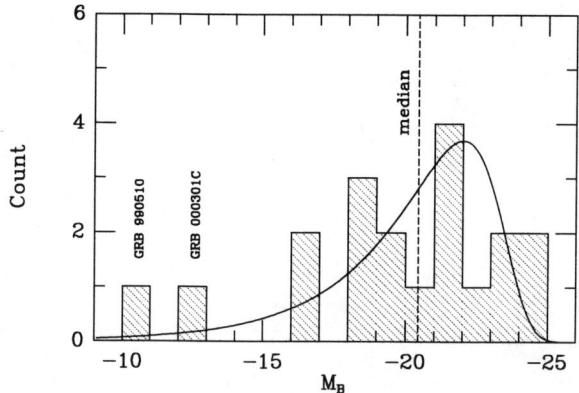

**FIGURE 1.** Histogram of estimated absolute B-band magnitudes of GRB host galaxies with known redshifts. These rest-frame magnitudes were computed from the observed R-band magnitudes by approximating the galaxy spectra as $f_\nu \propto \nu^{-1}$. The sample median is $M_B = -20.4$ mag. The solid curve is a heuristic model representing a luminosity-weighted Schechter function with $M_* = -23$ mag and $\alpha = -1.6$.

age FIR interferometers as providing the most decisive picture of the cosmic evolution of stellar energy density.

## GRB HOST GALAXIES: STRENGTHS

The host galaxies of GRBs offer a unique perspective into the SFR history of the Universe for the following reasons:

1. The existing data show excellent circumstantial evidence linking GRBs to massive stars (e.g. [8]). More to the point, every well-studied GRB so far has been identified with a host galaxy (Fig. 1).
2. GRBs are so bright that they are detectable to redshifts > 20 (should they exist; [9]). Thanks to the broad-band afterglow spectrum (X-ray through decimeter radio) not only can the host be accurately localized, but the redshift can also be obtained. Usually $z$ is obtained via optical spectroscopy, though with GRB 990705 [10] we now have a redshift from X-ray spectroscopy of the afterglow.
3. The immense dust-penetrating power of GRBs (only limited by Compton thick column densities) results in a sample of galaxies that is independent of the global dust properties.
4. Again thanks to the afterglow, a GRB host galaxy need only be detected via imaging (with $z$ via absorption spectroscopy) and this has resulted in the faintest luminosity distribution of star-forming galaxies with some hosts 12 magnitudes below $L_*$ (Fig. 1). In contrast, current state of the art optical/UV/NIR surveys reach $\sim L_*$.

## GRB HOST GALAXIES: POSSIBLE DRAWBACKS

It is clear that the sample of GRB hosts offers powerful diagnostics in our quest to decipher the SFR history of the Universe. However, it has two limitations. First, we have to assume that the GRB rate is linearly proportional to SFR. The circumstantial evidence for the association of GRBs with massive stars, and hence SFR is good [11, 12, 8, 13, 14]. Second, the GRB sample is quite small, especially when compared to the optical/UV sample.

However, the first problem is not as severe as one may think at first glance. All techniques used so far – optical/UV, ISOCAM, sub-mm and decimeter – require large extrapolations (and implicitly, constancy of spectral energy densities) to obtain the bolometric power. Converting this uncertain bolometric estimate to SFR requires detailed assumptions of the IMF of stars and the distribution of ISM in these distant galaxies. The severity of the second problem diminishes when one realizes that the number of *securely* identified sub-mm galaxies is, as of August 2001, only four! (M. Longair, talk at ESO Lighthouse conference).

## THE ORIGIN OF RADIO AND SUB-MM EMISSION FROM GALAXIES

Having argued that GRBs offer a unique perspective into the cosmic star formation, I now provide a short overview of the underlying sources and emission mechanisms of radio and sub-mm emission from galaxies. The radio luminosity from star-forming galaxies is a combination of synchrotron and thermal emission components, both directly related to the formation rate of massive stars via simple relationships [15]. This is simply due to the fact that radio synchrotron emission comes from electrons accelerated in supernova shocks, the end products of massive stars, and thermal emission comes from HII regions and is dominated by the most luminous (i.e. massive) stars. In addition, since the lifetime of massive stars is $\sim 10^7$ years, and the lifetime of the synchrotron emitting electrons is $\sim 10^8$ years, radio emission traces the instantaneous SFR [15].

Similarly, sub-mm (and FIR) emission traces star formation since it arises from star-light reprocessed by dust. In this case too the massive stellar population dominates the power output in the host, and therefore the amount of reprocessed radiation. Since the emission in the radio and sub-mm regimes is a tracer of the massive stellar population, it is not surprising that there is a simple relation between the radio and sub-mm luminosities of star forming galaxies. It turns out that this relation is sensitively dependent on redshift [16] [17] [18].

One complication to the preceding discussion is the possibility of emission from an obscured AGN, which will contribute to both the radio and sub-mm luminosities of the host. Observations with X-ray satellites can provide an estimate of the fraction of emission (if any) that arises from an active nucleus.

## RECENT DETECTIONS: GRB 980703 AND GRB 010222

Recently, we have detected the host galaxy of GRB 980703 in the radio [12], and the host of GRB 010222 in the sub-mm [13]. Fig. 2 shows the light-curve of the 8.46 GHz emission from GRB 980703; the flattening at $t > 350$ days can only be explained in terms of host emission. We detect similar levels of emission at 1.43 and 4.86 GHz. At a redshift of $z = 0.966$ these flux levels translate to an emitted luminosity at 1.43 GHz of $L_{em}(1.43) \approx 4.7 \times 10^{30}$ erg sec$^{-1}$ Hz$^{-1}$.

What can we learn about the host galaxy of GRB 980703 from the observed radio luminosity? First, the emitted 1.43 GHz luminosity immediately translates to a formation rate of stars more massive than 5 M$_\odot$, SFR(M > 5M$_\odot$) $\approx$ 90 M$_\odot$/yr, and a total star formation rate (using the Salpeter IMF) of $\approx$ 500 M$_\odot$/yr. Second, based on this luminosity and the radio/sub-mm relation, we find that this galaxy is an Ultra-Luminous Infra-Red Galaxy (ULIRG; see [19]), with $L_{FIR} \approx 10^{12}$ L$_\odot$. A comparison to the properties of radio-selected galaxies at $z \sim 1$ from a survey of the HDF [20], shows that the host of GRB 980703 is by no means an unusual galaxy. On the other hand, a comparison to the optically-derived SFR ($\sim$ 20 M$_\odot$/yr; [21]) shows that most of the star formation in this galaxy is obscured.

An alternative explanation for the radio emission is that it originates from an AGN. Surveys of the Hubble Deep Field (HDF), its flanking fields, and the Small Selected Area 13 (SSA13) have shown that approximately 20% of the radio sources are AGN [22, 23, 6].

We consider the AGN hypothesis unlikely based on optical spectroscopy. Optical spectra of the source obtained by Djorgovski et al. (1998) show no evidence for an unobscured AGN: high-ionization lines such as Mg II $\lambda$2799, [NeV]$\lambda$3346, and [NeV]$\lambda$3426 are absent, and the [OIII]$\lambda$4959 to H$\beta$ ratio is approximately 0.4, much lower than [OIII]/H$\beta$ > 1.3 for AGN [24]. In addition, AGN have redder colors for similar [OII] EW, relative to normal galaxies [25]. Using the spectrum of GRB 980703 we evaluate the color index, $(41 - 50) \equiv 2.5\log[f_v(5000)/f_v(4100)] \approx 0 \pm 0.1$; an AGN with the same [OII] EW would have a value > 0.3 [25].

However, it is not possible to rule out the existence of an obscured AGN. Future observations with XMM will

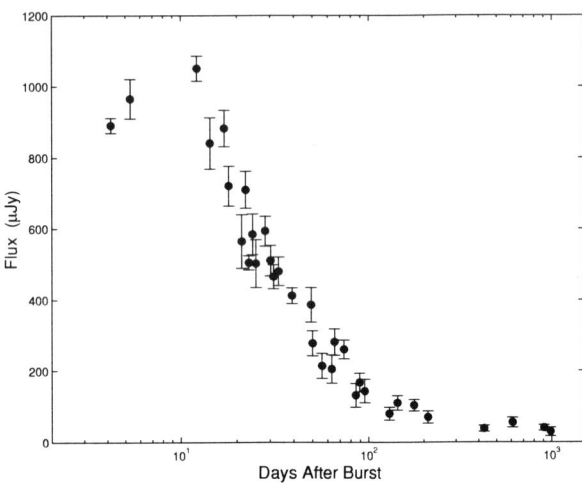

**FIGURE 2.** Radio light-curve at 8.46 GHz, showing the customary initial rise followed by decay of the afterglow of GRB 980703 (Berger et al. 2001). Observations from $\sim$ 100 – 300 days after the burst already show signs of flattening, due to the flux contribution from the host, while observations from $t > 350$ days directly probe the emission of the host. On these timescales, the afterglow contribution is negligible.

probe this possibility directly.

The high resolution afforded by VLA observations has shown that the GRB-host offset of GRB 980703 is negligible (see Fig. 3), indicating that the burst most probably took place within a nuclear starburst; in this case the resolution of the VLA allows a better offset determination than HST observations [8]. The nuclear starburst origin lends strong support to the collapsar model of GRBs.

The sub-mm detection of the host of GRB 010222 paints a similar picture (Fig. 4). The implied SFR is close to 1000 $M_\odot$/yr, and the FIR luminosity clearly indicates that this host is also a ULIRG [13].

## FUTURE PROSPECTS

Future radio/sub-mm/FIR studies of GRB hosts will be augmented by X-ray observations in order to assess the importance of obscured AGN in these host galaxies. There is some indication from studies of local ULIRGs [19] that high SFR is usually accompanied by some AGN activity. Thus, the advent of observatories such as the EVLA, SKA, SIRTF, and ALMA, in addition to XMM and Chandra will greatly increase our ability to study the properties of these hosts with greater sensitivity and resolution. For example, with a factor ten increase in resolution and a factor five increase in sensitivity over the current VLA, we will be able to probe scales of approximately 5 mas with the EVLA; for a galaxy at $z \sim 1$ this translates to a physical scale of 150 pc. In

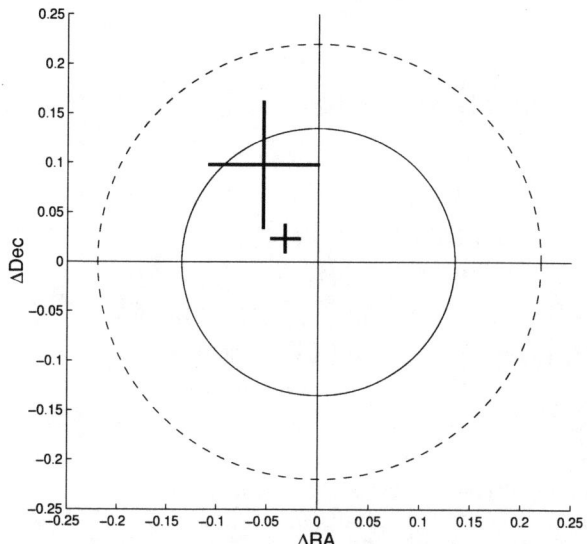

**FIGURE 3.** The weighted average GRB-host offset in RA and Dec from all VLA observations of GRB 980703 (small cross). The larger cross is the offset measurement from [8]. The solid circle designates the projected maximum source size from the radio observations, and the dashed circle is the optical size from [26]. Clearly the formation of massive stars is concentrated in the central region of the host, and the small offset of the burst from the host center indicates that GRB 980703 occurred in the region of maximum star formation [12]. This points to a link between GRBs and massive stars.

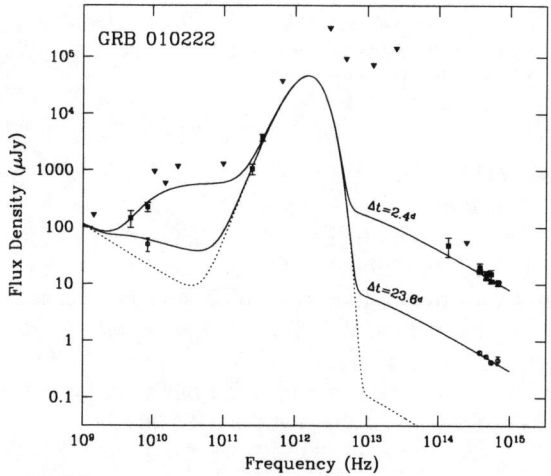

**FIGURE 4.** Spectral energy distribution of the host+afterglow emission from GRB 010222. The dashed line is the contribution of the host, and the solid line is the fading afterglow. The sub-mm flux densities at 250 and 350 GHz point to SFR~ 750 $M_\odot$/yr, and a ULIRG host galaxy. From [13]

addition, EVLA will detect galaxies with a total SFR as low as 50 $M_\odot$/yr at $z \sim 1$.

The potential of a GRB-selected galaxy sample is immense and unique. The dust-penetrating power of GRBs and their broad-band afterglow emission, offer a number of unique diagnostics: the obscured star formation fraction, the ISM within the disk, the local environment of the burst, and global and line-of-sight extinction, to name a few. In addition, GRBs allow us to select a wide range of galaxies independent of their emission propertied in any wavelength regime, and in addition they supply redshift information for these galaxies; the lack of accurate redshifts is one of the main problems of sub-mm studies of high-redshift galaxies.

It appears, therefore, that the numerous and detailed optical studies of GRB hosts are only the tip of the iceberg in our understanding of galaxies at high redshifts.

# REFERENCES

1. Madau, P., et al. 1996, MNRAS, 283, 1388.
2. Rauch, M. 1998, ARAA, 36, 267.
3. Storrie-Lombardi, L. J. & Wolfe, A. M. 2000, ApJ, 543, 552.
4. Adelberger, K. L. & Steidel, C. C. 2000, ApJ, 544, 218.
5. Ramaprakash, A. N., et al. 1998, Nature, 393, 43.
6. Barger, A. J., Cowie, L. L., & Richards, E. A. 2000, AJ, 119, 2092.
7. Peacock, J. A., et al. 2000, MNRAS, 318, 535.
8. Bloom, J. S., Kulkarni, S. R., & Djorgovski, S. G. 2001, submitted to AJ; astro-ph/0010176.
9. Lamb, D. Q. & Reichart, D. E. 2000, ApJ, 536, 1.
10. Amati, L., et al. 2000, Science, 290, 953.
11. Piro, L., et al. 2000, Science, 290, 955.
12. Berger, E., Kulkarni, S. R., & Frail, D. A. 2001, ApJ, 560, 652.
13. Frail, D. A., et al. 2001, Accepted to ApJ; astro-ph/0108436.
14. Reichart, D. E. 2001, Submitted to ApJL; astro-ph/0107546.
15. Condon, J. J. 1992, ARAA, 30, 575.
16. Carilli, C. L. & Yun, M. S. 1999, ApJ, 513, L13.
17. Carilli, C. L. & Yun, M. S. 2000, ApJ, 530, 618.
18. Dunne, L., Clements, D. L., & Eales, S. A. 2000, MNRAS, 319, 813.
19. Sanders, D. B. & Mirabel, I. F. 1996, ARAA, 34, 749+.
20. Haarsma, D. B., et al. 2000, ApJ, 544, 641.
21. Djorgovski, S. G., et al. 1998, ApJ, 508, L17.
22. Richards, E. A., et al. 1999, ApJ, 526, L73.
23. Richards, E. A. 2000, PASP, 112, 1001.
24. Rola, C. S., Terlevich, E., & Terlevich, R. J. 1997, MNRAS, 289, 419.
25. Kennicutt, R. C. 1992, ApJ, 388, 310.
26. Holland, S., et al. 2001, submitted to A&A. astro-ph/0103058.

# The Search for the Afterglow of the Dark GRB 001109[1]

J.M. Castro Cerón*, J. Gorosabel[†, **, ‡], A.J. Castro-Tirado[**, ‡], V.V. Sokolov[§],
V.L. Afanasiev[§], T.A. Fatkhullin[§], S.N. Dodonov[§], V.N. Komarova[§],
A.M. Cherepashchuk[¶], K.A. Postnov[¶], J. Greiner[∥], S. Klose[††], J. Hjorth[‡‡],
H. Pedersen[‡‡], E. Rol[§§], J. Fliri[¶¶], M. Feldt[***], G. Feulner[¶¶], M.I. Andersen[†††],
B.L. Jensen[‡‡], F.J. Vrba[‡‡‡], A.A. Henden[§§§, ‡‡‡] and G. Israelian[¶¶¶]

*Real Instituto y Observatorio de la Armada, Sección de Astronomía, 11.110 San Fernando-Naval (Cádiz) Spain
†Danish Space Research Institute, Juliane Maries Vej 30, 2100 Copenhagen Ø Denmark
**Instituto de Astrofísica de Andalucía, Apartado de Correos 3.004, 18.080 Granada Spain
‡Laboratorio de Astrofísica Espacial y Física Fundamental, Apartado de Correos 50.727, 28.080 Madrid Spain
§R.A.S. Special Astrophysical Observatory, Nizhnij Arkhyz, Karachai-Cherkessia, Russia 357147
¶Sternberg Astronomical Institute, MV Lomonosov State University, Moscow, Leninskie Gory, Russia 119899
∥Astrophysikalisches Institut, An der Sternwarte 16, 14482 Potsdam Germany
††Thüringer Landessternwarte Tautenburg, 07778 Tautenburg Germany
‡‡Astronomical Observatory, University of Copenhagen, Juliane Maries Vej 30, 2100 Copenhagen Ø Denmark
§§Astronomical Institute A. Pannekoek, Amsterdam Univ., Kruislaan 403, 1098 SJ Amsterdam The Netherlands
¶¶Universitäts-Sternwarte München, Scheinerstraße 1, 81679 München Germany
***Max Planck Institut für Astronomie, Königstuhl 17, 69117 Heidelberg Germany
†††Division of Astronomy, PO Box 3000, 90014 University of Oulu Finland
‡‡‡US Naval Observatory, Flagstaff Station, PO Box 1149, Flagstaff AZ 86002-1149 USA
§§§Universities Space Research Association
¶¶¶Instituto de Astrofísica de Canarias, c/. Vía Láctea s/n, 38.200 La Laguna (Tenerife) Spain

**Abstract.** We present optical and near IR follow up observations of the dark GRB 001109, 1–300 days after the event. No transient emission has been found for these wavelengths within this GRB's (Gamma Ray Burst) BeppoSAX 1' radius error box. Strong limits are set with $K' > 20$. A highly reddened galaxy, with variable radio emission, was found at $z = 0.398$. It is undergoing star formation, but seems to be unrelated to the event. We discuss the implications of these observations in the context of the dark GRB class.

## INTRODUCTION

The GRB 001109 was detected on 09.391169 UT November 2000 ($t_0$ hereafter) by the BeppoSAX [1] with a refined uncertainty of 2.5' [2, 3]. A BeppoSAX NFI (Narrow Field Instrument) observation at $t_0 + 16.5$ hrs detected a previously unknown source inside the 2.5' radius WFC (Wide Field Camera) error box [4]. The source, designated 1SAX J1830.1+5517, had R.A. (J2000) = $18^h30^m07.8^s$, Dec. (J2000) = $+55°17'56''$ (error radius = $50''$) and a 2–10 keV flux of $7.1 \pm 0.5 \times 10^{-13}$ erg cm$^{-2}$ s$^{-1}$.

A radiosource (dubbed VLA J1830+5518) was found within the NFI error box [5]. It seemed to decrease in brightness over a time span of 2 days [6], but further observations at the VLA for $\sim 390$ days failed to reveal a consistent decay. Flux variations were $\sim 10$–20% and its spectral slope is typical of extragalactic radio sources, thus ruling out this object as the GRB host [7].

## OBSERVATIONS

Target of Opportunity observations started at $t_0 + 9.1$ hrs. Table 1 displays the observing log. We performed aperture photometry using SExtractor [8] to study the contents of the BeppoSAX error box. The field was

---

[1] Based on observations made with telescopes at the Centro Astronómico Hispano Alemán (1.23 m + 3.50 m), the Roque de los Muchachos Observatory (NOT + WHT), the United States Naval Observatory (1.00 m) and the Russian Academy of Sciences's Special Astrophysical Observatory (6.05 m).

**TABLE 1.** Journal of observations of the GRB 001109 field

| Date UT | Telescope + Instrument | Filtre | Exposure Time (seconds) | Limiting Magnitude |
|---|---|---|---|---|
| 09.7708–09.8590/11/2000 | 1.23CAHA (CCD) | R | 7 × 500 | 20.9 |
| 09.7847–09.8854/11/2000 | 1.23CAHA (CCD) | B | 3 × 600 | 19.8 |
| 09.7848–09.7961/11/2000 | 4.20WHT (INGRID) | Ks | 750 | 19.9 |
| 09.7968–09.8081/11/2000 | 4.20WHT (INGRID) | H | 750 | 21.0 |
| 09.8083–09.8128/11/2000 | 4.20WHT (INGRID) | J | 300 | 21.3 |
| 09.8447–09.8845/11/2000 | 2.56NOT (StanCam) | I | 4 × 600 | 22.9 |
| 10.0876–10.1084/11/2000 | 1.00USNO (CCD) | I | 1 800 | 21.0 |
| 10.7618–10.8417/11/2000 | 1.23CAHA (CCD) | R | 9 × 500 | 20.9 |
| 10.7363–10.7883/11/2000 | 3.50CAHA (OMEGA Prime) | H | 10 × 300 | 20.5 |
| 11.8191–11.8281/11/2000 | 4.20WHT (INGRID) | H | 600 | 20.7 |
| 11.8292–11.8383/11/2000 | 4.20WHT (INGRID) | Ks | 600 | 19.4 |
| 11.8423–11.8514/11/2000 | 4.20WHT (INGRID) | J | 600 | 21.4 |
| 13.0560–13.0768/11/2000 | 1.00USNO (CCD) | I | 1 800 | 21.0 |
| 22.8278–22.8444/11/2000 | 2.56NOT (ALFOSC) | U | 2 × 600 | 24.0 |
| 22.1590–22.1938/11/2000 | 2.56NOT (ALFOSC) | B | 600 | 23.0 |
| 23.8035–22.8194/11/2000 | 2.56NOT (ALFOSC) | B | 2 × 600 | 23.5 |
| 26.7576–26.7618/11/2000 | 3.50CAHA (MOSCA) | R | 120 | 22.0 |
| 27.7514–27.7556/11/2000 | 3.50CAHA (MOSCA) | R | 180 | 22.3 |
| 22.1590–22.1938/05/2001 | 4.20WHT (PF) | B | 3 × 900 | 24.0 |
| 22.1951–22.2079/05/2001 | 4.20WHT (PF) | V | 3 × 450 | 23.5 |
| 29.1249–29.1795/05/2001 | 2.56NOT (ALFOSC) | U | 3 × 1 500 | 23.5 |
| 30.1249–30.1723/05/2001 | 2.56NOT (ALFOSC) | V | 900 + 300 | 23.5 |
| 31.0468–31.0548/05/2001 | 2.56NOT (ALFOSC) | V | 600 | 22.0 |
| 18.0361–18.0924/06/2001 | 4.20WHT (PF) | U | 5 × 900 | 23.5 |
| 30.0583–30.1361/06/2001 | 3.50CAHA (OMEGA Cass) | $K'$ | 120 × 60 | † |
| 01.0354–01.1181/07/2001 | 3.50CAHA (OMEGA Cass) | $K'$ | 120 × 60 | 21.0 |
| 24.8655–24.8828/07/2001 | 6.05SAO (SCORPIO) | R | 3 × 180 | 25.5 |
| 14.0524–14.0734/08/2001 | 2.56NOT (ALFOSC) | R | 600 + 900 | 23.8 |
| 14.9983–15.0223/08/2001 | 2.56NOT (ALFOSC) | R | 2 × 900 | 24.0 |
| 16.0571–16.1169/08/2001 | 2.56NOT (ALFOSC) | B | 4 × 1 200 | 25.0 |
| 16.9835–17.0570/08/2001 | 2.56NOT (ALFOSC) | U | 5 × 1 500 | 24.1 |
| 17.0148–17.0720/08/2001 | 2.56NOT (ALFOSC) | V | 5 × 900 | 24.5 |
| 17.0720–17.1148/08/2001 | 2.56NOT (ALFOSC) | I | 6 × 600 | 23.7 |

† The images from 30/6–01/07/2001 were coadded in just a single limiting magnitude, $K' = 21.0$.

**TABLE 2.** Photometric secondary standards in the field of the GRB 001109

| | RA (J2000) ($^h$ $^m$ $^s$) | Dec (J2000) (° ′ ″) | U | B | V | R | I |
|---|---|---|---|---|---|---|---|
| 1 | 18 29 52.55 | 55 16 37.8 | 18.62 ± 0.03 | 18.53 ± 0.08 | 17.95 ± 0.02 | 17.58 ± 0.02 | 17.25 ± 0.02 |
| 2 | 18 30 18.61 | 55 16 46.6 | 21.02 ± 0.17 | 19.57 ± 0.04 | 18.49 ± 0.02 | 17.79 ± 0.02 | 17.16 ± 0.02 |
| 3 | 18 30 02.94 | 55 17 03.2 | 19.24 ± 0.06 | 18.48 ± 0.07 | 17.55 ± 0.02 | 16.90 ± 0.02 | 16.36 ± 0.02 |
| 4 | 18 30 04.05 | 55 17 33.7 | 21.31 ± 0.17 | 19.99 ± 0.05 | 18.97 ± 0.02 | 18.16 ± 0.02 | 17.52 ± 0.02 |
| 5 | 18 29 48.91 | 55 19 20.5 | 20.26 ± 0.12 | 19.27 ± 0.06 | 18.33 ± 0.02 | 17.73 ± 0.02 | 17.25 ± 0.02 |
| 6 | 18 30 22.09 | 55 19 36.9 | 19.16 ± 0.06 | 19.45 ± 0.02 | 19.03 ± 0.07 | 18.77 ± 0.02 | 18.45 ± 0.02 |
| 7 | 18 30 20.65 | 55 19 40.7 | 20.63 ± 0.15 | 20.49 ± 0.08 | 20.05 ± 0.02 | 19.68 ± 0.03 | 19.27 ± 0.04 |
| 8 | 18 30 14.57 | 55 20 43.3 | 19.54 ± 0.08 | 18.62 ± 0.03 | 17.28 ± 0.02 | 16.33 ± 0.02 | 15.30 ± 0.02 |

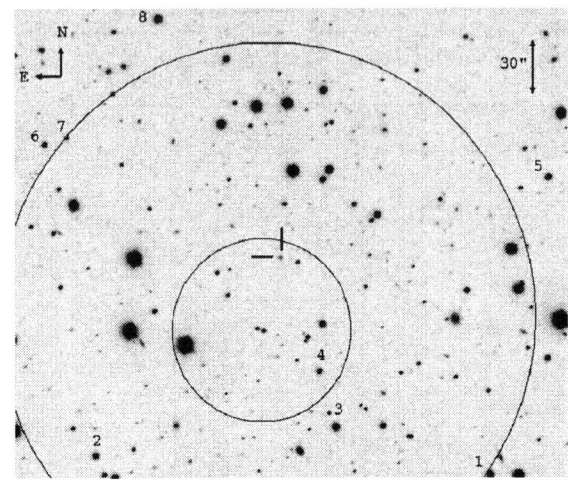

**FIGURE 1.** The contents of the BeppoSAX error box for the GRB 001109 field. This $R$ band image was taken with the 2.56NOT on 14.0524 UT August 2001. The source in between ticks is the galaxy coincident with VLA J1830+5518. The numbered stars are the secondary standards indicated in Table 2. The large circle represents the refined wide field camera error box [3] and the small one the NFI error circle [4]. The field of view covered by the figure is $5.1' \times 4.3'$. North is upwards and East is leftwards.

calibrated observing the Landolt field SA113 [9] in the *UBVRI* bands, at airmasses similar to that of the GRB field. Table 2 shows the positions and magnitudes of several secondary standards in the GRB field (see Fig. 1). Spectroscopic observations were made at the 3.50CAHA, 4.20WHT and 6.05SAO telescopes, the later (12 × 600 s exposures) with SCORPIO and a 300 lines/mm grating. The spectral resolution (FWHM) obtained was ∼ 20 Å and the effective wavelength coverage was 3 500–9 500 Å [10].

## RESULTS AND DISCUSSION

### Contents of the BeppoSAX NFI error box

No OA (Optical Afterglow) was detected in the first 1.23CAHA [11] and 2.56NOT frames ($R_{lim} > 20.9$ mag at 9.1 hrs and $I_{lim} > 22.9$ mag at 10.9 hrs after the GRB). Strong limits come from the deep near IR observations. The $H$ and $K'$ 3.50CAHA images have been compared to the $H$ and $Ks$ 4.20WHT [12]. We derived the following upper limits[2] for any near IR transient emission within the NFI error box: $K' > 19.9$, $H > 20.5$ and $J > 21.3$, ∼ 10 hrs after, all of them with a 3σ confidence level.

---

[2] We have assumed $K' = Ks$

## VLA J1830+5518

We performed astrometry on two different data sets to reveal the location of the VLA radiosource. For the first data set, 10 USNO A2.0 stars, not saturated on the 6.05SAO images, were used. The astrometrical uncertainty was found to be ∼ $0.5''$, including both, statistical and systematic errors [13]. For the second data set, an independent astrometric solution, based on 50 USNO A2.0 stars, was obtained using the coadded *I* band image taken at the 2.56NOT (see last entry in Table 1). It yielded a similar uncertainty ($0.57''$). Both astrometric solutions showed, independently, that the radiosource is located on the West outskirt of the brighter component (object A hereafter) of a double component system (see Fig. 2).

The spectral analysis revealed a double object[3] with spectra showing the continua and clearly identified Balmer breaks. Object A's redshift is $z = 0.398 \pm 0.002$, obtained from the identification of the Hα (6 563 Å) and O[III] (4 959 Å, 5 007 Å) emission lines (see Fig. 3 and [10]). Object B's redshift is $z = 0.3399 \pm 0.0005$, obtained from the identification of the Hα (6 563 Å) and Hβ (4 861 Å) emission lines [10]. If the VLA radiosource were indeed located in this galaxy and related to GRB 001109 then, the redshift of the GRB event would be $0.398 \pm 0.002$.

We have determined the flux distribution of the galaxy associated with VLA J1830+5518 by means of our *UBVRI* broad band photometric measurements together with the *JHKs* broad band measurements reported by [12]. The fluxes at the *UBVRIJHKs* passbands's wavelengths have been dereddened of Galactic extinction using a value of $E(B - V) = 0.04$ (DIRBE/IRAS dust maps; see [14]). The effective wavelengths and normalisations have been obtained by convolving the *UBVRIJHKs* filtres sensitivity profiles (plus the corresponding CCD efficiencies) with the spectrum of α-Lyrae. The *UBVRIJHKs* passbands's fluxes (measured in units of $2 \times 10^{-17}$ erg cm$^{-2}$ s$^{-1}$ Å$^{-1}$; see Fig. 4) correspond to the following values: $0.118 \pm 0.014$, $0.234 \pm 0.007$, $0.475 \pm 0.012$, $0.684 \pm 0.034$, $0.573 \pm 0.030$, $0.526 \pm 0.023$, $0.393 \pm 0.018$ and $0.315 \pm 0.014$, respectively. We have tried to reproduce the observed SED (Spectral Energy Distribution) using stellar population synthesis techniques [15] and leaving the extinction and the redshift as free parametres. The best fit is obtained with a dusty starburst galaxy SED at $z = 0.381$, with $A_V = 1.4$ mag and a starburst age of 0.25 Gyr (see Fig. 4).

---

[3] From the difference of the redshifts we calculated the separation of the two objects to be of the order of 200 Mpc thus, by "double object", we do not imply dynamical interaction.

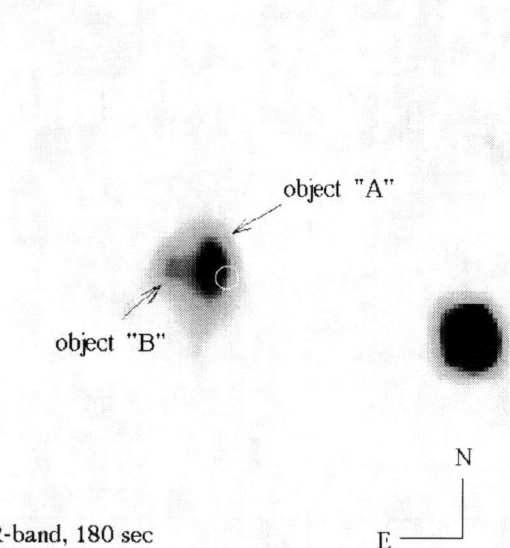

R-band, 180 sec

**FIGURE 2.** $20'' \times 20''$ $R$ band image of the galaxy coincident with VLA J1830+5518. The object is clearly extended and double. The centre of the circle marks the position of the radio source [5], R.A. (J2000) = $18^h30^m06.51^s$, Dec. (J2000) = $55°18'35.7''$. The size of the circle is $1''$.

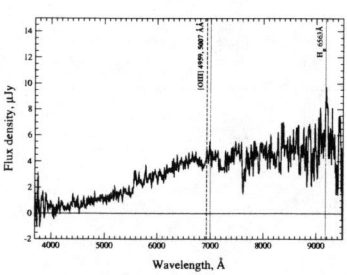

**FIGURE 3.** Optical spectrum of the galaxy coincident with VLA J1830+5518 obtained at the 6.05SAO. It shows restframe bright emission lines for H$\alpha$ (6563 Å) and O[III] (4959 Å, 5007 Å).

## The lack of an OA and its implication in the study of dark GRBs

The GRB 001109 resembles those GRBs observed in the past (970111, 970828, 991014, 000214, 000528 and 010214) for which afterglows at longer wavelengths were not found, in spite of intensive searches, at the position of the corresponding X ray afterglows detected by the BeppoSAX. The prompt, deep, near IR observations performed for the GRB 001109 suggest at least three possibilities:

*i*) some bursts are intrinsically faint in the near IR,

*ii*) a large extinction is present so, nothing is observed, even in the near IR,

*iii*) or this burst is shifted past $z = 17$.

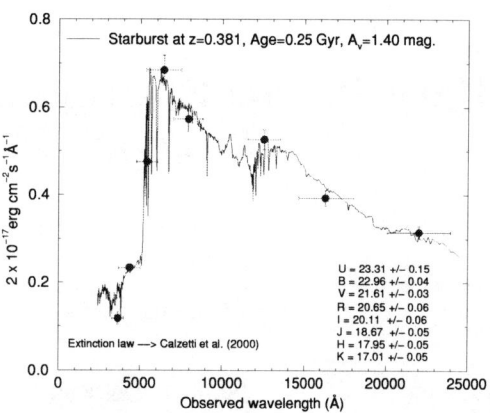

**FIGURE 4.** SED of the galaxy coincident with VLA J1830+5518. The solid line is the SED of a dusty starburst galaxy at $z = 0.381$, with $A_V = 1.4$ mag and an age of 0.25 Gyr. To construct the near IR part of the SED we used the homogeneous data taken with the 4.20WHT.

Should the latter be the case, the afterglow would be invisible, due to Lyman-$\alpha$ blanketing, in all the filtres used.

Rapid alerts provided by future space based missions (INTEGRAL, Swift ...) plus deep near IR follow up observations will allow one to discriminate among the above listed possibilities.

## REFERENCES

1. Boella, G. et al., *A&AS*, **122**, 299–307 (1997).
2. Gandolfi, G., *GCN*, **878**, 1+ (2000).
3. Gandolfi, G., *GCN*, **879**, 1+ (2000).
4. Amati, L. et al., *IAU Circular*, **7519**, 1+ (2000).
5. Taylor, G. B., Frail, D. A., and Bloom, J. S., *GCN*, **880**, 1+ (2000).
6. Rol, E. et al., *GCN*, **889**, 1+ (2000).
7. Berger, E., and Frail, D. A., *GCN*, **1168**, 1+ (2001).
8. Bertin, E., and S., A., *A&AS*, **117**, 393–404 (1996), http://terapix.iap.fr/soft/sextractor/.
9. Landolt, A. U., *AJ*, **104**, 340–371; 436–491 (1992).
10. Afanasiev, V. et al., *GCN*, **1090**, 1+ (2001).
11. Greiner, J. et al., *GCN*, **887**, 1+ (2000).
12. Vreeswijk, P. et al., *GCN*, **886**, 1+ (2000).
13. Sokolov, V., Fatkhullin, T., and Komarova, V., *GCN*, **1092**, 1+ (2001).
14. Schlegel, D. J., Finkbeiner, D. P., and Davis, M., *ApJ*, **500**, 525–553 (1998).
15. Bolzonella, M., Miralles, J. M., and Pelló, R., *A&A*, **363**, 476–492 (2000).

# Unveiling the Progenitors of Gamma-Ray Bursts through Observations of their Host Galaxies

Ranga-Ram Chary

*UCO/Lick Observatory, University of California, Santa Cruz, CA 95064*

**Abstract.** Analysis of the multi-wavelength broadband photometry between rest-frame ultraviolet and near-infrared wavelengths indicates that the extinction corrected star-formation rates per unit stellar mass of a small sample of gamma-ray burst (GRB) host galaxies are higher than those of prototypical, nearby starbursts. This result, the confirmed detection of the host of GRB980703 at radio wavelengths, and the tentative evidence in favor of a supernova light curve underlying the visible light transient associated with some GRBs provides evidence for a connection between stellar phenomenon and GRBs. Fitting population synthesis models to the multiband photometry of the host galaxies reveals the presence of a young stellar population with age less than 50 Myr in 4 out of 6 galaxies, albeit with large uncertainties. Determining the age of the stellar population in a large number of GRB host galaxies using this technique could be one of the more reliable ways of distinguishing between the collapsar model for GRBs and models that involve the merger of degenerate objects in a binary system.

## THE STARBURST-GRB CONNECTION

Detailed analysis of the cosmic infrared background and long wavelength galaxy counts by various groups have illustrated the large (>70%) contribution from infrared luminous galaxies ($L_{IR}=L(8-1000\,\mu m)>10^{11}\,L_\odot$) to the star-formation rate density at $z<3$ [3, 6]. Infrared luminous galaxies emit as much as 90% of their bolometric luminosity at far-infrared wavelengths indicating that dust reprocessing is extremely significant at $z\sim 1-3$. Most of these galaxies are inconspicuous at optical/UV wavelengths except that many of them show evidence of merger activity. This seems to suggest that tidally induced starbursts dominate the co-moving star-formation rate (SFR). However, the classification of a starburst galaxy based on its UV-determined SFR is ambiguous. For example M82, which is not an infrared luminous galaxy, is classified as a starburst galaxy based on stellar population synthesis model fits to its optical/near-infrared multiband photometry. These fits yield a young age for the stellar population, of order 10 Myr. While it's star-formation rate is about 7 $M_\odot$ yr$^{-1}$, no higher than that of M51 (the Whirlpool Galaxy), it's specific star-formation rate i.e. the star-formation rate per unit stellar mass is about an order of magnitude higher than M51, promoting it to the 'starburst' galaxy category. Thus, if GRBs were indeed associated with massive stars, almost all the hosts should be starburst galaxies with a large fraction of these being infrared luminous objects. To test this hypothesis it is necessary to obtain either a measure

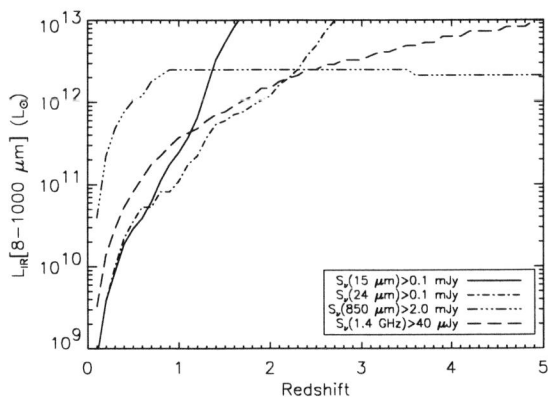

**FIGURE 1.** Minimum infrared luminosity that can be detected by various long wavelength observations as a function of redshift assuming the template spectral energy distributions of [3]. The sensitivity values adopted correspond to the limits of ISOCAM (15$\mu$m), SCUBA (850$\mu$m) and VLA/WSRT (21 cm). The 24$\mu$m sensitivity is the MIPS/SIRTF limit for a ~30 minute integration. The mid-infrared regime is the most sensitive to estimating dust-enshrouded star-formation at $z<2$. If we ignore contributions to the IR light from an AGN, $L_{IR}=10^{10}\,L_\odot$ corresponds to a star formation rate of 1.7 $M_\odot$/yr.

of the far-infrared luminosity or the extinction corrected specific SFR in the GRB host galaxies. Figure 1 shows the sensitivity of the deepest surveys at a variety of mid- and far-infrared wavelengths which allow a direct determination of $L_{IR}$ (see also [5]). Clearly, the mid-infrared regime (7−25 μm) is most sensitive to detecting infrared luminous galaxies at $z < 2.5$. However, even the deepest surveys at these wavelengths observe only galaxies with $L_{IR} > 10^{11} L_\odot$.

This is further illustrated by the fact that of the 16 GRB host galaxies that have been observed so far at 850 μm using the SCUBA instrument, only one of them (GRB010222) has been detected at late-times after the afterglow has faded completely [1]. The derived SFR for this object is comparable to the SFR inferred from the radio observations of GRB980703 [1] suggesting that it is an ultraluminous infrared galaxy (ULIG).

The low detection rate of GRB host galaxies in the submillimeter questions the validity of the association between star-formation and GRBs. However, at the typical SCUBA sensitivity of 3σ ∼3 mJy, only the brightest ULIGs that are more luminous than $L_{IR} \sim 4 \times 10^{12} L_\odot$ can be detected at $z > 1$. These hyper luminous galaxies comprise only a small subset of the entire ULIG population. Since the total contribution from ULIGs to the global star-formation history is at most 30% [3] the contribution from these hyper-luminous sources is less than 10%.

Thus, as inferred independently by Ramirez-Ruiz et al. [10], only about 10% of GRB hosts are likely to be detected by SCUBA at sensitivity limits of 3 mJy which is consistent with the small detection rate mentioned above and the small number of statistics. MIPS/SIRTF observations will be vital in deriving the far-infrared luminosity of a large number of GRB hosts.

Fortunately, at the faint end of the infrared luminosity function, a complementary technique which utilizes the UV-slope seems to measure the dust-obscured star-formation reasonably well [9]. At $L_{IR} < 4 \times 10^{11} L_\odot$, much of the dust obscuration is optically thin i.e. the far-infrared emission is dust reprocessed UV emission from the same part of the galaxy. However, in the more luminous sources the dust is concentrated so strongly that the UV opacity is very high in the neighborhood of the star-forming regions and so *all* the UV photons are re-radiated at longer wavelengths. In this case, the UV and far-infrared observations trace the low and high opacity regions of the same galaxy respectively and the UV-slope is only a lower limit to the amount of obscuration.

To search for signatures of a starburst in the GRB hosts using this technique, the multiband photometry on GRB host galaxies between rest-frame ultraviolet and near-infrared wavelengths were fit using the stellar population synthesis models of Bruzual & Charlot, taking into account internal dust extinction [4]. The total (obscured+UV) star-formation rate in these galaxies is not found to be unusually high (Figure 2). Although only 2 of the 6 host galaxies with accurate photometry have derived $L_{IR} > 10^{11} L_\odot$ and high resultant SFRs, the ratio

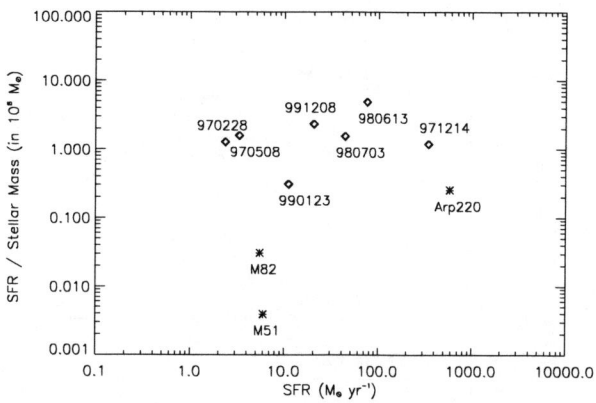

**FIGURE 2.** Total (obscured+unobscured) star-formation rates for selected gamma-ray burst host galaxies derived using the β-slope technique plotted against the ratio between the star formation rates and the midpoint of the range of stellar masses derived from fits to the multiband photometry. The uncertainty on the GRB971214 point is quite large because of the lack of good photometry at multiple wavelengths. The star-formation rates for the GRB hosts are lower limits, so the data points are likely to move higher and to the right. Estimates of the stellar mass typically have 95% confidence intervals that span an order of magnitude. Also plotted are the corresponding values for two prototypical starbursts Arp220 and M82 and the relatively quiescent Sbc galaxy M51. GRB hosts have star-formation rates per unit stellar mass much higher than local starbursts.

between the dust-obscured SFR and the unobscured SFR for the GRB hosts has an average value of ∼4. This is in agreement with the redshift dependent value of 3−7 that Chary & Elbaz [3] find for the ratio between the global comoving dust-obscured star-formation rate and the unobscured star-formation rate derived from the UV.

The near-infrared wavelengths constrain the stellar mass and age of the galaxy very well (Figure 3). Our analysis of the stellar population age and baryonic mass of 6 of the host galaxies with accurate photometry and redshifts reveal that all of them have high specific star-formation rates compared to local starbursts (Figure 2). The spectral energy distribution of 4 of them (GRB970228, GRB970508, GRB980613, GRB980703) are dominated by a young (age<50 Myr) stellar population which favors the collapsar model. However, 2 of them (GRB990123, GRB991208) have stellar populations of age ∼200 Myr suggesting that the merger of double degenerate systems might power some of the GRBs.

## FUTURE WORK

Various techniques have been suggested to constrain the progenitors of GRBs: studying the environments of

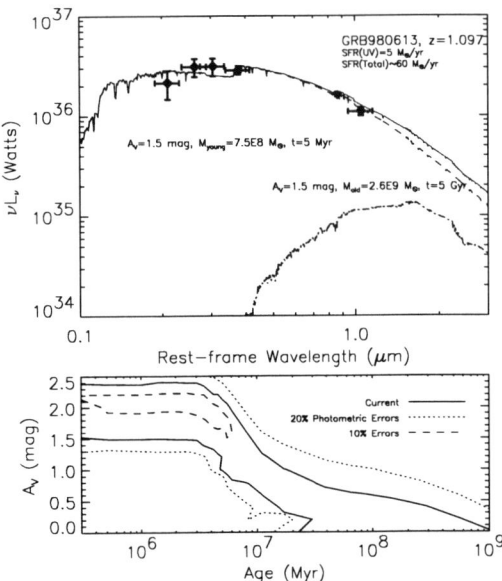

**FIGURE 3.** Optical/near-infrared photometry for the host of GRB980703 along with a template SED of a galaxy with an exponentially decaying starburst. The template consists of a young stellar population fit to the optical/UV photometry and the maximal contribution from an older stellar population which might have formed in an initial burst at $z \sim 10$ and has evolved over the 5 Gyr between $z = 10$ and $z = 1.097$. The solid black line is the sum of the two components. The luminosity of many GRB host galaxies are dominated by a young (10-50 Myr) stellar component which favors an origin of GRBs from collapsars. GRB host galaxies whose SEDs are dominated by an older stellar component would favor alternate mechanisms such as merging neutron stars. The lower plot shows the range of values for the extinction and age which result in SEDs that are within $2\sigma$ of the data points at B,V,R,I and K. The different contours illustrate the variation in the parameters with photometric uncertainty.

GRBs through spectral analysis of the X-ray afterglow [8], searching for signatures in the light curve of the optical transient associated with the GRB [2, 7] and studying the nature of the host galaxies. Of these, the latter is less direct but relatively simple and robust.

Associating GRBs with stellar phenomena necessitates that the host galaxies are either undergoing active star-formation or have been through epochs of substantial star-formation in the past. Since much of the high-redshift star-formation is obscured by dust, accurately measuring the SFR requires tracing the amount of internal extinction in the host galaxies.

Previous star-formation episodes can be constrained either by measuring the metallicity of the host galaxies or by determining the stellar mass. Estimates of galaxy masses derived by fitting template SEDs to the photometry between UV and visible wavelengths only provide lower limits since they trace the young stellar component which has a low mass to light ratio. Accurate masses can only be derived by including data at rest-frame $\lambda \sim 2\,\mu m$ which traces the old/cool stellar component. At high redshift, IRAC/SIRTF observations will be able to constrain the contribution from this component. Thus, by fitting population synthesis models to the rest-frame UV to near-infrared light one can estimate the mass fraction of gas that has gone through the most recent starburst, the internal extinction, the age of the starburst and fraction of galaxy mass incorporated in an old stellar population.

Determining the distribution of stellar ages in a large number of host galaxies will provide one of the more reliable ways to distinguish between models where massive stars evolve to become GRBs. In particular models which involve the production of GRBs through the core-collapse of isolated massive stars into a black hole/accretion disk system would take place on timescales comparable to the epoch of star-formation i.e. $\sim 10$ Myr. On the other hand merging double degenerate models suffer an evolutionary delay from the epoch of star-formation which is induced by the timescale for orbital decay of the binary system $>100$ Myr.

Thus, if a majority of the GRB hosts are found to have high specific SFRs (SFR/stellar mass) and young (age<10 Myr) stellar populations, this would be convincing evidence for 'collapsars' to be the progenitors of GRBs. However, most of the high redshift infrared luminous galaxies show the presence of a significant older stellar population that is dominating the rest-frame near-infrared light of the galaxy. This indicates that infrared luminous galaxies have undergone previous epochs of star-formation which could potentially provide an origin for double degenerate systems (Figure 3). As a result, if the majority of the hosts have a large old stellar population in addition to the young component, this would make it impossible to discriminate between the two models using this technique. A positional analysis of the burst locations on the host would then have to used as the discriminator due to the large kick velocities imparted to the degenerate objects from their supernovae.

# REFERENCES

1. Berger, E., 2002, these proceedings
2. Bloom, J. S., et al., 1999, Nature, 401, 453
3. Chary, R., & Elbaz, D., 2001, ApJ, 556, 562
4. Chary, R., Becklin, E. E., & Armus, L., 2002, ApJ, 566, 1
5. Elbaz, D., et al., 2002, A&A, in press (astro-ph/0201328)
6. Franceschini, A., et al., 2001, A&A, 378, 1
7. Galama, T. J., et al., 2000, ApJ, 536, 185
8. Galama, T. J., & Wijers, R. A., 2001, ApJ, 549, L209
9. Meurer, G. R., Heckman, T. M., Calzetti, D., 1999, ApJ, 521, 64
10. Ramirez-Ruiz, E., Trentham, N., & Blain, A. W., 2002, MNRAS, 329, 465

# VI. GAMMA-RAY BURSTS AND COSMOLOGY

# Gamma-Ray Bursts as a Probe of Cosmology

## Donald Q. Lamb

*Department of Astronomy & Astrophysics, University of Chicago, 5640 South Ellis Avenue, Chicago, IL 60637*

**Abstract.** We show that, if the long GRBs are produced by the collapse of massive stars, GRBs and their afterglows may provide a powerful probe of cosmology and the early universe.

## INTRODUCTION

There is increasingly strong evidence that gamma-ray bursts (GRBs) are associated with star-forming galaxies [1,2,3,4] and occur near or in the star-forming regions of these galaxies [2,3,4,5,6]. These associations provide indirect evidence that at least the long GRBs detected by BeppoSAX are a result of the collapse of massive stars. The discovery of what appear to be supernova components in the afterglows of GRBs 970228 [7,8] and 980326 [9] provides tantalizing direct evidence that at least some GRBs are related to the deaths of massive stars, as predicted by the widely-discussed collapsar model of GRBs [10,11,12,13,14]. If GRBs are indeed related to the collapse of massive stars, one expects the GRB rate to be approximately proportional to the star-formation rate (SFR).

## DETECTABILITY OF GRBS AND THEIR AFTERGLOWS

We have calculated the limiting redshifts detectable by BATSE and HETE-2, and by *Swift*, for the sixteen GRBs with well-established redshifts and published peak photon number fluxes. In doing so, we have used the peak photon number fluxes given in Table 1 of [15], taken a detection threshold of 0.2 ph s$^{-1}$ for BATSE and HETE-2 and 0.04 ph s$^{-1}$ for *Swift*, and set $H_0 = 65$ km s$^{-1}$ Mpc$^{-1}$, $\Omega_m = 0.3$, and $\Omega_\Lambda = 0.7$ (other cosmologies give similar results). Figure 1 displays the results. This figure shows that BATSE and HETE-2 would be able to detect half of these GRBs out to a redshift $z = 20$ and 20% of them out to a redshift $z = 50$. *Swift* would be able to detect half of them out to redshifts $z = 70$, and 20% of them out to a redshift $z = 200$, although it is unlikely that GRBs occur at such extreme redshifts. Consequently, if GRBs occur at very high ($z > 5$) redshifts (VHRs), BATSE has probably already detected GRBs

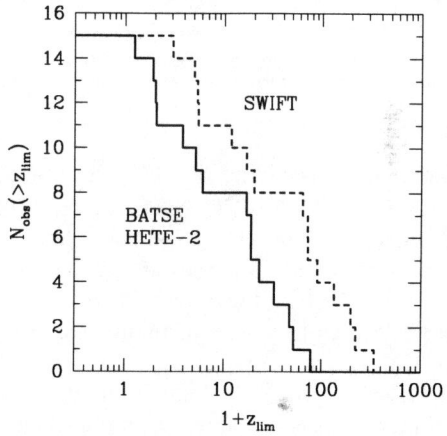

**FIGURE 1.** Cumulative distributions of the limiting redshifts at which the 15 GRBs with well-determined redshifts and published peak photon number fluxes would be detectable by BATSE and HETE-2, and by *Swift*.

at these redshifts, and HETE-2 and *Swift* should detect them as well.

The soft X-ray, optical and infrared afterglows of GRBs are also detectable out to VHRs. The effects of distance and redshift tend to reduce the spectral flux in GRB afterglows in a given frequency band, but time dilation tends to increase it at a fixed time of observation after the GRB, since afterglow intensities tend to decrease with time. These effects combine to produce little or no decrease in the spectral energy flux $F_\nu$ of GRB afterglows in a given frequency band and at a fixed time of observation after the GRB with increasing redshift:

$$F_\nu(\nu,t) = \frac{L_\nu(\nu,t)}{4\pi D^2(z)(1+z)^{1-a+b}}, \quad (1)$$

where $L_\nu \propto \nu^a t^b$ is the intrinsic spectral luminosity of the GRB afterglow, which we assume applies even at early times, and $D(z)$ is the comoving distance to the burst.

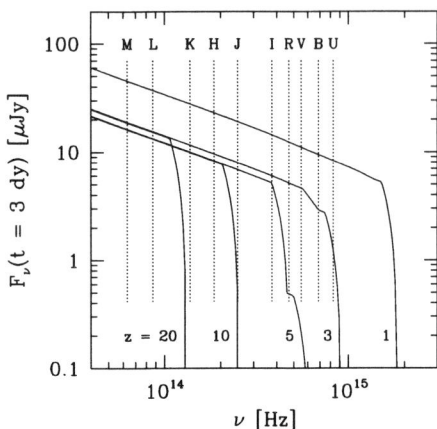

**FIGURE 2.** The best-fit spectral flux distribution of the early afterglow of GRB 000131, as observed one day after the burst, after transforming it to various redshifts, and extinguishing it with a model of the Lyα forest.

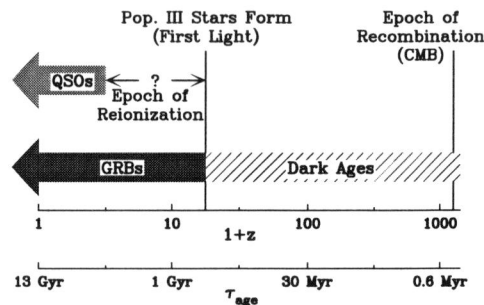

**FIGURE 3.** Cosmological context of VHR GRBs. Shown are the epochs of recombination, first light, and re-ionization. Also shown are the ranges of redshifts corresponding to the "dark ages," and probed by QSOs and GRBs.

Many afterglows fade like $b \approx -4/3$, which implies that $F_\nu(\nu, t) \propto D(z)^{-2}(1+z)^{-5/9}$ in the simplest afterglow model, where $a = 2b/3$ [16]. In addition, $D(z)$ increases very slowly with redshift at redshifts greater than a few. Consequently, there is little or no decrease in the spectral flux of GRB afterglows with increasing redshift beyond $z \approx 3$.

In fact, in the simplest afterglow model where $a = 2b/3$, if the afterglow declines more rapidly than $b \approx 1.7$, the spectral flux actually *increases* as one moves the burst to higher redshifts! An example of this is the afterglow of GRB 000131. Its peak flux $F_{peak}$ was in the top 5% of all BATSE bursts and the break energy $E_{break}$ in its spectrum was 164 keV, yet it occurred at a redshift $z = 4.50$. We have calculated the best-fit spectral flux distribution of the afterglow of GRB 000131 from [17], as observed three days after the burst, transformed to various redshifts. The transformation involves (1) dimming the afterglow, (2) redshifting its spectrum, (3) time dilating its light curve, and (4) extinguishing the spectrum using a model of the Lyα forest (for details, see [15]). Finally, we have convolved the transformed spectra with a top hat smearing function of width $\Delta\nu = 0.2\nu$. This models these spectra as they would be sampled photometrically, as opposed to spectroscopically; i.e., this transforms the model spectra into model spectral flux distributions.

Figure 2 shows the resulting spectral flux distribution. The spectral flux distribution of the afterglow is cut off by the Lyα forest at progressively lower frequencies as one moves out in redshift. Thus high redshift afterglows are characterized by an optical "dropout" [4], and VHR afterglows by a near infrared "dropout." We conclude that, if GRBs occur at very high redshifts, both they and their afterglows can be easily detected.

## GRBS AS A PROBE OF COSMOLOGY AND THE EARLY UNIVERSE

Theoretical calculations show that the birth rate of Pop III stars produces a peak in the SFR in the universe at redshifts $16 \lesssim z \lesssim 20$, while the birth rate of Pop II stars produces a much larger and broader peak at redshifts $2 \lesssim z \lesssim 10$ [18,19,20]. Therefore one expects GRBs to occur out to at least $z \approx 10$ and possibly $z \approx 15 - 20$, redshifts that are far larger than those expected for the most distant quasars.

Figure 3 places GRBs in a cosmological context. At recombination, which occurs at redshift $z = 1100$, the universe becomes transparent. The cosmic background radiation originates at this redshift. Shortly afterwards, the temperature of the cosmic background radiation falls below 3000 K and the universe enters the "dark ages" during which there is no visible light in the universe. "First light," which occurs at $z \approx 20$, corresponds to the epoch when the first stars form. Ultraviolet radiation from these first stars and/or from the first active galactic nuclei re-ionizes the universe. Afterward, the universe is transparent in the ultraviolet.

QSOs are currently the most powerful probes of the high redshift universe. GRBs have several advantages relative to QSOs as probes of cosmology. First, GRBs are expected to occur out to $z \approx 20$, whereas QSOs occur out to only $z \approx 5$. Second, very high redshift GRB afterglows can be 100 - 1000 times brighter at early times than are high redshift QSOs. This makes possible very sensitive high dispersion spectroscopy of the metal absorption lines and the Lyman α forest in the spectrum of the afterglows. Third, no "proximity effect" on intergalactic distances scales is expected for GRBs and their afterglows, in contrast to QSOs. Thus GRBs may be rel-

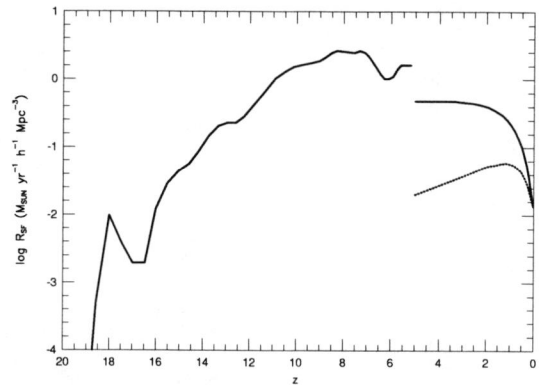

**FIGURE 4.** The cosmic SFR $R_{SF}$ as a function of redshift $z$. The solid curve at $z < 5$ is the SFR derived by [25]; the solid curve at $z \geq 5$ is the SFR calculated by [18] (the dip in this curve at $z \approx 6$ is an artifact of their numerical simulation). The dotted curve is the SFR derived by [24]. From [15].

atively "clean" probes of the intergalactic medium, the Lyman α forest, and damped Lyman α clouds, even in the vicinity of the GRBs.

The important cosmological questions that observations of GRBs and their afterglows may be able to address include the following:

- Information about the epoch of "first light" and the earliest generations of stars from merely the detection of GRBs at very high redshifts;

- Information about the growth of metallicity in the universe in the star-forming entities in which the bursts occur, in damped Lyman α clouds, and in the Lyman α forest from observations of the metal absorption line systems in the spectra of their afterglows;

- Information about the large-scale structure of the universe at VHRs from the clustering of the Lyman α forest lines and the metal absorption-line systems in the spectra of their afterglows; and

- Information about the epoch of re-ionization from the depth of the Lyman α break in the spectra of their afterglows.

Below we consider the first of these questions: the epoch of "first light" and the earliest generations of stars.

## GRBS AS A PROBE OF STAR FORMATION

Observational estimates [21,22,23,24] indicate that the SFR in the universe was about 15 times larger at a redshift $z \approx 1$ than it is today. The data at higher redshifts from the Hubble Deep Field (HDF) in the north suggests a peak in the SFR at $z \approx 1-2$ [24], but the actual situation is highly uncertain.

In Figure 4, we have plotted the SFR versus redshift from a phenomenological fit [25] to the SFR derived from submillimeter, infrared, and UV data at redshifts $z < 5$, and from a numerical simulation by [18] at redshifts $z \geq 5$. The simulations done by [18] indicate that the SFR increases with increasing redshift until $z \approx 10$, at which point it levels off. The smaller peak in the SFR at $z \approx 18$ corresponds to the formation of Population III stars, brought on by cooling by molecular hydrogen. Since GRBs are detectable at these VHRs and their redshifts may be measurable from the absorption-line systems and the Lyα break in the afterglows [4], if the GRB rate is proportional to the SFR, then GRBs could provide unique information about the star-formation history of the VHR universe.

We have calculated the expected number $N_*$ of stars as a function of $z$ assuming (1) that the GRB rate is proportional to the SFR[1], and (2) that the SFR is that given in Figure 4 (see [15] for details). The left panel of Figure 5 shows our results for $N_*(z)$ for an assumed cosmology $\Omega_M = 0.3$ and $\Omega_\Lambda = 0.7$ (other cosmologies give similar results). The solid curve corresponds to the star-formation rate in Figure 4; the dashed curve corresponds to the star-formation rate derived by [24]. Figure 5 shows that $N_*(z)$ peaks sharply at $z \approx 2$ and then drops off fairly rapidly at higher $z$, with a tail that extends out to $z \approx 12$. The rapid rise in $N_*(z)$ out to $z \approx 2$ is due to the rapidly increasing volume of space. The rapid decline beyond $z \approx 2$ is due almost completely to the "edge" in the spatial distribution produced by the cosmology. In essence, the sharp peak in $N_*(z)$ at $z \approx 2$ reflects the fact that the SFR we have taken is fairly broad in $z$, and consequently, the behavior of $N_*(z)$ is dominated by the behavior of the co-moving volume $dV(z)/dz$; i.e., the shape of $N_*(z)$ is due almost entirely to cosmology. The right panel in Figure 5 shows the cumulative distribution $N_*(>z)$ of the number of stars expected as a function of redshift $z$. The solid and dashed curves have the same meaning as in the upper panel. Figure 5 shows that for the particular SFR we have assumed, $\approx 40\%$ of all stars (and therefore of all GRBs) have redshifts $z > 5$.

## ESTIMATES OF THE GRB RATE

Is the GRB rate indeed proportional to the SFR (at least roughly)? We address this question in two ways. First, we

---

[1] This may underestimate the GRB rate at VHRs since it is generally thought that the initial mass function will be tilted toward a greater fraction of massive stars at VHRs because of less efficient cooling due to the lower metallicity of the universe at these early times.

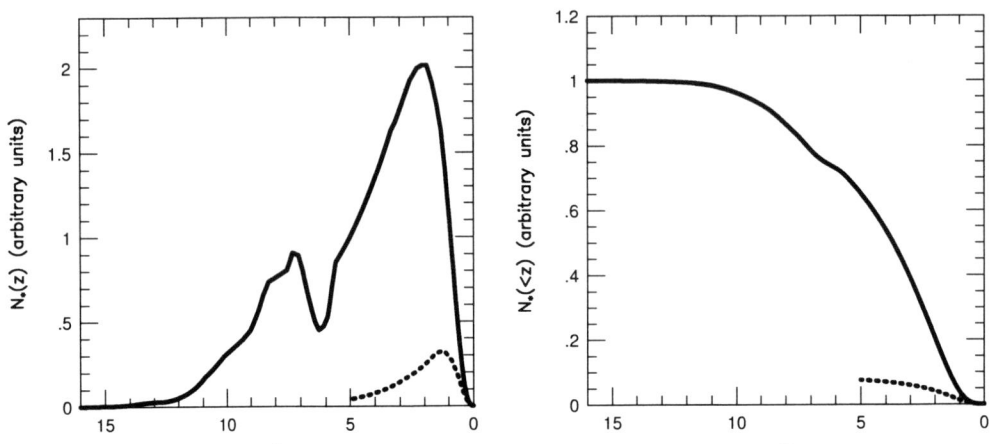

**FIGURE 5.** Left panel: The number $N_*$ of stars expected as a function of redshift $z$ (i.e., the SFR from Figure 4, weighted by the differential comoving volume, and time-dilated) assuming that $\Omega_M = 0.3$ and $\Omega_\Lambda = 0.7$. Right panel: The cumulative distribution of the number $N_*$ of stars expected as a function of redshift $z$. Note that $\approx 40\%$ of all stars have redshifts $z > 5$. The solid and dashed curves in both panels have the same meanings as in Figure 4. From [15].

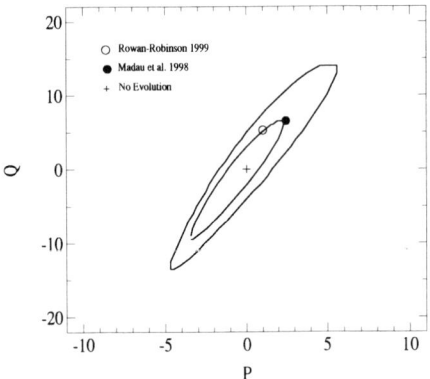

**FIGURE 6.** Credible regions for the GRB rate parameters $P$ and $Q$. The solid curves correspond to the 68% and 95% probability contours. Also shown are the (P,Q)-values corresponding to no space density evolution, the Madau et al. [24] SFR, and the Rowan-Robinson model fit to IR, optical and UV data [25]. From [26].

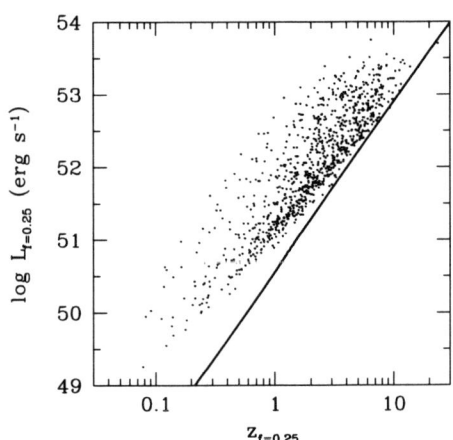

**FIGURE 7.** The joint redshift and luminosity distribution of the qualitatively acceptable redshift distribution (see Figure 3 in [27]). The diagonal solid line shows the 10% detection threshold of BATSE. From [27].

consider the sample of GRBs with known redshifts. We adopt a Bayesian approach and calculate the likelihood of the data, assuming a very general model for the GRB rate and a power-law model for the intrinsic GRB photon luminosity distribution [26]. We fit the model jointly to the peak fluxes and redshifts of the 14 GRBs with known $z$, and the 7 GRBs for which there are constraints on $z$. Figure 6 shows our preliminary results. These suggest that the SFR models lie at about a 68% excursion from the best-fit GRB rate model. Thus we find that, despite the qualitative differences that exist between the observed GRB rate and estimates of the SFR in the universe, current data are consistent with the actual GRB rate being approximately proportional to the SFR when observational selection effects are taken into account.

Second, we use the variability of the lightcurves of long GRBs to estimate their distribution as a function of intrinsic photon luminosity and redshift [27]. Figure 7 shows the resulting joint redshift and luminosity distribution. This distribution suggests that the GRB rate continues to increase at very high redshifts, and that the intrinsic luminosities of GRBs evolve with redshift.

## CONCLUSIONS

If the long GRBs are indeed produced by the collapse of massive stars, one expects GRBs to occur out to $z \approx 15 - 20$, redshifts that are far larger than those expected for the most distant QSOs. We have shown that both GRBs and their afterglows are easily detected out to these VHRs. GRBs can therefore give us information about the star-formation history of the universe, including the earliest generations of stars. The absorption-line systems and the Ly$\alpha$ forest visible in the spectra of GRB afterglows can be used to trace the evolution of metallicity in the universe, and to probe the large-scale structure of the universe at VHRs. Finally, measurement of the Ly$\alpha$ break in the spectra of GRB afterglows can be used to constrain, or possibly measure, the epoch at which reionization of the universe occurred.

## REFERENCES

1. Castander, F. J., & Lamb, D. Q. 1999, ApJ, **523**, 593
2. Fruchter, A. S., et al. 1999, ApJ, **516**, 683
3. Kulkarni, S. R., et al. 1998, Nature, **395**, 663
4. Fruchter, A. S. 1999, ApJ, **516**, 683
5. Sahu, K. C., et al. 1997, Nature, **387**, 476
6. Kulkarni, S. R., et al. 1999, Nature, **398**, 389
7. Reichart, D. E., 1999, ApJ, **521**, L111
8. Galama, T. J., et al. 2000, ApJ, **536**, 185
9. Bloom, J. S., et al. 1999, Nature, **401**, 453
10. Woosley, S. E. 1993, ApJ, **405**, 273
11. Woosley, S. E. 1996, in Gamma-Ray Bursts, eds. C. A. Meegan, R. D. Preece, & T. M. Koshut (New York: AIP), 520
12. Paczyński, B. 1998, ApJ, **494**, L45
13. MacFadyen, A. I., & Woosley, S. E. 1999, ApJ, **524**, 262
14. Wheeler, J. C., et al. 2000, ApJ, **537**, 810
15. Lamb, D. Q., & Reichart, D. E., 2000, ApJ, **536**, 1
16. Wijers, R. A. M. J., Rees, M. J., & Mészáros, P. 1997, MNRAS, **288**, L51
17. Andersen, M. I., et al. 2000, A&A, **364**, L54
18. Ostriker, J. P., & Gnedin, N. Y. 1996, ApJ, **472**, L63
19. Gnedin, N. Y., & Ostriker, J. P. 1997, ApJ, **486**, 581
20. Valageas, P., & Silk, J. 1999, A&A, **347**, 1
21. Gallego, J. 1995, ApJ, **455**, L1
22. Lilly, S. J., et al. 1996, ApJ, **460**, L1
23. Connolly, A. J. 1997, ApJ, **486**, L11
24. Madau, P., Pozzetti, L., & Dickinson, M. 1998, ApJ, **498**, 106
25. Rowan-Robinson, M. 1999, Ap&SS, **266**, 291
26. Weinberg, N., Graziani, C., Lamb, D. Q., and Reichart, D. E. 2001, in Proceedings of the Rome Workshop, in press (astro-ph/010759)
27. Reichart, D. E. and Lamb, D. Q. 2001, in Proceedings of the 20th Texas Symposium on Relativistic Astrophysics, in press (astro-ph/0103255)

# Measuring $\Omega_M$, $\Omega_\Lambda$ and the SFR with class III GRBs

Andreu Balastegui*, Pilar Ruiz-Lapuente[†*] and Ramon Canal[***]

*Departament d'Astronomia i Meteorologia, Universitat de Barcelona, Martí i Franqués 1, Barcelona 08028
[†]Max-Planck-Institut für Astrophysik, Karl-Schwarzschild-Strasse 1, 85740 Garching bei München, Germany
[**]Institut d'Estudis Espacials de Catalunya, Nexus Building, Gran Capità 2-4, Barcelona 08034, Spain

**Abstract.** New evidences of the existence of three classes of GRBs are presented. In addition, it is shown the potential of class III GRBs for measuring the cosmological parameters $\Omega_M$ and $\Omega_\Lambda$ by comparing observed log N-log P distributions with the theoretical ones via a $\chi^2$ test. The result is that all three classes obtained from a neural network classification algorithm by Balastegui, Ruiz-Lapuente & Canal [1] (2001; BRC hereafter) do fit the cosmological distributions, whereas the cluster analysis classification deviates from them. To derive the theoretical distributions of GRBs, rates proportional to various proposed star formation rates (SFR) are used, and three different SFRs [2] [3] [4] are compared. We also show that in order to obtain acceptable values of $\Omega_M$ and $\Omega_\Lambda$ for class III GRBs, some evolution in luminosity has to be assumed. The fact that class III GRBs fit a rate proportional to the SFR makes them good candidates to come from collapsars, while the shallower distribution of classes I and II (with z < 2) rather suggests that they result from NS-NS or NS-BH mergings, the difference from the distribution of class III arising for the time-delay of the merging with respect to the SFR.

## INTRODUCTION

It has recently been shown that some automatic classifier algorithms seem to favor a classification of GRBs in three classes [5], but those classifications have often been interpreted as arising from instrumental bias [6]. The most recent classification of GRBs comes from BRC, who applied a cluster analysis and a neural network algorithm to the BATSE Current GRB Catalog, in its version of 2000 September (http://www.batse.msfc.nasa.gov/batse/grb/catalog/current), which is the largest and most homogeneous catalog of GRBs up to now. It is the first time that a neural network algorithm has been used to classify GRBs, and the algorithm applied was the "Self-Organizing Map" [7], a completely unsupervised method. This method has the advantage over cluster analysis that it can deal with highly non-linear relationships.

Concerning whether this classification refers to a physical difference between classes of GRBs or is just a consequence of an instrumental bias, several clues pointing to the first reason are reported by BRC. The main clue arises from the fact that the sample of classes II and III together shows the trend that bursts with lower values of $\langle V/V_{max}\rangle$, and thus the more distant ones (based on the relationship between $\langle V/V_{max}\rangle$ and $z_{max}$ [8]), tend to be harder (i.e. have higher values of the hardness ratio $H_{32}$). When classes II and III are taken separately, it is shown that class II has completely lost this trend, and only class III shows a correlation between $\langle V/V_{max}\rangle$ and $H_{32}$. Based in the hardness-intensity correlation [9] BRC conclude that distant class III GRBs are brighter than closer ones, in other words, that there exists an evolution of the luminosity with redshift. This evolution should have been predicted theoretically, since if collapsars are the GRBs progenitors and, as it is known, lower metallicities allow to form more massive stars, then collapsars should have more energy available to power the GRB in the ancient Universe, when the metallicity was lower than today. Recently, Lloyd-Ronning, Fryer & Ramirez-Ruiz [10] (2001; LFR hereafter) have also found such luminosity evolution and they have even been able to derive a luminosity evolution law $L \propto (1+z)^{1.4\pm0.2}$. LFR found the luminosity evolution in a sample of 220 bursts with redshifts and luminosities deduced from a luminosity-variability correlation. From those 220 bursts 205 entered into the classification of BRC and 198 were classified as class III GRBs by the neural network algorithm. Therefore, the luminosity evolution found by LFR refers only to class III GRBs.

From the eight bursts with known redshifts that entered into the classification of BRC, seven were classified as class III and only one as class II. From that it can be inferred that class III GRBs are certainly of cosmological origin, this fact being less clear for classes I and II, although the inhomogeneity and high isotropy of their distributions do also point to that same origin. The preceding can be used as a test to reject a given classification, since the distribution of class III GRBs should be consistent with a cosmological distribution. Weinberg [11] (1972) described the distribution of measured fluxes from cosmological sources, and if the three class classification from BRC is correct, either all classes or at least class III should fit the theoretical log N-log P distribution, and that fitting could be used to measure $\Omega_M$ and $\Omega_\Lambda$ and to infer the behavior of the SFR from $z \sim 2$ up.

## METHODOLOGY

The procedure is to compare the theoretical cumulative log N-log P distribution with the observed one, and that is constructed from bins of constant width in logarithmic scale. For classes II and III, whose durations are longer than 1 s, the peak flux in the 1024 ms integration time is taken, whereas for class I GRBs, that contains bursts with less than 1 s of duration, the peak flux in the 64 ms integration time is taken. The distribution is corrected from efficiency, and only those bursts with $P_{1024} > 0.44$ $\gamma\,s^{-1}\,cm^{-2}$ are retained, because at low peak fluxes the efficiency is uncertain due to the atmospheric backscattering, and also those with $P_{64} > 1.1\,\gamma\,s^{-1}\,cm^{-2}$, where the efficiency is higher than 50%. The log N-log P distributions from the cluster analysis have 37, 16 and 43 bins for classes I, II and III respectively, whereas the neural network distributions have 33, 27 and 43 bins each.

The theoretical N(P) distribution comes from:

$$N(>P) = \frac{c}{H_0}\int_0^{z_{max}} \frac{4\pi}{1+z} n_z \times \quad (1)$$
$$\frac{D^2(z)}{\sqrt{\Omega_M(1+z)^3 + \Omega_k(1+z)^2 + \Omega_\Lambda}} dz \int_{L_{min}}^{L(z,P)} \Phi(L)dL,$$

where D(z) is the comoving distance, and $H_0 = 65$ km s$^{-1}$ Mpc$^{-1}$. Here a comoving burst rate, $n_z$, has to be inserted, as well as a luminosity function $\Phi(L)$. Neither of these two quantities is very well known, but if collapsars were GRBs progenitors, $n_z$ should closely follow the SFR, since very massive stars have lifetimes < 10 Myr. Five different rates are adopted: constant comoving rate; rates proportionals to three different SFRs; and the GRB rate derived by LFR (Fig. 1). The latter was derived up to $z \sim 10$, and as it is expected that the SFR decreases at very high z, a rate decreasing as $(1+z)^{-2}$ is adopted for z > 10. For the same reason we also tested two more SFRs, which coincide with the Steidel one up to z = 5 and z = 10, respectively, and decrease like the Madau SFR from these z up in both cases. These SFR are labelled Steidel(5) and Steidel(10) in Table 1.

On the other hand, several attempts to indirectly measure the luminosity function of GRBs have been reported

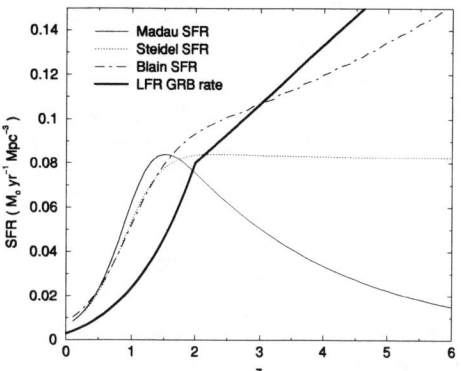

**FIGURE 1.** The different GRBs comoving rates used.

recently (e.g. LFR and [12]). Here three different luminosity functions are used: the standard candle case (equation (2)), and a broken power law for two cases (equation (3)): the one derived by LFR, with $\kappa_1=-1.5$ and $\kappa_2=-3$, and a symmetric power law, with $\kappa_1=2$ and $\kappa_2=-2$.

$$\Phi(L) = \delta(L - L_0) \quad (2)$$

$$\Phi(L) = C \times \begin{cases} \left(L/L_0\right)^{\kappa_1} & L_{min} \leq L < L_0 \\ \left(L/L_0\right)^{\kappa_2} & L_0 \leq L \leq L_{max} \end{cases} \quad (3)$$

Those luminosity functions are assumed to span two orders of magnitude, that is from $L_{min} = L_0/10$ to $L_{max} = 10L_0$. As stated above, class III GRBs do show luminosity evolution, and in that case we will, in addition, use the luminosity evolution law derived by LFR of $L \propto (1+z)^{1.4}$. In the case of no evolution, the farthest GRB observed, $z_{max}$, would be the one with luminosity $L_{max}$ and measured peak flux $P_{1024} = 0.44\,\gamma\,s^{-1}\,cm^{-2}$, since that is the minimum detectable peak flux. In the case of luminosity evolution, $L_{max}$ increases with redshift, and thus the farthest GRB could be at z > 100. Since that is not physically acceptable, a maximum redshift has to be introduced by hypothesis. Lamb & Reichart [13] suggested that GRBs should occur up to $z \sim 20$, and this is the limit introduced here.

Finally, a Band spectrum is adopted with $\alpha = -1$, $\beta = -2.25$ and break energy $E_b = 511$ keV. From this spectrum, S(E), L(z,P) in equation (1) comes from:

$$P(L,z) = \frac{\int_{(1+z)50}^{(1+z)300} S(E)dE}{4\pi D^2(z)(1+z)} \quad (4)$$

Our definition for the luminosity will be:

$$L = \int_{20}^{2000} ES(E)dE \quad (5)$$

Once a GRB rate and a luminosity function are chosen, three free parameters are left: $\Omega_M$, $\Omega_\Lambda$, and $L_0$. Finally, the theoretical log N-log P distribution is normalized so that N(>$P_{min}$) coincides with the observed one. There is no need then to adjust the proportionality between the SFR and $n_z$, but the number of degrees of freedom in the $\chi^2$ test will be the number of bins of the distribution minus one.

# RESULTS

The first thing that we want to know is whether all the classes derived by BRC are likely to be cosmological populations (as GRBs at large are). For that purpose, the cosmological parameters are fixed to $\Omega_M = 0.3$, and $\Omega_\Lambda = 0.7$, the values recently obtained from Type Ia supernovae [14]. We then look for the value of $L_0$ which minimizes $\chi^2$ and with that $L_0$ we calculate $\chi^2$, varying $\Omega_M$ and $\Omega_\Lambda$ to plot the lines of equal $\chi^2$ in the plane $\Omega_M$-$\Omega_\Lambda$. This way it can be seen how large is the area on the $\Omega_M$-$\Omega_\Lambda$ plane where the observed GRB log N-log P distribution is compatible with the theoretical one.

The result is that there is no combination of $n_z$ and luminosity function for which classes II and III from the cluster analysis classification are compatible with a cosmological distribution with probability higher than $10^{-8}$. In particular, for class II the best compatibility achieved is $10^{-29}$. On the other hand, for class I, compatibility levels of $\sim 90\%$ and $\sim 50\%$ are obtained in the case of GRBs being standard candles and $n_z$ proportional to the Madau and Steidel SFRs, respectively. Since it is known that at least class III GRBs are cosmological in origin, the above classification can be rejected.

In contrast, with the neural network classification all classes do fit very well the expected log N-log P distribution for a cosmological population. In the standard candle hypothesis, classes I and II fit well for all three SFRs used, because the maximum redshift of a GRB is $z_{max} \sim 1.90$ and the behavior of the corresponding $n_z$ is very similar up to $z \sim 2$. Class III only shows good compatibility with the Steidel SFR. Fig. 2 shows as an example the lines of equal $\chi^2$, in the $\Omega_M$-$\Omega_\Lambda$ plane, for class I GRBs with the standard candle hypothesis and $n_z \propto$ Steidel SFR. Each line is labeled with the probability of the theoretical and observed distributions being the same, instead of the value of $\chi^2$ arising from the comparison.

The region of good compatibility for class II is very similar to Fig. 2, while for class III it is narrower, that being due to the higher $z_{max}$ of class III GRBs, which is $z_{max}=4.45$ for $n_z \propto$ Steidel SFR and $L_0=1.7\cdot10^{52}$ erg s$^{-1}$.

Using a symmetric power law for the luminosity function, these results vary in that, for class III, good compatibilities are found only for the case of $n_z \propto$ Madau SFR, and that since then the bursts can have higher luminosities, the values of $z_{max}$ are also higher.

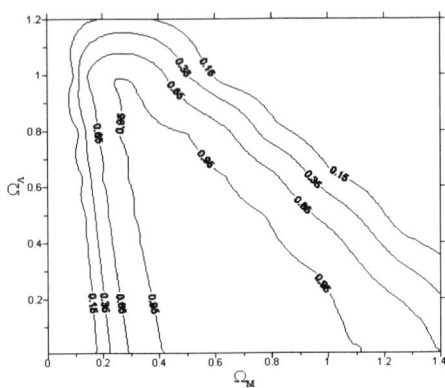

**FIGURE 2.** Equiprobability lines in the $\Omega_M$-$\Omega_\Lambda$ plane, for class I GRBs in the standard candle case and $n_z \propto$ Steidel SFR. $L_0 = 7.1\cdot10^{51}$ erg s$^{-1}$.

Up to now, what has been learnt is that classes I and II are closer populations of GRBs, with $z_{max}$ around 2 for standard candles and around 3 when adopting a luminosity function. From those GRBs alone it is not possible to discriminate among the different GRB rates proposed. On the other hand, class III GRBs are detected up to $z_{max}$ around 4.5 for standard candles and around 7 with a luminosity function, and only for them just some $n_z$ are permitted. The combinations of values of $\Omega_M$ and $\Omega_\Lambda$ for which good compatibility is achieved are also more restrictive than for classes I and II. It is possible, then, to measure $\Omega_M$, $\Omega_\Lambda$ and the SFR using class III GRBs. For that, the minimization of $\chi^2$ should not be made with just $L_0$ as a free parameter, but with $\Omega_M$ and $L_0$ jointly, and assuming $\Omega_\Lambda = \Omega_{total} - 1$, that is a flat Universe. As discussed above class III GRBs show luminosity evolution, and thus the luminosity function and evolution law calculated by LFR will be adopted in our procedure.

In Table 1 are shown the best values of $\Omega_M$, $\Omega_\Lambda$, and $L_0$ for each $n_z$ used. There are now two possible approaches to the problem: either to accept that the values for $\Omega_M$ and $\Omega_\Lambda$ from [14] are correct, and then derive the behavior of the SFR, or to assume instead that we have the correct $n_z$ and then derive $\Omega_M$ and $\Omega_\Lambda$.

From the first procedure, it would be concluded that the GRB rate is proportional to the Madau SFR, and since GRBs coming from collapsars trace the SFR, that would point to a SFR that peaks around $z \sim 1.5$ and starts to decline from there on. Even without assuming that $\Omega_M = 0.3$ and $\Omega_\Lambda = 0.7$, it would still be possible to reject the LFR GRB rate, because then the best values for $\Omega_M$ and $\Omega_\Lambda$ would imply an age of the Universe below 10 Gyr, which would be in conflict with the age of globular clusters. This also happens with the Blain SFR, for which best values for $\Omega_M$ are too high (not shown in the Table). It is also seen that, in order to obtain low

**TABLE 1.** Values of $\Omega_M$, $\Omega_\Lambda$ and $L_0$ that minimize $\chi^2$.

| $n_z$ | $\Omega_M$ | $\Omega_\Lambda$ | $L_0$ (erg s$^{-1}$) | $\chi^2$ |
|---|---|---|---|---|
| LFR | 1.38 | −0.38 | $1.2 \cdot 10^{52}$ | 17.85 |
| Madau | 0.30 | 0.70 | $1.1 \cdot 10^{52}$ | 18.19 |
| Steidel | 0.88 | 0.12 | $7.0 \cdot 10^{51}$ | 17.25 |
| Steidel(5) | 0.76 | 0.24 | $7.5 \cdot 10^{51}$ | 17.52 |
| Steidel(10) | 0.81 | 0.19 | $7.4 \cdot 10^{51}$ | 17.45 |

values for $\Omega_M$, the SFR should start decreasing at lower redshift. Even if the SFR started to decrease at z = 5, the best values for $\Omega_M$ and $\Omega_\Lambda$ would imply an age of the Universe of about 11.2 Gyr only, while the present uncertainty in the globular clusters ages spans 9-12 Gyr. Thus the SFR favored by this measure is the Madau one, and from considerations of the age of the Universe it can be concluded that the SFR should start to decrease somewhere between z ∼ 1.5 and z ∼ 5, contrarily to the GRB rate measure given by LFR, where the GRB rate increases up to at least z ∼ 10.

If, on the other hand, we had a measure of the luminosity function and the burst rate, which may happen in the near future thanks to satellites like *HETE-2* and *SWIFT*, it would be possible to introduce them in equation (1) and find the best values for $\Omega_M$ and $\Omega_\Lambda$. Taking $n_z \propto$ Madau SFR, and $L_0 = 1.1 \cdot 10^{52}$ erg s$^{-1}$, Fig. 3 shows the probabilities of compatibility of the cumulative log N-log P distributions with the pairs of values of $\Omega_M$, $\Omega_\Lambda$. It can be seen that the region with P > 15% almost coincides with that obtained from the BOOMERANG experiment [15].

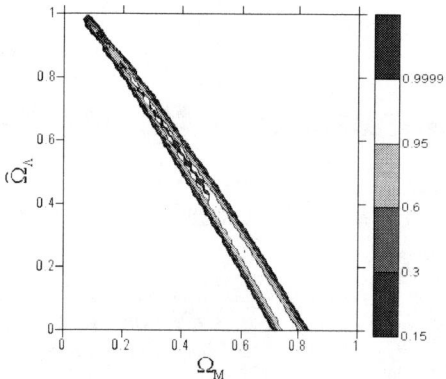

**FIGURE 3.** Equiprobability lines in the $\Omega_M$-$\Omega_\Lambda$ plane, for class III GRBs with the luminosity function and evolution law from LFR, $n_z \propto$ Madau SFR, and $L_0 = 1.1 \cdot 10^{52}$ erg s$^{-1}$.

but individual $T_{90}$ ranging between 1-100 s) makes them bad candidates to have mergings as progenitors, but in spite of having durations one order of magnitude higher than class I, their mean total fluence is only twice that of class I. They may thus be bursts with various short peaks separated in time by few seconds. The analysis of the lightcurves could then be the next step in the study of the classification by BRC, as far as class II is concerned.

Finally, it has been shown that the very high depth of class III GRBs, which are produced up to z ∼ 20, and whose detection up to this redshift is possibly due to the increase in luminosity with z, can be used to measure $\Omega_M$ and $\Omega_\Lambda$, and also to estimate the behavior of the SFR up to very high z.

## CONCLUSIONS

A test that should be passed by any subclassification of GRBs has been presented, that is, the classes must show a cosmological distribution, or at least those classes for which redshift measurements have been obtained. This requirement is guaranteed by the compatibility of the observed cumulative log N-log P distribution with the theoretical one. An analysis of the classification by BRC has shown that the neural network classification does fulfill that requirement while the cluster analysis classification does not. Further tests should be applied to this classification in order to check its physical basis.

As it was already shown in BRC, classes I and II are closer populations of GRBs than class III, and they have very similar $z_{max}$. GRBs of class III show luminosity evolution, as expected for collapsars, and the lower $z_{max}$ for classes I and II could be due to the time-delay of NS-NS and NS-BH mergings, which makes the mean redshift of mergings to be only 20-50% that of the collapsars. The durations of class II GRBs ($\langle T_{90} \rangle \sim 25$ s,

## REFERENCES

1. Balastegui A., Ruiz-Lapuente P., and Canal R., *MNRAS* **328**, 283-290 (2001) (BRC).
2. Madau P., and Pozzetti L., *MNRAS* **312**, L9-15 (2000).
3. Steidel C. C., Adelberger, K. L., Giavalisco, M., Dickinson, M., and Pettini, M., *ApJ* **519**, 1-17 (1999).
4. Blain A. W., Kneib, J. P., Ivison, R. J., and Smail, I., *ApJ* **512**, L87-90 (1999).
5. Mukherjee S., Feigelson E. D., Babu G. J., Murtagh F., Fraley C., and Raftery A., *ApJ* **508**, 314-327 (1998).
6. Hakkila J., Haglin D. J., Pendleton G. N., Mallozzi R. S., Meegan C. A., and Roiger R. J., *ApJ* **538**, 165-180 (2000).
7. Kohonen T., *IEEC Proceedings* 78, 1990, pp 1464-1480.
8. Mao S., and Paczyński B., *ApJ* **388**, L45-48 (1992).
9. Dezelay J. P. et al., *ApJ* **490**, L17-20 (1997).
10. Lloyd-Ronning N., Fryer C. L., and Ramirez-Ruiz E., *astro-ph/0108200* (2001) (LFR).
11. Weinberg S., *Gravitation and Cosmology*, Wiley, New York, 1972, pp. 451-457.
12. Schmidt, M., *ApJ* **552**, 36-41 (2001).
13. Lamb, D. Q., and Reichart, D. E., *ApJ* **536**, 1-18 (2000).
14. Perlmutter, S. et al., *ApJ* **517**, 565-586 (1999).
15. Melchiorri, A. et al., *ApJ* **536**, L63-66 (2000).

# Gamma-Ray Bursts and Cosmic Radiation Backgrounds

D. H. Hartmann[*], T. M. Kneiske and K. Mannheim[†] and K. Watanabe[**]

[*]*Department of Physics and Astronomy, Clemson University, Clemson, SC 29634-0978, USA*
[†]*Universität Würzburg, Am Hubland, 97057 Würzburg, Germany*
[**]*SSAI/LHEA, NASA/GSFC, Code 664, Greenbelt, MD 20771, USA*

**Abstract.** If gamma-ray bursts trace the cosmic star formation rate to large redshifts, their prompt and delayed emissions provide new tools for early universe cosmology. In addition to probing the intervening matter via absorption lines in the optical band, GRB continua also contribute to the evolving cosmic radiation background. We discuss the contribution of GRBs to the high-energy background, and the effect pair creation off low-energy background photons has on their observable TeV spectra.

## INTRODUCTION

Cosmic gamma-ray bursts (GRBs) have redshifts comparable to or perhaps even larger than those of quasars. Indeed, they are the most energetic explosions in the universe, with energies (uncorrected for beaming) of order $M_\odot c^2$. Their host galaxies are often sub-$L_*$, but actively forming stars at rates typical for galaxies in the early universe. The current paradigm associates GRBs with the formation of black holes in massive, rapidly rotating stars, or with the merger of compact star binaries. GRBs thus trace directly, or perhaps with a short delay, the cosmic star formation history, and may be the most easily detectable signposts of the first generation of stars (e.g., [1] and references therein). The redshifted gamma-ray flux from GRBs contributes to the evolving radiation background of the universe, as discussed in the next section, and at the same time serves as a probe of the cosmic radiation field through electron-positron pair creation absorption of their highest energy photons [2][3].

The cosmic microwave background (CMB) provides abundant soft photons for pair production of very high energy photons (in the TeV - PeV regime), but at lower energies (GeV-TeV regime) the target photons are optical and IR photons produced by stars and reprocessed by surrounding dust. Cosmic chemical evolution is intimately linked to the cosmic star formation history, and the present day extragalactic background light (EBL) provides a record of that history. Gamma-ray sources, such as GRBs and blazars, probe the evolution of this photon field through absorption effects at high energies (e.g., [2][3]). There are only three nearby active galaxies for which this absorption effect has been observed, Mrk 421, Mrk 501 (both at z = 0.03), and BL Lac (at z = 0.044). TeV emission from GRBs has only been reported for GRB970417a [4]. GRB power spectra ($\nu f_\nu$) typically peak at photon energies of a few hundred keV, but their power-law high energy emission may extend well into the GeV or even TeV regime. EGRET aboard the Compton Observatory has established that emission above 100 MeV is common, and in the case of GRB940217 a maximum photon energy of $E \sim 20$ GeV was determined [5].

Theoretical models (e.g., [6]) certainly suggest that GeV-TeV emission should be expected for a significant fraction of all bursts. The next generation GLAST experiment is expected to observe a large number of GRBs with spectral coverage up to 300 GeV. Ongoing improvements of ground-based experiments (VERITAS, HESS, HEGRA, MILAGRO, MAGIC, ...) lead to reduced sensitivities and thresholds, thus overlapping with space-based experiments. It will thus be possible to explore GRB spectra from the X-ray regime to the TeV regime, and the effects of propagation effects such as the above mentioned electron-positron pair creation must be taken into account.

To correct for $\gamma\gamma$ absorption, it is necessary to determine the cosmic evolution of the target photon distribution function, which we refer to as the metagalactic radiation field (MRF). In the third section of this paper we briefly describe our simulations of the evolving low-energy MRF, and demonstrate the extinction effect in the high energy part of GRB spectra. The gamma-ray horizon of the universe can perhaps be probed with GRBs, which would provide another powerful tool for the study of stellar evolution on the cosmic scale. GRB detections point to the onset of star formation in the universe, and their high energy spectra probe the production of light throughout the cosmic ages.

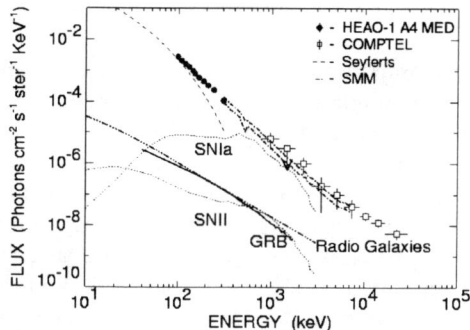

**FIGURE 1.** The observed gamma-ray background and estimated contributions from supernovae, radio galaxies, and gamma-ray bursts [7]–[9].

## THE GAMMA-RAY BACKGROUND

The unresolved cosmic gamma-ray background (CGB) from 10 KeV to 100 GeV is predominantly due to the superposition of three source populations (e.g., [7] and references therein): Seyfert galaxies, which dominate below $\sim 100$ keV; blazars, which dominate above a few MeV, and Type Ia supernovae, which fill the gap between the contributing active galaxies. The flux in the MeV regime, detected with COMPTEL and SMM, can be accounted for with nuclear line emission from the decay chain $^{56}\text{Ni} \rightarrow ^{56}\text{Co} \rightarrow ^{56}\text{Fe}$, escaping from expanding supernova ejecta (e.g. [8]). Despite their lower rate SNIa contribute the bulk of the CGB in the MeV regime due to their higher yields, and low mass envelopes. Core collapse supernovae (SNII) do contribute about 1% of the flux, as do radio galaxies such as Cen A ([9]). Figure 1 shows the observations and the contributions from various sources. The fact that SNIa so closely match the observed spectrum of the CGB places significant constraints on the cosmic star formation rate SFR(z). Most of observed flux is due to the integrated emission from supernovae that exploded at redshifts less than $\sim 1$, but a SFR that continues to rise as rapidly as observed between the present epoch and $z \sim 1$ would overproduce the CGB. The MeV data thus provide an independent constraint on the high-z star formation history of the universe.

Throughout the observable universe GRBs occur at significantly lower rates than supernovae, but their high $\gamma$-ray luminosity suggests a possible contribution to the CGB. Watanabe and Hartmann [9] used the BATSE data from the 4B catalog to estimate the contribution of observed GRBs (corrected for Earth blocking) to the CGB. As Figure 1 shows, GRBs compete with radio galaxies and SNII, but do not contribute a large portion of the observed background. Still, their role could be significant by reducing some of the deficit at 200-400 keV between the observed spectrum and the predicted flux from Seyfert galaxies.

## THE LOW-ENERGY BACKGROUND

Numerical simulations of hierarchical structure formation in a globally homogeneous Universe are tractable, but connecting the evolving gravitational structures to observable fluxes of electromagnetic radiation involves uncertain empirical descriptions of star formation, supernova feedback, and the dust-gas-star interplay. The necessary input comes from extensive observational campaigns, such as deep galaxy surveys, which measure the number of galaxies, morphological types, colors, fluxes, and distances in presumably representative solid angles out to large redshifts. The wealth of information derived from these observations can significantly complicate efforts to link theories of galaxy evolution and large scale structure formation. It is helpful to single out key quantities for which predictions can be compared with observations. One such quantity is the cosmic star formation rate (SFR) and its associated metagalactic radiation field (MRF). The MRF at $z = 0$ is commonly referred to as Extragalactic Background Light (EBL). The contribution of galaxies to the MRF is most significant between the far-infrared and the ultraviolet, while at longer wavelengths the 2.7 K microwave background (CMB) radiation from the big bang dominates.

In principle, the evolution of the MRF should be predictable from structure formation models [10] so that the observed MRF could be used to infer the role of AGNs, low surface brightness objects, and the decays of relic particles, or to single out global cosmological parameters. However, these models still rely on many uncertain parameters, and we are far from the ultimate goal of a first principles theory of the MRF. Here we compute the MRF directly from the global SFR inferred from tracers of cosmic chemical evolution, such as various Lyman $\alpha$ absorber systems, or from deep galaxy surveys. The spectral energy distribution (SED) for the globally averaged stellar population residing in galaxies can be estimated with population synthesis models [11] available for various input parameters, of which the initial mass function (IMF) and metallicity (Z) are the most important ones. Reprocessing by gas and dust is taken into account explicitly via some model of the evolution of the dust and gas content in galaxies, in combination with assumed dust properties derived from local observations in the Milky Way. The details of our modeling of the MRF are presented in [12].

Observational attempts to determine or constrain the present-day background face severe problems due to emissions from the Galaxy, which can introduce large systematic errors [13]. Nevertheless, studies with COBE have resulted in highly significant detections of a residual diffuse IR background, providing an upper bound on the MRF in the IR regime. Similarly, the cumulative flux from galaxies detected in deep HST or ISO exposures provide useful lower limits to the present-day MRF (e.g., [13]).

The method for calculating the MRF from a given SFR relies on an accurate knowledge of evolving stellar spectra and the reprocessing of star light in various dusty environments. Luminosity evolution of stellar populations is sensitive to the IMF, evolution of the mean cosmic metallicity, and the amount of interstellar extinction. Starting point of any model is the spectral energy density (SED) produced by a population of stars resulting from an instantaneous burst of star formation (commonly normalized to the mass of stars formed). Because star formation is an ongoing process with relatively short time scales of $10^{5-7}$ yrs, the starburst spectra can be directly convolved with the global SFR, $\dot{\rho}_*(z)$, to derive the evolution of the global luminosity density due to cosmic star formation. The SEDs are constructed from realistic stellar evolution tracks combined with detailed atmospheric models (e.g., [11]). The temporal evolution of the specific luminosity, $L_\nu(t)$ (in erg s$^{-1}$Hz$^{-1}$ per unit mass of stars formed) is then determined by the choices of IMF and the initial stellar metallicity.

From the population synthesis starburst models we obtain the comoving emissivity (luminosity density) at cosmic epoch $t$ from the convolution

$$\mathcal{E}_\nu(t) = \int_{t_m}^{t} L_\nu(t-t')\dot{\rho}_*(t')dt' \quad (\text{ergs}^{-1}\text{Hz}^{-1}\text{Mpc}^{-1}) \tag{1}$$

where $\dot{\rho}_*(t) = \dot{\rho}_*(z)$ is the star formation rate per comoving unit volume. Rewriting Eq. (1) in terms of redshift, $z = z(t)$, yields

$$\mathcal{E}_\nu(z) = \int_z^{z_m} L_\nu(t(z)-t(z'))\dot{\rho}_*(z')\left|\frac{dt'}{dz'}\right|dz', \tag{2}$$

where we assumed that star formation began at some finite epoch $z_m = z(t_m)$. For given evolution of the emissivity a second integration over redshift yields the energy density, or, after multiplication with $c/4\pi$, the comoving power spectrum of the MRF

$$P_\nu(z) = \nu I_\nu(z) = \nu \frac{c}{4\pi}\int_z^{z_m} \mathcal{E}_{\nu'}(z')\left|\frac{dt'}{dz'}\right|dz', \tag{3}$$

with $\nu' = \nu(1+z')/(1+z)$. Cosmological parameters enter through $dt/dz$, given by

$$\left|\frac{dt}{dz}\right| = \frac{1}{H_0(1+z)E(z)} \tag{4}$$

with an equation of state

$$E(z)^2 = \Omega_r(1+z)^4 + \Omega_m(1+z)^3 + \Omega_R(1+z)^2 + \Omega_\Lambda. \tag{5}$$

The term proportional to $\Omega_r$ takes into account the contribution from relativistic components such as the CMB and star light, although the latter would also require a new function describing the production of light as a function of time. The density parameter of this component is defined as $\Omega_r = u_r/\rho_{\text{crit}}c^2$, where $u_r$ refers to the relativistic energy density and $\rho_{\text{crit}}$ is the critical density of the universe; $\rho_{\text{crit}} = 3H_0^2/8\pi G = 10.54\ h^2$ keV/cm$^3$.

The average metallicty of gas in galaxies slowly increases with cosmic time, but the present-day value is not known precisely (e.g., [14]). We thus adopt an average extinction curve

$$A_\lambda = 0.68 \cdot E(B-V) \cdot R \cdot (\lambda^{-1} - 0.35) \tag{6}$$

with $R = 3.2$ and where $A_\lambda$ with $\lambda$ [$\mu$m] determines the absorption coefficient according to $g(\lambda) = 10^{-0.4 \cdot A_\lambda}$. Reemission by dust is calculated as the sum of three modified Planck spectra

$$L_\lambda^d(L_{bol}) = \sum_{i=1}^{3} c_i(L_{bol}) \cdot Q_\lambda \cdot B_\lambda(T_i) \tag{7}$$

where $Q_\lambda \propto \lambda^{-1}$. Two temperatures characterize warm and cold dust in galaxies, and one temperature is included to emulate a PAH component, which is also assumed to emit like a Blackbody. Dust in the ISM of the Milky Way is known to coexist at several different temperatures, determined by the distances from various heat sources. Hot dust has temperatures ranging from 50 K to 150 K-200 K when the dust is in equilibrated within HII regions, or near massive stars or compact accreting sources. Radiation from this dust component predominantly emerges in the mid-infrared and reprocesses only a small fraction of the stellar luminosity. Warm dust with temperatures between 25 K and 50 K corresponds to regions heated by the mean interstellar radiation field. Dust inside molecular clouds is shielded against high-energy radiation, and thus appears at low temperatures between 10 K and 25 K. Very cold dust at temperatures of 10 K or less can be present in the densest parts of molecular clouds or in outer regions of the galaxy where the flux of the interstellar radiation field has dropped to the value of the MRF.

The cosmic star formation rate density SFR(z) has been determined with different methods and for large set of input data. Many of these studies suggest that the original Madau curve [15] should be considered a lower limit, and that realistic rates could be larger by a factor 2−3 at all redshifts. A review of published SFR(z) functions shows that we do not yet understand systematic effects well enough to obtain a reliable estimate for

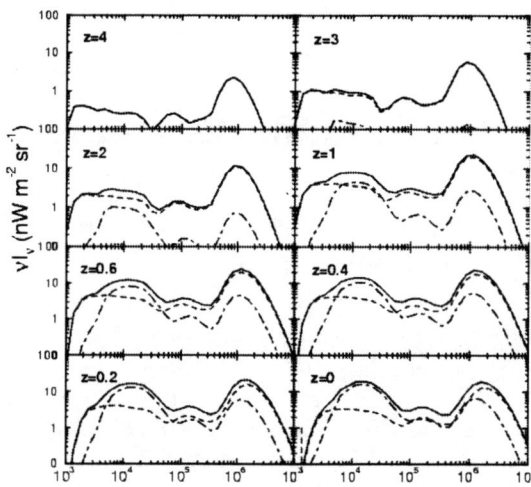

**FIGURE 2.** The evolving spectrum of the extragalactic background light. *Dashed* lines show the contribution of massive stars and *dot-dashed* lines the contribution of low mass stars

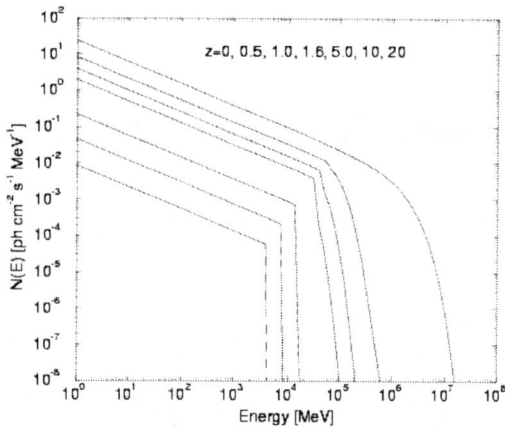

**FIGURE 3.** Using a power-law spectrum as a template (with an intrinsic cut-off above $E \sim 1$ TeV), the figure shows the effect of pair creation absorption for various redshifts.

SFR(z). This is especially true at redshifts beyond unity. This uncertainty enters in the final step of computing the MRF, the integration of the emissivity over cosmic time using Eq. (3). The evolution of the resulting MRF spectrum (comoving frame) is shown in Fig. 2 for several redshifts.

## ABSORBED TEV SPECTRA

Gamma-ray absorption due to $\gamma\gamma$-pair creation on cosmological scales depends on the line-of-sight integral of the evolving density of low-energy photons in the Universe, i.e. on the history of the diffuse, isotropic radiation field. Above we briefly discussed our semi-empirical MRF model, which is based on stellar light produced and reprocessed in evolving galaxies and calibrated with the EBL. The optical depth of the universe is given by

$$\tau_{\gamma\gamma}(E,z) = \int_0^z dz (\partial_z l)\, n(\varepsilon,z)\, \langle \sigma(E,z) \rangle \quad (8)$$

where E is the energy of the observed photon, n(ε,z) represents the evolving MRF, and the angle averaged cross section $\langle \sigma \rangle$ is of order of the Thomson cross section, $\sigma_T$. Using a power law spectrum as template (with an intrinsic absorption above 1 TeV) we show in Figure 3 how the line of sight optical depth to $\gamma\gamma$-pair creation affects the spectra. It is apparent that most GRB spectra, if they intrinsically extend into the TeV regime, are severely affected as soon as their redshifts exceed $z \sim 0.1$. If the GRB distribution traces the cosmic star formation history, we expect only a a small fraction of all bursts to be close enough for detectable TeV emission. Ongoing efforts to observe TeV emission from GRBs have so far only turned up one possible detection (GRB 970417a), which suggests either that GRBs do not commonly radiate in this regime, or that they do but are extinct by the opacity along the line of sight. If the latter interpretation is correct, TeV detections of GRBs (for example with ground based muon detectors; [16]) will be rare [2] but valuable probes of GRB physics and MRF evolution. TeV GRBs could significantly enhance the insights gathered from the limited set of TeV blazars (e.g., [17])

## REFERENCES

1. Hartmann, D. H., MacFadyen, A. I., and Woosley, S. E. in *Gamma-Ray Bursts*, eds. R. M. Kippen, R. S. Mallozzi, and G. J. Fishman, AIP 526, p. 653.
2. Mannheim, K., Hartmann, D. H., and Funk, B. 1996, ApJ 467, 532.
3. Salamon, M. H., and Stecker, F. W. 1998, ApJ 493, 547
4. Atkins, R., et al. 2000, ApJ 533, L119
5. Hurley, K., et al. 1994, *Nature* 372, 652
6. Zhang, B., and Meszaros, P. 2001, ApJ 559, 110
7. Watanabe, K., et al. 1999, ApJ 516, 285
8. The, L.-S. et al. 1993, ApJ 403, 32
9. Watanabe, K. and Hartmann, D. H. in *Gamma 2001*, eds. S. Ritz, N. Gehrels, and C. R. Shrader AIP 587, p. 442.
10. Somerville, R. & Primack, J.R. 1999 MNRAS 310, 1087
11. Bruzual, A. G., & Charlot, S. 1993, ApJ 405, 538
12. Kneiske, T. M., Mannheim, K., and Hartmann, D. H. 2002, A&A, submitted
13. Bernstein, R.A., Freedman, W., Madore, B.F. 2001, astro-ph/0112153
14. Pei, Y.C., Fall, S.M., & Hauser, M.G. 1999, ApJ 522, 604.
15. Madau, P. 1997, ApJ 475, 429.
16. Gupta, N. & Bhattacharjec, P. 2001, astro-ph/0108311
17. Aharonian, F. A., 2001, astro-ph/0112314

# Estimation of the Redshifts for Long Gamma-Ray Bursts

Zsolt Bagoly[*], István Csabai[†], Attila Mészáros[**], Peter Mészáros[‡], István Horváth[§],
Lajos G. Balázs[¶] and Roland Vavrek[¶]

[*]*Lab. for Information Technology, Eötvös University, Pázmány s. 1./A, H-1518 Budapest, Hungary*
[†]*Dept. of Physics for Complex Systems, Eötvös University, Pázmány s. 1./A, H-1518 Budapest, Hungary*
[**]*Astronomical Institute of the Charles University, V Holešovičkách 2, CZ-180 00 Prague 8, Czech Republic*
[‡]*Dept. of Astronomy, Pennsylvania State University 525 Davey Lab. University Park, PA 16802, USA*
[§]*Dept. of Physics, Bolyai Defense Academy, Box 12, H-1456 Budapest, Hungary*
[¶]*Konkoly Observatory, Box 67, H-1505 Budapest, Hungary*

**Abstract.**
It is known that the soft tail of gamma-ray bursts' spectra show excesses from the exact power-law dependence. In this article we show that this departure can be detected in the peak flux ratios of different BATSE DISCSC energy channels. This effect allows us to estimate the redshifts of bright long gamma-ray bursts given in the BATSE Catalog. For the 8 gamma-ray bursts, which have both BATSE DISCSC data and measured optical spectroscopic redshifts, the correlation between the true and estimated redshifts is remarkable, and the average error is $\Delta z \approx 0.5$. The method is similar to the photometric redshift estimation of galaxies in the optical range, hence we call it "gamma photometric redshift estimation". These redshifts for the remaining 857 long bright gamma-ray bursts are are up to $z \simeq (4-7)$. For the the faint long bursts - which should be up to $z \simeq 20$ - the redshifts hardly can be determined unambiguously from this method.

## INTRODUCTION

The gamma-ray bursts (hereafter GRBs) of the long subgroup are at high redshifts. The highest directly measured redshift is at $z = 4.5$ [1], but there are indirect considerations - based on BATSE data - predicting the existence of redshifts up to $z \simeq 20$ [2, 3, 4]. But this conclusion holds only in statistical sense, and there are known only a few cases when the observation with the BeppoSAX satellite [5] or other instruments [6] made possible to detect the afterglows and the measurement of redshifts using optical spectroscopy.

In [7] and [8] a linear relation between the intrinsic peak luminosities of GRBs and their so called "variabilities" was found. Similarly, [9] found a relation between the so called spectral lag and the peak-luminosity allowing to estimate the redshift of long GRBs.

In this article we present a new method of the estimation of the redshifts for the long GRBs. This procedure is in some sense similar to the optical observations of galaxies, where the number of objects with broad band photometric observations is much larger than the number of objects with measured spectroscopic redshifts [10]. Here we present a method that is quite similar to these methods (hence we call it as *gamma photometric redshift estimation*; GPZ for short). We utilize the fact that broadband fluxes change systematically, as characteristic spectral features redshift into, or out of the bands.

## GAMMA PHOTOMETRIC REDSHIFT ESTIMATION

To understand our method in this Section let us outline the general scheme of broadband observations, which is generally true both for the optical and gamma-ray region. Our considerations are general, and may be applied, e.g. in optical band, too, for optical sources.

Let us take two different instrumental channels defined by $E_4 > E_3$ and $E_2 > E_1$. Then one obtains

$$\frac{L_{4,3} - L_{2,1}}{L_{4,3} + L_{2,1}} = \frac{P_{4,3} - P_{2,1}}{P_{4,3} + P_{2,1}} = R(E_4, E_3, E_2, E_1, z), \quad (1)$$

where

$$L_{2,1} = \int_{E_1(1+z)}^{E_2(1+z)} L_E dE, \quad L_{4,3} = \int_{E_3(1+z)}^{E_4(1+z)} L_E dE, \quad (2)$$

and

$$P_{2,1} = \int_{E_1}^{E_2} P_E dE, \quad P_{4,3} = \int_{E_3}^{E_4} P_E dE. \quad (3)$$

Here $L_E dE$ denotes (in units photons/$(cm^2\ s)$) the luminosity, and $P_E dE$ is the observed flux (in units photons/$(cm^2\ s)$) for the energy band $[E, (E + dE)]$. $z$ is the redshift of GRB. $P_{4,3}$ and $P_{2,1}$, respectively, are the observed fluxes at the given channels.

Assume for the moment that one observes a pure power-law spectrum. This means that for the $L_E \propto E^{-\alpha}$ case $R = R(E_4, E_3, E_2, E_1, z)$ is not depending on $z$. In general case $R$, of course, will be a function of $z$.

In real situation, the incident spectrum measured by the detector is convolved with the detector's response function resulting the measured flux of the corresponding channel. For the channel with energy range $E_2 > E > E_1$ the measured flux $P_{1,2}$ is given by

$$P_{1,2} = \int_{E_1}^{E_2} P_E c(E) dE, \quad (4)$$

where $c(E)$ is the detector's response function. Hence, in general, $R$ is depending both on the spectrum and response function.

If on would know the restframe spectrum for a GRB one could calculate the theoretical $R$ as a function on $z$. This can be compared with $R$ obtained from broadband measurements ($R_{meas}$). The redshift, where $(R - R_{meas})^2$ is minimal, gives the estimated gamma photometric redshift.

Regarding this gamma photometric redshift estimation, the major problem comes from the fact that the spectra are changing quite rapidly with the time; the typical timescale for the time variation is $\simeq (0.5 - 2.5)\ s$ [11, 12]. Therefore, we will consider the spectra in the 320 ms time interval (i.e. in five 64 ms time intervals), with the peak flux at the middle of this interval.

In the following we will assume that the spectrum has the same shape around the time of the peak flux for all long bursts. We will test our assumption on GRBs, where spectroscopic redshifts are available.

## CALIBRATION

It is well-known [13] that the time-integrated average spectra of GRBs may be well approximated by a broken power-law; the break is at some energy $E_o$. Typical *restframe* energy for $E_o$ is above $\simeq 500\ keV$ [14, 15], but this varies for different GRBs.

Nevertheless the broken power-law spectrum is an approximation: first, because the break around $E_o$ may have a more complicated form (cf. [14, 15]; and, second, because at low *restframe* energies (around $\simeq 80 keV$) there may be essential departures from the power-law. This is the so called soft-excess, which is confirmed for $\simeq 15\%$ of GRBs on the high confidence level [16, 14,

15]; also for the remaining GRBs the soft-excess seems to occur, too [16].

Based on this, we construct our *template spectrum* in the following: let the spectrum be a sum of the Band's function [13] and a low energy power-law taking the form

$$\text{for } E \leq E_o\quad L_E = a(E/E_{cr})^{-\alpha} + a(E/E_{cr})^{-\beta} \quad (5)$$
$$\text{for } E \geq E_o\quad L_E = a_3(E/E_o)^{-\gamma} \quad (6)$$

,where $a_3 = a[(E_o/E_{cr})^{-\alpha}(E_o/E_{cr})^{-\beta}]$. In this spectrum there are six parameters, but the amplitude $a$ is for $R$ unimportant, and need not be specified. We choose the parameters so that the low energy cross-over is at $E_{cr} = 90\ keV$, $E_o = 500\ keV$, and the spectral indices are $\alpha = 3.2, \beta = 0.5$ and $\gamma = 3.0$ [14, 15].

There are several instruments [6] that observe GRBs and measure fluxes in the X-ray and gamma-ray band. To be able to cut out the time interval around the peak flux, we need data with reasonably good time resolution, so in our study we used the $64 ms$ resolution BATSE LAD DISCSC data from the public BATSE Catalog [17]. The 4 energies are defined by $E_1 = 25\ keV, E_2 = E_3 = 55\ keV, E_4 = 100\ keV$. Using the detector response matrices [18] one can calculate the observed counts for any incoming spectrum.

Before starting the detailed investigation of the fluxes, that one can get using the template spectra with the response matrices by Equ.(6), let us test the above trend in a simple way. Let us introduce the *peak flux ratio* (PFR hereafter) in the following way:

$$\text{PFR} = \frac{l_{34} - l_{12}}{l_{34} + l_{12}} \quad (7)$$

where $l_{ij}$ is the BATSE DISCS flux in energy channel $E_i < E < E_j$ integrated for 320 ms around the peak flux. For the above model spectrum this ratio should increase with $z$ with small redshift values. On Figure 1. we plot the theoretical PFR curves calculated from the above defined template spectrum using the averaged detector response matrices for the 9 bursts that have both BATSE data and measured redshifts.

If we redshift the template spectrum and apply the corresponding detector response matrix of the given burst, we can get a PFR value for any redshift. Figure 2 shows the theoretical curves together with the 7 observed GRB data (the DISCSC data for GRB 970828 are missing). We used the template spectrum defined in Eq. 6 and different response matrices corresponding to the observational conditions of the given burst.

There is a clear trend: as expected from the above considerations, PFR increases with increasing redshifts. Except for the supernova and the the upper limit the remaining 6 GRBs have a clearly increasing PFR with increasing $z$.

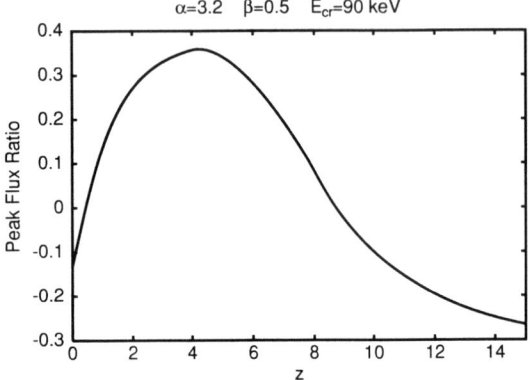

**FIGURE 1.** The theoretical PFR curves calculated from the model spectrum using the average detector response matrix.

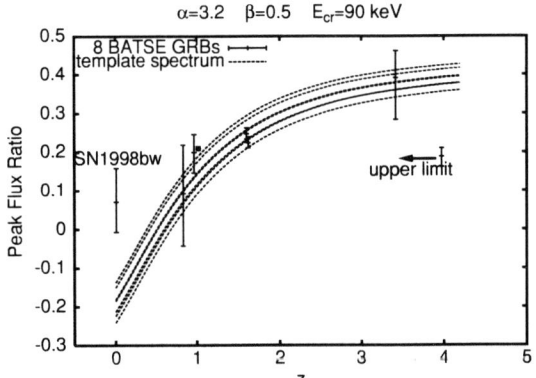

**FIGURE 2.** PFR for the 8 bursts that have both BATSE DISCS data and measured redshifts.

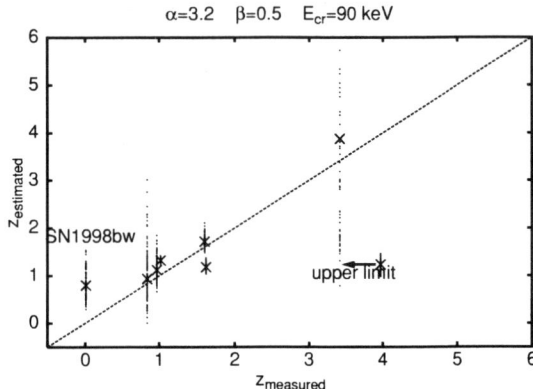

**FIGURE 3.** The measured spectroscopic redshift values compared with gamma photometric redshift estimation for the 8 considered GRBs.

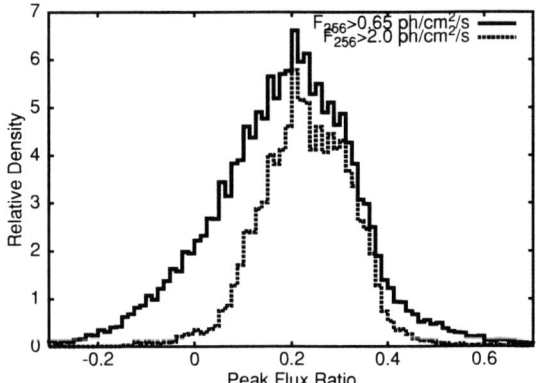

**FIGURE 4.** The PFR distribution of the long GRBs having DISCSC data.

In the plotted region the relation between $z$ and PFR is invertable, hence we can use it to estimate the *gamma photometric redshift* (GPZ) from PFR. In Figure 3 the measured spectroscopic redshift values are compared to GPZ values for the $n = 7$ considered GRBs. The errorbars show the effect of counts' Poisson noise only.

Leaving out the supernova and the burst that has redshift upper limit only, the estimation error is $\sigma_z = \sqrt{\sum_{i=1}^{n}(z_i^{spec} - z_i^{GPZ})^2} = 0.3$.

As the soft excess region redshifts out from the BATSE DISCS energy channels around $z \approx 4$, the theoretical curves converge to a constant value and start to decrease when the power-law breakpoint ($E_o$) redshifts in (i.e. around $z \simeq 4$). Hence, using only the 25-55 keV and 55-100 keV BATSE energy channels, this method can be used to estimate GPZ only in the redshift range $z \lesssim 4$; outside of this region the $z$ vs. PFR relation is non-invertable. Using further energy channels this confusion can be avoided; working out the details of the avoiding of this confusion is subject of our future work.

## ESTIMATION OF REDSHIFTS

Let us assume for a moment, that all the observed long bursts have $z < 4$. Then the estimation of $z^{GPZ}$ is straightforward for any GRB, which has calculable PFR from BATSE DISCS data. Using our method we calculate the redshift from the measured PFR values. Figure 4 shows the distribution of the measured PFR's of the long GRBs having DISCSC data. Figure 5 shows the distribution of the estimated redshifts.

The fact that the number of objects outside of the physically reasonable region ($-0.2 < $ PFR $ < 0.45$, see Figures 2. and 3.) is small suggests that the method can be valid not just for the 7 bursts with spectroscopic redshift where we could directly verify it. Also the fact that the density curve declines above $z \approx 2$ and almost vanishes around $z \approx 4$ suggests (but of course not proves) that the number of bursts with $z > 4$ aliased into the $z < 4$ region is probably small. The distribution has a peak value around $z \approx 1.5$.

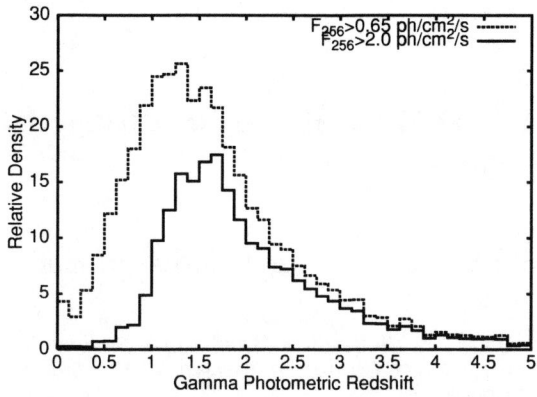

**FIGURE 5.** The gamma photometric redshift estimation distribution for the brighter half of the long GRBs having DISCSC data.

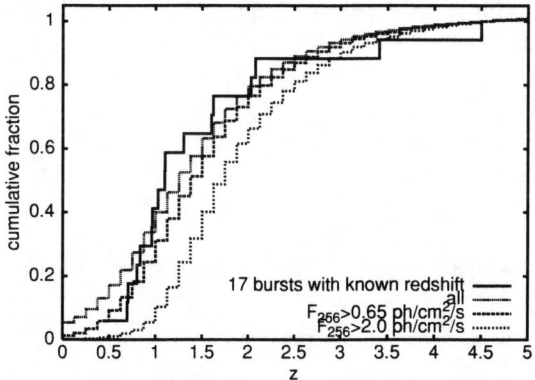

**FIGURE 6.** The cumulative redshift distribution of the 17 GRBs' with known redshift and the cumulative distributions from the gamma photometric redshift estimations with different peak fluxes.

Figure 6. shows the cumulative redshift distribution of the 17 GRBs' with known redshift and the cumulative distributions from the gamma photometric redshift estimations with different peak fluxes. the measured PFR's of the long GRBs having DISCSC data. There's a suprisingly good agreement between the measured and the GPZ distribution.

## ACKNOWLEDGMENTS

This research was supported in part through OTKA grants T024027 (L.G.B.), F029461 (I.H.) and T034549, NASA grant NAG5-2857, Guggenheim Foundation and Sackler Foundation (P.M.), Czech Research Grant J13/98:113200004 (A.M.).

## REFERENCES

1. Andersen, M. I., Hjorth, J., Pedersen, H., Jensen, B. L., Hunt, L. K., Gorosabel, J., Møller, P., Fynbo, J., Kippen, R. M., Thomsen, B., Olsen, L. F., Christensen, L., Vestergaard, M., Masetti, N., Palazzi, E., Hurley, K., Cline, T., Kaper, L., and Jaunsen, A. O., **A&A**, **364**, L54–L61 (2000).
2. Meszaros, P., and Meszaros, A., **ApJ**, **449**, 9+ (1995).
3. Meszaros, A., and Meszaros, P., **ApJ**, **466**, 29+ (1996).
4. Horvath, I., Meszaros, P., and Meszaros, A., **ApJ**, **470**, 56+ (1996).
5. Costa, E., Feroci, M., Frontera, F., Zavattini, G., Nicastro, L., Palazzi, E., Spoliti, G., di Ciolo, L., Coletta, A., D'Andreta, G., Muller, J. M., Jager, R., Heise, J., and in 't Zand, J., **IAU Circ.**, **6572**, 1+ (1997).
6. Klose, S., *Reviews of Modern Astronomy*, **13**, 129+ (2000).
7. Ramirez-Ruiz, E., and Fenimore, E. E., **ApJ**, **539**, 712–717 (2000).
8. Reichart, D. E., Lamb, D. Q., Fenimore, E. E., Ramirez-Ruiz, E., Cline, T. L., and Hurley, K., **ApJ**, **552**, 57–71 (2001).
9. Nemiroff, R. J., Marani, G. F., Norris, J. P., Bonnell, J. T., Meegan, C. A., and Hurley, K. C., "Gamma-Ray Burst Lensing Limits on Cosmological Parameters", in *AIP Conf. Proc. 526: Gamma-ray Bursts, 5th Huntsville Symposium*, 2000, pp. 663+.
10. Budavári, T., Csabai, I., Szalay, A. S., Connolly, A. J., Szokoly, G. P., Vanden Berk, D. E., Richards, G. T., Weinstein, M. A., Schneider, D. P., Benítez, N., Brinkmann, J., Brunner, R., Hall, P. B., Hennessy, G. S., Ivezić, Ž., Kunszt, P. Z., Munn, J. A., Nichol, R. C., Pier, J. R., and York, D. G., **AJ**, **122**, 1163–1171 (2001).
11. Ryde, F., and Svensson, R., **ApJ**, **512**, 693–698 (1999).
12. Ryde, F., and Svensson, R., **ApJ**, **529**, L13–L16 (2000).
13. Band, D. L., Ford, L. A., Matteson, J. L., Briggs, M., Paciesas, W., Pendleton, G., Preece, R., Palmer, D., Teegarden, B., and Schaefer, B., **ApJ**, **434**, 560–569 (1994).
14. Preece, R. D., Espley, J. R., and Briggs, M. S., "X-Ray Excesses in GRB Spectra", in *AIP Conf. Proc. 526: Gamma-ray Bursts, 5th Huntsville Symposium*, 2000, pp. 175+.
15. Preece, R. D., Briggs, M. S., Mallozzi, R. S., Pendleton, G. N., Paciesas, W. S., and Band, D. L., **ApJS**, **126**, 19–36 (2000).
16. Preece, R. D., Briggs, M. S., Pendleton, G. N., Paciesas, W. S., Matteson, J. L., Band, D. L., Skelton, R. T., and Meegan, C. A., **ApJ**, **473**, 310+ (1996).
17. Meegan, C. e. a., *Current BATSE Gamma-Ray Burst Catalog* (2000), URL http://www.batse.msfc.nasa.gov.
18. Pendleton, G. N., Paciesas, W. S., Briggs, M. S., Mallozzi, R. S., Koshut, T. M., Fishman, G. J., Meegan, C. A., Wilson, R. B., Harmon, A. B., and Kouveliotou, C., **ApJ**, **431**, 416–424 (1994).

# Determining the GRB (Redshift, Luminosity)-Distribution Using Burst Variability

Timothy Donaghy[*], Donald Q. Lamb[*], Daniel E. Reichart[†] and Carlo Graziani[*]

[*]*Department of Astronomy & Astrophysics, University of Chicago, 5640 South Ellis Avenue, Chicago, IL 60637*
[†]*Department of Astronomy, California Institute of Technology, 1201 East California Boulevard, MS 105-24, Pasadena, CA 91125*

**Abstract.** We use the possible Cepheid-like luminosity estimator for the long-duration gamma-ray bursts (GRBs) developed by Reichart et al. (2000) to estimate the intrinsic luminosity, and thus the redshift, of 907 long-duration GRBs from the BATSE 4B catalog. We describe a method based on Bayesian inference which allows us to infer the intrinsic GRB burst rate as a function of redshift for bursts with estimated intrinsic luminosities and redshifts. We apply this method to the above sample of long-duration GRBs, and present some preliminary results.

## INTRODUCTION

There is increasing evidence that gamma-ray bursts (GRBs) are due to the collapse of massive stars (see, e.g., [1] for a discussion of this evidence). If GRBs are indeed related to the collapse of massive stars, one expects the GRB rate to be roughly proportional to the star-formation rate (SFR). However, the observed redshift distribution of GRBs differs noticeably from that of the SFR: the observed GRB redshift distribution peaks at $z \approx 1$ and few bursts are observed beyond $z \sim 1.5$, while the SFR peaks at $z \approx 2$ and 10-40% of stars are thought to form beyond $z = 5$ (see, e.g., [2,3]).

However, observational selection effects play a key role in determining the observed redshift distribution of GRBs. Among these selection effects are the efficiencies with which burst redshifts can be determined by spectroscopic observations of the burst afterglow and/or the host galaxy of the burst. In addition, both of these methods of determining redshift require the identification of an optical afterglow of the GRB, and the detectability of the optical afterglow may be a function of redshift.

In this paper, we take a different approach. We use the possible Cepheid-like luminosity estimator for the long-duration GRBs developed by Reichart et al. [4] to estimate the intrinsic luminosity, and thus the redshift, of 907 long-duration GRBs from the BATSE 4B catalog [5]. This approach is free of the important and difficult-to-quantify observational selection effects that affect the redshift distribution of the GRBs with known redshifts. We assume a very general model for the GRB rate and a truncated power-law model for the intrin-

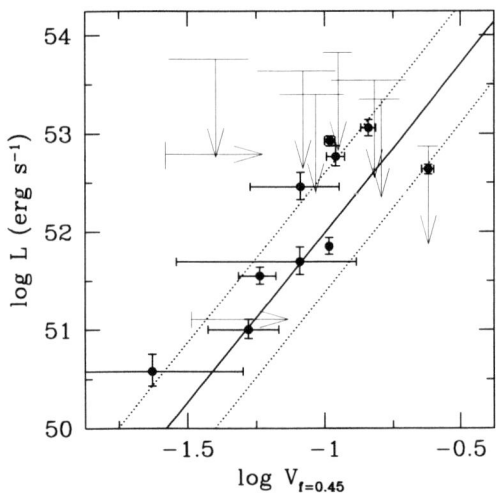

**FIGURE 1.** The variabilities $V$ and isotropic-equivalent peak photon energy luminosities $L$ of the 19 bursts for which some redshift information exists. The solid and dotted lines mare the center and 1-$\sigma$ widths of the best-fit model of these bursts in the $(\log L, \log V)$-plane.

sic GRB isotropic-equivalent photon luminosity distribution (in this work we assume that the amplitude and the power-law index of the isotropic-equivalent photon luminosity distribution does not evolve; we will relax this assumption in future work). Adopting a Bayesian approach, we calculate the likelihood of the data given the model, and convert it to a posterior distribution on the model parameters.

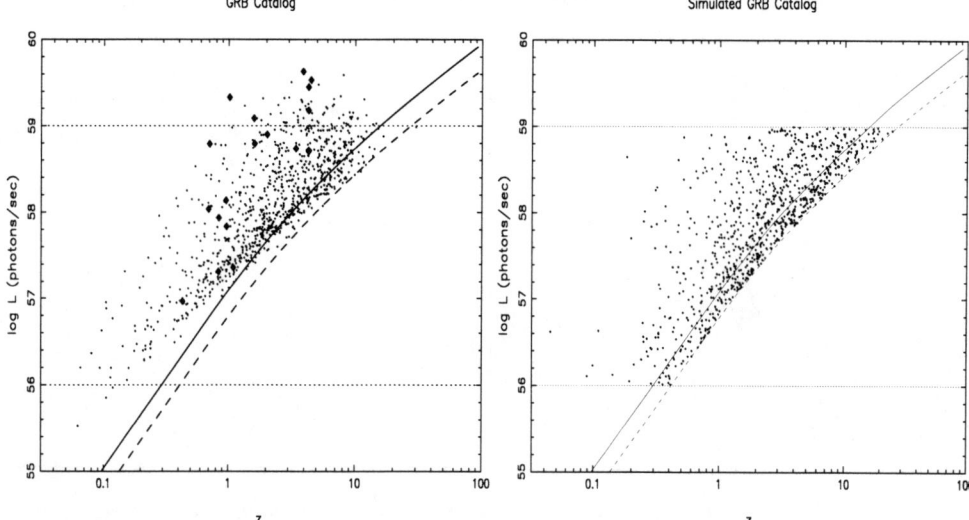

**FIGURE 2.** Left panel: distribution of the 907 long-duration BATSE bursts in the ($\log z$, $\log L$)-plane, as determined by our luminosity estimator. The solid and dashed curves represent the 90% and 10% detection thresholds of BATSE. The horizontal dotted lines show the values of $L_{\min}$ and $L_{\max}$ adopted in this study. The diamonds show the locations of the 19 GRBs with known redshifts used to calibrate the luminosity estimator. Right panel: a Monte Carlo simulation of the expected distribution of $\approx 900$ bursts with peak photon fluxes $P$ above the BATSE 90% detection threshold, assuming the maximum likelihood best-fit parameters for the model (see text).

## LUMINOSITY ESTIMATOR

A possible Cepheid-like luminosity estimator for gamma-ray bursts has been suggested by Ramirez-Ruiz and Fenimore [6] and developed further by Reichart et al. [4,7,8]. These authors have shown that there exists a correlation between a measure $V$ of the variability of the burst time history and the intrinsic isotropic-equivalent peak photon energy luminosity $L$ of the burst for the 19 GRBs for which some redshift information (either a redshift measurement or a redshift limit) exists. We show this correlation in Figure 1.

We apply this luminosity estimator to 907 bursts from the BATSE 4B catalog [5] that have durations $T_{90} > 10$ sec. We show the resulting redshift-luminosity distribution for these bursts in Figure 2 (see also [7] and [8]).

## MODEL

In this work, we assume that the rate of GRBs per unit redshift and luminosity is given by a separable function of redshift $z$ and intrinsic isotropic-equivalent photon number luminosity $L$. That is, we assume that the intrinsic photon number luminosity distribution of GRBs does not evolve with redshift; we will relax this assumption in future work. Then the rate of GRBs per unit redshift and luminosity can be written as [9]

$$\frac{dN}{dz\,dL_N} = \rho(z) f(L)\,, \quad (1)$$

where

$$\rho(z) = R_{\text{GRB}}(z;P,Q) \times (1+z)^{-1} \times 4\pi r(z)^2 (dr/dz) \quad (2)$$

is the rate of GRBs that occur at redshift $z$, and $r(z)$ is the comoving distance to the source.

We adopt a phenomenological model for the rate of GRBs that occur at redshift $z$ per unit comoving volume of the form proposed by Rowan-Robinson for the star formation rate [10]. In this model, the GRB rate is given by

$$R_{\text{GRB}}(z;P,Q) = R_0 \left(\frac{t(z)}{t(0)}\right)^P \exp\left[-Q\left(1-\frac{t(z)}{t(0)}\right)\right]. \quad (3)$$

where $P$, $Q$ and $R_0$ are model parameters, and $t(z)$ is the time since the Big Bang corresponding to the redshift $z$.

We take the intrinsic photon luminosity distribution of GRBs to be a truncated power-law,

$$\begin{aligned} f(L) &= \frac{1-\beta}{L_{\max}^{1-\beta} - L_{\min}^{1-\beta}} L^{-\beta} \\ &\quad \times \Theta(L-L_{\min})\Theta(L_{\max}-L). \end{aligned} \quad (4)$$

Thus the model has six parameters: $P$, $Q$, $R_0$, $\beta$, $L_{\min}$, and $L_{\max}$.

In this work we assume a flat universe with $\Omega_M = 0.3$ and $\Omega_\Lambda = 0.7$. Furthermore, we fix $L_{\min}$, and $L_{\max}$ to be constants (see Figure 2), reducing the number of free parameters in our model to four.

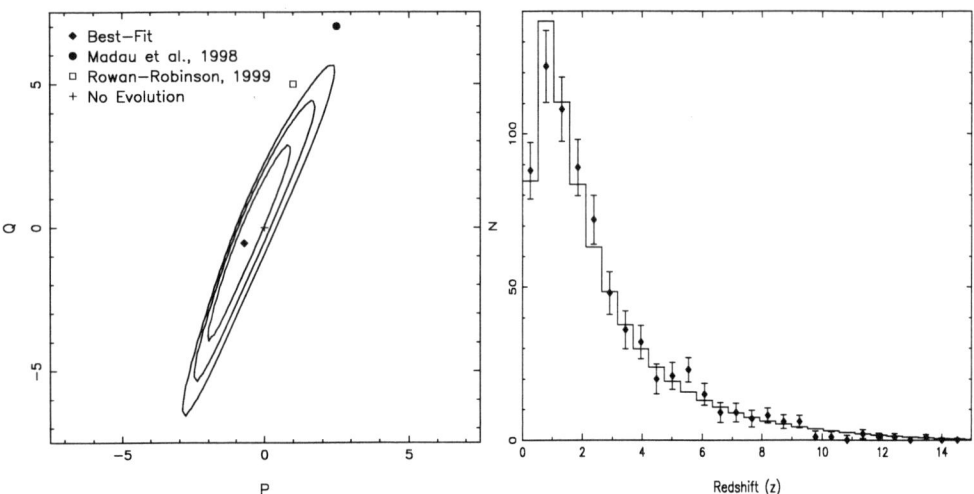

**FIGURE 3.** Left panel: the best-fit point in the $(P,Q)$-plane at the maximum likelihood best-fit values of the normalization $R_0 = 3.46$ of the GRB rate and the power-law index $\beta = 1.52$ of the GRB luminosity distribution. The contours shown correspond to $\Delta \log \mathcal{L} = 150$. Right panel: a comparison of the expected and observed number of bursts with $P$ above the BATSE 90% detection threshold for the maximum likelihood best-fit model.

## LIKELIHOOD FUNCTION

The likelihood function for GRBs that are Poisson in time is given by [9]

$$\mathcal{L} = \exp\left\{-\int dz\,dP\,\mu(z,P)\varepsilon(z,P)\right\} \prod_{i=1}^{N} \mu(z_i,P_i), \quad (5)$$

where the event likelihood,

$$\begin{aligned}
\mu(z_i,P_i) &= \int_0^\infty dL\,\rho(z_i)f(L)\delta\left(P_i - \frac{L}{4\pi r(z_i)^2(1+z_i)^\alpha}\right) \\
&= \rho(z_i) \times f\left(4\pi r(z_i)^2(1+z_i)^\alpha P_i\right) \\
&\quad \times 4\pi r(z_i)^2(1+z_i)^\alpha, \quad (6)
\end{aligned}$$

is the expected number of events observed within $dz\,dP$ of redshift $z_i$ and peak photon number flux $P_i$. The quantity $\alpha$ is the burst spectral index, which we take to be equal to 1.0, a value that is typical of GRBs [11]. By an application of Bayes' Theorem, we regard $\mathcal{L}$ as an (unnormalized) probability distribution on the model parameters. We estimate the best-fit parameters in our model by maximizing this likelihood function over the parameter space.

We take the BATSE observing efficiency $\varepsilon(z,P)$, which appears in the exponential factor in the likelihood function, to be $\theta(P - P_{th})$, where $P_{th} = 0.4$ photons cm$^{-2}$ sec$^{-1}$ is the 90% BATSE detection threshold. The product of the event likelihoods $\mu(z_i,P_i)$ runs over the 907 BATSE bursts considered in this study.

## RESULTS

Each $(z,L)$-point in the left panel of Figure 2 corresponds to the maximum value of a probability distribution for that particular burst. Thus each point has statistical and systematic errors that are associated with it and that are not displayed in the figure. Since the estimate of the redshift $z$ is calculated from the estimate of the peak photon luminosity $L$ (and the peak observed photon number flux $P$), the errors in $z$ and $L$ are completely correlated; that is, the uncertainty lies along curves of constant $P$ (i.e., they are diagonals) in the $(z,L)$-plane. Furthermore, these uncertainties are not symmetric; they are skewed toward low $L$ at low $z$ and toward high $L$ at high $z$ [7,8].

In this preliminary work, we neglect these statistical and systematic errors, and use only the best-fit values in our analysis. Calculating the likelihood of this data, given the model, and converting it to a posterior distribution on the model parameters, we find the maximum likelihood parameters for this data set to be: $P = -0.70, Q = -0.55$, $R_0 = 3.46$ and $\beta = 1.52$. The right panel of Figure 2 shows a Monte Carlo simulation of the distribution of $\approx 900$ bursts whose peak photon fluxes $P$ are above the BATSE 90% detection threshold for the maximum likelihood best-fit parameters.

The left panel of Figure 3 shows the best-fit point and contours of $\Delta \log \mathcal{L} = 150$ in the $(P,Q)$-plane at the maximum likelihood best-fit values of the normalization $R_0 = 3.46$ of the GRB rate and the power-law index $\beta = 1.52$ of the GRB luminosity distribution. The left panel of Figure 3 thus represents a two-dimensional slice through the four-dimensional parameter space of

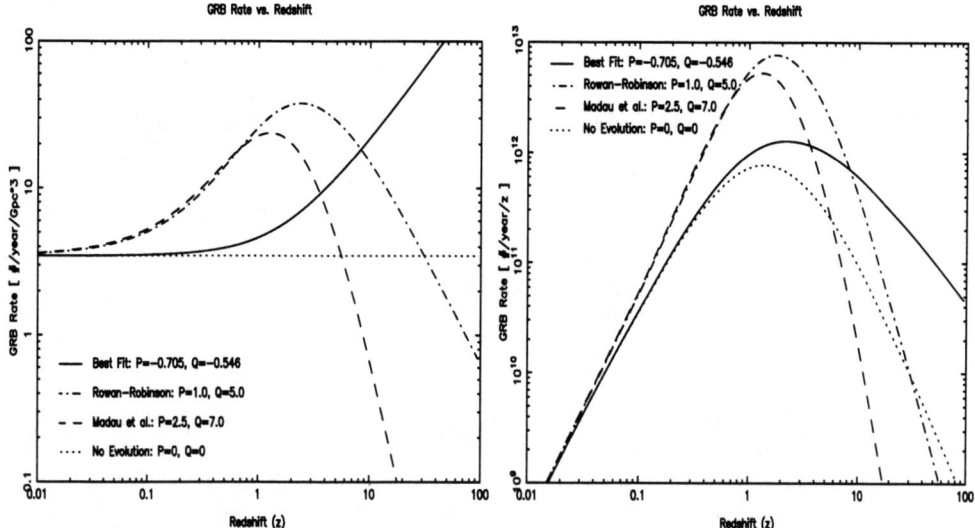

**FIGURE 4.** Left panel: the best-fit GRB rate per unit comoving volume (solid line). Right panel: the GRB rate per unit redshift $z$. Estimates of the star formation rate as a function of redshift $z$ made by Madau et al. (1998) and Rowan-Robinson (1999) are shown in both panels for comparison, as is the no evolution with redshift $z$ model ($P = Q = 0$). We emphasize that we have not taken into account the statistical and systematic errors in the redshifts $z$ and intrinsic peak photon luminosities $L$ derived from the variability measure $V$, and therefore cannot quote meaningful confidence regions for our best-fit parameters.

the model. The contours shown in Figure 3 do not correspond to credible regions because we have not taken into account the statistical and systematic errors in the redshifts $z$ and intrinsic peak photon luminosities $L$ derived from the variability measure $V$. However, taking the preliminary results at face value, the best-fit model appears to be consistent with no evolution but to be inconsistent with the Rowan-Robinson [10] and the Madau et al. [12] expressions for the star formation rate as a function of $z$.

The right panel of Figure 3 shows a comparison of the observed and expected numbers of bursts with $P$ above the BATSE 90% detection threshold for the maximum likelihood best-fit model. These distributions thus correspond to projections onto the $z$-axis of the burst distributions in the left and right panels of Figure 2, respectively.

The left panel of Figure 4 shows the best-fit GRB rate per unit comoving volume, while the right panel shows the GRB rate per unit $z$. Estimates of the star formation rate as a function of $z$ made by Madau et al. (1998) and Rowan-Robinson (1999) are shown for comparison, as is the no evolution model ($P = Q = 0$).

Applying the intrinsic peak luminosity estimator of Ramirez-Ruiz and Fenimore [5] to the brightest 220 long-duration bursts in the BATSE 4B catalog [11], Lloyd-Ronning et al. [13] find that the GRB rate increases with increasing redshift ($R_{\rm GRB} \propto (1+z)^3$ until $z \sim 2$ and $\propto (1+z)^1$ from $z \sim 2$ until $z \sim 10$), and that the mean luminosity of the GRBs also increases with increasing redshift ($<L> \propto (1+z)^{1.4\pm0.2}$). The results found in the present preliminary work differ somewhat from these conclusions. But we again emphasize that (1) the model that we have used in the present preliminary study does not allow for the evolution of $L$ with $z$, and (2) we have not taken into account the statistical and systematic errors in the redshifts $z$ and intrinsic peak photon luminosities $L$ derived from the variability measure $V$, and therefore cannot quote meaningful confidence regions for our best-fit parameters. In future work, we will relax the assumption of no evolution and will take into account the statistical and systematic errors.

## REFERENCES

1. Lamb, D. Q. 2000, Phys. Reports, **333**, 505
2. Gnedin, N. Y., & Ostriker, J. P. 1997, ApJ, **486**, 581
3. Valageas, P., & Silk, J. 1999, A&A, **347**, 1
4. Reichart, D., et al. 2001 ApJ, **552**, 57
5. Paciesas, W. S. et al. 1999, ApJS, **122**, 465
6. Ramirez-Ruiz, E. & Fenimore, E. E. 2000, ApJ, submitted (astro-ph/0004176)
7. Reichart, D. E. and Lamb, D. Q. 2001a, in Gamma-Ray Bursts in the Afterglow Era, ed. E. Costa, F. Frontera, and J. Hjorth (Springer-Verlag; Berlin), 233
8. Reichart, D. E. and Lamb, D. Q. 2001b, in Relativistic Astrophysics, AIP Conference Proceedings No. 586, ed. J. C. Wheeler and H. Martel (AIP: New York), p. 599
9. Weinberg, N., Graziani, C., Lamb, D. & Reichart, D., 2001, in Gamma-Ray Bursts in the Afterglow Era, ed. E. Costa, F. Frontera, and J. Hjorth (Springer: Berlin), 252
10. Rowan-Robinson, M., 1999, Ap&SS, **266**, 291
11. Mallozzi, R. S., et al. 1996, ApJ, **471**, 636
12. Madau, P., et al. 1998, ApJ, **498**, 106
13. Lloyd-Ronning, N. M. et al. (astro-ph/0108200)

# What are Luminosity Indicators Telling Us?

Nicole M. Lloyd-Ronning*, Chris L. Fryer[†] and Enrico Ramirez-Ruiz**

*Canadian Institute for Theoretical Astrophysics
[†]Los Alamos National Labs
**Institute of Astronomy, Cambridge

**Abstract.** We find that there exists a significant correlation between GRB luminosity and redshift in the sample of 220 Gamma-Ray Burst (GRB) that have redshifts derived from the luminosity-variability relationship [1]. In particular, we find that the relation between luminosity and redshift can be parameterized as $L \propto (1+z)^{1.4\pm\sim0.5}$. We discuss the possible reasons behind this correlation, which could result from either energy or jet opening angle evolution with redshift. In addition, we use non-parametric statistical techniques to independently estimate the distributions of the luminosity and redshift of bursts, accounting for the luminosity evolution (in contrast to previous studies which have assumed that the luminosity function is independent of redshift). Most significantly, we find a co-moving rate density of GRBs that continues to increase to $(1+z) > 10$. From this estimate of the GRB rate density, we then use the population synthesis codes of [2] to estimate the star formation rate at high redshifts, based on different progenitor models of GRBs. We find that no matter what the progenitor or population synthesis model, the star formation rate increases or remains constant to very high redshifts ($z > 10$).

## INTRODUCTION

Recently, several authors [1, 3] have discovered so-called "standard candles" - that is, relationships between luminosity and some directly measureable quantity - in a sample of GRBs that have independently measured redshifts. For example, based on eight bursts with spectroscopically measured redshifts, Fenimore and Ramirez-Ruiz ([1]; hereafter FRR) found a correlation between the bursts' luminosities $L$ (modulo a beaming factor) and the variability $V$ (or "spikiness") of their light curves. Reichart et al. [4] confirmed this relation using a different measure of variability. The physical nature of this relationship is probably related to a variation in the bulk Lorentz factor along the line of sight (see, e.g., [5, 6]).

## GAMMA-RAY BURST LUMINOSITIES AND REDSHIFTS

This relationship can be used to derive GRB luminosities and therefore redshifts. The data we use in this analysis is from Table 2 of FRR, which contains redshifts and luminosities from the empirically determined $L - V$ relationship for 220 BATSE bursts. From the burst time profile, they calculate a measure of variability (see their equation 2), obtain the luminosity from the empirical relation, and then solve for a redshift given the observed flux (and a cosmological model).

The FRR sample is chosen to have peak fluxes above the threshold $f_{256} = 1.5 \text{ph/cm}^2/\text{s}$, which defines a limiting luminosity as a function of redshift (see, e.g., [7] for more details). The data and their truncation limit as defined by this equation are shown in Figure 1, where the solid squares denote those bursts with measured redshifts; the horizontal and verticle lines are relevant for our statistical methods (described briefly below). The truncation places an important restriction on the amount of information we can obtain from the data. Because our goal is to learn about the *parent* (or intrinsic) distributions of $L$ and $z$, we must account for the trunction limit in some way. Fortunately, because the threshold is so well defined, we can use previously developed non-parametric techniques to gain information on the underlying distributions from the observed distributions and knowledge of the truncation limit.

There are simple, straightforward ways to estimate the parent distribution of truncated samples of data using maximum likelihood arguments. These methods are based on ideas first put forth by Lynden-Bell [9] and then further developed by Efron & Petrosian [8]. These non-parametric statistical techniques used a well defined truncation criterion (and the assumption that the observed sample is the one most likely to *be* observed) to estimate the correlation (if any) between the relevant variables, and their underlying parent distributions. The basic idea

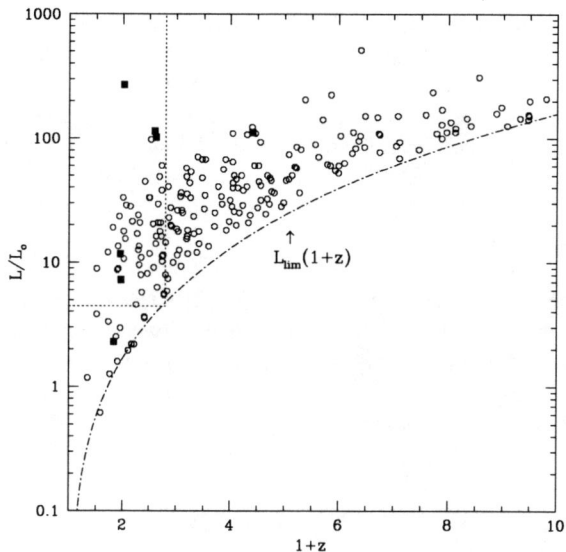

**FIGURE 1.** Luminosity vs. redshift for the GRBs in our sample.

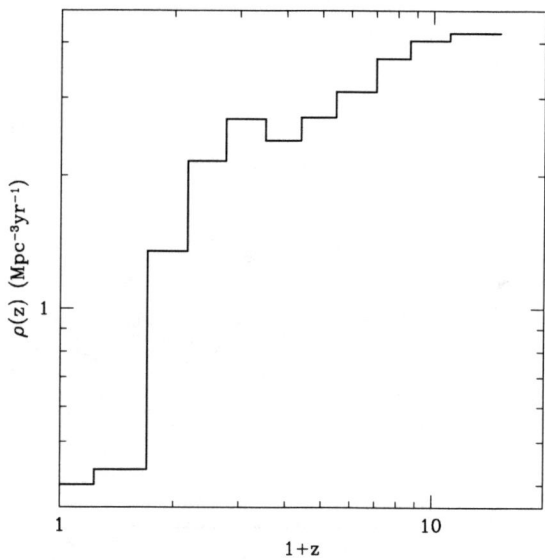

**FIGURE 2.** GRB co-moving rate density as a function of redshift.

involves defining an "eligible set" for each point by making truncations parallel to the axes (dotted lines in Figure 1), and then assigning a particular weight to each point based on the number in its eligible set. Correlations are computed via rank statistics, by only comparing points within eachother's observable limits (i.e. in the eligible set). A more detailed explanation of the techniques can be found in [7, 9, 8].

## Evolution of the GRB Luminosity Function

We apply the statistical techniques described above to the 220 FRR bursts shown in Figure 1. We find that there is a significant ($> 10\sigma$) correlation between the underlying parent distributions of luminosity and redshift in this sample, *after accounting for the truncation*. In other words, the null hypothesis that parent luminosity distribution is independent of redshift is rejected at the $10\sigma$ level. The functional form of the correlation can be conveniently parameterized by $\lambda(Z) \propto (1+z)^{1.4\pm0.5}$, in the sense that $L' = L/\lambda(Z)$ is uncorrelated with redshift (see [7] for a discussion of the error due to the scatter in the L-V relation).

If the redshifts from the $L - V$ relation are indeed valid and the GRB luminosity function therefore evolves with redshift, this has significant physical implications. Two possible causes for a luminosity-redshift correlation could be:

1) <u>Energy Evolution</u>. There are a number of observational and theoretical arguments that the stellar initial mass function (IMF) was "top heavy" in the earlier stages of the universe. This means that GRB progenitors would have on average a larger energy budget (e.g. their mass) and therefore possibly be more luminous at high redshifts. This simple "more mass = more energy" argument ignores many of the subtleties associated with GRB production, however; nonetheless, it is shown to be valid in the case of at least some progenitors [10].

2) <u>Jet Opening Angle Evolution</u> On the other hand, it has been suggested that the GRB energy output is in fact approximately constant, but the jet opening angle varies greatly from burst to burst [11], [12]. This implies that the "luminosity" evolution we find is actually evolution of the jet opening angle, where $1/\Omega \propto (1+z)^{1.4}$. This means that the average GRB opening angle is smaller at higher redshifts, and could be an indication of - among many other factors that determine the jet opening angle - the evolution of rotation rates of the progenitor.

## GRB Co-moving Rate Density

Since we have an estimate of the correlation present between $L$ and $Z$, we can independently compute the distributions of the uncorrelated variables $L'$ and $Z$. From the differential number distribution, we can derive a co-moving rate density of gamma-ray bursts through the

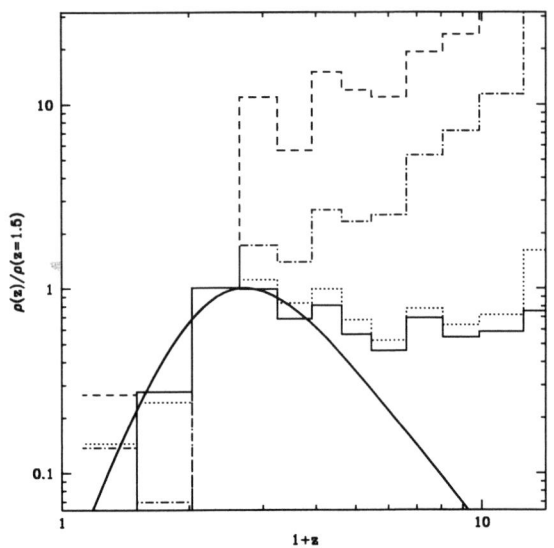

**FIGURE 3.** Star formation rate based on various progenitors for GRBs. See text for full explanation.

relation:

$$\rho(Z) = \frac{dN}{dZ}(1+z)\left(\frac{dV}{dZ}\right)^{-1}, \quad (1)$$

where $V$ is volume. We have plotted the density distribution in Figure 2 for $\Omega_\Lambda = 0.7$, $\Omega_m = 0.3$, and $H_o = 65 km/s/Mpc$, where we have arbitrarily normalized the curve so that $\rho(Z) = 1$ at $Z = 1$. Note that the density sharply rises as $\sim (1+z)^3$ to a redshift of about 2, very slightly declines, and then rises again after that as $\sim (1+z)^1$ to at least a redshift of 10.

## STAR FORMATION RATE

Most gamma-ray burst models are connected to the formation of massive stars and for some mechanisms, it is likely that the gamma-ray burst rate traces the SFR in the universe [2]. But, to extract a star formation rate at high redshifts from the gamma-ray burst distribution, we must not only know which progenitor(s) produce gamma-ray bursts, but how the evolution of these progenitors (e.g. due to metallicity effects, an evolving IMF, etc.) varies with increasing redshift. Although deriving an exact star formation rate is impossible at this time because of uncertainties (such as rotation rates of massive stars or neutron star kicks), with a few assumptions we can place some interesting limits on the star formation history using gamma-ray bursts. The details of this calculation can be found in [2, 7].

Figure 3 shows the co-moving rate density derived for both merger and collapsar progenitor models. We have assumed a particular metallicity dependence of the mass loss rate and parametrize metallicity evolution according to [13]. The solid and dotted lines in Figure 3 correspond to with and without the effects of a flattening IMF in the *collapsar model*. This provides a lower-limit for the slope of the SFR. The dashed and dot-dashed lines correspond, respectively, to the same IMF scenarios (with and without IMF flattening) for the *compact merger models*. These merger models provide a rough upper limit for the SFR derived from the rate of GRBs (He-merger and supranova GRBs will lie somewhere between the collapsar and compact merger burst models). For reference, we have superposed on the plot the star formation rate (SFR) from [14] (dark solid line).

## CONCLUSIONS

If the luminosity-variability relation of FRR [1] provides valid redshifts and luminosities to GRBs, then we can conclude that: 1) There exists a significant correlation between the luminosity of a GRB and its redshift, which can be parameterized by $L \propto (1+z)^{1.4 \pm \sim 0.5}$. 2) The comoving rate density of GRBs increases as $\rho \sim (1+z)^3$ until a redshift $1+z \sim 3$, and then continues to increase as $\sim (1+z)^1$ until a redshift of $(1+z) \sim 10$, beyond which our sample truncates. 3) From the GRB redshift distribution, we have used the population synthesis codes of Fryer et al. (1999) to compute an estimate of the star formation rate at high redshifts. We find that the star formation rate increases or at least remains constant to very high redshifts, no matter what the progenitor model. We also find that because our GRB rate density increases to high redshifts, merger models - which tend to overproduce GRBs at low redshifts - cannot be the sole source for all GRBs.

## REFERENCES

1. Fenimore, E.E. & Ramirez-Ruiz, E. 2000, *ApJ*, submitted
2. Fryer, C.L., Woosley, S.E. & Hartmann, D.H. 1999, *ApJ*, 526, 152
3. Norris, J.J., Marani, G.F, Bonnell, J.T. 2000, *ApJ*, 534, 248
4. Reichart, D.E., et al. 2000, *ApJ*, 552, 57
5. Kobayashi, S., these proceedings
6. Salmonson, J., these proceedings
7. Lloyd-Ronning, N.M., Fryer, C.L., Ramirez-Ruiz, E.
8. Efron, B. & Petrosian, V. 1992, *ApJ*, 399, 345
9. Lynden-Bell, D. 1971, MNRAS, 155, 95
10. Zhang, W. & Fryer, C.L. 2001, *ApJ*, 550, 357
11. Frail, D., et al., *ApJ*, in press
12. Panaitescu, A. & Kumar, P. 2001, *ApJ*, 554, 667
13. Pei Y.C., Fall, S.M., & Hauser, M.G. 1999, *ApJ*, 522, 604
14. Blain, A.W. 2001, in the Ringberg Proceedings

# Consequencs of a Dependence of GRB Properties on Local Metallicity

Enrico Ramirez-Ruiz[*], Andrew W. Blain[†] and Davide Lazzati[*]

[*]*Institute of Astronomy, Madingley Road, Cambridge CB3 0HA, England*
[†]*Department of Astronomy, California Institute of Technology, Pasadena, CA 91125, USA*

**Abstract.** We report a correlation between the isotropic equivalent energy of GRBs and their position offset from their host galaxies. This is possibly due to a dependence of the end point of massive stellar evolution on metallicity. If confirmed in further host observations, this correlation will both complicate interpretation of GRBs as tracers of cosmic star formation, and potentially allow a new probe of the astrophysics in high-redshift galaxies.

## THE BIRTH PLACE OF GAMMA-RAY BURSTS

Evidence is accumulating that GRBs are intimately linked with the deaths of massive stars. For the long-burst afterglows localized so far, the host galaxies show signs of the ongoing star formation activity necessary for the presence of young, massive progenitor stars [1, 2, 3]. The physical properties of the afterglows, their locations in host galaxies [4], iron line features, and evidence for supernova components several weeks after the burst [5, 6]. strongly support the idea that the most common GRBs are linked to the collapse of massive stars.

The circumburst medium provides a natural laboratory for studying GRBs. Stars that readily shed their envelopes have short jet-crossing times and are more likely to produce a GRB. Stars with less radiative mass loss retain a hydrogen envelope, in which a poorly collimated jet is likely to lose energy and fail to breaking out of the star [7]. Finding useful diagnostics for the progenitors is simplified if the metallicity of and physical conditions in the local ISM influences the evolution of the progenitor. GRBs occur close to the birth sites of their short-lived progenitors, and so their evolution is likely to be affected only by local properties of the host galaxy [8].

Here, we show that bursts located closer to the center of their parent galaxies have smaller isotropic equivalent energies (or broader jets), and so progenitors in inner and outer galactic locations may be intrinsically different (see Figure 1). We suggest that this could be the outcome of abundance gradients in the host galaxy.

## ELEMENT ABUNDANCES FROM THE LOCAL UNIVERSE TO HIGH REDSHIFTS

An exciting recent development in observational cosmology has been the extension of studies of abundances from the local Universe to high redshifts. The dependence of metallicity on environment appears to be stronger than on the redshift of formation: galaxies selected using the same techniques have metallicities rather independent of redshift, and old stars are not necessarily metal-poor [9]. Chemical abundances within different galaxies depend strongly on luminosity and environment [9]. From the center to the outermost 10 kpc, metallicity typically decreases by a factor of ten. A comparable change in metallicity only occurs over a range of a factor of a thousand in luminosity (see Fig. 5 of [9]). This is a much greater range of luminosity than displayed by moderate-redshift GRB host galaxies, which usually have magnitudes $R \approx 25$. These host galaxies are UV-bright [10], and so may exhibit comparable abundance gradients to their local counterparts. Drawing inferences about GRB hosts from local galaxies is difficult, however, since both merging and secular evolution are likely to be important and will complicate a direct comparison. Nonetheless, a direct association between abundance gradients in GRB hosts and in local galaxies could be responsible for the correlation presented in Figure 1.

Low-metallicity stars, which are likely to be more prominent in the outskirts of the galaxy, are smaller and have less mass loss than their metal-rich counterparts. Both properties inhibit the loss of angular momentum [7], and so low-metallicity stars are likely to be rotating rapidly. Equatorial accretion may thus be delayed and

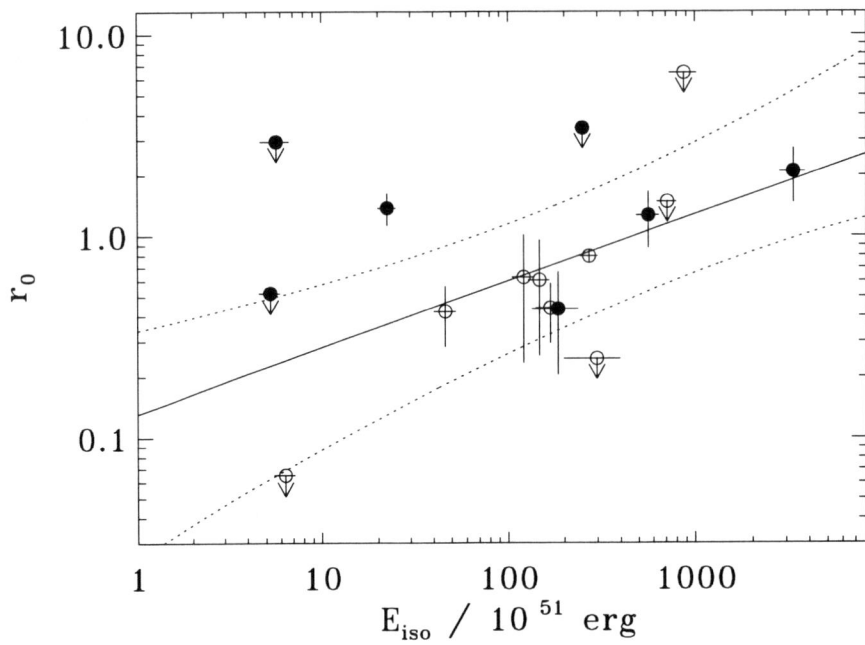

**FIGURE 1.** The projected observed offset of GRBs from their parental galaxy as a function of the burst isotropic equivalent energy. The center of the assigned host is determined as the centroid of the brightest component of the host system. The fractional isophotal offsets $r_0$ are the observed offsets $R_0$ normalized by the host half-light radius. Solid and dotted lines mark the center and $1\sigma$ widths of the best-fit model distributions parameters. The filled circles are bursts that occur in the most irregular, possibly merging galaxies, while the empty circles are bursts with more regular hosts. There is a tentative trend: the inner most bursts seem to be less energetic.

a funnel may be produced along the rotation axis. For higher rotational velocities this evacuated region will be more collimated, reducing the jet opening angle. Furthermore, for a given mass-loss rate, the lower the metallicity, the higher both the WR stellar mass, and the mass threshold for the removal of the hydrogen envelope by stellar winds. These effects all increase the mass of the helium core and favor black hole formation [7, 11]. If there are abundance gradients in the hosts, then the likely metallicity dependence of both black-hole formation and rotation suggests that GRBs in outer galactic locations may be more energetic (greater helium core mass) or less collimated (faster rotation) than those close to the galactic center. In the local Universe, regular galaxies are found to have steeper abundance gradients than those with complex morphologies [12]. Indeed, it is reassuring that the most regular GRB host galaxies (shown as open circles in Fig. 1) firmly support the trend between $E_{\rm iso}$ and $r_0$. Note that the scatter in Fig. 1 can be due to the dependence of metallicity on luminosity. A more detailed analysis of the underlying reasons for the correlation requires a large and unbiased sample of GRBs hosts, and knowledge of both the underlying GRB and afterglow luminosity functions.

## THE EFFECTS OF METALLICITY

What are the potential effects of a significant dependence of GRB luminosity, as detected by unextinguished $\gamma$-ray photons, on their location in the host galaxy, which could reflect the metallicity of their progenitors? The most significant is a potential offset between the true star-formation rate and that traced by GRB. If GRBs in outlying, low-metallicity environments and in low-mass galaxies are more luminous, then they are likely to be overrepresented in GRB samples, and especially in the bright BATSE catalog, as compared with those in high-metallicity environments.

The radial dependence of metallicity $Z$ in low-redshift spiral [12] and elliptical galaxies [13], is $Z \propto \exp(-1.9R/R)$, while the dependence of metallicity at fixed radius on enclosed mass $M_{\rm enc}$ in spiral galaxies derived from Fig. 4 of [13] is $Z \propto M_{\rm enc}^{\simeq -0.5}$. These functions both depend strongly on radius. Therefore, it is likely that local environmental effects will overcome global enrichment effects [9], but that there will be a gradual increase in the typical luminosity of GRBs with increasing redshift (see [14]).

Low-mass galaxies are likely to have statistically lower metallicities and thus contain more luminous GRBs than high-mass galaxies. As galaxy mass is expected to build up monotonically through mergers, then it is possible that the highest-redshift GRBs could be systematically more luminous due to the lower mass of their hosts, perhaps by a factor of 2–3 at $z \simeq 3$. This effect is likely to be more significant than, but in the same direction as, the global increase in metallicity with cosmic time.

The most luminous GRBs of all could be associated with metal-free Population-III stars; however, their very high redshifts would make examples difficult to find even in the *Swift* catalog of hundreds of bursts.

Star-formation activity is likely to be enhanced in merging galaxies. In major mergers of gas-rich spiral galaxies, this enhancement takes place primarily in the inner kpc, as bar instabilities drive gas into the core [15]. Metallicity gradients in the gas are likely to be smoothed out, both by mixing prior to star formation, and by SN enrichment during the burst of activity. GRB luminosities could thus be suppressed in such well-mixed galaxies, making GRBs more difficult to detect in these most luminous objects, in which a significant fraction of all high-redshift star formation is likely to have occurred. Shocks in tidal tails associated with merging galaxies are also likely to precipitate the formation of high-mass stars, yet as tidal tails are likely to consist of relatively low-metallicity gas, it is perhaps these less intense sites of star-formation at large distances from galactic radii that are more likely to yield detectable GRBs.

For star formation taking place in both merging and quiescent high-redshift galaxies, there should thus be a bias in favor of detecting GRBs at a greater projected distance from the host galaxy than the mean radius of the star-formation activity. Hence, based on the correlation shown in Figure 1, we predict that the radial distribution of a large sample of GRBs around their host galaxies should be considerably more extended than the signatures of star-formation regions within the host, such as blue colors, location of H$\alpha$ emission, intense radio emission etc. This might have the unfortunate consequence of making GRBs more difficult to use as clean markers of high-redshift star-formation activity. Detailed observations of the astrophysics of individual GRB host galaxies may be essential before a large sample of bursts can be interpreted. More optimistically, the astrophysics of star formation in high-redshift galaxies could perhaps be studied using the intrinsic properties of a well-selected population of GRB with deep, resolved host galaxy images.

If confirmed in detailed studies, a metallicity selection effect for GRBs may be able to explain the differences between the star-formation rate inferred from observations of galaxies [16, 17], which tend not to increase with redshift beyond $z \simeq 2$, and the rate inferred from GRB counts assuming a variability–luminosity relation [18, 14], which continues to increase to the highest redshifts. This increase may reflect a bias to detecting high-redshift GRBs in more numerous, low-mass, low-metallicity high-redshift galaxies.

Another test of the effect could be provided by a comparison of the luminosity function of GRB host galaxies with that of the total galaxy luminosity function over the same redshift range. If there is a bias towards the discovery of GRBs in low-metallicity regions, then the GRB host galaxy luminosity function should be biased to low luminosities by an increasing amount as redshift increases.

We thank G. Denicoló, M. Pettini, M. J. Rees, C. Tout, and N. Trentham for useful comments and suggestions. ERR acknowledges support from CONACYT, SEP and the ORS foundation. AWB thanks the Raymond & Beverly Sackler Foundation for financial support.

## REFERENCES

1. Kulkarni, S. R. et al., *Nature*, **395**, 663 (1998).
2. Fruchter, A. S. et al., *ApJ*, **520**, 54 (1999).
3. Berger, E., Kulkarni, S. R., Frail, D. A., *ApJ*, **560**, 652 (2001).
4. Bloom, J. S., Kulkarni, S. R., Djorgovski, S. G., *AJ*, **123**, 1111 (2002).
5. Bloom, J. S. et al., *Nature*, **401**, 453 (1999).
6. Reichart, D. E., *ApJ*, **521**, L111 (1999).
7. MacFadyen, A. I., Woosley, S. E., *ApJ*, **524**, 262 (1999).
8. Ramirez-Ruiz, E., Lazzati, D., Blain, A. W., *ApJ*, **565**, L9 (2002)
9. Pettini, M., *Proc. of the 'The promise of First', Pilbratt, Cernicharo, Heras & Prusti eds., ESA SP-460* (2001).
10. Trentham, N., Ramirez-Ruiz, E., Blain, A. W., *MNRAS in press (astro-ph/0204350)* (2002).
11. Ramirez-Ruiz, E., Dray, L., Madau, P., Tout, C. A., *MNRAS*, **327**, 829 (2001).
12. Zaritsky, D., Kennicut, R. C., Huchra J. P., *ApJ*, **420**, 87 (1994).
13. Henry, R. B. C., Worthey, G., *PASP*, **111**, 919 (1999).
14. Lloyd-Ronning, N. M., Fryer, C. L., Ramirez-Ruiz, E., *ApJ in press (astro-ph/0108200)* (2002).
15. Mihos, J. C., Hernquist, L., *ApJ*, **431**, L9 (1994).
16. Steidel, C. C., Adelberger, K. L., Giavalisco, M., Dickinson, M., Pettini, M., *ApJ*, **519**, 1 (1999).
17. Blain, A. W., Smail, I., Ivison, R. J., Kneib, J.-P., 1999, *MNRAS*, **302**, 632 (1999).
18. Fenimore, E. E., Ramirez-Ruiz, E., *ApJ submitted (astro-ph/0004176)* (2001).

# Determining the Gamma-Ray Burst Rate as a Function of Redshift

Nevin Weinberg[*], Carlo Graziani[†], Donald Q. Lamb[†] and Daniel E. Reichart[*]

[*]*Department of Astronomy, California Institute of Technology, Mail Code 105-24, 1201 East California Boulevard, Pasadena, CA 91125*
[†]*Department of Astronomy & Astrophysics, University of Chicago, 5640 South Ellis Avenue, Chicago, IL 60637*

**Abstract.** We exploit the 14 gamma-ray bursts (GRBs) with known redshifts $z$ and the 7 GRBs for which there are constraints on $z$ to determine the GRB rate $R_{\rm GRB}(z)$, using a method based on Bayesian inference. We find that, despite the qualitative differences between the observed GRB rate and estimates of the SFR in the universe, current data are consistent with $R_{\rm GRB}(z)$ being proportional to the SFR.

## INTRODUCTION

There is increasing evidence that GRBs are due to the collapse of massive stars (see, e.g., [1] for a discussion of this evidence). If GRBs are indeed related to the collapse of massive stars, one expects the GRB rate to be roughly proportional to the SFR. However, the observed redshift distribution of GRBs differs noticeably from that of the SFR: the observed GRB redshift distribution peaks at $z \approx 1$ and few bursts are observed beyond $z \sim 1.5$, while the SFR peaks at $z \approx 2$ and 10-40% of stars are thought to form beyond $z = 5$ (see, e.g., [2,3,]).

However, observational selection effects play an important role in determining the observed redshift distribution of GRBs. The important question is therefore whether or not the discrepancy between the observed GRB redshift distribution and the redshift dependence of the SFR is entirely due to selection effects; i.e., is the GRB rate roughly proportional to the SFR after taking observational selection effects into account? We address this question in this paper.

## METHOD

We adopt a Bayesian approach. We calculate the likelihood of the data given the model, and convert it to a posterior distribution on the model parameters. We assume a very general model for the GRB rate, and a power-law model for the intrinsic GRB photon luminosity distribution (we assume that the amplitude and the power-law index of the photon luminosity distribution does not evolve; we relax this assumption in future work). We determine the efficiency with which BeppoSAX and the IPN detect GRBs as a function of peak photon flux $P$ by comparing the BeppoSAX and IPN peak photon flux distributions to that of BATSE. We fit the model jointly to the peak fluxes and redshifts of the 14 GRBs with known $z$, and the 7 GRBs for which there are constraints on $z$.

We write the rate of GRBs that occur per unit redshift and luminosity as

$$dN/dz\,dL_N = \rho(z) f(L_N), \quad (1)$$

where

$$\rho(z) = R_{\rm GRB}(z;P,Q) \times (1+z)^{-1} \times 4\pi r(z)^2 (dr/dz) \quad (2)$$

is the rate of GRBs that occur at redshift $z$,

$$R_{\rm GRB}(z;P,Q) = \left[\frac{t(z)}{t(0)}\right]^P \exp\left[Q\left(1 - \frac{t(z)}{t(0)}\right)\right] \quad (3)$$

is the rate of GRBs that occur at redshift $z$ per unit comoving volume (see [5]), $t(z)$ is the elapsed time since the Big Bang, and

$$f(L_N) = L_N^{-\beta} \times \Theta(L_N - L_{\min})\Theta(L_{\max} - L_N) \quad (4)$$

is the intrinsic photon luminosity distribution of GRBs. Thus the model has five parameters: $P$, $Q$, $\beta$, $L_{\min}$, and $L_{\max}$.

We write the efficiency with which GRBs with known redshifts are found as $\varepsilon(z,P) = \varepsilon_z(z) \times \varepsilon_{\rm ST}\varepsilon_P(P)$, where $\varepsilon_z(z)$ is the efficiency with which the redshifts of GRBs are determined from optical observations once they are detected by a γ-ray burst instrument. We take $\varepsilon_z(z) = 1$

Figure 1 compare the best-fit models of $\varepsilon_P$ and the cumulative peak flux distributions of BATSE, the IPN, and BeppoSAX, respectively. Figure 2 shows the best-fit $\varepsilon_P$ for each of the three experiments.

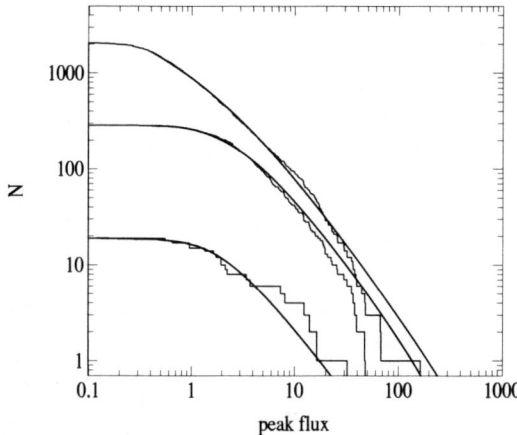

**FIGURE 1.** Comparison of the best-fit models to the differential distribution of peak fluxes of BATSE, IPN and BeppoSAX bursts, and the cumulative peak flux distributions for the three experiments.

for GRBs whose redshifts were determined by detection of an absorption-line system in the optical afterglow of the burst and $\varepsilon_z(z) = \Theta(1-z) + \Theta(z-2.5)$ for GRBs whose redshifts were determined by measuring emission lines in the spectra of the host galaxy. The latter expression accounts qualitatively for the difficulty in measuring redshifts when the $H_\alpha$ and O[II] emission lines from host galaxies do not lie in the visible spectrum. The quantity $\varepsilon_{ST}$ is the "stereo-temporal" efficiency that accounts for limitations of exposure in time and solid angle and $\varepsilon_P(P)$ is the efficiency with which BATSE, the IPN and BeppoSAX detect GRBs as a function of peak flux $P$.

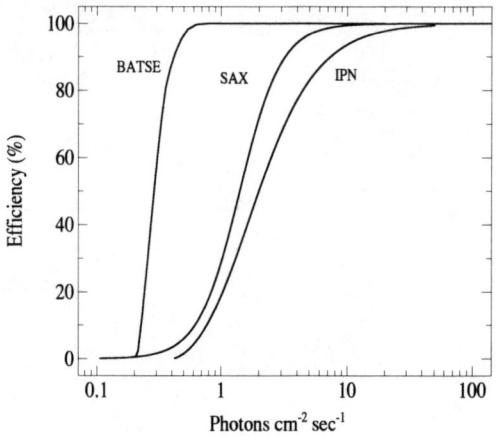

**FIGURE 2.** Efficiency $\varepsilon(P)$ with which BATSE, IPN and BeppoSAX detect GRBs as given by the best-fit models to the differential distribution of peak fluxes of BATSE, IPN and BeppoSAX bursts.

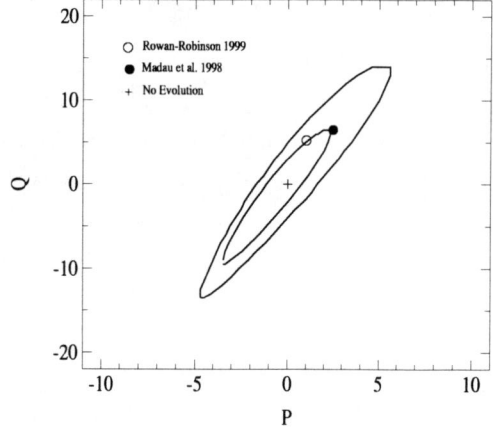

**FIGURE 3.** Probability contours in the (P,Q)-plane for the GRB rate parameters $(P,Q)$ found from fitting to the 14 GRBs with known $z$ and the 7 GRBs with constraints on $z$. The solid curves correspond to the 68% and 95% probability contours. Also shown are the (P,Q)-values corresponding to no space density evolution (+), the Madau et al. SFR [4], and the Rowan-Robinson phenomenological model fit to IR, optical and UV data [5].

The likelihood function is then given by

$$\mathcal{L} = \exp\left\{-\int dz\,dP\,\mu(z,P)\varepsilon(z,P)\right\} \prod_{i=1}^{N} \mu(z_i, P_i), \quad (5)$$

where

$$\mu(z,P) = \int_0^\infty dL_N\,\rho(z)f(L_N)\delta\left(P - \frac{L_N}{4\pi r(z)^2(1+z)^\alpha}\right)$$
$$= \rho(z) \times f\left(4\pi r(z)^2(1+z)^\alpha P\right) \times 4\pi r(z)^2(1+z)^\alpha$$
(6)

is the expected number of events observed within $dz\,dP$ of $(z,P)$. The quantity $\alpha$ is the burst spectral index, which we set equal to one in this work. By an application of Bayes' Theorem, we now regard $\mathcal{L}$ as an (unnormalized) probability distribution on the model parameters.

## RESULTS

Figure 3 shows 68% and 95% probability contours for the GRB rate parameters $P$ and $Q$. Also shown on the plots are the best-fit SFR models of Madau et al. [4] and Rowan-Robinson [5]. The SFR models lie at about a 68% excursion from the best-fit GRB rate model. Thus we find that, despite the qualitative differences that exist between the observed GRB rate and estimates of the SFR in the universe, current data are consistent with the actual GRB rate being approximately proportional to the SFR.

## REFERENCES

1. Lamb, D. Q. 2000, Phys. Reports, **333**, 505
2. Gnedin, N. Y., & Ostriker, J. P. 1997, ApJ, **486**, 581
3. Valageas, P., & Silk, J. 1999, A&A, **347**, 1
4. Madau, P., Pozzetti, L., & Dickinson, M. 1998, ApJ, **498**, 106
5. Rowan-Robinson, M. 1999, Ap&SS, **266**, 291

# VII. INSTRUMENTATION AND TECHNIQUES

## Current and Future Satellite Missions

## Ground-Based Instruments

## Techniques

# The Swift GRB MIDEX Mission

## N. Gehrels

*Mail Code 661, Goddard Space Flight Center, Greenbelt, MD 20771*

*On behalf of the Swift team*

**Abstract.**
Swift is a first-of-its-kind multiwavelength transient observatory for gamma-ray burst astronomy. It has the optimum capabilities for the next breakthroughs in determining the origin of gamma-ray bursts and their afterglows, as well as using bursts to probe the early Universe. Swift will also perform the first sensitive hard X-ray survey of the sky. The mission is being developed by an international collaboration and consists of three instruments, the Burst Alert Telescope (BAT), the X-ray Telescope (XRT), and the Ultraviolet and Optical Telescope (UVOT). The BAT, a wide-field gamma-ray detector, will detect 3-7 gamma-ray bursts per week with a sensitivity 5 times that of BATSE. The sensitive narrow-field XRT and UVOT will be autonomously slewed to the burst location in 20 to 70 seconds to determine 0.3-5.0 arcsec positions and perform optical, UV, and X-ray spectrophotometry. Strong education/public outreach and follow-up programs will help to engage the public and astronomical community. The Swift launch is planned for September 2003.

## INTRODUCTION

The discovery by BeppoSAX and ground observers [1, 2, 3] of afterglow from gamma-ray bursts (GRBs) has shown that they are cosmological, involving the most powerful explosions known. These explosions are thought to create super-relativistic blast-waves resulting in afterglow that fades from gamma-rays to radio. However, important information on the afterglow is lost by the $\sim$ 8 hour or longer delay between the initial burst and follow-up observations.

Swift is a multiwavelength observatory that exploits the afterglow characteristics of GRBs to make a comprehensive study of hundreds of bursts. It will determine the origin of GRBs, tell us how the blast wave interacts with its suroundings, and identify classes of bursts. Swift will also investigate how GRBs can be used to study the early Universe.

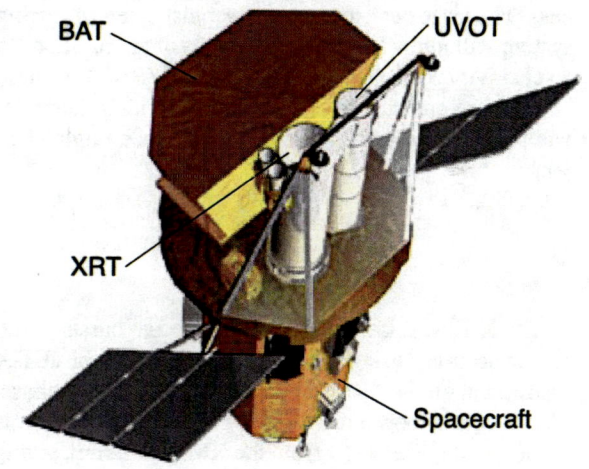

**FIGURE 1.** The Swift satellite.

## SWIFT INSTRUMENTS

The Swift instruments were carefully chosen for GRB discovery. It incorporates a wide-field GRB detector and two sensitive narrow-field telescopes for identifying and observing X-ray, UV, and optical afterglow (Figure 1).

### Burst Alert Telescope

The Burst Alert Telescope (BAT) covers the 15-150 keV energy band and will detect 150-300 bursts per year (the range is due to uncertainties in the number of weak GRBs). The GRB detector has a CdZnTe (CZT) detector array with an area of 5200 cm$^2$ and a coded aperture mask covering 1.4 sr of the sky. The mask is positioned one meter away from the detectors and will provide posi-

**TABLE 1.** BAT parameters.

| | |
|---|---|
| Energy Range | 15-150 keV |
| Aperture | Coded mask |
| Detecting Area | 5200 cm$^2$ |
| Detector | CdZnTe (CZT) |
| Detector Operation | Photon counting |
| Field of View | 1.4 sr (half-coded) |
| Detection Elements | 256 modules of 128 elements |
| Detector Element Size | $4 \times 4 \times 2$ mm$^3$ |
| Telescope PSF | 17 arcmin |

**TABLE 2.** XRT parameters.

| | |
|---|---|
| Energy Range | 0.2-10 keV |
| Telescope | JET-X Wolter 1 |
| Detector | EPIC CCD |
| Effective Area | 110 cm$^2$ @ 1.5 keV |
| Detector Operation | Photon counting, integrated imaging, and timing |
| Field of View | $23.6 \times 23.6$ arcmin |
| Detection Elements | $600 \times 600$ pixels |
| Pixel Scale | 2.36 arcsec |
| Telescope PSF | 15 arcsec HPD @ 1.5 keV |
| Sensitivity | $2 \times 10^{-14}$ erg cm$^{-2}$ s$^{-1}$ in $10^4$ seconds |

**TABLE 3.** UVOT parameters.

| | |
|---|---|
| Wavelength Range | 170 nm - 600 nm |
| Telescope | Modified Ritchey-Chrétien |
| Aperture | 30 cm diameter |
| F-number | 12.7 |
| Detector | Intensified CCD |
| Detector Operation | Photon counting |
| Field of View | $17 \times 17$ arcmin |
| Detection Elements | $2048 \times 2048$ pixels |
| Telescope PSF | 0.9 arcsec @ 350nm |
| Colors | 6 |
| Sensitivity | $B = 24$ in white light in 1000 s |
| Pixel Scale | 0.5 arcsec |

tions of 1-4 arcmin accuracy depending on burst brightness. The large detector area and sophisticated triggering system will allow BAT to detect bursts of all durations to a sensitivity 5 times better than BATSE (BAT threshold $\sim 10^{-8}$ erg cm$^{-2}$ for a 1 sec GRB). The instrument development is led by NASA's Goddard Space Flight Center.

## X-Ray Telescope

The X-Ray Telsecope (XRT) will locate bursts to 5.0 arcsec accuracy using flight-spare optics from the JET-X instrument on the Spectrum X mission. The mirror has a 15 arcsec half-power diameter at 1.5 keV. The detector is a 600 square pixel CCD from the XMM program, giving a FOV of 24 arcmin square in an energy range of 0.2-10.0 keV. The XRT has twice the effective area ($\sim 110$ cm$^2$ @ 1.5 keV) relative to the BeppoSAX X-ray telescope and four times better angular resolution. The instrument is being developed by Penn State University, University of Leicester, and Osservatorio Astronomico di Brera.

## UV/Optical Telescope

The UV/Optical Telescope (UVOT) is a 30 cm diameter modified Ritchey-Crétien equipped with an image intensified CCD covering 170-600 nm. It has a FOV 17 arcmin square and is based closely on the design of the XMM Optical Monitor (OM). The UVOT is able to reach $m_B = 24$ in 1000 s (open filter). A filter wheel provides 6 colors, two grisms and a $4\times$ magnifier. The optical point spread function of the telescope is 0.9 arcsec allowing for excellent astrometry. By registering the field against foreground stars, the UVOT will provide $< 0.3$ arcsec positions. The instrument is being developed by Penn State University and Mullard Space Science Laboratory.

## SWIFT MISSION

The strategy of the Swift mission is to slew to each new GRB position as soon as possible and to follow the GRB afterglows as long as they are visible. To observe the earliest phase of the afterglow, new BAT positions will trigger an autonomous spacecraft slew followed by a programmed sequence of observations with the XRT and UVOT.

The initial GRB position is normally determined by the BAT, but positions can also be uploaded from other satellites through a real-time TDRSS uplink. Either case will trigger the spacecraft software to plan and execute an autonomous slew. All calculations of slew path and pointing constraints will be done on-board. Figure of Merit (FoM) software will determine when to slew to a new position. The FoM has a flexible design that can accommodate more focused studies of specific GRB questions as the mission progresses.

Each of the three Swift instruments rapidly produces alert messages after a GRB is detected. To ensure prompt delivery, these messages are sent through a real-time TDRSS downlink to the ground, and routed immediately to the GRB Coordinates Network (GCN) [4] for delivery to the community.

When Swift is not engaged in prompt observations of the most recent bursts, it will follow a schedule uploaded from the ground each working day and as needed. This schedule will provide for long term follow-up of GRB

**TABLE 4.** Swift mission characteristics.

| |
|---|
| Autonomous slew decision capability |
| Fast Slew - 50° in < 60 s |
| Low Earth Orbit |
| 22° Inclination |
| Launch Vehicle: Delta 7320 with 3 meter fairing |
| Mass: 1500 kg |
| Power: 1000 W |
| Launch Date: September 2003 |

afterglows and other science. The PSU Mission Operational Center (MOC) will be capable of generating a new schedule in < 2 hours.

## SWIFT SCIENCE

Recent GRB discoveries have shown that X-ray, optical, and radio afterglows exist, continuing for days after the bursts, but fading quickly ($t^{-1}$ to $t^{-2}$ is typical). Better data on faster time scales for many more bursts is needed. (See [5] for discussion of requirements for future GRB missions.) The Swift mission provides the capability to answer the following four key science questions: What are the progenitors of GRBs? How does the blast-wave evolve and interact with its surroundings? Are there different classes of bursts with unique physical processes at work? What can GRBs tell us about the early Universe?

### GRB Progenitors

To determine the origin of GRBs, three parameters are needed: the total energy released, the nature of the host galaxy (if one exists), and the location within the host galaxy. The Swift mission is optimized to measure all three of these for hundreds of bursts.

Obtaining the energetics requires a reliable redshift measurement. Ideally this should be done independently for both the afterglow and any host galaxy to check that there has not been a chance coincidence [6]. The UV grisms and filters of Swift can make redshift determinations by searching for the Ly-$\alpha$ cutoff in the UV and eliminate the $1.3 < z < 2.5$ deadband of current observations [6] during the early phase of the afterglow. In addition, time varying optical, UV and X-ray lines and edges are expected within the first hour following a burst from the illumination of the immediate (100 pc) environment by the initial event [7, 8]. The rapid response of Swift will enable a search for predicted X-ray lines (see next section) and again provide a direct redshift measure from the afterglow.

The UVOT will obtain < 0.3 arcsec positions by using background stars to register the field, providing a unique host galaxy ID and allowing later comparison with HST fields to determine the position within the galaxy.

There will probably be events where no optical afterglow is detected because of dust extinction surrounding the site of the GRB. The position from the XRT will then be crucial. By obtaining 5.0 arcsec positions, the XRT will enable unique identification of candidate host galaxies down to $m_R \sim 26$. Follow-up observations with Chandra made within a couple of days for a selection of these events will give sub-arcsec positions within the Swift 5.0 arcsec error circle.

### Blast-wave Interactions

Afterglow is thought to be produced by the interaction of an ultra-relativistic blast-wave with the interstellar or intergalactic medium. The blast-wave model [9] predicts a series of stages as the wave slows. A key prediciton is a break in the spectrum that moves from the gamma to optical band, and is responsible for the power law decay of the source flux. This break moves through the X-ray band in a few seconds, but takes up to 1000 s to reach the optical. Thus observations within the first 1000 s in the optical and UV are crucial to see this early phase. While it now seems likely that all GRBs have an X-ray afterglow, not all have a bright optical afterglow (at least after several hours). This may be due to optical extinction, but it is also possible that in some cases the optical (and X-ray) afterglow is present but decays much more rapidly [10] and is a function of the density of the local environment [11]. Prompt high-quality X-ray, UV and optical observations over the first minutes to hours of the afterglow (inaccessible without Swift) are crucial to resolve this question. Continuous monitoring is important since model-constraining flares can occur in the decaying emission.

Star forming regions are embedded in large columns of neutral gas and dust. The presence of extinction can be readily determined by multi-band photometry in the optical and IR. The simultaneous detection of high X-ray absorption, coupled with photometric E (B-V) measurements with Swift, will determine whether dust and gas are present. Continuous monitoring over the first few hours to days will indicate whether dust is building up (due to condensation out of an expanding hot wind) or disappearing (due to ablation and evaporation).

### Classes of GRBs

While some evidence of sub-classes has been obtained (e.g. bimodal duration distribution, possible correlation of hardness and logN-logP shape, short bursts having

$V/V_{max}$ consistent with a Euclidean distribution), it is not clear if these are real differences or, rather, the result of the distribution fuction of GRB properties such as beaming angle, density of the local medium, or initial energy injection. Swift data will determine locations, redshifts, and afterglow properties of the different classes and thus allow physical understanding of their nature.

If there are classes of GRBs that are the signal of conventional supernova explosions (e.g., refs. [12, 13]), the UVOT will provide unique and unprecedented coverage of the optical and UV light curve during the early stage.

## Bursts as Astrophysical Tools

Since GRBs are the most luminous objects in the Universe, they provide a unique opportunity to probe the intergalactic medium (IGM) and the ISM of the host galaxies via measurment of absorption along the line of sight [14]. Depending on evolution, GRBs might originate from redshifts up to $\sim 15$ and have a median redshift $> 2$, larger than that of any other observable population. By rapidly providing both accurate positions and optical brightness, Swift will enable the immediate follow-up of those GRBs bright enough for high resolution optical absorption line spectroscopy at redshifts large enough to study the reionization of the IGM [15]. This information on the high-z Ly-$\alpha$ forest will be unique because there are currently no known bright ($m < 17$) galaxies or quasars at $z > 6.3$ [14].

## GROUND SYSTEM AND DATA ANALYSIS

A layered data analysis approach will be used to achieve rapid dissemination of Swift results and data to the community. The most urgently needed results, namely GRB positions, are produced on the spacecraft. Quicklook results, including optical finding charts and multiwavelength light curves, are produced in the Penn State Mission Operations Center (MOC) in near real-time and distributed using the GCN. Definitive standard products, including spectra, multi-band light curves, and images, will be made into production FITS files.

All the Swift data will be processed at the Swift data center at Goddard and will be made available to the general public through the HEASARC in the US and data centers in the UK and Italy. The end result will be easy access for the entire community to a broad range of timely information on GRBs.

## ORGANIZATION

Swift is the result of an international collaboration with GSFC, Penn State University (PSU) and institutions in the USA, United Kingdom, and Italy. The responsibilities of the various institutions are listed in Table 5.

**TABLE 5.** Swift mission responsibilities and institutions.

| Responsibility | Institution |
|---|---|
| Mission Management | GSFC |
| XRT | PSU, LU, OAB |
| UVOT | PSU, MSSL |
| BAT | GSFC, LANL |
| Ground System | GSFC |
| Mission Operations | PSU |
| Data Centers | GSFC, ASI, LU |
| EPO | SSU, PSU, GSFC |
| GRB Follow-up Coordination | UCB |

GSFC = Goddard Space Flight Center
PSU = Penn State University
LU = Leicester University
OAB = Osservatorio Astronomico di Brera
MSSL = Mullard Space Science Laboratory
LANL = Los Alamos National Laboratory
ASI = Italian Space Agency
SSU = Sonoma State University
UCB = University of California, Berkeley

## REFERENCES

1. Costa, E., et al., *Nature*, **387**, 783 (1997).
2. Van Paradijs, J., et al., *Nature*, **386**, 686 (1997).
3. Frail, D.A., et al., *Nature*, **389**, 261 (1997).
4. Barthelmy, S. et al., *Gamma-Ray Bursts: 4th Huntsville Symposium*, edited by C. Meegan, R. Preece, and T. Koshut, AIP Conference Proceedings 428, American Insitute of Physics, New York, 1998, p 139.
5. Gehrels, N. and Macomb, D., *Cosmic Explosions: Tenth Astrophysics Conference*, edited by S. Holt and W. Zhang, AIP Conference Proceedings 522, American Institude of Physics, New York, 2000, p 227.
6. Hogg, D. W. & Fruchter, A. S., *ApJ*, **520**, 54, 1999
7. Perna, R. & Loeb, A., *ApJ*, **501**, 467, 1998
8. Mészáros, P. & Rees, M. J., *ApJL*, **502**, L105, 1998
9. Mészáros, P. and Rees, M., *ApJL*, **418**, L59 (1993).
10. Groot, P. J., et al., *ApJL*, **502**, L123, 1998
11. Piran, T., "Gamma-Ray Bursts–The Second Revolution", *Frontiers Science Series 23: Black Holes and High Energy Astrophysics*, edited by H. Sato and N. Sugiyama, Frontiers in Science Series Number 23, Universal Academic Press, 1998, p 217.
12. Bloom, J. S., et al., *Nature*, **401**, 453, 1999
13. Woosley, S. E., Eastman, R. G. & Schmidt, B. P., *ApJ*, **516**, 788, 1999
14. Lamb, D.Q. and Reichart, D.E., *ApJ*, **536**, 1 (2000).
15. Miralda-Escudé, J., *ApJ*, **501**, 15 (1998).

# The GLAST Burst Monitor

C. Meegan*, G. Lichti†, M. Briggs**, R. Diehl†, G. Fishman*, R. Kippen**,
C. Kouveliotou*, A. von Kienlin†, W. Paciesas**, R. Preece** and V. Schönfelder†

*NASA/Marshall Space Flight Center, Mail Code SD50, Huntsville, AL 35812*
†*Max Planck Institute for Extraterrestrial Physics, Garching, Germany*
**University of Alabama in Huntsville, Huntsville, AL 35899*

**Abstract.** The Gamma Ray Large Area Space Telescope (GLAST), scheduled for launch in 2006, comprises a Large Area Telescope (LAT) and a GLAST Burst Monitor (GBM). The LAT is a pair telescope with unprecedented sensitivity in the 20 MeV to 300 GeV energy range. The GLAST Burst Monitor consists of an array of NaI and BGO scintillation detectors operating in the 10 keV to 25 MeV range and covering a wide field of view. The GBM will enhance LAT observations of GRBs by extending the spectral coverage into the range of current GRB databases, and will provide a trigger for repointing the spacecraft to observe delayed emission from bursts outside the LAT field of view.

## INTRODUCTION

The Gamma Ray Large Area Space Telescope (GLAST) is NASA's next major observatory dedicated to gamma ray astronomy. The primary instrument is the Large Area Telescope (LAT), which detects gamma rays in the ∼20 MeV to ∼300 GeV range by tracking the electron-positron pairs using an array of silicon strip detectors. The LAT builds upon and greatly extends the capabilities of the Energetic Gamma Ray Experiment Telescope (EGRET) on the Compton Gamma Ray Observatory. One of the science goals of the LAT is the study of gamma-ray bursts (GRBs). EGRET detected high energy emission from several bursts, including a remarkable observation of delayed emission [1], but the relatively small effective area and the high dead time per event were significant limiting factors. With its high effective area (over 10 times EGRET) and very low dead time per event, the LAT will make significant improvements in the observations of high energy emission from bursts.

It was recognized early in the GLAST definition phase that burst observations would be greatly enhanced if the observatory included an instrument capable of detecting GRBs over a wide field of view in the traditional GRB energy range. The GBM satisfies this need within the necessarily stringent cost, mass, and power allocations. The primary goal of the GBM is to extend the spectral coverage of GRBs from the LAT range down to ∼10 keV. A secondary goal is to provide on-board coarse locations of intense bursts that are outside the LAT field of view. This will allow repointing the observatory to allow the LAT to observe any extended high energy emission.

GBM is a collaborative effort involving scientists at the Marshall Space Flight Center, the University of Alabama in Huntsville and the Max Planck Institute for Extraterrestrial Physics.

## HARDWARE DESCRIPTION

GBM has similarities to the Burst and Transient Source Experiment (BATSE) on CGRO. GBM comprises twelve sodium iodide (NaI) scintillation detectors and two Bismuth Germanate (BGO) scintillation detectors. The NaI detectors are 5" diameter by 0.5" thick, and each is viewed by a single photomultiplier tube. They are sensitive over the energy range ∼10 keV to ∼1 MeV and are used to determine burst locations by measuring the relative rates on detectors with different orientations. The BGO detectors are 5" diameter by 5" thick, and each is viewed by two photomultiplier tubes to provide redundancy and improve light collection. They cover the energy range of ∼150 keV to ∼25 MeV.

In the baseline plan (Figure 1), the twelve NaI detectors are oriented in four banks equally spaced in azimuth. The three detectors at each azimuth have their normals at zenith angles of 30°, 60°, and 90°. The BGO detectors are placed on opposite sides of the LAT. With this configuration, any burst within 120 degrees of the viewing axis (+Z direction) will illuminate at least three NaI detectors and one BGO detector, enabling a computation of the burst location and providing full spectral cover-

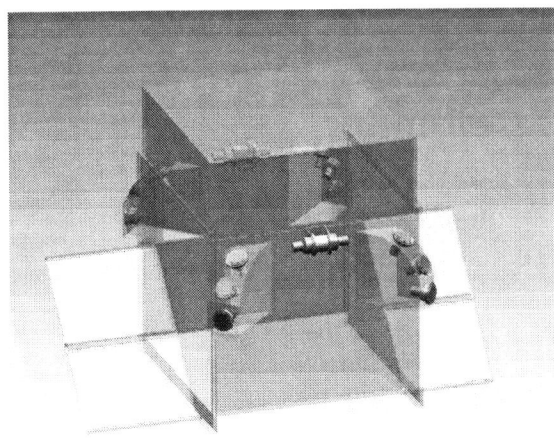

**FIGURE 1.** Baseline detector configuration

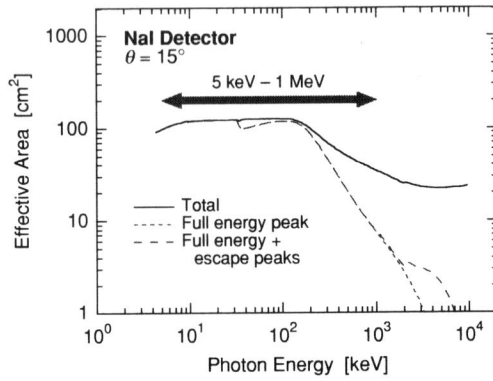

**FIGURE 2.** Effective Area of the NaI Detectors

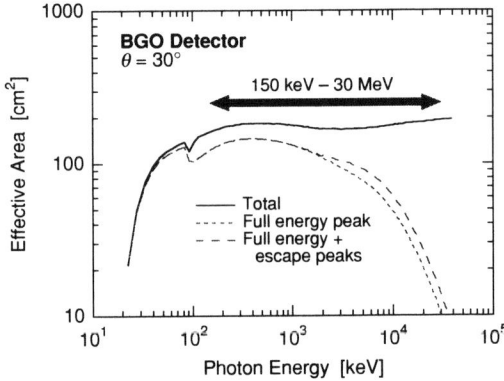

**FIGURE 3.** Effective Area of the BGO Detectors

age. The current operations scenario calls for the LAT to always be pointed away from the earth, so it is not necessary for the GBM to have high sensitivity beyond zenith angles of 120°.

Event data from each detector are sent to the Data Processing Unit (DPU). The DPU digitizes the data, constructs the various data types, detects bursts, computes burst locations, and provides the electrical interface to the spacecraft.

GBM will detect bursts in the same manner as BATSE, i.e., as statistically significant increases in the rates of two or more NaI detectors. Rates will be tested on time scales as short as 16 ms. When a burst is detected, a coarse location will be computed on-board and transmitted to the ground for distribution via the GRB Coordinates Network (GCN). If GBM detects a particularly intense burst that is outside the LAT field, there is an option to automatically repoint the spacecraft to obtain LAT observations of delayed emission. There is also a possibility of catching the prompt emission in cases of short slews to long-duration bursts. Sufficient burst data will be transmitted in near real-time to permit more accurate locations to be computed on the ground, both automatically and with a scientist in the loop.

## PERFORMANCE

Table 1 presents a list of the performance requirements and goals for the GBM, with comparisons to BATSE. On-board locations need only be accurate enough to allow the observatory to repoint to place a burst within the LAT field, which is greater than 30° in radius. The systematic error on ground computed locations is expected to be comparable to what was achieved on BATSE, although statistical errors will be larger due to the smaller sensitive area.

The threshold for on-board burst detection will be about 0.6 photons cm$^{-2}$ s$^{-1}$, which will provide 150–200 burst triggers per year. The threshold for detecting bursts in ground-based searches is expected to be about 0.35 photons cm$^{-2}$ s$^{-1}$.

Figures 2 and 3 show the effective area as a function of energy for the NaI and BGO detectors, respectively.

Figure 4 shows a simulated deconvolved spectrum of a strong burst. The crosses represent data from an NaI detector, the filled circles represent data from a BGO detector, and the squares represent data from the LAT. The combination of the GBM and the LAT will provide good measurements of the peak energy and the high and low energy spectral indices for many bursts.

## DATA PRODUCTS

GBM has two primary modes of operation. When not acquiring burst data, GBM transmits background information using two data types: CSPEC, with 128 channel energy resolution and 8 s temporal resolution, and CTIME, with 8 channel energy resolution and 0.256 s

**TABLE 1.** GBM Instrument Requirements

| Parameter | Requirement | Goal | BATSE |
|---|---|---|---|
| Energy Range | 10 keV – 25 MeV | 5 keV – 30 MeV | 10 keV – 1.8 MeV (LAD)<br>15 keV – 30 MeV (SD) |
| Energy Resolution | 20% FWHM at 511 keV | – | ~ 20 FWHM at 511 keV |
| Time resolution | 10 $\mu$s | 2 $\mu$s | 2 $\mu$s |
| On-board GRB locations | 15° accuracy (1$\sigma$ radius) within 2 s | 10° within 1 s | none |
| Rapid ground GRB locations | 5° accuracy (1$\sigma$ radius) within 5 s | 3° within 1 s | 10° within 5 s;<br>3° within 20 min. |
| Final GRB locations | 3° accuracy (1$\sigma$ radius) within 1 day | – | 3° within a few days |
| GRB sensitivity | 0.5 photons cm$^{-2}$ s$^{-1}$ (peak flux 50-300 keV) | 0.3 photons cm$^{-2}$ s$^{-1}$ (peak flux 50-300 keV) | 0.1 photons cm$^{-2}$ s$^{-1}$ (peak flux 50-300 keV) |
| Field of View | 8 steradians | 10 steradians | 12.6 (4$\pi$) steradians |
| Deadtime | <10 $\mu$s/count | <3 $\mu$s/count | ~10 $\mu$s/count |

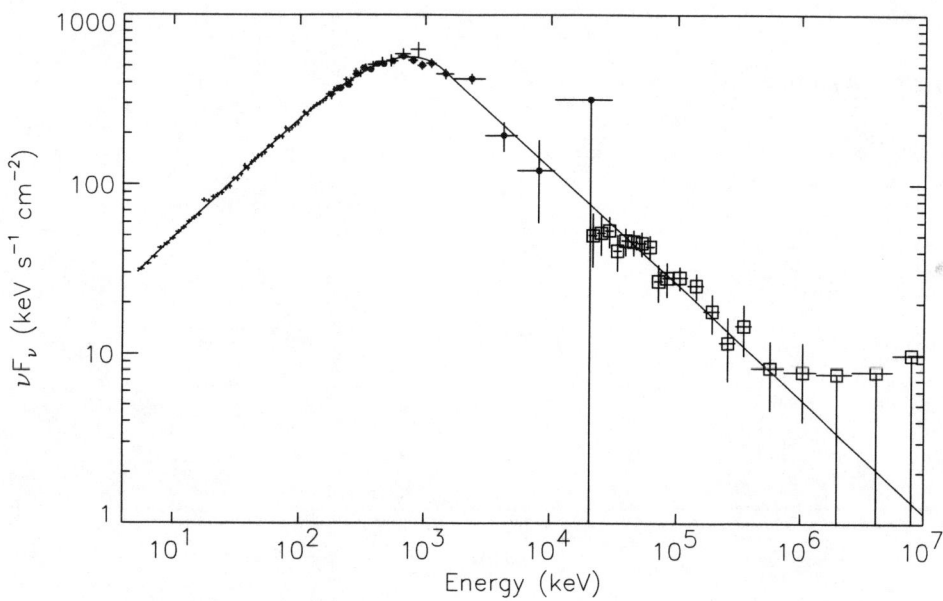

**FIGURE 4.** Simulated GLAST Spectrum of a Strong Burst

temporal resolution. When a burst occurs, a total of at least 250,000 counts are tagged with time and energy, which includes some pre-trigger events that are stored in a ring buffer.

Whenever a burst is detected on-board, either by the GBM or the LAT, the spacecraft will establish a real-time link to provide rapid notification to the ground for dissemination via the GCN.

## SUMMARY

GBM will significantly enhance the GLAST capability to study bursts by extending the energy range several decades. GBM is well-suited to scientific investigations of other sources, such as discrete transient sources, solar flares, and soft gamma repeaters. These studies take on added significance if GBM is the only operating wide field hard X-ray instrument in the 2006 to 2011 timeframe.

## REFERENCES

1. Hurley, K., et al., *Nature* **372**, 652(1994).

# The Current Performance of the Third Interplanetary Network

K. Hurley*, T. Cline[†], I. Mitrofanov**, E. Mazets[‡], S. Golenetskii[‡], F. Frontera[§], E. Montanari[¶], C. Guidorzi[¶] and M. Feroci[‖]

*UC Berkeley Space Sciences Laboratory, Berkeley, CA 94720-7450*
[†]*NASA GSFC, Code 661, Greenbelt, MD 20771*
**IKI, 117810, Profsouznaya 84/32. GSP-7, Moscow, Russia*
[‡]*Ioffe Physical-Technical Institute, St. Petersburg, 194021, Russia*
[§]*Istituto Tecnologie e Studio Radiazioni Extraterrestri, CNR, Via Gobetti 101, 40129, Bologna, Italy*
[¶]*Universita di Ferrara, via Paradiso, 12, 44100 Ferrara, Italy*
[‖]*Istituto di Astrofisica Spaziale, C.N.R., via Fosso del Cavaliere, Rome I-00133, Italy*

**Abstract.**
The 3rd Interplanetary Network (IPN) has been operating since April 2001 with two distant spacecraft, Ulysses and Mars Odyssey, and numerous near-Earth spacecraft, such as BeppoSAX, Wind, and HETE-II. Mars Odyssey is presently in orbit about Mars, and the network has detected approximately 30 cosmic, SGR, and solar bursts. We discuss the results obtained to date and use them to predict the future performance of the network.

## INTRODUCTION

The 3rd IPN began with the launch of Ulysses in November 1990. Ulysses is in a heliocentric orbit roughly perpendicular to the ecliptic, with perihelion of about 1.5 AU and an aphelion of about 5 AU. Until 1992, the network had Ulysses and Pioneer Venus Orbiter (PVO) as its distant points, and utilized many near-Earth spacecraft such as the Compton Gamma-Ray Observatory (CGRO) as its third point. PVO entered the atmosphere of Venus in 1992, and it was to be replaced in the IPN by NASA's Mars Observer, which was lost during insertion into Martian orbit. Finally, in 1999, the network was completed by the X- and Gamma-Ray Spectrometer experiment aboard the Near Earth Asteroid Rendezvous (NEAR) spacecraft. In this configuration, the IPN operated quite successfully until NEAR landed on the asteroid Eros in February 2001. Figure 1 summarizes some of the results. The Mars Odyssey mission, launched in April of that year, contains two experiments which have been modified for gamma-ray burst detection, the Gamma-Ray Spectrometer (GRS) and the High Energy Neutron Detector (HEND). As the GRS will not be turned on permanently until the spacecraft completes its aerobraking maneuver, we will discuss the results obtained with HEND.

## SOME IPN RESULTS TO DATE

HEND has an effective area of about 40 cm$^2$ of CsI and operates in the energy range above 40 keV. It transmits data continuously with a time resolution of 0.25 s, so that "triggering" is done with ground software. In addition, whenever a burst is recorded by Konus, the BeppoSAX GRBM, Ulysses, or HETE-II, HEND data are extracted for the appropriate crossing window and transmitted to Goddard and Berkeley for analysis. HEND was been on for a large fraction of the cruise phase, and has detected many cosmic, solar, and soft gamma repeater (SGR) events, some of which are listed in table 1.

This table is neither exhaustive, since we have not been able to analyze all the data yet, nor representative, since there were numerous solar flare particle events which raised the backgrounds and reduced the sensitivities of some IPN experiments. (Solar activity will be decreasing in the years to come.) Thus in about 4.5 months, HEND detected 21 confirmed

TABLE 1. A partial list of HEND solar, cosmic, and SGR bursts.

| Date | Seconds | Type | HEND | Ulysses | HETE | Konus | SAX |
|---|---|---|---|---|---|---|---|
| 010508 | 47828 | Cosmic | Yes | Yes | No | Yes | No |
| 010517 | 85894 | Cosmic | Yes | Yes | No | Yes | Yes |
| 010523 | 17059 | Cosmic | Yes | Yes | No | Yes | No |
| 010605 | 17145 | Solar | Yes | No | No | Yes | No |
| 010607 | 53723 | Cosmic | Yes | Yes | No | Yes | No |
| 010625 | 51802 | SGR1900+14 | Yes | Yes | No | Yes | No |
| 010627 | 04172 | Cosmic | Yes | Yes | No | No | No |
| 010628A | 03914 | Cosmic | Yes | Yes | Yes | Yes | No |
| 010628B | 68418 | Cosmic | Yes | Yes | No | Yes | No |
| 010701A | 02789 | Cosmic | Yes | Yes | No | Yes | No |
| 010701B | 07102 | Cosmic | Yes | Yes | No | Yes | No |
| 010702 | 12848 | SGR1900+14 | Yes | Yes | Yes | Yes | Yes |
| 010703 | 73842 | Cosmic | Yes | Yes | No | Yes | Yes |
| 010706 | 29689 | Cosmic | Yes | Yes | No | Yes | No |
| 010710 | 84642 | Cosmic | Yes | Yes | No | Yes | Yes |
| 010723 | 63602 | Cosmic | Yes | Yes | No | Yes | No |
| 010725 | 61288 | Cosmic | Yes | Yes | No | Yes | No |
| 010726 | 05392 | Cosmic | Yes | Yes | No | Yes | No |
| 010804 | 72805 | Cosmic | Yes | Yes | No | Yes | Yes |
| 010805 | 54213 | Solar | Yes | Yes | No | Yes | No |
| 010807 | 59183 | Solar | Yes | Yes | No | Yes | No |
| 010821A | 48423 | Cosmic | Yes | Yes | No | Yes | No |
| 010821B | 78900 | Solar | Yes | Yes | No | Yes | No |
| 010827 | 23676 | Solar | Yes | Yes | No | Yes | No |
| 010830 | 74159 | Solar | Yes | Yes | No | Yes | No |
| 010831A | 38298 | Solar | Yes | Yes | No | No | No |
| 010831B | 81626 | Solar | Yes | Yes | No | Yes | No |
| 010913 | 71191 | Cosmic | Yes | Yes | No | Yes | No |
| 010921 | 18552 | Cosmic | Yes | Yes | Yes | Yes | Yes |

cosmic or SGR events. HEND is now on almost continuously, but in an eccentric orbit which leads to a variable background rate. The orbit is gradually being circularized by aerobraking. In figure 1 we show an example of a HEND burst. Although the GRS has been on only sporadically during cruise, it too has detected several bursts.

## THE FUTURE OF THE IPN

The Mars Odyssey mission will collect data for at least one Martian year. The nominal end of mission will be around early 2004. Ulysses is currently funded through at least the start of 2004, with a slight extension possible. However, the decay of the radioactive power system probably precludes operation past the end of 2004. The Wind mission is funded through 2002 at least. The BeppoSAX is presently funded until April 2002, but it may be extended until it re-enters. HETE-II is funded through 2002 at least, and possibly for two years beyond.

The INTEGRAL mission is due to be launched into a highly elliptical orbit in October 2002. It will therefore either replace or complement the near-Earth spacecraft, depending on whether they remain operational or have been terminated by that time. Finally, Swift should be launched in late 2003. Thus, we expect to maintain a viable three-spacecraft IPN for a minimum of about 2.5 years.

## EXPECTED RESULTS

This new IPN is similar in many respects to one which was in operation when NEAR was in the network. Between December 1999 and January 2001, that IPN localized 57 GRBs and circulated their positions to the wide astronomical community via the GRB Coordinates Network (GCN). Of the 25 bursts which were followed up with long-wavelength observations, counterparts were found for 9. The success rate, 9/25 or 36%, is consistent with the overall success rate for bursts, which is about 40%.

There are factors which will both increase and decrease the performance of the new IPN compared to the old one. We first list the "increase" factors.

1. In the old network, the Earth-Ulysses distance

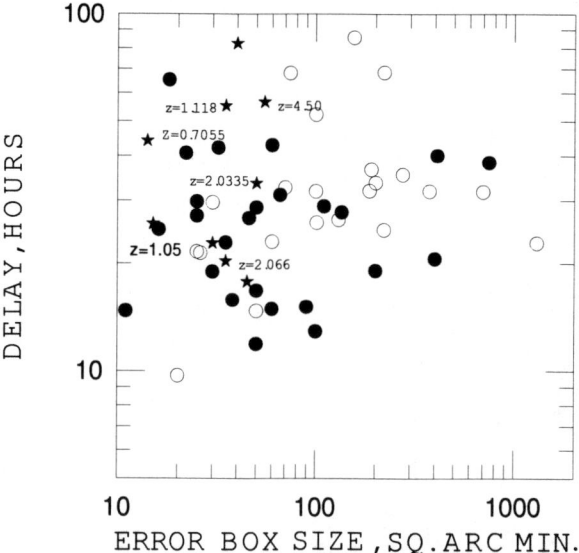

**FIGURE 1.** Fifty-seven GRBs detected by the IPN with NEAR. Each burst is characterized by the size of the error box and the delay to obtain it. The stars represent GRBs for which counterparts were identified; six of them have had their redshifts measured. The filled circles represent bursts which were followed up in the radio and/ or optical range, but for which no counterparts were identified. The hollow circles represent those events which were not followed up. In slightly over a year of operation, the database on GRB counterparts increased by 50% due to the IPN results alone.

varied between 2.3 and 4.4 AU, while the Earth-NEAR distance varied between 0.8 and 2.1. In the new network, the Earth-Ulysses distance will vary between 2 and 6 AU, while the Earth-Mars distance will vary between 0.45 and 1.8 AU. Thus the average baselines, and hence the annulus widths, will be slightly better.

2. In the old network, the NEAR XGRS experiment had a time resolution of 1 second. In the new network, the time resolution will be between 0.031 s (for GRS) and 0.250 s (for HEND), resulting in a gain in accuracy of the cross-correlations and thus narrower annuli.

3. In the old network, the XGRS detector had a lower energy threshold of 150 keV. The other experiments in the network have energy ranges of 25-150 keV. Since GRB time histories are energy-dependent, this led to increased uncertainty in the cross-correlations and thus wider annuli. HEND operates in an energy range which is better matched to those of the other IPN instruments.

4. Based on the number of bursts observed with HEND during the cruise phase (21 in 4.5 mo), and taking into account the facts that HEND was not on continuously, and that all the data have not been examined, the burst detection rate, 4.7/month, was greater than that of the old network.

The following are "decrease" effects:

1. Once MO is in a circular orbit, HEND will have an unobstructed field of view of about 2 pi sr, or a factor of about 2 less than in the cruise phase. This decrease should be offset by two other factors, however. First, bursts arriving from the anti-planet hemisphere should still be detectable in some cases, since they will either penetrate the material behind the detectors, and/or backscatter off the Martian atmosphere. Second, GRS will be on continuously, and due to its different detection efficiency, energy range, and field of view, it will detect some bursts that HEND does not.

2. The future tracking efficiencies (that is, the length of time between telemetry passes) of some spacecraft in the network are unknown. This applies particularly to Konus-Wind and to Mars Odyssey. Longer periods between downlinks translate to later error boxes, which must be searched more deeply for counterparts.

All these effects are difficult to quantify, but our expectation is nevertheless that the performance of this network will closely resemble that of its predecessor.

## CONCLUSIONS

In the 2.5 year MO mission, we expect to detect and localize over 100 GRBs and circulate their positions rapidly to astronomers around the world for multiwavelength counterpart studies. This should lead to the observation of numerous afterglows and the determination of many GRB redshifts.

## ACKNOWLEDGMENTS

KH is grateful for IPN support under JPL Contract 958056.

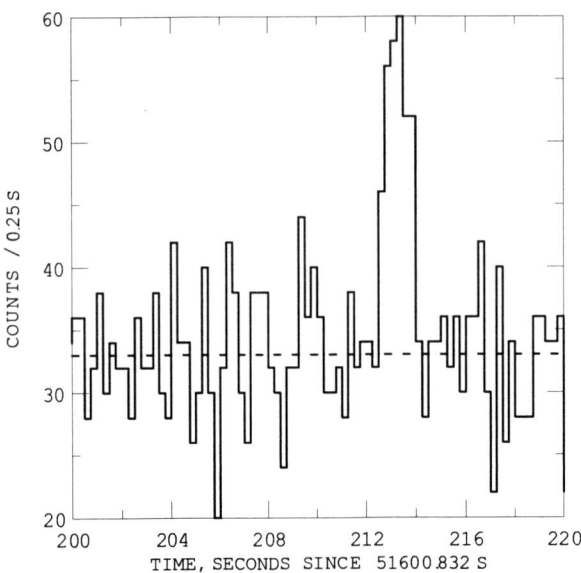

**FIGURE 2.** The time history of GRB010625 as observed by HEND in the 50-3000 keV energy range. The dashed line shows the background. This event was from SGR1900+14 and had a fluence of approximately $3 \times 10^{-6} \text{erg cm}^{-2}$

# Proposed Next Generation GRB Mission: EXIST

J. Grindlay[1], N. Gehrels[2], F. Harrison[3], R. Blandford[3], G. Fishman[4],
C. Kouveliotou[4], D.H. Hartmann[5], S. Woosley[6], W. Craig[7] and J. Hong[1]

*1. Harvard, 2. NASA/GSFC, 3. Caltech, 4. NASA/MSFC, 5. Clemson Univ., 6. UC Santa Cruz, 7. LLNL*

**Abstract.** A next generation Gamma Ray Burst (GRB) mission to follow the upcoming *Swift* mission is described. The proposed Energetic X-ray Imaging Survey Telescope, *EXIST*, would yield the limiting (practical) GRB trigger sensitivity, broad-band spectral and temporal response, and spatial resolution over a wide field. It would provide high resolution spectra and locations for GRBs detected at GeV energies with *GLAST*. Together with the next generation missions *Constellation-X, NGST* and *LISA* and optical-survey (*LSST*) telescopes, *EXIST* would enable GRBs to be used as probes of the early universe and the first generation of stars. *EXIST* alone would give ~10-50" positions (long or short GRBs), approximate redshifts from lags, and constrain physics of jets, orphan afterglows, neutrinos and SGRs.

## INTRODUCTION

Gamma-ray bursts (GRBs) are the most luminous events since the Big Bang. Current models for at least the "long" GRBs favor their origin in the collapse of massive stars. As such, their study with the most powerful telescopes can map the universe back to the very first stars, believed to be very massive. GRBs from such large redshifts will require a next generation telescope to follow the *Swift* mission, currently planned for launch in 2003. The proposed Energetic X-ray Imaging Survey Telescope, *EXIST*, can achieve these objectives as the Next Generation GRB mission.

We describe the current mission concept for *EXIST* and then some of the GRB and associated science that could be conducted with a nominal 5y mission.

## *EXIST* MISSION CONCEPT

The primary mission requirements for a next generation GRB mission are: 1. very large area, for maximum GRB sensitivity; 2. large field of view (FOV), for collection of a large GRB sample and sensitivity to rare events, including low-luminosity nearby GRBs possibly observed off-axis (e.g. SN1998bw); 3. high angular resolution and fine positional determination (10") for (near-) real time optical identifications; and 4. broad energy band coverage, with imaging response down to ~10 keV for high z GRBs and up to ~600 keV to extend beyond the ~300 keV $\nu F\nu$ energy peak and maximize sensitivity to broad energy-dependent lags (hard-soft) which may allow measures of GRB luminosity-distance.

Broad-band (~10-600 keV) hard x-ray imaging over a wide FOV is best conducted with a coded aperture telescope with an imaging detector capable of fine position resolution (to record the coded mask shadow), high Z stopping power for good high energy sensitivity in moderate detector thickness, and compact mounting and tiling capability for extension to a very large area total detector array. Coded aperture imagers are background limited and so record signal to noise $S/N \sim (A_{det} \cdot T/B)^{0.5}$, for a given detector area $A_{det}$ recording background B (cts cm$^{-2}$sec$^{-1}$) for a source observed for time T(sec). Optimum imaging sensitivity requires systematic variations on the detector (e.g., due to gain and non-uniform background variations) to be effectively averaged, which can best be achieved by continuously scanning the detector-telescope across the sky. Since for the wide-field (>>10°) imaging needed for GRB sample statistics, the background is dominated by the diffuse cosmic flux (primarily over ~20-200 keV) recorded in the FOV of size $\theta \times \theta$, then $B \sim \theta^2$. Similarly, the exposed $A_{det}$, available integration time T, and recorded B are each proportional to the angular width $\theta$ of the FOV in the scan direction. Thus, for a scan at orbital rate $d\varphi/dt \sim 4°$ min$^{-1}$, the expected $S/N \sim \theta^{0.5}$ for GRBs with duration $T_b \sim \theta(d\varphi/dt)^{-1} \sim$ 3-10min (e.g. long GRBs, or possibly high-z GRBs) and S/N ~ independent of $\theta$ for typical GRBs ($T_b <$ 1min). Since the total GRB sample $N_b \sim \theta^2$, large $\theta$ is optimum; this also maximizes the persistent source sensitivity.

These general considerations, as well as the primary goal to conduct a hard x-ray imaging survey which extends ROSAT sensitivity to >100keV but with all sky coverage each orbit [1], yield the preliminary *EXIST* mission concept outlined below. The baseline implementation is for a Free Flyer mission although a version studied originally [1,2] could be mounted on the International Space Station.

## Mission Implementation Overview

Three telescopes, each with a 60° x 75° fully-coded FOV, are mounted on a base spacecraft as shown in Figure 1a to form a combined 180° x 75° fan beam which images the full sky each orbit. Each telescope is constructed (Figure 1b) of a 3 x 3 array of actively collimated (CsI) overlapping FOV (60° x 50°) sub-telescope modules, each read out with arrays of Cd-Zn-Te (CZT) detectors (5mm thick; 1.3mm pixels) which view the sky through a curved coded aperture.

**Table 1:** *EXIST* **Mission Parameters**

| | |
|---|---|
| Energy range | 10-600 keV |
| FOV | 180° x 75° (fully coded) |
| | ~5 steradians (partial coded) |
| Angular Resolution | 2-5' (10-50"source locations) |
| Energy/Temporal Resolution | 1-3%; 2 μsec |
| Sensitivity (5σ, ≤1y) | ~0.05 mCrab (10-100 keV) |
| | ~0.5 mCrab (>200 keV) |
| Telescopes, Detectors | Coded aperture, 8 m² CZT |
| Pointing, Aspect | ~1° stability, 5"knowledge |
| Mass, Power, TM | 8500kg, 1500W, 1.5Mbs |
| Launch, Cost (incl. Ops) | Delta IV, $330M (w/ cont.) |

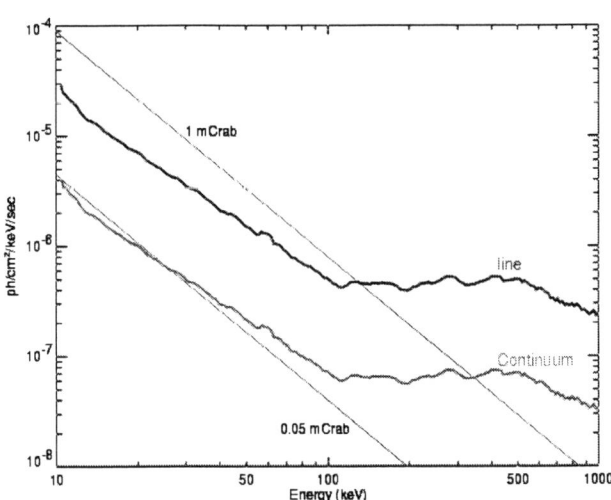

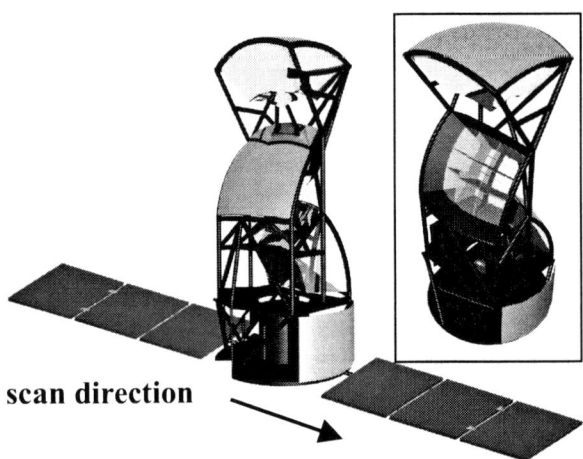

**FIGURE 1. a)** *EXIST* telescopes (3) on spacecraft, zenith pointed along orbital scan direction, **b)** cutaway view showing imaging CZT detector arrays and active collimator.

**FIGURE 2.** *EXIST* survey sensitivities. GRB sensitivities are ~50mCrab for an assumed 10sec duration burst.

Parameters for the mission are summarized in Table 1 and estimated continuum and line sensitivities (5σ, ≤1y, depending on source orbital latitude) are shown in Figure 2. After the first year all-sky survey, or at any time for Targets of Opportunity, the mission would be operated as an Observatory, with an active guest observer program for pointed observations with the central telescope while the outer two telescopes continue the Survey. GRBs are imaged in the wide combined field (5sr) regardless of Survey or Observatory mode. The active collimator and rear shield (1cm and 2cm thick CsI, respectively) will be pulse-height-analyzed to extend GRB spectral coverage up to ~1-3MeV. Burst positions are derived to ~1-3' within 10sec for rapid transmission to ground and other observatories, and to within ~10-50" in ground analysis of the full data within ~3hours.

Primary Survey science is the study of obscured AGN and black holes on all scales. Details of the full mission and science will be given in a later paper.

# GRB SCIENCE FROM EXIST

*EXIST* is a Next Generation Burst Observatory. With sensitivity to weak events a factor ~20 below BATSE and ~3 – 10 below *Swift* (given the extended low and high energy band of *EXIST*), it should provide ~10 – 50" locations for 2 – 3 GRBs a day. With its large instantaneous field of view it can study both low luminosity nearby GRB events (like SN1998bw) as well as the brightest events most likely to be observed by gravitational wave and neutrino detectors. *EXIST* will be on orbit at an amazing time, when *NGST* will be studying the high-redshift universe and *Contellation-X* will enable high resolution spectra of x-ray afterglows pinpointed by *EXIST* to be measured. *LISA* and *LIGO2* will be providing the first sensitive gravitational wave detections, and IceCube, Auger, and other high-energy neutrino and ultra-high energy cosmic ray detectors will be operating. The coincidence of these capabilities with *EXIST* will provide opportunities to search for the first massive stars in the universe to very high redshifts and open up non-electromagnetic channels of GRB energy release for observation – challenging our theories of relativistic shocks and testing physics from special relativity to neutrino masses and couplings.

## First Massive Stars to *EXIST*

Nearly 2/3 of all GRBs, the "long bursts", explode at significant cosmological distances. With its high sensitivity, EXIST can detect GRBs at high redshift (z~10-20; cf. Figure 3), enabling the first direct search

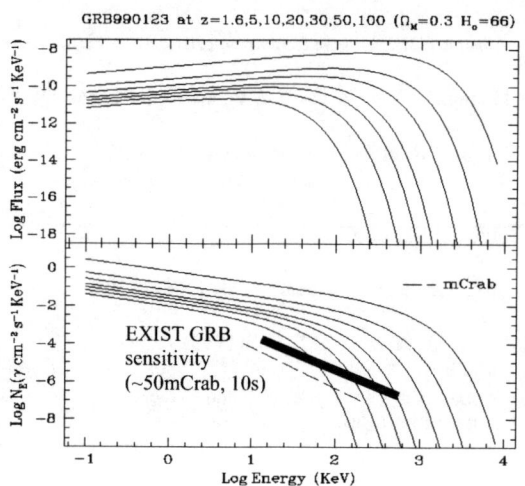

**FIGURE 3.** GRB sensitivity vs. z (adapted from [3]).

for the initial generation of stars (Pop III) that were likely very massive. Their resulting epoch of black hole (BH) formation would, given increasing evidence for long burst (>2sec) GRB production by hypernovae in Collapsars [4, 5], produce an epoch of GRBs. Spectroscopy of their optical afterglows, detectable in the IR with NGST, would enable mapping cosmic structure back to the "dark ages" [6]. This requires a rapid estimate of GRB redshift, which could be provided by the "photometric redshifts" to be derived from the observed relation [7] (cf. Figure 4) between GRB hard-soft lags and absolute luminosity.

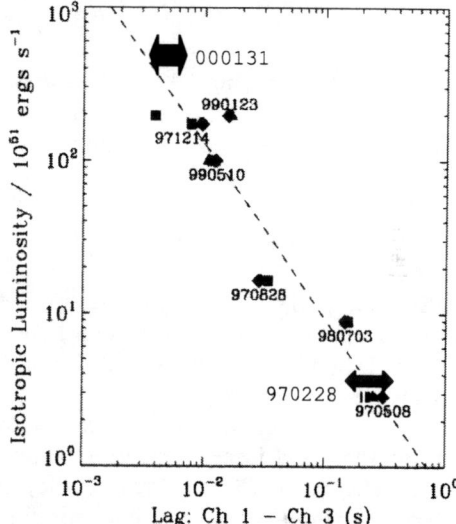

**FIGURE 4.** Luminosity-lag relation for GRBs (adapted from [7]). The observed correlation, $L_{pk} \propto \Delta\tau^{-1.14}$, may be understood as a kinematic effect from beaming [8].

Application of this technique to GRBs at high z requires the very large collection area for optimum statistics as well as the broad energy band coverage and good spectral resolution of *EXIST*.

The large GRB sample with approximate redshifts collected by *EXIST* will constrain the star formation rate to large z, complementing *NGST*. The *EXIST* deep sample will also provide possible GRB locations and redshifts for "orphan afterglows", probably found with SDSS [9] and expected in quantity from LSST in the *EXIST* era. Since orphans may preferentially be off-axis GRBs [10], they are likely to be x-ray bright and thus possibly related to the x-ray flashes (XRFs) which appear to form a spectral extension to GRBs [11]. Alternatively, the XRFs may be GRBs from massive (~300M$_\odot$) Pop III stars at z ~ 10 undergoing collapse to BHs, which also would produce background high energy (~$10^5$ GeV) neutrinos detectable with IceCube [12].

## Highest Energy Particles to *EXIST*

The relativistic shocks driven into the interstellar medium after a GRB explosion can produce high-energy neutrinos in several components with energies ~10 GeV, ~100 TeV and ~$10^{18}$ eV on timescales of <10s. Theory [13] predicts that several tens of muon induced neutrino events/year should be detectable with the IceCube neutrino telescope in coincidence with GRBs. Detection of the high energy neutrinos would not only test the shock acceleration mechanism, but would also imply GRBs are sources of Ultra High Energy Cosmic Rays (UHECR). A GRB at 100 Mpc (for which the wide-field, high sensitivity *EXIST* trigger is needed) producing 100 TeV neutrinos also allows a test of neutrino mass mixing five orders of magnitude more sensitive than solar neutrinos.

## Short GRBs and SGRs

Short (<2sec) GRBs are still unidentified with afterglows or hosts and are thus without luminosities or redshifts. They are consistent with arising in NS-NS (or perhaps NS-BH) mergers [5]. *EXIST* should detect ~$10^3$ over a 3 year mission with fluxes above the BATSE threshold and locate these to ~10" positional accuracy, sufficient to determine whether they are also (like long GRBs) associated with star formation regions in galaxies. If they arise from NS-NS mergers, or from accretion induced collapse of WDs [14], they may preferentially arise in globular cluster systems. Studies of the light curve variability enabled by the high statistics of *EXIST* can constrain short GRB jets (testing AIC), as it has apparently for long GRBs [15].

The soft gamma-ray repeaters (SGRs) are detected in the Galaxy (3) and LMC (1) and are associated with NSs, probably magnetars [16]. The Giant Flares detected from two of these reach highly super-Eddington luminosities which *EXIST* could detect with its all-sky monitoring and imaging out to ~3-10Mpc. This would survey SGRs and the incidence and activity of magnetars throughout the Local Group.

## CONCLUSIONS

A Next Generation GRB mission to conduct the highest sensitivity direct study of GRBs would culminate the steady advance in sensitivity, resolution (spatial and spectral) and broad sky coverage, from BATSE through HETE-2 and *Swift*. The proposed *EXIST* mission would achieve this. It would provide the high sensitivity-resolution spectra and images for high energy GRBs from *GLAST* and afterglows from *Constellation-X*, *NGST*, *LSST*, and very large ground-based telescopes. As recommended by the Decadal Survey, *EXIST* could be launched by 2010 to open a new frontier in the study of the most extreme cosmic phenomena and the first stellar objects.

## REFERENCES

1. Grindlay, J.E. et al., "EXIST: A High Sensitivity Hard X-ray Imaging Sky Survey Mission for ISS" in *Proc. 5$^{th}$ Compton Symposium*, edited by M. McConnell and J. Ryan, AIP Conf. Proc., **510**, 784 (2000).

2. Grindlay, J.E. et al., "EXIST: The Ultimate Spatial/Temporal Hard X-ray Survey", in *Gamma 2001*, eds. S. Ritz, N. Gehrels, and C. Shrader, AIP Conf. Proc., **587**, 899 (2001).

3. Hartmann, D., MacFadyen, A. and Woosley, S., "The Most Distant Gamma-ray Bursts", in *Proc. 5$^{th}$ Huntsville Symposium*, eds. R. Kippen, R. Mallozzi and G. Fishman,, AIP Conf. Proc., **526**, 653 (2000).

4. MacFadyen, A. and Woosley, S., *Ap. J.*, **524**, 262 (1999).

5. Narayan, R., Piran, T. and Kumar, P., *Ap. J.*, **557**, 949 (2001).

6. Lamb, D. and Reichert, D.E., *Ap. J.*, **536**, 1 (2000).

7. Norris, J., Marani, G. and Bonnell, J., *Ap. J.*, **534**, 248 (2000).

8. Salmonson, J. and Galama, T., astro-ph/0112298 (2001).

9. Vanden Berk, D. et. al.., astro-ph/0111054 (2001).

10. Huang, Y., Dai, Z. and Lu, T., astro-ph/0112469 (2001).

11. Kippen. R. et. al., these proceedings (2002).

12. Schneider, R, Guetta, D. and Ferrara, A., astro-ph/0201342 (2002).

13. Waxman, E. and Bahcall, J., *Ap.J.*, **541**, 707.

14. Usov, V., *Nature*, **357**, 472 (1992).

15. Kobayashi, S., Ryde, F. and MacFadyen, A., astro-ph/0110080 (2001).

16. Hurley, K. "Soft Gamma Repeaters in Review", in *Proc. 5$^{th}$ Huntsville Symposium*, eds. R. Kippen, R. Mallozzi and G. Fishman,, AIP Conf. Proc., **526**, 763 (2000).

# ECLAIRs: A Microsatellite to Observe the Prompt Optical and X-Ray Emission of Gamma-Ray Bursts

## Didier Barret

*Centre d'Etude Spatiale des Rayonnements, CNRS/UPS, 31028 Toulouse Cedex 04, France*

**Abstract.** The prompt γ-ray emission of Gamma-Ray Bursts (GRBs) is currently interpreted in terms of radiation from electrons accelerated in internal shocks in a relativistic fireball. On the other hand, the origin of the prompt (and early afterglow) optical and X-ray emission is still debated, mostly because very few data exist for comparison with theoretical predictions. It is however commonly agreed that this emission hides important clues on the GRB physics and can be used to constrain the fireball parameters, the acceleration and emission processes and to probe the surroundings of the GRBs. *ECLAIRs* is a microsatellite devoted to the observation of the prompt optical and X-ray emission of GRBs. For about 100 GRBs yr$^{-1}$, *independent of their duration*, *ECLAIRs* will provide high time resolution high sensitivity spectral coverage from a few eV up to ∼ 50 keV and localization to ∼ 5" in near real time. This capability is achieved by combining wide field optical and X-ray cameras sharing a common field of view ($\gtrsim$ 2.2 steradians) with the coded-mask imaging telescopes providing the triggers and the coarse localizations of the bursts. Given the delays to start ground-based observations in response to a GRB trigger, *ECLAIRs* is unique in its ability to observe the early phases (the first ∼ 20 sec) of all GRBs at optical wavelengths. Furthermore, with its mode of operation, *ECLAIRs* will enable to search for optical and X-ray precursors expected from theoretical grounds. Finally *ECLAIRs* is proposed to operate simultaneously with *GLAST* on a synchronous orbit. This combination will ensure broad band spectral coverage from eV to GeV energies for the GRBs detected by the two satellites, *ECLAIRs* further providing their accurate localization to enable follow-up studies. *ECLAIRs* relies upon an international collaboration involving theoretical and hardware groups from Europe and the United States. In particular, it builds upon the extensive knowledge and expertise that is currently being gained with the *HETE-2* mission.

## INTRODUCTION

GRBs occur at cosmological distances and are the most violent explosive phenomena presently observed in the Universe. For the strongest GRBs, up to ∼ 10$^{54}$ erg (assuming isotropic emission) can be radiated in the γ-ray domain on a very short timescale (from milliseconds to ∼ 100 seconds). In the currently favored models, GRBs are associated with the collapse of massive stars (collapsar, [35, 24]) or mergers of two compact stars (two neutron stars or a neutron star and a black hole, e.g., [13]) (see e.g., [25, 8] for recent reviews). Since GRBs are observable across the whole Universe, if they are indeed linked to the ultimate stages of massive star evolution, their redshift distribution should reveal the formation rate of massive stars up to very high redshifts, thus making GRBs effective cosmological probes (see e.g., [29, 18]).

In both the collapsar and the merger models, the final product is a stellar mass black hole and a rapidly rotating torus, from which the energy can be extracted via magnetohydrodynamic processes. The energy release in a very small volume produces a relativistic fireball with a Lorentz factor of at least a few hundred (e.g., [32]). When the relativistic flow decelerates in the interstellar medium (ISM), a forward and a reverse external shock are produced. The forward external shock can account for the afterglow emission observed at radio, optical and X-ray wavelengths ([12] for a recent review).

Whereas the physics of the afterglow is relatively well understood, the origin of the prompt emission is still debated, especially in X-rays and optical. In the relativistic fireball model, the Lorentz factor of the wind is supposed to be variable so that successive shells of plasma have large relative velocities leading to the formation of internal shocks ([28, 9, 4]). In that model, the non-thermal γ-ray emission is associated with either synchrotron or inverse Compton emission of electrons accelerated in those shocks. In X-rays and optical, the picture is not as clear, mostly because there are not as many high quality observations available in that energy range compared to the γ-ray domain. There is however growing observational and theoretical evidence that this energy domain contains critical information on the GRB physics, the nature of the progenitors, the way the initial bulk energy is converted into electromagnetic radiation. A better understanding of the GRB physics is required to test models predicting that GRBs may be sources of ultra high-energy cosmic rays, neutrinos, gravitational waves (see [27] for a recent review). This understanding is also required if one intends to make GRBs a reliable tool for cosmology and for the study of the Universe at very high redshifts.

*ECLAIRs* is a microsatellite proposed to the French Space Agency (CNES). It is specifically devoted to the observation of the prompt optical and X-ray emission of GRBs. In the next section, we briefly describe the main scientific objectives of *ECLAIRs* emphasizing on the area where it will bring an outstanding contribution. We then present the science payload and mission concept. Finally we emphasize on the complementarity of *ECLAIRs* with two other missions (*GLAST* and *SWIFT*) supposed to fly simultaneously with *ECLAIRs*.

## SCIENTIFIC OBJECTIVES

With *ECLAIRs* we wish to use, for the short and long duration GRBs, the prompt optical and X-ray emission to probe 1) the physics at work during the event and 2) the surrounding of the burst to get insights on their origin. In addition, thanks to its instrumental capabilities, with *ECLAIRs*, we will investigate the existence of optical and X-ray precursors expected from theoretical grounds, and whose presence would put unprecedented constraints on any GRB models.

As far as the prompt emission is concerned, as discussed above, very few data exist in optical and X-rays in contrast to the $\gamma$-ray domain. In the optical, so far only one GRB has been detected (by the ROTSE automated telescope; GRB990123, [1]), and for a few others, upper limits are available for the late part ($\geq 10-20$ second after the onset) of the events (e.g. [5]). In X-rays, the situation is slightly better, mostly thanks to the *GINGA* (e.g., [21]) and *Beppo-SAX* satellites [15].

### What can be learned from the prompt optical emission?

It has suggested that the prompt optical emission as the one observed in GRB990123 may be associated with electrons accelerated in the reverse external shock [33]. The strength of the optical emission depends on various parameters, but can in principle yield constraints on the wind initial Lorentz factor and the interstellar medium density [33, 17]. Alternatively, the prompt optical emission could arise from the forward shock of the blast wave when it propagates in the pre-accelerated and pair-loaded environment [3]. This emission can also give constraints on the radiation process itself; i.e., synchrotron versus Inverse Compton emission; a much stronger optical flash is expected in the Inverse Compton scenario (e.g., [10]).

The reasonable question to ask is why the prompt optical emission has so far been observed from only one GRB (GRB990123, [1]). The poor location accuracy of GRB detectors (e.g., BATSE), the delays in getting the finalized positions to the ground, the limited observing efficiency of automated optical telescopes, their response time, all conspire to make sensitive (below mag $\sim 14$) and truly simultaneous observations of GRBs over the whole event almost impossible. For ROTSE, the shortest response time that has been achieved is $\sim 10$ seconds [16]. *ECLAIRs* will not face any of these problems as the optical and X-ray cameras will operate continuously over a common field of view. Optical coverage will thus be granted for all types of bursts, independently of their duration, before, during and even after the event. This unique capability will also offer the opportunity to study the transition between the prompt and early afterglow phases at similar wavelengths.

Extinction by dust in the host galaxy may naturally prevent the detection of the prompt optical emission. This is the argument used to explain the lack of optical emission in some afterglows, otherwise detected in X-rays and in radio. Dust extinction is not unexpected if GRBs are associated with massive star formation. Djorgovski et al. (2001) have however found that the *maximum* fraction of optical afterglows hidden by dust is $\sim 50\%$. This is an upper limit, as some optical afterglows may have been missed for various reasons: very high redshifts, rapid decline rate, intrinsic faintness.

*ECLAIRs* will seek optical emission down to magnitude $\sim 15$ (R band, 8 sec) for all GRBs. The properties of the prompt optical emission will be correlated with the properties of the afterglow optical emission, thus providing complementary constraints on the relativistic flow parameters and the surroundings of the event.

### What can be learned from the prompt X-ray emission?

Let us now consider the prompt X-ray emission. The good correlation between the temporal behaviour of the prompt X-ray and $\gamma$-ray emissions suggests that it is also produced in internal shocks. In X-rays however, there might be additional contributions from the reverse shock [9], the forward shock [3] or from the photosphere of the fireball [19]. With its excellent sensitivity, *ECLAIRs* will observe the prompt X-ray emission of all GRBs, allowing detailed time-resolved X-ray spectroscopy to be performed. These studies will set constraints on the Lorentz factor of the wind, its baryon loading, the emission mechanism and the relative contribution of the various shock regions in the overall emission.

The importance of observing the prompt X-ray emission has recently been reinforced by the discoveries of X-ray spectral features in *Beppo-SAX* observations: e.g., a transient absorption edge in GRB990705 [2] and a transient emission feature in GRB990712 [14] (spectral features are also observed in the X-ray afterglows, e.g., [26]). Well before these discoveries, it was predicted that effects of photo-electric absorption and Compton scattering from the circum-burst material should lead to observable changes in the intrinsic GRB spectrum, with the introduction of absorption cutoffs and features such as K-edges and emission lines [20]. In principle, these features can be used to determine the density and composition of the ISM in the immediate vicinity of the GRB, the GRB redshift and possibly the nature of the GRB progenitor. For instance, the transient absorption edge observed from GRB990705 was satisfactorily modeled with photo-electric absorption by a medium with a large iron abundance, which could have been left there by a supernova event which occurred about 10 years before the burst [2]. Similarly the transient emission feature seen in GRB990712 was shown to be consistent with thermal emission of a baryon-loaded expanding fireball when it becomes optically thin [14]. The above interpretations are however made difficult by the limited statistical quality of the data. With its improved sensitivity and good time and spectral resolution, *ECLAIRs* will be able to observe the prompt emission of GRBs over the whole event and for all types of events.

## What are the short bursts ? What about the X-ray precursors ?

GRBs display a bimodal distribution in durations; the border is around 2 seconds with about 25% of the GRBs with durations less than that value. This distribution seems to correlate with spectral hardness; the shortest GRBs have on average harder spectra [11]. It seems therefore plausible that the two distributions represent two distinct, although quite similar, physical phenomena. Extremely short GRBs may be due to primordial black hole evaporation, short GRBs to merging neutron stars, and the long ones to collapsars (see e.g., [25]). So far, due to observational limitations, afterglows have only been identified for the long duration GRBs and very little is known about the short GRBs. *ECLAIRs* will have the unique capability to observe both short and long duration GRBs. These observations will thus provide clues to the following questions: How does the multi-wavelength prompt emission of the short GRBs compare with those of the long GRBs? How does the prompt emission relate to the afterglow properties? What is the redshift distribution of short GRBs? Answering these questions will help in assessing whether the short duration GRBs are of different nature than the long ones.

By its mode of operation, *ECLAIRs* will also enable us to search and study X-ray and optical precursors. X-ray precursors have already been observed ([21, 15]), arising between 10 and 100 seconds before the main event. Several models have been put forward to explain these X-ray precursors (or soft excesses) (e.g., [24, 22, 19]), all making some specific predictions which require further data to be tested. What is the spectrum of the X-ray precursor? How does it evolve with time? How do its properties relate to the properties of the main event? These are some of the questions which will be addressed by *ECLAIRs*. In addition, Paczyński (2001) recently pointed out that there are theoretical reasons to expect strong optical flashes preceding GRBs (e.g., [3]), the detection of which would put stringent constraints on the range of parameters for GRB models.

## THE *ECLAIRS* MISSION CONCEPT

The *ECLAIRs* mission concept results from the scientific goals described above and is optimized under the stringent constraints of a microsatellite: 50 kilos, 50 Watts and a total volume of $\sim 60$ cm $\times$ 60 cm $\times$ 30 cm (length, width, height) available for the science payload. A technical assessment study of *ECLAIRs* was carried out by CNES in December 2001. This study showed that *ECLAIRs* was feasible as a microsatellite, albeit with not much margins. *ECLAIRs* is now lining up for a selection at the end of 2002. This clearly leaves some time to work on the optimization of the science payload. To help us in this task, an international Science Advisory Committee (SAC) was set up for the mission. The science payload described below accounts for the results from the CNES study and for the advises received from the SAC.

### The science payload

The science payload consists of three sets of instruments (see Table 1). The Large Area X-ray Telescope (E-LAXT), the Soft X-ray Cameras (E-SXC), and the Wide Field Optical Cameras (E-WFOC) (see Fig 1). It will be provided by a consortium of institutes which have developed a considerable expertise along the preparation and operation of missions, such as *HETE-2* and INTEGRAL. The US contribution to *ECLAIRs* will be the subject of a SMEX/MOO proposal to NASA in 2002.

**TABLE 1.** The ECLAIRs science payload consisting of three instruments: the ECLAIRs Large Area X-ray Telescope (E-LAXT), the ECLAIRs Soft X-ray Cameras (E-SXC), the ECLAIRs Wide Field Optical Cameras (E-WFOC). Two options are considered for E-LAXT: Silicon and higher density detectors (e.g. CdZnTe). Si would cover the energy range 3-50 keV whereas CdZnTe would cover from 5 keV to 150 keV.

|  | E-LAXT | E-SXC | E-WFOC |
|---|---|---|---|
| Band pass | $\sim$ 3–50/5-150 keV | 0.4–15 keV | 500-700 nm |
| Number of units | 2 | 6 (3 pairs) | 4 |
| Offset angle | $\pm 10°$ | $-28°, 0, +28°$ | $\pm 25°$ |
| Mass (kg) | 14 | 11 | 14 |
| Power (instrument + electronics) (W) | 18 | 8 | 16 |
| Field of view (one unit, FWZR) | $\sim 120° \times 120°$ | $\sim 53° \times 53°$ | $\sim 50° \times 50°$ |
| Positionning accuracy | $\sim 0.5°$ | 5" | 5" |
| Number of GRBs yr$^{-1}$ (total) | $\sim 100$ | $\sim 100$ | ? |
| Limiting mag. (R) (S/N=8) | ... | ... | 14.8, 17.4 (8, 1000 sec) |

## ECLAIRs - Large Area X-ray Telescope

The Large Area X-ray Telescope (E-LAXT) is made of two identical conventional 2D coded-mask imaging telescopes, with offset looking directions. The mask is located 15 cm above the detector. The detector is a pixel semiconductor detector. In the baseline, each pixel was a 2 mm thick Si PIN diodes of 1cm$^2$. The mask cells match the pixel size. Large area Si PIN detectors with their associated low-noise low-power front-end electronics are currently developed at CESR as part of an R&T program funded by CNES. At low power, the expected noise level should result in an energy resolution of $\sim 1$ keV (at 6 keV, -40C) making possible a low energy threshold of $\sim 3$ keV. The thickness of the diodes ensures an energy coverage up to $\sim 40$ keV: matrix of Si PIN diodes are therefore a possible detector solution for *ECLAIRs*.

However, as suggested by the SAC, there are alternatives to Silicon for the E-LAXT detector: CdTe, CdZnTe. Both would extend the energy range of *ECLAIRs* in the hard X-ray range, which would help for the detection of GRBs. CdZnTe, as the ones developed for *SWIFT* or AXO [7] have excellent performances for a mW/cm$^2$ ratio similar to Silicon. In addition, using strip readout techniques they have been demonstrated to work down to 5 keV (e.g. [7]). Considering CdZnTe, the detector of one E-LAXT could have an effective area of $25 \times 25$ cm$^2$, covered with pixels of $5 \times 5$ mm$^2$ and 2 mm thickness. The imaging system would have an angular resolution of 2 degrees and a positioning accuracy of $\sim 0.5$ degree. Due to the stringent mass constraints on a microsatellite, the mask and shielding will be effective only below $\sim 50$ keV. Whereas the trigger will be obtained at energies above $\sim 50$ keV, the position of the GRB will be derived from the images reconstructed below that energy. Using a simplified model for the E-LAXT and the Log(N)/Log(P) curve derived from the BATSE 4B catalog [23], we have estimated the rate of GRBs localized by the two E-LAXT units to be larger than 100 GRB yr$^{-1}$.

## ECLAIRs - Soft X-ray Cameras

The Soft X-ray Cameras (E-SXC) for *ECLAIRs* are based upon the successfully-flown *HETE-2* design [30] (see also these proceedings). The operating principle is that of a coded-mask imager, in which a 1-D coded mask is rigidly suspended above an X-ray charge-coupled device (CCDID-34). The E-SXC assembly is made of 6 camera modules, covering a field of view of 2.7 sr. The CCDID-34 (3K$\times$6K array; 10$\mu$m square pixels, 20") has an overall size of 30 mm $\times$ 60 mm and is currently in production at MIT Lincoln Laboratory. It improves over the CCID-20 used for *HETE-2* by a greater energy coverage (0.4-15 keV versus 0.8-10 keV), a better time resolution (0.25 sec versus 1 sec), and a better quantum efficiency (sensitivity of $\sim 400$ mCrab, 1 sec, 4$\sigma$). The E-SXC will provide 5" burst localizations (at S/N =8). About 100 GRBs yr$^{-1}$ should be detected in the 6 units.

## ECLAIRs - Wide-Field Optical Cameras

The Wide-Field Optical Cameras (E-WFOC) for *ECLAIRs* are derived from the star camera units successfully flown on *HETE-2* [30]. The large field of view is achieved by four such cameras. The limiting magnitude in R is 14.8 (8 s at S/N=8) for one E-WFOC. Each of the four modules utilizes a moderately fast, well-corrected optical lens (focal length of 80 mm, f/0.9) coupled to a 2$\times$2 array of MIT CCID-34 sensors, resulting in a hybrid focal plane with 6K$\times$ 6K pixels; each pixel is 10$\mu$m x 10$\mu$m (25.8"). The integration time is 2 seconds.

To achieve the light weight and low power required for *ECLAIRs*, the drive and readout electronics, as well as the digital frame buffer memory, for the E-SXC and E-WFOC instruments will be combined to the maximum degree possible.

The operating mode for the E-WFOC relies upon digitizing and storing successive 300 MB image frames in a four stage deep buffer, requiring a total of 1.2 GB SRAM.

variability so far.

We are currently investigating near-infrared cameras (NIRC), 1-2 microns) as an alternative to the E-WFOC. There are three main advantages of considering NIRC. First, in NIR, the extinction is smaller than in the optical. This means that GRBs produced in dusty regions and not visible in the optical might become detectable in NIR. Second, NIRC would have the potential to observe higher z events, due to the Lyman alpha break. Finally, the discovery space is much larger in NIR than in the optical, as the NIR sky has never been searched for variability so far.

A detailed study is now required to determine whether cooled NIRC with good sensitivity, large field of view can be accommodated on *ECLAIRs* under the stringent constraints on power, mass, etc. of a microsatellite.

## Implementation of the mission

The baseline for *ECLAIRs* is an equatorial 550 km (low inclination) orbit for a low, stable background, low radiation damage to the CCD, and for the download of the science data to be possible with a single ground station. This orbit could be achieved by various launchers, as for example a PEGASUS from the Marshall Islands or from Alcantara. We plan to reuse the *HETE-2* segment. In particular, one of the 3 S band stations (Singapore, Kwajalein, or Cayenne) will be converted to X band. For the alert system, we plan to use the *HETE-2* network of 12 VHF stations located along the equator. As will be discussed below, *ECLAIRs* is proposed to fly simultaneously with *GLAST*; therefore an ideal launch date would be around the end of 2006. The lifetime of the mission is foreseen for 5 years for a maximum synergy with *GLAST*.

## Operational considerations

As far as the attitude control is concerned, the instruments will point in the anti-earth direction during night time, and look at the pole during day time. This way the operating temperature of the instruments will be kept low (below -50 C). The triggering system of *ECLAIRs* is relatively simple. The on-board computer monitors continuously the count rates in E-LAXT. When a transient event is detected, a signal is sent to the E-SXC and E-WFOC for the most recent data to be stored in a dedicated memory. Two images from E-LAXT are then reconstructed before and during the event. From the difference of the two images, the rough position (∼ 0.5 deg accuracy) of the event is obtained and sent out to the ground and to the secondary instruments which use this position to obtain the final more accurate position. After ∼ 30 seconds, the final position (5" accuracy) is transmitted to the ground. During the next passage to the ground station a high rate X-band communication (16 Mbits/s) allows the whole data set associated with the event to be downloaded.

The mission and science operations will be performed with the help of CNES control center in Toulouse. The mission and operation center which may be combined will be responsible for receiving the data from CNES, generating the spacecraft commands on a weekly basis, monitoring the health of the spacecraft and science payload, recovering the attitude, and for the quick-look anal-

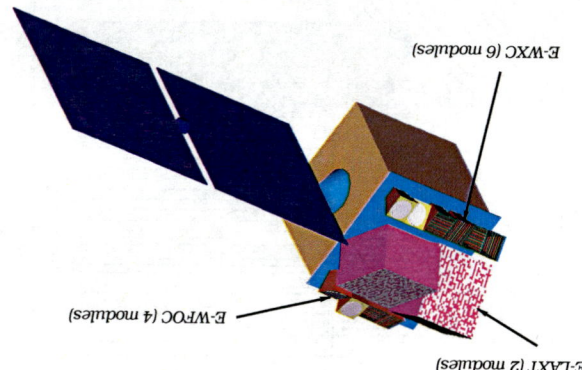

**FIGURE 1.** The *ECLAIRs* science payload on a microsatellite Myriade spacecraft. The three sets of instruments which are shown fit within the geometrical constraints imposed to the science payload.

In response to triggers from the E-LAXT or E-SXC, we will select 4° × 4° regions-of-interest (ie 512 x 512 sub-arrays) from this large buffer, centered on the suspected burst coarse localization, for transfer into an optical burst memory. In addition, neighborhoods of twenty-five stars, extending out to 64×64 pixels (27'×27'), will also be stored as astrometric and photometric references. The accumulation of 500 frames (=1000 sec), each with burst and reference star data, will reside in 377MB of SRAM, and require 3 minutes to downlink during an X-band contact with an *ECLAIRs* ground station. Shift-and-add summation of the digitized, two-second resolution CCD data in ground processing will permit the E-WFOC to achieve an ultimate limiting sensitivity of R=17.4 (1000s at S/N=8). Centroiding will result in bright optical transient localizations accurate to ± 2", even in the presence of spacecraft pointing drift (assumed to be 4"/s, 3σ, as specified for the Myriade spacecraft). For long term optical monitoring, we will also be able to downlink 45 full image frames per day (whole field of view at full angular resolution, every ∼ 30 minutes), each containing more than 2.5 million star images. Downlinking of the full frame data will require 13 GB/day.

ysis of the data for rapid distribution of the GRB final positions. In addition it will be responsible for the data archives and the education and public outreach program. The center will likely be provided by a consortium of institutes including the Geneva observatory, the Strasbourg observatory and the Leicester University X-ray group.

## A MISSION COMPLEMENTARY TO GLAST AND SWIFT

*GLAST*[1] is scheduled to be launched in March 2006 into a low earth orbit. The satellite will carry 2 instruments: the Large Area Telescope (LAT), which will observe emission from 20 MeV to 200 GeV, and the Gamma-ray Burst Monitor (GBM), which will detect transients from 20 keV to 20 MeV. The LAT detector is 50 times more sensitive than it's predecessor, *EGRET*. While only a few GRBs were detected by *EGRET*, *GLAST* is expected to observe nearly 200 GRBs per year. These bursts will also be detected by the GBM, so that the spectrum will be measured over 7 orders of magnitude. Unfortunately, only for the brightest bursts the positions derived by the LAT will be accurate enough to be used for follow up observations. Provided that *ECLAIRs* and *GLAST* can remain on a similar orbit (adjustment and maintenance of the orbit can indeed be achieved through the chemical propulsion system of the microsatellite) *ECLAIRs* would greatly enhance the *GLAST* science, by both extending the spectra to lower energies (down to $\sim 2$ eV) and by improving the localizations in near real-time to enable follow-up observations in the afterglow regime. Given that the GBM has a FOV much larger than *ECLAIRs*, all GRBs seen by *ECLAIRs* will be also detected by the GBM, thus providing for $\sim 100$ GRB yr$^{-1}$, spectral coverage from about 7 decades in energy, and for those detected by the LAT ($\sim 80$) over 11 decades in energy! This broad band spectral coverage will enable discrimination between the various radiation processes proposed for the multi-wavelength GRB emission, including those, yet to be tested, put forward for the GeV emission (inverse Compton, Synchrotron emission, see e.g., [3,4,6]). This will open a completely new window on the GRB physics, setting for the first time real constraints on models predicting that GRBs are sources of ultra-high energy cosmic rays and neutrinos. Furthermore, the ability to locate the *GLAST*-GRBs precisely, making possible the identification of the host galaxies and the measure of the redshifts will enable the systematics of the GeV emission to be studied, and the *GLAST*-GRBs to be compared, as a class of events, to the GRBs detected by satellites operating at lower energies (*Beppo-SAX*, *HETE-2* and *SWIFT*). Finally, for those GRBs for which the redshift will be determined, cut-offs in the observed GeV spectrum can be used to infer the level of ultra-violet to infrared background light which is a direct tracer of star and Galaxy formation in the early Universe [31]. The complementarity between *GLAST* and *ECLAIRs* is best illustrated in Fig. 2 where the observing energy range is plotted against the observing time window of the events. *ECLAIRs* was presented at the last *GLAST* GRB working group and received strong support.

*SWIFT*[2] is a NASA mission dedicated to the study of the GRB afterglows. It should be launched in Fall 2003, with a nominal on-orbit lifetime of 3 years. It will carry three instruments: The Burst Alert Telescope (BAT) covering the 10 to 150 keV range, and two Narrow Fields Instruments (NFIs); the X-Ray Telescope (XRT, 0.2-10 keV) and the UV Optical Telescope (UVOT, 170-650 nm). The observing strategy of *SWIFT* is to point

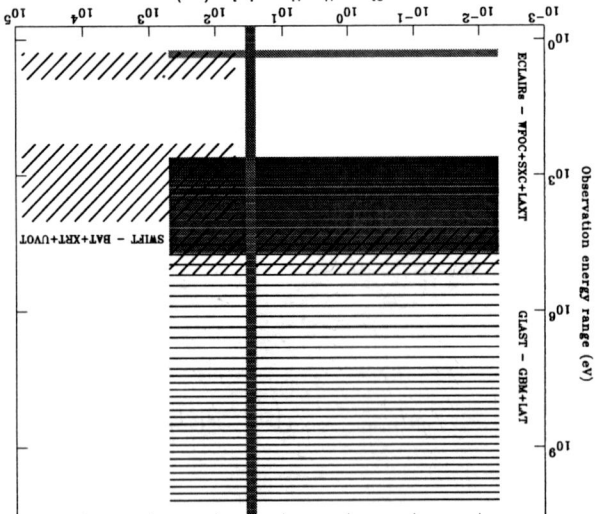

**FIGURE 2.** Comparing *ECLAIRs* (filled regions) with *GLAST* (horizontal lines) and *SWIFT* (tilted lines). The time window of the observations is given on the X axis whereas the Y axis represents the energy range of the instruments. As can be seen, the combination of *ECLAIRs* and *GLAST* would provide spectral coverage over 11 decades in energy. Note also the complementarity between *ECLAIRs* and *SWIFT*; *ECLAIRs* is focused on the prompt optical/X-ray emissions whereas *SWIFT* is designed for the afterglow emission in the same energy range. The mean GRB duration is also shown for indication (vertical box).

---

[1] http://www-glast.stanford.edu/       [2] http://swift.sonoma.edu/

the NFIs after the detection of a GRB in the BAT. This strategy clearly means that *SWIFT* will miss the early X-ray and optical emission of all GRBs. The time to point the NFIs to the direction of the GRB should range between 20 and 70 seconds, with a mean value of 50 sec. Its ability to observe the precursors and activity during the burst will thus make *ECLAIRs* a very complementary mission to *SWIFT* (see Fig. 2).

## CONCLUSIONS

Fortunately GRBs are extremely bright events which can easily be detected and studied with an instrumentation matching the stringent mass and power constraints of a microsatellite. GRBs have been proved to be highly complex phenomena whose understanding requires multi-wavelength observations of the prompt and afterglow phases and follow-up ground-based observations to determine their host galaxies and their redshifts. *ECLAIRs* will thus bring a significant contribution to a better understanding of GRBs by providing high sensitivity observations of the prompt optical/X-ray emission and accurate localization of more than 100 gamma-ray burst per year.

## ACKNOWLEDGMENTS

It is a real pleasure for me to thank all the members of the *ECLAIRs* collaboration for their interest in the mission and their support: J.L. Atteia, M. Boër, A. Beloborodov, A. Castro-Tirado, T. Courvoisier, F. Daigne, J.P. Dezalay, B. Dingus, M. Ehanno, P. Goldoni, P. Guillout, J.M. Hameury, G. Henri, K. Hurley, P. Jean, G. Jernigan, J.P. Kneib, D. Lamb, P. Mandrou, A. Marcowith, F. Martel, L. Michel, R. Mochkovitch, C. Motch, J.P. Osborne, M. Pakull, G. Pelletier, J. Poutanen, V. Reglero, G. Ricker, J. Rodrigo, A. Short, R. Svensson and M. Ward.

A special thank also to C. Meegan and N. Gehrels for their strong support on behalf of the GLAST collaboration.

I also wish to thank the members of the *ECLAIRs* Science Advisory Committee for their very valuable inputs which will help in optimizing the mission: J.L. Atteia, A. Castro-Tirado, F. Frontera, J. Hjorth, N. Kawai, S. Kulkarni, C. Meegan, R. Mochkovitch, T. Piran, G. Ricker, B. Stern, S. Woosley.

I am very grateful to G.K. Skinner, J.L. Atteia and G. Ricker for their continuous help and support.

## REFERENCES

1. Akerlof, C. et al. 1999, Nat, 398, 400.
2. Amati, L.et al. 2000, Science, 290, 953
3. Beloborodov A. M., 2002, ApJ, 565, 808
4. Beloborodov, A., 2000, ApJL, 539, L25
5. Boër M., et al. 2001, A&A, 378, 76
6. Böttcher, M. & Dermer, C. D. 1998, ApJL, 499, L131
7. Budtz-Joergensen, C. et al. 2001, Astrophysics & Space Science 276, 281
8. Castro-Tirado, A. J. 2001, ESA-SP Conf. Proc. in press, astro-ph/01021222
9. Daigne, F. & Mochkovitch, R. 2000, A&A, 358, 1157.
10. Daigne, F. & Mochkovitch, R. 1998, MNRAS, 296, 275.
11. Dezalay, J. P. et al. 1996, ApJL, 471, L27
12. Djorgovski, S. G. et al. 2001, , to appear in: Proc. IX Marcel Grossmann Meeting, eds. V. Gurzadyan, R. Jantzen, and R. Ruffini, Singapore: World Scientific, in press, astro-ph/0106574
13. Eichler D. et al. 1989, Nat. 340, 126
14. Frontera, F.et al. 2001, ApJL, 550, L47
15. Frontera, F. et al. 2000, ApJS, 127, 59
16. Kehoe, R. et al. 2001, ApJ, 554, 159
17. Kobayashi, S. 2000, ApJ, 545, 807
18. Lamb D. Q., Reichart D. E., 2000, Ap J, 536, 1
19. Mészáros, P. & Rees, M. J. 2000, ApJ, 530, 292
20. Mészáros, P. & Rees, M. J. 1998, ApJL, 502, L105
21. Murakami, T., et al. 1991, Nat, 350, 592
22. Nakamura, T. 2000, ApJL, 534, 159
23. Paciesas, W., et al. 1999, ApJS, 122, 465
24. Paczyński, B. 1998, ApJL, 494, L45
25. Piran T., 1999, Phys. Rep., 314, 575.
26. Piro, L. 1999, Proc.s of "X-Ray Astronomy '99:Stellar Endpoints, AGN and the Ddiffuse X-ray Background, astro-ph/0001436
27. Postnov K.A. 2001, published in Proc. XI Int. School Particles and Cosmology, Baksan Valley, astro-ph/0107122
28. Rees, M. & Mészáros P. 1994, ApJL, 430, L93
29. Ramirez-Ruiz E., Fenimore E. E. & Trentham N. 2000, Conference on Cosmology and Particle Physics, Verbier, Switzerland, eds. J. Garcia-Bellido, R. Durrer, and M. Shaposhnikov, astro-ph/0010588
30. Ricker, G., 2001, American Astronomical Society Meeting 198, 35.04 (see also this meeting)
31. Salamon, M. H. & Stecker, F.W., 1998, ApJ, 493, 547
32. Sari, R. & Piran, T. 1997, MNRAS, 287, 110
33. Sari, R. & Piran T. 1999, ApJ, 520, 641.
34. Waxman, E. 1997, ApJL, 485, L5
35. Woosley, S. 1993, ApJ, 405, 273.

# The *Swift* X-Ray Telescope

D. N. Burrows*, J. E. Hill*, J. A. Nousek*, A. Wells[†], A. Short[†], M. Turner[†], O. Citterio**, G. Tagliaferri** and G. Chincarini**

*Penn State University, 525 Davey Lab, University Park, PA 16802*
[†]*University of Leicester*
**Osservatorio Astronomico di Brera*

**Abstract.** The *Swift* Gamma-Ray Burst Explorer is designed to make prompt multiwavelength observations of Gamma-Ray Bursts (GRBs) and their afterglows. The X-ray Telescope (XRT) provides key capabilities that permit *Swift* to determine GRB positions with several arcsecond accuracy within 100 seconds of the burst onset. We present an overview of the XRT and its capabilities. The XRT is designed to observe GRB afterglows covering 7 orders of magnitude in flux in the 0.2-10 keV band, with completely autonomous operation. Accurate GRB positions are determined within seconds of target acquisition and are sent to the ground for distribution over the GCN. The XRT can also measure redshifts of GRBs if they have Fe line emission or other spectral features.

## INTRODUCTION

The *Swift* Gamma-ray Burst (GRB) Explorer, scheduled for launch in September 2003, is designed to make prompt multiwavelength observations of Gamma-Ray Bursts and afterglows. *Swift* is a NASA/GFSC mission with major contributions from Penn State University (PSU), the University of Leicester (UL; UK), the Mullard Space Sciences Lab (MSSL; England), and the Osservatorio Astronomico di Brera (OAB; Italy). *Swift* is a highly autonomous mission designed to respond very rapidly to new GRBs, obtaining accurate positions and spectroscopy in UV, X-ray, and Gamma-ray bands within minutes of the burst discovery. *Swift* has three scientific instruments. The Burst Alert Telescope (BAT), built by NASA/Goddard Space Flight Center, is a wide-angle coded aperture mask imager with CZT detectors, and provides arcminute burst positions. The X-ray Telescope (XRT), built by PSU/UL/OAB, provides arcsecond GRB positions and X-ray spectroscopy and lightcurves of afterglows. The UV/Optical Telescope (UVOT), built by PSU/MSSL, provides sub-arcsecond positions, finding charts, and redshift measurements.

The X-ray Telescope (XRT) provides key capabilities that permit *Swift* to determine GRB positions with several arcsecond accuracy within 100 s of the burst onset. The XRT is designed to observe GRB afterglows covering over seven orders of magnitude in flux in the 0.2-10 keV band, with completely autonomous operation. GRB positions are determined within seconds of target acquisition, and accurate positions sent to the ground for distribution over the GRB Coordinate Network. The XRT can also measure redshifts of GRBs with Fe line emission.

## OVERALL XRT DESIGN

The XRT is a Wolter Type II focusing telescope with a CCD detector, providing good angular resolution (15 arcseconds Half-Power Diameter at 1.5 keV) and moderate energy resolution (70 eV at 200 eV) from 0.2 – 10 keV. A schematic view of the instrument is shown below. The mirror is provided by OAB, the Focal Plane Camera is provided by Leicester, and the rest of the instrument is provided by PSU.

The scientific goals of the XRT instrument are: determine GRB position with 5 arcsecond accuracy within 100 s of the burst, transmit to ground immediately for distribution on GCN; obtain X-ray spectroscopy of the afterglow, measure redshift if possible; measure lightcurve of afterglow

The XRT instrumentation includes a spare mirror from the Spectrum X-Γ/JET-X instrument and a CCD detector developed for the XMM/EPIC instrument. A schematic diagram of the XRT is shown in Figure 1.

## XRT OPTICS

The mirror is a 12 shell Wolter I optic made by OAB and Medialario, and shown in Figure 2. The shells are

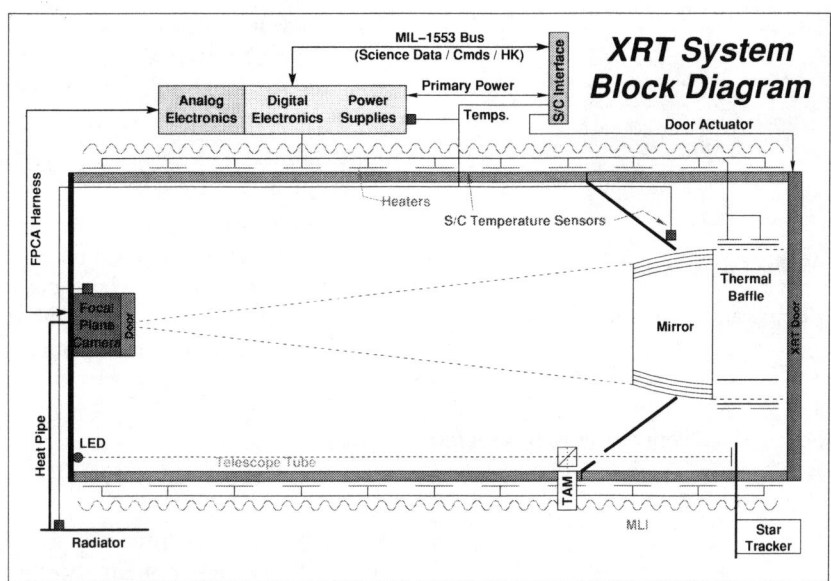

**FIGURE 1.** Schematic diagram of the XRT.

**FIGURE 2.** XRT Mirrors

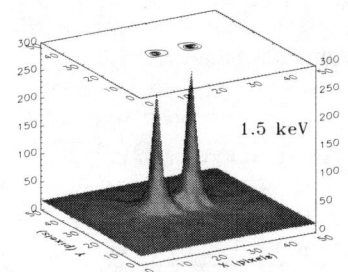

**FIGURE 3.** XRT Point Spread Function

gold-plated Ni. The mirror has excellent imaging performance, with a measured PSF of 15 arcseconds Half-Energy Width at 1.5 keV. The mirror was calibrated at the Panter facility in 1996, and re-calibrated in 2000 with identical results. Figure 3 shows the image of two point sources separated by 20 arcseconds, made at 1.5 keV at the Panter facility.

The excellent imaging performance of these mirrors is critically important to achieving the prime goal of the XRT: arcsecond GRB positions. Calibration data and simulations both show that the XRT can position GRB afterglows to better than 1 arcsecond in detector coordinates for reasonable GRB intensities (>100 mCrab). The centroiding algorithm that will be used on the XRT employs an iterative center-of-gravity algorithm applied to detected X-ray events in a single image to generate centroids accurate to about 0.2 pixels in a single exposure.

Equally important to obtaining accurate positions is the our knowledge of the boresight of the XRT with respect to the spacecraft attitude. The *Swift* star trackers are mounted on the side of the XRT in order to minimize errors between the star tracker reference system and the XRT boresight, and the XRT telescope tube includes a set of heaters designed to maintain a stable thermal environment and avoid alignment errors due to thermal variations. Finally, the XRT includes a Telescope Alignment Monitor designed to measure and correct for any remaining internal alignment drifts.

## FOCAL PLANE CAMERA

The Focal Plane Camera Assembly (FPCA; Figure 4) is provided by the University of Leicester. The design is dominated by a rather massive proton shield designed to minimize radiation damage to the CCD. The CCD operates at −100C and is cooled by a Peltier cooler, with heat dumped through a heat pipe to an anti-solar radiator. Four $^{55}$Fe sources illuminate the corners of the detector with 5.9 keV X-rays to provide a continuous monitor of

**FIGURE 4.** XRT Focal Plane Camera Assembly

the detector performance.

The FPCA includes a baffled vent designed to vent the camera during the ascent phase. The FPCA door is based on designs from JET-X and EPIC.

## XRT ELECTRONICS

The XRT instrument interface to the spacecraft is via a MIL-STD-1553 bus and redundant +32VDC power lines. The XRT door is opened directly by the spacecraft control system. All other instrument functions are controlled by telecommands processed by the instrument CPU. Analog and digital electronics are in a common box divided into separate compartments to minimize noise. The CCD clock drive signals are produced by the CCD Clock Sequencer and Clock Driver boards. Video signals from the CCD are processed by the Signal Chain board, and the digitized CCD pixel data are returned to the ICP for processing.

The XRT Electronics Engineering Unit is currently undergoing testing of the engineering model electronics boards in the *Swift* laboratory at PSU.

## XRT DETECTOR MODES

The XRT supports three readout modes: Imaging, Timing, and Photon-Counting:

> **Imaging:** accumulates piled-up image of target (comparable to optical CCD image) during initial burst observation, gives position and broad-band photometry for Fx< 26 Crabs.
>
> **Timing:** Measurement of source light-curve and spectum with time resolution better than 1 msec. Sub-modes:
>   – *Photodiode:* integrates charge over entire CCD for 0.1 ms time res., no position res.
>   – *Windowed Timing:* readout of central 100 columns, 1 ms time res., 1-D position res.
>
> **Photon counting:** Provides spectroscopy, photometry (2.5 s resolution), and images with full spatial and spectral resolution and maximum sensitivity for weak sources.

Each observation begins with an Imaging mode exposure, which is used to obtain the GRB position. The instrument then switches into Photodiode mode, which offers the highest time resolution but no position resolution. Further mode switches occur automatically based on detected count rate, with mode switch points chosen to provide the maximum science return for each flux range.

## XRT DATA DISTRIBUTION

Following the first three months of checkout, all XRT data will be made available to the public within hours of receipt. XRT positions, images, and raw spectra will be issued through the GRB Coordinate Network on this schedule:

> Positions: within 100 s of burst
> Postage stamp image: within 500 s of burst
> Raw spectrum: within 1000 s of burst

Fully processed data products from the XRT, including event files in standard OGIP formats, are expected to be available via the Internet within 12 hours of each observation.

Data centers in the US (HEASARC), UK (Leicester) and Italy (Rome/Milano) will be responsible for distributing data through mirror archive sites, and for supporting users.

The XRT site at *http://www.swift.psu.edu/xrt* provides further instrumental details.

## ACKNOWLEDGMENTS

This work is supported in the US by NASA contract NAS5-00136; in the UK by funding from PPARC; and in Italy by funding from ASI. We gratefully acknowledge the contributions to the XRT from the entire XRT team at Penn State, University of Leicester, and Osservatorio di Brera; from our subcontractors at Southwest Research Institute (electronics), Swales Aerospace (engineering support), ATK (telescope tube), and Starsys Research (telescope door); and by our colleagues at NASA/Goddard Space Flight Center.

# The Trigger Algorithm for the Burst Alert Telescope on Swift

E. E. Fenimore[*], David Palmer[*], Mark Galassi[*], Tanya Tavenner[*], Scott Barthelmy[†], Neil Gehrels[†], Ann Parsons[†] and Jack Tueller[†]

[*]*Los Alamos National Laboratory, Los Alamos, NM USA*
[†]*Goddard Space Flight Center, Greenbelt, MD USA*

**Abstract.** The Swift Burst Alert Telescope (BAT) is a huge (5200 cm$^2$) coded aperture imager that will detect gamma-ray bursts in real time and provide a location that the Swift satellite uses to slew the optical and x-ray telescopes. This huge size is a challenge for the on-board triggering: trends as small as 1% over af second is equivalent to a 1 $\sigma$ statistical variation in 1 second. There will be 3 types of triggers, 2 based on rates and one based on images. The first type of trigger is for short time scales (4 msec to 64 msec). These will be traditional triggers (single background) and we check about 25,000 combinations of time-energy-focal plane subregions per second. The second type of trigger will be similar to HETE: fits to multiple background regions to remove trends for time scales between 64 msec and 64 seconds. About 500 combinations will be checked per second. For these rate triggers, false triggers and variable non-GRB sources will be rejected by requiring a new source to be present in an image. The third type of trigger works on longer time scales (minutes), and will be based on routine images that are made of the field of view.

## INTRODUCTION

The Burst Alert Telescope (BAT) on Swift is a large (5200 cm$^2$) CZT-based coded aperture imager. BAT's primary role on the Swift satellite is to detect when a gamma-ray burst (GRB) starts, quickly locate it, and direct the Swift satellite to point the optical and x-ray telescopes at the source. The BAT trigger must not only detect the occurrence of a GRB, but it also must select the time periods to form an image. This requires the triggering code to identify a range of times (the "background" period) when there is no apparent emission from the GRB and a range of times (the "foreground" period) where the GRB probably will produce the strongest image.

The BAT triggering code has three types of triggers. Two of these are "rate" triggers based on statistically significant increases in the counting rate in the focal plane (or a portion of the focal plane), and one is an "image" trigger based on new significant sources found in images of the field of view (FOV). The rate triggers are divided into the "short" rate triggers (with foreground periods of less than or equal to 64 msec) and "long" rate triggers (with foreground periods larger than or equal to 64 msec). The image triggers are intended for longer periods of time (from 64 sec to many minutes).

One goal is to explore the widest possible parameter space. As such, as many triggers as possible will be run simultaneously until the flight computer is nearly saturated. Thus, special attention will be paid to the CPU usage.

It is not hard to design a triggering code that responds to GRBs. The real challenge in the triggering code is to avoid false triggers. This is a special problem with BAT because its huge size means that a very slight trend can appear to be a significant increase in the count rate: a 1% change in the count rate over 1 sec appears to be a 1 $\sigma$ variation. Thus, the chief danger is false triggers due to trends in the background, uninteresting variations of sources in the FOV, or minor configuration changes (such as automatic gain adjust) that produces the appearance of a statistically significant increase. Our chief defense against such false triggers is that we will form an image and the trigger will only proceed to slewing the satellite if there is a clear point source (thus eliminating events associated with particle variations) that is not in the direction of known variable sources.

## SHORT RATE TRIGGERS

Running many short time scales through a triggering code can require most of the CPU time. Fortunately, the background counting rate of BAT is not expected to change on short time scales (i.e., less than a few seconds). Thus, for the short time scales we will use simple traditional triggers where there is a single background period of fixed duration before the foreground period. This is the type of trigger that was used on all GRB experiments from Vela to BATSE.

The short trigger looks for statistically significant increases in the count rate on five time scales: 4, 8, 16, 32, and 64 msec. This is done for nine different regions of the focal plane (four quadrants, the left half, right half, top half, bottom half, and the full focal plane) and for four energy ranges. Thus, there are 36 combinations of focal plane regions and energy ranges. Within each second there are 250 4-msec samples to check, 250 8-msec samples to check (assuming the foreground periods are checked at all 4-msec phases), 125 16-msec samples (assuming the foreground periods are checked at all 8-msec phases), 62 32-msec samples, and 32 64-msec samples.

Overall, there are more than 25,000 short trigger samples to check every second. The calculational effort is optimized by having the code responsible for reading the photons from the focal plane search for the maximum number of counts at each time scale and region-energy combination. Every 256 msec, the short triggering code is sent 180 samples: the maximum counts seen in the 5 time scales and the 36 region-energy combinations. The triggering code only has to check the maximum sample that occurred within each time scale-region-energy combination, not every observed sample. By having the code that ingests the photons identify the maxima, we can effectively check 25,000 samples a second with just 720 actual trigger calculations.

All of the short trigger calculations for a particular set of 180 samples uses the same background rates based on a 1.024 sec period. These background rates are determined by the long trigger algorithm. Let $C_{i,k}$ be the maximum counts observed on the $2^k$ msec time scale in the $i^{\text{th}}$ region-energy combination. Let $B_i$ be the counts observed in 1024 msec for the $i^{\text{th}}$ region-energy combination. The short trigger "score" is effectively the $\sigma^2$ of the net signal relative to the expected statistical variation. (We use $\sigma^2$ to avoid taking square roots: one can more easily find the maximum of $\sigma^2$ than the maximum of $\sigma$ and both methods will point to the same sample.) The precise definition of the short trigger score is:

$$S = \frac{(C_{i,k} - B_i 2^{k-10})^2}{C_{i,k} + \sigma_{\min}^2} . \tag{1}$$

Here, $\sigma_{\min}^2$ is a commandable control variable to ensure that there is a minimum variance when the counts are small. A trigger is declared if $S$ is greater than a threshold, $\sigma_{\text{threshold}}^2$. Each of the triggers for the 180 combinations are controlled by three commandable variables: an enable/disable, $\sigma_{\text{threshold}}^2$, and $\sigma_{\min}^2$.

## LONG RATE TRIGGERS

To avoid trends on longer time scales, one must fit a function to the background and remove the trend. This is the technique pioneered by the HETE GRB trigger [2, 3]. The long rate triggers are much more complicated than the short rate triggers and are based on time series with 64 msec time resolution. To cover as wide as parameter space as possible, the triggering code must be very efficient. In previous trigger algorithms (such as HETE), most of the CPU usage is for forming sums of counts. For each long trigger, one needs to have the sum of counts in several background periods (probably involving a few hundred 64 msec samples) plus the counts in foregrounds ranging from 64 msec in duration up to perhaps 64 sec. The trick to an efficient triggering algorithm is to store the integral sum of counts, not the counts within samples. The BAT triggering code maintains 36 time series in circular buffers (for the 9 detector regions and 4 energy ranges). Each time series consists of the integer sum of the counts from when the instrument was turned on up to time $T$:

$$I_i(T) = \sum_0^T C_i . \tag{2}$$

To obtain the sum of counts within a particular time period (say $T_1$ to $T_2$), one needs only a single arithmetic step, $I_i(T_2) - I_i(T_1)$, rather than a sum over the samples between $T_1$ and $T_2$. (Eventually, the integer sum will overflow the register capability of the computer. If the difference is negative, one just adds the maximum range of the computer registers.)

A second integral sum adds up the number of invalid samples for each of the 36 region-energy combinations. For example, if there is a configuration change in one of the quadrants, a 1 is added to the corresponding invalid integral sum. Whenever we seek the number of counts from a period of time that includes that sample, we also check that the difference of the integral sum of invalids that cover that same duration is zero. For every time that the high voltage is off or something else disables the detectors, a 1 is added to the sum of invalid samples. The trigger code runs all the time and each criterion turns on as soon as the corresponding difference in the sum of invalid samples is zero. Thus, criteria that require fewer samples turn on as soon as they are ready, increasing the on-time for the triggers.

Each long rate trigger is controlled by about 30 commandable parameters. These parameters define the relative times of several background periods, the time, duration, and amount to step the phase for the foreground period, the degree of the polynomial to fit, a minimum variance ($\sigma_{\min}$, needed for low count rates), a systematic noise level ($\beta$, needed for high count rates), a threshold for declaring a trigger ($\sigma_{\text{threshold}}^2$), several parameters that control the CPU usage, and an enable/disable parameter. The background regions can either be all before the foreground samples (an extrapolation) or can bracket the foreground sample (an interpolation). The CPU usage is controlled in several ways. One can specify which tick of the 64 msec clock that each criterion is evaluated. This ensures that the CPU usage is evenly spread out. There is a CPU usage monitor and if a commandable level is exceeded, triggers will autonomously turn themselves off for a commandable period of time to maintain an acceptable CPU usage.

To obtain the long rate trigger score, one first finds the counts and variance on the counts in the foreground period ($C_{fore}, \sigma^2_{fore}$). Second, one fits a function (constant, linear, $2^{nd}$ order) to the background samples. Third, one integrates the fit function during the foreground sample to find the expected background rate ($C_{back}, \sigma^2_{back}$). (We use "model variance", so $\sigma^2_{fore}$ is actually based on $C_{back}$.), Finally, the trigger score is calculated as:

$$S = \frac{(C_{fore} - C_{back})^2}{\sigma^2_{fore} + \sigma^2_{back} + \sigma^2_{min} + \beta^2 C^2_{back}}. \quad (3)$$

Here, $\sigma^2_{min}$ provides a minimum variance to protect against very low counts and $\beta$ protects against systematic effects at high count rates. When the count rate is very high, the variations will no longer be Poissonian. The presence of $\beta$ converts the trigger score from a signal-to-noise criterion to a percentage difference criterion. The units on $\beta$ is percent change per effective $\sigma$. At high count rates, the $C^2_{back}$ term will dominate over the $\sigma^2$ terms, and the trigger score becomes

$$S = \left[ \frac{\frac{C_{fore} - C_{back}}{C_{back}}}{\beta} \right]^2. \quad (4)$$

For example, if the foreground has a 5% net increase over the background and $\beta = 0.025$ (i.e., 2.5%), then the trigger score will be $\approx 2^2$. If one did not have the $\beta$ term, a 5% net increase would be about 5 $\sigma$ (assuming the BAT background rate) and the trigger score would be $5^2$.

The huge size of BAT means that even a small change in a non-transient, but variable source will appear to be a statistically significant change in the count rate. Sources such as Cyg X-1, Her X-1, and Sco X-1 could easily produce consistent triggers forcing us to raise our thresholds. To guard against this, we will implement a new concept which we call a "veto" trigger. We will use a coded aperture technique based on "URA-tagging" [1] to effectively deconvolve the mask pattern in real time. Each photon is assigned a probability that it came from a designated source location based on whether it could reach the detector through the mask pattern. The sum of the probabilities becomes that source's strength at that time. Although developed for URA patterns, teh mask tagging also works for random patterns such as BAT. Some of the trigger criteria will process the mask-tagged time histories in the same way that the 36 region-energy combinations are processed (i.e., they will be stored as integral sums like equation 2). If the trigger criterion exceeds a threshold, a (commandable) set of regular triggers are disabled for that time period. For example, we might have veto triggers that have foreground durations of 1, 5, and 10 sec that are applied to the tagged time series for Cyg X-1. It any of those veto triggers exceed their thresholds, we disable the 1, 5, and 10 sec long rate criteria. We will have the capability to track up to three nask-tagged sources for the veto system.

We plan space for about 500 long rate trigger criteria. The parameters that control the CPU usage ensures that we will be able to run as many as possible.

## IMAGE TRIGGERS

To search for GRBs or other transients on long time scales, we will form images of the FOV and search for new objects. The flight software forms an on-board encoded image every 8 seconds which are used as the background images for the long rate triggers. Those images are combined together on three different time scales (perhaps 64 sec, 10 minutes, half an orbit) and the on-board image deconvolves the mask pattern. (The on-board image process consists of a non-iterative clean to remove known bright sources, a mask deconvolution, and a back project analysis to refine the location.) We search each such image for significant sources and eliminate known sources using an on-board table. Sources that exceed a commandable threshold are declared sources suitable for Swift to slew to.

## SUMMARY

The BAT instrument on Swift will have trigger software that can explore a wide parameter space (4 msec to orbital time scales, 4 different energy ranges, subregions of the FOV). All types of triggers (short, long, and imaging) will use on-board imaging to detect false triggers or known steady sources. About 16,000 commandable "knobs" can be used to optimize its performance. Of course to avoid massive confusion, only a few of those will actually be adjusted on orbit. The software will autonomously adjust the number of criteria to maximize the use of the available CPU power.

## REFERENCES

1. [Fenimore 1987] Fenimore, E. E., 1987, Applied Optics 26, 2760.
2. [Fenimore & Galassi 2001] Fenimore, E. E., & Galassi, M., Proceedings of the 2000 Rome Conference on GRBs.
3. [Tavenner 2002] Tavenner, T. 2002, these proceedings.

# Lobster Eye: New Approach to Monitor GRBs in X-Rays

R. Hudec[a], A. Inneman[b], L. Pína[c], V. Hudcová[a]

[a]Astronomical Institute, CZ-251 65 Ondrejov, Czech Republic
[b]Centre of Advanced X-ray Technologies, Reflex, Prague, Czech Republic
[c]Czech Technical University, Faculty of Nuclear Science, Prague, Czech Republic

**Abstract.** The detection of X-ray afterglows for the large majority of GRBs allows a novel X-ray experiment to be considered to monitor and to detect GRBs based on their X-ray emission. Such an experiment could provide not only a superior localization accuracy if compared with gamma ray instruments, but probably also a higher rate based on the theoretical assumption that the beaming in gamma rays is narrower if compared with lower energies. We describe and discuss the LOBSTER EYE X-ray telescope project including results of the development and tests of the telescope prototype. Considerations for a space experiment on a small scientific satellite of a Nadezhda type are also presented and discussed. The scientific aspects of such experiment regarding the GRB study and statistics are addressed.

## WHY WIDE FIELD TELESCOPES?

The lobster-eye (LE) geometry X-ray optics offer an excellent opportunity to achieve very wide fields of view (1 000 square degrees and more) while the widely used classical Wolter grazing incidence mirrors are limited to roughly 1 deg FOV.

Wide field X-ray telescopes with imaging optics are expected to represent an important tool in future space astronomy projects, especially those for deep monitoring and surveys in X-rays over a wide energy range. Wide field X-ray optics has been suggested in 70ies by Schmidt (orthogonal stacks of reflectors) and by Angel (array of square cells) but has not been constructed yet. Up to 180 deg FOV may be achieved [1].

## SCHMIDT OBJECTIVES

The device consists of a set of flat reflecting surfaces. The plane reflectors are arranged in an uniform radial pattern around the perimeter of a cylinder of radius R. X-rays from a given direction are focussed to a line on the surface of a cylinder of radius R/2. The 1D dimensional focusing device offers a wide field of view, up to maximum of 2_ with the coded aperture. Two such systems in sequence, with orthogonal stacks of reflectors, form a double-focusing device.

## ANGEL OBJECTIVES

The full lobster-eye optical grazing incidence X-ray objective consists of numerous tiny square cells located on the sphere and is similar to the reflective eyes of macruran crustaceans such as lobsters. The field of view can be made as large as desired. It is possible to achieve good efficiency for photon energies up to 10 keV and/or even more if additional coatings are applied. Spatial resolution of a few seconds of arc over the full field is possible, in principle, if very small reflecting cells can be fabricated.

## LE TELESCOPE PROTOTYPES

The large prototype of the Schmidt geometry represents one module and consists of two perpendicular arrays of double-sided X-ray reflecting flats (36 and 42 double-sided flats 100 x 80 mm each). The flats are 0.3 mm thick and gold-coated. The microroughness is below 1 nm. The focal distance is 400 mm from the midplane. The FOV of one module is about 6.5 degrees. More such modules may create an array with substantially larger FOV. The optical and X-ray tests indicate performance close to those calculated and expected (e.g. by ray tracing).

The advanced mini-Schmidt prototypes are based on 0.1 mm thick glass plates 23 x 23 mm, gold coated,

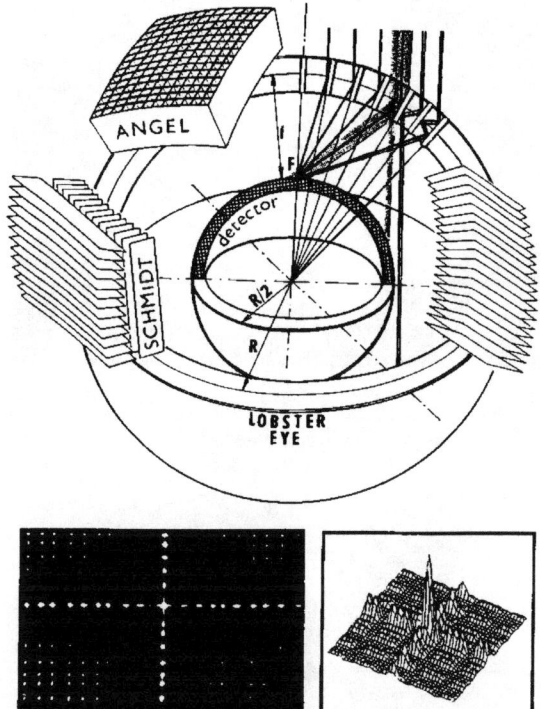

**FIGURE 1.** Schematic arangement of Schmidt and Angel X-ray objectives together with an image of distant point-like source and computer ray-tracing.

**FIGURE 2.** 2D Schmidt X-ray telescope prototype (100 x 80 mm plates) illuminated by laser beam.

spaced at 0.3 mm. 60 such plates are used for one module, the double focusing device is created by two such modules. The aperture/length ratio is ~80, the reflecting surface microroughness amounts to 0.2 – 0.5 nm. The FOV of the module is 2.5 deg.

For the Angel geometry, numerous square cells of very small size (about 1x1 mm or less at lengths of order of tens of mm, i.e. with the size/length ratio of 30 and more) are to be produced. This demand can be also solved by modified innovative replication technology. First test modules with LE Angel cells have been succesfully produced. First linear test module has 47 cells 2.5 x 2.5 mm, 120 mm long (i.e. size/length ratio of almost 50), surface microroughness 0.8 nm, f = 1.3 m. Second test module is represented by a L-shaped array of 2x18 = 36 cells of analogous dimension. The third test module with 6 x 6 = 36 cells is finished recently. The surface microroughness of the replicated reflecting surfaces is better than 1 nm. An innovative technique for production of 120 x 120 mm sized modules with large number of 3 x 3 mm cells, 120 mm long, is under development.

## THE TELESCOPES AND GRB X-RAY AFTERGLOWS

Almost every GRB is accompanied by a X-ray afterglow. The expected rate of GRBs is 1 per day, however the theoretical prediction assumes larger beaming angle in X-rays if compared with gamma rays, hence the actual rate of X-ray afterglows is expected to be larger than the rate of GRBs. The sensitivity of LE telescopes is sufficient enough to detect the X-ray GRB afterglows: for pointed observations, limits better than $10^{-14}$ erg sec$^{-1}$cm$^{-2}$ (0.5 – 3 keV) can be obtained ($10^{-12}$ erg sec$^{-1}$cm$^{-2}$ for daily observations).

The localization accuracy of the LE telescopes is of order of 1 arcmin, substantially exceeding the recent localization accuracy of most gamma ray instruments (2 deg and more). The LE telescopes are expected to provide a substantial contribution to the science and statistics of GRBs

## MORE SCIENCE WITH LE TELESCOPES

The additional science of LE X-ray telescopes includes supernova explosions, high energy binary sources, AGNs, blazars, X-ray novae, X-ray flares on stars, X-ray transients, cataclysmic variables etc. The use of LE telescopes will allow these objects to be detected and studied by sky patrol monitoring.

# DISCUSSION

The use of very wide field X-ray imaging system could be without doubts very valuable for many areas of X-ray and gamma-ray astrophysics.

Results of analyses and simulations of lobster-eye X-ray telescopes have indicated that they will be able to monitor the X-ray sky at an unprecedented level of sensitivity, an order of magnitude better than any previous X-ray all-sky monitor. Limits as faint as $10^{-12}$ erg cm$^{-2}$ s$^{-1}$ for daily observation in soft X-ray range are expected to be achieved, allowing monitoring of all classes of X-ray sources, not only X-ray binaries, but also fainter classes such as AGNs, coronal sources, cataclysmic variables, as well as fast X-ray transients including gamma-ray bursts and the nearby type II supernovae. For pointed observations, limits better than $10^{-14}$ erg sec$^{-1}$cm$^{-2}$ (0.5 to 3 keV) could be obtained, sufficient enough to detect X-ray afterglows of GRBs.

The various prototypes of both Schmidt as well as Angel arrangements have been produced and tested successfully for the first time, demonstrating the possibility to construct these lenses by innovative but feasible technologies. This makes the proposals for space projects with very wide field lobster eye optics possible.

# SATELLITE TEST EXPERIMENT

Recently we are exploiting the possibility to fly the test Lobster Eye X-ray telescope onboard the small Czech scientific satellite based on the Russian Nadezhda satellite bus expected to be provided as the refund of the Russian debt to the Czech Republic. The satellite project developed by the Institute of Atmospheric Physics of the Academy of Sciences of the Czech Republic will focus on the geophysics and upper atmosphere research, but there is a space to place two small LE units. These units should contain 10 x 10 Mini Schmidt LE modules each, covering thus the FOV of 25 x 25 degrees. As the focal detector, a modified X-ray sensitive pixel detector is under consideration.

# FUTURE STEPS TOWARD THE REAL LOBSTER EYE TELESCOPE

Both Schmidt and Angel arrangements

- application of additional layers to extent the energy range to higher energies

**FIGURE 3.** Advanced 2D Schmidt X-ray objectives based on 23 x 23 plates, 100 micron thick.

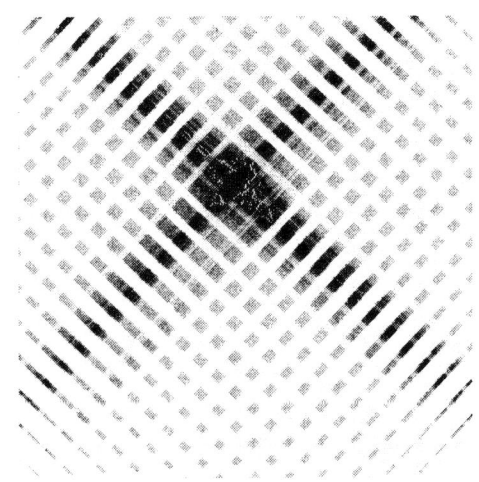

**FIGURE 4.** The mini Schmidt 2D prototypes module X-ray focal sport image (image area 12.3 x 12.3 mm; the measured gain is 570 for 8 keV).

- improving the surface quality (microroughness, slope errors) to further improve the reflectivity and the angular resolution
- construction of larger or multiple modules to achieve a larger FOV of order of at least 1 000 square deg or more
- further reduction of the cell apertures (Angel) and/or spacing and plate thickness (Schmidt) and to enhance the length/aperture ratios to achieve a better angular resolution

# ACKNOWLEDGEMENTS

We acknowledge the support provided by the Grant Agency of the Czech Republic, grant 102/99/1546.

# REFERENCES

1. Inneman A., Hudec R. and Pina L. Proc. SPIE 4138, 94-104, 2000.

# Afterglow Studies with the Swift UV/Optical Telescope

S. D. Hunsberger[*], P. W. A. Roming[*], J. A. Nousek[*] and K.O. Mason[†]

[*]*Department of Astronomy & Astrophysics, Pennsylvania State University*
[†]*Mullard Space Sciences Laboratory, University College London*

**Abstract.** The Swift Explorer mission, scheduled for launch in 2003, will be a unique observatory for multi-wavelength studies of GRBs. Within 50 seconds of a GRB detection by the Burst Alert Telescope (BAT), the X-Ray Telescope (XRT) and UV/Optical Telescope (UVOT) will begin their collection of data for afterglow studies. Within minutes, UVOT will provide a finding chart to accurately determine the position of the GRB afterglow. Because UVOT must autonomously execute its observing program, flight software will include several automated observing sequences. Using predictions of theoretical models, we produce expected science data products such as images and light curves based on execution of a "standard" GRB automated observing sequence.

## UVOT SUMMARY

The UVOT is a 30cm modified Ritchey-Chrétien telescope. An 11-position filter wheel contains six broadband filters (Fig. 1), two low-resolution grisms (Fig. 2), a 4x magnifier, plus open and blocked positions. Photons register on a microchannel plate intensified CCD allowing UVOT to operate in a photon counting mode. The telescope is also capable of producing images by accumulating photon counts into 10s tracking frames which are aligned and combined onboard. Imaging data provides a positional accuracy of 0.3".[3]

| UVOT Characteristics | |
|---|---|
| Telescope | Modified Ritchey-Chrétien |
| Aperture | 30cm diameter |
| Detector | Intensified CCD |
| Operation | Photon Counting |
| Field-of-View | 17' × 17' |
| Resolution | 2048 × 2048 pixels |
| PSF | 0.9" (FWHM @ 350nm) |
| Wavelength | 170 – 650nm |
| Filters | 11 (including blocked position) |
| Sensitivity | B=24 in white light in 1000s |
| Pixel Scale | 0.5" per pixel |

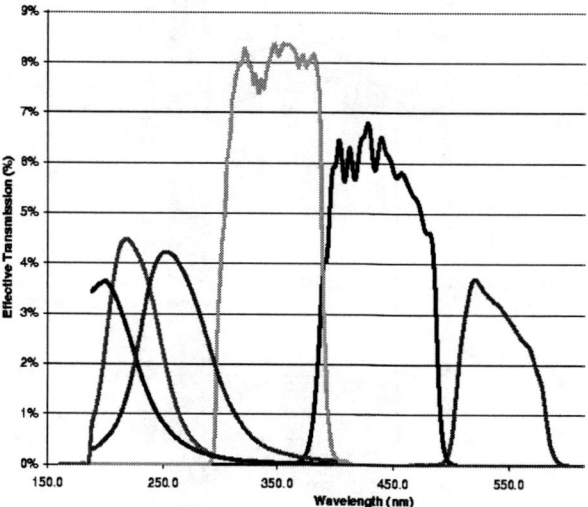

**FIGURE 1.** The graph shows total system throughput for each broadband filter (3 UV bands, U, B, and V). The cutoff below 180nm is due to a lack of measured data.

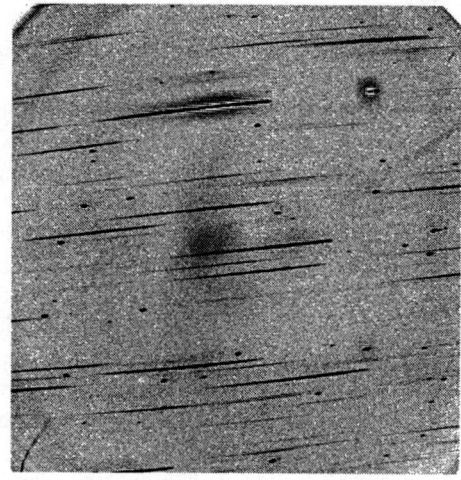

**FIGURE 2.** UV and optical grisms cover wavelength ranges 170 – 300nm and 300 – 650nm. The figure shows a CCD image produced by the optical grism on XMM-OM.

# SCIENCE OBSERVATIONS

A science exposure is specified by operating mode, observing windows, filter selection, and exposure time. Valid operating modes are Event Mode (photon counting with 10ms timestamps), Image Mode (imaging), or Event/Image (both). There are separate windows for collecting event and image data. Observing window sizes are constrained by CPU resources and telemetry rates.

There are two types of observations considered by UVOT: autonomous and pre-planned. On orbit, there will be a "set" of several autonomous observing sequences that will be updated as we learn more about the early evolution of GRB afterglows. Similarly, there will be a large set of pre-planned observing sequences. Pre-planned targets are objects that have been selected for follow-up observation by the mission operations team.

# AUTONOMOUS OBSERVING

One possible GRB observing sequence is presented in Fig. 3.

*Slew* — When a GRB is detected by the BAT, the spacecraft (S/C) slews to the burst position. The average slew time is 50 seconds. During this interval, no data is collected by UVOT.

*Settle* — During settling, the target is moving within the UVOT field-of-view. We utilize this opportunity to collect the earliest UV photons. Photon events are recorded over the entire CCD frame and data from the S/C attitude control system allows software on the ground to identify the GRB photons.

*Finding Chart* — UVOT begins collecting both image data and photon events within an $8' \times 8'$ window centered on the GRB position. The image from a 100s exposure in V provides a finding chart (Fig. 4) for ground observers. Due to telemetry constraints on the real-time downlink, a source list is relayed to the ground for image reconstruction within 270s. All image and event data is transmitted during routine ground contact.

*10 Second Exposures* — UVOT executes a series of 10s exposures, cycling through all broadband filters ten times. During these short exposures, UVOT operates only in Event Mode.

*100 Second Exposures* — UVOT executes a series of 100s exposures, cycling through all broadband filters five times. UVOT operates simultaneously in both modes.

*1000 Second Exposures* — UVOT executes a series of 1000s exposures, cycling once through all broadband filters. During these long exposures, UVOT operates only in Image Mode. This series of observations occurs at least 2 hours after GRB detection because the observing sequence is interrupted by Earth occultation (Fig. 5).

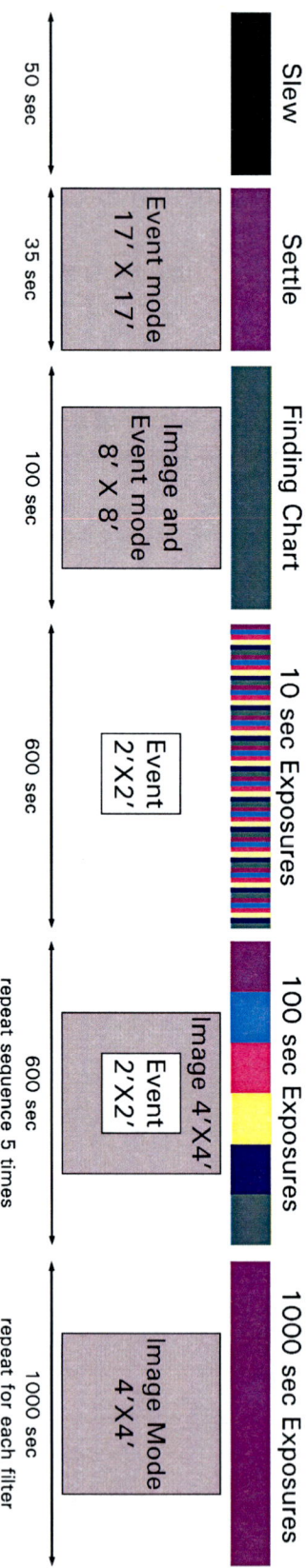

**FIGURE 3.** Autonomous Observing Timeline

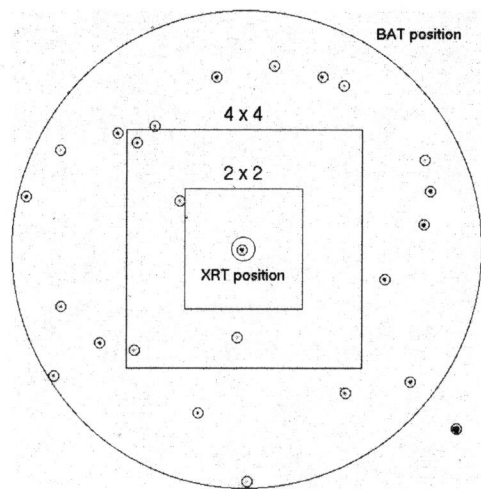

**FIGURE 4.** In this $8' \times 8'$ simulated image, the smallest circles highlight stellar objects including a GRB located near the center. The image also shows the BAT and XRT positional error circles with respect to Image and Event Mode windows used for afterglow observations.

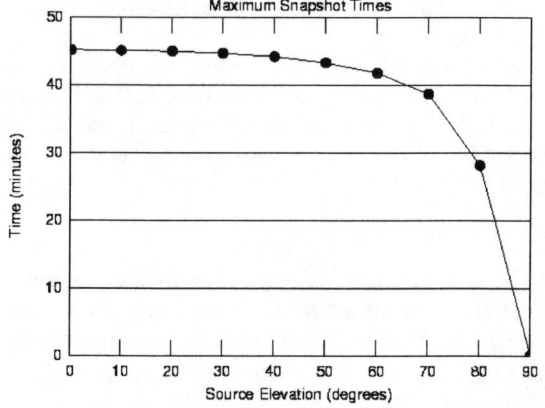

**FIGURE 5.** Because bright sources can damage its detector, UVOT cannot point within 30° of the Earth limb. Depending on the GRB position with respect to the S/C, time from burst to occultation can vary dramatically. The graph above plots maximum possible observing time as a function of elevation above the Swift observatory's orbital plane.

## SIMULATED RESULTS

Theoretical models are used as input to a UVOT simulator. Results presented focus on 3 variations of the dissipative fireball model [1]: forward blast wave (a1), reverse shock (a2), and wind model (b2). "In type 'a' models, the initial energy input is impulsive. As relativistic ejecta interact with the surrounding medium, much of the energy is radiated as external and reverse shocks. In type 'b' models, the energy input is continuous over some period of time, e.g., a relativistic wind. The resultant radiation is due to internal shocks within the wind." [2] The peak optical flux and decay rates will be very different for each of these cases.

Preliminary results of computer simulations demonstrate that these differences should be detectable in UVOT data collected during the early phases of autonomous observing. Fig. 6 shows resultant V-band light curves for each model assuming the automated sequence in Fig. 3 is executed and the maximum snapshot time is available.

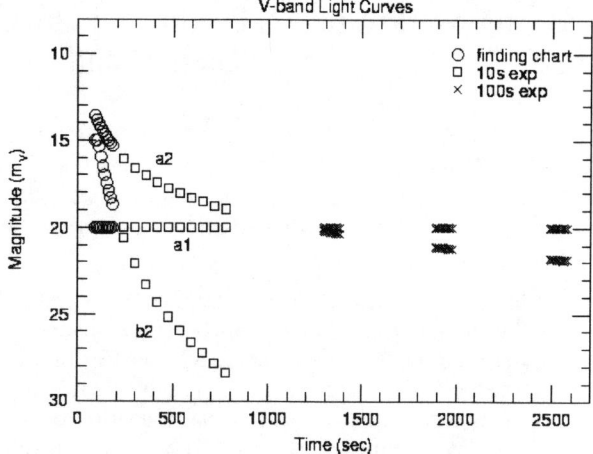

**FIGURE 6.** GRB Light Curves

## ACKNOWLEDGEMENTS

We wish to acknowledge members of the UVOT team at Penn State University, Mullard Space Sciences Lab, and Southwest Research Institute. Funding for this project is provided by NASA grant NAS5-00136 for PSU and by PPARC for MSSL.

## REFERENCES

1. Mészáros, P. and Rees, M. J., *The Astrophysical Journal* **476**, 232-237 (1997).
2. Roming, P., Hunsberger, S. D., Nousek, J., and Mason, K., "The Swift Ultra-Violet/Optical Telescope", in *GAMMA 2001: Gamma-Ray Astrophysics 2001*, edited S. Ritz et al., AIP Conference Proceedings 587, 2001, pp. 791-795.
3. Roming, P. W. A., et al., "The Ultra-Violet/Optical Telescope of the Swift MIDEX Mission", in *X-Ray & Gamma-Ray Instrumentation for Astronomy XI*, edited by K. A. Flanagan and O. H. W. Siegmund, SPIE Proceedings Series 4140, 2000, pp. 76-86.

# Searches for Hard X-Ray Gamma-Ray Burst Afterglows with the BAT on Swift

Hans A. Krimm[*,†], Louis M. Barbier[*], Scott D. Barthelmy[*], Ardeshir Eftekharzadeh[***], Edward E. Fenimore[‡], Neil Gehrels[*], Derek D. Hullinger[***], Craig Markwardt[***], David M. Palmer[‡], Ann M. Parsons[*], Hideki Ozawa[*,†], Jack Tueller[*] and Georg Weidenspointner[*,†]

[*]Code 661, NASA Goddard Space Flight Center, Greenbelt, MD 20771, USA
[†]Universities Space Research Association, 7501 Forbes Blvd., Suite 206, Seabrook, MD 20706, USA
[**]Department of Physics, University of Maryland, College Park, MD 20742, USA
[‡]Los Alamos National Laboratory, P.O. Box 1663, Los Alamos, NM 87545, USA

**Abstract.** The Burst Alert Telescope (BAT) on the Swift gamma ray burst mission will continue to observe the fields of all detected gamma-ray bursts for several days after the prompt emission has faded. Utilizing first event-by-event data, then one minute and later five minute survey accumulations, the BAT will be extremely sensitive to the hard X-ray afterglow known to be associated with many bursts.

This data will cover the crucial transition of the afterglow from rapid variability to the smoothly decaying power law in time and will extend observations of the tails of individual bursts to longer time scales than have been achievable so far. Since Swift is sensitive to short duration GRBs, we will also be able to determine whether hard X-ray afterglows are associated with short GRBs. The BAT will provide high resolution spectra of burst afterglows, allowing us to study in detail the time evolution of GRB spectra.

## INTRODUCTION

The Burst Alert Telescope [1, 2] on the Swift gamma-ray burst mission [3, 4] is a coded aperture telescope which provides the gamma-ray burst triggers for the mission. BAT will locate several hundred bursts per year to an accuracy of better than 4 arc minutes. Upon receipt of a valid BAT trigger the Swift satellite will rapidly slew to point the narrow field instruments, the X-Ray Telescope (XRT) and the Ultra-Violet / Optical Telescope (UVOT), toward the gamma-ray burst. The BAT will simultaneously provide follow-up hard X-ray and gamma-ray observations of burst afterglows. In addition to detecting and observing gamma-ray bursts, the BAT will provide a hard X-ray survey of unparalleled sensitivity [5].

## AFTERGLOW OBSERVATIONS

BAT and the Swift narrow-field instruments will be sensitive to afterglows during the largely unexplored time between $10^2$ and $2 \times 10^4$ seconds and, for up to the first hour, BAT will extend the energy range of afterglows up to 150 keV. The 5 keV energy resolution down to 15 keV of the BAT will enable us to study in detail the time evolution of afterglow spectra and overall spectral properties. This, combined with timing information, will give us new criteria by which to distinguish subclasses of GRB afterglows.

BAT will record event by event data with 0.1 msec time resolution for 10 minutes after the burst, and this data will allow us to study afterglow time variability with unprecedented precision. This data is collected even during the slew, so the burst light curve will include the transition from prompt to afterglow emission and will show any evidence of pulsations or flares.

After 10 minutes, the BAT will then switch to 1-minute survey mode for $\sim 30$ minutes and then to normal 5-minute survey mode. Even after the narrow field instruments are slewed to another target it is quite possible that the GRB will still be in the BAT field of view. The burst and afterglow data products are outlined below. For a more detailed description of BAT data products see [6].

### Burst afterglow data products

- 10 minutes of event by event data (detector number, time and energy for every photon)
- Several hours of masked-tagged data [See description below]
- approximately 30 minutes of one minute survey images (80 energy bins)
- Several hours of five minute survey images (80 energy bins)
- Sky images for each survey interval and from event by event data
- Light curve for entire burst and afterglow at appropriate time resolution
- Source counts spectra from event by event and survey data
- Response matrices for reconstruction of image and spectral data

There have been more than thirty GRBs for which X-ray afterglows have been observed. In Figure 1 it can be seen that BAT would have been sensitive to 6 of the 13 bursts shown, at times between 1 minute and 1 hour after the burst. Note that the GRB afterglow light curves are quite sensitive to the power law indices, which are poorly known. Furthermore, the afterglow curves shown should be taken as lower limits since fireball models predict a harder spectrum at earlier times, putting more photons into the BAT energy range than have been calculated using this simple model.

Since the XRT and UVOT will be observing the afterglow coincidentally with the BAT, it will be possible to extend the energy power law down to 0.2 keV and to compare the UV and X-ray fluxes as a function of time.

Although a number of afterglows have been detected by other instruments, Swift will be the first to be sensitive to GRBs shorter than a few seconds and will be able to study the afterglow from short bursts.

## MASKED TAGGED RATES

A powerful tool available in a coded aperture system such as BAT is a masked tagged rate. During each observation BAT will collect these rates for a commandable list of three field sources. Masked tagged rates are calculated on board by back projecting source photons through the mask onto the array and comparing rates in masked and unmasked detectors. This is done at 4 msec time resolution and allows monitoring of bright source behavior on time scales far shorter than the normal survey intervals.

## TRANSIENT ALERTS

BAT will serve as a hard X-ray transient monitor, in addition to its primary goal of discovering GRBs. For on-board imaging triggers on timescales > 90 seconds, transient alert and transient position messages will be sent out via TDRSS. Ground software will generate an email message (or a page if after hours) to a BAT on-call scientist who will examine the data and determine if the transient alert is genuine and worthy of a message sent out through the GRB Coordinates Network (GCN)[24, 25, 26]. This delay of up to an hour is required to assure data quality and reject false alerts.

BAT data will be examined on the ground for transients at a high sensitivity. But since this is part of the survey process, there will be a greater time delay. This analysis is done by measuring the flux of sources in a BAT transient catalog on time scales from 5-45 minutes. This part of the survey is also sensitive to unknown transients. For every transient found and vetted by the BAT team, a GCN alert will be sent out.

Timely BAT detection of transients will trigger numerous target of opportunity observations on other satellites. These observations will greatly increase the scientific impact of Swift transient detections.

BAT will be sensitive to Soft Gamma Repeaters because of its excellent short-burst trigger. While the SGR bursts are shorter than the Swift slew time, the XRT and UVOT will perform sensitive searches immediately after the burst for X-ray and optical counterparts and will likely be on-target when following bursts occur.

The Swift observatory has the capability for rapid reaction science, using the TDRSS uplink. This will provide a unique ability to respond in a matter of minutes with sensitive gamma ray, X-ray, UV, and optical observations to most events on the sky. This includes targets of opportunity for X-ray transients, pulsar glitches, outbursts from dwarf novae, and stellar flares. The BAT is more than ten times as sensitive as BATSE as a monitor, and BAT detections will initiate many of the targets of opportunity. As with any new observational capability, the potential for serendipitous science return is high.

## REFERENCES

1. Barthelmy, S., "The Burst Alert Telescope (BAT) on the Swift Mission" (2002), these proceedings.
2. Barthelmy, S., "Burst Alert Telescope (BAT) on the Swift MIDEX mission", in *X-Ray and Gamma-Ray Instrumentation for Astronomy XI, The Swift Mission*, edited by K. A. Flanagan and O. H. Siegmund, SPIE Proceedings 4140, 2000, pp. 50–63.
3. Barthelmy, S., "Swift: A Gamma Ray Burst MIDEX", in *Gamma-Ray Astrophysics*, edited by S. Ritz, N. Gehrels, and C. R. Shrader, AIP Conference Proceedings 587,

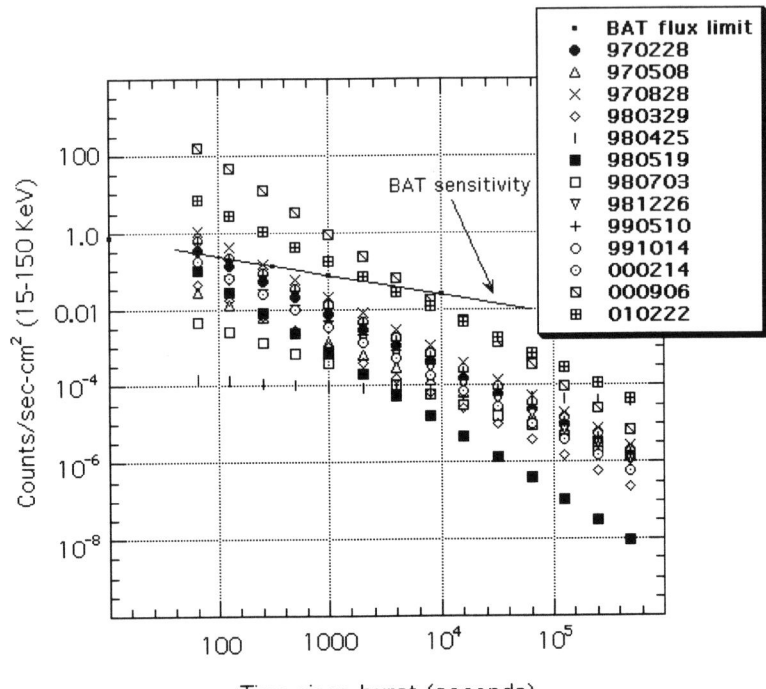

**FIGURE 1. BAT afterglow sensitivity.** BAT sensitivity ($5\sigma$, $\Delta t/t = 1$) compared to the 13 GRBs with published X-ray afterglow detections (see reference list below). Quoted (2-10 keV) flux was converted to BAT (15-150 keV) cts/sec-cm² using *grmc*, a photon tracking monte carlo program, and assuming a continuation of the lower energy X-ray spectral power laws. The flux at the time of the observations was extrapolated back to t=1 sec using the best fit light curve power laws from the X-ray observations. The BAT statistical sensitivity was calculated using grmc. References: GRB970228: [7, 8], GRB970508: [9], GRB97082: [10], GRB970828: [11], GRB980329: [12], GRB980425: [13, 14], GRB980519: [15], GRB980703: [16], GRB981226: [17], GRB990510: [18], GRB991014: [19], GRB000214: [20], GRB000926: [21, 22], GRB010222: [23]

Baltimore, Maryland, 2001, pp. 781–790.
4. Gehrels, N., "The Swift GRB MIDEX Mission" (2002), these proceedings.
5. Krimm, Hans A. et al., "Swift Burst Alert Telescope Hard X-Ray Monitor and Survey", in *Gamma-Ray Astrophysics*, edited by S. Ritz, N. Gehrels, and C. R. Shrader, AIP Conference Proceedings 587, Baltimore, Maryland, 2001, pp. 796–800.
6. Palmer, D.M. et al., "The Swift Burst Alert Telescope Data Products" (2002), these proceedings.
7. Wijers, R. A., Rees, M. J., and Meszaros, P., *Mon. Not. R. Astron. Soc*, **288**, L51–L56 (1997).
8. Frontera, F. et al., *Ap. J.*, **493**, L67–L70 (1998).
9. Piro, L. et al., *Astron. Astrophys*, **331**, L41–L44 (1998).
10. Yoshida, A. et al., *Astron. Astrophys, Suppl. Ser.*, **138**, 433–434 (1999).
11. Yoshida, A. et al., *Ap. J.*, **557**, L27–L30 (2001).
12. in't Zand, J.J.M. et al., *Ap. J.*, **505**, L119–L122 (1998).
13. Pian, E. et al., *Astron. Astrophys, Suppl. Ser.*, **138**, 463–464 (1999).
14. Pian, E. et al., *Ap. J.*, **536**, 778–787 (2000).
15. in't Zand, J.J.M. et al., *Ap. J.*, **516**, L57–L60 (1999).
16. Vreeswijk, P.M. et al., *Ap. J.*, **523**, 171–176 (1999).
17. Frontera, F. et al., *Ap. J.*, **540**, 697–703 (2000).
18. Pian, E. et al., *A. & A.*, **372**, 456–462 (2001).
19. in't Zand, J.J.M. et al., *Ap. J.*, **545**, 266–270 (2000).
20. Antonelli, L.A. et al., *Ap. J.*, **545**, L39–L42 (2000).
21. Piro, L. et al., *Ap. J.*, **558**, 442–447 (2001).
22. Harrison, F.A. et al., *Ap. J.*, **559**, 123–130 (2001).
23. in't Zand, J.J.M. et al., *Ap. J.*, **559**, 710–715 (2001).
24. Barthelmy, S., and Cline, T., "GCN: A Status Report" (2002), these proceedings.
25. Barthelmy, Scott et al., "The GRB Coordinates Network (GCN): A Status Report", in *Gamma-Ray Bursts 4th Huntsville Symposium*, edited by C. A. Meegan, R. D. Preece, and T. M. Koshut, AIP Conference Proceedings 428, Huntsville, Alabama, 1997, pp. 99–103.
26. Barthelmy, S., Cline, T., and Butterworth, P., "GRB Coordinates Network (GCN): A Status Report", in *Gamma-Ray Astrophysics*, edited by S. Ritz, N. Gehrels, and C. R. Shrader, AIP Conference Proceedings 587, Baltimore, Maryland, 2001, pp. 213–217.

# The Development of GRAPE, a Gamma Ray Polarimeter Experiment

M.L. McConnell*, J.R. Ledoux*, J.R. Macri* and J.M. Ryan*

*Space Science Center, University of New Hampshire, Durham, NH 03824*

**Abstract.**
The measurement of hard X-ray polarization in γ-ray bursts (GRBs) would add yet another piece of information in our effort to resolve the true nature of these enigmatic objects. Here we report on the development of a dedicated polarimeter design with a relatively large FoV that is capable of studying hard X-ray polarization (50-300 keV) from GRBs. This compact design, based on the use of a large area position-sensitive PMT (PSPMT), is referred to as GRAPE (Gamma-RAy Polarimeter Experiment). The feature of GRAPE that is especially attractive for studies of GRBs is the significant off-axis polarization response (at angles greater than 60°). For an array of GRAPE modules, current sensitivity estimates give minimum detectable polarization (MDP) levels of a few percent for the brightest GRBs.

## INTRODUCTION

In recent years, largely as a result of the observation of several X-ray, optical and radio afterglows, there has developed a growing consensus that classical γ-ray bursts (GRBs) are at cosmological distances. Such great distances imply that a typical GRB releases $10^{51} - 10^{53}$ ergs or more within a time span of several seconds.

The general picture that has emerged is one that describes the GRB phenomenom in terms of a relativistic fireball model [e.g., 1]. One common feature of many of these models is that the energy release takes the form of jets that are directed along the rotation axis of the system. Several indirect arguments have been used to argue that such jets are required to explain the observations. For example, the shapes of the observed light curves of some of the observed optical afterglows (e.g., GRB 990510) have been used as supporting evidence for jets related to the external shocks. Since the energy budget of a given GRB depends heavily on assumptions about the extent to which the flow is jet-like, determining the reality and nature of jets in GRBs is becoming an important goal of future observations.

Optical polarization measurements, and the theoretical studies that have been motivated by such measurements, have provided a better insight into the nature of the GRB phenomena. The optical studies, however, probe only the external shock region. In the context of the canonical fireball model, measurements of the hard X-ray polarization during the prompt phase of the GRB promise to provide a similar probe of the internal shock region. Since the outgoing flow at the internal shock is expected to be more tightly collimated than the flow at the external shock (resulting from a continuous spreading of the jet as it progresses outward through the fireball), one can perhaps expect a somewhat higher level of hard X-ray polarization (assuming that it arises from synchrotron emission) during the prompt phase of the GRB.

In contrast to the classical GRBs, the prevailing model of soft γ-ray repeaters (SGRs) involve emission from the vicinity of magnetars, neutron stars with magnetic fields in excess of $10^{14}$ G, with the energy release triggered by massive neutron star crustquakes. Baring [2] suggested that the softness of the events can be attributed to photon splitting in the extremely intense magnetic fields. The photon splitting process degrades the high energy γ-ray photons to hard X-ray energies [3]. One by-product of photon splitting is that the reprocessed photons would exhibit a polarization level of 25% [2]. Polarization measurements in the 50–300 keV energy range could therefore provide a test of the importance of photon splitting in SGRs.

## A MODULAR POLARIMETER DESIGN

The basic physical process used to measure linear polarization of hard X-rays (50–300 keV) is Compton scattering. The measurement is based on the fact that the incident photons tend to be scattered at right angles to the incident electric field vector. A Compton scatter po-

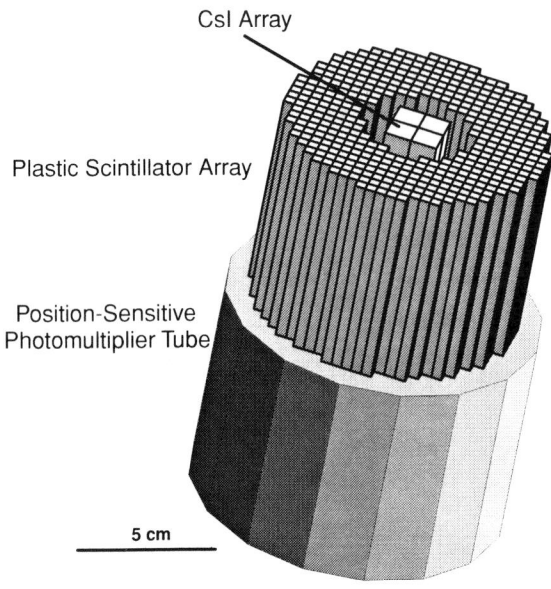

**FIGURE 1.** Schematic diagram of a polarimeter module.

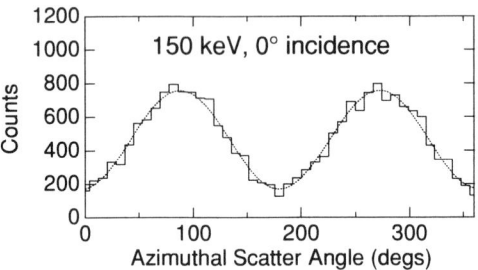

**FIGURE 2.** The simulated polarimetric response of a GRAPE module. Polarized 150 keV photons vertically incident on the module show an asymmetry in the azimuthal distribution of interactions in the plastic array. The data here have been corrected for intrinsic asymmetries in the module geometry.

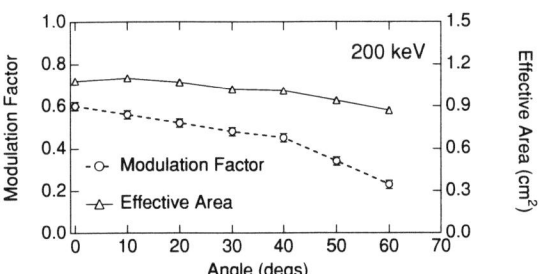

**FIGURE 3.** The effective area and modulation factor as a function of incidence angle for 200 keV photons (from simulations).

larimeter consists of two detectors (or detector arrays) that are used to determine the energies of both the scattered photon and the scattered electron. One detector (the *scattering detector*) provides the medium for the Compton interaction to take place. This detector must be designed to maximize the probability of a single Compton interaction with a subsequent escape of the scattered photon. The primary purpose of the second detector (the *calorimeter*) is to absorb the full energy of the scattered photon.

We have developed a modular design that places an entire device on the front end of a single 5-inch diameter position-sensitive PMT (PSPMT) [4, 5, 6]. An array of plastic scintillator elements (each 5 cm long with a cross section of 5 mm × 5 mm) provides the spatial resolution in the scattering elements. The spatial resolution is necessary for acurately reconstructing the scattering geometry. The array is arranged in the form of an annulus with an outside diameter of 10 cm. A 2 × 2 array of 1 cm CsI scintillators is positioned within a well at the center of the array. A multi-anode PMT (MAPMT) provides an independent readout of each CsI element (one element per anode). Independent readouts provide the necessary timing information used to define a Compton scatter event. Figure 1 shows a schematic view of a GRAPE module.

To be recorded as a polarimeter event, an incident photon Compton scatters from one (and only one) of the plastic scattering detectors into the central calorimeter. The incident photon energy can be determined from the sum of the energy losses in both detectors and the azimuthal scattering angle (used to define the polarimetric response) can be determined by the relative locations of the plastic and CsI detectors involved in the event. When the polarimeter is arranged so that the incident flux is parallel to the symmetry axis, unpolarized radiation will produce an axially symmetric coincidence rate. If the incident radiation is linearly polarized, then the coincidence rate will show an azimuthal asymmetry whose phase depends on the position angle of the incident radiation's electric vector and whose magnitude depends on the degree of polarization. Figure 2 shows the simulated polarimetric response of a GRAPE module to vertically incident polarized radiation at an energy of 150 keV. The amplitude of the modulation pattern serves to define the level of polarization. The angle at which the modulation pattern is a minimum corresponds to the polarization angle. Figure 3 shows that the module maintains a very good polarimetric response out to very large incidence angles (beyond 60°). This makes the GRAPE design especially appealing for GRB studies.

We anticipate that this design would be used in the context of an array of polarimeter modules (a bunch of GRAPEs). Table 1 shows the sensitivity of an array of 36 modules, based on Monte Carlo simulations of this design and reasonable estimates of the background at balloon altitudes. The sensitivity is presented in terms of

**TABLE 1.** GRB Polarization Sensitivity (50–300 keV)

| Fluence (ergs cm$^{-2}$) | MDP T = 10 s | MDP T = 100 s | Observed Rate ($N$ > Fluence) |
|---|---|---|---|
| $1 \times 10^{-4}$ | 2.8% | 2.9% | 1 per 320 days |
| $5 \times 10^{-5}$ | 3.9% | 4.4% | 1 per 80 days |
| $1 \times 10^{-5}$ | 9.3% | 13.7% | 1 per 10 days |
| $5 \times 10^{-6}$ | 14.0% | 24.4% | 1 per 6 days |
| $3 \times 10^{-6}$ | 19.4% | 38.6% | 1 per 4 days |
| $1 \times 10^{-6}$ | 43.2% | — | 1 per 2 days |

**FIGURE 4.** The components of the laboratory science model, showing (left to right) the PSPMT, the array of plastic scintillators, and the MAPMT/CsI assembly that fits into the central well of the plastic array.

both the minimum detectable polarization (MDP) and the frequency of observed events above a given fluence level. Note that, for a given fluence level, the MDP is lower for shorter events due to the larger signal-to-background integrated over the event duration. These results suggest that a balloon borne GRAPE would achieve a sensitivity level of a few percent for the strongest GRBs.

## LABORATORY SCIENCE MODEL

We have assembled a laboratory science model (SM) for the purpose of validating the GRAPE design. The components of the SM are shown in Figure 4. The current configuration uses a charge division readout for the PSPMT that provides an event location using only four channels (two readouts for $X$ and two readouts for $Y$). During the summer of 2001, we carried out an LED mapping of the PSPMT spatial response using the four channel readout. The LED was stepped through a sequence of locations with a fixed grid spacing of 2.5 mm. The array of measured grid locations is shown in Figure 3. Note the distortions of the grid pattern that are evident near the edge of the PSPMT.

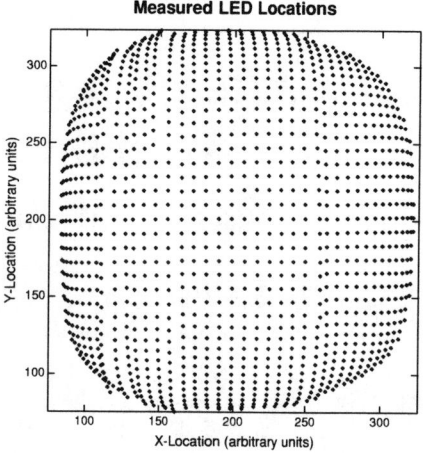

**FIGURE 5.** Plot of the measured (relative) positions of the LED calibration points.

We have recently started collecting data with the fully assembled science model using a partially polarized beam that is generated by scattering 662 keV photons from a $^{137}$Cs source off a block of plastic scintillator. This particular arrangement also provides for the electronic tagging of polarized photons. A preliminary set of uncorrected data shows evidence for a polarization signature. Further processing and analysis of these data are in progress. In the coming months we plan to complete the analysis of laboratory SM data and make detailed comparisons with Monte Carlo simulations. This effort will provide the basis for continued development of a future spacecraft or balloon payload.

## ACKNOWLEDGMENTS

This work is currently supported by NASA grant NAG5-5324.

## REFERENCES

1. Piran, T., *Phys. Rept.* **333**, 529-533 (2000).
2. Baring, M., *ApJ Letters* **440**, L69-L72 (1995).
3. Baring, M., *MNRAS* **262**, 20-26 (1993).
4. McConnell, M. L., Forrest, D. J., Macri, J., McClish, M., Osgood, M., Ryan, J. M., Vestrand, W. T., and Zanes, C., *IEE Trans. Nucl. Sci.* **45**, 910-914 (1998).
5. McConnell, M. L., Macri, J. R., McClish, M., Ryan, J., Forrest, D. J., and Vestrand, W. T., *IEE Trans. Nucl. Sci.* **46**, 890-896 (1999).
6. McConnell, M. L., Macri, J. R., McClish, M., and Ryan, J., *SPIE Conf. Proc.* **3764**, 70-78 (1999).

# Afterglow and Transient Astronomy With the Swift GRB Explorer

J. A. Nousek[*], M. M. Chester[*], F.E. Marshall[†] and the Swift Team

[*]*The Pennsylvania State Univ., Dept. of Astronomy & Astrophysics, 525 Davey Lab., University Park, PA 16802*
[†]*Goddard Space Flight Center, Code 661, Greenbelt, MD 20771*

**Abstract.** Swift is a multi-wavelength observatory designed for the autonomous detection and immediate follow-up of gamma-ray bursts and their afterglows. Following launch in late 2003, Swift's Burst Alert Telescope (BAT) will detect 100s of GRBs per year, and autonomously maneuver sensitive UV/optical and X-ray telescopes onto the burst within 10 to 75 seconds. GRB and X-ray positions and a UV/optical finding chart will be rapidly distributed through the GCN to promote ground-based observations. Afterglows will be monitored by Swift for days to weeks. All data will be converted into standard FITS formats and rapidly made available to the community from data centers in the US, Italy, and the UK.

Swift has two additional capabilities. A hard X-ray survey will be performed with the BAT, with daily pointings covering 80% of the sky and immediate follow-up of X-ray transients. Swift will also function as a Target of Opportunity observatory for sources requiring fast response and repeated short-term (days) or longer-term (months) monitoring. Swift will be operated so that its capabilities are available to the entire community, including a public database of GRB results from other instruments.

## THE SWIFT MISSION

The discovery by BeppoSAX of afterglow from gamma ray bursts (GRBs) revolutionized our understanding of these extraordinary events. We have identified X-ray and optical counterparts with typical redshifts measured in the range of $z \sim 0.5\text{-}4.5$. The inferred energies suggest that GRBs are the most powerful and relativistic explosions known. These afterglows are thought to be produced by a super-relativistic blast wave fading down from gamma rays to radio. A panchromatic approach is now essential for the next phase of discovery.

The Swift MIDEX satellite, selected by NASA for launch in late 2003, is a multiwavelength observatory that exploits the afterglow characteristics to localize ~1000 bursts to accuracies of a few arc seconds. In addition it will measure afterglow light curves and gamma-ray, X-ray and optical/UV spectra starting only seconds after the burst is discovered.

The Swift mission concept utilizes a broad wavelength coverage offered by a state-of-the-art suite of instruments, chosen to efficiently carry out the localization and spectral measurements. GRBs are discovered by the Burst Alert Telescope (BAT). The BAT has sensitivity five times better than the BATSE instrument on CGRO (within the BAT's 1.4 sr field of view.) As soon as the BAT processing electronics detects a new burst, it generates a request to maneuver the highly agile Swift spacecraft to point the co-aligned X-ray Telescope (XRT) and UV/Optical Telescope (UVOT) at the newly discovered GRB. Rapid BAT processing and a responsive spacecraft will result in making X-ray to optical measurements of the GRB counterpart within 20 seconds (for 10% of the BAT bursts) and 75 seconds (for 90% of the BAT bursts).

## OBSERVING OPPORTUNITIES

The Swift mission will offer multiple opportunities for carrying out rapid response observations of GRB afterglows and X-ray transients. In addition to the autonomous response by Swift in reacting to new GRBs, Swift can also be directed via uploaded command to carry out Target of Opportunity (ToO) observations.

These ToOs should be targets of extraordinary scientific importance. The Swift team policy is to consider requests from the entire astronomical community through an open Web form request. If the request meets pre-defined thresholds of criticality in both scientific importance and time-sensitivity, the

ToO request will be reviewed by the Swift PI, and if approved, will implemented as soon as possible by the Mission Operations (Ops) team.

ToO commands to the spacecraft will be transmitted via the TDRSS link, which is continuously available with a latency of less than one hour. Uploads of these commands are designed to be extremely efficient and simple to implement. The Ops team will only need to supply the celestial position of the target, a numerical value termed 'Merit' (which establishes the relative priority of the request), and select which of several pre-designed observing mode sequences will be used. After that point the on-board Figure-of-Merit algorithm responds to the ToO request in exactly the same manner as a new GRB detected by the BAT.

## Extended Observing Opportunities

In addition to ToOs, Swift will allow repeated observations of the same target (when scientifically justified). Examples of this include: multiple observations of ToOs (where the ToO mechanism is used repeatedly), monitoring of variable intensity sources (using a set of 15 positions which are chosen as look directions while the BAT is between burst detections), and multi-wavelength campaigns (based on Swift's excellent flexibility and complement of instruments).

## OBSERVING PLANNING

The BAT and XRT are fully autonomous. The BAT operates in two modes, one for GRB detection, the other for GRB observation following detection. The XRT initiates its mode selection based on the BAT trigger or ToO message, and thereafter dynamically adjusts its observing mode based on detected source counting rate.

The UVOT Observing Modes are selectable sequences based on pre-planned steps through various filter wheel positions. For GRB observations a standard program of filter or grisms will be cycled through for a nominal 10,000 second observing cycle. For ToOs or observations uploaded through the standard mission planning process the Science Ops Team will select the sequence.

The UVOT will suffer damage to its photocathode if it views bright sources (brighter than $8^{th}$ magnitude) for extended times. The UVOT will autonomously protect itself through both an on-board star catalog (supplemented by ephemeredes predicting the position of bright planets and asteroids) and a detector level safing system which automatically protect the instrument.

The spacecraft is planned to have a $96^m$ orbit at an inclination of 20°. Viewing constraints enforced by the spacecraft exclude pointing the instruments toward the Sun, Earth, Moon or a small area in the Ram direction. (The Ram direction is the local orientation of spacecraft motion with respect to the residual atmosphere around the satellite. Energetic ions impinging down the telescope tubes of the XRT and UVOT could erode the instrument filters.) The instruments will also shut down during the enhanced radiation environment of the South Atlantic Anomaly.

The result of these constraints is that observations will necessarily be broken into segments of 40 to 90 minutes. In order to segregate data for analysis, it will be tagged by identifiers including the target, observing sequence number and observing segment. Processing will automatically reassemble data for targets and observing segments.

After Quick-look data products are reviewed at the Penn State based MOC, standard data processing will be performed at the GSFC based Swift Data Center. These pipeline-based products will be distributed to the scientific community through the High Energy Astrophysics Science Archive Center (for the U.S.) with mirror processing and distribution capabilities at the Italian Swift Analysis Center (in Milan and Rome) and the United Kingdom Data Centre (in Leicester). An additional site in Japan is under consideration.

## SERENDIPITOUS SCIENCE

While the BAT is monitoring the sky for GRBs, it is also collecting gamma-ray/hard X-ray photons from the non-GRB sources within the BAT's field of view. These photons can be interpreted to produce an all-sky map of the 10-150 keV energy range, and derive a catalog of point sources which will be approximately 30 times more sensitive than the HEAO A-4 survey.

Swift observations between GRB follow-ups will be chosen to assure that at least 80% of the sky will fall within the BAT field of view. This will allow the BAT to also survey the sky for X-ray transients.

## SWIFT SUMMARY CAPABILITIES

The primary Swift data objective is development of a new set of fully observed GRBs, with localized source counterparts in the X-ray and UV/optical range for hundreds of bursts within a three year mission life.

Swift achieves this goal by being the first mission with the capability to point sensitive and accurate X-ray and UV/optical telescopes at newly discovered GRBs within minutes of the burst. This rapid localization of the burst and identification of the counterpart with arcsecond accuracy will be sent to the ground immediately. Later analysis on the ground, will result in sub-arcsecond relative positions for all the bursts with detectable UV/optical counterparts. The X-ray spectroscopy, and filter and grism spectroscopy will also measure redshifts to many bursts, even without ground follow-up.

The dedicated nature of the Swift satellite will also enable multwavelength uniform afterglow studies on timescales ranging from minutes to weeks, as we follow afterglows to the limit of detection by the Swift instruments.

## ACKNOWLEDGMENTS

We acknowledge support from the many scientists and engineers of the Swift team who are working hard and long to assure the success of Swift at Penn State, Goddard Space Flight Center and across the world. Funding at Penn State is supplied by NASA contract NAS5-00136.

# High Resolution Spectroscopy of the X-Ray Emission of GRBs by IMBOSS on the ISS

L. Colasanti, L. Piro, L. Pacciani, E. Costa, G. Gandolfi, P. Soffitta*, F. Gatti, D. Pergolesi, M. Razeti, R. Vaccarone, G. Testera, M. Pallavicini[†], A. Ferrari, E. Trussoni, M. Orio**, D. Mc Cammon, T. Sanders, M. Galeazzi[‡] and A. Szymkowiak, S. Porter, R. Kelley[§]

*Istituto di Astrofisica Spaziale e Fisica Cosmica, CNR, Roma, Italy*
[†]*INFN-Univerista' di Genova, Italy*
**Osservatorio Astronomico di Torino, Italy*
[‡]*University of Wisconsin, Madison, USA*
[§]*GSFC-NASA, USA*

**Abstract.** The IMBOSS (Interstellar/Intergalactic Medium and gamma-ray Burst Observatory and Spectroscopy Survey) is an experiment proposed to fly on the ISS (International Space Station), in order to perform an all-sky survey to study the diffuse X-ray emission and to measure the spectra of Gamma-Ray Bursts (GRB) with high energy resolution in the 0.1-10 keV energy range. In a 3-year mission, the experiment will detect about 20 to 40 GRBs. In several events, we can perform high resolution spectroscopy of the iron emission lines and absorption edges. Such components have been observed by BeppoSAX and Chandra. The measurements of these features would provide a direct diagnostic of the physical and kinematical state of the medium surrounding a GRB. Furthermore, they would supply information about the origin of the progenitors of the GRBs (probably a supernova explosion of massive stars) and their site formation (possibly a star-forming region).

## TECHNICAL DESCRIPTION

The three main parts of the experiments are: the collimators and the detectors; the dewar and the cryogenic insert; the external analog and digital electronics.

There will be two *detectors* with different fields of view. The first one, with a collimator of $10° \times 10°$, will be used to make a survey of the diffuse X-ray emission produced by Interstellar and Intergalactic Media. The second detector, with a collimator with a minimum field of view of $40° \times 40°$ and a final goal of $60° \times 60°$, will be used to observe and achieve high resoluton spectroscopy of the X-ray emission of Grbs. Each detector will be an array of 36 silicon bolometers, with HgTe absorber, a total area of 1.44 cm$^2$ and a resolution of 4-10 eV.

The *dewar* will contain 100-120 L of superfluid helium (at 1.3 K), surrounding the magnet with a maximum field of 4 T. The *cryogenic insert* will include the paramagnetic salt pill of ADR refrigerator (at operating temperature of 60 mK), the mechanical thermal switch, 5 IR filters.

The elements of *external analog* and *digital electronics* are: J-Fet preamplifiers operating at 120 K; external low noise amplifiers with a gain of 20000; antialiasing filters; the analog to digital conversion (up to 100 kHz) of the signals, with a 14 bit resolution; the control logic for the refrigeration cycle; the interface to the ISS (cryostat power line, telemetry and telecommand interface).

## HOW MANY GRBS?

We used the BeppoSAX data to build up a Log N(>S)-Log S. Taking into account the observed spectral shape (a power-law with photon index 1.5) and a typical absorption column density (due to the our Galaxy) of $2 \times 10^{20}$ cm$^{-2}$), about 180 cts (for t=10 s) from the diffuse component, assuming a signal to noise ratio of 5$\sigma$, with a field of view of $40° \times 40°$, in a three-year mission we can detect 18 GRBs. Assuming a field of view of $60° \times 60°$, we have about 250 counts from the background; in this case, we can detect 42 GRBs.

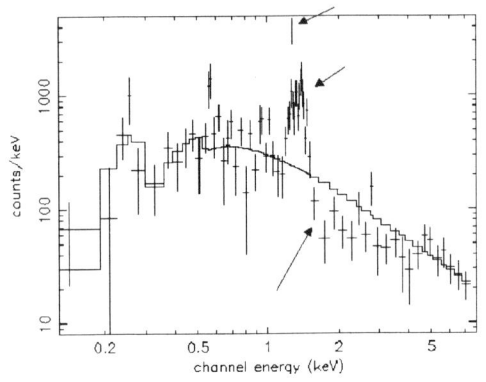 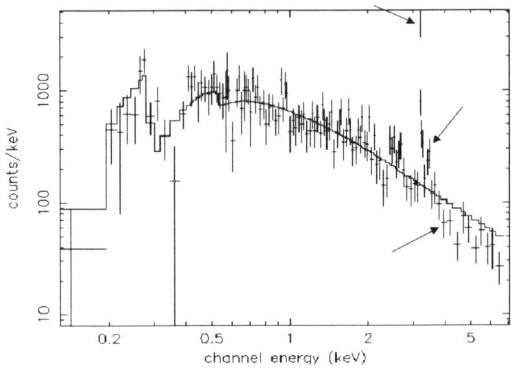

**FIGURE 1.** *Left*: Simulated spectrum for a field of view of $40° \times 40°$. *Right*: Simulated spectrum for a field of view of $60° \times 60°$. The deviations from the best-fit power-law (continuous histogram line), indicated by arrows, represent the three added components (see text).

## IRON FEATURES AND SIMULATED SPECTRA

The observation of iron emission lines and absorption edges is an important tool to understand the origin of GRBs and the nature of their progenitors and to investigate the physical and kinematical state of the medium surrounding a GRB. If they arise from a supernova explosion of a massive star, we expect to observe such features.

The most significant detections of emission lines are those in GRB 970508 [1] and GRB 991216 [2]. In GRB 990705 [3] BeppoSAX detected an iron absorption edge.

On the basis of the above observed features, we have simulated some spectral emissions of GRBs, adding emission lines and absorption edge. The first simulation (figure 1 *left*) is for a field of view of $40° \times 40°$, assuming a fluence of $4.8 \times 10^{-6}$ erg/cm$^2$ (5 GRBs in 3 years). To the GRB emission (continuous line), we have added two emission lines (E=6.4 keV, E.W.=0.3 keV, $\sigma_L$=0 eV and E=6.9 keV, E.W.=4 keV, $\sigma_L = 0.4$ keV in the restframe) and an absorption edge (E=7.1 keV and $\tau$=1.5) at redshift z=4. In the second simulation (figure 1 *right*), we assumed a field of view of $60° \times 60°$, a fluence $S=8.9 \times 10^{-6}$ erg/cm$^2$ for the GRB emission, two emission lines ($E = 6.4$ keV, $E.W. = 0.3$ keV, $\sigma_L = 0$ eV and $E = 6.9$ keV, $E.W. = 1$ keV, $\sigma_L = 0.4$ keV in the restframe) and an absorption edge ($E = 7.1$ keV and $\tau =1.5$ in the restframe) at redshift z=1.

emission lines by the determination of their width $\sigma_L$. A narrow line is generated from a medium at rest in space. Therefore it should be associated to stellar forming regions. Broad lines are index of moving medium, probably pre-ejected from the supernova explosion of a massive star. The line sensitivity of IMBOSS will allow us to observe features in 15 GRBs (at least) and broad lines in about 7 GRBs, assuming a field of view of $60° \times 60°$. By an high resolution spectroscopy of X-ray emission of GRBs, we can investigate the state and the nature of medium that surrounds them. In this way, it is possible to draw important information about the origin of GRBs and the nature of their progenitors.

## REFERENCES

1. Piro et al., ApJ 514, L73, 1999
2. Piro et al., Science 290, 889, 2000
3. Amati et al., Science 290, 953, 2000

## CONCLUSION

The above simulated spectra show that we can perform a high resolution spectroscopy of iron emission lines and absorption edges. We can discriminate the nature of

# Performance of the Swift X-ray Telescope (XRT) Mirror/Detector Combination

A.D. Short[a], R.M. Ambrosi[a], I.B. Hutchinson[a], R. Willingale[a], A.F. Abbey[a], A.A. Wells[a], J.E. Hill[b], D.N. Burrows[b], G. Tagliaferri[c], O. Citterio[c]

*a Physics & Astronomy Department, University of Leicester, UK*
*b Department of Astronomy & Astrophysics, Penn State University, US.*
*c Osservatorio Astronomico di Brera, Milano, Italy*

**Abstract.** The Swift XRT [4] uses a 12 shell grazing incidence mirror developed for the JET-X telescope and an open electrode CCD detector developed for XMM-Newton [9]. The mirror has recently undergone re-characterization at the Panter facility and the flight detector is currently being calibrated at the University of Leicester. We present key performance characteristics of the mirror/detector combination, in particular, the point spread function (PSF) and centroiding capability and the impact of background events, as well as the telescope effective area and spectral response.

## INTRODUCTION

At present, GRB afterglow is rarely acquired less than 8 hours after the burst and since the decay is rapid (typically $t^{-1.2}$) this means that most of the afterglow emission is being missed. This 'data gap' will be addressed in 2003 with the launch of Swift. The Swift X-ray telescope is under construction and calibration will begin in the spring of 2002, however preliminary tests of the flight optics with a flight like detector have already been carried out and the baseline instrument response has been constructed through simulation. The results of this work are presented here.

## THE SWIFT OBSERVATORY

Swift is due for launch in the autumn of 2003. It will detect GRBs at a rate of approximately one per day and the primary objective is to report the location of each burst to the astronomical community within ~5 minutes. The second objective is to measure the X-ray and UV-optical light curves for a period of hours or days and the third objective is to study spectral evolution as the afterglow decays.

Swift will carry three instruments. The wide field, gamma ray Burst Alert Telescope (BAT) provided by Goddard Space Flight Centre has coded aperture optics and Cd(Zn)Te detectors. It will monitor the sky over 2sr and detect gamma ray bursts, reporting their position to the spacecraft with arc-minute resolution. The Spacecraft will then slew autonomously to point two co-aligned narrow field instruments. The X-ray Telescope (XRT) will locate each burst with arc-second resolution passing the location to the UV/optical telescope, which will search the XRT error box and identify new sources with sub arc-second resolution using on-board star charts. The design of the UV/optical telescope provided by MSSL in collaboration with Penn State University (PSU) is based on the Optical Monitor aboard XMM-Newton.

### The Swift X-ray Telescope

The Swift X-ray telescope is being provided through a collaboration between Penn State University, the University of Leicester and the Osservatorio Astronomico di Brera. PSU are providing the telescope tube and telescope door, and are responsible for system engineering, electronics and software. Brera are providing the X-ray mirror [1], which is a flight spare unit from JET-X. The University of Leicester is providing the focal plane camera [4,8], the telescope alignment monitor and the electron deflector, as well as supporting telescope integration and alignment, qualification, calibration, spacecraft integration and post launch operations.

## POINT SPREAD FUNCTION & CENTROIDING

The XRT mirror and a flight like CCD were tested together at the Panter facility in July 2000 [1,2,6]. The results confirm an on-axis, in-focus PSF Half Energy Width (HEW) of 16" and a corresponding centroiding capability better than 1" for a 100 mCrab source.

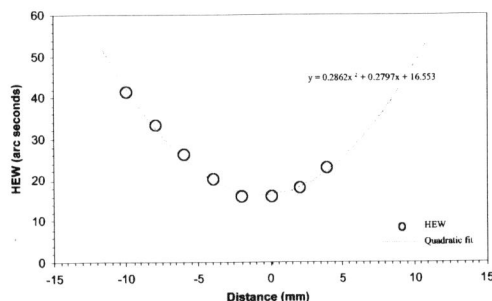

**FIGURE 1.** PSF HEW as a function of 'out of focus position'.

'Out of focus' measurements indicate that a 2mm error in focal length can be tolerated without compromising the centroiding accuracy and without increasing the HEW by more than 2" (figure 1). Off axis measurements confirm that there is little variation in HEW up to 8' from the centre of the CCD (figure 2). This is well matched to the error circle of the Swift Burst Alert Telescope (BAT) which will locate GRBs within a 4' radius.

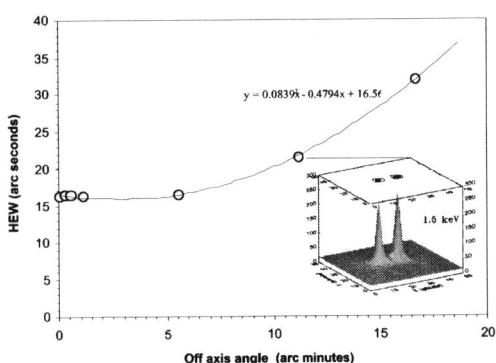

**FIGURE 2.** PSF HEW as a function of off axis angle

The XRT is required to obtain centroids from the first frame obtained on each new source, but centre of gravity (C of G) centroiding methods are highly sensitive to the presence of background events [6]. Their effect may be reduced by event recognition, decreasing the sampling area around the PSF or by employing more sophisticated centroiding algorithms such as matched filters.

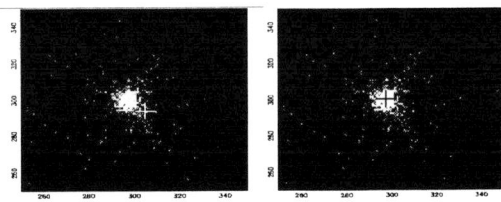

**FIGURE 3.** Centre of gravity (left) and matched filter (right) centroids (crosses). Background event 200 pixels away.

Figure 3 shows centroid locations calculated using simple C of G and matched filter techniques for a single frame containing a background event 200 pixels from the PSF. The C of G technique introduces an error of ~20" (requirement <1") and the matched filter technique is expected to be too processor intensive for the flight algorithm. The flight software will therefore employ event recognition and an iterative C of G algorithm.

## RE-DISTRIBUTION & EFFECTIVE AREA

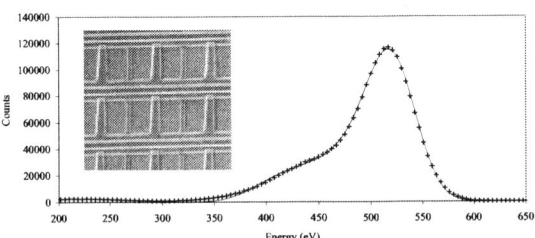

**FIGURE 4.** Spectral re-distribution. Calibration data (points) and analytical model (line). A-symmetric surface losses are significant due to 'open' electrode structure (inset)

The XRT CCD has an open electrode structure [9] (figure 4 inset) which increases the low energy quantum efficiency. However, low energy X-rays absorbed close to the electrode structure suffer a degree of charge loss which is a function of both interaction depth and position within the pixel. The spectral re-distribution below ~1keV is therefore non gaussian and detailed modelling of the CCD response has been conducted in order to generate the CCD component of the response matrix [7,8](figure 5 inset). An example of measured and best-fit analytical model re-distribution is given in figure 4.

By combining the CCD response (described above) with the mirror effective area and the optical filter transmission across the energy band, the preliminary XRT response matrix has been generated. This will be updated as the calibration progresses, but the current XRT effective area is shown in figure 5.

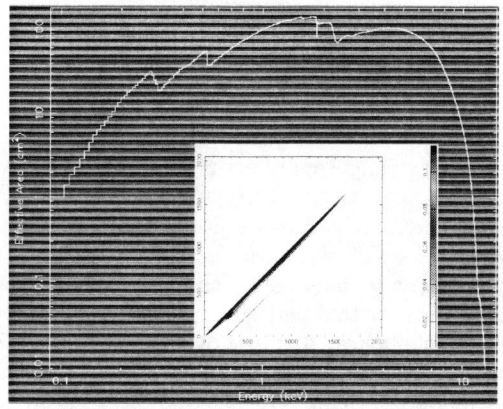

**FIGURE 5.** Preliminary total effective area and response matrix (inset) of the Swift XRT

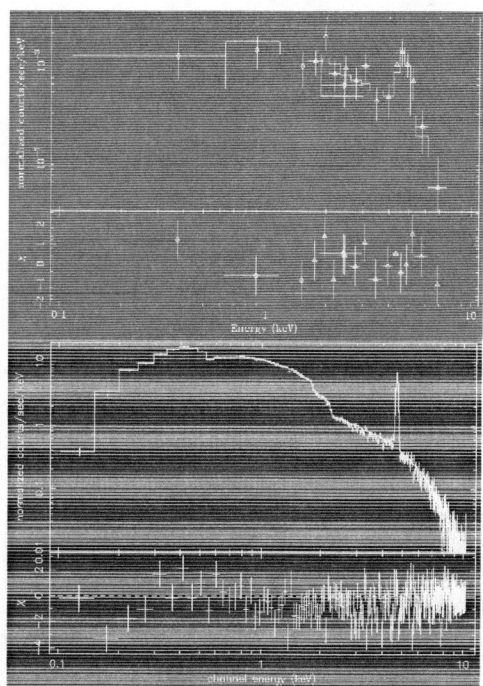

**FIGURE 6.** GRB 000214 X-ray afterglow. Fit is absorbed power law ($\Gamma=2.2$, nH=$5.5\times10^{20}$) plus red shifted Fe line at 4.7keV Top panel: 51ks BeppoSAX observation commencing 12 hours after the burst. Fe line chance probability =0.27%, equivalent width ~2keV Bottom panel: Simulated 3.6ks Swift observation commencing 2 hours after the burst

## SIMULATED SWIFT XRT GRB SPECTRUM

The upper panel of figure 6 shows a BeppoSAX GRB X-ray afterglow spectrum from a 15 hour observation which began 12 hours after the burst [3]. A red shifted Fe line was detected with a luminosity of $9\times10^{-6}$ photons cm$^{-2}$s$^{-1}$. The continuum effective flux (2-10keV) was $2.9\times10^{-13}$ erg cm$^{-2}$s$^{-1}$. Assuming an afterglow decay rate of $t^{-1.2}$ we may conservatively estimate that a one hour Swift observation within the first hours after the burst would encounter an effective flux >1000 times greater. This is simulated in the lower panel. The most striking difference is the sharpness of the line. In fact the resolution of the XRT (~100eV @ 5keV) is similar to the predicted intrinsic line width. Table 1 gives parameter constraints from the real and simulated data.

**TABLE 1. Parameter Constraints**

| Parameter | Value | 90% conf. range (observation) | 90% conf. range (simulation) |
|---|---|---|---|
| nH (cm$^{-2}$) | $5.5\times10^{20}$ | $8.2\times10^{20}$ | $0.3\times10^{20}$ |
| $\Gamma$ | 2.2 | 0.6 | 0.03 |
| Fe E (keV) | 4.7 | 0.4 | 0.01 |

## FUTURE WORK

The assumption here that the spectrum does not evolve during the burst is entirely illustrative. Simulations of this type will be conducted using models of X-ray afterglow generation and evolution. These will establish the sensitivity of Swift observations to detect lines and edges, to distinguish between line emission mechanisms and hence to determine line energy and red-shift.

## REFERENCES

1. Ambrosi, R., et al. 2001 SPIE, (in print)
2. Ambrosi, R., et al. 2001 NIM A (in print)
3. Antonelli, A., et al. 2000, ApJ, 545, L39
4. Burrows, D.N., et al. 2000 SPIE 4140, L64-75
5. Hiraga, J., et al. 2001 NIM A, 465, L384-393
6. Hutchinson, I., et al. 2001 NIM A (in print)
7. Short, A.D., et al. 1998 SPIE 3445, L13-27
8. Short, A.D., et al. 2001 NIM A, (in print)
9. Turner, M.J.L., et al. 2000 A&A 365, L27-L35

# The ROTSE-IIIa Telescope System

D. Smith[*], C. Akerlof[*], M. C. B. Ashley[†], D. Casperson[**], G. Gisler[**], R. Kehoe[*], S. Marshall[‡], K. McGowan[**], T. McKay[*], M. A. Phillips[†], E. Rykoff[*], W. T. Vestrand[**], P. Wozniak[**] and J. Wren[**]

[*]*2477 Randall Laboratory, University of Michigan, 500 E. Univeristy Ave., Ann Arbor, MI, 48109, USA*
[†]*School of Physics, University of New South Wales, Sydney 2052, Australia*
[**]*Los Alamos National Laboratories, Los Alamos, NM, 87545, USA*
[‡]*Lawrence Livermore National Laboratory, 7000 East Avenue, Livermore, CA, 94550, USA*

**Abstract.**
We report on the current operating status of the ROTSE-IIIa telescope, currently undergoing testing at Los Alamos National Laboratories in New Mexico. It will be shipped to Siding Spring Observatory, Australia, in first quarter 2002. ROTSE-IIIa has been in automated observing mode since early October, 2001, after completing several weeks of calibration and check-out observations. Calibrated lists of objects in ROTSE-IIIa sky patrol data are produced routinely in an automated pipeline, and we are currently automating analysis procedures to compile these lists, eliminate false detections, and automatically identify transient and variable objects. The manual application of these procedures has already led to the detection of a nova that rose over six magnitudes in two days to a maximum detected brightness of $m_R \sim 13.9$ and then faded two magnitudes in two weeks. We also readily identify variable stars, includings those suspected to be variables from the Sloan Digital Sky Survey. We report on our system to allow public monitoring of the telescope operational status in real time over the WWW.

## TELESCOPE STATUS

ROTSE-IIIa is a 0.45-m robotic reflecting telescope which is managed by a fully-automated system of interacting daemons within a Linux environment. The telescope has an f-ratio of 1.9, yielding a field of view (FOV) of $1.85 \times 1.85$ degrees. The control system is connected via a TCP/IP socket to the Gamma-Ray Burst Coordinate Network (GCN), and a flexible scheduler daemon plans observation sequences including sky patrols, targeted monitoring programs, and fast ($< 10$ s) responses to GRB alerts. Upon receipt of a GRB alert over the GCN, we begin a program of 10 5-s, 10 20-s, and 80 60-s exposures (with a read-out time of 7 s between images). We also automatically schedule blocks of 30 60-s follow-up exposures. These blocks are spaced at ever-increasing intervals. The system began its testing run on October 11, 2001, at the ROTSE-I site at Los Alamos National Laboratory. ROTSE-IIIa will be shipped to Siding Springs Observatory, Australia, upon completion of the testing phase.

ROTSE-IIIa currently uses unfiltered CCDs, although we have included a slot in the mechanical design that will allow for the insertion of a filter. We currently calibrate ROTSE magnitudes against the USNO R-band magnitudes, and we include a constant offset of 0.3 magnitudes to convert these numbers to a V-band magnitude. This necessarily introduces unknown systematic errors for objects with atypical spectra. As shown in Figure 1, ROTSE-IIIa can reach 17th magnitude in a 5-s exposure, 17.5 in 20-s, and 18.5 in a 60-s exposure. Longer exposures are not practical due to saturation of the sky. Multiple images can be co-added to reach $\sim$ 19th magnitude.

Figure 2 shows the dispersion of calibrated stellar magnitudes over time (corrected to V band). Plotted are the standard deviations vs. the mean magnitude of $\sim$ 3000 objects detected in at least 10 of 20 observations of a single field. The intensity of bright stars varies by less than 1%, while the RMS deviation rises to $\sim$ 30% for the dimmest stars ROTSE-IIIa can detect.

To measure the astrometric accuracy of our analysis, we compare the relative positions of objects found in both the USNO A2.0 catalog as well as in the above-mentioned 20 ROTSE-IIIa observations. The transformation of the CCD coordinates to celestial coordinates is achieved through a third-order polynomial warp. Most bright stars ($m_V < 14$) can be localized to better than 0.3 arcsec, or one-tenth of a pixel, and the bulk of the faintest objects stay within two-thirds of a pixel.

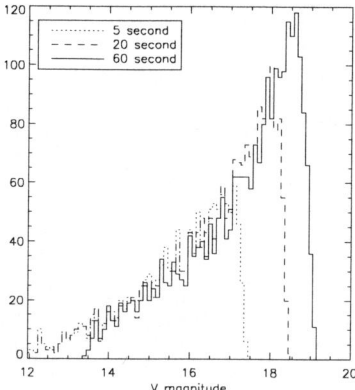

**FIGURE 1.** Number of objects per magnitude bin for a random ROTSE-IIIa field at three exposure times. Magnitudes were derived by calibrating to the USNO R-band and then applying a constant correction factor of 0.3 mag to convert to an approximate V-band.

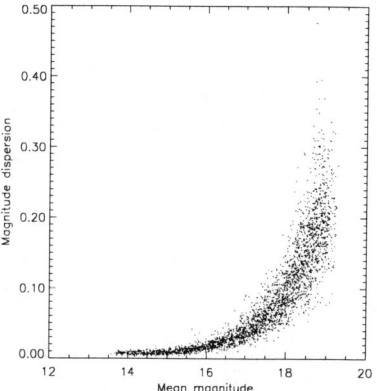

**FIGURE 2.** Photometric scatter in 20 60-s observations of 3,000 objects over a two-week interval as a function of V-band magnitude. The images have been corrected using a relative photometry algorithm. Root mean square deviations vary from $\sim 1\%$ at 14th magnitude to $\sim 10-30\%$ at the magnitude limit.

## DATA ANALYSIS PIPELINE

We are implementing software to automatically analyze all images recorded with the ROTSE-III telescopes. The full analysis pipeline for each telescope will be installed on a computer at each site, but during testing we are copying most of the data to the University of Michigan for easier control. All images are automatically dark- and flat-field corrected as soon as they are recorded. Immediately after each corrected image file is written, it is processed into an object list with the SExtractor package [1], using an aperture 5 pixels in diameter, to identify all source candidates within the FOV. The list of these candidate objects is compared against the USNO A2.0 catalog using a triangle-matching routine to compile a calibrated list of R-band magnitudes and celestial locations for these sources. Calibrated object lists also allow diagnostic parameters to be measured, such as the astrometric accuracy and the focus quality. These steps have been implemented on an on-site computer and the system produces a calibrated object list for each image within 45 s from the closing of the shutter.

Once more than one calibrated list for a given field is available, these lists are compiled into a "match structure"; a data structure that enables us to filter out objects that do not appear at the same location in sequential observations. The telescope aspect is shifted through a small, random vector between observations. This enables the elimination of hot CCD pixels and cosmic ray events. We then apply a relative photometry algorithm to stabilize the magnitude estimates and calibrate the systematic errors. This reduces the scatter in light curves for stable bright objects to $< 1\%$ (Fig. 2) and allows us to reliably identify variable sources. We can also flag known distractions such as asteroids and "masked" stars (stars that appear in ROTSE images but are too close to bright stars for sensitive cataloged surveys to resolve). Once this procedure is automated, we anticipate that the system will be able to report an arcsecond position for a bright ($m_R > 17$), variable object within five minutes of the receipt of a burst trigger via a GCN alert.

## DISCOVERIES

Calibrated match structures can easily be searched for previously unknown objects. For our first testing of the procedures involved, we constructed these structures for 20 epochs over two weeks for $\sim 500$ sq. deg. of sky patrols along the Equatorial strips of the Sloan Digital Sky Survey Early Data Release. Variable stars can be identified through application of the techniques developed for ROTSE-I [2]. In Figure 3, we show the folded light curve for a typical RR Lyra star in a ROTSE-IIIa field. The best-fit period of 0.64 d was found using a cubic spline method [3]. This method provides a best-fit period and error estimate for a variable star light curve, as well as a spline-interpolated approximation for the source light curve.

Non-periodic variables can also be identified. Figure 4 shows the light curve for a new transient that was not detected on Oct 11.424, 2001 (with a limiting magnitude of 18.2), but was easily visible at a magnitude of 14.0 on Oct 13.291, 45 h later. Clouds prevented observations on the night of Oct 12. The transient then faded two magnitudes over the subsequent two weeks until proximity to the waxing moon interfered with further monitoring.

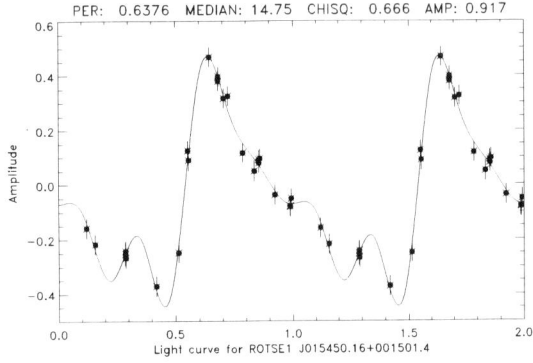

**FIGURE 3.** Folded light curve in calibrated R-band for an identified variable star in one of the ROTSE-IIIa sky patrol fields. An estimate of a systematic error of 4% has been added in quadrature to the statistical errors. The mean magnitude of 14.75 has been subtracted.

The best-fit power law index for the fading light curve is ~ 0.9, but the shape of the curve is clearly not consistent with a single power law. With a mean magnitude of 15 and an RMS deviation around that mean of 0.66, this source stood out starkly against the typical behavior of other stars (Fig. 2). The ease with which this source was detected is a positive indication of the potential for using ROTSE-IIIa to identify and track new transients.

## REALTIME WWW MONITORING

The telescope operating system writes all values of its status variables to a file once every minute. These variables include the current pointing direction, the name of the most recent image, whether the enclosure roof is open or closed, the current velocity of the mount along its two axes, and whether or not any alarms are active. This file is copied back to the University of Michigan by a cron job and parsed into an HTML file that can be viewed in real time over the WWW at http://www.rotse.net. Also included in this display are a thumbnail of the most recent image and graphs of the diagnostic parameters derived from the calibrated object lists. This interface allows interested parties to monitor burst response and telescope status in near-real-time.

## CONCLUSIONS

ROTSE-IIIa is working well in its first testing phase at Los Alamos. Operation and image processing has been automated, and image analysis is being automated at the time of this writing. When full automation is implemented, we will be able to report an arcsecond

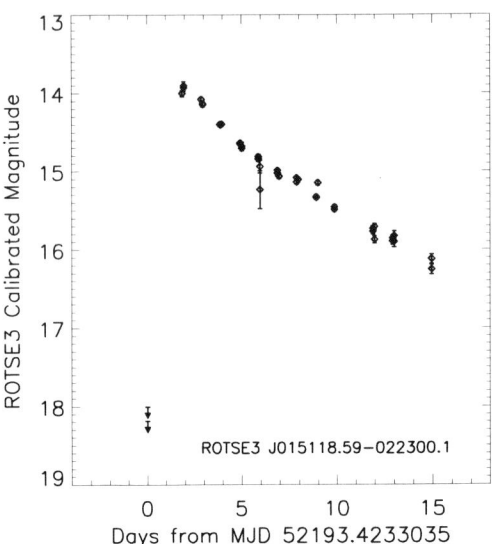

**FIGURE 4.** Light curve in the calibrated R-band for a transient nova discovered in the first two weeks of automated sky patrols by ROTSE-IIIa. Arrows indicate the limiting magnitudes of two images taken 1.8 d before the first detection at $m_R = 14.00 \pm 0.04$ on Oct 13.291, 2001. The source was easily detectable for the following two weeks, until proximity to the waxing moon interfered with continued monitoring.

position for a bright ($m_R > 17$), rapidly variable ($\Delta m > 10\%$ within the first minute) object within five minutes of receipt of a burst trigger. Limiting magnitudes under good conditions in 60-s exposures can approach 19th magnitude, and we anticipate gaining further sensitivity at the (darker) permanent site in Australia.

## ACKNOWLEDGMENTS

D. Smith is supported by NSF fellowship 00-136. Work performed at LANL is supported by NASA SR&T through Department of Energy (DOE) contract W-7405-ENG-36 and through internal LDRD funding. Work performed at the University of Michigan is supported by NASA under SR&T grant NAG5-5101, the NSF under grants AST 97-03282 and AST 99-70818, the Research Corporation, the University of Michigan, and the Planetary Society. Work performed at LLNL is supported by NASA SR&T through DOE contract W-7405-ENG-48.

## REFERENCES

1. Bertin, E., and Arnouts, S., *A&AS*, **117**, 393–404 (1996).
2. Akerlof, C., et al., *AJ*, **119**, 1901–1913 (2000).
3. Akerlof, C., et al., *ApJ*, **436**, 787–794 (1994).

# Monitoring of the Prompt GRB Afterglow with the REM Telescope

S. Covino*, F. Zerbi[†], G. Chincarini[†], G. Ghisellini[†], M. Rodonó**, L.A. Antonelli[‡], P. Conconi[†], G. Cutispoto**, E. Molinari[†], L. Nicastro[§] and E. Palazzi[¶]

*Brera Astronomical Observatory, via E. Bianchi 46, 23807, Merate (LC), Italy.
[†]Osservatorio Astronomico di Brera, via Bianchi 46, 23807 Merate (LC), Italy
**Osservatorio Astronomico di Catania, Via S.Sofia, 78, 95123 Catania, Italy
[‡]Osservatorio Astronomico di Roma, Via di Frascati 33, 00040 Monte Porzio Catone (ROMA), Italy
[§]CNR-IFCAI, Via Ugo La Malfa 153, 90146 Palermo, Italy
[¶]CNR-TESRE, via P. Gobetti 101, 40129 Bologna, Italy

**Abstract.** In these pages we present REM (Rapid Eye Mount), a fully robotized fast slewing telescope equipped with a high throughput NIR (Z', J, H, K) camera and an Optical slitless spectrograph (ROSS) optimized for the monitoring of the prompt afterglow of Gamma Ray Bursts. Covering the NIR domain REM can discover objects at extremely high red-shift and trigger large telescopes to observe them when they are still bright. The synergy between REM-IR cam and ROSS makes REM a powerful observing tool for any kind of fast transient phenomena.

## INTRODUCTION

Many fascinating questions remain open about GRB origin and structure. Some of these questions should be answered by collecting data in the early phases of the afterglow in a frequency domain as wide as possible. Space borne observatories (like HETE II or Swift) have been developed to detect the burst and monitor the afterglow at higher energies. These satellites have also the capability to send a triger to facilities on the ground which can take care of the afterglow monitoring at lower frequencies, e.g. optical, IR, radio.

An important role is played in this context by small robotic facilities that can collect in a timely manner relevant information and act as an inter-mediate step to activate Target of Opportunity procedure at telescopes with larger area. REM is one of these, the only one foreseen so far to monitor the near infrared region as well as to sample intensively the optical continuum.

The REM telescope is a Ritchey-Chretien system with a 60 cm f/2.2 primary and a overall f/8 focal ratio mounted in an alt-azimuth mount in order to provide stable Nasmyth focal stations, suitable for fast motions. At the first focal station a dichroic, working at 45 degrees in the f/8 convergent beam, will split the beam to feed the REM-IR camera and the ROSS Spectrograph. Both the instuments have the same field of view ($10 \times 10$ arcmin$^2$, matched to the most common errorbox of gamma ray detectors) but with different plate scale (1.2 as/pixel for the IR arm and 0.55 as/px for the optical arm). A complete description of the REM assembly is given in [1].

REM is designed to provide the coordinates of the NIR transient in less than 1 minutes upon trigger reception and colors in less than 5 minutes. Both these informations will be made available to the public immediately via the usual means. Curves and Spectra will be available at a later stage since they will be reduced off-line.

## SCIENCE WITH REM

Although we have only one example of prompt optical/IR emission during the first minutes of the afterglow (GRB 990123), we can estimate the typical early magnitudes by extrapolating back in time the light curves of known afterglows, (figure 1). The IR Camera on REM is expected to reach magnitude H=15.5, 16.04 and 17.11 with exposure times of 5, 30 and 600 seconds respectively (S/N=5). With the ROSS spectrograph a V=14 point-like source is recorded better than $10\sigma$ in 1 sec exposures.

The above numbers suggest that we can detect the IR afterglow during the first 2-4 hours even with an exposure time of 5 seconds. This will allow the study of the light curve in great detail and the detection of possible (even if short) variations from the smooth power-law be-

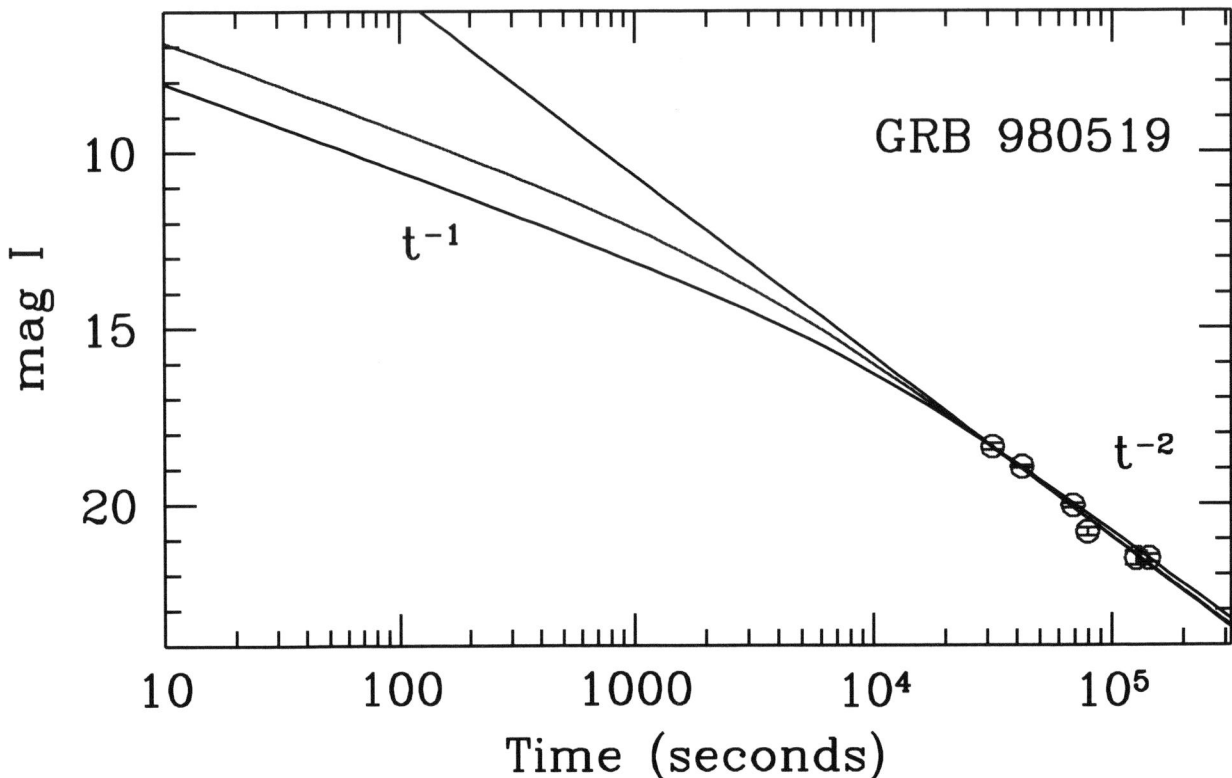

**FIGURE 1.** The optical light curve of GRB 980519, with 3 possible extrapolation towards earlier times. Should the early afterglow behave "normally" (i.e. with a $t^{-1}$) we would have evidence of beaming

havior and the definition of any possible break. Increasing the exposure time (after the initial phases) to 10 minutes, we can follow typical bursts up to 12 hours, after which larger telescopes can take over.

One of the main goals of REM is to detect burst at high-z for which the Ly-break dumps all the light at optical wavelengths. Ly-$\alpha$ absorption falls in the REM NIR camera wavelength range for sources with red-shift between 8 and 15, i.e. any burst in this range can still be detected by REM NIR camera and its position determined with great accuracy. A highly absorbed burst might as well dump the optical photons out but the color-color techniques can discriminate between an absorbed and a high-z burst in real time. Large telescopes (such as VLT) can then point at the target while it is still bright enough for high-dispersion spectroscopic observations. This appears to be the only way to obtain high quality spectra for large redshift (even $z > 10$) objects.

Even in the case of the absorbed burst the role of REM can be crucial. Indeed the infrared light is much less absorbed, and therefore an IR transient can be detected even if the optical is not. In addition, by combining the two (IR and optical) REM scientific outputs it is possible to estimate the amount of absorption and then the characteristics of the circumburst environment.

It is known that a lack of dust absorption, thought to be associated with star-forming regions, has been reported in the spectra of some bursts. A possible explanation for such a lack is that dust grains are sublimated by the prompt optical/UV emission (see Fig. 2). In this case, an *IR flash* should be observed to start before the optical one, as the IR radiation can penetrate unabsorbed in the cloud while the higher energy photons progressively clean out the dust. The observed IR fluence before the detection of the optical flash would greatly constrain the amount of dust in the cloud.

Beside the most promising results highlighted above the two instruments on-board REM will provide other useful information for GRB science. The borad band spectral coverage (0.45-2.3 $\mu$m) will for instance help in understanding if comptonization dominates on synchrotron in the early phase of the afterglow. In addition such a broad band coverage will allow to compare bursts at different redshifts at the same comoving frequency.

## REFERENCES

1. Zerbi F.M., Chincarini G., Ghisellini G. et al., 2002, AN 322, 275

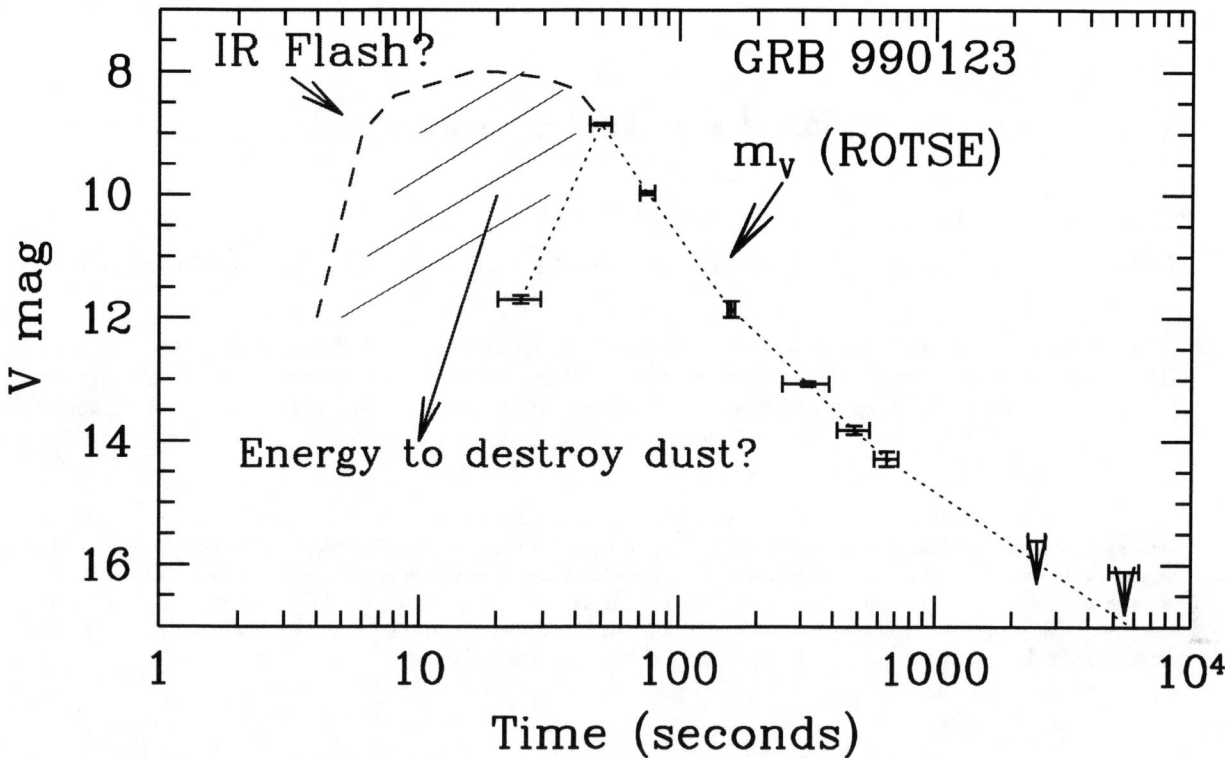

**FIGURE 2.** The optical flash of GRB 990123 as seen by ROTSE. Part of the optical UV photons could have been absorbed by dust, in the first part of the emission. After dust has evaporated, the line of sight become extinction free. Since IR photons are much less absorbed by dust, they could pass nearly unabsorbed, resulting in a more prompt emission.

# BART – Recent Status

M. Jelínek[a], R. Hudec[a], P. Kubánek[b], M. Nekola[a], J. Soldán[a], I. Stoklasová[a,c], M. Topinka[a,c], R. Smída[a,c], L. Svéda[a,c], F. Hroch[a,d], J. Polcar[a,d], A. J. Castro-Tirado[e]

[a] Astronomical Institute, Academy of Sciences of the Czech Republic, CZ-251 65 Ondrejov, Czech Republic
[b] Department of Informatics, Faculty of Mathematics and Physics, Charles University, Prague, Czech Republic
[c] Astronomical Institute, Faculty of Mathematics and Physics, Charles University, Prague, Czech Republic
[d] Masaryk University, Brno, Czech Republic
[e] LAEFF – INTA, Madrid, Spain, also IAA, Granada, Spain

**Abstract.** The Burst Alert Robotic Telescope is now in routine operation at the Ondrejov Observatory. The recent configuration includes a 25 cm optical tube and a 6.4 cm wide field camera with attached CCD cameras. The results of the test operation of the system as well as the system capabilities and scientific results obtained are presented and discussed. The cosmic ray statistics, background events statistics, methods of rapid data analyses as well as specific problems of reductions and evaluation of wide-field CCD images are also addressed.

## INTRODUCTION

The BART telescope (Burst Alert Robotic Telescope) is in routine operation at the Ondrejov Observatory since March 2001. It is a remotely controlled small aperture telescope with attached wide field camera and it serves as a dedicated optical telescope for gamma-ray bursts and high energy astrophysics. The BART represents a low-cost device based on commercially available parts with specially designed software suitable for networking and duplications.

## SCIENCE WITH BART

- optical follow-up observations of GRBs (HETE2, BeppoSAX, INTEGRAL, ...) with automated rapid response ( ~ 0.5 min depending on position)
- dense optical follow-up observations of GRBs, astrometry and photometry
- simultaneous and quasisimultaneous optical data for satellite observations and campaigns
- secondary science: photometry and optical monitoring of selected triggers (X-ray stars, AGNs, blazars, QSOs, SNe, CVs, targets of opportunity triggers, ....)

## TECHNICAL SOLUTION

- optical tube: Meade LX200 Schmidt-Cassegrainn telescope, aperture 25 cm
- LX200 Meade mount, 8 degrees/sec
- attached wide-field camera: Meopta lens, aperture 6.4 cm, FOV 5 degrees diameter (identical with INTEGRAL OMC Test Device)
- SBIG ST9E and ST8 CCD cameras
- resulting limiting magnitudes 18.5 for optical tube and 15.5 for the wide-field camera
- controller software written in Python (http://www.python.org), a high level scripting language with objects features
- cameras driver written in C, both cameras can shoot and download the images simultaneously
- control software runs on a Pentium PC under Linux RedHat 6.2 operating system
- user access is provided by simple web-based interface, obtained images can be downloaded by FTP
- users can create their own scripts, which controls the telescope movement and cameras exposition during automated observation

**FIGURE 1. Left.** The central part of the Orion constellation above the local horizon taken with the BART Wide Field Camera (identical with the INTEGRAL OMC Test Device).

**FIGURE 2. Right.** The optical system of the BART robotic telescope.

## RECENT STATUS

The system is in routine operation, more than 4 000 test images have been taken. The WF camera proved to be important for covering HETE triggers error boxes. With this camera, we can cover the entire HETE box with limiting magnitude of 15 even for large boxes (up to 3 degrees radius). Further attempts are done to improve the guidance and tracking capabilities of the LX200 mount. For more details see the BART web page: http://lascaux.asu.cas.cz

## BART AND BOOTES

BOOTES: Burst Observer and Optical Transient System-modified and advanced BART system, is a 30 cm aperture Meade LX200 optical tube with SBIG ST8 CCD camera and various WF cameras in routine operation in El Arenosilo, Spain. The 2nd station is in preparation at the 200 km distance [1].

## EXPERIENCE GAINED

Nearly 4000 CCD images were taken with NF and WF instruments during the test period. We have focused on the study of problems specific for WF CCD imaging and related data analyses (object recognition and classification, astrometry, photometry...) and also on searches for OTs over a wide FOV and on related study of false triggers caused by other phenomena.

We have analysed the Cosmic Rays (CR) since a fraction of them may mimic real OT images. The mean rate of CR on BOOTES ST-8 CCD cameras has been found to be 0.1 $min^{-1}$ $cm^{-2}$, nearly 50% point-like, a large fraction (but not all – short focal distances!) can be eliminated by the PSF fitting. The CR rate has been found to be highly non-isotropic in time.

The methods of OT/GRB analyses in real time have been exploited: e.g. creating a full sky coverage

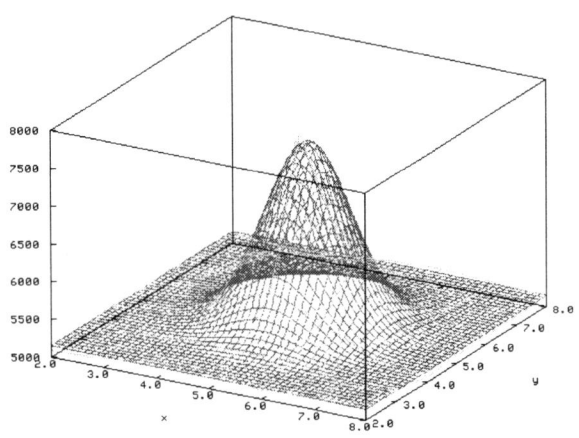 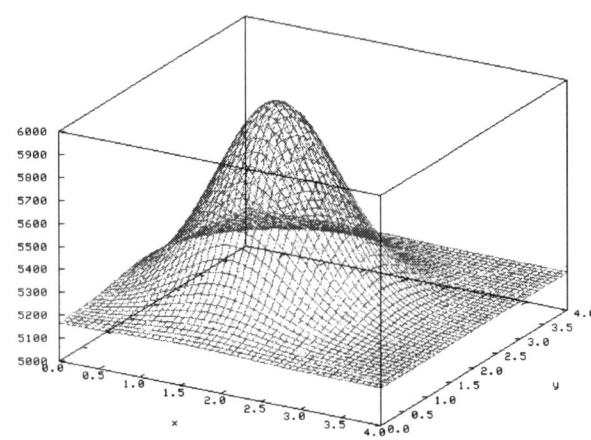

**FIGURE 3. Left.** PSF Fitting of a star image. **Right.** PSF Fitting of an OT candidate found on a WF CCD image (SBIG ST-8 CCD camera).

on HDD and catalogue (2125 fields on the sky, false detections eliminated by multiple images). This allows rapid real-time searches for OTs without need for a second comparison image.

The BART participates in various projects (e.g. BL Lac monitoring), every clear night service for GRB (HETE) alerts (but no GRB HETE trigger with known position in the visible and clear sky at night so far) since March 2001.

The further aspects include interdisciplinary project, as well as the University-Academy collaboration, a wide participation/teaching of students, and tests of innovative methods.

## CONCLUSION

There are 2 (and soon 3) new small aperture automated telescopes with attached wide-field cameras for rapid response to GRB satellite triggers with limiting magnitudes 18-19. The systems are also suitable to provide dense observations (photometry) of fading optical counterparts to GRBs, as well as of secondary science triggers of various physical nature such as X-ray stars, CVs, AGNs, QSOs, blazars, SNe, targets of opportunity, etc. as well as to provide simultaneous and quasisimultaneous optical observations for satellite observations and campaigns.

## ACKNOWLEDGEMENTS

We acknowledge the support provided by the Grant Academy of the Czech Republic (grant 205/99/0145), by the Ministry of Education of the Czech Republic (Project ES02) and by the ESA PRODEX (Contract 14527/00/NL).

## REFERENCES

1. Castro-Tirado, A.J. et al., in Gamma-Ray Bursts, edited by R.M. Kippen et al, AIP Conference Proceedings 526, Melville, New York, 2000, 260.

**FIGURE 4.** Example of the BART WF = INTEGRAL OMC TD image centered at the North America Nebula. The device provides numerous test images to test the various INTEGRAL OMC software packages.

# Analyses of GRBs on Astronomical Emulsions

R. Hudec[a], V. Hudcová[a], F. Krolupper[b], P. Kroll[c]

[a]*Astronomical Institute, Academy of Sciences of the Czech Republic, CZ-251 65 Ondrejov, Czech Rep.*
[b]*Dept. of Informatics, Faculty of Mathematics and Physics, Charles University, Prague, Czech Rep.*
[c]*Sonneberg Observatory, Sonneberg, Germany*

**Abstract.** The millions of archival astronomical plates located at various observatories represent a unique database for various scientific projects including the GRB analyses. They can easily provide thousands of exposures for any celestial position and reach monitoring intervals of more than one year of continuous monitoring. The very recent efforts to digitize the archival plates and the corresponding software development significantly facilitate the extraction of unique scientific data from archival records and related reductions and analyses. We discuss the ways in which this material can be used for GRB related studies and analyses, with examples of results obtained. We also address the possibility to search for additional GRB/collapsar events for the host galaxies/star forming regions where GRB have been observed using archival data and discuss the related constraints and scientific aspects.

## INTRODUCTION

The operation of HETE, BeppoSAX and RXTE satellites has provided a list of nearly 100 reliably and precisely positioned gamma ray bursts (GRB). The typical localization accuracy amounts to roughly 1 ... 20 armin, and is even essentially better if there have been X-ray, optical and/or radio afterglows found.

For a large fraction of these GRBs, low energy counterparts-afterglows have been found including optical (see [1] for a review). The optical afterglows (OAs) of GRBs have been found to peak at roughly 18-23 mag and then decreasing slowly according to a power law over a time period of weeks to months. In one case, the direct optical emission peaking at about mag 9 and lasting for about 1 min has been found [2].

These magnitudes are inside the sensitivity limits of numerous plate collections, hence the archival plate databases may be considered as a data source for optical GRB analyses.

## THE ARCHIVAL PLATES

There are nearly 3 millions astronomical archival plates located at different observatories. These plates represent a unique database for various scientific projects including the GRBs – afterglows analyses [3].

These archives can easily provide thousands of exposures for any celestial position, reaching monitoring intervals of up to few years of continuous monitoring – i.e. tens of thousands of hours [4].

The recent efforts to digitize the plates and the corresponding software development significantly facilitate the extraction of unique scientific data from archival records and related reductions and analyses.

## THE UDAPAC PROJECT

The recent UDAPAC Project represents the first effort to create an European Plate Centre with related facilities, staff, software and expertise to extract scientific information from archival astronomical plates [5]. The UDAPAC plate centre is located at the Royal Observatory in Brussels, Belgium, with many other scientists involved
(http://midasf.oma.be/~fido/ovid.html).

## THE SOFTWARE DEVELOPMENT

So far, the data recorded on archival plates were accessible only by special procedures. The recent wide digitisation of plate collections offers significantly easier access by computers. However, there is still a gap between the digitised archive and the scientific use. Special software is required to fill this gap.

We have developed new algorithms to access data on digitised plates and have tested these techniques in trial sets of digitised plates from the Sonneberg Observatory sky patrol archive. The new algorithm is based on the flood method. This method, applied to the digitised photographic plates, is able to reveal the star images for further analyses. This method has been tested on a set of digitised Sonneberg Sky Patrol plates with very promising results if compared with other methods. The flood method is based on a similar idea as the watershed method and is reasonably simple and quick.

## WHY TO USE ARCHIVAL PLATES FOR GRB-AFTERGLOWS ANALYSES

- to search for underlying GRB hosts and/or peculiar objects
- the search for possible activity (including recurrence)
- to provide reliable statistics of background events
- to search for orphan afterglows (with observable optical but not gamma ray emission)
- to search for OTs and OAs of GRBs upon their optical emission independently on satellites – and also in historical data

The suggested association of GRBs and star forming regions can account for activity recurrence due not by the recurrent physical model of the central engine, however by the apparently recurrent (but not identical) events in the same galaxy/star forming region.

## THE SENSITIVITY LIMITS OF PLATES

The limiting magnitudes depend on particular telescope/emulsion combination and typically amount to mag 12 – 17 but there are also plate collections going as deep as mag 23. This is sufficient to detect OT and OA emission related to GRBs. Example: The 18 000 archival plates taken by the UKSTU in Siding Springs, Australia, have typical limiting magnitudes of 19...23 and field of view of 6.4 x 6.4 deg$^2$ and are hence well suited to search for optical emissions related to GRBs including faint OAs.

**TABLE 1.** Examples of GRBs investigated on UKSTU plates

| GRB | OA yes/no | Number of plates | Optical activity |
|---|---|---|---|
| 970228 | y | 2 | n |
| 970402 | n | 13 | y[1)] |
| 970616 | n | 9 | n |
| 980109 | n | 8 | n |
| 980326 | y | 14 | n |
| 980425 | SN? | 25 | n |
| 980515 | ? | 27 | n |
| 981226 | ? | 7 | n |
| 990506 | n | 14 | n |
| 990510 | y | 6 | n |
| 990705 | y | 251 | y |
| 990712 | y | 15 | n |
| 990908 | ? | 54 | n |

Note: [1)] variable star BL Circini.

## THE EXAMPLES OF GRB PLATE ANALYSES

The study based on the use of UKSTU plates may serve as an example. We have analyzed altogether 700 high quality plates covering the positions of 15 BeppoSAX-RXTE precisely positioned GRBs in the UKSTU (United Kingdom Schmidt Telescope Unit) plate archive located at the Royal Observatory Edinburgh, UK. Their limiting magnitudes were in the range 19 ... 23, depending both on the observing conditions, exposure times as well as filter used. The Table 1 gives few examples.

Since the positional errors of the triggers studied were usually small, of order of 1 arcmin and/or below, the GRBs localizations were investigated both by the Zeiss blinkmicroscope as well as by the plate microscope. This proved to be more effective than the time-consuming scanning of the whole plates. Moreover, the false objects may be much effectively eliminated on original plates if compared with digitized files where the 3rd dimension (along the line of sight) is lost. The goal for the study was to search for any kind of optical activity (including possible burst recurrence and light variability of underlying objects/hosts) at the positions of GRB triggers at times before and after GRBs.

Although the recent results on GRBs do not seem to provide any kind of support for recurrence of the triggers, they still cannot rule out this completely. The

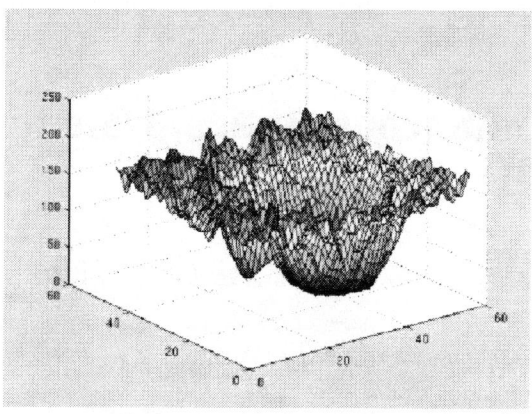 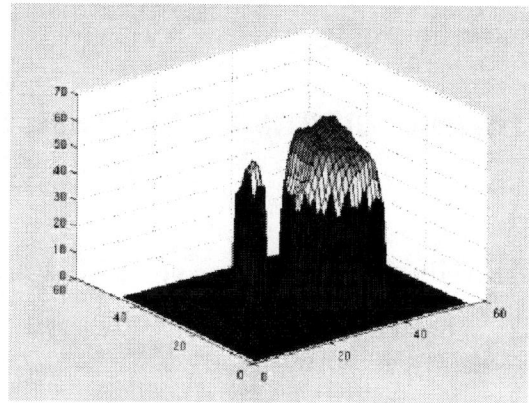

**FIGURE 1.** The digitized star images and their response to the flood method. Faint star with large response of 100 drops in maximum.

BATSE GRB catalog (4B) shows trigger overlaps consistent with statistical expectations for non-recurrent triggers but these studies are affected by large errors of events detected by BATSE and they are also unable to detect any kind of repetitions with time scales equal or exceeding the operational time of CGRO which was about 8 years at the time of the study.

The recent GRB results relate GRBs to faint and distant host galaxies and there are indications that GRBs are related to star-forming regions [6]. This scenario however do not fully exclude the possibility of events recurrence on a long time basis. The recurrent trigger would then simply represent an another event in the same galaxy.

It should be also noted that the recurrence and/or non-recurrence of the triggers could be crucial in understanding their physical nature which still - despite of recent progress in GRB investigations - remains hidden. This strengthens the value of searches for event repetitions since confirmation of recurrence could strictly support one group of theoretical models explaining GRBs and definitely eliminate other one. In star-forming scenario, the estimated apparent recurrence would provide unprecedented possibility to estimate the star forming rate.

## CONCLUSION

Despite of negative results found so far, this study provides valuable experience for using high quality deep optical archival data to search for OAs and OTs of GRBs in general i.e. independently on satellite experiments. The recent wide digitization od plate collections will allows further analyses to be carried out much more effectively.

## ACKNOWLEDGEMENTS

This study has been supported by the grant provided by the Grant Agency of the Czech Republic No. 205-99-0145, by the project KONTAKT ES002 provided by the Ministry of Education and Youth of the Czech Republic and also by the Academic Link between the Astronomical Institute Ondrejov and the University of Westminster, Harrow, UK provided by the British Council in Prague. We also acknowledge the support provided by the UKSTU staff at the ROE.

## REFERENCES

1. McNamara B. J. and Harrison T. E., Nature 396, 233-236, 1999.
2. Akerlof C.W. and McKay T.A., GCN Circular No. 205, 1999.
3. Hudec R., Acta Historica Astronomiae 6, 127-131, 1999a.
4. Hudec R., Acta Historica Astronomiae 6, 28-40, 1999b.
5. de Cuyper J.-P. et al., Astronomical Data Analysis Software and Systems X, ASP Conference Proceedings, Vol. 238. Edited by F.R. Harnden, Jr., Francis A. Primini, and Harry E. Payne. San Francisco: Astronomical Society of the Pacific, ISSN: 1080-7926, 2001, p. 125.
6. Djorgovski, S. G., Kulkarni, S. R., Bloom, J. S., Goodrich, R., Frail, D. A., Piro, L. and Palazzi, E., ApJ 508, L17-L20, 1998.
7. Bloom J. S., Djorgovski S. G., Kulkarni, S. R. and Frail, D. A., ApJ 507, L25-L28, 1998.
8. Hurley K. et al., IAU 6966,1998a.
9. Kouveliotou C. et al. IAUC 6944, 1998b.
10. Woods P. M. et al. ApJ 519, 1999, L139-L142.
11. Axelrod T., Mould J. and Schmidt B., GCN Circular No. 408, 1999.
12. Gandolfi G. et al., GCN Circular No. 406, 1999.

# Simultaneous and Quasisimultaneous Optical Data for GRBs

R. Hudec[a], I. Stoklasová[a,b], M. Jelínek[a], R. Smída[a,b], L. Svéda[a,b], P. Kroll[c]

[a]*Astronomical Institute, Academy of Sciences of the Czech Republic, CZ-251 65 Ondrejov, Czech Republic*
[b]*Astronomical Institute, Charles University, Prague, Czech Republic*
[c]*Sonneberg Observatory, Sonneberg, Germany*

**Abstract.** The photographic all-sky patrol network operated by the Ondrejov Observatory and based on 11 stations in the Czech Republic still represents one of very few possibilities to get real-time simultaneous optical data (no delay with respect to the GRB emission) and even pre-burst data for GRBs. The system can reach the limiting magnitude for real time optical data down to mag 12. Capabilities of the system as well as selected scientific results of related GRB analyses (including real time optical data) are presented and discussed. Analogous optical system but based on 4K x 7K CCD camera has been developed at the Sonneberg Observatory – the details are also presented and discussed.

## INTRODUCTION

The time 0 sec after the GRB onset can be never achieved by alert systems. Also, the alert systems will never achieve coverage for times before the GRB triggers. On the other hand, there are theoretical expectations that optical flashes may preceed GRBs [1]. Both these time coverages can be easily achieved by either photographic or CCD sky monitors.

## EN: PHOTOGRAPHIC ALL SKY MONITORING

- European Fireball Network, EN, 11 Stations in the Czech Republic

- Simultaneous optical data for various projects, complete sky monitoring (180 degrees diameter field of view)

- Optics: Fish-Eye Objective F-Distagon 3.5/30

- Detector: Planfilm FOMAPAN 400 ASA or 100 ASA (panchromatic emulsion) 90 x 120 mm, sky diameter 80 mm

- Typical exposure time: 3 hrs for guided cameras, whole night for fixed cameras

- 2 stations equipped with guided and fixed cameras

- 9 stations equipped with fixed cameras

- Sensitivity for brief 1 sec triggers 2-3 mag, for stars up to mag 12

- Response limited to the red light above 400 nm

## PREFERENCES

- Large sky coverage (full visible hemisphere)

- Large fraction of observation time: 2 400 to 6 000 sr.h for one station/year

- Multiplicity of data to eliminate background triggers easily

- Classification of detected triggers by parallax

- Simultaneous and pre-burst optical data (limits) for GRBs

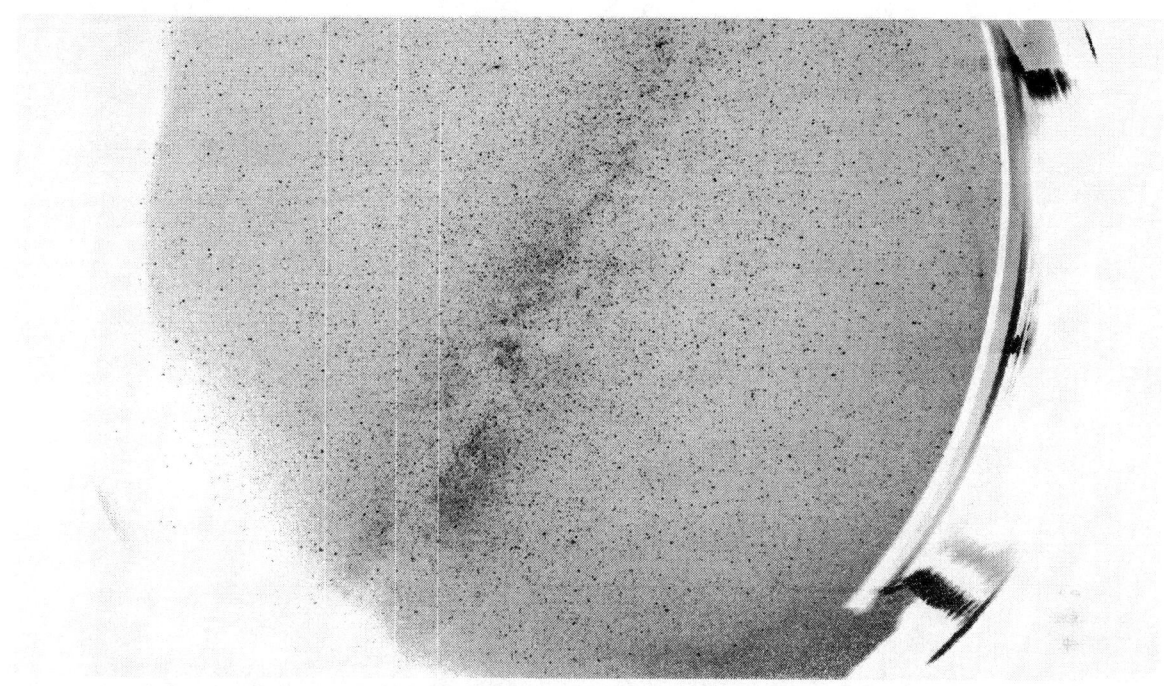

**FIGURE 1.** The image taken by the Sonneberg All Sky 7k x 4k CCD camera, 10 min exposure, 4 x 4 binning.

The network will be soon operated as a fully remote controlled network - without any human assistance. The access to the plate data will be soon facilitated by the new high quality flatbed CCD film scanner with optical resolution of 3 000 dpi connected to a powerful computer/graphic station.

## EXAMPLE OF RECENT RESULTS

The real time and pre-burst optical data have been analysed e.g. for GRB000926.

For this trigger which occurred at roughly 23 h 50 m, simultaneously taken pointed Ondrejov EN exposures covering the time interval 23 h 22 m – 03 h 03 m have been analysed. The trigger position close to the horizon was the cause for reduced sensitivity limit (magnitude 8 for stars). No optical (red) emission has been confirmed for this trigger. The very faint object at the trigger position is just 0.2 mag above the plate fog and hence cannot be confirmed as a real object, and probably represents a grain cumulating. The previous exposure was taken between 19 h 59 m and 23 h 21 m, i.e. ends nearly 29 minutes before the GRB trigger. No optical activity has been revealed down to limiting magnitude for stars 10 in this case.

## SUMMARY OF GRB RELATED RESULTS

No optical emission above mag 5 (1 sec duration assumed) or mag 13 (full exposure time) or Lg/Lo> 100...300 has been detected for a few GRBs. The faintest limit (320) exists for GRB 830313 [2], [3]. No optical emission above magnitudes 0...3 (1 sec duration assumed) or 4-11 (full exposure time) or Lg/Lo>0.1 ... 10 has been detected for many (~140) GRBs (mostly from the 4B catalog).

An Optical Transient (OT) was detected on the plate taken ~7 h after the GRB790929 inside its error box [4].

## THE SONNEBERG ALL-SKY CCD CAMERA

The all-sky camera based on analogous optical system (Zodiak 3.5/30 mm) as the Czech EN network but with a 7k x 4k CCD camera OES MM7k4k (Philips chip 7168x4096 pixels 12 microns earch, 16 bit, binning 4x4, readout time 60 sec) instead of film detector has been tested at the Sonneberg Observatory. During the tests, the camera has achieved a 9 mag sensitivity limit for 5 min exposure. The system is expected to be operated routinely since the beginning of 2002.

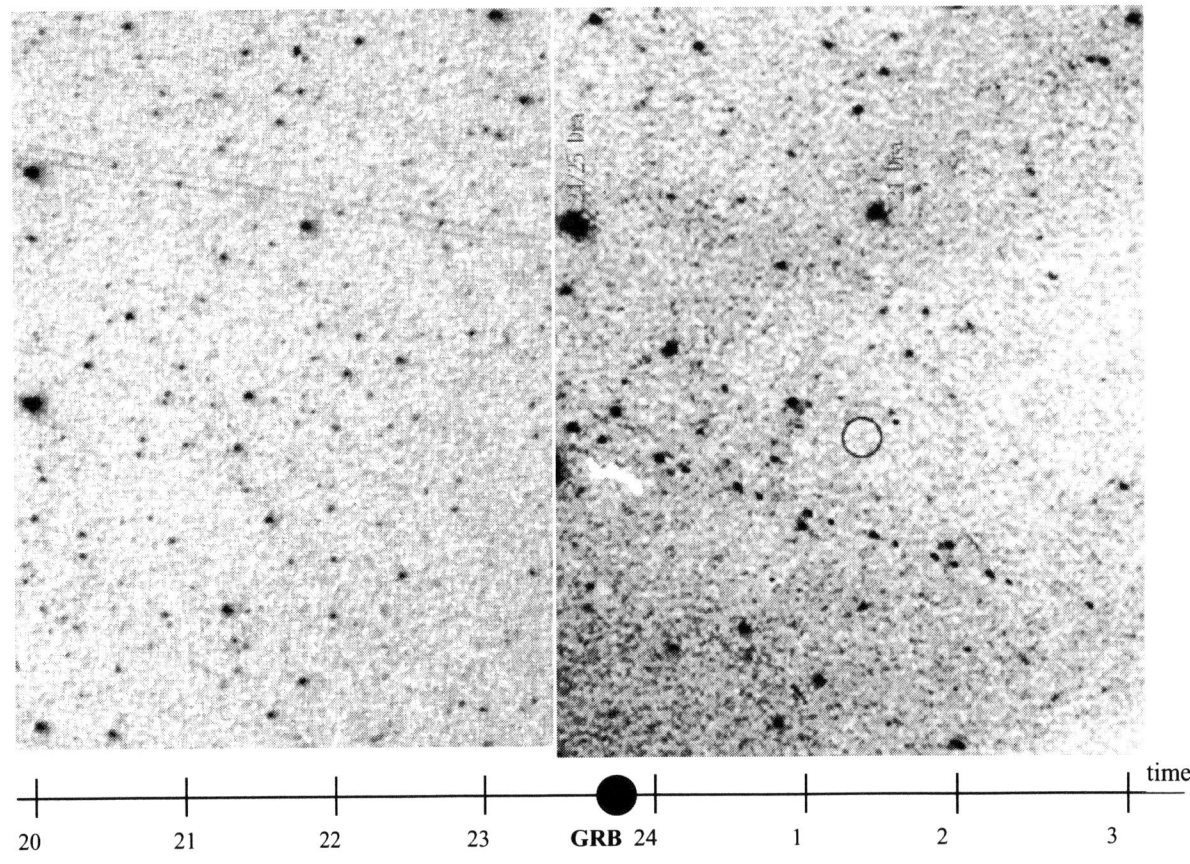

**FIGURE 2.** The area of the GRB000926 on the plates of the EN network. Left: pre-burst image (end of exposure 29 minutes before the GRB trigger), limiting magnitude 10. Right: simultaneous image, limiting magnitude 8. The position of the GRB is indicated by a circle. Both images contain airplane trails/lights.

## CONCLUSION

The all-sky sky monitors operated at the Ondrejov and Sonneberg Observatories are able to provide valuable optical real-time and pre-burst data for GRBs.

The recent analyses provide valuable limits for simultaneous optical emission of GRBs, for a 1 min duration, these limits are between mag 3 and 6. Better limits are expected to be achieved for triggers occurring at clear nights high above the local horizon.

The All Sky CCD Camera tested at the Sonneberg Observatory is expected to achieve limits of order of mag 9 for 1 min duration, hence capable to detect OTs analogous to the prompt optical emission of GRB990123.

## ACKNOWLEDGEMENT

The Czech GRB analyses on all-sky images are supported by the grant provided by the Grant Agency of the Academy of Sciences of the Czech Republic, grant IA3003206, by the grant 205/99/0145 provided by the Grant Agency of the Czech Republic and by the Project KONTAKT ES002 provided by the Ministry of Education and Youth of the Czech Republic.

## REFERENCES

1. Paczynski, B. et al., 2001, astro-ph/0108522.
2. Hudec, R., 1993, AA 270, 151.
3. Hudec, R., 1993, Astrophys. Lett., 28, 359.
1. Borovi_ka, J., et al., 1992, AA 258, 379.

# VHE Observations of GRB with Milagro

J.E. McEnery[†] for the Milagro Collaboration

[†] *Physics Department, University of Wisconsin, Madison WI 53706*

**Abstract.** The Milagro gamma-ray observatory employs a novel water Cherenkov detector to observe extended air showers produced by high energy particles impacting the earth's atmosphere. The detector consists of a large pond instrumented with an array of 723 photomultiplier tubes. The instrument operates 24 hours a day and continuously observes the entire overhead sky (∼2 sr). Because of its wide field of view and high duty cycle Milagro is uniquely capable of searching for gamma-ray bursts. Milagro can play a role in the extension of the measured spectrum of prompt and afterglow emission to VHE energies (>500 GeV). Detection of VHE counterparts would place powerful constraints on GRB mechanisms, and because of their attenuation from pair production on background IR fields, provide an additional estimate of the source redshift when optical lines cannot be detected. More than 20 GRB have occured within the field of view of Milagro since observations began in January 2000. We describe the results of a search for VHE counterparts to these GRB.

## INTRODUCTION

Some of the most important contributions to our understanding of gamma-ray bursts have come from observations of afterglows over a wide spectral range. This has allowed detailed modeling of gamma-ray burst afterglow properties both as a function of time and a function of wavelength. However, because of the very short duration, far less is known about the broadband spectra during the prompt phase of gamma-ray bursts. Almost all gamma-ray bursts are detected in the energy range between 20 keV and 1 MeV. A few gamma-ray bursts have been observed at energies above 100 MeV by EGRET indicating that the spectra of gamma-ray bursts extends out to at least 100s of MeV. However, it is unknown how far in energy gamma-ray burst spectra extend, or indeed, whether there may be a second higher energy component of emission, similar to that seen in other objects observed at TeV energies.

## THE MILAGRO DETECTOR

Milagro is a ground-based gamma-ray observatory sensitive at energies around a few TeV. With a large field of view of >2 steradians, high duty cycle (>90%) and large effective area, Milagro is ideally suited to making VHE observations of gamma-ray bursts. Milagro consists of 723 photomultiplier tubes (PMTs), submerged below the surface of a large, covered pool of water. The PMTs are placed on a grid with 2.8 m spacing in each of two layers, at 1.5 m and 7 m below the surface. A VHE gamma-ray interacts with the Earth's atmosphere to produce an extensive air-shower. The relativistic charged particles in the shower which reach ground level radiate Cherenkov light in the water. An event is recorded when ∼60 PMTs sense light within a ∼200ns of one another. The resulting trigger rate is around 1500 Hz. The relative arrival times of the shower front at photomultiplier tubes on the top layer are used to determine the origin (on the sky) of the particle or gamma ray initiating the shower to within ∼ 0.75°. The lower layer of photomultiplier tubes is used to identify and reject hadron-induced showers which dominate the data.

The expected performance of the Milagro detector has been simulated using CORSIKA [1] to model air-shower development in the atmosphere, and GEANT [2] to model the detector response. The sensitivity and energy threshold of Milagro are strong functions of zenith angle. This is because showers which originate closer to the horizon pass through more atmosphere and are attenuated. We have investigated the sensitivity and energy response of Milagro for a rudimentary binned analysis in direction with no background rejection due to gamma-hadron differentiation. Figure 1 shows the median energy for gamma-rays detected in Milagro as a function of zenith angle assuming an $E^{-2.4}$ differential photon spectrum. Also shown is the sensitivity to bursts on timescales of 1, 10 and 100 seconds as a function of zenith angle. Due to the rapidly falling sensitivity and in-

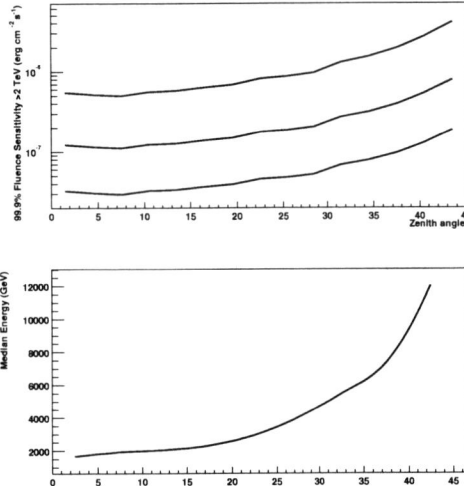

**FIGURE 1.** (top) The sensitivity of Milagro as a function of zenith angle for 1, 10 and 100 second bursts. (bottom) the median energy of gamma-rays which trigger Milagro as a function of zenith angle

creasing energy threshold we restrict our search for VHE counterparts to bursts that are within 45° of the zenith at Milagro. While the peak sensitivity of Milagro to a typical VHE spectrum of $E^{-2.4}$ is at energies around a few TeV, it is important to note that there is significant sensitivity at lower energies, and thus to spectra with cutoffs at lower energies. This is illustrated in Figure 2 which shows the distribution of energies for gamma-rays which trigger Milagro.

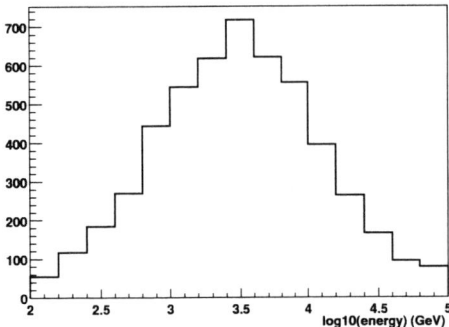

**FIGURE 2.** Distribution of energies of gamma-rays, from an $E^{-2.4}$ differential photon spectrum, which trigger Milagro

## THE GAMMA-RAY BURST SAMPLE

The very high data rate in Milagro makes it prohibitively expensive to archive all raw data. Instead the data are processed online and only the reconstructed arrival times

**TABLE 1.** Limits on VHE fluence (above 2 TeV) on 26 bursts which were within the field of view of Milagro.

|  | Dur(s) | Zenith ang. | 99% fluence limit (erg cm$^{-2}$) |
|---|---|---|---|
| GRB000113 | 370 | 20.8 | 6.89e-06 |
| GRB000212 | 8 | 2.21 | 1.12e-06 |
| GRB000226 | 10 | 31.5 | 2.39e-06 |
| GRB000301c | 14 | 37.5 | 1.9e-06 |
| GRB000302 | 120 | 31.9 | 2.3e-06 |
| GRB000317 | 550 | 6.39 | 1.6e-06 |
| GRB000330 | 0.2 | 30.0 | 8.5e-07 |
| GRB000331 | 55 | 38.3 | 6.6e-06 |
| GRB000408 | 2.5 | 31.1 | 2.8e-06 |
| GRB000508 | 30 | 34.1 | 3.0e-06 |
| GRB000615 | 10 | 39.0 | 8.9e-07 |
| GRB000630 | 20 | 33.2 | 1.5e-06 |
| GRB000727 | 10 | 40.8 | 2.6e-06 |
| GRB000730 | 7 | 19.2 | 4.1e-07 |
| GRB000926 | 25 | 15.9 | 1.2e-06 |
| GRB001017 | 10 | 42.1 | 2.1e-06 |
| GRB001018 | 31 | 31.8 | 1.4e-06 |
| GRB001019 | 10 | 19.5 | 8.45e-07 |
| GRB001105 | 30 | 8.6 | 9.5e-07 |
| GRB010104 | 2 | 19 | 3.2e-07 |
| GRB010115 | 5 | 41 | 2.9e-06 |
| GRB010220 | 150 | 26.9 | 1.489e-06 |
| GRB010613 | 150 | 24 | 2.9e-06 |
| GRB010921 | 12 | 10.3 | 5.9e-07 |
| GRB011130 | 30 | 33.6 | 1.7e-06 |
| GRB011212 | 80 | 33.0 | 3.3e-06 |

and directions of the showers are archived. The quality of the reduced data being stored will improve as analysis techniques are refined. It is advantageous to store the complete data for interesting regions of the sky and/or interesting periods of time. This allows more sensitive analyses to be carried out at a later date.

We respond to a GRB notification from the Gamma-ray burst Coordinated Network (GCN) by storing complete data on all events from 30 minutes before the GRB trigger to 2 hours after the trigger. Since Milagro began full operation in January 2000, there have been 26 GRB detected by satellite experiments within its field of view.

We have searched for an VHE excess from each of these bursts during the duration reported by BATSE, HETE, IPN or BeppoSAX. In cases where more than one duration was reported, we searched over the longest duration. For some bursts the position was not well known. In these cases we tiled the GRB error region with an array of 1.2 degree bins centered on a 0.1x0.1 degree grid. The bin with largest positive fluctuation from background was used to calculate a VHE upper limit or flux. No significant VHE excesses were found. In table 1 we list the 99% fluence upper limit above 2 TeV for each burst assuming a differential photon spectral index of $E^{-2.4}$. This energy was chosen as it is close to the median energy, so the quoted fluence limit is relatively insensitive to the assumed spectral index.

## GRB010921

TeV photons are absorbed by pair production with infrared background photons. The importance of this process depends on the energy of the TeV photon and the distance to the GRB. The limits on VHE emission presented in this paper, while interesting, do not constrain the emission spectrum of the bursts in a straightforward way. Without knowing the distance to the burst we cannot distinguish whether the upper limit is telling us that the emitted TeV flux was low, or if it has simply been absorbed in intergalactic space.

An interesting exception to this is GRB010921. This burst has a measured redshift of 0.45 [3]. The upper energy cutoff due to absorption on the IR background is uncertain, but is likely to be somewhere between about 100 GeV and 300 GeV [4]. We have crudely approximated the effect of IR absorption as a sharp cutoff at 150 GeV. In this case, we find a upper limit on the fluence at the 99% confidence level of $5.6 \times 10^{-5}$ erg cm$^{-2}$.

## CONCLUSIONS AND FUTURE PROSPECTS

In the two years since Milagro began full operation, observations have been made of 26 gamma-ray bursts. No evidence for a prompt VHE counterpart was found for any of these bursts. High energy gamma-rays are attenuated by interactions with the intergalactic infrared background. Thus the limits presented in this paper can constrain either the distance to the burst or the level of VHE flux. Our observations of GRB010921 represent, for the first time, a limit on the VHE flux which constrains emission from the GRB and not simply its distance from us.

The sensitivity of Milagro compares favorably to that of other instruments which make prompt observations of GRB. Figure 3 shows the sensitivity of Milagro (>1 TeV), GLAST (>100 MeV) and EGRET (>100 MeV). The data points show the distribution of fluence and duration seen by the BATSE instrument.

## ACKNOWLEDGMENTS

This work is supported by the Department of Energy Office of High Energy Physics, the National Science Foundation, the LDRD program at Los Alamos National Laboratory, the University of California, the Institute of Geophysics and Planetary Physics, the Research Cor-

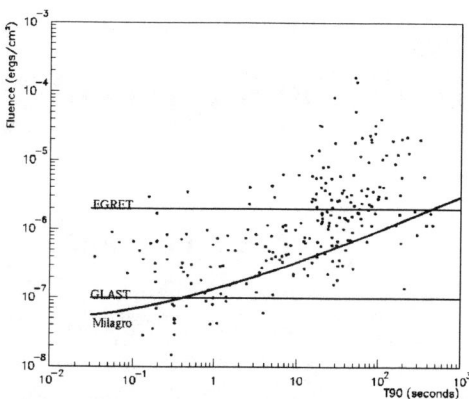

**FIGURE 3.** Sensitivity of Milagro, EGRET and GLAST superimposed on the distribution (black dots) of fluence and duration seen by BATSE.

poration, and the California Space Institute. We would also like to recognise the hard work of Scott Delay and Michael Schneider, without whom these data would not exist.

## REFERENCES

1. Heck, D., et al., *CORSIKA: A Monte Carlo Code to Simulate Extensive Air Showers*, FZKA 6019, Forschungszentrum Karlsruhe, 1998.
2. CERN-Application-Software-Group, *CERN W*, 5013, version 3.21, 1994.
3. Ricker, G., et al., GCN 1096, Tech. rep., GCN 1096 (2001).
4. Primack, J., *Frascati Physics Series* (2002), in press (astro-ph/0201119).

# Rapid Notification of TeV Transients with the Milagro Telescope

Miguel F. Morales[†] for the Milagro Collaboration

[†]*Santa Cruz Institute for Particle Physics, University of California, Santa Cruz, CA 95064*

**Abstract.** The Milagro telescope is a wide field of view TeV observatory that is ideally suited for the observation of transients in the 300 GeV-10 TeV band. Milagro has a very large field of view (about 2 sr), nearly continuous operation, and a fluence sensitivity that is comparable to satellite experiments. Milagro is currently performing real time searches for TeV transients, with the goal of prompt notification and multi-wavelength observation of any detected TeV transients.

## THE MILAGRO OBSERVATORY

The Milagro detector is centered around a large reservoir of water located high in the Jemez mountains of New Mexico[1]. The reservoir (affectionately know as "The Milagro Pond") is approximately 4800 $m^2$ in size and is located at an altitude of 2600 m (8600 ft.). This reservoir is covered with a light tight cover, and instrumented with 723 photo tubes arranged in two layers (see figure 1).

When a very high energy (VHE) photon enters Earth's atmosphere, it pair produces and starts a cascade of particles that travels as a relativistic pancake through the atmosphere. This relativistic pancake is known as an extensive air shower (EAS), and passes through the light tight cover and into the Milagro Pond. The relativistic particles in the EAS front exceed the speed of light in water and emit Cherenkov radiation, primarily in the blue and near ultra-violet (charged particles in the shower front immediately Cherenkov radiate, whereas secondary gamma rays pair produce in the water, and the daughter particles radiate). The Cherenkov radiation is then detected by the photo tubes. By careful timing of the detected Cherenkov light, the direction of the incident EAS front (and thus the original photon) can be reconstructed to 0.75° on average.

Milagro's use of Extensive Air Showers offers several compelling features. Because EAS penetrate through the atmosphere, observation can be continuous. Milagro observes day and night and is immune to the weather. Shower reconstruction depends on timing, not collimation, bestowing Milagro with a very wide field of view (∼2 sr). Additionally, TeV EAS fronts have large diameters, so the effective area of Milagro is considerably larger than the physical area of the pond for much of its energy range.

The cosmic ray background is a large problem for all EAS detectors, and our long duration signals are background dominated. There are characteristic differences between gamma ray and cosmic-ray initiated showers which can be exploited to reject a portion of the cosmic ray flux. We are currently using the bottom layer of tubes to reject cosmic ray showers. This is an active area of both hardware and analysis research (see the last section on future directions for the Milagro detector).

The water Cherenkov technique used by Milagro is unique among EAS detectors, and allows us to achieve sensitivity to below 300 GeV, good background rejection, and a small angular resolution. Milagro's unique abilities are enabling the first comprehensive search for TeV transients.

## REAL TIME TRANSIENT SEARCHES WITH MILAGRO

The study of gamma-ray bursts has been revolutionized by multi-wavelength observations. While afterglow observations at energies below a few keV tell us about the environment and development of the GRB fireball, observations (or lack thereof) of TeV emission can set strong constraints on the basic production mechanism of gamma-ray bursts. For example, if the observed spectral peak at BATSE energies is due to synchrotron radiation, one expects a prompt second peak in the TeV due to inverse Compton scattering[2]. The exact shape, timing, and amplitude of this second peak is highly model dependent, and can set strong constraints on the production mechanism of GRBs. It is this TeV emission peak that

**FIGURE 1.** This photograph shows the Milagro reservoir (80 m x 60 m) from under the light tight cover. Clearly visible are the photo tubes arranged on a 3 m grid. The upper layer of phototubes is under 1.5 m of water, while the second layer is under 7 m of water.

Milagro is uniquely able to observe.

Unfortunately, TeV photons are absorbed over cosmological distances by pair production with IR background photons. This absorption limits the distance to which Milagro can observe GRBs to a red shift significantly less than 1, though the exact cutoff depends on the cosmological model. Though this may become of powerful tool for measuring the IR background, it hampers our ability to see most GRBs. If we were limited to doing coincidence observations with satellites, there would be very few GRBs for which we could place direct limits on the TeV emission (especially with the current low rate of GRB detections). However, Milagro is a very sensitive detector in its own right, and can search the sky for TeV transients with very little sensitivity degradation (see figure 2). An independent search for transient TeV signals has the added benefit of not biasing the search with satellite selection criteria.

The fluence sensitivity of Milagro near 1 TeV is comparable to the best GRB satellites in the keV-MeV range (again see figure 2), and extends from below 300 GeV to above 10 TeV. For the past year, Milagro has been searching the sky for transients of 250 microseconds to 40 seconds duration. This effort has recently been augmented by a new analysis that is being used for the 40 second to 2 hour time region [3]. During the past year of operation, no GRBs of 250 microseconds to 40 seconds duration have been seen by Milagro (see figure 3), but this is not unexpected due to the limited distance to which Milagro is sensitive.

Both of these analyses are now running and allow Milagro to provide prompt GCN notification of any TeV

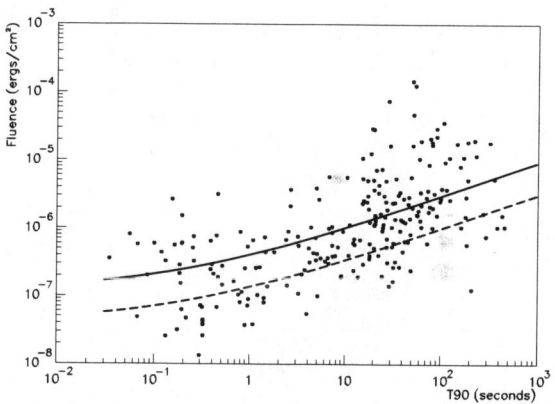

**FIGURE 2.** This plot shows a sample of BATSE bursts (block dots) plotted as X-ray fluence versus burst duration. The lines show the 5 sigma detection threshold of the Milagro detector for fluence at TeV energies. The dashed line is for an externally triggered GRB, while the solid line represents a blind search. Note that the trials penalty for a blind search does not make a large difference in the detection threshold, because a small increase in the number of signal photons leads to a large decrease in the Poisson probability. For example, doubling the flux of a 5 $\sigma$ detection changes it to a 10 $\sigma$ detection, or a factor of more than $10^{-16}$ in probability.

transients with a resolution of a few thenths of a degree. Any transients seen with Milagro will be ideal candidates for multi-wavelength observations because of their proximity (z<1), good localization, and the prospect of detailed observation of both the emission mechanism and the developing fireball.

## FUTURE DIRECTIONS

The Milagro Observatory is still under construction. The portion of the Milagro detector in the reservoir is complete – and is the basis for all of the data shown here – but is only the central piece of a larger detector. We are in the process of deploying ~170 outrigger detectors around the central reservoir in order to complete the Milagro telescope. The final Milagro detector will have significantly better sensitivity than the current partial version.

The outrigger detectors are cylindrical water filled cisterns approximately 2.4 m in diameter and 1 m deep. Each cistern is lined with reflective Tyvek, equipped with a single photo tube, and placed on a 15 m grid around the central Milagro reservoir.

These outriggers significantly increase the size of the Milagro detector. The outriggers allow us to contain the full size of most EAS, giving us sparse sampling across the entire shower front and detailed timing and calorimetry on the portion of the EAS incident on the central Milagro pond. This combination of large scale sparse information and detailed information on one area will improve our angular resolution, our energy resolution, and allow for more sophisticated background rejection techniques.

## ACKNOWLEDGMENTS

This research was supported in part by the National Science Foundation (Grant Numbers PHY-9722617, -9901496, -0070927, -0070933, -0070968, -0096256), the U. S. Department of Energy Office of High Energy Physics, Los Alamos National Laboratory, the LDRD program at LANL, the University of California, the Institute of Geophysics and Planetary Physics, The Research Corporation, and the California Space Institute. The author would also like to acknowledge the support of the NASA Graduate Student Research Fellowship and the Mellon Foundation.

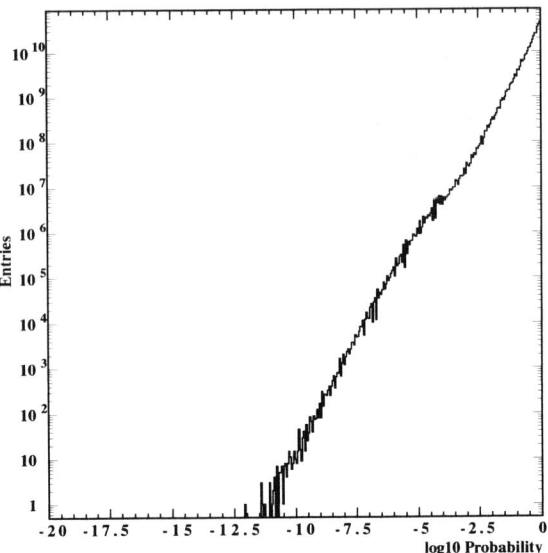

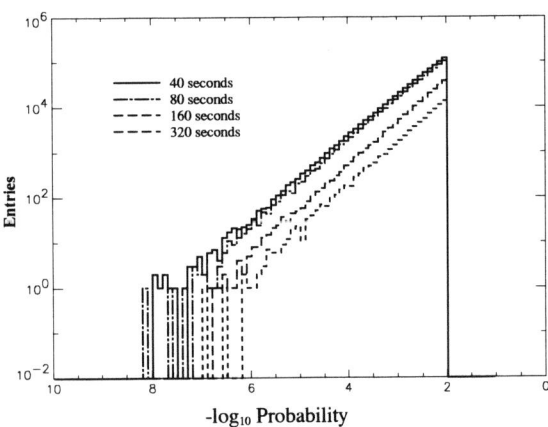

**FIGURE 3.** In Milagro, the real-time transient search is divided between two independent analyses. The binned analysis covers the time scale of $250\,\mu s - 40$ s, and has been running for one year. It has recently been joined by the weighted analysis (see [3]), which covers the $40$ s $- 2$ h region. The figures above show examples of the probability distributions observed by the two analyses. These distributions are consistent with the expected background fluctuation distributions; any signals would appear as isolated points of very low probability. The top figure shows the observed probability distribution for the binned analysis on a time scale of 10 seconds, for 200 days of Milagro data. The kink in the distribution at $10^{-4}$ is due to a transition between sparse and fine sampling by the search algorithm. The lower figure shows the observed probability distributions for 4 separate time scales of the weighted analysis, collected over 1 day. Events with a probability greater than $10^{-2}$ are not recorded. No TeV transients have been observed, but both analyses are currently running and will provide prompt notification of any signals to the GRB community.

## REFERENCES

1. The Milagro Collaboration, *Status of the Milagro Gamma Ray Observatory*, Hamburg: 27th International Cosmic Ray Conference, 2001, astro-ph/0110513.
2. Dermer, C.D., Böttcher, M., Chiang, J., *ApJ*, **537**, 255-260 (1999).
3. Miguel F. Morales for the Milagro Collaboration, "Discovery Mode Search Techniques For Gamma-Ray Telescopes," *These Proceedings*, Paper 14.11

# Discovery Mode Search Techniques For Gamma-Ray Telescopes

Miguel F. Morales[†] for the Milagro Collaboration

[†]*Santa Cruz Institute for Particle Physics, University of California, Santa Cruz, CA 95064*

**Abstract.** Like many wide field gamma-ray telescopes, the Milagro observatory has a highly variable point spread function (PSF), dependent upon the characteristics of the incoming photon. Because of the large variations in the PSF, a binned transient search technique is not optimal, and maximum likelihood can be computationally unfeasible. I have expanded upon the Gaussian weighting technique [1] to develop a transient search method that is sensitive, model independent, and computationally tractable. This method is currently being used to look for TeV transients from 40 seconds – 2 hours duration with the Milagro detector.

## THE PROBLEM

My goal in developing the weighted analysis technique was to perform a near optimal discovery mode search for point sources and transients in the Milagro data set. Milagro, like many wide field gamma-ray telescopes, has a highly variable resolution that depends on the characteristics of the individual photon. In an optimal bin analysis, one looks at the PSF of the detector and chooses a bin size that maximizes the expected signal to noise. These analyses are fast and work very well for detectors with a set PSF of near Gaussian shape. However, in wide field gamma-ray observatories the PSF can vary by more than an order of magnitude from one photon to the next. In binned analyses all photons are treated as equal, and the quality of an event is ignored. This can seriously degrade the significance of a signal and the flux limit of a detector.

The maximum likelihood technique is able to use the photon by photon PSF information, and is the standard tool of choice for dealing with variable PSF detectors. Unfortunately, maximum likelihood can be difficult to implement in a way that it is both computationally efficient and model independent. Because we don't know what the characteristics of a TeV transient signal will be, we must also make sure that our discovery mode search is not biased by an assumed signal type. The weighted analysis technique uses all of the available information, is computationally fast, and model independent.

## THE METHOD

For a gamma-ray telescope every event has two associated probabilities, the probability that it was a gamma ray (as opposed to a cosmic ray - $P(\gamma)$), and the PSF, or probability that the initial photon came from the direction determined by the reconstruction algorithm. Both of these probabilities can vary considerably from one photon to the next, and the distributions are determined by the characteristics of the detector (for an example, see the next section where we characterize the Milagro detector).

We can make a map of the sky that exactly represents our knowledge by placing at each point the probability that a photon came from that location. This is equivalent to putting the product $(dPSF/d\Omega)P(\gamma)$ onto the sky map. Our sky map is then the probability density map and represents our complete knowledge of the dataset.

So at any given point, the cumulative photon probability density (or weight) is:

$$w = \sum_{i}^{all\,events} \frac{dPSF_i(r_i)}{d\Omega} P_i(\gamma)$$

Where $i$ indexes all of the events, and $r_i$ is the angular distance of the reconstructed event direction from the point in question. In a discovery mode search, we want to determine the probability that a given weight is a fluctuation of the background. There are two independent effects which contribute to variations in the weight. First there are fluctuations in the number of events observed (and thus used in the sum) compared to the expected number of events. Second, there are fluctuations in the weight due to unusually close events (PSF term), or from

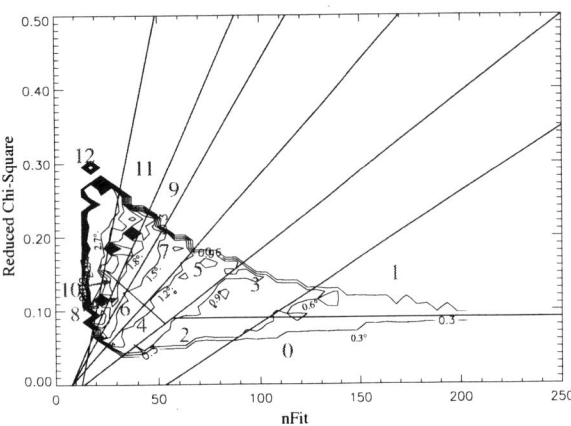

**FIGURE 1.** Contour plot of the PSF width as a function of the fit variables nFit and reduced chi-square. Overplotted are the 13 regions of similar PSF.

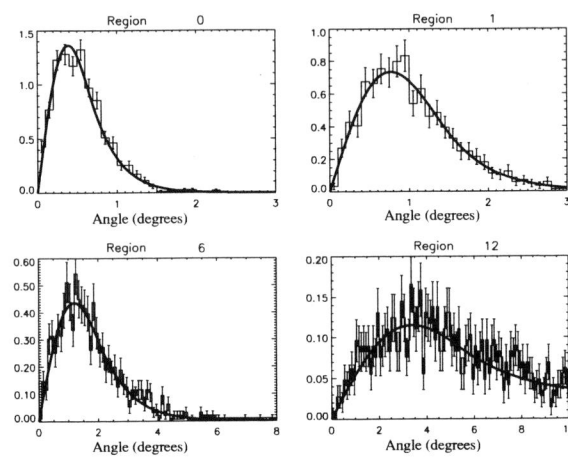

**FIGURE 2.** The radial PSFs for 4 of the 13 regions indicated in figure 1, with the associated fits.

unusually gamma-like events (gamma term). The effects of weight variations and number variations can be separated by looking independently at the fluctuations in the average weight and the number of events observed. The question then becomes, given the distribution of average weights and expected number of events, what is the probability of observing a particular average weight and number of events? Mathematically this can be written as:

$$P(w_{avg}, N_{obs} | \ P(w_{avg}|N_{obs}), N_{exp}) =$$
$$P(w_{avg} \ | \ P(w_{avg}|N_{obs})) \ P(N_{obs}|N_{exp})$$

The first term on the right is just the probability of getting a certain average weight (easily measured from our background), and the second term is straight Poisson statistics. Given the weight at a point, and the measured distribution of weights seen in the background, the probability that a given weight comes from the background distribution can be easily calculated.

For a full sky search, the sky map only needs to be sparsely sampled. The smallest PSF sets the spatial scale for variations, so sampling on a scale smaller than the smallest PSF retains all of the information. This spacing turns out to be only a little smaller than the bin size in an optimal binned analysis. Additionally, because we are sampling at points, tiling problems associated with binning a spherical sky never occur and separate sky maps can be directly summed. This can be a significant calculational advantage when searching over many time scales.

## MILAGRO DETECTOR CHARACTERISTICS

In order to use the weighted analysis technique (or maximum likelihood), the characteristics of the detector must be understood. For Milagro, the PSF correlates well with the reconstruction variables nFit and reduced chi-square. The Milagro reconstruction algorithm performs an iterative chi-square fit where tubes with times far from the fit are removed or added in each iteration. nFit is the number of tubes used in the final fit, and the reduced chi-square is the true reduced chi-square times $\sim 0.1$ (for historical reasons). Figure 1 is a contour plot of the PSF width as a function of nFit and reduced chi-square. In this work, this space is divided into the 13 regions shown in the figure, and the PSF of the detector is found for each of these regions. Events are categorized by which region they fall into and the corresponding PSF is used. Figure 2 shows the Monte Carlo distributions for gamma rays in 4 selected regions, and the corresponding fits. In order to avoid noise due to the low statistics of our Monte Carlo data set, these fits are used to determine the PSF used in the analysis.

The probability that an event was initiated by a gamma ray can be determined in a similar manner. The Milagro background rejection is currently based on the "compactness" parameter. Compactness characterizes the distribution of light in the lower layer of phototubes, and shows signficantly different behavior for gamma and proton initiated showers[2]. For the weighted analysis technique we want to determine the probability that a given shower was initiated by a gamma ray. For each region, we can use Monte Carlo simulations to determine the compactness distributions for both protons and gamma rays. Figure 3 shows the compactness distributions and gamma

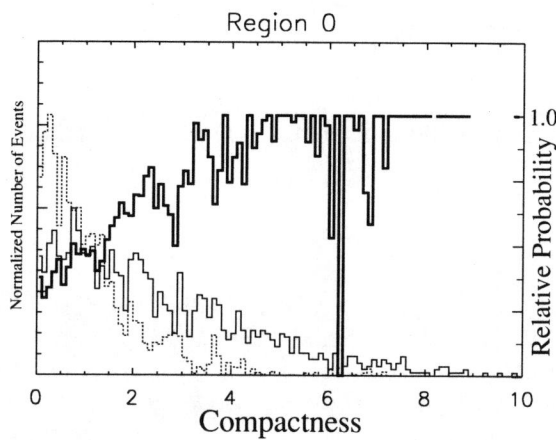

**FIGURE 3.** This figure shows the compactness histograms and gamma probability distributions for an example region. The compactness histograms for gamma and proton initiated showers are shown by the thin solid and dashed lines, respectively. The solid thick line shows the fraction of events which are gamma rays as a function of the compactness parameter

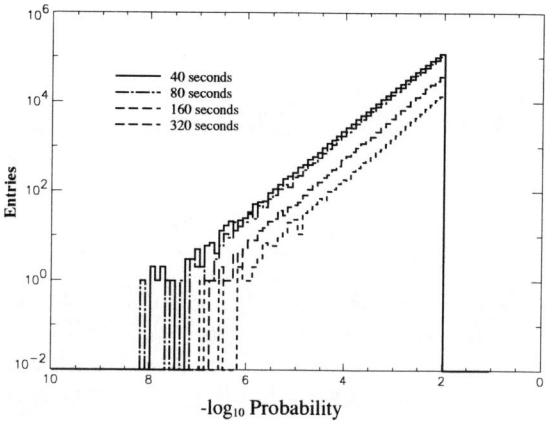

**FIGURE 4.** Histograms of the observed probability distributions for one day of TeV transient searches at 4 different time scales. These distributions are consistent with the expected background fluctuation distributions; any signals would appear as isolated points of very low probability. The weighted analysis method is currently being used for real time transient searches with the Milagro observatory.

probability for a representative region. The gamma probability distribution is fit to avoid noise from the low statistics of our Monte Carlo data set, and for computational reasons we currently use only showers with a gamma probability greater than 0.5.

## IMPLEMENTATION

The weighted analysis technique is amenable to a number of approximations that can significantly speed up the analysis. First, the PSF can be truncated to some set distance; for Milagro 4 degrees is the largest radius used. The current implementation also does not calculate the distance from the photon to each point, but instead uses a table of approximate PSF values that are copied onto the sky map. Separate signal and background maps can be directly added to form new maps, greatly increasing the speed of analyzing additional time scales. On current hardware (dual ~1Ghz Pentium 4, standard memory) the current algorithm can process ~2000 events/second while analyzing 9 time scales between 40 seconds and 2 hours.

Recently, this technique has been used to search through the Milagro dataset in real time for transients of 40 seconds to 2 hours duration. Example background distributions for these searches are shown in figure 4. No significant burst has been detected to date, but rapid notification of any observed TeV transients will be distributed to the GRB community.

## ACKNOWLEDGMENTS

This research was supported in part by the National Science Foundation (Grant Numbers PHY-9722617, -9901496, -0070927, -0070933, -0070968, -0096256), the U. S. Department of Energy Office of High Energy Physics, Los Alamos National Laboratory, the LDRD program at LANL, the University of California, the Institute of Geophysics and Planetary Physics, The Research Corporation, and the California Space Institute. The author would also like to acknowledge the support of the NASA Graduate Student Research Fellowship and the Mellon Foundation.

## REFERENCES

1. Woodhams, M.D., *Scintillator Data Analysis for the JANZOS Array*, M.A. Thesis, University of Auckland, 1989, pp. 33-42.
2. Sinnis, C., for the Milagro Collaboration, "Background Rejection in the Milagro Gamma Ray Observatory," Proceedings 27th ICRC, Hamburg 2001, **OG187**

# Optical Transient Monitor (OTM) for BOOTES Project

P. Páta[1], M. Bernas[1], A.J. Castro-Tirado[2,3], R. Hudec[4]

[1] *Czech Technical University, Faculty of Electronic Engineering, Dep. of Radioelectronics, Prague, Czech Republic*
[2] *Laboratorio de Astrofísica Espacial y Física Fundamental (INTA), Villafranca del Castillo, Madrid, Spain*
[3] *Instituto de Astrofísica de Andalucía (CSIC), Granada, Spain*
[4] *Astronomical Institute, Academy of Sciences of the Czech Republic, Ondrejov, Czech Republic*

**Abstract.** The Optical Transient Monitor (OTM) is a software for control of three wide and ultra-wide filed cameras of BOOTES (Burst Observer and Optical Transient Exploring System) station. The OTM is a PC based and it is powerful tool for taking images from two SBIG CCD cameras in same time or from one camera only. The control program for BOOTES cameras is Windows 98 or MSDOS based. Now the version for Windows 2000 is prepared. There are five main supported modes of work. The OTM program could control cameras and evaluate image data without human interaction.

## INTRODUCTION

The project BOOTES (Burst Observer and Optical Transient Exploring System) [4] is a system for sky monitoring and optical transient of GRB searching. The project has two parts, BOOTES-1 and BOOTES-2. First one is located in south Spain (Mazagón) and is operating since July 1998. Second station is in eastern part of south Spain (La Mayora). Both stations are in full operation July 2001.

The BOOTES stations has three MEADE based telescopes and wide, ultra-wide CCD cameras. Short Nikkor lenses are attached to SBIG CCD cameras – ST-7, ST-8 and newer version ST-9. Cameras are controlled with OTM (Optical Transient Monitor) software. This program has been especially developed for BOOTES project.

## OTM DESIGN

Presented software Optical Transient Monitor (OTM) is a controlling program for the BOOTES project [3,4]. The program controls the wide field cameras ST8 (Santa Barbara Instruments Group) for continuous monitoring of the same fields of view as BeppoSAX, HETE-2 or INTEGRAL. There are two versions of OTM. The older is MSDOS based program and it requires the acceleration board Photomate 20 with two signal processors TMS320 C40 (Texas Instruments) [1]. This version has been in operation at BOOTES-1 station (El Aeronosillo, INTA, Spain) since July 1998. The second version, which is designed as a full 32 bit application for W95/W98, has been in operation for one year (since June 2000). There acceleration board Photomate 20 is recommended but not required. Detail hardware requirements of 32 bit version of OTM program are shown in Table 1.

## COMMUNICATION MODES

The OTM system supports both the manual and full automatic regime (i.e. without interaction with operator). Each step of the image processing from data acquisition to its evaluation in the manual mode can be managed separately according to user's requirements. The second mode allows setting up the system so that the cameras would start and finish the data acquisition automatically every night. The real time image processing can be carried out in the MSDOS version only. The important information are available during observation in a form of table containing temperatures of CCDs and air, system state, log file, etc. The cameras can be controlled separately or synchronously. There are two algorithms of image capturing. The main difference between them is the fact, that the both cameras work parallel despite the second algorithm, where one camera takes exposures while the second is read out at the same time.

**Table 1** Hardware requirements of 32 bits version OTM program.

| System requirements |
| --- |
| MS Windows 95, 98, ME |
| Version for MS Windows NT or 2000 is prepared |
| PC Pentium 70 MHZ or better |
| 64 MB RAM memory |
| One extended parallel card (for two cameras mode) |
| SBIG CCD cameras ST-8 compatible |
| SVGA resolution |

## IMAGE PROCESSING FUNCTIONS

There are implemented many useful functions for evaluation image data in OTM program. The Photomate 20 card is used in MSDOS version as an accelerated image processing board independence on main PC.

Supported functions are:

1. Automatic compensation of dark frame and flat fielding.

2. Polynom fit of the background value using histogram, iteration method or least square fit.

3. Searching for objects - elimination of optic's defects, CCD's and false objects.

4. Building of photometric table of found stars and others objects.

5. Comparison of the new table with previous tables from the same camera. Construction of light curves for each object in the image.

6. Pseudocorrelation method - finding of transformation matrix, translation vector.

7. Differential photometry - automatic or manual selection of standard stars.

8. Searching for new objects based on change of brightness.

9. Comparison of actual results from both cameras - elimination of suspicion and false objects.

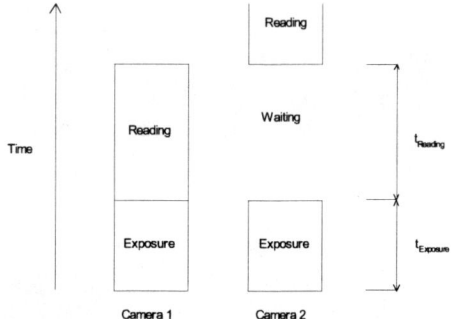

**Figure 1** Serial mode of data acquisition.

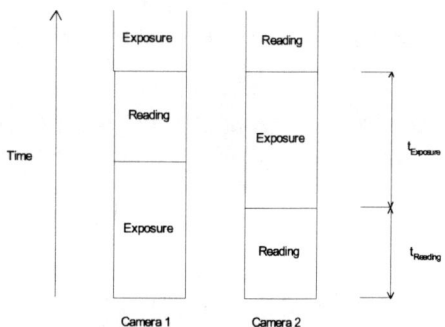

**Figure 2** Full parallel mode of data acquisition.

## STATUS OF BOOTES OTM MONITOR

OTM program monitors many important values of BOOTES status during observation without human interaction:
1. Status of precooling phase of CCDs.
2. UT time during each observation step.
3. CCD cameras status – temperature CCDs, Air, power of cooling, TCP-IP and SAMBA connection.
4. Description of OTM cycle phase.
5. Telescope position – taken from TCM (Telescope Control Monitor) via TCP-IP.
6. PC clock synchronizing with timer computer with GSM or TCM.
7. Recorded status to OTM form and OTM.LOG file in image directory.
8. Saving images in FITS, DAT, SBIG and compressed format.

## ACKNOWLEDGMENTS

This work has been supported by the research program No. **J04/98:212300014** "Research in the Area of Information Technologies and Communications" of the Czech Technical University in Prague. A part of this research work has been partially supported by the grant IG CTU 02111113 "The Scientific Image Data Optical Processing" of the Czech Technical University in Prague.

## REFERENCES

1. Bernas M., Páta P., Hudec R., Rezek T., Castro-Tirado A., "Optical Transient Monitor" In: Gamma Ray Bursts, *AIP Conference Proceedings,* New York: American Institute of Physics, **pp. 864-868**, (1998).

2. Páta P., "Formulation of Basic Criterions for Real-time Recognition of Some Objects in Astronomical Images", Poster 97, **EEC 25**, FEE CVTU in Prague, (1997).

3. Castro-Tirado, A.J. et al., "The Burst Observer and Transient Exploring System (BOOTES)" *Third Huntsville Symposium on Gamma-Ray Bursts*, ed: Kouveliotou, C. et al., AIP Conf. Proc. 384, p. 814 (1996).

4. Castro-Tirado, A.J. et al., "The Burst Observer and Optical Transient Exploring System (BOOTES)" *Astronomy and Astrophysics Suppl. Ser.,* 138, 583 (1999).

5. Castro-Tirado, A.J. et al., "First results of the Burst Observer and Optical Transient Exploring System (BOOTES-1)" *The Fifth Hunstville Gamma-ray Burst Symposium*, eds: Kippen, M. et al., AIP 526, p. 260 (2000).

# Rapid Identification of Optical Afterglows: Bright Prospects

Paul A. Price*, Brian P. Schmidt* and Tim S. Axelrod*

*Research School of Astronomy and Astrophysics, Australian National University, Australia*

**Abstract.** We have developed an automated system for the rapid observation and identification of gamma-ray burst (GRB) afterglows. We expect to observe GRB localisations from the High Energy Transient Explorer within 2 minutes of the GRB, and automatically identify an optical transient within 10 minutes. Such early observations are expected to yield new information on the GRB phenomenon, constrain possible progenitor models and allow rapid spectroscopic follow-up.

## INTRODUCTION

The optical afterglows of GRBs are usually identified at $R \sim 18$–$20$ mag, 3 hours or more after the burst. Projecting their observed decay back in time, we expect that they should be $R \sim 14$–$15$ mag, 5 to 6 minutes after the burst. At such bright magnitudes, they will be easily identified, allowing study of the early decay, and enabling spectroscopy (currently done with 8–10 metre telescopes) by 2 metre class telescopes.

## ROBOTIC TELESCOPE

The automation of the Great Melbourne Telescope (GMT) for the Southern Edgeworth Kuiper-belt Survey creates a powerful instrument for observations of GRB afterglows. We have taken advantage of this ability, and initiated an Australian GRB follow-up program. The GMT is connected to the GRB Coordinates Network through the internet, and receives GRB localistions within seconds of detection by by the High Energy Transient Explorer (HETE-2). We estimate that it will be possible to observe GRB error-circles less than 2 minutes (slew-time limited) from detection.

## AUTOMATIC IDENTIFICATION

Once observations have been made of a GRB error box, an automatic identification algorithm processes the observations, comparing them with the Digitised Sky Survey. Sources present on both the red and blue GMT images but not on the DSS are identified as candidates for the optical afterglow and listed on a web page for human inspection within 10 minutes.

After a second epoch observation, the program automatically subtracts the two epochs. Combination of the blue and red subtracted images provides an image clean of cosmic rays and other artifacts from which variable objects can be readily identified. These objects are also displayed on the web page of candidates, along with the coordinates and an estimated brightness.

## PROSPECTS

Regular identification of afterglows at early times (minutes after the GRB) should invigorate the field of GRBs, enabling a new set of observational tests for these enigmatic objects and providing the ability to better use them as cosmological probes. With the dual-beam MACHO imager on the GMT, we will be able to determine the spectral slope in the optical and so study the early afterglow in unprecedented detail, and for the first time distinguish between evolution in an homogenous ISM and a wind-stratified medium by observing the evolution of the cooling frequency and the sychrotron peak. From these, we should be able to infer the progenitor population from the properties of the circum-burst medium. Further, identification of a bright afterglow will enable high S/N spectra of high-redshift objects to be obtained from 2 metre class telescopes.

Using the robotic GMT, we have already identified the optical afterglows of two GRBs (000911 and 001007), observed at 1.4 and 3.5 days after the GRB, respectively. The afterglow of GRB 000911 was also observed with the VLA, and from broad-band modelling of this multi-wavelength data [3], it appears to be characterised by an hard electron energy distribution, $p \sim 1.5$. Further, the identification of this afterglow enabled stringent limits on the presence of an underlying supernova to be made [2].

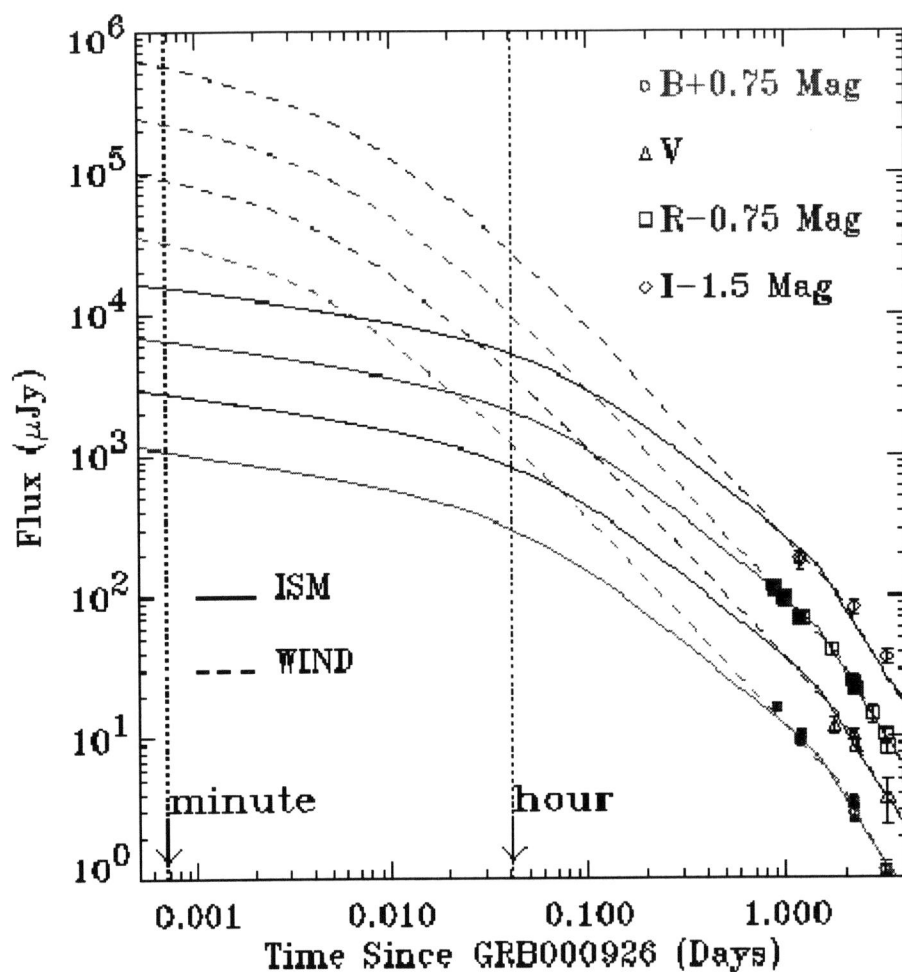

**FIGURE 1.** Predicted early-time light curves of GRB000926, based on broad-band modelling (Harrison et al. [1], courtesy S. Yost). Early optical observations would distinguish for the first time between the homogeous ISM and wind environments.

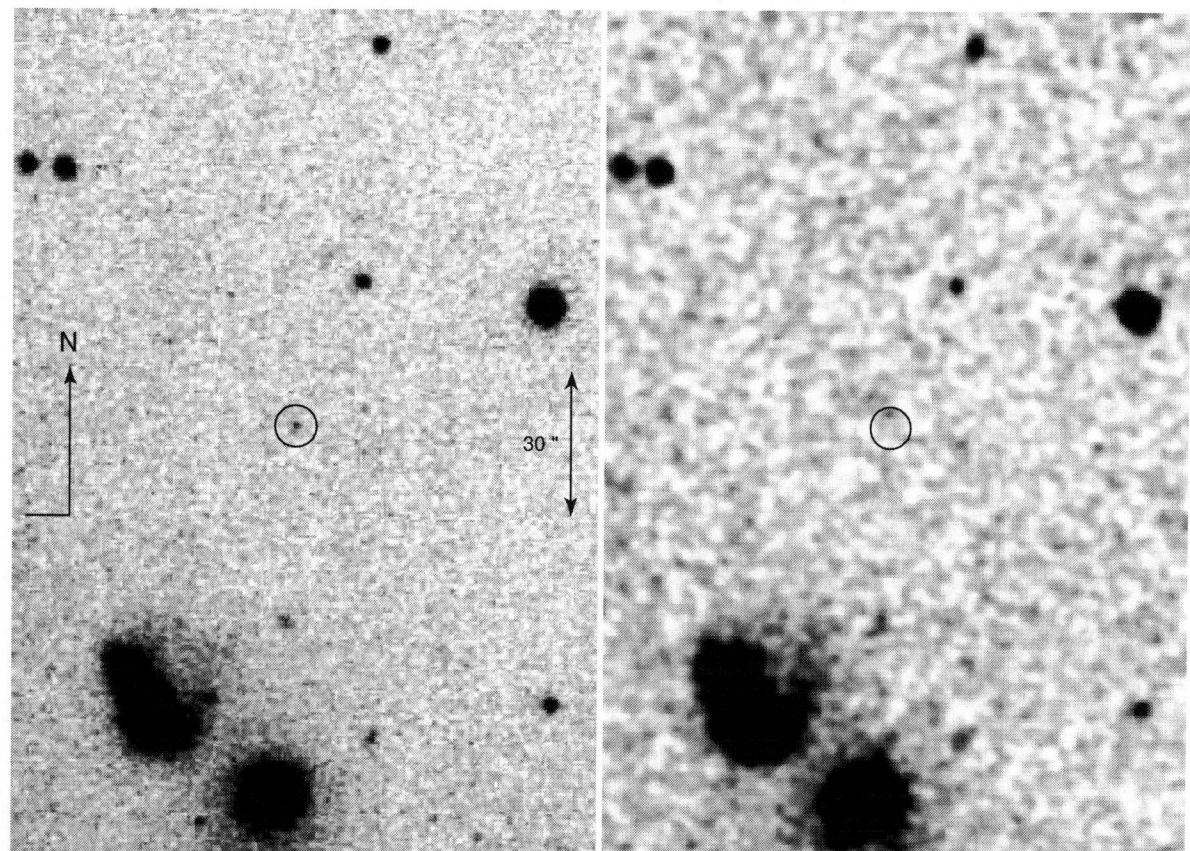

**FIGURE 2.** Identification of the optical afterglow of GRB 000911. Left: $R_M$ image from the 50-inch GMT at MSO. Right: Corresponding image from the DPOSS $F$-plate. The location of the optical afterglow is circled. From Price et al. [3].

## ACKNOWLEDGEMENTS

We thank Rachel Moody and Jon Smillie for their part in the automation of the Great Melbourne Telescope.

## REFERENCES

1. Harrison, F.A., et al., 2001, ApJ, 559, 123.
2. Lazzati, D., et al., 2001, A&A, 378, 996.
3. Price, P.A., et al., 2001, ApJ (submitted), astro-ph/0110315.

# Secondary Science with ROTSE Data

J. Štrobl[a,b], R. Hudec[a], M. Jelínek[a], V. Šimon[a], F. Hroch[a,c], C. Akerlof[d], ROTSE team[d]

[a]*Astronomical Institute Academy of sciences of the Czech Republic, CZ-251 65 Ondřejov, Czech Republic*
[b]*Astronomical Institute, Charles University, Prague, Czech Republic*
[c]*Masaryk University, Brno, Czech Republic*
[d]*University of Michigan, Ann Arbor, Michigan, USA*

Abstract. The ROTSE optical GRB follow-up instrument offers an excellent possibility for a secondary science with the data obtained within the sky monitoring. We present and discuss the results of a project of analysing two selected ROTSE monitoring fields with the goal to study the long-term behaviour of the objects located inside. The method developed and tested can be applied in a general way to study light changes of astrophysical objects of various types within the limiting magnitude of the ROTSE device.

## INTRODUCTION

The WF (Wide-Field) CCD systems are very commonly used in recent automated observing experiments like ROTSE, LOTIS, BOOTES, BART, TAROT etc. These WF projects come with very special and specific problems, which embarrass the reduction process and following scientific research. This study attempts to find solution of these problems and find nearly comfortable automatic reduction process and test this process on real sky monitoring data from project ROTSE (Akerlof et al. 2000ab, McKay et al. 2000). The main goal of the study was to propose and to test procedures allowing to use the ROTSE monitor data for secondary science such as study of variable astrophysical objects by generating and analyzing their light curves.

## WIDE-FIELD CCD IMAGING

The WF imaging has especially the following problems:
- difficult manipulation with data (one CCD picture has a big size)
- overlapping of stars (high areal density of stars)
- computing speed (really many objects in large FOV)
- problems with calibration (non constant extinction in large FOV)

## THE PROCEDURE TESTED

For the basic reduction the significantly modified software pack "munipack" (http://www.ian.cz/munipack/) by F. Hroch was used. Most of these modifications should occur in the next version of this software. The other phases of reduction process are realised by a set of quite simple utilities programmed in C. Everything is connected with the small set of shell scripts, which make the use of the whole package easy and comfortable. The whole process can be divided into following steps: (1) The basic (instrumental) calibration (application of dark frame, flat field), (2) The stars identification and photometry process, (3) Matching founded stars with catalogue, (4) Photometric calibration.

First three parts are performed with adapted utilities from package "munipack". The last one (photometric calibration) is based on the calibration of instrumental magnitude regarding to the UBV system by fitting of magnitudes of constant stars to catalogue values.

The reduction process is already functional, though it is still in development.

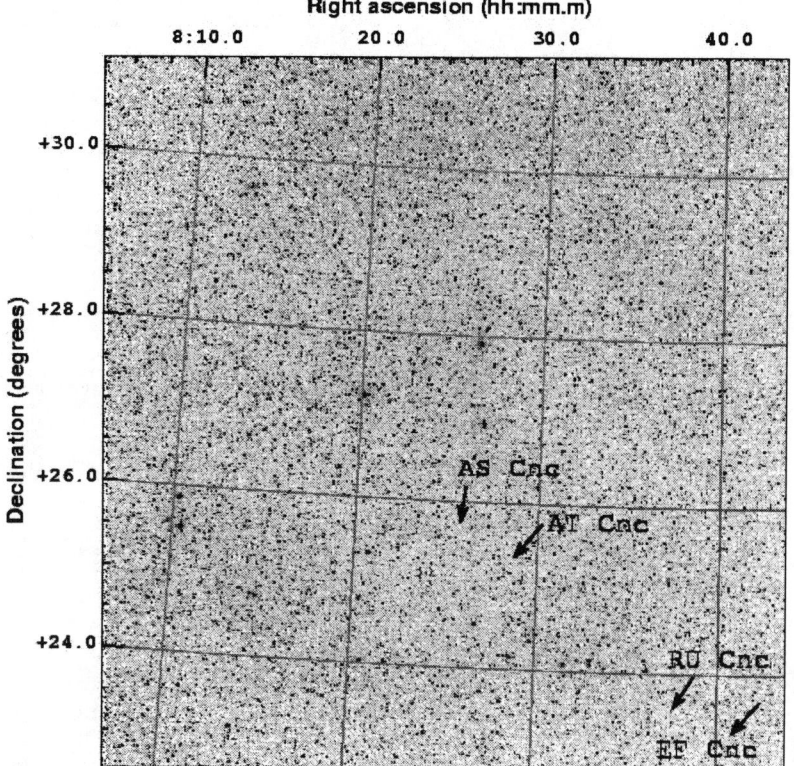

**FIGURE 1.** The ROTSE field 54b with indicated variable stars whose light curves are shown as examples.

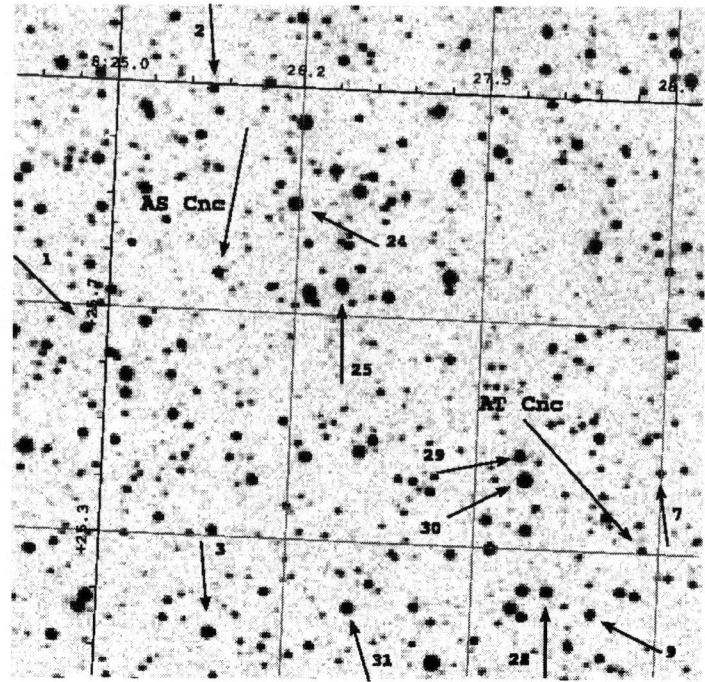

**FIGURE 2.** The surroudings of stars AS Cnc and AT Cnc on ROTSE CCD image.

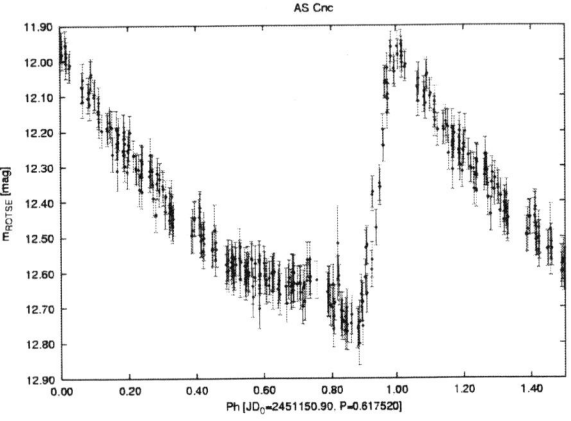
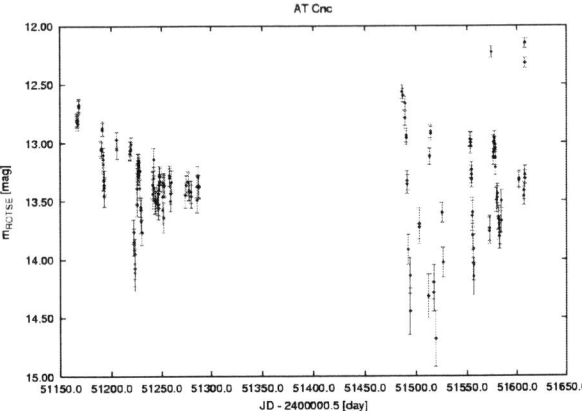

**FIGURE 3. Left.** The generated light curve (orbital) of AS Cnc. **Right.** The generated light curve of AT Cnc.

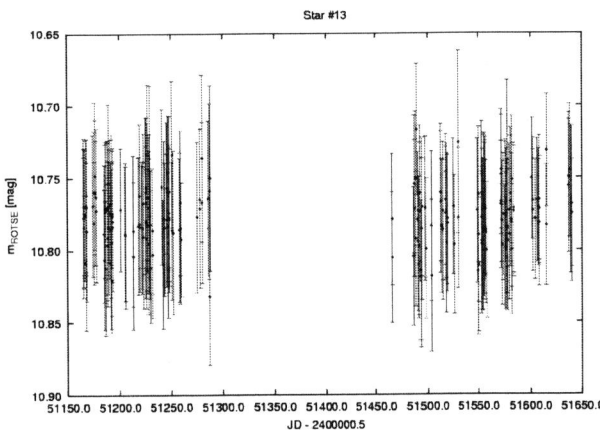
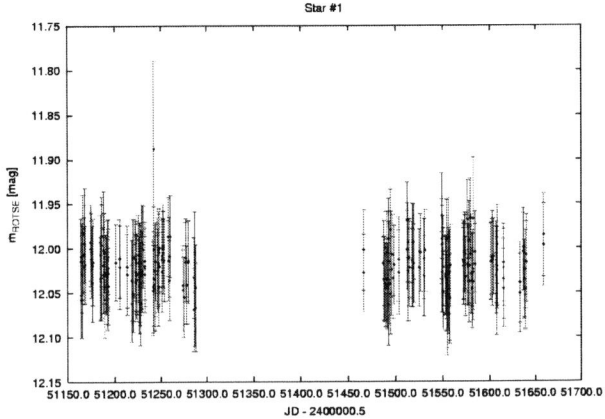

**FIGURE 4.** The light curves of constant stars

## EXAMPLES OF RESULTS

We present examples of results from processing of the field 54b of the ROTSE survey.

This field has FOV approximately 8 deg. Our processing detected about 20000 stars on each picture. The whole process of all data (about 300 pictures) has taken tens of hours (on computer with processor Intel Celeron 566MHz).

## CONCLUSION

The ROTSE WF CCD monitoring data represent a unique database for study of variable astrophysical objects. We have tested procedures dedicated to create light curves for all objects inside the FOV of particular fields. The accuracy of photometric measurements represents few hundreds of magnitude, hence real light changes of order of 0.1 mag can be revealed and analysed.

## ACKNOWLEDGEMENTS

This study has been made as Diploma Thesis of Jan Strobl, student of Charles University in Prague.

The project utilizes data obtained by the Robotic Optical Transient Search Experiment. ROTSE is a collaboration of Lawrence Livermore National Lab, Los Alamos National Lab, and the University of Michigan (www.umich.edu/~rotse). The work has been also supported by the grant provided by the Grant Agency of the Czech Republic, 205/99/0145.

## REFERENCES

1. Akelof, C. et al.: www.umich.edu/~rotse
2. Akerlof, C. AJ 119 (2000), 1901A.
3. Akerlof, C. et al.: ApJ 532 (2000), L25.
4. McKay, C. et al.: AASM 197 (2000), 4621.

# Searching for Optical Transients in Real-Time: The RAPTOR Experiment

W.T. Vestrand, K. Borozdin, S. P. Brumby, D. Casperson, E. Fenimore,
M. Galassi, G. Gisler, K. McGowan, S. Perkins, W. Priedhorsky, D. Starr,
R. White, P. Wozniak, and J. Wren

*Los Alamos National Laboratory, Los Alamos, NM, 87545, USA*

**Abstract.** A rich, but relatively unexplored, region in optical astronomy is the study of transients with durations of less than a day. We describe a wide-field optical monitoring system, RAPTOR, which is designed to identify and make follow-up observations of optical transients in real-time. The system is composed of an array of telescopes that continuously monitor about 1500 square degrees of the sky for transients down to about $12^{th}$ magnitude in 60 seconds and a central fovea telescope that can reach $16^{th}$ magnitude in 60 seconds. Coupled to the telescope array is a real-time data analysis pipeline that is designed to identify transients on timescales of seconds. In a manner analogous to human vision, the entire array is mounted on a rapidly slewing robotic mount so that the fovea of the array can be rapidly directed at transients identified by the wide-field system. The goal of the project is to develop a ground-based optical system that can reliably identify transients in real-time and ultimately generate alerts with source locations to enable follow-up observations with other, larger, telescopes.

## INTRODUCTION

A surprising fact about modern optical astronomy is that the nightly variation of the optical sky is largely unmonitored [1]. The fact that spectacular celestial transients are being missed was clearly demonstrated by the detection of an optical transient associated with a Gamma Ray Burst (GRB) at redshift z=1.6 [2]. That cosmological optical transient reached an astounding peak apparent magnitude of 9---making it the most luminous optical source ever measured ($M_V$=-36.4). However, without the real-time position provided by a high-energy satellite that cued robotic optical telescopes to slew to the correct position, the remarkable transient, which was observable potentially even with binoculars, would have been missed.

There are reasons to suspect the existence of celestial optical transients that cannot be found through sky monitoring by high-energy satellites. For example, it has been suggested that there may be a class of orphan transients that are detectable as off-axis optical emission from beamed GRBs [3,4]. It has also been suggested that optical transients could be precursors to GRBs [5]. But an equally exciting possibility is that there are new, as yet undiscovered, classes of rapid optical transients that are completely unrelated to high-energy transients.

In this paper we briefly discuss the RAPTOR (**Rap**id **T**elescope for **O**ptical **R**esponse) project at Los Alamos National Laboratory. The primary goal of the project is to construct an autonomous robotic system for monitoring the sky that is capable of both independently discovering optical transients and instantly following-up to verify the discovery.

## THE RAPTOR SYSTEM: AN ANALOGUE OF HUMAN VISION

As predators, we humans have evolved a highly sophisticated vision system for both imaging and change detection. The human eye has a wide-field, low-resolution, imager (rod cells of the retina) as well as a narrow-field, high-resolution imager (cone cells of the fovea) [6]. Both eyes send image information to a powerful real-time processor, the brain, running "software" for the detection of interesting targets. If a target is identified, both eyes are rapidly slewed to place the target on the central fovea imager for detailed "follow-up" observations with color sensitivity and higher spatial resolution. Human vision also employs two spatially separated eyes viewing the same scene both to eliminate image faults like "floaters" and to extract distance information about objects in the scene.

The system concept for RAPTOR is best understood as an analogue of human vision. RAPTOR employs two primary telescope arrays (RAPTOR A and B) that are separated by a distance of 20 miles to provide stereoscopic imaging. Each telescope array simultaneously images the same 1500 square-degree field with a wide-field imager and a central 16 square-degree with a narrow-field "fovea" imager. Real-time processors instantly analyze images from RAPTOR A and B and the positions of interesting transients are fed back to the mount controllers with instructions to point the fovea telescopes at the transient. The two fovea cameras then image the transient with higher spatial resolution and at a faster cadence to gather light curve information. Each fovea camera also images the transient through a different filter to provide color information. Altogether, the RAPTOR system therefore acts as a closed loop system that autonomously identifies and makes detailed follow-up observations of optical transients in real-time.

## The RAPTOR Wide-Field Telescopes

The RAPTOR A and B wide-field imaging arrays are each composed of four Canon 85mm f/1.2 lenses with CCD cameras at the focal planes. The cameras are thermo-electrically cooled Apogee AP-10 cameras, which employ a 2Kx2K format Thomson 7899M CCD chip with 14-micron pixels. Each camera of the array covers a 19.5°x19.5° field with a single-pixel spatial resolution of 34 arcseconds. Together the mosaic of the four fields for each array enables simultaneous monitoring of approximately 1500 square-degrees. The limiting magnitude of this wide-field system is $m_R \sim 12$ for a sixty-second exposure (see Figure 1).

In the center of each wide-field array is a fovea telescope (see Figure 2). It is composed of a large 400mm focal length telephoto lens with a 5.6-inch objective diameter and, at least initially, an AP-10 CCD camera. In this configuration the fovea cameras will cover a 4°x4° field-of-view and have nearly five times the spatial resolution of the wide-field array. Further, to provide color information about the transient, the fovea camera for RAPTOR A will have a Johnson I filter and the RAPTOR B camera will employ an R filter.

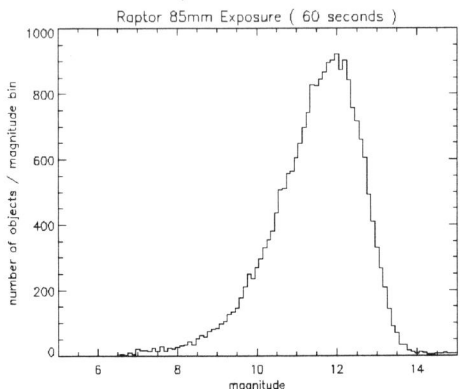

**FIGURE 1.** The number of objects detected per magnitude bin by the wide-field array in a random mosaic tile. The exposure had a sixty-second duration and was taken at the RAPTOR site with higher sky brightness. The limiting magnitude for same array is approximately 0.5 magnitude deeper at the darker site.

Both RAPTOR telescope arrays are mounted on rapidly slewing, robotic, mounts. The mounts were designed to have the capability to place the fovea camera on the transient and to begin follow-up observations within a few seconds no matter where the transient occurred in the 1500 square-degree monitoring field. Tests of the completed mounts indicate that they can accelerate at rate of up to 400 degrees/sec$^2$ and reach speeds of 200 degrees/sec. In practice this means that the arrays can be slewed from horizon to horizon in 1.5 seconds and, after waiting another second for mount and telescope vibrations to damp, begin imaging. To our knowledge, the RAPTOR mounts are the swiftest ever constructed for astronomical purposes.

## The Real-Time Pipeline

A key enabling aspect of the RAPTOR system is the fast analysis pipeline that is designed to identify transients in real time. Immediately upon completion of each exposure, the raw images are combined with flat-field and dark frames to form corrected images. Sources are then extracted using a modified version of the SExtractor package [7] to form a source-object file for each corrected image. Using the Tycho star catalog as a reference, the extracted source-object files are then image registered to calculate the source coordinates and the relative photometry is derived to form a calibrated object file for each image. The entire process of calibrated list extraction is accelerated to take less than 10 seconds and runs in parallel for all ten of the array cameras.

**FIGURE 2.** A RAPTOR wide-field telescope array. The 5.6-inch central "fovea" telescope is surrounded by an array of four 85mm lenses with large format CCD cameras that constitute the wide-field monitoring system. The entire array is mounted on a rapidly slewing telescope mount that can point the "fovea" telescope at the location of any transient identified by the wide-field system in one second.

To identify astrophysical transients, the calibrated object files from pairs of consecutive RAPTOR A mosaic images and the simultaneous RAPTOR B pair are compared. To minimize the number of false positives generated by hot CCD pixels and cosmic ray hits, the pointing directions are slightly dithered between each exposure. Only sources that are detected at the same sky position in both images of the pair are identified as potential celestial sources. This list of potential sources is then compared to an internal source catalog at each telescope site. If the transient is identified in each mosaic image of the pair then it is flagged as a potential astrophysical transient. The two lists of candidate transients from each array are then compared. If the transient is present in the image pair from both arrays and no parallax shift is observed, then the transient is identified as a real astrophysical transient and an alert is generated. The position of the transient is then fed back to the mount and telescope arrays are rapidly slewed to point the fovea telescopes at the transient to begin follow-up observations.

## MINING THE SKY IN REAL-TIME

The RAPTOR project is developing hardware and software that will enable autonomous robotic searches of the optical sky in real-time. Further, the RAPTOR system is designed to filter out the myriad of non-celestial signals that can mimic celestial transients in order to allow robust identification of real celestial transients. Our goal is to monitor a sizable fraction of the sky and to generate alerts with source locations that will enable follow-up observations with other telescopes. Such a system will open an exciting new area of discovery space. The record of the sky collected by RAPTOR will also be a powerful resource for exploring sky variability.

## ACKNOWLEDGMENTS

Internal Laboratory Directed Research and Development funding supports the RAPTOR project at Los Alamos National Laboratory.

## REFERENCES

1. Paczynski, B., *Pub. Astron. Soc. of the Pacific*, **112**, 1281-1283 (2000).

2. Akerlof, C., et al., *Nature*, **389**, 400-402 (1999).

3. Katz, J. I., *Astrophys. Journal*, **432**, L107-L109 (1994).

4. Meszaros, P, and Rees, M.J., *Astrophys. Journal*, **476**, 232-237 (1997).

5. Paczynski, B., astro-ph/0108522 (2001).

6. Hubel, D.H., *Eye, Brain, and Vision*, New York: W. H. Freeman and Company, 1995, pp. 33-59.

7. Bertin, E., and Arnouts, S., *Astron. Astrophys. Suppl. Ser.*, **117**, 393-404 (1996).

# A System for Photon-Counting Spectrophotometry of Prompt Optical Emission from Gamma-Ray Bursts

W.T. Vestrand, K. Albright, D. Casperson, E. Fenimore, C. Ho, W. Priedhorsky, R. White, and J. Wren

*Los Alamos National Laboratory, Los Alamos, NM, 87545, USA*

**Abstract.** With the launch of HETE-2 and the coming launch of the Swift satellite, there will be many new opportunities to study the physics of the prompt optical emission with robotic ground-based telescopes. Time-resolved spectrophotometry of the rapidly varying optical emission is likely to be a rich area for discovery. We describe a program to apply state-of-the-art photon-counting imaging technology to the study of prompt optical emission from gamma-ray bursts. The Remote Ultra-Low Light Imaging (RULLI) project at Los Alamos National Laboratory has developed an imaging sensor which employs stacked microchannel plates and a crossed delay line readout with 200 picosecond photon timing to measure the time of arrival and positions for individual optical photons. RULLI detectors, when coupled with a transmission grating having 300 grooves/mm, can make photon-counting spectroscopic observations with spectral resolution that is an order of magnitude greater and temporal resolution three orders of magnitude greater than the most capable photon-counting imaging detectors that have been used for optical astronomy.

## INTRODUCTION

Charged Coupled Detectors (CCDs) are currently the pre-eminent detectors for optical astronomy. Since their introduction into astronomy in the late 1970's, their dramatically higher quantum efficiency (at least a factor of 50 in the red), ease of calibration due to their highly linear response (plates have a very non-linear response), and intrinsically digital nature, have given CCDs important advantages over photographic plates. Modern CCD imagers, with spatial resolution comparable to the best photographic plates and pixel numbers approaching $10^8$, have made photographic techniques obsolete. However, despite their considerable advantages as imaging detectors for optical astronomy, CCDs have an important limitation--- they cannot count individual optical photons. Further, their readout time and readout noise is a limiting factor when employing short integration times to study rapid astronomical phenomena.

The ideal imaging detector for exploring the properties of rapidly varying optical emission would be an imager with photon counting capability and good spectral resolution. Here we report on a system under development at Los Alamos National Laboratory (LANL) that is designed to make photon-counting spectrophotometric observations and briefly discuss application of the technology to prompt optical emission from gamma-ray bursts.

## A PHOTON-COUNTING SPECTROPHOTOMETER

The Remote Ultra-Low Light Imaging (RULLI) project is a very successful program at LANL for the development of optical photon-counting imagers. The RULLI imagers are composed of a light sensitive photocathode, a stack of three microchannel plates (MCPs), and a crossed delay line (CDL) readout, all hermetically sealed in a vacuum tube (Figure 1). When used for the detection of optical photons, they employ an S-20 photocathode, which has a quantum efficiency that allows effective detection of optical photons up to wavelengths of about 750 nanometers (Figure 2). Each photoelectron ejected by the photocathode generates a cascade in the MCP stack that produces a gain of ~$10^7$. The emergent electron cloud spreads over about 10 wire pairs in the tightly wound CDL grid to allow 2-D centroiding of the event to a spatial resolution of about 70 microns on the 40 mm diameter active area. This sensor tube is connected to electronics that provide absolute timing accuracy for the photon arrival times to better than 200 picoseconds. The maximum event rate for the sensor and electronics is about $10^6$ events/second. This combination of spatial resolution and unprecedented ability to detect individual photons with extremely high time resolution has led to important applications in time-resolved imaging of cloud scattering and literal 3-D ranging and mapping of solid objects [1,2].

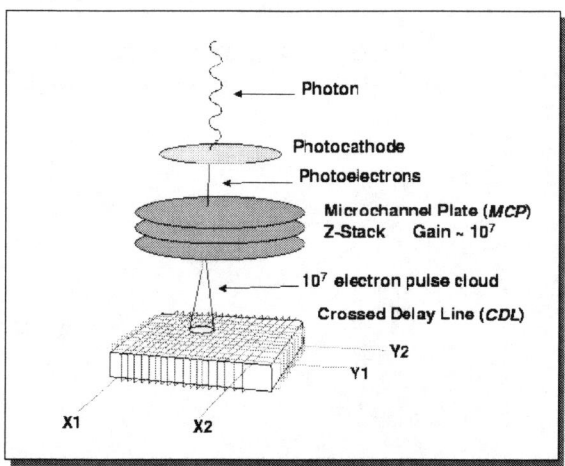

**FIGURE 1.** The internal components of a RULLI imager.

Application of RULLI technology has the potential to revolutionize the field of photon-counting spectrophotometry in astrophysics. A RULLI detector, when coupled with a simple transmission grating having 300 grooves per mm, can make photon-counting spectroscopic observations with spectral resolution that is an order of magnitude greater and temporal resolution three orders of magnitude greater than the best existing cryogenic detectors. The RULLI detector can also handle incident photon rates that are an order of magnitude higher, and therefore has a greater dynamic range than the cryogenic detectors. Further, since they have essentially no noise, RULLI detectors have the ability to operate at room temperature and have an observing sensitivity that is only limited by the natural sky background and the available statistics. By eliminating the power, weight, complexity, and cost issues associated with cryogenic cooling, RULLI technology is more attractive for many applications.

At LANL we are currently incorporating one of these RULLI detectors into a system designed for photon-counting spectrophotometry of rapidly varying astronomical objects. Our spectrophotometry system employs a 30-cm F7 Ritchey-Chretien telescope with a two-element field flattener that yields a spot size smaller 20 microns across the central 40mm diameter image circle. The nominal field-of-view for the telescope, when employing the full 40mm image circle, is 1.1 degrees---a size that is well matched for covering a typical HETE-2 error box. Our ruggedized telescope also incorporates Invar rods to minimize temperature induced focus variations and an interferometrically matched front window constructed of fused silica to seal the tube from dust and moisture. To obtain spectral dispersion, a transmission grating is located between the field flattener and a sensor-mounting flange that can accommodate either a conventional Apogee AP-10 CCD camera or a RULLI detector. This configuration, when used in first order with a grating of 300 grooves/mm, yields a spectral resolution of $R=\Delta E/E \sim 700$ at 600 nm with the AP-10 CCD camera and R~110 with a RULLI detector.

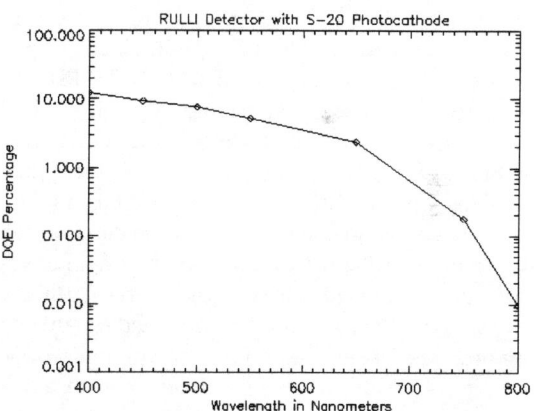

**FIGURE 2.** The measured Detection Quantum Efficiency (DQE) as a function of photon wavelength for a RULLI imager employing a S-20 photocathode.

To study the prompt optical emission from gamma ray bursts (GRBs), the entire spectrometry system will be mounted on a robotic rapidly slewing mount at Fenton Hill Observatory, which is located about 20 miles west of LANL in the Jemez mountains at an altitude of 8,600 feet. The rapid telescope mount, which can slew from horizon to horizon and settle in less than 3 seconds, is connected via socket to the GCN network for prompt response to GRB alerts.

# SPECTROPHOTOMETRY OF GRBS

The ability of high-energy satellites to provide rapid, accurate, GRB positions in recent years led to the discovery of fading radio, optical, and X-ray afterglows at the locations of the GRBs. Optical spectroscopy of those afterglows taught us that gamma-ray bursts are more amazing than almost everyone had dared to speculate---they are the largest explosions since the big bang and contribute in a substantial way to the entire energy output of the universe. Further, for the gamma rays to escape the intense internal radiation fields requires the emitting material to be driven to Lorentz factors of 100 to 1000. Such extreme ultra-relativistic bulk flows occur nowhere else in the Universe. The extreme ultra-relativistic nature of the outflow means that the complex history of interactions and deceleration of the bulk flow, which occurs on the timescale of a day in the plasma frame, is carried by emission that arrives at Earth within the span of a minute. Unfortunately the observations of afterglows taken hours later therefore tell us little about the details of these cataclysmic events.

On January 23, 1999, a ROTSE telescope at LANL detected a short-lived luminous optical transient from the bright GRB 990123 [3]. This observation showed that prompt optical emission could also be generated during the initial GRB outburst. Most models assume that the optical emission from GRBs is synchrotron radiation generated by energetic electrons in the burst outflow. By noting that the optical and gamma ray intensities were not correlated, many authors argued that the emission in the two bands originated in different regions and suggested that the prompt optical outburst is generated as a reverse shock that traverses the relativistic outflow. Such a reverse shock can only occur once in a given GRB, and indeed, the ROTSE data seem to show a single peak leading to power-law decay. On the other hand, there are models that predict rapid optical variations were present that should be correlated with the gamma ray fluctuations that were as short as tens of milliseconds in this event [4]. The ROTSE light curve during the burst proper is composed of only three points (Figure 3), and is too highly under-sampled to definitively distinguish between models. More bursts, measured with far better temporal resolution, are needed.

Application of our photon-counting RULLI technology to this problem would have a major scientific impact. It would allow time-resolved optical spectroscopy of the evolving prompt emission and would represent an opportunity to measure the unique physics of the ultra-relativistic flow. During the rapid deceleration that occurs within the first minute or so of the gamma-ray burst, there are major parameters that define the physics of the flow: the Lorentz factor, the ambient density, the fraction of the energy in mass of the particles, and the fraction of the energy in the magnetic field. By measuring the spectral shape of the emission and how it evolves with our RULLI-based spectrophotometer, each of these parameters can be determined [5]. Thus, with our detailed photon-counting spectroscopy, we will determine the physical conditions of the only known place in the universe to have ultra-relativistic motion since the Big Bang.

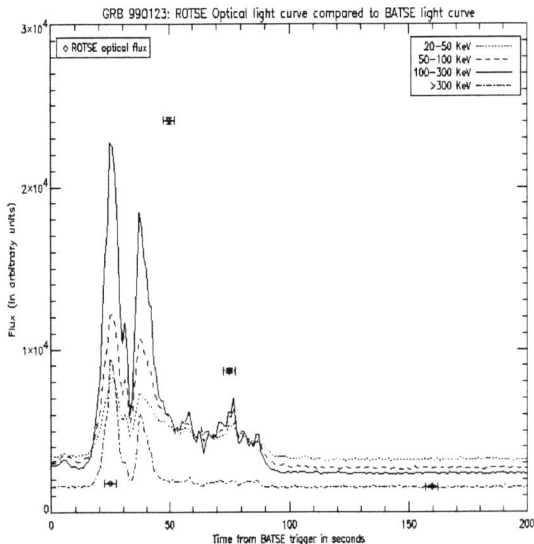

**FIGURE 3**. The gamma ray and optical lightcurves for prompt emission from GRB 990123. The optical lightcurve during this prompt phase is composed of only three broadband CCD measurements. RULLI will give us time-resolved spectral measurements comparable to those in the gamma rays and provide powerful tests of GRB models.

# REFERENCES

1. Priedhorsky, W. et al., *Appl. Optics* **35**, 441-452 (1996).

2. Ho, C., et al. *Appl. Optics* **38**, 1833-1840 (1999).

3. Akerlof, C., et al., *Nature*, 398, 400-401 (1999).

4. Liang, E.P. et al., *ApJ*, 519, L21-24 (1999).

5. Wijers, R. and Galama, T., *ApJ*, 525, 177 (1999).

# Recent Developments in the BOOTES Experiment

A. de Ugarte Postigo*, T.J. Mateo Sanguino[†], J.M. Castro Cerón**, P. Páta[‡],
M. Bernas[‡], M. Jelinek[§], R. Hudec[§], S. McBreen[¶], J.Á. Berná[∥], C.E. García Dabó*,
J. Gorosabel[††], [‡‡], [§§], J.M. Más-Hesse[§§], T. Soria[¶¶], B.A. de la Morena[†], J. Torres
Riera***, A.J. Castro-Tirado[‡‡]

*Facultad de Ciencias Físicas, Universidad Complutense de Madrid, Madrid Spain
[†]Centro de Experimentación del Arenosillo (INTA), Mazagón (Huelva) Spain
**Real Instituto y Observatorio de la Armada, Sección de Astronomía, San Fernando (Cádiz) Spain
[‡]Czech Technical Univ., Faculty of Electronic Engineering, Dept. of Radioelectronics, Prague Czech Republic
[§]Astronomical Institute, Academy of Sciences of the Czech Republic, Ondřejov Czech Republic
[¶]Department of Experimental Physics, University College Dublin, Dublin Ireland
[∥]Dpto. de Física, Ing. de Sist. y Teo. de la Señal, Univ. de Alicante, San Vicente del Raspeig (Alicante) Spain
[††]Danish Space Research Institute, Copenhagen Ø Denmark
[‡‡]Instituto de Astrofísica de Andalucía (CSIC), Granada Spain
[§§]Laboratorio de Astrofísica Espacial y Física Fundamental (INTA), Villafranca del Castillo (Madrid) Spain
[¶¶]Estación Experimental de la Mayora (CSIC), Costa del Algarrobo (Málaga) Spain
***División de Ciencias del Espacio y Tecnologías Electrónicas (INTA), Torrejón de Ardoz (Madrid) Spain

**Abstract.** BOOTES, the **B**urst **O**bserver and **O**ptical **T**ransient **E**xploring **S**ystem, is mostly a Spanish-Czech international collaboration that works to fill the space that there actually exists in the rapid variability Astronomy. It is specially aimed towards the detection and study of the optical transients that are generated in conjunction with the elusive GRBs (Gamma Ray Bursts). Since 1998 BOOTES has provided follow ups for more than 60 of these phenomena with the first of its observatories; the most important results obtained so far are the detection of an OT (Optical Transient) in the GRB 000313 error box and the non detection of optical emission simultaneous to the GRB 010220 event. With the recent inauguration of a second observatory BOOTES multiplies its science capabilities.

## INTRODUCTION

BOOTES [1] saw its first light in 1998. A pioneer robotic observatory for OT (Optical Transient) follow ups [2] at that time, it is also considered part of the ground segment for the European Space Agency's INTEGRAL (INTErnational Gamma RAy Laboratory) satellite [3], dedicated to High Energy Astrophysics and to be launched in the Autumn of 2002.

Although it was first thought of as a tool to perform rapid searches of GRB (Gamma Ray Burst) optical counterparts, BOOTES has also participated in other studies like meteor storms campaigns. Additionally, it has accumulated a significantly large database of sky images to allow the study of objects which exhibit short lived variations.

In this paper we present an update of our *technical developments*, namely a new observatory which gives BOOTES improved observing capabilities and that has been recently inaugurated, and a communication system for remote control of the observatory through a GSM (Groupe Spécial Mobile) mobile phone that has been already implemented. Additionally, we will describe *selected recent observations* and conclude with *future* perspectives.

## TECHNICAL DEVELOPMENTS

BOOTES-1, the main BOOTES observatory, is located in Mazagón (Huelva), a very dark sky area in southern Spain, and is hosted by the INTA (Instituto Nacional de Técnica Aeroespacial) in its "Centro de Experimentación del Arenosillo". It has twin domes (BOOTES-1A and BOOTES-1B) that shelter two Schmidt-Cassegrain telescopes and several wide field cameras. Following complementing schemes, all instruments carry out systematic explorations of the sky each night [4, 5, 6].

A second observatory (BOOTES-2) has been recently inaugurated 240 km East of BOOTES-1, at Málaga's "La Mayora", an experimentation centre property of the CSIC (Consejo Superior de Investigaciones Científicas). BOOTES-2 is both, identical and uses the same instrumentation as BOOTES-1B. Observing simultaneously the same field of the sky from both locations BOOTES has a stereoscopic view. Thus, using parallax, it can discriminate against near Earth detected sources up to a distance of $10^6$ km.

One of the technical aspects that has been developed in BOOTES is a remote information and control system that, using GSM mobile communication tools, allows for full remote control of the observatories without needing any additional staff [7].

BOOTES had already a meteorological station which was equipped with software capable of deciding whether the weather conditions were optimal for astronomical observation or not. This station controlled the aperture of the domes. The new remote information and control system adds the following functions:

1. On demand reports of the observatories's status, covering the position of the domes and complete meteorological information.
2. Alerts of system malfunctions such as problems while opening or closing the domes, adverse weather conditions or system breakdown due to power failures, etc.
3. Manual interaction with the system to open or close the domes at any moment.

In order to achieve these capabilities, a communication protocol between a personal computer and a GSM modem via RS 232 (Recommended Standard) has been developed. It is based on the SMS (Short Message Service) standard (version 03.40) and programmed in the LabView environment.

## RESULTS

In this section we highlight two of the most interesting BOOTES's GRB follow up observations. They nicely examplify its capabilities. Additionally we present a series of images obtained in a meteor campaign as a brief example of some of the colateral scientific work carried out by BOOTES.

Since 1998 BOOTES has responded to more than 60 GCN (GRB Coordinates Network) triggers with different results [4, 5, 6]. The majority of the follow ups have been negative detections despite the prompt response, thus ratifying the elusiveness of these phenomena. Here we present two of the most significant GRB error zones exploration results:

### GRB 000313: an OT in the follow up images

On $13^{th}$ March 2000 BOOTES responded to a GCN alert [8] and started exposing at the coordinates given, 4 minutes after the GRB was detected by the BATSE (Burst and Transient Source Experiment). The image was 5 minutes long with an $I$ filtre, going down to $13^{th}$ magnitude.

A $9^{th}$ magnitude source, not present in the Digital Sky Survey, was found in the said image. After the first one and, because of the big error box given by BATSE, BOOTES started a mosaic around this field, so no other image was taken at the original position until 55 minutes had elapsed since the trigger. By then further imaging revealed that the source had faded away. The results are discussed in [9].

### GRB 010220: simultaneous observation of the GRB error zone

This is the first time that a GRB error box is imaged by a CCD (Charge Coupled Device) camera prior to the beginning of the burst as well as during and after it [6, 10]. The first of the images that we present in [6, 10] began 52 seconds before the gamma ray event registered on $20^{th}$ February 2001 and covered the first 8 seconds of it. The second one covered from 68 to 128 seconds after the trigger.

No OT was detected in any of these images. This can be attributed to the high extinction ($A_V$ = 3.3 magnitudes) in this direction, at only 1.38° from the galactic plane.

## Additional science

BOOTES collaborates with the "Sociedad de Observadores de Meteoros y Cometas de España" in the study of meteor showers, including double station observations and evolution of meteor trails in the upper atmosphere. See [11] for a description of the double station observational set up and the results obtained during the Perseids campaign in the summer of 1999. Fig. 1 shows the evolution of a bright meteor trail observed on $18^{th}$ November 2000 during the Leonids campaign.

## FUTURE

At this moment the BOOTES team is already working on introducing new instrumentation and analysis procedures to increase the productivity of its observatories. This will serve to improve the science that can be made with them

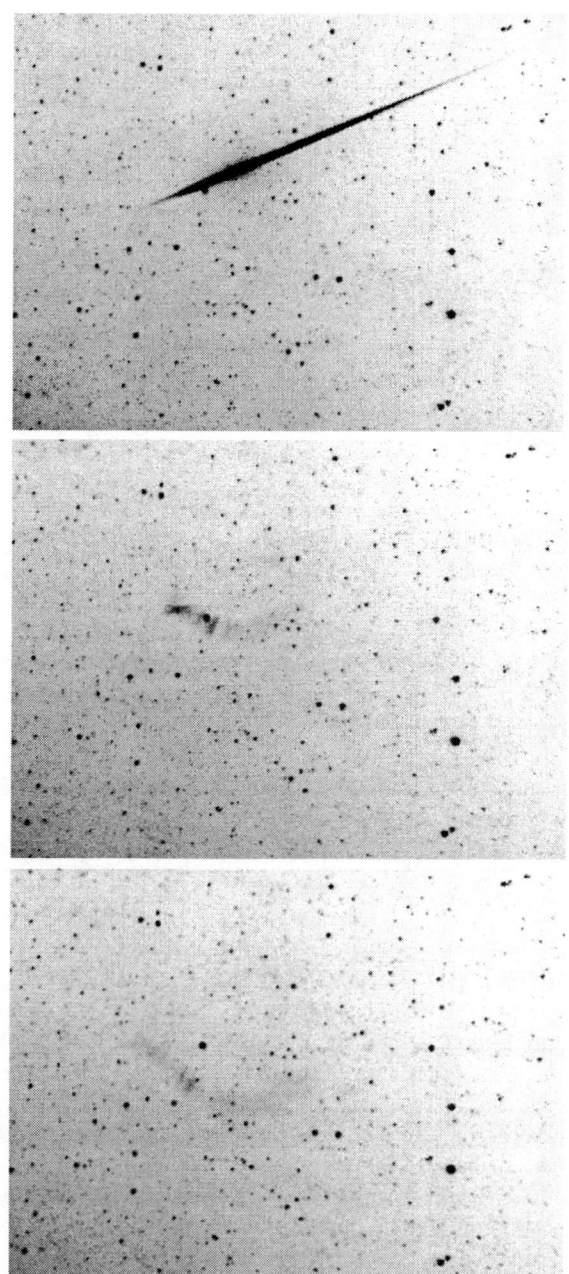

**FIGURE 1.** Evolution of a Leonid trail during the 2000 storm.

towards filling the gap that there still exists in the rapid variability Astronomy.

## ACKNOWLEDGMENTS

We thank the INTA's "División de Ciencias del Espacio y Tecnologías Electrónicas" for their support through project IGE 4900506. JMCC, JÁB and JG acknowledge research grants from Spain's "Ministerio de Ciencia y Tecnología", Spain's "Generalidad Valenciana" through its "Oficina de Ciencia y Tecnología" (Project GV01-361, "Estudio de atmósferas estelares y envolturas en estrellas calientes") and the European Commission respectively. This work is partially supported by Spain's "Comisión Interministerial de Ciencia y Tecnología" grant ESP95-0389-C02-02. The Czech contribution is supported by the Ministry of Education and Youth of the Czech Republic, projects ES02 and ES36 and by the ESA Prodex Project 14527, as well as by the Grant Agency of the Czech Republic, grant 205/99/0145.

## REFERENCES

1. Castro-Tirado, A. J., Hudec, R., and Soldán, J., "B.O.O.T.E.S.", in *Proceedings of the 3rd Huntsville Symposium, held in Huntsville, AL USA, 25–27 October 1995*, edited by C. Kouveliotou et al., American Institute of Physics, Woodbury, 1996, vol. 384 of *AIP Conference Proceedings*, pp. 814+.
2. Castro-Tirado, A. J. et al., *A&AS*, **138**, 583–585 (1999).
3. Hudec, R. et al., "BART, BOOTES and OMC: monitoring of AGNs-blazars", in *Blazar Monitoring towards the Third Millennium; Proceedings of the OJ-94 Annual Meeting 1999, held in Torino, Italy, 19–21 May*, edited by C. M. Raitieri et al., Osservatorio Astronomico di Torino, Pino Torinese, 1999, pp. 131–133.
4. Castro Cerón, J. M. et al., "Search for gamma ray burst quasi simultaneous optical emission with BOOTES-1", in *Highlights of Spanish Astrophysics II; Proceedings of the 4th Scientific Meeting of the Spanish Astronomical Society (SEA), held in Santiago de Compostela, Spain, 11–14 September 2000*, edited by J. Zamorano et al., Kluwer Academic Publishers, Dordrecht, 2001, pp. 37–40.
5. Castro Cerón, J. M. et al., "Two years of gamma ray burst follow up observations with BOOTES-1", in *Exploring the Gamma-Ray Universe; Proceedings of the Fourth INTEGRAL Workshop, held in Alicante, Spain, 4–8 September 2000*, edited by Á. Giménez et al., SP-459, European Space Agency – Agence spatiale européene, Noordwijk, 2001, pp. 407–410.
6. Castro Cerón, J. M. et al., "Gamma-Ray Burst Follow Up Observations with BOOTES in 1998–2000", in *Gamma-Ray Burst in the Afterglow Era; Proceedings of the International Workshop, held in Rome, Italy, 17–20 October 2000*, edited by E. Costa et al., ESO Astrophysics Symposia XIX, Springer-Verlag, Berlin, 2001, pp. 53–55.
7. Mateo Sanguino, T. J., *Sistema de control GSM de un observatorio astronómico*, Universidad de Granada, 18.071 Granada España, 2001, Proyecto de Fin de Carrera – Ingeniería Superior.
8. Castro-Tirado, A. J. et al., *GCN*, **612**, 1+ (2000).
9. Castro-Tirado, A. J. et al., *A&A – submitted* (2002).
10. Castro-Tirado, A. J. et al., *GCN*, **957**, 1+ (2001).
11. Trigo-Rodríguez, J. M. et al., *WGN, Journal of the International Meteor Organisation*, **28**, 120–125 (2000).

# An Update on the GRB ToolSHED Project Status

Jon Hakkila*, David J. Haglin†, Richard J. Roiger†, Timothy W. Giblin*, William S. Paciesas** and Charles A. Meegan‡

*Department of Physics and Astronomy, College of Charleston, Charleston, SC
†Department of Computer and Information Sciences, Minnesota State University, Mankato, MN
**Department of Physics and Astronomy, University of Alabama in Huntsville, AL
‡NASA NSSTC, Huntsville, AL

**Abstract.** The GRB ToolShed is an online suite of induction-based machine learning and statistical tools designed for gamma-ray burst classification and cluster analysis. The ToolSHED also includes a large preprocessed gamma-ray burst database. We report on the current status of the ToolSHED.

## INTRODUCTION

Development continues on the GRB ToolSHED [1]. Our purpose in designing the ToolSHED is to supply the astronomical community with an online gamma-ray burst database and a suite of machine learning algorithms to aid in the study of this database. We have previously demonstrated the usefulness of machine learning algorithms in the study of gamma-ray burst data (*e.g.* [2] [3]). Our project is in the third year of a three-year NASA Applied Information Science Research (AISR) grant (which funds us to construct the tool), and in the first year of a three-year NSF Research in Undergraduate Institutions (RUI) grant (which supports database augmentation).

## SPECIFIC FEATURES

The GRB ToolSHED, physically located at Minnesota State University, Mankato, can be found electronically at http://grb.mnsu.edu/grbts/. A mirror site at the College of Charleston is planned. The ToolSHED has many features already implemented and several additional ones to be implemented in the near future.

### Database

The primary database will be the BATSE 5B catalog (other BATSE catalogs are currently supported). Afterglow data is expected soon.

The individual attributes for the BATSE data include: BATSE basic (location) data, trigger data and instrumental triggering criteria, flux/fluence data, and duration data. In addition, attributes pertaining to the internal luminosity function [4] (see Hakkila et al., this conference), duty cycle [5], color-color diagrams (*e.g.* [6]), gamma-ray burst pulse data, and redshift and luminosity data from bursts with afterglows. User-provided GRB attributes and data are also supported.

### Machine Learning

There are two major categories of machine learning models: *supervised*, and *unsupervised*. The supervised learning models look for correlations among a data set capable of "predicting" the value of a certain *output* attribute. That is, the data is searched for ways to predict the value of a certain attribute based on values of all other attributes. Generally, higher prediction accuracies are considered more useful. Some supervised learning models require the output attribute to be numeric, others require it to be *categorical* (e.g. long or short).

The unsupervised learning models look for ways to clump the instances together so that they can be described with simple "cut-off" rules. Considering the attribute space to be $n$-dimensional, a cutting plane in this $n$-dimensional space might describe two separate clumps of instances: those on one side versus those on the other side. Note that unsupervised learning models cannot provide a name for the clumps; they can only provide a listing of instances in each group along with a "rule" for placing instances into those groups.

There are currently several machine learning algorithms that are available with the ToolSHED. For supervised learning, there is a java-based version of C4.5 (called J48), M5′, Linear Regression, K*, Naïve

Bayes, and a Back-Propagation Neural Network. For unsupervised learning, Estimation Maximization (EM), KMeans, Cobweb and a Kohonen Neural Network are available. Soon a tool capable of doing either supervised or unsupervised learning called ESX will be included.

When the ToolSHED asks the user for a selection of learner models, it presents them with a list of models along with the input attribute and output attribute (if any) restrictions imposed by that model.

## Data Visualization

There are currently no direct visualization tools available from the ToolSHED. However, users may download data directly to a spreadsheet for producing either graphs within the spreadsheet or input to some other graphing tool such as IDL. A direct connection to using ION for producing graphs within the ToolSHED is planned.

## A SESSION WALK-THROUGH

To help the potential user visualize a complete mining session, a sequence of screen shots are presented. The first screen gives the layout of the GRB ToolSHED. There is a navigation menu along the left side of the screen that is capable of expanding and minimizing menu selections within the organization of folders. Session-based functions are provided along the top of the screen. Before a user logs in, most of the selections of the navigation menu are displayed in italics but are not "clickable."

Once logged in, the user can navigate to various functions. One typical function is that of "Select" to select a subset of the Gamma-Ray Burst data. The user may have the web application save a certain selection for later retrieval. This feature is indicated in the navigation menu as *remember current extraction* and *reload previous extraction*. Figure 1 shows an example of selecting certain columns or attributes of the data.

To carry out a data mining session, one of the many learner models available within the application must be chosen (see Figure 2). The navigation menu has an option for selecting a learner model. The menu shows a partial list of the available learner models, along with some characteristics of each model.

Once a model has been selected, additional information specific to that model will be requested. After invoking J48, for example, the user will see a request for some tuning parameter values for the learner model along with suggested default values (see Figure 3). The output attribute is a pull-down selection with every categorical attribute in the users data extraction.

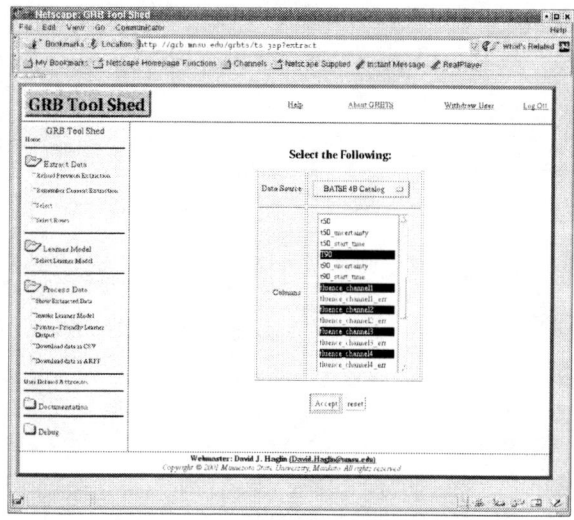

**FIGURE 1.** Selecting data attributes for study.

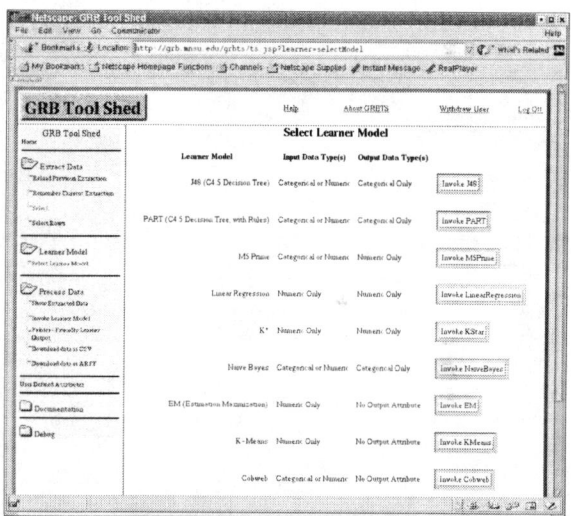

**FIGURE 2.** Selecting a classifier.

And finally, results of a J48 supervised learning session is shown to the user (See Figure 4). Since J48 is a decision tree learner model, the outcome of a mining session is a tree structure:

```
hr321 <= 3.426546: long
(590.77/82.0) hr321 > 3.426546:
short (177.23/55.23)
```

## CONCLUSIONS

Although the GRB ToolSHED is still under construction, the project is approaching the point where pre-beta testers are needed. Please contact any co-authors of this

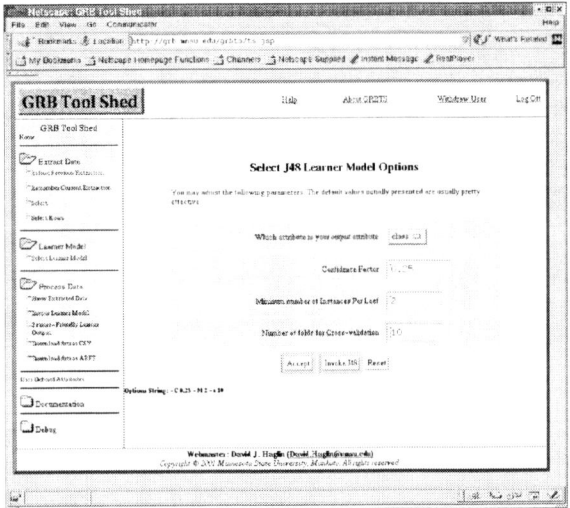

**FIGURE 3.** Selecting options for J48.

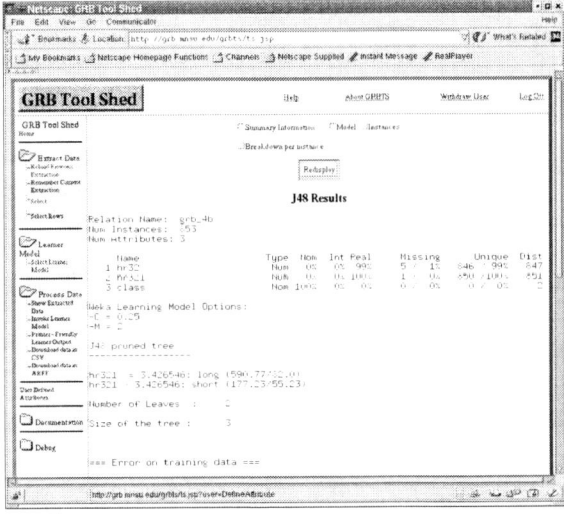

**FIGURE 4.** Results of a J48 supervised classification session.

paper if you are interested in participating.

It has been exciting to realize that until recently, when a new mining experiment was posed, the way to conduct that experiment was to manually construct a data set with the exact data in question in the proper format for a specific data mining tool. This project has "emerged" in that now when a new mining experiment is posed, the fastest way to get that experiment set up is to use the GRB ToolSHED.

## ACKNOWLEDGMENTS

We would like to gratefully acknowledge NASA support under grant NRA-98-OSS-03 (the Applied Information Systems Research Program) and NSF grant AST-0098499 (Research in Undergraduate Institutions).

## REFERENCES

1. Haglin, D. J., Roiger, R. J., Hakkila, J., Pendleton, G. N., and Mallozzi, R., "A GRB Tool Shed", in *AIP Conf. Proc. 526: Gamma-ray Bursts, 5th Huntsville Symposium*, 2000, pp. 877+.
2. Hakkila, J., Haglin, D. J., Pendleton, G. N., Mallozzi, R. S., Meegan, C. A., and Roiger, R. J., *The Astrophysical Journal*, **538**, 165–180 (2000).
3. Roiger, R. J., Hakkila, J., Haglin, D. J., Pendleton, G. N., and Mallozzi, R. S., "Unsupervised Induction and Gamma-Ray Burst Classification", in *Proceedings of the Fifth Huntsville Gamma-Ray Burst Symposium*, edited by M. Kippen, R. S. Mallozzi, and G. J. Fishman, AIP Conference Proceedings 526, American Institute of Physics, New York, 2000, pp. 38–42.
4. Horack, J. M., and Hakkila, J., *The Astrophysical Journal*, **479**, 371+ (1997).
5. Hakkila, J., Haglin, D. J., Pendleton, G. N., Mallozzi, R. S., Meegan, C. A., and Roiger, R. J., *The Astrophysical Journal*, **538**, 165–180 (2000).
6. Kouveliotou, C., Paciesas, W. S., Fishman, G. J., Meegan, C. A., and Wilson, R. B., *Astronomy and Astrophysics Supplement Series*, **97**, 55–57 (1993).

# VIII. SOFT GAMMA REPEATERS

# The Effects of Burst Activity on Soft Gamma Repeater Pulse Properties and Persistent Emission

Peter M. Woods

*Universities Space Research Association*
*National Space Science and Technology Center*
*320 Sparkman Dr., Huntsville, AL 35805*

**Abstract.**
Soft Gamma Repeaters (SGRs) undergo changes in their pulse properties and persistent emission during episodes of intense burst activity. SGR 1900+14 has undergone large flux increases following recent burst activity. Both SGR 1900+14 and SGR 1806−20 have shown significant changes in their pulse profile and spin-down rates during the last several years. The pulse profile changes are linked with the burst activity whereas the torque variations are not directly correlated with the bursts. Here, we review the observed dynamics of the pulsed and persistent emission of SGR 1900+14 and SGR 1806−20 during burst active episodes and discuss what implications these results have for the burst emission mechanism, the magnetic field dynamics of magnetars, the nature of the torque variability, and SGRs in general.

## INTRODUCTION

Soft Gamma Repeaters (SGRs) are an exotic class of high energy transient, very likely isolated, strongly magnetized neutron stars or "magnetars." For periods of days to months, SGRs can be found in burst active states where they emit anywhere from a handful to several hundred bursts. Typically, the bursts last ∼0.1 sec and have energy spectra (E >25 keV) that can be modeled as a power-law convolved with an exponential. At lower energies, however, this empirical model fails to fit the spectrum [1]. The burst energies follow a power-law number distribution up to $\sim 10^{42}$ ergs ($dN/dE \propto E^{-5/3}$ [2, 3]), consistent with a so-called self-organized critical system (e.g. earthquakes, Solar flares, etc. [4]) where the burst energy reservoir greatly exceeds the energy output within any given burst. On two occasions, more energetic bursts or giant flares were recorded from SGR 0526−66 on 1979 March 5 [5] and SGR 1900+14 on 1998 August 27 [6, 7, 8, 9]. Each of these extraordinary events had a bright ($\sim 10^{44}$ ergs s$^{-1}$), spectrally hard initial spike followed by a softer, several minute long tail showing coherent pulsations at 8 and 5 s, respectively. More recently, an intermediate flare ($\sim 10^{43}$ ergs) lasting 40 s was recorded from SGR 1900+14 on 2001 April 18 [10].

All SGRs are associated with persistent X-ray counterparts; three of them have quiescent luminosities $\sim 10^{34}$ ergs s$^{-1}$, while the quiescent flux level of SGR 1627−41 has not yet been determined [11]. The spectra of three SGRs can be modeled with a power-law (photon indices ∼ 2−3.5); SGR 1900+14 requires an additional blackbody component ($kT$ ∼0.5 keV [12, 13, 14]). Two SGRs show low-amplitude pulsations in their persistent emission. The frequency of these pulsations is increasing rapidly, consistent with the interpretation of an underlying strongly magnetized neutron star [15]. For a more comprehensive review of the properties of SGRs, see [16].

During the last few years, changes in the X-ray emission properties of SGRs have been noted during episodes of burst activity [13, 17]. Through studying the transient effects imparted upon SGRs (or the lack thereof) during times of burst activity, we have gained deeper insight into the nature of the burst mechanism and the SGR systems in general. Here, we review the observed influence of burst activity on SGR pulse properties and persistent X-ray emission, limiting ourselves to the two SGRs that show pulsations in their X-ray emission, namely SGR 1900+14 and SGR 1806−20.

## PERSISTENT AND PULSED FLUX

Changes in the flux of SGRs was first noted in SGR 1900+14 following the giant flare of August 27 [18, 19, 20, 12]. Following this discovery, a compilation of persistent and pulsed flux measurements over several years (Figure 1 [13]) revealed that, in general,

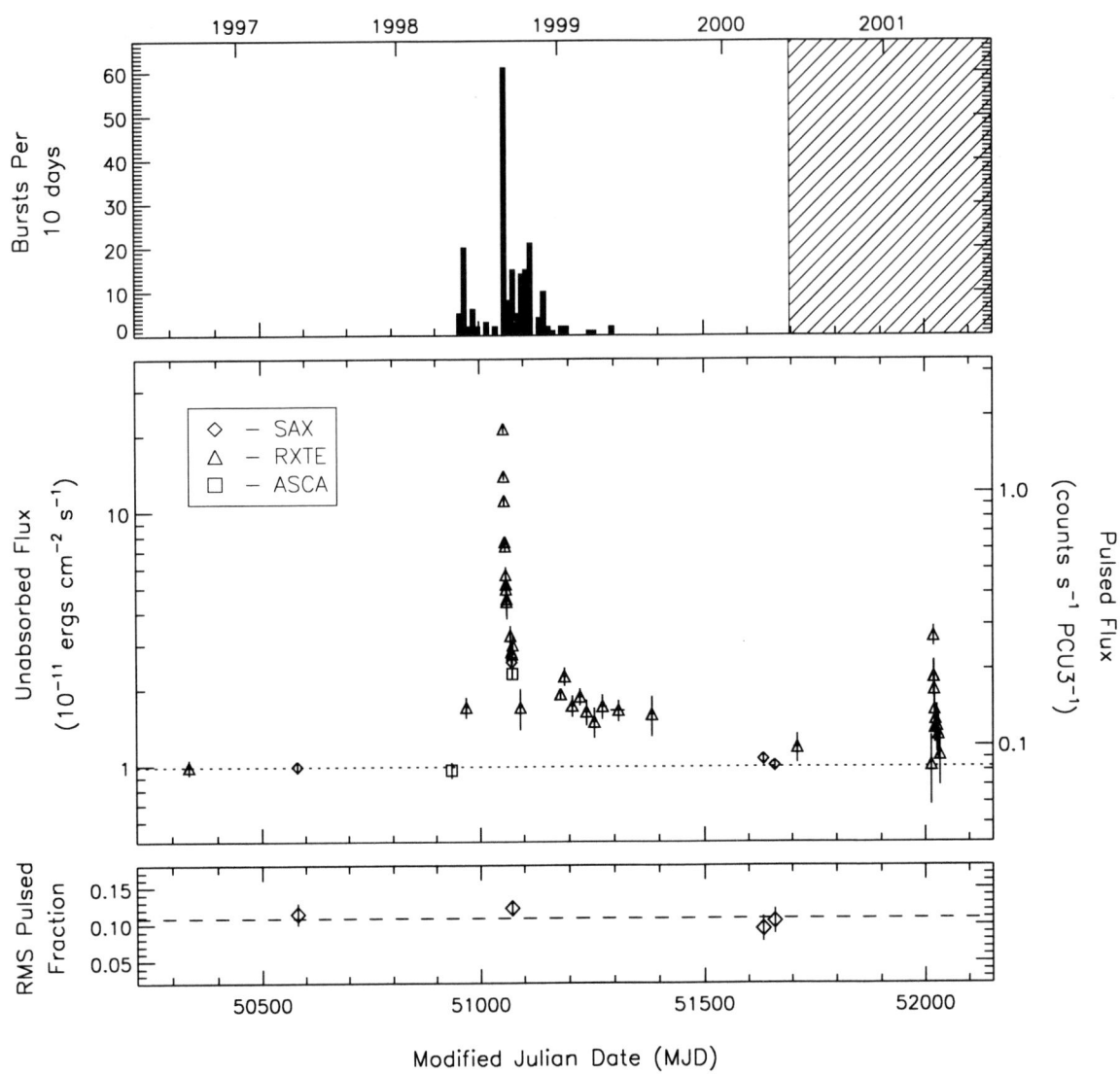

**FIGURE 1.** *Top panel* – Burst rate history of SGR 1900+14 as observed with BATSE. *Middle panel* – Persistent/Pulsed flux history of SGR 1900+14 covering 4.5 years. The left vertical scale is unabsorbed 2–10 keV flux and the right is pulsed flux in units of counts s$^{-1}$ PCU3$^{-1}$. The dotted line marks the nominal quiescent flux level of this SGR. Note that the spike in the pulsed flux shortly after MJD 52000 coincides with a burst active episode not covered by the BATSE monitoring (*top*). See text for further details. *Bottom panel* – Pulse fraction of SGR 1900+14 (2–10 keV) as measured within the four BeppoSAX observations using the MECS instruments. The dashed line marks the mean RMS pulsed fraction ($f_{RMS} \sim 0.11$).

there is an excellent correlation between burst activity (top) and enhancements in the persistent/pulsed flux from this SGR (middle). We have found that the pulse fraction (bottom) is consistent with remaining constant at most epochs despite changes in the persistent flux. It is by assuming that this fraction remains constant at all times that we can plot both the pulsed flux (*RXTE* PCA) and the persistent flux (*BeppoSAX* and *ASCA*) on the same scale. We note, however, that there are exceptions to this rule when the pulse fraction has increased for short periods of time (see below).

We have found that the brightest pulsed/persistent flux excess seen in Figure 1 is directly linked with the August 27 flare. The excess decays approximately as a power-law in time ($F \propto t^{-0.7}$) following the giant flare (Figure 2 [13]), qualitatively similar to GRB afterglows. In order to avoid confusion between the two phenomena, we will refer to the excesses in SGRs as X-ray tails rather than afterglows hereafter. The spectrum (0.1–10 keV) of the X-ray tail at ~19 days after the flare was found to be exclusively non-thermal [12].

value. The pulsed fraction increases above the quiescent level (11% RMS) up to ~20% during this tail (Figure 4 [21]), and the phase of the pulsations do not shift during the tail.

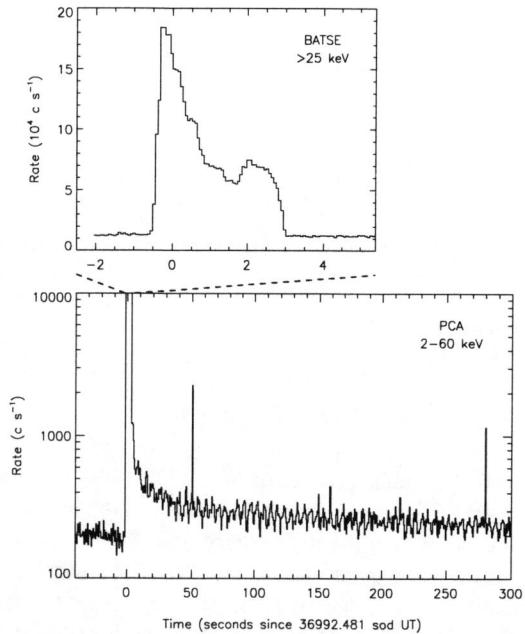

**FIGURE 3.** The energetic burst of 1998 August 29 from SGR 1900+14 as seen with BATSE (top panel) and the RXTE PCA (bottom panel).

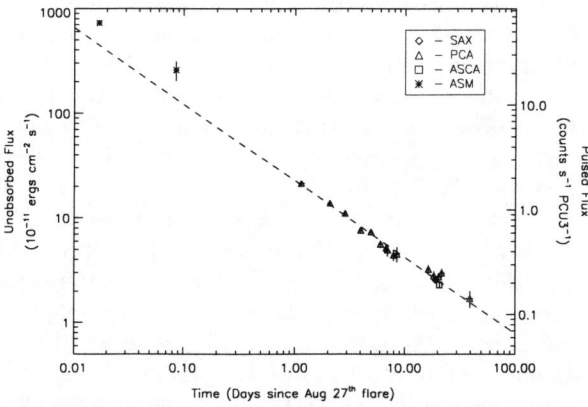

**FIGURE 2.** The flux decay following the 1998 August 27 flare from SGR 1900+14. The reference time is the beginning of the flare as observed in soft γ-rays. The dotted line is a fit to the RXTE/PCA, BeppoSAX, and ASCA data only (i.e. the ASM data are not included in the fit). The slope of this line is $-0.713 \pm 0.025$.

For SGR 1900+14, there are now four X-ray tails that can be linked with specific bursts or flares. The second of these events was recorded on 1998 August 29 (Figure 3 [17]). This burst had a high gamma-ray fluence and an X-ray tail whose bolometric flux decayed approximately as a power-law in time. The spectrum of this tail softens with time. Formally, the spectrum is equally well fit by a power-law plus a blackbody or a thermal bremsstrahlung, each with interstellar attenuation [17]. However, the bremsstrahlung model yields a column density ~5 times larger than the measured column from the persistent emission whereas the two component model fit yields a column consistent with the persistent emission

The last two bursts with X-ray tails were detected on 2001 April 18 and April 28. Spectral analysis of the April 18 burst tail is presented in [14, 22, 23]. The April 28 event is discussed in greater detail elsewhere in this volume [21]. During each of these events, the pulse fraction was found to increase during the tail [24, 21]. The spectrum of the April 28 tail is a cooling blackbody [21], dissimilar to the 1998 August tails which each required a power-law component.

Even though we have detected just four X-ray tails following energetic bursts from SGR 1900+14, we find significant differences between them. First, there are varying levels of thermal and non-thermal emission within the tails. Also, the pulse fraction increases by up to a factor ~3 in one case (April 28) and not at all in another (August 27). An interesting trend which arises from this small set of X-ray tails is that the pulse fraction enhancement in the separate X-ray tails appears to correlate with the magnitude of the thermal contribution to the X-ray flux. That is, tails with the highest relative blackbody flux show the largest increase in pulse fraction.

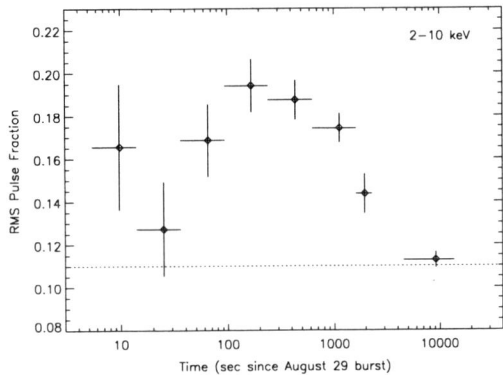

**FIGURE 4.** The evolution of the 2–10 keV pulse fraction during the X-ray tail following the burst of 1998 August 29. The dotted line denotes the average pulse fraction observed during quiescence.

## PULSE PROFILES

Currently, the pulse profiles of both SGR 1900+14 and SGR 1806−20 are very nearly sinusoidal (i.e. they show very little power at the higher harmonics). This has not always been the case, however, as both SGRs have shown significant changes in their pulse profiles during the last several years. The most notable of which was the dramatic change in the pulse profile of SGR 1900+14 during the tail of the giant flare of August 27 [6, 7, 8, 9].

Forty seconds after the onset of the August 27 flare, 5.16 s coherent gamma-ray pulsations at high amplitude emerged. Initially, the pulse profile was complex, having four distinct maxima per rotation cycle. Toward the end of the flare, the pulse profile was significantly more sinusoidal (Figure 5 – middle row). The same qualitative behavior was observed in the persistent X-ray emission from SGR 1900+14. In all observations prior to 1998 August 27, the pulse profile was complex having significant power at higher harmonics (Figure 5 – top row). For all observations after August 27 through early 2000, the pulse profile remained relatively simple (Figure 5 – bottom row). Hence, the pulse profile change observed at gamma-ray energies during the tail of the August 27 flare translated to the persistent emission from this SGR in a sustained manner (i.e. for years after the August 27 X-ray tail had disappeared) [13].

By the middle of the year 2000, the pulse profile began to show slightly more power in the higher harmonics [25]. The next observations of SGR 1900+14 took place in the hours and days following the April 18 intermediate flare. At some point between the latter half of 2000 and the days directly after the April 18 flare, the pulse profile simplified in shape [24]. As with the August 27 flare, the direction of the pulse profile change here was the same in that power at the higher harmonics lessened following the flare [25].

A systematic study of the temporal and spectral evolution of the pulse profile of SGR 1900+14 using exclusively *RXTE* PCA observations has recently been completed [25]. In this study, we show that there is a significant energy dependence of the pulse profile, particularly when the profile was complex in shape (i.e. prior to 1998 August 27). Moreover, the pulsed flux spectrum during the X-ray tail of the August 27 flare becomes harder with time.

The evolution of the pulse profile of SGR 1806−20 is also presented in [25]. We find significant temporal evolution in the pulse profile of this SGR from 1996 November to 1999 January. Due to the sparseness of the observations, however, we cannot determine the exact time of this change, nor the timescale over which it progressed to better than 2.3 years.

## PULSE TIMING

Coherent pulsations from the persistent emission of SGR 1806−20 were discovered within an *RXTE* PCA observation from 1996 November [15]. From archival observations, it was found that the spin frequency of this SGR was decreasing rapidly, indicative of a strongly magnetized neutron star spinning down via magnetic braking [15]. The spin frequency history of this SGR now extends from 1993 through 2001 (Figure 6 [26]). We have found that at all times, the SGR has been spinning down, but the rate of spindown shows substantial variability. In fact, the measured spin-down torque on this SGR has been found to vary by up to a factor ∼4. Unlike the flux variability of SGR 1900+14, the torque variations seen in SGR 1806−20 do not correlate with the burst activity [26].

Pulsations from the X-ray counterpart of SGR 1900+14 were discovered during an ASCA observation in 1998 April [27], shortly before the SGR entered an intense, sustained burst active interval [28]. Similar to SGR 1806−20, subsequent observations showed that this SGR was spinning down rapidly and irregularly [19, 29]. The spin frequency history of this SGR now extends from 1996 through 2001 (Figure 7 [26]). As with SGR 1806−20, the variations in torque do not directly correlate with the burst activity from this SGR with one notable exception, the giant flare of August 27.

From an earlier compilation of pulse frequency measurements between 1996 September and 1999 January, we showed that the spin down of this SGR showed small variations, yet remained constant on average for timescales longer than about a month at nearly all epochs

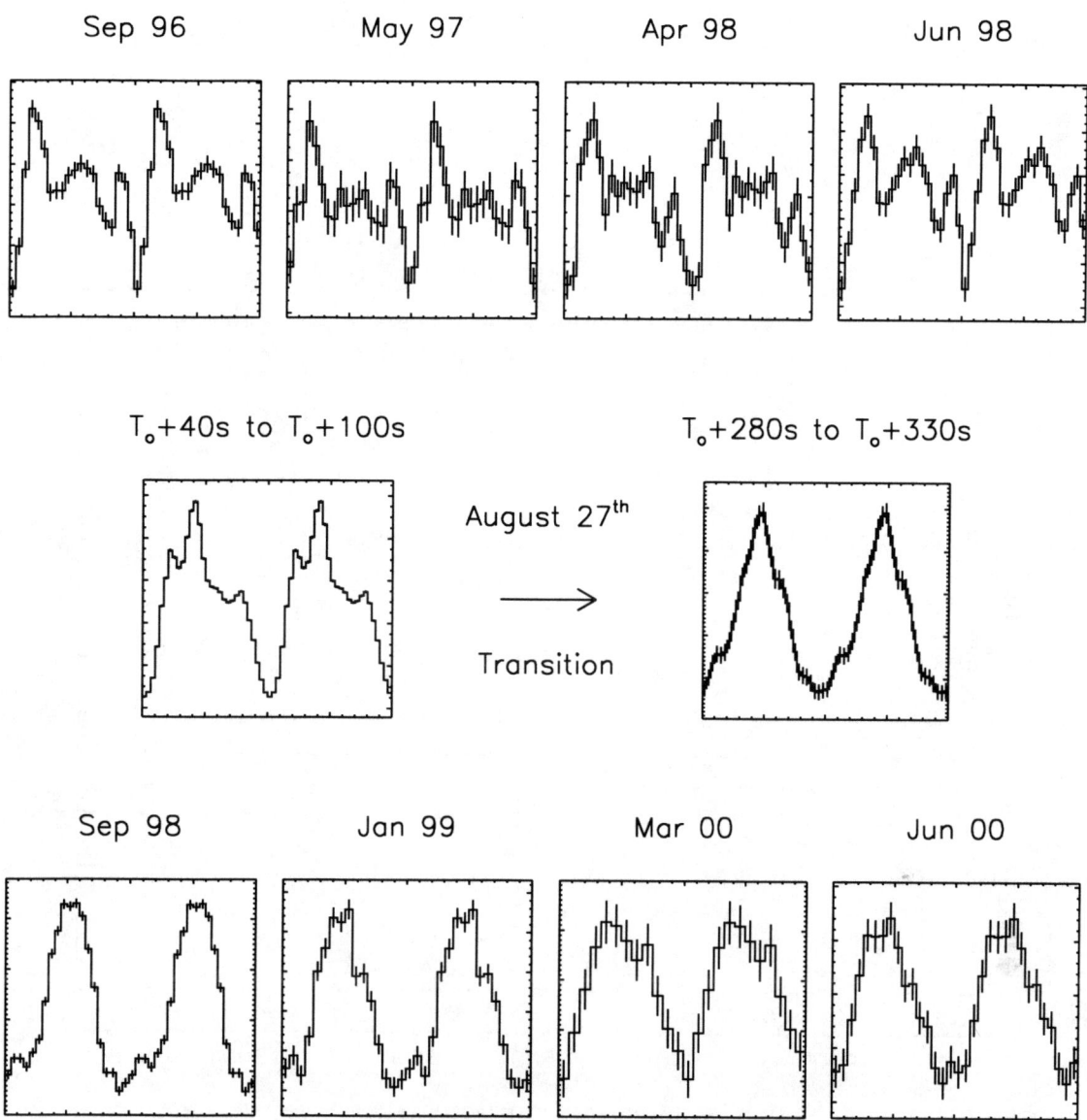

**FIGURE 5.** Evolution of the pulse profile of SGR 1900+14 covering 3.8 years. All panels display two pulse cycles and the vertical axes are count rates with arbitrary units. The two middle panels were selected from Ulysses data (25–150 keV) of the August 27$^{th}$ flare. Times over which the Ulysses data were folded are given relative to the onset of the flare ($T_o$). See text for further details. The top and bottom rows are integrated over the energy range 2–10 keV. From top-to-bottom, left-to-right, the data were recorded with the RXTE, BeppoSAX, ASCA, RXTE, RXTE, RXTE, BeppoSAX, and RXTE.

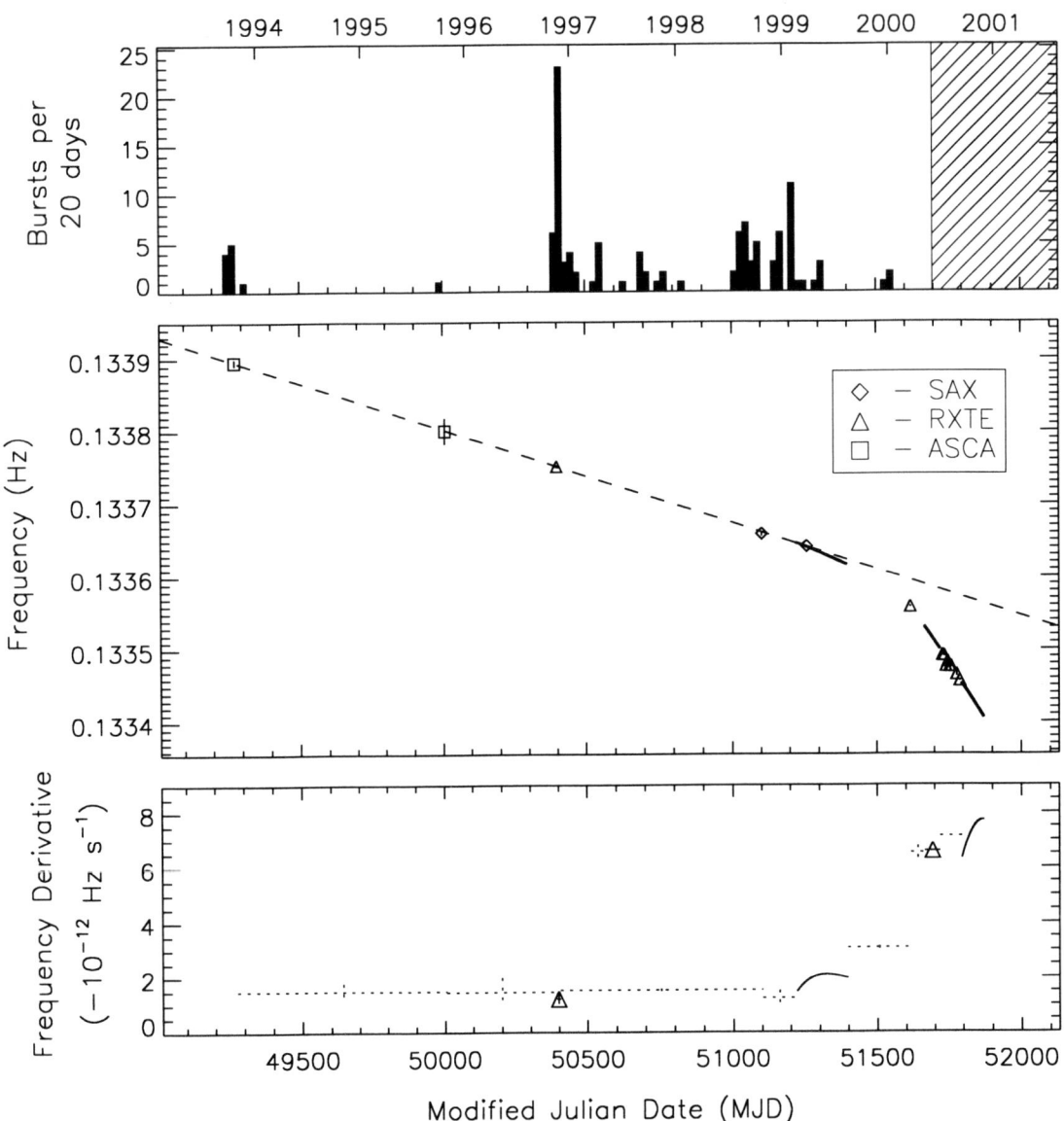

**FIGURE 6.** *Top* – Burst rate history of SGR 1806−20 as observed with BATSE. The hashed region starts at the end of the *CGRO* mission. *Middle* – The frequency history of SGR 1806−20 covering 7.1 years. Plotting symbols mark individual frequency measurements and solid lines denote phase-connected timing solutions. The dashed line marks the average spin-down rate prior to burst activation in 1998. *Bottom* – The frequency derivative history over the same timespan. Dotted lines denote average frequency derivative levels between widely spaced frequency measurements. Solid lines mark phase-coherent timing solutions and triangles mark instantaneous torque measurements, both using *RXTE* PCA data.

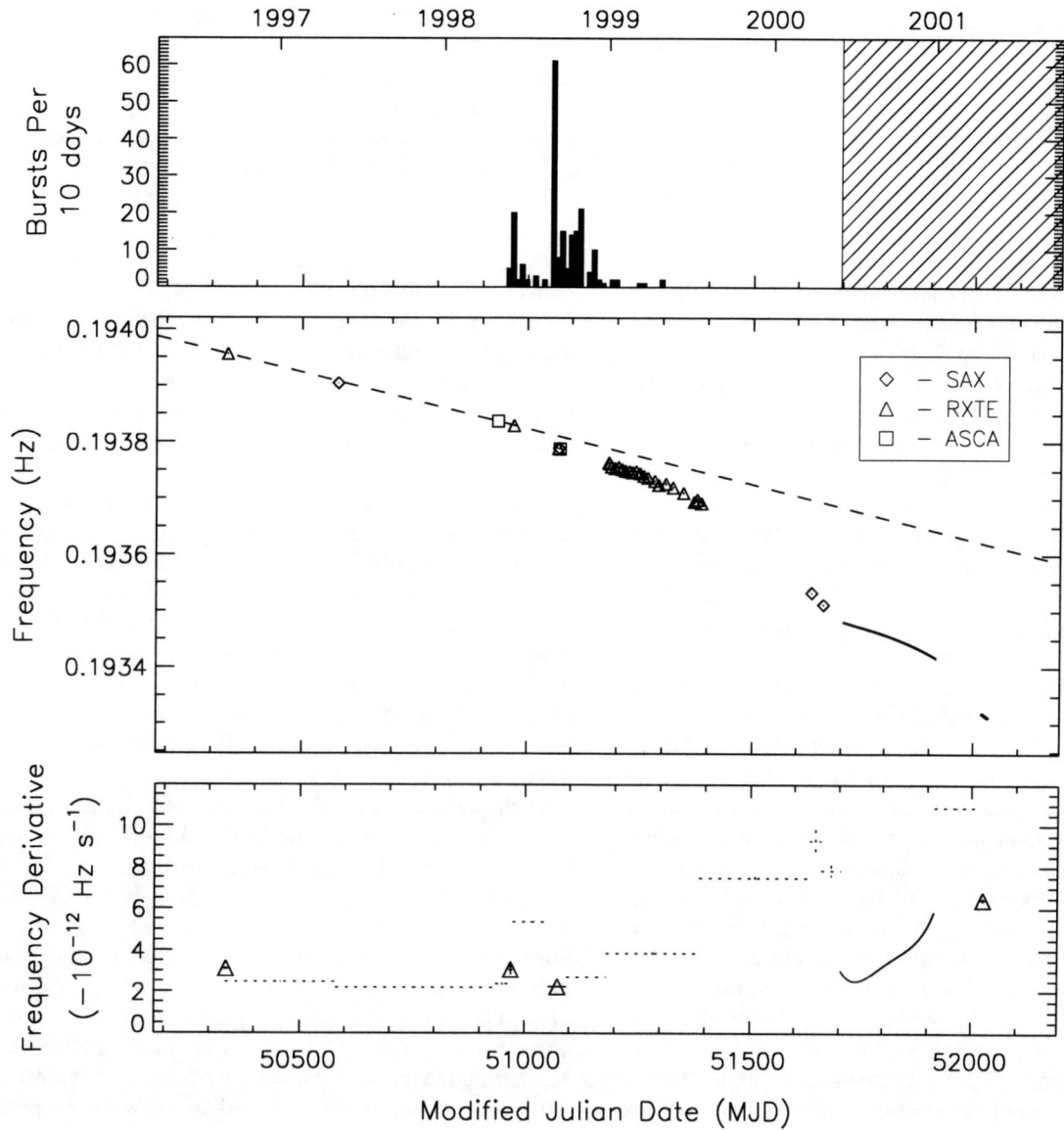

**FIGURE 7.** *Top* – Burst rate history of SGR 1900+14 as observed with BATSE. The hashed region starts at the end of the *CGRO* mission. *Middle* – The frequency history of SGR 1900+14 covering 4.7 years. Plotting symbols mark individual frequency measurements and solid lines denote phase-connected timing solutions. The dashed line marks the average spin-down rate prior to burst activation in 1998. *Bottom* – The frequency derivative history over the same timespan. Dotted lines denote average frequency derivative levels between widely spaced frequency measurements. Solid lines mark phase-coherent timing solutions and triangles mark instantaneous torque measurements, both using *RXTE* PCA data.

[29]. The lone exception to this rule was an 80 day interval during the middle of 1998 where the average spin-down rate nearly doubled. Contained within this 80 day interval was the August 27 flare and we argued that this spin-down anomaly was likely linked to the flare [29]. Subsequently, Palmer [30] showed that the phase of the pulsations in gamma-rays was offset from a backward extrapolation of the X-ray pulse train recorded during the days following the flare, supporting our earlier inference that the star had spun down rapidly during/after the giant flare. The conclusion of rapid spin down depends critically on the energy dependence of the pulse profile. More recent observations of gamma-ray pulsations during the 2001 April 18 flare have shown that there is very little change in the pulse profile with energy for this source [24], substantiating the claim that the phase offset in the August 27 flare was due to a sudden change in torque, perhaps from a relativistic particle outflow from the stellar surface. Gamma-ray observations during the flare [9] and a transient radio nebula discovered following the flare [31] provide independent evidence for the existence of a particle outflow.

As illuded to in the previous paragraph, we searched for a similar effect in the aftermath of the April 18 intermediate flare. We found that unlike the August 27 flare, the phase of the gamma-ray pulsations matched nicely with the X-ray ephemeris [24]. A fortuitous X-ray monitoring observation 4 days prior to the flare shows that there was a short timescale change in the spin ephemeris somewhere between April 14 and April 18, however, the brevity of the April 14 observation precluded us from constraining the manner in which this change occurred.

Finally, we note that although the August 27 flare likely did alter the spin down of SGR 1900+14, its impact was very small relative to the much larger variations observed during $\sim 5$ years of monitoring. So, in general, the direct effects of burst activity are insignificant to the overall torque noise in each of these SGRs. For a more complete discussion of the torque variability in these two SGRs, as well as a quantitative analysis of the torque noise, see [26].

## DISCUSSION

We have summarized the recent observations of dynamic behavior in the persistent and pulsed emission from SGR 1900+14 and SGR 1806−20. Now, we will discuss what constraints these observations place on the models for the SGRs, in particular the magnetar model.

The magnetar model postulates that the SGRs are young neutron stars with super-strong magnetic fields ($B \sim 10^{14} - 10^{15}$ G). It is the decay of this strong field which powers both the burst and persistent emission [32, 33]. The steady X-ray emission is generated by persistent magnetospheric currents and low-level seismic activity within and beneath the stellar surface. The burst emission is due to the build up of stress in the stellar crust from the evolving magnetic field and the eventual release of this stress when the crust fractures. To date, this model provides the most accurate description of the persistent, pulsed, and burst properties of SGRs.

Currently, only four clear X-ray tails have been detected, all from SGR 1900+14. As mentioned earlier, there is a potential correlation between the relative abundance of thermal emission in the tail and the enhancement of the pulse fraction. This correlation, if proved correct with the detection and analysis of several more SGR tails, would provide a strong argument for heating of a localized region on the neutron star during bursts. Since the pulse fraction increases during some of these tails, the flux enhancement must be anisotropic about the star. In the cases of the August 29 and April 28 burst tails, the location of the heating is also constrained. In each of these events, we have precise pulse phase information prior to, during, and after the tail. For both bursts, the phase of the pulsations during the tail does not shift relative to the pulse phase prior to the burst. This requires that the localized region on the neutron star with the largest relative flux enhancement is the same region giving rise to the persistent X-ray pulse peak (e.g. the polar cap).

With regards to the magnetar model, localized heating of the polar cap requires that the fracture site of the burst be at the same location. The peak of the August 29 flare in gamma-rays lags behind the centroid of the pulse profile peak in X-rays by $\sim 0.1$ cycle, although the burst light curve (duration $\sim 3.5$ s) covers a large fraction of a pulse cycle spanning the pulse valley. The phase alignment of the April 28 burst is yet to be determined. Measuring the phase alignment of the April 28 burst and detailed modeling of the expected lag between the peak in the X-ray pulse profile and the rise and/or peak of the burst are required before one can determine whether or not these two bursts fit with the picture of a localized fracture region near the polar cap.

The dramatic change in the pulse profile of SGR 1900+14 in conjunction with the giant flare requires a substantial change in the magnetic field of the neutron star [13, 34]. In the magnetar model, there are at least two possible ways this can happen. One possibility is that a twist in the magnetosphere is generated following the flare, driving a persistent current which produces an optically thick scattering screen at some substantial distance ($\sim 10 R_*$) from the stellar surface. In this model, the surface field geometry remains complex at all times. The pulse profile, however, simplifies when the scattering screen is present (i.e. after the flare). The scattering screen must have the properties of redistributing the radiation in phase, but not in energy in order to account for the reemergence of the blackbody component after the August 27 tail fades away [35]. The decay of this magnetospheric twist is believed to be several years. An alternative scenario involves restructuring of the surface magnetic field geometry. In this picture, the field geometry is complex prior to the flare and relaxes to a more dipolar structure following the event giving rise to the observed change in pulse profile.

Thompson, Lyutikov & Kulkarni [35] recently investigated each of these scenarios in detail, noting advantages and disadvantages for each model. In this work, they have identified further observational tests involving the energy spectrum of the emission before and after the flare. Simulations of the expected behavior [36] and an analysis of the spectral evolution of SGR 1900+14 [37] are currently underway to work towards resolving this issue.

Unlike the flux variability, the torque enhancements in these systems do not correlate with the burst activity. In the context of the magnetar model, the abscence of a direct correlation between these two parameters has strong implications for the underlying physics behind each phenomenon. The magnetar model postulates that the bursting activity in SGRs is a result of fracturing of the outer crust of a highly magnetized neutron star. Furthermore, the majority of models proposed to explain the torque variability in magnetars invoke crustal motion and/or low-level seismic activity [38, 39, 34]. Since there is no direct correlation between the burst activity and torque variability, then either (*i*) the seismic activities leading to each observable are decoupled from one another, or (*ii*) at least one of these phenomena is *not* related to seismic activity [26]. Simultaneous spectral information from imaging X-ray telescopes (e.g. *BeppoSAX*, *Chandra*, and *XMM-Newton*) complimentary to the torque measurements obtained with the *RXTE* PCA would be useful in determining the nature of the torque variabilitity in these SGRs.

## ACKNOWLEDGMENTS

I thank the many collaborators who have contributed to the results discussed here. I would like to thank Chryssa Kouveliotou for many useful discussions and a careful reading of the manuscript. I acknowledge my own support from the Long Term Space Astrophysics program (NAG 5-9350).

## REFERENCES

1. Olive, J.-F., et al., "FREGATE observations of a strong burst from SGR 1900+14", in *Woods Hole 2001 GRB Conference*, edited by R. Vanderspek, AIP, New York, 2002.
2. Cheng, B., et al., *Nature*, **382**, 518 (1996).
3. Göğüş, E., et al., *ApJ*, **558**, 228 (2001).
4. Bak, P., Tang, C., and Wiesenfeld, K., *Phys. Rev. A*, **38**, 364 (1988).
5. Mazets, E., et al., *Nature*, **282**, 587 (1979).
6. Hurley, K., et al., *Nature*, **397**, 41 (1999).
7. Mazets, E., et al., *Astron. Lett.*, **25**, 635 (1999).
8. Feroci, M., et al., *ApJ*, **515**, L9 (1999).
9. Feroci, M., Hurley, K., Duncan, R., and Thompson, C., *ApJ*, **549**, 1021 (2001).
10. Guidorzi, C., et al., *GCN Circ. 1041* (2001).
11. Kouveliotou, C., et al., *in preparation* (2002).
12. Woods, P., et al., *ApJ*, **518**, L103 (1999).
13. Woods, P., et al., *ApJ*, **552**, 748 (2001).
14. Kouveliotou, C., et al., *ApJ*, **558**, L47 (2001).
15. Kouveliotou, C., et al., *Nature*, **393**, 235 (1998).
16. Hurley, K., "The 4.5±0.5 Soft Gamma Repeaters in Review", in *Gamma-Ray Bursts: 5th Huntsville Symp.*, edited by R. Kippen, R. Mallozzi, and G. Fishman, AIP 526, New York, 2000, p. 763.
17. Ibrahim, A., et al., *ApJ*, **558**, 237 (2001).
18. Remillard, R., Smith, D., and Levine, A., *IAU Circ. 7002* (1998).
19. Kouveliotou, C., et al., *ApJ*, **510**, L115 (1999).
20. Murakami, T., et al., *ApJ*, **510**, L119 (1999).
21. Lenters, G., et al., "An Extended Burst Tail from SGR 1900+14 with a Thermal X-ray Spectrum", in *Woods Hole 2001 GRB Conference*, edited by R. Vanderspek, AIP, New York, 2002.
22. Feroci, M., et al., *in preparation* (2002).
23. Fox, D., et al., *ApJ submitted* (2002).
24. Woods, P., et al., *in preparation* (2002).
25. Göğüş, E., et al., *ApJ*, **submitted** (2002).
26. Woods, P., et al., *ApJ*, **in press** (2002).
27. Hurley, K., et al., *ApJ*, **510**, L111 (1999).
28. Hurley, K., et al., *ApJ*, **510**, L107 (1999).
29. Woods, P., et al., *ApJ*, **524**, L55 (1999).
30. Palmer, D., *astro-ph/0103404* (2001).
31. Frail, D., Kulkarni, S., and Bloom, J., *Nature*, **398**, 127 (1999).
32. Thompson, C., and Duncan, R., *MNRAS*, **275**, 255 (1995).
33. Thompson, C., and Duncan, R., *ApJ*, **473**, 322 (1996).
34. Thompson, C., et al., *ApJ*, **543**, 340 (2000).
35. Thompson, C., Lyutikov, M., and Kulkarni, S., *ApJ*, **in press** (2002).
36. Thompson, C., et al., *in preparation* (2002).
37. Woods, P., et al., *in preparation* (2002).
38. Thompson, C., and Blaes, O., *Phys. Rev. D*, **57**, 3219 (1998).
39. Harding, A., Contopoulos, I., and Kazanas, D., *ApJ*, **525**, L125 (1999).

# An Extended Burst Tail from SGR 1900+14 with a Thermal X-Ray Spectrum

Geoffrey T. Lenters[*], Peter M. Woods[†], Johnathan E. Goupell[*], Chryssa Kouveliotou[**], Ersin Göğüş[‡], Kevin Hurley[§], Dmitry Frederiks[¶] and Sergey Golenetskii[¶]

[*]*Hope College, Department of Physics & Engineering, Holland, MI 49422*
[†]*USRA, National Space Science and Technology Center, Huntsville, AL 35805*
[**]*MSFC/IPA, National Space Science and Technology Center, Huntsville, AL 35805*
[‡]*Dept. of Physics, University of Alabama in Huntsville, Huntsville, AL 35899*
[§]*University of California, Berkeley, Space Sciences Laboratory, Berkeley, CA 94720–7450*
[¶]*Ioffe Physical-Technical Institute, St. Petersburg, 194021, Russia*

**Abstract.**
The Soft Gamma Repeater, SGR 1900+14, entered a new phase of activity in April 2001 initiated by the intermediate flare recorded on April 18. Ten days following this flare, we discovered an abrupt increase in the source flux which was later attributed to a high fluence burst from SGR 1900+14 recorded by other spacecraft (*Ulysses* and *KONUS*) while the SGR was Earth-occulted for *RXTE*. We present here spectral and temporal analysis of both the burst of April 28 and the long X-ray tail following it. We show that the tail spectrum is exclusively thermal, indicative of a cooling hotspot on the neutron star surface. From the combined spectral and temporal results, we place constraints on the burst mechanism and its effects on the persistent emission.

## INTRODUCTION

Soft Gamma Repeaters (SGRs) are a small class of astrophysical objects (four confirmed sources and a fifth candidate source) discovered by their emission of bright bursts of soft gamma-rays. A brief summary of the salient properties of SGRs appears in these proceedings [1], and a more comprehensive review appears in [2].

In this paper we focus on SGR 1900+14 which has been the most prolific burst emitter during the last several years. The most energetic event during the recent activity was a giant flare recorded on 1998 August 27 [3, 4, 5, 6]. Following the flare, the persistent X-ray flux increased by more than a factor of ~20 and decayed rapidly as a power-law over at the next 40 days [7]. Two weeks into the decay phase, the X-ray emission was found to have a non-thermal (powerlaw) spectrum [7]. In addition to this event, two other high-fluence bursts have shown extended X-ray tails. These two events were recorded on 1998 August 29 [8] and more recently on 2001 April 18 [9, 10, 11, 12].

Here we present results on the temporal decay and spectral evolution of the tail of a burst recorded on 2001 April 28, ten days after the intense April 18 event. Using data acquired with the *Rossi X-ray Timing Explorer* (RXTE) Proportional Counter Array (PCA), we fit several spectral models to the tail flux to constrain the radiative mechanism that governs its emission. We discuss the implications of these results on the current models for SGR emissions.

## 2001 APRIL 28 BURST

The reactivation of SGR 1900+14 in 2001 April began with the detection of a energetic (~$10^{43}$ ergs) intermediate flare on April 18. This activity triggered a series of ToO observations of the source with several satellites (*BeppoSAX*, *RXTE*, *Chandra*). An observation of the SGR with the *RXTE* PCA on April 28, 10 days following the intermediate flare, revealed an intriguing discontinuity in the source flux between spacecraft orbits (see Figure 1). The trend was similar to the one observed during the 1998 August 29 event [8], where the decaying tail extended over two *RXTE* orbits, and thus, strongly indicative of burst activity during Earth occultation of the source. A search in *Ulysses* and *KONUS* Wind data revealed a relatively intense burst on 2001

April 28, recorded approximately 121 seconds *before* SGR 1900+14 emerged from Earth-occultation at a higher flux level in the PCA.

## Gamma-ray Observations

On 2001 April 28, *Ulysses* and *KONUS* Wind observed a burst whose triangulated position (an annulus) contained SGR 1900+14. Its time history was unusual in that it had a duration of ∼2 s (Figure 1), compared to the more common, short duration SGR bursts. The *KONUS* energy spectrum was well-fit by a power-law times an exponential function ($\frac{dN}{dE} \propto E^{-0.5} \exp(-\frac{E}{kT})$) with $kT = 14.6$ keV. We estimated the (15 − 100 keV) source fluence and peak flux (over 0.016 s) to be $1.4 \times 10^{-5}$ ergs cm$^{-2}$ and $1.0 \times 10^{-5}$ ergs cm$^{-2}$ s$^{-1}$, respectively. The corresponding burst energy is ∼ $3.3 \times 10^{41}$ ergs (assuming a source distance for SGR 1900+14 of 14 kpc; [13]).

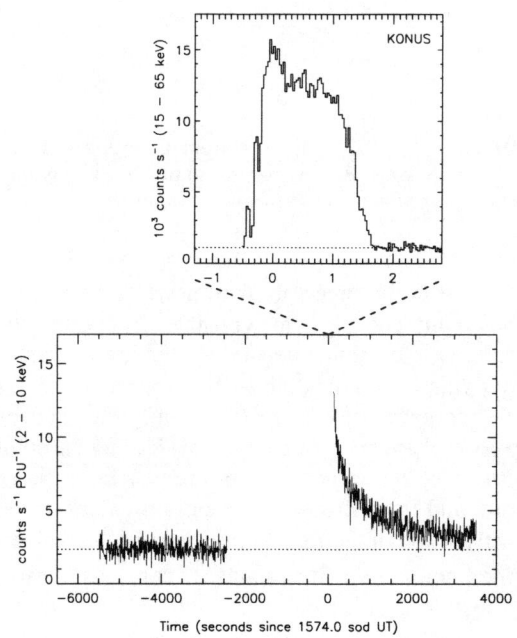

**FIGURE 1.** *Top Panel: KONUS* 15−65 keV time history of the 2001 April 28 burst from SGR 1900+14. The background level is indicated with a dotted line. *Bottom Panel: RXTE*/PCA 2−40 keV lightcurve with the nominal background denoted by the dotted line.

## X-ray Observations

### Temporal Analysis

We have performed an extensive temporal analysis of the 2001 April 28 *RXTE*/PCA tail data. Using the *ftool* PCABACKEST, we subtracted the nominal background responsible for the slow orbital variations visible in the lightcurve. We then subtracted the contribution from the quiescent SGR flux, other discrete X-ray sources in the field-of-view, and the Galactic ridge [14], all of which are assumed to be constant over this 0.2 day interval. We estimated the average of the residual count rate during the two *RXTE* orbits prior to the burst and subtracted this value (shown as the dotted line in Figure 1) from the orbit containing the burst tail.

The tail counts were divided into three energy bands (2−5, 5−10, and 10−20 keV) and binned logarithmically. A single power-law function (i.e. $F \propto t^{-\alpha}$) was then fit to each lightcurve yielding decay indices ($\alpha$) of 0.64(1), 0.83(1), and 0.91(3), respectively. We observe a significant increase of the decay index with energy, indicating spectral softening of the tail with time. However, we note that the $\chi^2$ values for the three bands are large which suggests that a single power-law model is not an accurate representation of the decay.

Next, we studied the pulse properties of SGR 1900+14 during the X-ray tail. The precise ephemeris was determined elsewhere [12] from the complete *RXTE*/PCA data taken during the 2001 April reactivation. We folded seven time intervals during the decaying burst tail as well as the two orbits prior to the burst for comparison, all using 2−10 keV data.

The level of the RMS pulse fraction (2−10 keV) increased directly after the April 18 flare to ∼18% [5, 10, 12] and subsequently decayed to 14% by May 1 [10, 12]. From these data, we infer a pulse fraction of ∼15% for the pre-April 28 burst PCA data. Under this assumption, we can determine the background level during the two pre-burst orbits on April 28 and thereby measure the pulse fraction during the tail. The RMS pulse fraction during the seven intervals of the tail, the quiescent pulse fraction level (dotted line), and the inferred level (dashed line) are shown in Figure 2. We find that the pulse fraction increases significantly after the burst to ∼33%.

We find no evidence for a systematic shift in phase of the pulse maximum or the slope of the phases (i.e. pulse frequency) relative to the pre-defined ephemeris, i.e., the pulse phase remained steady during the tail. We find that the pulse shape is constant, showing only marginal variability (i.e. low-level phase noise).

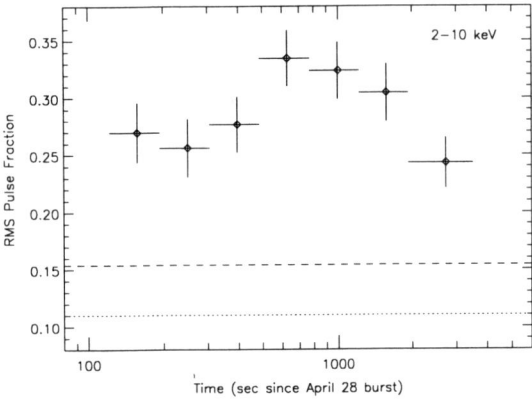

**FIGURE 2.** The evolution of the 2–10 keV pulse fraction during the X-ray tail following the burst of 2001 April 28. The dashed line marks the pulse fraction inferred from imaging telescopes. The dotted line denotes the average pulse fraction observed during quiescence.

## Spectral Analysis

We removed the Galactic ridge component, quiescent SGR flux, and other discrete X-ray sources for the spectral analysis of the tail emission as follows. We first subtracted the background from the source spectra of the two orbits prior to the burst; the background spectrum was estimated using PCABACKEST. We then fit the residual spectrum (2–30 keV) with a multicomponent function (blackbody + power-law + Gaussian line), which was sufficient to model the observed residual spectrum. The blackbody temperature and absorption were fixed at the measured values for SGR 1900+14 and all other parameters were allowed to vary in the fit. Using this model, we obtained a good fit to the data ($\chi^2_\nu = 1.15$ for 51 degrees of freedom). Please note that this spectrum is *not* entirely the quiescent spectrum of the SGR, but is rather the sum of the SGR spectrum with the Galactic ridge and other discrete X-ray sources in the PCA field-of-view.

Further, we used the instrumental background (as estimated with PCABACKEST) together with the spectrum defined above as the "background" for our spectral analysis of the tail flux. More specifically, we defined a two part model, where one component is the presumed tail spectrum and the second (fixed) part is the remaining background as defined above. We have fit several models to the time-integrated tail: bremsstrahlung, power-law, and blackbody models (each with interstellar absorption) yielding $\chi^2$ values of 70.54, 119.6, and 21.03, respectively; all having 29 degrees of freedom. Clearly, the simple blackbody model yields the best fit to the data (Figure 3).

To study the spectral evolution of the tail, we divided

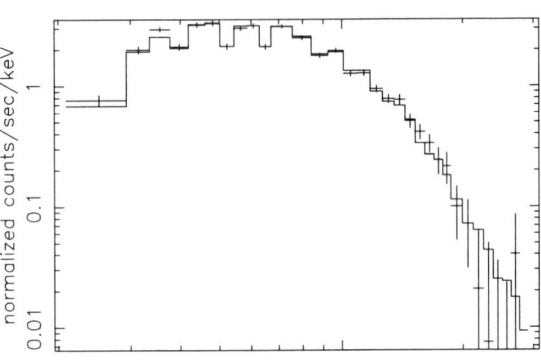

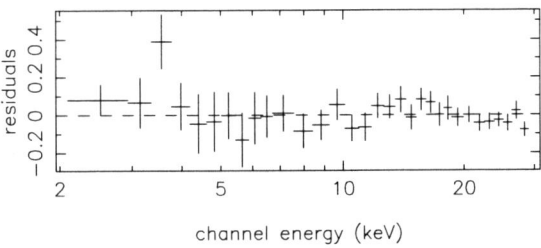

**FIGURE 3.** The *RXTE* PCA spectrum of the X-ray tail of the 2001 April 28 burst (900 s exposure) fit to a blackbody model. The residuals to the fit are shown in the bottom panel.

the data into eight intervals with nearly an equal number of counts and fit them with the blackbody model. We find that the column density ($N_H$) does not vary significantly across the tail, while the temperature ($kT$) decreases monotonically. To better quantify the changes in $kT$, we simultaneously fit all intervals forcing the column density to be the same in each interval. Both the temperature and normalization (i.e. emitting area) decrease modestly through the tail (Figure 4). The $N_H$ value determined here ($N_H = 3(1) \times 10^{\times 10^{22}}$ cm$^{-2}$) is consistent with the quiescent value [7].

## DISCUSSION

We have detected, for the first time, a cooling, exclusively thermal X-ray tail following an SGR burst. During the $>10^3$ s long tail, we measure a 35% decrease in the blackbody temperature and a 50% reduction in the emitting area. At all times, the inferred emitting area encompasses a small fraction of the neutron star surface. These observations provide strong evidence for surface heating following this burst from SGR 1900+14.

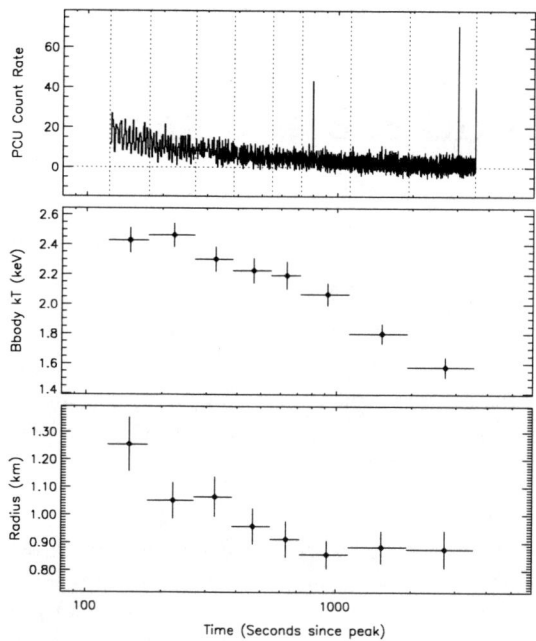

**FIGURE 4.** The parameters of the simple blackbody fit as a function of time for the 2001 April 28 decay. The top panel contains the light curve, the middle panel shows the blackbody temperature, and the bottom panel shows the radius of the emitting area. The vertical dotted lines in the top panel denote the time intervals over which spectra were fit and the horizontal dotted line is at zero counts per second. The radius was determined assuming a distance of 14 kpc to the source [13].

The pulse properties of the tail further constrain the location of the heating on the stellar surface following this burst. The increase in pulse fraction *and* the stability of the phase of the pulsations during the X-ray tail suggests that the location of the heating on the stellar surface must be the same location that produces the peak in the quiescent X-ray pulse profile (e.g. the polar cap). If a different area of the neutron star were heated enough to alter the pulse fraction, as is the case here, the pulse shape would change and the RMS pulse fraction would be expected to *decrease*.

In the magnetar model, it is postulated that the build up of magnetic stresses within the neutron star crust are sufficient to overcome the tensile strength of the crust and crack it [15]. Fracturing the stellar crust perturbs field lines and ultimately leads to a trapped pair-photon fireball. Thompson & Duncan [15] considered the likelihood of heating of the neutron star surface by the resultant trapped fireball suspended above the fracture site. For shallow heating, they estimated the cooling timescale to be equivalent to the burst duration. For deep heating of the crust, the cooling will be longer due to a larger conduction time of the heat to the surface. [16] have shown that the temporal evolution of this cooling could resemble a powerlaw in time with a decay index $\sim -0.8$. This picture fits nicely with the cooling thermal components observed in this X-ray tail. One requirement of this model with regards to this burst is that the burst fracture site be located at one of the magnetic poles. We are currently investigating the phase alignment of the burst in gamma-rays with the X-ray pulse profile.

## ACKNOWLEDGMENTS

GL acknowledges support from NASA's Michigan Space Grant Consortium seed grant program as well as Hope College. PMW and CK acknowledge support from the Long Term Space Astrophysics program (NAG 5-9350).

## REFERENCES

1. Woods, P., "The Effects of Burst Activity on Soft Gamma Repeater Pulse Properties and Persistent Emission", in *these proceedings*, 2002.
2. Hurley, K., "The 4.5±0.5 Soft Gamma Repeaters in Review", in *Gamma-Ray Bursts: 5th Huntsville Symp.*, edited by R. Kippen, R. Mallozzi, and G. Fishman, AIP 526, New York, 2000, p. 763.
3. Hurley, K., et al., *Nature*, **397**, 41 (1999).
4. Feroci, M., et al., *ApJ*, **515**, L9 (1999).
5. Feroci, M., Hurley, K., Duncan, R., and Thompson, C., *ApJ*, **549**, 1021 (2001).
6. Mazets, E., et al., *Astron. Lett.*, **25**, 635 (1999).
7. Woods, P., et al., *ApJ*, **552**, 748 (2001).
8. Ibrahim, A., et al., *ApJ*, **558**, 237 (2001).
9. Guidorzi, C., et al., *GCN Circ. 1041* (2001).
10. Kouveliotou, C., et al., *ApJ*, **558**, L47 (2001).
11. Feroci, M., et al., *in preparation* (2002).
12. Woods, P., et al., *in preparation* (2002).
13. Vrba, F., et al., *ApJL* (2000).
14. Valinia, A., and Marshall, F., *ApJ* (1998).
15. Thompson, C., and Duncan, R., *MNRAS*, **275**, 255 (1995).
16. Thompson, C., Lyutikov, M., and Kulkarni, S., *ApJ*, **in press** (2002).

# The Environments of SGRs: A Brief & Biased Review

Stephen S. Eikenberry

*Astronomy Department, Cornell University, Ithaca, NY USA*

**Abstract.**
I review some recent developments in our understanding of the environments of soft gamma-ray repeaters. I pay particular attention to the apparent association betwwen SGR 1900+14 and SGR 1806-20 and embedded clusters of stars.

## INTRODUCTION

The environments of soft gamma-ray repeaters (SGRs) can provide many important insights into these intriguing and unusual objects. First, the environment of an SGR may provide clues to its origins, including possible progenitor stars, the age of the SGR, its space velocity, etc. Such environmental clues may be gleaned from a surrounding supernova remnant, nearby stars, and unusual ISM features, among other things. Secondly, by studying the impact of the SGR on its surrounding ISM (or vice versa) we can also hope to learn more about the evolution of SGRs.

In this review, I will first go over what we knew (or thought we knew) about SGR environments previously. I will then go over our current state of knowledge regarding the environments of SGRs which are well-localized at the time of the Woods Hole 2001 meeting: SGR 0525-66, SGR 1900+14, and SGR 1806-20.

## WHAT WE USED TO "KNOW"

Prior to about 1998, the community had a reasonably clear (though not universally accepted) picture of SGR environments. The three well-localized SGRs were all "known" to be associated with supernova remnants. SGR 0525-66 lies inside the supernova remnant N49 in the LMC [1]. SGR 1900+14 was localized to within a few arcminutes of G42.8+0.6 (Hurley, et al. [2]; Vasisht, et al. [3]), and also apparently coincident with an unusual double MIa star system [4]. Finally, SGR 1806-20 was found to lie very close to the apparent SNR G10.0-0.3 [2]. This supposed SNR had a bright, variable radio core apparently associated with an X-ray source (Vasisht et al. [5]; Kulkarni & Frail [6]), indicating a possible plerionic SNR similar to the Crab Nebula. Subsequent observations revealed a bright heavily-absorbed infrared star at the radio core position [7], which was then found to be a candidate luminous blue variable (LBV) star based on its IR spectrum [8]. Due to the extreme rarity of both LBVs and SGRs, a chance superposition of these two seemed very unlikely and led to the conclusion that the SGR must be related to the LBV. As I describe below, we now recognize that many of these associations are either coincidental or due to some more complicated underlying associations.

## SGR 0525-66

Since the giant outburst of March 1979, SGR 0525-66 has been associated with the Large Magellanic Cloud [9]. More recently, Rothschild, et al. [1] found an X-ray source within the $\gamma$-ray error box, which lies inside the contours of the N49 SNR. Simple analyses show that the chance projection of the SGR within this SNR is < 1%, even making rather stringent assumptions [10]. Thus, this SGR/SNR association seems robust.

SGR 0525-66 and its host SNR also have the lowest extinction of any known SGR, with $A_V \sim 1$ mag. More recently, Kaplan et al. [11] have obtained HST optical observations of the area surrounding SGR 0525-66. They find several stars within the X-ray positional error box, as well as some apparent bright "lumps" from the SNR itself. The stars are consistent with G/K main sequence stars. It is interesting to note, in comparison with the other well-localized SGRs, that there are **no luminous stars nearby**.

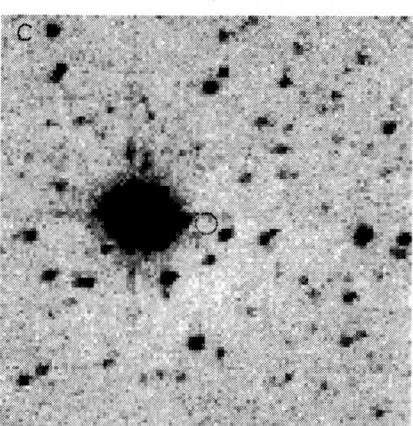

**FIGURE 1.** Infrared K-band image of the field near SGR 1900+14 taken near the outburst of 1998 (after Eikenberry & Dror [13]). The small circle indicates the radio positional error box for SGR 1900+14. The cluster of stars including the luminous M-star pair are seen off to the left. North is up and east is to the left in this image.

## SGR 1900+14

The γ-ray IPN localization for SGR 1900+14 initially led to an apparent association with an X-ray source [2]. This in turn led to an initial optical/IR identification of SGR 1900+14 with an unusual pair of M-type supergiant stars [4]. The relative scarcity of such pairings made a coincidental alignment of the SGR with them moderately unlikely.

However, the 1998 major outburst of SGR 1900+14 produced, among other things, a radio transient which Frail et al. [12] localized to ∼ 1-arcsec precision. This position was shown to be inconsistent with the M-star pair [13] (see Figure 1), and no variable source was found from IR observations during the 1998 flare (Oppenheimer et al. [14]; Eikenberry & Dror [13]).

More recently, Vrba et al. [15] have shown that the M-star pair are in fact the two brightest members of an entire cluster of supergiant and giant stars. This dense cluster has a reddening consistent with the X-ray absorption towards SGR 1900+14, and, at an assumed distance of $\sim 10-15$ kpc, lies a projected distance of ∼ 1 pc from the SGR. The same arguments apply for the association with the SGR and the cluster – the odds of having two such unusual objects coincidentally aligned on the sky are low. One possible interpretation is that SGR 1900+14 was "born" in the cluster of stars, and either left in its progenitor stage or was ejected by a supernova kick (or both). If we assume this scenario and a "standard" pulsar kick velocity of a few hundred km/s, then a separation of 1 pc implies an age for SGR 1900+14 of $< 10^4$ years – consistent with several models for SGR activity.

## SGR 1806-20

As noted above, SGR 1806-20 was thought to be associated with a luminous blue variable (LBV) star which lies at the time-variable (in both flux and morphology) core of the radio nebula G10.0-0.3 (Vasisht et al. [5]; Kulkarni & Frail [6]). However, the recent Inter-Planetary Network (IPN) localization of SGR 1806-20 provides a position inconsistent with that of the LBV star and radio core [16]. Furthermore, Gaensler et al. [10] argue that G10.0-0.3 is not a supernova remnant at all, but is rather powered by the tremendous wind of the LBV star. Infrared observations of the field of SGR 1806-20 reveal that the LBV star is not alone, but appears to be part of a cluster of embedded, hot, luminous stars [17], and the IPN position for SGR 1806-20 is consistent with membership in that cluster. Recently, Eikenberry et al. [18] have used near-infrared photometry and spectroscopy to conclude that this cluster contains what may be the most luminous star in the Galaxy (the LBV star), at least one Wolf-Rayet star of type WCL, and at least two blue "hypergiants" of luminosity class Ia+. These properties make the cluster resemble a somewhat smaller and older version of the "super" star cluster R136 [19], making the potential association with SGR 1806-20 even more intriguing.

*Chandra* observations of SGR 1806-20 have provided a sub-arcsecond localization of the SGR in this crowded field (Kaplan et al. [20]; Eikenberry et al. [21]). The localization matches the IPN position for SGR 1806-20, and completely excludes a direct association of the SGR with the LBV and radio core. This seems to confirm the conclusion of Gaensler et al. [10] that G10.0-0.3 is in fact not a SNR, but is instead a radio nebula powered by the mass-loss wind of the LBV star at its core. Figure 2 shows J-, H-, and K-band infrared images taken with OSIRIS on the CTIO 4-m telescope of the region near SGR 1806-20, along with the 90% positional error circle. As noted above, SGR 1806-20 lies in the direction of an unusual embedded cluster of massive, luminous young stars (Fuchs et al. [17]; Eikenberry et al. [18]), with a distance of $14.5 \pm 1.4$ kpc and a reddening of $A_V = 29 \pm 2$ mag (Corbel et al. [22]; Eikenberry et al. [18]). The fact that stars D and E in Figure 2 have $J - K = 5.0$ mag indicates that they are members of this cluster ($E_{J-K} = 5.0$ mag for $A_V = 29$ mag), and thus that SGR 1806-20 lies within the radial extent of the cluster on the sky. Furthermore, the X-ray absorption towards SGR 1806-20 is $\sim 5-6 \times 10^{22}$ cm$^{-2}$ (see above, and Mereghetti et al. [23]), which is consistent with the extinction towards the cluster. Thus, it seems likely that SGR 1806-20 is also a member of this massive star cluster at a distance of $14.5 \pm 1.4$ kpc.

The IR colors (and upper limits) of the two candidate counterparts to SGR 1806-20 are consistent with both of them being stellar members of the star cluster ($J - K =$

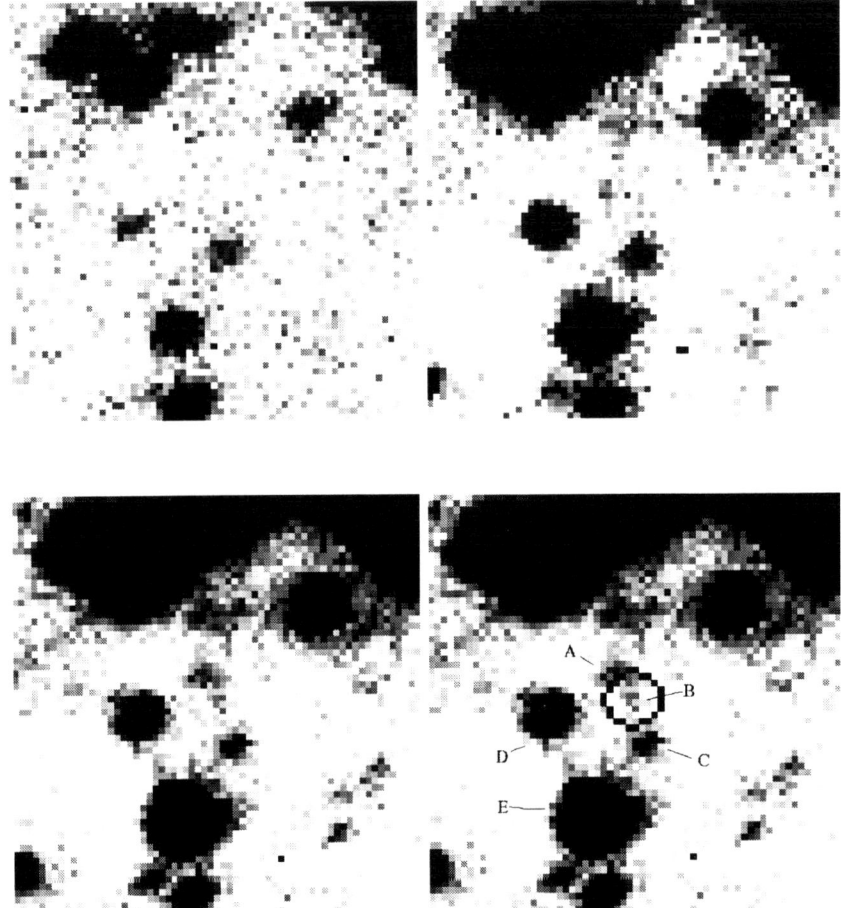

**FIGURE 2.** Infrared images of the field of SGR 1806-20 (as in Eikenberry et al. [21]). Images shown are: J-band (top left); H-band (top right); K-band (bottom left); K-band with stars labelled (bottom right). North is up and east is to the left. The circle indicates the 0.7-arcsec-diameter X-ray error circle from *Chandra*. The bright stars at the top of the images are near the cluster center.

5.0 mag, $H - K = 2.0$ mag). Note that the brighter star just outside the error circle ("C" in Figure 2) appears to be a foreground star ($J - K = 3$ mag), confirming that it is not a likely counterpart to the SGR. However, as can be seen in Figure 2, the field of SGR 1806-20 is highly crowded in the IR by both foreground/background objects and cluster members, and the simple fact that two stars lie within the 90% confidence error circle and another just outside it shows that the probability of a chance coincidence of unrelated IR objects is high. Thus, we cannot conclude definitely that *either* of the possible counterparts is actually related to SGR 1806-20. If both candidates are in fact members of the cluster, we can estimate their absolute magnitudes to be $M_K = -2.3$ mag (A) and $M_K = -0.4$ mag (B). For Star A, this is consistent with stars of luminosity matching a B1V or K3III star, and is inconsistent with any stars of luminosity class I. After correcting for the $H - K = 2$ mag differential extinction toward the cluster, the intrinsic color of $(H - K)_{intrins} = 0.8 \pm 1.2$ mag is essentially consistent with all stellar spectra earlier than late M, and does not significantly constrain the classification. For Star B, the absolute magnitude is consistent with stars of luminosity matching a B8V star, and is inconsistent with any stars of luminosity class III or higher. It is important to note that no observations have yet probed the distribution of stars in the cluster with masses below that of late B main sequence stars. Thus, it is possible that there are even further stars within the error circle at significantly lower mass/luminosity, unless the cluster mass distribution shows a sharp lower cutoff.

One particularly intriguing aspect of the association between SGR 1806-20 and the star cluster is that a neutron star progenitor went supernova *before* the stars currently observed in the cluster. Since more massive stars evolve to the supernova stage more rapidly, *if* the progenitor of SGR 1806-20 formed at the same time as the currently observed massive stars, its mass must have been greater. However, the mass estimate for the LBV is $> 200 M_\odot$ [18], and several of the other stars are likely to have masses in the range of $\sim 50 - 100 M_\odot$ [18]. While recent theories have predicted that very massive stars may produce neutron star remnants due to the effect of envelope loss via dense stellar winds, the upper limit on their masses is $\sim 80 M_\odot$, with higher mass stars producing massive black hole remnants. Thus, it seems unlikely that the SGR 1806-20 progenitor formed at the same time as the currently observed massive stars in the cluster.

Alternately, SGR 1806-20 may have formed prior to these stars – in fact, Kaplan et al. [20] suggest that the supernova event that produced SGR 1806-20 may have triggered the star formation activity that produced the massive stars in the cluster. However, the massive stars we observe in the cluster are evolved, with ages of $\sim 10^6$ yr to reach the LBV and Wolf-Rayet stages of their lives. If the SGR 1806-20 supernova event led to the birth of these stars, then SGR 1806-20 is much older than the $\sim 10^3 - -10^4$ yr typically considered for magnetars. Alternately, SGR 1806-20 may simply be taken as evidence for prior massive star formation at this location. While at least one supernova occurred here, perhaps it was not the first, and an earlier supernova event triggered the formation of the currently observed massive stars.

## SOME CONCLUSIONS

Here I present some of the conclusions of the above discussions:

- Of the well-localized SGRs, only SGR 0525-66 seems to have a clear association with a SNR.
- SGR 1900+14 is not *directly* associated with the MIa double star system.
- SGR 1806-20 is not *directly* associated with the LBV star near it, nor is there a SNR evident in this neighborhood
- Both SGR 1806-20 and SGR 1900+14 are in/near dense clusters of massive stars. These are rare enough that a chance superposition is low.
- The association of SGRs with such clusters explain the apparent, but eventually false, association with some of the particular massive stars in the clusters.
- SGR 0525-66 is *not* associated with any apparent cluster of massive stars.
- Th environments of SGRs are giving some very interesting, if somewhat mixed, signals regarding the origins and evolutions of these objects.

## REFERENCES

1. Rothschild, R. E.; Kulkarni, S. R.; Lingenfelter, R. E 1994, *Nature*, 368, 432
2. Hurley, K., Sommer, M., Kouveliotou, C., Fishman, G., Meegan, C., Cline, T., Boer, M., Niel, M. 1994, *ApJ*, 431, L31
3. Vasisht, G.; Kulkarni, S. R.; Frail, D. A.; Greiner, J. 1994, *ApJ*, 431, L35
4. Vrba, F. J., Luginbuhl, C. B., Hurley, K. C., Li, P., Kulkarni, S. R., van Kerkwijk, M. H., Hartmann, D. H., Campusano, L. E., Graham, M. J., Clowes, R. G., Kouveliotou, C., Probst, R., Gatley, I., Merrill, M., Joyce, R., Mendez, R., Smith, I., Schultz, A. 1996, *ApJ*, 468, 225
5. Vasisht,G., Frail,D.A., Kulkarni,S.R. 1995, *ApJ*, 440, L65
6. Kulkarni,S.R & Frail,D.A. 1995, *Nature*, 365, 33
7. Kulkarni, S. R., Matthews, K., Neugebauer, G., Reid, I. N., van Kerkwijk, M. H., Vasisht, G. 1995, *ApJ* 440, L61
8. van Kerkwijk, M. H., Kulkarni, S. R., Matthews, K., Neugebauer, G. 1995, *ApJ* 444, L33

9. Mazets,E.P. et al. 1979, PaZh, 5, 307
10. Gaensler,B.M., Slane,P.O., Gotthelf,E.V., Vasisht,G. 2001, *ApJ*, 559, 963
11. Kaplan, D. L., Kulkarni, S. R., van Kerkwijk, M. H., Rothschild, R. E., Lingenfelter, R. L., Marsden, D., Danner, R., Murakami, T. 2001, *ApJ*, 556, 399
12. Frail, D. A., Kulkarni, S. R., Bloom, J. S. 1999, *Nature*, 398, 127
13. Eikenberry, S.S. & Dror,D.H. 2001, *ApJ*, 537, 429
14. Oppenheimer, B. R., Bloom, J. S., Eikenberry, S. S., Matthews, K. 1998, *IAUC 6933*
15. Vrba,F.J. et al. 2000, *ApJ*, 533, L17
16. Hurley,K., Kouveliotou,C., Cline,T., Mazets,E., Golenetskii,S., Frederiks,D.D., van Paradijs,J. 1999b, *ApJ*, 523, L37
17. Fuchs,Y., Mirabel,F., Chaty,S., Claret,A., Cesarsky,C.J., Cesarsky,D.A. 1999, *A & A*, 350, 891
18. Eikenberry,S.S. et al., 2002, *ApJ*, submitted.
19. Massey,P. & Hunter,D.A. 1998, *ApJ*, 493, 180
20. Kaplan, D. L., Fox, D. W., Kulkarni, S. R., Gotthelf, E. V., Vasisht, G., Frail, D. A. 2002, *ApJ*, 564, 935
21. Eikenberry, S. S., Garske, M. A., Hu, D., Jackson, M. A., Patel, S. G., Barry, D. J., Colonno, M. R., Houck, J. R. 2001, *ApJ*, 563, L133
22. Corbel,S. et al., 1997, *ApJ*, 478, 624
23. Mereghetti,S., Cremonesi,D., Feroci,M., Tavani,M. 2000, *A & A*, 361, 240

# Mid-infrared observations of the SGR 1900+14 error box

Sylvio Klose*, Bringfried Stecklum*, Dieter H. Hartmann[†], Frederick J. Vrba**, Arne A. Henden[‡] and Aurore Bacmann[§]

*Thüringer Landessternwarte, 07778 Tautenburg, Germany
[†]Clemson University, Clemson, South Carolina, SC 29634-0978
**US Naval Observatory, Flagstaff, AZ 86002-1149
[‡]USRA/US Naval Observatory, Flagstaff, AZ 86002-1149
[§]Astrophysikalisches Institut und Universitäts-Sternwarte, 07745 Jena, Germany

**Abstract.** We report on mid-infrared observations of the compact stellar cluster located in the proximity of SGR 1900+14, and the radio/X-ray position of this soft-gamma repeater. Observations were performed in May and June of 2001 when the bursting source was in an active state. At the known radio and X-ray position of the SGR we did not detect transient mid-IR activity, although the observations were performed only hours before and after an outburst in the high-energy band.

## INTRODUCTION

Recent deep, high-resolution multi-wavelength timing observations of SGRs led to significant progress in our understanding of these enigmatic sources [1, 2]. Key goals in current SGR studies include the identification of their counterparts at long-wavelengths and a better understanding of their past and future evolutionary states (e.g., [3]). Of particular interest is their possible relation to the class of anomalous X-ray pulsars [4].

From the point of view of ground-based astronomy, among the known four (perhaps five) soft gamma-ray repeaters [1] SGR 1900+14 has the advantage that it (or better, its error box) is observable from the northern as well as the southern hemisphere. This increases the opportunities to monitor this source in the optical/infrared bands whenever it is in an active state. During its recent activity cycle in spring/summer 2001 we observed the SGR 1900+14 error box with the ESO 3.6-m telescope using the newly commissioned TIMMI 2 mid-infrared camera. The campaign covered the position shortly after and before an outburst in the high-energy band.

## OBSERVATIONS

TIMMI 2 (Thermal Infrared Multi Mode Instrument) is a thermal infrared camera designed for direct imaging at 5, 10, and 17 microns. This instrument is the successor of TIMMI 1 which was decommissioned in 1999. TIMMI 2 uses a 240 × 320 pixel AsSi BIB detector [5], the scale is 0.3 arcsec per pixel for observations in the $N$ band. The field of view is $96'' \times 72''$. An overview of TIMMI 2 is given in [6], and some early observational results are presented in [7].

Our first observing run was performed on May 22, about 1 month after the giant gamma-ray flare detected from SGR 1900+14 on April 18 [8], and also about 1 month before the detection of the next high-energy outburst from this source [9]. A second observing run, again using TIMMI 2 at the ESO 3.6-m telescope, was carried out on June 28, now only about 9 hours after and 6.5 hours before a gamma-ray outburst [9, 10]. All observations were performed using the N11.9 filter. N11.9 is a narrow-band filter centered at about 11.6 microns (FWHM of ∼ 1.2 microns). We selected this filter, because it offers the highest sensitivity [5].

## RESULTS AND DISCUSSION

With a flux density limit of about 3 mJy our observations represent the deepest mid-infrared observations of the SGR 1900+14 field performed to date. Furthermore, with a delay of only 9 hours after a high-energy outburst our observations probe the SGR environment at a time when it could still be affected by the energy input from the burst. However, we do not detect any mid-IR flux that would indicate an energizing interaction between the burst and the circum-burster medium.

Basically, there are two issues that can be addressed with our observations. The first one is the relevance

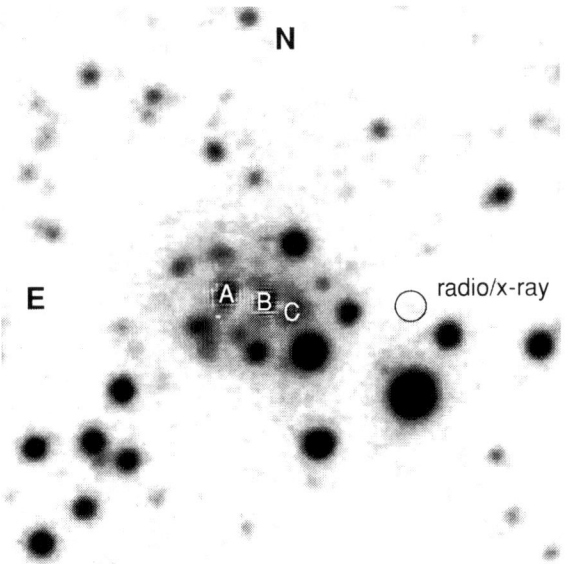

**FIGURE 1.** Deep *I*-band image of the compact stellar cluster in the proximity of SGR 1900+14 with the position of its two most prominent members, two nearly identical M5 supergiants, indicated as A,B. In order to identify the cluster the M5 supergiants and star C have been psf subtracted. Also indicated is the putative radio/X-ray position of SGR 1900+14 suggested in [11, 12]. Adapted from [13]. To provide a scale, star B is separated by 11.8 arcsec from this radio position [14].

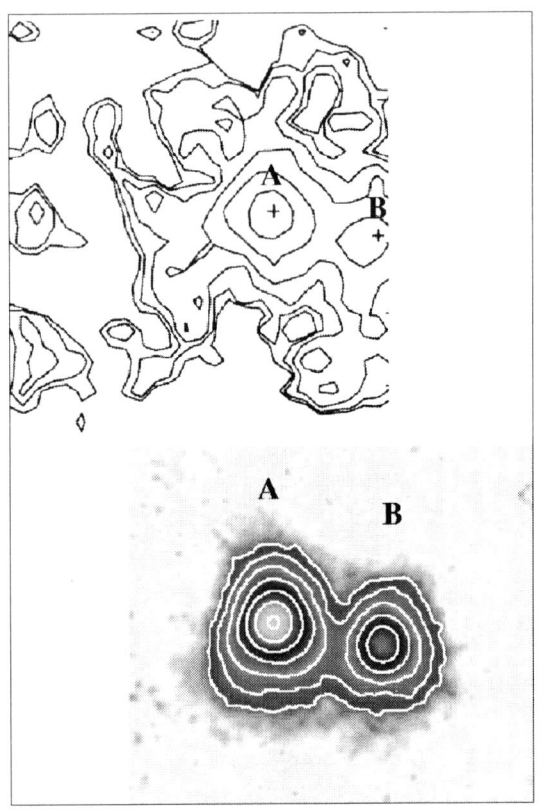

**FIGURE 2.** Part of van Paradijs' published countour plot of their mid-infared observations with TIMMI 1 (top; [15]) compared to our TIMMI 2 image (May 2001; bottom). The stars A and B refer to Fig 1. Contour levels correspond to 28, 56, 112, 224, 448 and 896 mJy per square arcsec (top panel) and also include 14 mJy in the bottom panel. The point-spread function of the 3.6-m telescope is known to deviate slightly from circular.

of the compact stellar cluster of high-mass stars seen in the proximity (Fig. 1) to the X-ray/radio position of SGR 1900+14 [13]. The second issue concerns the implications of the non-detection of the SGR in the mid-IR for models of the burst source and its immediate environment.

## The compact stellar cluster

To our knowledge, the only mid-infrared observations of the SGR 1900+14 error box to date were performed by van Paradijs et al. in July 1995 [15]. These authors used the ESO 3.6-m telescope equipped with the TIMMI 1 camera. At the time of their observations neither a radio transient nor an underlying compact stellar cluster was known, so their main focus was the bright M5 supergiants discovered in the arcmin-sized SGR error box ([16, 17]; Fig. 1).

Based on the earlier observations, our primary attention focused on potential evidence for long-term variability in the interstellar medium surrounding the bursting source, including any gas associated with the M5 supergiants (Fig. 2). Van Paradijs et al. measured a flux density of stars A and B in the *N* band of about 1.64 and 0.86 Jy, respectively [15]. On our May 22 image we measure a flux density in the N11.9 filter of 2.11 and 0.90 Jy, respectively. Since the spectrum of these stars is rapidly rising in the mid-infrared (see figure 5 in [17]), these measurements are not in conflict with each other. Furthermore, they are in agreement with recent observations of this region by the MSX satellite during its Galactic Plane survey (Fig. 3; see the MSX database at [18]): For the D-band image taken at 12.13 microns we measure a total flux density of the unresolved A+B stars of 2.97 Jy.

A comparison between our two observing runs seems to reveal a slight short-term variability of star B. We consider this not surprising, however, since supergiants are known to exhibit some optical variability [17]. Spots on the stellar surface, for example, could be responsible for variations of a cool supergiant [19]. Van Paradijs et al. found evidence for a possible component of extended diffuse emission surrounding the M5 supergiants [15]. Such a component is not apparent in our data, although

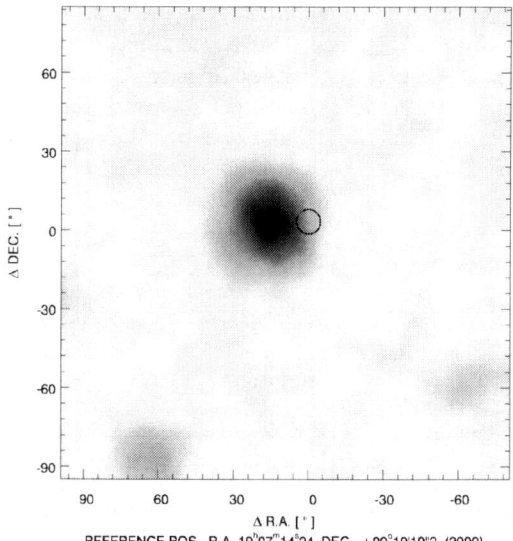

**FIGURE 3.** The SGR 1900+14 error box was also imaged by the Midcourse Space Experiment (MSX satellite) in 1996/97 [18]. Shown is the image obtained at 8.28 microns. Note that the MSX satellite did not resolve the stellar cluster. The image appears extended along the scan direction of the satellite.

our observations are more sensitive. The non-detection of a diffuse emission component is interesting because it constrains the amount of hot dust within the stellar cluster. Moreover, a large local flux of UV photons could produce emission from PAHs of which one feature falls into our chosen filter band. We do not detect this feature. Although we have not yet performed detailed simulations, we believe that this non-detection indicates that there are no strong UV sources within this cluster. This is in agreement with the non-detection of radio continuum emission from this cluster [20]. Obviously in this stellar cluster high-mass star-formation stopped $\gtrsim 10^7$ years ago.

The recent discovery of a compact stellar cluster underlying the M supergiants [13], and the discovery of a similar stellar cluster close to SGR 1820–20 [21], raises the question of what role such clusters might play in the formation of SGRs. In the case of SGR 1900+14 there is an extensive debate in the literature of whether the cluster and the SGR are indeed physically related or whether they are located at very different distances in the Galactic Plane and only appear to be connected by random projection. Recently, the debate has focused on the different interstellar extinction measured towards the M supergiants on the one hand [17], i.e. the most luminous members of the stellar cluster, and toward the quiescent X-ray counterpart of the SGR on the other hand [20]. Since the former is several magnitudes higher than the latter, this seems to argue against a physical relationship between the cluster and the SGR.

In principle, mid-infrared observations could be used to constrain the distance of the stellar cluster, because they are less affected by interstellar extinction. Assuming that star C in Fig. 1 is either a member of this stellar cluster or seen in front of this cluster, we can place a lower limit on the distance of the cluster by the fact that we do not detect this star on our frames. The spectral type of this star was determined by Vrba et al. to be M2 III [17]. Its predicted $N$-band luminosity [22, 23] together with our non-detection of it places this star at a distance of $d \gtrsim 8$ kpc, in agreement with constraints deduced from earlier optical obervations [17]. A similar derivation of an upper limit on the cluster distance based on our observations is more uncertain. Although the spectral type of the supergiants seems to be relatively secure (M5; [17, 24]), there is still the question of their absolute luminosity in the $N$ band. Although theoretical $V - N$ colors are available in the literature, the observational basis of mid-infrared observations of supergiants is still very small. This makes it difficult to estimate the uncertainty that can be attributed to the choice of a certain $N$-band luminosity for these stars.

Although we cannot provide a strong upper limit on the distance of the stellar cluster based on our mid-infrared observations, we draw attention to the following issue concerning the extinction measured toward the M supergiants [17]. In light of the recent discovery of an underlying compact stellar cluster [13] the possibility remains that the measured extinction is the sum of ordinary interstellar extinction and intrinsic extinction within the compact stellar cluster. The observed scatter in the $I - J$ colors of the cluster stars [13] significantly exceeds their uncertainties, indicating that several stars in this cluster suffer significant intracluster extinction. Not only that, but the spread of the $(I - J)$ color of the cluster stars takes up about one half of the $\Delta(I - J)$ between the supergiants A and B and the least reddened star at about $I - J \approx 6.7$ (table 1 in [13]). Naturally, there is no *ad hoc* reason to assume that the M supergiants, the brightest beacons of this cluster, are by chance located in front of the stellar cluster. We discuss this point further elsewhere [25].

## The SGR

Naturally, our hope was a detection of the SGR in the mid-IR with observations placed so closely before *and* after a high-energy outburst. However, at the position of the radio transient associated with the 1998 August 27 burst [11] as well as the quiescent X-ray source discovered in the SGR error box [12] we do not detect the burster. The fact that there is no detection bears on the unknown nature of the SGR environment. Because of the possible presence of fossil accretion disks around SGRs, and their potential AXP relatives, we modeled the spectral energy distribution (SED) of such a disk following Perna et al. [26]. At first glance, mid-infrared observations are very promising for detecting such disks since their SEDs can peak in the infrared. However, even if we include 10 to 20 magnitudes of extinction ($A_V$) toward SGR 1900+14, compared to deep near-infrared observations [20] our flux density limit does not provide a strong constraint on any persistent accretion disk. We can report, however, that if any such a disk exists around SGR 1900+14 then hours before and after a high-energy burst its flux density in the $N$ band is less than 3 mJy.

## CONCLUDING REMARKS

The issue of whether or not SGR 1900+14 and the compact stellar cluster its proximity are physically related is crucially linked to the very uncertain distances. Are these objects located at the same distance or not? In this particular case, this question could be answered if a better understanding of the measured extinction toward the M supergiants and toward the SGR is achieved. The former might profit from deeper mid-infrared observations, the latter might gain from the recent discovery [27] of a persistent dust-scattered X-ray halo around the quiescent X-ray counterpart of the SGR, which could lead to a direct measurement of the extinction by the scattering dust along the line of sight [28].

The non-detection of any signal from the SGR might not be surprising if there is no accretion disk at all around the burster, as indicated by the observations performed to date [20]. But should one expect to detect any non-gamma-ray signal from the burster within hours of a high-energy outburst? This question remains to be addressed by further theoretical studies, but we note that there are now two cases where mid-IR observations were performed only a few hours after a SGR outburst (the other case is SGR 1820–20 [21]) and no signal from the burster or the ambient interstellar medium was detected down to a flux density limit of a few mJy. Future observations need to further push the sensitivity limit, and also sample more closely in time. Truly simultaneous coverage would require robotic observations, similar to those used in the search for prompt optical emission [29, 30].

## ACKNOWLEDGMENTS

The authors thank J. Greiner and U.R.M.E. Geppert (both AIP Potsdam, Germany) for valuable comments on the manuscript. This research made use of data products from the Midcourse Space Experiment. Processing of the data was funded by the Ballistic Missile Defense Organization with additional support from NASA Office of Space Science. This research has also made use of the NASA/IPAC Infrared Science Archive, which is operated by the Jet Propulsion Laboratory, California Institute of Technology, under contract with the National Aeronautics and Space Administration.

## REFERENCES

1. Hurley, K., in *Gamma-Ray Bursts*, edited by R. M. Kippen, R. S. Mallozzi, and G. J. Fishman, AIP Conf. Proc. 526, American Institute of Physics, New York, 2000, p. 763.
2. Ibrahim, A. I. et al., *Astrophys. J.*, **558**, 237 (2001).
3. Kaplan, D. L. et al., *Astrophys. J.*, **556**, 399 (2001).
4. Thompson, C. et al., *Astrophys. J.*, **543**, 340 (2000).
5. TIMMI 2 users manual, http://www.ls.eso.org/lasilla/Telescopes/360cat/timmi/html/manual.html.
6. Relke, H. et al., *SPIE* Vol. **4009**, 440.
7. Stecklum, B., in *The Origins of stars and planets: The VLT view*, edited by J. Alves et al., ESO, Garching 2001, in press.
8. Guidorzi, C. et al., GCN #1041 (2001).
9. Ricker, G. et al., GCN #1073 (2001).
10. Ricker, G. et al., GCN #1074 (2001).
11. Frail, D. A., Kulkarni, S. R., and Bloom, J. S., *Nature* **398**, 127 (1999).
12. Fox, D. W. et al., astro-ph/0107520 (2001).
13. Vrba, F. J. et al., *Astrophys. J.*, **533**, L 17 (2000).
14. Vrba, F. J. et al., *Gamma-Ray Bursts*, edited by R. M. Kippen, R. S. Mallozzi, and G. J. Fishman, AIP Conf. Proc. 526, American Institute of Physics, New York, 2000, p. 809.
15. van Paradijs, J. et al., *Astron. Astrophys.*, **314**, 146 (1996).
16. Hartmann, D. H. et al., in *Workshop on High Velocity Neutron Stars*, edited by R. E. Rothschild, and R. E. Lingenfelter, AIP Conf. Proc. 366, American Institute of Physics, New York, 1996, p. 84.
17. Vrba, F. J. et al., *Astrophys. J.*, **468**, 225 (1996).
18. see: http://www.ipac.caltech.edu/ipac/msx/
19. Tuthill, P. G., Haniff, C. A., Baldwin, J. E., *MNRAS* **285**, 529 (1997).
20. Kaplan, D. L. et al., astro-ph/0107519 (2001).
21. Fuchs, Y. et al., *Astron. Astrophys.*, **350**, 891 (1999).
22. Wainscoat, R. J. et al., *Astrophys. J. Suppl. Ser.* **83**, 111 (1992).
23. Ducati, J. R. et al., *Astrophys. J.*, **558**, 309 (2001).
24. Guenther, E., Klose, S., and Vrba, F. J., in *Gamma-Ray Bursts*, edited by R. M. Kippen, R. S. Mallozzi, and G. J. Fishman, AIP Conf. Proc. 526, American Institute of Physics, New York, 2000, p. 825.
25. Klose, S. et al., in preparation (2003).
26. Perna, R., Hernquist, L., and Narayan, R., *Astrophys. J.*, **541**, 344 (2000).
27. Kouveliotou, C. et al., *Astrophys. J.*, **558**, L 47 (2001).
28. Predehl, P., and Klose, S., *Astron. Astrophys.*, **306**, 283 (1996).
29. Akerlof, C. et al., *Astrophys. J.*, **542**, 251 (2000).
30. Park, H. S. et al., *Astron. Astrophys. Suppl. Ser.*, **138**, 577 (1999).

# Search for Optical Activity of SGR 1806-20 at the SAO 6-m Telescope

G. Beskin*, V. Debur*, A. Panferov*, I. Panferova*, V. Plokhotnichenko*, A. Pozanenko[†], V. Loznikov[†], M. Boer**, J.-L. Atteia**, A. Klotz** and G. Ricker[‡]

*SAO RAS, Karachai-Cherkessia, Russia
[†]IKI RAS, Moscow, Russia
**CESR/CNRS, Toulouse, France
[‡]MIT Center for Space Research, Cambridge, MA, USA

**Abstract.** The field of SGR 1806-20 has been observed at the 6-meter telescope of the Special Astrophysical Observatory during a period of SGR gamma-ray activity. No variable emission in the range $2\,\mu s$ - $10\,s$ was found. The technique and observation details are presented.

## OBSERVATIONS

The field of SGR 1806-20 was observed with the 6-meter telescope of the Special Astrophysical Observatory during its gamma-ray activity. Observations were carried out with the Panoramic Photometer–Polarimeter (PPP) with fine time resolution at the primary telescope focus on June 20th, 2001 (approximately 2 days after HETE trigger [1], and 2.8 days before next outbursts [2, 3]), and on August 22nd, 2001 (15 hours after HETE trigger [4]). Details of the observations are presented in Tables 1 and 2. The image of the field observed in June is presented in Figure 1.

The main part of the PPP is a Positional Sensitive Detector (PSD), which consists of a vacuum tube with a standard S20 photocathode, a set of micro-channel plates and a charge-distribution measuring anode. The pixel size of the detector is $0''.21$, the FOV is a circle of about $1'.5$ and time resolution (dead time) is $2\,\mu s$ [5].

The MANIA acquisition system measures the arrival time of detected photons with an accuracy of 20 ns and sends the data to a PC for subsequent storage [6].

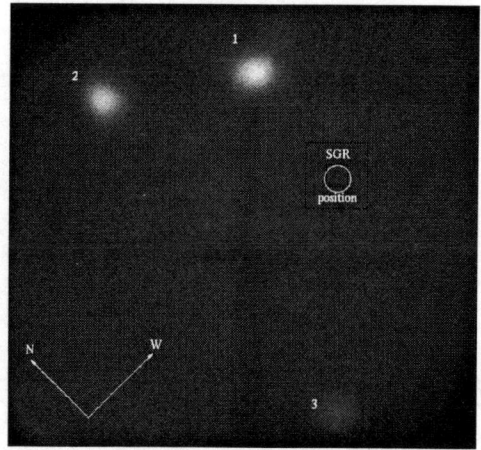

**FIGURE 1.** Observed field of SGR 1806-20. Numbered stars are identified in the Table 2

## RESULTS

All registered photons found in the set of square boxes measuring $6''.5 \times 6''.5$ covering the IPN source position [7] were processed by a special MANIA program. The MANIA algorithm is based on a comparison of the distribution of photon arrival intervals with a Poisson distribution. No variable emission was found in 300 sq. arcsec region centered on the IPN source position. The upper limit for a variable component relative to the background level (152 photons/s) in the range from $2\,\mu s$ to $10\,s$ calculated for level of confidence of 99% is shown in Figure 2. Detailed analyses, including improved position of SGR [8] and a Fourier analysis of the photon list, are in progress.

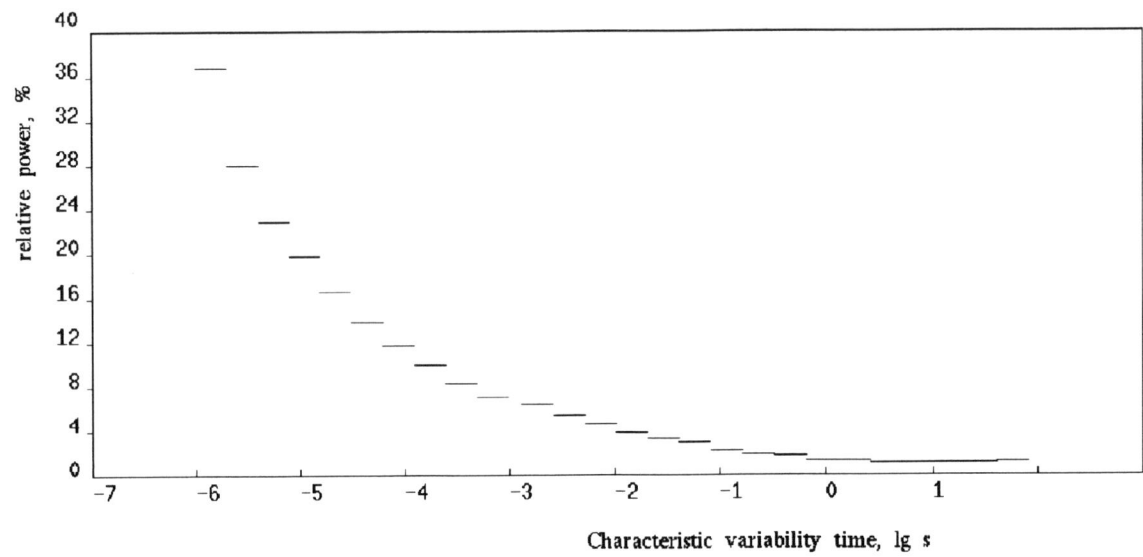

**FIGURE 2.** Upper limit for relative power of variable emission

**TABLE 1.** Observation details

| Date | Start Time (UT) | Exposure Time (s) | Zenith Distance | Filter |
|---|---|---|---|---|
| 20.06.2001 | $19^h58^m$ | 4544.6 | $67.°1$ | B |
| 22.08.2001 | $20^h31^m$ | 800.33 | $77.°7$ | B |

**TABLE 2.** USNO-A2.0 stars in the field of SGR 1806-20

| N | R.A. J2000 * | DEC. J2000 | Bmag | Rmag |
|---|---|---|---|---|
| 1 | $18^h08^m39.^s42$ | $-20°24'11.''74$ | 14.6 | 13.7 |
| 2 | $18^h08^m41.^s13$ | $-20°24'02.''29$ | 15.5 | 14.6 |
| 3 | $18^h08^m41.^s47$ | $-20°25'12.''61$ | 16.5 | 15.9 |

* Monet D.G., 1998, BAAS 30, Vol.4, 120.03

## ACKNOWLEDGMENTS

These observations were performed in the program of Rapid follow-up optical observations of cosmic Gamma-Ray Bursts. The authors would like to thank the 6-meter telescope time allocation committee for granting the observation time.

This study was partially supported by the Russian Ministry of Science, Russian Foundation of Basic Researches (grant 01-02-17857), Federal Programs "Astronomy" and "Integration" and Science-Education Center "Cosmion".

## REFERENCES

1. Ricker, G., et al., 2001, GCN GRB observation report 1068
2. Ricker, G., et al., 2001, GCN GRB observation report 1070
3. Hurley, K., et al., 2001, GCN GRB observation report 1072
4. Ricker, G., et al., 2001, GCN GRB observation report 1089
5. Beskin, G.M., et al., 1999, Report SAO RAS, 269
6. Beskin, G.M., Komarova, V.N., Neizvestny, S., et al., 1997, Experimental Astronomy, 7, 413
7. Hurley, K., et al., 1999, Ap.J., 523, L37
8. Kaplan, D.L., et al., 2001, astro-ph/0108195

# Symposium Participants

Akerlof, Carl UNIVERSITY OF MICHIGAN — cakerlof@umich.edu
Almohd, Abdalla ALABAYATT OBSERVATORY — ASEEL_00@YAHOO.COM
Aloy, Miguel-Angel MAX-PLANCK-INSTITUT FUER ASTROPHYSIK — maa@mpa-garching.mpg.de
Anfimov, Dmitriy SPACE RESEARCH INSTITUTE — dima@cgrsmx.iki.rssi.ru
Atteia, Jean-Luc C.E.S.R — atteia@cesr.fr
Bagoly, Zsolt EÖTVÖS UNIVERSITY — bagoly@ludens.elte.hu
Balastegui, Andreu UNIVERSITY OF BARCELONA U.B. — abalaste@am.ub.es
Band, David GLAST GSFC — dband@lheapop.gsfc.nasa.gov
Barraud, Celine IAP — barraud@hotmail.com
Barthelmy, Scott NASA-GSFC — scott@lheamail.gsfc.nasa.gov
Berger, Edo CALTECH — ejb@astro.caltech.edu
Bonnell, Jerry USRA/GSFC — bonnell@grossc.gsfc.nasa.gov
Briggs, Michael UNIVERSITY OF ALABAMA IN HUNTSVILLE — Michael.Briggs@msfc.nasa.gov
Burrows, David PENN STATE UNIVERSITY — burrows@astro.psu.edu
Burud, Ingunn SPACE TELESCOPE SCIENCE INSTITUTE — burud@stsci.edu
Butler, Nathaniel MIT — nrbutler@space.mit.edu
CastroCerón, José María REAL INST. Y OBSERVATORIO DE LA ARMADA — josemari@alumni.nd.edu
Cline, Thomas GODDARD SPACE FLIGHT CENTER — cline@apache.gsfc.nasa.gov
Clingempeel, Richard AIIR — oerlicon@hroads.net
Connaughton, Valerie UNIVERSITY OF ALABAMA IN HUNTSVILLE — valeriec@hiwaay.net
Costa, Enrico INSTITUTO DI ASTROFISICA SPAZIALE - CNR — costa@ias.rm.cnr.it
Covino, Stefano BRERA ASTRONOMICAL OBSERVATORY — covino@merate.mi.astro.it
Crew, Geoffrey MIT CENTER FOR SPACE RESEARCH — gbc@space.mit.edu
Daigne, Frederic MAX-PLANCK-INSTITUT FUER ASTROPHYSIK — daigne@mpa-garching.mpg.de
de Val Borro, Miguel STOCKHOLM OBSERVATORY — miguel@astro.su.se
Dezalay, Jean-Pascal CENTRE D'ETUD SPATIALE DES RAYONNEMENTS — dezalay@cesr.fr
Dingus, Brenda UNIVERSITY OF WISCONSIN — dingus@physics.wisc.edu
Donaghy, Timothy UNIVERSITY OF CHICAGO — quinn@midway.uchicago.edu
Doty, John MIT — jpd@space.mit.edu
Dullighan, Allyn MIT — allyn@space.mit.edu
Eikenberry, Stephen CORNELL UNIVERSITY — sse2/2cornell.edu
Farewell, Jean MIT CENTER FOR SPACE RESEARCH — farewell@space.mit.edu
Fenimore, Ed LANL — efenimore@lanl.gov
Feroci, Marco ISTITUTO DI ASTROFISICA SPAZIALE - CNR — feroci@ias.rm.cnr.it
Fox, Derek CALTECH — derekfox@astro.caltech.edu
Fruchter, Andrew SPACE TELESCOPE SCIENCE INSTITUTE — fruchter@stsci.edu
Fryer, Chris LOS ALAMOS NATIONAL LABORATORY — fryer@lanl.gov
Galassi, Mark LOS ALAMOS NATIONAL LABORATORY — rosalia+woodshole@galassi.org
Gandolfi, Giangiacomo INSTITUTO DI ASTROFISICA SPAZIALE - CNR — gandolfi@ias.rm.cnr.it
Gehrels, Neil NASA/GSFC — gehrels@lheapop.gsfc.nasa.gov
Ghirlanda, Giancarlo SISSA/ISAS - TRIESTE - ITALY — ghirland@sissa.it
Giblin, Timothy DEPT. OF PHYSICS & ASTRON., THE COLL OF CHARLESTON — giblint@cofc.edu
Gonzalez Maria, Magdalena UNIVERSITY OF WISCONSIN-MADISON — magda@titus.physics.wisc.edu
Goodrich, Bob THE W. M. KECK OBSERVATORY — rgoodrich@keck.hawaii.edu
Granot, Jonathan INSTITUTE FOR ADVANCED STUDY — granot@ias.edu
Graziani, Carlo UNIVERSITY OF CHICAGO — carlo@oddjob.uchicago.edu
Hakkila, Jon COLLEGE OF CHARLESTON — hakkilaj@cofc.edu
Hardee, Philip UNIVERSITY OF ALABAMA — hardee@athena.astr.ua.edu
Hardtke, Rellen UNIVERSITY OF WISCONSIN - MADISON — rellen@alizarin.physics.wisc.edu
Hartmann, Dieter CLEMSON UNIVERSITY — hartmann@grb.phys.clemson.edu
Heger, Alexander DEPT. OF ASTRONOMY AND ASTROPHYSICS, UCSC — 1@2sn.org
Heise, John SRON SPACE RESEARCH NETHERLANDS — j.heise@sron.nl
Henden, Arne USRA/USNO — aah@nofs.navy.mil
Hirshfeld, Alan UNIVERSITY OF MASSACHUSETTS — ahirshfeld@umassd.edu
Hudec, Rene ASTRONOMICAL INSTITUTE ONDREJOV — rhudec@asu.cas.cz
Hullinger, Derek UNIVERSITY OF MARYLAND — derek@milkyway.gsfc.nasa.gov
Hunsberger, Sally PENNSYLVANIA STATE UNIVERSITY — sdh@astro.psu.edu
Hurley, Kevin UC BERKELEY — khurley@sunspot.ssl.berkeley.edu
Jernigan, Garrett UC BERKELEY — jgj@ssl.berkeley.edu
Kaneko, Yuki UNIVERSITY OF ALABAMA IN HUNTSVILLE — kanekoy@email.uah.edu
Kawai, Nobuyuki TITECH — nkawai@hp.phys.titech.ac.jp
Khasawneh, Abeer Mansour ALBALKAA APPLAID UNIVERSITY — aseel_00@yahoo.com
Kippen, Marc NATIONAL SPACE SCIENCE & TECH. CENTER — marc.kippen@msfc.nasa.gov
Klose, Sylvio THURINGER LANDESSTERNWARTE TAUTENBURG — klose@tls-tautenburg.de
Kniffen, Donald NASA HEADQUARTERS — donk@annapolis.net

| Name & Affiliation | Email |
|---|---|
| Kobayashi, Shiho OSAKA UNIVERSITY | shiho@vega.ess.sci.osaka-u.ac.jp |
| Kocevski, Daniel RICE UNIVERSITY | kocevski@rice.edu |
| Kocharovsky, Vladimir INSTITUTE OF APPLIED PHYSICS RAS | kochar@appl.sci-nnov.ru |
| Konigl, Arieh UNIVERSITY OF CHICAGO | arieh@jets.uchicago.edu |
| Koshut, Tom OFFICE OF CONGRESSMAN BUD CRAMER | tom.koshut@mail.house.gov |
| Kosugi, George SUBARU TELESCOPE | george@naoj.org |
| Kouveliotou, Chryssa NASA/MSFC | chryssa.kouveliotou@msfc.nasa.gov |
| Krimm, Hans A. USRA/GSFC | krimm@milkyway.gsfc.nasa.gov |
| Kulkarni, Shrinivas CALTECH | srk@astro.caltech.edu |
| Kumar, Pawan INST. FOR ADVANCED STUDY | pk@ias.edu |
| Lamb, Don UNIVERSITY OF CHICAGO | lamb@oddjob.uchicago.edu |
| Lazzati, Davide INSTITUTE OF ASTRONOMY | lazzati@ast.cam.ac.uk |
| Lee, Brian FERMILAB | bclee@fnal.gov |
| Lenters, Geoffrey HOPE COLLEGE | lenters@hope.edu |
| Levine, Alan MIT CENTER FOR SPACE RESEARCH | aml@space.mit.edu |
| Liang, Edward RICE UNIVERSITY | liang@rice.edu |
| Lichti, Giselher MAX-PLANCK-INST. FUER EXTRATERRESTRISCHE PHYSIK | grl@mpe.mpg.de |
| Litvak, Maxim SPACE RESEARCH INSTITUTE | max@cgrsmx.iki.rssi.ru |
| Lloyd-Ronning, Nicole CANADIAN INST. FOR THEORETICAL ASTROPHYSICS | nicole@urania.stanford.edu |
| MacFadyen, Andrew UCSC/CALTECH | andrew@ucolick.org |
| Martel, Francois ESPACE | fm@space.mit.edu |
| McBreen, Brian UNIVERSITY COLLEGE DUBLIN | Bran.McBreen@ucd.ie |
| McBreen, Sheila UNIVERSITY COLLEGE DUBLIN | smcbreen@bermuda.ucd.ie |
| McConnell, Mark UNIVERSITY OF NEW HAMPSHIRE | Mark.McConnell@unh.edu |
| McEnery, Julie UNIVERSITY OF WISCONSIN | mcenery@titus.physics.wisc.edu |
| Meegan, Charles NASA/MSFC | charles.meegan@msfc.nasa.gov |
| Miller, Warner LOS ALAMOS NATIONAL LABORATORY | wam@lanl.gov |
| Mirabal Nestor COLUMBIA UNIVERSITY | abulafia@astro.columbia.edu |
| Mochkovitch, Robert INSTITUT D'ASTROPHYSIQUE DE PARIS | mochko@iap.fr |
| Monnelly, Glen MIT | monnelly@space.mit.edu |
| Morales, Miguel F. UNIVERISTY OF CALIFORNIA SANTA CRUZ | mmorales@scipp.ucsc.edu |
| Norris, Jay NASA/GODDARD SPACE FLIGHT CENTER | norris@groax0.gsfc.nasa.gov |
| Nousek, John PENN STATE UNIVERSITY | nousek@astro.psu.edu |
| Oksanen, Arto AAVSO | arto.oksanen@jklsirius.fi |
| Olive, Jean-Francois CESR | olive@cesr.fr |
| Paciesas, William UNIVERSITY OF ALABAMA IN HUNTSVILLE | bill.paciesas@msfc.nasa.gov |
| Palmer, David LOS ALAMOS NATIONAL LABORATORY | palmer@lanl.gov |
| Panaitescu, Alin PRINCETON UNIVERSITY | adp@astro.princeton.edu |
| Pangia, Michael GEORGIA COLLEGE & STATE UNIVERSITY | mpangia@gcsu.edu |
| Park, Hye-Sook LLNL | hpark@llnl.gov |
| Perna, Rosalba HARVARD UNIVERSITY | rperna@cfa.harvard.edu |
| Petrosian, Vahe' STANFORD UNIVERSITY | vahe@astronomy.stanford.edu |
| Phillips, Andre UNIVERSITY OF NEW SOUTH WALES | a.phillips@unsw.edu.au |
| Piran, Tsvi HEBREW UNIVERSITY | tsvi@phys.huji.ac.il |
| Piro, Luigi INSTITUTO DI ASTROFISICA SPAZIALE - CNR | prio@ias.rm.cnr.it |
| Pizzichini, Graziella TESRE/CNR | pizzichini@tesre.bo.cnr.it |
| Preger, Barbara IAS/CNR-CIFS | preger@ias.rm.cnr.it |
| Price, Paul RESEARCH SCHOOL OF ASTRONOMY AND ASTROPHYSICS, ANU | pap@mso.anu.edu.au |
| Price, Aaron AAVSO | aaronp@aavso.org |
| Prigozhin, Gregory MIT | gyp@space.mit.edu |
| Ramirez-Ruiz, Enrico INSTITUTE OF ASTRONOMY | enrico@ast.cam.ac.uk |
| Reichart, Dan CALTECH | der@astro.caltech.edu |
| Rhoads, James STSCI | rhoads@stsci.edu |
| Ricker, George MIT | grr@space.mit.edu |
| Rossi, Elena INSTITUTE OF ASTRONOMY | emr@ast.cam.ac.uk |
| Rosswog, Stephan PHYSICS AND ASTRONOMY, UNIVERSITY OF LEICESTER | sro@star.le.ac.uk |
| Ruffert, Maximilian UNIVERSITY OF EDINBURGH | m.ruffert@ed.ac.uk |
| Ryde, Felix STANFORD UNIVERSITY | felix@ahoor.stanford.edu |
| Sakamoto, Takanori TOKYO INSTITUTE OF TECHNOLOGY | sakamoto@tithp1.hp.phys.titech.ac.jp |
| Salmonson, Jay LAWRENCE LIVERMORE NATIONAL LABORATORY | salmonson@llnl.gov |
| Sari, Re'em CALTECH | sari@tapir.caltech.edu |
| Savcheva, Antonia MIT | savcheva@mit.edu |
| Schaefer, Bradley UNIVERSITY OF TEXAS AT AUSTIN | schaefer@astro.as.utexas.edu |
| Schilling, Govert FREELANCE ASTRONOMY WRITER | goverts@yahoo.com |
| Shirasaki Yuji NASDA | yshirasa@oasis.tksc.nasda.go.jp |
| Short, Alexander UNIVERSITY OF LEICESTER | adts@star.le.ac.uk |
| Six, Frank MSFC | frank.six@msfc.nasa.gov |
| Smith, Donald UNIVERSITY OF MICHIGAN | donaldas@umich.edu |
| Smith, Ian RICE UNIVERSITY | ian@spacsun.rice.edu |
| Takagi, Ryo TOKYO INSTITUTE OF TECHONOLOGY | rtakagi@th.phys.titech.ac.jp |

| | |
|---|---|
| **Tavenner, Tanya** LANL | tanya@lanl.gov |
| **Thompson, Christopher** CANADIAN INST. FOR THEORETICAL ASTROPHYSICS | thompson@cita.utoronto.ca |
| **Totani, Tomonori** PRINCETON UNIVERSITY OBSERVATORY | ttotani@astro.princeton.edu |
| **van Putten, Maurice** MIT | vmp@math.mit.edu |
| **Vanden Berk, Daniel** FERMILAB | danvb@fnal.gov |
| **Vanderspek, Roland** MIT | roland@space.mit.edu |
| **Vestrand, Tom** LOS ALAMOS NATIONAL LABORATORY | vestrand@lanl.gov |
| **Villasenor, Jesus Noel** MIT | jsvilla@space.mit.edu |
| **Vlahakis, Nektarios** THE UNIVERSITY OF CHICAGO | vlahakis@jets.uchicago.edu |
| **Wang, Xiaohu** HARVARD UNIVERSITY | xwang@cfa.harvard.edu |
| **Wijers, Ralph** DEPT. OF PHYSICS AND ASTRONOMY | rwijers@mail.astro.sunysb.edu |
| **Williams, George** STEWARD OBSERVATORY | gwilliams@as.arizona.edu |
| **Woods, Peter** USRA/NSSTC | Peter.Woods@msfc.nasa.gov |
| **Woosley, Stan** ASTRONOMY DEPARTMENT | woosley@ucolick.org |
| **Yonetoku, Daisuke** ISAS | yonetoku@astro.isas.ac.jp |
| **Yost, Sarah** CALTECH | yost@srl.caltech.edu |
| **Zhang, Bing** PENN STATE UNIVERSITY | bzhang@astro.psu.edu |
| **Zhang, Weiqun** UNIVERSITY OF CALIFORNIA, SANTA CRUZ | zhang@ucolick.org |

# AUTHOR INDEX

## A

Abazajian, K., 349
Abbey, A. F., 511
Ables, E., 366
Adelman, J., 349
Afansiev, V. L., 424
Afonso, J., 357
Akerlof, C., 514, 544
Albright, K., 550
Amati, L., 387
Ambrosi, R. M., 511
Andersen, M. I., 357, 424
Annis, J., 349
Antonelli, L. A., 517
Aptekar, R. I., 143
Ashley, M. C. B., 514
Atteia, J.-L., 3, 17, 42, 59, 63, 73, 82, 88, 111, 583
Axelrod, T. S., 541
Ayal, S., 313
Azzibrouck, G., 107

## B

Bacmann, A., 579
Bagoly, Z., 137, 163, 446
Balastegui, A., 438
Balázs, L. G., 137, 163, 446
Band, D. L., 140
Barbier, L. M., 500
Barraud, C., 59, 73, 82, 88
Barret, D., 481
Barthelmy, S. D., 101, 143, 366, 491, 500
Beacom, J. F., 349
Belczynski, K., 417
Belyanin, A. A., 159
Berger, E., 393, 420
Berná, J. Á., 553
Bernas, M., 538, 553
Beskin, G., 583
Blain, A. W., 457
Blandford, R., 477
Boer, M., 17, 107, 583
Boettcher, M., 295

Borgonovo, L., 264
Borozdin, K., 547
Bradt, H., 276
Brady, T., 38
Braga, J., 107
Briggs, J. W., 349
Briggs, M. S., 176, 244, 248, 273, 379, 469
Brinkmann, J., 349
Brumby, S. P., 547
Burrows, D. N., 488, 511
Butler, N., 17, 45, 63, 70, 73, 82, 88, 111
Butterworth, P. S., 143

## C

Campana, S., 393
Canal, R., 438
Casperson, D., 514, 547, 550
Castro Cerón, J. M., 387, 424, 553
Castro-Tirado, A. J., 357, 424, 520, 538, 553
Celotti, A., 270, 393
Chary, R.-R., 428
Chen, B., 349
Cherepashchuk, A. M., 424
Chester, M. M., 506
Chincarini, G., 488, 517
Christensen, L., 357
Cimatti, A., 393
Citterio, O., 488, 511
Cline, T. L., 42, 143, 366, 473
Colasanti, L., 509
Conconi, P., 517
Connaughton, V., 273, 379
Costa, E., 123, 387, 509
Cotin, F., 17
Couteret, J., 17
Covino, S., 393, 517
Craig, W., 477
Crew, G. B., 3, 17, 33, 38, 42, 49, 66, 70, 73, 82, 88, 91, 101, 107
Csabai, I., 349, 446
Cutispoto, G., 517

## D

Daigne, F., 289, 299
Debur, V., 583
de la Morena, B. A., 553
Della Valle, M., 393
Derishev, E. V., 159, 292
De Rújula, A., 319
de Ugarte Postigo, A., 553
de Val Borro, M., 264
Dezalay, J.-P., 17, 59, 73, 82, 88
Diehl, R., 469
Dill, R., 33, 38
Dingus, B. L., 240, 267
di Serego Alighieri, S., 393
Dodonov, S. N., 424
Doi, M., 346
Donaghy, T. Q., 25, 59, 82, 111, 450
Doty, J. P., 3, 17, 33, 38, 42, 45, 49, 66, 70, 73, 82, 88, 91, 107
Dullighan, A., 63

## E

Eftekharzadeh, A., 500
Ehanno, M., 17
Eikenberry, S. S., 574
Evrard, J., 17

## F

Fatkhullin, T. A., 424
Feldt, M., 424
Fenimore, E. E., 3, 25, 59, 82, 94, 97, 111, 117, 491, 500, 547, 550
Feroci, M., 387, 473
Ferrari, A., 509
Feulner, G., 424
Fishman, G. J., 273, 469, 477
Fliri, J., 424
Foster, R., 107
Francis, J., 38
Frederiks, D. D., 143, 570
Freismuth, T. M., 147
Frontera, F., 123, 357, 387, 473
Fruchter, A., 357
Fryer, C. L., 199, 454
Fugazza, D., 393

Fynbo, J. U., 357

## G

Galassi, M., 3, 25, 59, 82, 94, 97, 111, 114, 117, 491, 547
Galeazzi, M., 509
Gandolfi, G., 387, 509
García Dabó, C. E., 553
Gatti, F., 509
Gehrels, N., 366, 465, 477, 491, 500
Ghirlanda, G., 270
Ghisellini, G., 169, 270, 393, 411, 517
Giblin, T. W., 144, 147, 179, 273, 379, 556
Giommi, P., 387
Gisler, G., 514, 547
Golenetskii, S. V., 143, 473, 570
González, M. M., 267
Gorosabel, J., 357, 424, 553
Goupell, J. E., 570
Granot, J., 327
Graziani, C., 3, 25, 59, 76, 79, 82, 94, 97, 111, 114, 450, 460
Greiner, J., 357, 424
Grindlay, J., 477
Guidorzi, C., 387, 473

## H

Haardt, F., 393
Haglin, D. J., 144, 147, 179, 556
Hakkila, J., 144, 147, 176, 179, 556
Hanlon, L., 280
Hardtke, R., 150
Harrison, F., 477
Hartmann, D. H., 366, 442, 477, 579
Harvanek, M., 349
Hatsukade, I., 25
Heger, A., 185, 214
Heise, J., 229, 244
Henden, A. A., 349, 424, 579
Hennessy, G. S., 349
Hill, J. E., 488, 511
Hjorth, J., 357, 424
Ho, C., 550
Hong, J., 477
Horváth, I., 137, 163, 446

Hroch, F., 360, 520, 544
Hudcová, V., 494, 523
Hudec, R., 338, 360, 363, 494, 520, 523, 526, 538, 544, 553
Huffman, G., 38
Hullinger, D. D., 500
Hunsberger, S. D., 497
Hurley, K., 3, 17, 42, 73, 82, 107, 111, 153, 349, 366, 473, 570
Hutchinson, I. B., 511
Hyvönen, H., 352

## I

Il'inskii, V. N., 143
Inneman, A., 494
in 't Zand, J. J. M., 244, 387
Israel, G. L., 393
Israelian, G., 424
Ivezic, Z., 349

## J

Janka, H.-T., 193
Jelínek, M., 520, 526, 544, 553
Jensen, B. L., 357, 424
Jernigan, J. G., 3, 49, 91

## K

Kaneko, Y., 267
Kaper, L., 357
Kawai, N., 3, 25, 59, 63, 82, 94, 97, 107, 111, 114, 117, 346, 383
Kawasaki, W., 346
Kehoe, R., 349, 514
Kelley, R., 509
Kent, S., 349
Kippen, R. M., 176, 244, 379, 469
Kissel, S., 33
Kleinman, S., 349
Klose, S., 357, 424, 579
Klotz, A., 583
Kneiske, T. M., 442
Kobayashi, S., 260
Kocevski, D., 130, 156, 237, 286, 295
Kocharovsky, V. V., 159, 292
Kocharovsky, Vl. V., 159, 292
Komarova, V. N., 424
Königl, A., 166, 327
Kouveliotou, C., 273, 342, 357, 469, 477, 570
Krimm, H. A., 500
Kroll, P., 523, 526
Krolupper, F., 523
Kron, R., 349
Krzesinski, J., 349
Kubánek, P., 520
Kulkarni, S., 393
Kumar, P., 305

## L

Lagrange, D., 17
Lamb, D. Q., 3, 25, 59, 76, 82, 94, 97, 111, 114, 349, 415, 433, 450, 460
Lazzati, D., 169, 206, 335, 393, 411, 457
Ledoux, J. R., 503
Lee, B. C., 349
Lee, W. H., 217
Lenters, G. T., 570
Lestrade, J.-P., 73
Levine, A. M., 49, 276
Liang, E., 130, 156, 237, 286, 295
Lichti, G., 469
Lindsay, K., 366
Lloyd-Ronning, N. M., 252, 454
Long, D., 349
Loznikov, V., 583
Lupton, R., 349

## M

MacFadyen, A. I., 202, 206, 226, 260
Macri, J. R., 503
Mallén-Ornelas, G., 357
Mallozzi, R. S., 248
Manchanda, R., 107
Mannheim, K., 442
Markwardt, C., 500
Marshall, F. E., 506
Marshall, S., 514
Masai, K., 383
Masetti, N., 338, 357, 363, 387
Más-Hesse, J. M., 553

Mason, K. O., 497
Mateo Sanguino, T. J., 553
Matsuoka, M., 3, 25, 59, 82, 94, 111, 117
Mazets, E. P., 143, 473
McBreen, B., 280
McBreen, S., 280, 553
Mc Cammon, D., 509
McConnell, M. L., 503
McEnery, J. E., 529
McGowan, K., 514, 547
McKay, T., 349, 514
McMillan, R., 349
Meegan, C. A., 176, 379, 469
Meegan, D. J., 144
Meegan, W. S., 556
Mészáros, A., 137, 163, 446
Mészáros, P., 137, 292, 446
Mitrofanov, I., 473
Miyazaki, S., 346
Mizumoto, Y., 346
Mochkovitch, R., 289, 299
Moilanen, M., 352
Molinari, E., 517
Møller, P., 357
Monnelly, G., 33, 42, 49, 66, 70, 107
Montanari, E., 387, 473
Morales, M. F., 532, 535
Murakami, T., 383

## N

Namiki, M., 383
Neilsen, E. H., 349
Nekola, M., 520
Nemiroff, R., 366
Newberg, H. J., 349
Newman, P. R., 349
Nicastro, L., 517
Niel, M., 17
Nitta, A., 349
Nousek, J. A., 488, 497, 506

## O

Ogasawara, R., 346
Oksanen, A., 352
Olive, J.-F., 17, 59, 73, 82, 88
Orio, M., 509

Ozawa, H., 500

## P

Pacciani, L., 509
Paciesas, W. S., 144, 147, 179, 248, 469, 556
Palazzi, E., 357, 517
Pallavicini, M., 509
Palmer, D. M., 491, 500
Pal'shin, V. D., 143
Palunas, P., 349
Panaitescu, A., 305
Panferov, A., 583
Panferova, I., 583
Pangia, M. J., 283
Park, H. S., 366
Parsons, A. M., 491, 500
Páta, P., 538
Pedersen, H., 357, 424
Pendleton, G. N., 176
Pereira, W., 366
Perez-Ramirez, D., 366
Pergolesi, D., 509
Perkins, S., 547
Perna, R., 411, 417
Phillips, M. A., 514
Pian, E., 357
Pina, L., 494
Piran, T., 313
Piro, L., 372, 387, 509
Pizzichini, G., 3, 107, 338, 363
Plokhotnichenko, V., 583
Polcar, J., 360, 363, 520
Porter, S., 509
Postnov, K. A., 424
Pozanenko, A., 583
Preece, R. D., 176, 244, 248, 267, 273, 283, 379, 469
Preger, B., 97
Price, P. A., 56, 355, 393, 403, 541
Priedhorsky, W., 547, 550
Prigozhin, G., 33, 49, 70, 91

## Q

Quilligan, F., 280

## R

Ramirez-Ruiz, E., 169, 172, 206, 217, 393, 454, 457
Razeti, M., 509
Rees, M. J., 335
Reichart, D. E., 349, 403, 450, 460
Remillard, R., 276
Ricker, G. R., 3, 17, 33, 42, 49, 66, 70, 73, 82, 91, 107, 111, 583
Ríkal, V., 360
Roberts, J., 38
Rodonó, M., 517
Roiger, R. J., 144, 147, 179, 556
Rol, E., 342, 357, 424
Roming, P. W. A., 497
Rossi, E., 335
Rosswog, S., 220
Rouaix, G., 17
Ruffert, M., 193
Ruiz-Lapuente, P., 438
Ryan, J. M., 503
Ryde, F., 260, 264, 286
Rykoff, E., 514

## S

Sakamoto, T., 25, 59, 82, 94, 111, 114, 117
Salamanca, I., 357
Salmonson, J. D., 134, 223
Sanders, T., 509
Saracco, P., 393
Schmidt, B. P., 541
Schneider, D. P., 349
Schönfelder, V., 469
Sheets, T. B., 143
Shirasaki, Y., 3, 25, 59, 82, 94, 97, 111, 117
Short, A. D., 488, 511
Šimon, V., 338, 544
Smída, R., 520, 526
Smith, D. A., 276, 514
Smith, I. A., 342
Snedden, S., 349
Soderberg, A. M., 172
Soffitta, P., 387, 509
Sokolov, V. V., 424
Soldán, J., 520

Soria, T., 553
Souleille, P., 17
Sprague, A. J., 147
Stallworth, A. D., 147
Starr, D., 547
Stecklum, B., 579
Stella, L., 393
Stoklasová, I., 520, 526
Stoughton, C., 349
Štrobl, J., 544
Svéda, L., 520, 526
Svensson, R., 264
Szymkowiak, A., 509

## T

Tagliaferri, G., 488, 511
Takada, T., 346
Takagishi, K., 25
Takahashi, D., 25, 94, 114
Tamagawa, T., 3, 25, 59, 63, 82, 94, 97, 111, 114, 117
Tanvir, N., 342, 357
Tavenner, T., 25, 59, 82, 97, 114, 491
Testera, G., 509
Tichá, J., 363
Tichy, M., 363
Tilanus, R. P. J., 342
Topinka, M., 363, 520
Torii, K., 25, 59, 82, 94, 111, 114, 117
Torres Riera, J., 553
Totani, T., 346
Treyer, M. A., 357
Trussoni, E., 509
Tucker, D. L., 349
Tueller, J., 491, 500
Turner, M., 488

## U

Urata, Y., 346

## V

Vaccarone, R., 509
Vanden Berk, D. E., 349
van den Heuvel, E., 357

Vanderspek, R., 3, 17, 25, 33, 38, 42, 49, 66, 70, 73, 82, 88, 91, 97, 101, 107, 111, 114, 117
van Paradijs, J., 273
van Putten, M. H. P. M., 210
Vavrek, R., 163, 446
Vedrenne, G., 3, 17
Vestrand, W. T., 514, 547, 550
Villasenor, J. N., 3, 33, 42, 49, 66, 70, 91, 107
Vlahakis, N., 166
von Kienlin, A., 469
Vrba, F. J., 424, 579
Vreeswijk, P. M., 342, 357

## W

Watanabe, K., 442
Watson, D., 280
Weidenspointner, G., 500
Weinberg, N., 460
Wells, A. A., 488, 511
White, R., 547, 550
Wijers, R. A. M. J., 273, 342, 357, 396
Wilhite, B., 349
Williams, G. G., 366
Willingale, R., 511
Wilson, J. R., 223
Wolf, C., 357
Woods, P. M., 244, 561, 570
Woosley, S. E., 3, 185, 214, 226, 477
Wozniak, P., 514, 547
Wren, J., 349, 514, 547, 550

## Y

Yamauchi, M., 25
Yanny, B., 349
Yasuda, N., 346
Yonetoku, D., 383
York, D., 349
Yoshida, A., 3, 25, 59, 63, 82, 94, 97, 107, 111, 114, 117, 346, 383
Yost, S. A., 369
Young, K. C., 147

## Z

Zerbi, F., 517
Zhang, B., 331
Zhang, W., 185, 226